14

555
Counters
Synchronous

"ALL WORK AND NO PLAY?"

Ask
your instructor
to contact Kim Harris
toll-free at
1-800-347-7707
to ask about our
group purchase
discount!

Student Computer Disks* let you have fun testing your knowledge of topics covered in the best-seller ***DIGITAL: SYSTEMS, LOGIC AND APPLICATIONS***!

And because the ***2-disk set*** is designed specifically to accompany your textbook, you'll be able to make the most of your study time!

Available exclusively from Delmar Publishers, the Student Disks include:

- Randomly generated ***multiple choice questions/answers*** for drill and practice! Keyed to sections in the text!
- ***Interactive exercises*** (covering topics such as Number Conversion, Logic Gates, Boolean Algebra, K-Mapping, and J-K Flip-Flops) to reinforce important concepts! At least 1 per chapter!
- ***Troubleshooting simulation programs*** to provide opportunities for hands-on practice with specific logic and digital systems!

Use the handy order form on the back to purchase your Student Disks today!
Supply is limited; be sure to act quickly -- At $9.95, these disks won't last long!

*For use with IBM or IBM-compatible personal computers.

❒ **YES,** Please send me the outstanding

Student Disks

to accompany **Digital: Systems, Logic and Applications.**

QUANTITY (# of sets)	TITLE	UNIT PRICE (2-disk set)	TOTAL
	IBM Diskettes **# 3119-3**	**$9.95**	
		Sub Total	
		Add State, Local Taxes	
		Total Order	

CHECK ONE:

❒ Payment in full is enclosed for $ ____________

❒ Charge my order to (circle one): VISA MASTERCARD

Card Number ____________________

Expiration Date ____________ Signature ____________________________

NAME ____

ADDRESS ____

CITY __________________ STATE __________ ZIP __________

SCHOOL/INSTITUTION ____

Note: Your order must be accompanied by your signature, plus full payment or complete credit card information.

Mail to: Delmar Publishers Inc.®
2 Computer Drive, West
Box 15015
Albany, New York 12212-5015

All sales for educational software are final once the package seal is broken. We will gladly replace any defective disk for a $2.50 service charge. Prices subject to change without notice. Please allow 4-6 weeks for delivery.

DIGITAL SYSTEMS, LOGIC AND APPLICATIONS

Thomas A. Adamson

Delmar Publishers Inc.®

NOTICE TO THE READER

Publisher and author do not warrant or guarantee any of the products described herein or perform any independent analysis in connection with any of the product information contained herein. Publisher and author do not assume, and expressly disclaim, any obligation to obtain and include information other than that provided to them by the manufacturer.

The reader is expressly warned to consider and adopt all safety precautions that might be indicated by the activities described herein and to avoid all potential hazards. By following the instructions contained herein, the reader willingly assumes all risks in connection with such instructions.

The publisher and author make no representations or warranties of any kind, including but not limited to, the warranties of fitness for particular purpose or merchantability, nor are any such representations implied with respect to the material set forth herein, and the publisher and author take no responsibility with respect to such material. The publisher and author shall not be liable for any special, consequential or exemplary damages resulting, in whole or in part, from the readers' use of, or reliance upon, this material.

DEDICATION: To my wife, Judy Adamson,
whose continued encouragement, trust
and love made another book possible.

COVER PHOTOS: 88100 RISC microprocessor photo courtesy of Motorola Inc.
Computer screen photos by Tom Carney

Delmar Staff
Associate Editor: Cameron Anderson
Editing Manager: Gerry East
Production Editor: Eleanor Isenhart
Design Coordinator: Susan C. Mathews
Publications Coordinator: Karen Seebald

For information, address Delmar Publishers Inc.,
2 Computer Drive West, Box 15-015,
Albany, New York 12212

Printed in the United States of America
Published simultaneously in Canada
by Nelson Canada
A Division of International Thomson Limited

10 9 8 7 6 5 4 3 2

Library of Congress Cataloging-in-Publication Data

Adamson, Thomas A., 1936–
Digital : systems, logic, and applications,

Includes index.
1. Digital electronics. 2. Microprocessors.
3. Logic circuits. I. Title.
TK7868.D5A29 1989 621.391'6 88-25806
ISBN 0-8273-3112-6
ISBN 0-8273-3113-4 (instructor's guide)
ISBN 0-8273-3114-2 (lab manual)

CONTENTS

PREFACE

Digital: Systems, Logic and Applications is written for the reader who is interested in learning digital logic as a means of understanding microprocessor-based systems. It presents to the beginner, the prerequisite material required for advanced microprocessor courses.

The text assumes no prior knowledge of electronics or computer programming and may be used by electronic technology majors in the first semester of instruction with a concurrent DC electrical circuits course. Nonelectronics majors may also use this text by omitting the electronic application section of each chapter without loss of continuity.

When you complete this text you will be well prepared to tackle any microprocessor course or industrial electronics course. This text contains the latest concepts in standard digital circuits including programmable logic devices (such as PROMs, PLAs and PALs) and the new ANSI/IEEE Standard 91-1984 logic symbols as well as a rigorous treatment of Boolean algebra, K-mapping, interfacing and data transmission.

Every chapter contains an electronic application section for those who will be involved in the troubleshooting and prototyping of digital integrated circuits. These sections include TTL and CMOS integrated circuits, data sheet interpretation, signature analysis, CMOS/TTL interfacing, GPIB, A/D and D/A conversion as well as the use of logic analyzers in the troubleshooting and analysis of digital circuits.

Each chapter also contains a microprocessor application section which shows you how to immediately utilize the concepts just learned in a practical digital system environment. This gives the user of this text a tremendous advantage upon entrance into a microprocessor, robotics and/or industrial technology course.

Even though each chapter has a microprocessor application section, the focus of the text is not exclusively on computers. Equal emphasis is placed on those principles that apply to industrial automation, communications and process control. This approach gives the user of this text the necessary broader base in this exciting and rapidly expanding field.

This is an excellent text for a two-year college curriculum and may also be used as an introductory course in a four-year engineering technology or computer science program. It is suitable for private technical schools as well as industrial training programs.

The text contains many detailed illustrations, tables, and worked-out examples and it is easy to read and understand. With such an abundance of pedagogical features much of the material may be covered rapidly. As such this text may be used as a one-semester course by omitting some topics or presented at a more leisurely pace for a two-semester course.

Features

Each chapter of the text contains the following important learning features:

Chapter Objectives. A set of specific statements outlining exactly what the reader can expect to gain from the chapter.

Chapter Introduction. A discussion of the chapter material to follow containing historical antidotes and other topics of interest designed to stimulate reader interest and motivation.

Key Terms. A list of new terminology that the reader will find in the chapter. Each of these terms will be defined in the chapter the first time it is used.

Running Glossary. This ensures that each new term is defined when it is first introduced.

Chapter Sections. Each chapter consists of small sections that introduce a single new concept.

Review Questions. Each chapter section contains a set of review questions with answers. This gives readers an opportunity to immediately check their understanding of the section just completed.

Example Problems. Every new concept presented is immediately followed by an example problem showing the reader how to apply the new material.

Microprocessor Application. Every chapter contains a section that illustrates how the material of that chapter is used in a microprocessor-based system. Thus the reader sees an immediate application of all new material.

Electronic Application. Every chapter contains a section that illustrates the electrical aspects of digital circuits. These sections may be used by the student who has a basic electricity background. For nonelectronic majors, this section may be omitted with no loss of continuity in the material.

Troubleshooting Simulation. Floppy disks are available to support the text. They contain computer simulated troubleshooting, test questions and interactive exercises of digital devices, circuits and systems. This provides the student an opportunity to do hands on practice in developing troubleshooting skills in literally thousands of possible problems for specific digital logic and systems.

Summary. A concise summary is presented at the end of each chapter to serve as a quick review. Each summary statement is keyed to the section where material was presented.

Chapter Self-Test. A student self-test with answers is included at the end of every chapter. This helps to immediately point out any weak spots before proceeding to new material.

Chapter Problems. Chapter problem sets are keyed to specific sections within the chapter. The problem set contains graded problems as follows:

- Basic Concepts
- Applications
- Troubleshooting
- Analysis and Design

Ancillary Materials

Student Disks (for use with IBM® or IBM-compatible personal computers)

The disks are keyed to each chapter of the text by providing the following:

- An interactive troubleshooting and demonstration program for each chapter as described in the corresponding chapter.
- A set of scrambled multiple choice distractors and single correct answers for each section of each chapter.
- A demonstration and testing program for each chapter that contains an interactive simulation of a key feature of the chapter (such as interacting with a JK flip-flop).
- A permanent record, automatically kept on each student disk, maintains the following information concerning the drill and practice programs, the interactive tests, and the troubleshooting simulations:

Section Scores	Student Name
Test Scores	Student ID Number
Chapter Running Scores	Student Section

The **Student Disks** may be purchased by individuals for $9.95.

However, **GROUP PURCHASE DISCOUNTS ARE AVAILABLE** for qualified adopters of the DIGITAL: SYSTEMS, LOGIC AND APPLICATIONS text!

Contact Kim Harris, toll-free at 1-800-347-7707 to:

1. *Obtain free master disks,* with permission to make as many copies as you need for your class(es)

- OR -

2. *Purchase multiple copies* of the disks at a special, low price

Instructor's Disk

Each instructor is provided with an instructor's disk that interacts with the student disks in the following manner:

- The instructor's disk contains software that allows the recording of the permanent records on the student disks that shall include:

Section Score Information	Student Name
Complete Test Score Information	Student ID Number
Complete Chapter Score Information	Section Number

- The information from the student disks may be displayed on the screen and copies made on an attached printer.
- Analysis may be for a single student or a class of students or from a single chapter to a range of chapters.

Instructor's Guide

An Instructor's Guide is available that contains detailed solutions to all Chapter Problems.

Laboratory Manual

A coordinated laboratory manual is available with the text. Experiments in the lab manual may be performed with or without a standard digital trainer. Each experiment contains an important troubleshooting section.

Overhead Transparency Reprints

These contain key figures from the text from which overhead transparencies may be made by the instructor for classroom use. This is a valuable time saving tool for presenting complex digital logic circuits for discussion and analysis.

Very few instructors present the class material in exactly the same order as the material appears in the book. This is done not only from a personal preference point but to accommodate specific groups of students. The structure of this book lends itself to a variety of approaches to this important subject.

When you complete this book, you will understand the prerequisite information that is required of the electronic technologist and others who are or will be responsible for the operation, programming, maintenance and design of digital and microprocessor-based systems.

Acknowledgments

No successful book is published without the talent and skill of many dedicated people. I would like to thank the staff of Delmar Publishers and the reviewers for making this book possible.

Digital: Systems, Logic and Applications is more than just a book; it is an educational system. The Instructor's Guide, done by Professor Charles Goodspeed, is one of the best prepared instructor's guides available with any text on this subject. A software instructional and monitoring system was made possible through the foresight of one of the most outstanding editors in the business, Cam Anderson. The coordinated Laboratory Manual, Wall Charts, Study Guide, and other materials are the result of proposals by Mark Huth and Cam Anderson.

Extensive reviews, done by fellow teaching professionals, represent hundreds of hours of work, and numerous suggestions, revisions and improvements to this important first edition. Each contributed a substantial portion of their knowledge to make the material now in this text. My heartfelt thanks to each:

Paul Bierbauer—DeVry (Chicago)
Pearly Cunningham—Allegheny Community College
Charles Goodspeed—Hartnell Community College
Bill Hames—Midlands Technical College
Peter Holsberg—Mercer County Community College
Rudy Lucas—DeVry (Chicago)
Bob Martin—North Virginia Community College

Don B. LeFavour—Columbus State Community College
John L. Morgan—DeVry (Dallas)
Chuck Parnell—Cochise Community College
Pat Smith—Texas State Technical Institute
Joseph M. Sweet—Pennco Tech
Steve Sink—Davidson County Community College
Parker Tabor—Greenville Technical College
Ross Thompson—DeVry (Alberta)
Steven Waterman—DeVry (Lombard)
Steven Yelton—Cincinnati Technical College

T. Adamson

Data sheets from National Semiconductor Corporation have been reprinted with permission and carry the following Life Support Policy statement.

LIFE SUPPORT POLICY

NATIONAL'S PRODUCTS ARE NOT AUTHORIZED FOR USE AS CRITICAL COMPONENTS IN LIFE SUPPORT DEVICES OR SYSTEMS WITHOUT THE EXPRESS WRITTEN APPROVAL OF THE PRESIDENT OF NATIONAL SEMICONDUCTOR CORPORATION. As used herein:

1. Life support devices or systems are devices or systems which, (a) are intended for surgical implant into the body, or (b) support or sustain life, and whose failure to perform, when properly used in accordance with instructions for use provided in the labeling, can be reasonably expected to result in a significant injury to the user.
2. A critical component is any component of a life support device or system whose failure to perform can be reasonably expected to cause the failure of the life support device or system, or to affect its safety or effectiveness.

National Semiconductor Corporation
2900 Semiconductor Drive
P.O. Box 58090
Santa Clara, CA 95052-8090
Tel: (408) 721-5000
TWX: (910) 339-9240

National Semiconductor GmbH
Westendstrasse 193-195
D-8000 Munchen 21
West Germany
Tel: (089) 5 70 95 01
Telex: 522772

NS Japan Ltd.
4-403 Ikebukuro,
Toshima-ku,
Tokyo 171, Japan
Tel: (03) 988-2131
FAX: 011-81-3-988-1700

National Semiconductor Hong Kong Ltd.
Southeast Asia Marketing
Austin Tower, 4th Floor
22-26A Austin Avenue
Tsimshatsui, Kowloon, H.K.
Tel: 3-7231290, 3-7243645
Cable: NSSEAMKTG
Telex: 52996 NSSEA HX

National Semicondutores Do Brasil Ltda.
Av. Brig. Faria Lima, 830
8 Andar
01452 Sao Paulo, SP. Brasil
Tel: (55/11) 212-5066
Telex: 391-1131931 NSBR BR

National Semiconductor (Australia) PTY, Ltd.
21/3 High Street
Bayswater, Victoria 3153
Australia
Tel: (03) 729-6333
Telex: AA32096

CHAPTER 1

Introduction to the Digital World

OBJECTIVES

After completing this chapter, you should be able to:

- ☐ Explain the purpose of this book.
- ☐ Recognize how the book is organized.
- ☐ Follow suggestions on how to use this book.
- ☐ Identify what you should know when you finish this book.
- ☐ Understand some important introductory facts about the digital world.
- ☐ Know a short history of the development of microprocessors and microcomputers.
- ☐ Distinguish between troubleshooting and instrumentation.

INTRODUCTION

Welcome! You are about to begin one of the most exciting journeys you have ever taken in your life. In using digital logic, you will discover the fundamental principles of computers, robots, automation, and all the many other new technology developments.

When you finish this book, you will know more about the foundations of the new technologies than did graduate engineers of the last generation. With this acceleration of knowledge, can you imagine what your children will be learning if they choose to follow your career choice? Perhaps you may be one of those who will contribute new knowledge or discover new ways of passing it on to others. In any case, your choice to learn this material will launch you into one of the most exciting and challenging careers imaginable. Your choice is a commendable one. Stick with it. Time goes by no matter what you are doing, so spend the time learning this material and be prepared for your future.

KEY TERMS

Chip (IC)
Digital
Analog
Microcomputer
Troubleshooting
Instrumentation

1–1 WHY THIS BOOK?

Discussion

This section presents a brief reason of why this book exists. You should know that before investing the amount of time in it that mastery of its subject will require. Here you will learn about very technical material in nontechnical terms.

Reasons Why

This book exists to prepare you for an eventual course in *microprocessor technology*. What this means is that this book is not an end in itself; it is a first stepping stone that begins a great journey into a vast and changing realm of knowledge. This book will help prepare you for the following courses:

MICROPROCESSORS
ROBOTICS
INDUSTRIAL AUTOMATION
COMPUTER MAINTENANCE AND REPAIR
DIGITAL COMMUNICATIONS
COMPUTER INTERFACING
BIOMEDICAL ELECTRONICS

This list is just a small example of the importance of the information contained in this book.

Historical Perspective

Not too long ago, books on this subject required an extensive knowledge of electronics circuits. Before starting, students had to know how transistors (or even vacuum tubes) worked. This, however, is no longer the case. In the "good old days" engineers and technicians had to construct their own logic circuits, where the design and troubleshooting of individual circuit components, such as the transistor, resistor, and other separate electronics devices, were required.

Today, this is no longer true. You do not need to know electronics to get started in this subject. You will need to know electronics if your career goal is the actual design or troubleshooting of digital systems. A knowledge of electronics is no longer required to start this subject because all of these once discrete electronic components are now placed inside a single object known as an integrated circuit and called a **chip** or **IC**.

Chip (IC):
An entire circuit manufactured on a single piece of electronics material. Some chips are no larger than an eraser on a pencil.

Some typical chips are shown in Figure 1–1.

As you will see later in this text, each of the digital logic circuits presented are contained in a chip. When you design or troubleshoot, you will be using chips. To start your introduction to this field of knowledge, it is only necessary to know what to expect from

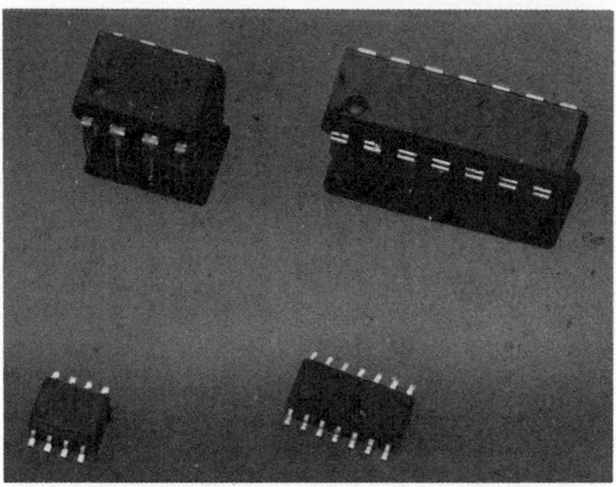

Figure 1–1 Typical chips. *Copyright* by Motorola, Inc. Used by permission.

the chip when you apply certain conditions to the chip. This cause and effect relationship does not require a knowledge of electronics. It does require a knowledge of the chip and digital logic. That is what you will learn here.

Conclusion

This section presented the reason for this book. Here you saw that the material contained in this text is to prepare you for future courses. Check your understanding of this section by trying the following review questions.

1–1 Review Questions

1. State the main purpose of this book.
2. Name three subject areas that require the information in this book.
3. Is it necessary to know electronics before starting this book? Explain.
4. What do you first need to know about digital ICs when beginning this book?

1–2 HOW THIS BOOK IS ORGANIZED

Discussion

The information in this section is very important. Since you will be spending many hours with this book, it is important that you make your time with it as efficient as possible. This book is organized to convey highly technical and complex information in the easiest possible manner. Understanding how the book is organized will help you take advantage of its many useful learning features.

Each Chapter

Every chapter starts with a list of *objectives* that lists those items you will learn from the chapter. This is followed by an *introduction* that explains the purpose of the chapter. Next, a list of *key terms* is presented. These are new words that are all defined in the chapter and contained in the glossary. All new terms are clearly defined immediately after the first time they are presented.

Each chapter is divided into *sections*. Each section starts with a *discussion* that gives you the general idea of what is in the section and ends with a *conclusion* that reviews what you should have gotten out of the section. Every section in the chapter presents *review questions* that contains questions primarily intended for classroom discussion of key topics.

All chapters contain three unique and important sections:

- Microprocessor application
- Electronic application
- Troubleshooting simulation

The *microprocessor application* sections of each chapter present material that will directly prepare you for a microprocessor course. In some cases, this section will present an application of the digital material presented in the chapter, while in other cases both types of information are covered.

The *electronic application* section requires some knowledge of electronics. This section presents the background information you will need in order to make electrical measurements and predictions in digital systems for the purpose of maintaining and repairing such systems.

The *troubleshooting simulation* section presents information about the interactive troubleshooting program for that chapter contained on the computer disk included with this book. The disk contains a unique interactive troubleshooting exercise that simulates real digital circuit problems. Here you have an opportunity to gain valuable troubleshooting experience that grows as you progress through each chapter. Because of the nature of these troubleshooting examples, the disk offers you literally thousands of different troubleshooting problems in a variety of digital circuits and systems.

Every chapter contains a *summary*. This is very useful to read just before a quiz or job interview. It summarizes the key points introduced in the chapter. They can also serve as an easy way to get an in-depth overview of the chapter material before starting the chapter.

At the end of each chapter is a *self-test*. This is an important learning and job preparation feature. It is an important learning feature because the answers are right there to give you immediate feedback. It is an important job preparation feature because the questions are divided into the different ways of asking written questions that you may encounter on a written exam for a job interview. It is good to go over all of these after completing this course and just before that first all-important job interview.

The *chapter problems* are all keyed to each section of the chapter. Thus, if you have problems answering a particular question, you may quickly refer to the section in the chapter that covered that particular material. In addition, these questions contain a unique *applications* section that asks you specific questions about important microprocessor applications, a *troubleshooting and instrumentation* section that addresses the electrical characteristics of digital devices, and an *analysis and design* section that invites you to use your new knowledge in an applied setting.

Laboratory Manual

A coordinated laboratory manual is available with this text.

Interactive Disk

The student disks contain interactive exercises that relate directly to the material presented in each chapter of this book. There are ten chapters, and ten major sections on the disks. Each disk section contains the following:

- Troubleshooting Simulation
  This contains a troubleshooting simulation program that allows you to interact with the computer. Here you have an opportunity to develop valuable troubleshooting skills that you can apply in a laboratory setting. These programs are divided into two modes. The first mode is the demonstration mode that shows how the program works and guides you in developing your troubleshooting skills. The second mode is the testing mode. Here the computer will place a problem in a simulated circuit for you to find. The program will confirm if your analysis is correct, give you a score and show you the correct analysis if needed.
- Multiple Choice Questions
  These are questions that are related directly to each chapter section. When you do these, the computer will tell you immediately if your answer is correct and refer you back to the specific chapter section of the book. These questions are excellent for checking your understanding of chapter material, preparing for a test, or to use with the book just before a job interview.
- Interactive Exercises
  The third part of each chapter section on your disk contains a unique interactive section. Here, a key feature of the chapter is automated in such a manner that you may interact with it. This allows you to observe the behavior of new concepts under conditions imposed by you. These interactive exercises are divided into two modes: the demo mode, which takes you on a guided tour of the simulation and then a testing mode. The testing mode gives you an opportunity to test your understanding in a variety of new situations.

Instructor's Disk

Your instructor is supplied with a special computer disk. This disk allows the instructor to check the progress you have made on your disk. Thus, when you are using the student disk that comes with your book, you are not working alone. What you do, how you respond and areas in which you experience difficulty are all revealed to the instructor through the use of the special instructor's disk. Through this system, your instructor is always there to give you guidance and direction. Some instructors may wish to assign certain areas of the disk as homework or quiz assignments. Others may require certain scores as a prerequisite to laboratory exercises. There are so many different ways this learning system may be used. Check with your instructor on how this will be used in your school situation.

Conclusion

This section presented the organization of the book. Here you saw the key features and why they were useful. Check your understanding of this section by trying the following review questions.

1–2 Review Questions

1. State what is contained in every *section* of each chapter.
2. What material is contained at the beginning of each chapter?
3. Name three key sections contained in each chapter. State their purposes.
4. How are the end-of-chapter problems arranged?

1–3 HOW TO USE THIS BOOK

Discussion

Traditionally only electronic majors took a course that required the knowledge presented in this text. However, this is no longer the case. Because of the use of the microprocessor and digital systems in almost every field of technology, from auto mechanics to building construction, more technology majors outside of the field of electronics require the information presented here. The auto mechanic, building contractor, health technologist, and technical programmer, as well as the electronics technician, are expected to have a background in this material.

This section will give you some suggestions for the different ways you may use this book in a class environment or on your own. Keep in mind that your instructor may select an arrangement of material that is not suggested in this section. The suggestions here are merely to serve as a rough guide to the versatility of this book.

Traditional Method

The traditional method of presenting digital logic is to require that the student have completed a course in basic DC/AC electrical circuits and be currently enrolled in an electronic devices course. Normally these more traditional courses do not introduce any microprocessor terminology until the end of the digital course or until the beginning of the microprocessor course. Using the book in this method would require that the electronic application sections of each chapter be covered and the microprocessor application sections be deferred until the end of the course or used in the first part of the microprocessor course. A concurrent digital lab is offered and the experiments may be followed in any order suggested by your instructor.

Electronics Technology Major—No Electronics Background

This book may be used as a first course in electronics technology. It is suggested that the student also be concurrently enrolled in a traditional DC/AC circuits course. The electronic application sections may be deferred until later, depending upon the pace of the DC/AC course. It is suggested that the microprocessor applications section be covered in each chapter. The reason is that microprocessor courses continue to become more advanced. Many of the newer texts that present the 16- and 32-bit machines assume at least a fundamental understanding of the basic processes inherent in a microprocessor. Presenting these sections concurrently with the more traditional digital material will give the student a tremendous advantage in the microprocessor course.

The digital laboratory should be offered concurrently with this approach. It is suggested that the lab exercises be taken in sequence, since the first few do not require a prior knowledge of electronic equipment.

Computer Science Major—Nonelectronics Majors

This book may be used by nonelectronics majors, such as computer science students, who require a knowledge of the hardware and logic circuits that are used in microprocessor-based systems. In this case, all sections are presented except the troubleshooting and instrumentation sections.

Special emphasis should be placed on the microprocessor applications section of each chapter. Usually in this setting, the digital laboratory is omitted.

Industrial Training

The microprocessor sections should be covered if the training will eventually lead to microprocessor-based systems. Students in this category may not have a laboratory situation available. Under these conditions, it is suggested that class demonstrations be used.

Conclusion

The important point of this section was that the material in this book is now required by almost every area of technology. This book is designed to serve you in ways not offered by more traditional textbooks on the subject. It will not only serve you well in the classroom but also as a useful review and reference before that important first job interview. Check your knowledge of this section by trying the following review questions.

1–3 Review Questions

1. State the change that has taken place concerning people who need to know the information in this book.
2. State the purpose of the electronic application section.
3. State the purpose of the microprocessor application section.

1–4 WHAT YOU WILL KNOW

Discussion

This section will give you an idea of what you will know when you finish this book. In fairness to yourself, you should have an idea of what you will get in return for the time and effort you are willing to invest.

Building Blocks of Automation

Computers, robots, and other such systems are made of essentially the same kind of computing components. For example, a satellite orbiting a distant planet uses the same kind of computing circuits as does your home computer or an automated production line. This book explains how these computing parts work. Consider these computing parts as the building blocks of such systems.

As you will see, thousands of these building blocks can be produced inexpensively in a chip. How to identify these chips and the significance of the interconnections between them is contained in this book. The opportunity for you to learn how to measure their conditions and isolate the cause of improper system operation is available as well.

Analysis

Sections in this book contain information concerning the relationship between symbolic figures and the actual building block itself. You will have the opportunity to learn ways of simplifying the construction of the interconnections between these building blocks. Here, the analytical tools, invented by many people and tested over several lifetimes of work are available for you to adopt as your own in a matter of a few hours.

Real Things

It is one thing to analyze these building blocks on paper, but, to a technician, it is equally important to analyze real building blocks to ensure that they perform according to their intended purposes. In this book there is the opportunity to learn about the real building blocks and their limitations.

If you require an understanding of the electronics of these building blocks, then that is available in the electronic application sections. If you do not require this background, then these sections may be omitted without any loss of continuity. If you do require an electrical understanding of these devices, then chances are you will be taking a laboratory and using the coordinated laboratory manual available with this text. In this situation, you will have the opportunity to gain valuable experience in the actual construction, measurement, and design of these building blocks.

Preparation for Future Courses

In a standard sequence of courses for electronics technology or a related specialty, this book serves as the major text for the prerequisite course for microprocessor technology, robotics, computer maintenance and repair, or similar studies. Because of this, the microprocessor application section in every chapter will help you build that important foundation necessary for those future courses.

Employment Preparation

If one of the motivating factors for your willingness to invest the time and effort required to learn this material is to eventually obtain gainful employment, then you will find that many of the learning features in this book will help prepare you for that important first job interview. The chapter summaries are handy to skim over just before the interview. The chapter self-test questions contained at the end of each chapter present an opportunity for you to develop test-taking skills in different kinds of written question formats. As you will see, this book is a good reference for use on the job as well as in future courses.

Conclusion

This section presented the knowledge and skills you will have the opportunity to acquire as you progress through this book. This information was presented in a nontechnical manner so that, no matter what your background before starting this material, you would have a good idea of the benefits available to you. Check your understanding of this section by trying the following review questions.

1–4 Review Questions

1. State what a computer, robot, and automated assembly line have in common in relation to the subject of this book.
2. Explain what skills you would have the opportunity to acquire as an electronics major using this book in a classroom setting.
3. Can this book help for future courses? future employment? Explain.

1–5 THE WORLD OF DIGITAL

Discussion

This section presents some important introductory concepts. These concepts focus on the relationship between the *discrete* and the *continuous*. You can think of the discrete as a set of dots on a sheet of paper, and the continuous as a line connecting these dots.

Comparing the Two

An ON–OFF switch for your living room light is an example of a discrete operation. The light can be in only one of two possible states (conditions), ON or OFF; no other condition is allowed.

A dimmer switch connected to your living room light is an example of the continuous. There is a continuous range of brightness presenting countless possible states (conditions). As a matter of fact, you may think of a dimmer switch as the combining of digital and analog. The ON/OFF action of the dimmer is the digital part while the continuous part is the changing for different light levels.

Concerning the subject of this book, operations that are discrete (have only two conditions) are referred to as **digital,** while operations that are continuous (have countless possible states) are referred to as **analog.**

Digital:
In the context of this book, refers to devices that have discrete conditions.

Analog:
In the context of this book, refers to devices that have a continuous range of possible conditions.

As an example, consider a new car. You control the degree of pressure (an analog quantity) you put on the accelerator—but the reading on your speedometer may be digital; that is, it indicates discrete values of the car's speed, not continuous ones. A digital speedometer may show, for example, two fives to indicate a speed of 55 miles per hour instead of a needle changing its position to indicate an increase/decrease in speed.

Analog Control

Many modern buildings contain automated interior lighting systems that attempt to use outside lighting inside the building, thus reducing the cost of artificial lighting. Such a system measures the amount of indoor light and adjusts the amount of artificial light ac-

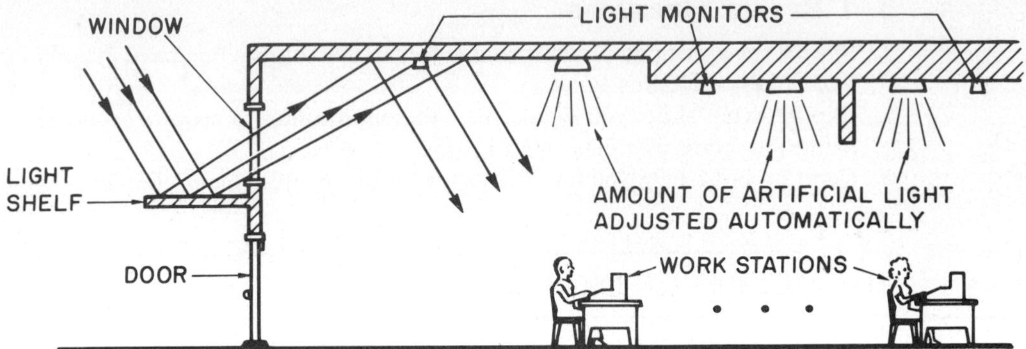

Figure 1–2 Example of analog control.

cordingly. This is an example of the *analog control* of an automated system. See Figure 1–2. Other examples of analog would be a thermostat adjustment, or using a ruler or adjusting the focus on your camera.

Digital Control

The same modern building with its automatic light control has a security system that can detect when a door is open or closed. This is an example of a *digital control*. See Figure 1–3. Other examples of digital would be your car's odometer, a television channel selector, or adjusting the shutter speed on your camera.

The Revolution

Figure 1–4 compares the historical analog object with its digital counterpart. This figure dramatically illustrates the influence of the discrete on our prior dependency of the continuous.

The Computer

There are two major classes of computers. One is the digital computer and the other the analog computer. The most popular class of computers today is the digital computer. The digital computer produces exact answers (discrete values) while the analog computer gives approximate ones. There are special circuits that will take analog information and convert it to digital. Not surprisingly these circuits are called *analog-to-digital converters* and are

Figure 1–3 Example of digital control.

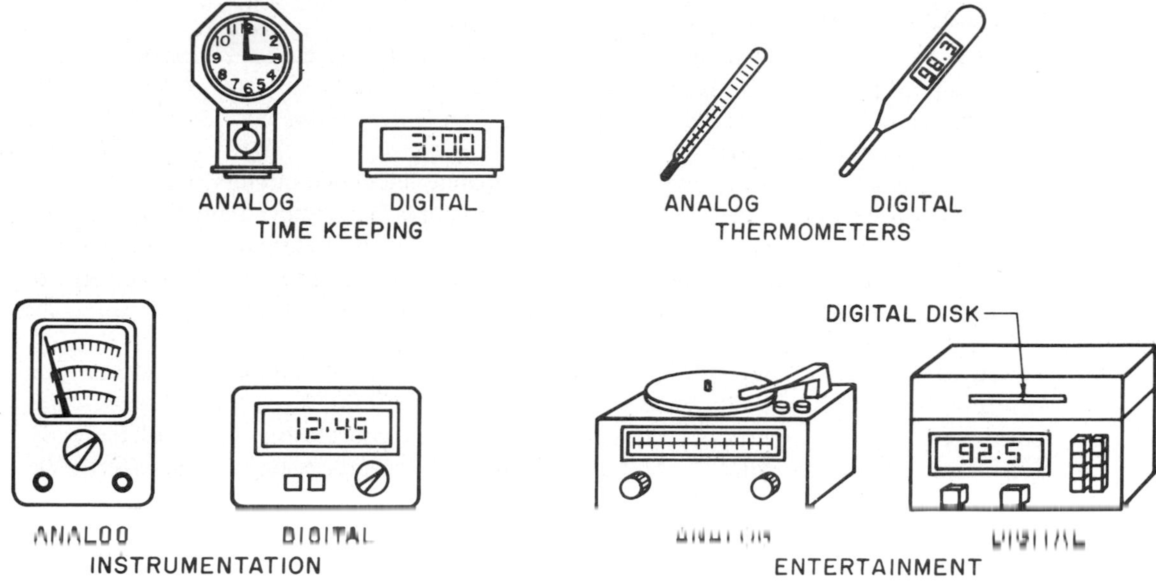

Figure 1–4 The influence of digital.

present in this book. These circuits, for example, would be used to convert a patient's blood pressure reading into digital information. Thus, along with other information about the patient, the digital computer may be used as a monitoring and diagnostic tool. Conversely, there are circuits that convert digital information into analog. As you probably suspect these circuits are called *digital-to-analog converters*. An example of this would be the adjustment of the air pressure in a space station or the temperature in a green house.

Conclusion

This section introduced you to the difference between *digital* and *analog*. Here you saw some examples of the two and an application. In the next section, you will have the opportunity to learn about the exciting world of microprocessors. For now, check your understanding of this section by trying the following review questions.

1–5 Review Questions

1. Give an example of a *discrete* operation and a *continuous* operation.
2. Define *digital* and *analog*.
3. Give an example of a digital device and an analog device.

MICROPROCESSOR APPLICATION

Discussion

This section gives you some interesting background information about microprocessors. A formal definition of a microprocessor will be presented in a later chapter. For now, this background information will set the stage for this remarkable device.

The Beginning

In 1969, the Japanese calculator company Busicom asked the American company Intel to design a portable calculator that would consist of about 50,000 transistors. A former Stanford University research assistant and physicist, Ted Hoff, headed up the design team at Intel to work on this project.

The original Busicom design required 12 separate integrated circuits, which Hoff rejected as too expensive. His approach was to use one integrated circuit that relied on software (a built-in program to cause switching between the transistors to take place in such a way that one circuit could replicate several different circuits). This concept of having small circuits perform many different processes produced the name *micro* (small) *processor*.

In 1970 Federico Faggin designed the 4004 microprocessor chip in nine months. The consumer cost of this revolutionary chip was around $200.00. Even though this chip is crude by today's standards, its computational powers came close to those of the early room-size ENIAC computer! Today, you can buy a microprocessor that far exceeds the capabilities of the original and costs less than a six-pack of soda pop.

Taking the Ball and Running

It's ironic that one of the first users of the microprocessor, the calculator company Busicom, is no longer in business. However, many other companies followed Intel's example and started producing their own version of the chip. However it wasn't until 1975 that a computer was developed that used the microprocessor. This achievement was through the efforts of Ed Roberts, who produced the first **microcomputer.**

Microcomputer:
A computer that uses a microprocessor as its processing unit.

The first microcomputer produced by Roberts was the Altair 8008. Roberts manufactured these computers in his company Micro Instrumentation Telemetry Systems (MITS). They sold in kit form for slightly less than $400.00 and were purchased primarily by a small core of young computer hobbyists.

Around this time, a young college student, Steve Jobs, left his one-semester attempt at Reed College in Oregon to work for Nolan Bushnell, who founded a then small video game company called Atari. This led Jobs to frequent the meetings of computer hobbyists, who used Ed Roberts' Altair 8008 as well as their own designs. At the meeting of a group who called themselves the Home Brew Computer Club, Jobs met a technician employed by the Hewlett-Packard computer company by the name of Steve Wozniak.

Up to this time, the public had an image of computers as giant machines that possessed magical abilities and could only be operated by a very special and gifted group of people called programmers. The practice was that no one could learn programming unless they were well versed in mathematics and had a solid background in the construction of the computer. In 1976, Steve Jobs convinced Wozniak (Woz) to leave Hewlett-Packard, and with $1 300.00 cash between them they developed the first computer specifically targeted for the general public—a venture that no established computer company would dare take. As founders of Apple Computer®, these two young men turned this idea into sales of over

$139,000,000.00 in just three years after the first Apple was sold. At this time, the larger, established computer companies started making their own personal computers.

What's Happening Now

Today the chip and its associated technology have reached into almost every aspect of human enterprise. These devices are now contained in almost all home appliances, from dishwashers and television sets to your personal telephone. They are used by law inforcement agencies, the medical profession, educational institutions, government, and entertainment, as well as the more traditional use in microcomputers, robots, and automated control systems. It wasn't long ago that the list of human activities that did not use this technology was much longer than those that did. Today, the opposite is the case.

Figure 1–5 shows a glove that interacts with a computer. This is part of a system that produces another world for the user in conjunction with a head-mounted monitor with a position and orientation sensor. Working through a computer, this system keeps track of hand and finger movements and, with a microphone wired for voice recognition, causes the user to be completely immersed in a new environment. The user issues instructions to the computer by pointing, talking, gesturing, and actually handling graphics images. Imagine playing a video game or learning how to repair a complex robot using such a system.

The Future

As you will discover in this text, the computers of today perform their tasks by a *sequence* of processes. The trend is to perform many processes at the same time. This is called *parallel processing*. A parallel computer, called the Connection Machine, is capable of

Figure 1–5 Interface for advanced computing. *Courtesy* of Jon Brenneis in *Scientific American.*

Figure 1–6 A parallel processing board. *Courtesy* of Jon Brenneis in *Scientific American*.

performing 65,536 processes at once. In Figure 1–6 each of the 64 integrated circuits is a processor that has roughly the power of a VAX 11/750 minicomputer.

One of the major advantages of the computer over that of the human brain is speed. However, even with this, traditional computers can take several hours to recognize the face of a friend (or enemy), a task that the human brain can do very quickly. Even though individual brain cells operate at a much lower speed than computer circuits, the brain is capable of processing thousands of tasks at the same time. Thus, when your brain processes an image it does it all at once, while for the traditional computer it does it one part at a time, like looking through a keyhole. The parallel computing process may change that.

Conclusion

This section presented some important background information on the microprocessor. Here you had an opportunity to see its beginnings, where it is now, and where it may be in the future. Check yourself out on this section by trying the following review questions.

1–6 Review Questions

1. State the main event that prompted the design of a microprocessor.
2. Name the people who designed the first microprocessor.
3. Who produced the first commercial microcomputers?
4. What was the philosophy of Steve Jobs concerning the marketing of microcomputers?
5. State one of the main differences between today's personal computers and the advanced machines now being built.

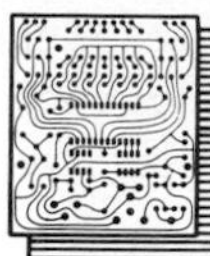

ELECTRONIC APPLICATION

Discussion

This section gives an introduction to the concepts of troubleshooting and instrumentation. These sections usually require an understanding of basic electrical circuits. In most of the chapters in which they appear, they offer valuable background information that is required for the repair and maintance of these systems.

Troubleshooting

The area of **troubleshooting** requires an understanding of the major causes of improper operation of electrical systems.

★ **Troubleshooting:**
The act of locating and repairing a fault that causes improper operation in an electronic circuit.

Problems may be categorized in the following areas.

- *Complete Failures*
  This is usually the easiest kind of problem to correct. A complete failure means that the entire electrical system is inoperative. A complete failure may occur because the power cord to the equipment is not connected to the wall outlet, or the ON-OFF switch may be faulty, or the electrical power supply within the system is defective.
- *Poor System Performance*
  This area can be more difficult to analyze. Poor performance means that the system is not operating within the performance limits of its original design. Thus a computer may not perform all of its functions properly, just some of them. Poor system performance may be due to a weak component within the system, or a particular section of the system may be faulty. A knowledge of system operation and proper instrumentation is necessary to locate errors due to poor system performance.
- *Tampered Equipment*
  An old saying among technicians is, "If it works, don't fix it!" Tampered equipment is equipment that has been incorrectly changed for the purpose of experiment, modification, or attempted repairs. This can leave the system inoperative or performing poorly.
- *Intermittent Fault*
  This problem is usually the most difficult to repair. An intermittent is an inconsistent fault that causes the system to be inoperative or perform poorly. The difficulty of finding the cause of an intermittent is in keeping the problem consistent. Intermittents can be caused by mechanical defects, temperature changes, or erratic electrical behavior of components and connections.
- *Massive Traumas*
  These system failures may cause more than one part of the system to fail. A massive trauma is a system failure caused by an outside intrusion into the system that results in an inoperative system, poor system performance, or intermittents. Examples of

massive traumas are fire, smoke, dropped equipment, water immersion, lightning damage, and applications of voltages with incorrect polarities or values. A common massive trauma with personal computers is the accidental spilling of liquids such as coffee or soda pop on the system.

Instrumentation

Instrumentation involves the action of making, recording, and interpreting measured information.

Instrumentation:
The use of equipment to measure electrical and logical quantities for the purpose of keeping these quantities within prescribed limits to ensure proper system operation.

Instrumentation involves the following processes:

- *A decision is made to make the measurement*
  This decision may be made after the technician determines that a system fault exists. This determination usually results from experience with the system to be measured.
- *A measurement procedure is selected*
  The procedure may consist of making sure that all system operating controls have been properly set, or it may be a good visual inspection. The technician may "feel" for the presence or absence of heat, based on what the technician has experienced as normal. The technician's sense of smell can detect the odor of a burning resistor or disk drive motor. However you must be cautious of intentionally inhailing fumes from faulty electrical equipment. These fumes may be harmful.
- *The measurement is conducted*
  Here the actual measurement is made. It is important to record what the measurement indicated. There is another old saying: "The lightest pencil mark is better than the world's best memory." There are usually good reasons for old sayings.
- *The data are analyzed*
  The process of analyzing the data consists of comparing the recorded measurement to some standard. This step is very important. From previous measurements, experienced technicians usually know what the standard should be. If the equipment being measured comes with complete documentation, then electrical values printed on schematics or measurement tables are used as the standard. If none of these are available, then making the same measurement on an identical working system, if available, will yield a standard of reference. Without any of these standards, the technician may have to perform a mathematical or logical analysis.
- *Act on the analysis*
  If a discrepancy exists between the measured data and the standard, then some action is required, such as troubleshooting to correct the fault, reporting the discrepancy to others, or simply giving the system to someone else for final repair.

Conclusion

This section presented an overview of the major areas concerning troubleshooting and instrumentation. The technician must understand the various classes of system performance that require attention and the different approaches to determining a course of action. Check your understanding of this section by trying the following review questions.

1–7 Review Questions

1. How would you define troubleshooting? instrumentation?
2. What is the easiest category to troubleshoot? What is the most difficult?
3. Before making a measurement, what must you be sure of?
4. State a good habit when making measurements.

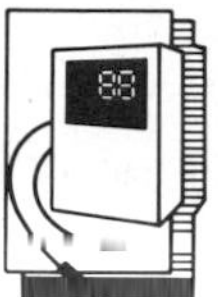

TROUBLESHOOTING SIMULATION

Discussion

The troubleshooting simulation program for this chapter introduces the use of the *logic probe*. Here you will have an opportunity to develop basic troubleshooting skills in finding the location of an open. The program is self-explanatory and allows you to see the circuit under normal operating conditions in order to gain experience in using the logic probe. Then, when you're ready, the random number generator inside the computer will place a "bug" in the circuit. It will be up to you to find where it is. Once you complete your analysis, the computer will present a possible list of faulty conditions. If your analysis is incorrect, the computer will show you where the problem is. If your analysis is correct, your score is recorded and the computer will enter another problem for you.

This troubleshooting simulation program is referred to as a *smart system*. This means that it simulates the behavior of a master technician in its analysis of the trouble. This is a great learning opportunity for you, since you are simulating sitting elbow to elbow with a master troubleshooter!

SUMMARY

- The main purpose of this book is to prepare you for a course in microprocessor technology.
- Material presented in this book does not *require* a background in electronics.
- Most of the circuits used for digital systems are constructed with integrated circuits.
- Every chapter has an introduction to give you an overview of the chapter, a set of objectives, and key terms.
- Each chapter of this book is divided into sections. Each section starts with a discussion, and ends with a conclusion and some section review questions.
- All new terms introduced in the book are defined immediately after they are presented.
- All chapters in this book contain a section on microprocessor application and electrical application.
- Every chapter ends with a summary, chapter self-test, and problems.

- All chapter problems are divided according to the section where the information was presented. These chapter problems also contain a microprocessor applications section as well as troubleshooting and instrumentation and analysis and design sections.
- A coordinated laboratory manual is available with this text.
- The material contained in this text is now required by many different technology disciplines.
- It isn't necessary to have completed a traditional DC/AC circuits course, but for those who need the troubleshooting and instrumentation sections with this book, such a course should be taken concurrently.
- There are several approaches as to how the material in this text may be presented in a classroom setting, depending upon the overall program goal and technology discipline.
- This text offers you the opportunity to know the fundamental principles involved in computer-type logic circuits.
- This book presents methods of analyzing digital or two-state devices as opposed to analog or variable-state devices.
- There are many applications for analog as well as digital control of automated systems.
- The first microprocessor was designed to be used in a Japanese calculator in 1970.
- The first computer to use the microprocessor was the Altair 8008, manufactured by MITS in 1975.
- The first successful microcomputer made available for sale to the general public was produced by Steven P. Jobs and Stephen Wozniak, who developed and marketed the Apple Computer®.
- The microprocessor chip is currently used in a variety of system applications from music to law enforcement as well as the more traditional applications of computers, robotics, and automation.
- The trend in the development of automated digital systems is toward *parallel* rather than *sequential* processing of information.
- Troubleshooting requires an understanding of the major causes of improper operation of electrical systems.
- Complete system failure is usually easier to repair than an intermittent fault.
- Instrumentation involves the action of making, recording, and interpreting measured information.
- One of the important items in making measurements is to record them so that they do not have to be made again.

CHAPTER SELF-TEST

I. TRUE/FALSE

Answer the following questions true or false.

1. This book is designed to prepare you for a course in microprocessor technology.

2. In the past, books on this subject required an extensive background in basic electronics.
3. An extensive background in electronics is not necessary for this text because no mathematics is used or required.
4. Each chapter represents the smallest unit of learning in the design of this book.
5. Each chapter in this text includes a microprocessor applications section.

II. MULTIPLE CHOICE

Answer the following questions by selecting the most correct answer.

6. One of the major sections contained in every chapter of this book is
   (A) electronic application.
   (B) advanced mathematical analysis.
   (C) laboratory experiments.
   (D) none of the above.
7. The best way to use this book
   (A) is to cover every section in the sequence presented.
   (B) is to select only those topics that interest you.
   (C) is to ensure that you have an electronics background before starting.
   (D) depends upon the goals of the program within which the book is used.
8. The section(s) in each chapter that may require some knowledge of electronics are the
   (A) microprocessor applications section.
   (B) electronic application section.
   (C) sections on logic analysis and simplifying.
   (D) All of the above are correct.
9. In which subject(s) would your major most likely be if you were using this book in a classroom setting without a concurrent laboratory?
   (A) electronics technology
   (B) industrial automation technology
   (C) computer science
   (D) None of the above are correct.
10. The purpose of the microprocessor application section is to
    (A) teach you how to program a computer.
    (B) help you learn enough about them so you can omit a microprocessor course.
    (C) prepare you for your first microprocessor course.
    (D) have a section that can be used for lab experiments.

III. MATCHING

Match the phrase on the right to its corresponding phrase on the left.

11. Complete failure	(A) Spilled coffee
12. Poor system performance	(B) Unplugged computer
13. Tampered equipment	(C) User-modified model
14. Intermittent fault	(D) Keyboard bad sometimes
15. Massive trauma	(E) None of these

IV. FILL-IN

Fill in the blanks with the most correct answer(s).

16. The word ______ refers to devices that have discrete conditions.
17. The word ______ refers to devices that have a continuous range of possible conditions.
18. An ON–OFF light switch is an example of ______ control.
19. A light dimmer switch is an example of ______ control.
20. The first microprocessor chip was designed to be used in a/an ______.

V. OPEN-ENDED

Answer the following questions as indicated.

21. Who developed the first computer to use a microprocessor? What was the intended market for this computer?
22. Define troubleshooting and instrumentation.
23. State the difference between *sequential* processing and *parallel* processing. Which method resembles the mechanism used by the human brain?
24. State the first three steps in instrumentation.
25. State the five areas of troubleshooting.

Answers to Chapter Self-Test

1] T 2] T 3] F 4] F 5] T 6] A 7] D 8] B 9] C 10] C
11] B 12] A, C 13] C 14] D 15] A 16] digital 17] analog
18] digital 19] analog 20] calculator 21] Ed Roberts, computer hobbyist.
22] Troubleshooting is the act of locating and repairing in an electrical circuit a fault that causes improper operation of the circuit. Instrumentation is the use of equipment to measure electrical and logical quantities for the purpose of keeping these quantities within prescribed limits to ensure proper system operation.
23] Sequential processing is doing one process at a time. Parallel processing is doing several processes at the same time. Parallel processing resembles the mechanism used by the human brain.
24] The first three steps in instrumentation are deciding to make the measurement, selecting a measurement procedure, conducting the measurement.
25] Complete failure. Poor system performance. Tampered equipment. Intermittent fault. Massive traumas.

CHAPTER PROBLEMS

Basic Concepts

Section 1–1

1. State the primary purpose for this book.
2. Explain why a background in electronics is not necessary in order to *start* using this book.
3. What is a chip?

Section 1–2

4. State how each chapter in this book is organized.
5. What is a chapter section? How are these sections organized?
6. What is the purpose of the Key Terms at the beginning of each chapter?
7. Give some of the uses of the chapter summary and self-test.
8. State the purposes of the microprocessor applications section and the electrical application section.
9. Explain how the end-of-chapter problems are arranged.
10. State the purpose of the troubleshooting simulation section of each chapter.

Section 1–3

11. State the suggested method of using this book by an electronics major with no electronics background.
12. What is the suggested method of using this book by a nonelectronics major, such as a computer science student?
13. When should a concurrent laboratory be taken with this book? Explain.
14. When should the microprocessor applications sections be emphasized? Explain.

Section 1–4

15. State what computers, robots, and other such systems have in common.
16. What is the advantage of using a chip?
17. What does this book offer in terms of analytical tools?
18. What sections of this text present the electrical characteristics of the chips used in digital?
19. List some of the future courses this book will help prepare you for.
20. Name some of the features of this book that can directly help you on your first job interview.

Section 1–5

21. Give an example of a controlling device that has only two possible conditions.
22. Give an example of a controlling device that has a range of possible conditions.
23. What is the difference between a discrete operation and a continuous operation?
24. Define *digital* as used in this book. Also define *analog*.
25. Give an example of the analog control of a system.
26. Give an example of the digital control of a system.
27. List some of the items that used to be analog and are now available in digital.

Applications

The questions in this section refer to the microprocessor application section of the chapter.

Section 1–6

28. What was the main reason for the design of the first microprocessor chip?
29. Name those involved in the design of the first microprocessor chip.
30. What computer was the first to use a microprocessor? Name its primary market.
31. State the major difference in the marketing strategy for the computer developed by Jobs and Wozniak.

32. Explain the differences and advantages of parallel processing over that of sequential processing.

Troubleshooting and Instrumentation

The questions in this section refer to the electronic application section of the chapter.

Section 1–7

33. Explain what the word *troubleshooting* means as used in this text.
34. List the five major areas that may account for poor system performance.
35. During troubleshooting, what type of problem is usually the easiest to repair? the most difficult? Explain.
36. Explain what the word *instrumentation* means as used in this text.
37. List the five major processes used in instrumentation.

Analysis and Design

Questions in this section concern themselves with using the technical information presented in this chapter in a setting that requires careful analysis or asks for a creative design. Since this chapter did not contain any technical information, no questions are presented.

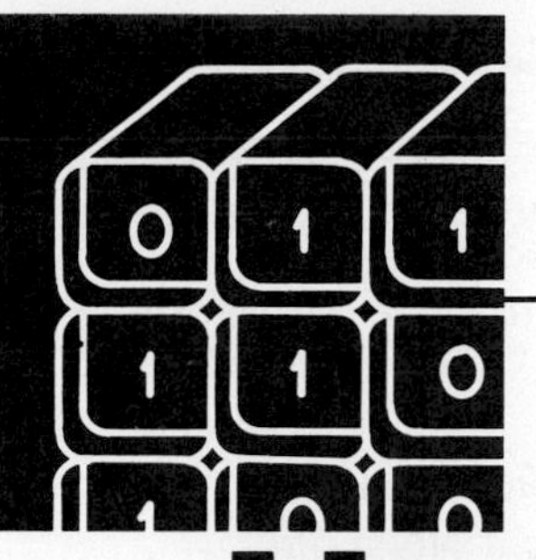

CHAPTER 2

Numbers Used by Microprocessors

OBJECTIVES

After completing this chapter, you should be able to:

- ☐ Understand the meaning of the base of a number system.
- ☐ Use binary numbers to represent any number.
- ☐ Explain the methods of converting between binary and decimal numbers.
- ☐ Understand the purpose of binary-coded decimal (BCD).
- ☐ Convert between BCD and decimal numbers.
- ☐ Understand the reason for a 16-symbol number system.
- ☐ Convert between hexadecimal and binary numbers.
- ☐ Convert between hexadecimal and decimal numbers.
- ☐ Use codes to represent other symbols.
- ☐ Understand the organization method used by the computer to store information.
- ☐ Understand the use of hardware to represent 1's and 0's.
- ☐ Understand how digital instruments are used to create and test for 1's and 0's.

INTRODUCTION

This chapter sets the foundation for the rest of the book. The very essence of digital logic is the idea of representing information by one of two conditions: ON or OFF. The reason for this is to allow the use of an inexpensive and reliable method for developing automated systems, such as the computer or modern assembly line. You will discover new ways of representing numbers. While doing so, you will begin to unlock the mysteries of the digital computer. Here, in this chapter, are its most fundamental secrets.

KEY TERMS

Base
Positional Number Systems
Place Value
Powers of 10
Binary Numbers
Bit
Base Subscript
Least Significant Bit (LSB)
Most Significant Bit (MSB)
Word
Nibble
Byte
Light Emitting Diode (LED)
Keypad
8421 Code
Hexadecimal Number System
Alphanumeric Codes
American Standard Code for Information Interchange (ASCII)
Complement
One's Complement Notation
Two's Complement Notation
Address
Data
Memory Size
High-Order Nibble
Low-Order Nibble
Parity
Odd Parity
Even Parity
Parity Bit

2–1 YOUR EVERYDAY NUMBER SYSTEM

Discussion

The numbers you use to balance your checkbook, write down a phone number, or address a letter are so familiar that they seem to be the only "right" way of doing things. In this section, you will have the opportunity to take a closer look at these "everyday" numbers. The reason why you are doing this is to develop a terminology with a familiar set of numbers that can be used to explain a not-so-familiar set of numbers.

How Many Symbols

The "everyday" number system contains 10 different symbols:

0 1 2 3 4 5 6 7 8 9

The 10 symbols can be used to represent any number:

12   4 680   932 569 042 567

With the use of a symbol called the decimal point, very small numbers can be represented:

0.12   0.002 3   0.000 452 068 608

These two methods can be combined to show precise values:

12.03 4 689.05 12 168.034 028

Because the number system has 10 symbols, it is said to be to the **base** 10 and is called the decimal number system.

Base:
Number of characters used in each position of a number system. Sometimes called the radix.

It surprises many people that there are number systems that use more than 10 symbols. Our everyday number system is called a **positional number system.**

Positional Number Systems:
A systematic method for representing values, where any value may be represented as a sequence of multiples of successive powers of a given base. For example, in decimal notation:

$$3241 = 3 \times 10^3 + 2 \times 10^2 + 4 \times 10^1 + 1 \times 10^0$$

All of these other number systems can also represent numbers as large or as small as you please. Believe it or not, the computer—deep down inside where outsiders can't see what is going on—uses a number system to the base 2! That's only two symbols.

Place Value

Taking a closer look at the familiar base 10 number system will help you learn some terminology and techniques for learning number systems that use a different base. Recall that you can represent any value number with only 10 symbols in the base 10 number system. The reason why you can do that is because of the concept of **place value.**

Place Value:
The representation of quantities by a positional value system. For example, the symbol 3 represents different values, depending upon its placement with respect to the decimal point: 3.0, 30.0, 0.3.

The symbols used in the decimal number system are effectively given different values depending on their location with respect to the decimal point. Each position to the left or to the right represents a change in value by a factor of 10 (this is sometimes referred to as the weight of the number). This concept is illustrated in Figure 2–1. The following example illustrates.

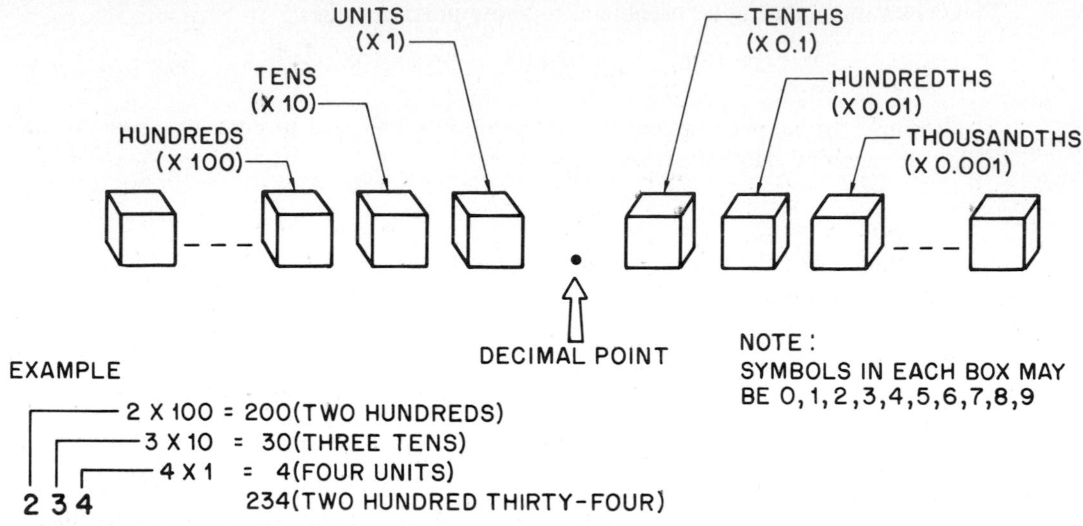

Figure 2–1 Positional value representation of decimal numbers.

Example 2–1

State the place value for each digit of the following decimal numbers:

(A) 42 (B) 0.31 (C) 2 013.045

Solution

(A) 4 = Tens place, 2 = Units place.

(B) 3 = Tenths place, 1 = Hundredths place.

(C) 2 = Thousands place, 0 = Hundreds place, 1 = Tens place, 3 = Units place, 0 = Tenths place, 4 = Hundredths place, 5 = Thousandths place

Representing the Base

Another method for representing a decimal number is by using **powers of 10** notation.

Powers of 10:

The power of 10 is expressed as an exponent of the base 10. The exponent indicates the number of decimal places to the left or right of the decimal place.

Powers of 10 notation is an abbreviated way of representing large or small numbers. For large numbers:

$$10^2 = 10 \times 10 = 100$$
$$10^3 = 10 \times 10 \times 10 = 1\,000$$
$$10^4 = 10 \times 10 \times 10 \times 10 = 10\,000$$

For small numbers:

$$10^{-2} = \frac{1}{10^2} = \frac{1}{10 \times 10} = \frac{1}{100} = 0.01$$
$$10^{-3} = \frac{1}{10^3} = \frac{1}{10 \times 10 \times 10} = \frac{1}{1\,000} = 0.001$$
$$10^{-4} = \frac{1}{10^4} = \frac{1}{10 \times 10 \times 10 \times 10} = \frac{1}{10\,000} = 0.000\,1$$

Thus, the place value of each decimal number can be represented as a power of its base. You should note that, by definition, $10^0 = 1$. This concept is illustrated in Figure 2–2.

Thus, the value of any number of a given number system can be represented as the sum of the positional values. In general terms,

$$S_X \times B^N + \cdots + S_2 \times B^1 + S_1 \times B^0 + S_1 \times B^{-1} + S_2 \times B^{-2} + \cdots + S_X \times B^{-N}$$

where: S_X = value of symbol
B = base of the number system
N = exponent of the place value

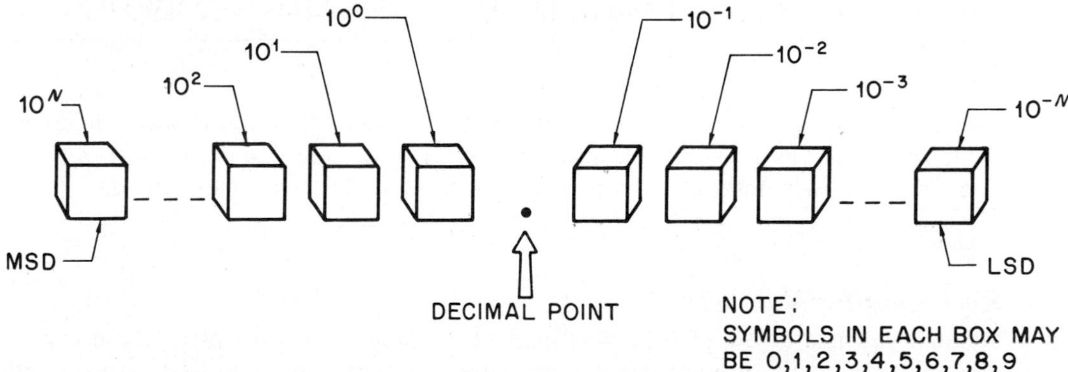

Figure 2–2 Representation of positional value using the base of the number.

Example 2–2

Represent the following decimal numbers, using powers of 10 notation:

(A) 12 (B) 340 (C) 1 450.025

Solution

(A) $12 = 1 \times 10^1 + 2 \times 10^0$

(B) $340 = 3 \times 10^2 + 4 \times 10^1 + 0 \times 10^0$

(C) $1\ 450.025 = 1 \times 10^3 + 4 \times 10^2 + 5 \times 10^1 + 0 \times 10^0 + 0 \times 10^{-1} + 2 \times 10^{-2} + 5 \times 10^{-3}$

Conclusion

This section presented a formal view of the "everyday" number system, the decimal system. You discovered other ways of representing this number system. The skills you learned here can now be applied to the analysis of other number systems. Test your understanding of this section by trying the following review questions.

2–1 Review Questions

1. How many symbols does the "everyday" number system contain?
2. State the base of our "everyday" number system.
3. Give the meaning of the term *place value*. State an example.
4. Represent 6 258.2 in powers of 10 notation.

2–2 THE NUMBER SYSTEM OF DIGITAL—BINARY NUMBERS

Discussion

This section shows you how to represent numbers in exactly the same way a microcomputer does. The "language" of all microprocessors is the language of ON and OFF. Here you will see how numbers can be represented by using this two-state method.

Representing Numbers

Assume you had a row of four cups marked as shown in Figure 2–3. These cups can represent part of the insides of a computer because they can be viewed as being either ON or OFF. A toothpick will be put in a cup to indicate it is ON. Using the value marked on each cup, you can represent different numbers. This is demonstrated by the following example.

Figure 2–3 Four marked cups.

Example 2–3

Determine the values represented by each set of cups in Figure 2–4.

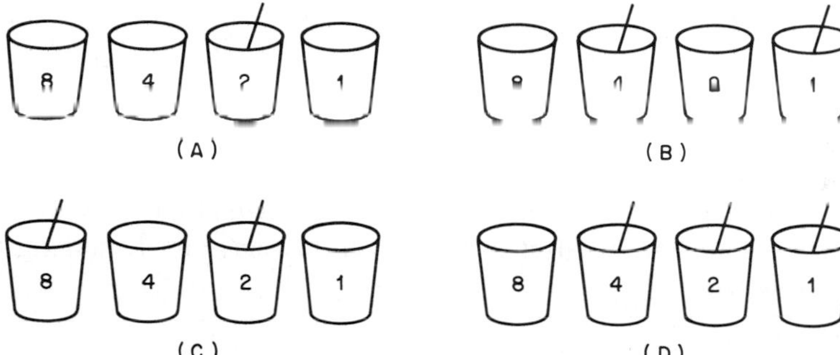

Figure 2–4

Solution

The general rule for getting the decimal value represented by the cups is to get the total value of all cups that are ON.

(A) Since there is a toothpick in only the cup marked "2," it is the only one that is ON. Hence, the number being represented is 2.

(B) Here, there are two cups that are ON—the ones marked "4" and "1." Hence, the value represented is 4 + 1 = 5.

(C) Again, two cups are ON—the "8" and the "2." This results in a value of 8 + 2 = 10.

(D) In this example, three cups are ON. Adding up their marked values produces 4 + 2 + 1 = 7.

Thus, the cups in each group represent the numbers shown in Figure 2–5.

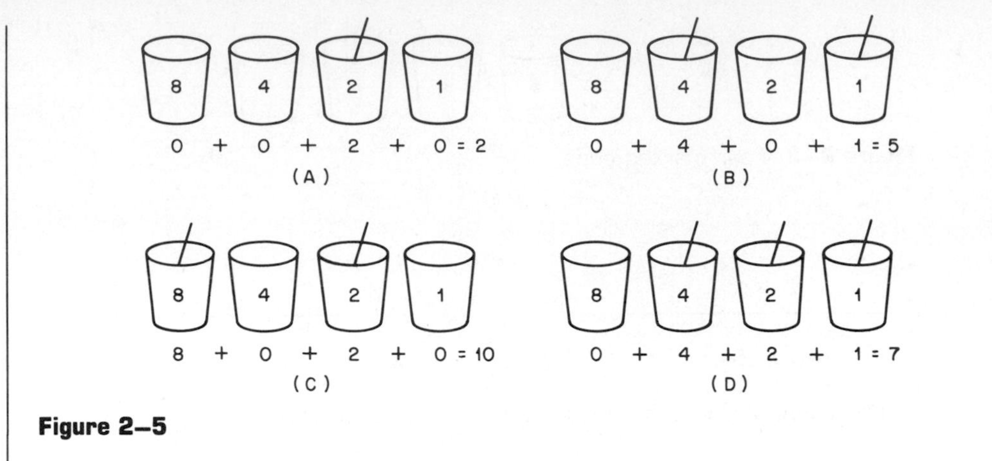

Figure 2–5

If you wanted to represent a number larger than 15 (this is a toothpick in each cup: 8 + 4 + 2 + 1 = 15), you would need another cup with a value of "16" marked on it (double the value of the last cup). Every time you add another cup to the system, its value becomes double that of the last cup. See Figure 2–6. In this manner, you can represent any whole number by using the concept of ON and OFF. With the four cups from the previous example, 16 different values can be represented, ranging from 0 to 15. This is shown in Figure 2–7.

Binary Numbers

The previous exercise introduced you to the way a microprocessor represents numbers—**binary numbers.**

Binary Numbers:
A base 2 number system consisting of two symbols (1 and 0) to represent all numerical values. This system is useful for giving numerical value to the two-state (ON/OFF) condition of digital systems.

Since the binary number system is a base 2 number system, it has only two symbols: 1 and 0. This corresponds to the ON and OFF system used by computers. It also is representative of the cups with their toothpicks. You can think of a cup as containing one **bit** of information.

Bit:
Abbreviation for binary digit—a unit of data in binary notation. Each of the 1's and 0's used in binary is called a bit.

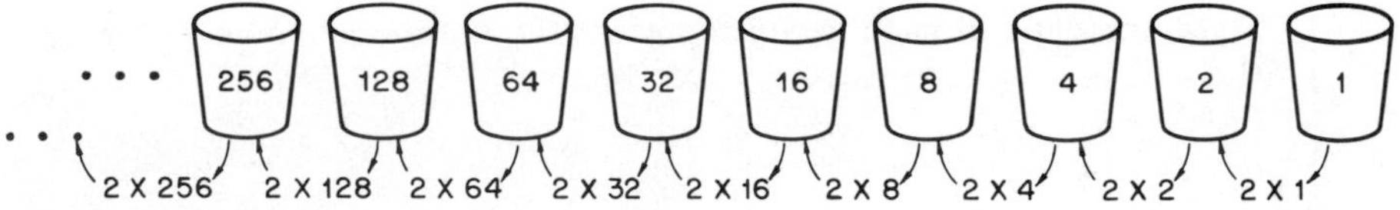

Figure 2–6 Adding more cups.

Figure 2–7 Representing 16 different values.

The base notation of binary numbers is

$$S_X \times 2^N + \cdots + S_3 \times 2^2 + S_2 \times 2^1 + S_1 \times 2^0 + S_1 \times 2^{-1} + S_2 \times 2^{-2} \cdots + S_X \times 2^{-N}$$

In order to distinguish one number system from the other, a **base subscript** is used.

Base Subscript:
For identification of the base of the number. Unless indicated otherwise, the base of a number is assumed to be 10. Base subscript notation of a base 10 number, for example, is 123_{10}, and for a base 2 number 1010_2.

Figure 2–8 shows the binary number representation of various numbers using the cup examples. The concept of **least significant bit** (LSB) and **most significant bit** (MSB) is also important in the study of these binary numbers.

Least Significant Bit (LSB):
In a binary number, it is the significant bit contributing the smallest quantity of the value of the number. For example, in 101_2, the rightmost 1 is the LSB.

Most Significant Bit (MSB):
In a binary number, it is the significant bit contributing the largest quantity of the value of the number. For example, in 101_2, the leftmost 1 is the MSB.

The values of the binary numbers shown in Figure 2–8 could also have been computed by using the binary base notation.

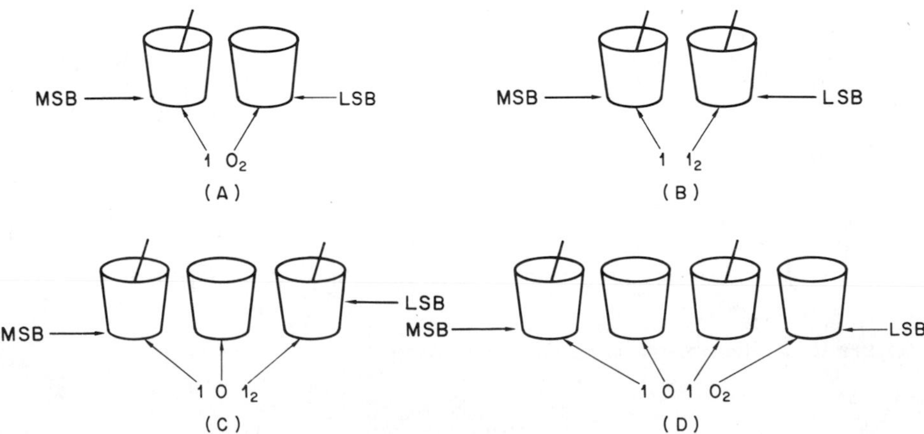

Figure 2–8 Binary number representation.

(A) $10_2 = 1 \times 2^1 + 0 \times 2^0 = 1 \times 2 + 0 \times 1 = 2 + 0 = 2_{10}$

(B) $11_2 = 1 \times 2^1 + 1 \times 2^0 = 1 \times 2 + 1 \times 1 = 2 + 1 = 3_{10}$

(C) $101_2 = 1 \times 2^2 + 0 \times 2^1 + 1 \times 2^0 = 1 \times 4 + 0 \times 2 + 1 \times 1 = 4 + 1 = 5_{10}$

(D) $1010_2 = 1 \times 2^3 + 0 \times 2^2 + 1 \times 2^1 + 0 \times 2^0$
$= 1 \times 8 + 0 \times 4 + 1 \times 2 + 0 \times 1 = 8 + 2 = 10_{10}$

Note from example (D) in Figure 2–8 that if base subscripts were not used, the binary number could easily be interpreted incorrectly as a decimal number, or the resulting decimal number could be mistaken for a binary number. Table 2–1 gives the decimal values of binary numbers up to four places.

Note the similarity between the binary number system and the cup system presented in Figure 2–7. This similarity is why the binary number system was chosen to express numerical values in digital systems. The place value of binary numbers is represented in

Table 2–1 BINARY NUMBER VALUES

Binary Number	Value
0000_2	0_{10}
0001_2	1_{10}
0010_2	2_{10}
0011_2	3_{10}
0100_2	4_{10}
0101_2	5_{10}
0110_2	6_{10}
0111_2	7_{10}
1000_2	8_{10}
1001_2	9_{10}
1010_2	10_{10}
1011_2	11_{10}
1100_2	12_{10}
1101_2	13_{10}
1110_2	14_{10}
1111_2	15_{10}

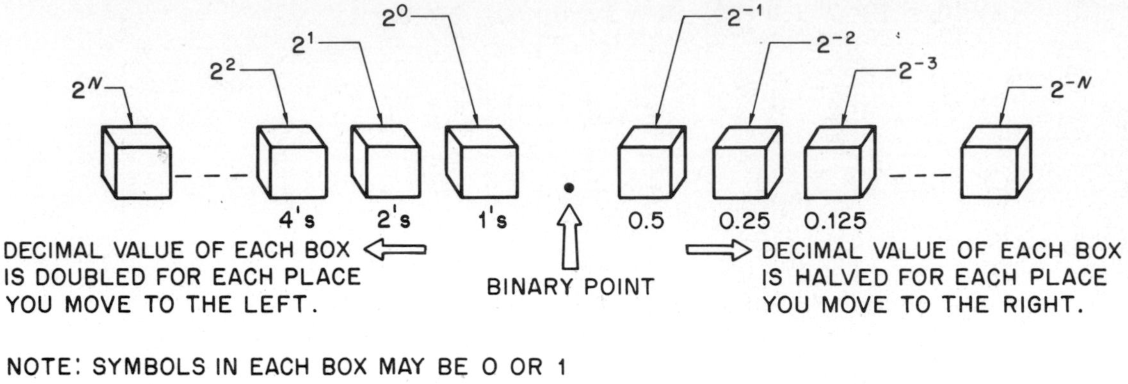
2^N
2^2
2^1
2^0
2^-1
2^-2
2^-3
2^-N
4's
2's
1's
0.5
0.25
0.125
DECIMAL VALUE OF EACH BOX IS DOUBLED FOR EACH PLACE YOU MOVE TO THE LEFT.
BINARY POINT
DECIMAL VALUE OF EACH BOX IS HALVED FOR EACH PLACE YOU MOVE TO THE RIGHT.
NOTE: SYMBOLS IN EACH BOX MAY BE 0 OR 1

Figure 2–9 Place value representation of binary numbers.

Figure 2–9. Using this system, you can represent any number. If you think of the cup-and-toothpick analogy, it becomes easy to figure out what value a binary number represents. This is demonstrated in the following example.

Example 2–4

Determine the decimal value represented by the following binary numbers:

(A) 10_2 (B) 11_2 (C) 101_2 (D) 1011_2

Solution

You can use the cups or base 2 notation. Both methods are shown in Figure 2–10.

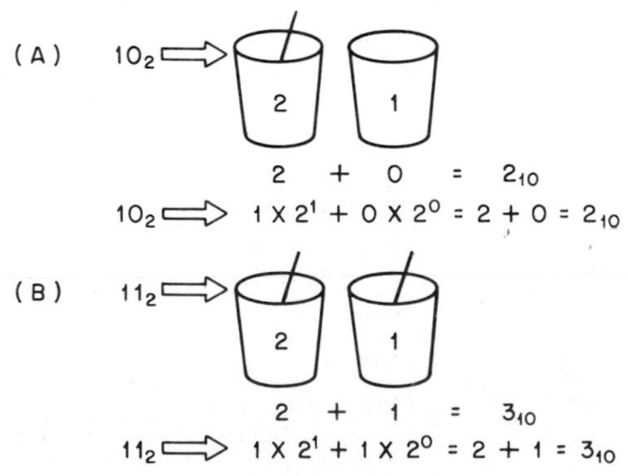
(A)
10_2
2
1
2 + 0 = 2_10
10_2 ⇒ 1 X 2^1 + 0 X 2^0 = 2 + 0 = 2_10
(B)
11_2
2
1
2 + 1 = 3_10
11_2 ⇒ 1 X 2^1 + 1 X 2^0 = 2 + 1 = 3_10

Figure 2–10

(C) 101_2 ⇒ 4 2 1

4 + 0 + 1 = 5_{10}

$101_2 \Rightarrow 1 \times 2^2 + 0 \times 2^1 + 1 \times 2^0 = 4 + 0 + 1 = 5_{10}$

(D) 1011_2 ⇒ 8 4 2 1

8 + 0 + 2 + 1 = 11_{10}

$1011_2 \Rightarrow 1 \times 2^3 + 0 \times 2^2 + 1 \times 2^1 + 1 \times 2^0 = 8 + 0 + 2 + 1 = 11_{10}$

Figure 2–10 Continued

Numbers Less Than 1

As you might suspect, binary numbers can be used to represent numbers that are less than 1. This is indicated by the binary number being to the right of the binary point (decimal point). For example, the binary number

$$0.101_2$$

in base two notation is

$$1 \times 2^{-1} + 0 \times 2^{-2} + 1 \times 2^{-3}$$

To evaluate the negative exponents

$$2^{-1} = \frac{1}{2^1} = \frac{1}{2} = 0.5_{10}$$

$$2^{-2} = \frac{1}{2^2} = \frac{1}{4} = 0.25_{10}$$

$$2^{-3} = \frac{1}{2^3} = \frac{1}{8} = 0.125_{10}$$

Thus,

$$1 \times 0.5 + 0 \times 0.25 + 1 \times 0.125 = 0.5 + 0.125 = 0.625_{10}$$

Table 2–2 lists some of the values for binary numbers less than 1.

Table 2–2 BINARY NUMBER VALUES (LESS THAN 1)

Binary Number	Decimal Value
0.1_2	0.5_{10}
0.01_2	0.25_{10}
0.001_2	0.125_{10}
0.0001_2	0.0625_{10}
0.00001_2	0.03125_{10}
0.000001_2	0.015625_{10}

Conclusion

In the next section, you will learn more about converting from binary numbers to decimal and from decimal to binary. In this section, you saw what a binary number was and how the ON and OFF conditions of computers could be used to represent binary numbers, and thus any number, as large or as small as you please. Test your understanding of this section by trying the following review questions.

2–2 Review Questions

1. What is a binary number? Give an example.
2. How many values can a binary number represent?
3. Explain if it is possible for a binary number to represent values less than 1.

2–3 CONVERTING BETWEEN BINARY AND DECIMAL NUMBERS

Discussion

You had some experience in interpreting the value of a binary number from the last section. In this section, you will learn some more rigorous methods of converting from binary to decimal and from decimal to binary.

Binary to Decimal Conversion

In the last section, you saw how to convert some binary numbers to their decimal equivalent. Here, you will see more examples of doing the same activity. Some helpful terminology concerning binary numbers is shown in Figure 2–11.

Word:
Set of bit patterns that occupies a specific storage location and is treated by the digital system as a unit and moved around the system as such.

Nibble:
Generally considered to be a 4-bit word.

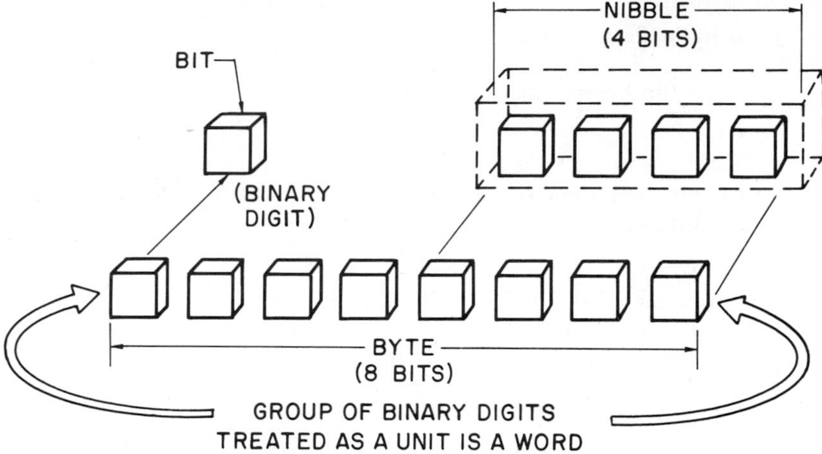

Figure 2–11 Binary number terminology.

Byte:
Generally considered to be an 8-bit word.

The following example illustrates how to use binary weights to evaluate a binary number. Notice the similarity between this and envisioning cups that are ON or OFF.

Example 2–5

Evaluate the binary number 10110111_2.

Solution

Determine the weight of each bit. Then find the total of the weights that have a bit of 1.

Binary weight:	2^7	2^6	2^5	2^4	2^3	2^2	2^1	2^0
Weight value:	128	64	32	16	8	4	2	1
Binary number:	1	0	1	1	0	1	1	1

Resulting sum: $128 + 0 + 32 + 16 + 0 + 4 + 2 + 1$
$= 183_{10}$

Note that the result is the same as

$1 \times 128 + 0 \times 64 + 1 \times 32 + 1 \times 16 + 0 \times 8 + 1 \times 4 + 1 \times 2 + 1 \times 1$
$= 183_{10}$

You can also use the same method for evaluating binary numbers that are less than 1, as illustrated by the following example.

Example 2–6

Evaluate the binary fraction 0.1001_2.

Solution

As before, determine the weight of each bit. Then find the total of the weights that have a bit of 1.

Binary weight:	2^{-1}	2^{-2}	2^{-3}	2^{-4}
Weight value:	0.5	0.25	0.125	0.0625
Binary fraction:	1	0	0	1

Resulting sum: $0.5 + 0 + 0 + 0.0625 = 0.5625_{10}$

Note that the result is the same as

$1 \times 0.5 + 0 \times 0.25 + 0 \times 0.125 + 1 \times 0.0625 = 0.5625_{10}$

The previous examples can be combined to evaluate a binary number that contains a fractional part. This is illustrated in Example 2–7.

Example 2–7

Evaluate the binary number 1011.01_2.

Solution

Use the same method as in the previous two examples.

Binary weight:	2^3	2^2	2^1	2^0	2^{-1}	2^{-2}
Weight value:	8	4	2	1	0.5	0.25
Binary number:	1	0	1	1	0	1

Resulting sum: $8 + 0 + 2 + 1 + 0 + 0.25 = 11.25_{10}$

Decimal to Binary Conversion

It is equally important to know how to convert from a decimal number to a binary number, because you need to know what the decimal numbers you want to enter into a digital system will look like in their binary form.

One method of converting a decimal number to a binary number is to work the problem backwards. Essentially you are asking yourself, what binary number converts to the given decimal number? There is a step-by-step process that you can follow for this method. Basically, the method goes like this: Suppose you wanted to convert a decimal number to its binary equivalent. The first thing to do is to recognize that each binary weight is double the value of the previous weight (for a whole number). Start by writing down the unit binary weight and proceed to the left, doubling this value until you come to the number that is larger than the decimal number you want to convert. This number is larger than the given number so don't use it. The next place to the right is the MSB. Starting with the MSB and going to the right, write a 1 in every place that will not cause you to exceed the value of the given number. If a weight causes you to exceed the value of the given number write a 0. This is illustrated in the following example.

Example 2–8

Convert the following decimal numbers to their binary equivalents:

(A) 5_{10} (B) 12_{10} (C) 23_{10} (D) 150_{10}

Solution

(A) Write down the binary weights starting at 1, and double to the left until a value larger than 5 is reached. Omit the 8 (it's larger than 5). Now use what you have to make 5.

8	4	2	1	
X	1	0	1	$= 101_2$

Thus $5_{10} = 101_2$

(B) Write down the binary weights. Omit the 16 (it's larger than 12).

16	8	4	2	1	
X	1	1	0	0	$= 1100_2$

Thus $12_{10} = 1100_2$

(C) Write down the binary weights. Omit the 32.

32	16	8	4	2	1	
X	1	0	1	1	1	$= 10111_2$

Thus $23_{10} = 10111_2$

(D) Write down the binary weights. Omit the 256.

256	128	64	32	16	8	4	2	1	
X	1	0	0	1	0	1	1	0	$= 10010110_2$

Thus $150_{10} = 10010110_2$

Another method for converting a decimal number into a binary is by taking the decimal number and repeatedly dividing it by 2. The remainder generated by each division produces the binary number. The first remainder is the LSB in the binary number. This method is illustrated in the following example.

Example 2–9

Using the divide-by-2 method, convert the following decimal numbers to their binary equivalents.

(A) 5 (B) 12 (C) 36

Solution

Refer to Figure 2–12.

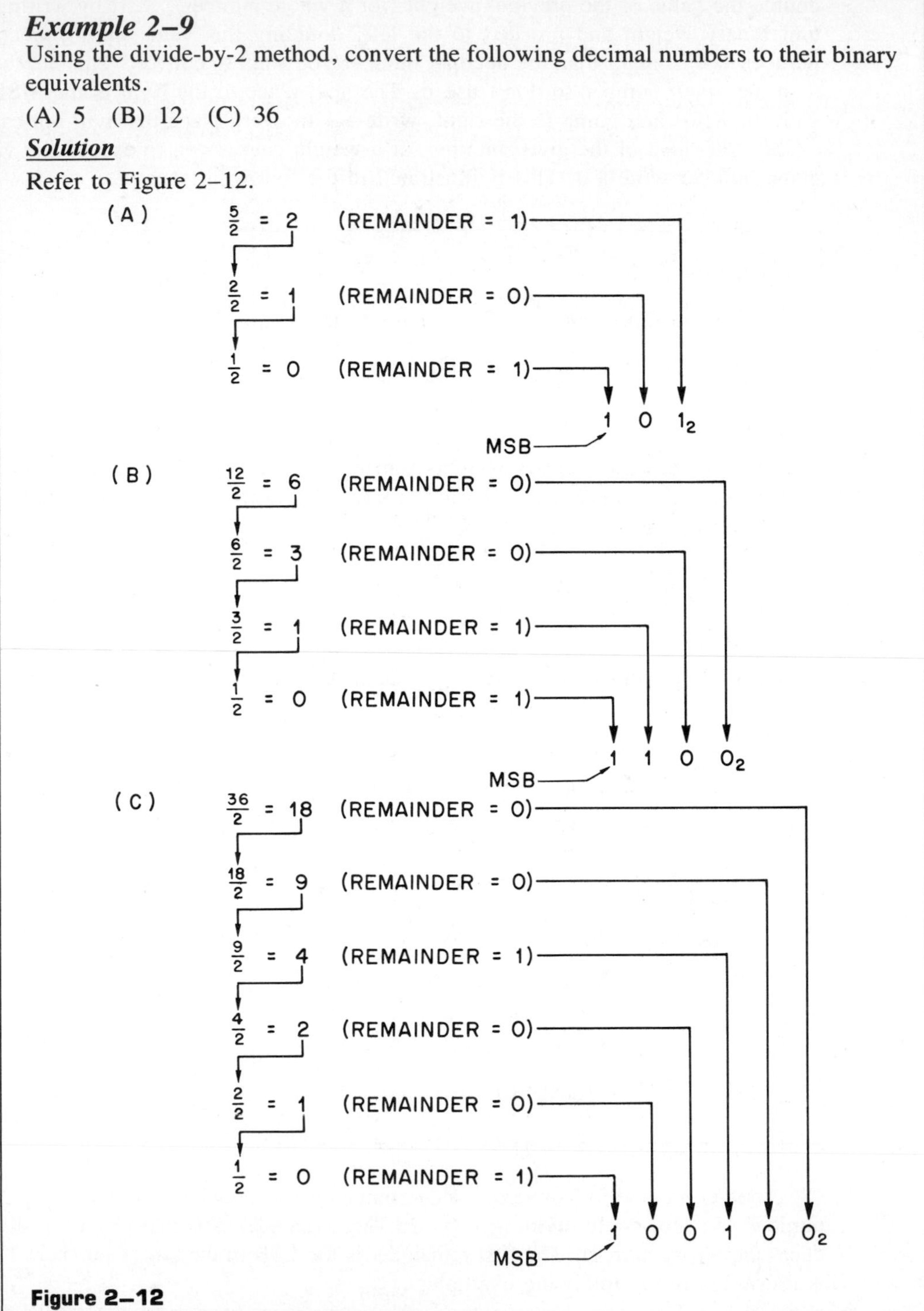

Figure 2–12

Converting from Decimal Fraction to Binary

You can convert a decimal fraction to a binary number by repeated multiplication by 2. When a 1 is generated to the left of the decimal point in the answer, this represents a binary 1; when there is no 1 generated, this represents a binary 0. The binary number is arranged starting with the MSB to the right of the binary point and going to the right. For example, convert 0.125_{10} to binary.

$$\begin{array}{rrl} & 0.125 & \\ & \times 2 & \\ \hline \text{0 to the left of the decimal point} \rightarrow & 0.250 & \rightarrow 0.0_2 \\ & \times 2 & \\ \hline \text{0 to the left of the decimal point} \rightarrow & 0.50 & \rightarrow 0.00_2 \\ & \times 2 & \\ \hline \text{1 to the left of the decimal point} \rightarrow & 1.0 & \rightarrow 0.001_2 \end{array}$$

To convert a number such as 25.38 to binary, you must use two different procedures. There is a whole number part (25) and the fractional part (.38). Each will be converted in a separate procedure.

Using the Scientific Calculator

There are many scientific calculators that will convert from binary to decimal and back. The limitation is quite severe, however, because most calculators of this type have a limited number of digits they can display. For example, a calculator that can display nine digits is limited to the following binary number:

$$111111111_2 = 511_{10}$$

Conclusion

You should now be at the point where you can convert between binary and decimal numbers. You will have many opportunities to practice doing this as you progress through this book. Test your understanding of this section by trying the following review questions.

2–3 Review Questions

1. Explain how to convert a binary number to decimal.
2. What is the meaning of the term “weight”?
3. State how to convert a decimal number to binary.

2–4 REPRESENTING DIGITAL ELEMENTS

Discussion

In this section, you will be introduced to the concept that an ON and OFF combination can represent numbers in ways that are different from binary numbers. Here you will also see how the ON and OFF property of switches and lights can be used to represent numbers.

Using Display Lights

Small lights such as the **light emitting diode (LED)** is used extensively in digital circuits to display the ON and OFF condition of the digital circuit. Typical LEDs and their construction are shown in Figure 2–13.

Light Emitting Diode (LED):
A solid-state device that emits light when the correct amount and polarity of electrical potential are applied.

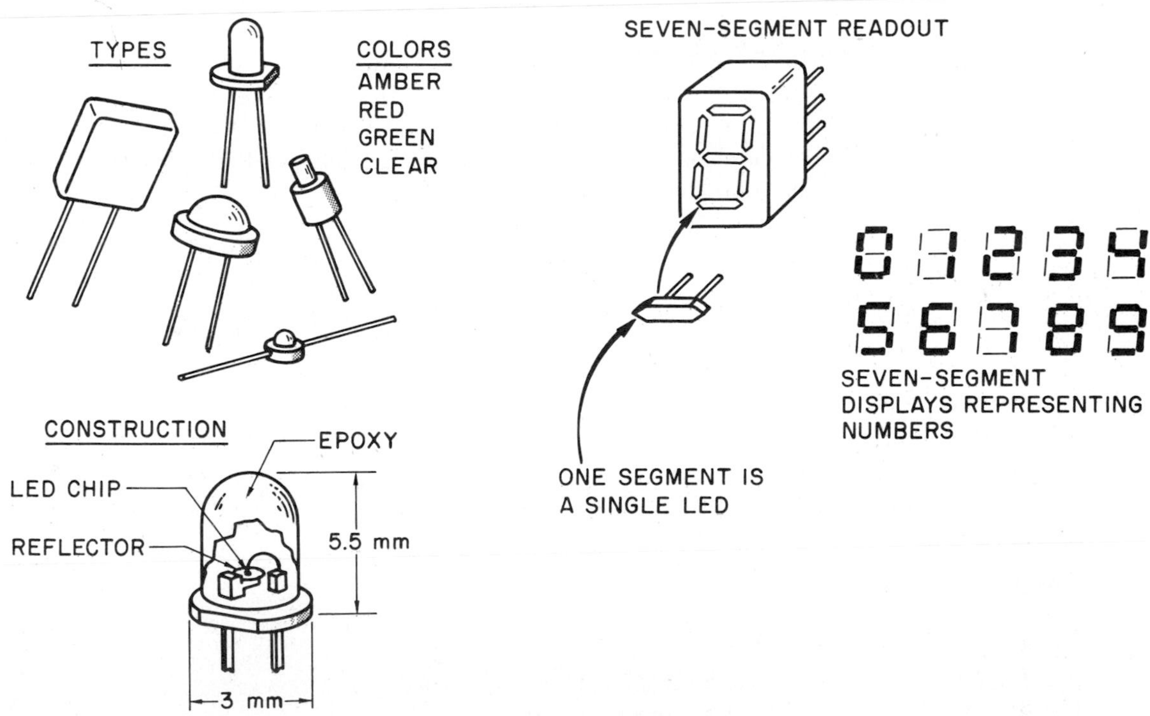

Figure 2–13 Types of LEDs and their construction.

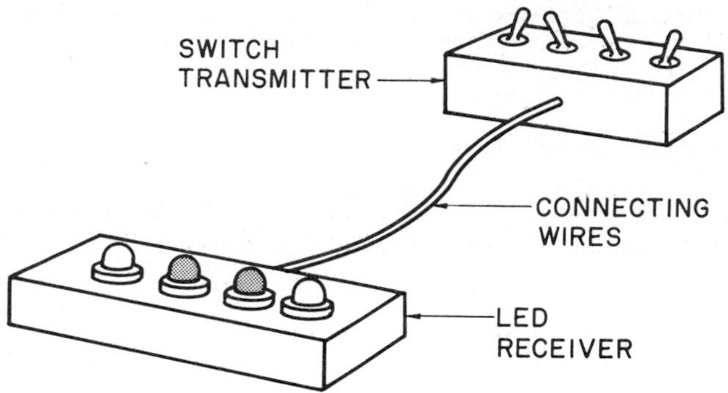

Figure 2–14 Binary number transmitter.

Figure 2–14 shows an application of LEDs, a binary number transmitter. The binary values are placed in the switches, and the LEDs read out the binary pattern.

Example 2–10

State the numerical values of each light pattern shown in Figure 2–15. Assume that an ON light represents a binary 1 and an OFF light a binary 0.

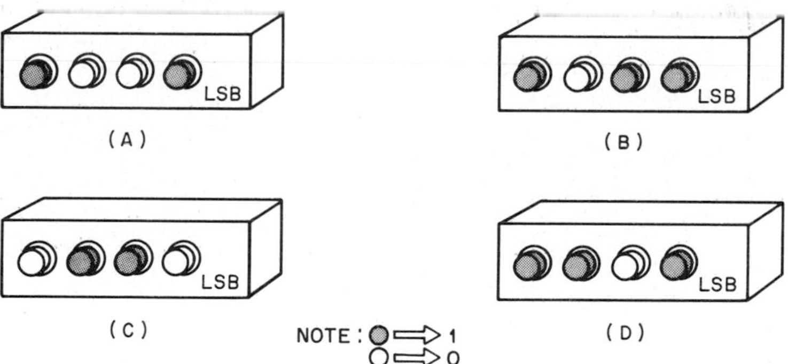

Figure 2–15

Solution

(A) $1001_2 = 8 + 1 = 9_{10}$

(B) $1011_2 = 8 + 2 + 1 = 11_{10}$

(C) $0110_2 = 4 + 2 = 6_{10}$

(D) $1101_2 = 8 + 4 + 1 = 13_{10}$

Numerical values may also be represented by switches, as demonstrated in the next example.

Example 2–11

Determine the values represented by the switch positions shown in Figure 2–16. Assume that an ON switch represents a binary 1 and an OFF switch a binary 0.

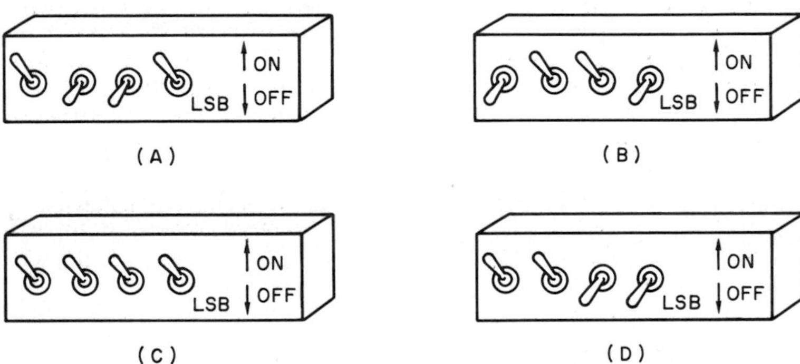

Figure 2–16

Solution

(A) $1001_2 = 8 + 1 = 9_{10}$

(B) $0110_2 = 4 + 2 = 6_{10}$

(C) $1111_2 = 8 + 4 + 2 + 1 = 15_{10}$

(D) $1100_2 = 8 + 4 = 12_{10}$

Using a Keypad

A **keypad** is used to enter values into the computer.

Keypad:
A small keyboard (or section of a keyboard) containing a smaller number of keys than a typewriter keyboard. It serves as one of the simplest input devices to a computer.

A typical 10-key keypad is shown in Figure 2–17. The keypad is used to enter numbers into the digital control system of a robot. As each number is entered, the LEDs indicate its value, and then the number is stored in the robot control unit. Only one key is pressed at a time (any value from 0 to 9), and the resulting value is converted to binary. When a number larger than 9 is to be entered, say 53_{10}, first press the key marked 5, then the key marked 3. This is the same way you enter numbers into your scientific pocket calculator.

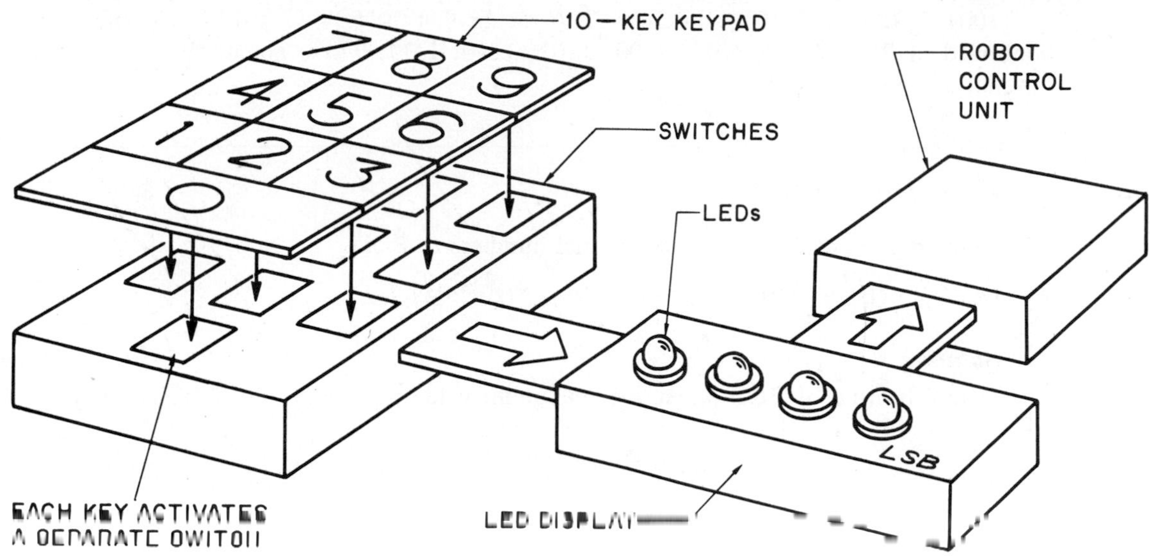

Figure 2–17 Ten-key input to robot control unit.

When the number 53_{10} is entered, the following binary numbers are stored in the control unit:

$$\text{Enter } 5 \rightarrow 0101$$
$$\text{Enter } 3 \rightarrow 0011$$

The final representation of 53_{10} will be 0101 0011. (The first group of 4 bits represents the 5, and the second group of 4 bits represents the 3.) The number that was just entered cannot be interpreted as a binary number. Take another look:

$$01010011 = 64 + 16 + 2 + 1 = 83_{10}$$

This is not the value of 53 that was originally entered! However, if the resulting group of 0's and 1's is not interpreted as a binary number but is treated as two groups of four digits, with each group representing a single decimal digit between 0 and 9, then no confusion will result. As you may suspect, such a system is used, which leads to the next topic.

Binary-Coded Decimal (BCD)

Binary-coded decimal (BCD) is a type of **8421 code** that uses 4 bits to represent a decimal digit from 0 to 9.

8421 Code:
The place value of each bit corresponds to the place value of a binary number.

$$2^3\ 2^2\ 2^1\ 2^0 = 8421$$

The main advantage of this code is that it is easy to convert between it and decimal numbers. The disadvantage is that the largest number allowed for the 4 bits in BCD is 9_{10}

$= 1001_{BCD}$ (note the BCD subscript). This means that the following 4-bit patterns are not allowed in BCD: 1010, 1011, 1100, 1101, 1110, 1111. These potentially useful representations are essentially wasted.

Example 2–12

Convert each of the following decimal numbers to BCD.

(A) 5_{10} (B) 9_{10} (C) 12_{10} (D) 346_{10} (E) 2047_{10}

Solution

Using the rule of BCD where the maximum value represented by 4 bits is 9, you get

(A) 5 → 0101
(B) 9 → 1001
(C) 1 2 → 0001 0010
(D) 3 4 6 → 0011 0100 0110
(E) 2 0 4 7 → 0010 0000 0100 0111

As you can see from this example, it is very easy to convert from decimal to BCD. It is just as easy to convert from BCD to decimal. Just start at the LSB of the BCD number, break it up into groups of 4 bits, and write the decimal equivalent of each group of 4 bits.

Example 2–13

Determine the decimal numbers represented by the following BCD codes:

(A) 10010011_{BCD} (B) 00110111_{BCD} (C)100000000001_{BCD}
(D) 100110011001_{BCD}

Solution

(A) 1001 0011 → 9 3
(B) 0011 0111 → 3 7
(C) 1000 0000 0001 → 8 0 1
(D) 1001 1001 1001 → 9 9 9

Conclusion

This section presented the concept that other number codes can be represented by ON and OFF conditions. In particular, you were presented the BCD code that allowed an easy conversion to and from decimal numbers. Test your new skills with the following review questions.

2–4 Review Questions

1. Describe an LED.
2. Explain how light patterns and the position of switches can represent numbers.
3. Describe a keypad.
4. Define BCD.
5. State the advantage and disadvantage of BCD.

2–5 MORE EFFICIENCY—HEXADECIMAL NUMBERS

Discussion

In the last section, you were introduced to BCD. Its main disadvantage was that it wasted six combinations of the 4 binary bits that it used (1010, 1011, 1100, 1101, 1110, and 1111 are not allowed in BCD). The reason for this was that the largest-valued key on the 10-key keypad was the number 9. If larger numbers were to be entered, more than one key had to be pressed, resulting in the creation of another set of 4 bits for every key press. This section shows how the limitations of BCD are overcome by using a different keypad while still preserving the ease of conversion of a keystroke to a binary code.

An Improved Keypad

To enter numbers easily into a digital system (such as a robot) and allow all of the possible combinations of 4 bits to be used, you need a 16-key keypad. See Figure 2–18. Observe that in this system, all combinations of the 4 bits are used, and a real binary number (not BCD) is produced. Since the keypad produces 16 distinct codes, it represents a number system that has base 16. Such a number system is called the **hexadecimal number system.**

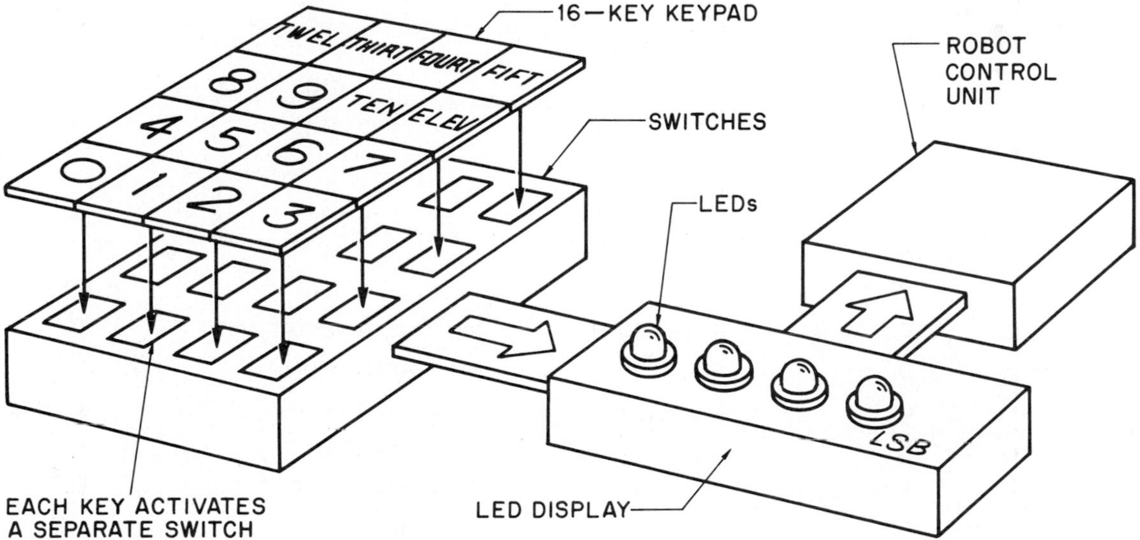

Figure 2–18 Sixteen-key keypad.

Figure 2–19 Typical hexadecimal keypad. *Courtesy* of Grayhill, Inc., La Grange, IL.

Hexadecimal Number System:
A number system with a base of 16. Thus, 16 symbols are used: 0–9, A, B, C, D, E, and F, which are sometimes referred to as hex numbers.

The general form of this number system is

Table 2–3 COMPARISON OF DIFFERENT SYSTEMS

Hexadecimal	Binary	BCD	Decimal
0	0000	0000	0
1	0001	0001	1
2	0010	0010	2
3	0011	0011	3
4	0100	0100	4
5	0101	0101	5
6	0110	0110	6
7	0111	0111	7
8	1000	1000	8
9	1001	1001	9
A	1010	0001 0000	10
B	1011	0001 0001	11
C	1100	0001 0010	12
D	1101	0001 0011	13
E	1110	0001 0100	14
F	1111	0001 0101	15

$$S_X \times 16^N + \cdots + S_3 \times 16^2 + S_2 \times 16^1 + S_1 \times 16^0 + S_1 \times 16^{-1} + \cdots + S_X \times 16^{-N}$$

A typical hexadecimal keypad is shown in Figure 2–19. The first 6 capital letters of the English alphabet represent the extra six symbols. This is a convenient arrangement because now, the conversion between keystroke and binary is very direct and the resulting binary code is a real binary number. Table 2–3 compares the different systems presented to this point.

Converting from Binary to Hexadecimal

Converting from binary to hexadecimal is a very straightforward process. Take any binary number and starting from the LSB, break it into groups of four. Once this is done, convert each group of four to its hexadecimal equivalent. The following example illustrates this procedure.

Example 2–14

Convert the following binary numbers to hexadecimal.

(A) 101101001010_2 (B) 11111011111_2 (C) 1011011011_2

Solution

(A) 1011 0100 1010	(B) 0111 1101 1111	(C) 0010 1101 1011
B 4 A	7 D F	2 D B

Note that on problem (B) one 0 was added to the left, and on problem (C) two 0's were added to the left to keep the binary numbers in groups of four.

Converting from Hexadecimal to Binary

To convert from hexadecimal to binary, simply do the opposite procedure of converting from binary to hexadecimal and replace each hexadecimal symbol with its corresponding four binary bits. The next example illustrates this procedure.

Example 2–15

Convert each of the following hexadecimal numbers to their binary equivalents.

(A) $2A_{16}$ (B) $C06_{16}$ (C) FFA_{16}

Solution

(A) 2 A	(B) C 0 6	(C) F F A
0010 1010	1100 0000 0110	1111 1111 1010

Using Hexadecimal Numbers

Hexadecimal numbers are a very efficient way of representing binary numbers. As such, you will find this base 16 number system very common in digital systems. You saw that a 16-key keypad could be used to enter numbers into a digital system. The hexadecimal number system also makes it easy to display the values of 4-bit binary numbers. The device used to do this is a seven-segment LED. See Figure 2–20. Observe that all 16 hexadecimal digits can be represented by the seven-segment display. With several seven-segment displays, very large binary numbers can be represented.

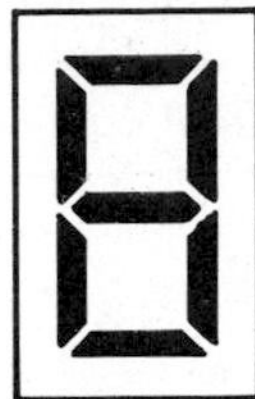

Figure 2–20 Seven-segment LED and corresponding readouts.

Example 2–16

What binary pattern is represented by the seven-segment displays of Figure 2–21? Determine the decimal equivalent number.

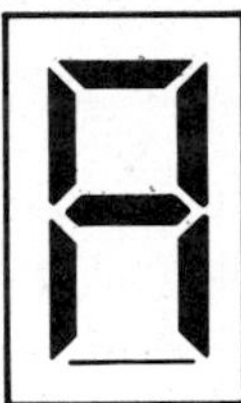
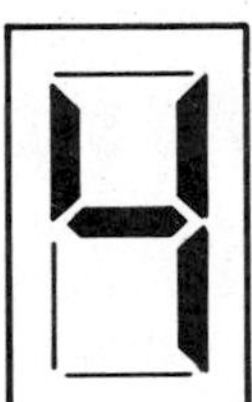
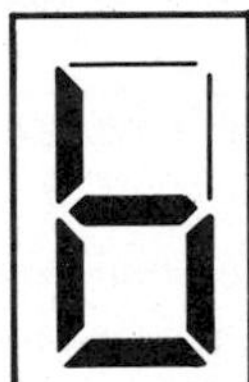
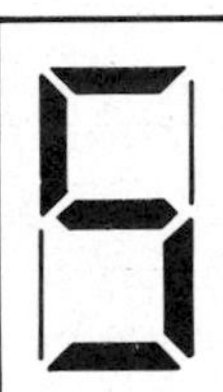

Figure 2–21

Solution

This is nothing more than converting from hexadecimal to binary:

A	4	B	5
1010	0100	1011	0101

The decimal equivalent is

$$32\,768 + 8\,192 + 1\,024 + 128 + 32 + 16 + 4 + 1 = 42\,165_{10}$$

Converting from Hexadecimal to Decimal

As you saw in the last example, one method of converting from hexadecimal to decimal was to first convert the hexadecimal number to binary and then convert the binary number to its decimal value. Scientific pocket calculators are also available to do the same thing—a decimal or hexadecimal number can be entered and the push of a button will do the conversion. These same calculators will also convert between hexadecimal and binary.

Another way of converting from hexadecimal to decimal is to multiply each hexadecimal digit by its weight and then take the sum of these products. This process is illustrated in the following example.

Example 2–17

Convert the following hexadecimal numbers to their decimal values.

(A) 36_{16} (B) $E2A_{16}$

Solution

(A) $36_{16} = 3 \times 16^1 + 6 \times 16^0 = 3 \times 16 + 6 \times 1 = 48 + 6 = 54_{10}$

(B) $E2A_{16} = E \times 16^2 + 2 \times 16^1 + A \times 16^0$
$= 14 \times 16^2 + 2 \times 16^1 + 10 \times 16^0$
$= 14 \times 256 + 2 \times 16 + 10 \times 1$
$= 3\,584 + 32 + 10 = 3\,626$

Converting Decimal to Hexadecimal

Figure 2–22 illustrates the process of converting from decimal to hexadecimal by repeated division of the base (16). This is similar to the process of converting decimal to binary.

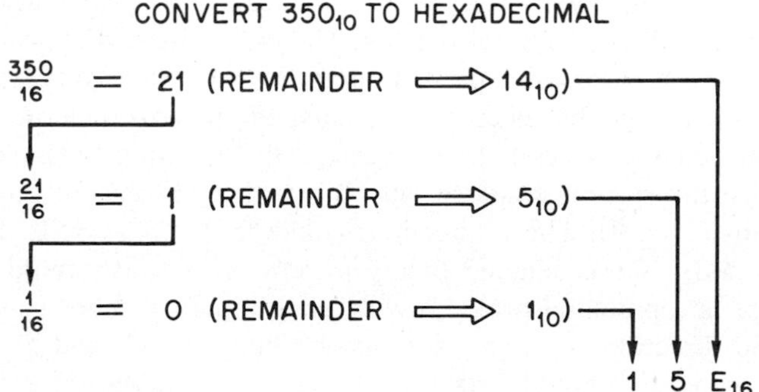

Figure 2–22 Process of converting from decimal to hexadecimal.

Conclusion

This section demonstrated the use of hexadecimal numbers and the ease they offered in conversions with binary numbers. In the next section, you will see how information other than numbers can be represented by a system of ON and OFF. For now, test your new skills with the following review questions.

2–5 Review Questions

1. Describe a hex keypad.
2. State the advantage of a hex keypad over that of a 10-key keypad.
3. Describe the hexadecimal number system.
4. Explain how to convert between the hexadecimal and binary number systems.
5. Explain how to convert between the hexadecimal and decimal number systems.

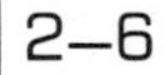

THE NEED FOR OTHER CODES

Discussion

This section presents codes that are developed for a particular application. You saw how the BCD code was created to make coding information from a 10-key keypad easier. There are instances where information must be gathered without human intervention—such as with robots, remote weather stations, and extraterrestrial sensors. In these cases hexadecimal numbers are not needed.

Need for Another Code

Figure 2–23 shows a representation of a remote weather station with a wind direction sensor. This system transmits wind direction information from a remote location to a central location where a binary display indicates the wind direction. The direction of the wind causes the weather vane to turn. It is attached to a shaft that causes a binary-coded disk to turn under three electrical sensors. The binary pattern is printed on the disk in the same manner as the wires on a printed circuit board. Three stiff wires pressed on the surface of the disk detect, by electrical currents, the pattern underneath them. The system is in turn connected to wires that can transmit the resulting binary code to a central location.

Since the system contains only 3 bits, it can transmit the information of only eight directions ($2^3 = 8$). These directions could be N, NE, E, SE, S, SW, W, and NW. More sensors and a more detailed binary pattern on the disk could produce greater accuracy, but there is a potential problem with using binary numbers, as shown in Figure 2–24. If the wind direction is making a transition between NE and N (binary 111_2 and 000_2), an erroneous reading could occur. In the case shown, the information being transmitted by

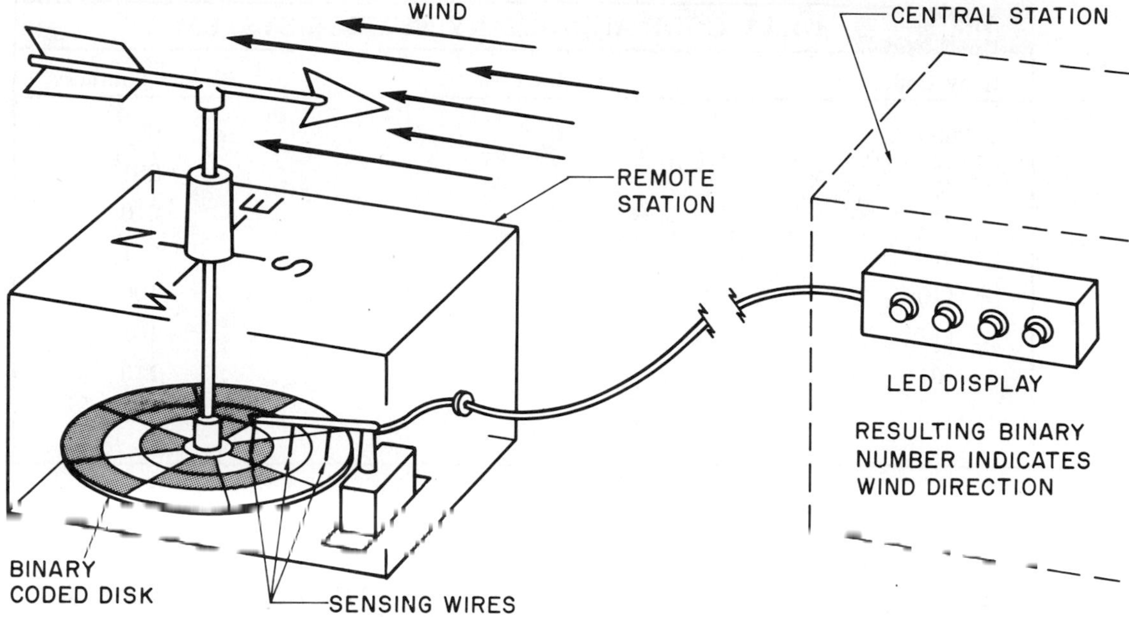

Figure 2–23 Remote weather station with wind direction sensor.

the sensing heads indicates a direction of 101_2 (W), which is incorrect. The problem arises because wires are not in perfect alignment. One possible way of reducing this error is by a more careful manufacturing process. However, another method is available that doesn't change with system use. This method simply uses an ON and OFF code that allows only one bit at a time to change. Doing this would eliminate the erroneous reading.

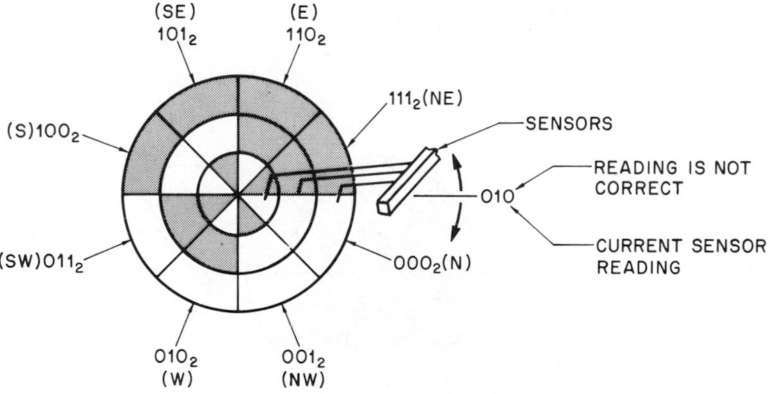

Figure 2–24 Potential problem with binary code on disk.

Table 2–4 GRAY CODE AND BINARY NUMBER SYSTEM

Gray Code	Decimal	Binary Numbers
000	0	000
001	1	001
011	2	010
010	3	011
110	4	100
111	5	101
101	6	110
100	7	111

The Gray Code

The gray code allows only one bit at a time to change. Table 2–4 compares this code with the binary number system. Note from the table that only one bit at a time changes with the gray code. Compare this to the binary code where more than one bit at a time can change. For example 011 to 100 is a change of 3 bits.

Figure 2–25 shows a sensor disk that has the Gray code etched on its surface. When this code is used, an erroneous reading will not be sent.

Excess-Three Code

The excess-3 code is similar to the BCD code. Like the BCD, the excess-3 code uses groups of 4 binary bits to represent a single decimal digit. What distinguishes it from BCD is that 3 is added to the decimal digit before it is converted to binary. This process is illustrated in the following example.

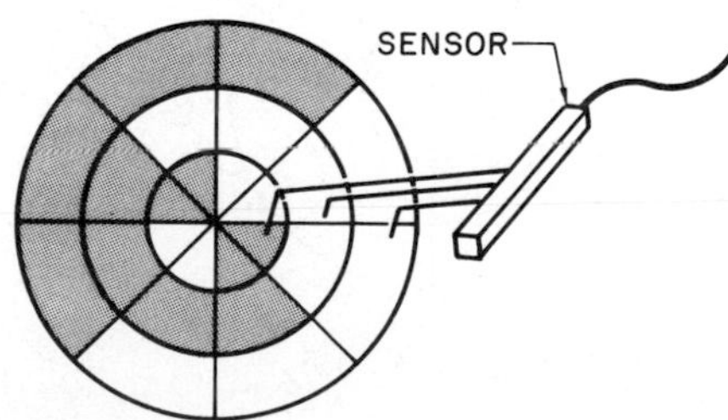

Figure 2–25 Coding disk using the Gray code.

Example 2–18

Convert the following decimal numbers to the excess-3 code.

(A) 4 (B) 12 (C) 120 (D) 789

Solution

(A)	4	(B)	1	2	(C)	1	2	0
	+3		+3	+3		+3	+3	+3
	(7)		(4)	(5)		(4)	(5)	(3)
	0111		0100	0101		0100	0101	0011

(D)	7	8	9
	+3	+3	+3
	(10)	(11)	(12)
	1010	1011	1100

Conclusion

In this section, you were introduced to two other important codes used in digital systems. Each code was developed for a unique application. The next section introduces you to the ON and OFF code that allows words as well as numbers to be used by digital systems. For now, test your understanding of this section by trying the following review questions.

2–6 Review Questions

1. State a disadvantage of the binary number system.
2. Explain how the disadvantage in Question 1 can be overcome.
3. Give the major advantage of the Gray code.
4. State how decimal numbers are converted to the excess-3 code.

2–7 USING LETTERS AND NUMBERS—THE ASCII CODE

Discussion

To communicate with digital systems, you need letters, numbers, and other symbols. In a word processor, for example, uppercase and lowercase letters, numbers, punctuation marks, the space between words, and the carriage return for starting a new line must all be stored in ON or OFF notation.

What is Needed

For this storage to be accomplished, the number of binary bits must be large enough so that a unique bit pattern for the 10 number symbols and the 26 letters exists. Thus, 26 + 10 = 36 different symbols are needed, so at least 6 bits must be used ($2^6 = 64$ different combinations). This leaves plenty of room for other symbols plus capital and lowercase letters. Codes that can handle both letters and numbers are called **alphanumeric codes.**

Alphanumeric Codes:
Codes that can represent letters of the alphabet, numerals, and other symbols such as punctuation marks and mathematical symbols.

The American Standard Code for Information Interchange

One standardized alphanumeric code is called the **American Standard Code for Information Interchange (ASCII).**

American Standard Code for Information Interchange (ASCII):
Pronounced "Askee." Standard data transmission code used to achieve compatibility between data devices. Has a total of 128 code combinations.

The ASCII code is found in almost every microcomputer system. It is a 7-bit code. All of the decimal digits start with 011 followed by their BCD equivalent. For example, the decimal number 5 is 011 0101. The letter A is represented by 100 0001. Table 2–5 shows the ASCII code. Many of the functions are carried over from the older teletype systems.

Example 2–19

Reading from left to right, determine the word represented by the following ASCII code:

100010011010011100111110100111101001100001110110001000010000100

Solution

Start by breaking up the code into groups of 7 bits each (it is convenient to break it further into its characteristic 3-bit 4-bit pattern). Then use the ASCII table for the translation.

100 0100	110 1001	110 0111	110 1001	111 0100	110 0001
D	i	g	i	t	a

110 1100	010 0001	000 0100
l	!	EOT (End of Transmission)

Table 2–5 AMERICAN STANDARD CODE FOR INFORMATION INTERCHANGE

	000	001	010	011	100	101	110	111
0000	NUL	DLE	SP	0	@	P	\`	p
0001	SOH	DC_1	!	1	A	Q	a	q
0010	STX	DC_2	"	2	B	R	b	r
0011	ETX	DC_3	#	3	C	S	c	s
0100	EOT	DC_4	$	4	D	T	d	t
0101	ENQ	NAK	%	5	E	U	e	u
0110	ACK	SYN	&	6	F	V	f	v
0111	BEL	ETB	'	7	G	W	g	w
1000	BS	CAN	(	8	H	X	h	x
1001	[illegible]	[illegible]	)	9	I	Y	i	y
1010	LF	SUB	*	:	J	Z	j	z
1011	VT	ESC	+	;	K	[	k	{
1100	FF	FS	,	<	L	\	l	¦
1101	CR	GS	–	=	M	]	m	}
1110	SO	RS	.	>	N	∧	n	~
1111	SI	US	/	?	O	—	o	DEL

ACK → Acknowledge; BEL → Bell; BS → Backspace; CAN → Cancel; CR → Carriage Return; DC_1–DC_4 → Direct Control; DEL → Delete; DLE → Data Link Escape; EM → End of Medium; ENQ → Enquiry; EOT → End of Transmission; ESC → Escape; ETB → End of Transmission Block; ETX → End Text; FF → Form Feed; FS → Form Separator; GS → Group Separator; HT → Horizontal Tab; LF → Line Feed; NAK → Negative Acknowledge; NUL → Null; RS → Record Separator; SI → Shift In; SO → Shift Out; SOH → Start of Heading; STX → Start Text; SUB → Substitute; SYN → Synchronous Idle; US → Unit Separator; VT → Vertical Tab.

Conclusion

This section presented the most common alphanumeric code in use today, the ASCII. You also had an opportunity to interpret a sample of the code. The next section demonstrates an application of what you have learned here to microprocessors and their associated circuits. Check your understanding of this section by trying the following review questions.

2–7 Review Questions

1. What is meant by an alphanumeric code?
2. What does ASCII stand for?
3. What can the ASCII code represent?
4. How is the ASCII code structured?

2–8 ARITHMETIC OPERATIONS

Discussion

Computers do all of their arithmetic using binary. Since you will need to know everything possible about how computers do things, you should know how to do arithmetic in binary. Since many digital systems are programmed using hexadecimal, you should also know something about doing arithmetic with hexadecimal numbers. You'll see how in this section.

Binary Counting

In order to count in binary, observe the following:

0_{10} 0000_2
1_{10} 0001_2 The units column alternates every other bit.
2_{10} 0010_2 The 2's column alternates every 2 bits.
3_{10} 0011_2
4_{10} 0100_2 The 4's column alternates every 4 bits.
5_{10} 0101_2
6_{10} 0110_2
7_{10} 0111_2
8_{10} 1000_2 The 8's column alternates every 8 bits.

Binary Addition

It is very easy to add using binary numbers. All you need to learn are the following rules:

$$\text{(A)}\ \begin{array}{r} 0_2 \\ +0_2 \\ \hline 0_2 \end{array} \quad \text{(B)}\ \begin{array}{r} 0_2 \\ +1_2 \\ \hline 1_2 \end{array} \quad \text{(C)}\ \begin{array}{r} 1_2 \\ +0_2 \\ \hline 1_2 \end{array} \quad \text{(D)}\ \begin{array}{r} 1_2 \\ +1_2 \\ \hline 10_2 \end{array} \quad \text{(E)}\ \begin{array}{r} 1_2 \\ 1_2 \\ +1_2 \\ \hline 11_2 \end{array}$$

This should make sense. In (A), 0 plus 0 is 0; in (B) and (C) 1 added to 0 is 1. In (D), 1 plus 1 is 0 carry the 1—and isn't the answer 10_2 equal to 2_{10}? Sure it is, and the answer checks. Next, look at (E). This says that 1 plus 1 plus 1 is 1 carry the 1. The result is 11_2, and isn't that equal to 3_{10}? It is, and this is one way of checking the binary addition, by a quick conversion to decimal.

Example 2–20

Add the following binary numbers:

$$\text{(A)}\ \begin{array}{r} 11_2 \\ +01_2 \\ \hline \end{array} \quad \text{(B)}\ \begin{array}{r} 101_2 \\ +10_2 \\ \hline \end{array} \quad \text{(C)}\ \begin{array}{r} 1101_2 \\ +1111_2 \\ \hline \end{array}$$

Solution

(A) $\begin{array}{r} 1 \\ 11_2 \\ +01_2 \\ \hline 100_2 \end{array}$ ← Carry the one Check: $\begin{array}{r} 3_{10} \\ +1_{10} \\ \hline 4_{10} \end{array}$

(B) $\begin{array}{r} 101_2 \\ +10_2 \\ \hline 111_2 \end{array}$ Check: $\begin{array}{r} 5_{10} \\ +2_{10} \\ \hline 7_{10} \end{array}$

(C) $\begin{array}{r} 111 \\ 1101_2 \\ +1111_2 \\ \hline 11100_2 \end{array}$ ← Carries Check: $\begin{array}{r} 13_{10} \\ +15_{10} \\ \hline 28_{10} \end{array}$

Binary Subtraction (One's Complement)

Many people are surprised to find that about the only arithmetic operation most microprocessors can do is addition. *They subtract by adding!* What you just read is not a misprint, but a very important statement: Most microprocessors *subtract by adding*. To do this trick used by microprocessors, you must know the meaning of the term **complement.**

Complement:
A reversion of the digital state. The complement of "OFF" is "ON," and the complement of "ON" is "OFF." Thus, the complement of 1 is 0, and the complement of 0 is 1.

One binary subtraction scheme uses what is called the **one's complement notation.**

One's Complement Notation:
A number in binary derived from another binary number resulting from the complement of each bit.

The complement of a binary number is found by complementing each bit of a binary number (if it is a 1, make it a 0; if a 0, make it a 1).

To subtract binary numbers by adding, first complement the number to be subtracted (the subtrahend). The two numbers are then added, and a 1 is added to the answer. Any final carry is added back to the answer. Here is an example:

$\begin{array}{r} 12_{10} \\ -4_{10} \\ \hline 8_{10} \end{array}$ $\begin{array}{r} 1100_2 \\ -0100_2 \\ \hline \end{array}$

	1100_2
complement →	$+1011_2$
add last carry to	10111_2
answer →	1
answer →	1000_2

As you can see, 12 − 4 = 8 and in binary, the same results are achieved by adding! Note that leading zeros must be placed in the subtrahend before complementing in order for the number to have as many bits as the minuend.

Example 2–21

Using one's complement notation, subtract the following binary numbers:

(A) $\begin{array}{r} 100_2 \\ -10_2 \\ \hline \end{array}$ (B) $\begin{array}{r} 1010_2 \\ -11_2 \\ \hline \end{array}$ (C) $\begin{array}{r} 10010110_2 \\ -10000_2 \\ \hline \end{array}$

Solution

Be sure to add leading zeros before complementing the subtrahend.

(A) $\begin{array}{r} 100_2 \\ -010_2 \\ \hline \end{array}$ complementing → $\begin{array}{r} 100_2 \\ +\ 101_2 \\ \hline 1001_2 \\ 1 \\ \hline 010_2 \end{array}$ add last carry back to ans → Check: $\begin{array}{r} 4_{10} \\ -2_{10} \\ \hline 2_{10} \end{array}$

(B) $\begin{array}{r} 1010_2 \\ -0011_2 \\ \hline \end{array}$ complementing → $\begin{array}{r} 1010_2 \\ +\ 1100_2 \\ \hline 10110_2 \\ 1_2 \\ \hline 111_2 \end{array}$ add last carry back to ans → Check: $\begin{array}{r} 10_{10} \\ -\ 3_{10} \\ \hline 7_{10} \end{array}$

(C) $\begin{array}{r} 10010110_2 \\ -00010000_2 \\ \hline \end{array}$ complementing → $\begin{array}{r} 10010110_2 \\ +11101111_2 \\ \hline 110000101_2 \\ 1 \\ \hline 10000110_2 \end{array}$ add last carry back to ans → Check: $\begin{array}{r} 150_{10} \\ -16_{10} \\ \hline 134_{10} \end{array}$

Binary Subtraction (Two's Complement)

The most popular way of subtracting is to add the **two's complement** of the subtrahend to the minuend.

Two's Complement Notation:
A value arrived at by taking the one's complement and adding 1 to the result.

To get the two's complement of any number, simply take its one's complement and add 1 to it. For example, to find the two's complement of 1010, take the one's complement 0101 and add 1 to get 0110.

Example 2–22

Find the two's complement of the following binary numbers:

(A) 0100 (B) 1100 (C) 10011001

Solution

Take the one's complement; then add 1 to the result.

(A) 0100	one's complement →	1011
	adding one →	+1
		1100
(B) 1100	one's complement →	0011
	adding one →	+1
		0100
(C) 10011001	one's complement →	01100110
	adding one →	+1
		01100111

There is a short-cut method of finding the two's complement of a binary number. Starting at the right and working toward the left, copy every digit up to the first 1 you come to. Then, complement every digit:

11000	two's complement → 01000
10110	two's complement → 01010
11111	two's complement → 00001

To subtract using the two's complement, simply add the two's complement to the minuend and ignore any carry.

Example 2–23

Subtract the following numbers, using two's complement notation:

(A) 1100	(B) 11011	(C) 1001001
−10	−1101	−0010

Solution

Add the required leading zeros to the subtrahend, take the two's complement, and add the numbers.

(A)	1100		1100	Check:	12_{10}
	−0010	two's complement →	+1110		-2_{10}
		ignore the carry →	1/1010		10_{10}
(B)	11011		11011	Check:	27_{10}
	−01101	two's complement →	+10011		-13_{10}
		ignore the carry →	1/01110		14_{10}
(C)	1001001		1001001	Check:	73_{10}
	−0010010	two's complement →	+1101110		-18_{10}
		ignore the carry →	1/0110111		55_{10}

Binary Multiplication

Many computers perform binary multiplication by repeated addition. For example, if you wanted to multiply 4 × 3, you would arrive at the answer by adding 4 to itself three times:

$$4 + 4 + 4 = 12$$

or, by adding 3 to itself four times:

$$3 + 3 + 3 + 3 = 12$$

If this problem were done in many digital computers, then the process would appear similar to the following:

```
  100 → first time
+ 100 → second time
 1000
+ 100 → third time
 1100 → answer
```

Binary Division

You can do division by repeated subtraction. You just keep subtracting the divisor from the dividend and count the number of times you can do this until the answer is zero. Anything that is left over is the remainder. As an example:

$$\frac{12}{4} = \begin{array}{rl} 12 & \\ -4 & \\ \hline 8 & \leftarrow 1 \\ -4 & \\ \hline 4 & \leftarrow 2 \\ -4 & \\ \hline 0 & \leftarrow 3 \text{ [3 is the answer.]} \end{array}$$

Since a microprocessor subtracts by adding, the same division process can be accomplished; thus for 12 ÷ 4:

$$\frac{12}{4} = \begin{array}{rrl} \text{convert to binary} \rightarrow & 1100 & \\ \text{two's complement of 4} \rightarrow & 1100 & \\ \hline & 1/1000 & \leftarrow 1_2 \\ & 1100 & \\ \hline & 1/0100 & \leftarrow 10_2 \\ & 1100 & \\ \hline & 1/0000 & \leftarrow 11_2 \end{array}$$

The final binary answer is 11_2 which is 3_{10}.

Counting in Hexadecimal

Since the hexadecimal number system is so common for many forms of programming, you should know how to use it to count (or you may think your computer is giving you a wrong answer when it isn't). Observe what happens in hexadecimal once you get to F. It's the same thing that happens in decimal once you get to its last symbol (9)—you indicate that you are starting over again by placing a 1 in the 10's column. In hexadecimal, you place a 1 in the 16's column:

1, 2, 3, 4, 5, 6, 7, 8, 9, A, B, C, D, E, F, 10, 11, 12, 13,
14, 15, 16, 17, 18, 19, 1A, 1B, 1C, 1D, 1E, 1F, 20, 21, 22,
23, 24, 25, 26, 27, 28, 29, 2A, 2B, 2C, 2D, 2E, 2F, 30 . . .

Adding Hexadecimal Numbers

Many pocket scientific calculators will add in hexadecimal. The process requires that you obey the following rules:

1. Always add two digits at a time, and think of the sum in decimal ($7_{16} = 7_{10}$ and $E_{16} = 14_{10}$).
2. If you find that the sum you get is 15_{10} or less, then write the resulting hexadecimal digit.
3. However, if you find that the sum is greater than 15_{10}, then write the amount that this exceeds 16_{10} and carry a 1 to the next column.

The following example demonstrates this process.

Example 2–24

Determine the sum of the following hexadecimal numbers:

(A) $14_{16} + 12_{16}$ (B) $84_{16} + 27_{16}$ (C) $5C_{16}$ $A2_{16}$ (D) $BE_{16} + CF_{16}$

Solution

(A) $14_{16} + 12_{16} = 26_{16}$

(Right column) $4_{10} + 2_{10} = 6_{10} = 6_{16}$
(Left column) $1_{10} + 1_{10} = 2_{10} = 2_{16}$

(B) $84_{16} + 27_{16} = AB_{16}$

(Right column) $4_{10} + 7_{10} = 11_{10} = B_{16}$
(Left column) $8_{10} + 2_{10} = 10_{10} = A_{16}$

(C) $5C_{16} + A2_{16} = FE_{16}$

(Right column) $12_{10} + 2_{10} = 14_{10} = E_{16}$
(Left column) $5_{10} + 10_{10} = 15_{10} = F_{16}$

(D) $BE_{16} + CF_{16} = 18D_{16}$

(Right column) $14_{10} + 15_{10} = 29_{10}$
$29_{10} - 16_{10} = 13_{10} =$ D with carry
(Left column) $11_{10} + 12_{10} + 1_{10} = 24_{10}$
$24_{10} - 16_{10} = 8_{10}$ with carry

Conclusion

This section gave you a real insight into the operations of a computer. It is really amazing what you can do by just adding! Here, you also saw how to count in binary and hexadecimal. These skills will be developed further in later chapters of the book. For now, test your new skills with the following review questions.

2–8 Review Questions

1. Give the binary counting sequence from 11000_2 to 11011_2.
2. State the rules for adding binary numbers.
3. Explain the meaning of the term *complement*.
4. Define the binary one's complement. Do the same for the two's complement.
5. Give the hexadecimal count from 7_{16} to 10_{16}.

MICROPROCESSOR APPLICATION

Discussion

This section will give you an idea of how information is stored in digital systems. You now know that all the information used by digital systems is in an ON or OFF condition. You and I interpret this as a 1 or 0 value. The question now is, how are all those complex things tucked away for use by a digital system?

How Memory Is Organized

Digital memory is actually thousands of tiny circuits that can each represent a 1 or a 0. It is convenient to think of these tiny memory circuits as organized in rows where each row contains a distinct piece of information (called a word). A very common arrangement is a row with a word size of 8 bits (a byte); however, for the discussion here, 4 bits (a nibble) will be used. This type of organization is illustrated in Figure 2–26.

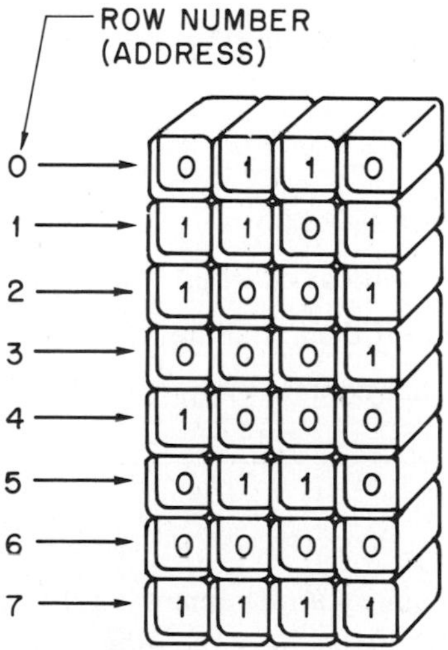

Figure 2–26 Memory organized in nibbles.

Figure 2–26 could represent a small computer memory that will hold eight rows of 4 bits each. Look at the very first row—it is called row 0! The 4 bits in row 0 are 0110. The second row (called row 1) contains the bits 1101. In digital memory, the row number is referred to as the **address,** and the bits inside are referred to as **data.** This means that address 0 contains the data 0110. Another way of saying this is that the data stored in memory location 0 is 0110, and the data stored at memory location 1 is 1101.

Address:
A label, name, or number identifying a location or unit where information is stored.

Data:
A general term that identifies basic elements of information that can be processed or produced by a digital system.

Example 2–25

Referring to Figure 2–26, answer the following questions.

(A) What is the data at address 3?

(B) Give the hexadecimal value of the data at address 7.

(C) What is the decimal value of the data at address 5?

(D) What address(es) contain 9_{16}?

Solution

(A) Data at address 3 → 0001.

(B) Hexadecimal value at address 7 → $1111_2 = F_{16}$.

(C) Decimal value of data at address 5 → $0110_2 = 6_{10}$.

(D) The address that contains 9_{16} → 1001_2 is address 2.

The preceding memory arrangement is called an 8 × 4 memory. Note that the number of addresses (rows) is given first; then the number of bits in each row. This is how the **memory size** is stated.

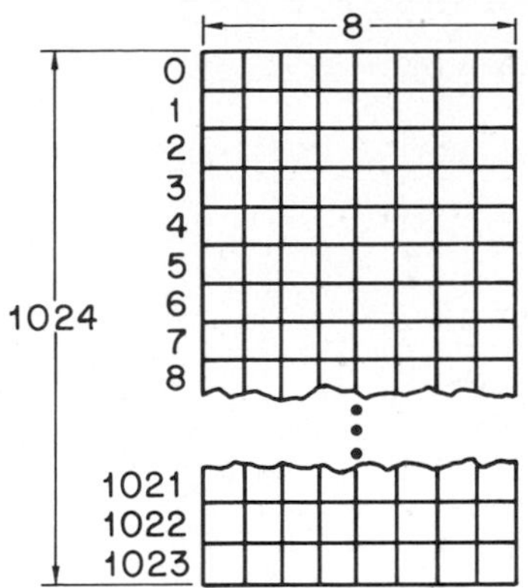

Figure 2–27 1024 × 8 ROM organizations.

Memory Size:
An indication of how bit patterns are accessed in a two-dimensional matrix. The notation is $A \times D$, where A gives the number of addresses (rows) and D gives the number of bits at each address (column or word size).

As an example, Figure 2–27 illustrates a ROM chip as a 1024 × 8 memory. It has 1024 addresses (memory locations) and contains 8 bits of data in each location. This means that its word size is 8. Other memory chips have word sizes of 16 (2 bytes) and 32 (4 bytes). The EA 4600 is a 4096 × 4 ROM; this means it has 4096 memory locations and a word size of 4.

What Is a K?

The letter *k* stands for kilo which means 1000. However, in digital the letter *K* stands for 1024. This is the reason. Look at the following binary progression (2^N, where $N = 0, 1, 2, \ldots$):

0, 1, 2, 4, 8, 16, 32, 64, 128, 256, 512, 1024, 2048, 4096, 8192, . . .

Notice that each number beyond 1024 is a multiple of 1024 (2 × 1024 = 2048; 4 × 1024 = 4096, etc.). It is therefore convenient to say that 1024 memory locations represents 1 K of memory and 2048 memory locations represents 2 K of memory. Thus 4 K of memory would mean 4096 memory locations and 16 K of memory would represent 16 × 1024 = 16 384 memory locations. It is common practice for industry to make memory chips with the number of addresses in multiples of 1024. There is a good reason for this as you will see in later chapters. For now, the important point is to be able to visualize what the terminology means.

Note that the amount of memory locations doesn't tell you the word size; it only tells you how many words can be stored.

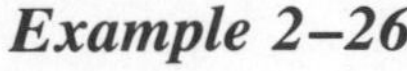

Example 2–26

Figure 2–28 illustrates different types of memory organizations. Give the size of each memory.

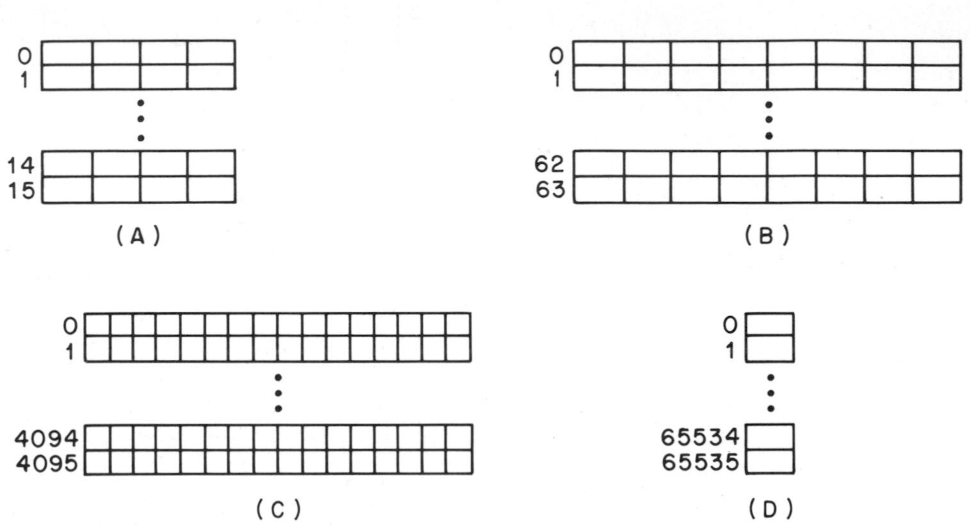

Figure 2–28

Solution

(A) There are 16 memory locations (0 through 15) with a word size of 4. Therefore the memory size is 16 × 4.

(B) This contains 64 memory locations (0 through 63) with a word size of 8. The memory size is 64 × 8.

(C) Here are 4096 memory locations (0 through 4095) with a word size of 16. First convert 4096 to K: 4096/1024 = 4K. Therefore the memory size is 4K × 16.

(D) For this one, the number of memory locations is 65 536 (0 through 65 535). Thus 65536/1024 = 64K. The size is therefore 64K × 1.

Memory Maps

Many times you will want to know the exact bit pattern contained in a section of memory. Recall that ROM is memory that keeps the same bit pattern. Say you wanted to know the bit pattern in a certain ROM chip. The manufacturer could supply you with a bit pattern, but all of these 1's and 0's could get confusing. Therefore hexadecimal numbers are usually used to represent the actual bit pattern located inside the memory. See Figure 2–29. Usually, the addresses of the memory map are also given in hexadecimal.

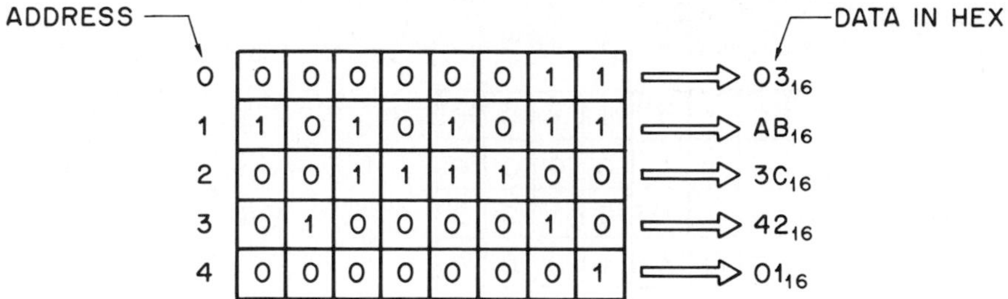

Figure 2–29 Using a memory map.

Example 2–27

Answer the following questions by referring to the memory map of Figure 2–30.

Address	Data	Address	Data
0	$3C_{16}$	8	09_{16}
1	00_{16}	9	$4C_{16}$
2	24_{16}	A	DF_{16}
3	18_{16}	B	EE_{16}
4	$B2_{16}$	C	$A3_{16}$
5	$C5_{16}$	D	$C4_{16}$
6	36_{16}	E	96_{16}
7	51_{16}	F	CD_{16}

Figure 2–30

(A) State the memory size in decimal.

(B) Convert the addresses to decimal and the bit patterns to binary—redraw the memory with the actual bit pattern.

Solution

(A) There are 15_{10} + 1 (the 0 location) = 16 memory locations. There must be enough bits in each memory location to represent two hexadecimal digits. It takes 4 bits to represent one hexadecimal digit, so this memory must have 8 bits in each location—an 8-bit word. The size of this memory is therefore 16 × 8.

(B) See Figure 2–31.

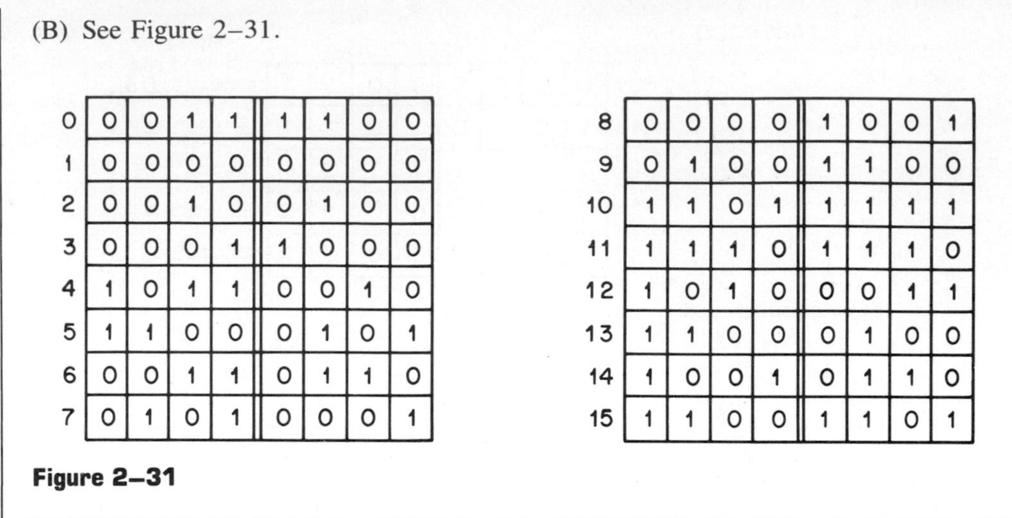

0	0	0	1	1	1	1	0	0
1	0	0	0	0	0	0	0	0
2	0	0	1	0	0	1	0	0
3	0	0	0	1	1	0	0	0
4	1	0	1	1	0	0	1	0
5	1	1	0	0	0	1	0	1
6	0	0	1	1	0	1	1	0
7	0	1	0	1	0	0	0	1

8	0	0	0	0	1	0	0	1
9	0	1	0	0	1	1	0	0
10	1	1	0	1	1	1	1	1
11	1	1	1	0	1	1	1	0
12	1	0	1	0	0	0	1	1
13	1	1	0	0	0	1	0	0
14	1	0	0	1	0	1	1	0
15	1	1	0	0	1	1	0	1

Figure 2–31

The first 4 bits, including the MSB, of an 8-bit word is the first nibble of data at a certain memory location. These 4 bits are called the **high-order nibble.** The second nibble of data at the same memory location is called the **low-order nibble.** As an example, the high-order nibble at memory location 0 of Figure 2–30 is 3_{16} (representing 0011_2), and the low-order nibble at the same memory location is C_{16} (representing 1100_2).

High-Order Nibble:
The first 4 bits, including the MSB, of an 8-bit word.

Low-Order Nibble:
The last 4 bits, including the LSB, of an 8-bit word.

Suppose you wanted a particular bit pattern placed inside a certain memory location. The bit pattern you selected would probably be entered as a hexadecimal number. In this case, there are two things you would need to enter: the address and the data.

The hexadecimal loader may be part of a microprocessor-based system used to control a robot or a factory assembly line. See Figure 2–32. There are 16 buttons on the hexadecimal loader, which represent the hexadecimal numbers. There is a seven-segment readout that indicates the hex value of the most recent button pressed. The unit will load bit patterns into a 16 × 8 memory (16 addresses with an 8-bit word size—a byte).

When the address button is pressed, the displayed value will activate an address; this then becomes the current activated address. When the HI button is pressed, the displayed value will be loaded into the high-order nibble at the current activated address. Pressing the LOW button will load the low-order nibble of the current activated address. In order to change the address, you must enter the hex value of the address and press the address button.

The basic idea of how to get a specific bit pattern into a specific memory location with a hexadecimal loader is as follows. Suppose you were using the hexadecimal loader of Figure 2–32, and you wanted to enter the bit pattern shown in Figure 2–33. It makes no difference in what order you address the memory, but just keep things simple. Start

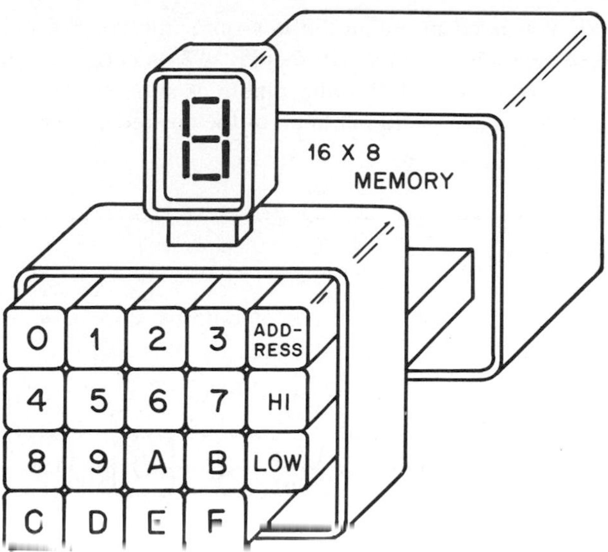

Figure 2–32 Typical hex loader used to program a robot.

with memory location 0 and proceed from there. Press button 0; this will cause the seven-segment readout to display a 0. The system now needs to know if the 0 is to be an address or data. Since this is to be an address, simply push the address button, and address 0 will be activated.

Now look at the bit pattern of Figure 2–33. The first memory location (address 0) is to have 0110 0011, which is 63_{16}. The high-order nibble is 6_{16}, and the low-order nibble is 3_{16}. So your next step is to press the 6 key and then the HI key. You will see the seven-segment LED light up with a 6 as soon as you push the 6 key. Since you already activated address 0, the bit pattern for 6_{16} will be entered into the high-order nibble at address 0 as soon as you push the HI key.

	HIGH				LOW			
0	0	1	1	0	0	0	1	1
1	1	0	1	1	0	1	0	0
2	1	1	1	1	1	1	0	0
3	1	0	1	0	0	0	0	0
4	0	0	0	1	1	1	1	0
5	1	0	1	1	0	1	0	0
6	0	0	0	0	1	1	0	1
7	1	1	1	0	1	1	1	1

Figure 2–33 Desired bit pattern to be put into memory.

Next, you need to put in the low-order nibble. So, enter 3 and press the LOW key. The seven-segment readout will display a 3 as soon as you press the 3 key. It will enter this bit pattern into the low-order nibble as soon as you press LOW.

This process is repeated until all of the desired bits are entered. The following example illustrates.

Example 2–28

A student pushes the sequence of buttons shown in Figure 2–34. Determine the address and bit pattern stored there.

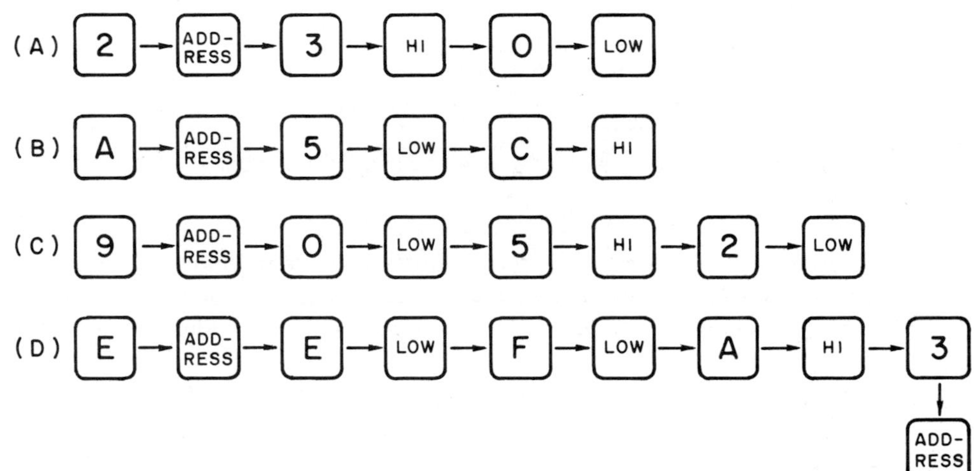

Figure 2–34

Solution

(A) The address is 2_{10}. The bit pattern is 30_{16} = 0011 0000.

(B) The address is 10_{10}. 5 was entered as the low nibble, C_{16} as the high nibble. Thus the bit pattern is $C5_{16}$ = 1100 0101.

(C) The address is 9_{10}. 0 was first entered into the low nibble, then 5 into the high nibble, and then 2 into the low nibble (the low nibble will contain the last pattern entered into it). Thus the pattern is 52_{16} = 0101 0010.

(D) The address is 14_{10}. E_{16} was entered into the low nibble, followed by F_{16} into the low nibble (low nibble now has F_{16}). Then A_{16} was entered into the high nibble. The resulting bit pattern is AF_{16} = 1010 1111. Lastly, address 3_{10} was activated but no data were entered.

The resulting bit pattern is shown in Figure 2–35.

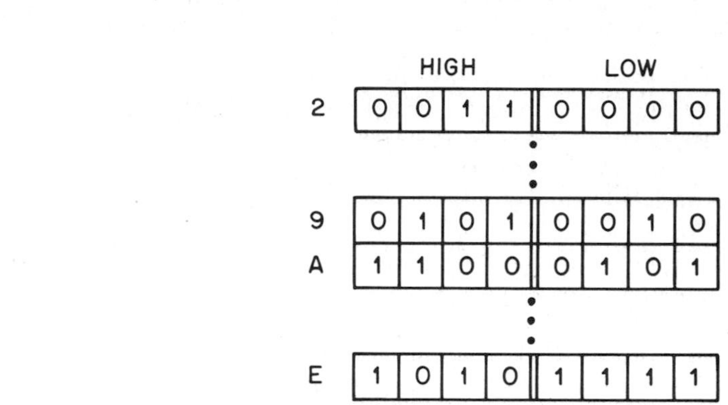

Figure 2–35

Parity

The storage and transmission of bits in and between systems can cause a bit to change accidentally. Thus, a 1 could become a 0, or a 0 a 1. There are systems that can check to see if this has happened. Here you will see two different methods of checking a binary pattern. One is called **odd parity,** and the other is called **even parity.**

Parity:
Relates to the maintenance of a count by keeping the same number of binary 1's in a computer word in order to be able to check that word's correctness.

Odd Parity:
Creating a word with an extra bit so that the number of 1's in the word adds up to an odd number.

Even Parity:
Creating a word with an extra bit so that the number of 1's in the word adds up to an even number.

No matter which type of parity you use, you must have an extra bit within the computer word. If you are using odd parity, this bit will be 0 or 1 in order to always make the number of bits come out odd. This **parity bit** can be placed at the beginning or at the end of the word. It is usually placed at the beginning.

Parity Bit:
An extra bit added to a word so that an accidental change of a single bit in the word can be detected.

Example 2–29

For the following words, the leftmost bit of an 8-bit word represents a parity bit. State the value of the 8-bit number if the following decimal numbers are to be stored by using odd parity.

(A) 6_{10} (B) 15_{10} (C) 24_{10}

Convert the resulting numbers into its hexadecimal equivalent for loading into a hexadecimal loader.

Solution

First convert each decimal number to its binary equivalent. Then count the number of 1's. If the number is odd, set the leftmost bit to 0. If the number is even, set the leftmost bit to 1.

(A) $6_{10} = 110_2$: even number of bits → 1000 0110 (Parity bit = 1)

(B) $15_{10} = 1111_2$: even number of bits → 1000 1111 (Parity bit = 1)

(C) $24_{10} = 11000_2$: even number of bits → 1001 1000 (Parity bit = 1)

Convert each number into its hexadecimal equivalent:

(A) 1000 0110 = 86_{16}

(B) 1000 1111 = $8F_{16}$

(C) 1001 1000 = 98_{16}

When using this system, the largest binary number that can be represented is $2^7 - 1 = 127_{10}$ = X111 1111 (the X represents the parity bit).

Example 2–30

Tom wants to store his name in ASCII code in memory, starting at memory location 7. He is using a hexadecimal loader of the type in Figure 2–32 and wishes to use even parity. Code what he is to do in hexadecimal and state each step.

Solution

First convert the name to ASCII (refer to Table 2–5).

T → 101 0100
o → 110 1111
m → 110 1101

Since the ASCII code uses 7 bits, an eighth bit can be added to the left as the parity bit. Converting to even parity gives

101 0100 → 1101 0100 → $D4_{16}$
110 1111 → 0110 1111 → $6F_{16}$
110 1101 → 1110 1101 → ED_{16}

Tom would then enter the keying sequence shown in Figure 2–36.

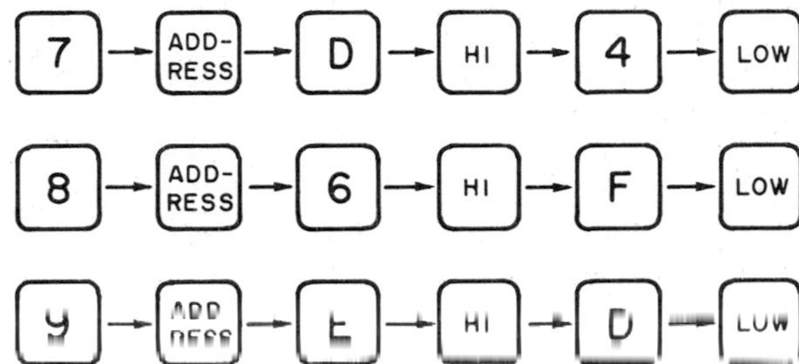

Figure 2–36

Conclusion

This concludes the microcomputer applications section of this chapter. What you have learned here is fundamental to understanding digital systems. You can see why learning binary and hexadecimal numbers is so important. The applications presented here will be built upon in the microcomputer application sections of other chapters. Test your understanding of this section by trying the following review questions.

2–9 Review Questions

1. Describe the notation used to indicate how digital memory is organized.
2. Explain the difference between an address and data.
3. Explain how the hexadecimal loader described in this section is used to place data at a given address.
4. Describe the difference between even parity and odd parity.

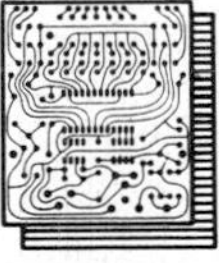

ELECTRONIC APPLICATION

Note: Electronic Application sections may require some knowledge of electronics. Non-electronics majors may omit these sections with no loss in continuity of the subject matter.

Discussion

Since the focus of this chapter was on the ON and OFF conditions of the computer, this section will introduce some of the more common devices and test equipment used to create and check the 1 and 0 world of digital systems.

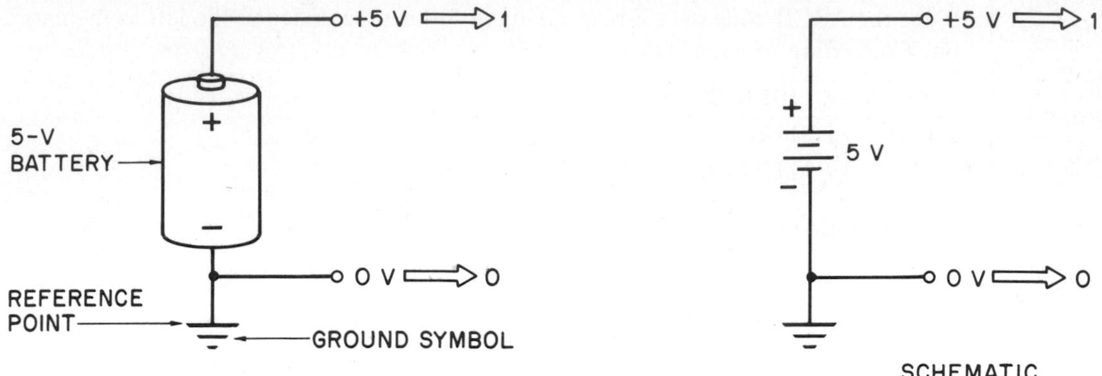

Figure 2–37 Method of representing 1's and 0's.

Sources

One of the most common systems for representing digital information is to use +5 V to represent a 1 and 0 V to represent a 0. This is illustrated in Figure 2–37. Note that for the purposes of this text, the ground symbol represents a 0 and the +5 V symbol represents a 1. In order to represent 1's and 0's electrically, programming devices, such as switches, and display devices, such as miniature lights, are needed.

Miniature Switches

Figure 2–38 illustrates some typical switches found in digital work. Figure 2–39 illustrates the schematic of several different types of switches used in digital systems along with their corresponding nomenclature.

Figure 2–38 Typical digital switches. *Courtesy* of AMP Incorporated, Harrisburgh, PA.

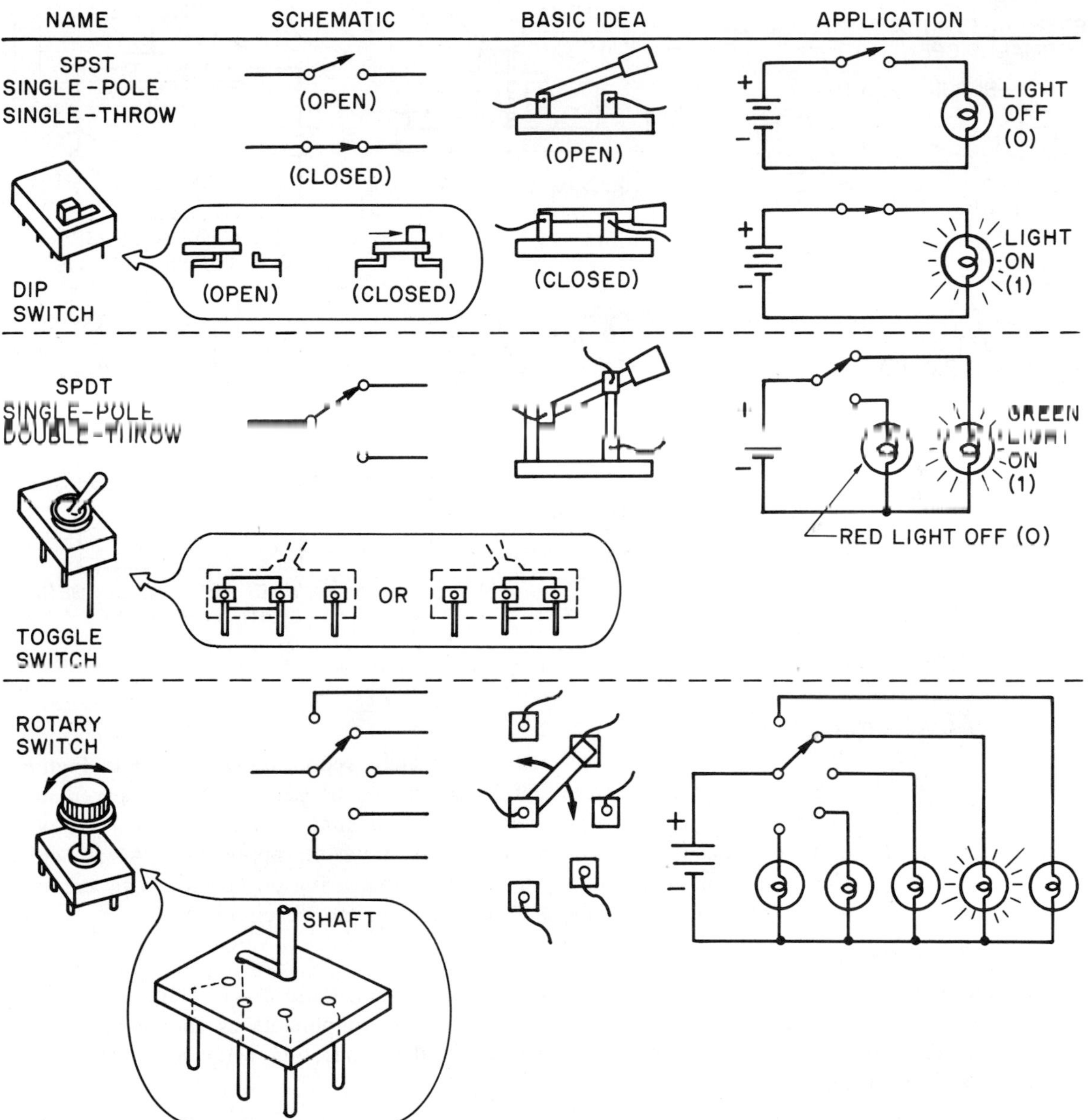

Figure 2–39 Schematic of different types of switches.

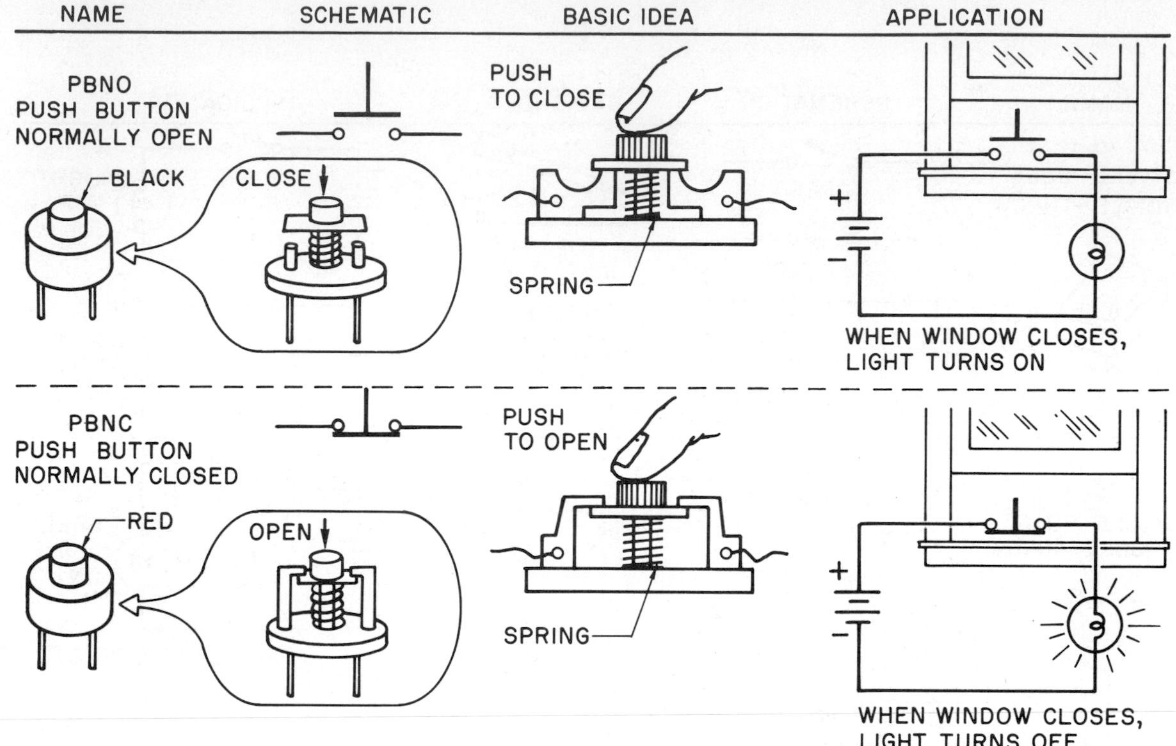

Figure 2–39 Continued

Light Emitting Diodes

An LED is an economical and simple method of visually representing a 1 or a 0. Figure 2–40 shows the LED connection. Note that the polarity of the LED must be observed and that a resistor must be placed in series with the LED to limit its current (usually about 220 Ω) for 5 V, but this value may vary depending upon the application. Along with switches, LEDs can be mounted on a test board without using any solder. This creates temporary connections used for testing and debugging electronic circuits.

Prototype Boards

Prototype boards come in a variety of sizes, as shown in Figure 2–41. The construction of these boards is similar. They are made to accept components as shown in Figure 2–42. When using these boards, never strip back the connecting wires so far that they could short against each other.

Wire-Wrapping

A prototype board is good for a quick "try-and-see" circuit. However, it is easy for wires to get pulled up or shorted. A more lasting method is to use wire-wrapping. This is an excellent method to use when a more mechanically stable digital circuit is desired or a few prototypes are to be constructed. The wire-wrapping technique is shown in Figure 2–43.

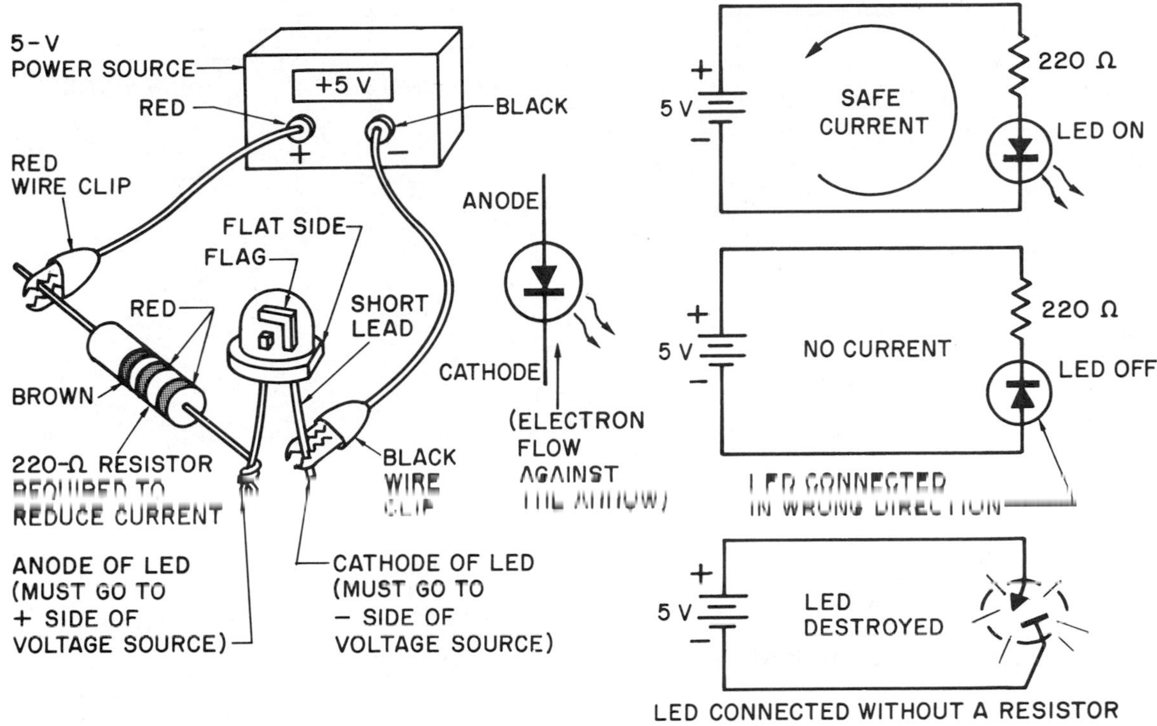

Figure 2–40 Connecting LEDs.

Figure 2–41 Representative prototype boards. *Courtesy* of Global Specialties.

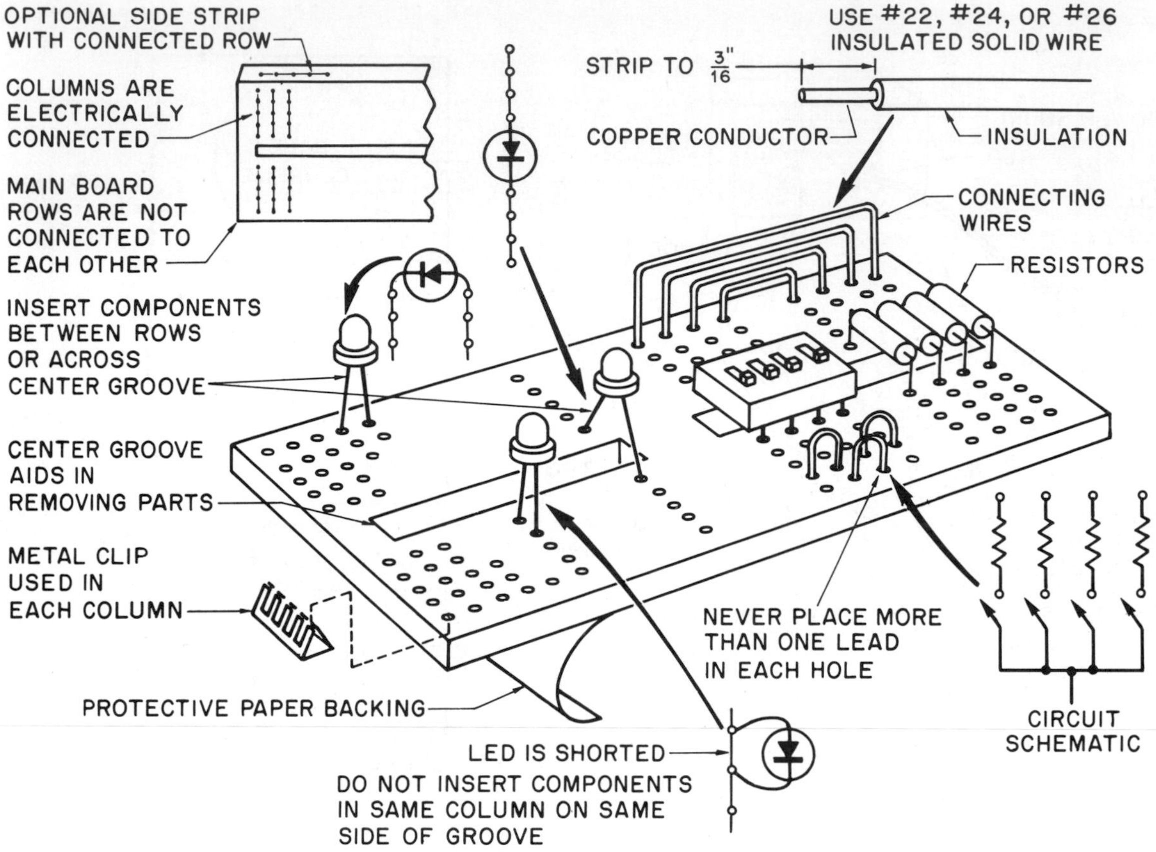

OPTIONAL SIDE STRIP
WITH CONNECTED ROW
COLUMNS ARE
ELECTRICALLY
CONNECTED
MAIN BOARD
ROWS ARE NOT
CONNECTED TO
EACH OTHER
INSERT COMPONENTS
BETWEEN ROWS
OR ACROSS
CENTER GROOVE
CENTER GROOVE
AIDS IN
REMOVING PARTS
METAL CLIP
USED IN
EACH COLUMN
PROTECTIVE PAPER BACKING
STRIP TO 3"/16
COPPER CONDUCTOR
USE #22, #24, OR #26
INSULATED SOLID WIRE
INSULATION
CONNECTING
WIRES
RESISTORS
NEVER PLACE MORE
THAN ONE LEAD
IN EACH HOLE
CIRCUIT
SCHEMATIC
LED IS SHORTED
DO NOT INSERT COMPONENTS
IN SAME COLUMN ON SAME
SIDE OF GROOVE

Figure 2–42 Using the prototype board for mounting parts.

COMPONENT SIDE
WIRE-WRAP SIDE
HANDLE FOR
TWISTING
WIRE-WRAP BIT
WIRE-WRAPPING
TOOL
HOLE FOR
WIRE-WRAP
TERMINAL
BOARD
WITH HOLES
WIRE
PART

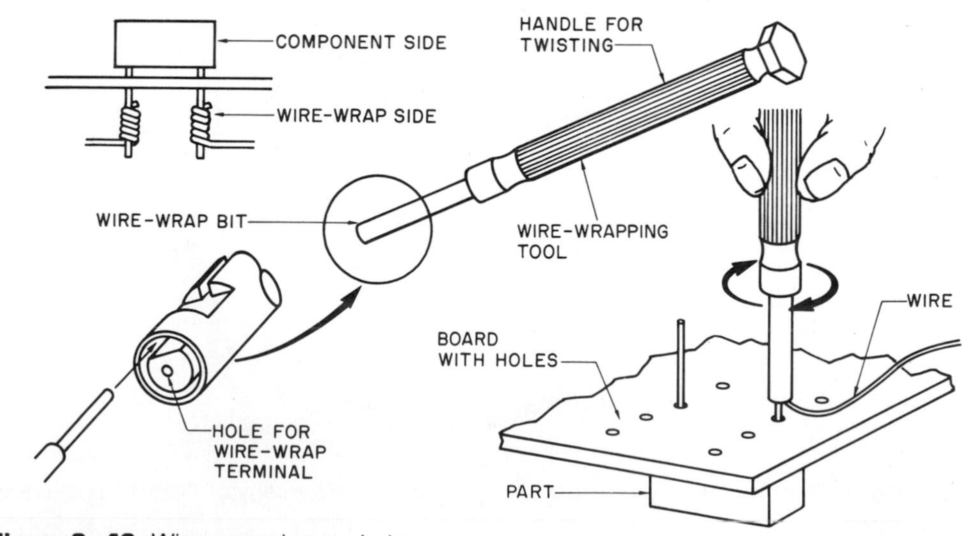

Figure 2–43 Wire-wrapping technique.

Figure 2–44 Typical digital probe. *Courtesy* of Global Specialties.

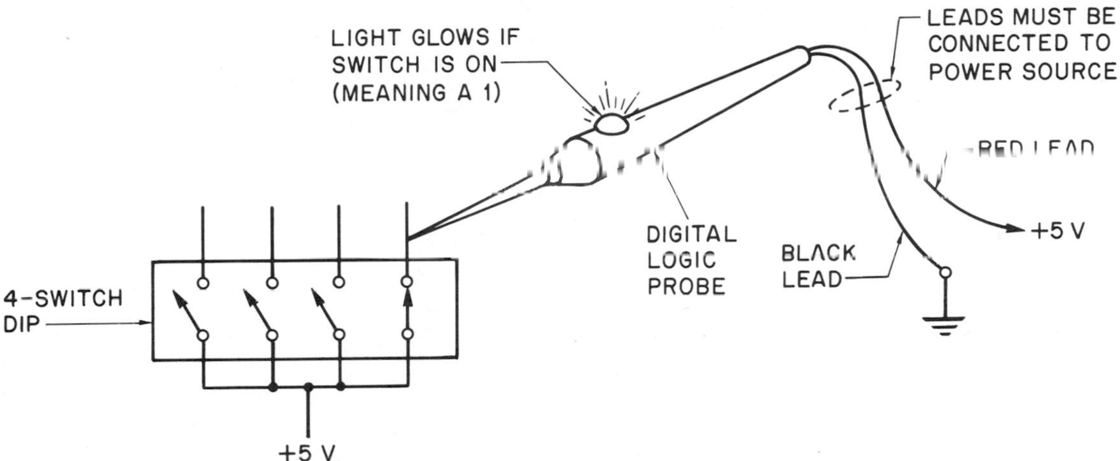

Figure 2–45 Digital probe testing a digital system.

Digital Probe

An instrument used to test if a 1 (+5 V) or a 0 (0 V or ground) is present in a circuit is called a *digital probe*. A typical digital probe is shown in Figure 2–44. Figure 2–45 shows a digital probe being used to check the condition of a simple binary circuit. Note that this circuit can represent 4 bits. The input device is four switches, and the output device could be four LEDs.

Digital Pulser

Where the digital probe looked for the presence of a 1 or a 0, a *digital pulser* can create them. A typical digital pulser probe is shown in Figure 2–46. Figure 2–47 shows a digital pulser in action. Here it is checking to make sure that an LED will indicate the presence of a 1 (+5 V). Since the pulser is on the continuous mode, the LED will be blinking ON and OFF about five times each second.

Figure 2–46 Typical digital pulser probe. *Courtesy* of Global Specialties.

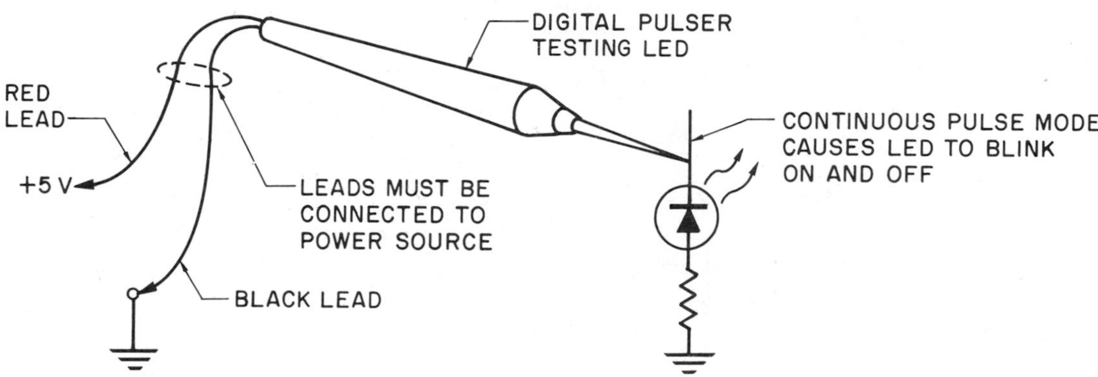

Figure 2–47 Digital pulser in action.

Conclusion

This section introduced you to the most fundamental hardware and instrumentation tools used by the digital technician. For those of you who will be responsible for the repair, maintenance, design, and prototyping of digital systems, these sections are as important as the theory of each chapter. Check your understanding of this section by trying the following review questions.

2–10 Review Questions

1. Describe the most common method of electrically representing a 1 or a 0.
2. Explain the difference between a PBNO switch and a PBNC switch.
3. Describe the use of a prototype board.
4. Describe wire-wrapping construction. When should it be used?

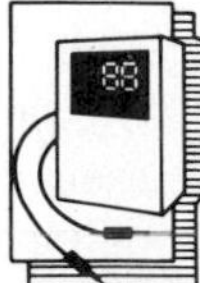

TROUBLESHOOTING SIMULATION

The troubleshooting simulation program for this chapter presents an opportunity for you to work with real hardware. Here you will see how a digital circuit is wired and then tested. When you are ready, the computer will wire the circuits for you. It is up to you to determine if they are wired correctly. This is an important introduction to the troubleshooting simulations that follow.

SUMMARY

- The decimal number system contains 10 symbols.
- Using positional notation, number system symbols can represent any numeric value.
- The base of a number system is the number of separate characters used by the number system.
- A number of any system may be represented by the sum of the products of the weighted characters and exponents of the system base.
- The ON and OFF states of digital systems can be interpreted as representations of numbers.
- A binary number is a number system to the base 2 and uses the symbols 0 and 1 to represent all numerical values.
- A base subscript is needed to indicate the base of the number being used.
- The scientific calculator can be used to convert between binary and decimal.
- The ON and OFF pattern of display lights and switches can be used to represent binary numbers.
- Keypads can be used to enter digital information into digital systems.
- Binary coded decimal (BCD) is an 8421 code that uses 4 bits to represent a decimal digit from 0 to 9.
- The advantage of BCD is the ease of conversion between decimal and BCD.
- The disadvantage of BCD is that six available combinations of 4 binary bits are wasted.
- The hexadecimal number system consists of 16 different symbols and is therefore to the base 16.
- The advantage of the hexadecimal number system is the ease of conversion between it and binary.
- The hexadecimal number system overcomes the disadvantage of BCD in that all 16 combinations of 4 binary bits are used.
- Hexadecimal values may be represented using seven-segment readouts.
- There are advantages for using other number codes to represent the ON and OFF conditions of digital systems other than binary numbers.
- The Gray code is useful in applications where the change of only one bit between values adds to the reliability of the system.
- The excess-3 code is another binary code.
- An alphanumeric code is a system that allows the ON and OFF states of a digital system to represent alphabetical, numerical, and special characters.
- The American Standard Code for Information Interchange (ASCII) is the most common alphanumeric code.

- Arithmetic operations, such as addition, subtraction, multiplication, and division, can be performed using binary numbers.
- Most digital systems capable of arithmetic processes perform all of these processes by addition.
- The one's or two's complement system can be used to perform subtraction by adding binary numbers.
- Computer memory can be visualized as storing 1's and 0's in an orderly fashion of rows and columns.
- An address is the location of a particular row of 1's and 0's.
- Data is the pattern of 1's and 0's stored at a given address.
- In binary numbers, a K stands for 1024 bits.
- A memory map is a convenient way of representing the bit patterns of memory using the hexadecimal number system.
- Parity is a method of testing the accuracy of a given bit pattern.
- It is common practice in digital systems to have an ON condition represented by +5 V and an OFF condition represented by 0 V or ground.
- Miniature switches are one way of generating a digital code.
- Miniature lights, such as LEDs, are one way of observing a digital code.
- Prototype boards are useful for easily assembling a temporary circuit for testing and observation.
- Wire-wrapping is a means of creating a more permanent type of circuit.
- The home, school, or industrial computer can be used to easily convert between number systems.

CHAPTER SELF-TEST

I. TRUE/FALSE

Answer the following questions true or false.

1. There are 10 symbols in the base 10 number system.
2. The term *base* means the largest or smallest value that can be represented by a number system.
3. Binary numbers are a way of representing the ON and OFF conditions on digital circuits.
4. The number 1011_2 is a binary number.
5. A bit represents four binary places.

II. MULTIPLE CHOICE

Answer the following questions by selecting the most correct answer.

6. The decimal equivalent of 1010_2 is
   (A) 1010_{10} (B) 10_{10} (C) $10 + 10 = 20$ (D) not defined
7. The decimal equivalent of 0.01_2 is
   (A) 0.25_{10} (B) 0.01_{10} (C) 0.5_{10} (D) not defined

8. A nibble is
   (A) a binary bit (B) 8 binary digits (C) 4 binary bits (D) none of these
9. The binary equivalent of 12_{10} is
   (A) 12_2 (B) 1100_2 (C) $0001\ 0010_2$ (D) not defined
10. A word can be the size of a
   (A) bit (B) nibble (C) byte (D) any of these if they are treated as a unit.

III. MATCHING

Match the items on the right to the correct statements on the left.

11. BCD	(A) 0001, 0011, 0010 → 1, 2, 3
12. Binary	(B) $6C_{16}$
13. Hex	(C) 0101 1001 = 59_{10}
14. Excess-3	(D) 1101_2
15. Gray code	(E) 1011 1100 = 89_{16}

IV. FILL-IN

Fill in the blanks with the most correct answer(s).

16. A 16 × 4 memory means it is organized into ______ rows with ______ bits in each row.
17. The location of data in a memory is referred to as the ______ .
18. In computer memory, a size of 12K means exactly ______ decimal memory locations.
19. A ______ ______ is an easy way of representing memory bit patterns using hex numbers.
20. The ______ bit is a way of ensuring that a bit pattern is correct.

V. OPEN-ENDED

Answer the following questions as indicated.

21. Convert the following binary numbers to decimal:
   (A) 0011_2 (B) 1100_2 (C) $1\ 1011_2$
22. Convert the following decimal numbers to binary:
   (A) 5_{10} (B) 14_{10} (C) 248_{10}
23. Convert the following numbers to BCD:
   (A) 18_{10} (B) 238_{10}
24. Convert the following hex numbers to binary:
   (A) $4F_{16}$ (B) $3CA_{16}$ (C) $620A_{16}$
25. Convert the following binary numbers to hex:
   (A) 1101_2 (B) 10110101_2 (C) 100100001101_2

Answers to Chapter Self-Test:

1] T 2] F 3] T 4] T 5] F
6] B 7] A 8] C 9] B 10] D
11] C 12] D 13] B 14] E 15] A
16] 16, 4 17] address 18] 12 × 1024 = 12 288
19] memory map 20] parity
21] (A) 3_{10} (B) 12_{10} (C) 27_{10}
22] (A) 0101_2 (B) 1110_2 (C)11111000_2

23] (A) $0001\ 1000_{BCD}$ (B) $0010\ 0011\ 1000_{BCD}$
24] (A) $0100\ 1111_2$ (B) $0011\ 1100\ 1010_2$ (C) $0110\ 0010\ 0000\ 1010_2$
25] (A) D_{16} (B) $B5_{16}$ (C) $90D_{16}$

CHAPTER PROBLEMS

Basic Concepts

Section 2–1:

1. Give the place value for each digit in the following decimal numbers:
   (A) 25 (B) 379 (C) 3 987 (D) 2 057 008.97 (E) 3.141 59
2. Represent the following decimal numbers using powers of 10 notation:
   (A) 36 (B) 0.25 (C) 489.3 (D) 34 900.002 5
   (E) 12 900 046.005 6
3. State the final value of each of the numbers written in powers of 10 notation:
   (A) $3\times10^2 + 0\times10^1 + 4\times10^0$
   (B) 5×10^{-3}
   (C) $4\times10^3 + 8\times10^2 + 3\times10^1 + 0\times10^0 + 2\times10^{-1} + 5\times10^{-2}$

Section 2–2:

4. Determine the values represented in each part of Figure 2–48.

Figure 2–48

5. Referring to Figure 2–49, determine the values represented by each section.

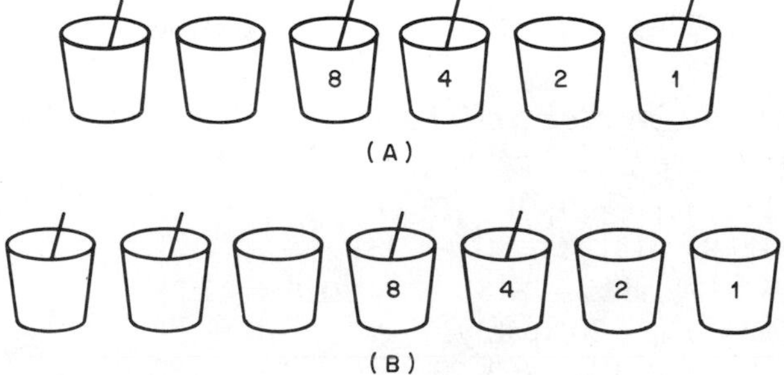

Figure 2–49

Section 2–3:

6. Convert each binary number to decimal:
   (A) 0101_2 (B) 1101_2 (C) 110110_2 (D) 1011001010_2
   (E) 1001001001_2 (F) 10101101011010_2
7. Convert the following decimal numbers to binary:
   (A) 8_{10} (B) 18_{10} (C) 86_{10} (D) 583_{10}
   (E) 999_{10} (F) $12\ 380_{10}$
8. Convert the following decimal numbers to binary (no more than 16 places):
   (A) 0.5_{10} (B) 0.625_{10} (C) $0.037\ 9_{10}$ (D) 0.001_{10}
   (E) $0.001\ 25_{10}$ (F) $0.000\ 062\ 5_{10}$
9. Convert the following binary numbers to decimal:
   (A) 10.10_2 (B) 1.010_2 (C) 1101.1011_2
   (D) 1010011.010110_2 (E) 10100110110.0100101_2
   (F) $11011110011001.000100010001_2$
10. Convert the following decimal numbers to binary (no more than 16 places):
   (A) 8.25_{10} (B) 19.025_{10} (C) 86.625_{10}
   (D) $168.003\ 5_{10}$ (E) $12\ 876.001\ 26_{10}$ (F) 3.141 59

Section 2–4:

11. Determine the binary values represented by each of the light patterns in Figure 2–50.

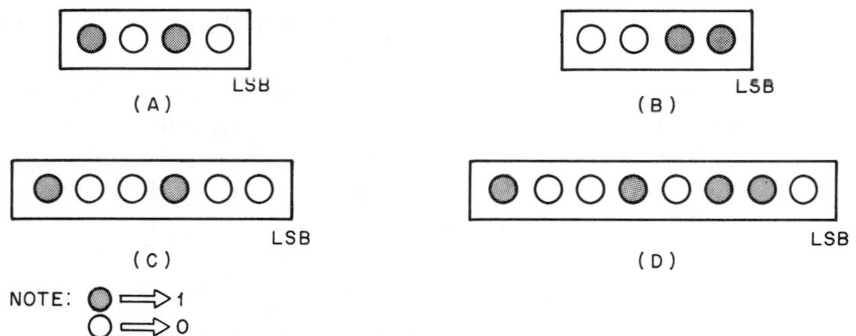

Figure 2–50

12. Convert each of the light patterns of Figure 2–50 to its decimal equivalent.
13. Give the binary pattern represented by each of the seven-segment readouts of Figure 2–51.

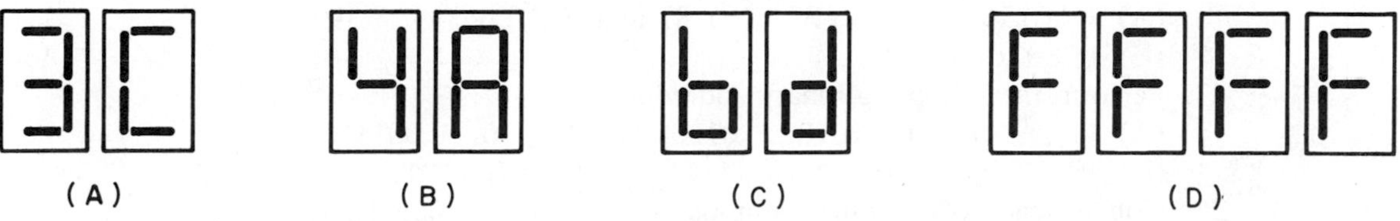

Figure 2–51

14. Convert the following decimal numbers to BCD:
    (A) 3_{10} (B) 84_{10} (C) 790_{10} (D) $6\,001_{10}$
    (E) $12\,124_{10}$ (F) $184\,030_{10}$
15. Convert the following BCD numbers to decimal:
    (A) 0100_{BCD} (B) $0100\,0100_{BCD}$ (C) $1001\,0111\,0011_{BCD}$
    (D) $0100\,0010\,0001\,0000_{BCD}$
    (E) $1000\,0101\,0111\,1001\,0100_{BCD}$

Section 2–5:

16. Convert the following hexadecimal numbers to binary:
    (A) 4_{16} (B) B_{16} (C) 38_{16} (D) DE_{16} (E) $D0C_{16}$
    (F) $35F0_{16}$ (G) $C5D6E7F9_{16}$
17. Convert the following binary numbers to hexadecimal:
    (A) 0000_2 (B) 1010_2 (C) 11010_2 (D) 1011101_2
    (E) 1001001001_2 (F) $1010011011001011 1_2$
    (G) 101100111010111_2
18. Convert the following hexadecimal numbers to decimal:
    (A) B_{16} (B) 10_{16} (C) 100_{16} (D) $C0_{16}$ (E) 450_{16}
    (F) $3CDE_{16}$ (G) $FFFF_{16}$
19. Convert the following decimal numbers to hexadecimal:
    (A) 1_{10} (B) 16_{10} (C) 256_{10} (D) $1\,024_{10}$
    (E) $16\,384_{10}$ (F) $65\,536_{10}$

Section 2–6:

20. Convert the following decimal numbers to their Gray code equivalents:
    (A) 2 (B) 1 (C) 5 (D) 3 (E) 7
21. Convert the following Gray code numbers to their binary equivalents:
    (A) 010 (B) 110 (C) 001 (D) 111 (E) 101
22. Convert the following decimal numbers to excess-3 code:
    (A) 6_{10} (B) 18_{10} (C) 268_{10} (D) 864_{10}
    (E) $2\,300_{10}$ (F) $58\,010_{10}$

Section 2–7:

23. Convert your initials to their ASCII code.
24. Convert your address to its ASCII code, include your zip code.
25. Decode the following ASCII code message. There are 7 bits per characters.
    000001010110011101111111010101000001100100110100111001000
    01000000110000101000000110011111011111101111111001000100000
    11010101101111111000100100000100101111

Section 2–8:

26. Do a binary count between the indicated decimal values:
    (A) 1 to 15 (B) 24 to 32 (C) 87 to 103 (D) 389 to 395
    (E) 1 280 to 1 287
27. Perform the following binary additions:
    (A) $11_2 + 10_2$ (B) $1010_2 + 0100_2$ (C) $11000_2 + 00111_2$
    (D) $1011010_2 + 1011_2$ (E) $111111_2 + 1_2$ (F) $1101101_2 + 101101_2$
28. Convert each of the binary numbers to its one's complement:
    (A) 0001_2 (B) 1110_2 (C) 1010_2 (D) 0101_2 (E) 101101_2
    (F) 101101101101_2 (G) 1011010111_2 (H) 111111_2

29. Subtract each of the binary quantities using the one's complement method:
(A) $101_2 - 10_2$ (B) $1100_2 - 11_2$ (C) $1010_2 - 1001_2$
(D) $101101_2 - 10011_2$ (E) $1011011011_2 - 11111_2$
30. Convert each of the binary numbers to its two's complement:
(A) 0000_2 (B) 1111_2 (C) 1010_2 (D) 0101_2 (E) 10110000_2
(F) 11011011011_2 (G) 1010011100011000_2
31. Subtract each of the binary numbers using the two's complement method:
(A) $101_2 - 11_2$ (B) $1010_2 - 101_2$ (C) $1111_2 - 1000_2$
(D) $101101_2 - 1101_2$ (E) $11110000_2 - 10110_2$
32. Multiply each of the following binary numbers using the process of repeated addition:
(A) $10_2 \times 10_2$ (B) $11_2 \times 11_2$ (C) $100_2 \times 11_2$
(D) $1010_2 \times 11_2$ (E) $1101_2 \times 100_2$
33. Divide the following binary numbers using the process of subtraction in the two's complement system:
(A) $100_2/10_2$ (B) $1000_2/100_2$ (C) $1010_2/101_2$
(D) $1100_2/10_2$
34. Continue the count in hexadecimal between the following hexadecimal values:
(A) 3_{16} to A_{16} (B) 8_{16} to 10_{16} (C) 19_{16} to 20_{16}
(D) 100_{16} to 110_{16} (E) $FFE6_{16}$ to $FFFF_{16}$
35. Add the following hexadecimal numbers:
(A) $4_{16} + 3_{16}$ (B) $A_{16} + 2_{16}$ (C) $C_{16} + 5_{16}$
(D) $10_{16} + 12_{16}$ (E) $DE_{16} + 38_{16}$ (F) $FE5A_{16} + DA_{16}$

Applications

36. Answer the following questions referring to Figure 2–52. Assume that this is a 10 × 8 memory.
(A) What is the decimal value of the data stored at address 11_2?
(B) If data at address 100_2 is BCD, what is its value in decimal?
(C) If even parity is being used, then what is represented at address 101_2 if it is in ASCII? (Use first bit as parity.)
(D) What is represented at address 101_2 if it is excess-3 code?

0	1	0	1	0	1	1	0	1
1	1	1	0	1	0	0	0	0
2	1	0	0	0	1	1	0	0
3	1	0	1	1	0	0	1	1
4	0	0	1	1	0	1	1	0
5	1	1	0	0	0	1	1	0
6	0	0	0	0	1	0	0	0
7	1	1	1	1	0	1	1	1
8	1	0	0	1	0	1	1	0
9	0	1	0	1	1	0	0	1

Figure 2–52

37. Determine the memory size of each of the representations in Figure 2–53.
38. Sketch the memories represented by the following:
    (A) 4 × 4 (B) 1K × 8 (C) 16K × 1 (D) 64K × 16
39. Refer to the memory map of Figure 2–54.
    (A) What is the size of the memory?
    (B) Convert the addresses to decimal and the bit pattern to binary—redraw the memory with the actual bit pattern.
40. Construct a memory map using hexadecimal numbers of the bit patterns in Figure 2–55.

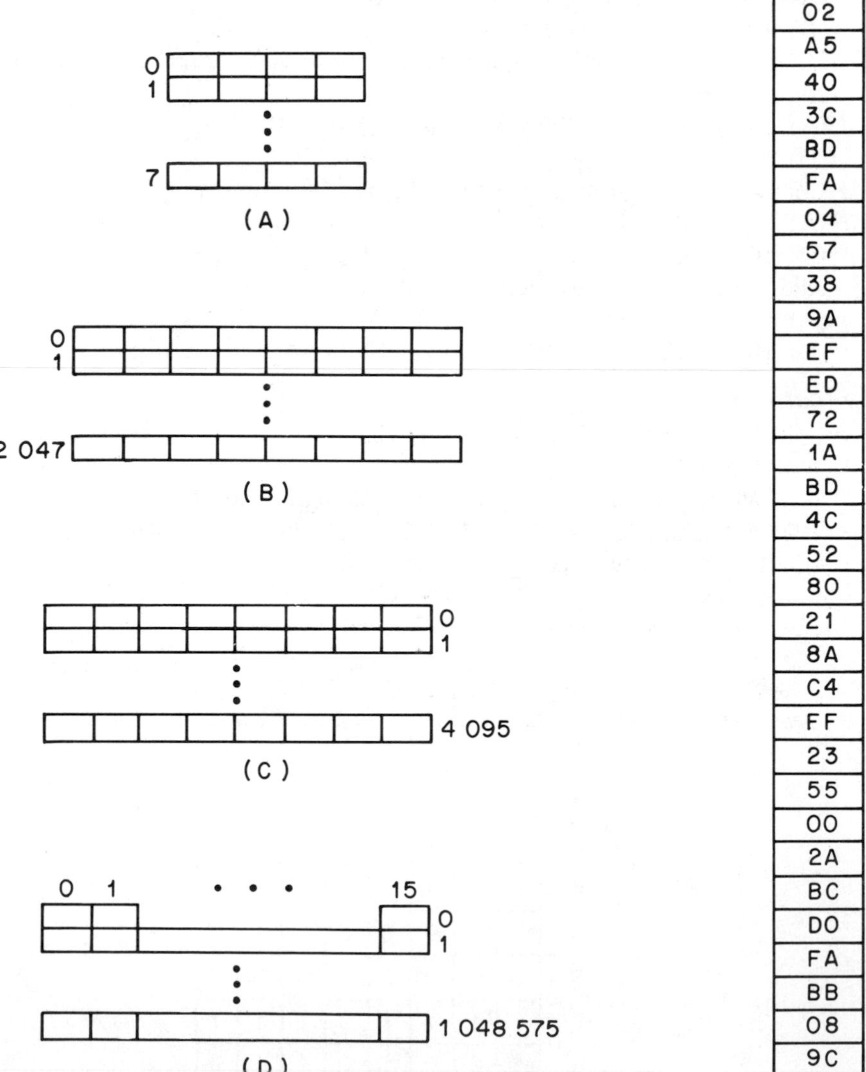

Figure 2–53

02
A5
40
3C
BD
FA
04
57
38
9A
EF
ED
72
1A
BD
4C
52
80
21
8A
C4
FF
23
55
00
2A
BC
D0
FA
BB
08
9C

Figure 2–54

0	0	1	0	0	1	1	0	1
1	1	1	0	1	1	0	1	1
2	0	0	0	1	0	0	1	0
3	1	1	1	0	1	1	0	0
4	0	0	1	0	1	0	1	1
5	1	1	0	1	0	1	1	0
6	1	1	1	1	1	1	0	1
7	1	0	1	0	0	0	0	0
8	0	1	0	0	1	0	0	0
9	1	1	0	1	1	1	0	1
A	0	1	1	0	1	1	0	1
B	0	0	0	1	0	0	1	0
C	1	1	0	1	1	1	1	1
D	1	0	1	0	0	1	1	1
E	0	0	1	1	1	1	0	0

Figure 2–55

41. Give the sequence of button presses you would use with the hexadecimal loader used in this chapter to load the memory in Problem 40 with the indicated bit patterns in Figure 2–55.
42. From the sequence of buttons pressed in Figure 2–56, determine what address and data are loaded.

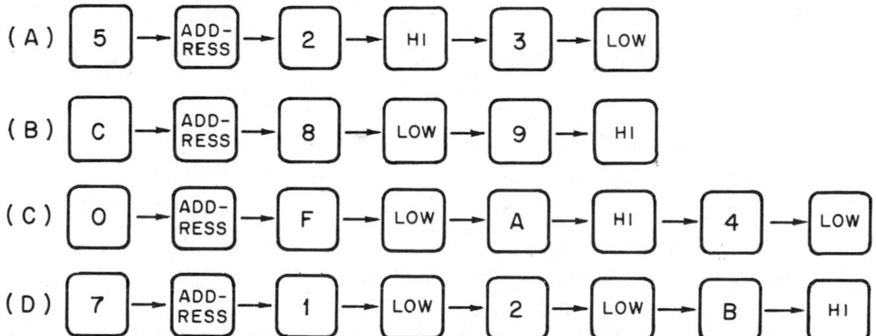

Figure 2–56

43. How would you use the hexadecimal loader (the button sequence) to load your initials into memory, starting at address 12_{16} and using (A) even parity? (B) What would be the sequence if the parity were odd?
44. Do the same as question 43 for your zip codes (both even and odd parities).

Troubleshooting

45. Name any errors in the schematics of Figure 2–57.

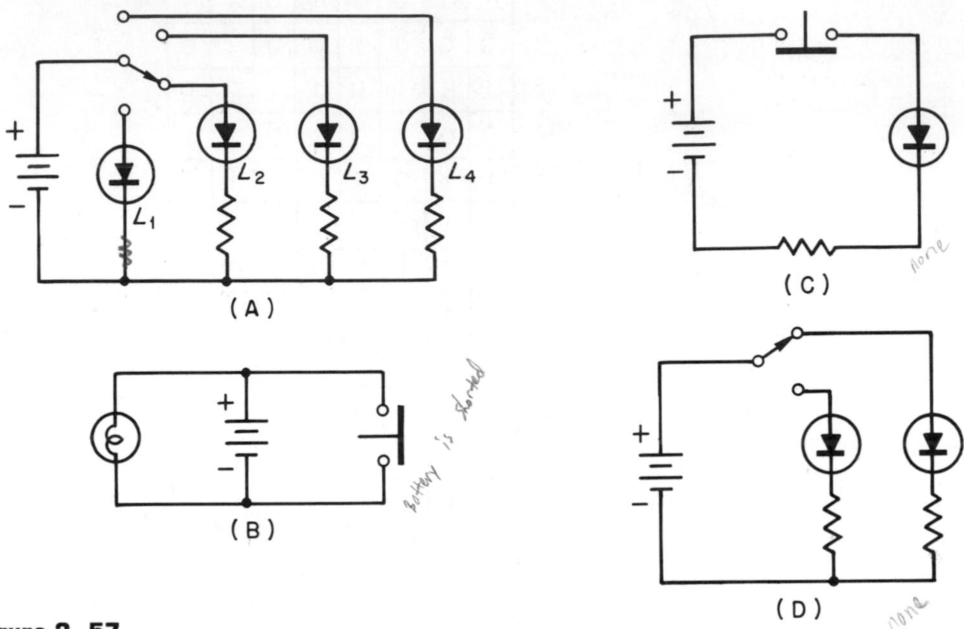

Figure 2–57

46. Locate the wiring errors in Figure 2–58.

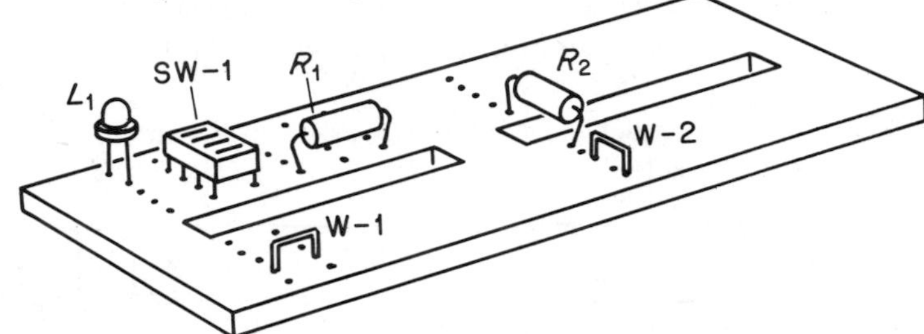

Figure 2–58

Analysis and Design

47. You are employed as a digital technician at a company that produces automated assembly lines for manufacturing space satellite equipment. The engineer you work for wants you to do a preliminary design on a position indicator that will transmit data on the position of a part. The position indicator is to indicate its direction every 22.5°. The part rotates 360°.
    Your Task: Make recommendations as to the general design of such a system and how it should be tested.
48. As a systems technician at a robotics manufacturing company, you are responsible for the assembly of the seven-segment readout used as part of the robot readout system.
    Your Task: Design a testing station that would test a seven-segment readout to ensure that each of its segments is working properly.

CHAPTER 3

The Building Blocks of Logic

OBJECTIVES

After completing this chapter, you should be able to:

- ☐ Explain ways of representing an ON or OFF condition.
- ☐ Understand the meaning of pulse waveforms.
- ☐ Explain the operation of the AND gate.
- ☐ Explain the operation of the OR gate.
- ☐ Explain the operation of the inverter.
- ☐ Explain some applications of gates.
- ☐ Explain basic operations using Boolean algebra.
- ☐ Understand how computers use logic functions.
- ☐ Identify the various types of chips available for logic circuits.
- ☐ Understand the process of identifying and using real logic circuits.

INTRODUCTION

This chapter shows the fundamental building blocks that make up the "brains" of any digital system. This consists of tiny microscopic circuits that serve as the building blocks of digital systems.

You will use the ON and OFF concepts presented in the last chapter in a new and exciting way. Here you will see how these basic building blocks can be used to control simple processes and decision making. These skills will then be used to connect these miniature circuits together to create digital systems that can control the complex operations required of computers and other powerful digital systems.

Here you will also learn about the different kinds of "chips" that contain these "thinking" circuits. You will learn how to identify them and use them. This is a very practical and essential part for those interested in the actual hardware used in the construction of digital systems.

Study this chapter carefully. The material presented here is the foundation for the rest of the book.

KEY TERMS

Graph
Axes
Pulse
Leading Edge
Training Edge
Waveform
Periodic
Nonperiodic
Square Wave
Truth Table
Logic Gate
Logic Symbol
Logic Diagram
Boolean Algebra
NOT Function
Boolean Multiplication
Boolean Addition
Microprocessor
Register
Accumulator
Digital Integrated Circuit
IC Family
CMOS
TTL
Subfamilies

3–1 OTHER WAYS OF THINKING ABOUT 1's AND 0's

Discussion

In the last chapter, you were introduced to the concept that an ON condition could be represented by a one and an OFF condition could be represented by a 0. Looking at ONs and OFFs this way allowed numbers to be represented using the binary number system. It also made for the development of useful codes that could represent other things, such as letters of the alphabet. Recall that you were also shown that a 1 could be represented by +5 volts and a 0 by 0 volts or ground.

This section will expand your concept of the ON and OFF conditions of digital circuits to prepare you for the dynamic conditions found in an operating digital system, such as the computer. You will also see that there are many other ways of expressing the concept of ON and OFF.

A Digital Graph

Look at the circuit in Figure 3–1. The circuit shows an LED connected through a current limiting resistor (recall the resistor is necessary to limit the LED current to a safe value). The LED is also connected to a single-pole, single-throw (SPST) switch and a 5-volt source. When the switch is open, no current flows and the LED is OFF. When the switch is closed, current flows and the LED goes ON.

Because these ON and OFF conditions will be analyzed using graphical techniques, it's important that you understand how to represent digital conditions graphically.

Graph:
A picture showing the relationship between variables. In electronics a graph can show the relationship between an electrical variable and time.

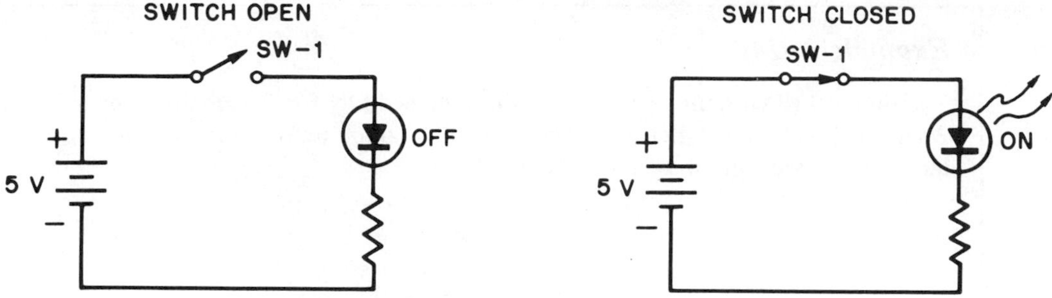

Figure 3–1 Digital LED circuit.

Figure 3–2 shows the graphical representation of the LED circuit when the switch is open and when the switch is closed. Note the following important points about the **axes** of these graphs:

1. The vertical axis indicates the ON or OFF condition of the LED.
2. The horizontal axis indicates the amount of time the LED has been in the indicated condition.

Axes:
The lines of a graph used to indicate the magnitude of the variables represented by the graph. A two-dimensional graph uses two mutually perpendicular lines as the axes. The vertical line is referred to as the *Y*-axis and the horizontal as the *X*-axis.

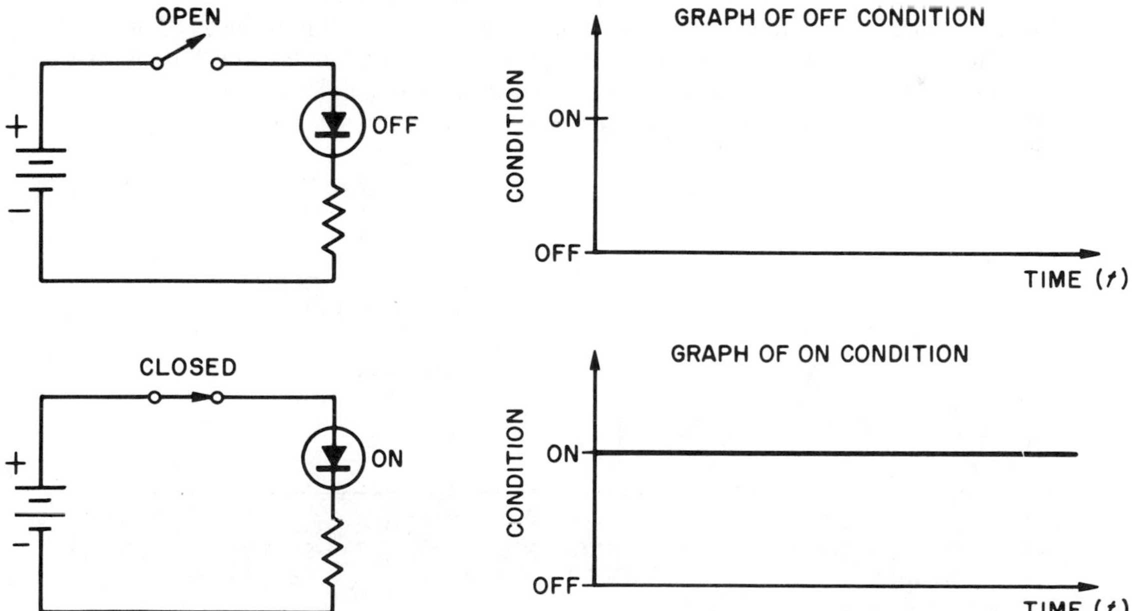

Figure 3–2 Graphical representation of ON and OFF circuits.

Example 3–1

The digital LED circuit of Figure 3–3 is switched ON for 2 seconds, then OFF for 2 seconds, and then switched back ON. Draw a graph representing this action. Start the graph at the time the LED is turned ON.

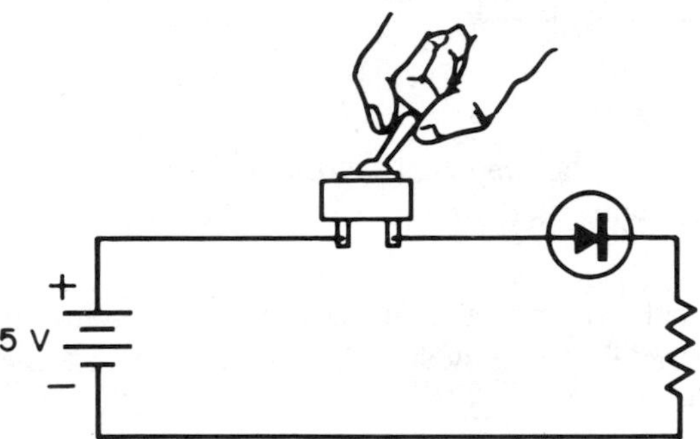

Figure 3–3

Solution

Draw the vertical axis of the graph to represent the condition of the LED and the horizontal axis to represent the amount of time the circuit is in the indicated conditions. Next, construct and label the graph to show the ON and OFF times of the LED. This is shown in Figure 3–4. Observe that no more of the graph can be drawn beyond 4 seconds because you were never told in the problem what happened after that point in time.

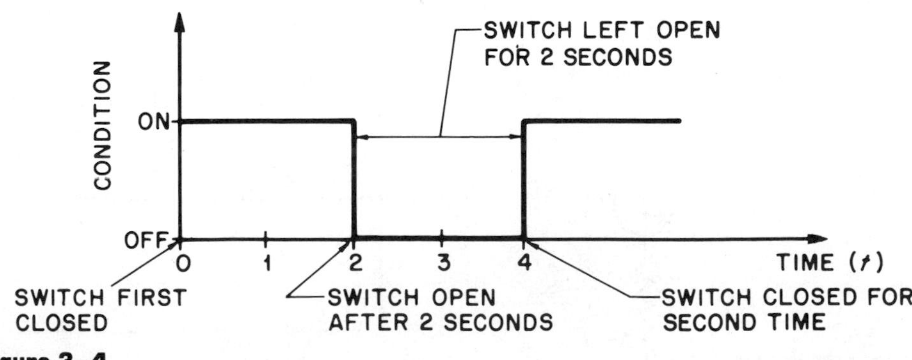

Figure 3–4

Example 3–2

Explain in words the meaning of the graph of Figure 3–5.

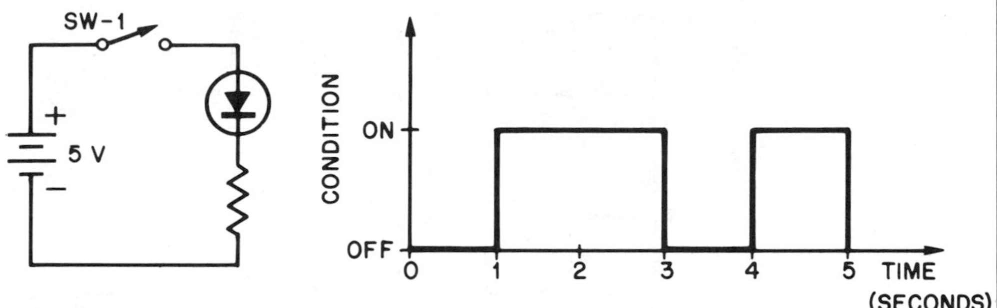

Figure 3–5

Solution

You read the graph by starting at zero time and observing the resulting condition as follows:

∗ 0–1 second

The LED is OFF. The LED stayed OFF for 1 second.

∗ 1–3 seconds

The LED is ON. The LED stayed ON for (3 − 1 = 2) 2 seconds.

∗ 3–4 seconds

The LED is OFF. The LED stayed OFF for (4 − 3 = 1) 1 second.

∗ 4–5 seconds

The LED is ON. The LED stayed ON for (5 − 4 = 1) 1 second.

This information is illustrated in Figure 3–6.

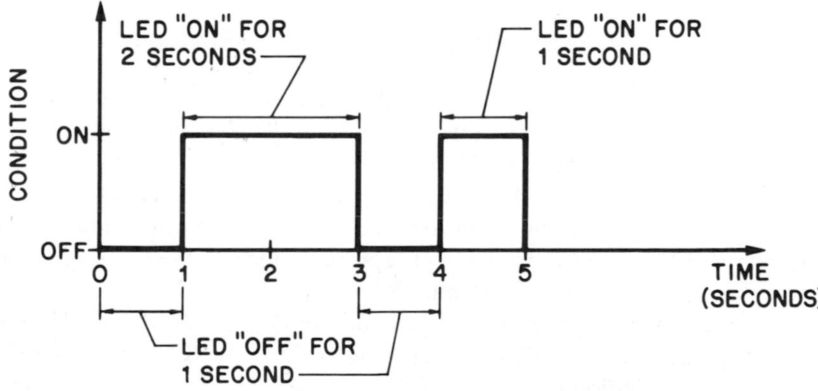

Figure 3–6

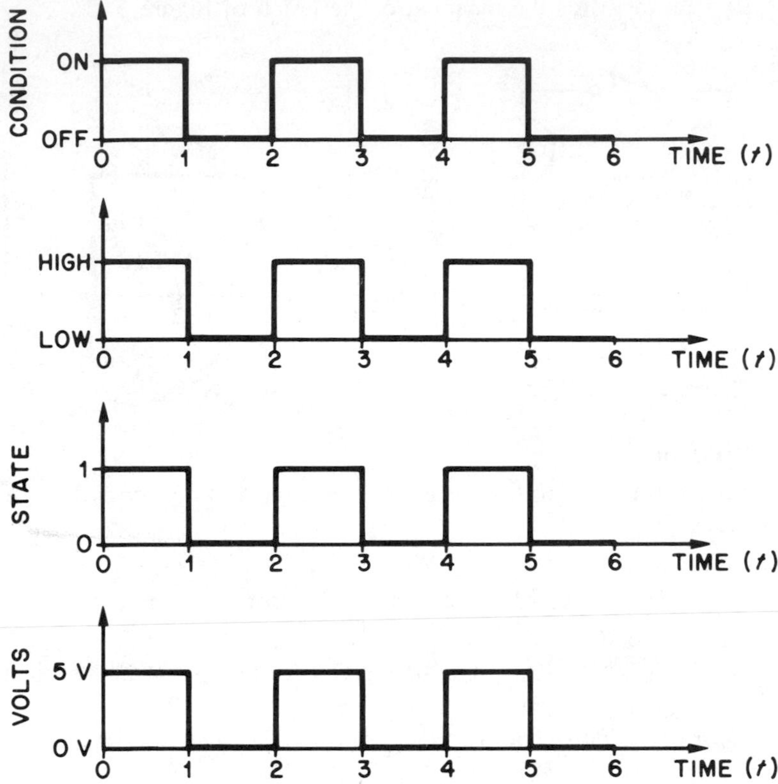

Figure 3–7 Digital graphs representing the same information.

Other Ways of Representing the Digital Graph

Observe that there are only two conditions that can be represented by the digital graph: ON or OFF—there are no values in between. Technicians and engineers usually refer to the ON condition as HIGH and the OFF condition as LOW. Recall that a 1 and 0 (as well as a +5 volts and 0 volts) can also be used to represent an ON or OFF condition. Figure 3–7 shows several digital graphs representing the same information. Note that even though the vertical axis used four different representations of the digital condition, the horizontal axis was always measured in time.

Because the graph of a digital circuit consists of a series of ups and downs, the term HIGH is used to mean ON or 1 and the term LOW is used to mean an OFF or 0. This important terminology is shown in Figure 3–8.

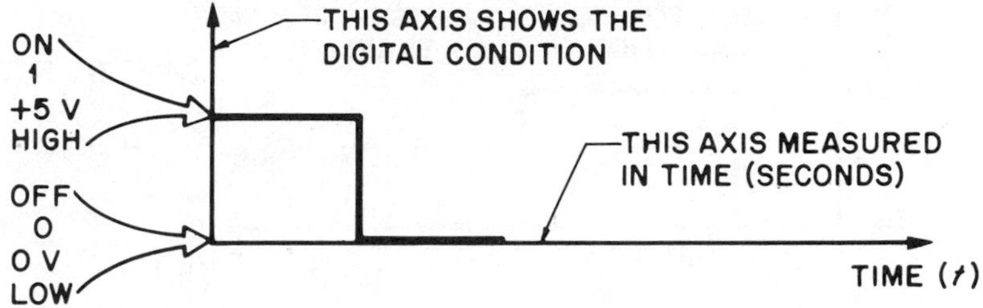

Figure 3–8 Important digital terminology.

Example 3–3

For the LED display shown in Figure 3–9, state the resultant outputs in terms of HIGH or LOW. Then express the result in binary and convert to hexadecimal.

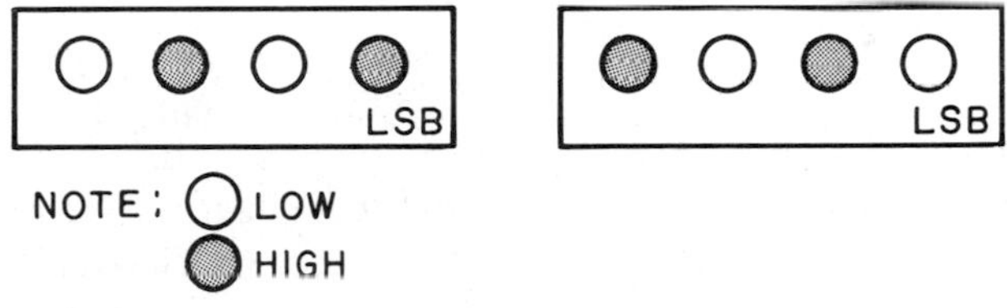

Figure 3–9

Solution

When an LED is ON, this represents a HIGH condition; when it is OFF, it represents a LOW condition. Thus the LED readout of Figure 3–9 is:

LOW-HIGH-LOW-HIGH	HIGH-LOW-HIGH-LOW
0101	1010
5	A

Hence the binary representation is:

$$01011010_2$$

and the hexadecimal representation is:

$$5A_{16}$$

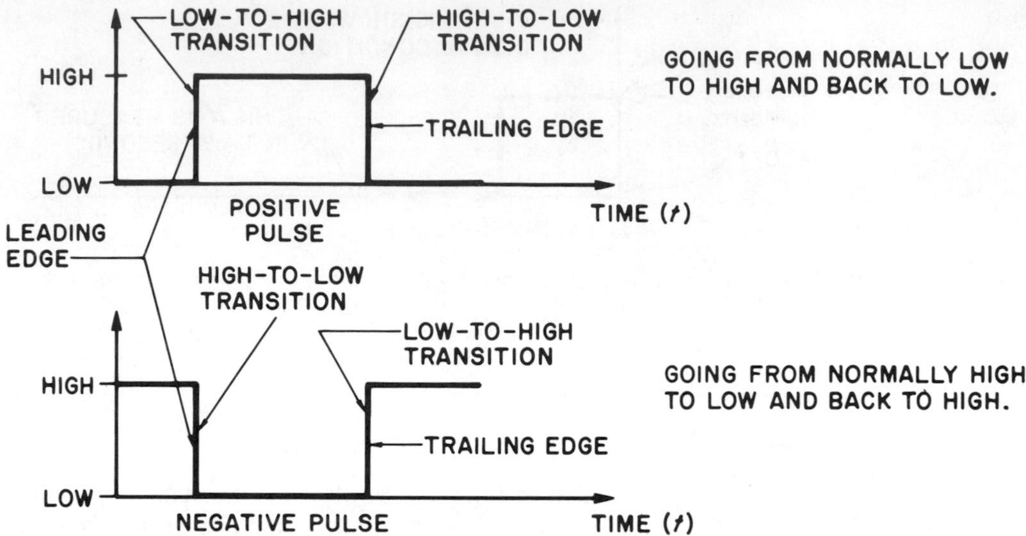

Figure 3–10 Pulse representation and terminology.

Pulses

The graphical representation of a digital condition can be referred to as a **pulse.** This is illustrated in Figure 3–10.

Pulse:
A momentary change in an electrical characteristic of a circuit—usually voltage or current.

Leading Edge: Occurs first in time
For a positive going pulse, it is the LOW-to-HIGH transition. For a negative going pulse, it is the HIGH-to-LOW transition.

Trailing Edge:
For a positive going pulse, it is the HIGH-to-LOW transition. For a negative going pulse, it is the LOW-to-HIGH transition.

Example 3–4

Sketch the resulting pulses from the actions of each of the switches in Figure 3–11.

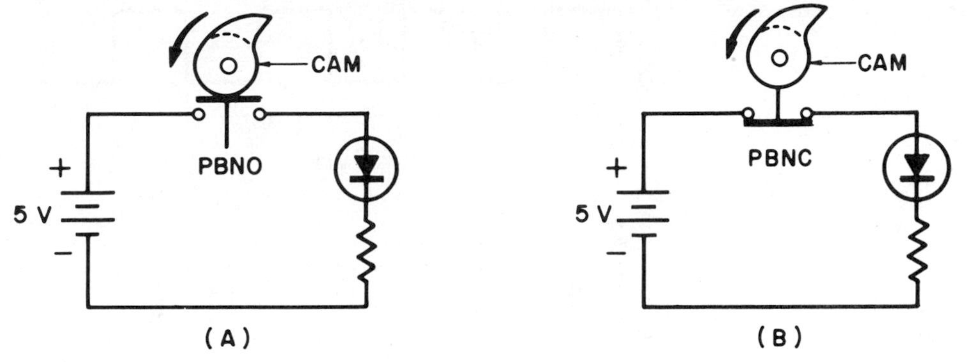

Figure 3–11

Solution

Circuit A contains a PBNO and its normal condition is open (which represents a LOW, OFF, or 0). The rotating cam will cause the switch to close for a short time causing *positive* pulses.

Circuit B contains a PBNC and its normal condition is closed (which represents a HIGH, ON, or 1). The rotating cam will cause the switch to open for a short time causing *negative* pulses.

These conditions are shown in Figure 3–12.

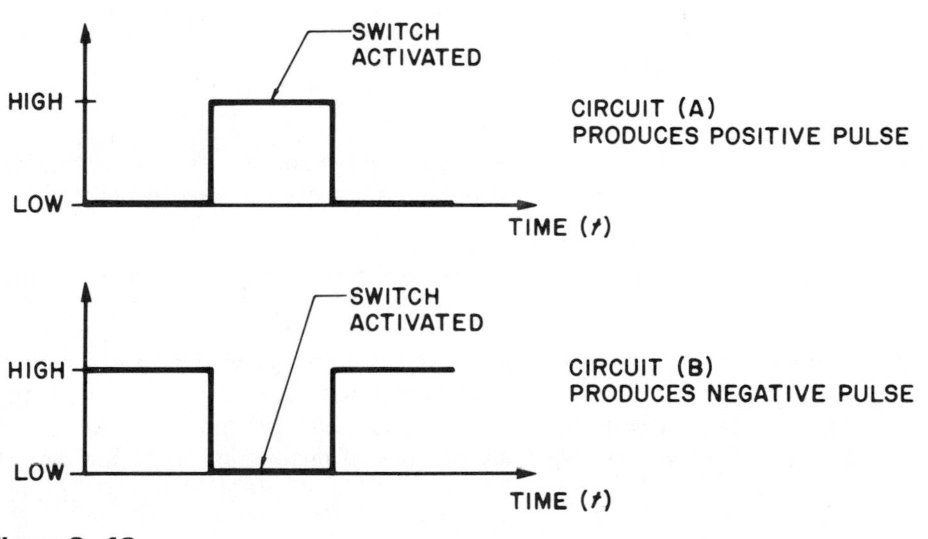

Figure 3–12

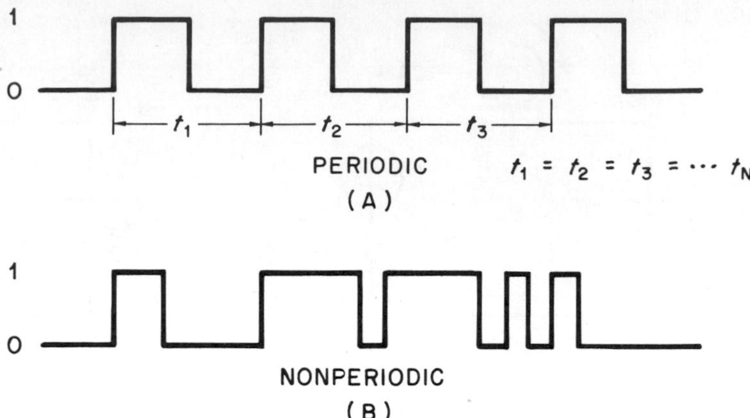

Figure 3–13 Types of pulse waveforms.

Waveforms

A series of pulses is called a pulse waveform or simply a **waveform.**

> **Waveform:**
> The graph of an electrical characteristic of a circuit. Usually the vertical axis represents voltage or current, and the horizontal axis is time.

There are two fundamental types of pulse waveforms: **periodic** and **nonperiodic.** These are shown in Figure 3–13.

> **Periodic:**
> A waveform that continually repeats itself.

> **Nonperiodic:**
> A waveform that is not periodic.

Note that the periodic pulse has equal time increments between its leading edges. A very common type of pulse waveform is referred to as a **square wave.**

> **Square Wave:**
> A periodic waveform where the ON time is equal to the OFF time.

Conclusion

The intention of this section was to broaden your understanding of methods used to represent digital conditions. Here you saw how to graphically represent logical conditions. The next section will utilize what you learned here and show you a digital "thinking" circuit. For now, test your understanding of this section by trying the following review questions.

3–1 Review Questions

1. Explain how to construct a graph that will represent the conditions of a digital circuit.
2. What does the horizontal axis of a digital graph always represent?
3. Give the other terms used for ON and OFF.
4. State what is meant by a digital pulse. Describe what is meant by leading edge and trailing edge.
5. Explain the meaning of a digital waveform. What do periodic and nonperiodic mean?

3–2 THE AND CONCEPT

Discussion

In this section, you will meet your first digital building block. It is one of three used in digital circuits.

ANDing

Observe the digital circuit of Figure 3–14. In order for the LED to be HIGH (ON), PB-1 must be closed AND PB-2 must be closed. No other combination will cause the LED to be HIGH. This is illustrated in Figure 3–15. Observe from the figure that all possible combinations of the two switches were considered, and the resulting condition of the LED was also given.

Truth Tables

A **truth table** is a handy tool for analyzing the conditions of digital circuits.

> **Truth Table:**
> A listing showing all possible input combinations of a digital circuit and the resultant output(s). Usually done in terms of any two-state symbolism such as 1's and 0's.

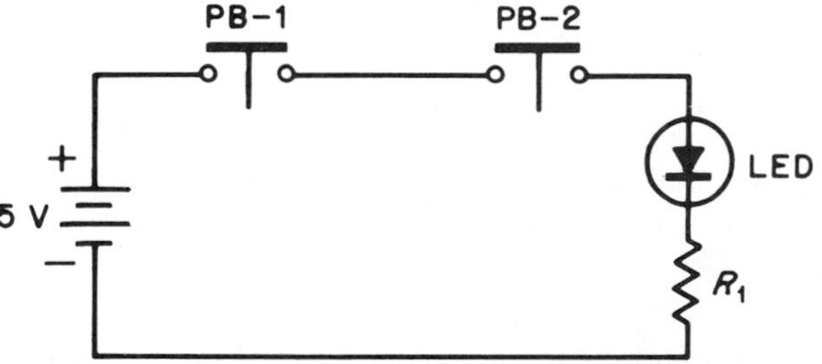

Figure 3–14 Digital AND circuit.

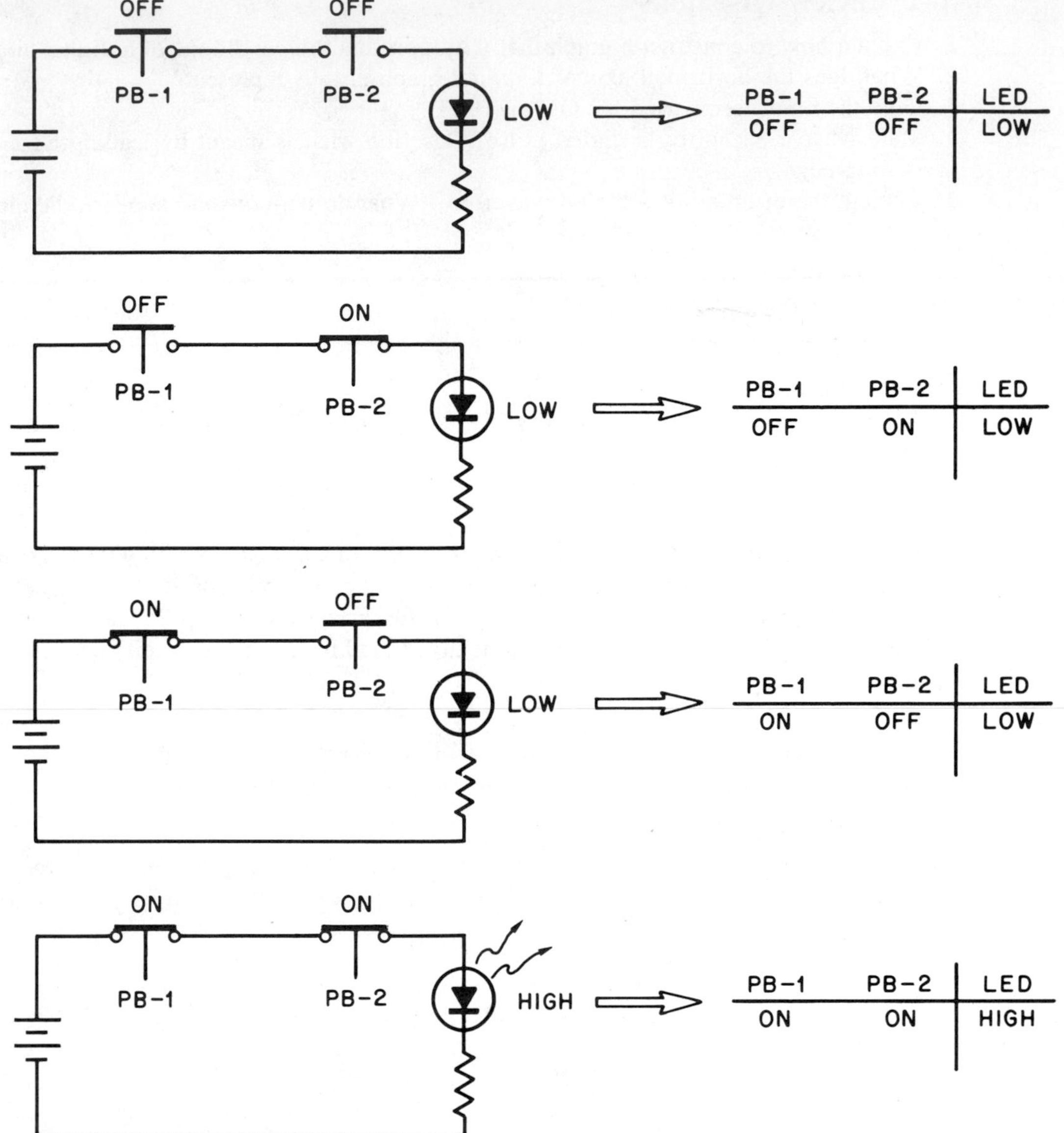

Figure 3–15 All possible combinations of AND circuit.

A truth table is simply a way of easily recording all possible conditions of a digital circuit. Figure 3–16 illustrates. Note that the circuit was divided into an input and an output. Observe that if the open and closed conditions of the switches and the ON and OFF conditions of the LED are converted to a 0 and a 1, then the input combination of the switches can assume a binary count while the output condition is given as 0 or 1.

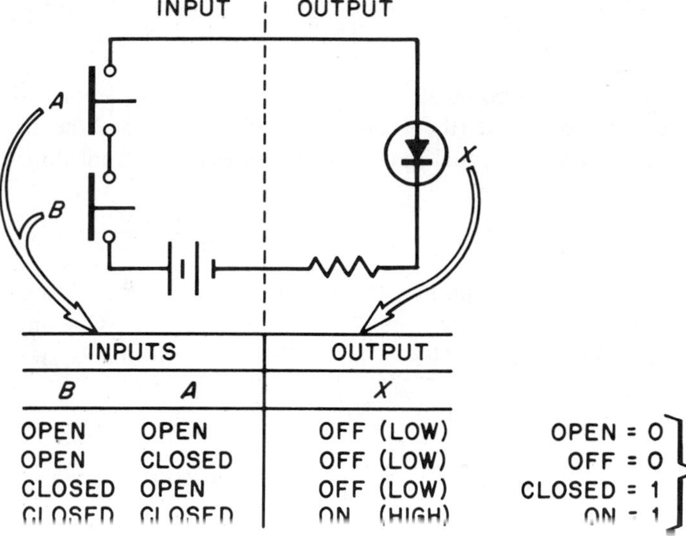

TRUTH TABLE

INPUTS		OUTPUT
B	A	X
0	0	0
0	1	0
1	0	0
1	1	1

Figure 3–16 Development of truth table.

AND Circuit		
B	*A*	*X*
0	0	0
0	1	0
1	0	0
1	1	1

The truth table shows that for a two-input AND circuit, the output is a 1 only when both the inputs are a 1. Under all other possible conditions, the output is a 0. There are four possible combinations for this truth table.

You can have AND circuits with as many inputs as you wish. The number of combinations for the inputs can be found from the relation

$$\text{Number of combinations} = 2^N$$

where: N = the number of inputs.

A binary count is used for the input to ensure that all possible combinations of the input have been considered. It may seem like a waste of time to develop a truth table for an AND circuit, since you know that the only time the output will be HIGH (a 1) is when all of the inputs are HIGH. However, this simple example is being used here so you can develop skills in creating truth tables. Later, these skills will be used to analyze more complex "thinking" circuits.

Example 3–5

A digital circuit is to be developed that will indicate when a back door and two side doors of a warehouse are closed (they can only be opened from the inside). An indicator is needed in order to signal that it is OK to exit the front door and lock the warehouse.

Solution

Use switches for the three doors and an LED for the output indicator. Indicate each of the three doors by the letters *A*, *B*, and *C*. The LED is to be ON only when *A* AND *B* AND *C* are closed (a 1). This is an AND condition and is shown in the circuit of Figure 3–17.

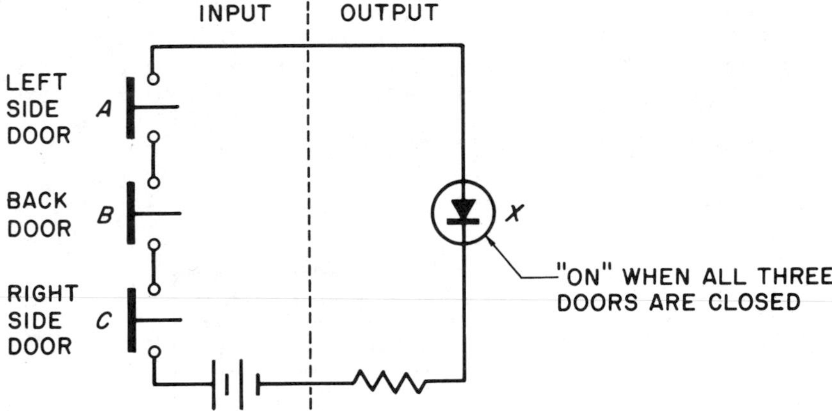

Figure 3–17

The truth table will have three inputs, giving $2^3 = 8$ different combinations. Construct the truth table by doing a binary count of the three inputs. This ensures that no possible input combination has been missed:

Truth Table

A	*B*	*C*	*X*
0	0	0	0
0	0	1	0
0	1	0	0
0	1	1	0
1	0	0	0
1	0	1	0
1	1	0	0
1	1	1	1

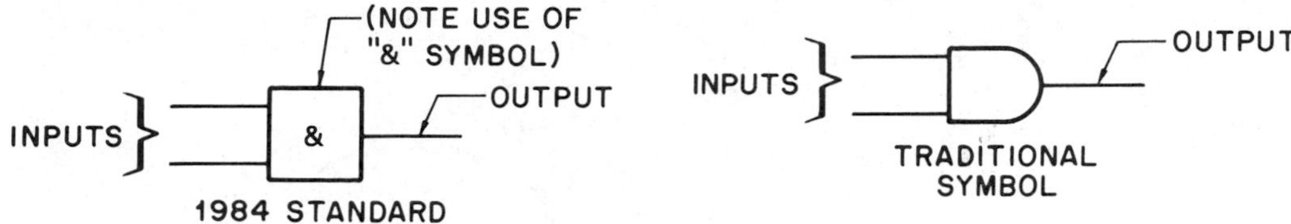

Figure 3–18 Logic symbols of AND gate.

Gates

A gate is a circuit that performs a "thinking function" for a digital circuit and is usually called a **logic gate.**

Logic Gate:

An electrical circuit that simulates a logic function.

There are three basic gates used in digital circuits. One of these is called the AND gate. Its **logic symbol** is shown in Figure 3–18.

Logic Symbol:

A drawing that represents a logic function.

The two lines on the left of the AND gate are called the inputs, and the line on the right is called the output. The two-input AND gate is shown in a test circuit in Figure 3–19 along with the resulting truth table.

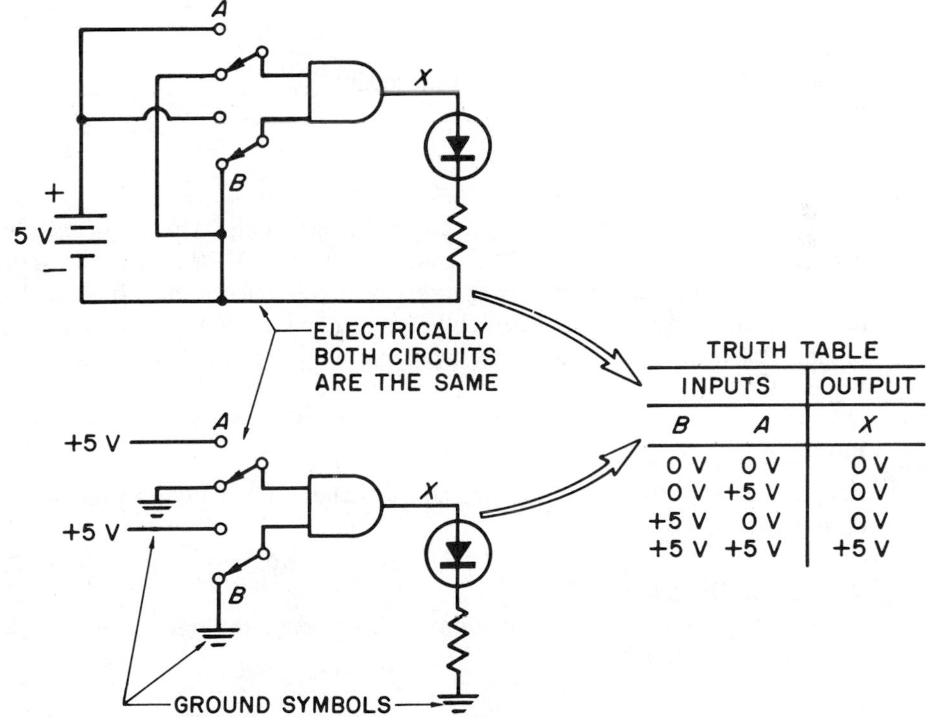

TRUTH TABLE

INPUTS		OUTPUT
B	*A*	*X*
0 V	0 V	0 V
0 V	+5 V	0 V
+5 V	0 V	0 V
+5 V	+5 V	+5 V

Figure 3–19 AND gate test circuit.

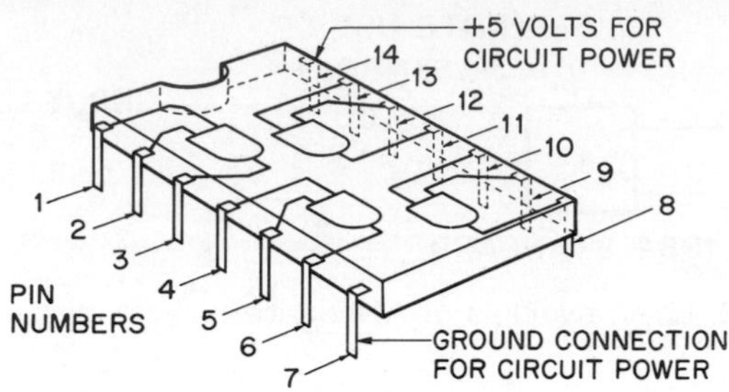

Figure 3–20 IC chip with four two-input AND gates.

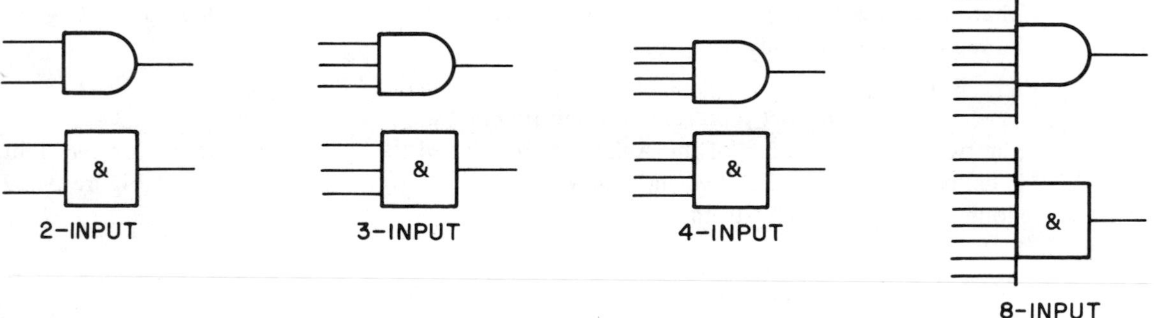

Figure 3–21 AND gates with different inputs.

The simplest kind of AND gate comes in an IC chip. The chip costs only a few cents and gives you four two-input AND gates. See Figure 3–20. As shown in the figure, pins 7 and 14 are used to supply power to all the AND gates in the chip. The chip is called a quad two-input AND gate. Theoretically, you can have AND gates with any number of inputs. Practically, AND gates are available with two, three, four, and eight inputs. See Figure 3–21.

Logic Diagrams

Rather than show a battery, LED, and switches, use a **logic diagram** to represent the digital circuit.

> **Logic Diagram:**
> The graphical representation of one or more logic functions and their interconnections.

When using a logic diagram, you should understand that the inputs and outputs of the IC will be either 0 (LOW) or 1 (HIGH). This is illustrated in Figure 3–22.

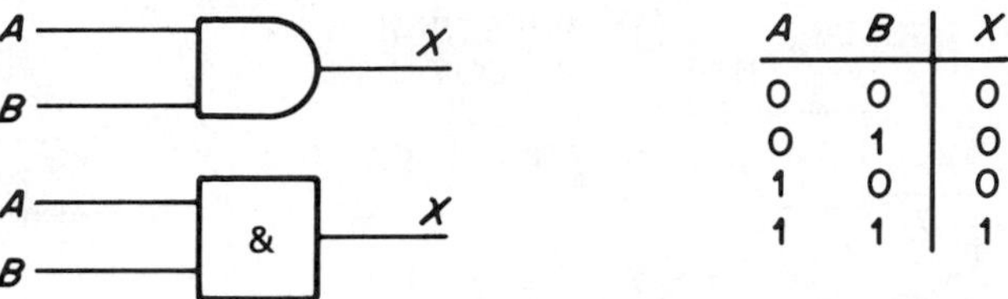

A	B	X
0	0	0
0	1	0
1	0	0
1	1	1

Figure 3–22 Logic diagram of AND gate with corresponding truth table.

Example 3–6

Redo the three-door problem of Example 3–5, using a logic diagram.

Solution

See Figure 3–23.

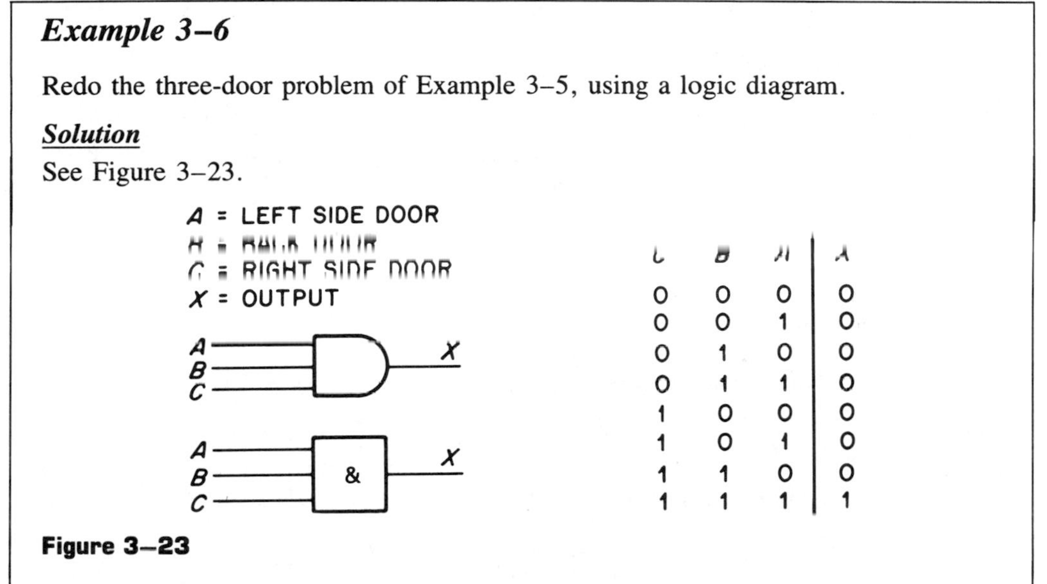

C	B	A	X
0	0	0	0
0	0	1	0
0	1	0	0
0	1	1	0
1	0	0	0
1	0	1	0
1	1	0	0
1	1	1	1

Figure 3–23

AND Gate Analysis

You will encounter AND gates that have pulse waveforms as their inputs. These logic circuits are used in complex digital systems such as computers, so it is important that you develop skills that will help you analyze them. As an example of such a condition, observe the action of the AND gate in Figure 3–24.

Look at the truth table for the logic circuit of Figure 3–24(A). Since the *B* input is always 1, there are only two possible input conditions. Because an AND gate will have a 1 output only when *all* inputs are a 1, then its output, under these conditions, will follow the *A* line input. Thus the digital signal is allowed to "pass" through the gate.

Observe the truth table of Figure 3–24(B). Again, since the *B* input always has the same input (this time a 0), the output will now always be LOW. This is indicated by the truth table, the output is always 0, no matter what the state of input line *A* becomes. Thus, this logic circuit has the effect of controlling the passage of pulses, determined by the condition of one of its input lines.

Multiple-Pulse Inputs

There are times when pulses are applied to both inputs of an AND gate. This is illustrated in the following example.

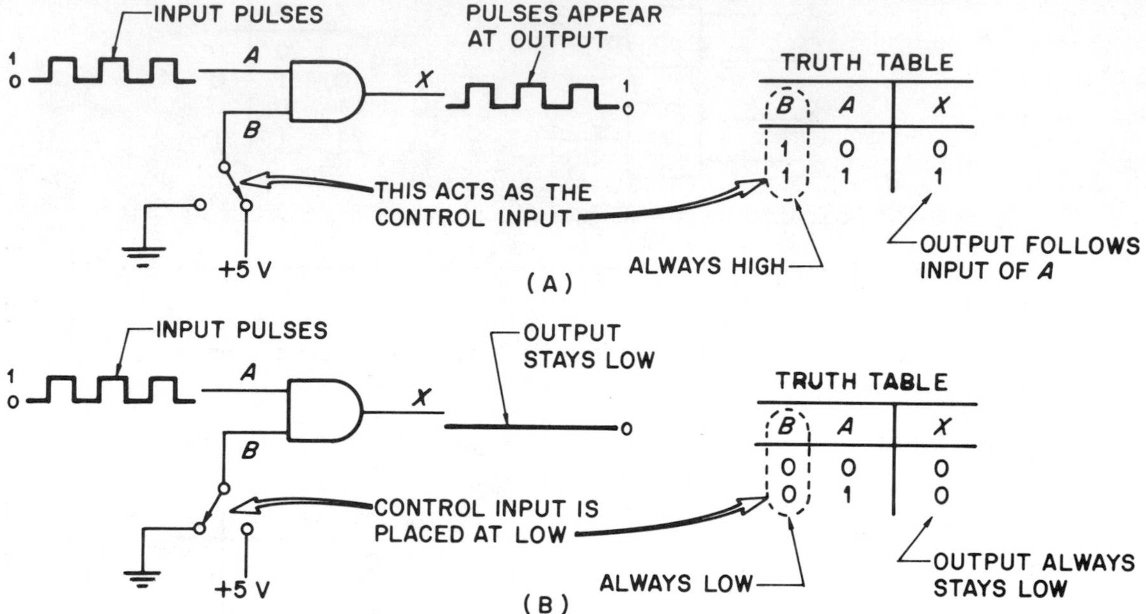

Figure 3–24 Using an AND gate to control the passing of pulses.

Example 3–7

Draw the resulting waveform for the logic circuit shown in Figure 3–25.

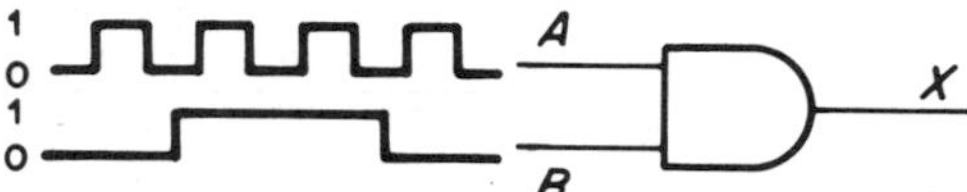

Figure 3–25

Solution

Analyzing the circuit as an AND gate—that is, the only time the output will be HIGH is when both inputs are HIGH. You can develop the output logic waveform by lining up the input pulses; every time both are HIGH, the output will be HIGH. Figure 3–26 shows how.

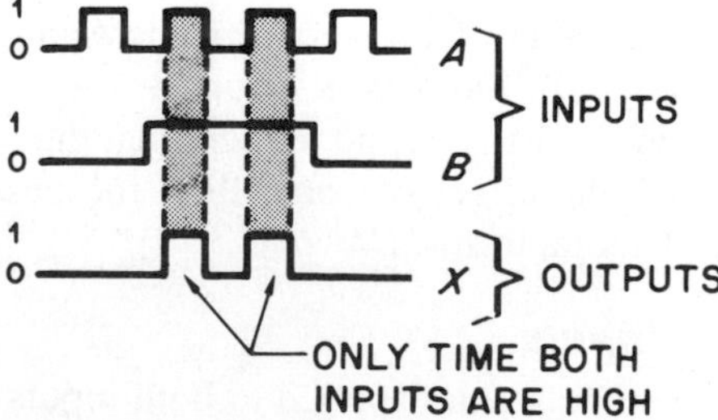

Figure 3–26

Example 3–8

Develop the output pulse waveform for each logic gate in Figure 3–27.

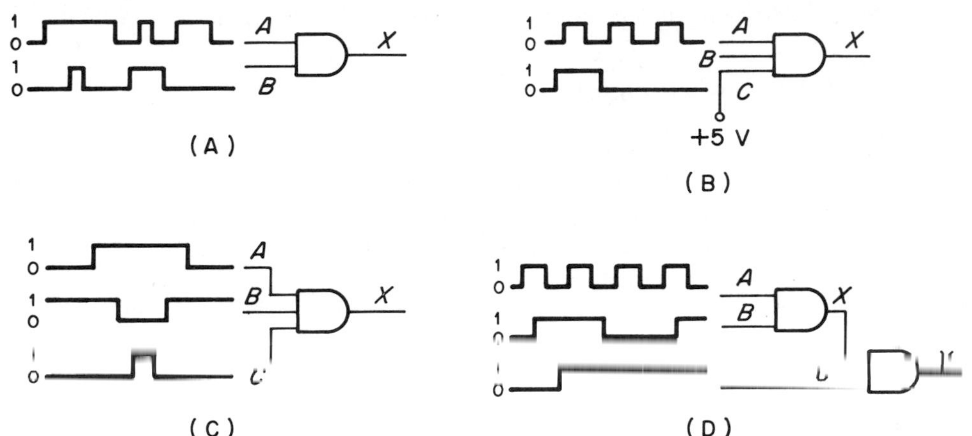

Figure 3–27

Solution

As before, analyze each circuit by observing when all inputs are HIGH. This is the only time the output will be HIGH. Refer to Figure 3–28 for the solutions.

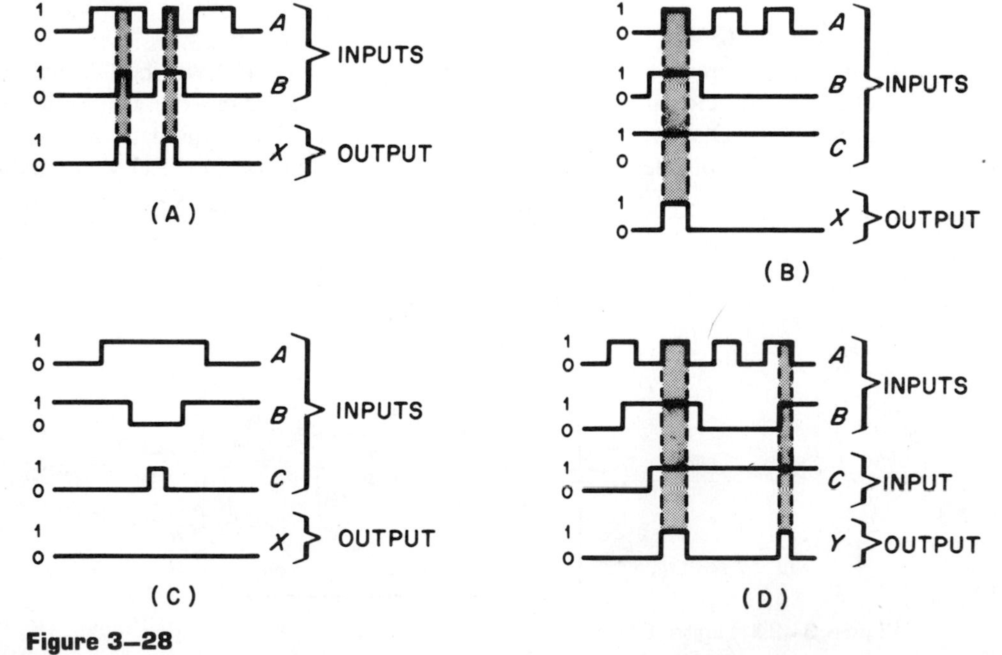

Figure 3–28

Conclusion

This section presented your first "thinking" circuit—the AND gate. You learned the meaning of the AND function, saw some applications, and had a chance to practice with some real AND gate logic circuits.

The next section presents the second of the three fundamental "thinking" digital circuits. Test your understanding of this section by trying the following review questions.

3–2 Review Questions

1. State what the AND function means.
2. Describe an electrical circuit consisting of two switches, a voltage source, an LED, and a resistor that demonstrates the AND function.
3. What is a truth table? How is it constructed?
4. What is an AND gate?
5. What is an IC chip called that consists of four AND gates where each gate has two inputs?

3–3 THE OR CONCEPT

Discussion

This section presents your second digital building block. You will see that the methods used for its analysis are the same as the methods used for analyzing the AND gate.

The OR Concept

In the digital circuit of Figure 3–29, observe that the LED will be ON when either PB-1 OR PB-2 are closed OR when both PB-1 and PB-2 are closed. This action is shown in Figure 3–30. Observe from the figure that all possible combinations of the two input switches were considered, and the resulting conditions of the LED were also given.

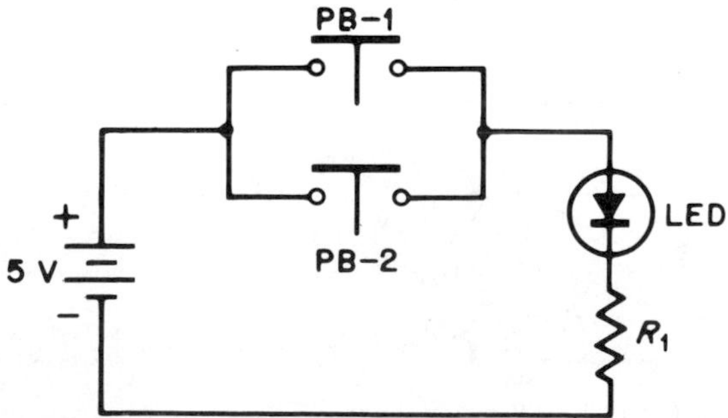

Figure 3–29 Digital OR circuit.

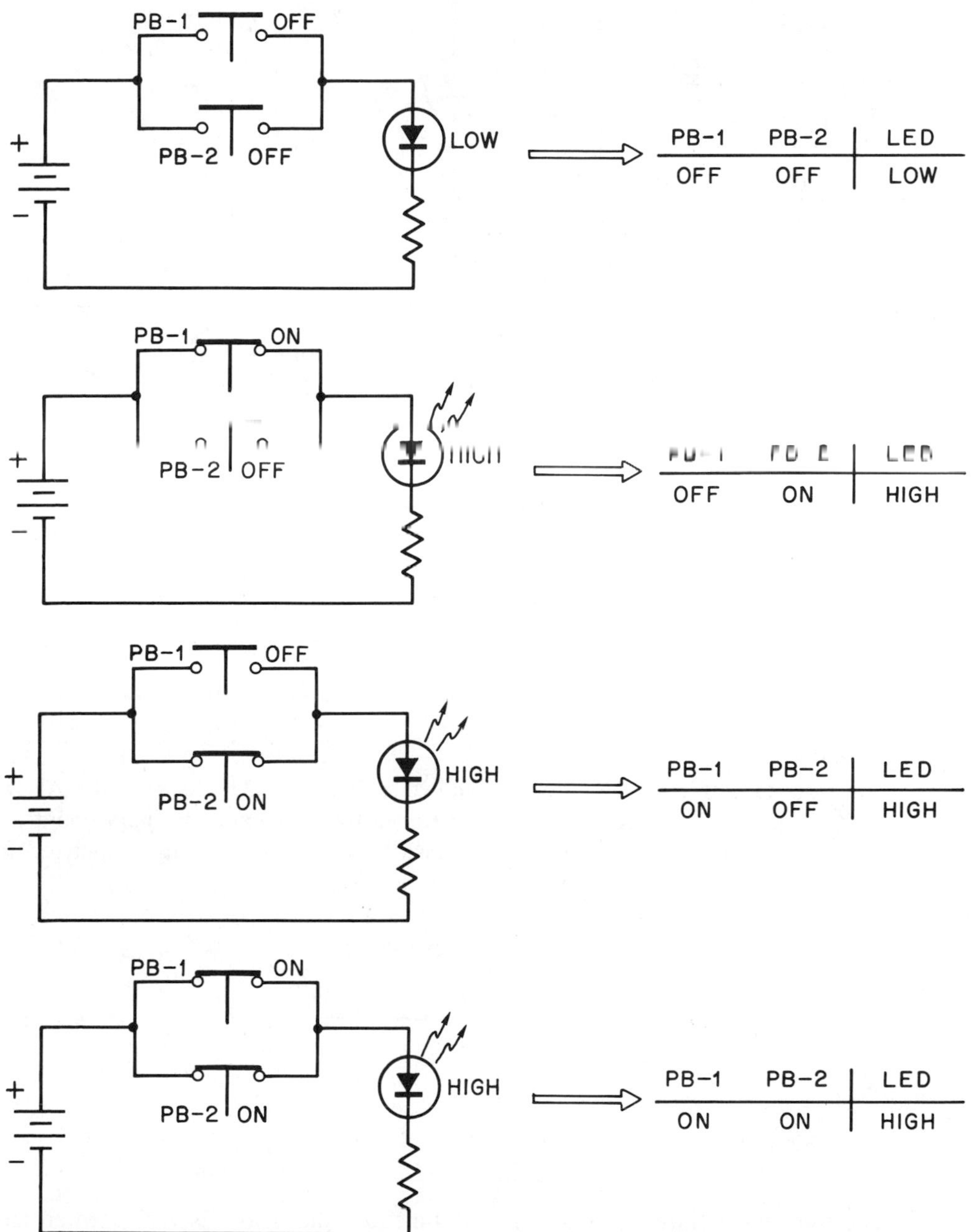

Figure 3–30 All possible combinations of OR circuit.

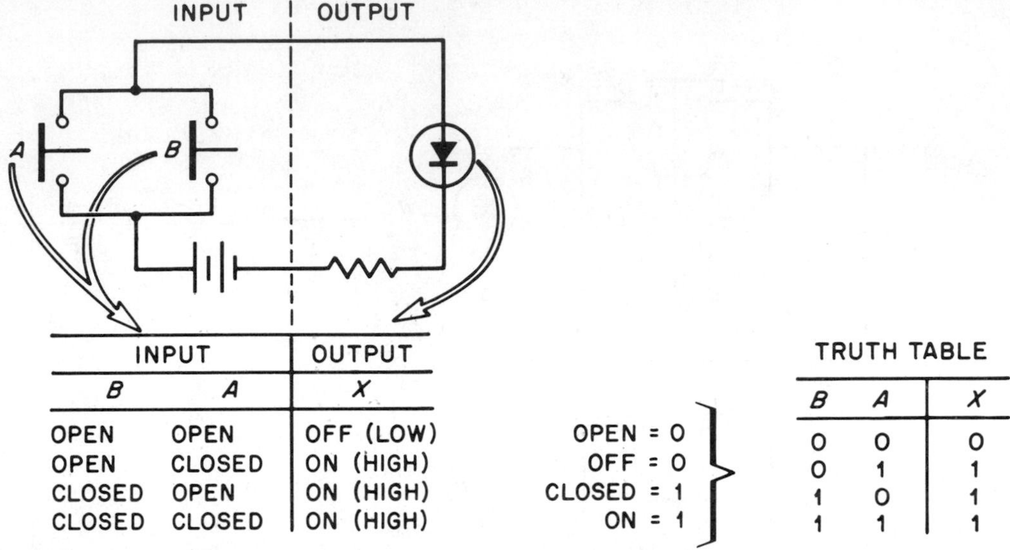

INPUT		OUTPUT
B	A	X
OPEN	OPEN	OFF (LOW)
OPEN	CLOSED	ON (HIGH)
CLOSED	OPEN	ON (HIGH)
CLOSED	CLOSED	ON (HIGH)

OPEN = 0
OFF = 0
CLOSED = 1
ON = 1

TRUTH TABLE

B	A	X
0	0	0
0	1	1
1	0	1
1	1	1

Figure 3–31 Development of truth table for the OR.

Truth Table

The truth table for an OR circuit is shown in Figure 3–31. As with the AND function, the truth table of an OR function is developed by performing a binary count of the input to ensure all combinations have been considered and the resulting output recorded.

OR Circuit		
B	*A*	*X*
0	0	0
0	1	1
1	0	1
1	1	1

The truth table shows that for the OR function, the output is HIGH when any input is HIGH. The OR function can have as many inputs as you like. However, as you will see, there are some practical limitations.

Example 3–9

A digital circuit is to be developed that will indicate if any of three hospital patients in the same room need service. The output is to be an LED located at the nurses' station. Construct the corresponding truth table.

Solution

Use switches for each of the patients and indicate them as switch *A*, *B* and *C*. Construct the circuit so that *A* OR *B* OR *C* will cause the LED to light. This will be an OR circuit. See Figure 3–32.

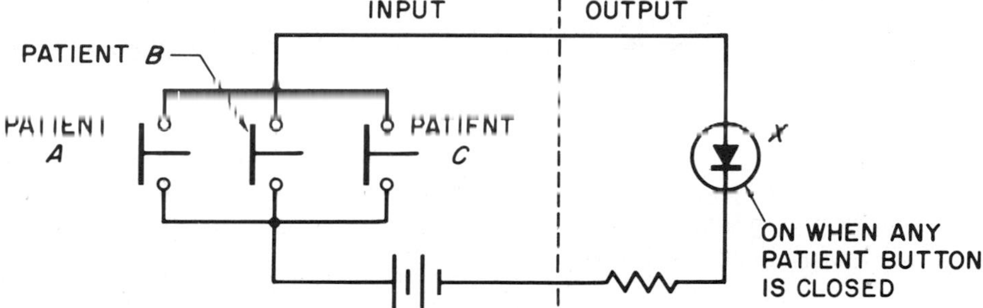

Figure 3–32

The truth table will have three inputs, giving $2^3 = 8$ different combinations. As with the AND function, construct the truth table by doing a binary count, which ensures that all possible combinations have been considered.

OR Truth Table

A	*B*	*C*	*X*
0	0	0	0
0	0	1	1
0	1	0	1
0	1	1	1
1	0	0	1
1	0	1	1
1	1	0	1
1	1	1	1

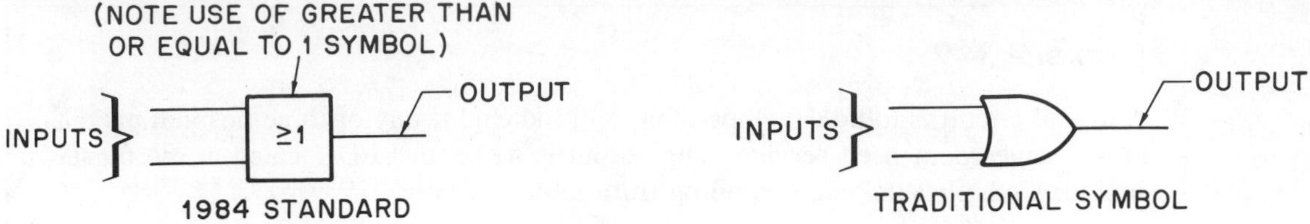

Figure 3–33 Logic symbols of an OR gate.

OR Gates

The logic symbols for an OR gate are shown in Figure 3–33. As with the AND gate, the two lines on the left of the OR gate are the inputs, and the single line on the right is called the output. The two-input OR gate is shown in a test circuit in Figure 3–34 along with the resulting truth table.

The simplest kind of OR gate comes in an IC chip. Like the AND gate chip, it costs only a few cents and gives you four two-input OR gates. See Figure 3–35. The chip is a 14-pin chip. Since it contains four two-input OR gates, it is called a quad two-input OR gate. Note that the same pins as with the quad two-input AND gate are used to supply power to the chip. Theoretically, you can have OR gates with any number of inputs. Practically speaking, OR gates are available with two, three, four, and eight inputs, as shown in Figure 3–36.

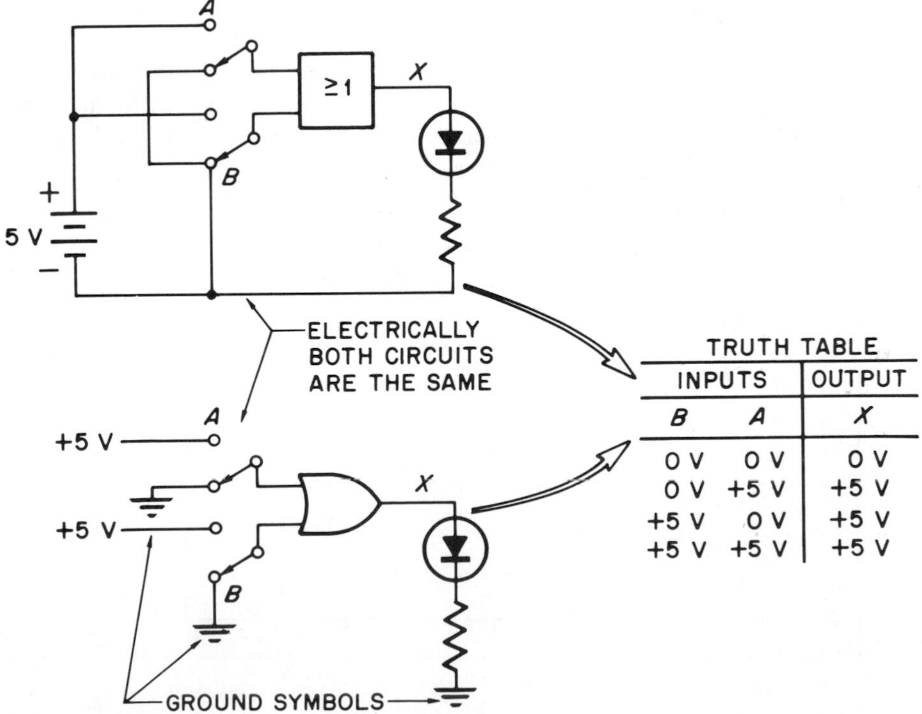

TRUTH TABLE		
INPUTS		OUTPUT
B	*A*	*X*
0 V	0 V	0 V
0 V	+5 V	+5 V
+5 V	0 V	+5 V
+5 V	+5 V	+5 V

Figure 3–34 OR gate test circuit.

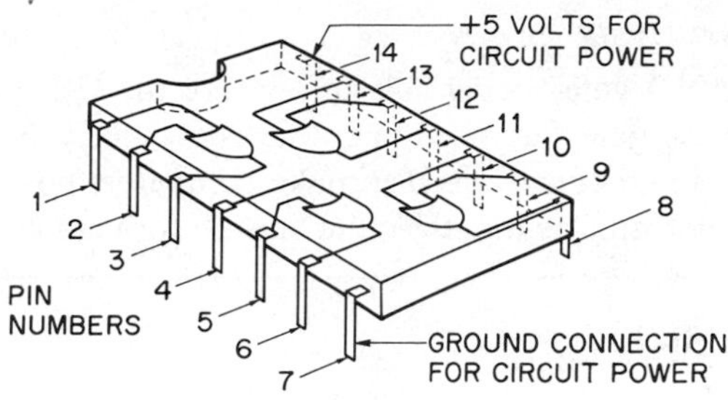

Figure 3–35 IC chip with two-input OR gates.

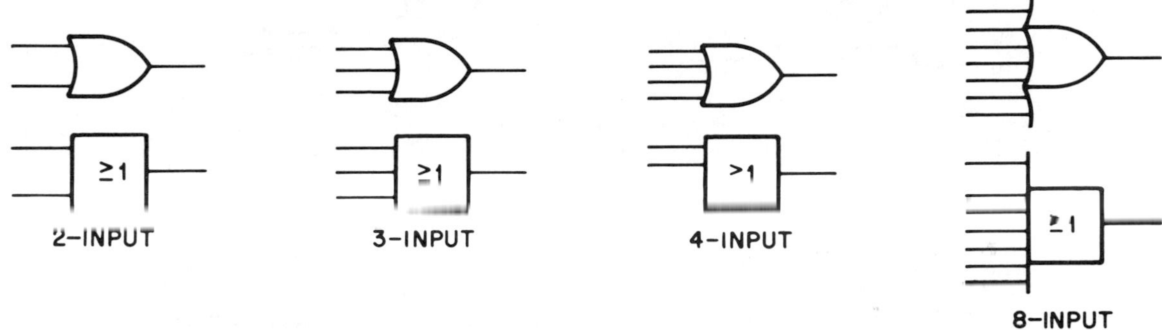

Figure 3–36 OR gates with different inputs.

Logic Diagrams

OR gates use logic diagrams just as AND gates do. This is illustrated in the following example.

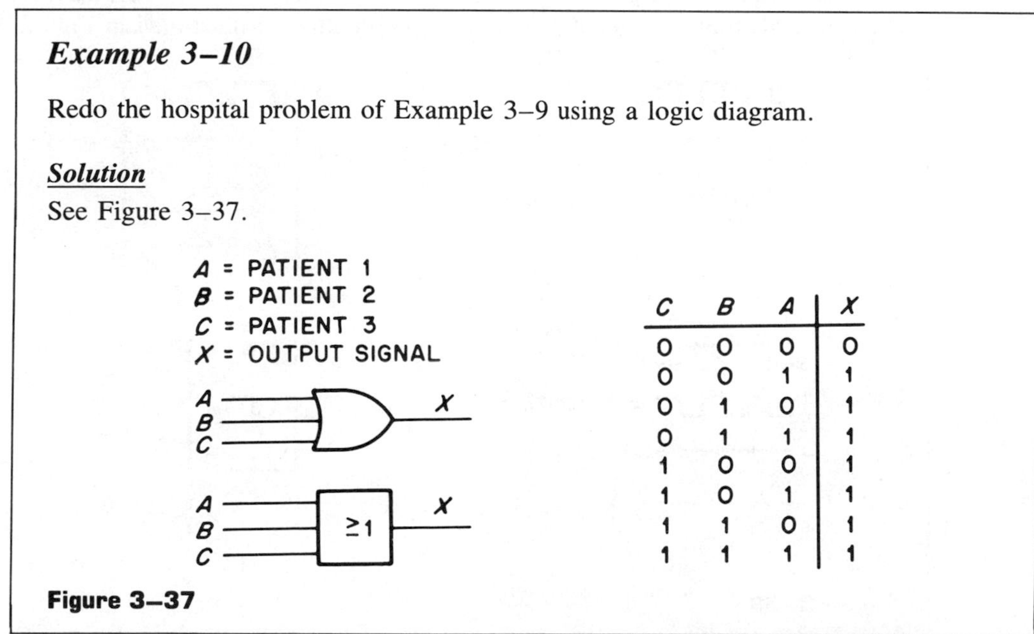

Example 3–10

Redo the hospital problem of Example 3–9 using a logic diagram.

Solution

See Figure 3–37.

C	*B*	*A*	*X*
0	0	0	0
0	0	1	1
0	1	0	1
0	1	1	1
1	0	0	1
1	0	1	1
1	1	0	1
1	1	1	1

Figure 3–37

OR Gate Analysis

As with the AND gate, being able to analyze the OR gate with pulse waveforms is very important. These gates are used in conjunction with AND gates in digital systems to perform complex logic operations that make computers possible. The analysis of the OR gate with pulse waveforms as input is quite straightforward as shown in the following example.

Example 3–11

Draw the resulting output waveform for each of the logic circuits in Figure 3–38.

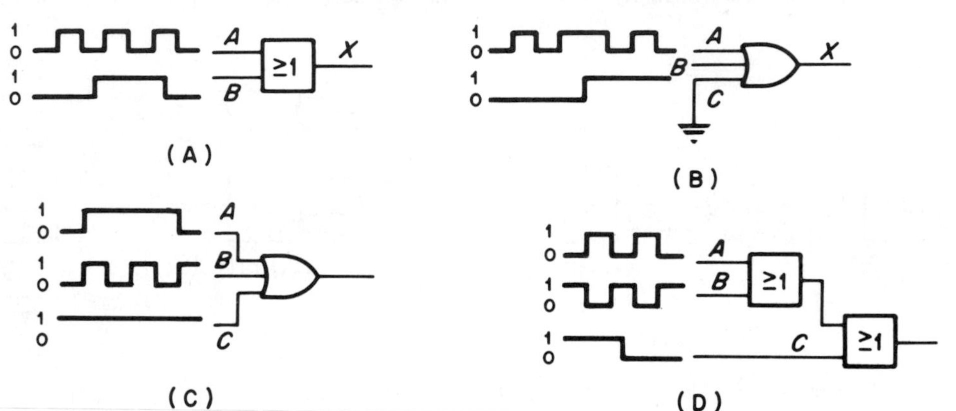

Figure 3–38

Solution

With an OR gate, the output is HIGH when any input is HIGH. This leaves you the task of analyzing each input and noting its condition as illustrated in Figure 3–39.

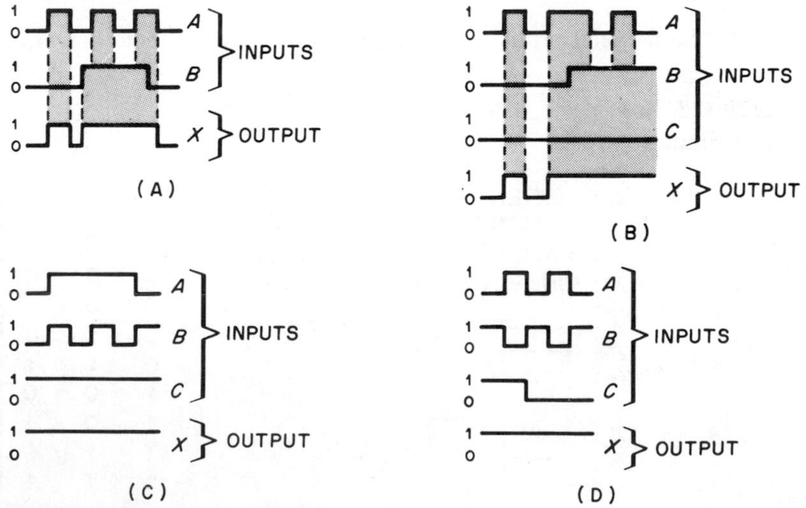

Figure 3–39

Conclusion

In this section, your second building block was presented: the OR gate. You learned what OR meant, saw an application, and had a chance to practice with some real OR gate logic circuits. The next section presents the last of the three fundamental logic circuits used by digital systems. Before starting the next section, test your understanding of this one by trying the following review questions.

3–3 Review Questions

1. State what the OR function means.
2. Describe an electrical circuit consisting of two switches, a voltage source, an LED, and a resistor which demonstrates the OR function.
3. What is an OR gate?
4. Describe the truth table for a two-input OR gate.
5. What is an IC chip called that consists of four OR gates where each gate has two inputs.

3–4 THE INVERTER

Discussion

Here, you will meet the last of the three fundamental building blocks used in digital systems. At first, it will seem as if this logic circuit doesn't do very much. But later, you will see how important and powerful a function this really is.

The Inverter Concept

Do you remember how to take the complement of a binary number? You changed all of the 1's to 0's and all of the 0's to 1's. Essentially, you inverted each bit. See Figure 3–40. Observe from the figure that in order for the LED to be HIGH the input switch must be OFF. When the input switch is ON, the LED is OFF. This is shown in the truth table in Figure 3–40. If you convert the ON and OFF switch notation and the HIGH and LOW LED notation to 1's and 0's, then you have the following truth table for an inverter.

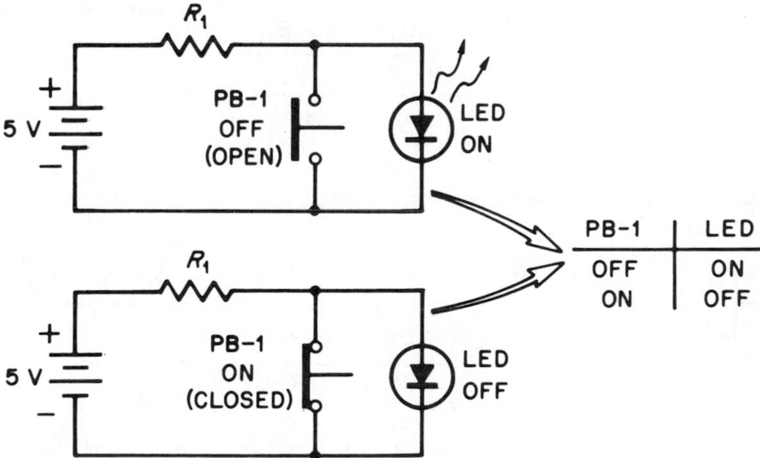

Figure 3–40 Digital inverter circuit.

Inverter	
Input *A*	**Output** *X*
0 1	1 0

This truth table uses the conventional notation where letters from the beginning of the alphabet (A, B, C, etc.) represent the inputs while letters from the end of the alphabet (X, Y, Z) represent the outputs.

Unlike the OR gate and the AND gate, an inverter has only one input. An inverter takes the logic level of the input and inverts it to the opposite logic level. As you can see from the truth table, if the input is a 0, the output will be a 1, and if the input is a 1, the output will be a 0.

The Inverter

Figure 3–41 shows the logic symbols of an inverter. The important point about the logic symbols for the inverter is the small circle or triangle on the output. The small circle or triangle, in logic circuit notation, means *invert* or *complement*.

There is a digital circuit called a *buffer* that has a symbol very similar to that of an inverter. However, unlike the inverter, a *buffer* does not invert the input logic level. Figure 3–42 illustrates. The small circle or triangle is not used on the buffer.

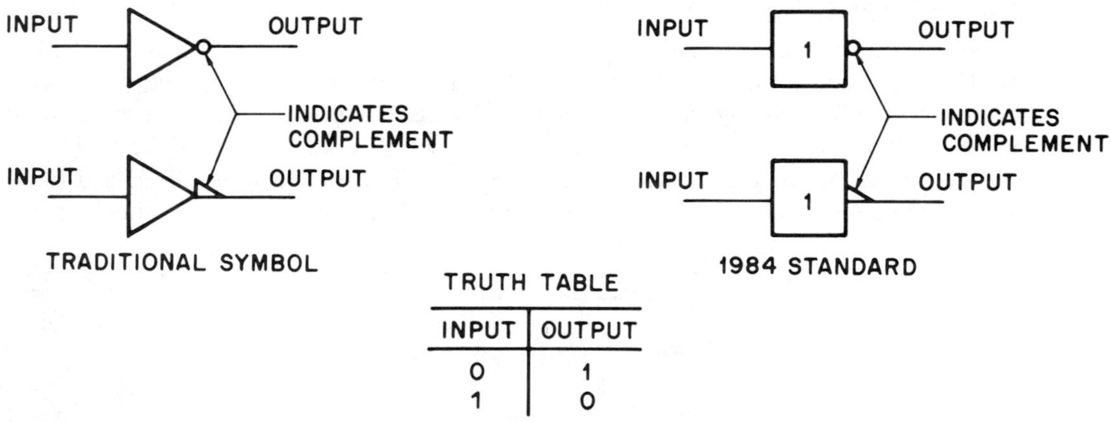

Figure 3–41 Logic symbols for an inverter.

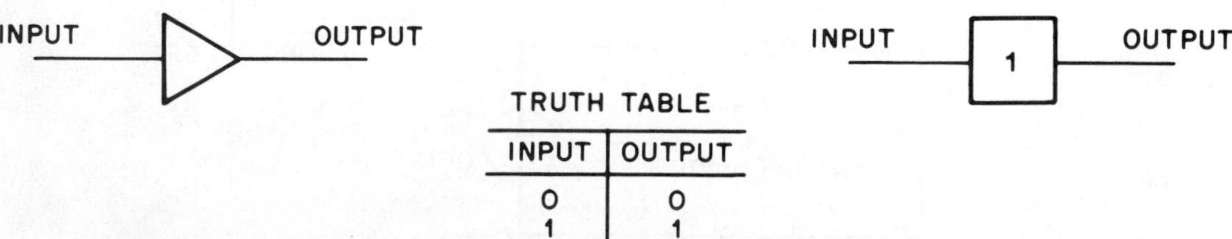

Figure 3–42 Logic symbol for the buffer with truth table.

The purpose of a buffer is to help increase the electrical power of the digital signal. You will learn more about its use as you study more details of digital circuits. What is important now is that you recognize what a buffer looks like and what it does.

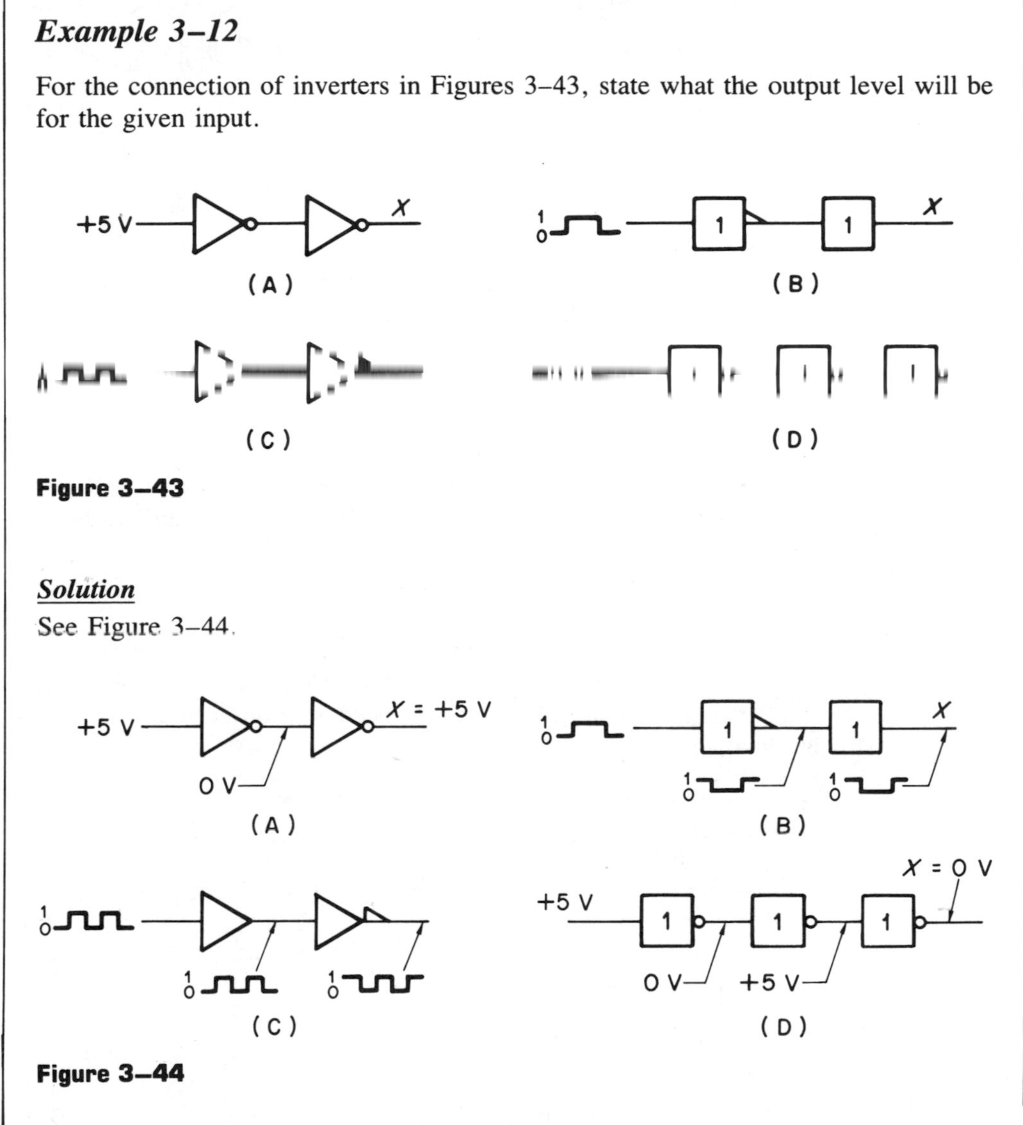

Example 3–12

For the connection of inverters in Figures 3–43, state what the output level will be for the given input.

Figure 3–43

Solution

See Figure 3–44.

Figure 3–44

Real Inverters

Figure 3–45 shows one of the simplest kinds of inverters. It is an IC chip that costs only a few cents and gives you six inverters. Because there are six inverters, it is called a *hex inverter*. The same figure also shows an IC chip consisting of six buffers. As you have probably already guessed, this chip is call a *hex buffer*.

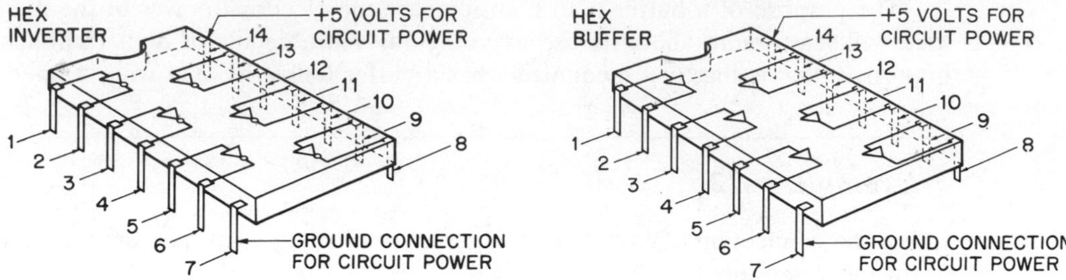

Figure 3–45 IC chips for inverters and buffers.

Inverter Applications

The following examples illustrate some inverter applications using some of the examples presented previously in the chapter.

Example 3–13

Using logic diagrams, show how you would construct a circuit that would activate a green LED if all three doors to the warehouse were closed and would activate a red LED if any door were open. You may be tempted to use an OR gate for the red LED, but there is a more efficient design.

Solution

This problem is similar to the warehouse example given in Section 3–2 and 3–3. There an AND gate was used to solve the problem. Again, an AND gate can be used to activate the green LED. You only need an inverter to activate the red LED. This is illustrated in Figure 3–46.

A = LEFT SIDE DOOR
B = BACK DOOR
C = RIGHT SIDE DOOR
X = GREEN LED
Y = RED LED

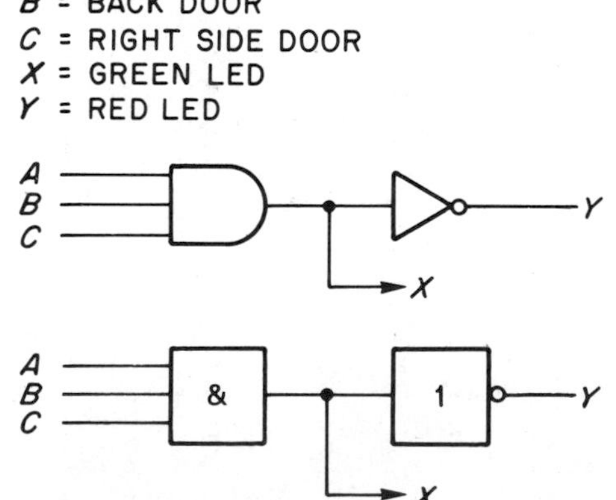

INPUTS			OUTPUTS	
C	B	A	X	Y
0	0	0	0	1
0	0	1	0	1
0	1	0	0	1
0	1	1	0	1
1	0	0	0	1
1	0	1	0	1
1	1	0	0	1
1	1	1	1	0

Figure 3–46

Example 3–14

Using logic diagrams, show how you would construct a circuit that would light a green LED if all of three patients in the same hospital room do not need service, and a red LED if any patient does need service.

Solution

This problem is similar to the one presented in the last section, where an OR gate was used. The same arrangement can be used here along with an inverter. Figure 3–47 shows how.

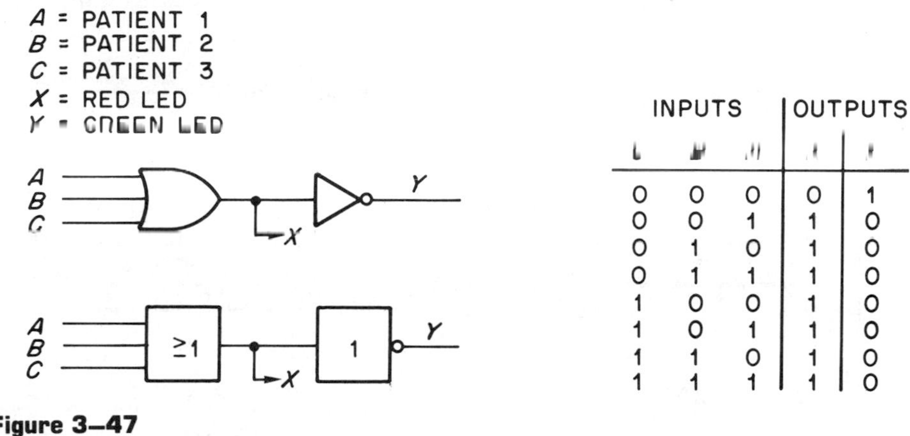

INPUTS			OUTPUTS	
[illegible]	[illegible]	[illegible]	[illegible]	[illegible]
0	0	0	0	1
0	0	1	1	0
0	1	0	1	0
0	1	1	1	0
1	0	0	1	0
1	0	1	1	0
1	1	0	1	0
1	1	1	1	0

Figure 3–47

Inverter Analysis

As you may suspect, inverter analysis with pulse waveforms is important in the study of digital systems. Just remember to complement each pulse as it passes through the inverter. The following example shows what to do.

Example 3–15

For the inverter circuits in Figure 3–48, sketch the resulting output waveforms.

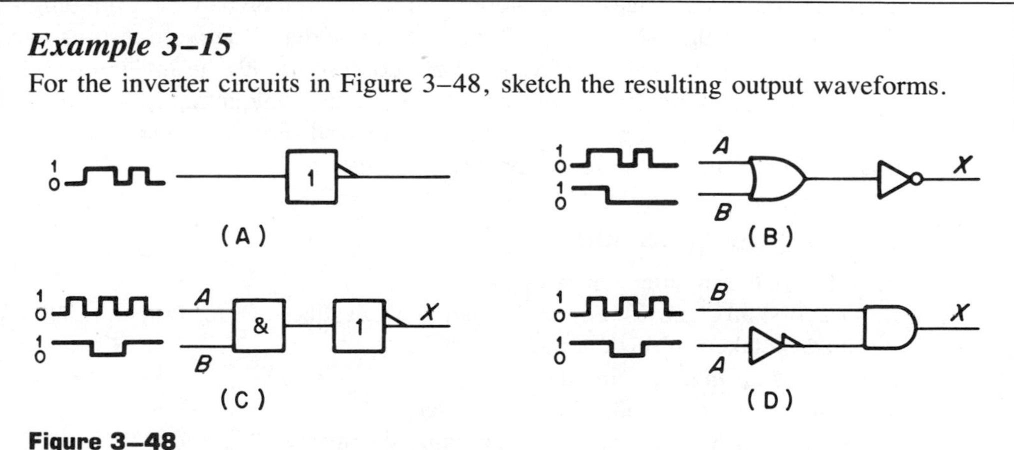

Figure 3–48

Solution

See Figure 3–49.

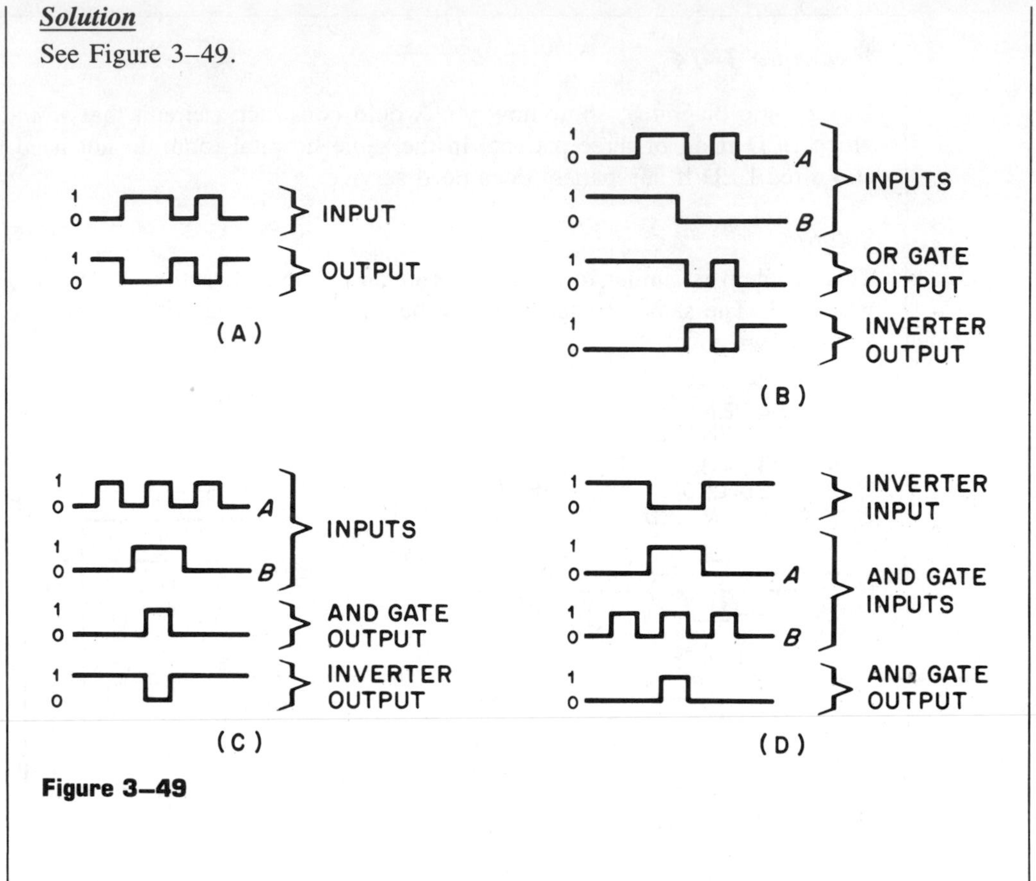

Figure 3–49

Conclusion

Congratulations! You have now worked with the three fundamental building blocks of all digital systems: the AND gate, OR gate, and inverter. You were also shown the buffer, but that was just to make sure you did not confuse it with the inverter.

In the next section, you will be introduced to a unique and powerful analyzing tool for these basic "thinking" parts of the digital world. For now, test your understanding of this section by trying the following review questions.

3–4 Review Questions

1. What is the inverter function?
2. Describe an electrical circuit that demonstrates the inverter function. What precautions must be taken in this circuit?
3. What is an inverter circuit?
4. Describe the truth table for an inverter.
5. What is an IC chip called that contains six inverters?

3–5 ANOTHER WAY—BOOLEAN ALGEBRA

Discussion

This section presents an important and powerful way of representing the "thinking" circuits of digital systems. Invented by a self-made mathematician, George Boole, in 1847, it is a very simple kind of algebra that has only two values: TRUE or FALSE. It is a very specialized way of converting logical statements into simple mathematical relationships. This allows logic problems to be expressed in a manner similar to ordinary algebra.

An engineer by the name of Claude E. Shannon, in 1938, showed how the algebra of Boole (now called **Boolean algebra**) could be used to analyze the two-state switching of relay circuits. Thus, you can see that the TRUE/FALSE states of Boolean algebra lend themselves readily to the ON/OFF conditions of digital circuits.

Boolean Algebra:
A two-value algebra similar in form to ordinary algebra.

[illegible]

There are three basic relationships in Boolean algebra (you will see that all of these relationships relate to the invert, AND, and OR gates you have just studied). The most basic Boolean relationship is the **NOT function.** In words, it states

$$X = \text{NOT } A$$

It means if A = TRUE, then

$$X = \text{NOT TRUE} = \text{FALSE}$$

and, if A = FALSE, then

$$X = \text{NOT FALSE} = \text{TRUE}$$

If you substitute a 1 for TRUE and a 0 for FALSE, then you have the Boolean statement for an inverter:

$$X = \text{NOT } A$$

If $A = 1$, then:

$$X = \text{NOT } 1 = 0$$

If $A = 0$, then:

$$X = \text{NOT } 0 = 1$$

See Figure 3–50.

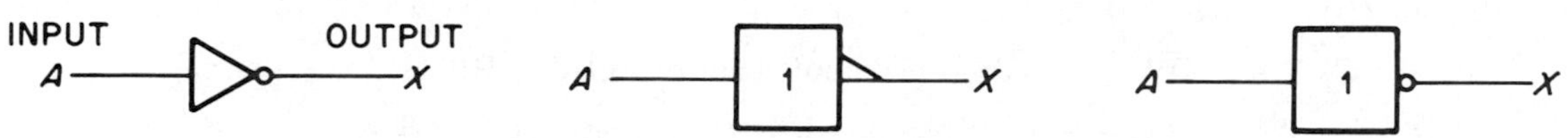

Figure 3–50 Logic symbols for the NOT expression.

NOT Function:
Complement or inversion of a variable. NOT TRUE is FALSE. NOT FALSE is TRUE.

The standard truth table can be used to represent the NOT operation:

NOT Operation	
A	**NOT *A*** ***X***
FALSE	TRUE
TRUE	FALSE

NOT Operation	
A	**NOT *A*** ***X***
0	1
1	0

In logic circuits, the *overbar* represents the NOT operation. This means that

$$X = \text{NOT } A$$

can be written as

$$X = \overline{A}$$

You read this as "X equals NOT A" or "X equals A bar" or "X equals the complement of A." Using this notation, you have
If $A = 0$, then

$$X = \overline{A} = \overline{0} = 1 \qquad \text{(NOT 0 is 1)}$$

and, if $A = 1$,

$$X = \overline{A} = \overline{1} = 0 \qquad \text{(NOT 1 is 0)}$$

Example 3–16

Solve the following Boolean expressions:

(A) $Z = \overline{C}$ for C = TRUE (B) Y = NOT B, for $B = 0$

(C) $T = \overline{D}$ for D = *LOW* (D) $W = \overline{E}$, for E = ground

Solution
Just remember that the NOT statement means the opposite, and Boolean algebra has only one of two possible answers.

(A) $Z = \overline{\text{TRUE}}$ = FALSE (NOT TRUE = FALSE)

(B) Y = NOT 0 = 1

(C) $T = \overline{\text{LOW}}$ = HIGH (the complement of LOW is HIGH)

(D) $W = \overline{\text{ground}}$ = +5 volts (the inversion of ground is +5 volts)

Boolean Multiplication

Multiplication in Boolean algebra simply represents the AND function. It does not mean multiply in the ordinary sense of the word. For example,

$$X = A \cdot B \qquad \text{means} \qquad X = A \text{ AND } B$$

This represents the Boolean expression for a two-input AND gate. See Figure 3–51.

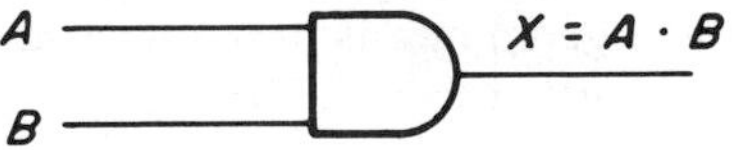

Figure 3–51 Logic symbol for Boolean multiplication.

Boolean Multiplication:
Boolean representation of the AND function.

The expression

$$Y = A \cdot B$$

is read "*Y* equals *A* and *B*" or "*Y* equals *A* dot *B*." It is the standard way of expressing an AND gate in Boolean algebra, and, as such, the truth table for the AND gate applies:

AND Operation		
A	*B*	*X*
FALSE	FALSE	FALSE
FALSE	TRUE	FALSE
TRUE	FALSE	FALSE
TRUE	TRUE	TRUE

AND Operation		
A	*B*	*X*
0	0	0
0	1	0
1	0	0
1	1	1

Example 3–17

Determine the results of the following Boolean equations:

(A) $X = A \cdot B$, where A = FALSE, B = TRUE

(B) $Z = C \cdot D$, where D = HIGH, C = HIGH

(C) $V = D \cdot A$, where $D = 1$, $A = 0$

(D) $Y = A \cdot F$, where $A = +5$ volts, $F = 0$ volts

Solution

Using the definition of the Boolean AND, you obtain

(A) X = FALSE AND TRUE = FALSE

(B) Z = HIGH and HIGH = HIGH

(C) V = 1 AND 0 = 0

(D) Y = +5 volts AND 0 volts = 0 volts

Figure 3–52 Logic symbol for Boolean addition.

Boolean Addition

Addition in Boolean algebra means the same as the OR function. It does not mean addition in the ordinary sense of the word. For example,

$$X = A + B \quad \text{means} \quad X = A \text{ OR } B$$

This represents the Boolean expression for a two-input OR gate. See Figure 3–52.

Boolean Addition:
Boolean representation of the OR function.

The expression

$$Y = A + B$$

is read as "Y equals A OR B." It is the standard way of expressing an OR gate in Boolean algebra, and, as such, the truth table for the OR gate applies.

OR Operation		
A	*B*	*X*
FALSE	FALSE	FALSE
FALSE	TRUE	TRUE
TRUE	FALSE	TRUE
TRUE	TRUE	TRUE

OR Operation		
A	*B*	*X*
0	0	0
0	1	1
1	0	1
1	1	1

Example 3–18

Determine the results of the following Boolean equations:

(A) $X = A + B$, where A = FALSE, B = TRUE

(B) $Z = C + D$, where D = HIGH, C = HIGH

(C) $V = D + A$, where $D = 1$, $A = 0$

(D) $Y = A + F$, where $A = +5$ volts, $F = 0$ volts

Solution

Using the definition of the Boolean AND, you obtain

(A) X = FALSE OR TRUE = TRUE

(B) Z = HIGH OR HIGH = HIGH

(C) V = 1 OR 0 = 1

(D) Y = +5 volts OR 0 volts = +5 volts

Conclusion

This section introduced you to the three elements of Boolean algebra. You learned that TRUE or FALSE were the only two allowed values of the algebra, and these states can also be represented by voltages and binary 1 or 0, as well as the terms HIGH or LOW. You also saw that the NOT function mathematically represented the inverter, that Boolean multiplication represented the AND gate, and that Boolean addition represented the OR gate.

In the next section, you will have an opportunity to apply these concepts in the analysis of real digital logic circuits. For now, test your understanding of this section by trying the following review questions.

3–5 Review Questions

1. State the values allowed in Boolean algebra.
2. Explain what the Boolean NOT statement means.
3. Define Boolean multiplication.
4. Define Boolean addition.
5. Why is Boolean algebra useful in the analysis of digital circuits?

3–6 BOOLEAN APPLICATIONS

Discussion

This section will show you the beginning steps in applying Boolean algebra to the analysis of logic gates. This will be the beginning of your skill development for gaining the knowledge of how digital systems work. Once you know how to analyze the "thinking" process of these machines, you will then be well prepared to tackle complex systems, such as computers and robotics.

Analyzing Logic Circuits

The three building blocks of digital systems are the inverter, the AND gate, and the OR gate. The three mathematical representations of these are the Boolean NOT, multiplication, and addition. These fundamental elements are used together to analyze, design, maintain, and repair digital systems. The following example gives some practice in developing these skills.

Example 3–19

For the logic circuits shown in Figure 3–53:

(A) Write the Boolean expression.

(B) Develop the truth table.

(C) Determine the output of each if $A = 1$ and $B = 0$.

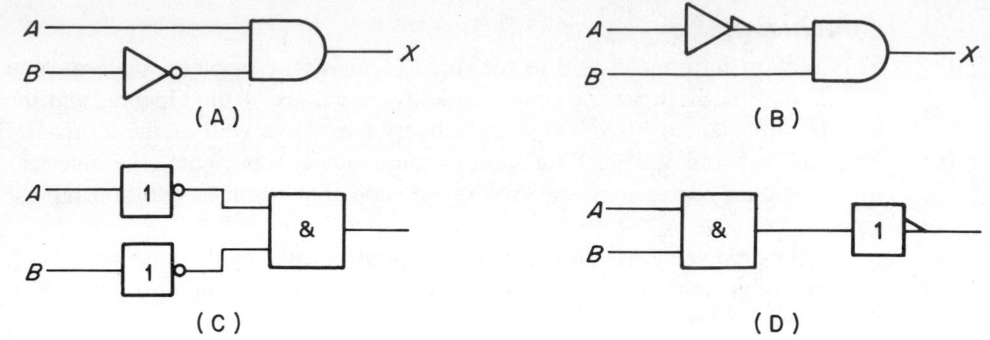

Figure 3–53

Solution

(A) Write the Boolean expression. For circuit A, the input to the AND gate is *A* AND NOT *B*. These are the two quantities that will be ANDed. Thus the Boolean expression is

$$X = A \cdot \overline{B}$$

For circuit B, this time the input to the AND gate is NOT *A* AND *B*. Thus, the Boolean expression is

$$X = \overline{A} \cdot B$$

For circuit C, in this case, both the *A* and *B* inputs are being inverted before they are ANDed. The resulting Boolean expression is

$$X = \overline{A} \cdot \overline{B}$$

For circuit D, here, *A* is ANDed with *B* and the whole result is inverted. Therefore,

$$X = \overline{A \cdot B}$$

(B) Developing the truth tables. For circuit A, start with a binary count of all possible combinations of the input:

Truth Table Circuit A		
A	***B***	
0	0	
0	1	
1	0	
1	1	

For this logic circuit, the *B* input is inverted. Show this next on the truth table:

Truth Table Circuit A		
A	***B***	$\overline{B}$
0	0	1
0	1	0
1	0	1
1	1	0

Now, observe from the Boolean expression for this circuit that the two quantities to be ANDed are A and NOT B.

$$X = A \cdot \overline{B}$$

Thus, on the resulting truth table, show the results of ANDing A and NOT B:

Truth Table Circuit A				
A	B	$\overline{B}$	X	
0	0	1	0	
0	1	0	0	
1	0	1	1	← (A = 1 AND $\overline{B}$ = 1)
1	1	0	0	

For circuit B, do the same process as before, but this time the A input is inverted.

Truth Table Circuit B				
A	B	$\overline{A}$	X	
0	0	1	0	
0	1	1	1	← ($\overline{A}$ = 1 AND B = 1)
1	0	0	0	
1	1	0	0	

For circuit C, both inputs are inverted:

Truth Table Circuit C					
A	B	$\overline{A}$	$\overline{B}$	X	
0	0	1	1	1	← ($\overline{A}$ = 1 AND $\overline{B}$ = 1)
0	1	1	0	0	
1	0	0	1	0	
1	1	0	0	0	

For circuit D, the inversion comes *after* the ANDing!

Truth Table Circuit D			
A	B	$A \cdot B$	$\overline{A \cdot B}$
0	0	0	1
0	1	0	1
1	0	0	1
1	1	1	0

(C) Determining the output for the condition that $A = 1$ and $B = 0$. For circuit A, the Boolean expression is $X = A \cdot \overline{B}$. Thus,

$$X = 1 \cdot \overline{0} = 1 \cdot 1 = 1$$

This can be verified by the truth table for circuit A:

Truth Table Circuit A				
	A	B	$\overline{B}$	X
	0	0	1	0
	0	1	0	0
(condition for $A = 1$ and $B = 0$) →	1	0	1	1
	1	1	0	0

For circuit B, the Boolean expression is $X = \overline{A} \cdot B$. Thus,

$$X = \overline{1} \cdot 0 = 0 \cdot 0 = 0$$

This can also be verified by the truth table for circuit B:

Truth Table Circuit (B)				
	A	B	$\overline{A}$	X
	0	0	1	0
	0	1	1	1
(condition for $A = 1$ and $B = 0$) →	1	0	0	0
	1	1	0	0

For circuit C, the Boolean expression is $X = \overline{A} \cdot \overline{B}$. Thus,

$$X = \overline{1} \cdot \overline{0} = 0 \cdot 1 = 0$$

Again, this can be verified by the truth table for circuit C:

Truth Table Circuit C					
	A	B	$\overline{A}$	$\overline{B}$	X
	0	0	1	1	1
	0	1	1	0	0
(condition for $A = 1$ and $B = 0$) →	1	0	0	1	0
	1	1	0	0	0

For circuit D, the Boolean expression is $X = \overline{A \cdot B}$. Thus,

$$X = \overline{1 \cdot 0} = \overline{0} = 1$$

Again, the truth table can be used:

Truth Table Circuit D			
A	B	$A \cdot B$	$\overline{A \cdot B}$
0	0	0	1
0	1	0	1
1	0	0	1
1	1	1	0

(condition for $A = 1$ and $B = 0$) → (row 3 of the table)

Note the methods available to you for predicting the output of a logic circuit—Boolean expressions and truth tables.

Developing Logic From Boolean Expressions

In the previous example, you were given various kinds of logic circuits and asked to develop a Boolean expression. Sometimes you will know the required Boolean expression and be required to construct a logic circuit to fulfill the Boolean conditions. The following example illustrates.

Example 3–20

Construct the logic circuit for each of the Boolean expressions:

(A) $Y = \overline{A} + B$

(B) $Y = A + \overline{B}$

(C) $Y = \overline{A + B}$

(D) $Y = \overline{A} + \overline{B}$

Solution

See Figure 3–54.

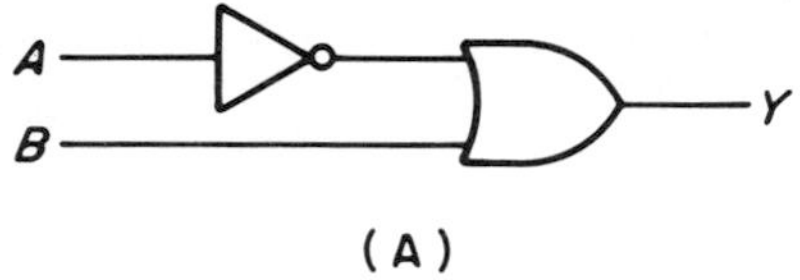

(A)

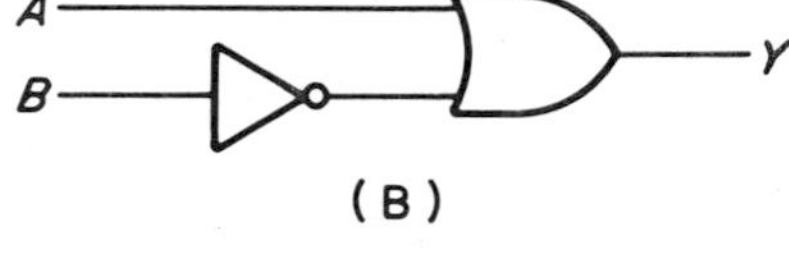

(B)

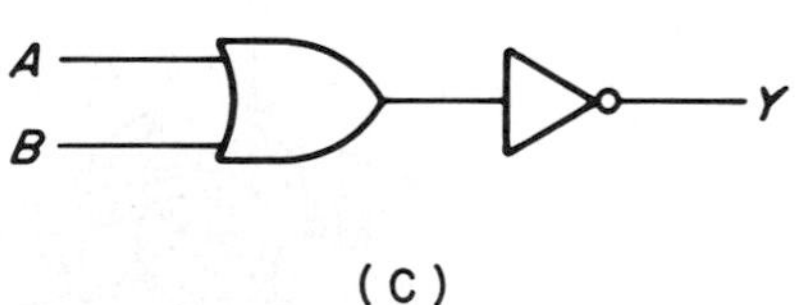

(C)

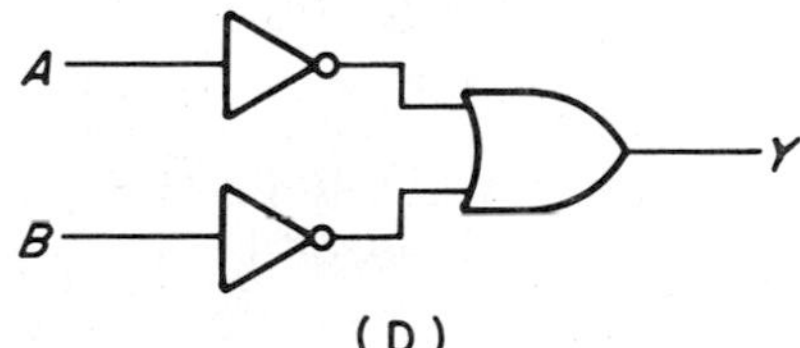

(D)

Figure 3–54

Developing Boolean Expressions from Truth Tables

Sometimes a certain condition is desired and, from the stated conditions, a digital circuit must be designed. An interim step in doing this is to take the given requirements and convert them to a truth table, and from the truth table construct a Boolean expression. Once you have the Boolean expression, you can convert it to its equivalent logic circuit. From that point, the circuit can actually be wired using available ICs.

Example 3–21

Develop the truth tables for the conditions shown in Figure 3–55.

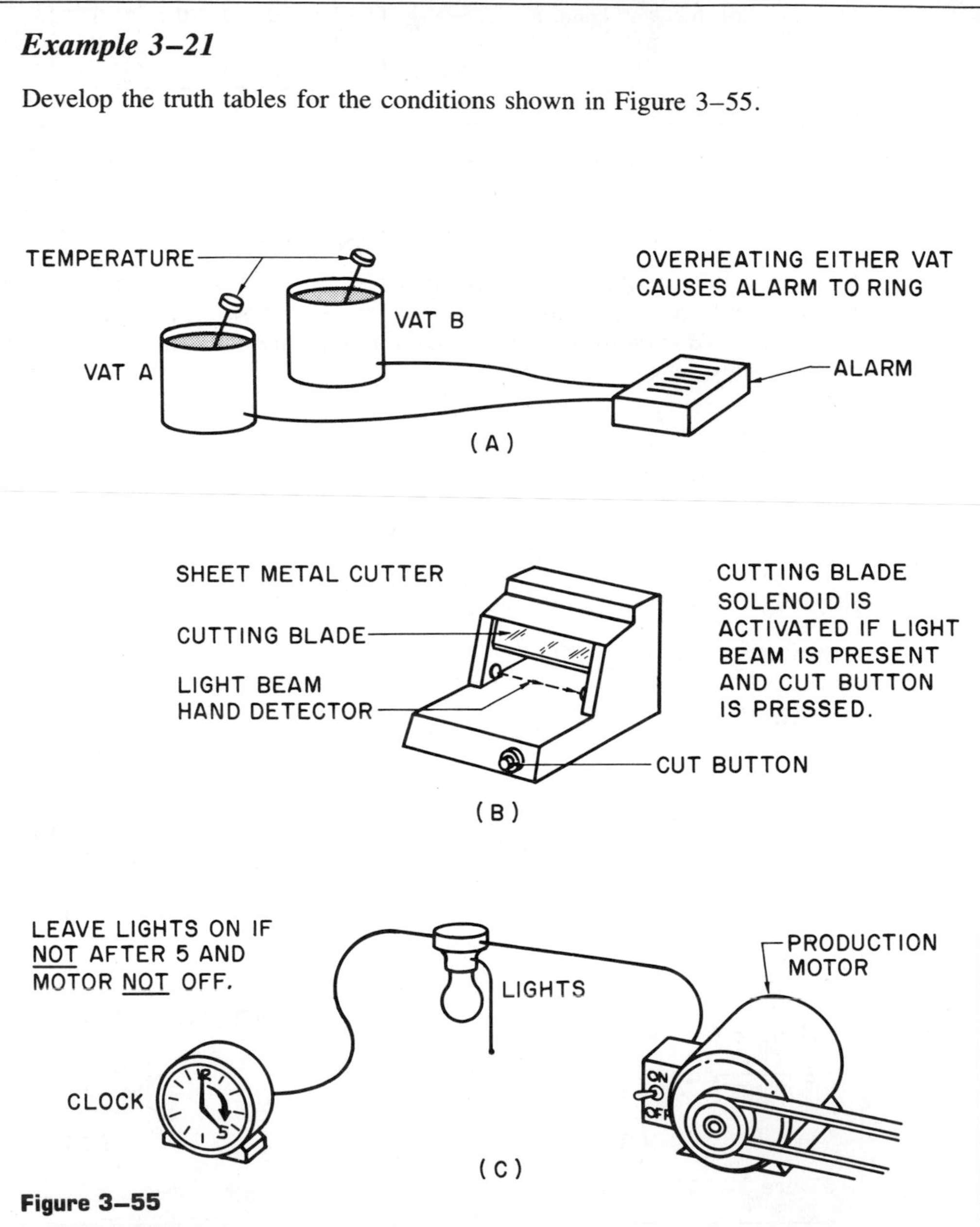

Figure 3–55

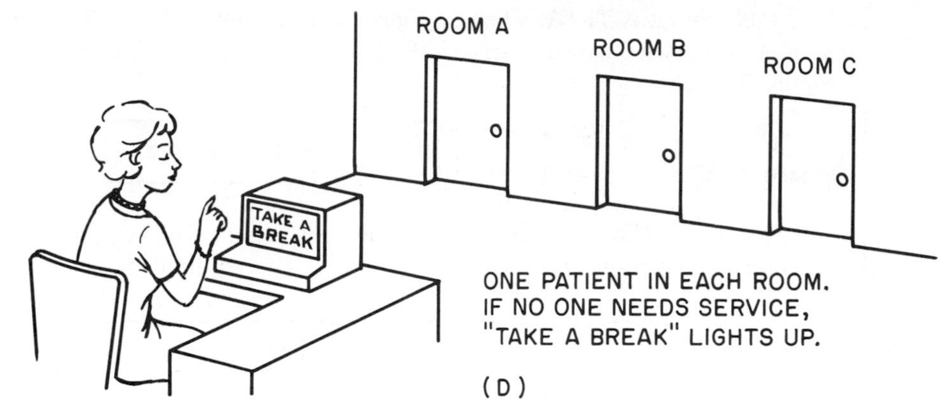
ROOM A
ROOM B
ROOM C
TAKE A BREAK
ONE PATIENT IN EACH ROOM.
IF NO ONE NEEDS SERVICE,
"TAKE A BREAK" LIGHTS UP.
(D)

Figure 3–55 Continued

Solution

System A. Here, you have two vats. The alarm will ring if any vat becomes overheated. Represent each vat by a letter (remember, use the first letters of the alphabet to represent an input) and the alarm by a letter (use one of the last letters of the alphabet to represent an output). Write the required conditions in a truth table, making sure to account for all possible combinations of the inputs. Do this by a binary count for the inputs:

Vat Conditions		
Vat A Overheated	**Vat B Overheated**	**Alarm X Activated**
NO	NO	NO
NO	YES	YES
YES	NO	YES
YES	YES	YES

You can now simplify the truth table by letting NO = 0 and YES = 1:

Vat Conditions		
Vat A Overheated	**Vat B Overheated**	**Alarm X Activated**
0	0	0
0	1	1
1	0	1
1	1	1

Since the output is TRUE when any input is true, you have the Boolean expression $X = A + B$, which results in an OR gate. The required logic circuit is shown in Figure 3–56.

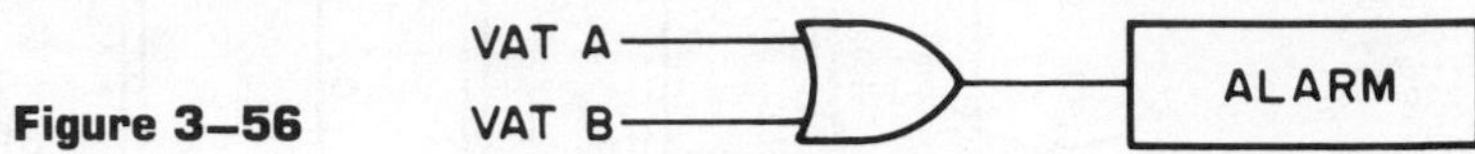

Figure 3–56

System B: The condition here is that the soleniod will be activated if the hand detector is active and the cut button is active. The required truth table is

Paper Cutter Conditions		
Hand Detector Active	**Cut Button Active**	**Solenoid Active**
NO	NO	NO
NO	YES	NO
YES	NO	NO
YES	YES	YES

Notice that the only time the cutting solenoid is allowed to activate and cut the paper is when the light beam is not broken and the cut button is pressed. Converting the NO to 0 and the YES to 1, you have

Paper Cutter Conditions		
Hand Detector Active	**Cut Button Active**	**Solenoid Active**
0	0	0
0	1	0
1	0	0
1	1	1

Letting the hand detector = A, the cut button = B, and the solenoid = X, write the Boolean expression for the only time the X output is a 1:

$$X = A \cdot B$$

The resulting required logic circuit is shown in Figure 3–57.

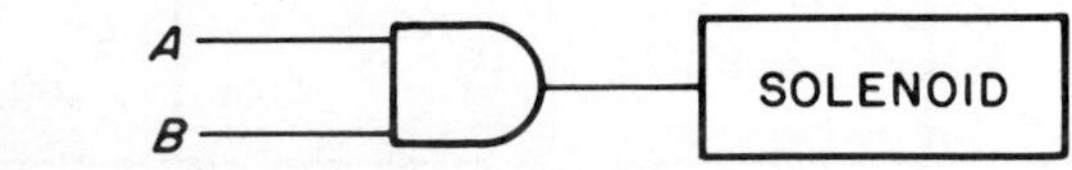

Figure 3–57

System C: This time, let "after 5:00 P.M." = A, "production motor off" = B, and "factory lights on" = X. A 1 = YES and a 0 = NO. The required truth table is

Factory Lights		
A	*B*	*X*
0	0	1
0	1	1
1	0	1
1	1	0

The Boolean expression for this table indicates that the output is FALSE only when both inputs are TRUE. This is an inverted AND expression:

$$X = \overline{A \cdot B}$$

The required logic circuit is shown in Figure 3–58.

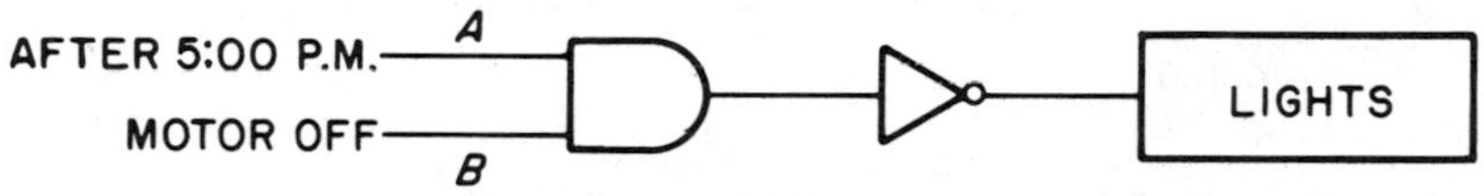

AFTER 5:00 P.M. A
MOTOR OFF B
LIGHTS

Figure 3–58

System D: Let room A = A, room B = B, and room C = C, and the "Take a Break Light On" = X. The required truth table is

Take a Break			
A	*B*	*C*	*X*
0	0	0	1
0	0	1	0
0	1	0	0
0	1	1	0
1	0	0	0
1	0	1	0
1	1	0	0
1	1	1	0

Write the Boolean expression for the only time the "Take a Break Light" is ON:

$$X = \overline{A + B + C}$$

The required logic circuit for this system is shown in Figure 3–59.

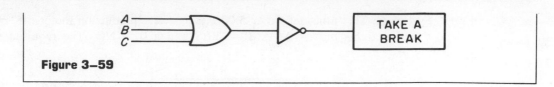

A
B
C
TAKE A BREAK

Figure 3–59

Conclusion

This section gave you a taste of using what you have learned up to this point in practical applications. Here, you applied gates, truth tables and Boolean algebra, in the analysis and applied design of actual digital systems. In the next section, you will see how a computer uses the fundamental Boolean operations to manipulate a computer word in many interesting and useful ways. Test your understanding of this section by trying the following review questions.

3–6 Review Questions

1. State the two ways to analyze an existing logic circuit.
2. In a truth table, what do the terms YES and NO represent?
3. What kind of logic circuit will give you a YES when any input condition is a YES?
4. What kind of logic circuit will give you a NO when all inputs are YES?
5. How do you ensure that all possible input conditions of a digital system have been considered?

MICROPROCESSOR APPLICATION

Discussion

A **microprocessor** is an integrated circuit that contains prewired circuits to cause the device to perform many different processes. It usually comes on a small 40-pin chip—hence its name, microprocessor. In this section, you will get a general idea of what's inside a microprocessor and some of the logic functions that all microprocessors can perform.

Microprocessor:
A single integrated circuit chip containing a fixed set of processes such as arithmetic, logic, and bit manipulation.

Why Use a Microprocessor?

The microprocessor is a separate chip for the memory presented in the last chapter. This is shown in Figure 3–60. Recall that a computer word is a group of ON/OFF bits treated as a unit. Each computer word is stored in a specific memory location called an *address*. A separate action referred to for now as an *address selector* selects one address at a time.

Memory by itself is very static. That is, once you store a word in memory, it stays there, and as long as electrical conditions allow, it will stay exactly the way you put it. It is the microprocessor that can copy a word from memory, do something with it, and even have it recopied into another memory location in the same form or in a different form. You will soon understand the digital circuits used to perform this process, but for now consider the very simple microprocessor shown in Figure 3–61. The microprocessor

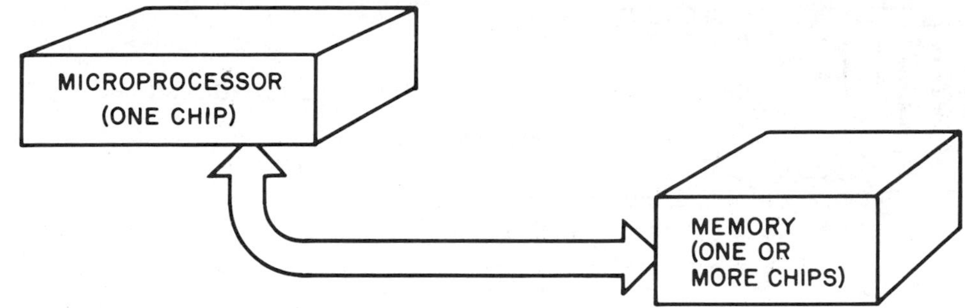

Figure 3–60 Microprocessor with memory.

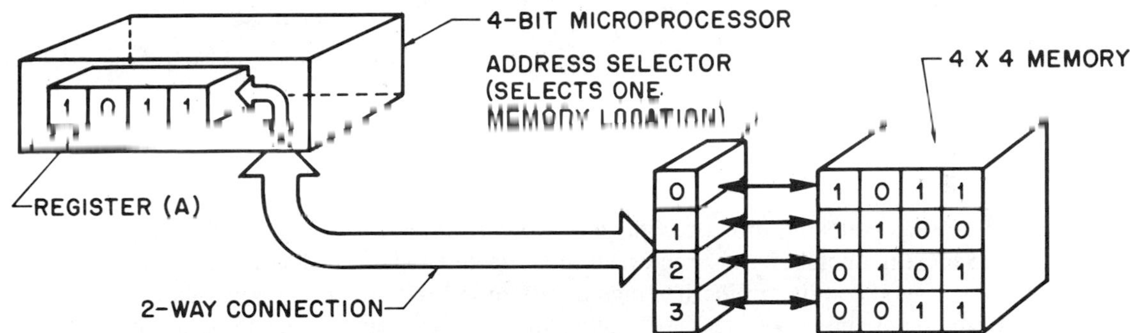

Figure 3–61 Very simple microprocessor with memory.

is connected to a single memory storage location. By using appropriate logic circuits in the address selector, it can take any word from the 4 × 4 memory and copy it into this single memory location inside the microprocessor. This storage location inside the microprocessor is called a **register.** Note that the microprocessor in Figure 3–61 contains one register while the memory contains four registers.

Register:
A storage place for one or more computer words.

Since the microprocessor under discussion has a 4-bit register, it is called a 4-bit microprocessor. Most modern microprocessors are 8-bit, 16-bit or even 32-bit microprocessors. The first microprocessor ever built was a 4-bit. It gave all the engineers and technicians of the day a chance to learn about microprocessors. You too will be given the same learning opportunity.

Looking Inside the Microprocessor

Suppose a seven-segment readout was connected through appropriate logic circuits to the microprocessor so that the contents of its 4-bit register could be displayed. This arrangement is shown in Figure 3–62.

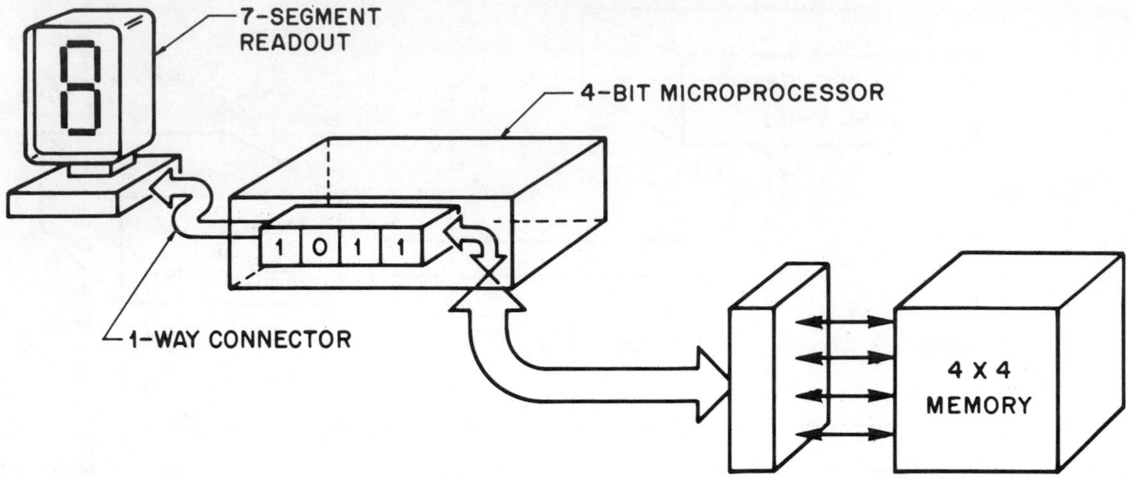

Figure 3–62 Adding a seven-segment readout.

When you work directly with a microprocessor, usually your view of it will be through hexadecimal numbers. It's important that you develop the skills in interpreting these numbers so you can determine if a digital system is performing its processes correctly. Some of the many processes that can be performed by a microprocessor are logic functions, such as NOT, AND, and OR.

The NOT Process

All microprocessors have a NOT process. This process consists of taking the word in its register and performing a complement on each bit. Usually, you will verify the process through hexadecimal numbers.

Example 3–22

A 4-bit microprocessor performs a NOT process on the word in its register. What will be the seven-segment readout after the NOT process if the original register contents were

(A) F_{16} (B) 0_{16} (C) A_{16} (D) 3_{16}

Solution

First, convert the hex number to its binary equivalent, then take the one's complement of the number (invert each bit of the word). Then convert the result back to a hex number.

(A) $F_{16} = 1111_2$.
Complementing each bit $\rightarrow 0000_2 = 0_{16}$.
Thus: NOT $F_{16} = 0_{16}$.

(B) $0_{16} = 0000_2$.
Complementing each bit → $1111_2 = F_{16}$.
Thus: NOT $0_{16} = F_{16}$.
Note that complementing the same word twice produces the original word:
NOT $F_{16} = 0_{16}$ and NOT $0_{16} = F_{16}$.

(C) $A_{16} = 1010_2$.
Complementing each bit → $0101_2 = 5_{16}$.
Thus: NOT $A_{16} = 5_{16}$.

(D) $3_{16} = 0011_2$.
Complementing each bit → $1100_2 = C_{16}$.
Thus: NOT $3_{16} = C_{16}$.

More inside the Microprocessor

All microprocessors have more than one internal register. In order to distinguish one register from the other, they are given names. The most active register in any microprocessor [illegible] particular process. The register that does this most of the time is called the **accumulator,** because that's where the results of most processes accumulate.

Accumulator:

A register inside a microprocessor. It is usually used for the intermediate storage of a word or the copying of a word to or from memory or another device external to the microprocessor. It is also used to temporarily store the result of an arithmetic, logical, or transfer process.

The second register for our 4-bit microprocessor will be referred to as the B register. This is illustrated in Figure 3–63.

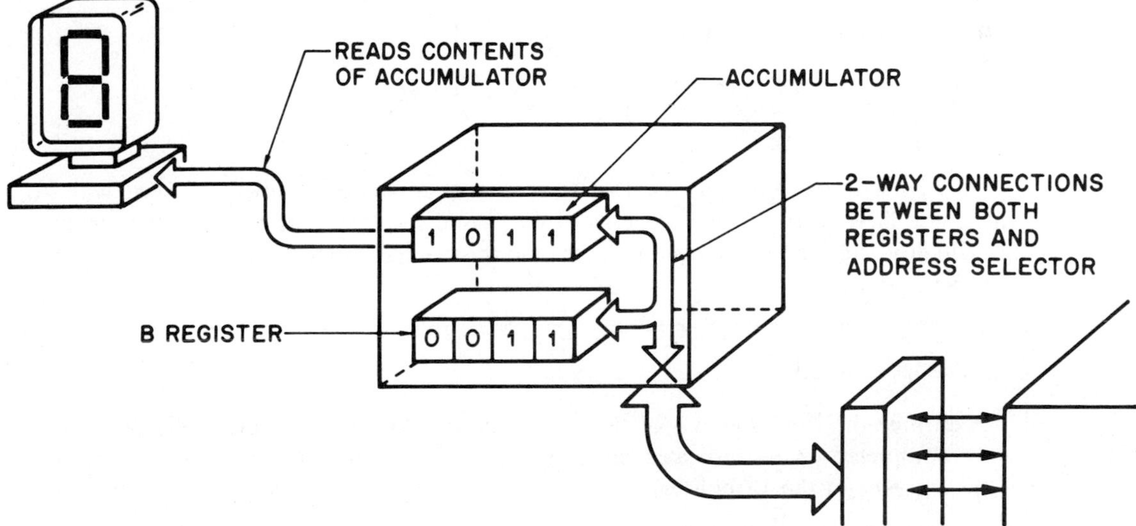

Figure 3–63 Four-bit microprocessor with accumulator and B register.

The AND Process

When a microprocessor ANDs two words, it ANDs each bit pair of each word (such as the two LSBs) and stores the results in the accumulator. For the microprocessor presented here, this means that the word in the accumulator will be ANDed with the word in the B register, and the results will be stored in the accumulator. This action is demonstrated in Figure 3–64.

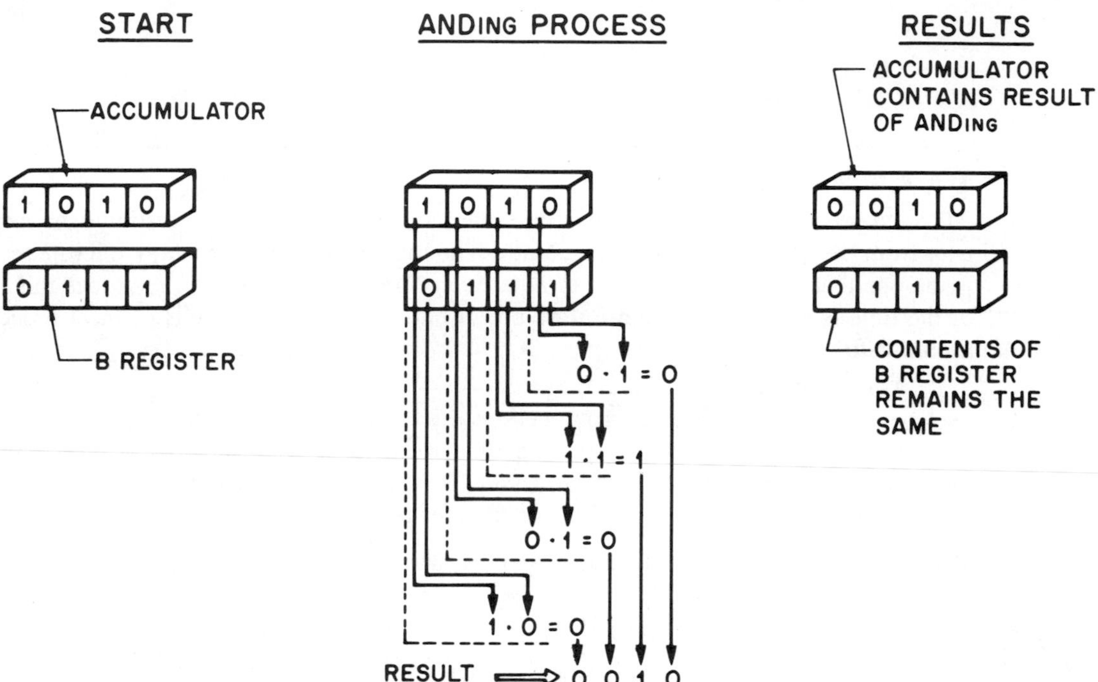

Figure 3–64 The AND process.

In Figure 3–64, you can see that the contents of the accumulator started with A_{16}, which was ANDed to 7_{16}, the contents of the B register. After the process, the result was 2_{16}, which was stored in the accumulator (and the original value in the accumulator of A_{16} was lost).

Example 3–23

Determine the final contents of the accumulator after the following AND processes are completed (A → indicates the contents of the accumulator and B → indicates the contents of the B register).

(A)	(B)	(C)	(D)
$A \rightarrow F_{16}$	$A \rightarrow F_{16}$	$A \rightarrow B_{16}$	$A \rightarrow A_{16}$
$B \rightarrow F_{16}$	$B \rightarrow 0_{16}$	$B \rightarrow 8_{16}$	$B \rightarrow 6_{16}$

Solution

First convert each hex number to its binary equivalent; then AND each bit pair. Convert the result back to hexadecimal.

(A) $F_{16} = 1111_2$
$F_{16} = 1111_2$
$1111_2 \leftarrow$ AND each bit pair
Thus: F_{16} AND $F_{16} = F_{16}$

(B) $F_{16} = 1111_2$
$0_{16} = 0000_2$
$0000_2 \leftarrow$ AND each bit pair
Thus: F_{16} AND $0_{16} = 0_{16}$

(C) $B_{16} = 1011_2$
$8_{16} = 1000_2$
$1000_2 \leftarrow$ AND each bit pair
Thus: B_{16} AND $8_{16} = 8_{16}$

(D) $A_{16} = 1010_2$
$6_{16} = 0110_2$
$0010_2 \leftarrow$ AND each bit pair
Thus: A_{16} AND $6_{16} = 2_{16}$

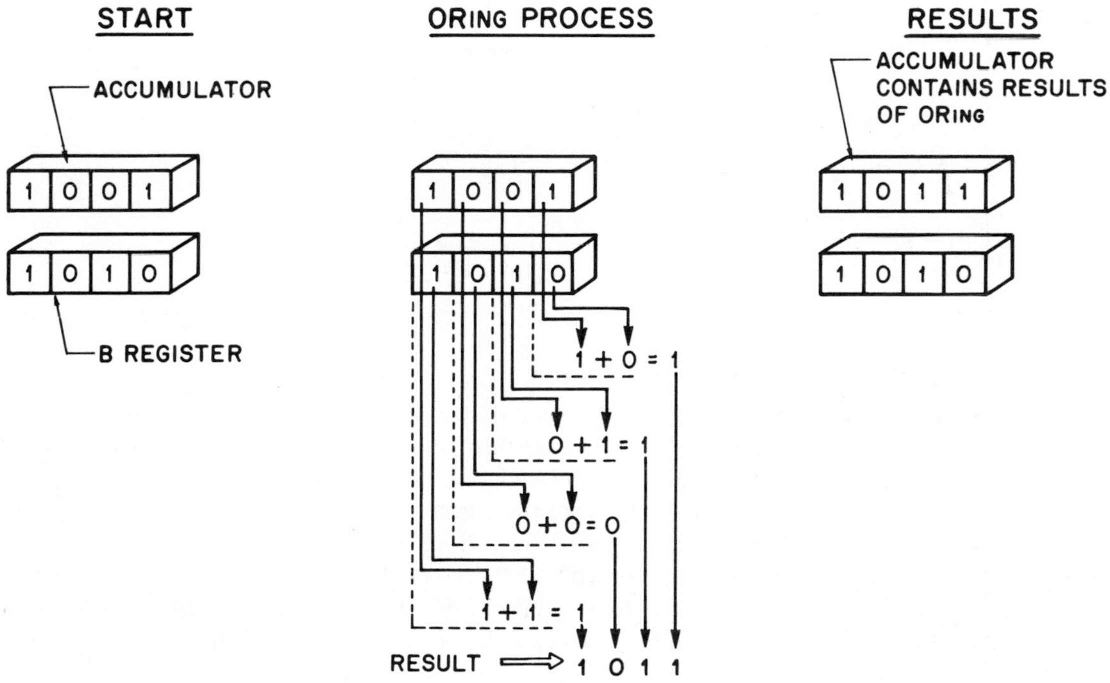

Figure 3–65 The OR process.

OR Process

The OR process for a microprocessor simply does a bit-by-bit ORing of each bit pair of the accumulator contents and the B register contents. The result is stored in the accumulator. Again, the original contents of the accumulator are lost. This process is illustrated in Figure 3–65. Expressing the values in Figure 3–65, in hex, you see that the accumulator started with a value of 9_{16} and the B register had A_{16}. After the OR process, the content of the accumulator was B_{16}.

Example 3–24

Determine the final contents of the accumulator after an OR process is performed.

(A) $A \rightarrow F_{16}$ $B \rightarrow F_{16}$ (B) $A \rightarrow F_{16}$ $B \rightarrow 0_{16}$ (C) $A \rightarrow A_{16}$ $B \rightarrow 6_{16}$

Solution

Convert each hex number to its binary equivalent. Then OR each bit pair of the accumulator and B register. Convert the resulting number back to hex.

(A) $F_{16} = 1111_2$
$F_{16} = 1111_2$
$1111_2 \leftarrow$ OR each bit pair
Thus: F_{16} OR $F_{16} = F_{16}$

(B) $F_{16} = 1111_2$
$0_{16} = 0000$
$1111_2 \leftarrow$ OR each bit pair
Thus: F_{16} OR $0_{16} = F_{16}$

(C) $A_{16} = 1010_2$
$6_{16} = 0110_2$
$1110_2 \leftarrow$ OR each bit pair
Thus: A_{16} OR $6_{16} = E_{16}$

A Look at 8-Bit Microprocessors

An 8-bit microprocessor has an 8-bit word size. This means that its accumulator and B register can both work with 8-bit words (a byte). It's convenient for an 8-bit microprocessor to work with a memory with an 8-bit word size as well. This arrangement is shown in Figure 3–66. As you can see from the figure, two hex numbers are now required to represent the contents of the accumulator. (A 16-bit microprocessor would require four hex numbers to represent the contents of its accumulator.)

The NOT, AND and OR process for an 8-bit microprocessor is still the same as it is for the 4-bit, you just need to compare 8 bits instead of 4. The following example demonstrates.

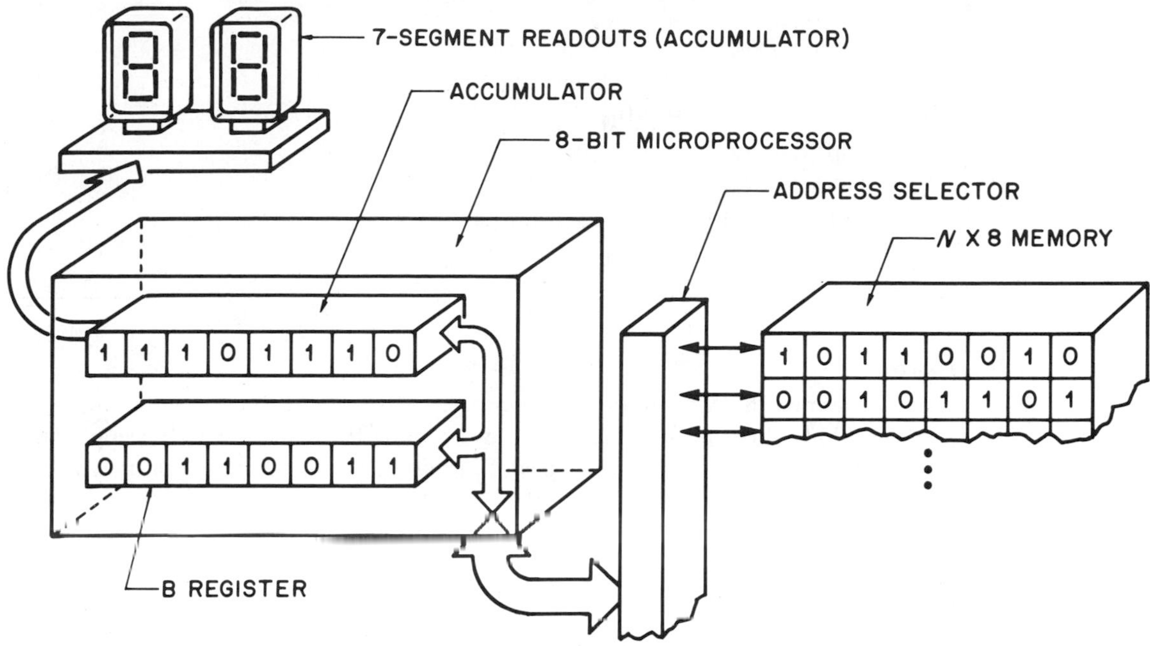

Figure 3–66 Eight-bit microprocessor—accumulator and B register.

Example 3–25

Determine the final contents of the accumulator if the following processes are performed on an 8-bit microprocessor:

(A) NOT FF_{16} (B) $C2_{16}$ OR 50_{16} (C) $A3_{16}$ AND $F0_{16}$ (D) 144_{10} AND 20_{10}

Solution

Follow the same procedure as with the 4-bit microprocessor. First convert the numbers from their base to binary; then do the bit-by-bit process. Convert the binary result back to the original base. Note that problem (D) is presented in base 10; therefore the answer should be given in base 10.

(A) $FF_{16} = 1111\ 1111_2$
Complementing each bit: $0000\ 0000_2 = 00_{16}$
Thus: NOT $FF_{16} = 00_{16}$

(B) $C2_{16} = 1100\ 0010_2$
$50_{16} = 0101\ 0000_2$
$1101\ 0010_2$ ← OR each bit pair
Thus: $C2_{16}$ OR 50_{16} = $D2_{16}$

(C) $A3_{16} = 1010\ 0011_2$
$F0_{16} = 1111\ 0000_2$
$1010\ 0000_2 \leftarrow$ AND each bit pair
Thus: $A3_{16}$ AND $F0_{16} = A0_{16}$

(D) $144_{10} = 1001\ 0000_2$
$20_{10} = 0001\ 0100_2$
$0001\ 0000_2 \leftarrow$ AND each bit pair
Thus: 144_{10} AND $20_{10} = 16_{10}$

Conclusion

This section presented the concept of a microprocessor. Here you saw how the microprocessor performed the three basic logic processes on its words. The electronic application section presents the two most common types of digital integrated circuits that you will be using in the lab. For now, test your understanding of this section by trying the following review questions.

3–7 Review Questions

1. State the purpose of a microprocessor.
2. Give the name of the most-used register in a microprocessor.
3. What is the main difference between 4-bit and 8-bit microprocessors?
4. Explain the NOT process of a microprocessor.
5. Explain the AND and the OR processes of a microprocessor.

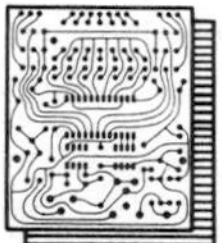

ELECTRONIC APPLICATION

Note: Electronic application sections may require some knowledge of electronics. Non-electronics major may omit these sections, with no loss in subject continuity.

Discussion

A **digital integrated circuit** (IC) is a complete circuit in a single chip.

> **Digital Integrated Circuit:**
> An electronic circuit where all of the components are produced on a single substance (usually silicon) and the circuit responds to a two-level (TRUE/FALSE or HIGH/LOW) electrical signal.

A major part of troubleshooting is understanding the "hardware" you will be working with. As such, you must observe operating precautions when using digital circuits, just as with any other electronic device. This section presents the most popular digital ICs you will be using in the lab, along with some of their operating characteristics and safety precautions.

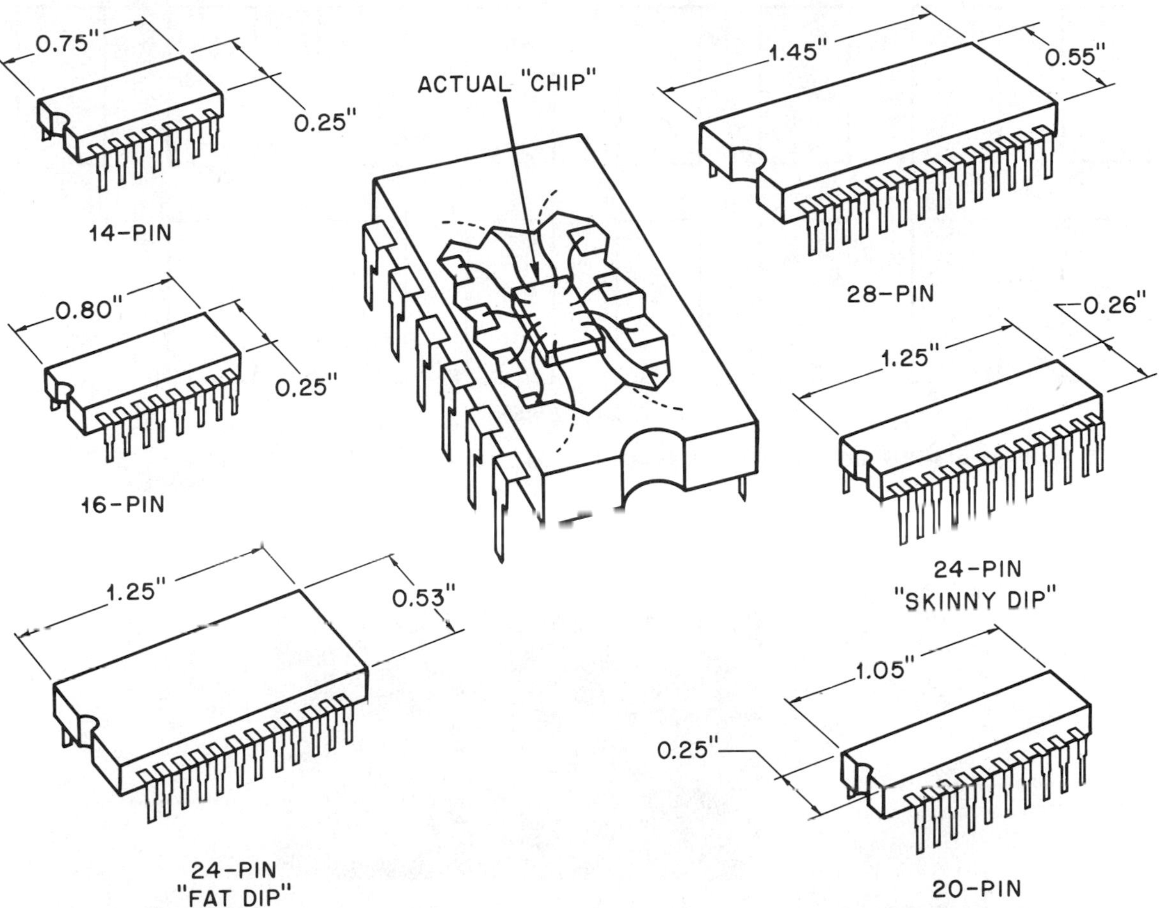

Figure 3–67 Typical digital ICs.

Typical Digital ICs

Figure 3–67 shows the fundamental construction of a digital IC and typical pin configurations with various sizes. Digital ICs come most commonly in packages that contain 8, 14, 16, and 24 pins, called a DIP (dual in-line package). Figure 3–68 shows the numbering scheme used on DIPs. Note that there are various ways of identifying which end contains pin 1.

Manufacturing Processes

Figure 3–69 is a photograph of various silicon wafers. These represent integrated circuits in final stages of manufacture. Each wafer is made from a thin slice of a silicon ingot. All the chips are made at the same time on the wafer by a process that includes photochemical etching. Each chip is then tested, cut apart by a diamond-tipped saw, and the functional ones are packaged by automated machinery (controlled by digital systems themselves manufactured by the same process).

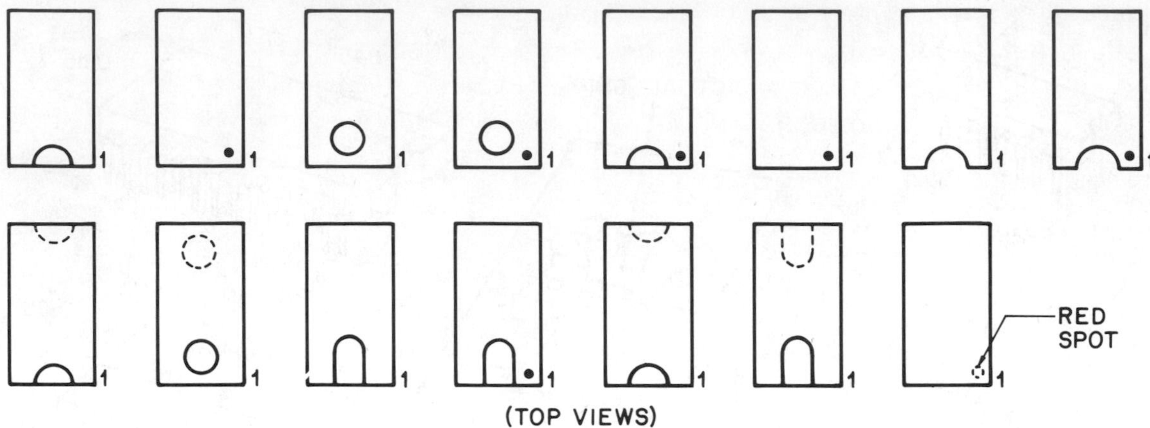

Figure 3–68 Locating pin numbers on DIPs.

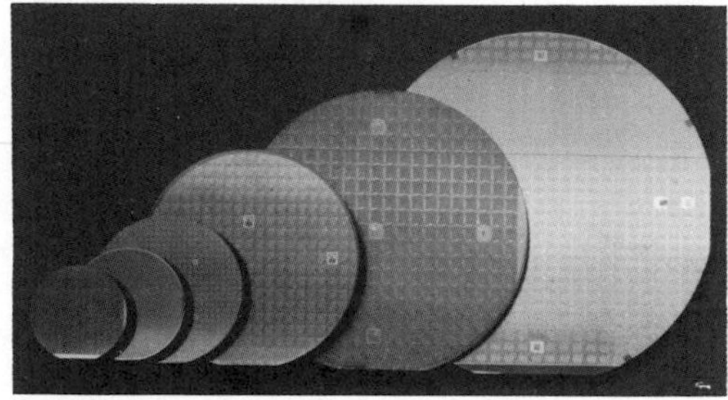

Figure 3–69 Silicon wafers. *Courtesy* of National Semiconductor Corporation.

Packaging Densities

Back in 1959, only one transistor could be fitted on a chip. Currently, more than a million transistors can be placed on a single chip. By the year 2000 it is expected that package densities will be more than a billion components per chip. Figure 3–70 shows the development of packaging densities over the years. Breakthroughs in new technologies promise even greater densities than are available with present wafer processes.

IC Labeling

Digital ICs are labeled on the top with a number that identifies its electrical characteristics and type of logic circuit. Many manufacturers include a date code. The system used to label one type of digital ICs is shown in Figure 3–71.

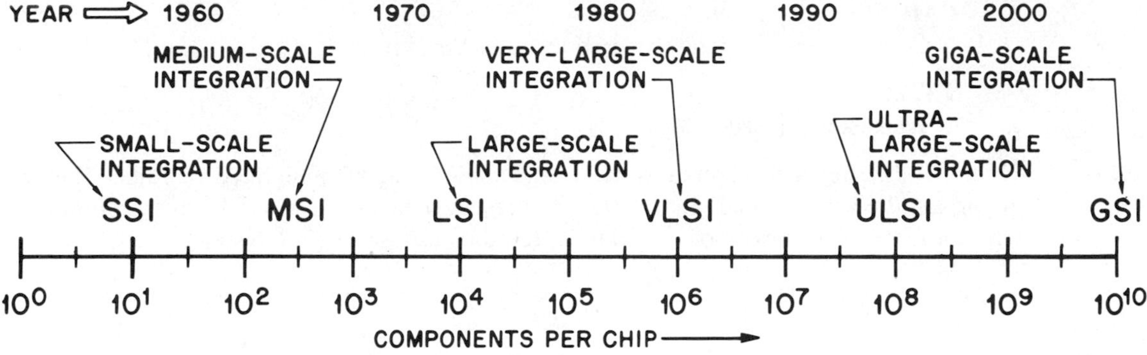

Figure 3–70 Number of components per chip.

Major Families

Several digital **IC families** are available for digital systems. The two major ones that you will be using in the lab are **TTL** and **CMOS.**

> **IC Family:**
> A group of integrated circuits manufactured using specific materials and processes.

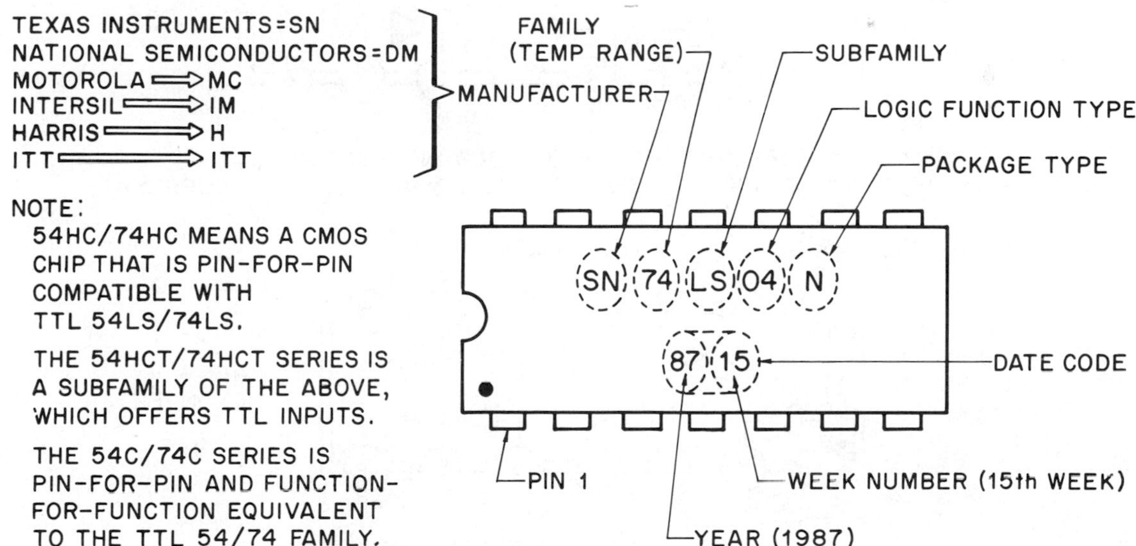

Figure 3–71 System used for labeling digital ICs.

CMOS:

Complementary metal-oxide semiconductor.

TTL:

Transistor-transistor logic.

Figure 3–72 shows representative TTL and CMOS digital integrated circuits. The two major families shown in Figure 3–72 are presented in this section. In this section, you will learn about the two most popular IC families used in digital systems.

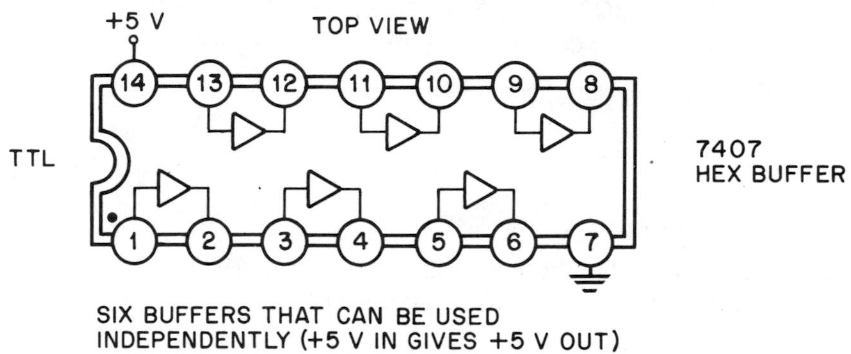

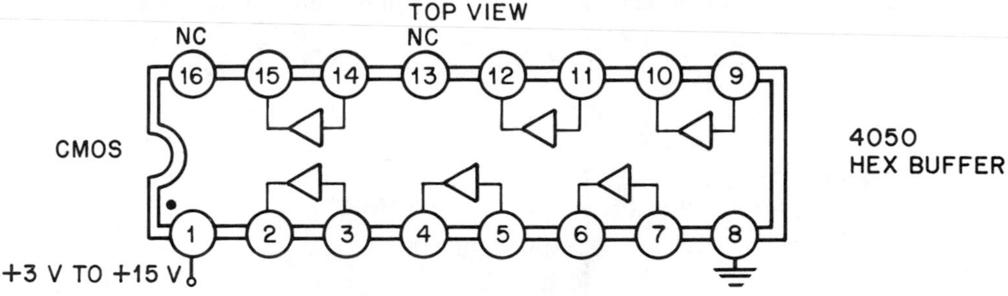

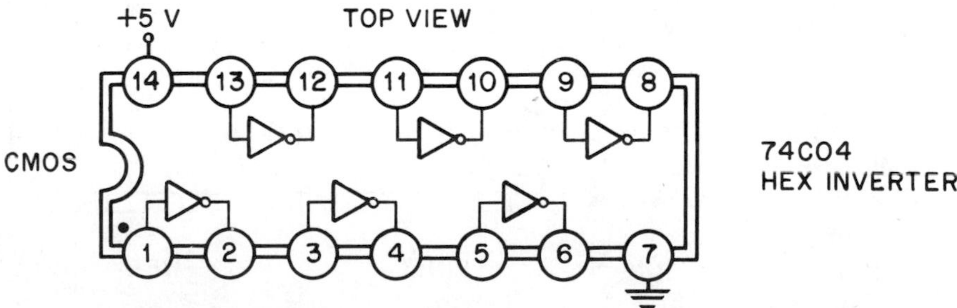

Figure 3–72 Representative TTL and CMOS integrated circuits.

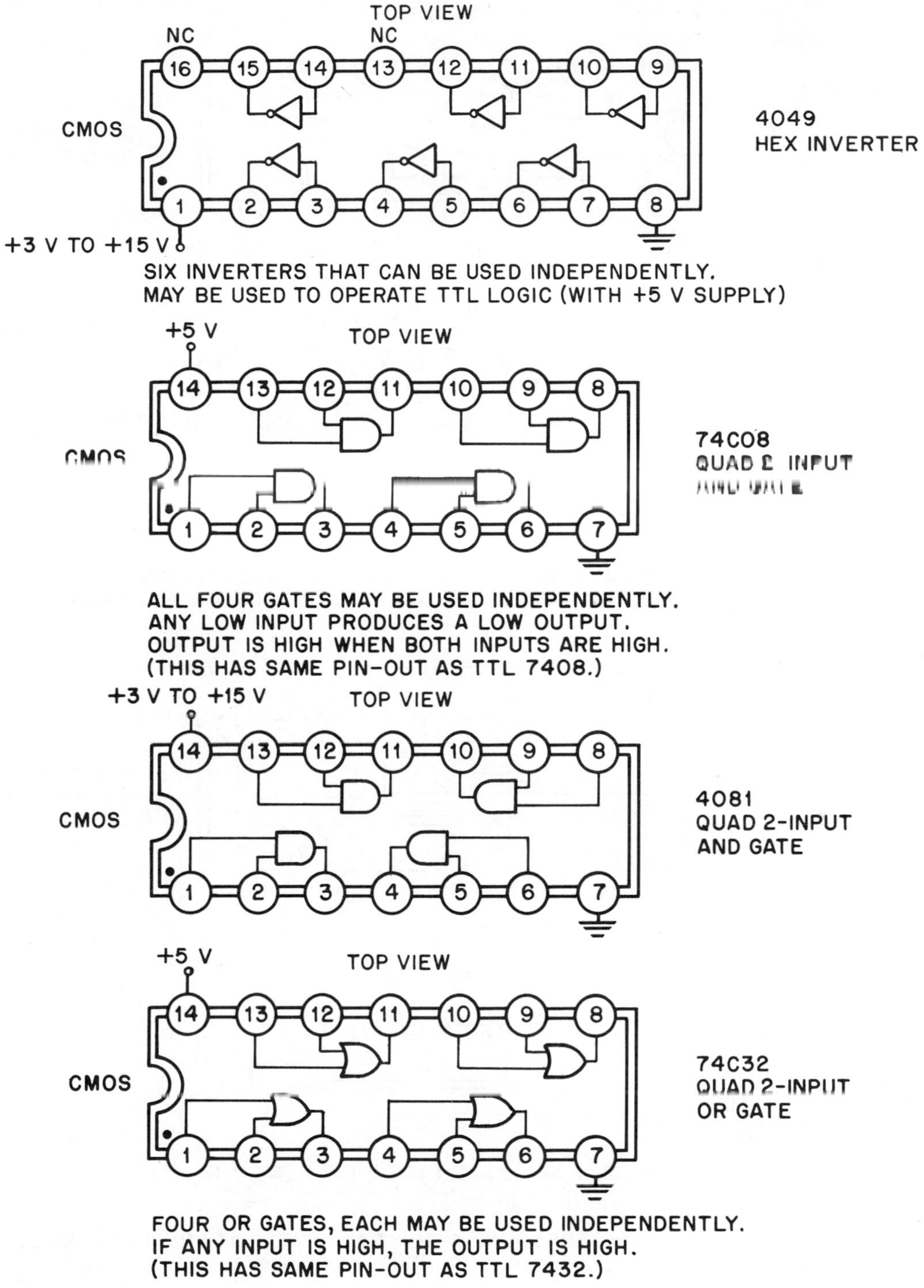

TOP VIEW

NC NC

CMOS

4049
HEX INVERTER

+3 V TO +15 V

SIX INVERTERS THAT CAN BE USED INDEPENDENTLY.
MAY BE USED TO OPERATE TTL LOGIC (WITH +5 V SUPPLY)

+5 V TOP VIEW

CMOS

74C08
QUAD 2 INPUT
AND GATE

ALL FOUR GATES MAY BE USED INDEPENDENTLY.
ANY LOW INPUT PRODUCES A LOW OUTPUT.
OUTPUT IS HIGH WHEN BOTH INPUTS ARE HIGH.
(THIS HAS SAME PIN-OUT AS TTL 7408.)

+3 V TO +15 V TOP VIEW

CMOS

4081
QUAD 2-INPUT
AND GATE

+5 V TOP VIEW

CMOS

74C32
QUAD 2-INPUT
OR GATE

FOUR OR GATES, EACH MAY BE USED INDEPENDENTLY.
IF ANY INPUT IS HIGH, THE OUTPUT IS HIGH.
(THIS HAS SAME PIN-OUT AS TTL 7432.)

Figure 3–72 Continued

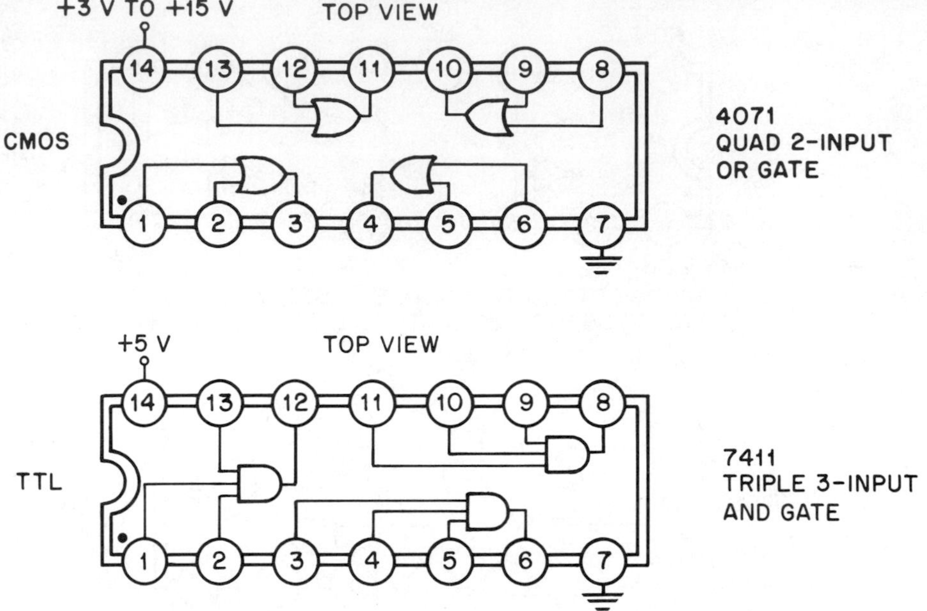

EACH AND GATE MAY BE USED INDEPENDENTLY. ANY LOW INPUT GIVES A LOW OUTPUT. WHEN ALL INPUTS ARE HIGH, OUTPUT IS HIGH.

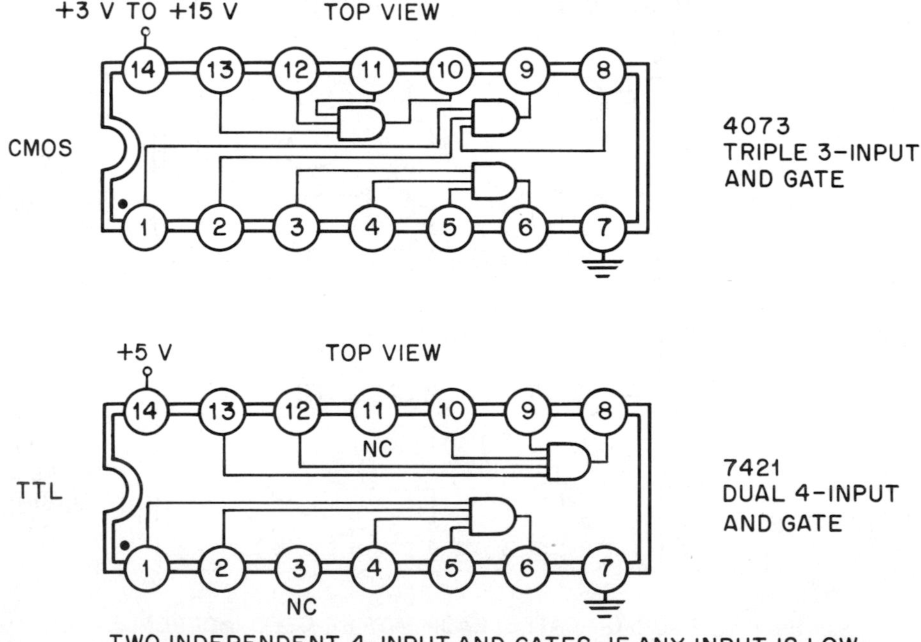

TWO INDEPENDENT 4-INPUT AND GATES. IF ANY INPUT IS LOW, OUTPUT IS LOW. OUTPUT IS HIGH WHEN ALL INPUTS ARE HIGH.

Figure 3–72 Continued

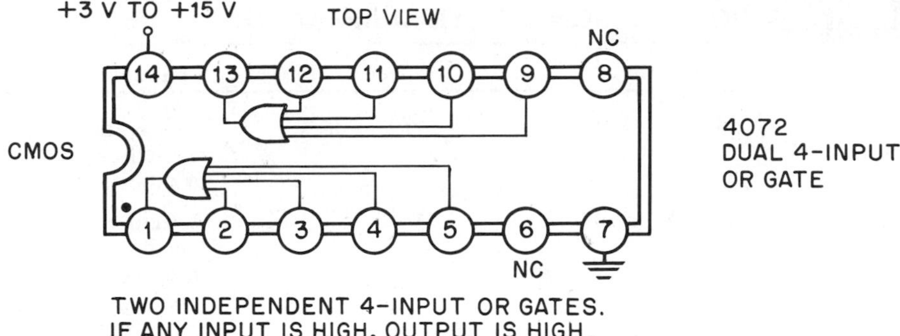

Figure 3–72 Continued

Transistor-Transistor-Logic

Transistor-transistor-logic (TTL) is historically the most inexpensive and readily available digital chip. It is powered from a +5-volt supply and uses more electrical power than the CMOS family. TTL has several **subfamilies** that have some important differences.

Subfamilies:
A group of ICs from the same family with similar electrical characteristics.

The major difference between the subfamilies of TTL is the 54XX series and the 74XX series (the X's will have numbers to represent a specific type of logic circuit, such as a hex inverter or a quad two-input AND gate).

The 54XX subfamily series has an operating temperature range of −55°C to +125°C, and the 74XX subfamily has a smaller operating temperature range from 0°C to 70°C. The 54XX series is made for rugged environmental use and generally costs more than an equivalent 74XX series. Table 3–1 compares the TTL subfamilies. What applies to the 74XX series shown in the table also applies to the 54XX. The speed of the subfamily refers to how fast the IC can respond to an input pulse: the greater the speed, the more digital operations per unit time that can be performed. This is important for systems that must do a vast number of calculations or must respond quickly to various conditions. The power is the amount of electrical power used by each gate in the chip. The more power used, the more energy the system will require. This may not be a problem for digital systems that can be "plugged into the wall," but it would be a major problem for portable digital systems, such as a digital watch.

Note: The supply voltage for the subfamilies in the table is +5 volts and usually designated by V_{CC}.

Design Considerations for TTL

The digital circuit designer must consider many variables in the final choice of an IC. Two major considerations are *speed* and *power* consumption. This is why it is important to use the exact subfamily when replacing damaged IC chips. Not doing so could cause a malfunction of the digital system because of slow response or excessive power demands. Generally speaking, the higher the switching speed is, the greater the power requirements are. This is shown in Table 3–2.

Table 3–1 COMPARISON OF TTL SUBFAMILIES

Subfamily	Designation	Speed	Power	Comments
Regular	54XX or 74XX	× 1	× 1	This is the oldest subfamily. It is used as a basis of comparison to other subfamilies.
Low-power	54LXX or 74LXX	× 1/3	× 1/10	A tremendous saving in power but the trade-off is a reduction in speed. Historically used where power was the main consideration.
Schottky	54SXX or 74SXX	× 3.3	× 2	An increase in switching speed with the trade-off of consuming more power.
Low-power Schottky	54LSXX or 74LSXX	× 1	× 1/5	Provides a reduction in power compared to standard TTL with the same switching speed.
Advanced Low-power Schottky	54ALSXX or 74ALSXX	× 2	× 1/10	One of the most advanced TTL families. Provides up to a 50% reduction in power and up to twice the data handling speed of the LS.
Advanced Schottky	54ASXX or 74ASXX	× 6.6	× 0.7	This family is produced primarily for very high speed operations. Power dissipation is slightly better than standard TTL.

Table 3–2 TTL SPEED AND POWER REQUIREMENTS

Fastest speed					Slowest speed
Type: AS	S	ALS	LS	TTL	L
Lowest power					Largest power
Type: L	ALS	LS	AS	TTL	S

TTL Unused Inputs

An unconnected TTL gate input acts as a HIGH because of the affect this open has on the electrical circuit inside the chip. Bccause of the chance of noise pick-up, unused TTL inputs should be connected as shown in the following table.

Gate	Recommended Connection
AND	To V_{CC} through a 1 kΩ resistor
OR	To ground
General	To a used input

Table 3–3 MAJOR CMOS CHARACTERISTICS

Subfamily	Designation	Comments
A	4XXXA	This is the old type of CMOS. It is the cheapest and historically had the most devices available. Speed is about 10 times slower then standard TTL.
B	4XXXB	This is an improvement over that of the A-type CMOS. There is a slight improvement in speed and some protection against static discharge. This subfamily is preferred over the older A-type.
TTL-compatible	74CXX	This is CMOS that is pin for pin compatible with TTL. For example, a 74C04 CMOS hex inverter is pin-for-pin compatible with the 7404 TTL hex inverter. Its major disadvantage is much slower speed over TTL
High-speed	74HCXX	An improved CMOS where the switching speed is equivalent to that of the low-power Schottky TTL. Probably the best choice overall of any digital IC.

Complementary Metal Oxide Semiconductor (CMOS)

The other major family of IC chips you will encounter in the lab is complementary metal-oxide semiconductor (CMOS). The main advantage over TTL, is a great reduction in power requirements. The major disadvantage of CMOS is that it is much slower than TTL. You must be careful in handling these chips because they are easily damaged by static electricity. Your body can accumulate static electricity, and the mere act of picking up one of these chips to place it in a circuit could damage it. The major subfamilies of CMOS and their characteristics are listed in Table 3–3.

Power Dissipation

The amount of power generated by a CMOS depends upon several factors. First, a CMOS circuit only uses significant power when switching (going from a HIGH to a LOW or back to a HIGH from a LOW). When there is no switching, CMOS hardly uses any power at all (only microwatts!). The other variable with CMOS is its power supply voltage (called V_{DD} and V_{SS}). The V_{SS} connection is connected to ground while V_{DD} is connected to the + voltage. The value of V_{DD} can vary from +3 volts up to +20 volts (depending on the CMOS). Higher voltages will cause a faster switching speed and the more power the unit will dissipate. As a general rule, the optimal switching speed is achieved at about +12 volts.

Another important factor in power dissipation is how quickly the logic level into the CMOS changes from HIGH to LOW or LOW to HIGH. Slowly changing logic levels should be avoided because they can cause a very large increase in power dissipation. Large power consumption can also be caused by unused CMOS inputs and outputs. Therefore all unused logic terminals of a CMOS must be electrically connected to ground or through a 1-kΩ resistor to $+V_{DD}$.

CMOS Precautions

Figure 3–73 summarizes the precautions to be taken when using CMOS.

Interfacing CMOS and TTL

CMOS and TTL digital ICs can be used together, provided you take into account some electrical considerations. These factors are presented later in this text.

MESFET

The MESFET (*me*tal *s*emiconductor *f*ield-*e*ffect *t*ransistor) is a new technology that promises higher speeds and greater packaging density than the MOSFET can. This family reduces the time it takes an electron to travel within its structure, thus increasing the switching speed. Its packaging density decreases the distance between conducting paths, which also contributes to reduced switching speed.

MODFET

The MODFET (*mod*ulated *d*oped *f*ield-*e*ffect *t*ransistor) promises even greater speed and packaging density of any IC process yet developed. Essentially it is manufactured from a

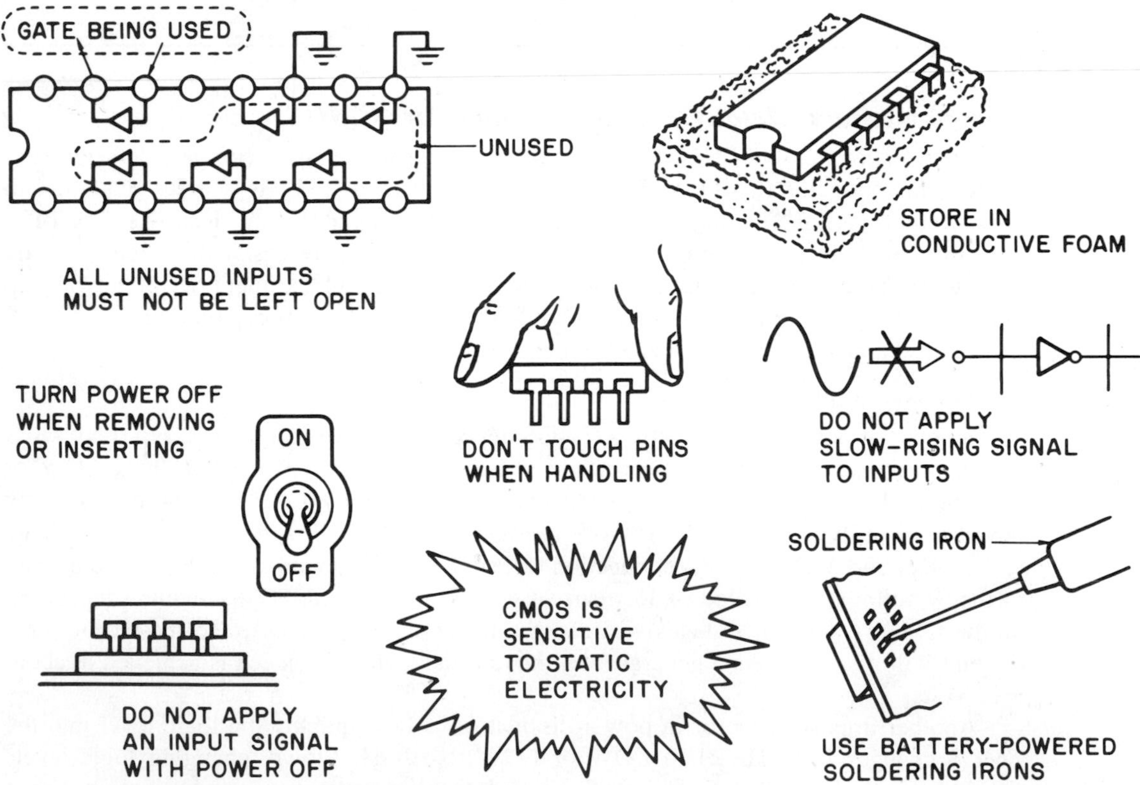

Figure 3–73 CMOS precautions.

thin layer of aluminum-gallium-arsenide that is deposited on a gallium arsenide substrate, which makes them faster than a MESFET, which is faster than a MOSFET. Because of the physical process of charge movement within this material, it is possible to make electrical connections in three dimensions rather than the flat two dimensions currently available. This three-dimensional construction will allow packaging densities of over 1 billion components per chip (called GSI for giga-scale integration).

Conclusion

This section presented the two major digital IC families you will encounter in a digital lab. Here you saw how ICs were coded, the subfamilies of TTL and CMOS, and some important considerations when using ICs. Test your understanding of this section by trying the following review questions.

3–8 Review Questions

1. Describe a digital IC.
2. What is an IC family? How is this different from a subfamily?
3. What is the main advantage of TTL?
4. What is the main advantage of CMOS?
5. Name two disadvantages of CMOS.

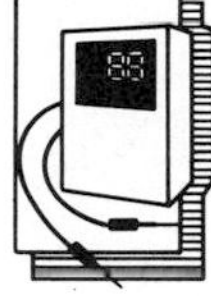

TROUBLESHOOTING SIMULATION

The troubleshooting simulation program for this chapter gives you an opportunity to troubleshoot logic gates. First, you will be able to test the condition of each gate under normal operating conditions. Then, when you are ready, the computer will place a problem in one of the gates. You will again be allowed to test all of the inputs and outputs of the gate to determine the most likely cause of the problem. The computer will then indicate if you are correct and generate a new problem for you.

SUMMARY

- An electronic graph can show the relationship between an electrical variable and time.
- A digital waveform is the graphical representation of a digital process.
- The AND concept means that all inputs must be ON before the output is ON.
- A truth table gives a listing of all conditions of a logic circuit.
- Gates perform the "thinking" function of digital circuits.
- The OR concept means that if any input is ON, the output will be ON.
- The inverter concept means that the output is always the opposite of the input.
- Boolean algebra is similar to "regular" algebra, except that it works with only two values: TRUE or FALSE.
- Boolean inversion is equivalent to the inverter.
- Boolean multiplication is equivalent to the AND gate.
- Boolean addition is equivalent to the OR gate.

- Boolean algebra is useful in the analysis and design of logic circuits.
- A microprocessor is a single IC chip that contains a built-in fixed set of processes, such as arithmetic, logic, and bit manipulation.
- All microprocessors contain the logic processes of NOT, AND, and OR.
- The two major digital IC families are TTL and CMOS.

CHAPTER SELF-TEST

I. TRUE/FALSE

Answer the following questions true or false.

1. The horizontal axis of a digital graph represents the variable of time.
2. A periodic waveform is one that is predictable.
3. A square wave is a periodic waveform that is ON for the same amount of time that it is OFF.
4. The AND function means that all inputs must be ON before the output is ON.
5. The OR function means that one input must be ON (but not all of them) before the output is ON.

II. MULTIPLE CHOICE

Answer the following questions by selecting the most correct answer.

6. A quad two-input AND gate is an IC chip that contains
   (A) two AND gates with four connections.
   (B) four AND gates with two inputs each.
   (C) two AND gates with two inputs each.
   (D) none of the above.
7. The Boolean expression for an AND gate followed by an inverter is
   (A) $X = A \cdot B$
   (B) $X = \overline{A} + \overline{B}$
   (C) $X = \overline{A \cdot B}$
   (D) $X = \overline{A} \cdot \overline{B}$
8. The Boolean expression for an OR gate followed by an inverter is
   (A) $X = A + B$
   (B) $X = \overline{A \cdot B}$
   (C) $X = \overline{A + B}$
   (D) $X = \overline{A} + \overline{B}$
9. The Boolean expression for an AND gate with each input containing an inverter is
   (A) $X = A \cdot B$
   (B) $X = \overline{A} + \overline{B}$
   (C) $X = \overline{A \cdot B}$
   (D) $X = \overline{A} \cdot \overline{B}$

10. The Boolean expression for an OR gate with each input containing an inverter is
(A) $X = A + B$
(B) $X = \overline{A \cdot B}$
(C) $X = \overline{A + B}$
(D) $X = \overline{A} + \overline{B}$

III. MATCHING

Match the correct truth table on the right to the stated logic function on the left.

11. AND
12. $Y = A + B$
13. $Y = \overline{A}$
14. OR
15. $Y = A \cdot B$

(A)

A	*B*	*Y*
0	0	0
0	1	1
1	0	1
1	1	1

(B)

A	*B*	*Y*
0	0	0
0	1	0
1	0	0
1	1	1

(C)

A	*Y*
0	1
1	0

IV. FILL-IN

Fill in the blanks with the most correct answer(s).

16. The 74LXX is a ______ ______ digital integrated circuit.
17. The advanced low-power Schottky TTL has a part designation of ______.
18. Generally speaking, low-power logic devices will have a ______ switching speed.
19. A group of ICs from the same family with similar electrical characteristics is called a ______-______.
20. The ______ process will convert FF_{16} to 00_{16}.

V. OPEN-ENDED

Answer the following questions as indicated.

21. State the actual IC you would use if you needed a three-input AND gate. What pin numbers are used for powering this device?
22. State the results you would expect when the following values are logically ORed:
(A) $FF_{16} + FF_{16}$ (B) $00_{16} + EF_{16}$ (C) $36_{16} + 75_{16}$

23. Determine the results you would expect if the following were logically ANDed: (A) $FF_{16} + FF_{16}$ (B) $00_{16} + EF_{16}$ (C) $36_{16} + 75_{16}$
24. Give the results of performing the NOT process on the following: (A) F_{16} (B) 0_{16} (C) $3C_{16}$
25. Sketch the internal structure of an 8-bit microprocessor, showing the flow of information between its accumulator, B register, memory, and readouts.

Answers to Chapter Self-Test

1] T 2] T 3] T 4] T 5] F 6] B 7] C 8] C 9] D 10] D
11] B 12] A 13] C 14] A 15] B 16] low power 17] 74ALSXX
18] low (slow) 19] subfamily 20] NOT
21] 7411 or 4073, pin 14 for V_{CC} and pin 7 for ground.
22] (A) FF_{16} (B) EF_{16} (C) 77_{16} 23] (A) FF_{16} (B) 00_{16} (C) 34_{16}
24] (A) 0_{16} (B) F_{16} (C) $C3_{16}$ 25] Figure 3–74

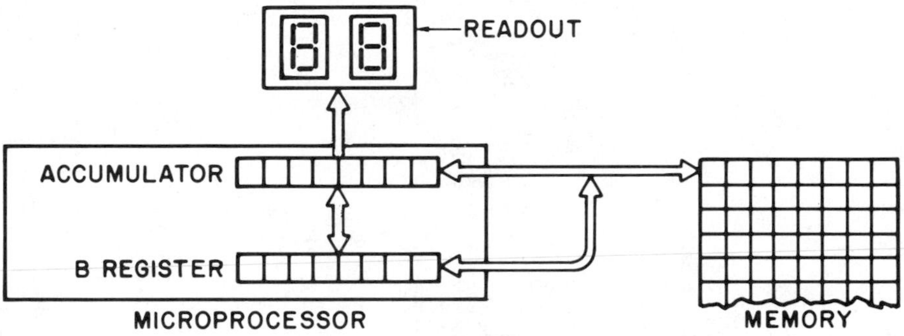

Figure 3–74

CHAPTER PROBLEMS

Basic Concepts

Section 3–1:

1. Look at the graph of Figure 3–75(A). How long is it HIGH? LOW?
2. For the graph of Figure 3–75(B), how many times does it go LOW? HIGH?
3. Explain in words the action of the graph of Figure 3–75(B).
4. State the action of the graph of Figure 3–75(A).

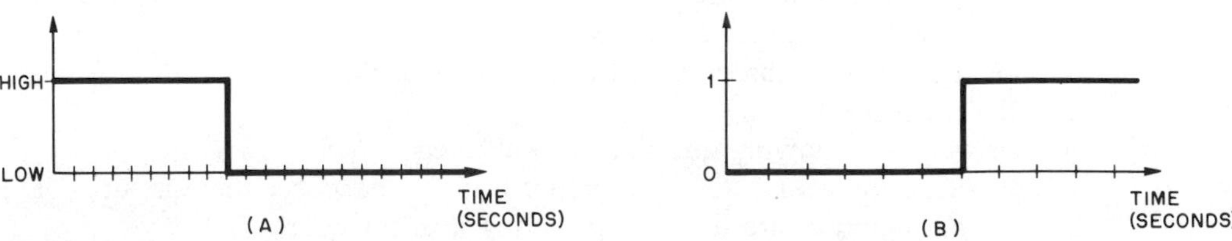

Figure 3–75

5. What units can be used for the vertical axis of a digital graph? for the horizontal axis?

6. Sketch a positive pulse and indicate the leading and trailing edges. Do the same for a negative pulse.
7. Construct a graph of the digital circuit shown in Figure 3–76(A).

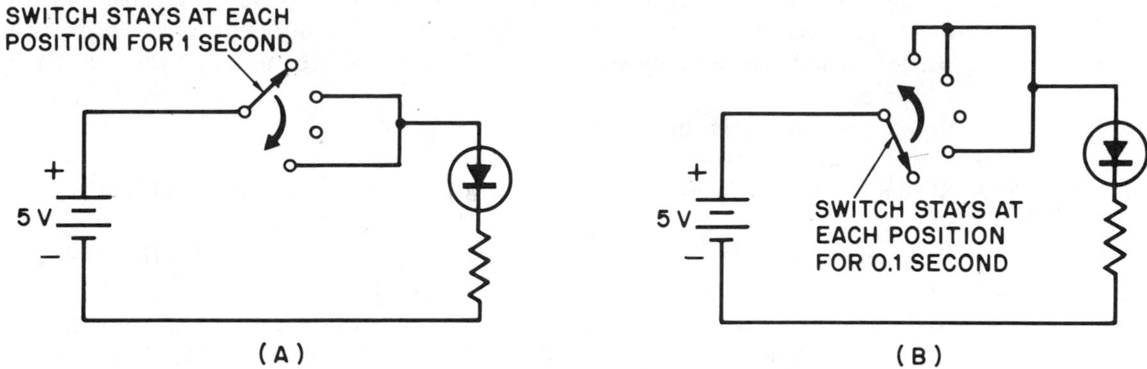

Figure 3–76

8. Sketch the resulting digital waveform of the digital circuit of Figure 3–76(B).
9. Which digital waveforms of Figure 3–77 are periodic?
10. Which waveforms of Figure 3–77 are nonperiodic?

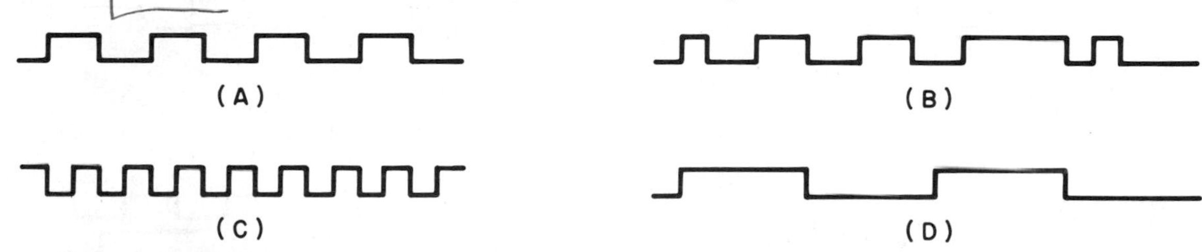

Figure 3–77

Section 3–2:

11. Construct a truth table for each logic circuit in Figure 3–78.
12. For each logic circuit of Figure 3–78, determine when the LED will be ON.

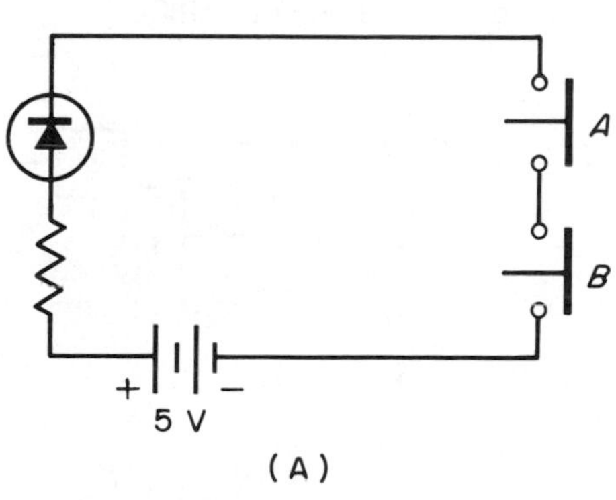

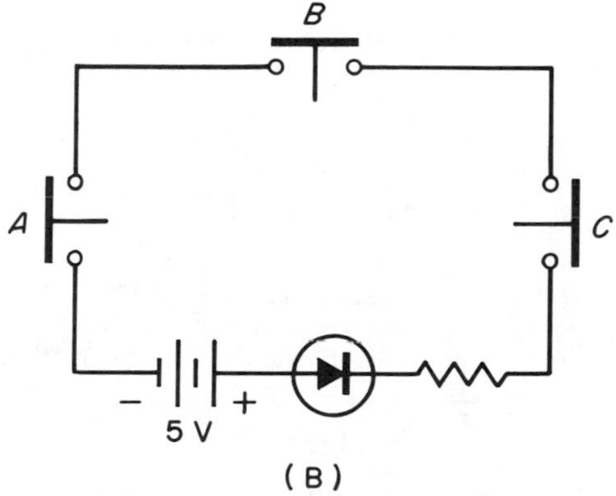

Figure 3–78

13. Develop a logic circuit using switches, a power source, and an LED to indicate the condition of three separate water sprinklers on a golf course. It is part of a water management system, and you need to know when all three are ON.
14. Using switches, a power source, and an LED, sketch a logic circuit that will indicate the condition of three traffic lights used at three different intersections as a part of an ambulance safety system. You are only required to know when all three are green.
15. Create the truth table for the logic circuit of problem 14.
16. Develop the truth table for the logic circuit of problem 13.
17. Sketch the two symbols for AND gates with the following number of inputs: (A) 2 (B) 3 (C) 4 (D) 8
18. Write the truth tables for each of the AND gates in problem 17 (A), (B), and (C).
19. Redo the logic circuit of problem 14 using AND gates.
20. Redo the logic circuit of problem 13 using AND gates.
21. Construct a logic diagram of how an AND gate can be used to control the passing of pulses.
22. Develop the output waveform for each logic arrangement of Figure 3–79.

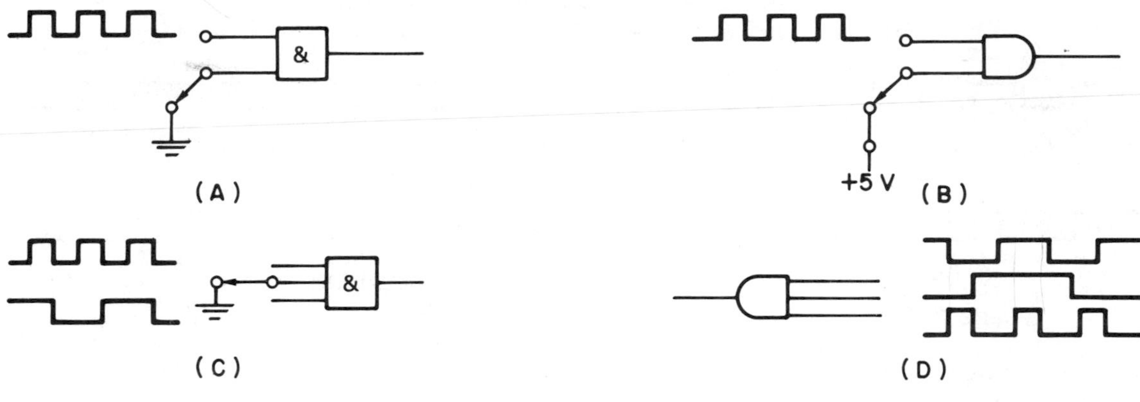

Figure 3–79

23. Sketch the output waveform for each logic arrangement of Figure 3–80.

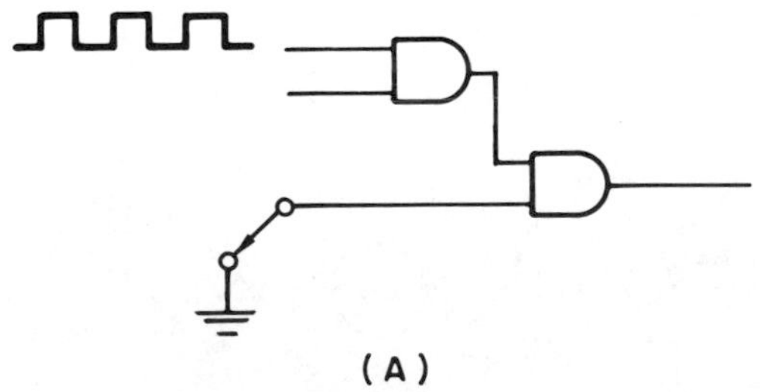

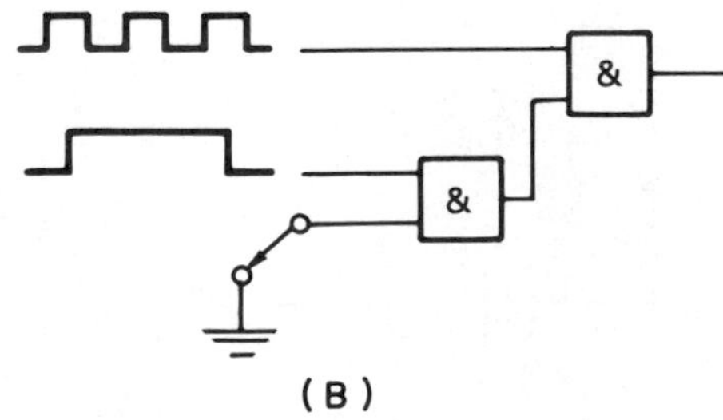

Figure 3–80

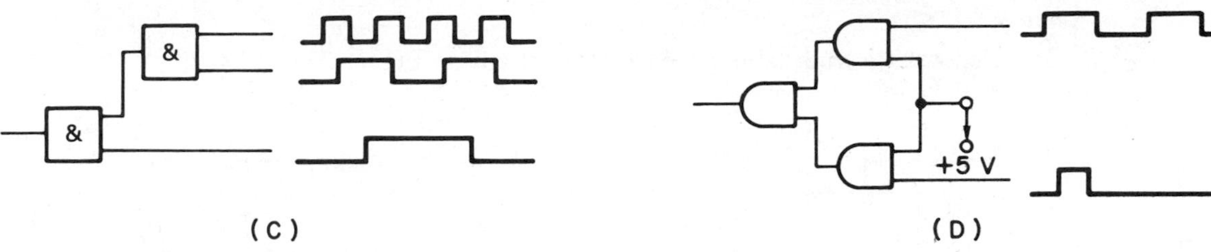

Figure 3–80 Continued

24. For the indicated input waveforms, determine what the missing waveform must be in each logic circuit of Figure 3–81 in order to achieve the given output waveform.

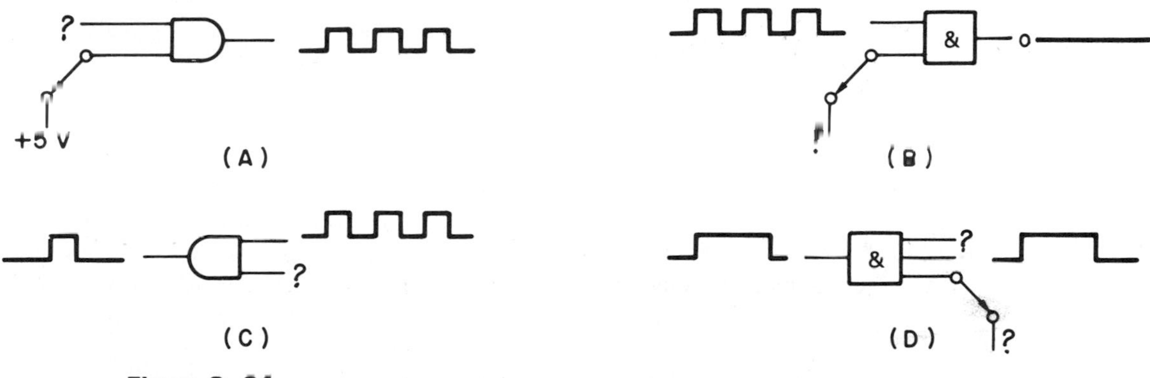

Figure 3–81

25. Determine the shape of the missing waveform in order to achieve the given output for each of the logic circuits of Figure 3–82.

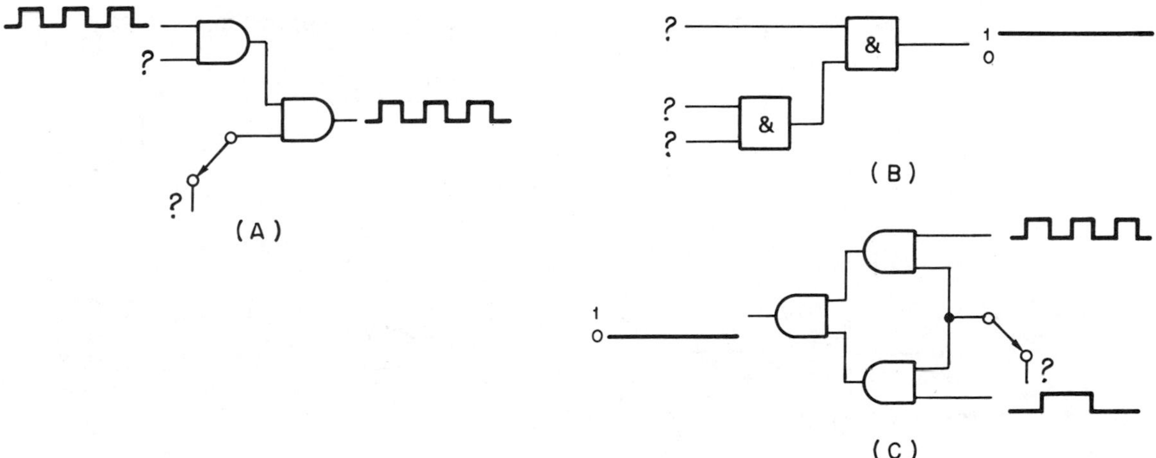

Figure 3–82

Section 3–3:

26. Construct a truth table for each logic circuit in Figure 3–83.

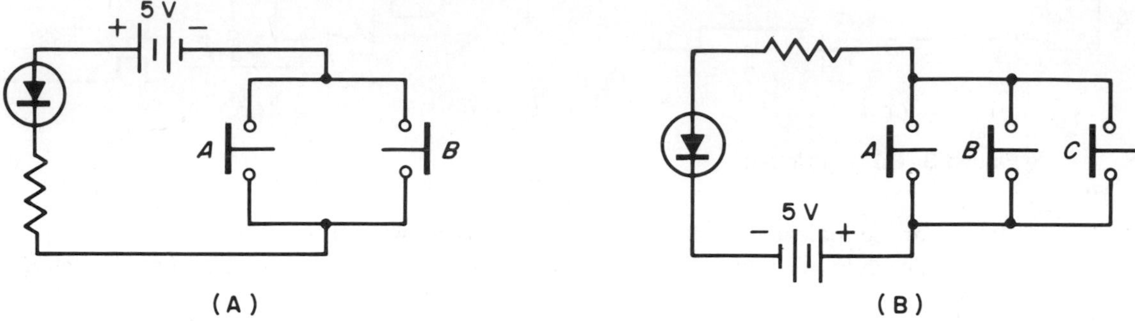

Figure 3–83

27. For each logic circuit of Figure 3–83, determine when the LED will be ON.
28. Develop a logic circuit using switches, a power source, and an LED to indicate the condition of three separate water sprinklers on a golf course. It is part of a water management system, and you need to know when all three are OFF.
29. Using switches, a power source, and an LED, sketch a logic circuit that will indicate the condition of three traffic lights used at three different intersections as a part of an ambulance safety system. You are only required to know when any one is red.
30. Create the truth table for the logic circuit of problem 29.
31. Develop the truth table for the logic circuit of problem 28.
32. Sketch the two symbols for OR gates with the following number of inputs: (A) 2 (B) 3 (C) 4 (D) 8
33. Write the truth tables for each of the OR gates in problem 17.
34. Redo the logic circuit of problem 29 using OR gates.
35. Redo the logic circuit of problem 28 using OR gates.
36. Develop the output waveform for each logic arrangement of Figure 3–84.

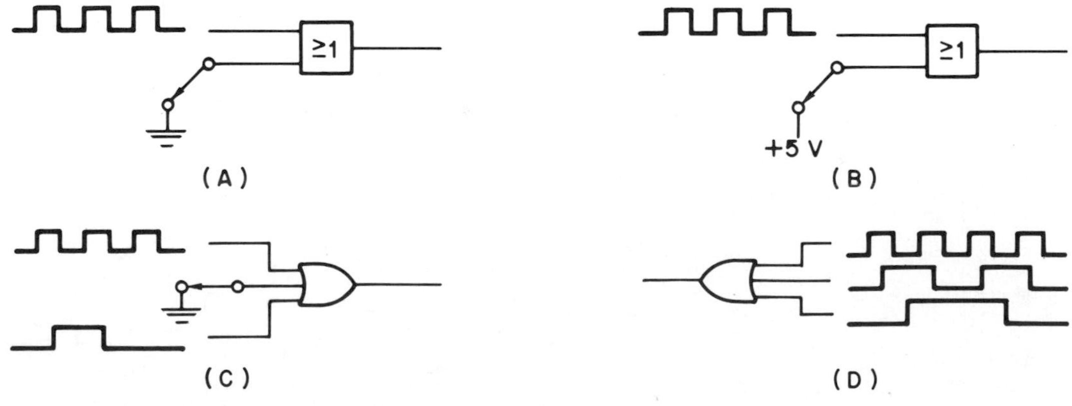

Figure 3–84

37. Sketch the output waveform for each logic arrangement of Figure 3–85.

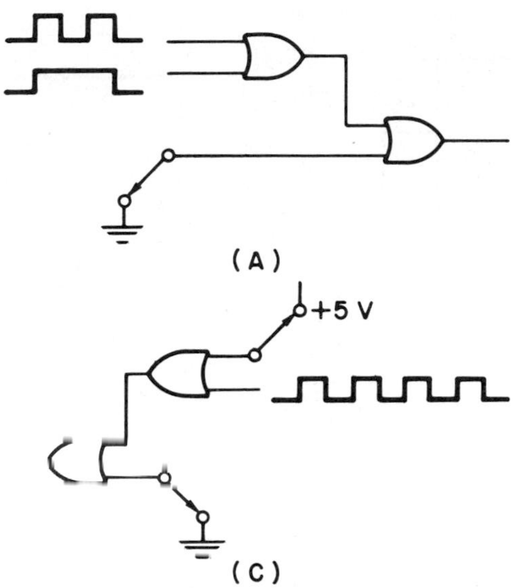

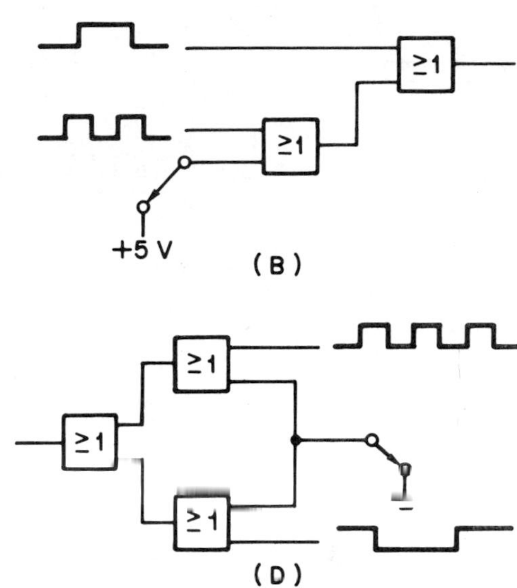

Figure 3–85

38. For the indicated input waveforms, determine what the missing waveform must be in each logic circuit of Figure 3–86 in order to achieve the given output waveform.

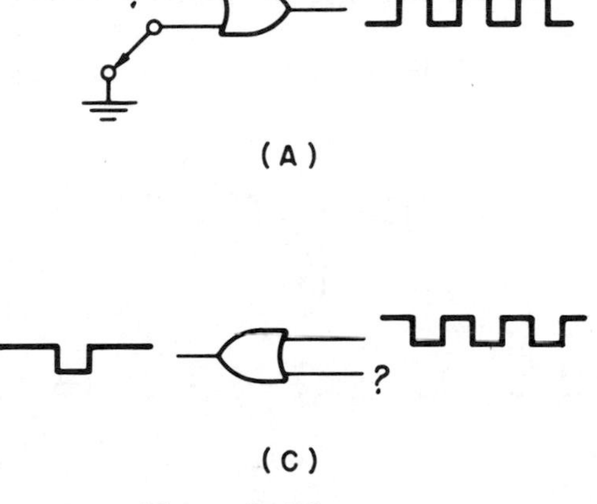

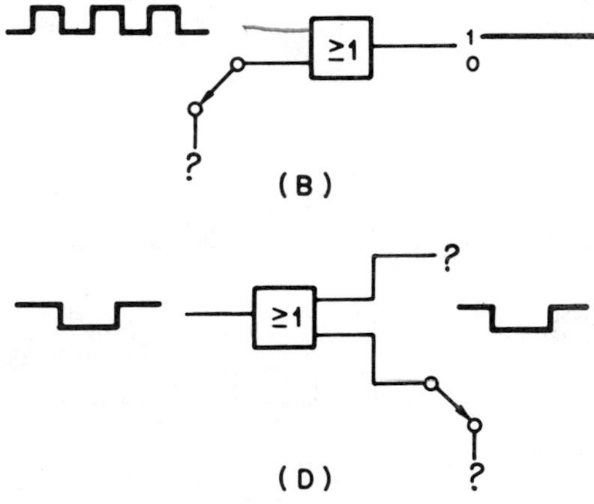

Figure 3–86

39. Determine the shape of the missing waveform in order to achieve the given output for each logic circuit of Figure 3–87.

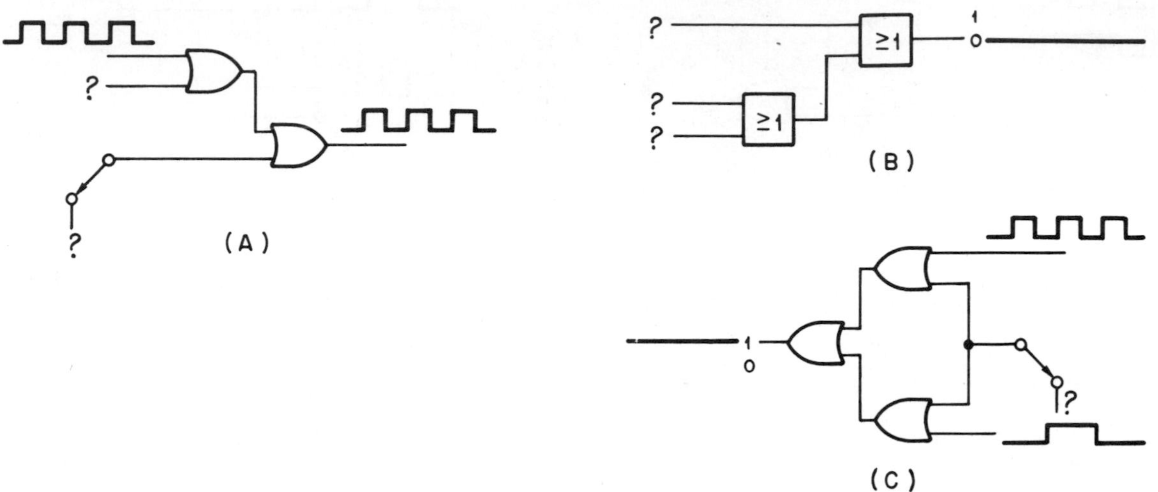

Figure 3–87

Section 3–4:

40. Draw a circuit using a power source, switch, resistor, and LED that will cause the LED to be ON when the switch is OFF.
41. Sketch the four different logic symbols used for an inverter.
42. Sketch the different logic symbols used for a buffer.
43. Construct the truth table for an inverter and for a buffer.
44. State the difference between an inverter and a buffer.
45. Sketch the pinout diagram for a hex inverter.
46. Using inverters where necessary, design a logic circuit that would light a green LED if three traffic lights were all green and a red LED if any traffic light were red.
47. Sketch the missing waveforms for the inverter and buffer circuits of Figure 3–88.

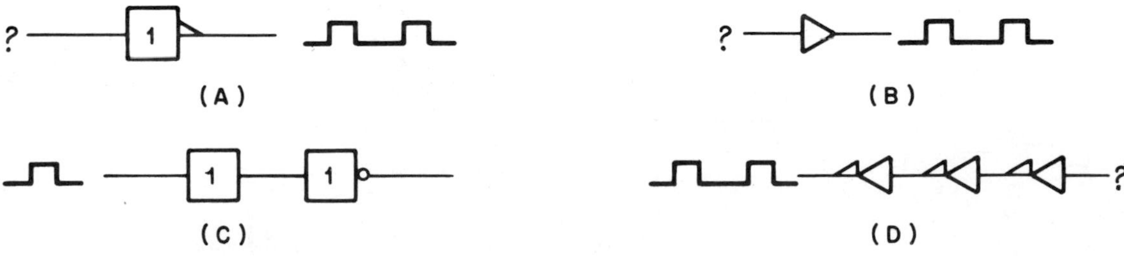

Figure 3–88

48. Sketch the missing waveforms for the inverter and buffer circuits of Figure 3–89.

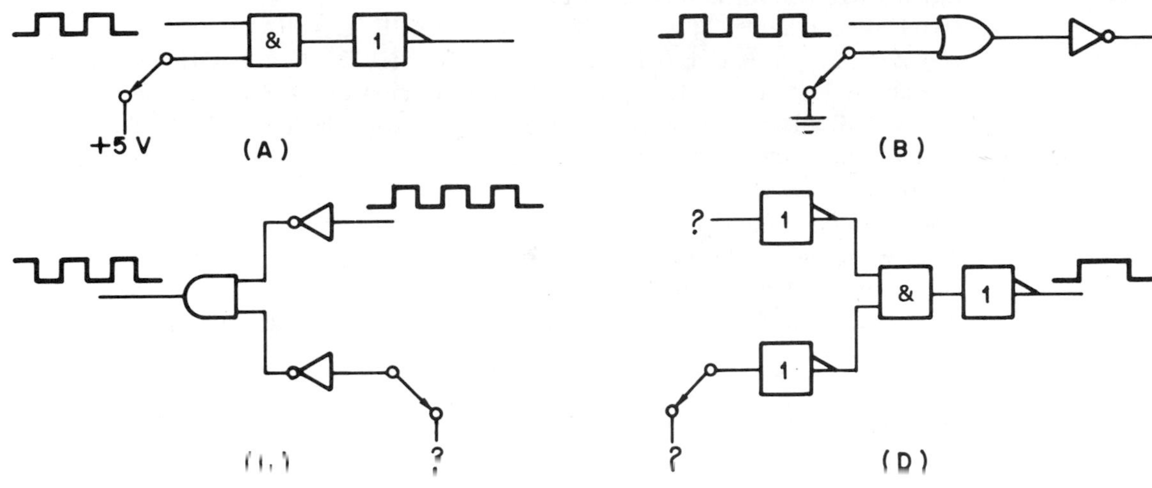

Figure 3–89

Section 3–5:

49. Solve the following Boolean expressions:
    (A) $X = \overline{A}$ for $A = 1$ (B) $Y = \overline{B}$ for $B =$ HIGH (C) $Z = \overline{C}$ for $C =$ FALSE
50. Solve the following Boolean expressions:
    (A) $Y = \overline{C}$ for $C = 0$ (B) $Z = \overline{A}$ for $A = +5$ volts
    (C) $X = \overline{B}$ for $B =$ ON
51. Construct the logic circuit that would perform each expression of problem 50.
52. Sketch the logic circuit for each Boolean expression of problem 49.
53. How is the Boolean operation that uses the multiplication symbol different from decimal multiplication.
54. Determine the values of the following Boolean expressions:
    (A) $Y = A \cdot B$ for $A =$ LOW, $B =$ HIGH
    (B) $Z = \overline{B \cdot C}$ for $B =$ ON, $C =$ ON
    (C) $X = \overline{A} \cdot \overline{B} \cdot \overline{C}$ for $A =$ ground, $B = +5$ volts, $C = +5$ volts
55. Solve the following Boolean expressions:
    (A) $X = \overline{B \cdot C}$ for $B = 1$, $C = 1$
    (B) $Y = A \cdot D \cdot C$ for $A =$ TRUE, $D =$ TRUE, $C =$ FALSE
    (C) $Z = \overline{A} \cdot \overline{B} \cdot \overline{C} \cdot \overline{D}$ for $A =$ HIGH, $B =$ HIGH, $C =$ HIGH, $D =$ LOW
56. Sketch the logic circuit for each Boolean expression in problem 55.
57. For each Boolean expression in problem 54, sketch the logic circuit that would perform that expression.
58. How is the Boolean operation that uses the addition symbol different from decimal addition.
59. Determine the value of the following Boolean expressions:
    (A) $Y = \overline{A} + B$ for $A =$ LOW, $B =$ HIGH
    (B) $Z = \overline{B + C}$ for $B =$ ON, $C =$ ON
    (C) $X = \overline{A} + \overline{B} + \overline{C}$ for $A =$ ground, $B = +5$ volts, $C = +5$ volts

60. Solve the following Boolean expressions:
    (A) $X = \overline{B + C}$ for $B = 1$, $C = 1$
    (B) $Y = \overline{A} + \overline{D} + \overline{C}$ for A = TRUE, D = TRUE, C = FALSE
    (C) $Z = \overline{A} + \overline{B} + \overline{C + D}$ for A = HIGH, B = HIGH, C = HIGH, D = LOW
61. Sketch the logic circuit for each Boolean expression in problem 60.
62. For each Boolean expression in problem 59, sketch the logic circuit that would perform that expression.

Section 3–6:

63. For the logic circuits of Figure 3–90:
    - Write the Boolean expression.
    - Develop the truth table.
    - Determine the output of each if $A = 0$ and $B = 1$, and if there is a C input, $C = 1$.

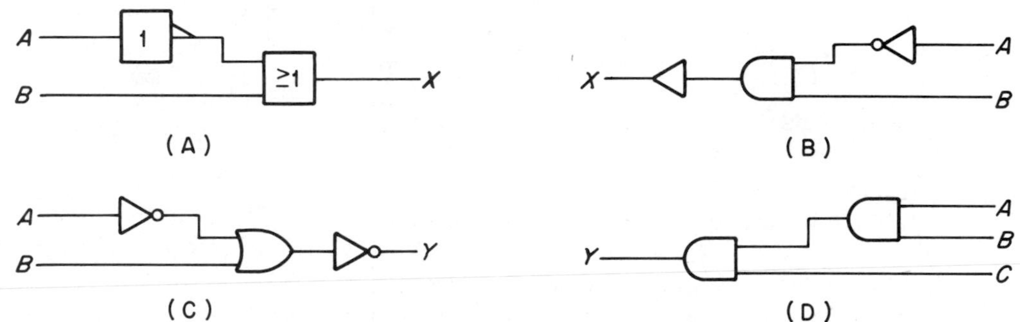

Figure 3–90

64. For the logic circuits of Figure 3–91:
    - Write the Boolean expression.
    - Develop the truth table.
    - Determine the output of each if A = HIGH, B = HIGH, and if there is a C input, C = LOW.

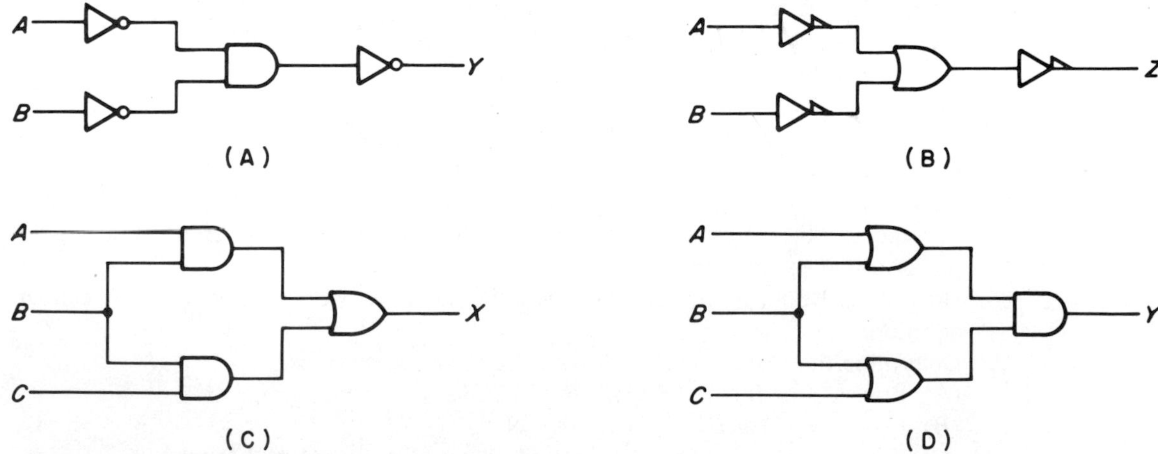

Figure 3–91

65. Create the truth table for the required conditions of Figure 3–92.

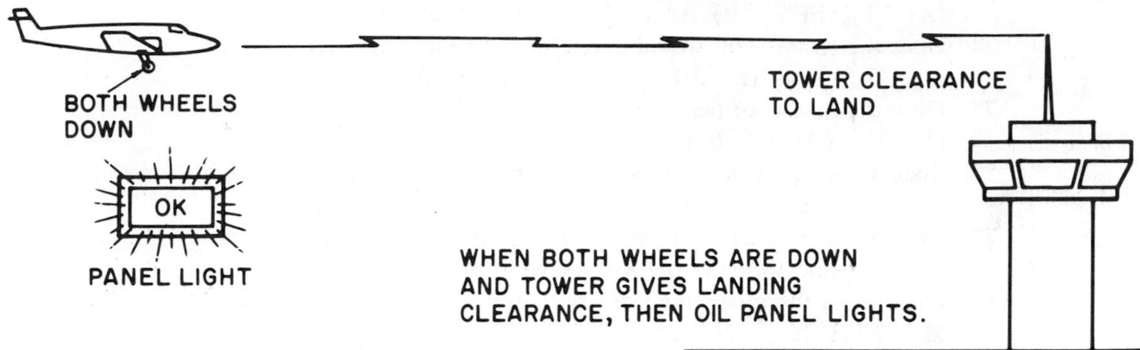
BOTH WHEELS DOWN
OK
PANEL LIGHT
TOWER CLEARANCE TO LAND
WHEN BOTH WHEELS ARE DOWN AND TOWER GIVES LANDING CLEARANCE, THEN OIL PANEL LIGHTS.

Figure 3–92

66. For the given conditions of Figure 3–93.
    (A) Develop the truth table that will turn on the computer correctly.
    (B) Develop a truth table for all terminals that are OFF.
    (C) Develop a truth table for all terminals in use.

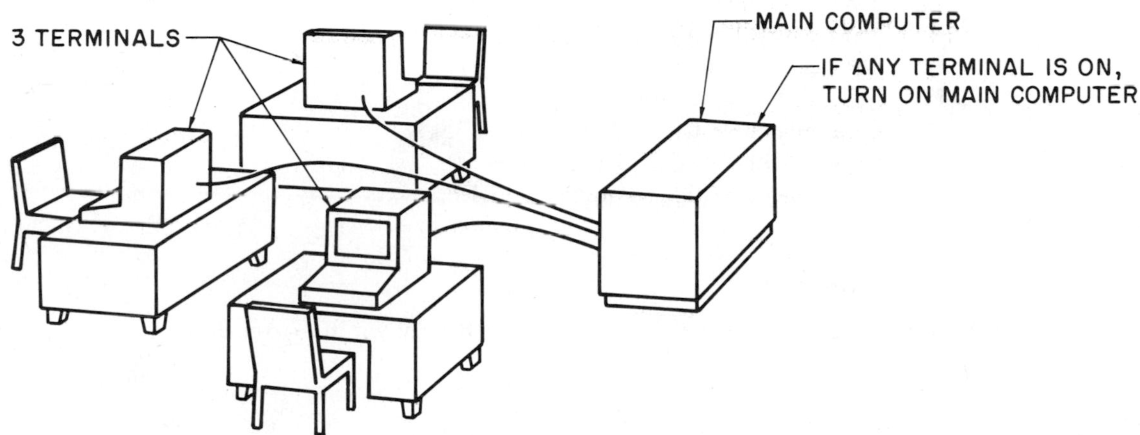
3 TERMINALS
MAIN COMPUTER
IF ANY TERMINAL IS ON, TURN ON MAIN COMPUTER

Figure 3–93

67. Draw the logic circuit to implement the conditions of problem 66.
68. For problem 65, construct the logic circuit required to implement the given conditions.

Applications

69. Describe the following processes performed by a microprocessor:
    (A) NOT (B) AND (C) OR
70. (A) What is a microprocessor? (B) Sketch an 8-bit microprocessor, indicating the accumulator, B register, and flow of data between the microprocessor, memory, and readouts.
71. State the results you would expect when the following values are logically ORed:
    (A) $EF_{16} + FE_{16}$ (B) $AB_{16} + 02_{16}$ (C) $90_{16} + 88_{16}$
72. Give the results of performing an OR process on the following:
    (A) $49_{16} + 05_{16}$ (B) $C0_{16} + 48_{16}$ (C) $09_{16} + 53_{16}$

73. Determine the results you would expect if the following were logically ANDed:
(A) $EF_{16} \cdot FE_{16}$ (B) $BA_{16} \cdot 20_{16}$ (C) $75_{16} \cdot 36_{16}$
74. Give the results you would expect if the following were ANDed:
(A) $73_{16} \cdot 37_{16}$ (B) $A0_{16} \cdot 0A_{16}$ (C) $FF_{16} \cdot 00_{16}$
75. Give the results of performing the NOT process on the following:
(A) 0_{16} (B) F_{16} (C) $C3_{16}$
76. State what you would expect as a result of the NOT process from the following:
(A) C_{16} (B) 5_{16} (C) 48_{16}
77. Determine the final contents of the accumulator for each of the given initial conditions if the following processes were performed on the original microprocessor contents in the order indicated:
    - NOT the accumulator
    - OR accumulator and B register

    (A) Accumulator → FC_{16}, B register → 05_{16}
    (B) Accumulator → 00_{16}, B register → 00_{16}
    (C) Accumulator → 46_{16}, B register → $3C_{16}$
78. Determine the final contents of the accumulator if the given processes were performed on each of the microprocessor contents in the order indicated:
    - AND accumulator and B register
    - NOT the accumulator
    - OR accumulator and B register
    - NOT the accumulator

    (A) Accumulator → CD_{16}, B register → 35_{16}
    (B) Accumulator → FF_{16}, B register → 00_{16}
    (C) Accumulator → 00_{16}, B register → FF_{16}

Electronic Application

79. Indicate the pin numbers on each of the ICs shown in Figure 3–94.

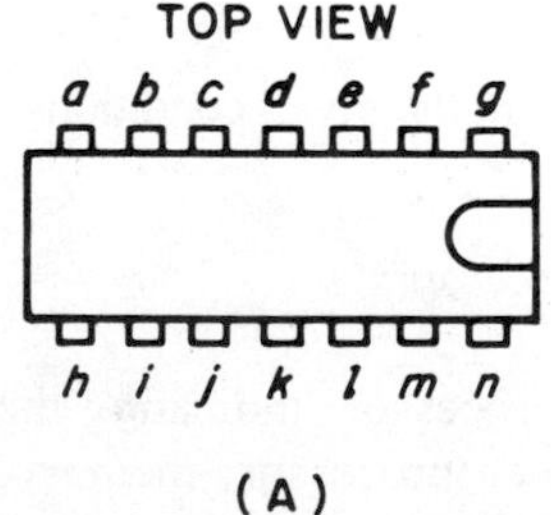

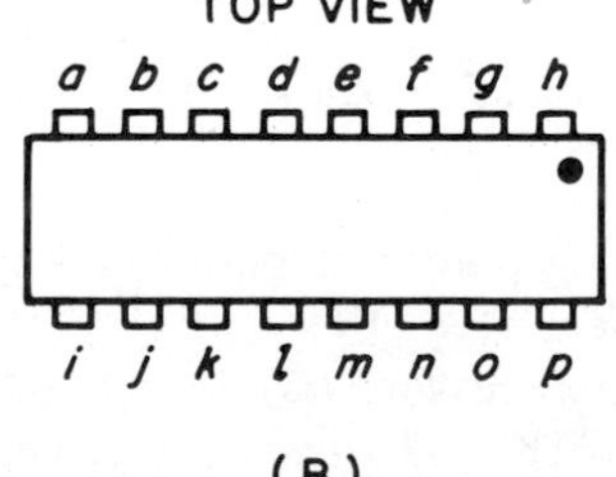

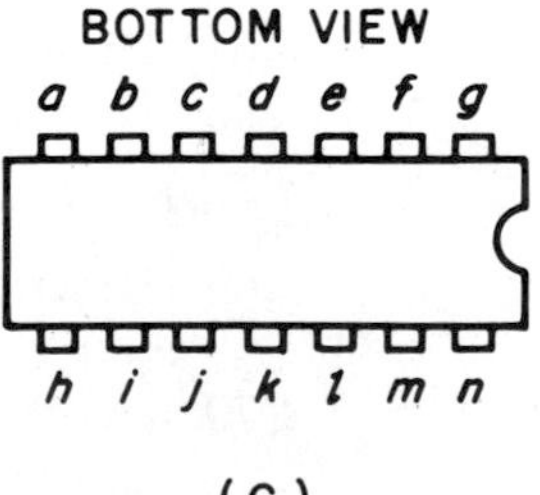

Figure 3–94

80. From the IC labels shown in Figure 3–95, determine the following:
    - Manufacturer
    - Week and year of manufacture
    - Operating temperature range
    - Subfamily type
    - Logic function

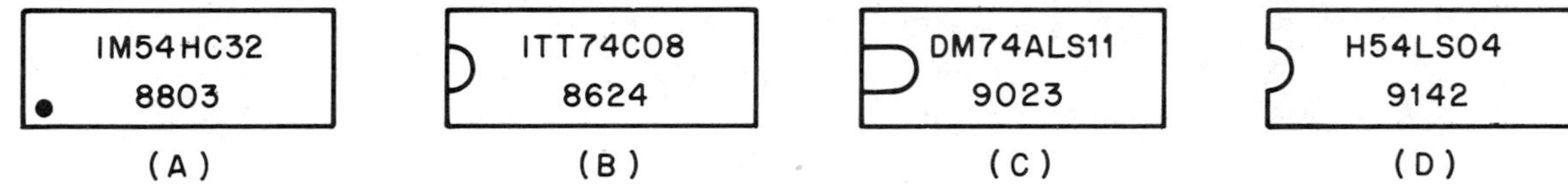

Figure 3–95

81. You are using a logic probe to troubleshoot the logic chips shown in Figure 3–96. From the indicated readings, determine if there is a problem and, if so, the most likely cause.

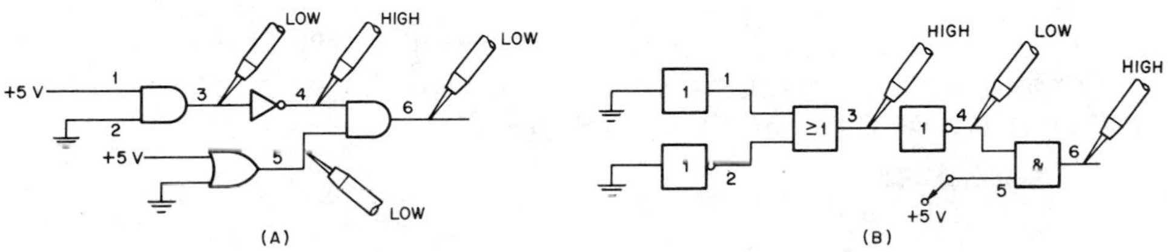

Figure 3–96

82. The logic chip connections shown in Figure 3–97 may contain a problem. From the indicated logic probe readings, determine if there is a problem and, if so, the most likely cause.

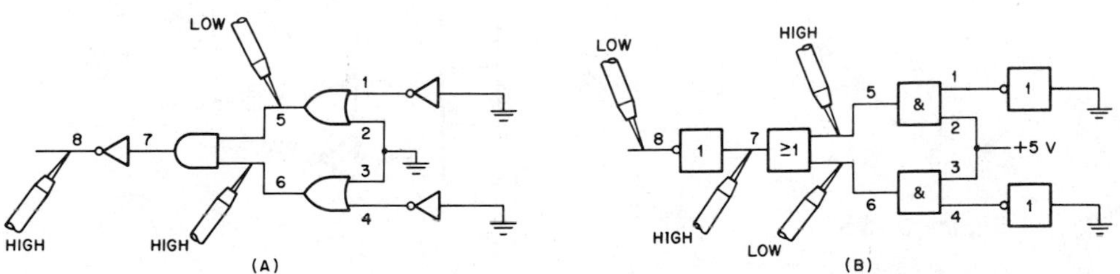

Figure 3–97

Analysis and Design

83. The output of the accumulator of the 4-bit microprocessor of Figure 3–98 is connected to the input of the indicated logic circuit. Starting with the initial contents of the accumulator and B register, explain what process(es) you would have it perform in terms of NOT, OR, and AND in order to do the following in the indicated sequence. (Note that all the motors are now OFF.)
    - Turn all motors ON.
    - Turn every even-numbered motor OFF.
    - Turn all motors OFF.
    - Turn every odd-numbered motor OFF.
    - Turn all motors OFF.

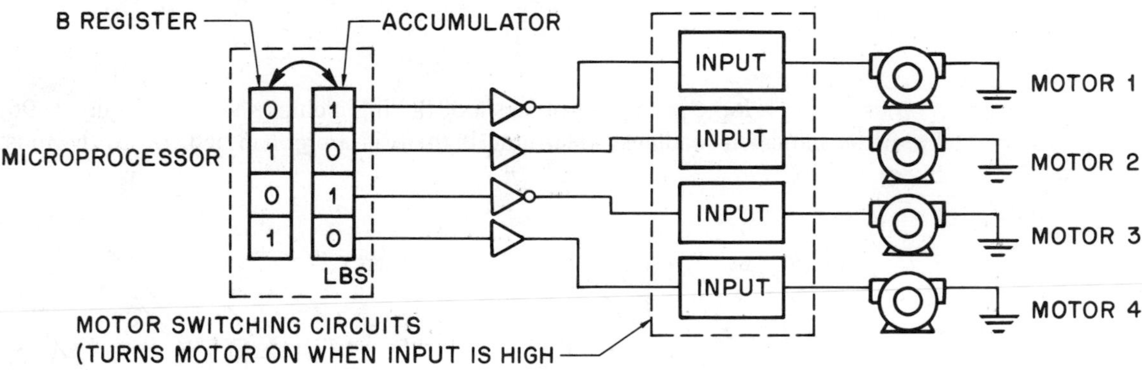

Figure 3–98

Hint: Logic may also have to be done on the contents of the B register.

84. Using standard logic gates, sketch how you would actually wire the logic circuit of Figure 3–99. Start with the four outputs of the accumulator. Develop the truth table and the Boolean expression for the digital circuit.

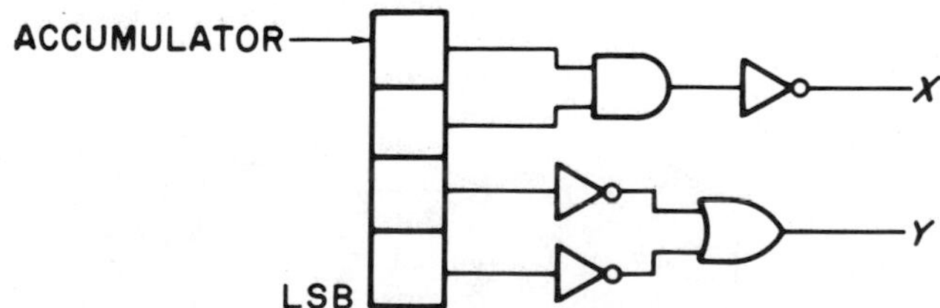

Figure 3–99

CHAPTER 4

Logic Analysis and Simplification

OBJECTIVES

After completing this chapter, you should be able to:

- ☐ Know the logic symbols and understand the logical operation of the
  NOR GATE (NOT OR)
  NAND GATE (NOT AND)
- ☐ Apply De Morgan's theorems in the simplification and analysis of logic circuits.
- ☐ See the relationships of Boolean laws and how they apply to logic gates.
- ☐ Understand Boolean rules and derived relations and how they apply to logic networks.
- ☐ Know the logic symbols and understand the logical operation of the
  XOR GATE (EXCLUSIVE OR)
  XNOR GATE (EXCLUSIVE NOR)
- ☐ Apply the concept of active logic levels.
- ☐ Understand the methods of simplifying logic diagrams using analysis simplification.
- ☐ Create any logic gate using the NAND gate.
- ☐ Create any logic gate using the NOR gate.
- ☐ Understand the stored program concept and how to apply it to a 4-bit microprocessor.
- ☐ Appreciate the differences in voltage levels for digital logic gates.
- ☐ Understand the reason for a noise margin and how it is achieved.
- ☐ Understand the current requirements of TTL digital ICs and the meaning of fan-out.

INTRODUCTION

The gates you studied in the last chapter along with the introduction to Boolean algebra will serve as a foundation for this chapter. Here you will learn about the power and versatility of combining the AND and OR gates with the inverter. These are the logic circuits used by real hardware. You will also learn the important techniques of gate simplification. This is a useful skill for understanding the versatility offered by logic systems.

Boolean algebra will be introduced as an analysis tool for logic circuits. The concepts of this algebra along with a very useful theorem are essential analytical tools for the logic designer and the logic troubleshooter.

This chapter will also show you the versatility of the NAND gate and the NOR gate. You will see how you can use either gate to create any logic condition you choose. You will gain a deeper understanding of the underlying principles of microprocessors. These principles are common to all microprocessors. In the electronic application section you will discover the real truths about the voltage levels and current requirements of digital ICs.

KEY TERMS

NOR Gate
Bubbled (Input/Output)
NAND Gate
De Morgan's First Theorem
De Morgan's Second Theorem
Exclusive OR Gate
Exclusive NOR Gate
Active Logic Level
Black Box
Instruction Set
Stored Program Concept
Instruction
Data
Worse-Case Conditions
Output Profile
Indeterminate Range
Input Profile
Electrical Noise
Noise Margin
Conventional Current
Current Sink
Current Source
Load
Unit Load
Fan-Out

4–1 THE NOR GATE

Discussion

Here you will discover a very practical application for the inverter. This section will demonstrate the operation of combining the OR gate with the inverter. This combination may not at first seem very practical, but have patience and learn this first section well. The applications will quickly follow.

Inverting the OR Gate

Figure 4–1 shows the output of an OR gate connected to an inverter. The truth table for such a combination follows.

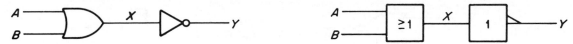

Figure 4–1 OR gate connected to an inverter.

Inputs A	Inputs B	OR output X	Inverted Y
0	0	0	1
0	1	1	0
1	0	1	0
1	1	1	0

Observe from the truth table the effects that the inverter has on the final output (Y output). The output of this combination is TRUE only when both inputs are FALSE. Such a combination is so widely used in digital systems that it has its own name and logic symbol. It is called a NOR gate (for NOT OR gate).

NOR Gate:
A logic gate formed by inverting the output of an OR gate. The output is HIGH when all inputs are LOW.

The logic symbols and truth table for a NOR gate are shown in Figure 4–2.

Example 4–1

Determine the output conditions for the following NOR gate input conditions:
(A) $A = 1, B = 1$ (B) $A = 0, B = 0$ (C) $A = 0, B = 1$

Solution

Utilizing the NOR gate truth table, you find

(A) $A = 1, B = 1; X = 0$

(B) $A = 0, B = 0; X = 1$

(C) $A = 0, B = 1; X = 0$

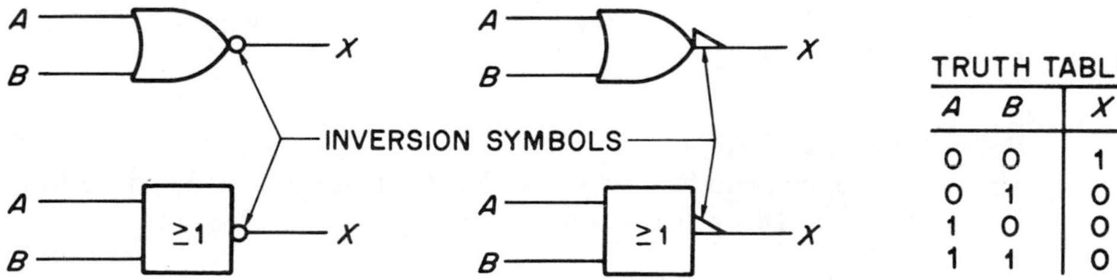

Figure 4–2 NOR gate.

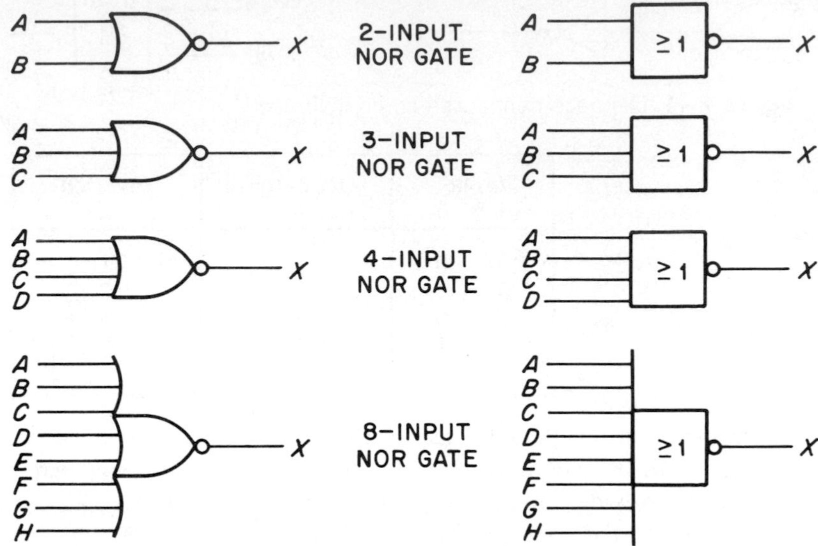

Figure 4–3 Kinds of NOR gates.

Multiple Inputs

NOR gates come with multiple inputs, as shown in Figure 4–3. The truth table for a three-input NOR gate follows.

Inputs A	B	C	Output X
0	0	0	1
0	0	1	0
0	1	0	0
0	1	1	0
1	0	0	0
1	0	1	0
1	1	0	0
1	1	1	0

Again, observe that the only time the output of a NOR gate is TRUE is when all inputs are FALSE. Typical IC NOR gates are shown in Figure 4–4.

Boolean Expressions

The Boolean expression for a two-input NOR gate is developed by inverting the expression for the OR gate. This process is shown in Figure 4–5. Thus, the Boolean expression for the NOR gate is

$$X = \overline{A + B}$$

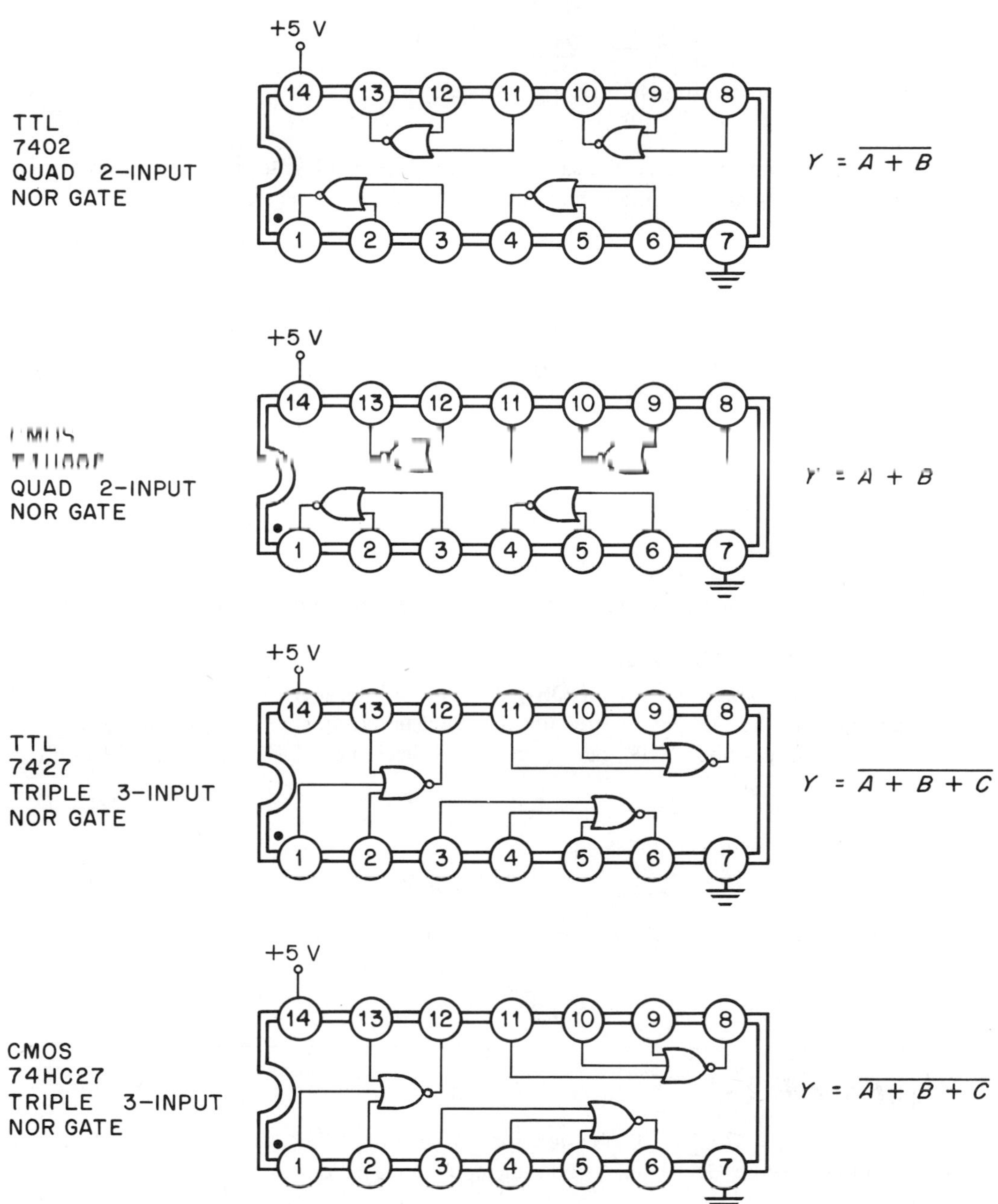

Figure 4–4 TTL and CMOS NOR gates.

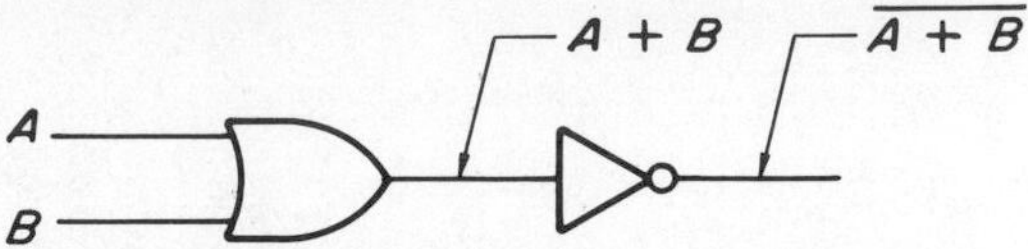

Figure 4–5 Developing NOR Boolean expression.

This expression must give the same results as the NOR gate truth table. To verify this, substitute all possible combinations into the Boolean expression:

For $A = 0$, $B = 0$:

$$X = \overline{0 + 0} = \overline{0} = 1$$

For $A = 0$, $B = 1$:

$$X = \overline{0 + 1} = \overline{1} = 0$$

For $A = 1$, $B = 0$:

$$X = \overline{1 + 0} = \overline{1} = 0$$

For $A = 1$, $B = 1$:

$$X = \overline{1 + 1} = \overline{1} = 0$$

Again, the output of the NOR gate is TRUE only when all inputs are FALSE.

Multiple-Input Expressions

The Boolean expression for a NOR gate with any number of inputs is developed by first starting with the OR gate expression and then inverting the whole expression.

For a three-input NOR gate, start with a three-input OR,

$$A + B + C$$

and invert the *whole* expression:

$$\overline{A + B + C}$$

Thus, the Boolean expression for a three-term NOR gate is

$$X = \overline{A + B + C}$$

Example 4–2

Develop the Boolean expression for a four-input NOR gate. Demonstrate that the output will be HIGH only when all inputs are LOW.

Solution

Start with the Boolean expression for a four-input OR gate:

$$A + B + C + D$$

Invert the *entire* expression:

$$\overline{A + B + C + D}$$

Thus, the Boolean expression for a four-input NOR gate is

$$X = \overline{A + B + C + D}$$

To show the output condition when all inputs are LOW, set each input variable to 0 and solve the Boolean expression:

$$X = \overline{0 + 0 + 0 + 0}$$
$$X = \overline{0} = 1$$

Closer Investigation

Consider the gates shown in Figure 4–6. It's important to note that the results of the two logic circuits are *different*. This can be demonstrated by developing a truth table for each one.

Inversion *after* ORing			
A	B	$A + B$	$\overline{A + B}$
0	0	0	1
0	1	1	0
1	0	1	0
1	1	1	0

Inversion *before* ORing				
A	B	$\overline{A}$	$\overline{B}$	$\overline{A} + \overline{B}$
0	0	1	1	1
0	1	1	0	1
1	0	0	1	1
1	1	0	0	0

Note that the results between the two logic arrangements are very different. This means, and this is important, that

$$\overline{A + B} \neq \overline{A} + \overline{B}$$

The two expressions are not equal because the results of their truth tables are not equal for the same set of input conditions.

The following example uses **bubbled inputs** in place of inverters. This is another way of representing an inversion on a gate input.

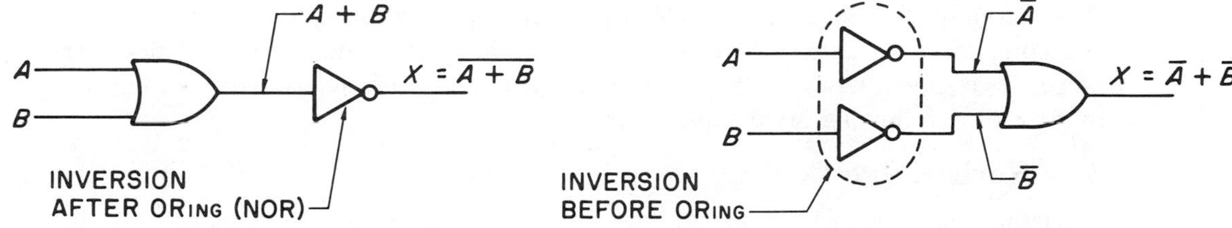

Figure 4–6 Comparison of inversion functions.

Bubbled (Input/Output):
A method of showing a logic inversion.

Example 4–3

Demonstrate that the logic circuits shown in Figure 4–7 are not equivalent.

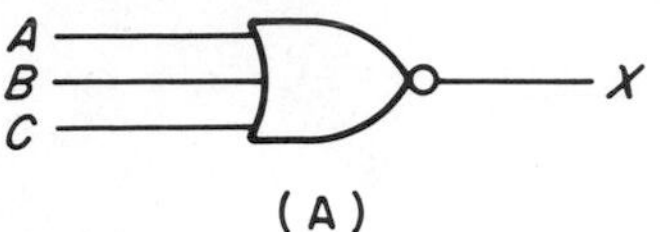

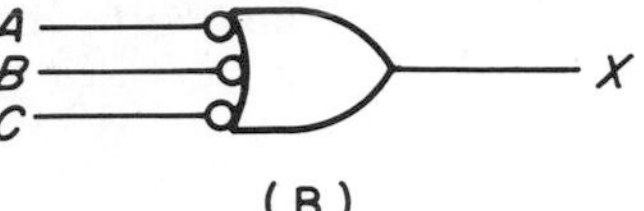

Figure 4–7

Solution

Observe that the gate of Figure 4–7(A) is the standard symbol for a three-input NOR gate. The logic gate of Figure 4–7(B) has its inputs inverted *before* the ORing takes place. Develop a truth table for each case and see if the resulting outputs are identical for all possible input conditions:

Circuit A NOR				
A	B	C	X	$\overline{X}$
0	0	0	0	1
0	0	1	1	0
0	1	0	1	0
0	1	1	1	0
1	0	0	1	0
1	0	1	1	0
1	1	0	1	0
1	1	1	1	0

Circuit B Invert before OR						
A	B	C	$\overline{A}$	$\overline{B}$	$\overline{C}$	$\overline{A} + \overline{B} + \overline{C}$
0	0	0	1	1	1	1
0	0	1	1	1	0	1
0	1	0	1	0	1	1
0	1	1	1	0	0	1
1	0	0	0	1	1	1
1	0	1	0	1	0	1
1	1	0	0	0	1	1
1	1	1	0	0	0	0

Thus, the circuits are not equivalent because their truth tables are not identical.

Conclusion

This section presented the NOR gate, its real hardware, truth table, Boolean expression, and comparison to inversion before ORing. In the next section, you will discover the properties of inverting the output of an AND gate. Test your understanding of this section by trying the following review questions.

4–1 Review Questions

1. Define, in terms of basic gates, the logical construction of a NOR gate.
2. State the only time the output of a NOR gate is TRUE.

3. Give the Boolean expression for a three-input NOR gate.
4. Are the following expressions equal: $\overline{A + B}, \overline{A} + \overline{B}$?
5. Explain how you would demonstrate your answer to question 4.

4–2 THE NAND GATE

Discussion

This section presents the operation of combining the AND gate with an inverter. Again, you may not see immediate practical applications for this logic combination, but you will find that these gate arrangements are indeed very versatile.

Inverting the AND Gate

Figure 4–8 shows the output of the AND gate connected to an inverter. The truth table for such a combination follows:

Inputs *A*	*B*	AND output *X*	Inverted *Y*
0	0	0	1
0	1	0	1
1	0	0	1
1	1	1	0

Observe from the truth table the effects of the inverter on the final output. The output of this combination is FALSE only when *all* inputs are TRUE.

This is another useful logic combination used in digital systems. It too has a unique name and logic symbol called the **NAND** (from NOT AND) **gate.** Its logic symbols are shown in Figure 4–9.

Figure 4–8 AND gate connected to an inverter.

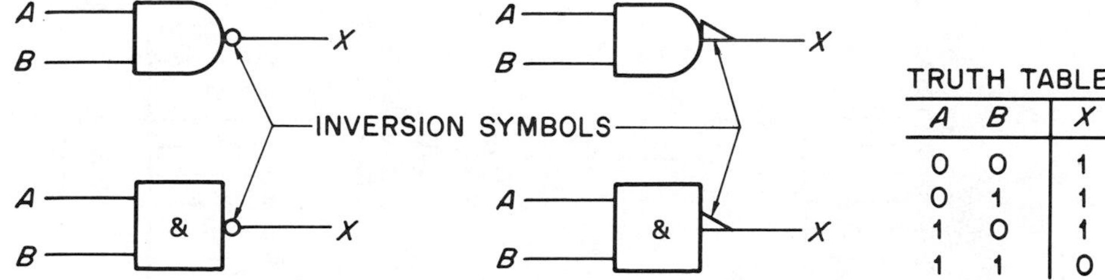

Figure 4–9 Logic symbols of NAND gates.

NAND Gate:

A logic circuit formed by inverting the output of an AND gate. The output is low only when all inputs are high.

Example 4–4

Determine the output conditions for the following NAND gate inputs.

(A) $A = 1, B = 1$ (B) $A = 1, B = 0$ (C) $A = 0, B = 0$

Solution

Utilizing the NAND gate truth table, you find

(A) $A = 1, B = 1; \quad X = 0$

(B) $A = 1, B = 0; \quad X = 1$

(C) $A = 0, B = 0; \quad X = 1$

Multiple Inputs

NAND gates come with multiple inputs, as shown in Figure 4–10. The truth table for a three-input NAND gate follows:

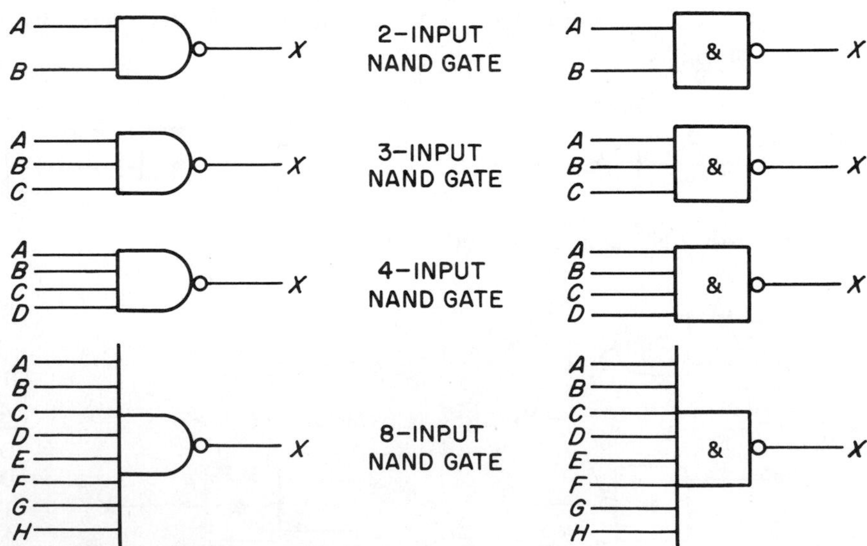

Figure 4–10 Multiple-input NAND gates.

Inputs A	B	C	Output X
0	0	0	1
0	0	1	1
0	1	0	1
0	1	1	1
1	0	0	1
1	0	1	1
1	1	0	1
1	1	1	0

Again, observe that the only time the output of a NAND gate is FALSE is when all inputs are TRUE.

Real NAND Gates

Three real NAND gates are shown in Figure 4–11. Note from Figure 4–11 that both TTL and CMOS are available and pin-for-pin compatible.

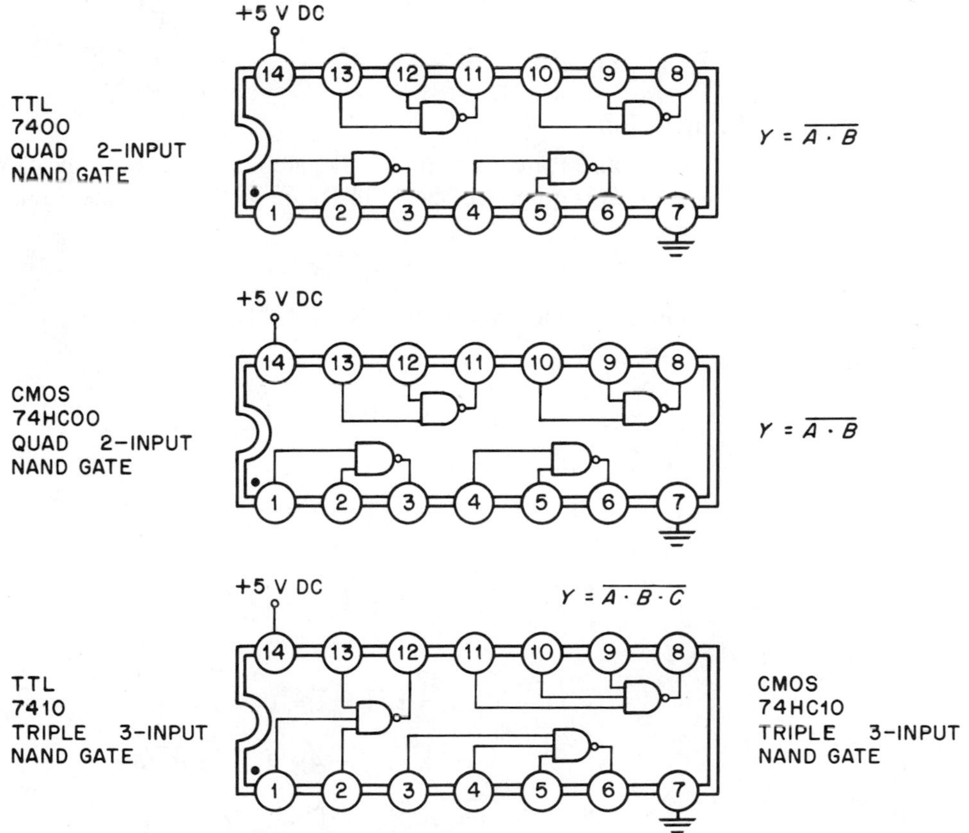

Figure 4–11 TTL and CMOS NAND gates.

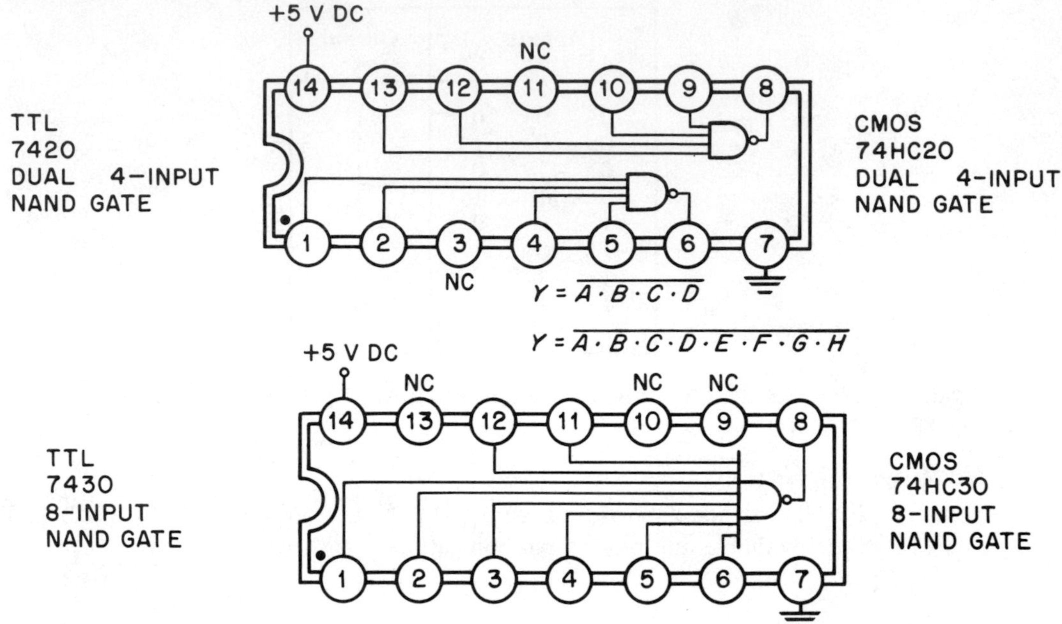

Figure 4–11 Continued

Boolean Expression

The Boolean expression for a two-input NAND gate is developed in the same manner as that of the NOR gate. This process is shown in Figure 4–12. Thus, the Boolean expression for the NAND gate is

$$X = \overline{A \cdot B}$$

This expression must give the same results as the NAND gate truth table:

For $A = 0$, $B = 0$:

$$X = \overline{0 \cdot 0} = \overline{0} = 1$$

For $A = 0$, $B = 1$:

$$X = \overline{0 \cdot 1} = \overline{0} = 1$$

For $A = 1$, $B = 0$:

$$X = \overline{1 \cdot 0} = \overline{0} = 1$$

For $A = 1$, $B = 1$:

$$X = \overline{1 \cdot 1} = \overline{1} = 0$$

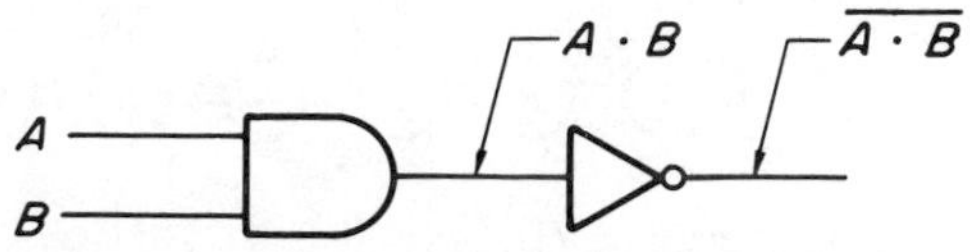

Figure 4–12 Developing NAND Boolean expression.

Observe that the output of the NAND gate is FALSE only when *all* inputs are TRUE.

Multiple-Input Expressions

The development of Boolean expressions for a NAND gate with multiple inputs uses the same process as with the NOR gate. For a three-input NAND gate, start with a three-input AND gate:

$$A \cdot B \cdot C$$

Invert the *whole* expression (this is done because inversion takes place *after* ANDing):

$$\overline{A \cdot B \cdot C}$$

Thus, the Boolean expression for a three-input NAND gate is

$$X = \overline{A \cdot B \cdot C}$$

Example 4–5

Develop the Boolean expression for a four input NAND gate. Demonstrate that the output is LOW only when all inputs are HIGH.

Solution

Start with the Boolean expression for a four-input AND gate:

$$A \cdot B \cdot C \cdot D$$

Invert the *entire* expression:

$$\overline{A \cdot B \cdot C \cdot D}$$

Thus, the Boolean expression for a four-input NAND gate is

$$X = \overline{A \cdot B \cdot C \cdot D}$$

To show the output condition when all inputs are true, set each variable to 1 and solve the Boolean expression:

$$X = \overline{1 \cdot 1 \cdot 1 \cdot 1} = \overline{1} = 0$$

If any input is LOW, the output will be HIGH.

Closer Investigation

As with the NOR gate, you should analyze the NAND gate and similar logic circuitry to be sure you understand the logical operation. Consider the gate combinations shown in Figure 4–13. Again, it is important to note that the two logic circuits produce two different

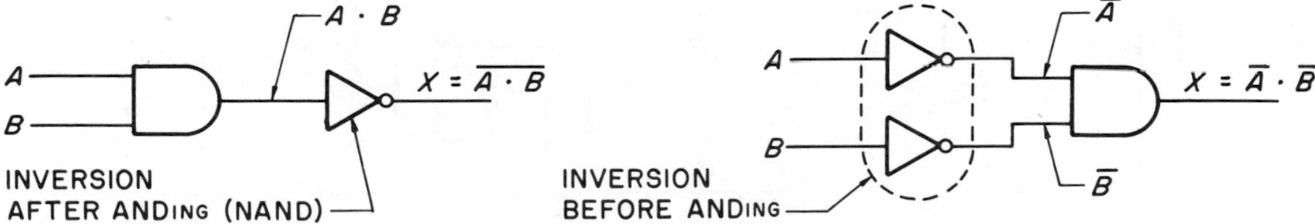

Figure 4–13 Comparison of inversion functions (AND gates).

Boolean expressions. This is demonstrated by the development of a truth table for each one.

Inversion *after* ANDing			
A B	$A \cdot B$	$\overline{A \cdot B}$	
0 0	0	1	
0 1	0	1	
1 0	0	1	
1 1	1	0	

Inversion *before* ANDing		
A B	$\overline{A}$ $\overline{B}$	$\overline{A} \cdot \overline{B}$
0 0	1 1	1
0 1	1 0	0
1 0	0 1	0
1 1	0 0	0

The results of the two truth tables are not equal. Many students assume that both logic circuits are equal. They are not equal because the results of their truth tables are not the same under identical input conditions. This means that

$$\overline{A \cdot B} \neq \overline{A} \cdot \overline{B}$$

Example 4–6

Demonstrate that the logic circuits shown in Figure 4–14 are not equivalent.

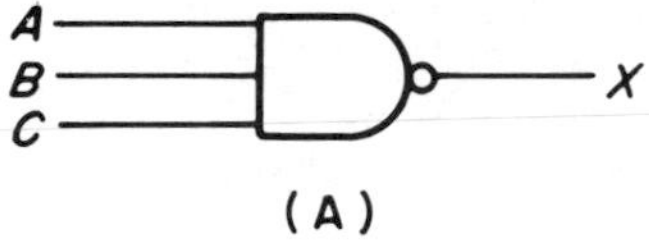

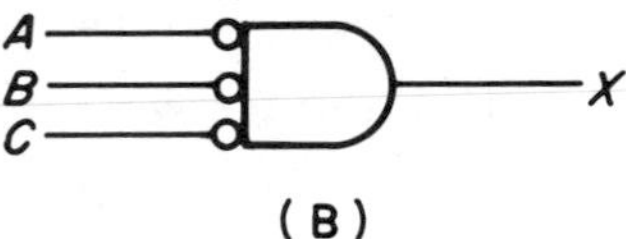

Figure 4–14

Solution

The logic circuit shown in Figure 4–14(A) is a NAND gate because inversion takes place *after* the ANDing. The logic arrangement of Figure 4–14(B) has inversion *before* ANDing.

The test for equality or inequality can be done by developing truth tables for both and comparing their outputs under identical input conditions:

Circuit A NAND		
A B C	$A \cdot B \cdot C$	$\overline{A \cdot B \cdot C}$
0 0 0	0	1
0 0 1	0	1
0 1 0	0	1
0 1 1	0	1
1 0 0	0	1
1 0 1	0	1
1 1 0	0	1
1 1 1	1	0

Circuit B Invert *before* ANDing		
A B C	$\overline{A}$ $\overline{B}$ $\overline{C}$	$\overline{A} \cdot \overline{B} \cdot \overline{C}$
0 0 0	1 1 1	1
0 0 1	1 1 0	0
0 1 0	1 0 1	0
0 1 1	1 0 0	0
1 0 0	0 1 1	0
1 0 1	0 1 0	0
1 1 0	0 0 1	0
1 1 1	0 0 0	0

Since the outputs are not equal for identical input conditions, the two logic circuits are not equal.

Conclusion

You have completed your introduction to NOR and NAND gates. In the next section, you will learn a very useful relationship concerning these logic gates. Try the following review questions.

4–2 Review Questions

1. Define, in terms of basic gates, the logical construction of a NAND gate.
2. State the only time the output of a NAND gate is low.
3. Give the Boolean expression for a three-input NAND gate.
4. Are the following expressions equal: $\overline{A \cdot B}$, $\overline{A} \cdot \overline{B}$?
5. Explain how you would demonstrate your answer to question 4.

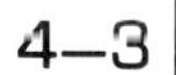

USEFUL RELATIONSHIPS—DE MORGAN'S FIRST THEOREM

Discussion

Initially, much of Boolean algebra was ignored until applied to the analysis of logic circuits. Augustus De Morgan was one exception. He felt that George Boole had made a profound achievement in the development of his new algebra. De Morgan spotted two important relationships of Boole's new algebra. Today, these are called De Morgan's first and second theorems. You already have enough information from the first two sections to describe what De Morgan did. But since he was the first to write about it, he deservedly gets the credit.

De Morgan's First Theorem

The first important logical relationship made by De Morgan is shown in Figure 4–15.

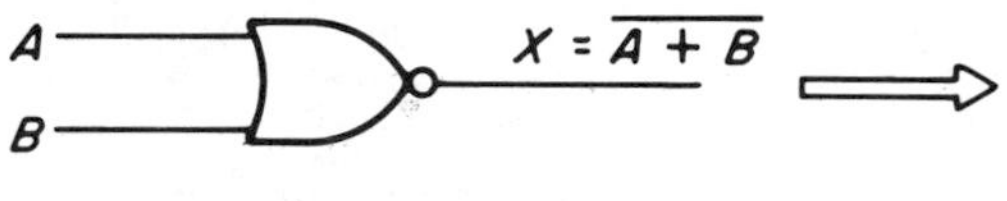

$X = \overline{A + B}$

A	B	X
0	0	1
0	1	0
1	0	0
1	1	0

(A)

$X = \overline{A} \cdot \overline{B}$

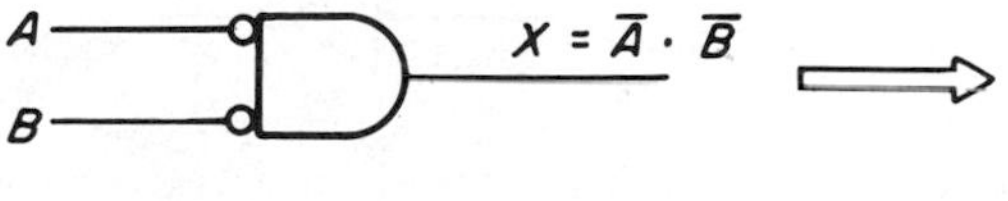

A	B	X
0	0	1
0	1	0
1	0	0
1	1	0

(B)

Figure 4–15 De Morgan's first theorem.

De Morgan's First Theorem:
The Boolean relationship showing that a NOR gate is logically equivalent to an AND gate with inverted inputs.

From Figure 4–15, you can see that the truth table for the NOR gate is identical to the truth table for the AND gate with inverted inputs. You have already studied both of these logic arrangements. The Boolean expression of De Morgan's first theorem is:

$$\overline{A + B} = \overline{A} \cdot \overline{B} \quad \text{or} \quad \overline{A} \cdot \overline{B} = \overline{A + B}$$

In the first equation, the left member represents a NOR gate, and the right member represents an AND gate with inverted inputs. Using this relationship, you can convert a logic operation to practical hardware and also simplify complex logic diagrams. A more general statement of the theorem is that the complement of a sum equals the product of the complement of each term or

$$\overline{A + B + \cdots Z} = \overline{A} \cdot \overline{B} \cdots \overline{Z}$$

This concept is shown in the following example:

Example 4–7

State De Morgan's first theorem for a three-input NOR gate. Draw the equivalent logic diagrams using both AND and OR gates with inverters and develop the truth table for each.

Solution

First, draw the logic circuits and from this develop the Boolean expressions. Then develop the truth tables for each. If the outputs are the same for each under identical conditions, then they are equal. Figure 4–16 shows the equivalent circuits.

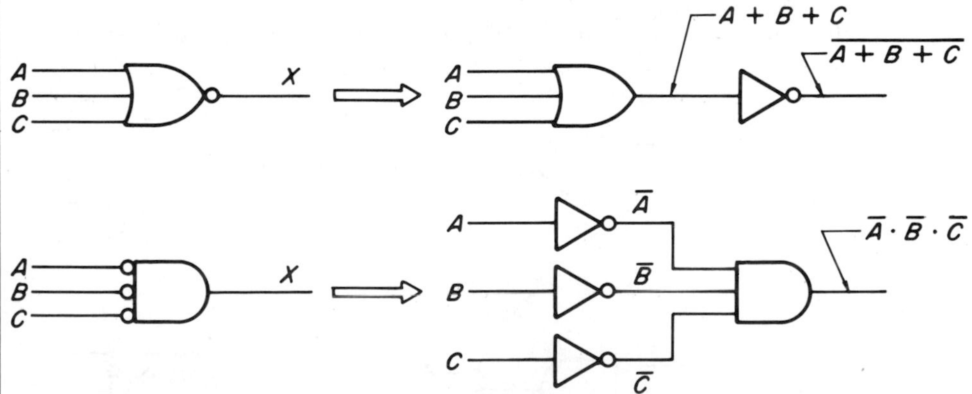

Figure 4–16

The Boolean expression for the three-input NOR gate is

$$X = \overline{A + B + C}$$

The Boolean expression for the AND gate with three inverted inputs is

$$X = \overline{A} \cdot \overline{B} \cdot \overline{C}$$

From the Boolean expressions develop the truth tables for each logic circuit:

Three-Input NOR			
A	B	C	$\overline{A + B + C}$
0	0	0	1
0	0	1	0
0	1	0	0
0	1	1	0
1	0	0	0
1	0	1	0
1	1	0	0
1	1	1	0

AND with Three Inverted Inputs						
A	B	C	$\overline{A}$	$\overline{B}$	$\overline{C}$	$\overline{A} \cdot \overline{B} \cdot \overline{C}$
0	0	0	1	1	1	1
0	0	1	1	1	0	0
0	1	0	1	0	1	0
0	1	1	1	0	0	0
1	0	0	0	1	1	0
1	0	1	0	1	0	0
1	1	0	0	0	1	0
1	1	1	0	0	0	0

Since the resulting outputs of each truth table are identical for all possible input combinations, the following Boolean expressions are identical.

$$\overline{A + B + C} = \overline{A} \cdot \overline{B} \cdot \overline{C}$$

Applying De Morgan's First Theorem

You will find many applications of De Morgan's two theorems. The following example gives an application of his first theorem.

Example 4–8

Sketch the resulting output waveform you would expect from the logic circuit of Figure 4–17.

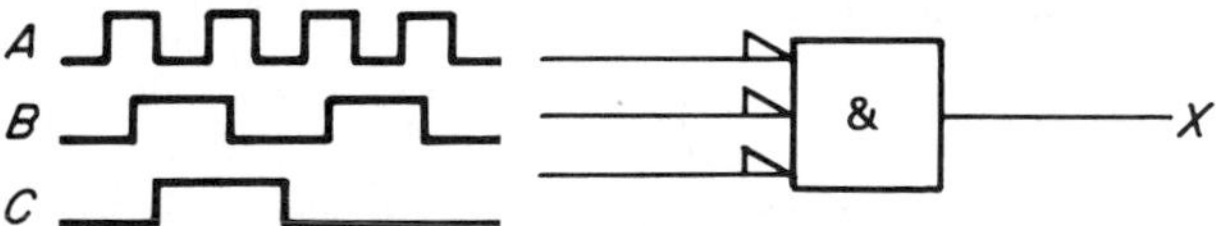

Figure 4–17

Solution

Two approaches to the solution of this circuit will be shown: a direct method, and a method using De Morgan's first theorem.

(A) Direct method

First, sketch the inverted signals as they would actually appear to the input of the AND gate. This process is shown in Figure 4–18.

(B) Applying De Morgan's first theorem

By applying De Morgan's first theorem, the logic circuit can be simplified. First, write the Boolean expression for the AND gate with the three inverted inputs:

$$X = \overline{A} \cdot \overline{B} \cdot \overline{C}$$

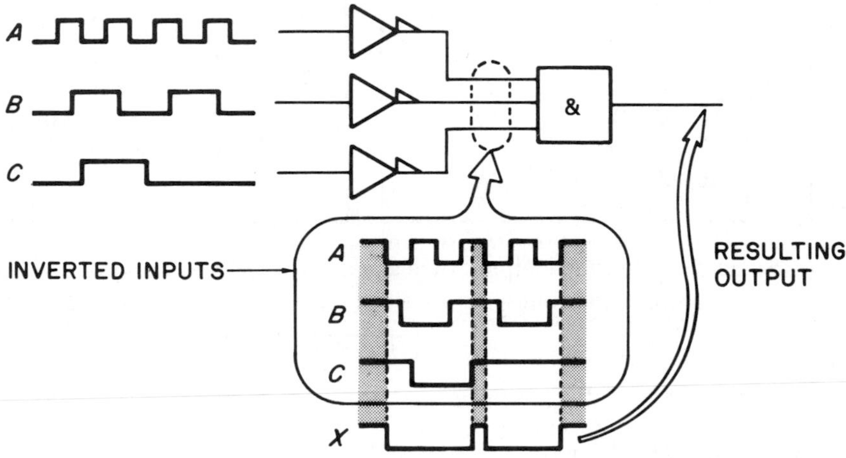

Figure 4–18

Then apply De Morgan's first theorem:

$$\overline{A} \cdot \overline{B} \cdot \overline{C} = \overline{A + B + C}$$

The result is a three-input NOR gate. The waveform analysis becomes much easier if the original circuit is analyzed as a NOR gate rather than as an AND gate with inverted inputs. Just recall that the only time the output of a NOR gate is HIGH is when all inputs are LOW. The analysis is shown in Figure 4–19.

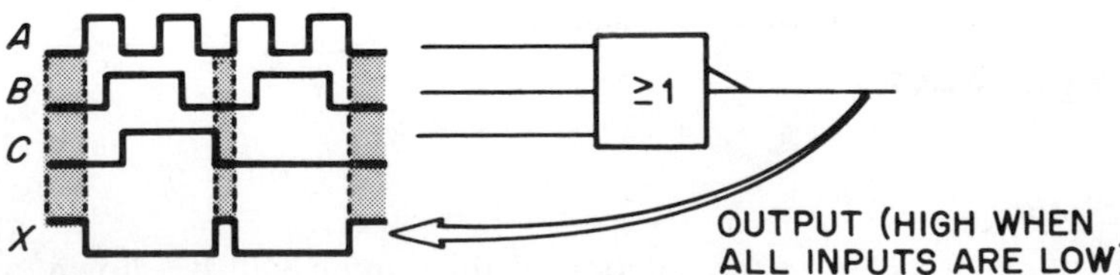

Figure 4–19 Application of De Morgan's first theorem.

Conclusion

This section presented De Morgan's first theorem. Here you saw how to develop its Boolean expression and use truth tables to prove the theorem. As you may suspect, the next section presents De Morgan's second theorem. Understanding how to apply both of these important theorems will open new horizons in your ability to analyze digital logic circuits. Test your understanding of this section by trying the following review questions.

4–3 Review Questions

1. State, in words, De Morgan's first theorem.
2. Explain how you could prove De Morgan's first theorem.
3. Give the equivalent logic circuit of a four-input AND gate with inverted inputs.

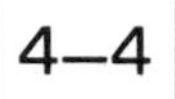

4–4 ANOTHER USEFUL RELATIONSHIP—DE MORGAN'S SECOND THEOREM

Discussion

Here you will see De Morgan's second theorem and how to apply it. Armed with those two powerful theorems, you will be well equipped for analyzing practical logic circuits.

De Morgan's Second Theorem

De Morgan's second important logical relationship is shown in Figure 4–20. Observe from the figure that the truth tables for both logic arrangements are the same under identical conditions. Thus, **De Morgan's second theorem** states

$$\overline{A \cdot B} = \overline{A} + \overline{B} \quad \text{or} \quad \overline{A} + \overline{B} = \overline{A \cdot B}$$

In the first equation, the left member represents a NAND gate, and the right member is an OR gate with inverted inputs.

De Morgan's Second Theorem:
A NAND gate is logically the same as an OR gate with inverted inputs.

A, B → $X = \overline{A \cdot B}$ ⟹

A	B	$\overline{A \cdot B}$
0	0	1
0	1	1
1	0	1
1	1	0

A, B → $X = \overline{A} + \overline{B}$ ⟹

A	B	$\overline{A}$	$\overline{B}$	$\overline{A} + \overline{B}$
0	0	1	1	1
0	1	1	0	1
1	0	0	1	1
1	1	0	0	0

Figure 4–20 De Morgan's second theorem.

As with De Morgan's first theorem, you can use this relationship to convert a logic operation to practical hardware and also simplify complex logic diagrams. A more general statement of De Morgan's second theorem is that the complement of a product is equal to the sum of the complements of each term or

$$\overline{A \cdot B \cdot \cdots Z} = \overline{A} + \overline{B} + \cdots \overline{Z}$$

This is illustrated in the following example.

Example 4–9

Develop De Morgan's second theorem for a three-input NAND gate. Draw the equivalent logic diagrams and develop the truth table for each.

Solution

As with the development of the three-input NOR gate in the last section, draw the logic circuits, develop the truth tables, and compare their outputs.

Figure 4–21 shows the equivalent circuits. The Boolean expression for the three-input NAND gate is

$$X = \overline{A \cdot B \cdot C}$$

The Boolean expression for the three-input OR gate with inverted inputs is

$$X = \overline{A} + \overline{B} + \overline{C}$$

From the Boolean expressions, develop and compare the outputs of the truth tables for each.

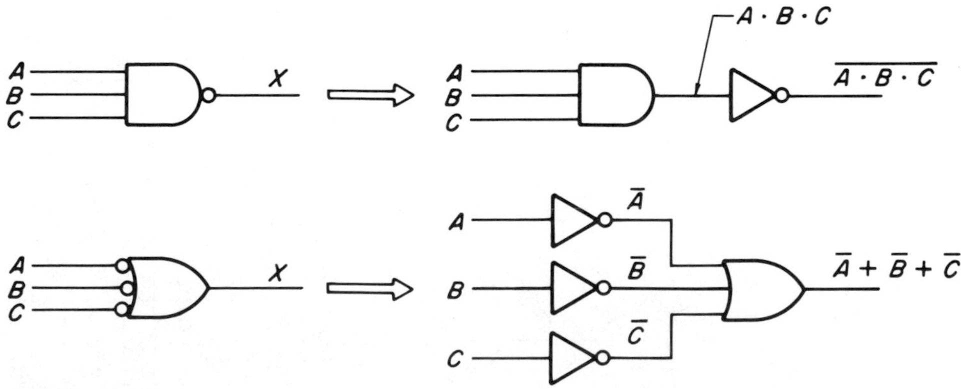

Figure 4–21

Circuit A NAND			
A	B	C	$\overline{A \cdot B \cdot C}$
0	0	0	1
0	0	1	1
0	1	0	1
0	1	1	1
1	0	0	1
1	0	1	1
1	1	0	1
1	1	1	0

Circuit B Invert *before* ORing						
A	B	C	$\overline{A}$	$\overline{B}$	$\overline{C}$	$\overline{A} + \overline{B} + \overline{C}$
0	0	0	1	1	1	1
0	0	1	1	1	0	1
0	1	0	1	0	1	1
0	1	1	1	0	0	1
1	0	0	0	1	1	1
1	0	1	0	1	0	1
1	1	0	0	0	1	1
1	1	1	0	0	0	0

Since the output of each truth table is the same for identical input conditions, then the two logic circuits are the same.

$$\overline{A \cdot B \cdot C} = \overline{A} + \overline{B} + \overline{C}$$

Applying De Morgan's Second Theorem

The process of applying De Morgan's second theorem is similar to the application of the first theorem, as illustrated in the following example.

Example 4–10

Sketch the resultant output waveform you would expect for the logic circuit of Figure 4–22.

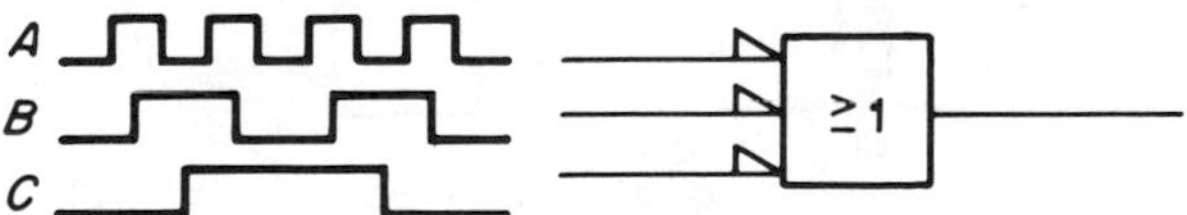

Figure 4–22

Solution

As before, two approaches will be demonstrated.

Direct method:

The logic circuit is redrawn into its component parts as shown in Figure 4–23.

Using De Morgan's second theorem

In this case, apply De Morgan's second theorem by recognizing that the logic circuit is a three-input OR gate with inverted inputs.

$$\overline{A} + \overline{B} + \overline{C}$$

Applying De Morgan's second theorem results in

$$\overline{A} + \overline{B} + \overline{C} = \overline{A \cdot B \cdot C}$$

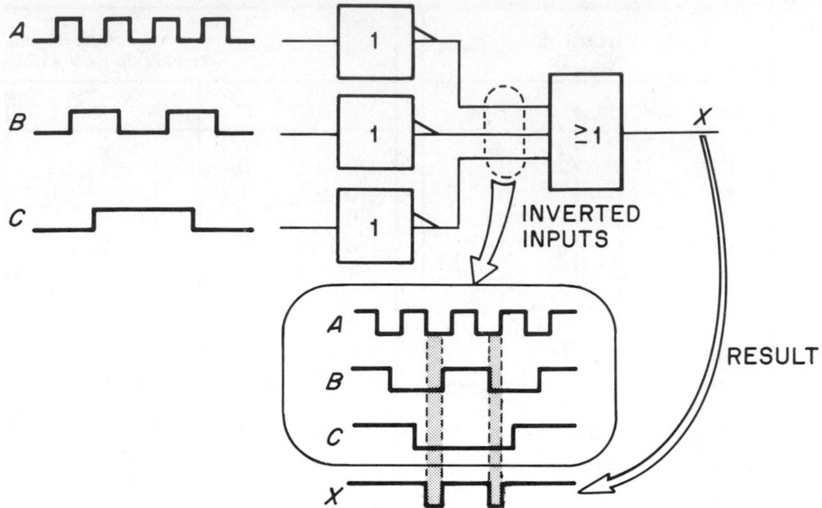

Figure 4–23

This leads to a three-input NAND gate. All that is necessary now is to recall that the only time the output of a NAND gate is LOW is when all inputs are HIGH. The process is illustrated in Figure 4–24.

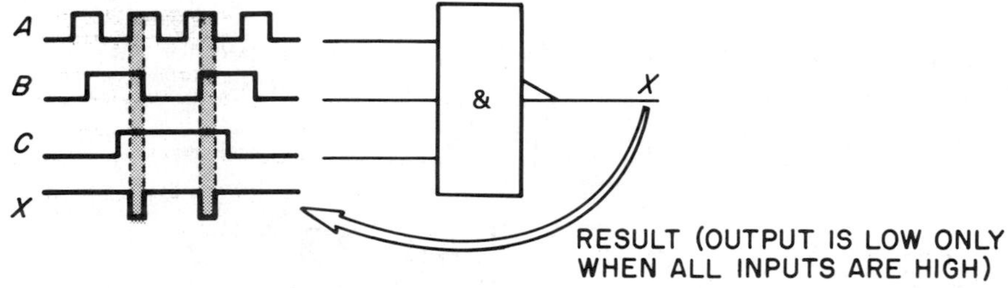

Figure 4–24 Application of De Morgan's second theorem.

Conclusion

De Morgan's theorems may be applied to any number of inputs. This completes your introduction to De Morgan's two important theorems. The next section demonstrates how they are applied. Test your understanding of this section by trying the following review questions.

4–4 Review Questions

1. State, in words, De Morgan's second theorem.
2. Explain how you could prove De Morgan's second theorem.
3. Give the equivalent logic circuit of a four-input OR gate with inverted inputs.

4–5 BOOLEAN EXPRESSIONS

Discussion

Boolean algebra is used in the design, analysis, and repair of digital systems. You have had a taste of working with Boolean algebra. In this section, you will be more involved in working with this important subject.

Precedence of Operations

Precedence of operations describes the order in which Boolean operations are to be performed. Normally, Boolean operations proceed from left to right. Boolean multiplication precedes addition. As an example, the expression $A \cdot B + C$, means that A is ANDed with B, and then this quantity is ORed with C.

Parentheses are used to indicate a different order of operations, and all operations inside them are performed first. In the expression $A \cdot (B + C)$, B and C are first ORed, and the result is ANDed with A. Parentheses are not needed if removing them does not change the order of operations: $(A + B) + C = A + B + C$. The use of parentheses in Boolean algebra is shown in Figure 4–25.

Boolean Laws

Up to this point, you have been working with many different types of logic gates and intuitively applying some Boolean laws that are formalized in Table 4–1. Many of these laws should be familiar to you from ordinary algebra.

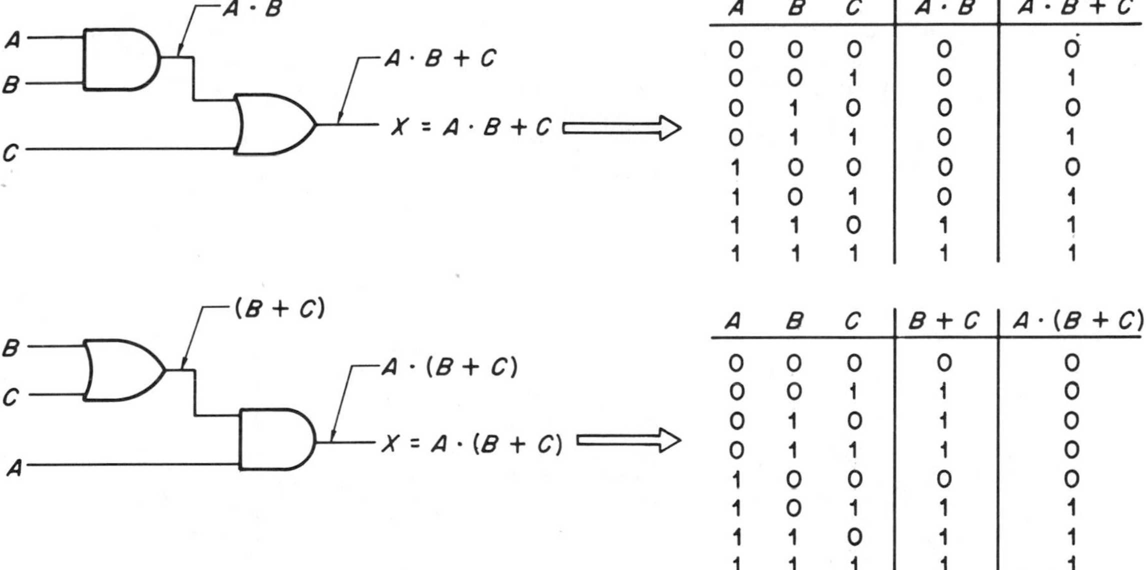

A	B	C	A · B	A · B + C
0	0	0	0	0
0	0	1	0	1
0	1	0	0	0
0	1	1	0	1
1	0	0	0	0
1	0	1	0	1
1	1	0	1	1
1	1	1	1	1

A	B	C	B + C	A · (B + C)
0	0	0	0	0
0	0	1	1	0
0	1	0	1	0
0	1	1	1	0
1	0	0	0	0
1	0	1	1	1
1	1	0	1	1
1	1	1	1	1

Figure 4–25 Demonstration of Boolean precedence.

Table 4–1 BOOLEAN LAWS

Name	Boolean Expression	Meaning	Figure
Commutative laws	[Addition] $A + B = B + A$	It makes no difference in which order variables are ORed; the result will be the same.	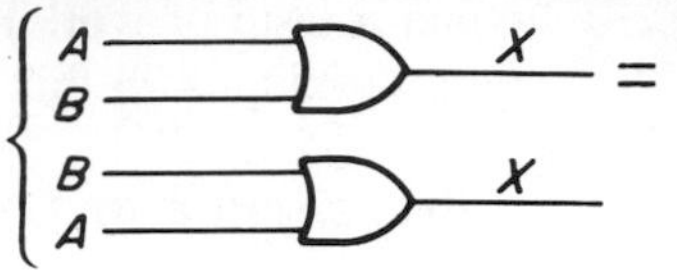 A, B → X = B, A → X
	[Multiplication] $A \cdot B = B \cdot A$	It makes no difference in which order variables are ANDed; the results will be the same.	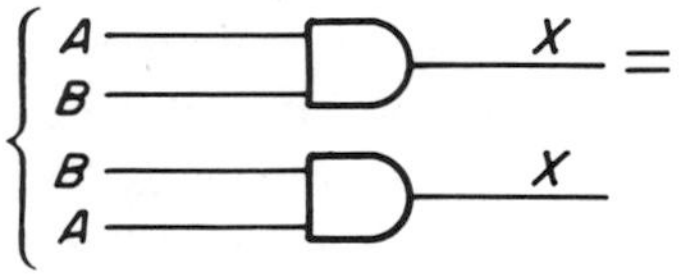 A, B → X = B, A → X
Associative laws	[Addition] $A + (B + C) =$ $(A + B) + C$	It makes no difference how variables are grouped when they are ORed; the results will be the same	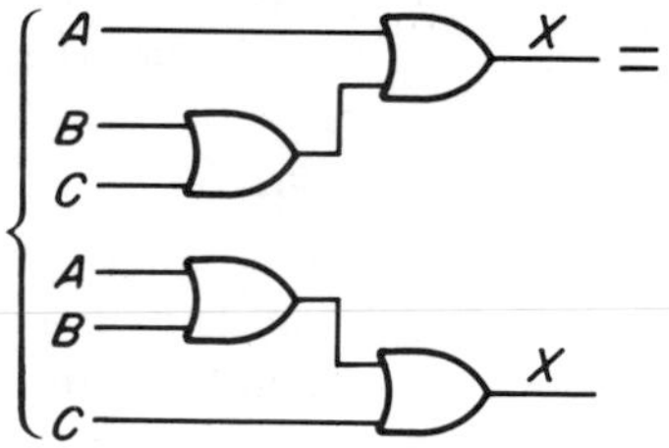 A, B, C → X = A, B, C → X
	[Multiplication] $A \cdot (B \cdot C) =$ $(A \cdot B) \cdot C$	It makes no difference how variables are grouped when they are ANDed; the results are the same.	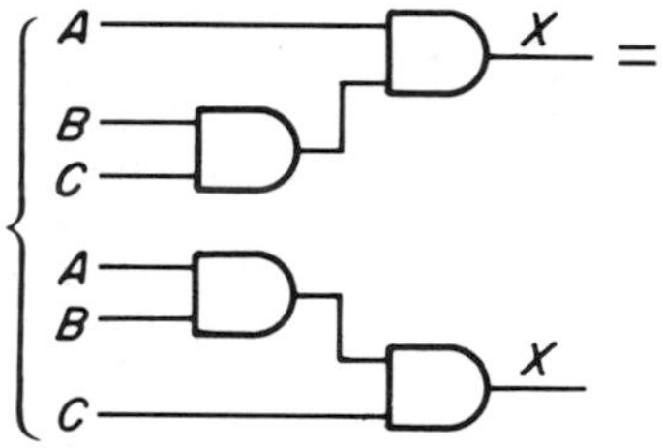 A, B, C → X = A, B, C → X
Distributive law	$A \cdot (B + C) =$ $A \cdot B + A \cdot C$	ORing variables that are then ANDed with a single variable is the same as ANDing each variable with the single variable and ORing the results.	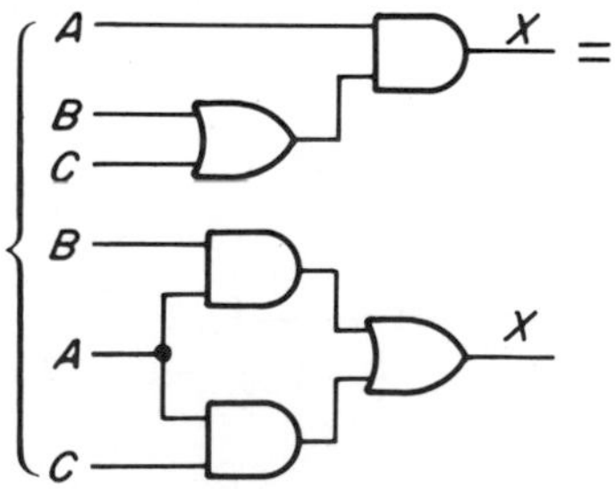 A, B, C → X = B, A, C → X

Applications

The advantages of these Boolean laws are very practical. They can help you simplify logic circuits. This is illustrated in the following example.

Example 4–11

Use the distributive law to simplify the logic circuit in Figure 4–26.

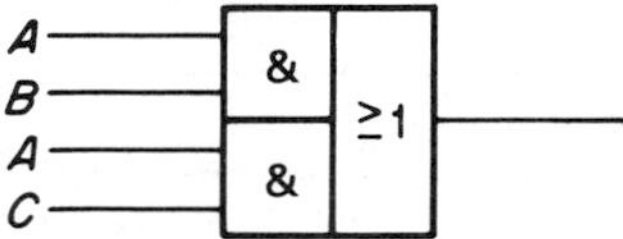

Figure 4–26

Solution

When using Boolean algebra to simplify a logic circuit, it is helpful if you write the Boolean expression for the output of each gate. This is illustrated in Figure 4–27. Applying the distributive law, you find

$$A \cdot B + A \cdot C = A \cdot (B + C)$$

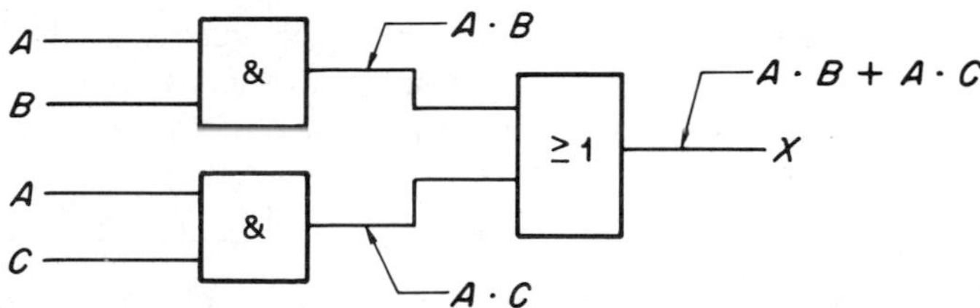

Figure 4–27

In Boolean algebra, the dot may be omitted; ANDing is still implied. Thus, you can write

$$AB + AC = A(B + C)$$

The result is shown in Figure 4–28. Note, that the same result is achieved with one fewer gate.

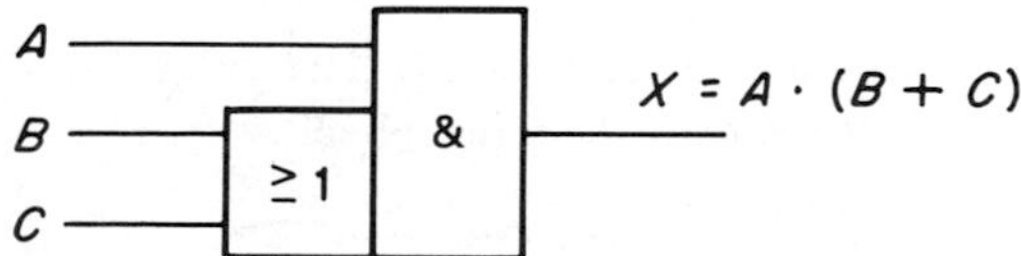

Figure 4–28

Example 4–12

Prove that the logic circuit of Figure 4–27 is equal to the logic circuit of Figure 4–28.

Solution

To prove this, simply develop a truth table for both systems and show that the output is the same for identical input conditions.

From Figure 4–27					
A	B	C	AB	AC	$AC + AB$
0	0	0	0	0	0
0	0	1	0	0	0
0	1	0	0	0	0
0	1	1	0	0	0
1	0	0	0	0	0
1	0	1	0	1	1
1	1	0	1	0	1
1	1	1	1	1	1

From Figure 4–28				
A	B	C	$B + C$	$A(B + C)$
0	0	0	0	0
0	0	1	1	0
0	1	0	1	0
0	1	1	1	0
1	0	0	0	0
1	0	1	1	1
1	1	0	1	1
1	1	1	1	1

As can be seen from the two truth tables, the outputs are the same for identical input conditions. Thus, even though the number of gates used is not the same, the results are. Therefore, the extra gate is not needed. What greater practical application could any algebra have?

The elimination of one gate by the application of Boolean algebra may not sound like such a saving, but in the example it represented a 33% reduction of hardware. If you consider a digital system with thousands of gates, the application of Boolean algebra can make a significant reduction of hardware. This results in a saving of power as well as increasing the system reliability.

Boolean Rules

There is a set of useful rules in Boolean algebra that will give you even more power in the analysis of digital systems. As with the Boolean laws, there are many practical applications for these rules. For example, one Boolean rule is

$$A + 1 = 1$$

Its meaning is shown in Figure 4–29. The rules for Boolean algebra, along with their implementations, are shown in Figure 4–30.

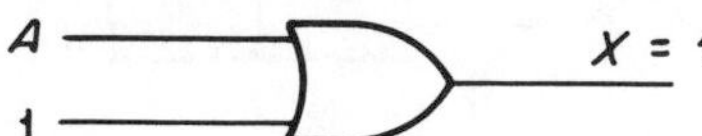

Figure 4–29 Illustration of $A + 1 = 1$

RULE	LOGIC REPRESENTATION	MEANING	RESULTS
(1) $A + 0 = A$	0, A → OR gate → X $A + 0 = A$ $0 + 0 = 0$ $1 + 0 = 1$ A | X 0 | 0 1 | 1	THE GATE OUTPUT WILL ALWAYS BE EQUAL TO THE VARIABLE INPUT.	A ○——○ X LOGIC CAN BE REPLACED BY A SINGLE CONNECTION FROM *A* TO *X*.
(2) $A + 1 = 1$	1, A → OR gate → X $A + 1 = 1$ $0 + 1 = 1$ $1 + 1 = 1$ A | X 0 | 1 1 | 1	THE GATE OUTPUT WILL ALWAYS BE EQUAL TO 1, NO MATTER WHAT THE VALUE OF *A*.	1 ○——○ X LOGIC CAN BE REPLACED BY A SINGLE CONNECTION OF *X* TO 1.
(3) $A \cdot 0 = 0$	0, A → AND gate → X $A \cdot 0 = 0$ $0 \cdot 0 = 0$ $1 \cdot 0 = 0$ A | X 0 | 0 1 | 0	THE GATE OUTPUT WILL ALWAYS BE EQUAL TO 0, NO MATTER WHAT THE [illegible]	0 ○——○ X LOGIC CAN BE REPLACED BY A SINGLE CONNECTION [illegible]
(4) $A \cdot 1 = A$	1, A → AND gate → X $A \cdot 1 = A$ $0 \cdot 1 = 0$ $1 \cdot 1 = 1$ A | X 0 | 0 1 | 1	THE GATE OUTPUT WILL ALWAYS BE EQUAL TO THE VARIABLE *A*.	A ○——○ X LOGIC CAN BE REPLACED BY A SINGLE CONNECTION OF *X* TO *A*.
(5) $A + A = A$	A, A → OR gate → X $A + A = A$ $0 + 0 = 0$ $1 + 1 = 1$ A | X 0 | 0 1 | 1	THE GATE OUTPUT WILL ALWAYS BE EQUAL TO THE VARIABLE *A*.	A ○——○ X LOGIC CAN BE REPLACED BY A SINGLE CONNECTION OF *X* TO *A*.
(6) $A + \overline{A} = 1$	A, $\overline{A}$ → OR gate → X $A + \overline{A} = 1$ $0 + 1 = 1$ $1 + 0 = 1$ A | X 0 | 1 1 | 1	THE GATE OUTPUT WILL ALWAYS BE EQUAL TO 1, NO MATTER WHAT THE VALUE OF *A*.	1 ○——○ X LOGIC CAN BE REPLACED BY A SINGLE CONNECTION OF *X* TO 1.
(7) $A \cdot A = A$	A, A → AND gate → X $A \cdot A = A$ $0 \cdot 0 = 0$ $1 \cdot 1 = 1$ A | X 0 | 0 1 | 1	THE GATE OUTPUT WILL ALWAYS BE EQUAL TO THE VARIABLE *A*.	A ○——○ X LOGIC CAN BE REPLACED BY A SINGLE CONNECTION OF *X* TO *A*.
(8)(a) $\overline{\overline{A}} = A$	A → inverter → inverter → X $\overline{\overline{A}} = A$ $\overline{\overline{0}} = \overline{1} = 0$ $\overline{\overline{1}} = \overline{0} = 1$ A | X 0 | 0 1 | 1	THE OUTPUT OF A DOUBLE INVERSION IS EQUAL TO THE INPUT.	A ○——○ X INVERTERS CAN BE REPLACED BY A SINGLE CONNECTION OF *X* TO *A*.

Figure 4–30 Rules of Boolean algebra.

RULE	LOGIC REPRESENTATION	MEANING	RESULTS
(8)(b) EVEN NUMBER OF BARS $\overline{\overline{A}} = A$	EVEN NUMBER OF INVERTERS 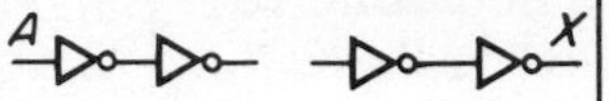 A → X EVEN $\overline{\overline{A}} = A$ A | X 0 | 0 1 | 0	THE OUTPUT OF AN EVEN NUMBER OF INVERTERS IS EQUAL TO THE VALUE OF THE INPUT.	 A ——— X INVERTERS CAN BE REPLACED BY A SINGLE CONNECTION OF X TO A.
(9) ODD NUMBER OF BARS $\overline{\overline{\overline{A}}} = \overline{A}$	ODD NUMBER OF INVERTERS 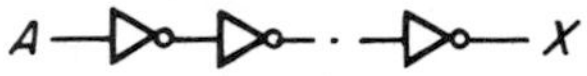 A → X ODD $\overline{\overline{\overline{A}}} = \overline{A}$ $\overline{0} = 1$ $\overline{1} = 0$ A | X 0 | 1 1 | 0	THE OUTPUT OF AN ODD NUMBER OF INVERTERS IS EQUAL TO A SINGLE INVERSION.	 A —▷o— X INVERTERS CAN BE REPLACED BY A SINGLE INVERTER.
(10) $A \cdot \overline{A} = 0$	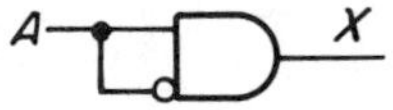 A → X $A \cdot \overline{A} = 0$ $0 \cdot 1 = 0$ $1 \cdot 0 = 0$ A | X 0 | 0 1 | 0	THE GATE OUTPUT WILL ALWAYS BE EQUAL TO 0, NO MATTER WHAT THE VALUE OF A.	0 ——— X LOGIC CAN BE REPLACED BY A SINGLE CONNECTION OF X TO 0.

Figure 4–30 Continued

Example 4–13

Express the logic circuit of Figure 4–31 as a Boolean statement. Simplify the resulting Boolean statement, using the rules of Boolean algebra. Then redraw the resulting logic circuit.

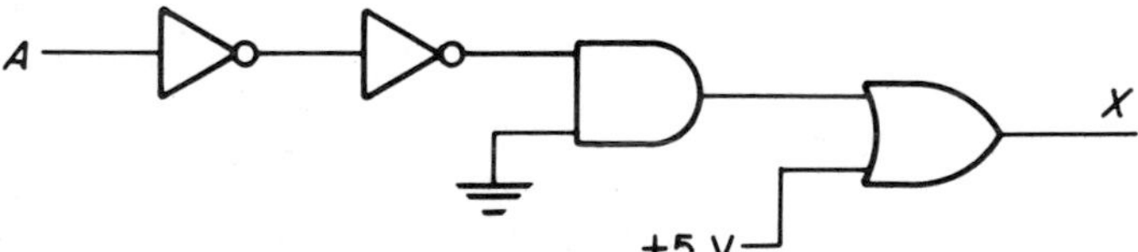

Figure 4–31

Solution

Figure 4–32 shows the steps involved in simplifying the given circuit. The resulting expression is

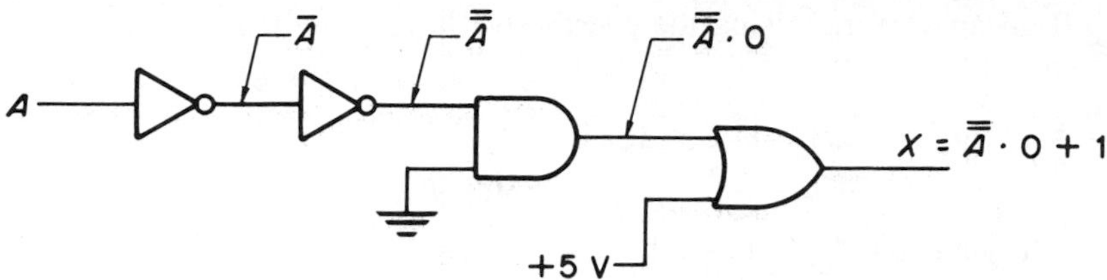

Figure 4–32

$$X = \overline{\overline{A}} \cdot 0 + 1$$

Using rule 8 [$\overline{\overline{A}} = A$], you have

$$\overline{\overline{A}} \cdot 0 + 1 = A \cdot 0 + 1$$

Using rule 3 [$A \cdot 0 = 0$] gives

$$A \cdot 0 + 1 = 0 + 1$$

This is an OR statement, where

$$0 + 1 = 1$$

Therefore, the original logic circuit can be replaced by a single connection from a HIGH input to the X output as shown in Figure 4–33.

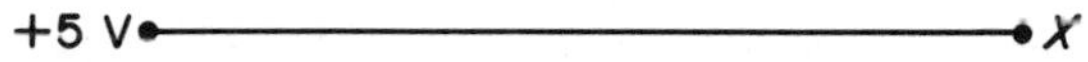

+5 V ——————————— X

Figure 4–33

Derived Boolean Rules

There are three other Boolean rules that can be derived from the first 10. The first of these is illustrated in Figure 4–34. This relationship can also be derived by using the rules of

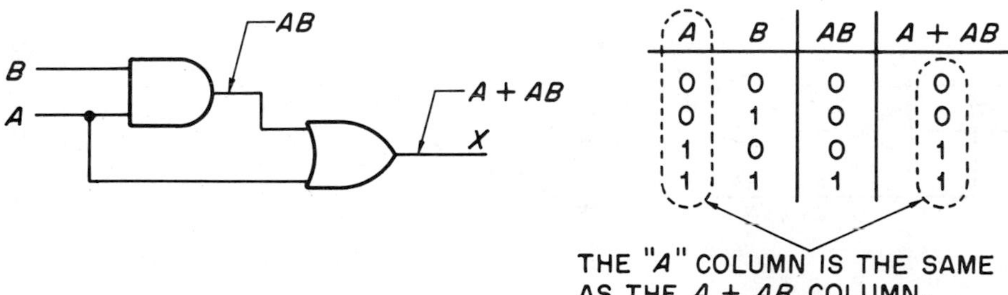

Figure 4–34 Illustration of the first derived rule.

Boolean algebra. Using the distributive law, you obtain

$$A + AB = A(1 + B)$$

Applying rule 2 $[A + 1 = 1]$ gives

$$A(1 + B) = A \cdot 1$$

Applying rule 4 $[A \cdot 1 = A]$ gives

$$A \cdot 1 = A$$

Thus, the first derived rule is

$$A + AB = A$$

Factoring Boolean Expressions

Factoring applies to Boolean algebra as well as ordinary algebra. Consider the following application of the distributive law:

$$(A + B)(C + D) = AC + BC + AD + BD$$

This means that if you started with the expression

$$AA + AB + AB + BB$$

it could be factored into

$$AA + AB + AB + BB = (A + B)(A + B)$$

The ability to factor Boolean expressions can be an aide in simplification. This is illustrated in working with the second derived Boolean rule.

More Derived Boolean Rules

The next derived Boolean rule is shown in Figure 4–35. Again, the relationship can also be derived using the rules of Boolean algebra. Using the first derived rule $[A + AB = A]$, you obtain

$$A + \overline{A}B = (A + AB) + \overline{A}B = A + AB + \overline{A}B$$

Factoring gives

$$A + AB + \overline{A}B = A + B(A + \overline{A})$$

Applying rule 6, $A + \overline{A} = 1$, gives

$$A + B(A + \overline{A}) = A + B$$

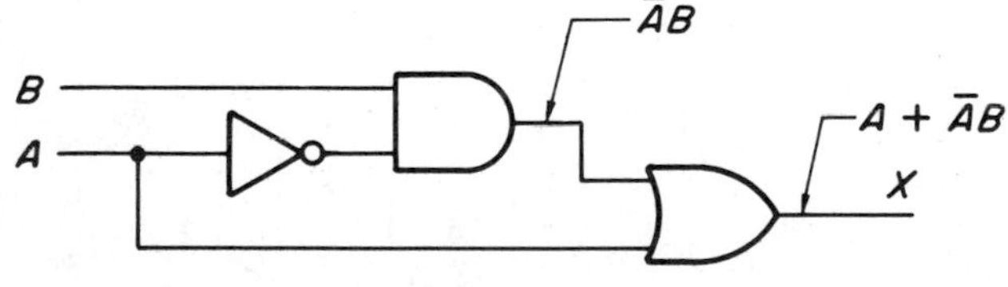

A	B	$\overline{A}$	$\overline{A}B$	$A + \overline{A}B$
0	0	1	0	0
0	1	1	1	1
1	0	0	0	1
1	1	0	0	1

THE RESULT OF THIS TRUTH TABLE IS THE SAME AS AN OR GATE. THUS: $A + \overline{A}B = A + B$

Figure 4–35 Second derived Boolean rule.

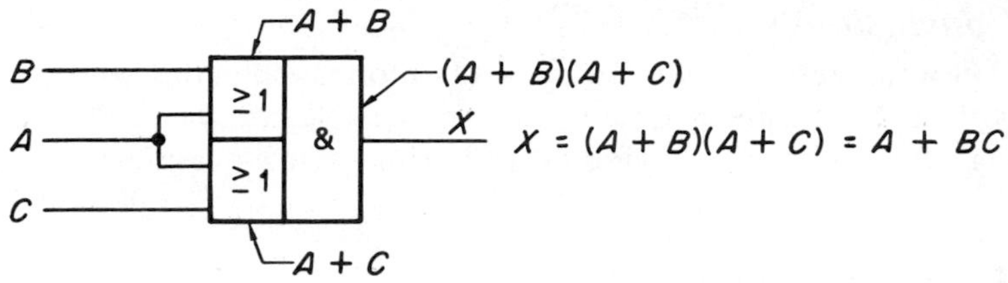

Figure 4–36 Illustration of third derived rule.

Thus,

$$A + \overline{A}B = A + B$$

The third derived rule is illustrated in Figure 4–36.

By analysis of the resulting truth table, the output is TRUE, when A OR the quantity B AND C are TRUE.

A	B	C	$(A + B)$	$(A + C)$	$(A + B)(A + C)$	$B\,C$	$A + B\,C$
0	0	0	0	0	0	0	0
0	0	1	0	1	0	0	0
0	1	0	1	0	0	0	0
0	1	1	1	1	1	1	1
1	0	0	1	1	1	0	1
1	0	1	1	1	1	0	1
1	1	0	1	1	1	0	1
1	1	1	1	1	1	1	1

The same results can be achieved by applying the rules of Boolean algebra. Using the distributive law, you have

$$(A + B)(A + C) = AA + AC + AB + BC$$

Applying rule 7 $[A \cdot A = A]$ gives

$$AA + AC + AB + BC = A + AC + AB + BC$$

Using the distributive law again, you have

$$A + AC + AB + BC = A(1 + C + B) + BC$$

Using rule 2 $[A + 1 = 1]$, anything ORed with 1 is 1. Thus,

$$A(1 + C + B) + BC = A + BC$$

Hence, we have

$$(A + B)(A + C) = A + BC$$

Conclusion

You were presented with much useful information in this section. In the next section, you will be shown applications of what you have learned here. For now, test your understanding of this section by trying the following review questions.

4–5 Review Questions

1. What is precedence of operations?
2. State the commutative laws of Boolean algebra.
3. State the associative laws of Boolean algebra.
4. Under what conditions is the output of an AND gate always equal to its input?
5. What generalization can you make about the number of inverters used in a logic circuit?

4–6 EXCLUSIVE LOGIC

Discussion

This section presents the last two fundamental gate units. These have a wide range of useful applications. They can serve as the building blocks for digital computer logic circuits that perform arithmetic operations. They serve equally well in automated systems and robotics. Your understanding of these two units is important in order to gain a complete understanding of complex digital systems.

Exclusive OR Gates

You already know the meaning of the OR function:

$$Y = A + B$$

Here, Y is TRUE when A or B is TRUE or *both* are TRUE. There are times, however, when you only want one thing or the other to happen, but not both. For example, you may have the option of making a purchase in a department store by using a credit card or cash. This is a situation where you would normally choose one method of purchase or the other, but not both. This situation is different from the OR logic condition you have been working with up to this point.

The situation where only one of two choices is allowed (but not both) is logically referred to as an **Exclusive OR** and is abbreviated XOR.

Exclusive OR Gate:
A two-input gate in which the output is TRUE when only one input is TRUE, and FALSE when both inputs are TRUE or both inputs are FALSE.

XOR Truth Table

The truth table for XOR and OR functions are shown along with their respective Boolean statements:

XOR		
A	*B*	*X*
0	0	0
0	1	1
1	0	1
1	1	0

$$X = A \oplus B$$

OR		
A	*B*	*X*
0	0	0
0	1	1
1	0	1
1	1	1

$$X = A + B$$

Note that the XOR operation has a circle around the + sign in order to distinguish it from the OR operation. Also note from the truth tables that the XOR function is TRUE only when the inputs are *different*. This is not the case with the OR function. The logic symbols for the XOR gate are illustrated in Figure 4–37. Typical XOR ICs are shown in Figure 4–38.

XOR Boolean Expression

Consider the Boolean expression

$$Y = \overline{A}B + A\overline{B}$$

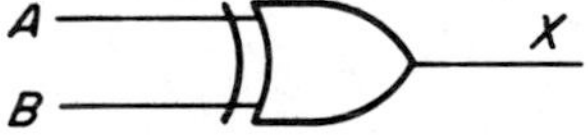

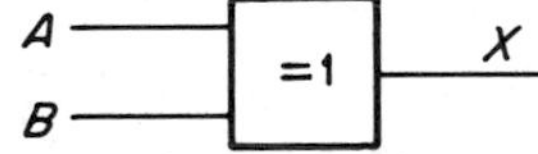

Figure 4–37 Exclusive OR logic symbols.

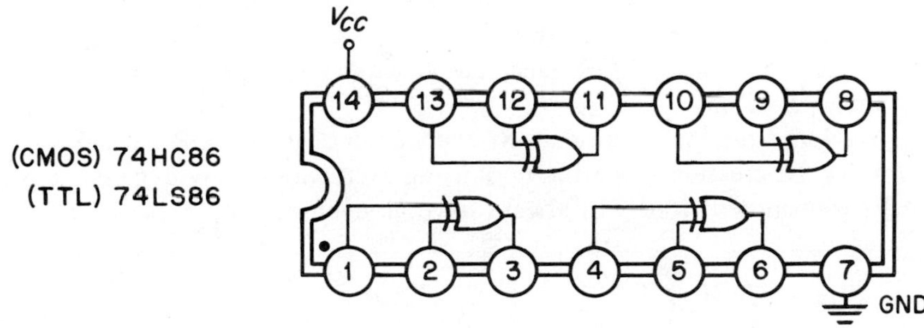

Figure 4–38 Typical XOR ICs.

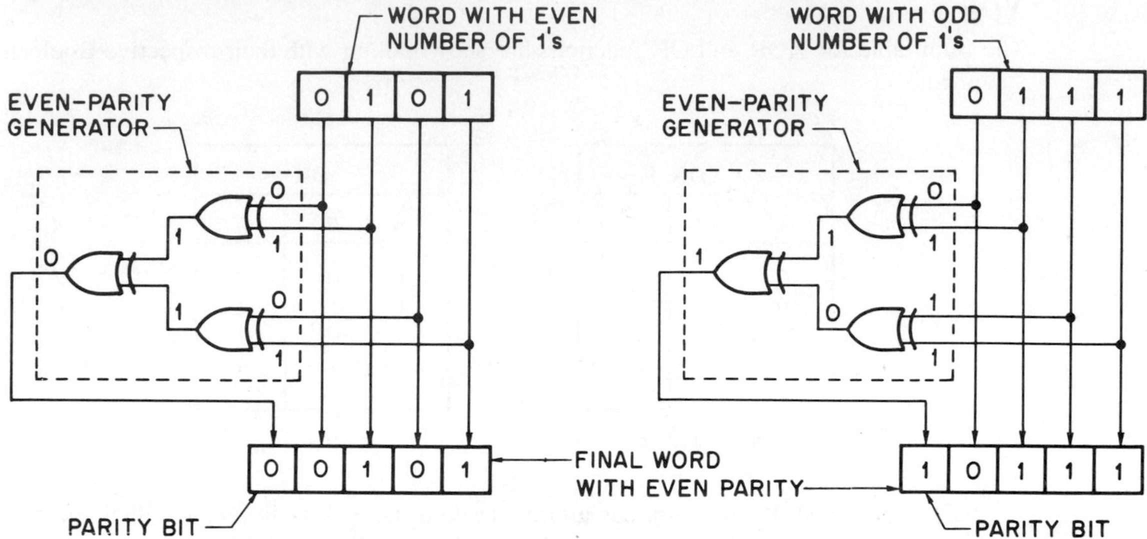

Figure 4–39 Two conditions for the even-parity generator.

The truth table for this expression is

A	B	$\overline{A}B$	$A\overline{B}$	Y
0	0	0	0	0
0	1	1	0	1
1	0	0	1	1
1	1	0	0	0

The results of this truth table are identical to the results of the XOR truth table. This relationship results in the Boolean identity

$$\overline{A}B + A\overline{B} = A \oplus B$$

XOR Applications

A direct application of XOR gates is parity generation. Recall that parity checking is a method used for catching errors in the transfer of digital information. Figure 4–39 shows a logic circuit using XOR gates that generate *even parity* for a 5-bit word where the MSB is the parity bit.

Note from the figure that the connection of three XOR gates ensures the proper parity bit for the creation of a 5-bit word using even parity. In both cases, as shown in the figure, the transmitted word will always have an even number of 1's.

Example 4–14

Develop the truth table for the logic circuit shown in Figure 4–40.

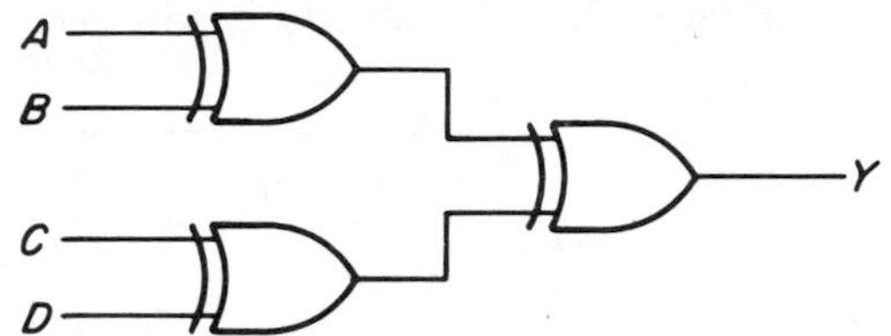

Figure 4–40

Solution

You should recognize the circuit as an even-parity bit generator. The truth table is

A	*B*	*C*	*D*	*A* ⊕ *B*	*C* ⊕ *D*	*Y*
0	0	0	0	0	0	0
0	0	0	1	0	1	1
0	0	1	0	0	1	1
0	0	1	1	0	0	0
0	1	0	0	1	0	1
0	1	0	1	1	1	0
0	1	1	0	1	1	0
0	1	1	1	1	0	1
1	0	0	0	1	0	1
1	0	0	1	1	1	0
1	0	1	0	1	1	0
1	0	1	1	1	0	1
1	1	0	0	0	0	0
1	1	0	1	0	1	1
1	1	1	0	0	1	1
1	1	1	1	0	0	0

Observe from the table that the only time the output *Y* is high is when there is an *odd* number of 1's in the input.

Exclusive NOR Gates

An **Exclusive NOR gate** (XNOR) is formed by connecting the output of an Exclusive OR gate to an inverter. This is shown in Figure 4–41.

Exclusive NOR Gate:

A two-input logic gate where the output is TRUE only if the logic inputs are the same (both TRUE or both FALSE). Otherwise the gate output is FALSE.

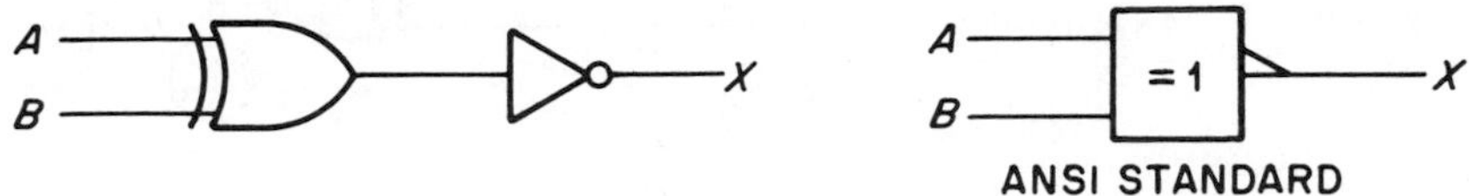

Figure 4–41 Construction of XNOR gate.

The truth table for an XNOR gate along with its Boolean expression is

A	B	X
0	0	1
0	1	0
1	0	0
1	1	1

$$X = \overline{A \oplus B} \quad \text{or} \quad X = AB + \overline{A}\,\overline{B}$$

As the truth table shows, when both inputs are HIGH or both inputs are LOW, the output is HIGH. The results of an XNOR gate are TRUE only when both inputs are the same. Note that the Boolean expression for the XNOR is the NOT of an XOR gate. This is what you would expect, since the gate is formed by adding an inverter to the output on an XOR gate.

XNOR Boolean Expression

The Boolean expression for an XNOR gate is

$$Y = \overline{A \oplus B}$$

You can construct a truth table of this expression to show that it is identical to that of the XNOR definition. You can also use Boolean algebra to show that if you invert the output of an XNOR gate, then you again wind up with an XOR.

Start with the XNOR expression:

$$Y = \overline{A \oplus B}$$

Invert:

$$\overline{Y} = \overline{\overline{A \oplus B}} = A \oplus B$$

Application of XNOR

The XOR gate was used as an even-parity generator. You can use the XNOR gate for an even-parity detector. This is a logic circuit that will detect if even parity is received during the transmission of a digital word. Figure 4–42 shows an even-parity detector for a 5-bit word (where again the fifth bit is the parity bit). Observe from the figure that the XNOR

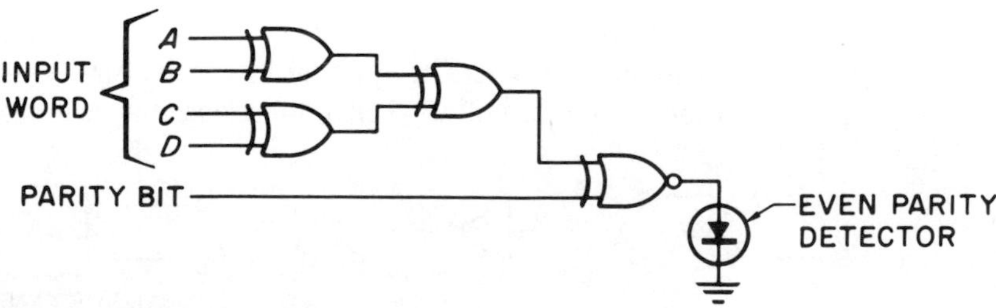

Figure 4–42 Even-parity detector for 5-bit word.

gate is used on the output to indicate if even parity has been received. The LED will be activated only when there is an even number of 1's on the input.

Example 4–15

Construct the truth table for the logic diagram of Figure 4–43.

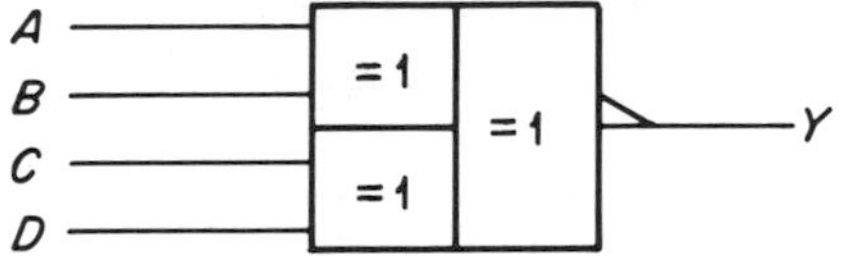

Figure 4–43

Solution

This circuit has two XOR gates connected to an XNOR gate. The resulting truth table is:

A	B	C	D	$A \oplus B$	$C \oplus D$	Y
0	0	0	0	0	0	1
0	0	0	1	0	1	0
0	0	1	0	0	1	0
0	0	1	1	0	0	1
0	1	0	0	1	0	0
0	1	0	1	1	1	1
0	1	1	0	1	1	1
0	1	1	1	1	0	0
1	0	0	0	1	0	0
1	0	0	1	1	1	1
1	0	1	0	1	1	1
1	0	1	1	1	0	0
1	1	0	0	0	0	1
1	1	0	1	0	1	0
1	1	1	0	0	1	0
1	1	1	1	0	0	1

Observe from this truth table that the logic circuit output is the inverse of the logic circuit output for the even-parity generator. Thus, this circuit can function as an odd-parity generator. You can check this for yourself. Note that if you count the number of 1's in the first four columns along with the bit of the last (*Y*) column, there will always be an odd number of 1's. Thus, the bit in the last column could serve as a parity bit.

Conclusion

This section presented two very important logic circuits: the XOR and the XNOR. In the next section, you will see how manufacturers of digital systems use alternative forms of gate representation. Test your understanding of this section by trying the following review questions.

4–6 Review Questions

1. Describe the difference between the OR function and the XOR function.
2. State when the output of an XOR gate will be low.
3. Give an application of XOR gates.
4. Explain the basic logic construction of an XNOR gate.
5. What is the logical symbol for the XOR process?

4–7 OTHER WAYS OF REPRESENTING LOGIC GATES

Discussion

Many manufacturers of digital systems utilize a slightly different variation of the standard logic symbols in their logic circuit diagrams. As you will see in this section, this is done because it makes the logic of a digital system easier to follow.

De Morgan Revisited

Recall that De Morgan's theorems allowed you to represent logic gates in two different ways. As an example, an OR gate with inverted inputs was logically identical to a NAND gate:

$$\overline{A} + \overline{B} = \overline{AB}$$

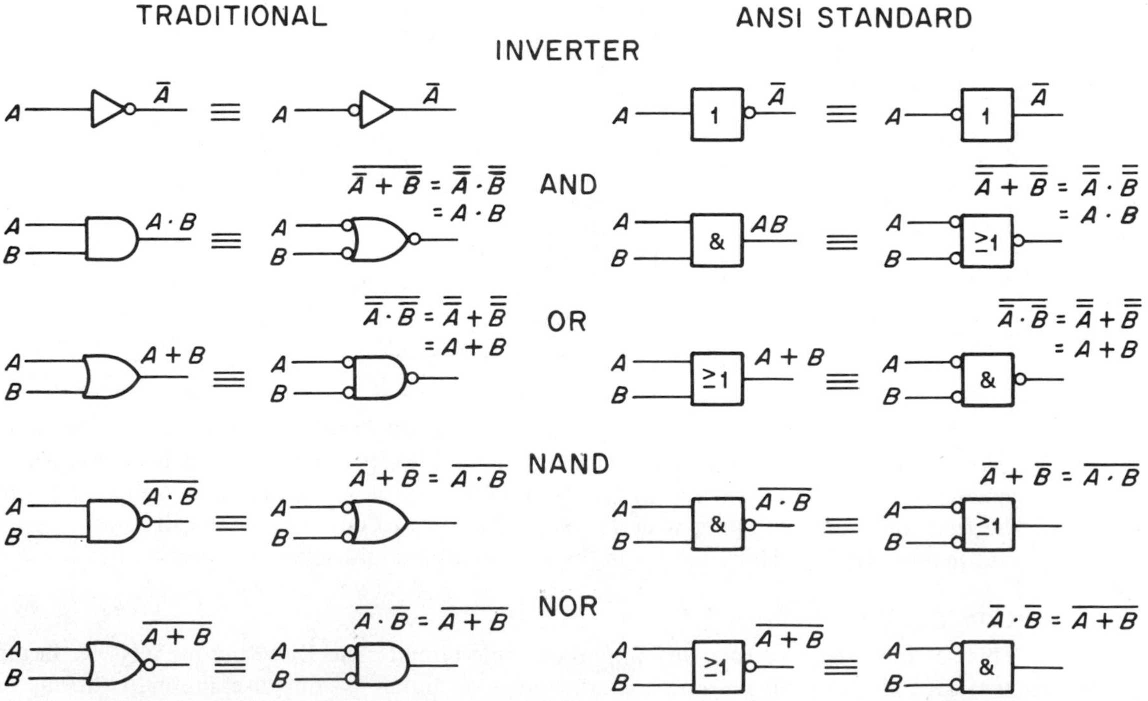

Figure 4–44 Duality of logic representation.

All logic gates, including the inverter had a similar duality. This is illustrated in Figure 4–44. There are several important points concerning the relationships of these equivalent gate symbols. First, note that the equivalent logic symbols all have bubbled inputs. Also note that with the exception of the inverter, all the equivalent gate symbols are opposites—if the original uses an AND symbol, its dual will use an OR symbol. Figure 4–44 shows the equivalent logic symbols for only two inputs, but this relationship is valid for any number of inputs. The important point to emphasize is that every gate can be expressed in terms of its standard symbol or its equivalent alternative symbol depending upon its application in a logic circuit. For example, the NAND gate is equivalent to the OR with inverted inputs.

Example 4–16

The logic circuit in Figure 4–45 uses alternate logic symbols. Redraw the circuit using simplified logic symbols. Write the resulting Boolean expression.

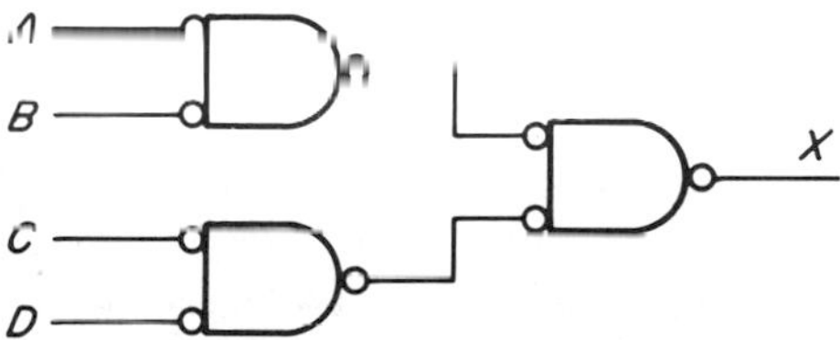

Figure 4–45

Solution

Converting each gate into its simplified symbol yields the logic diagram of Figure 4–46.

STEPS IN SOLUTION

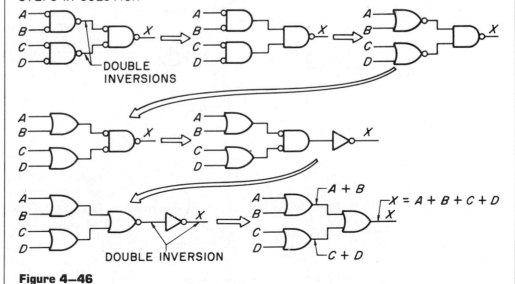

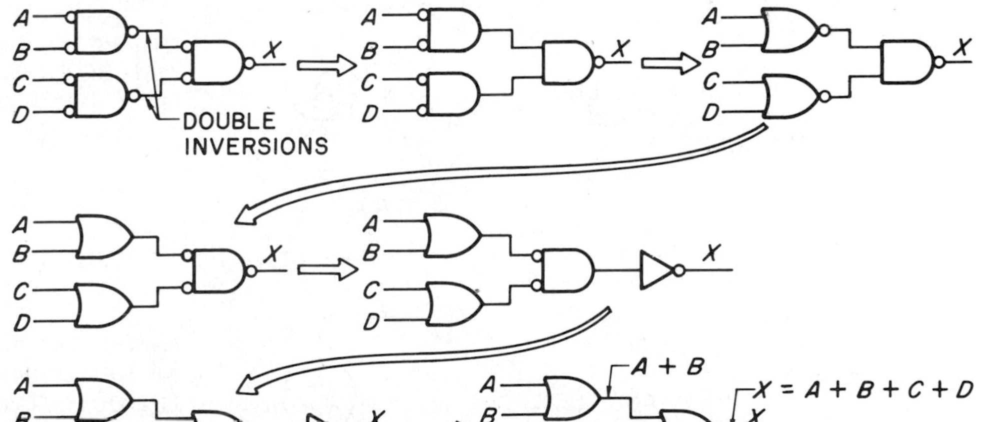

Figure 4–46

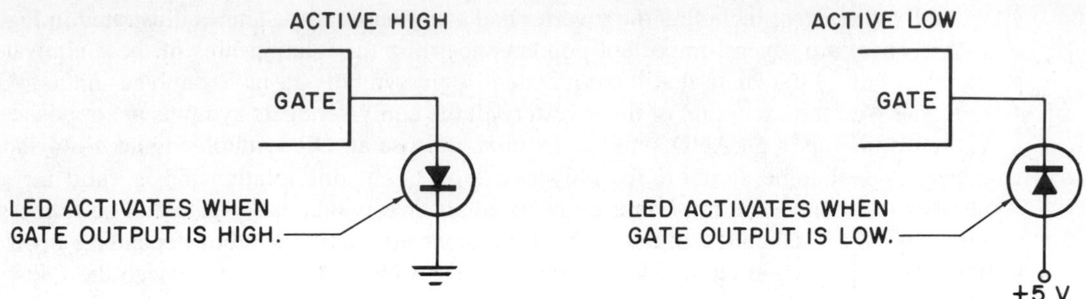

Figure 4–47 Concept of active logic levels.

Active Logic Level

There are two kinds of **active logic levels**: active LOW and active HIGH. This concept is illustrated in Figure 4–47.

Active Logic Level:
The logic state (HIGH or LOW) used to activate a logic or other device.

As an example of how the concept of an active logic level can be helpful in emphasizing which of its states will activate a device, consider Figure 4–48. It is important to recognize from the figure that in both cases you have an inverter activating an LED. Recall that the operation of LEDs was introduced in the electronic application section of Chapter 2. However, in the first case, Figure 4–48(A), the LED is activated when the inverter output is LOW, and in the other case, Figure 4–48(B), the LED is activated when the output of the inverter is HIGH. It must be emphasized that in both cases, the same inverter

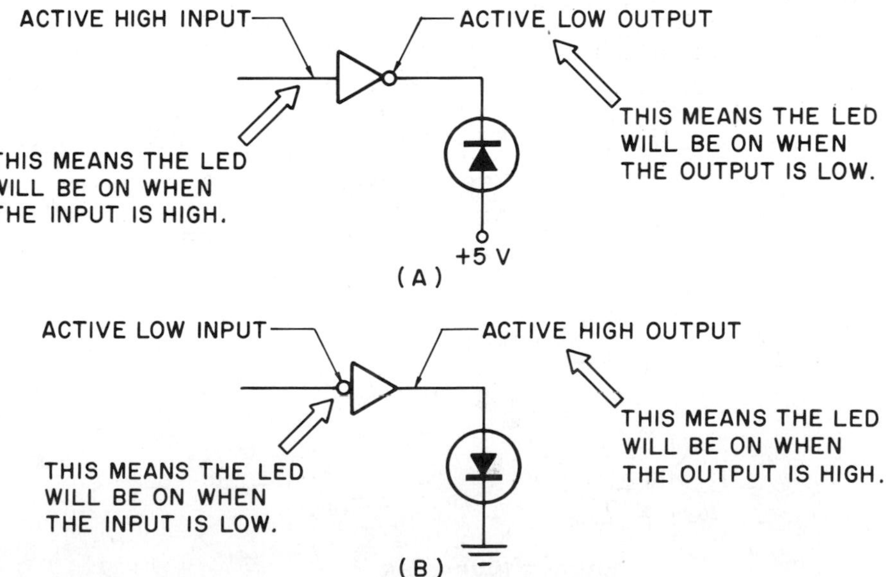

Figure 4–48 Active levels used for emphasis.

is being used; only the logic symbol is different to emphasize if a LOW or HIGH output level activates the LED!

Example 4–17

Construct the logic diagram of a three-input NAND gate to emphasize an active HIGH output.

Solution

The simplified logic symbol for a NAND gate uses an inverter symbol on the output. To indicate an active HIGH output, use the equivalent OR gate symbol with inverted inputs, as shown in Figure 4–49.

Figure 4–49 Active HIGH output NAND gate.

Conclusion

This section presented the useful concept of active logic levels. Keep in mind that many manufacturers use this alternative form of logic symbol in an attempt to make logic analysis of their circuits easier. In the next section, you will discover a technique of using this concept for making gate network analysis even easier. Test your understanding of this section by trying the following review questions.

4–7 Review Questions

1. Give the primary use for alternative forms of logic symbols.
2. What is meant by an active logic level?
3. Explain the difference between active HIGH and active LOW.
4. What generalization can be made about simplified gate symbols and their equivalents?

4–8 ANALYSIS SIMPLIFICATION

Discussion

This section presents a technique that will help you analyze the logic of logic gates. Basically it uses the concept presented in the last section, with a new twist. Once you understand this technique, you will have a very useful analysis tool at your command.

Basic Idea

Suppose you were testing the output of the logic circuit of Figure 4–50 for a HIGH or LOW output. From the diagram, you see an OR gate operated by two AND gates. If either or both of the inputs to the OR are HIGH, then the output (*X*) will be HIGH. This means if *A* AND *B* are HIGH or *C* AND *D* are HIGH, then the circuit output will be HIGH.

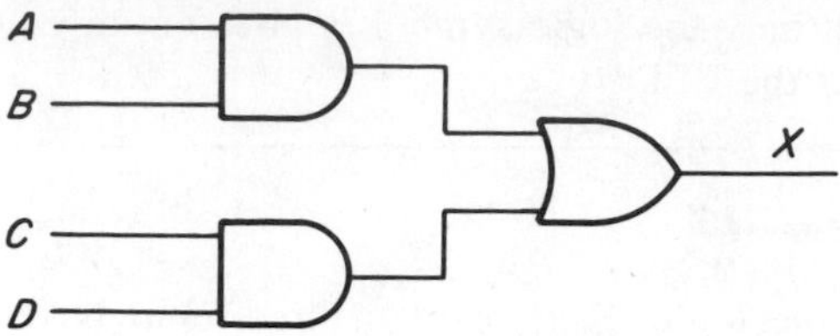

Figure 4–50 An easily analyzed logic circuit.

Now if you were testing the output of the logic circuit of Figure 4–51 for a HIGH or LOW condition, it is not immediately evident at what levels the inputs (*A*, *B*, *C*, *D*) must be. Why does this circuit look more complex then the one in Figure 4–50? The reason is the inversion bubbles.

In order to feel comfortable in analyzing this circuit, most technicians would first construct a truth table:

Inputs				Output
A	*B*	*C*	*D*	*X*
0	0	0	0	0
0	0	0	1	0
0	0	1	0	0
0	0	1	1	1
0	1	0	0	0
0	1	0	1	0
0	1	1	0	0
0	1	1	1	1
1	0	0	0	0
1	0	0	1	0
1	0	1	0	0
1	0	1	1	1
1	1	0	0	1
1	1	0	1	1
1	1	1	0	1
1	1	1	1	1

Analysis of this truth table reveals that the output is HIGH only when *A* AND *B* OR *C* AND *D* or both are HIGH. Thus, the logic circuit of Figure 4–51 is logically the same as the previous easier to analyze logic circuit of Figure 4–50. It took some time (and work) to see this relationship. There is another way.

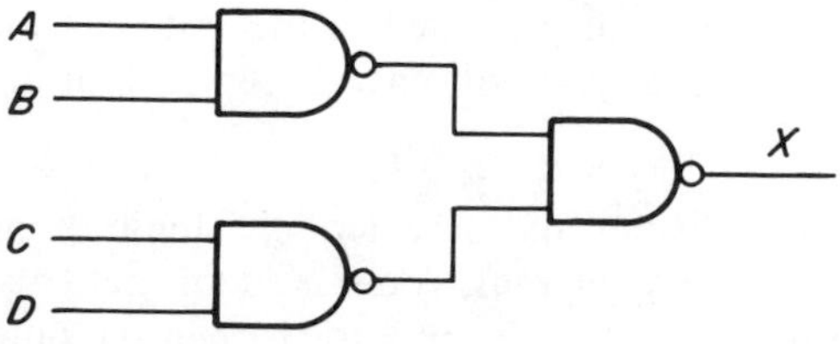

Figure 4–51 A not so easily analyzed logic circuit.

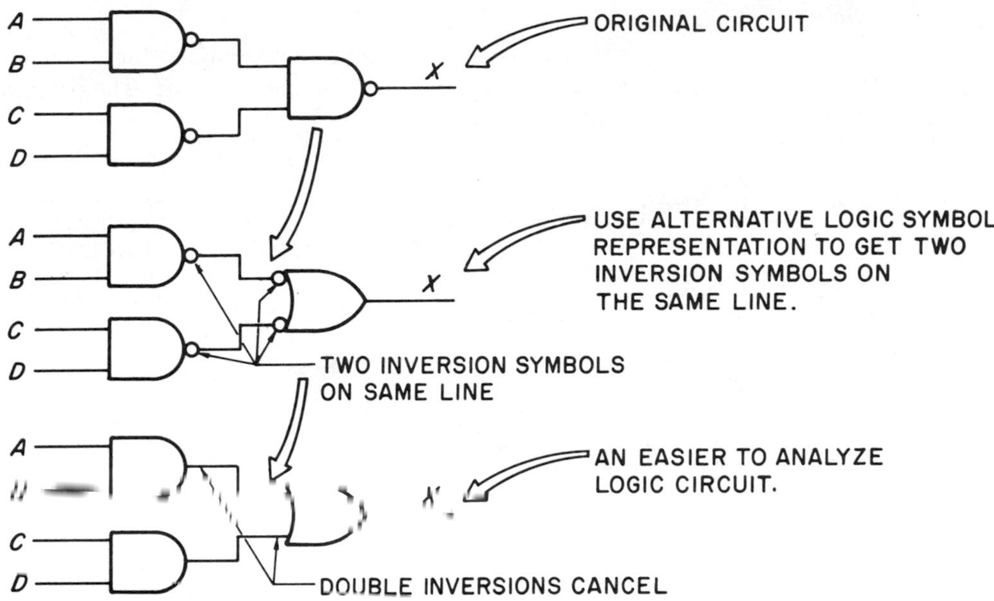

Figure 4–52 Simplifying logic circuits for analysis.

Circuit Conversion

Figure 4–52 shows how a logic circuit with inversion bubbles can be transformed into an easier to analyze logic circuit. The third circuit of this figure, known as AND/OR logic, is a common operation in logic. It is simpler to do this circuit using NAND gates.

Essentially what you do to simplify the analysis of gate networks is to use alternative logic symbols in those places that would create a *double inversion* on the same connecting line. The reason for doing this is that the double inversion will cancel, and thus the connecting line can be redrawn without them.

This technique does not simplify *all* of the possible logic combinations you will ever encounter, but knowing how to do it does make the life of a digital technician easier most of the time. As you will see, this technique can also serve you in other useful and practical ways.

Example 4–18

Simplify the logic circuit of Figure 4–53 for analysis.

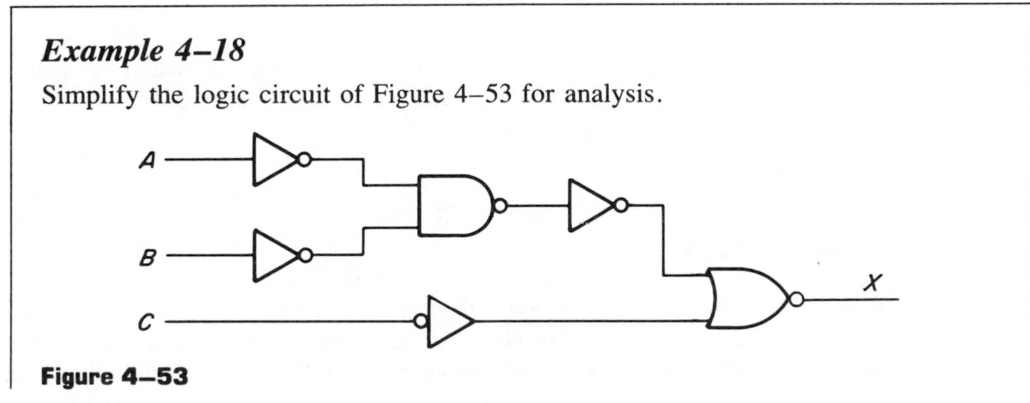

Figure 4–53

Solution

Try to make connecting lines contain two inversion symbols. Since a double inversion cancels, these can be replaced by a single connecting line. The process of analysis simplification is shown in Figure 4–54.

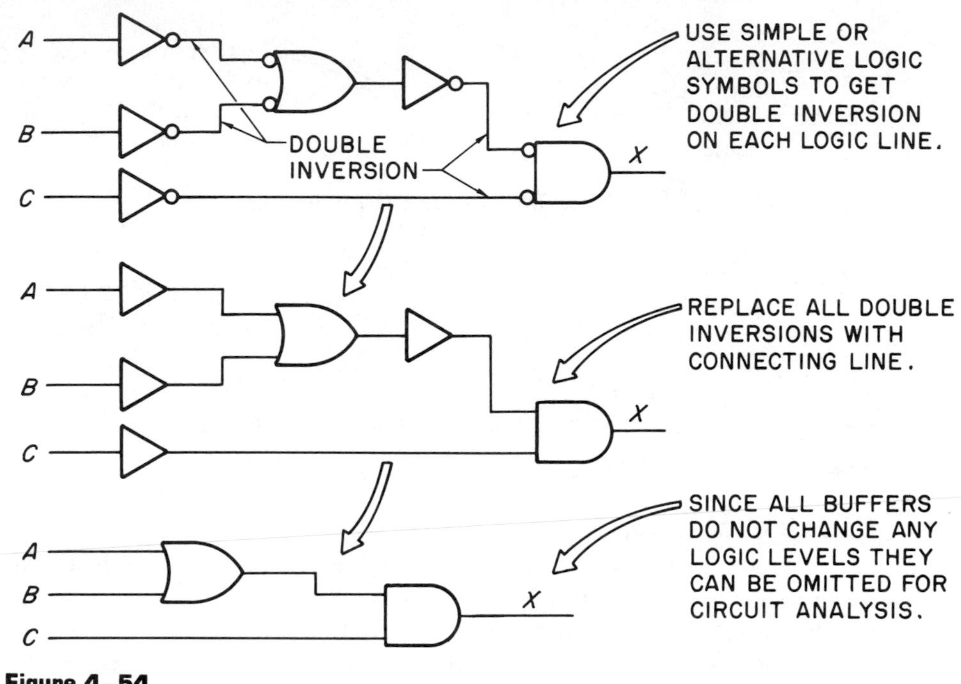

Figure 4–54

Note that the simplified logic diagram achieved in Figure 4–54 is much easier to analyze than the one in Figure 4–53. You can see that the output will be HIGH when *A* OR *B* AND *C* are HIGH.

The Other Way Around

This same technique can be "worked backwards," so to speak, when designing logic circuits that will actually use the NAND/NOR ICs rather than the AND/OR variety. This technique is illustrated by the following example.

Example 4–19

A technician has sketched a logic diagram for a required circuit using AND/OR gates. Your company has only NAND/NOR gates in stock. Convert the AND/OR gate circuit of Figure 4–55 to an equivalent NAND/NOR logic combination.

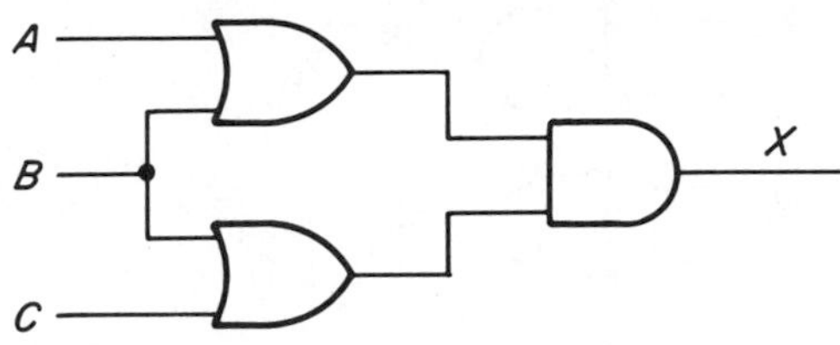

Figure 4–55

Solution

Convert each logic symbol into its equivalent symbol and replace all double inversion lines with a straight logic connection. This process is shown in Figure 4–56.

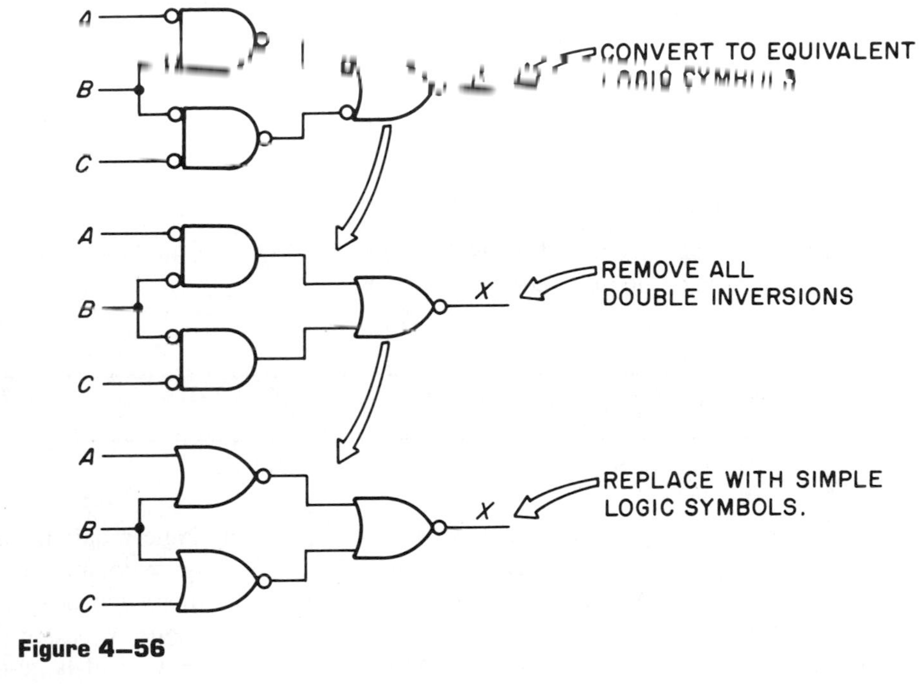

Figure 4–56

Note, just as the AND/OR combination is easily done with NAND gates, the OR/AND combination is also easily done with NOR gates.

Conclusion

This section introduced you to some of the techniques that will help you in the analysis of logic circuits. You also saw an example of how to convert an AND/OR logic circuit into a NAND/NOR circuit. The next two sections present a method of using NAND/NOR gates to represent *any* kind of logic you choose (including AND/OR). Thus, you will begin to see the reason for the popularity of the NAND/NOR logic circuits. Check your progress with the following review question.

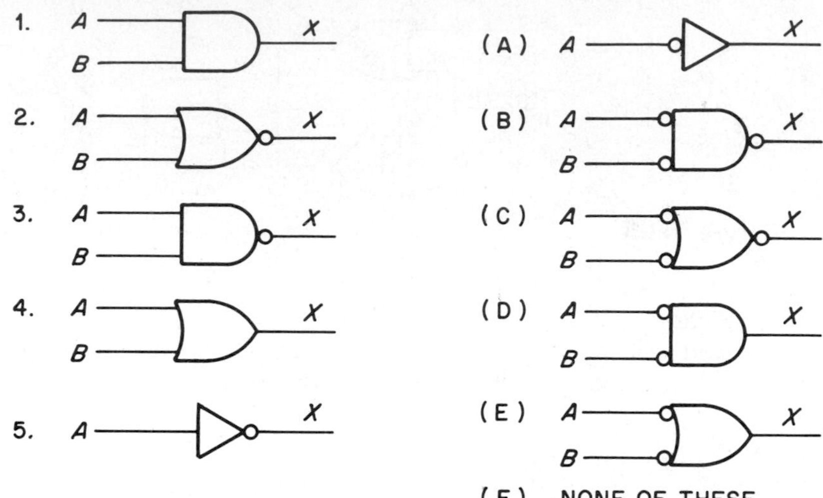

Figure 4–57

4–8 Review Questions

Using the gate circuits of Figure 4–57, match the logic circuits on the left to the corresponding equivalent logic circuits on the right.

4–9 THE VERSATILE NAND GATE

Discussion

In this section, you will discover how you can create *any* logic requirement using only NAND gates. Because of this versatility, the NAND gate is quite popular. This versatility is referred to as the universality of the NAND gate. The more you deal with practical digital circuits, the more you will see this economical technique of circuit design used. The reason this can be cost saving to the manufacturer is because of large-quantity discounts for ordering large amounts of the same item. There is also a reduced cost when fewer *different* parts must be handled, inspected, inventoried, stored, and distributed within the assembly area. Also, with the increasing popularity of automated assembly, the use of identical parts reduces the cost of special tooling and programming.

Equivalent Circuits

Figure 4–58 shows how NAND gates can be used to simulate any logic gate you may require. The following discussion will present the construction of each type of equivalent gate illustrated in Figure 4–58. You will see how the equivalent circuit is achieved by the application of two techniques: logic circuit simplification (covered in the last section) and Boolean algebra.

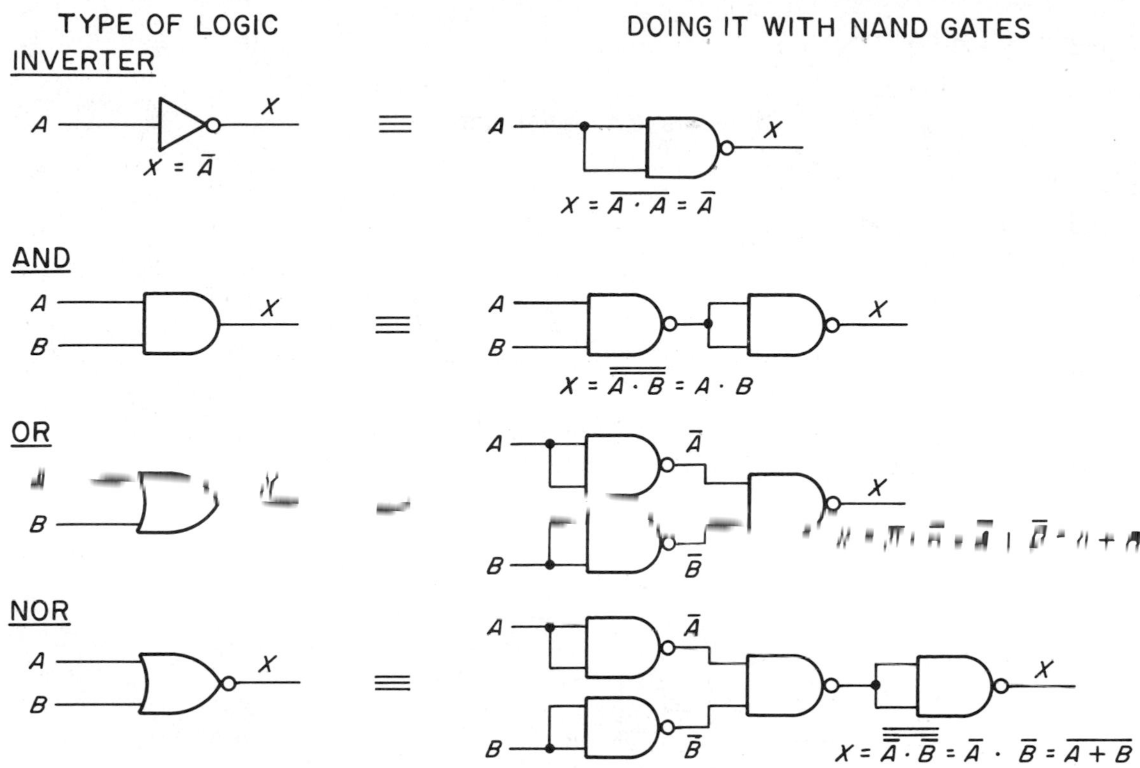

Figure 4–58 Equivalent logic circuits made with NAND gates.

The Inverter

Observe from Figure 4–58 that an inverter can be constructed from a NAND gate with its inputs tied together. This can be expressed as

$$X = \overline{AA}$$

Since $\overline{AA} = \overline{A}$, you have

$$X = \overline{A}$$

The last statement is the Boolean expression for an inverter. No logic circuit simplification will be used with the inverter because the Boolean algebra is so straight-forward.

The AND Gate

The development of the AND gate from two NAND gates is shown in Figure 4–59. The Boolean algebra analysis is demonstrated in Figure 4–60. Observe from the figure that effectively you have a NAND gate followed by an inverter. As shown by the Boolean algebra development, this inversion process results in the logical equivalent of an AND gate.

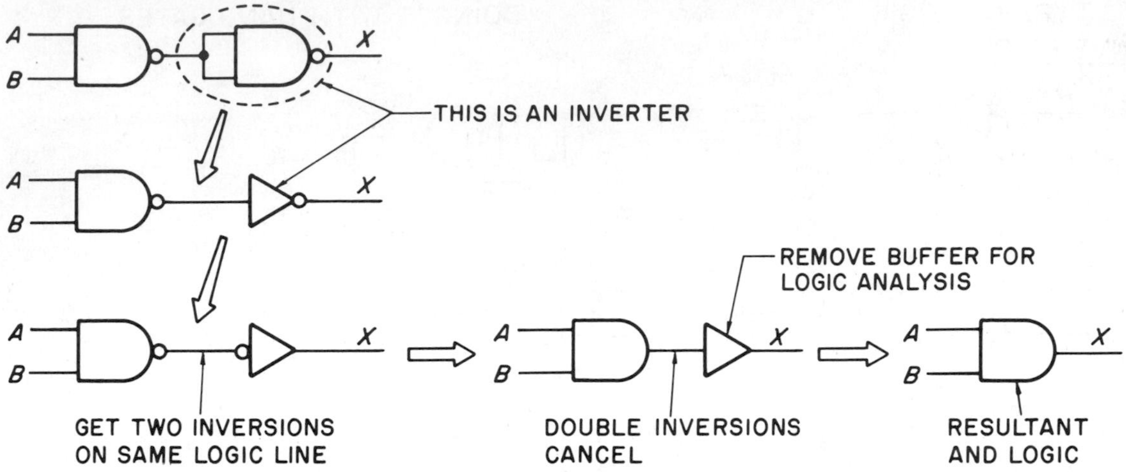

Figure 4–59 AND gate development from NAND gates.

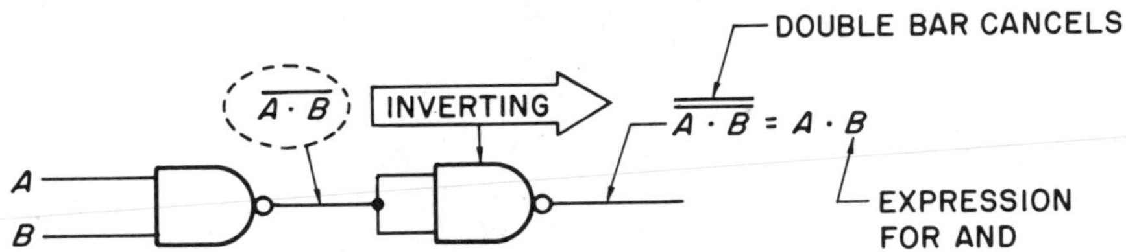

Figure 4–60 Boolean algebra development of AND from NAND.

The OR Gate

The construction of an OR gate from NAND gates is demonstrated by the gate simplification technique shown in Figure 4–61. As shown in the figure, the reduction of the NAND gates to inverters feeding an equivalent NAND gate circuit (the OR with inverted inputs) results in an OR gate with buffer inputs. However, from a logical analysis standpoint, since the buffer has no affect on the logic, it can be omitted.

The Boolean algebra analysis of this equivalency is demonstrated in Figure 4–62. Note that this time De Morgan's theorem was called into action. Effectively you are using De Morgan's theorem when you transfer from a NAND gate symbol to its equivalent OR gate with inverted inputs.

The NOR Gate

Figure 4–63 shows the development of the NOR gate from NAND gates. Observe that this equivalency uses more NAND gates than any other of this series. Upon close inspection, you should see that the only difference between the NOR equivalent and the OR equivalent is the addition of another NAND gate connected as an inverter on the output of an OR equivalent circuit. The Boolean development of the NOR gate from NAND gates is shown in Figure 4–64.

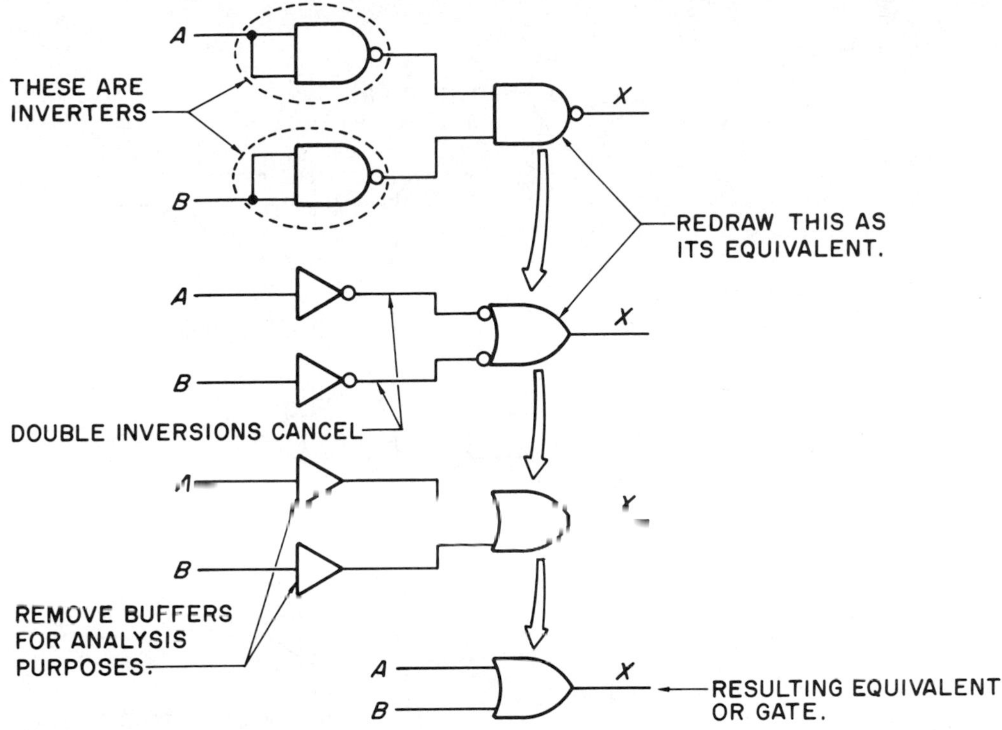

Figure 4–61 OR gate from NAND gates.

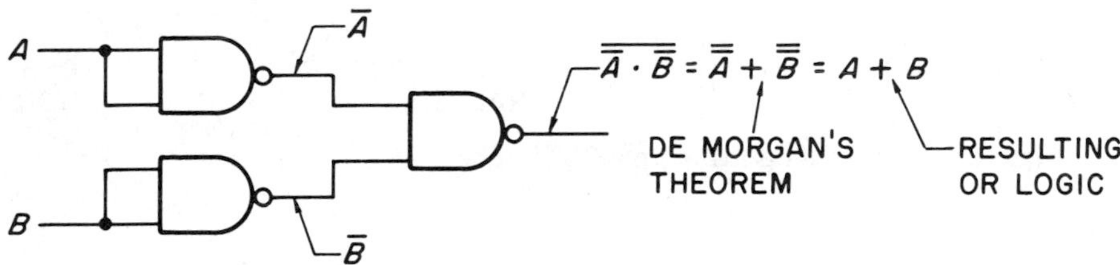

Figure 4–62 Boolean algebra development of OR from NAND.

De Morgan's theorem is again used to demonstrate the proof. Note that this circuit would use a whole IC (a quad two-input NAND gate) just to construct one NOR gate. Such construction may be hard to justify economically.

Conclusion

This section demonstrated how you could develop any kind of a gate circuit with the use of just NAND gates. In the next section, you will see how to do the same thing using the

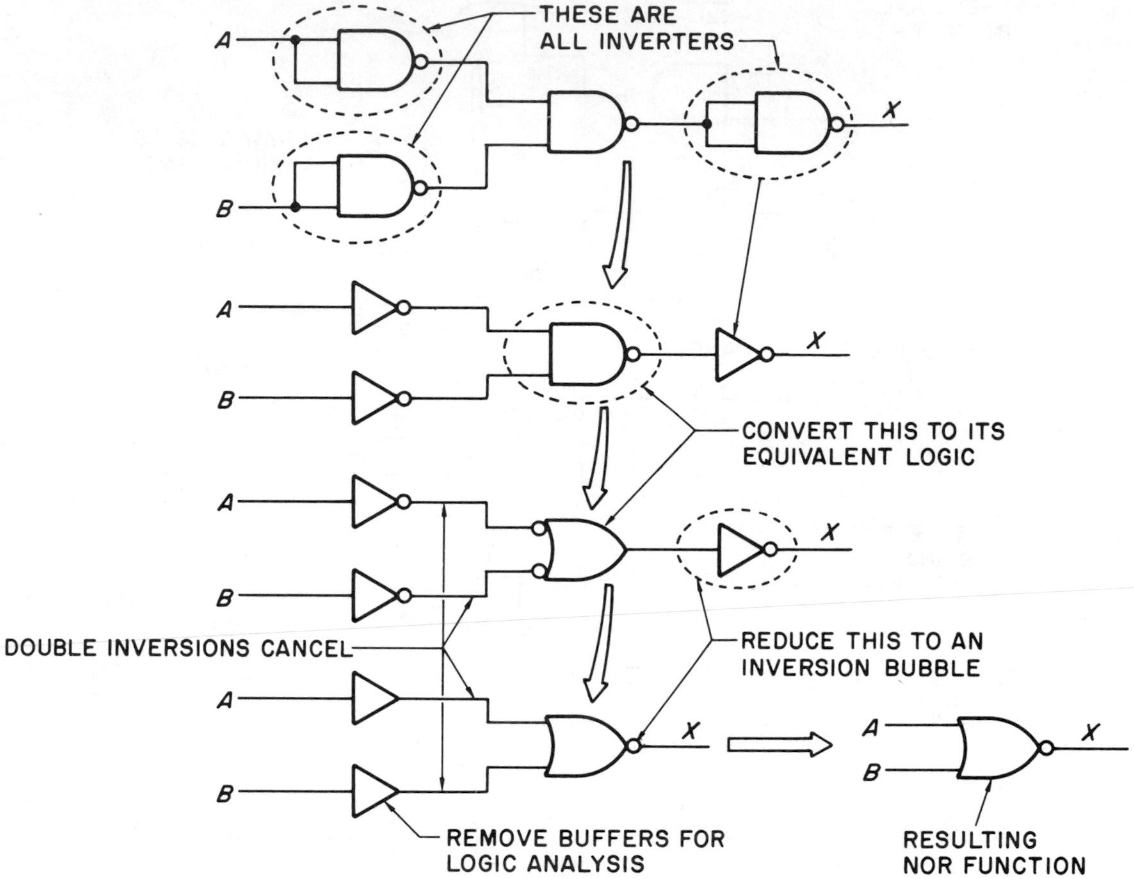

Figure 4–63 NOR gate from NAND gates.

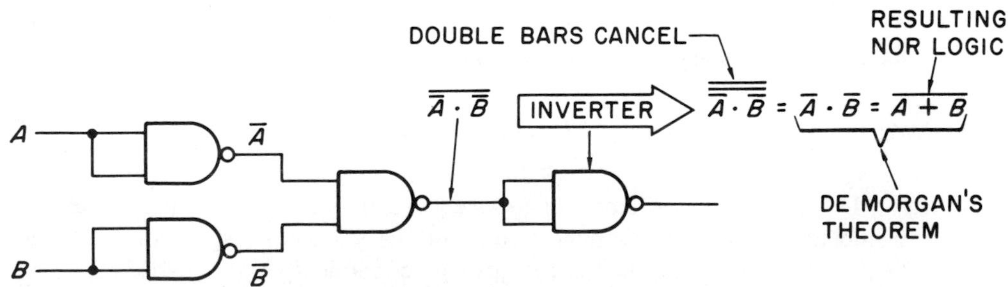

Figure 4–64 Development of NOR gates from NAND gates.

versatile NOR gate. For now, test your understanding of this section by trying the following review questions.

4–9 Review Questions

1. Explain how you would create an inverter from NAND gates.
2. Present the algebraic proof that you can create an OR gate from NAND gates.
3. Which equivalent gate circuit requires the greatest number of NAND gates? How many NAND gates does it require?
4. When is De Morgan's theorem used in the logic gate simplification?

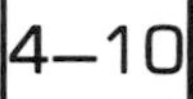

4–10 THE VERSATILE NOR GATE

Discussion

This section on the NOR gate is similar to the last section on the NAND gate. Here you will see how to use the NOR gate to implement any logic circuit you choose. Again, this speaks for the popularity of the NOR gate. This versatility is referred to as the universality of the NOR gate. Once you complete this section, you will have a strong practical and theoretical foundation in the fundamental building blocks of all digital systems: the logic gate.

Equivalent Circuits

Figure 4–65 shows how NOR gates can be used to simulate any logic gate you may require. The following discussion will present the construction of each type of equivalent gate illustrated in Figure 4–65. As with the NAND gate, you will see how the equivalent circuit is achieved by the application of two techniques: logic circuit simplification (covered in the last section) and Boolean algebra.

The Inverter

Observe from Figure 4–65 that an inverter can be constructed from a NOR gate with its inputs tied together. This can be expressed as

$$X = \overline{A + A}$$

Since $\overline{A + A} = \overline{A}$, then

$$X = \overline{A}$$

The last statement is the Boolean expression for an inverter. No logic circuit simplification will be used with the inverter because the Boolean algebra is so straight-forward.

The OR Gate

The construction of an OR gate from NOR gates is demonstrated by the gate simplification technique shown in Figure 4–66. The Boolean algebra analysis is demonstrated in Figure 4–67. Observe from the figure that effectively you have a NOR gate followed by an inverter. As shown by the Boolean algebra development, this inversion process results in the logical equivalent of an OR gate.

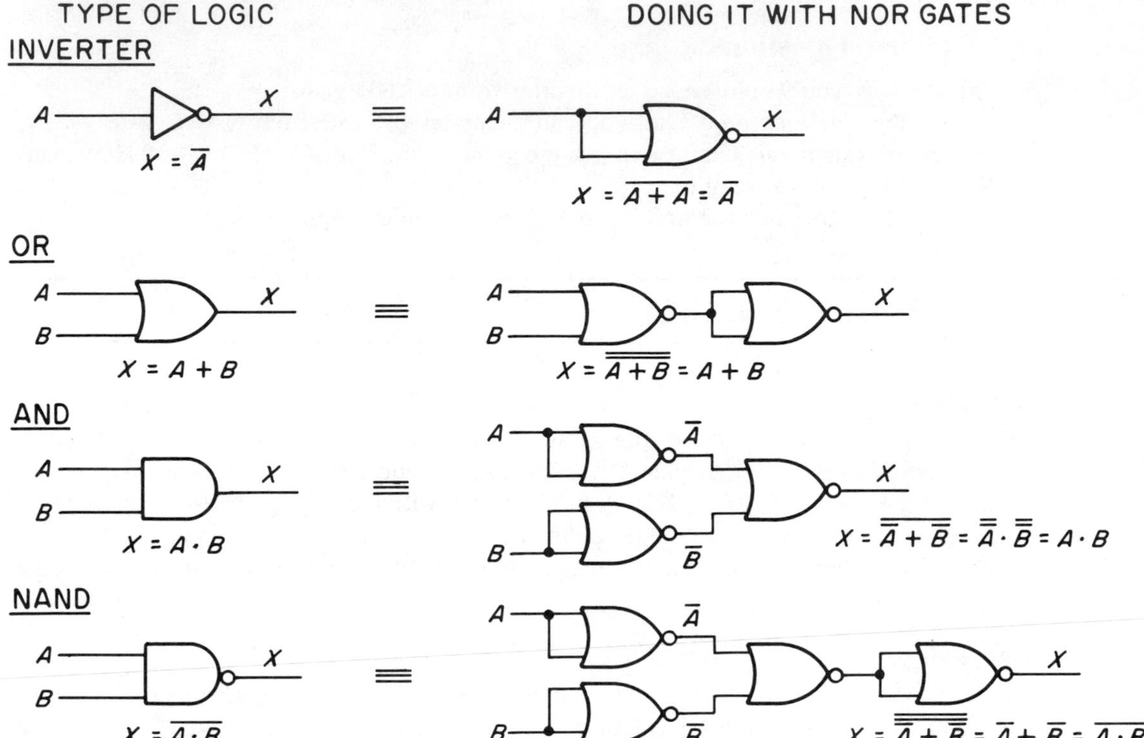

Figure 4–65 Equivalent logic circuits made with NOR gates.

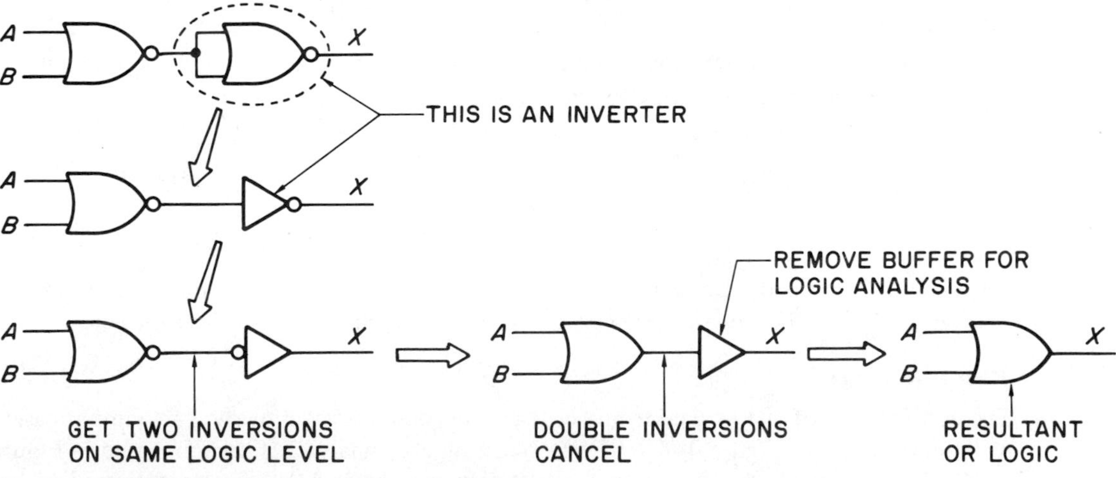

Figure 4–66 OR gate from NOR gates.

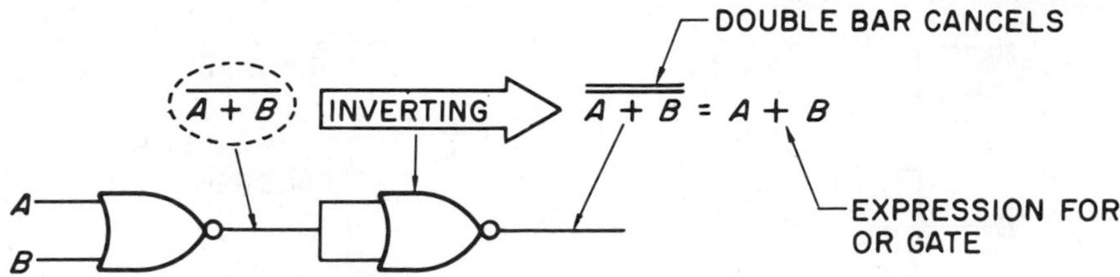

Figure 4–67 Boolean algebra development of OR from NOR.

The AND Gate

The development of the AND gate from three NOR gates is shown in Figure 4-68. As shown in the figure, the reduction of the NOR gates to inverters feeding an equivalent NOR gate circuit (the AND with inverted inputs) results in an AND gate with buffer inputs. However, from a logical analysis standpoint, the buffer circuits can be omitted.

The Boolean algebra analysis is demonstrated in Figure 4-69. Note that De Morgan's theorem was called into action. Effectively you are using De Morgan's theorem when you transfer from a NOR gate symbol to its equivalent AND gate with inverted inputs.

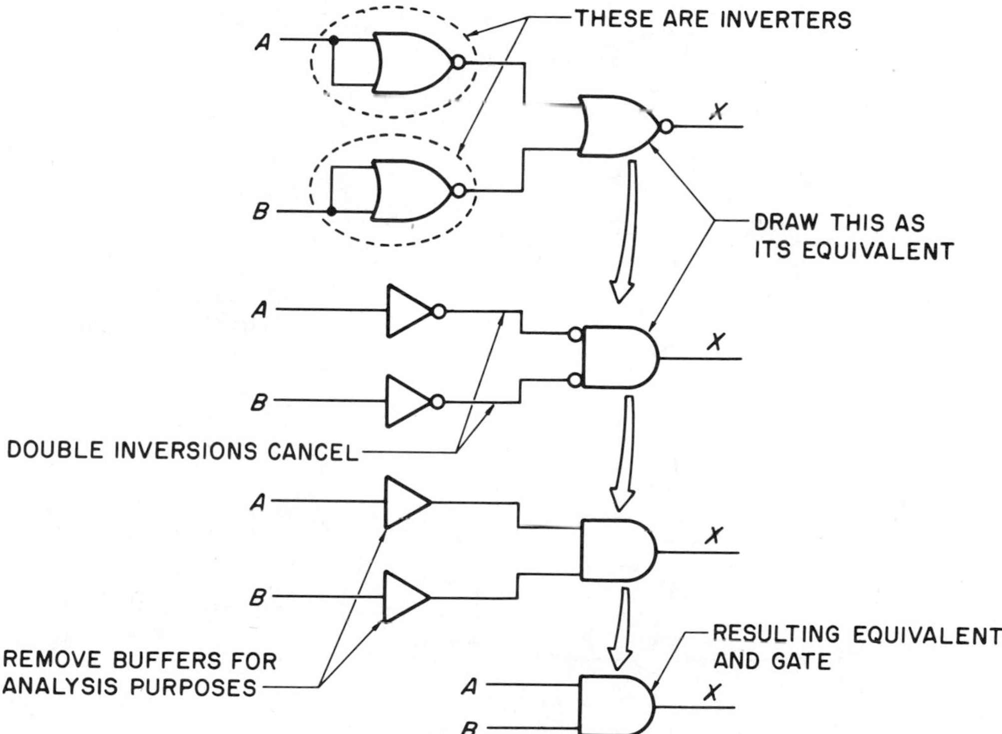

Figure 4–68 AND gate development from NOR gates.

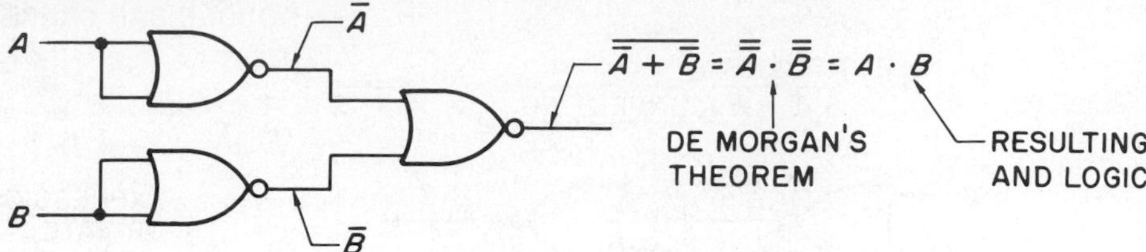

Figure 4–69 Boolean algebra development of AND from NOR.

The NAND Gate

Figure 4–70 shows the development of the NAND gate from NOR gates. Observe that this equivalency uses more NOR gates than do any other of this series. Upon close inspection, you should see that the only difference between the NAND equivalent and the AND equivalent is the addition of another NOR gate connected as an inverter on the output of an AND equivalent circuit. The Boolean development of the NAND gate from NOR gates is shown in Figure 4–71.

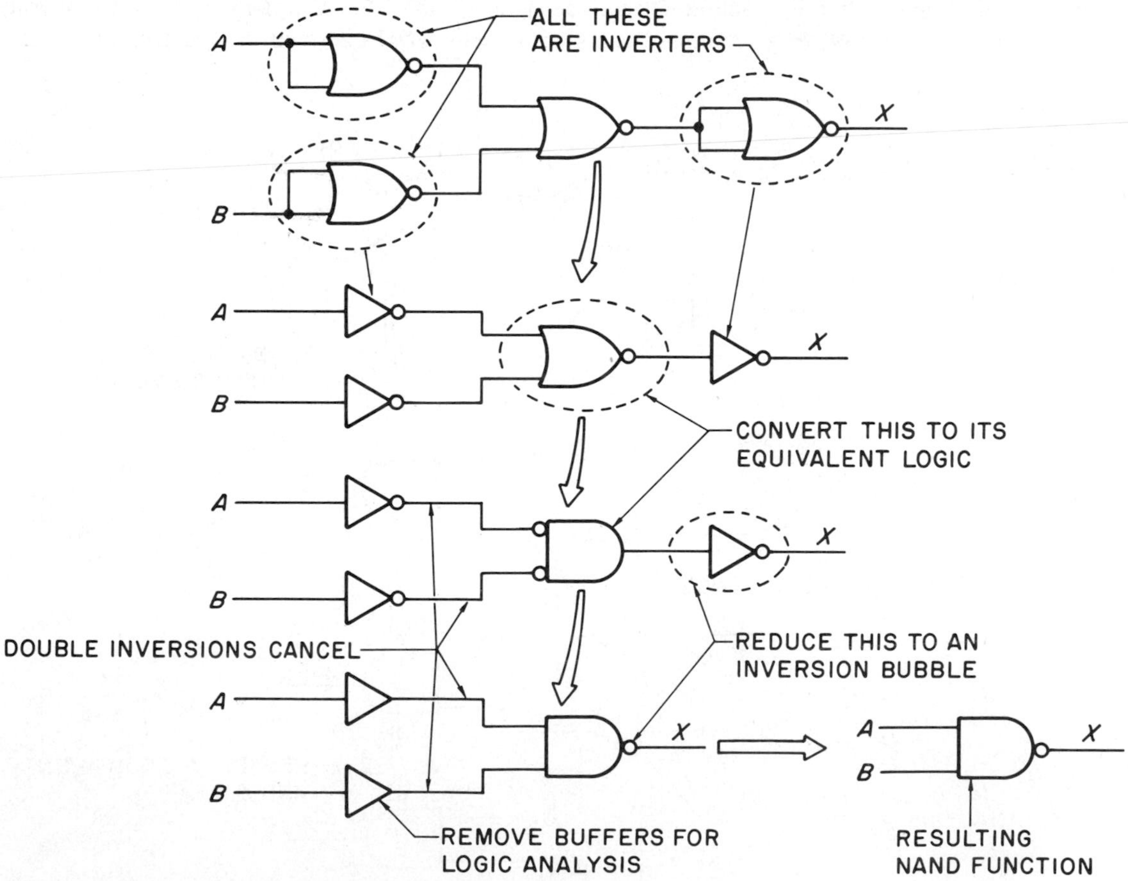

Figure 4–70 NAND gate from NOR gates.

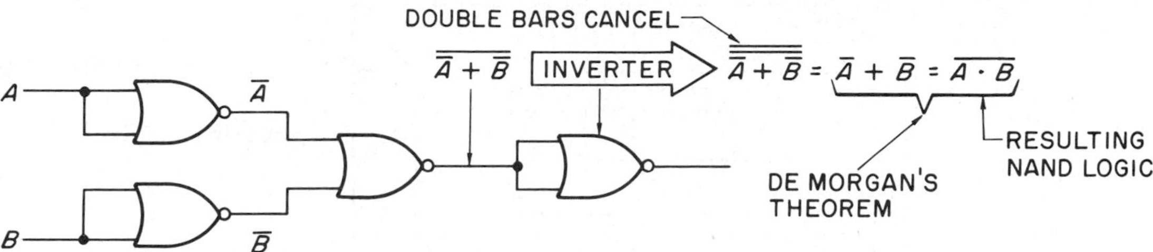

Figure 4–71 Development of NAND gates from NOR gates.

De Morgan's theorem is again used to demonstrate the proof. Note that this circuit would use a whole IC (a quad two-input NOR gate) just to construct one NAND gate. Such construction may be hard to justify economically.

Conclusion

You should now be able to construct any kind of basic logic function if you have a supply of NOR gates and/or NAND gates. This versatility has made almost obsolete the straight forward AND/OR gate logic circuits. On the one hand, the AND/OR logic circuits were easier to analyze, but the economic advantage of NAND/NOR gate design flexibility won the day. Test your understanding of this section by trying the following review questions.

4–10 Review Questions

1. Explain how you would create an inverter from NOR gates.
2. Present the algebraic proof that you can create an AND gate from NOR gates.
3. Which equivalent gate circuit requires the most NOR gates? How many NOR gates does it require?
4. Is there any basic logic circuit that cannot be constructed using NAND/NOR gate logic? Explain?

MICROPROCESSOR APPLICATION

Discussion

Recall the Chapter 3 section on microprocessor applications. You learned what some of the logical operations of a microprocessor actually did. There you learned about the NOT, AND, and OR processes.

In this section, you will be introduced to an important concept about microprocessors. This new concept will represent another major step toward your understanding of real microprocessors used in digital systems.

Another Register and a Black Box

The 4-bit microprocessor presented in the last chapter will be expanded upon here. Figure 4–72 shows the new expansions. As you can see from the figure, the new additions are another 4-bit register called the data register and a **black box** called the *process logic*.

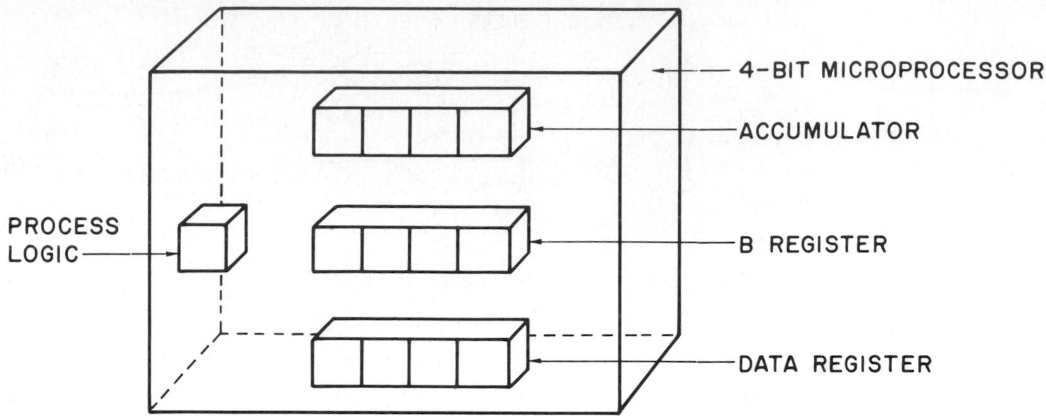

Figure 4–72 Expansions to the 4-bit microprocessor.

Black Box:
A term used in electronics to denote a complex circuit where the circuit details are not necessary for input and output analysis. The importance is placed on what the circuit does as a part of a total system.

Table 4–2 summarizes the function of all of the internal parts of the new microprocessor, including their interconnections.

Table 4–2 FUNCTIONS AND CONNECTIONS OF 4-BIT MICROPROCESSORS

Section Name	What It Does	Connections
4-bit accumulator	Stores the results of logic operations.	Connects to and from B register and data register.
B register	Stores a word that can be used with the accumulator for logic operations.	Connects to and from the accumulator and data B register.
Data register	Stores a word that can be used to control the process logic.	Connects to and from the accumulator, B register, and memory. Also contains connections to the process logic.
Process logic	Is capable of doing the following: 1. Copy bits between registers. 2. Copy bits between memory and data register. 3. Do logic operations with bits in the accumulator and B register. 4. Controls the transfer of data between all internal registers and memory.	Controls the flow of words between the microprocessor registers and between the data register and memory.

Recall from Chapter 3, the B register is unique to the hypothetical microprocessor used in this text. Some microprocessors use a register called the B register with similar functions of the B register used here. As you can see from this table, the addition of the data register and the process logic gives the new 4-bit microprocessor many options. The interconnections between the various sections and memory are illustrated in Figure 4–73. As shown in the figure, the process logic controls the copying of bit patterns (words) between the internal registers of the microprocessor and the data register and memory.

For now, think of the process logic as consisting of many logic gates connected in such a fashion to each other and the other registers that when the ON/OFF switch is first turned ON, the following process *will always take place*.

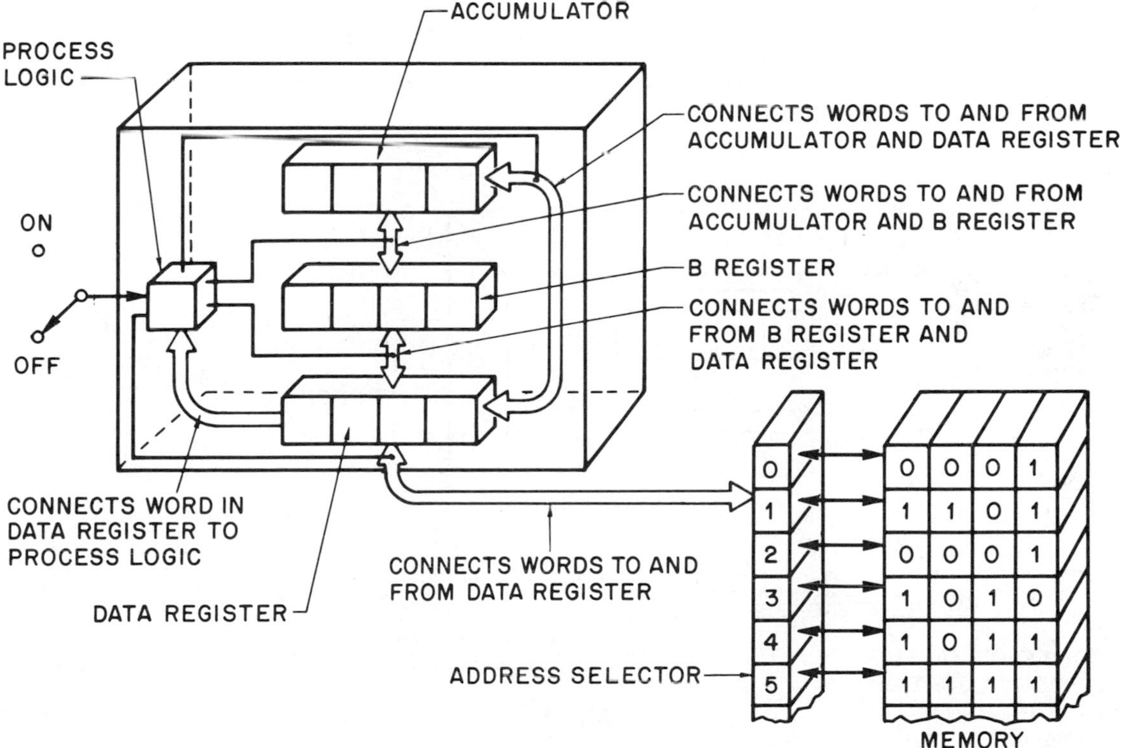

Figure 4–73 Four-bit microprocessor interconnections.

Process for Turning Microprocessor ON	
Event	**Process**
ON/OFF switch first turned ON	1. Bit pattern in first memory location will be copied into the data register. 2. The copied bit pattern in the data register will cause the logic circuits of the process logic to do another process. 3. The new process done by the process logic depends upon the bit pattern in the data register.

Figure 4–74 is a graphic illustration of the first step that happens when the microprocessor is first turned ON. The very next thing that will happen is that the bit pattern contained in the data register (which was copied from the very first memory location) will cause the process logic to do another process. What will the process be? It will depend upon two things: the actual bit pattern now in the data register, and how the logic circuits inside the process logic were constructed at the factory. In order for you to know what effect specific bit patterns found in the data register will have on the process logic, you need a list of these effects from the factory that built the microprocessor. Such a list is called an **instruction set.**

Instruction Set:
A listing of the bit patterns that will cause the microprocessor to perform a specific process.

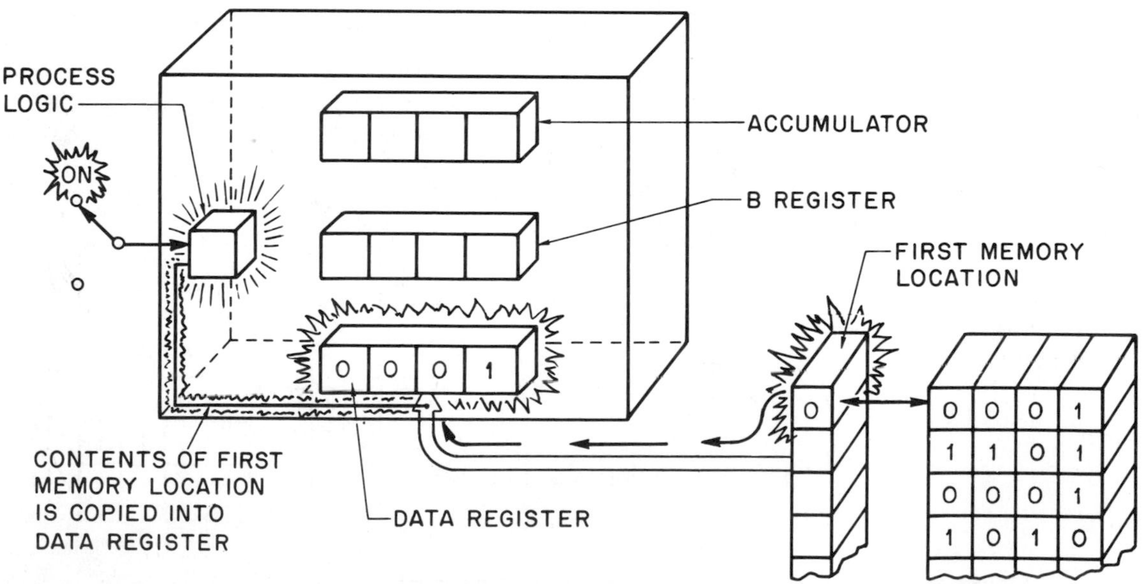

Figure 4–74 First process when first turned ON.

Table 4–3 4-BIT MICROPROCESSOR INSTRUCTION SET

Bit Pattern in Data Register	Hex Code	Causes the Process Logic to Perform the Following Processes
0 0 0 0	0_{16}	Copy the contents of the next memory location into the data register.
0 0 0 1	1_{16}	Copy the contents of the next memory location into the data register. Copy the contents of the data register into the accumulator.
0 0 1 0	2_{16}	Copy the contents of the next memory location into the data register. Copy the contents of the data register into the B register.
0 0 1 1	3_{16}	Copy the contents of the B register into the accumulator.
0 1 0 0	4_{16}	Copy the contents of the accumulator into the B register.
0 1 0 1	5_{16}	NOT the contents of the accumulator.
0 1 1 0	6_{16}	AND the contents of the accumulator with the B register and store the result in the accumulator.
0 1 1 1	7_{16}	OR the contents of the accumulator with the B register and store the result in the accumulator.
1 0 0 0	8_{16}	XOR the contents of the accumulator with the B register and store the result in the accumulator.
1 0 0 1	9_{16}	Copy the contents of the accumulator into the data register. Copy the contents of the data register into the next memory location.
1 0 1 0	A_{16}	Copy the contents of the B register into the data register. Copy the contents of the data register into the next memory location.
1 0 1 1	B_{16}	Increase the value of the accumulator by 1.
1 1 0 0	C_{16}	Decrease the value of the accumulator by 1.
1 1 0 1	D_{16}	Add the contents of the B register to the accumulator and store the results in the accumulator.
1 1 1 0	E_{16}	Set all the bits in the accumulator to 0.
1 1 1 1	F_{16}	Stop all further processing.

For the educational model 4-bit microprocessor used here, the instruction set is given in Table 4–3.

The significance of Table 4–3 is illustrated in Figure 4-75. As you can see, the bit pattern copied into the data register when the microprocessor is first turned ON determines what process the process logic will cause to happen next. For example, if the word in memory location 0 (the first memory location) were 0001, then the process logic would cause the word in the next memory location (address 1) to be copied into the data register

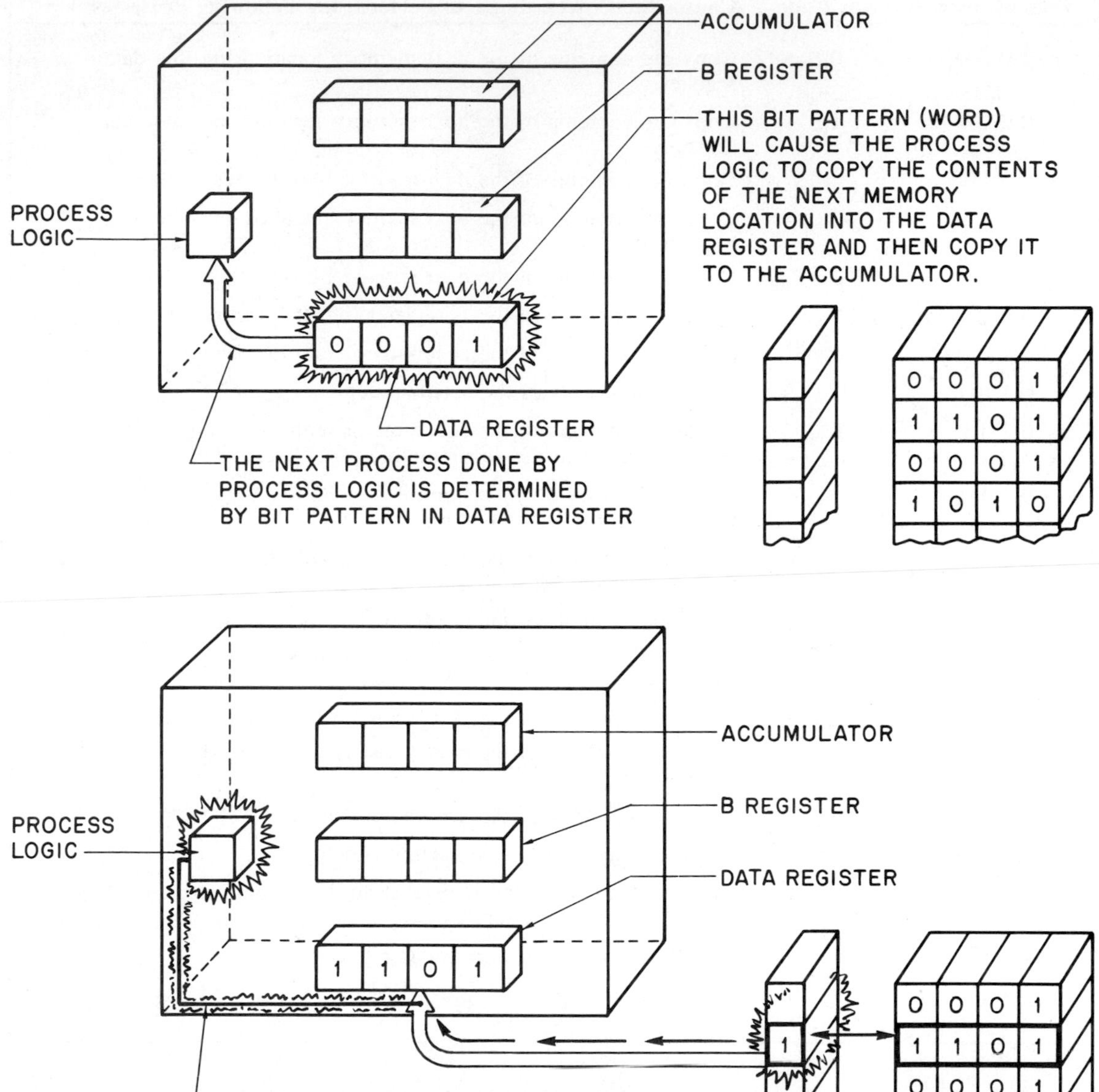

Figure 4–75 Example process with 4-bit microprocessor.

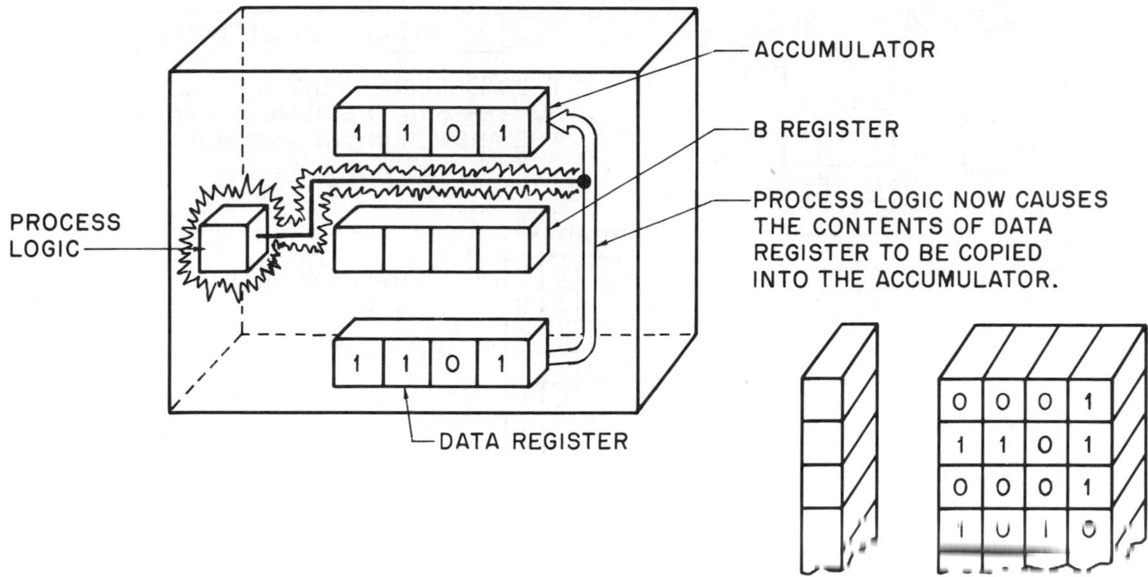

Figure 4–75 Continued

and then to be copied from the data register into the accumulator. This is what Table 4–2 predicts will happen (the microprocessor instruction set).

What happens next? What does the process logic cause the microprocessor to do when it is finished with a process from the instruction set? It always does the following:

> Go to the next memory location and copy its contents into the data register. This new bit pattern will not be another instruction.

GET and DO (FETCH and EXECUTE)

The process logic causes the microprocessor to do a continuous process of GET and DO. This is referred to as FETCH and EXECUTE. This means that it first GETs (FETCHES) a bit pattern from memory, then it DOes (EXECUTES) what the bit pattern indicates from the instruction set built into it at the factory. When finished with the process indicated by the bit pattern it got from memory, the process logic causes another FETCH and EXECUTE. That is, the data register now FETCHES the bit pattern from the *next* location in memory and EXECUTES what this new bit pattern indicates from the instruction set built into it at the factory. The only time this FETCH and EXECUTE process stops is when the bit pattern 1111 (meaning halt all processing) appears as an instruction in the data register.

Look at the sequence of processes shown in Figure 4–76. You will observe a microprocessor doing a program. The set of bit patterns in the memory is called a *program*. The microprocessor gets its name from the fact that it can do many small (micro) predictable processes. The order and number of times these processes are done depends upon the bit patterns stored in memory. That's all there is to the basic concept behind all microprocessors. The differences are that for a 4-bit microprocessor only $2^4 = 16$ different

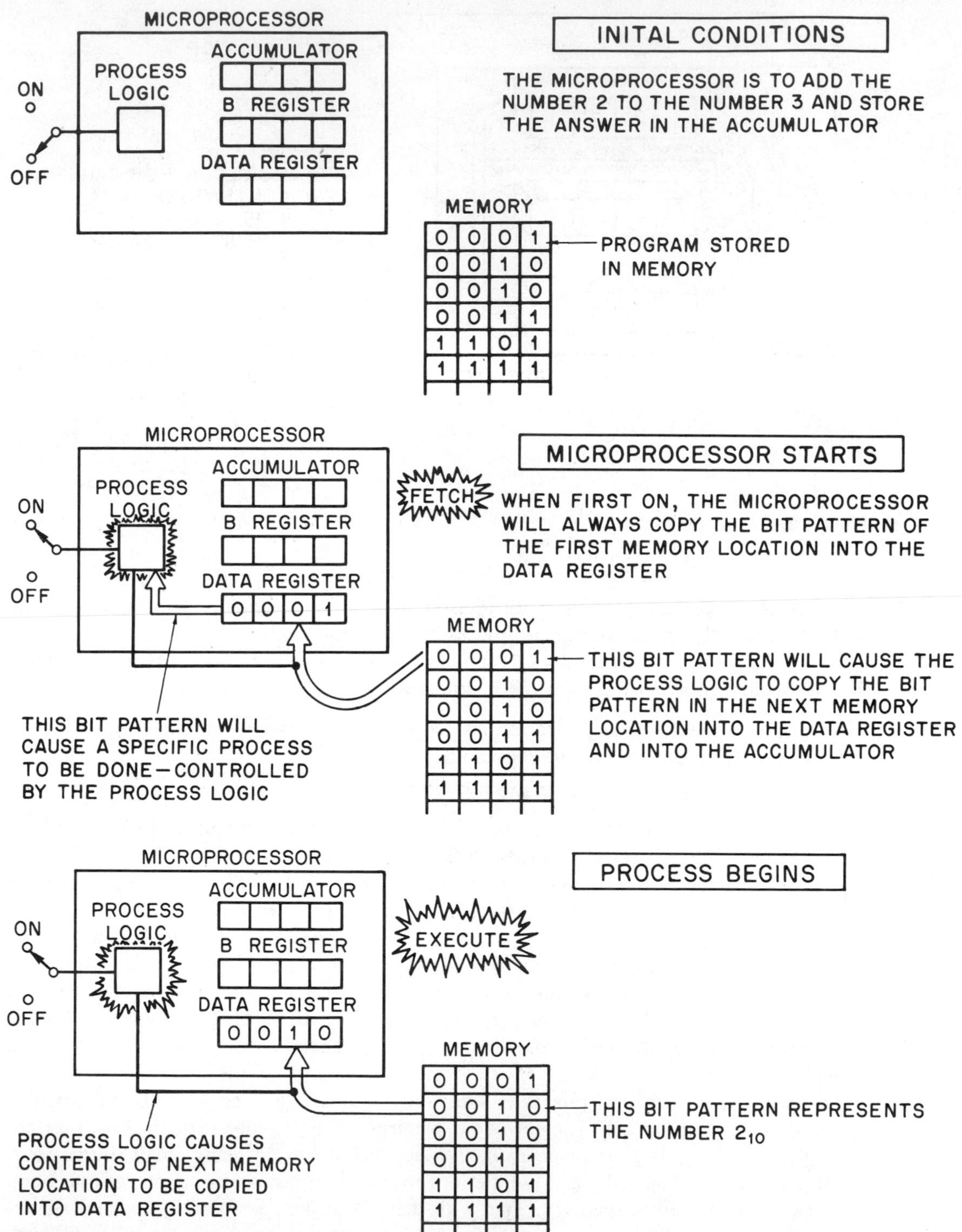

Figure 4–76 A typical set of processes.

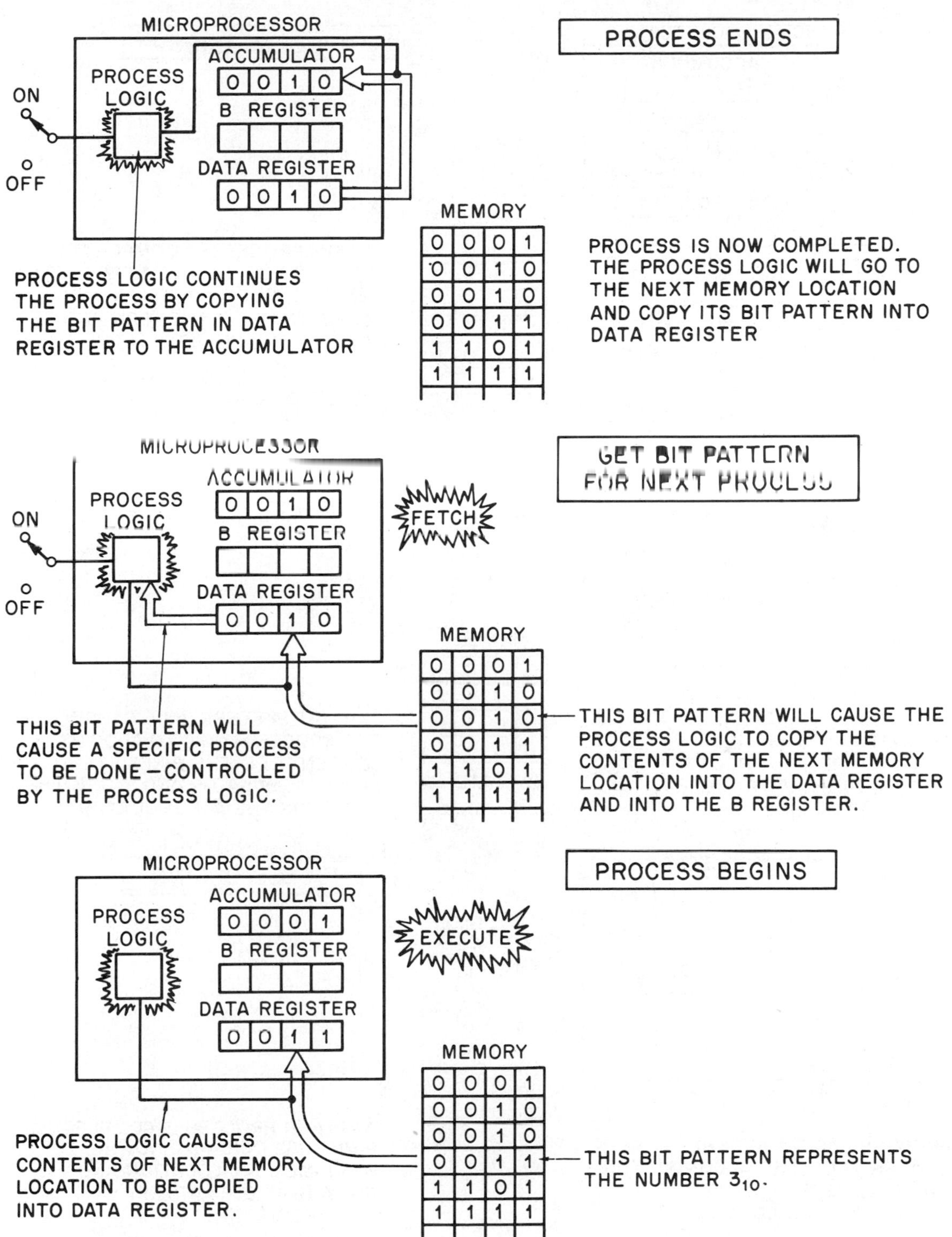

Figure 4–76 Continued

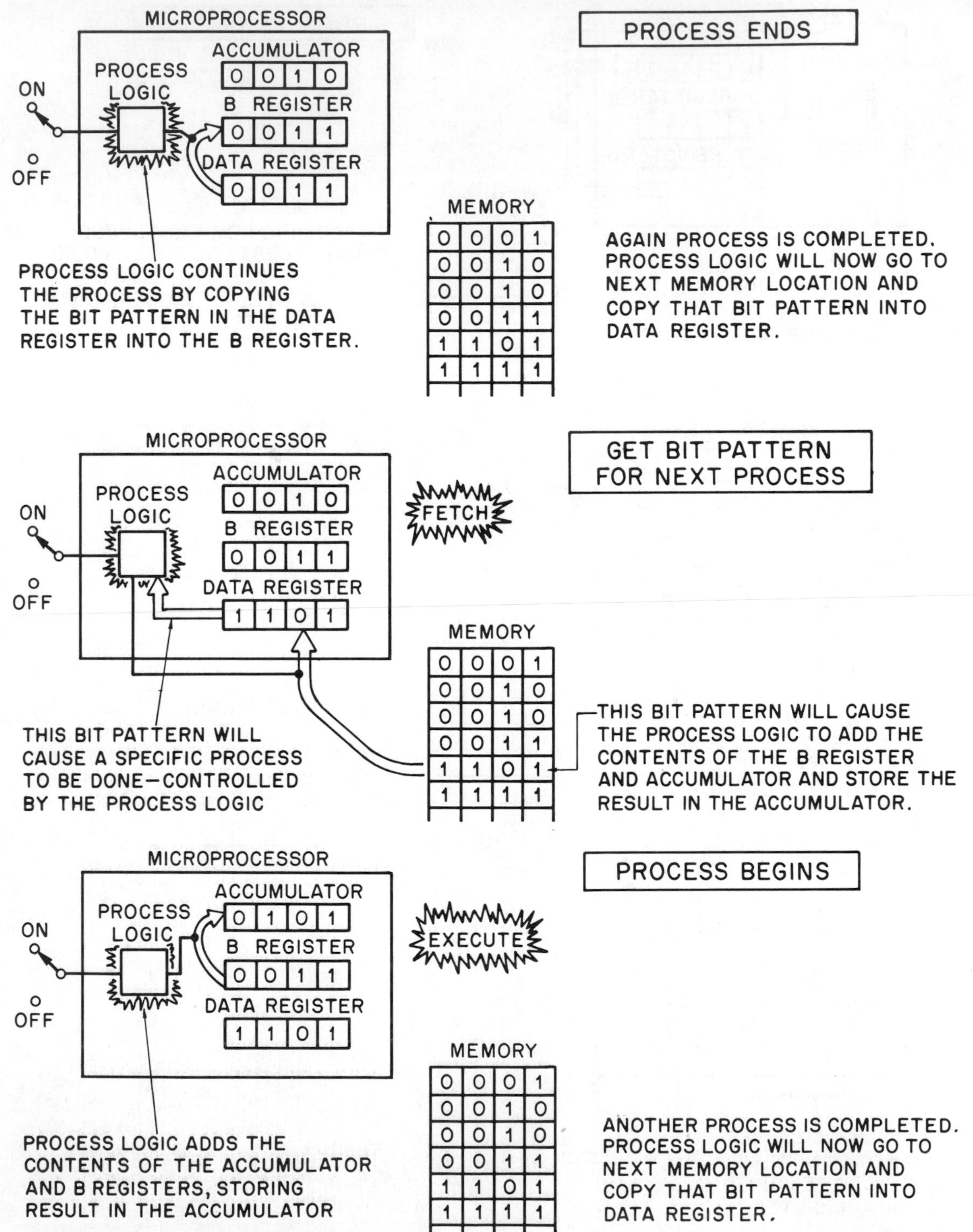

Figure 4–76 Continued

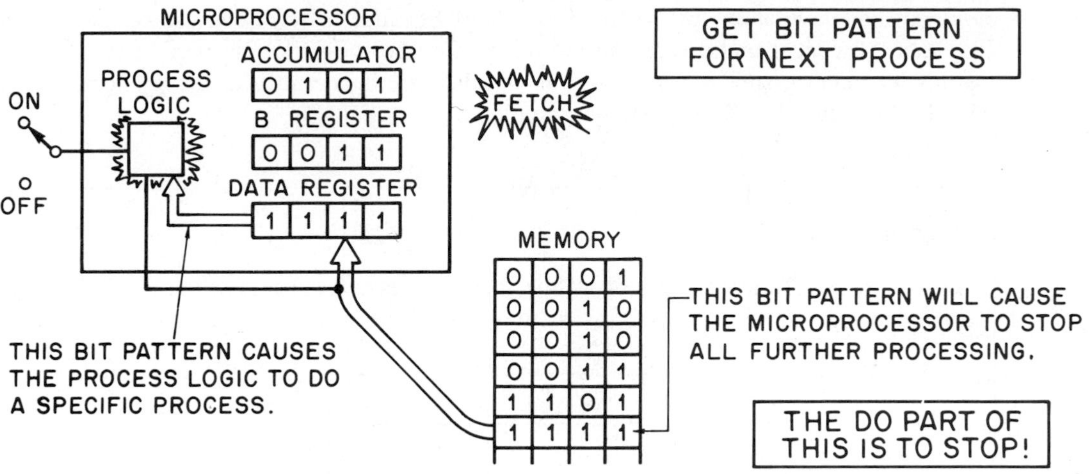

Figure 4–76 Continued

processes can be performed. For an 8-bit microprocessor, $2^8 = 256$ processes are possible, and for a 16-bit microprocessor, $2^{16} = 65\,536$ processes are possible!

Stored Program

What you have just witnessed is called the **stored program concept.** This means that the *instructions* are stored in memory right along with the data.

Stored Program Concept:
The concept used for programming digital systems where the instructions are included in memory with the data.

In this example, the bit patterns are represented by hex numbers 1_{16}, 2_{16} ... However, because many computers do not accept subscripts, the capital letter H is placed after the number to indicated hex numbers, for example, 1H, 2H, 3H ... As you could see from the example program, there is a difference between an instruction and data. An **instruction** is a bit pattern that causes the microprocessor to DO a specified process. **Data** is a bit pattern that something happens to within the process.

Instruction:
A bit pattern that causes the microprocessor to perform a predictable process.

Data:
A bit pattern that is used in a process.

You may be asking the question: "How does the microprocessor know the difference between an *instruction* and *data*?" (If you're not asking the question to yourself, maybe you should be.) The answer is, *it doesn't*! When it is first turned ON, the microprocessor *always takes the first bit pattern it sees to be an instruction*. This means, that *you* must make sure that the very first bit pattern in memory—for our simple 4-bit machine—is an instruction of something you want it to do. Every time the microprocessor completes a process, it goes to the next memory location and treats the bit pattern there as an *instruc-*

tion, no matter what the bit pattern looks like! Because instructions use bit patterns just like data, you must make sure that the microprocessor does not get them mixed up. You do this by making sure that instructions and data are in memory in the proper sequence. This is illustrated by the following examples.

Example 4–20

Determine the final contents of all the microprocessor registers when the following program is executed (indicate instructions and data):

Address	Contents
0	2_{16}
1	5_{16}
2	1_{16}
3	A_{16}
4	6_{16}
5	F_{16}
6	5_{16}

Solution

Start with the assumption that the microprocessor will be turned ON and copy the contents of the first memory location (address 0) into the data register. Since this is the FETCH part of the FETCH-EXECUTE process, the contents of this memory location will be used as an instruction. By referring to the instruction set of the 4-bit microprocessor (Table 4–3), you have $2_{16} = 0010_2$, which causes the contents of the next memory location (address 1) to be copied into the data register and then to be copied from the data register into the B register. This is the same effect as if the contents of the next memory location were simply copied into the B register. This process continues and is best understood by commenting on the previous program as follows:

Address		Contents	Comments	Type
FETCH	0	2_{16}	Copy the contents of the next memory location into the B register	Instruction
EXECUTE	1	5_{16}	This bit pattern will now be copied in the B register (0101_2)	Data
FETCH	2	1_{16}	Copy the contents of the next memory location into the accumulator	Instruction

EXECUTE	3	A_{16}	This bit pattern will now be copied into the accumulator (1010_2)	Data
FETCH/EXECUTE	4	6_{16}	AND the contents of the B register with the accumulator, store the results in the accumulator (0101 AND 1010 = 0000)	Instruction
FETCH/EXECUTE	5	F_{16}	Stop all further processing	Instruction
	6	5_{16}	Ignored	

The final contents will be
Accumulator → $0000_2 = 0_{16}$
B register → $0101_2 = 5_{16}$
Data register → 1111_2 F_{16}

Note that in order for you to understand what processes would happen, the instruction set for the microprocessor was needed. The program didn't do much more than make the contents of the registers predictable based upon the program in memory. For the simple 4-bit microprocessor, the important point is that if you are given a bit pattern in memory and the instruction set for the microprocessor using that memory, you can predict what the microprocessor will do.

In the next example, the program causes the microprocessor to change some memory.

Example 4–21

Determine the final contents of all the microprocessor registers and memory when the following program is executed (indicate instructions and data):

Address	Contents
0	1_{16}
1	8_{16}
2	2_{16}
3	4_{16}
4	9_{16}
5	0_{16}
6	F_{16}

Solution

Again, start with memory location 0 and refer to the instruction set. Then comment on what the program will actually do.

Address		Contents	Comments	Type
FETCH	0	1_{16}	Copy the contents of the next memory location into the accumulator	Instruction
EXECUTE	1	8_{16}	This bit pattern will now be copied in the accumulator (1000_2)	Data
FETCH	2	2_{16}	Copy the contents of the next memory location into the B register	Instruction
EXECUTE	3	4_{16}	This bit pattern will now be copied into the B register (0100_2)	Data
FETCH	4	9_{16}	Copy the contents of the accumulator into the next memory location	Instruction
EXECUTE	5	0_{16}	The contents of this memory will be changed to a copy of what is now in the accumulator (1000_2)	Data
FETCH/EXECUTE	6	F_{16}	Stop all further processing	Data

The final contents will be
Accumulator → $1000_2 = 8_{16}$
B register → $0100_2 = 4_{16}$
Data register → $1111_2 = F_{16}$
Memory:

Address	Contents	
0	1_{16}	
1	8_{16}	
2	2_{16}	
3	4_{16}	
4	9_{16}	
5	8_{16}	← (This has been changed.)
6	F_{16}	

Conclusion

In this section, you were introduced to an important concept concerning what a microprocessor is and what it does. In the microprocessor applications section of the next chapter, you will have more practice with this concept. For now realize that a microprocessor merely does predetermined small processes based upon bit patterns stored in memory. In order to know what these processes are, you must have a copy of the instruction set for that microprocessor. Test your understanding of this section by trying the following review questions.

4-11 Review Questions

1. State how a microprocessor got its name.
2. Explain the purpose of an instruction set for a microprocessor.
3. Describe the stored program concept.
4. What is the first thing a microprocessor does when turned ON?
5. What is the difference between an instruction and data? Do they appear differently in memory? Explain.

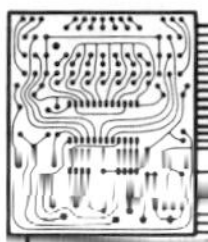

ELECTRONIC APPLICATION

Discussion

When troubleshooting digital ICs, you need to know something about the real voltage values you should be getting with logic ICs. Up to this point, you have been told that the voltage value for a HIGH state is +5 volts, and the voltage value for a LOW state is 0 volts. In actual circuits, these values are not achieved for reasons to be presented in this section. Here you will learn about the actual voltage levels you will encounter in the construction, troubleshooting, and servicing of digital systems.

Inverter Output Analysis

Assume you had an inverter, and you were measuring the voltage output under the conditions shown in Figure 4–77. In Figure 4–77 (A), an ideal voltage reading would be 0 volts, because with a HIGH input you would get a LOW output from the inverter. However, because of slight variations in the internal construction of the integrated circuits and because of the voltage and current demands of the circuits themselves, these are not the actual values you will read. The same is true when reading the HIGH output condition of

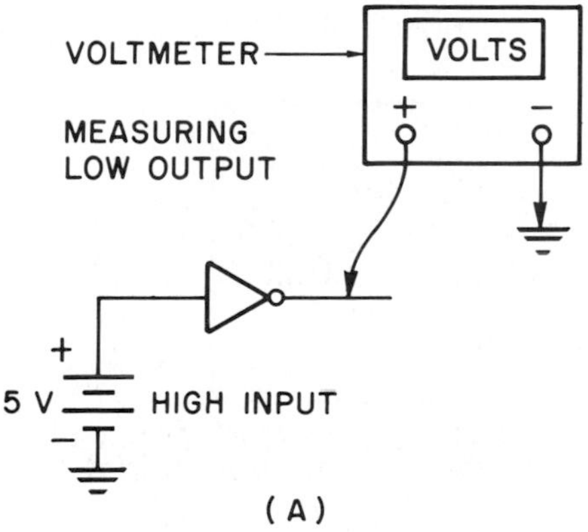

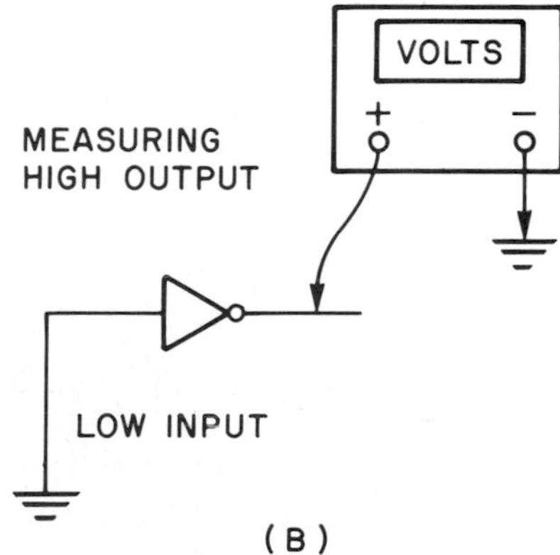

Figure 4–77 Measuring output voltage logic levels.

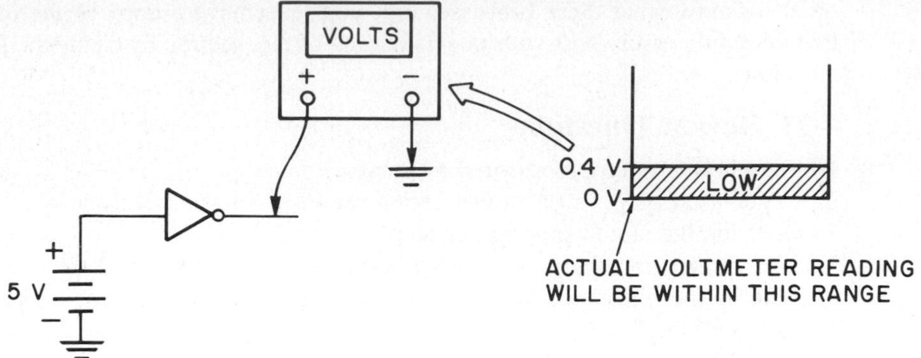

Figure 4–78 Output voltage range for a TTL LOW.

Figure 4–77 (B). Thus, manufacturers of digital ICs specify what they call **worse-case conditions.**

Worse-Case Conditions:
Circuit parameters (voltage, current, etc.) are measured under the worst conditions of voltage and temperature.

What the manufacturer does guarantee is the worst-case output voltage for a LOW condition and a HIGH condition. For the TTL inverter example, it is 0.4 volts. This is shown in Figure 4–78.

This means if you read the output of this TTL inverter when it is LOW, any reading between 0 volts and 0.4 volts would be considered OK. By the same token, manufacturers guarantee a *range* of voltages for a TTL output HIGH. This is illustrated in Figure 4–79. What this means is if you read the output voltage of this TTL inverter when its output is HIGH, any voltage reading from +2.4 volts to +5 volts is considered OK. These important facts can be represented by a diagram that digital IC manufacturers refer to as the **output profile** of an IC.

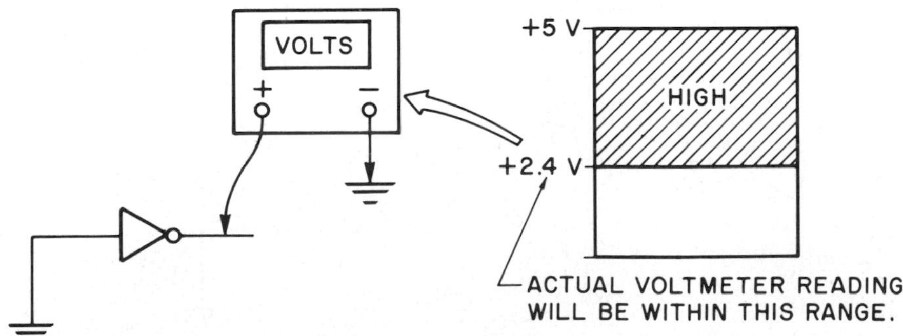

Figure 4–79 Output voltage range for a TTL HIGH.

Output Profile:
The specific range of logic HIGH and logic LOW output voltages for a l device.

The output profile diagram for the TTL inverter used in this discussion is shown in Figure 4–80. Observe from Figure 4–80 that there is a range of output voltages (from 0.4 volts to 2.4 volts) called the **indeterminate range.** For comparison, note the output profile of a typical CMOS.

Indeterminate Range
A range of voltage values for a digital IC that indicates the circuit is not functioning according to the manufacturer's specifications.

What this means is, if you read an output voltage from the TTL IC that is *inside* the indeterminate range, then you have a circuit problem.

Circuit Notation

The notations V_{OH} and V_{OL} are standard notations used in digital IC specification sheets. V_{OH} means "voltage output HIGH," and V_{OL} means "voltage output LOW."

Inverter Input Voltage

For practical design and troubleshooting, you must also know what to expect when measuring the input voltages of a TTL digital IC. Figure 4–81 shows measurements being conducted of the input voltage levels for the example inverter.

Recall that for an inverter it takes a HIGH input for the output to be LOW, and a LOW input for the output to be HIGH. What you need to know as a technician is *how much below* 5 volts can you decrease the input and still have a LOW on the output. Again,

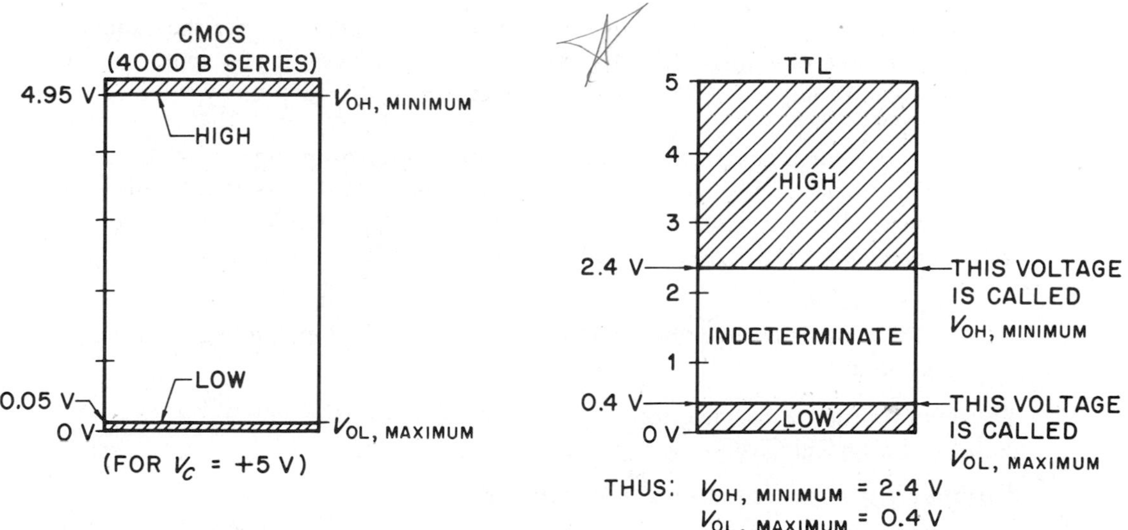

Figure 4–80 Output profile for TTL and CMOS inverters.

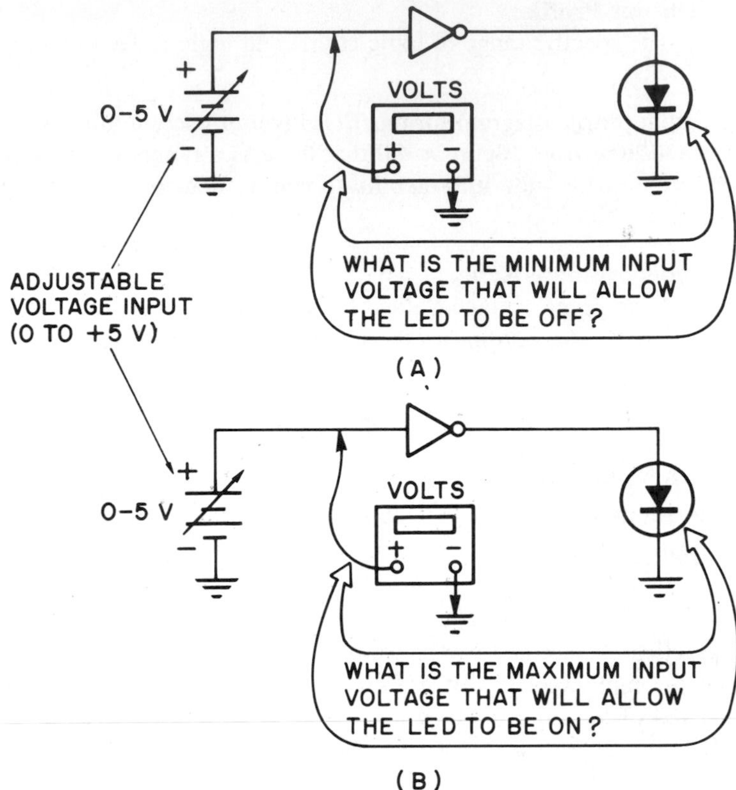

Figure 4–81 Measuring TTL input HIGH and input LOW.

manufacturers will specify this value. The actual expected range for the example TTL inverter is shown in Figure 4–82. What this actually means is, if you apply a voltage for a HIGH input of anywhere between +2 volts and +5 volts, then this will be accepted as a logic HIGH.

The range for a guaranteed logic LOW input is shown in Figure 4–83. This means that if you actually apply an input voltage anywhere from 0 volts to 0.8 volts, then the manufacturer guarantees that this will be accepted as a logic LOW. The range of readings for the input of a digital circuit is called the **input profile.**

Input Profile:
The specified range of logic HIGH and logic LOW input voltages for a digital circuit.

The input profile for the inverter discussed here is shown in Figure 4–84. Note the comparison of a typical CMOS input profile.

Comparing Input and Output Profiles

You may have been wondering why the input and output profiles for the TTL inverter are different. The reason is best understood by first considering the problems you would have

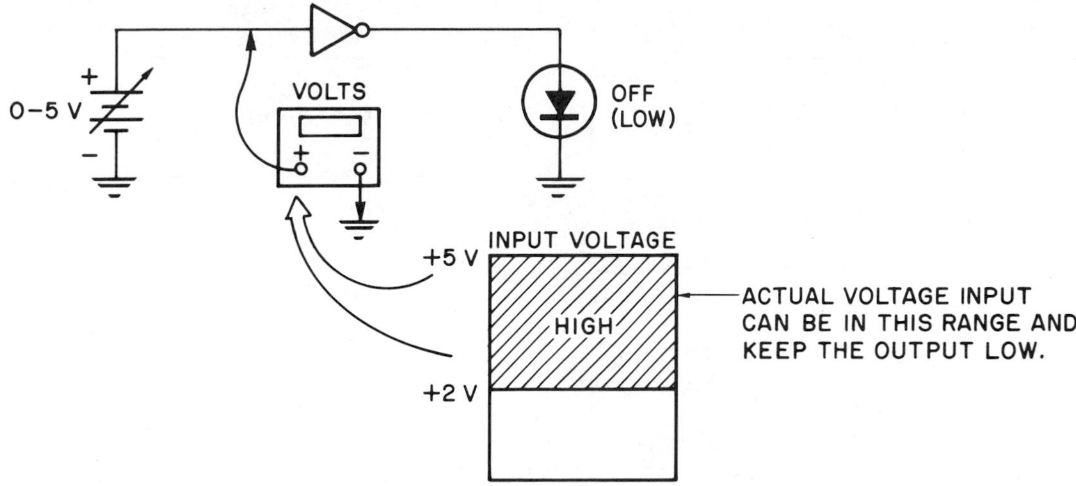

Figure 4–82 Actual input HIGH voltage range for TTL inverter.

if both input and output profiles were identical. Consider the situation of Figure 4–85. Assume that the maximum *output* voltage for a logic LOW is 0.4 volts ($V_{OL,max} = 0.4$ V), and that the maximum *input* voltage for a logic LOW is also 0.4 volts ($V_{IL,max} = 0.4$ V). If this were the case, then, as shown in the figure, it would be possible for the output of the first inverter to be as much as 0.4 V, for a logical LOW.

Under these conditions, there is a danger of the second inverter changing its logic level due to the effects of **electrical noise.**

Electrical Noise:
Undesirable and unpredictable electrical signals caused by natural phenomena and manufactured equipment.

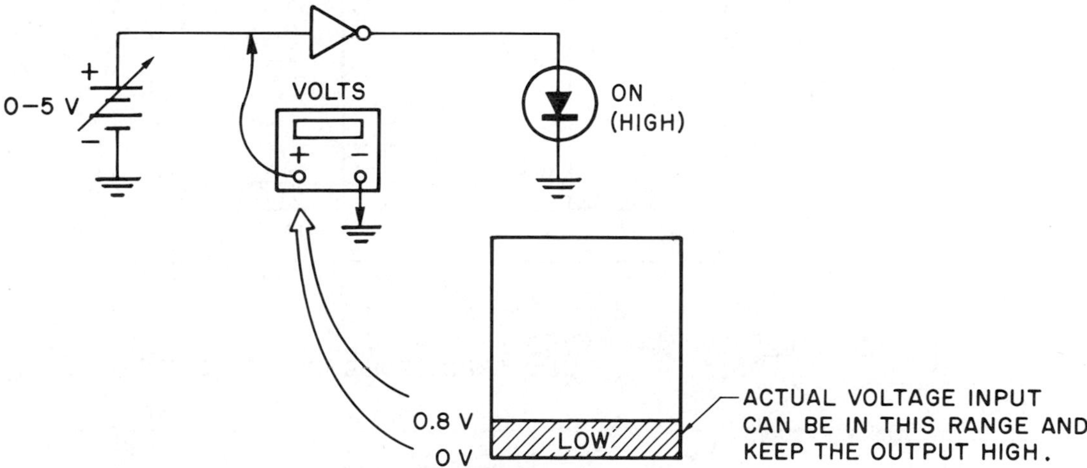

Figure 4–83 Actual input LOW voltage range for TTL inverter.

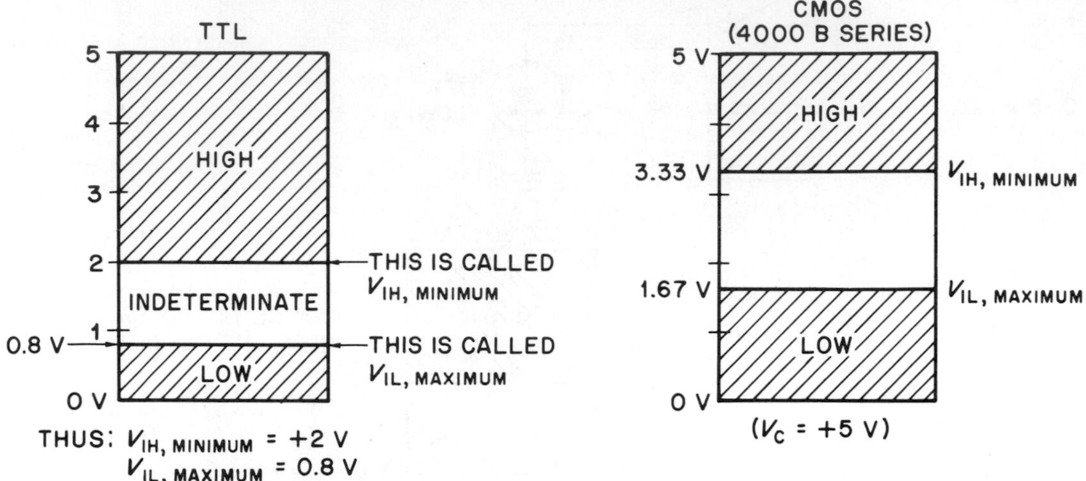

Figure 4–84 Input profile for TTL and CMOS inverters.

Electrical wires, such as used for connections on printed circuit boards, act as receiving antennas of electrical noise, where they are converted to small electrical signals that can easily mix with the digital signals in the digital system. As shown in Figure 4–85, if the input and output profiles were to be identical, then the *slightest* increase in voltage due to noise, above the 0.4 volts, would cause the input voltage to the second inverter to be

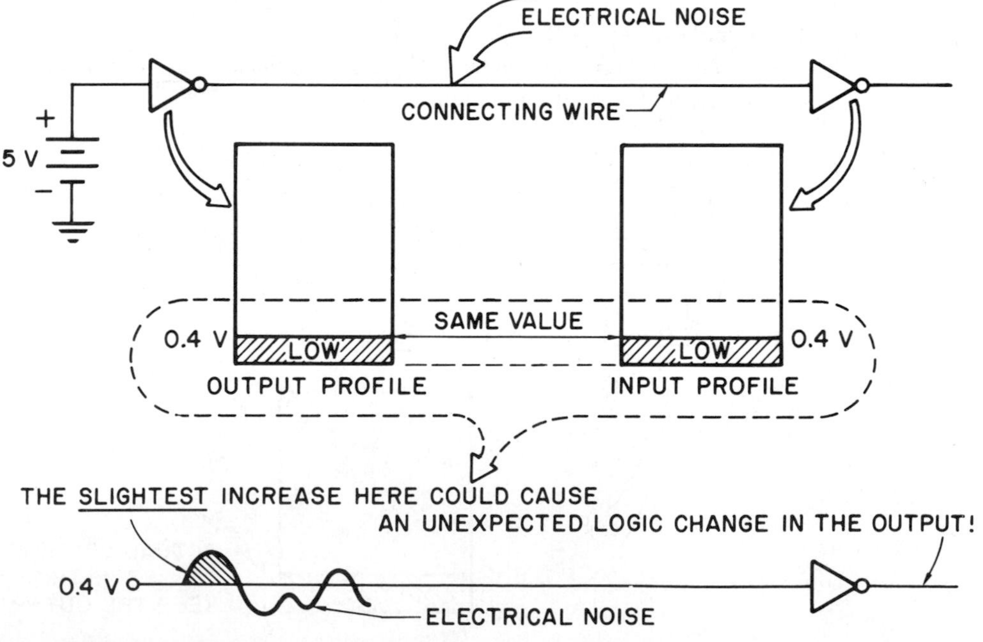

Figure 4–85 Undesirability of identical profiles for logical LOW.

pushed up into the indeterminate region. It is in this region that the manufacturer does not guarantee the effect on the output logic level. The output could be HIGH or LOW or even change rapidly between these two levels when the input voltage is in the indeterminate region.

The same type of problem is encountered if the profiles were identical and the output of the first inverter was at a logic HIGH. This condition is illustrated in Figure 4–86. Again, as the figure shows, there is the possibility for the second inverter to unexpectedly change its logic level due to unpredictable electrical noise.

The problem in both cases is that there is no margin for voltage level changes *caused by noise*. What is needed is a method of minimizing the effects of electrical noise in digital systems, not only for TTL inverters, but also for all other logic gates (AND, OR, NAND, etc). Their input and output profiles must allow for a margin of noise that will not cause an unexpected change in any logic level.

Noise Margin

Consider the two inverters in Figure 4–87. As shown, having an output profile different from an input profile can reduce the possibility of unexpected changes in logic levels due to electrical noise. As you can see from the figure, because the guaranteed maximum *output* voltage for a logic LOW is 0.4 volts *and* the guaranteed maximum *input* voltage that will *still be accepted* as a logic LOW is 0.8 volts, this creates a safety margin.

This same safety margin is achieved for a logic HIGH as shown in Figure 4–88. Again, you can see that a margin of safety has been created to greatly reduce the chances

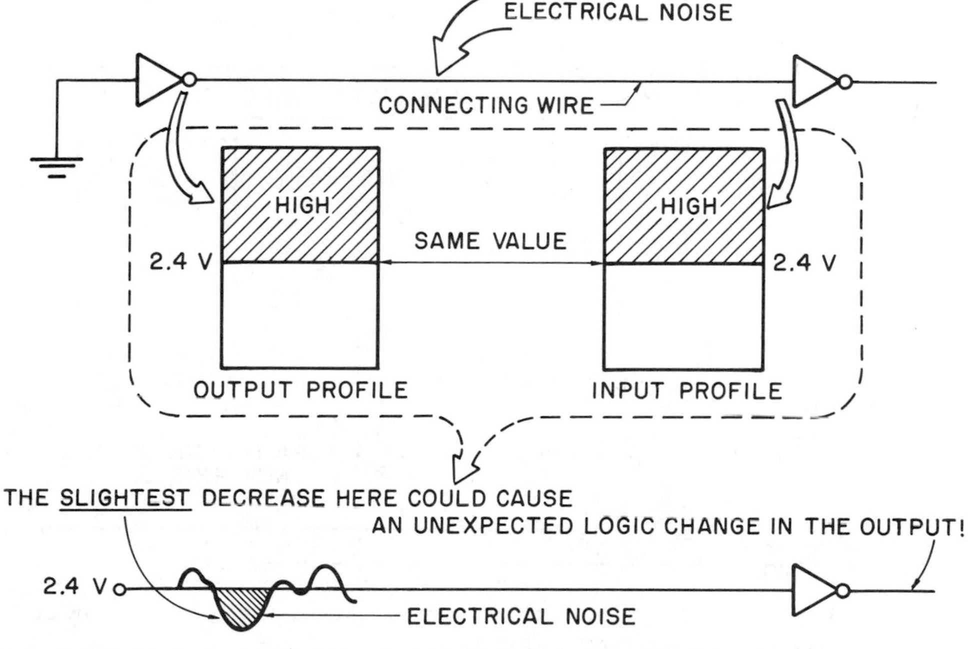

Figure 4–86 Undesirability of identical profiles for a logical HIGH.

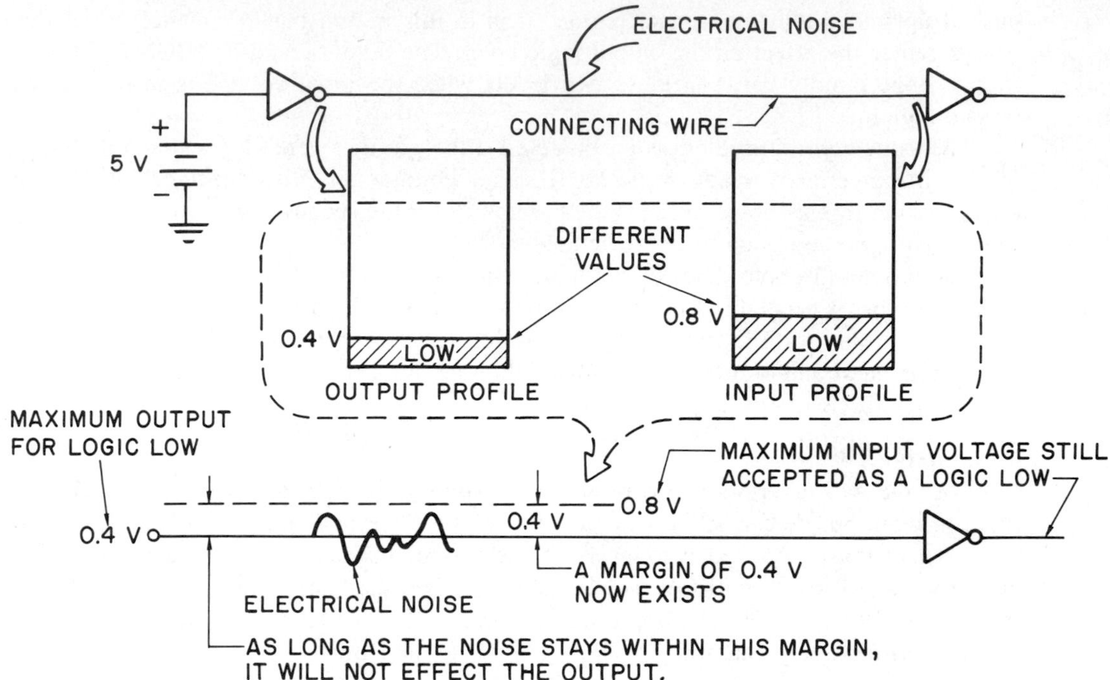

Figure 4–87 Desirability of different profiles for a logic LOW.

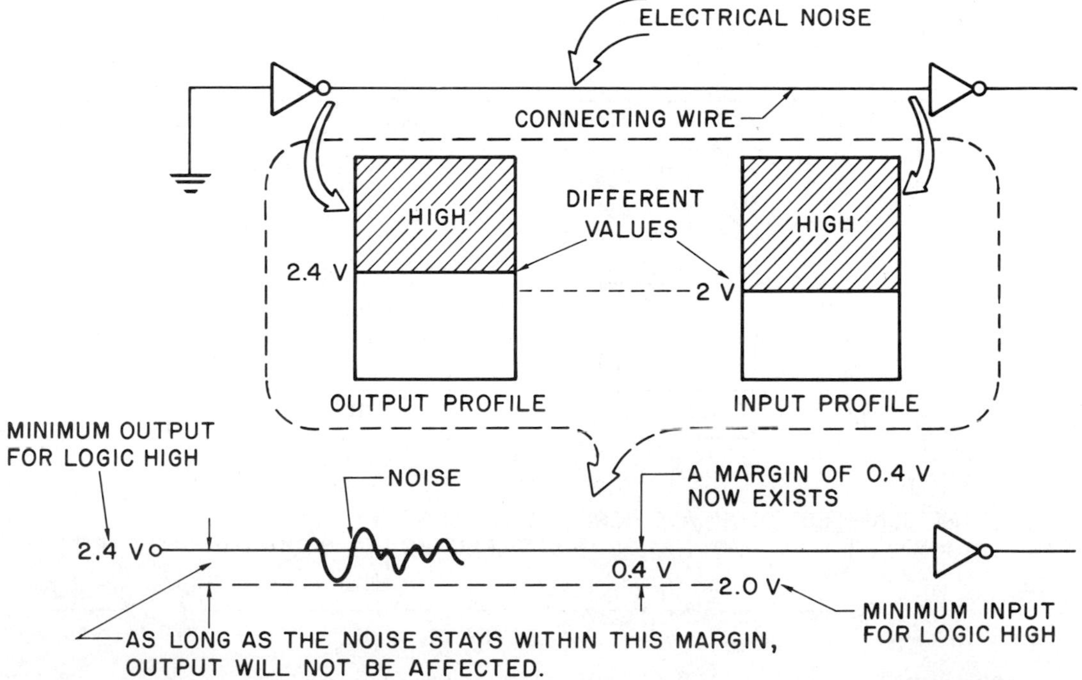

Figure 4–88 Desirability of different profiles for a logic HIGH.

of an unexpected change in logic level due to noise. This is achieved because the guaranteed lowest *output* voltage for a logic HIGH ($V_{OH,min}$) is 2.4 volts, whereas the guaranteed lowest *input* voltage for a logic HIGH ($V_{IH,min}$) is 2 volts. Thus, again, a safety margin of 0.4 volts has been achieved.

Both of these safety margins are referred to as the **noise margin.**

Noise Margin:
The voltage difference between the input and output profiles of a logic circuit.

The noise margin is expressed mathematically as

$$V_H = V_{OH,min} - V_{IH,min}$$
$$V_L = V_{OL,max} - V_{IL,max}$$

where V_H = logic HIGH noise margin
V_L = logic LOW noise margin

The noise margin is sometimes referred to as the *noise immunity* of a logic circuit.

Example 4–22

The input/output profiles for a 54HC04 CMOS hex inverter are shown in Figure 4–89. Compute the noise margins for this device.

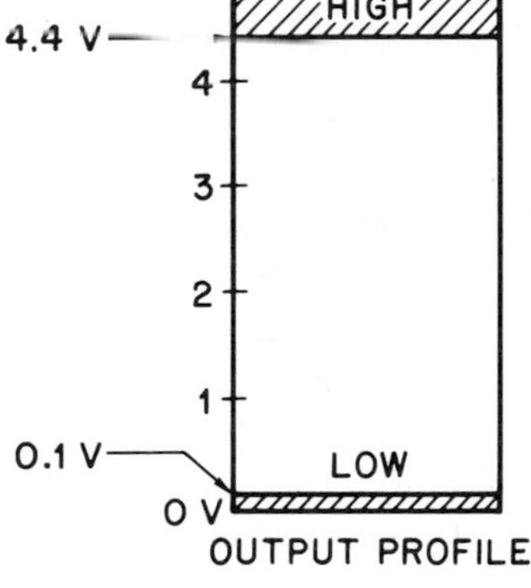

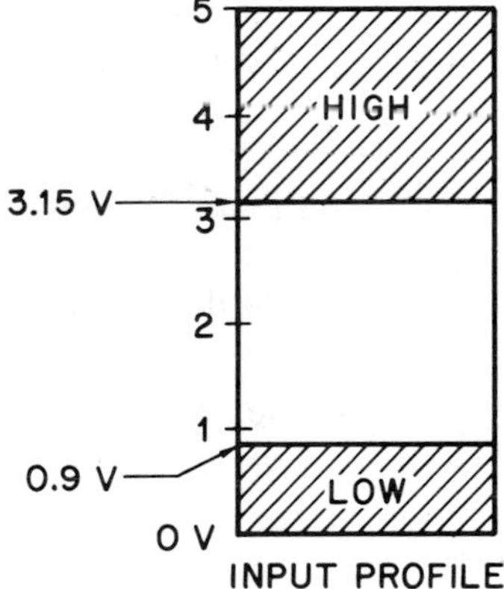

Figure 4–89

Solution

Using the relationship for noise margin, you have

$$V_H = V_{OH,min} - V_{IH,min}$$
$$V_H = 4.4\text{ V} - 3.15\text{ V} = 1.25\text{ V}$$
$$V_L = V_{IL,max} - V_{OL,max}$$
$$V_L = 0.9\text{ V} - 0.1\text{ V} = 0.8\text{ V}$$

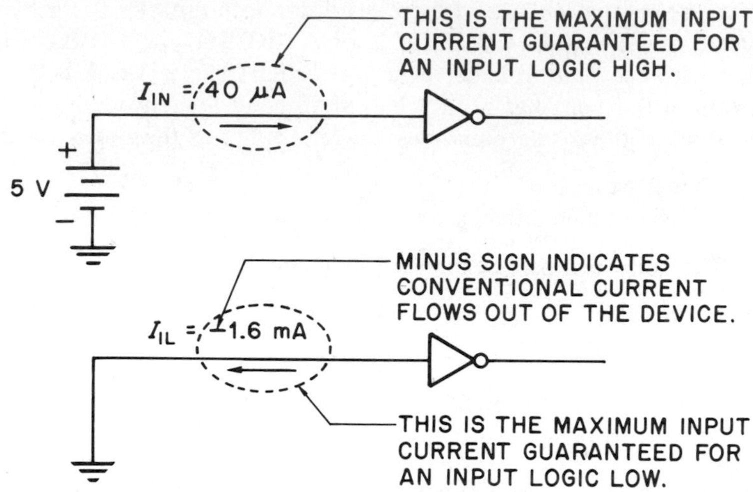
THIS IS THE MAXIMUM INPUT CURRENT GUARANTEED FOR AN INPUT LOGIC HIGH.
I_IN = 40 μA
5 V
+
−
MINUS SIGN INDICATES CONVENTIONAL CURRENT FLOWS OUT OF THE DEVICE.
I_IL = −1.6 mA
THIS IS THE MAXIMUM INPUT CURRENT GUARANTEED FOR AN INPUT LOGIC LOW.

Figure 4–90 TTL input currents.

Current Requirements

Besides knowing what the actual input and output *voltages* must be for chip families logic levels, the technician and digital designer must also know the practical *current* requirements. **Conventional current** flow will be used here because it is used on data sheets in industry.

Conventional Current:
Current flow where the direction is taken to be from the positive voltage potential to the negative. This is opposite to the direction of electron flow.

Current flow does take place between logic devices. The amount and direction of current that flows between them differ, depending on the logic level of the device. These conditions are shown in Figure 4–90. The amount of current for a logical LOW input is much larger than for a logical HIGH input. For the conditions shown, when a logical HIGH is applied, conventional current flows into the logic device, and therefore it is said to act as a **current sink.**

Current Sink:
A device is said to act as a current sink when it accepts current flow into it from an external source.

When the input logic level is LOW, conventional current flows out of the device. Under these conditions, the device is said to act as a **current source.**

Current Source:
A device is said to act as a current source when it delivers current flow to an external connection.

Example 4–23

The current characteristics of a standard TTL low-power Schottky NAND gate are shown in Figure 4–91. Determine the values of its source and sink currents.

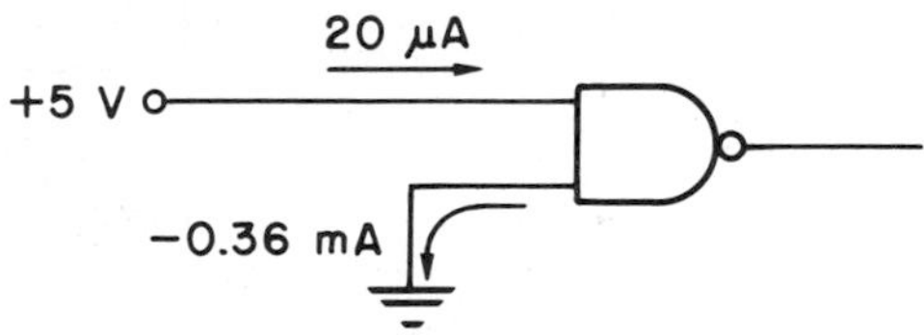

Figure 4–91

Solution

Current flowing *out of* a device is called the source current and is given a minus sign to indicate this. Current flowing *into* a device is called the sink current. Thus for the TTL NAND gate of this example,

$$I_{IL} = -0.36 \text{ mA}$$

and

$$I_{IH} = 20 \ \mu\text{A}$$

Again, observe that the negative sign indicates conventional current flow out of the device.

Output Current Characteristics

The output current characteristics of TTL logic are different from the input current characteristics. This must be the case if the output of a TTL logic device is to logically control several other logic devices. This requirement is illustrated in Figure 4–92.

What is happening in Figure 4–92 is very important to the designer as well as to the technician. The two output NAND gates are acting as a **load** for the inverter. This is usually presented as a **unit load (UL).**

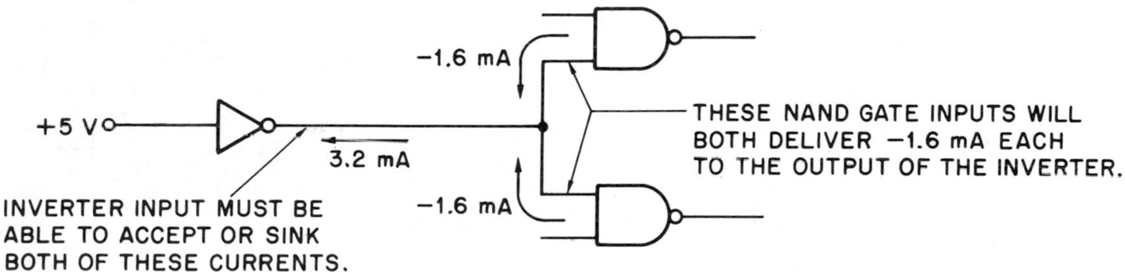

Figure 4–92 Output current effects from multiple connections.

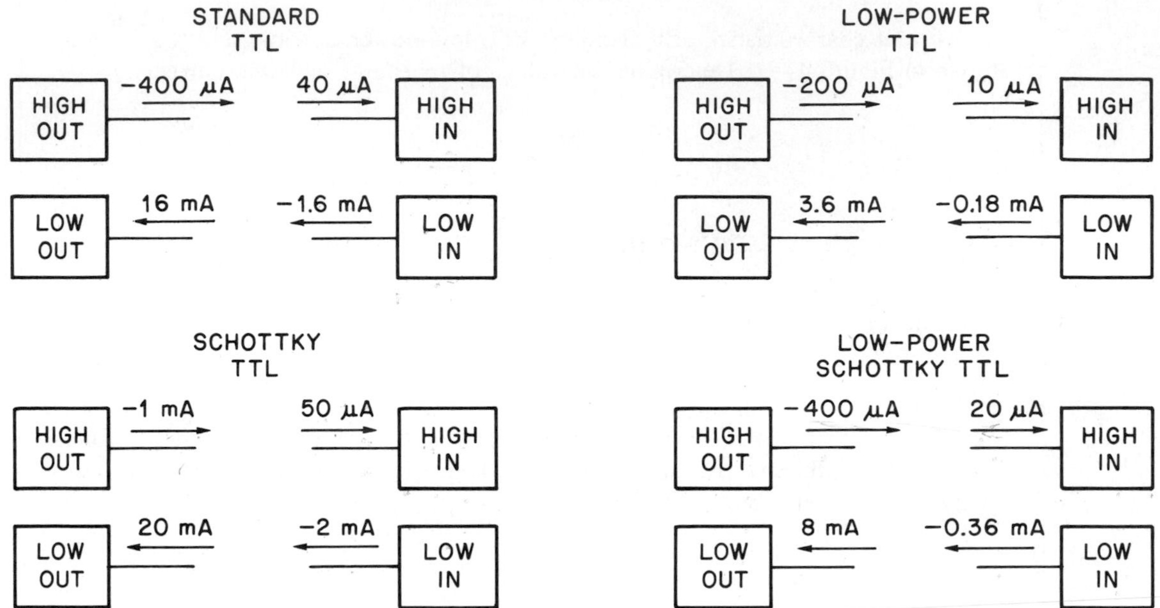

Figure 4–93 Current requirements for various logic subfamilies.

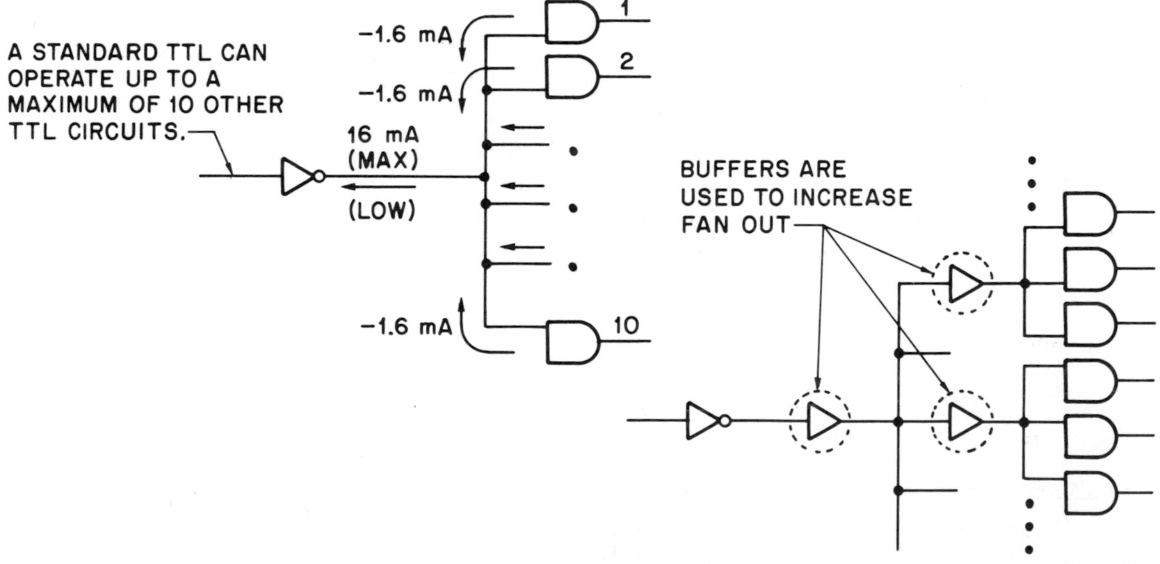

Figure 4–94 Current considerations of standard TTL.

Load:
A device or devices that consume electrical energy from a given source.

Unit Load (UL):
A gate input represents a unit load to a gate output within the same family.

Since the inputs of the NAND gates are at a logic low, they must each source −1.6 mA, and the inverter output must be able to sink at least twice this amount (3.2 mA). This requirement of the output of one device having to accommodate the summation of all the currents from devices it is operating creates a very important practical problem. The question that arises is, how many logic circuits can be connected from the output of one?

The answer is that the limit is set by the maximum amount of current a logic device can source or sink on its output compared to the amount sourced or sinked by the inputs connected to it. This amount varies between logic subfamilies as shown in Figure 4–93. As you can see, for a standard TTL logic circuit, the maximum number of logic devices that can be logically controlled by one of them is 16 mA/1.6 mA = 10. This limitation, and how to overcome it, is shown in Figure 4–94. The maximum number of logic devices that can be controlled by a single logic device is called its **fan-out.**

Fan-Out:
The maximum number of logic devices that can be controlled by the output of a single logic device.

Conclusion

This section presented the important practical details of real logic circuits. You are now in a position to appreciate that the logic symbols used to represent real logic circuits are greatly simplified where the voltages and currents between them are not considered for logical analysis purposes. However, when it comes time to actually design or repair such a digital system, then the voltages and currents are of utmost importance. You were introduced here to important background information needed for the troubleshooting and maintenance of digital systems. Test your understanding of this section by trying the following review questions.

4–12 Review Questions

1. Are the logic levels of TTL circuits +5 volts for HIGH and 0 volts for LOW? Explain.
2. What are input output profiles, as applied to digital logic circuits.
3. Are the input profiles the same as the output profiles for logic circuits? If not, explain what purpose this serves.
4. Explain current sourcing and current sinking. State what the minus sign for current indicates when used with logic circuits.
5. Discuss the meaning of fan-out. Why is this important?

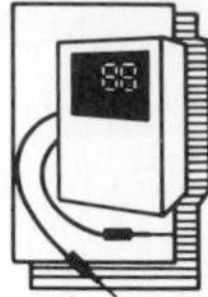

TROUBLESHOOTING SIMULATION

The troubleshooting simulation program for this chapter gives you an opportunity to develop your troubleshooting skills with NAND and NOR gates. The program presents working gates that you may test. A problem is placed in the gate. By using logic analysis you then determine the problem. The computer will indicate if your analysis is correct. This simulation prepares you for troubleshooting gate networks.

SUMMARY

- The NOR gate is logically an OR gate followed by an inverter.
- The output of a NOR is LOW when any input is HIGH.
- The NAND gate is logically an AND gate followed by an inverter.
- The output of a NAND gate is LOW when all inputs are HIGH.
- De Morgan's first theorem is useful in relating the logic of a NOR gate to that of an AND gate with inverted inputs.
- De Morgan's second theorem is useful in relating the logic of a NAND gate to that of an OR gate with inverted inputs.
- There is a *precedence* of operations in Boolean algebra such that Boolean multiplication is done before Boolean addition.
- Parentheses are used in Boolean algebra to change the order of operations; all work inside the parentheses is done first.
- Boolean rules are useful in the analysis of logic networks.
- Boolean equations can be factored in the same manner as "regular" algebraic equations.
- An Exclusive OR gate (XOR) has a TRUE output only when the inputs are different. For all other conditions, its output is FALSE.
- An Exclusive NOR gate (XNOR) has a TRUE output only when the inputs are the same. For all other conditions, its output is FALSE.
- An application of an XOR and XNOR gate is that of a parity generator and checker.
- Because of De Morgan's theorems, logic circuits may be presented in two different ways. This technique may aid in circuit simplification.
- Some manufacturers will draw their gate logic to indicate an active HIGH or active LOW.
- Many logic circuits using NAND/NOR gates may be simplified to AND/OR gates for logic analysis.
- NAND gates can be used to generate any type of logic circuit.
- NOR gates can be used to generate any type of logic circuit.
- A microprocessor contains process logic that is constructed at the factory to perform specific processes that are controlled by bit patterns.
- Instructions and data are contained together in memory in the form of bit patterns.
- A program is a set of bit patterns stored in memory that will cause a predetermined set of processes by the microprocessor.
- Digital ICs have input and output voltage profiles that are designed in such a way as to minimize the effects of electrical noise.

- Digital ICs have different input and output current requirements. This limits the number of logic circuits that can be operated from a single logic circuit.
- Buffers can be used to increase the fan-out capabilities of digital ICs.

CHAPTER SELF-TEST

I. TRUE/FALSE

Answer the following questions true or false.

1. A NOR gate can be constructed from an OR gate with inverted inputs.
2. A NAND gate can be constructed from an AND gate with an inverted output.
3. The output of a NOR gate will be HIGH when any of its inputs are HIGH.
4. The output of a NAND gate will be LOW when all of its inputs are HIGH.
5. A NAND gate followed by an inverter will produce an AND function.

II. MULTIPLE CHOICE

Answer the following questions by selecting the most correct answer.

6. The Boolean expression $X = \overline{A + B + C}$ represents a
   (A) three-input OR gate
   (B) three-input NOR gate
   (C) three-input NAND gate
   (D) single-output NAND gate
7. The Boolean expression for a two-input NAND gate is
   (A) $A + B$
   (B) $\overline{A}\overline{B}$
   (C) $\overline{AB}$
   (D) None of the above
8. De Morgan's first theorem is
   (A) $X = \overline{A} \cdot \overline{B}$
   (B) $\overline{A + B} = \overline{A} \cdot \overline{B}$
   (C) $\overline{A + B} = \overline{A \cdot B}$
   (D) $\overline{A} \cdot \overline{B} = \overline{A \cdot B}$
9. Applying De Morgan's first theorem to the statement $Y = \overline{A + B + C}$ yields
   (A) $Y = \overline{A} + \overline{B} + \overline{C}$
   (B) $Y = \overline{A \cdot B \cdot C}$
   (C) $Y = A \cdot B \cdot C$
   (D) None of the above
10. De Morgan's second theorem is
   (A) $\overline{A \cdot B} = \overline{A} + \overline{B}$
   (B) $\overline{A \cdot B} = \overline{A + B}$
   (C) $\overline{A} + \overline{B} = \overline{A} \cdot \overline{B}$
   (D) None of these

III. MATCHING

In Figure 4–95, match the logic diagrams on the right to those on the left.

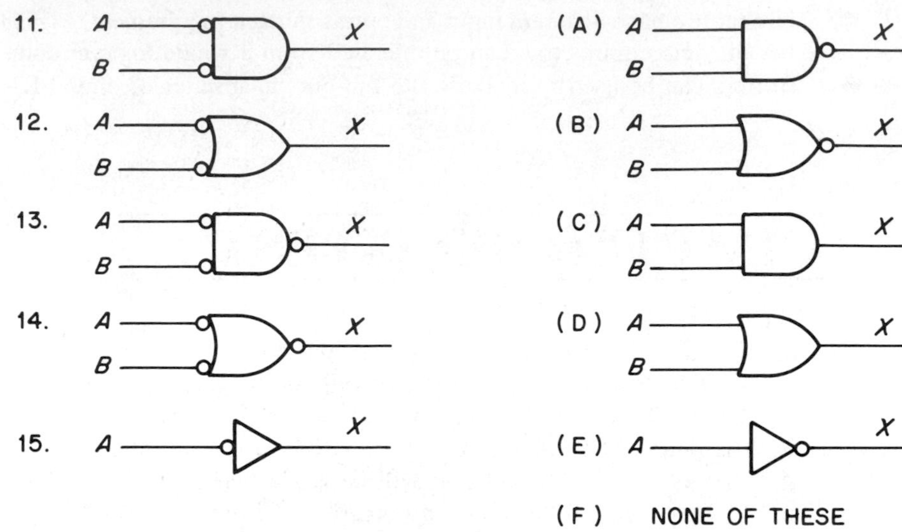

Figure 4–95

IV. FILL-IN

Fill in the blanks with the most correct word(s) or statements.

16. The output of a/an _______ gate will be TRUE only when the two inputs have different logic levels.
17. When error checking is achieved by counting the number of 1's in the word, this is called _______ checking.
18. The bit patterns that cause the process logic to perform a specific process is called the _______ set.
19. An AND gate can be constructed from a NAND gate with a/an _______ output.
20. The _______ register stores a word that can be used to control the process logic of the 4-bit microprocessor.

V. OPEN-ENDED

Answer the following questions as indicated.

21. Simplify the following logic statements:
    (A) $A \cdot 1$ (B) $A + 1$ (C) $A + A$ (D) $A \cdot \overline{A}$ (E) $A + \overline{A}$
22. Simplify the following logic statements:
    (A) $A + AB$ (B) $A + \overline{A}B$ (C) $(A + B)(A + C)$
23. Construct the following logic circuits using nothing more than NAND gates:
    (A) inverter (B) OR (C) NOR
24. Simplify the logic circuit of Figure 4–96.

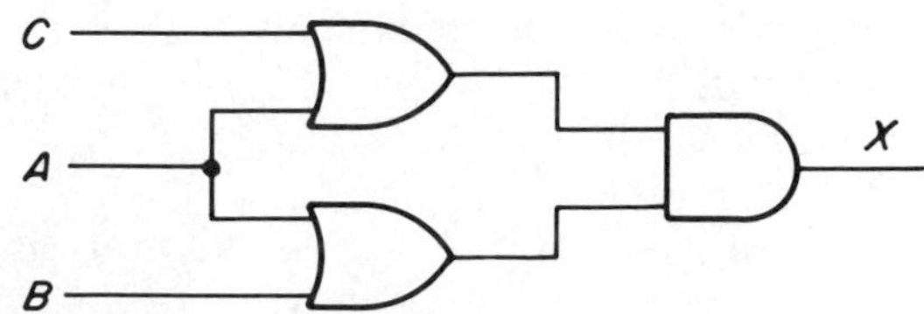

Figure 4–96

25. State what the contents of all the microprocessor registers will be when the following program is executed (use the 4-bit microprocessor instruction set).

Memory Location	Data
0	0001
1	1111
2	0010
3	1010
4	1011
5	1111

Answers to Chapter Self-Test

1] F 2] T 3] F 4] T 5] T 6] B 7] C 8] B 9] D 10] A
11] B 12] A 13] D 14] C 15] E 16] Exclusive OR
17] parity 18] instruction 19] inverted 20] data
21] (A) A (B) 1 (C) A (D) 0 (E) 1 22] (A) A (B) $A + B$
(C) $A + BC$ 23] Figure 4–97

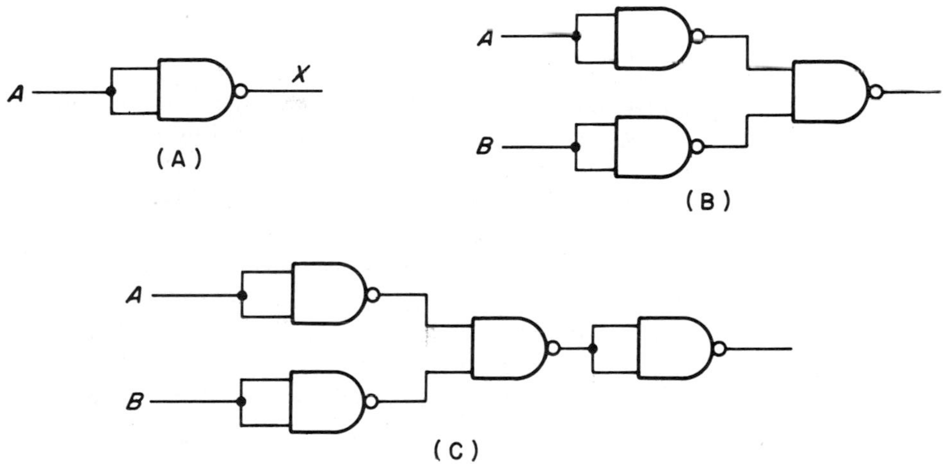

Figure 4–97

24] Figure 4–98

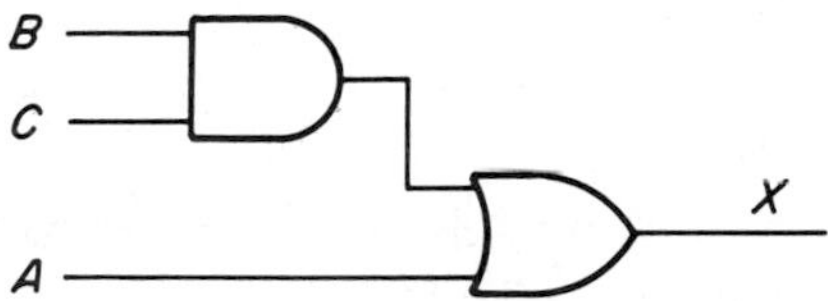

Figure 4–98

25] accumulator → 0000, B register → 1010, data register → 1111

CHAPTER PROBLEMS

Basic Concepts

Section 4–1:

1. Determine the output conditions for the following NOR gate input conditions:
   (A) $A = 1, B = 0$ (B) $A = 0, B = 0$ (C) $A = 1, B = 1, C = 0$
2. Assume the following inputs to the NOR gate network of Figure 4–99. Determine the resulting output conditions:
   (A) $A = 0, B = 0, C = 0$ (B) $A = 1, B = 1, C = 1$
   (C) $A = 1, B = 0, C = 1$
3. Construct the truth table for the NOR gate logic of Figure 4–99.

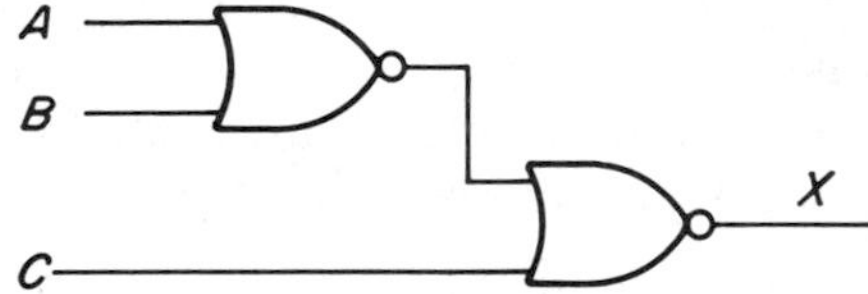

Figure 4–99

4. Using the diagram of a TTL 7402 quad two-input NOR gate presented in Section 4–1, sketch how you would actually construct the logic diagram shown in Figure 4–99.
5. Write the Boolean expression for a (A) two-input NOR gate, (B) three-input NOR gate, (C) four-input NOR gate.
6. Determine the output of each logic diagram in Figure 4–100 for the following conditions:
   (A) $A = 1, B = 0$ (B) $A = 1, B = 1$ (C) $A = 0, B = 0$

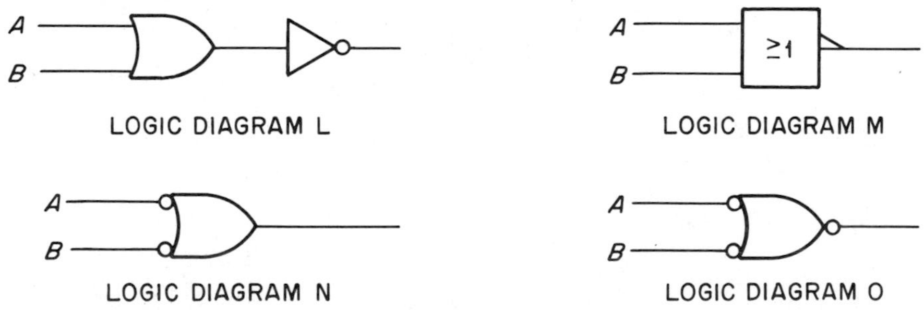

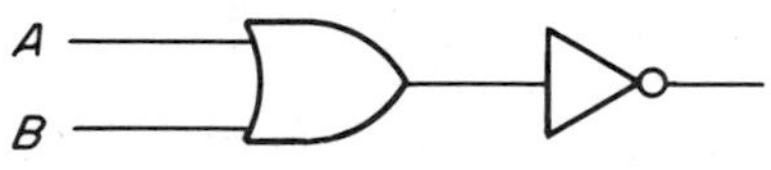

LOGIC DIAGRAM L

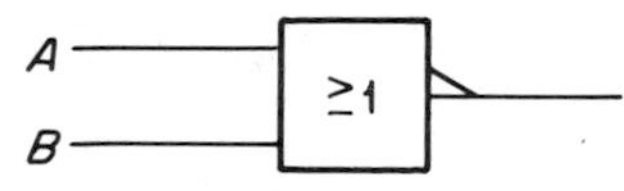

LOGIC DIAGRAM M

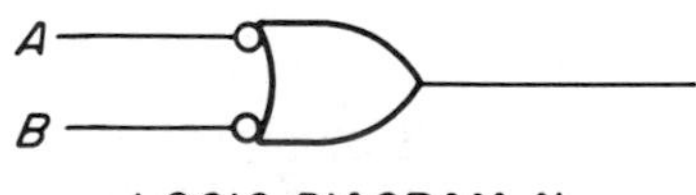

LOGIC DIAGRAM N

A
B

LOGIC DIAGRAM O

Figure 4–100

7. Construct the truth table for each logic diagram of Figure 4–100.
8. Prove that an OR gate with inverted inputs is not logically the same as a NOR gate.

Section 4–2:

9. Determine the output conditions for the following NAND gate input conditions:
   (A) $A = 1, B = 0$ (B) $A = 0, B = 0$ (C) $A = 1, B = 1, C = 0$

10. Assume the following inputs to the NAND gate network of Figure 4–101. Determine the resulting output conditions:
(A) $A = 0, B = 0, C = 0$ (B) $A = 1, B = 1, C = 1$
(C) $A = 1, B = 0, C = 1$
11. Construct the truth table for the NAND gate logic of Figure 4–101.

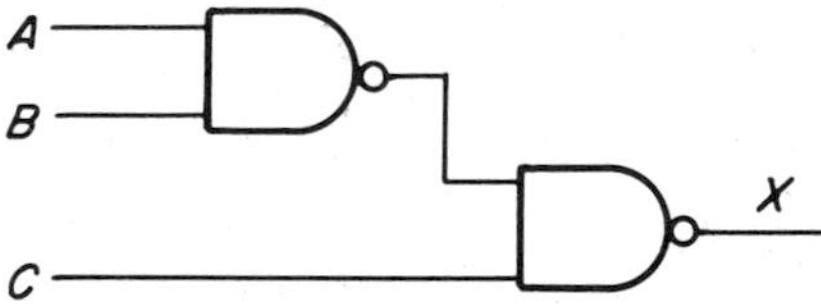

Figure 4–101

12. Using the diagram of a TTL 7400 quad two-input NAND gate shown in Section 4–2, sketch how you would actually construct the logic diagram shown in Figure 4–101.
13. Write the Boolean expression for a (A) two-input NAND gate, (B) three-input NAND gate, (C) four-input NAND gate.
14. Determine the output of each logic diagram in Figure 4-102 for the following conditions:
(A) $A = 1, B = 0$ (B) $A = 1, B = 1$ (C) $A = 0, B = 0$
15. Construct the truth table for the each logic diagram of Figure 4–102.

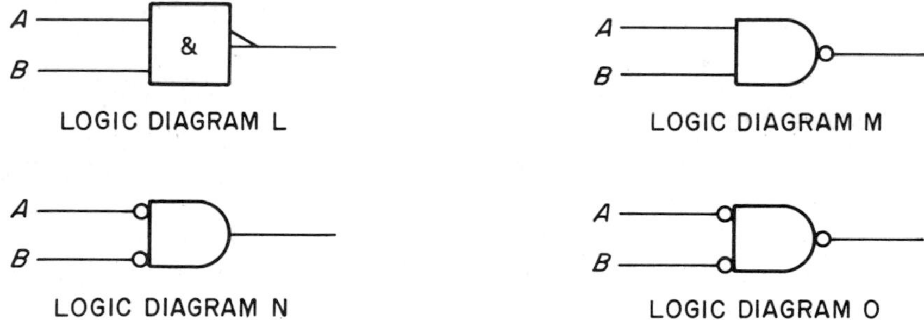

Figure 4–102

16. Prove that an AND gate with inverted inputs is not logically the same as a NAND gate.

Section 4–3:

17. Using logic diagrams and truth tables, prove De Morgan's first theorem.
18. State De Morgan's first theorem.
19. Convert the following logic expressions into NOR gates, using De Morgan's first theorem:
(A) $\overline{A} \cdot \overline{B}$ (B) $\overline{A} \cdot \overline{B} \cdot \overline{C}$ (C) $\overline{(A + B)} \cdot \overline{(C + D)}$

Section 4–4:

20. Using logic diagrams and truth tables, prove De Morgan's second theorem.
21. State De Morgan's second theorem.
22. Convert the following logic expressions into NAND gates, using De Morgan's second theorem:
    (A) $\overline{A} + \overline{B}$ (B) $\overline{A} + \overline{B} + \overline{C}$ (C) $\overline{AB} + \overline{CD}$

Section 4–5:

23. Explain what is meant by (A) *precedence* of operations as it applies to Boolean algebra. (B) Give an example.
24. Develop the Boolean expression and truth table for the logic network of Figure 4–103(A).

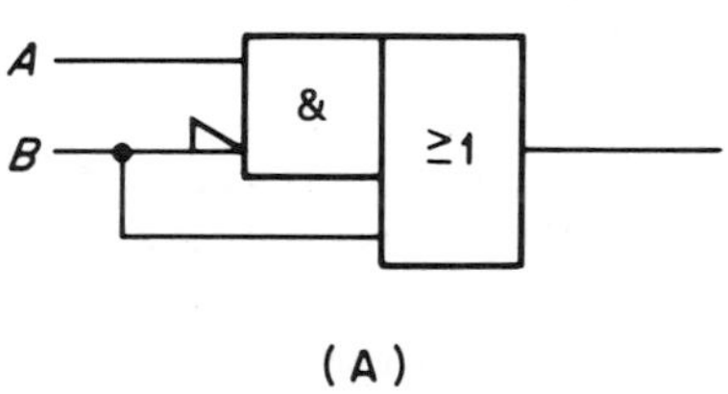

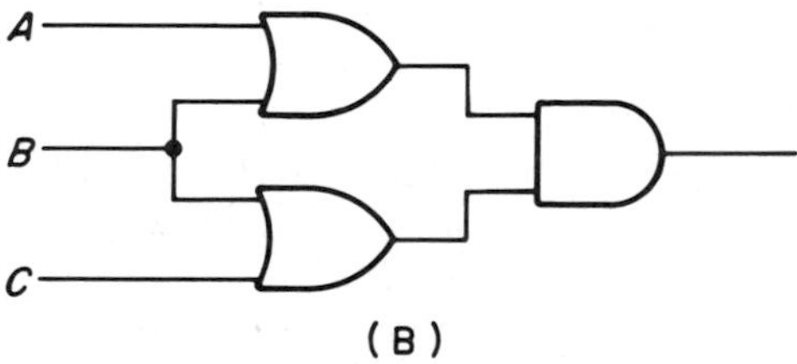

Figure 4–103

25. Develop the Boolean expression and truth table for the logic network of Figure 4–103(B).
26. Develop the Boolean expression and truth table for each logic network in Figure 4–104.

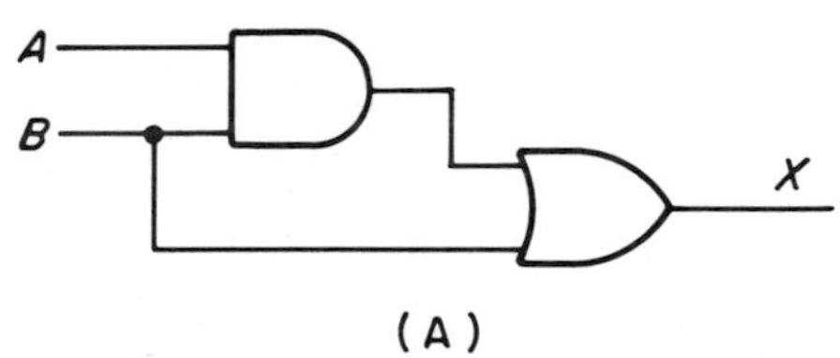

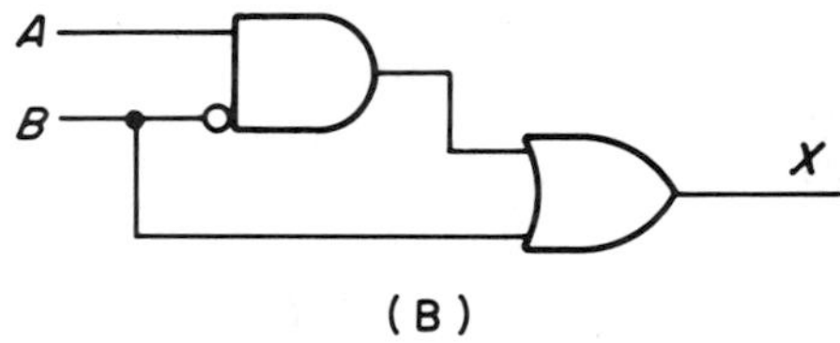

Figure 4–104

27. Use Boolean algebra to simplify the logic networks of Figure 4–105(A) and (B).

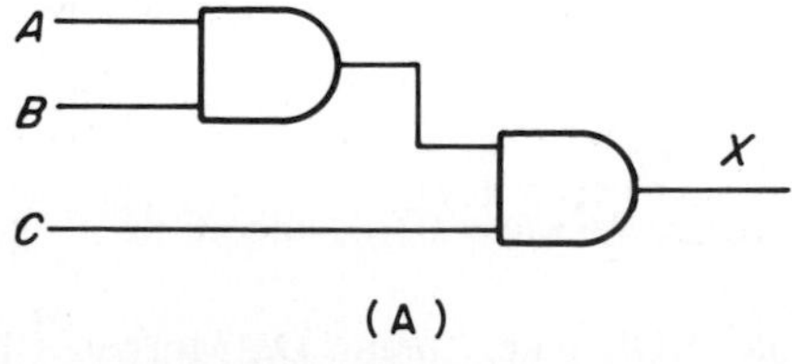

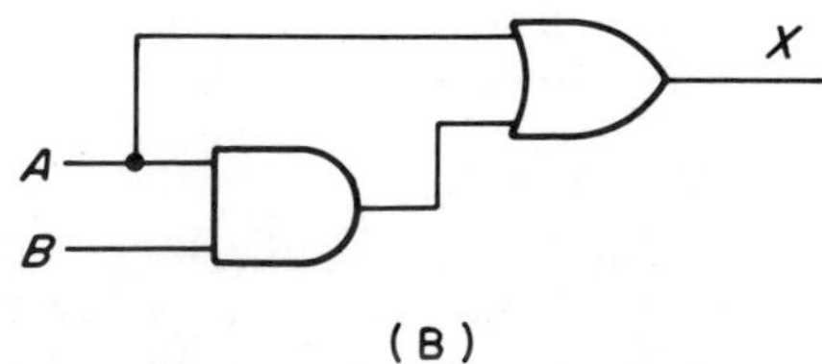

Figure 4–105

28. Use Boolean algebra to simplify the logic networks of Figure 4–105(C) and (D).

Figure 4–105 Continued

29. Use a truth table to show that the simplified logic network you produced for problem 28 is the same as the original ones.
30. Using Boolean rules, simplify the following logic statements:
    (A) $X = A \cdot 1$ (B) $Y = \overline{A} + A$ (C) $X = A \cdot A$ (D) $Y = A + 0$
31. Simplify the following logic statements, using Boolean rules:
    (A) $X = \overline{\overline{A}}$ (B) $Y = A \cdot 1 \cdot A$ (C) $X = A + A + 1$ (D) $Y = A \cdot A \cdot 0$
32. Construct the logic diagram that demonstrates the simplified relationship for each Boolean statement of problem 30. (HINT: Use a direct wire connection where possible.)
33. For each Boolean statement of problem 31, construct the logic diagram that demonstrates the simplified statement. (HINT: Use a direct wire connection where possible.)
34. Using Boolean algebra, simplify the diagrams of Figure 4–106(A) and (B).

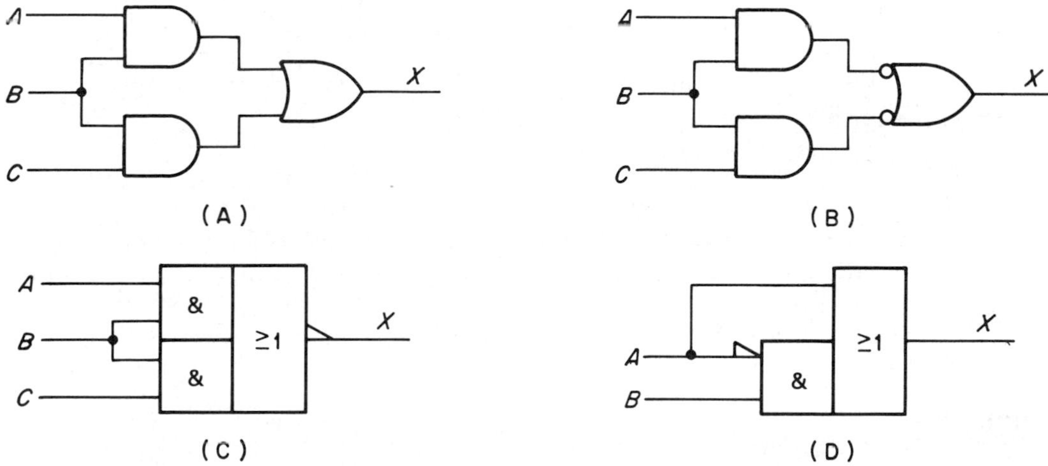

Figure 4–106

35. Simplify the logic diagrams of Figure 4–106(C) and (D).

Section 4–6:

36. (A) Construct the truth table for an XOR gate. (B) What general statement can you make when the output will be TRUE?

37. Determine the type of parity used for the parity generator of Figure 4–107.

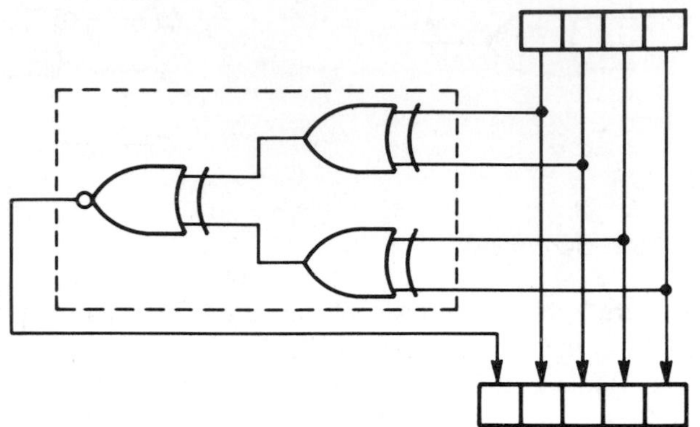

Figure 4–107

38. Develop a truth table for the logic diagram of Figure 4–108(A).
39. Develop a truth table for the logic diagram of Figure 4–108(B).

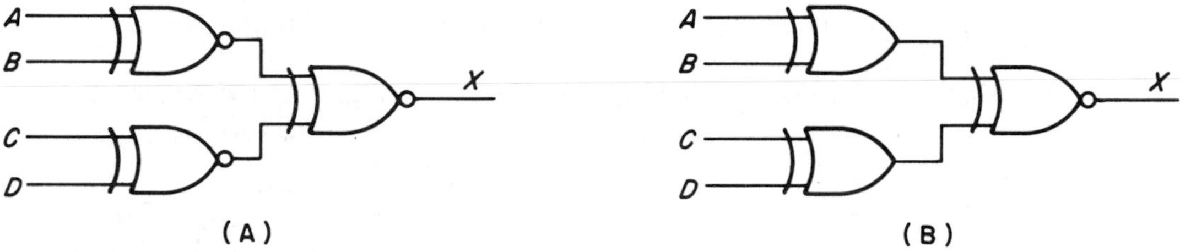

Figure 4–108

40. Construct the logic diagram of an odd-parity generator. (4-bit word)
41. Construct the logic diagram of an odd-parity checker. (4-bit word plus parity bit.)
42. Write the Boolean expressions for an XOR gate and an XNOR gate. Explain the differences.
43. Develop the output waveform for the logic diagram of Figure 4–109(A).

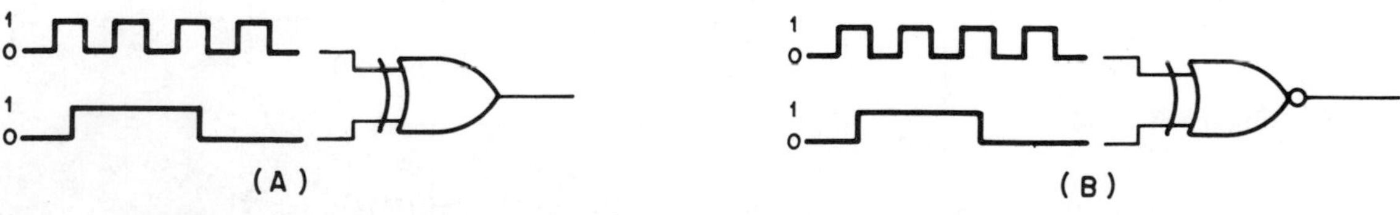

Figure 4–109

44. Develop the output waveform for the logic diagram of Figure 4–109(B).

Section 4–7:

45. Simplify the logic diagrams of Figure 4–110(A) and (B) into standard AND/OR gates.

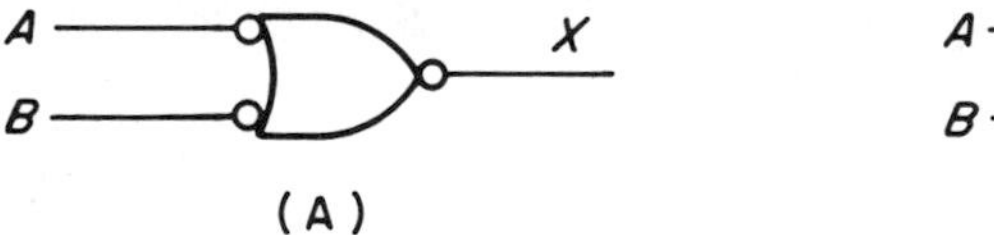

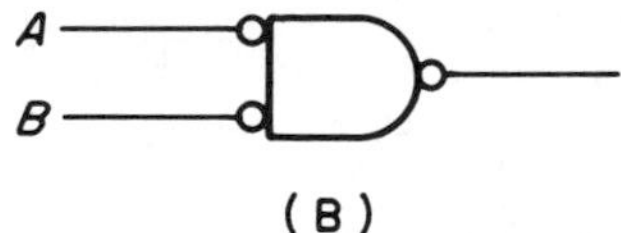

Figure 4–110

46. Simplify the logic diagrams of Figure 4–110(A) and (B) into standard NAND/NOR gates as presented in Sections 4–1 and 4–2. State the actual ICs you would use to implement the logic.
47. Redraw the logic diagrams of Figure 4–111(A) and (B) so that the output indicates an active LOW.

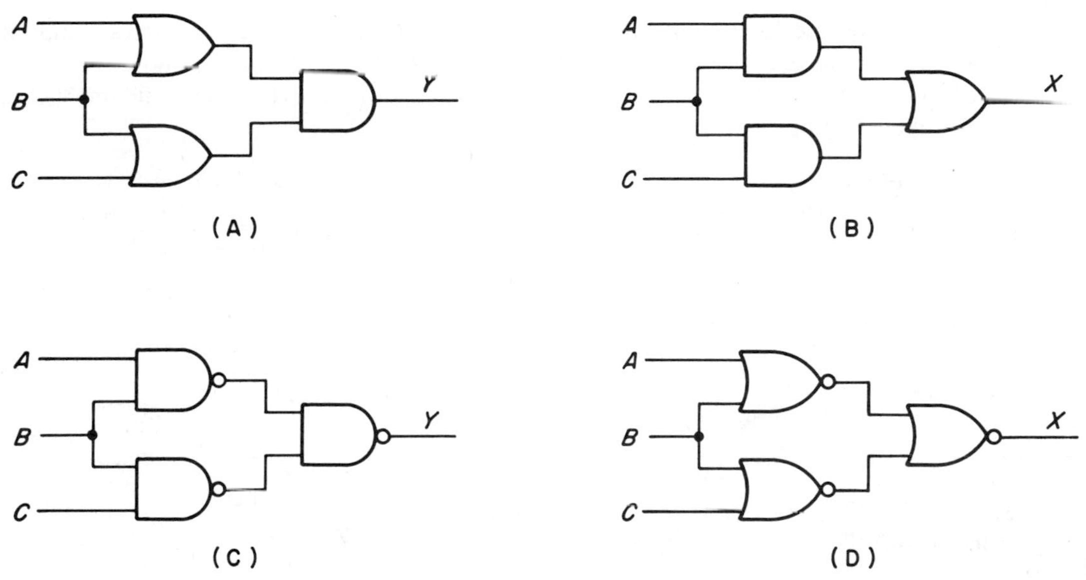

Figure 4–111

48. Redraw the logic diagrams of Figure 4–111(C) and (D) so that the output indicates an active HIGH.

Section 4–8:

49. Redraw the logic diagrams of Figure 4–112(A) and (B) so they are as easy to analyze as possible. Use analysis simplification.

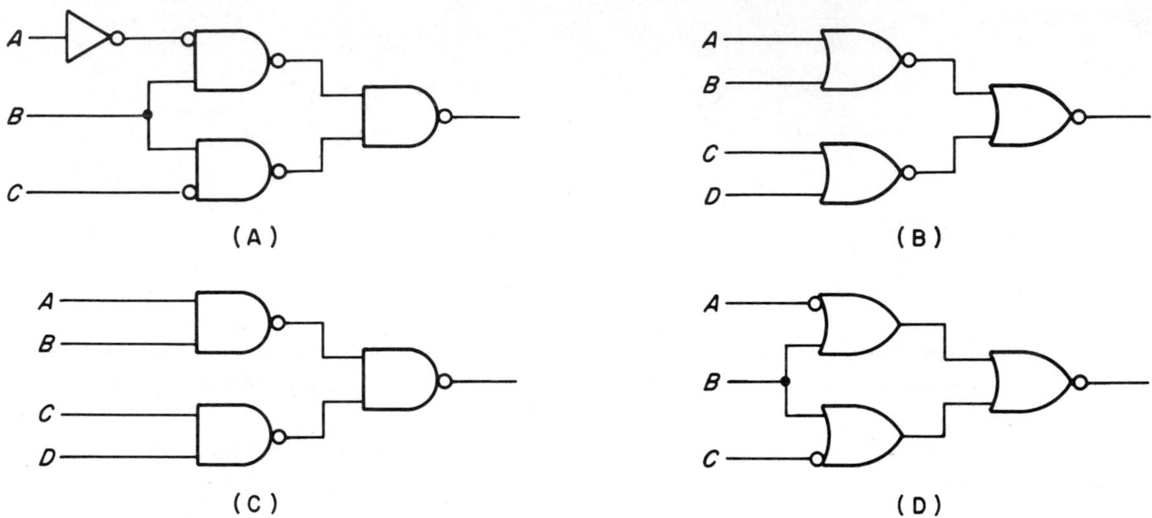

Figure 4–112

50. Redraw the logic diagrams of Figure 4–112(C) and (D), using analysis simplification, so they are as easy as possible to analyze.
51. Convert the AND/OR diagrams of Figure 4–113(A) and (B) to standard NAND/NOR logic.

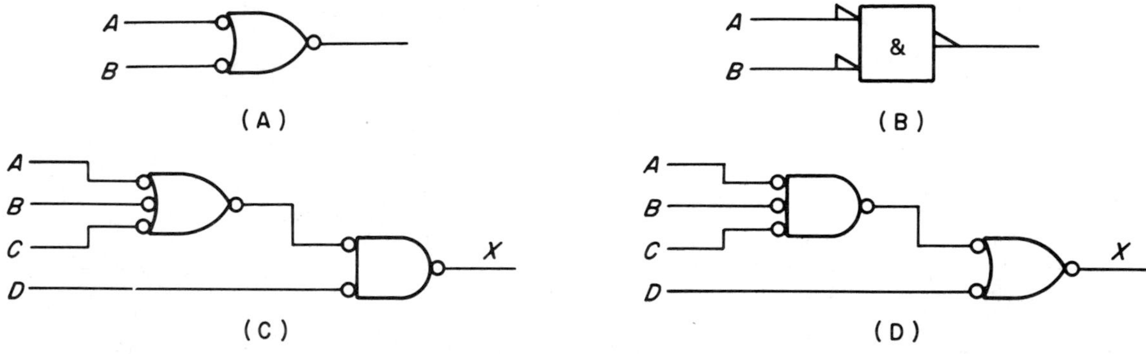

Figure 4–113

52. Convert the AND/OR diagrams of Figure 4–113(C) and (D) to standard NAND/NOR logic.

Section 4–9:

53. Develop the following gates, using only NAND gates:
(A) AND (B) OR (C) NOR

54. Show the Boolean development of each of the gates in problem 53.

Section 4–10:

55. Develop the following gates, using only NOR gates:
    (A) OR (B) AND (C) NAND
56. Show the Boolean development of each gate in problem 55.

Applications

57. State the functions of the following parts of the 4-bit microprocessor presented in this chapter:
    (A) accumulator (B) B register (C) data register (D) process logic
58. Sketch a diagram to illustrate the interconnections between the parts of the 4-bit microprocessor of question 57.
59. Explain what process always takes place when the microprocessor is first turned on. Why does this happen?
60. Explain the difference between FETCH and EXECUTE processes for the 4-bit microprocessor.
61. Which processes from the instruction set of the 4-bit microprocessor require a one-step process (an example would be the STOP instruction)?
62. Which processes from the instruction set of the 4-bit microprocessor change the contents of the accumulator?
63. State which processes from the instruction set of the 4-bit microprocessor change the contents of memory.
64. State the final contents of the internal registers of the 4-bit microprocessor and memory if the following set of instructions were executed:

(A) Memory Location	Data	(B) Memory Location	Data
0	0001	0	0010
1	0011	1	1111
2	0010	2	0011
3	0011	3	0101
4	1100	4	0111
5	0110	5	1111
6	1010		
7	1001		
8	0000		
9	1111		

65. Assume that the content of the accumulator is 0101 and of the B register is 1010. State what the final contents of all the internal registers of the 4-bit microprocessor and memory will be when the following programs are executed:

(A) Memory Location	Data	(B) Memory Location	Data
0	1000	0	0101
1	1101	1	0110
2	1001	2	1101
3	0000	3	1100
4	1111	4	1001
		5	0000

(continued)

(B) Memory Location	Data
6	1010
7	0000
8	1111

Troubleshooting

66. What would you conclude if you made the following measurements on the output of a LS TTL NAND gate?
(A) 4.2 V (B) 2.3 V (C) 0.3 V (D) 1.4 V
67. What would you conclude if you made the following measurements on the input of a LS TTL NOR gate?
(A) 3.8 V (B) 1.9 V (C) 0.4 V (D) 4.0 V
68. The output of a NAND gate is connected to the inputs of two NOR gates. If these are LS TTL circuits, what is the amount and direction of currents you would expect if the output of the NAND gate were (A) HIGH, (B) LOW?
69. The output of a LS TTL inverter is connected to the input of five NAND gates. Determine the amount and direction of the current on the output of the inverter when its output is (A) HIGH, (B) LOW.
70. Explain the meaning of *fan-out*.

Analysis and Design

71. The output of the accumulator of the 4-bit microprocessor presented in this chapter is connected to the input of the indicated logic circuit of Figure 4-114. Write a program (indicate the memory location and data) that would cause the following activation of relays for a robot control system. Use the instruction set for the 4-bit microprocessor presented in this chapter.
    - Deactivate all relays.
    - Activate all odd-numbered relays.
    - Deactivate all odd-numbered relays.
    - Activate all even-numbered relays.
    - Activate one relay at a time starting with relay 1.

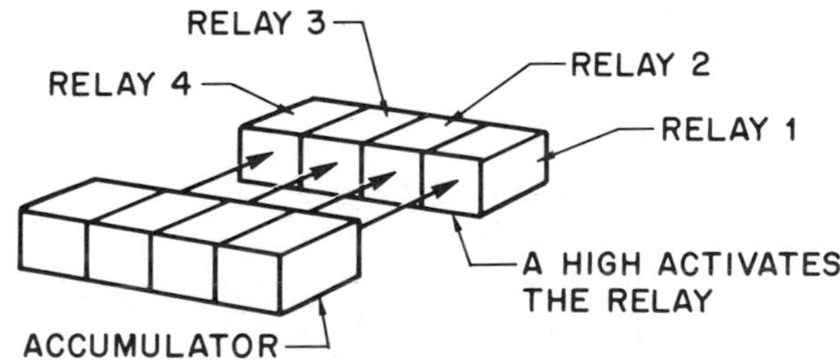

Figure 4–114

72. Develop a program using the instruction set of the 4-bit microprocessor presented in this chapter that will cause the contents of the accumulator to be the complement of the contents of the B register. The contents of the B register are determined by the contents of the third memory location.

CHAPTER 5

Analysis and Design of Combinational Networks

OBJECTIVES

After completing this chapter, you should be able to:

- ☐ Develop Boolean expressions from gate networks.
- ☐ Develop gate networks from Boolean expressions.
- ☐ Develop Boolean expressions from truth tables, using the sum of products method.
- ☐ Develop Boolean expressions from truth tables, using the product of sums method.
- ☐ Develop logic networks from Boolean expressions.
- ☐ Simplify Boolean expressions, using mapping techniques.
- ☐ Work with mapping techniques that use the sums of products.
- ☐ Work with mapping techniques that use the product of sums.
- ☐ Design logic systems utilizing the previous methods.
- ☐ Interpret digital IC specification sheets.
- ☐ Understand the fundamental concepts of assembly language programming, using the 4-bit microprocessor as the programming model.

INTRODUCTION

In this chapter you will learn how to take an idea and convert it into an actual logic circuit that will implement that idea. Of course, the idea must be expressible in terms of ON/OFF logic, and must be precisely predictable. But then, this covers the ideas of computer, robotics, automation, and other digital system applications.

Here you will learn techniques for expressing Boolean equations directly from logic diagrams and truth tables. The ability to do this will help you simplify the design and construction of logic systems. This chapter will use all of the information you have learned to this point—here you will learn how to fine tune it.

The microprocessor section develops the most important concept that is basic to all microprocessors—how programs are stored and used in memory. You will have an opportunity to see how a microprocessor performs its microprocesses.

In the electronic application section you will discover many of the important details that make the difference between a reliable digital system and one that only works sometimes. This is an important chapter—let's get started.

KEY TERMS

Logic Levels
Empty Term
Sum of Products
Product of Sums
Bus
Karnaugh Map
Pair Simplification
Quad Simplification
Octet Simplification
Redundant Group
Don't Cares
Mnemonic Code
Assembler
Assembly Language
Language Level
Machine Language

5–1 DEVELOPING BOOLEAN EXPRESSIONS FROM GATE NETWORKS

Discussion

You have had a sampling of developing Boolean expressions from a given gate network. This section will help you develop practical techniques for more complex logic circuits.

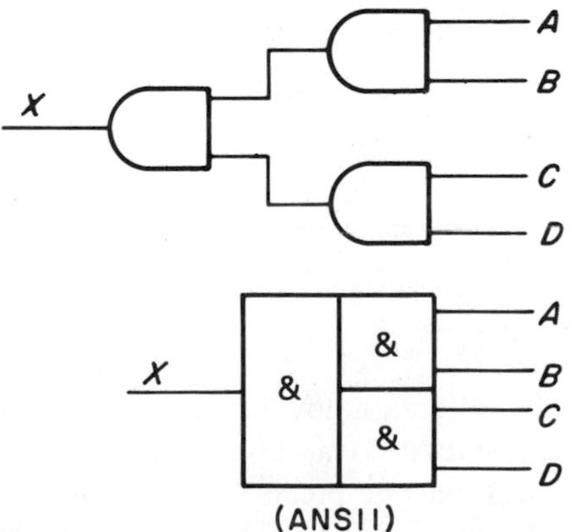

Figure 5–1 Typical logic network for analysis.

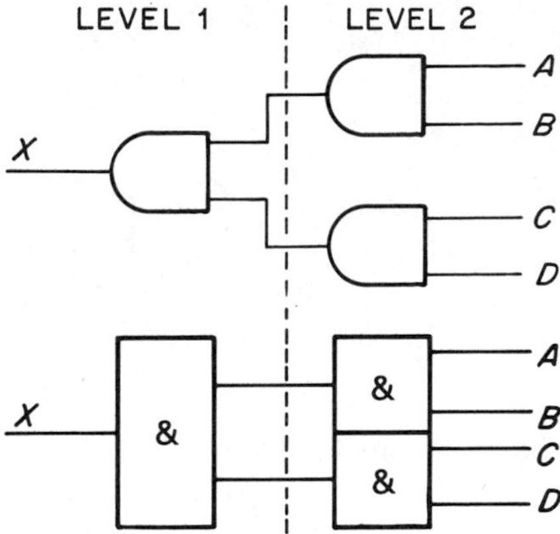

Figure 5–2 Logic levels for logic network.

Basic Idea

The basic idea is this: given a gate network (logic circuit), develop a Boolean expression that represents the logic of the network. An approach to this is to first start your analysis by sectioning the logic circuit into **logic levels.**

Logic Levels:
The sectioning of a logic circuit for analysis purposes, where the first level is the output gate, the second level consists of the gates feeding the output gate, the third level is the logic gates feeding the second level, and so on.

As an example, consider the logic network of Figure 5–1. The logic levels for the circuit of Figure 5–1 are shown in Figure 5–2.

Once the levels have been determined, you can start your analysis either at the outermost level (level 1) or the innermost level (level 2 in this example). Both methods will be demonstrated, first starting from the outermost level. As you can see from Figure 5–3, level 1 contains an AND gate, and its output expression will be the product of *two* terms.

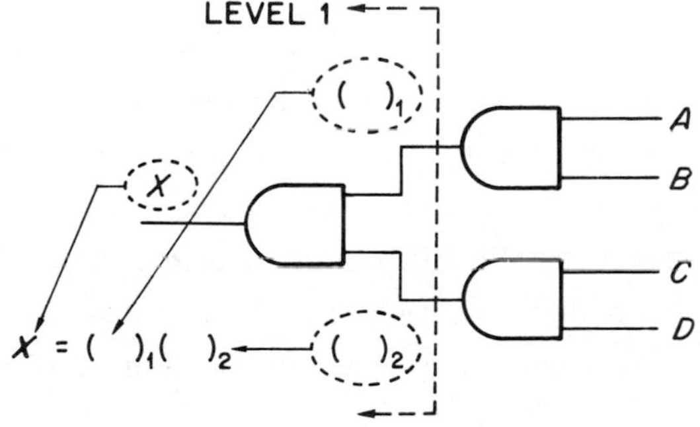

Figure 5–3 Starting analysis at the outer level.

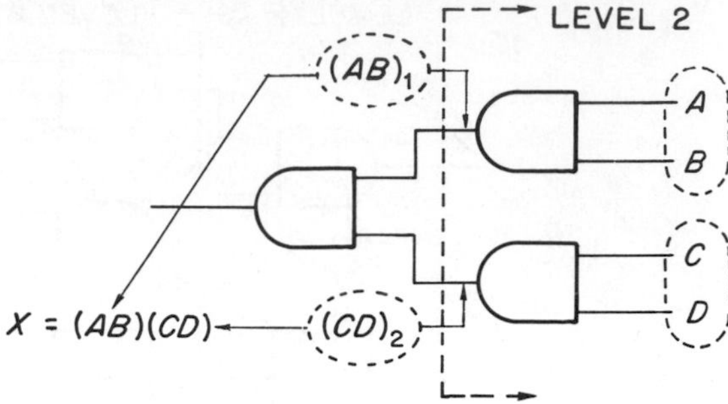

Figure 5–4 Filling in the empty terms.

Since these terms have not yet been evaluated, they are called **empty terms** and are designated by subscripts:

$$(\,)_N$$

where: N = number of the term

Empty Term:
Part of a Boolean expression used in the evaluation of a larger Boolean expression where the contents of the smaller expression have not yet been evaluated.

The next step in the analysis is to go to the next level (level 2 in this case) and evaluate each of the empty terms for level 1. This is illustrated in Figure 5–4. Note that the last step consists of "filling" the empty terms with the evaluations from the next level and removing the subscripts. The resulting expression is now the Boolean equation for the given logic circuit. The process is outlined in Table 5–1.

Table 5–1 EMPTY TERMS METHOD OF DEVELOPING BOOLEAN EXPRESSIONS FROM LOGIC NETWORKS

Step	Procedure
1	Separate the logic network into logic levels.
2	Develop the output expression for the first level using empty terms.
3	Go to the next level and develop the empty terms.
4	Repeat step 3 until there are no more levels.
5	Fill in all of the empty terms in the output expression—remove subscripts.
6	The resulting expression is the Boolean equation that represents the logic of the logic network.

The following example illustrates the same process with another logic circuit.

Example 5–1

Develop the Boolean expression for the logic network of Figure 5–5 using the method of empty terms.

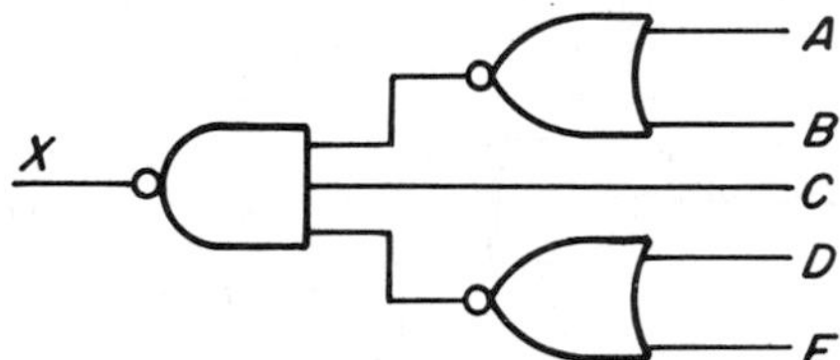

Figure 5–5

Solution

Use the steps outlined in Table 5–1 for the development of the Boolean expression for a given logic network. The process is demonstrated in Figure 5–6.

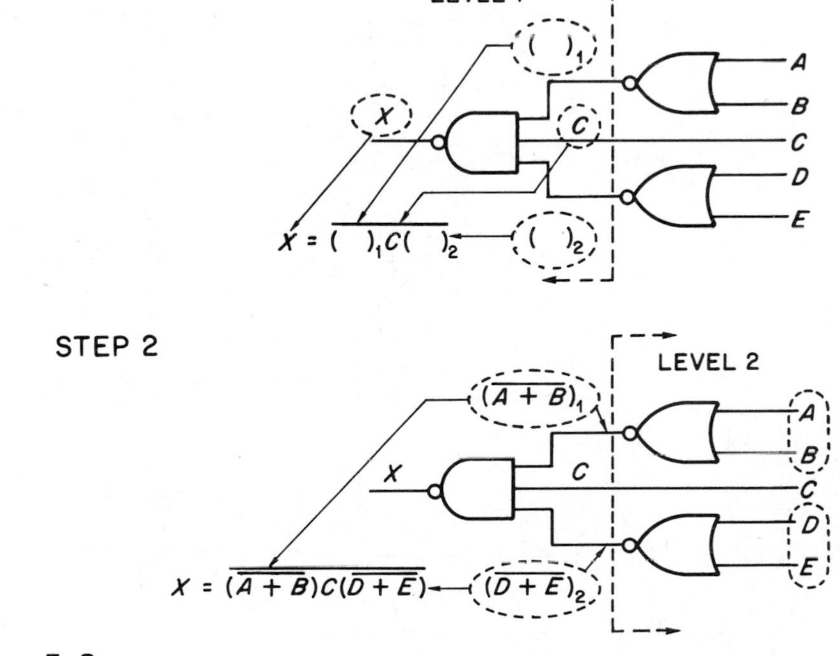

Figure 5–6

The next example demonstrates an application of empty terms analysis with three logic levels.

Example 5–2

Develop the Boolean expression for the logic network shown in Figure 5–7.

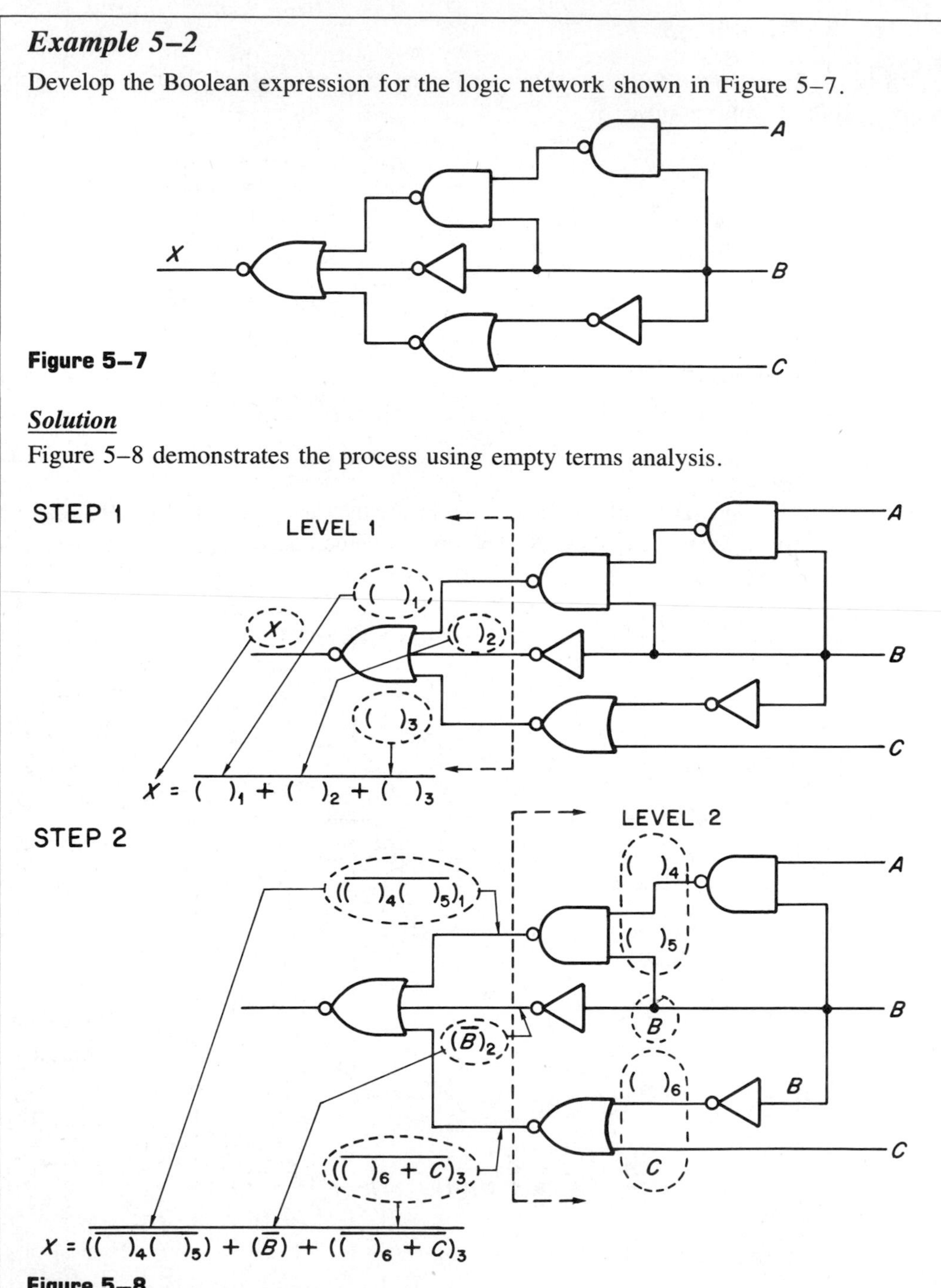

Figure 5–7

Figure 5–8

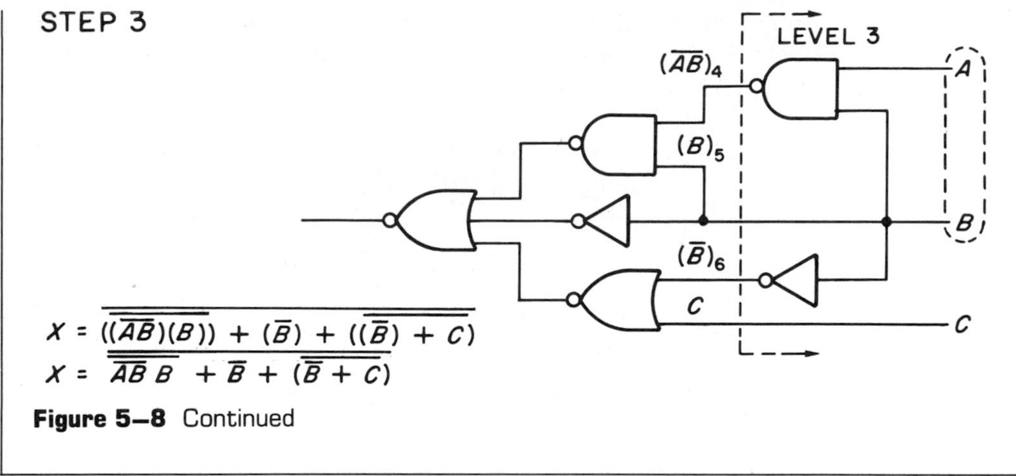

Figure 5–8 Continued

You will see later in this chapter techniques for simplifying the resulting Boolean expression. For now, the important skill to develop is the creation of the Boolean expression. Once you have gained confidence in doing that, you will be able to better focus your attention on the process of manipulating the resulting equation.

The Other Method

The other method for developing a Boolean expression for a given logic network is outlined in Table 5–2.

Table 5–2 GATE-BY-GATE ANALYSIS OF LOGIC NETWORKS

Step	Procedure
1	Separate the logic network into logic levels.
2	Develop the Boolean expressions for each logic gate in the deepest level.
3	Go to the next lowest level and develop the Boolean expression for each gate at that level.
4	Continue the process in step 3 until all levels are completed.
5	The resulting expression is the Boolean expression for the logic of the gate network.

As an example in using the gate-by-gate analysis technique, consider the development of the Boolean expression for the logic network of Figure 5–9. The process outlined in Table 5–2 is illustrated in Figure 5–10.

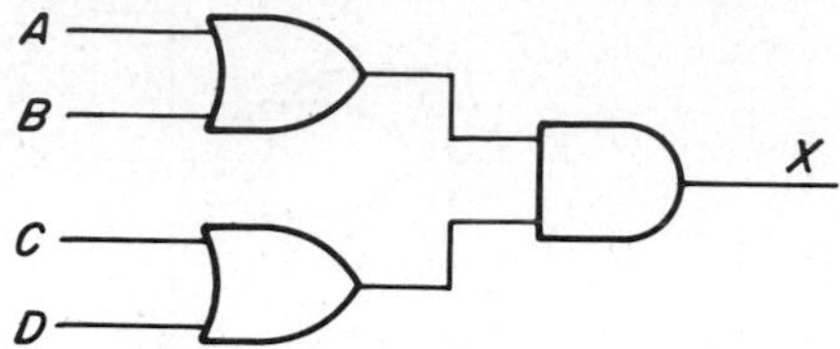

Figure 5–9 Logic network for gate-by-gate analysis.

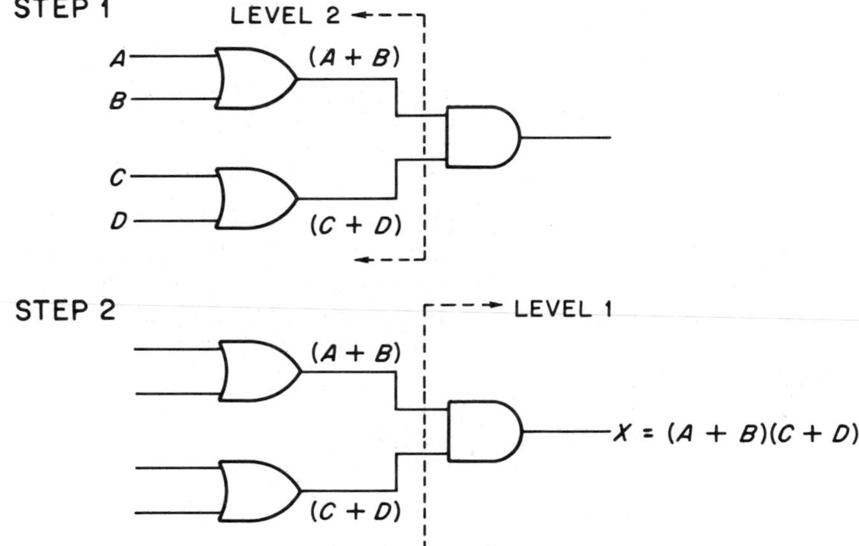

Figure 5–10 Gate-by-gate analysis of given logic network.

The gate-by-gate analysis technique is demonstrated in the following two examples.

Example 5–3

Develop the Boolean equation that represents the logic network of Figure 5–11.

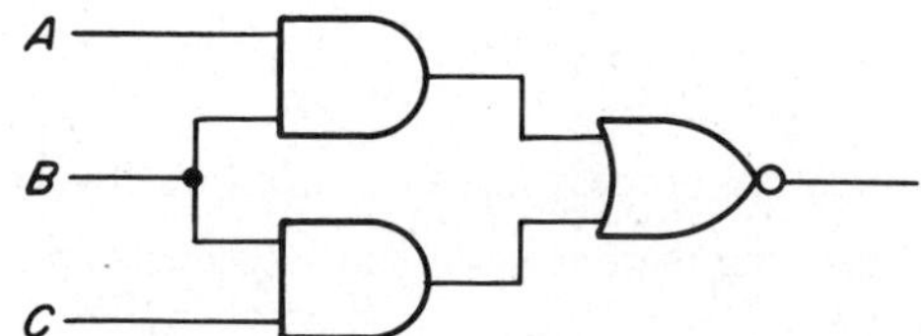

Figure 5–11

Solution

Use the steps outlined in Table 5–2 for gate-by-gate analysis. The process is demonstrated in Figure 5–12.

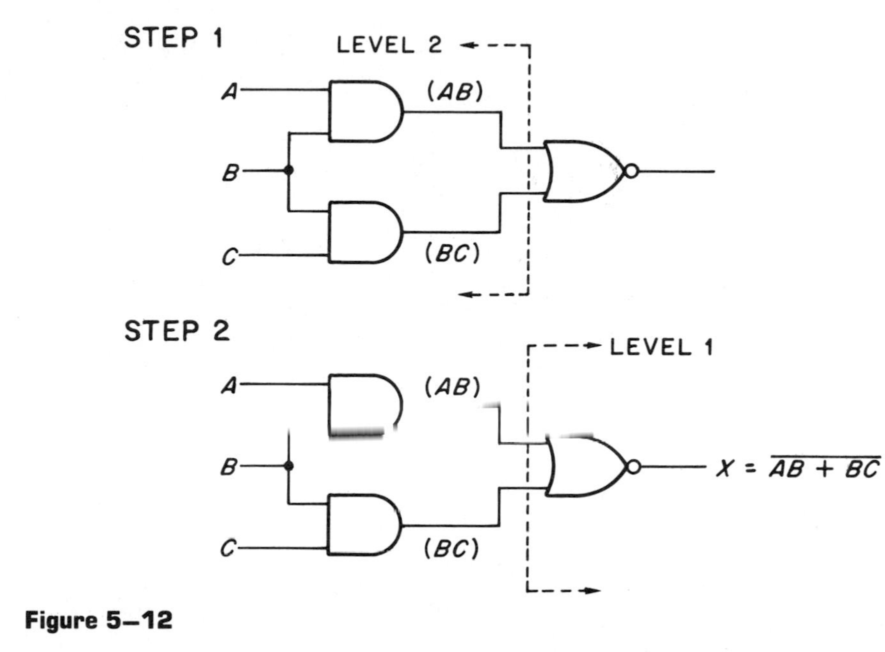

Figure 5–12

The next example demonstrates the gate-by-gate analysis technique for a network of three levels.

Example 5–4

Develop the Boolean expression that represents the logic network of Figure 5–13.

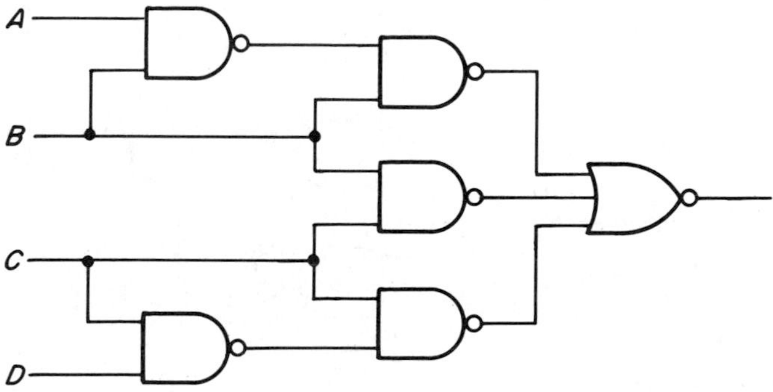

Figure 5–13

Solution

The process, using the gate-by-gate analysis technique, is illustrated in Figure 5–14.

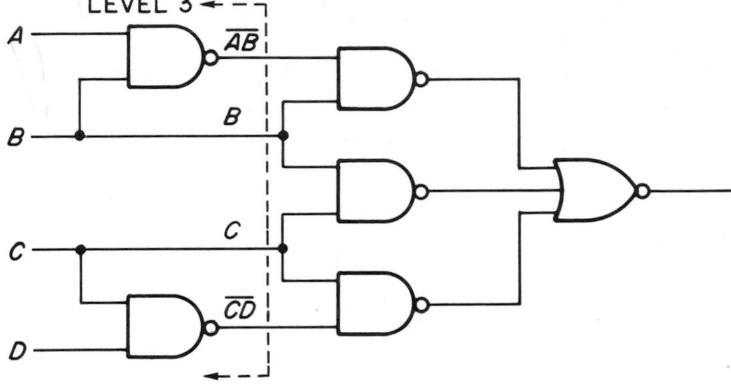

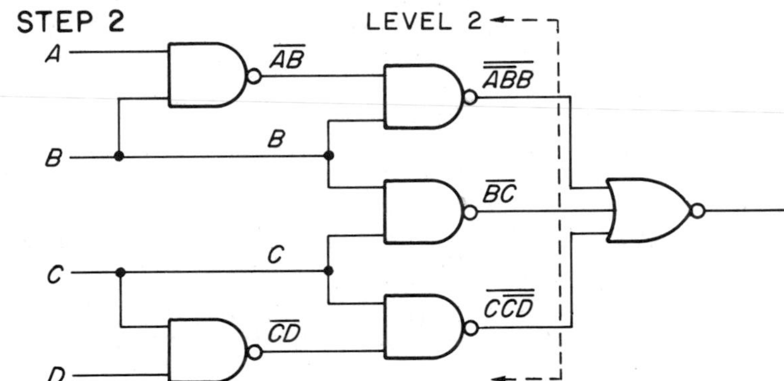

STEP 3

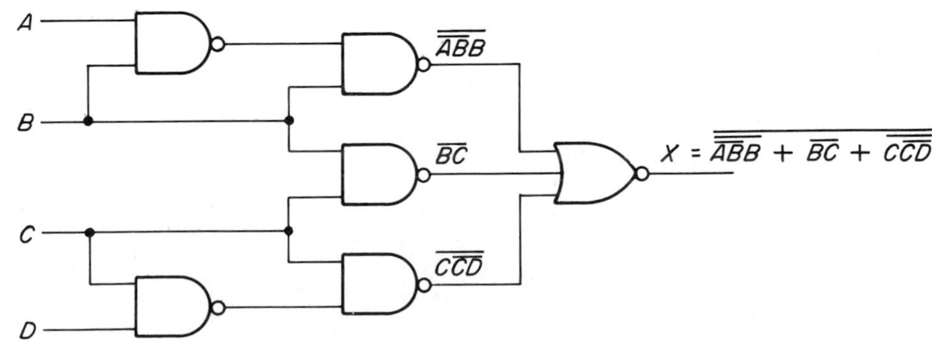

Figure 5–14

Conclusion

In this section you saw two methods that lead to developing a Boolean expression from a given logic network. The next section shows two methods for developing the Boolean expression that represents a given truth table. Check your understanding of this section by trying the following review questions.

5–1 Review Questions

1. What is meant by a *logic level*?
2. Explain the process of analyzing a logic network by the method of empty terms.
3. State the Boolean expression for the logic network of Figure 5–15(A).
4. Develop the Boolean expression for the logic network of Figure 5–15(B).

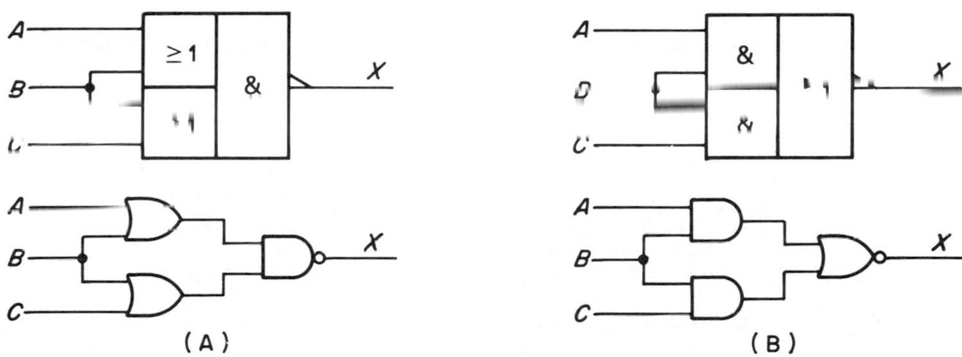

Figure 5–15

5–2 DEVELOPING BOOLEAN EXPRESSIONS FROM TRUTH TABLES

Discussion

In this section, you will learn new techniques in developing a Boolean expression from a given truth table. These techniques are important, because one of the usual first steps in logic design is the development of a truth table. The next important step in the design process is to develop the Boolean expression that represents the truth table. From this, as you will see later in this chapter, it is then possible to minimize the number of logic gates actually needed to implement the required truth table conditions.

Types of Boolean Equations

There are two general types of Boolean equations used in this section for the analysis of truth tables. One type is called the **sum of products** (SOP) form, and the other is the **product of sums** (POS) form.

Sum of Products:

A Boolean expression that consists of the ORing of terms that are ANDed.

Product of Sums:

A Boolean expression that consists of the ANDing of terms that are ORed.

To illustrate, the Boolean expression

$$X = AB + CD$$

is an example of a sum of products (SOP) expression. The Boolean expression

$$X = (A + B)(C + D)$$

is an example of a product of sums (POS) expression. The difference between these forms is illustrated in Figure 5–16. Recall that a Boolean product is the ANDing of logical variables, and a Boolean sum is the ORing of logical variables.

(A) SUM OF PRODUCTS

$X = (\cdot\ \cdot) + (\cdot\ \cdot) + (\cdot\ \cdot)$

BOOLEAN PRODUCTS (OR SINGLE TERMS) SEPARATED BY SUMS

EXAMPLE: $X = ABC + \bar{A}B + C$

(B) PRODUCT OF SUMS

$X = (+\ +)(+\ +)(+\ +)$

BOOLEAN SUMS (OR SINGLE TERMS) SEPARATED BY PRODUCTS

EXAMPLE: $X = (A + B + C)(\bar{A} + B)C$

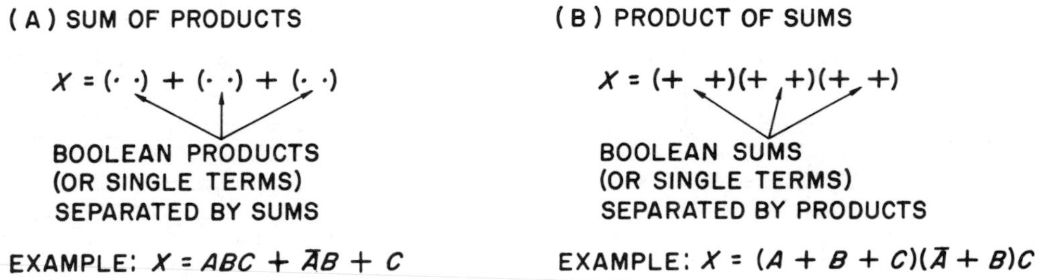

Figure 5–16 Difference between SOP and POS.

Example 5–5

Determine which of the Boolean expressions are POS or SOP.

(A) $AB + CD$

(B) $(A + B)(C + D)$

(C) $ABC + AB + B$

(D) $ABC + \bar{A}BC + A\bar{B}C$

(E) $(\bar{A} + B + \bar{C})(A + B + C)(\bar{A} + \bar{B} + C)$

Solution

By the definition of SOP and POS, you have

(A) SOP (B) POS (C) SOP (D) SOP (E) POS

Both types of Boolean expressions are commonly used in the development and analysis of logic networks. First, the use of SOP will be introduced.

Sum of Products

The sum of products (SOP) form of a Boolean expression can be used to develop a Boolean expression from a given truth table. As an example, suppose you were given the following truth table and asked to develop a Boolean expression that represented it.

A	*B*	*C*	*X*
0	0	0	0
0	0	1	0
0	1	0	1
0	1	1	1
1	0	0	1
1	0	1	0
1	1	0	0
1	1	1	0

As you can see, there are only three conditions when the output will be TRUE. These conditions are

$$A(\text{FALSE}) \text{ AND } B(\text{TRUE}) \text{ AND } C(\text{FALSE})$$

OR

$$A(\text{FALSE}) \text{ AND } B(\text{TRUE}) \text{ AND } C(\text{TRUE})$$

OR

$$A(\text{TRUE}) \text{ AND } B(\text{FALSE}) \text{ AND } C(\text{FALSE})$$

What has just been expressed in words is an SOP expression. You can reduce this to a Boolean expression by signifying $\overline{A}$ as A(FALSE), $\overline{B}$ as B(FALSE), and $\overline{C}$ as C(FALSE). You will then have a sum of products expression:

$$X = \overline{A}B\overline{C} + \overline{A}BC + A\overline{B}\,\overline{C}$$

It's convenient to show where on the original truth table each of these terms originated:

A	*B*	*C*	*X*	
0	0	0	0	
0	0	1	0	
0	1	0	1	← $\overline{A}B\overline{C}$
0	1	1	1	← $\overline{A}BC$
1	0	0	1	← $A\overline{B}\,\overline{C}$
1	0	1	0	
1	1	0	0	
1	1	1	0	

To demonstrate that the resulting Boolean expression does indeed represent the previous truth table, substitute each of the given conditions into the SOP expression:

A	*B*	*C*	$\overline{A}\cdot B\cdot \overline{C} + \overline{A}\cdot B\cdot C + A\cdot \overline{B}\cdot \overline{C}$
0	0	0	$\overline{0}\cdot 0\cdot \overline{0} + \overline{0}\cdot 0\cdot 0 + 0\cdot \overline{0}\cdot \overline{0} = 1\cdot 0\cdot 1 + 1\cdot 0\cdot 0 + 0\cdot 1\cdot 1 = 0 + 0 + 0 = 0$
0	0	1	$\overline{0}\cdot 0\cdot \overline{1} + \overline{0}\cdot 0\cdot 1 + 0\cdot \overline{0}\cdot \overline{1} = 1\cdot 0\cdot 0 + 1\cdot 0\cdot 0 + 0\cdot 1\cdot 0 = 0 + 0 + 0 = 0$
0	1	0	$\overline{0}\cdot 1\cdot \overline{0} + \overline{0}\cdot 1\cdot 0 + 0\cdot \overline{1}\cdot \overline{0} = 1\cdot 1\cdot 1 + 1\cdot 1\cdot 0 + 0\cdot 0\cdot 1 = 1 + 0 + 0 = 1$
0	1	1	$\overline{0}\cdot 1\cdot \overline{1} + \overline{0}\cdot 1\cdot 1 + 0\cdot \overline{1}\cdot \overline{1} = 1\cdot 0\cdot 0 + 1\cdot 1\cdot 1 + 0\cdot 0\cdot 0 = 0 + 1 + 0 = 1$
1	0	0	$\overline{1}\cdot 0\cdot \overline{0} + \overline{1}\cdot 0\cdot 0 + 1\cdot \overline{0}\cdot \overline{0} = 0\cdot 1\cdot 1 + 0\cdot 0\cdot 0 + 1\cdot 1\cdot 1 = 0 + 0 + 1 = 1$
1	0	1	$\overline{1}\cdot 0\cdot \overline{1} + \overline{1}\cdot 0\cdot 1 + 1\cdot \overline{0}\cdot \overline{1} = 0\cdot 0\cdot 0 + 0\cdot 0\cdot 1 + 1\cdot 1\cdot 0 = 0 + 0 + 0 = 0$
1	1	0	$\overline{1}\cdot 1\cdot \overline{0} + \overline{1}\cdot 1\cdot 0 + 1\cdot \overline{1}\cdot \overline{0} = 0\cdot 1\cdot 0 + 0\cdot 1\cdot 0 + 1\cdot 0\cdot 1 = 0 + 0 + 0 = 0$
1	1	1	$\overline{1}\cdot 1\cdot \overline{1} + \overline{1}\cdot 1\cdot 1 + 1\cdot \overline{1}\cdot \overline{1} = 0\cdot 1\cdot 0 + 0\cdot 1\cdot 1 + 1\cdot 0\cdot 0 = 0 + 0 + 0 = 0$

This demonstration verifies that the given SOP Boolean expression does accurately represent the given requirements of the truth table.

To summarize the procedure (SOP from truth table):

1. Inspect the truth table for each output of a 1.
2. For each output of a 1, make a product term of each input variable, complementing those with a 0 condition.
3. When step 2 is completed for the entire truth table, write a Boolean expression that is the sum of all the product terms.

Example 5–6

Using the SOP method, develop a Boolean expression for each of the following truth tables:

(A)

A	*B*	*C*	*X*
0	0	0	1
0	0	1	0
0	1	0	0
0	1	1	1
1	0	0	1
1	0	1	0
1	1	0	0
1	1	1	1

(B)

A	*B*	*C*	*X*
0	0	0	0
0	0	1	1
0	1	0	1
0	1	1	0
1	0	0	0
1	0	1	1
1	1	0	1
1	1	1	0

(C)

A	*B*	*C*	*X*
0	0	0	0
0	0	1	1
0	1	0	0
0	1	1	1
1	0	0	0
1	0	1	1
1	1	0	0
1	1	1	1

(D)

A	*B*	*C*	*X*
0	0	0	1
0	0	1	1
0	1	0	1
0	1	1	0
1	0	0	0
1	0	1	1
1	1	0	1
1	1	1	1

Solution

First, write each product term for each TRUE output condition on the given truth table. Then OR each of these terms to produce the SOP Boolean expression.

(A)

A	B	C	X	
0	0	0	1	$\leftarrow \overline{A}\,\overline{B}\,\overline{C}$
0	0	1	0	
0	1	0	0	
0	1	1	1	$\leftarrow \overline{A}BC$
1	0	0	1	$\leftarrow A\overline{B}\,\overline{C}$
1	0	1	0	
1	1	0	0	
1	1	1	1	$\leftarrow ABC$

The resulting SOP expression is

$$X = \overline{A}\,\overline{B}\,\overline{C} + \overline{A}BC + A\overline{B}\,\overline{C} + ABC$$

(B)

A	B	C	X	
0	0	0	0	
0	0	1	1	$\leftarrow \overline{A}\,\overline{B}C$
0	1	0	1	$\leftarrow \overline{A}B\overline{C}$
0	1	1	0	
1	0	0	0	
1	0	1	1	$\leftarrow A\overline{B}C$
1	1	0	1	$\leftarrow AB\overline{C}$
1	1	1	0	

The resulting SOP expression is

$$X = \overline{A}\,\overline{B}C + \overline{A}B\overline{C} + A\overline{B}C + AB\overline{C}$$

(C)

A	B	C	X	
0	0	0	0	
0	0	1	1	$\leftarrow \overline{A}\,\overline{B}C$
0	1	0	0	
0	1	1	1	$\leftarrow \overline{A}BC$
1	0	0	0	
1	0	1	1	$\leftarrow A\overline{B}C$
1	1	0	0	
1	1	1	1	$\leftarrow ABC$

The resulting SOP expression is

$$X = \overline{A}\,\overline{B}C + \overline{A}BC + A\overline{B}C + ABC$$

(D)

A	B	C	X	
0	0	0	1	$\leftarrow \overline{A}\,\overline{B}\,\overline{C}$
0	0	1	1	$\leftarrow \overline{A}\,\overline{B}C$
0	1	0	1	$\leftarrow \overline{A}B\overline{C}$
0	1	1	0	
1	0	0	0	
1	0	1	1	$\leftarrow A\overline{B}C$
1	1	0	1	$\leftarrow AB\overline{C}$
1	1	1	1	$\leftarrow ABC$

The resulting SOP expression is

$$X = \overline{A}\,\overline{B}\,\overline{C} + \overline{A}\,\overline{B}C + \overline{A}B\overline{C} + A\overline{B}C + AB\overline{C} + ABC$$

Example 5–7

Using the SOP, develop the Boolean expression for the following truth table:

A	B	C	D	X
0	0	0	0	1
0	0	0	1	0
0	0	1	0	0
0	0	1	1	1
0	1	0	0	1
0	1	0	1	0
0	1	1	0	0
0	1	1	1	0
1	0	0	0	0
1	0	0	1	0
1	0	1	0	0
1	0	1	1	1
1	1	0	0	0
1	1	0	1	0
1	1	1	0	1
1	1	1	1	0

Solution

First write the product expression for each TRUE term:

A	B	C	D	X	
0	0	0	0	1	$\leftarrow \overline{A}\,\overline{B}\,\overline{C}\,\overline{D}$
0	0	0	1	0	
0	0	1	0	0	
0	0	1	1	1	$\leftarrow \overline{A}\,\overline{B}CD$
0	1	0	0	1	$\leftarrow \overline{A}B\overline{C}\,\overline{D}$
0	1	0	1	0	
0	1	1	0	0	
0	1	1	1	0	
1	0	0	0	0	
1	0	0	1	0	
1	0	1	0	0	
1	0	1	1	1	$\leftarrow A\overline{B}CD$
1	1	0	0	0	
1	1	0	1	0	
1	1	1	0	1	$\leftarrow ABC\overline{D}$
1	1	1	1	0	

The resulting sum of products expression is

$$X = \overline{A}\,\overline{B}\,\overline{C}\,\overline{D} + \overline{A}\,\overline{B}CD + \overline{A}B\overline{C}\,\overline{D} + A\overline{B}CD + ABC\overline{D}$$

Again, don't worry yet about simplifying the resulting Boolean expression. You will see how to do that later in this chapter. What is important now is that you develop your analysis skills one step at a time. Once you have built confidence in one level, then move on to the next.

Product of Sums

The product of sums (POS) is another method for developing a Boolean expression from a given truth table. It is important that you know both of these methods because the ability to use either one will allow you to use the easiest analysis method for a given truth table.

In the SOP method, you were concerned with the TRUE terms. In the POS method, you will be concerned with the FALSE terms. You will see that there are advantages to both systems, depending upon the desired outcome of the given truth table.

As an example, consider the following truth table.

A	B	C	X
0	0	0	0
0	0	1	1
0	1	0	1
0	1	1	1
1	0	0	0
1	0	1	1
1	1	0	1
1	1	1	0

Now, concentrate on only the FALSE terms. The output is FALSE for three conditions. What is needed now is to write a Boolean expression for these three conditions. This can be done as follows (POS from truth table).

1. Inspect the truth table for each output of a 0.
2. For each output of a 0, make a sum term of each input variable, complementing those with a 1 condition.
3. When Step 2 is completed for the entire table, write a Boolean expression that is the product of all the sum terms.

Thus for the given truth table:

A	*B*	*C*	*X*	
0	0	0	0	$\leftarrow A + B + C$
0	0	1	1	
0	1	0	1	
0	1	1	1	
1	0	0	0	$\leftarrow \overline{A} + B + C$
1	0	1	1	
1	1	0	1	
1	1	1	0	$\leftarrow \overline{A} + \overline{B} + \overline{C}$

If these terms are ANDed and any one term is 0, then the result will be 0.

The resulting product of sums expression is therefore

$$X = (A + B + C)(\overline{A} + B + C)(\overline{A} + \overline{B} + \overline{C})$$

You can demonstrate that this Boolean statement represents the FALSE conditions of the truth table in the same manner as you did for the SOP expression:

A	*B*	*C*	$(A + B + C)\ (\overline{A} + B + C)\ (\overline{A} + \overline{B} + \overline{C})$
0	0	0	$(0 + 0 + 0)\ (\overline{0} + 0 + 0)\ (\overline{0} + \overline{0} + \overline{0}) = (0)\cdot(1 + 0 + 0)\cdot(1 + 1 + 1)$ $= (0)\cdot(1)\cdot(1) = 0$
0	0	1	$(0 + 0 + 1)\ (\overline{0} + 0 + 1)\ (\overline{0} + \overline{0} + \overline{1}) = (1)\cdot(1 + 0 + 1)\cdot(1 + 1 + 1)$ $= (1)\cdot(1)\cdot(1) = 1$
0	1	0	$(0 + 1 + 0)\ (\overline{0} + 1 + 0)\ (\overline{0} + \overline{1} + \overline{0}) = (1)\cdot(1 + 1 + 0)\cdot(1 + 0 + 1)$ $= (1)\cdot(1)\cdot(1) = 1$
0	1	1	$(0 + 1 + 1)\ (\overline{0} + 1 + 1)\ (\overline{0} + \overline{1} + \overline{1}) = (1)\cdot(1 + 1 + 1)\cdot(1 + 0 + 0)$ $= (1)\cdot(1)\cdot(1) = 1$
1	0	0	$(1 + 0 + 0)\ (\overline{1} + 0 + 0)\ (\overline{1} + \overline{0} + \overline{0}) = (1)\cdot(0 + 0 + 0)\cdot(0 + 1 + 1)$ $= (1)\cdot(0)\cdot(1) = 0$
1	0	1	$(1 + 0 + 1)\ (\overline{1} + 0 + 1)\ (\overline{1} + \overline{0} + \overline{1}) = (1)\cdot(0 + 0 + 1)\cdot(0 + 1 + 0)$ $= (1)\cdot(1)\cdot(1) = 1$
1	1	0	$(1 + 1 + 0)\ (\overline{1} + 1 + 0)\ (\overline{1} + \overline{1} + \overline{0}) = (1)\cdot(0 + 1 + 0)\cdot(0 + 0 + 1)$ $= (1)\cdot(1)\cdot(1) = 1$
1	1	1	$(1 + 1 + 1)\ (\overline{1} + 1 + 1)\ (\overline{1} + \overline{1} + \overline{1}) = (1)\cdot(0 + 1 + 1)\cdot(0 + 0 + 0)$ $= (1)\cdot(1)\cdot(0) = 0$

As you can see from the analysis, the POS method does an equally good job of predicting the required conditions of a given truth table.

The following two examples illustrate the POS method for developing a Boolean expression from truth tables.

Example 5–8

Using the POS method, develop the Boolean expression for the following truth tables:

(A)

A	*B*	*C*	*X*
0	0	0	1
0	0	1	0
0	1	0	0
0	1	1	1
1	0	0	1
1	0	1	0
1	1	0	0
1	1	1	1

(B)

A	*B*	*C*	*X*
0	0	0	0
0	0	1	1
0	1	0	1
0	1	1	0
1	0	0	0
1	0	1	1
1	1	0	1
1	1	1	0

(C)

A	*B*	*C*	*X*
0	0	0	0
0	0	1	1
0	1	0	0
0	1	1	1
1	0	0	0
1	0	1	1
1	1	0	0
1	1	1	1

(D)

A	*B*	*C*	*X*
0	0	0	1
0	0	1	1
0	1	0	1
0	1	1	0
1	0	0	0
1	0	1	1
1	1	0	1
1	1	1	1

Solution

(A)

A	*B*	*C*	*X*	
0	0	0	1	
0	0	1	0	$\rightarrow (A + B + \overline{C})$
0	1	0	0	$\rightarrow (A + \overline{B} + C)$
0	1	1	1	
1	0	0	1	
1	0	1	0	$\rightarrow (\overline{A} + B + \overline{C})$
1	1	0	0	$\rightarrow (\overline{A} + \overline{B} + C)$
1	1	1	1	

The resulting Boolean expression is

$$X = (A + B + \overline{C})\,(A + \overline{B} + C)\,(\overline{A} + B + \overline{C})\,(\overline{A} + \overline{B} + C)$$

(B)

A	B	C	X	
0	0	0	0	$\rightarrow (A + B + C)$
0	0	1	1	
0	1	0	1	
0	1	1	0	$\rightarrow (A + \overline{B} + \overline{C})$
1	0	0	0	$\rightarrow (\overline{A} + B + C)$
1	0	1	1	
1	1	0	1	
1	1	1	0	$\rightarrow (\overline{A} + \overline{B} + \overline{C})$

The resulting Boolean expression is

$$X = (A + B + C)(A + \overline{B} + \overline{C})(\overline{A} + B + C)(\overline{A} + \overline{B} + \overline{C})$$

(C)

A	B	C	X	
0	0	0	0	$\rightarrow (A + B + C)$
0	0	1	1	
0	1	0	0	$\rightarrow (A + \overline{B} + C)$
0	1	1	1	
1	0	0	0	$\rightarrow (\overline{A} + B + C)$
1	0	1	1	
1	1	0	0	$\rightarrow (\overline{A} + \overline{B} + C)$
1	1	1	1	

The resulting Boolean expression is

$$X = (A + B + C)(A + \overline{B} + C)(\overline{A} + B + C)(\overline{A} + \overline{B} + C)$$

(D)

A	B	C	X	
0	0	0	1	
0	0	1	1	
0	1	0	1	
0	1	1	0	$\rightarrow (A + \overline{B} + \overline{C})$
1	0	0	0	$\rightarrow (\overline{A} + B + C)$
1	0	1	1	
1	1	0	1	
1	1	1	1	

The resulting Boolean expression is

$$X = (A + \overline{B} + \overline{C})(\overline{A} + B + C)$$

Example 5–9

Using the POS method, develop the Boolean expression for the following truth table:

A	B	C	D	X
0	0	0	0	1
0	0	0	1	1
0	0	1	0	1
0	0	1	1	1
0	1	0	0	1
0	1	0	1	1
0	1	1	0	0
0	1	1	1	1
1	0	0	0	1
1	0	0	1	1
1	0	1	0	1
1	0	1	1	0
1	1	0	0	1
1	1	0	1	1
1	1	1	0	0
1	1	1	1	1

Solution

Write a sum term for each FALSE result. Then form the products of these terms:

A	B	C	D	X	
0	0	0	0	1	
0	0	0	1	1	
0	0	1	0	1	
0	0	1	1	1	
0	1	0	0	1	
0	1	0	1	1	
0	1	1	0	0	$\rightarrow (A + \overline{B} + \overline{C} + D)$
0	1	1	1	1	
1	0	0	0	1	
1	0	0	1	1	
1	0	1	0	1	
1	0	1	1	0	$\rightarrow (\overline{A} + B + \overline{C} + \overline{D})$
1	1	0	0	1	
1	1	0	1	1	
1	1	1	0	0	$\rightarrow (\overline{A} + \overline{B} + \overline{C} + D)$
1	1	1	1	1	

The resulting Boolean expression is

$$X = (A + \overline{B} + \overline{C} + D)(\overline{A} + B + \overline{C} + \overline{D})(\overline{A} + \overline{B} + \overline{C} + D)$$

Which Method?

You may be asking yourself which of the two methods (POS or SOP) is best to use when developing a Boolean expression. The answer is, use the method that produces the fewest terms. For example, consider the following two truth tables:

Table A

A	B	C	X
0	0	0	1
0	0	1	0
0	1	0	0
0	1	1	0
1	0	0	1
1	0	1	0
1	1	0	0
1	1	1	0

Table B

A	B	C	X
0	0	0	1
0	0	1	1
0	1	0	0
0	1	1	1
1	0	0	1
1	0	1	1
1	1	0	0
1	1	1	1

The Boolean expression for either table may be expressed by using POS or SOP. However, if the SOP method is used for Table A, the Boolean expression becomes

$$X = (\overline{A}\,\overline{B}\,\overline{C}) + (A\overline{B}\,\overline{C})$$

If the POS expression is used for the same table, it is

$$X = (A + B + \overline{C})(A + \overline{B} + C)(A + \overline{B} + \overline{C})(\overline{A} + B + \overline{C})(\overline{A} + \overline{B} + C)(\overline{A} + \overline{B} + \overline{C})$$

Now consider Table B. For this table, the Boolean expression for the SOP method is

$$X = (\overline{A}\,\overline{B}\,\overline{C}) + (\overline{A}\,\overline{B}C) + (\overline{A}BC) + (A\overline{B}\,\overline{C}) + (A\overline{B}C) + (ABC)$$

while the POS method produces

$$X = (A + \overline{B} + C)(\overline{A} + \overline{B} + C)$$

Conclusion

In this section you discovered two methods of developing a Boolean expression from a given truth table. The Boolean expressions arrived at by this method are not necessarily the most efficient form, but you will soon discover methods for simplifying them. In the next section, you will see how to develop logic networks that represent a given Boolean expression. You can see that you are on your way to creating real digital systems that can replicate ideas. Check your understanding of this section by trying the following review questions.

5–2 Review Questions

1. State the meaning of the term *sum of products* as it applies to a Boolean expression.
2. State the meaning of the term *product of sums* as it applies to a Boolean expression.
3. Explain how a *sum of products* expression is developed from a truth table.
4. Explain how a *product of sums* expression is developed from a truth table.
5. Explain how to know whether to use POS or SOP in developing a Boolean expression from a truth table.

5–3 LOGIC NETWORKS FROM BOOLEAN EXPRESSIONS

Discussion

You now have the ability to develop Boolean expressions from a given logic diagram or truth table. In this section, you will discover techniques for developing logic networks from a given Boolean expression. You will then be equipped with another link in the chain of taking an idea and making it a reality in the world of digital systems.

Two General Forms

Every logic circuit has a truth table. As you learned in the last section, you could develop a SOP or POS Boolean expression from *any* truth table. What this means is that any logic circuit may have its logic expressed in a POS or SOP Boolean expression.

In this section, you will learn the techniques for converting SOP and POS Boolean expressions into equivalent AND/OR logic networks and how to convert these networks to the more practical NAND/NOR networks using real digital integrated circuits.

Sum of Products Method

Given any SOP form of a Boolean expression, you can convert it to a logic circuit, using AND/OR gates. The process is demonstrated in Figure 5–17. The step-by-step process is listed in Table 5–3.

As an example, consider the following SOP expression:

$$X = A + BC + DE\overline{F}$$

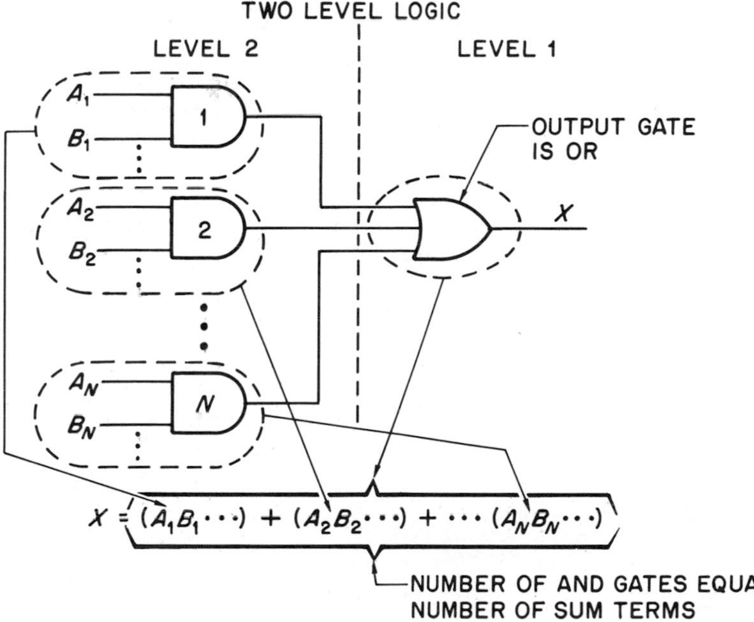

Figure 5–17 Converting SOP expression to logic gates.

Table 5–3	STEPS FOR CONVERTING SOP TO GATE LOGIC
Step	**Procedure**
1	Recognize that the resultant logic network will consist of two levels.
2	The first level will be an OR gate.
3	The number of inputs to the OR gate is equal to the number of product terms.
4	Each input to the OR gate is an AND gate or a single variable.
5	The number of inputs to each AND gate equals the number of variables in each product.
6	The inputs to each AND gate are the actual variables of each product.

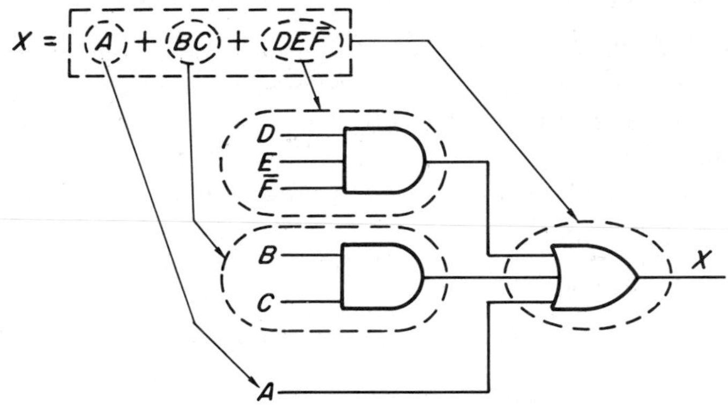

Figure 5–18 Example of SOP conversion to logic network.

Figure 5–18 graphically demonstrates the conversion process. The process is further demonstrated by the following example.

Example 5–10

Develop the equivalent AND/OR logic network for each of the following SOP expressions:

(A) $AB + \overline{A}B$
(B) $ABC + \overline{A}B + \overline{B}C$
(C) $A + B + \overline{B}C + A\overline{B}CD$

Solution

Figure 5–19 illustrates the resulting gates using the steps outlined in Table 5–3.

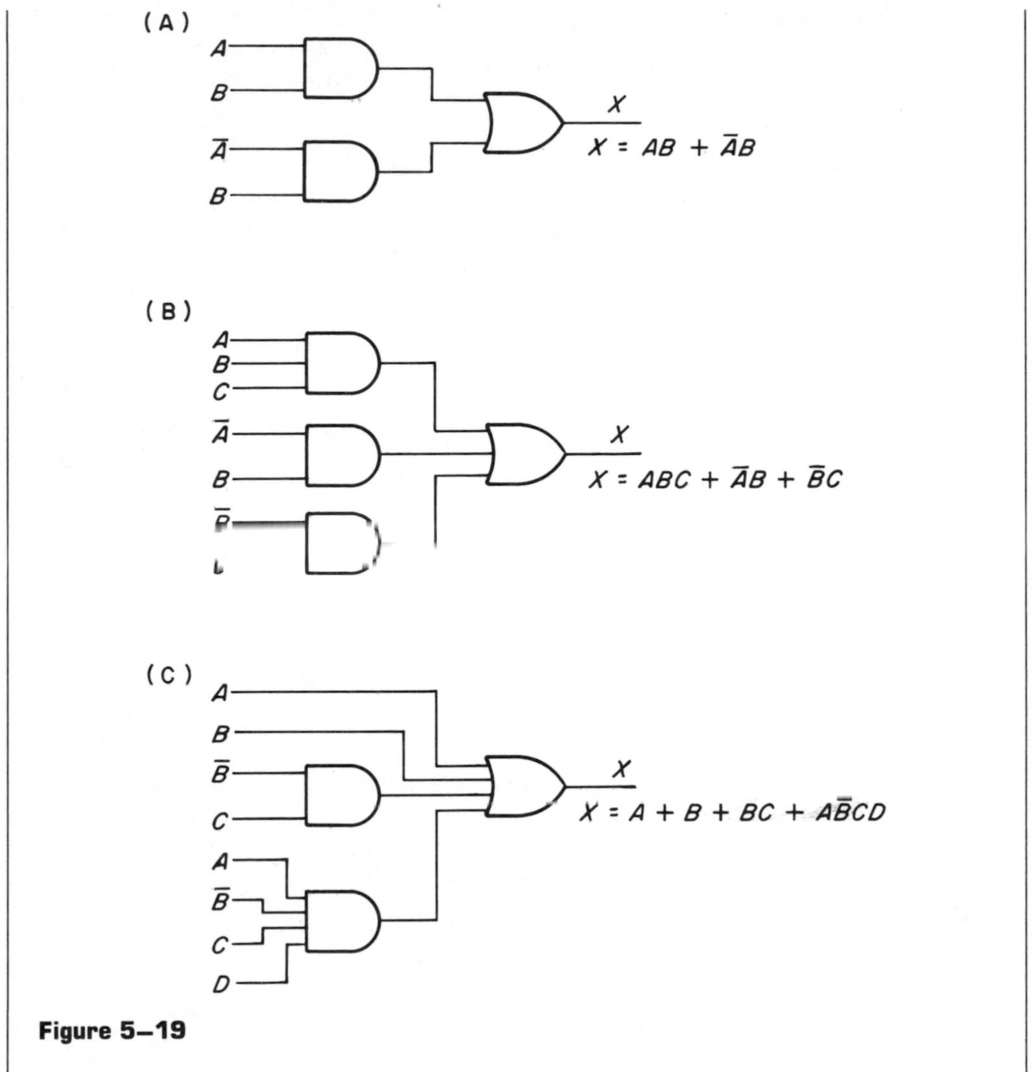

Figure 5–19

Practical Logic Gates

As you saw from the previous examples, the conversion from SOP to a logic circuit is a straightforward process. However, to actually implement the circuit using AND/OR logic is not practical. For instance, the last gate network of Example 1 required a four-input OR gate. Such a gate is not available in an integrated circuit. A more practical approach is to convert the AND/OR logic into its equivalent NAND/NOR logic. This process is also quite straightforward.

Recall the graphical techniques for simplifying logic networks, where part of the process consisted of replacing a double inversion with a single connection. This means that you can also do the opposite, that is, replace a single connection with a double inversion.

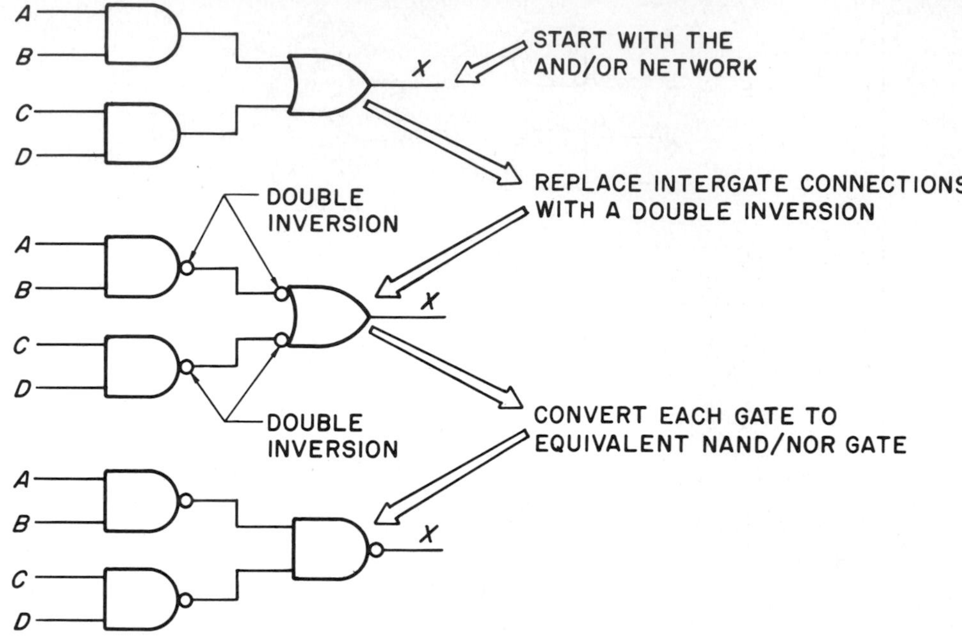

Figure 5–20 Converting from AND/OR to NAND/NOR.

Doing this with an AND/OR gate network produces a surprising result. This is illustrated in Figure 5–20.

As you can see from the figure, the resulting logic network consists solely of NAND gates! This is another reason for their popularity—it is a very practical means of implementing a Boolean SOP into real logic gates.

Example 5–11

Convert the following Boolean expressions to their equivalent NAND/NOR gates:

(A) $AB + C$

(B) $\overline{A}B + AB + AC$

(C) $ABC + A\overline{B}C + ABC + C\overline{D}$

Solution

First convert the expressions to their AND/OR equivalent forms. Then use the double-bubble interlogic connection procedure to convert them to NAND/NOR logic. The solutions are illustrated in Figure 5–21.

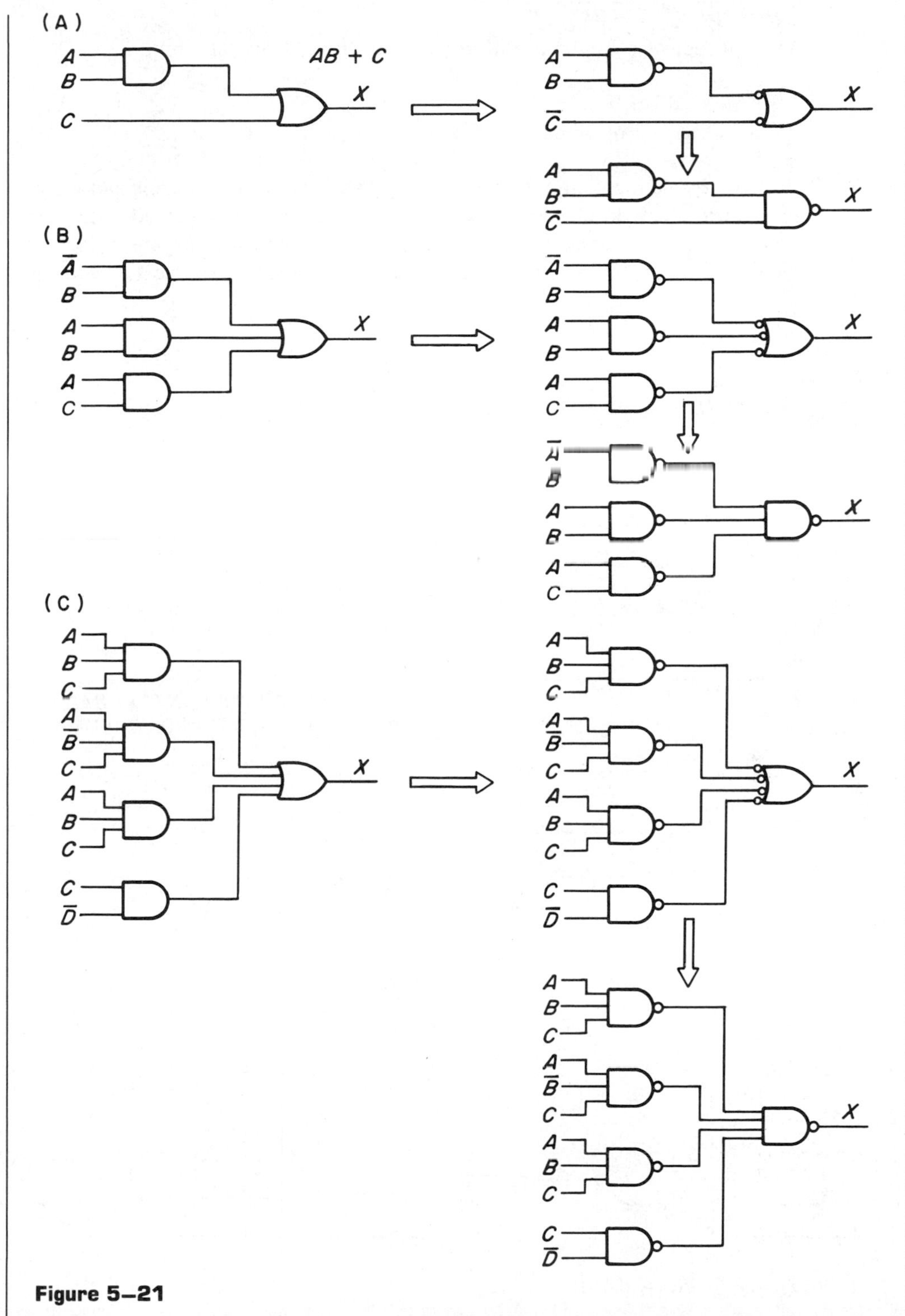
(A)
AB + C
A
B
C
X
(B)
$\overline{A}$
B
A
B
A
C
X
(C)
A
B
C
A
$\overline{B}$
C
A
B
C
C
$\overline{D}$
X

Figure 5–21

Concept of a Bus

The concept of a **bus** is useful in the realization of digital circuits.

Bus:

A group of wires in a digital system, treated as a unit.

In many of the logic circuits presented up to this point, logic terms and their complements have been used. In a practical sense, this is usually implemented with a group of wires that contains the variable and its complement, as shown in Figure 5–22. As you can see from the figure, a bus creates a convenient and practical source for Boolean variables.

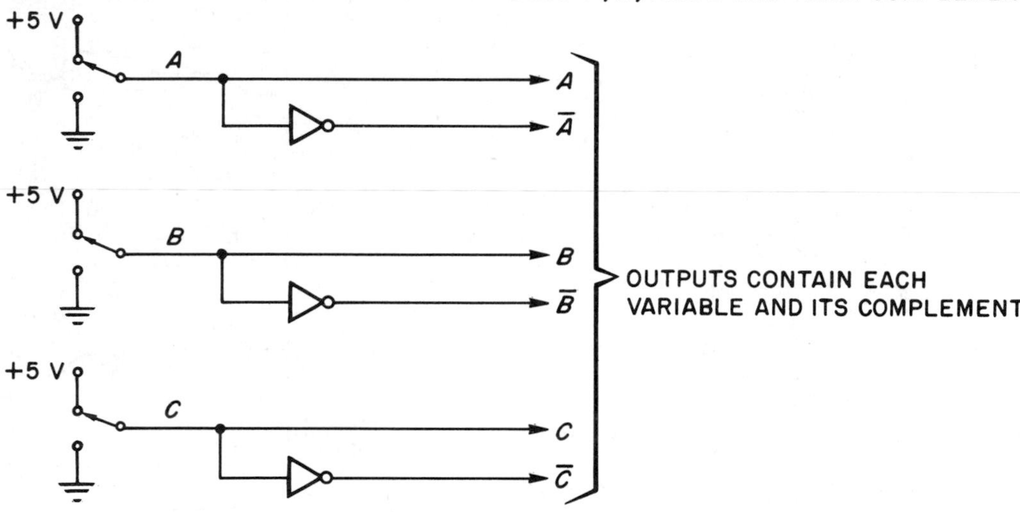

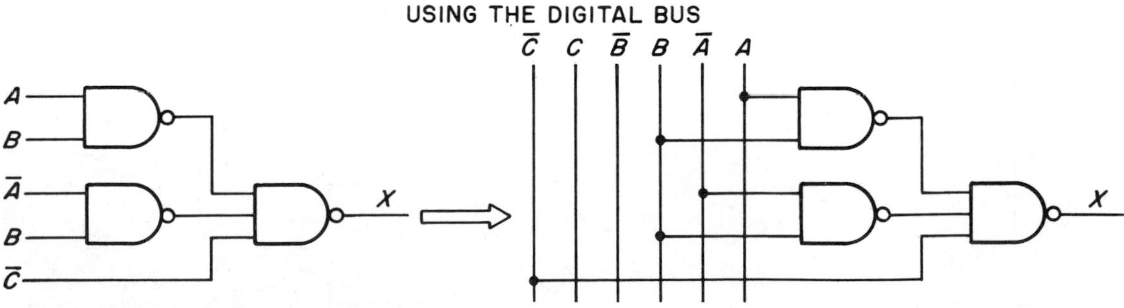

Figure 5–22 A digital bus.

Gates from POS

Converting a POS expression to its logic circuit equivalent is a straightforward process. This is illustrated in Figure 5–23 and Table 5–4 outlines the process.

As an example, consider the following SOP expression:

$$X = A(B + C)(D + E + \overline{F})$$

Figure 5–24 graphically demonstrates the conversion process. This process is further illustrated by the following examples.

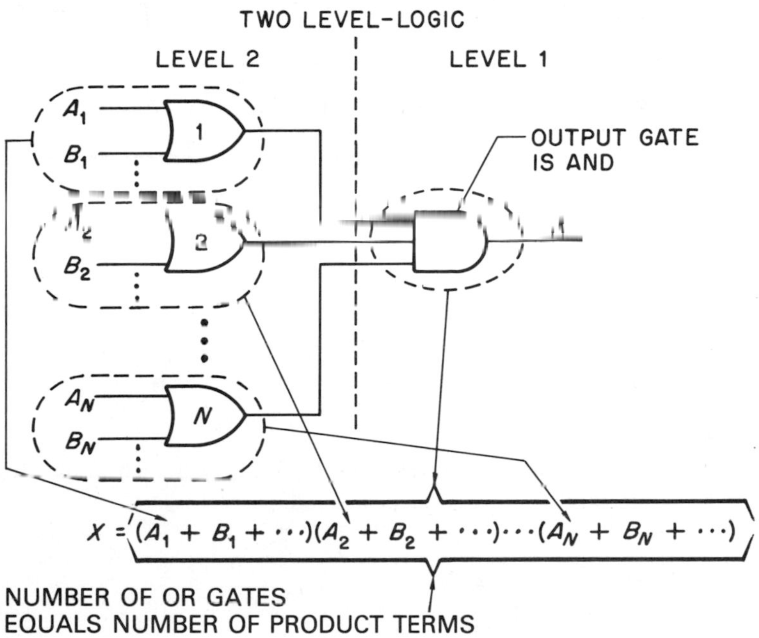

Figure 5–23 Converting POS into AND/OR logic.

Table 5–4	**STEPS FOR CONVERTING POS TO GATE LOGIC**
Step	**Procedure**
1	Recognize that the resultant logic network will consist of two levels.
2	The first level will be an AND gate.
3	The number of inputs to the AND gate is equal to the number of sum terms.
4	Each input to the AND gate is an OR gate or a single variable.
5	The number of inputs to each OR gate equals the number of variables in each sum.
6	The inputs to each OR gate are the variables of each sum.

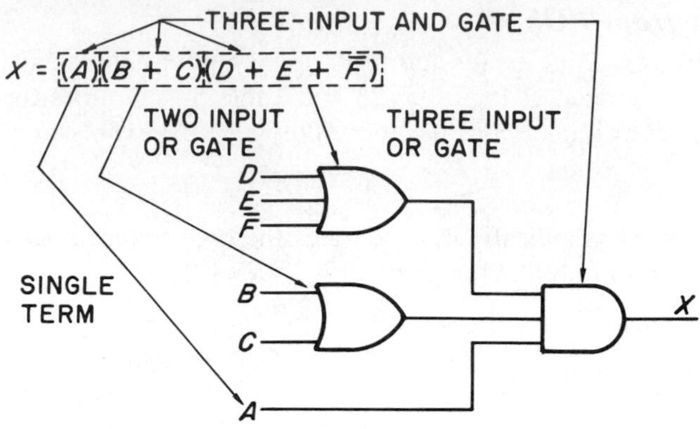

Figure 5–24 Example of conversion of POS to AND/OR logic.

Example 5–12

Develop the equivalent AND/OR logic for each of the following POS expressions:

(A) $(A + B)(\overline{A} + B)$

(B) $(A + B + C)(\overline{A} + B)(\overline{B} + C)$

(C) $AB(\overline{B} + C)(A + \overline{B} + C + D)$

Solution

Figure 5–25 illustrates the resulting logic networks using the steps outlined in Table 5–4. Note that the bus concept is used.

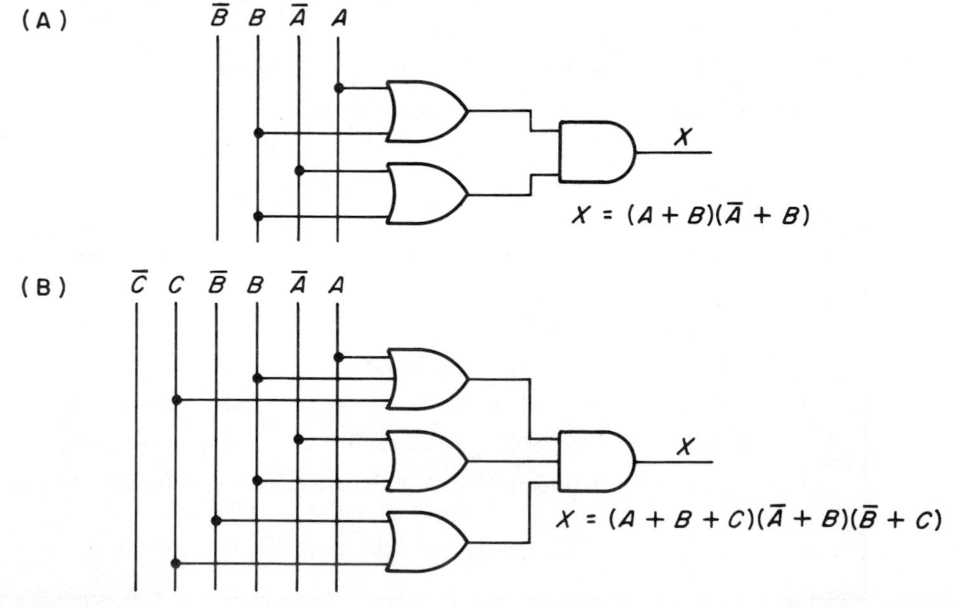

Figure 5–25

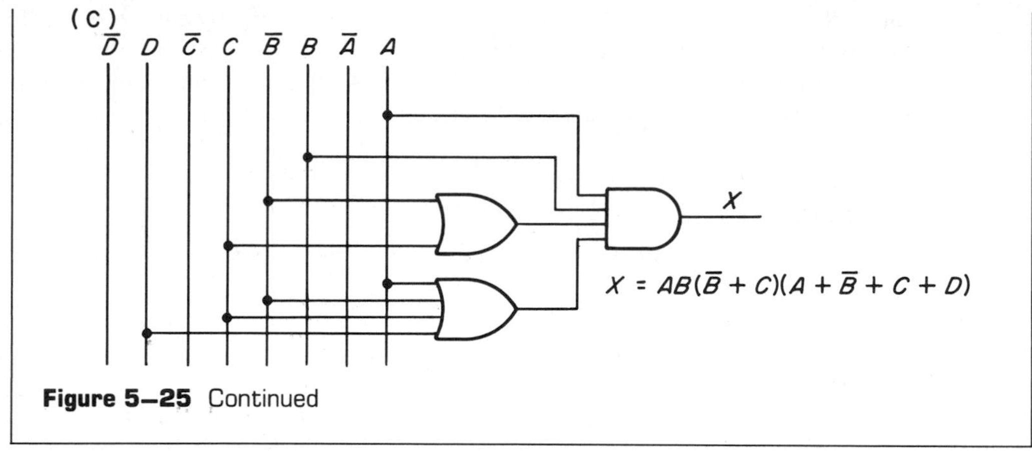

Figure 5–25 Continued

More Practical Circuits

As with the SOP, the conversion from POS to a logic network is again a straightforward process. However, you still need to substitute practical NAND/NOR logic for implementing the actual circuit.

Fortunately, the same process can be used here for conversion to NAND/NOR logic as was used for the SOP logic. The process is illustrated in Figure 5–26. Converting a POS logic circuit to NAND/NOR logic leaves you with a resulting logic network that consists only of NOR gates. Contrast this with the SOP logic that resulted in all NAND

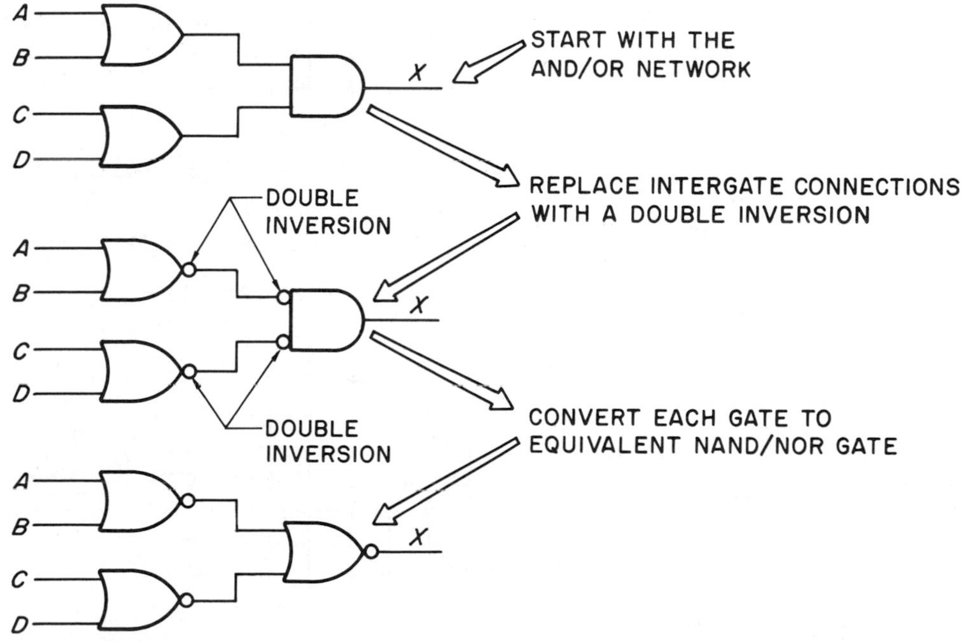

Figure 5–26 Converting from AND/OR to NAND/NOR.

gates. Since you can create a SOP or POS expression from any truth table, it follows that any truth table can therefore be represented by all NAND gates or all NOR gates. This is as you would expect, since, you may recall, NAND gates alone can be used to represent any logic function, and so can NOR gates.

Example 5–13

Convert the following Boolean expressions to their equivalent NAND/NOR gates:

(A) $(A + B)C$

(B) $(\overline{A} + B)(A + B + C)(B + \overline{C})$

(C) $\overline{A}B + A\overline{B}$

(D) $ABC + \overline{A}BC + A\overline{B}C$

Solution

First convert the expression to its AND/OR equivalent. Then use the double-bubble interlogic conversion technique to convert each AND/OR combination to its equivalent NAND/NOR combination. See Figure 5–27.

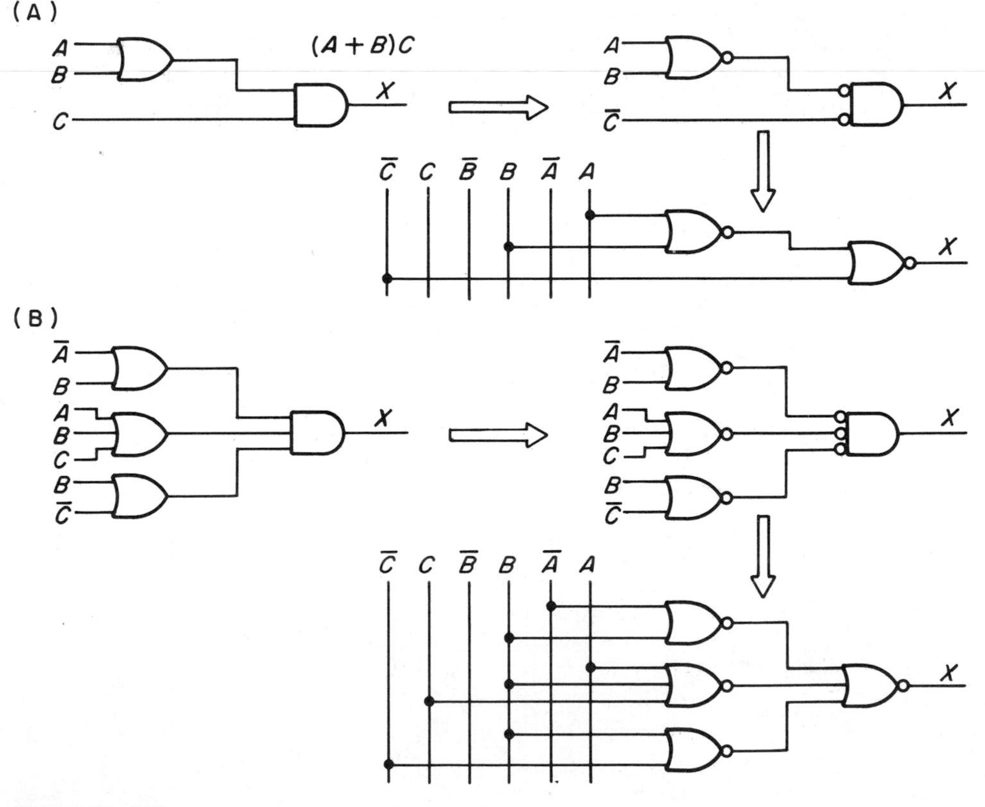

Figure 5–27

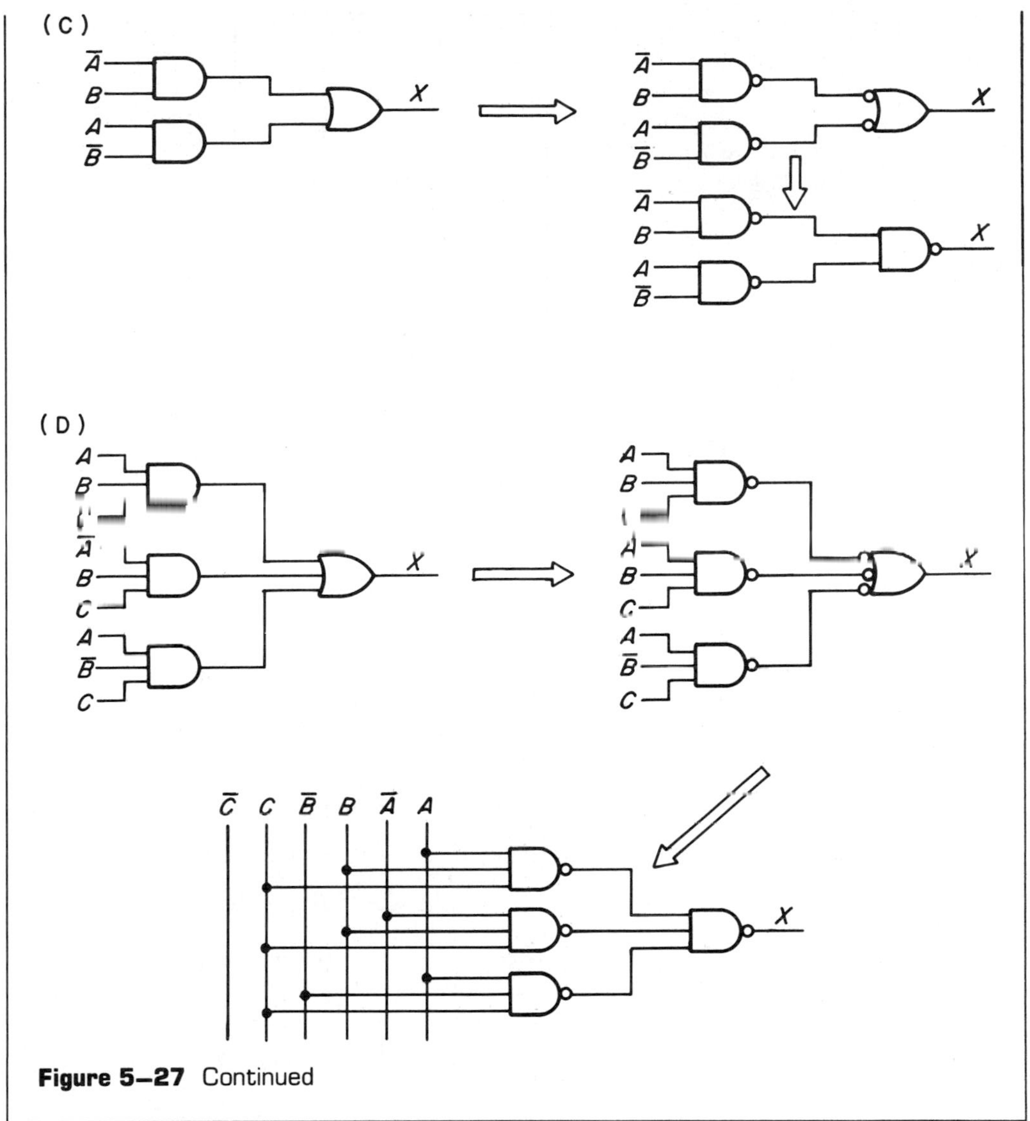

Figure 5–27 Continued

The following example illustrates the duality of NAND/NOR logic and reinforces the concept that you can replicate the logic of any truth table using *only* NAND gates or *only* NOR gates.

Example 5–14

Show how you would construct a logic circuit to implement the conditions of the following truth table, using (A) NAND gates, (B) NOR gates.

A	B	C	X
0	0	0	1
0	0	1	0
0	1	0	1
0	1	1	0
1	0	0	1
1	0	1	0
1	1	0	1
1	1	1	0

Solution

For using NAND gates, develop the SOP form of the Boolean expression for the given truth table:

A	B	C	X	
0	0	0	1	$\rightarrow \overline{A}\,\overline{B}\,\overline{C}$
0	0	1	0	
0	1	0	1	$\rightarrow \overline{A}B\overline{C}$
0	1	1	0	
1	0	0	1	$\rightarrow A\overline{B}\,\overline{C}$
1	0	1	0	
1	1	0	1	$\rightarrow AB\overline{C}$
1	1	1	0	

Thus, the SOP Boolean expression becomes

$$X = \overline{A}\,\overline{B}\,\overline{C} + \overline{A}B\overline{C} + A\overline{B}\,\overline{C} + AB\overline{C}$$

For NOR gates, develop the POS form of the Boolean expression:

A	B	C	X	
0	0	0	1	
0	0	1	0	$\rightarrow A + B + \overline{C}$
0	1	0	1	
0	1	1	0	$\rightarrow A + \overline{B} + \overline{C}$
1	0	0	1	
1	0	1	0	$\rightarrow \overline{A} + B + \overline{C}$
1	1	0	1	
1	1	1	0	$\rightarrow \overline{A} + \overline{B} + \overline{C}$

Thus, the POS Boolean expression becomes

$$X = (A + B + \overline{C})(A + \overline{B} + \overline{C})(\overline{A} + B + \overline{C})(\overline{A} + \overline{B} + \overline{C})$$

The resultant gate networks for each of these expressions is shown in Figure 5–28.

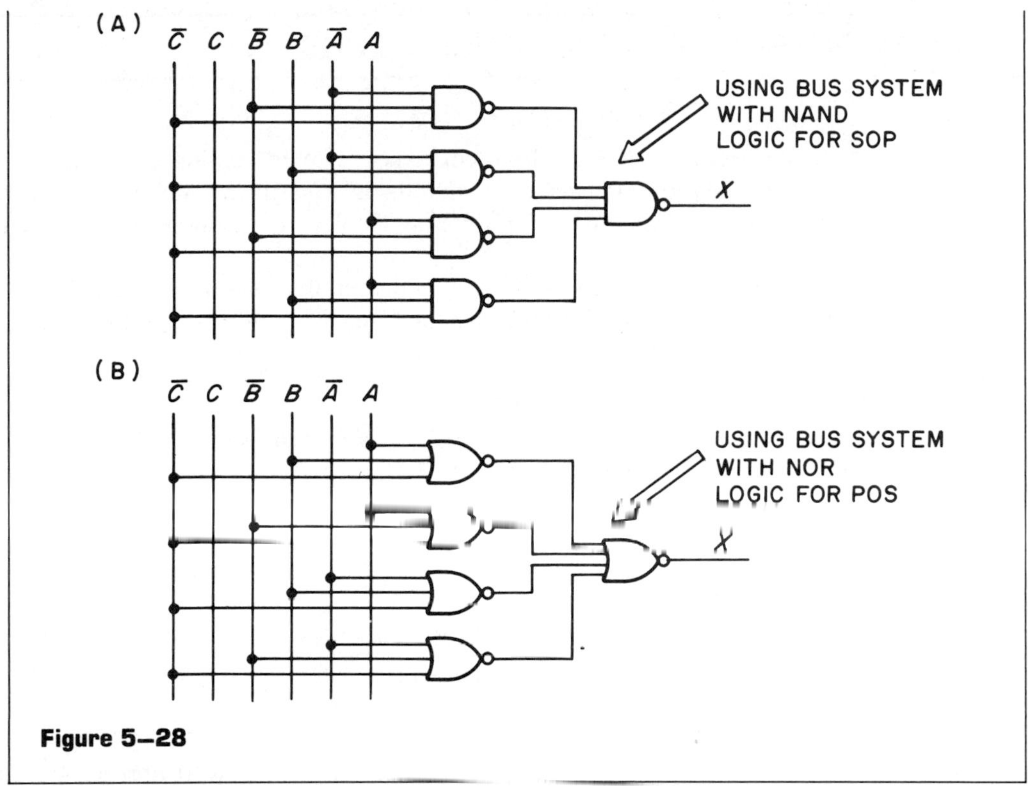

Figure 5–28

The previous logic circuits will implement the given truth table. They *do not represent the simplest logic networks possible*—you will see a method of doing that in the next section. The important point is that you can create a real logic network that will implement the requirements of any truth table.

Conclusion

In this section, you discovered how to develop a logic circuit that would implement the logic requirements of any truth table. The two approaches presented here were quite straightforward. By this, your only task was to develop a logic circuit that would "do the job." However, many of these logic circuits can be simplified, with a resulting reduction in their complexity. That's the topic of the next section. Check your understanding of this section by trying the following review questions.

5–3 Review Questions

1. State the type of AND/OR gate logic used to implement a POS Boolean expression.
2. Give the type of AND/OR logic used to implement an SOP Boolean expression.
3. What kind of NAND/NOR logic does a POS Boolean expression require?
4. For an SOP Boolean expression, state the kind of NAND/NOR logic that will represent it.
5. Explain what is meant by the statement that any logic network can be represented by all NAND gates or all NOR gates. Why is this possible?

5–4 LOGIC SIMPLIFICATION TECHNIQUES

Discussion

Up to this point, you have not been concerned with attempting to reduce the number of logic gates needed to reproduce the requirements of a given truth table, so that you could concentrate on the techniques of developing Boolean expressions and logic circuits from truth tables.

In this section, you will discover a technique that produces a visual graphic display of a given logic condition. This is much the same as a truth table, except that it will allow you to eliminate logic redundancy.

Logic Network Simplification

Consider the logic network of Figure 5–29. As you can see from the figure, replacing a whole logic network by a single piece of wire obviously saves money and greatly increases the total system reliability. The question is, what are the techniques involved for the reduction of a logic network into its simplest form?

One method of reducing logic gates is by using Boolean algebra, as was done in Figure 5–29. As you can see, using Boolean algebra for the reduction of logic circuits requires a familiarity with Boolean expressions that is gained by experience and some intuition.

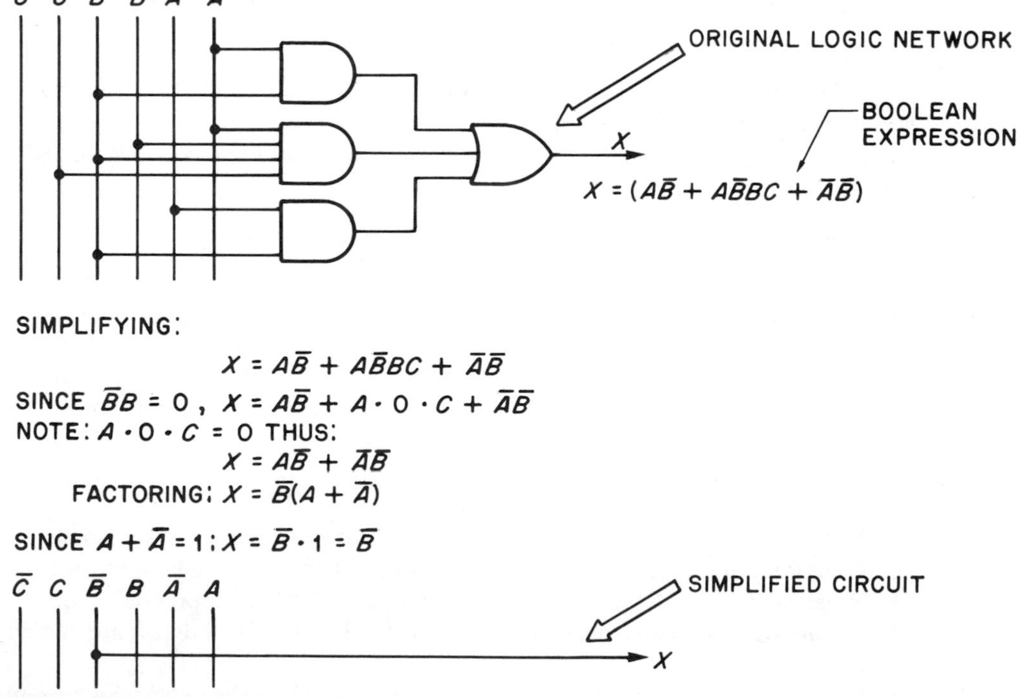

Figure 5–29 Example of logic network simplification.

There is another method of gate network simplification that is so straightforward it is as predictable as a computer program. Essentially, you can think of this method as a graphical program that has a very specific set of rules that can help you reduce the number of logic gates within a given logic network.

Karnaugh Mapping

A **Karnaugh map** (K-map) is a graphical representation of a given logic condition. Like a truth table, it represents all of the logic conditions for a given set of Boolean variables. Figure 5–30 shows how it works.

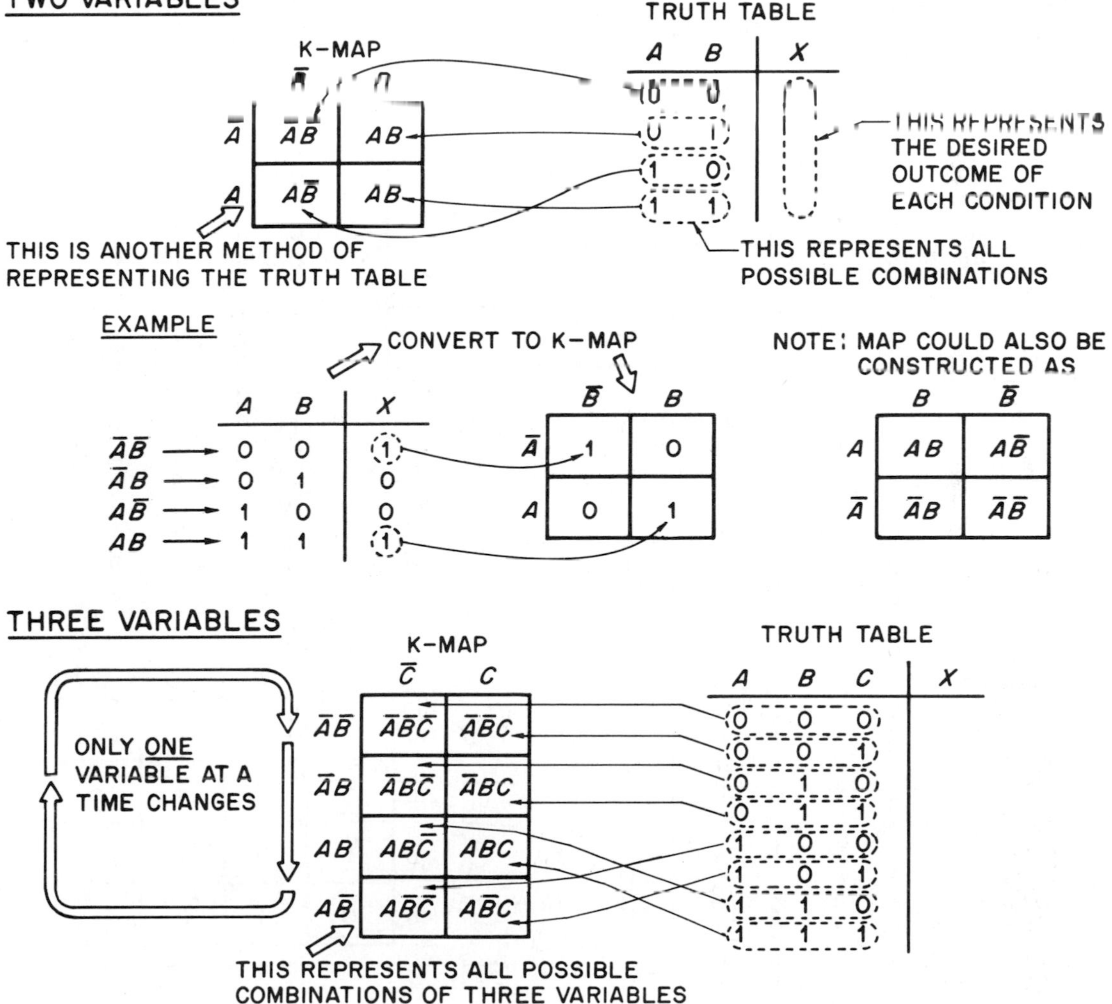

Figure 5–30 Construction of K-maps.

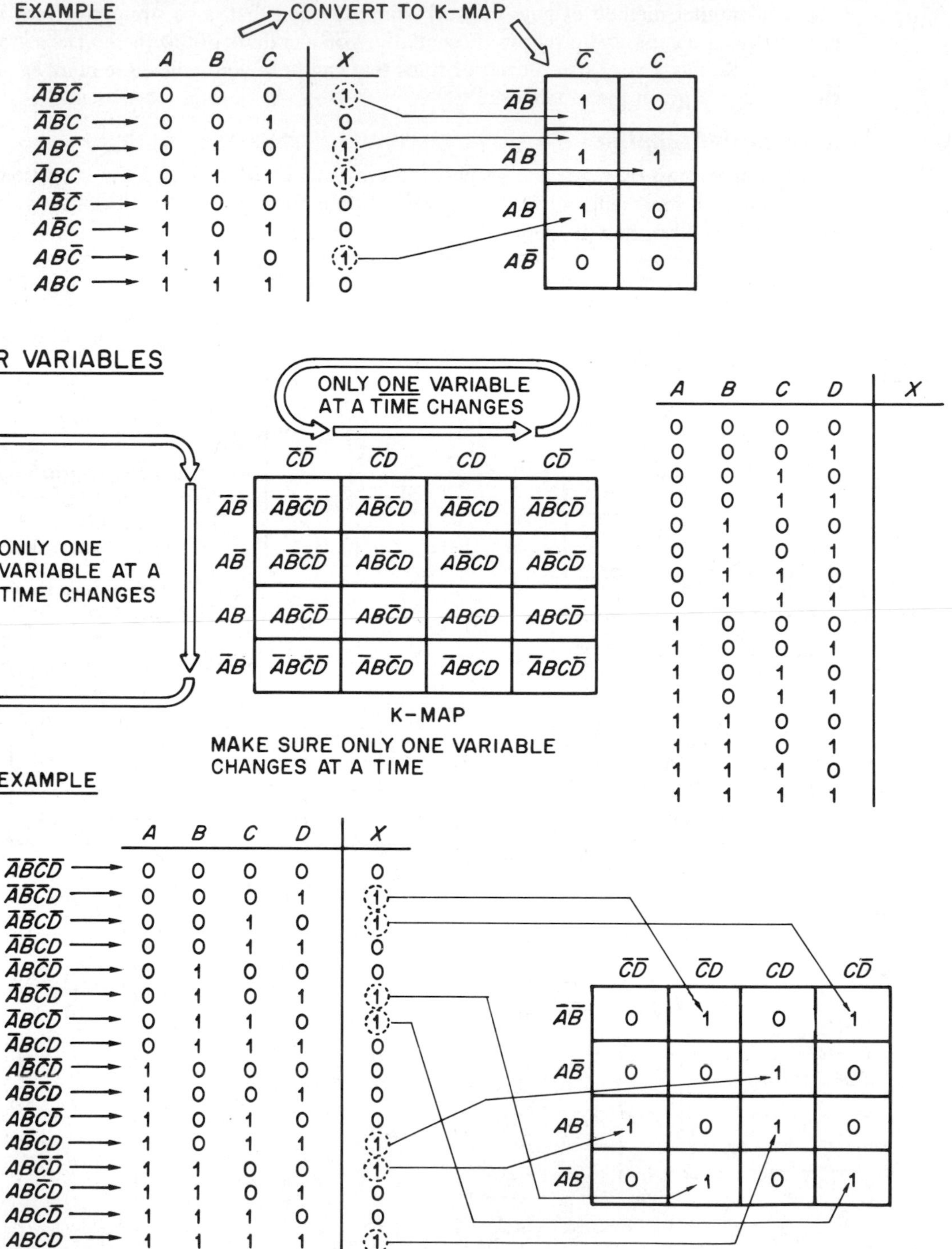

EXAMPLE
CONVERT TO K-MAP
A B C X
$\bar{A}\bar{B}\bar{C}$ 0 0 0 1
$\bar{A}\bar{B}C$ 0 0 1 0
$\bar{A}B\bar{C}$ 0 1 0 1
$\bar{A}BC$ 0 1 1 1
$A\bar{B}\bar{C}$ 1 0 0 0
$A\bar{B}C$ 1 0 1 0
$AB\bar{C}$ 1 1 0 1
ABC 1 1 1 0
$\bar{C}$ C
$\bar{A}\bar{B}$ 1 0
$\bar{A}B$ 1 1
AB 1 0
$A\bar{B}$ 0 0
FOUR VARIABLES
ONLY ONE VARIABLE AT A TIME CHANGES
ONLY ONE VARIABLE AT A TIME CHANGES
K-MAP
MAKE SURE ONLY ONE VARIABLE CHANGES AT A TIME
EXAMPLE

Figure 5–30 Continued

Karnaugh Map:
A graphical method of simplifying logic networks.

Observe the following about the construction of a K-map:

1. All combinations of the variables are represented by a box inside the K-map.
2. Along the sides of a K-map (as you move from column to column; row to row; or from the end back to the beginning) only one variable at a time is allowed to change from its complemented to its uncomplemented form or from its uncomplemented to its complemented form.
3. For a given variable combination that is TRUE, that cell inside the K-map representing the variable combination is given a 1, whereas for a variable combination that is FALSE, its corresponding cell is given a 0.
4. The K-map contains the same information as a truth table.

Example 5–15

Determine which of the truth tables are correctly represented by the K-maps of Figure 5–31.

(A)

A	B	X
0	0	0
0	1	1
1	0	1
1	1	0

	$\overline{B}$	B
$\overline{A}$	0	1
A	1	0

(B)

A	B	C	X
0	0	0	1
0	0	1	0
0	1	0	1
0	1	1	1
1	0	0	0
1	0	1	0
1	1	0	0
1	1	1	1

	$\overline{C}$	C
$\overline{A}\overline{B}$	1	0
AB	0	1
$\overline{A}B$	1	1
$A\overline{B}$	0	0

(C)

A	B	C	X
0	0	0	0
0	0	1	1
0	1	0	0
0	1	1	1
1	0	0	1
1	0	1	0
1	1	0	1
1	1	1	0

	$\overline{B}\overline{C}$	$\overline{B}C$	BC	$B\overline{C}$
$\overline{A}$	0	1	1	0
A	1	0	0	1

Figure 5–31

Solution

See Figure 5–32.

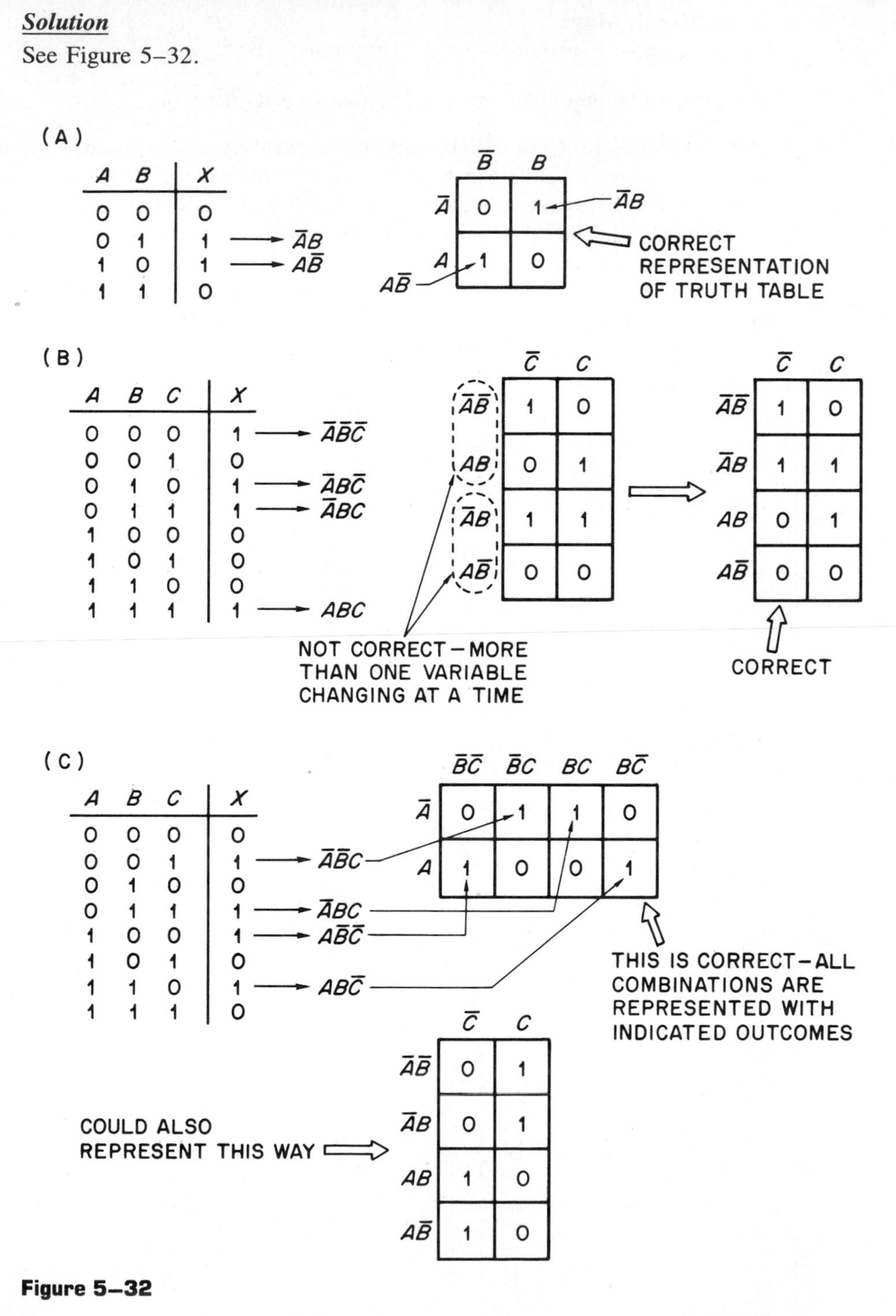

Figure 5–32

Pair Simplification

The purpose of representing a given logic condition with a K-map is to help reduce the number of redundant logic gates. One method of doing this is by **pair simplification.** It is traditional to construct K-maps with complemented input variables in the upper left corner of the map. Doing this makes it easier to go down the list of inputs starting with 0000, from the truth table, thus reducing the chance of error when transferring this information to the K-map. However, this is not a requirement as long as the variables for each column and row have only one condition at a time change. This is illustrated by the following K-map.

	CD	$C\overline{D}$	$\overline{C}\,\overline{D}$	$\overline{C}D$
AB				
$A\overline{B}$				
$\overline{A}\,\overline{B}$				
$\overline{A}B$				

Pair Simplification:
In K-map, looking for cells that are either vertically or horizontally adjacent for the purpose of eliminating one variable from the final Boolean expression represented by the K-map.

To illustrate pair simplification, consider the following K-map:

	$\overline{C}\,\overline{D}$	$\overline{C}D$	CD	$C\overline{D}$
$\overline{A}\,\overline{B}$	0	0	0	0
$\overline{A}B$	0	0	0	0
AB	0	1	1	0
$A\overline{B}$	0	0	0	0

This K-map represents the SOP term

$$Y = AB\overline{C}D + ABCD$$

However, using Boolean algebra and factoring, you have

$$Y = ABD(\overline{C} + C)$$

Since $(\overline{C} + C) = 1$ and $ABD \cdot (1) = ABD$, the equation reduces to one term:

$$Y = ABD$$

Notice that the C term has been eliminated. Figure 5–33 shows three ways a similar factor could be eliminated from different K-maps.

The following rules can be used to simplify a K-map with vertical or horizontal adjacent pairs:

1. Look for any horizontally or vertically adjacent pairs.
2. Observe which variable represented by the pair is represented in both the complemented and uncomplemented forms. This is the variable that will be eliminated from the final term.
3. A pair represents one term of an SOP expression.

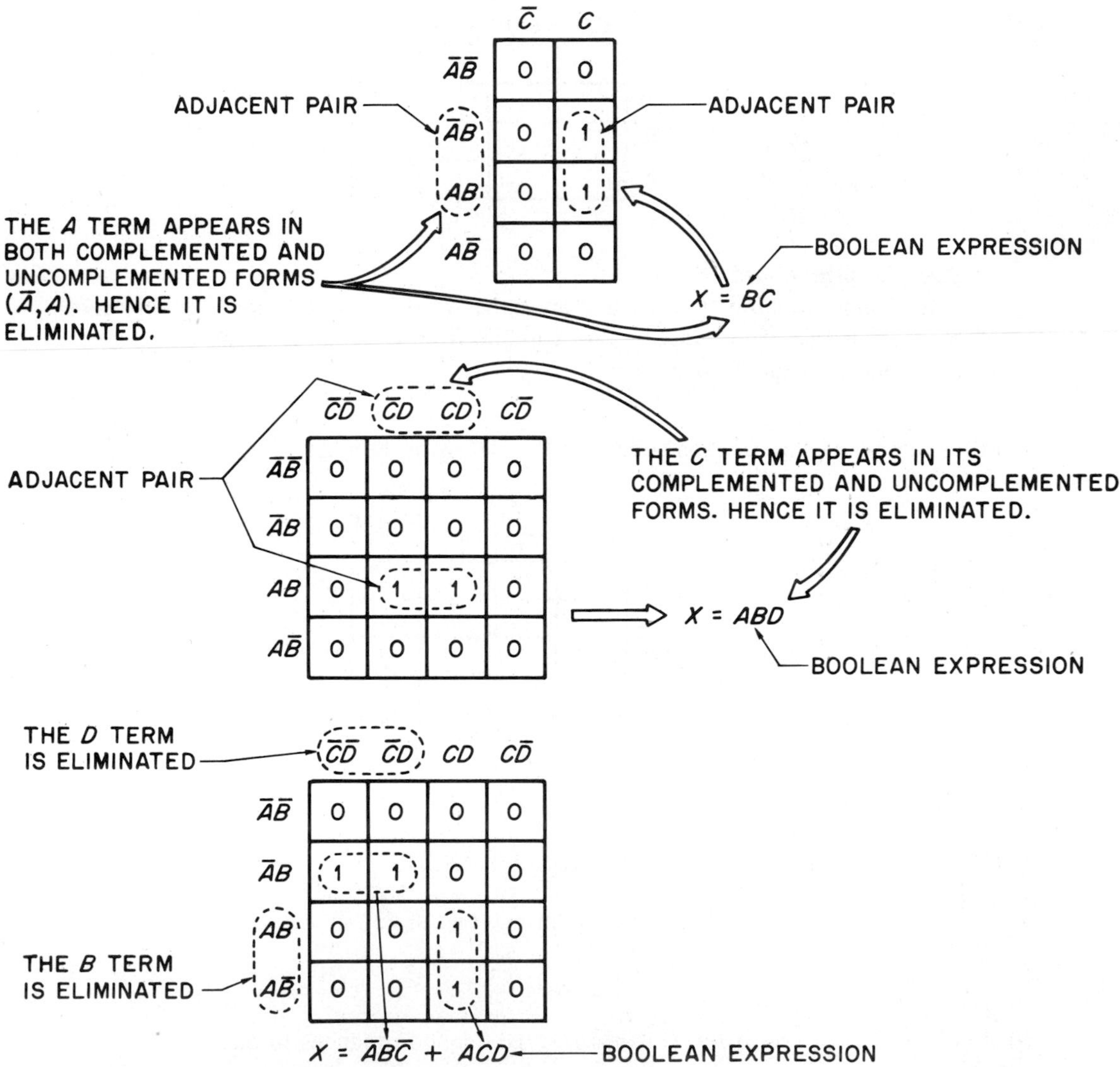

Figure 5–33 Reduction with K-map pairs.

Example 5–16

Construct the simplified SOP expression from the following truth table, using K-map techniques.

A	*B*	*C*	*D*	*X*
0	0	0	0	1
0	0	0	1	1
0	0	1	0	0
0	0	1	1	0
0	1	0	0	0
0	1	0	1	0
0	1	1	0	1
0	1	1	1	1
1	0	0	0	0
1	0	0	1	1
1	0	1	0	0
1	0	1	1	0
1	1	0	0	0
1	1	0	1	1
1	1	1	0	0
1	1	1	1	0

Solution

First develop each product term for every TRUE condition from the truth table:

A	*B*	*C*	*D*	*X*	
0	0	0	0	1	$\rightarrow \overline{A}\,\overline{B}\,\overline{C}\,\overline{D}$
0	0	0	1	1	$\rightarrow \overline{A}\,\overline{B}\,\overline{C}D$
0	0	1	0	0	
0	0	1	1	0	
0	1	0	0	0	
0	1	0	1	0	
0	1	1	0	1	$\rightarrow \overline{A}BC\overline{D}$
0	1	1	1	1	$\rightarrow \overline{A}BCD$
1	0	0	0	0	
1	0	0	1	1	$\rightarrow A\overline{B}\,\overline{C}D$
1	0	1	0	0	
1	0	1	1	0	
1	1	0	0	0	
1	1	0	1	1	$\rightarrow AB\overline{C}D$
1	1	1	0	0	
1	1	1	1	0	

Next, express the results in the form of a K-map. Circle all adjacent pairs and write the resulting Boolean SOP expression. This process is illustrated in Figure 5–34.

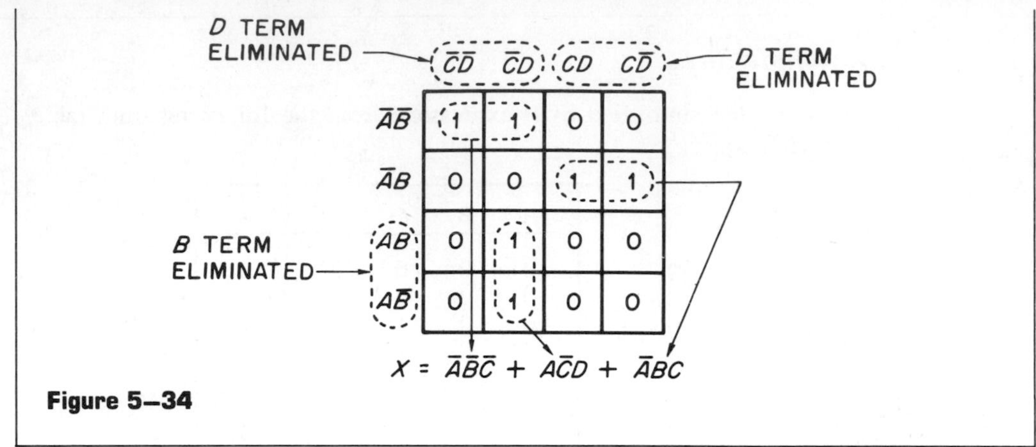

Figure 5–34

Because of K-mapping, you have reduced an SOP expression from a six-product term to a three-product term. This represents a great saving in gates when the gates are actually connected.

Rolling the Map

Recall that the correct construction of the K-map means that you have only one variable change at a time, from column to column or row to row. Think of rolling a K-map so

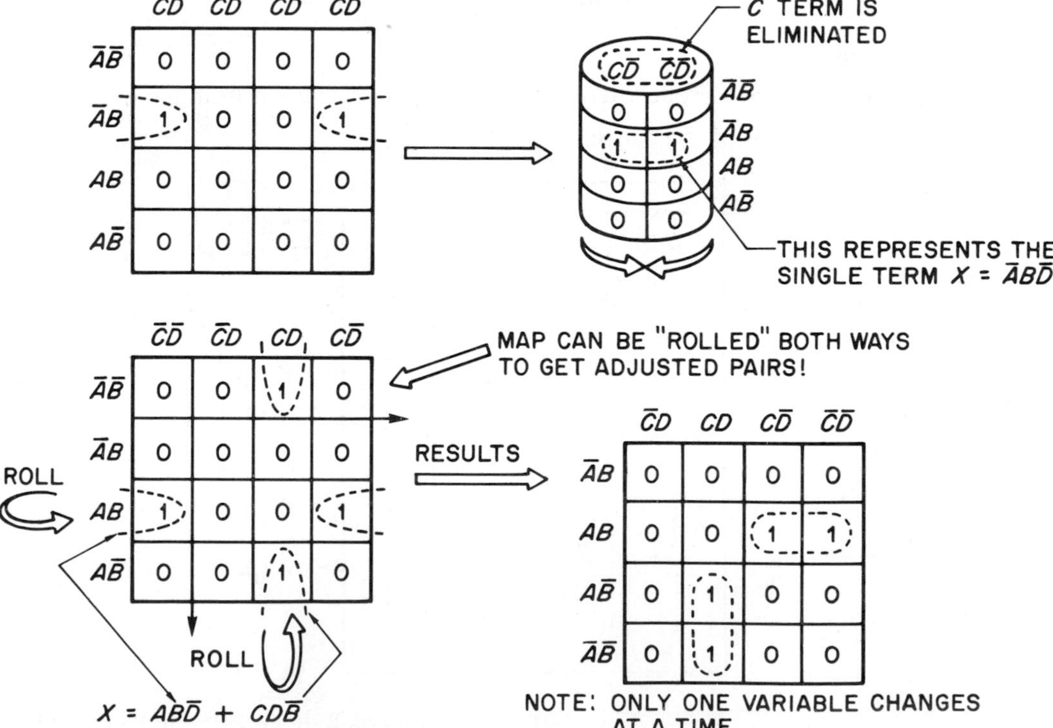

Figure 5–35 Rolling the K-map for adjacent pairs.

that its sides touch. You can do a vertical roll or a horizontal roll. You do this in order to look for vertically or horizontally adjacent 1's. This is demonstrated in Figure 5–35.

Example 5–17

Develop the simplified SOP expression from the following truth table, using K-mapping.

A	*B*	*C*	*X*
0	0	0	1
0	0	1	0
0	1	0	0
0	1	1	1
1	0	0	1
1	0	1	0
1	1	0	0
1	1	1	1

Solution

Write each product term from the truth table.

A	*B*	*C*	*X*	
0	0	0	1	$\rightarrow \overline{A}\,\overline{B}\,\overline{C}$
0	0	1	0	
0	1	0	0	
0	1	1	1	$\rightarrow \overline{A}BC$
1	0	0	1	$\rightarrow A\overline{B}\,\overline{C}$
1	0	1	0	
1	1	0	0	
1	1	1	1	$\rightarrow ABC$

Develop the corresponding K-map as shown in Figure 5–36.

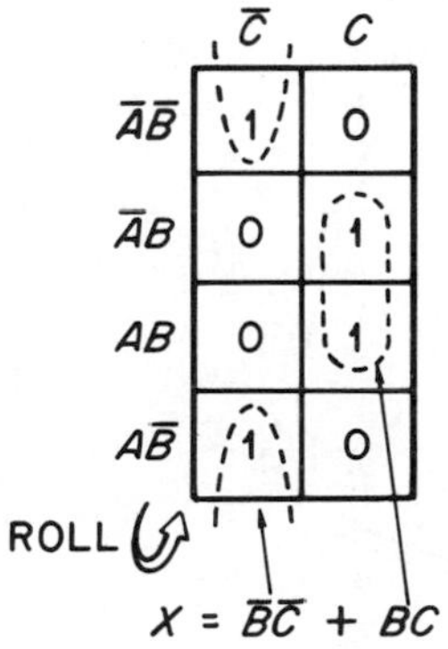

	$\overline{C}$	C
$\overline{A}\overline{B}$	1	0
$\overline{A}B$	0	1
AB	0	1
$A\overline{B}$	1	0

ROLL

$X = \overline{B}\overline{C} + BC$

Figure 5–36

Simplifying with Quads

Another method of simplifying logic networks is by looking for horizontally or vertically adjoining 1's to form a group of four. This is called **quad simplification.**

> **Quad Simplification:**
> Treating a group of four vertically or horizontally adjacent cells of a K-map as a single Boolean term.

To illustrate quad simplification, consider the following K-map:

	$\overline{C}\,\overline{D}$	$\overline{C}D$	CD	$C\overline{D}$
$\overline{A}\,\overline{B}$	0	0	0	0
$\overline{A}B$	0	1	1	0
AB	0	1	1	0
$A\overline{B}$	0	0	0	0

Taking each 1 of the K-map as a single-product term, you have the resulting SOP Boolean expression:

$$X = \overline{A}B\overline{C}D + \overline{A}BCD + AB\overline{C}D + ABCD$$

This expression can be simplified by using Boolean algebra techniques:

Factoring gives
$X = \overline{A}BD(\overline{C} + C) + ABD(\overline{C} + C) = \overline{A}BD + ABD$
Factoring gives
$X = BD(\overline{A} + A) = BD$

Thus, four adjacent 1's are equivalent to a single-product term consisting of only *two* variables. The process is illustrated in Figure 5–37.

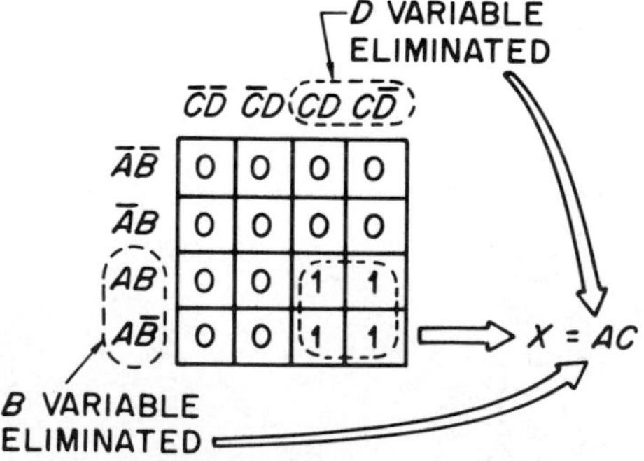

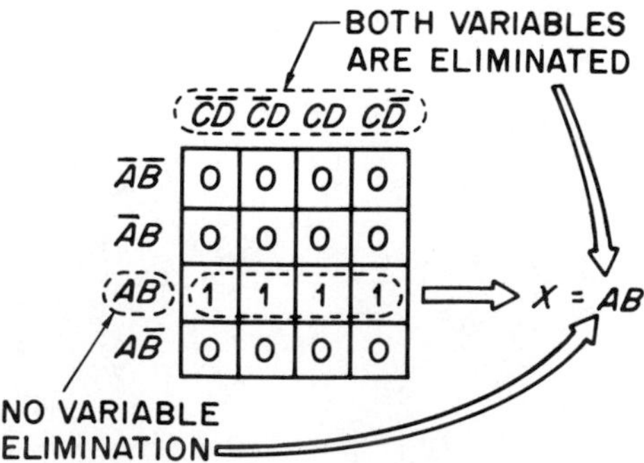

Figure 5–37 Examples of quads.

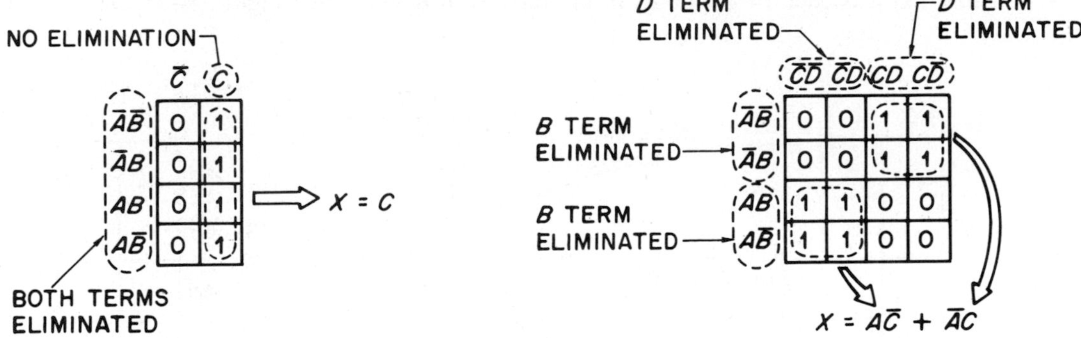

Figure 5–37 Continued

As you can see, a quad in a K-map results in a great simplification of the final number of terms in the resulting Boolean expression. This, of course, means fewer gates in the final implementation of the original truth table.

Example 5–18

Develop the simplified SOP expression for the following truth table.

A	*B*	*C*	*X*
0	0	0	1
0	0	1	1
0	1	0	0
0	1	1	0
1	0	0	1
1	0	1	1
1	1	0	0
1	1	1	0

Solution

First, develop each product term as shown:

A	*B*	*C*	*X*	
0	0	0	1	$\rightarrow \overline{A}\,\overline{B}\,\overline{C}$
0	0	1	1	$\rightarrow \overline{A}\,\overline{B}C$
0	1	0	0	
0	1	1	0	
1	0	0	1	$\rightarrow A\overline{B}\,\overline{C}$
1	0	1	1	$\rightarrow A\overline{B}C$
1	1	0	0	
1	1	1	0	

Next, express each of the terms in the form of a K-map. See Figure 5–38.

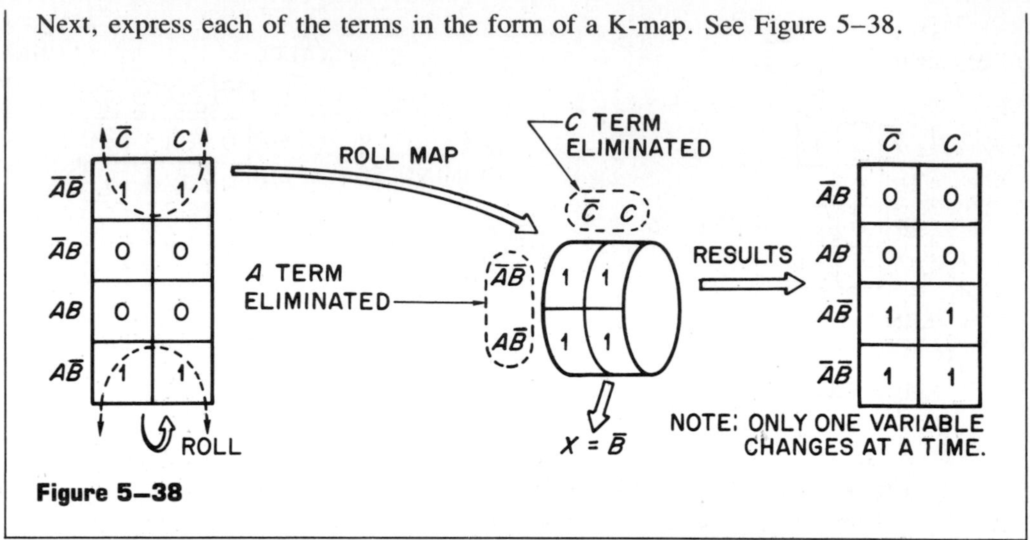

Figure 5–38

Overlapping Groups

You may use the same cell more than once in K-map simplification. This important concept is shown in Figure 5–39.

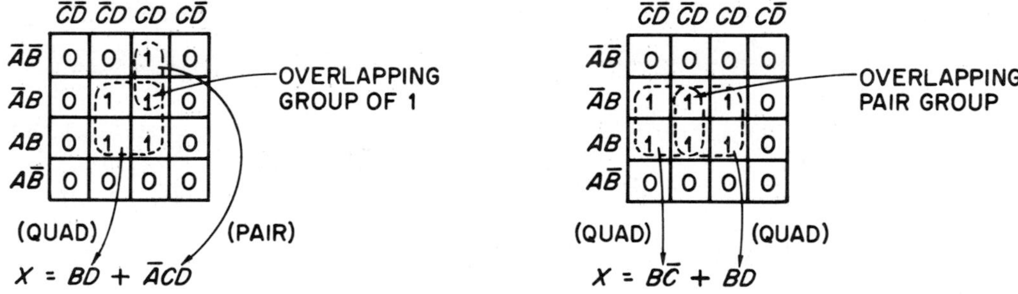

Figure 5–39 Examples of overlapping groups.

Example 5–19

Develop the simplified SOP expression using K-mapping techniques for the following truth table.

A	B	C	D	X
0	0	0	0	1
0	0	0	1	1
0	0	1	0	1
0	0	1	1	0
0	1	0	0	0
0	1	0	1	0
0	1	1	0	0
0	1	1	1	0
1	0	0	0	1
1	0	0	1	0
1	0	1	0	1
1	0	1	1	1
1	1	0	0	0
1	1	0	1	0
1	1	1	0	0
1	1	1	1	0

Solution

As before, write the product terms:

A	B	C	D	X	
0	0	0	0	1	$\rightarrow \overline{A}\,\overline{B}\,\overline{C}\,\overline{D}$
0	0	0	1	1	$\rightarrow \overline{A}\,\overline{B}\,\overline{C}D$
0	0	1	0	1	$\rightarrow \overline{A}\,\overline{B}C\overline{D}$
0	0	1	1	0	
0	1	0	0	0	
0	1	0	1	0	
0	1	1	0	0	
0	1	1	1	0	
1	0	0	0	1	$\rightarrow A\overline{B}\,\overline{C}\,\overline{D}$
1	0	0	1	0	
1	0	1	0	1	$\rightarrow A\overline{B}C\overline{D}$
1	0	1	1	1	$\rightarrow A\overline{B}CD$
1	1	0	0	0	
1	1	0	1	0	
1	1	1	0	0	
1	1	1	1	0	

Next, express each product term in a K-map. See Figure 5–40.

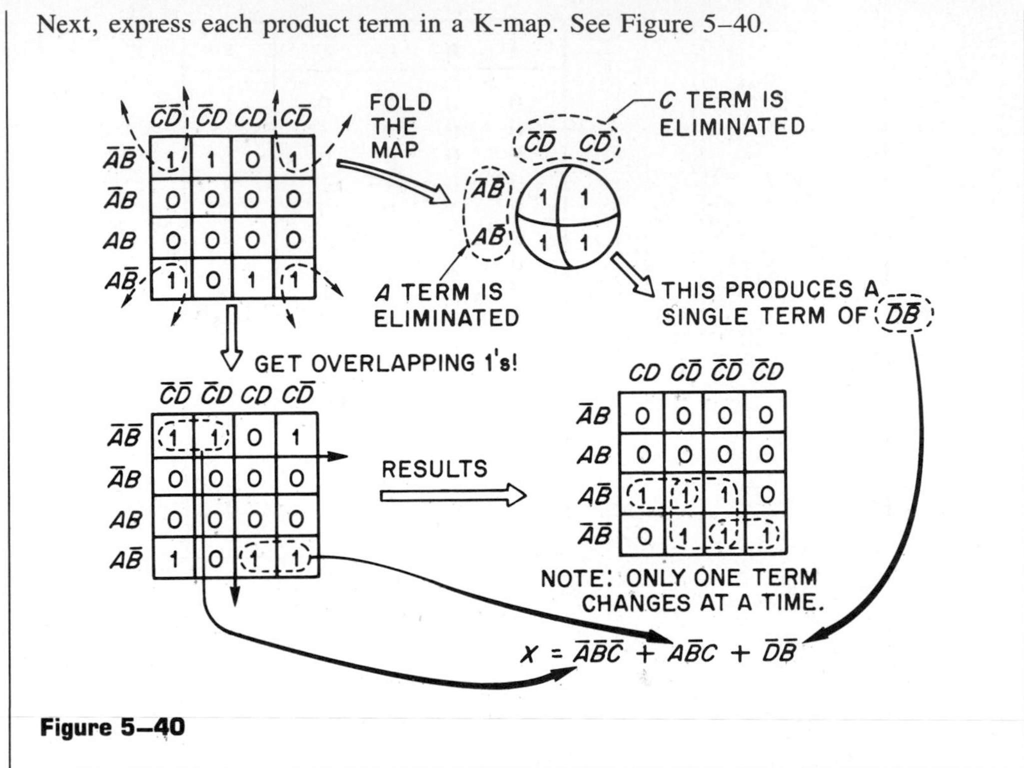

Figure 5–40

Simplifying with Octets

You will find that if you can get *eight* vertically and/or horizontally adjoining cells in a K-map, you can reduce this to a Boolean expression with only one variable. This process is called **octet simplification.**

Octet Simplification:
Treating a group of eight horizontally and/or vertically adjacent cells as a single term.

To illustrate octet simplification, consider the following K-map:

	$\overline{C}\,\overline{D}$	$\overline{C}D$	CD	$C\overline{D}$
$\overline{A}\,\overline{B}$	1	1	1	1
$\overline{A}B$	1	1	1	1
AB	0	0	0	0
$A\overline{B}$	0	0	0	0

Expressing this K-map as a Boolean SOP expression, you have

$$X = \overline{A}\,\overline{B}\,\overline{C}\,\overline{D} + \overline{A}\,\overline{B}\,\overline{C}D + \overline{A}\,\overline{B}CD + \overline{A}\,\overline{B}C\overline{D} + \overline{A}B\overline{C}\,\overline{D} + \overline{A}B\overline{C}D + \overline{A}BCD + \overline{A}BC\overline{D}$$

Using Boolean algebra techniques and factoring gives

$$X = \overline{A}\,\overline{B}\,\overline{C}(\overline{D} + D) + \overline{A}\,\overline{B}C(\overline{D} + D) + \overline{A}B\overline{C}(\overline{D} + D) + \overline{A}BC(\overline{D} + D)$$

Since $(\overline{D} + D) = 1$, one variable is eliminated:

$$X = \overline{A}\,\overline{B}\,\overline{C} + \overline{A}\,\overline{B}C + \overline{A}B\overline{C} + \overline{A}BC$$

Factoring again gives

$$X = \overline{A}\,\overline{B}(\overline{C} + C) + \overline{A}B(\overline{C} + C)$$

This eliminates another variable:

$$X = \overline{A}\,\overline{B} + \overline{A}B$$

One last factoring produces

$$X = \overline{A}(\overline{B} + B)$$

which reduces to the final result:

$$X = \overline{A}$$

Thus, an *octet* will reduce eight product terms to a single variable! Quite a saving of digital ICs!

Example 5–20

Develop a simplified expression using K-mapping for the following truth table.

A	*B*	*C*	*D*	*X*
0	0	0	0	1
0	0	0	1	1
0	0	1	0	0
0	0	1	1	0
0	1	0	0	1
0	1	0	1	1
0	1	1	0	1
0	1	1	1	1
1	0	0	0	1
1	0	0	1	1
1	0	1	0	0
1	0	1	1	0
1	1	0	0	1
1	1	0	1	1
1	1	1	0	1
1	1	1	1	1

Solution

As before, write each of the product terms from the truth table:

A	B	C	D	X	
0	0	0	0	1	$\rightarrow \overline{A}\,\overline{B}\,\overline{C}\,\overline{D}$
0	0	0	1	1	$\rightarrow \overline{A}\,\overline{B}\,\overline{C}D$
0	0	1	0	0	
0	0	1	1	0	
0	1	0	0	1	$\rightarrow \overline{A}B\overline{C}\,\overline{D}$
0	1	0	1	1	$\rightarrow \overline{A}B\overline{C}D$
0	1	1	0	1	$\rightarrow \overline{A}BC\overline{D}$
0	1	1	1	1	$\rightarrow \overline{A}BCD$
1	0	0	0	1	$\rightarrow A\overline{B}\,\overline{C}\,\overline{D}$
1	0	0	1	1	$\rightarrow A\overline{B}\,\overline{C}D$
1	0	1	0	0	
1	0	1	1	0	
1	1	0	0	1	$\rightarrow AB\overline{C}\,\overline{D}$
1	1	0	1	1	$\rightarrow AB\overline{C}D$
1	1	1	0	1	$\rightarrow ABC\overline{D}$
1	1	1	1	1	$\rightarrow ABCD$

Next, place each of the product terms in a K-map. See Figure 5–41.

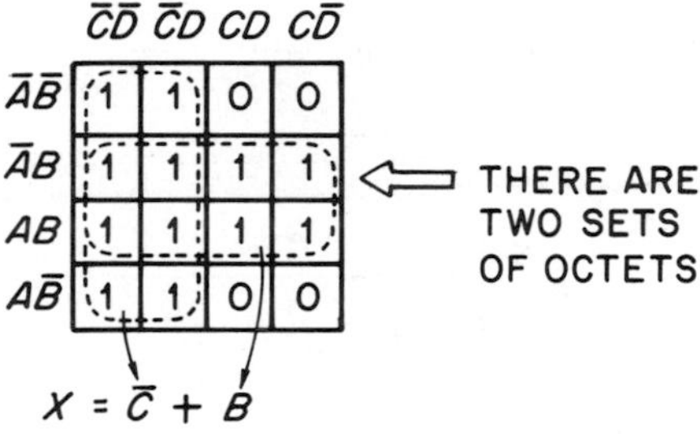

Figure 5–41

The previous example clearly demonstrates the power of the K-map simplifying logic networks.

Conclusion

In this section, you were introduced to the concept of K-mapping. In the next section, you will learn some time-saving tricks. Test your understanding of this section by trying the following review questions.

5–4 Review Questions

1. Explain the purpose of using K-mapping.
2. State the four rules for the development of a K-map.
3. Explain the meaning of *pair, quad,* and *octet* simplification.
4. What is meant by *folding* the K-map?
5. What is meant by *overlapping* cells in a K-map?

5–5 K-MAP TRICKS

Discussion

In the last section, you were introduced to K-mapping. There you learned some of the "tools of the trade" in the simplification of logic networks. In this section, you will learn more valuable K-mapping skills.

Redundant Groups

There are instances in K-mapping where a grouping of adjacent cells will overlap. An unnecessary product term can be created by this grouping. Such a group of cells is called a **redundant group.**

> **Redundant Group:**
> In K-mapping, a group of cells that are already used by other groups in logic network simplification.

Figure 5–42 shows an example of a redundant group. When the K-map contains possible redundant groups, you should include every 1 in the least number of largest groupings you can create. The following example illustrates this technique.

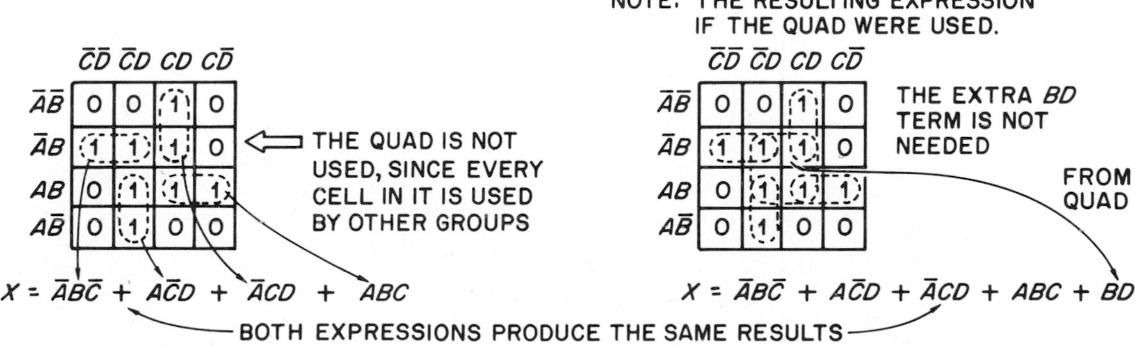

Figure 5–42 Illustration of a redundant group.

Example 5–21

Using K-mapping techniques, develop the Boolean expression for the following truth table.

A	*B*	*C*	*D*	*X*
0	0	0	0	1
0	0	0	1	1
0	0	1	0	0
0	0	1	1	0
0	1	0	0	1
0	1	0	1	1
0	1	1	0	1
0	1	1	1	1
1	0	0	0	0
1	0	0	1	0
1	0	1	0	0
1	0	1	1	0
1	1	0	0	1
1	1	0	1	1
1	1	1	0	0
1	1	1	1	0

Solution

First, write the product terms and then express them in K-map form.

A	*B*	*C*	*D*	*X*	
0	0	0	0	1	$\rightarrow \overline{A}\,\overline{B}\,\overline{C}\,\overline{D}$
0	0	0	1	1	$\rightarrow \overline{A}\,\overline{B}\,\overline{C}D$
0	0	1	0	0	
0	0	1	1	0	
0	1	0	0	1	$\rightarrow \overline{A}B\overline{C}\,\overline{D}$
0	1	0	1	1	$\rightarrow \overline{A}B\overline{C}D$
0	1	1	0	1	$\rightarrow \overline{A}BC\overline{D}$
0	1	1	1	1	$\rightarrow \overline{A}BCD$
1	0	0	0	0	
1	0	0	1	0	
1	0	1	0	0	
1	0	1	1	0	
1	1	0	0	1	$\rightarrow AB\overline{C}\,\overline{D}$
1	1	0	1	1	$\rightarrow AB\overline{C}D$
1	1	1	0	0	
1	1	1	1	0	

The resulting K-map is shown in Figure 5–43. Note the redundant group of two adjacent cells that are shared by three other groups.

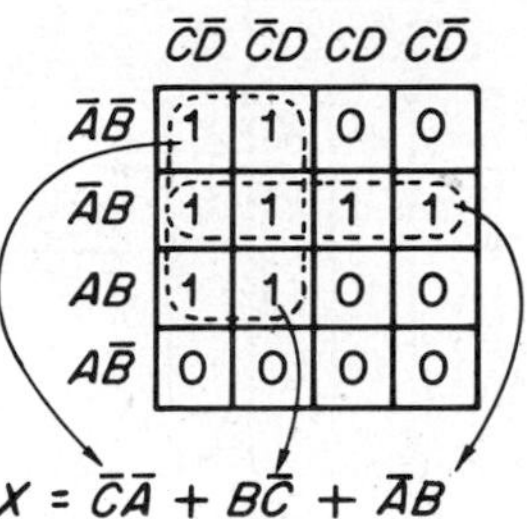

Figure 5–43

Don't Cares

There are times when certain outputs of a truth table are not used by any part of a digital system. For example, a given logic system may only be interested in logic conditions from 1000_2 to 1111_2. Then all the conditions from 0000_2 to 0111_2 would be **don't cares** and are represented on the output of a truth table as an X.

Don't Cares:
A logic condition whose outcome has no effect on the rest of the system.

A sample truth table where the first eight combinations are not used by the logic circuit follows. This truth table represents a condition where only when $A = 1$ is there a desired outcome. When $A = 0$, it makes no difference what the outcome is. This could represent a condition that the A variable represents if an assembly line is on or off and the B, C, and D variables are other conditions taking place on the assembly line. Thus if the assembly line is shut down ($A = 0$), it makes no difference what the other three conditions on the assembly line are.

A	*B*	*C*	*D*	*X*
0	0	0	0	X
0	0	0	1	X
0	0	1	0	X
0	0	1	1	X
0	1	0	0	X
0	1	0	1	X
0	1	1	0	X
0	1	1	1	X
1	0	0	0	1
1	0	0	1	0
1	0	1	0	1
1	0	1	1	0
1	1	0	0	0
1	1	0	1	1
1	1	1	0	0
1	1	1	1	1

(X → don't care)

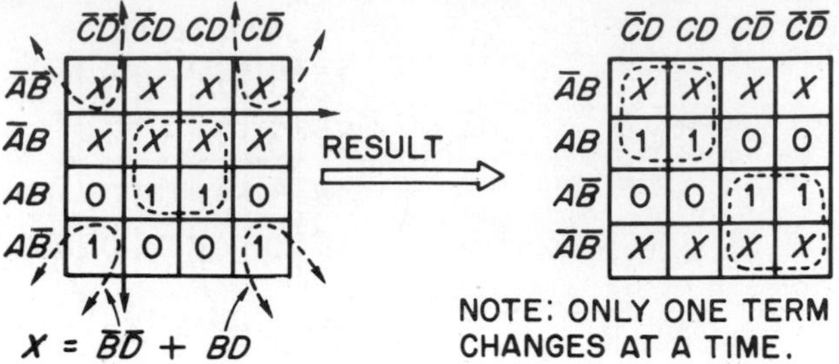

Figure 5–44 Using *Don't Cares* in a K-map.

Figure 5–44 shows the K-map of the previous truth table. Note that the don't cares may be treated as either a 1 or a 0. In this case, selected don't cares are treated as 1's for the purpose of simplifying the resultant expression as much as possible.

Example 5–22

Use K-mapping techniques to develop the Boolean expression for the following truth table.

A	B	C	X
0	0	0	0
0	0	1	1
0	1	0	1
0	1	1	0
1	0	0	0
1	0	1	X
1	1	0	X
1	1	1	X

Solution

Write the product expressions (including those for the don't cares) and enter them in the K-map. Indicate which is a 1 and which is a don't care. See Figure 5–45. You should recognize the resulting expression as the logic of an XOR gate.

A	B	C	X	
0	0	0	0	
0	0	1	1	→ $\bar{A}\bar{B}C$
0	1	0	1	→ $\bar{A}B\bar{C}$
0	1	1	0	
1	0	0	0	
1	0	1	X	→ $A\bar{B}C$
1	1	0	X	→ $AB\bar{C}$
1	1	1	X	→ ABC

	$\bar{C}$	C
$\bar{A}\bar{B}$		1
$\bar{A}B$	1	
AB	X	X
$A\bar{B}$		X

$X = \bar{C}B + C\bar{B}$

Figure 5–45

Using Product of Sum Terms

Recall that when the output conditions of a truth table had more 1's than 0's it was easier to use POS rather than SOP. The same idea is true for developing and analyzing a K-map.

You know that the 1's on a K-map each represent a product of an SOP Boolean expression. What you may not have realized is that the 0's represent the *sums* of the POS Boolean expression. This idea is illustrated in Figure 5–46. As you can see from the figure, the variable of each sum term is the complement of the K-map variable. With this in mind, you can use the sum terms with K-map pairs, quads, and octets in the simplification of Boolean expressions.

For example, using the truth table of Figure 5–47, the development of a Boolean expression for the product terms as well as the sum terms is illustrated. The Boolean expression derived from the product form would require two logic gates to implement (an AND and OR). However, as you can see from the figure, the expression for the sums requires only one gate. If both of these expressions represent all conditions of the same truth table, then how can they appear different and seem to require a different number of gates to perform the same function? Recall that the Boolean algebra rules presented in the last chapter showed that

$$A + \bar{A}B = A + B$$

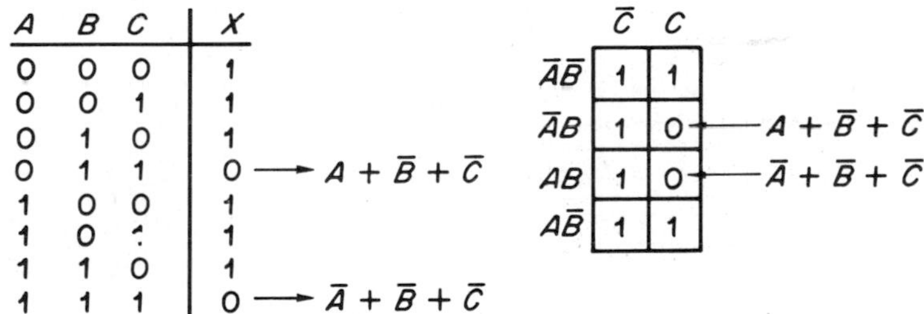

Figure 5–46 Product of sums K-map representation.

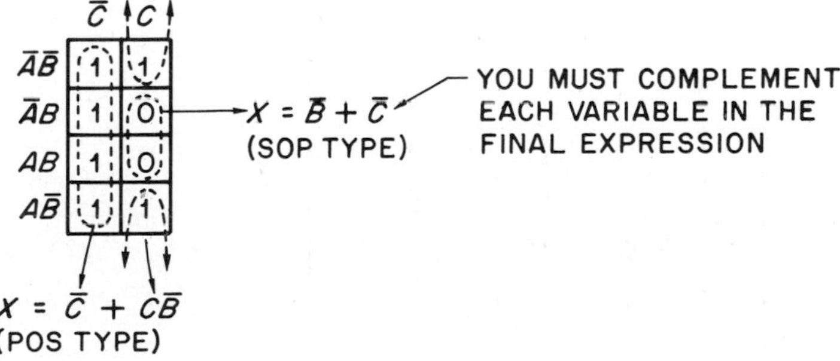

Figure 5–47 Developing sum of terms from a K-map.

Thus, upon closer inspection, you can see that

$$X = \overline{C} + C\overline{B} = \overline{C} + \overline{B}$$

which is identical to that arrived at by using the sum expression represented by each cell in the K-map that contained a 0.

Example 5–23

Use the POS terms for the K-map analysis of the following truth tables.

(A)

A	*B*	*X*
0	0	0
0	1	1
1	0	0
1	1	1

(B)

A	*B*	*C*	*X*
0	0	0	0
0	0	1	1
0	1	0	1
0	1	1	0
1	0	0	0
1	0	1	1
1	1	0	1
1	1	1	0

(C)

A	*B*	*C*	*D*	*X*
0	0	0	0	1
0	0	0	1	1
0	0	1	0	1
0	0	1	1	1
0	1	0	0	0
0	1	0	1	0
0	1	1	0	0
0	1	1	1	0
1	0	0	0	1
1	0	0	1	1
1	0	1	0	1
1	0	1	1	1
1	1	0	0	1
1	1	0	1	0
1	1	1	0	1
1	1	1	1	0

Solution

Refer to Figure 5–48.

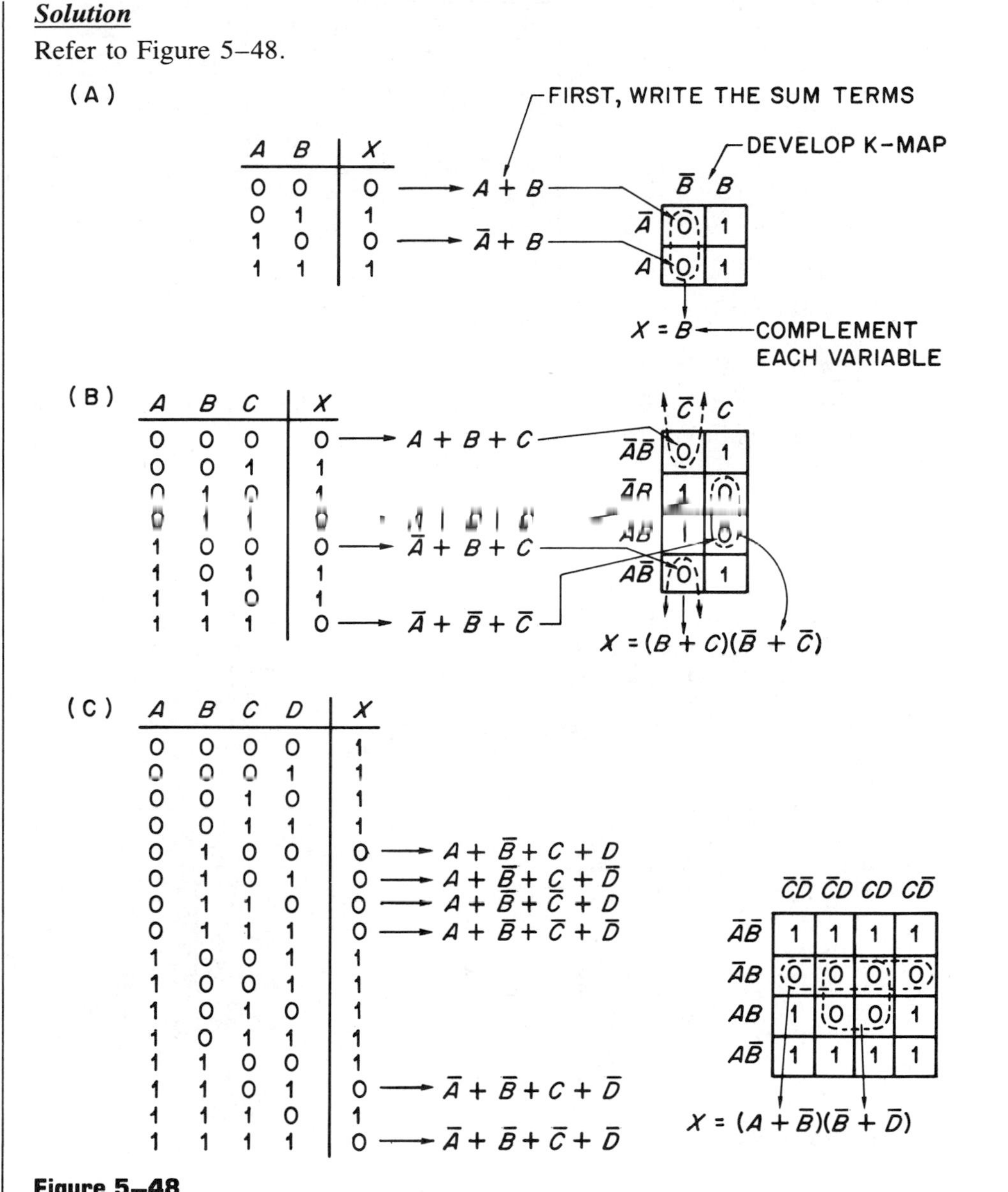

Figure 5–48

Example 5–24

Solve the following truth table by both the POS and SOP methods. Demonstrate that both produce the same logic.

A	*B*	*C*	*X*
0	0	0	1
0	0	1	1
0	1	0	0
0	1	1	1
1	0	0	1
1	0	1	1
1	1	0	0
1	1	1	1

Solution

Refer to Figure 5–49. Since both expressions are equal, their logic diagrams would be as shown in Figure 5–50.

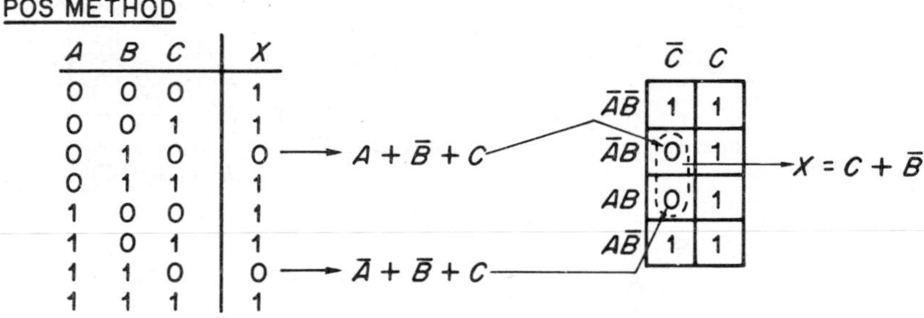

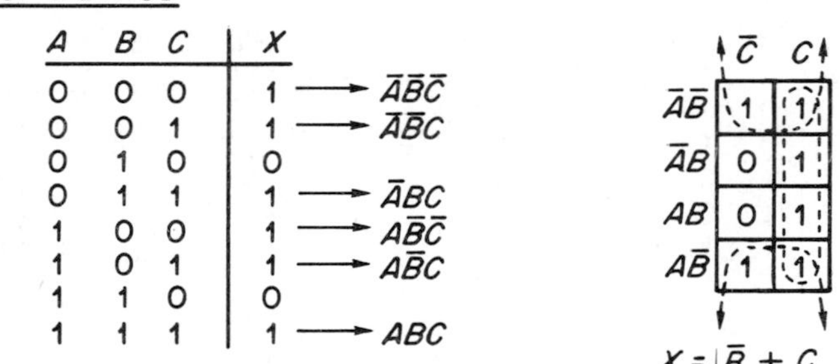

Figure 5–49

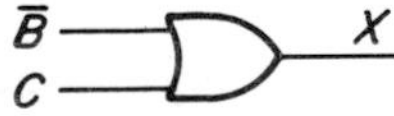

Figure 5–50

Conclusion

In this section you saw how to simplify logic diagrams by avoiding redundant K-map terms and by taking advantage of don't care conditions. You also saw how to use the POS form for K-map simplification, and compared this with the SOP form.

In the next section, you will be presented with some design problems that will start with a concept and be implemented with real integrated circuits. Check your understanding of this section by trying the following review questions.

5–5 Review Questions

1. What is a *redundant group* in K-mapping?
2. What is a *don't care* condition?
3. Explain how don't cares can be used in K-map simplification techniques.
4. Explain the technique used for developing POS terms from a K-map.
5. Which requires fewer steps, a POS or SOP K-map simplification?

5–6 LOGIC DESIGN

Discussion

The method of logic design presented in this section will use all of the techniques presented up to this point. The main idea of logic design is to have a systematic method that will allow you to replicate a logical concept using commercially available digital integrated circuits (ICs) in an economical and efficient manner. You are now at the point where you can take an idea and convert it into a logic circuit using discrete logic gates.

Steps of the Design Process

Table 5–5 shows the steps that will be used in the design of a digital system using NAND/NOR gate logic.

Table 5–5 STEPS IN GATE LOGIC DESIGN

Item	Reason
Write the problem.	Everyone involved in the design process will be working toward the same solution.
Ensure that there is a need to design something at all.	Often, no design is necessary. A "problem" may not require logic network for its solution.
Ensure that gate logic is the best solution.	The most common method of implementing digital logic design is through software or programmable logic devices (to be presented in the next chapter).
Reduce the problem to a truth table.	Clearly see the logic requirements of the problem.
Develop the K-map using POS or SOP.	Simplify the resulting Boolean expression.
Use Boolean algebra techniques to minimize the number of gates.	K-mapping itself does not guarantee the fewest gates. You may need to apply some Boolean algebra techniques.
Implement using economical and reliable components.	NAND/NOR ICs are the most readily available IC types with a proven record of performance.

Implementing the Process

A classical design problem in digital IC courses is that of a majority vote indicator. As a technician, you have been assigned to design a digital system that will indicate when the majority of a four-member board of directors for your company all vote in favor of the same issue. Each board member has a switch to indicate his or her secret ballot, and the outcome is indicated by a single green LED. Since your company manufactures NAND/NOR gates and no other product, the directors naturally want this done using the company product.

First Step:

State the problem in writing:
Given: Four input switches, each with an ON/OFF output.
Requirements: LED lights when three or more switches are ON.

Second Step:

Ensure that there is a need for the design.
In this case there is, since the board members want to show another use for the company product.

Third Step:

Ensure that gate logic is the best solution.
For this problem it is the only solution allowed.

Fourth Step:

Reduce the problem to a truth table.
Do this by representing each input switch by the variables A, B, C and D. Indicate the outcome by X. The truth table is

A	B	C	D	X
0	0	0	0	0
0	0	0	1	0
0	0	1	0	0
0	0	1	1	0
0	1	0	0	0
0	1	0	1	0
0	1	1	0	0
0	1	1	1	1
1	0	0	0	0
1	0	0	1	0
1	0	1	0	0
1	0	1	1	1
1	1	0	0	0
1	1	0	1	1
1	1	1	0	1
1	1	1	1	1

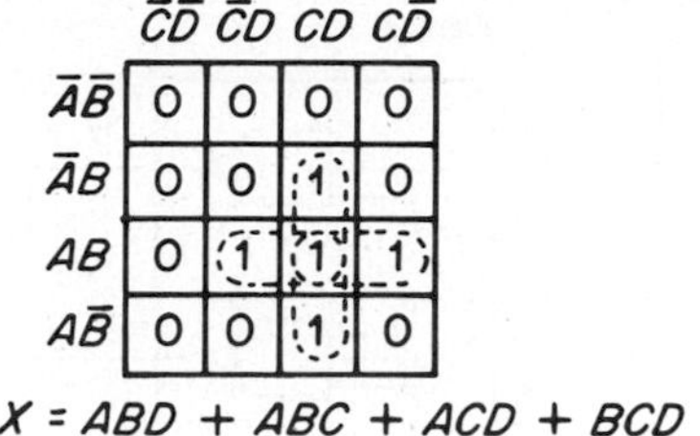

$\overline{C}\overline{D}$ $\overline{C}D$ CD $C\overline{D}$
$\overline{A}\overline{B}$ 0 0 0 0
$\overline{A}B$ 0 0 1 0
AB 0 1 1 1
$A\overline{B}$ 0 0 1 0
$X = ABD + ABC + ACD + BCD$

Figure 5–51 K-map for voting problem.

Fifth Step:

Develop the K-map using SOP or POS techniques.
Since there are fewer 1's on the output than 0's, the SOP techniques will be used. This is shown in Figure 5–51.

Sixth Step:

Use Boolean algebra techniques to minimize the number of gates.

Seventh Step:

Implement using economical and reliable components.
The implementation process is shown in Figure 5–52.

As you can see from the figure, you have two design choices. You can use two triple three-input NAND ICs with one dual four-input NAND or one triple three-input with one dual four-input NAND. The latter is the better choice because it requires only two gates. The final gates that would be used are shown in Figure 5–53.

An Industrial Design Problem

Design a logic network, using the minimum number of discrete NAND/NOR gates, that will implement the following requirements: A production company wants to have a bell ring under the following conditions:

1. It's after five o'clock and equipment is shut down.
2. Equipment is shut down and production is completed.
3. Both of the previous conditions.

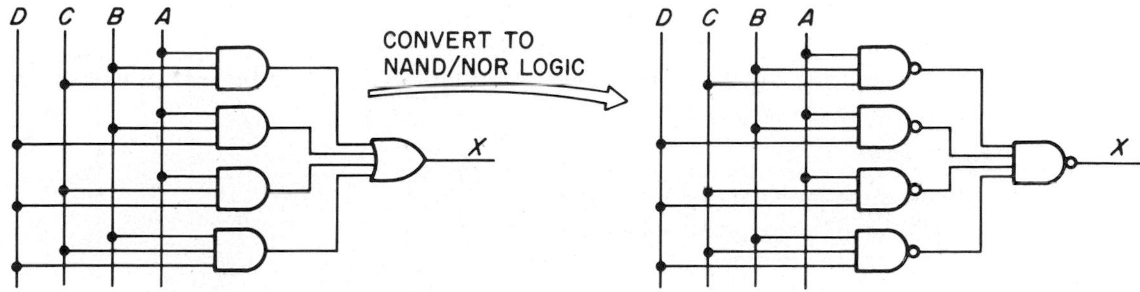

Figure 5–52 Implementing gates for voting problem.

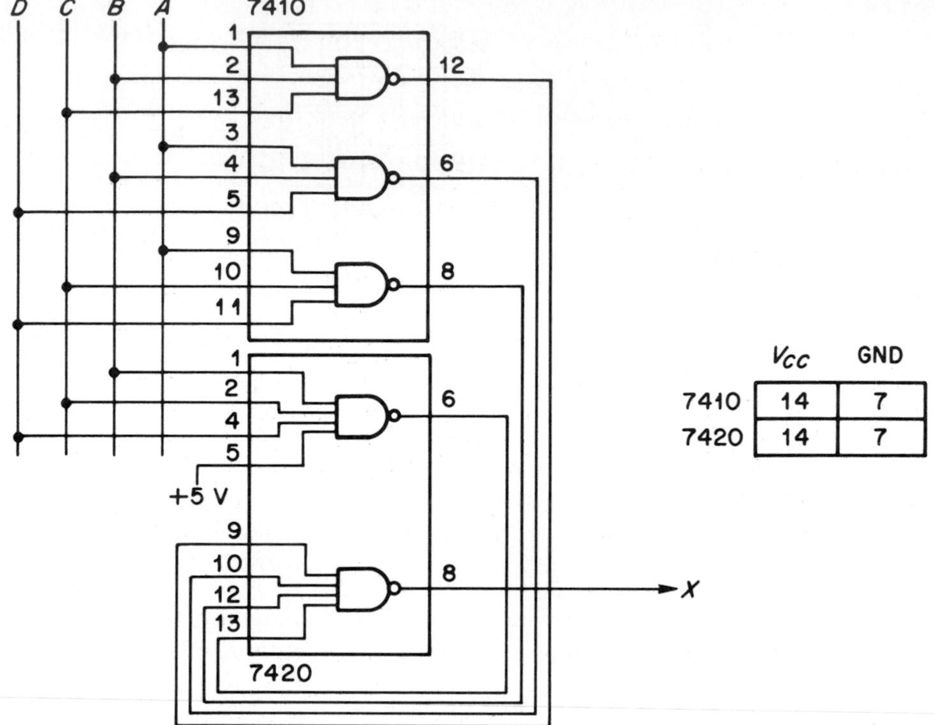

Figure 5–53 Final implementation of voting problem.

First Step:

State the problem in writing:
The problem is already in writing.

Second Step:

Ensure that there is a need for the design.
Assume that the design is needed as an exercise in using gate logic.

Third Step:

Ensure that gate logic is the best solution.
Gate logic may not be the best solution. But for this point in the text, it will be used to illustrate this application.

Fourth Step:

Reduce the problem to a truth table.
Each variable is represented as follows:
A = after five o'clock
B = equipment shut down
C = production completed

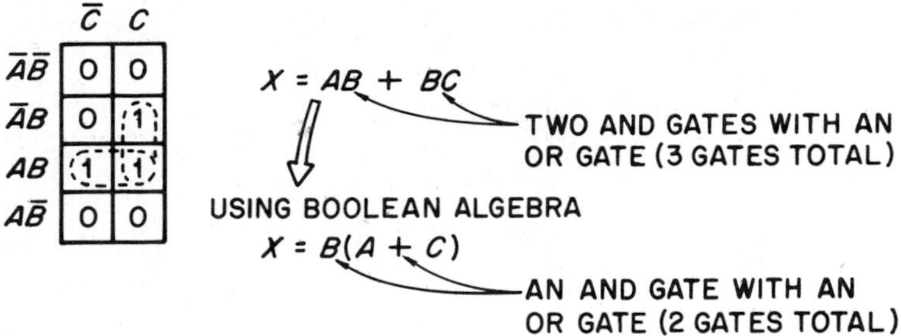

Figure 5–54 K-map for industrial problem.

A	*B*	*C*	*X*
0	0	0	0
0	0	1	0
0	1	0	0
0	1	1	1
1	0	0	0
1	0	1	0
1	1	0	1
1	1	1	1

Fifth Step:

Develop the K-map using SOP or POS techniques.

Since there are fewer 1's on the output than 0's, the SOP techniques will be used. This is shown in Figure 5–54.

Sixth Step:

Use Boolean algebra techniques to minimize the number of gates.

Seventh Step:

Implement using economical and reliable components.

The implementation process is shown in Figure 5–55. The actual logic circuit that would be used is one quad two-input NOR gate, as shown in Figure 5–56.

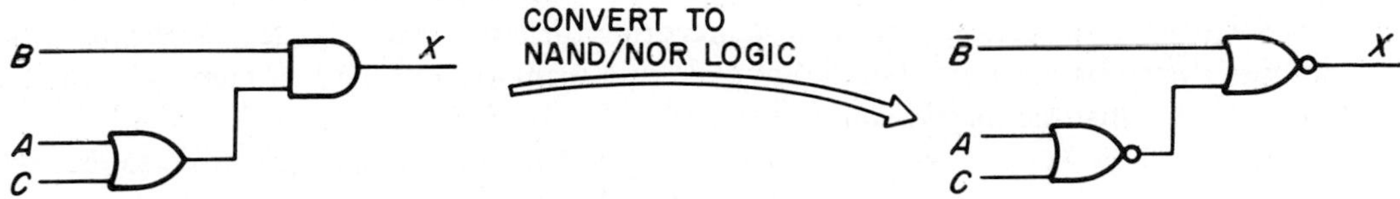

Figure 5–55 Implementing gates for industrial problem.

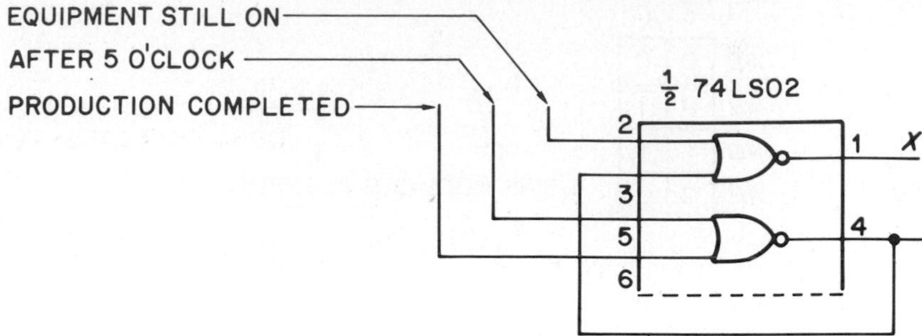

Figure 5–56 Final implementation of industrial problem.

Conclusion

This section showed the steps involved in the design of a digital logic system using gate networks. Here you saw two separate examples that demonstrated the implementation of real circuits from ideas. The analysis tools you have developed up to this point will serve you well in all aspects of digital logic networks. Check your understanding of this section by trying the following review questions.

5–6 Review Questions

1. State the steps used in the design of a digital logic system.
2. Are gates always necessary in the implementation of a logic design? Explain.
3. Why is Boolean algebra used as a part of the design process?
4. When developing the K-map, should you use SOP or POS terms?
5. What kind of gates are used for the final implementation of the problem? Explain.

MICROPROCESSOR APPLICATION

Discussion

In the microprocessor section of the last chapter you learned about an *instruction set* for the 4-bit microprocessor. There, you saw how a particular bit pattern was used to invoke a predictable set of processes. In this section, you will see how to construct a type of programming documentation that must be understood by digital technicians in order to troubleshoot and analyze digital systems.

Basic Idea

When instructions are written by a programmer to be inputted into memory, 1 and 0 codes are not used. The reason is that they are prone to error. Many times the hexadecimal number system is used to input digital code into memory, but the hex number itself does not indicate what the *instruction* will do.

In computer work, a short word usually consisting of three letters, is used to represent an instruction. These words are called **mnemonics,** and they are an abbreviation for what process the instruction would cause the microprocessor to perform.

Mnemonic Code:

A technique to assist the human memory. A mnemonic code resembles the word that describes the process that will be performed by the microprocessor.

As an example, the mnemonic LDR stands for the instruction for the 4-bit microprocessor that would load the data register with the contents of the next memory location.

Mnemonic Code

The mnemonic code for the hypothetical 4-bit microprocessor introduced in the last chapter is presented in Table 5–6. Note that the hex number is given under the bit pattern. In computer programming, hex numbers are identified by using the capital letter H following the number. The reason for doing this is because most computer programs will not accept subscripts. Thus 3H means 3_{16}.

As an example, one of the programs presented in the microprocessor section of the last chapter follows.

Address	Contents
0	2_{16}
1	5_{16}
2	1_{16}
3	A_{16}
4	6_{16}
5	F_{16}
6	5_{16}

Unless you have committed all of the binary codes of the instruction set to memory, it is not easily evident what this program is to do. You must refer to the instruction set for the microprocessor. However, with a little practice, it becomes clearer what the program is to do. In mnemonics the previous program is

Address	Operation
0	LDB 5
2	LDA A
4	ANA
5	HLT

This program should be easier to read. This is what it means:

LDB 5 → Load the number 5 into the B register.
LDA A → Load the number A_{16} into the accumulator.
ANA → AND the accumulator and B register; store the result in the accumulator.
HLT → Stop all further processing.

Note that the addressing in the program that uses mnemonics is somewhat different from the original program. It appears that some memory locations are being skipped. However, all memory locations are really being used as shown in Figure 5–57.

Table 5–6 FOUR-BIT MICROPROCESSOR INSTRUCTION SET

Mnemonic and bit pattern		Causes the process logic to perform the following processes:
NOP	0 0 0 0 0H	1. Copy the contents of the next memory location into the data register.
LDA	0 0 0 1 1H	1. Copy the contents of the next memory location into the data register. 2. Copy the contents of the data register into the accumulator.
LDB	0 0 1 0 2H	1. Copy the contents of the next memory location into the data register. 2. Copy the contents of the data register into the B register.
TBA	0 0 1 1 3H	1. Copy the contents of the B register into the accumulator.
TAB	0 1 0 0 4H	1. Copy the contents of the accumulator into the B register.
NTA	0 1 0 1 5H	1. NOT the contents of the accumulator.
ANA	0 1 1 0 6H	1. AND the contents of the accumulator with the B register and store the result in the accumulator.
ORA	0 1 1 1 7H	1. OR the contents of the accumulator with the B register and store the result in the accumulator.
XOR	1 0 0 0 8H	1. XOR the contents of the accumulator with the B register and store the result in the accumulator.
TAM	1 0 0 1 9H	1. Copy the contents of the accumulator into the data register. 2. Copy the contents of the data register into the next memory location.
TBM	1 0 1 0 AH	1. Copy the contents of the B register into the data register. 2. Copy the contents of the data register into the next memory location.
INR	1 0 1 1 BH	1. Increase the value of the accumulator by 1.
DCR	1 1 0 0 CH	1. Decrease the value of the accumulator by 1.
ADD	1 1 0 1 DH	1. Add the contents of the B register to the accumulator and store the results in the accumulator.
CLA	1 1 1 0 EH	1. Set the bits in the accumulator to 0.
HLT	1 1 1 1 FH	1. Stop all further processing.

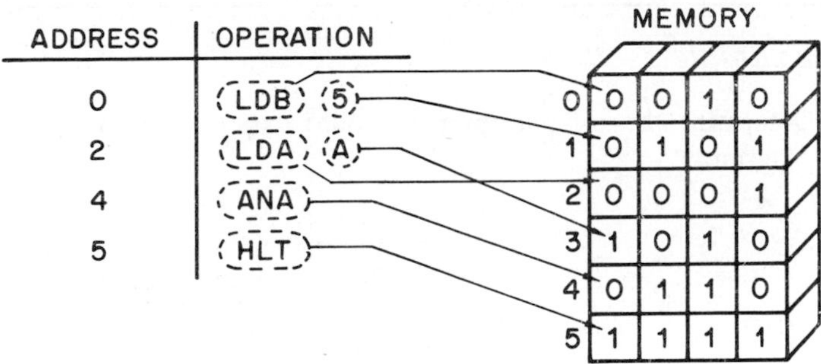

Figure 5–57 Address contents for sample program.

Understanding What the Mnemonics Mean

It is helpful to see why a certain three-letter code was used for each mnemonic. This is explained in Table 5–7.

Table 5–7 MEANING OF MNEMONICS FOR 4 BIT MICROPROCESSOR

Mnemonic	Meaning
NOP	No operation (This operation will cause the contents of the next memory location to be copied into the data register, but nothing else will happen.)
LDA	Load accumulator from memory.
LDB	Load B register from memory.
TBA	Transfer B register into accumulator.
TAB	Transfer accumulator into B register.
NTA	NOT the contents of the accumulator.
ANA	AND the contents of the accumulator with B register (result will be in the accumulator).
ORA	OR the contents of the accumulator with B register (result will be in the accumulator).
XOR	XOR the contents of the accumulator with B register (result will be in the accumulator).
TAM	Transfer the contents of the accumulator to memory.
TBM	Transfer the contents of the B register to memory.
INR	Increment the accumulator.
DCR	Decrement the accumulator.
ADD	Add the contents of the accumulator and B register (result will be in the accumulator).
CLA	Clear the accumulator (set contents to zeros).
HLT	Halt the processor.

Example 5–25

Determine the bit pattern you would expect to find in memory for the following program:

Address	Operation
0	LDB 3
2	CLA
3	XOR
4	HLT

Solution

The memory contents are shown in Figure 5–58.

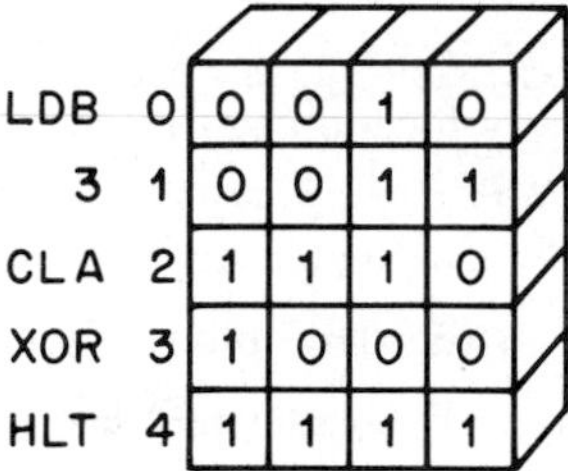

Figure 5–58

Assembler and the Language

Most microcomputers are capable of having their microprocessors programmed using mnemonics. A program is needed that will convert these mnemonics into the bit pattern that is understood by the microprocessor. The program that does this is called an **assembler.**

Assembler:
A computer program that operates on the symbolic input of instructions and translates them to bit patterns that are understood by the microprocessor.

Thus, when you write a program using mnemonics, it is called **assembly language** programming.

Assembly Language:
Programming instructions using mnemonics.

Usually, an assembly language program will contain comments. The comments are for clarification by the programmer and are not used by the microprocessor at all. For example, the following is an assembly language program that contains comments. It is standard convention to use a semicolon to indicate that a comment is to follow:

Address	Operation	
0	LDB 3	; Put 3 into the B register.
2	CLA	; Set accumulator to zero.
3	XOR	; XOR accum and B register.
4	HLT	; Stop the program.

As you can see from the *assembly program,* the addition of comments makes it easier to follow what the program is causing the microprocessor to do.

Language Levels

At this point, you should know three **levels** of programming: assembly, hexadecimal, and binary.

Language Level:
How close the programming code is to the actual bit pattern used by the microprocessor. The closer the programming code is to that of the microprocessor, the lower the level of the programming language level.

As an example, the lowest-level language is the actual binary number that represents the microprocessor instruction. The next higher-level language is hexadecimal numbers, and the next highest is assembly language, which uses mnemonics. When you use a bit pattern to program a microprocessor, this is called **machine language** programming.

Machine Language:
The actual bit pattern used by the microprocessor. Writing a program using this bit pattern is called machine language programming.

Many times, using hexadecimal numbers for programming a microprocessor is called *machine language* programming. This is not really the language of the machine (the microprocessor) because a logic circuit or program is required to translate the hex numbers into the actual bit pattern used by the microprocessor.

As a technician or designer of digital systems, you must be able to translate between assembly language, hex, and machine languages. It is important that you can look at a bit pattern in memory and, with access to the instruction set, develop the hex code or assembly code. You must also be able to take a given assembly or hex code and from that determine what the corresponding bit pattern in memory should be. The following examples demonstrate the requirements.

Translating to and from Machine Code

The following examples present some practical problems in the translation of assembly language into machine code and hex. They also demonstrate doing the converse—that is, given a bit pattern in memory, translate that into assembly language.

Example 5–26

Using the instruction set presented in this section for the 4-bit microprocessor, show what the actual bit pattern in memory would be for the following assembly language program.

Address	Operation
0	CLA ; Set accumulator to zero.
1	TAB ; Copy accum into B reg.
2	INR ; Increment accumulator.
3	ADD ; Add accum and B reg.
4	HLT ; Stop processing.

Solution

See Figure 5–59.

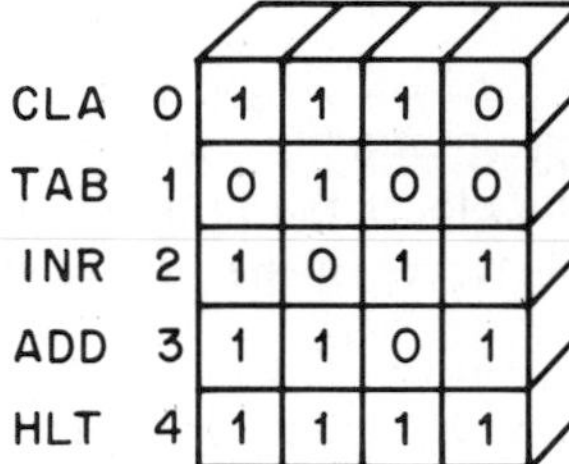

Figure 5–59

Example 5–27

Develop the assembly language program for the bit pattern shown in Figure 5–60.

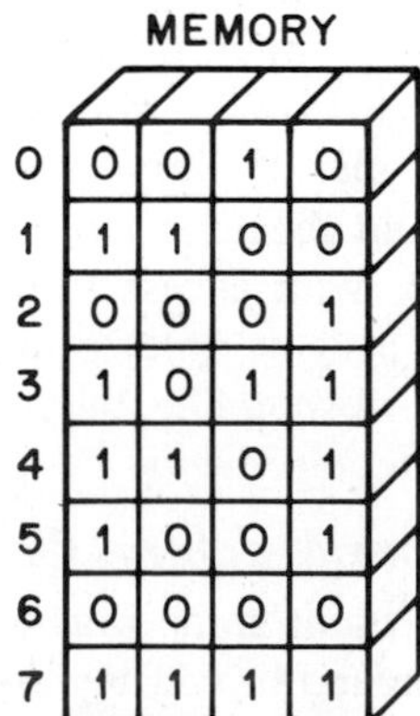

Figure 5–60

Solution

Refer to Figure 5–61.

ADDRESS	OPERATION
0	LDB C; PUT C_{16} INTO B REG.
2	LDA B; PUT B_{16} INTO ACCUM.
4	ADD ; ADD C_{16} + B_{16}
5	TAM ( ); STORE ANSWER
7	HLT

NOTE: MEMORY LOCATION 6 RESERVED FOR STORING ACCUMULATOR

Figure 5–61

Example 5–28

Check the following bit pattern in memory with the assembly program presented in Figure 5–62 and determine if there is a problem. If there is, state the problem(s).

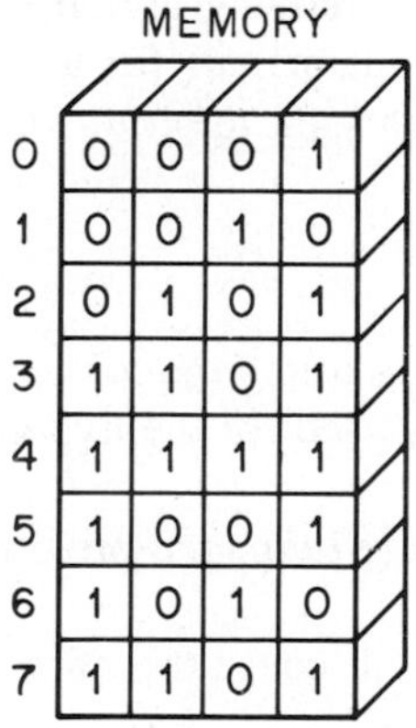

MEMORY

0	0	0	0	1
1	0	0	1	0
2	0	1	0	1
3	1	1	0	1
4	1	1	1	1
5	1	0	0	1
6	1	0	1	0
7	1	1	0	1

ADDRESS	OPERATION
0	LDA 9
2	LDB A
4	NTA
5	ADD
6	HLT

Figure 5–62

Solution

See Figure 5–63.

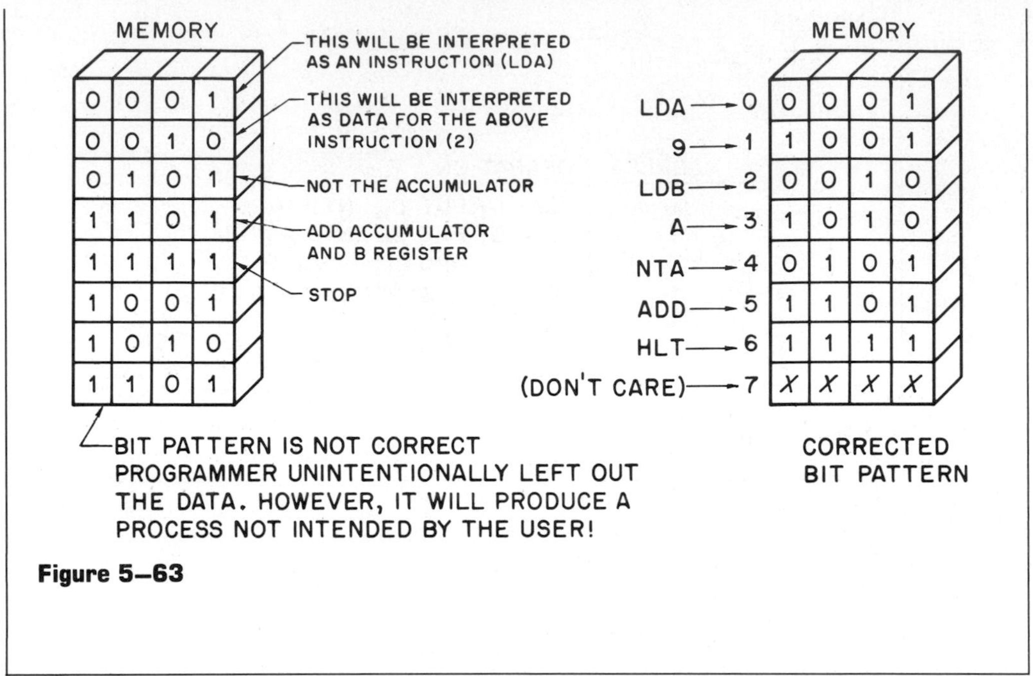

Figure 5–63

Conclusion

You have come a long way in the application of digital technology to microprocessors. Remember, this is the reason why you are studying digital techniques in the first place. It is important that you understand the fundamentals for all aspects of a digital system. Test your understanding of this section by trying the following review questions.

5–7 Review Questions

1. Why use *mnemonics* when programming a microprocessor?
2. Give an example of a mnemonic that is used for the hypothetical 4-bit microprocessor presented in this section.
3. State the difference between an *assembly language* program and a *machine language* program.
4. Describe what the following program will cause the microprocessor to do:

   0 LDA 7
   2 LDB 3
   4 ADD
   5 HLT

5. State what the contents of the accumulator and B register will be when the program in question 4 is executed.

DM54ALS04A/DM74ALS04A Hex Inverters

General Description

This device contains six independent gates each of which performs the logic INVERT function.

Features

- Switching Specifications at 50 pF.
- Switching Specifications Guaranteed Over Full Temperature and V_{CC} Range.
- Advanced [illegible] Schottky [illegible] Process
- Functionally and Pin For Pin Compatible with Schottky and Low Power Schottky TTL Counterpart.
- Improved AC Performance Over Schottky and Low Power Schottky Counterparts.

Absolute Maximum Ratings (Note 1)

Supply Voltage	7V
Input Voltage	7V
Operating Free Air Temperature Range	
DM54ALS	−55°C to 125°C
DM74ALS	0°C to 70°C
Storage Temperature Range	−65°C to 150°C

Note 1: The "Absolute Maximum Ratings" are those values beyond which the safety of the device can not be guaranteed. The device should not be operated at these limits. The parametric values defined in the [illegible] table are not guaranteed at the absolute [illegible] define the conditions for actual device operation.

Connection Diagram

Dual-In-Line Package

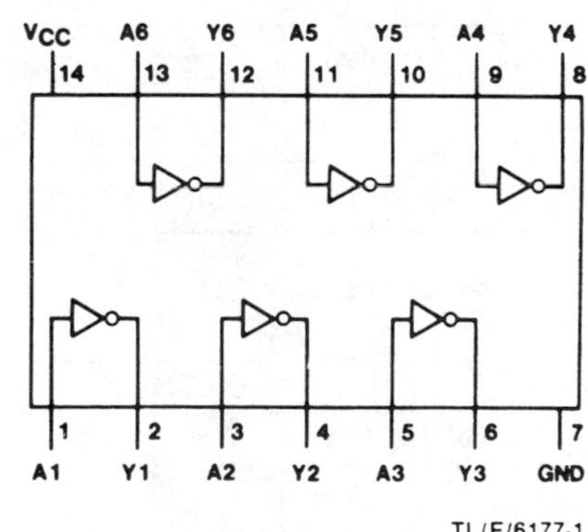

TL/F/6177-1

54ALS04A (J) 74ALS04A (J,N)

Function Table

$Y = \overline{A}$

Input	Output
A	Y
L	H
H	L

H = High Logic Level
L = Low Logic Level

Figure 5–64 First sheet of specifications for hex inverter. *Reprinted* with permission of National Semiconductor Corporation.

National holds no responsibility for any circuitry described. National reserves the right at any time to change without notice said circuitry.

Recommended Operating Conditions

Parameter	DM54ALS04A			DM74ALS04A			Unit
	Min	Nom	Max	Min	Nom	Max	
Supply Voltage, V_{CC}	4.5	5	5.5	4.5	5	5.5	V
High Level Input Voltage, V_{IH}	2			2			V
Low Level Input Voltage, V_{IL}			0.8			0.8	V
High Level Output Current, I_{OH}			−0.4			−0.4	mA
Low Level Output Current, I_{OL}			4			8	mA

Electrical Characteristics over recommended operating free air temperature range.

All typical values are measured at $V_{CC} = 5V$, $T_A = 25°C$.

Symbol	Parameter	Conditions		Min	Typ	Max	Unit
V_{IK}	Input Clamp Voltage	$V_{CC} = 4.5V$, $I_I = -18$ mA				−1.5	V
V_{OH}	High Level Output Voltage	$I_{OH} = -0.4$mA $V_{CC} = 4.5$ to 5.5V		$V_{CC}-2$			V
V_{OL}	Low Level Output Voltage	$V_{CC} = 4.5V$	54/74ALS $I_{OL} = 4$ mA		0.25	0.4	V
			74ALS $I_{OL} = 8$ mA		0.35	0.5	V
I_I	Max High Input Current	$V_{CC} = 5.5V$, $V_{IH} = 7V$				0.1	mA
I_{IH}	High Level Input Current	$V_{CC} = 5.5V$, $V_{IH} = 2.7V$				20	μA
I_{IL}	Low Level Input Current	$V_{CC} = 5.5V$, $V_{IL} = 0.4V$				−0.1	mA
I_O	Output Drive Current	$V_{CC} = 5.5V$	$V_O = 2.25V$	−30		−112	mA
I_{CC}	Supply Current	$V_{CC} = 5.5V$	Outputs High		0.65	1.1	mA
			Outputs Low		2.4	4.2	mA

Switching Characteristics over recommended operating free air temperature range (Note 1).

All typical values are measured at $V_{CC} = 5V$, $T_A = 25°C$.

Parameter	Conditions	DM54ALS04A			DM74ALS04A			Unit
		Min	Typ	Max	Min	Typ	Max	
T_{PLH}, Propagation delay time. Low to high level output	$V_{CC} = 4.5$ to 5.5V $R_L = 500\ \Omega$, $C_L = 50$ pF.	3		14	3		11	ns
T_{PHL}, Propagation delay time. High to low level output		2		12	2		8	ns

Note 1: See Section 1 for test waveforms and output load.

Figure 5–65 Second sheet of data sheet. *Reprinted* with permission of National Semiconductor Corporation.

ELECTRONIC APPLICATION

Discussion

It is important for the technician and digital designer to know how to interpret the specifications sheets for digital ICs. These specifications are prepared by the manufacturers of the ICs and present important operating conditions concerning each type of logic IC. The examples used here are from the Logic Databook by National Semiconductor Corporation, a manufacturer of digital ICs.

Overview of the Data Sheets

Refer to Figure 5–64. There is the first sheet of specifications for the hex inverter. Note that the sheet is divided into five major parts:

1. General Description: Describes what the IC contains.
2. Features: This is the manufacturer's "brag" area. Here, the manufacturer will list the outstanding features of its IC.
3. Absolute Maximum Ratings: Read the footnote here (Note 1). It's important to understand that you are not to operate the device at these ratings!
4. Connection Diagram: Here you see how the pin numbers of the chip relate to the actual logic circuits on the inside.
5. Function Table: This gives the logic relations between the input(s) and the output(s). Here you see the Boolean expression along with a corresponding truth table.

The second sheet of the data sheet is shown in Figure 5–65. It is divided into three major areas:

1. Recommended Operating Conditions: These are the values of voltage and current that you should use with this IC.
2. Electrical Characteristics: What is listed here are typical values you would expect to get from the IC under specific operating conditions.
3. Switching Characteristics: These characteristics specify how long it takes the output to respond to an input. See Figure 5–66.

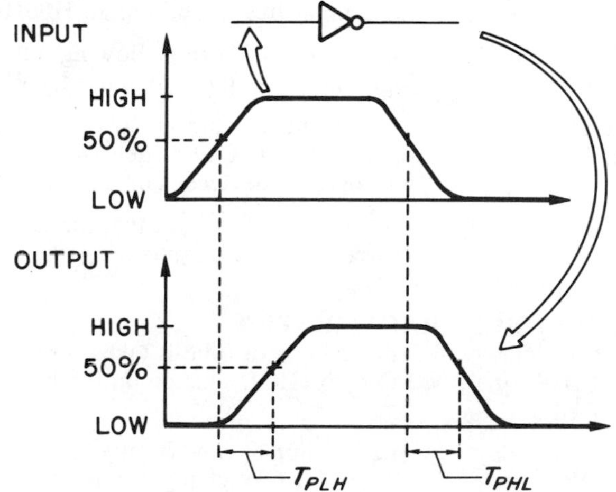

Figure 5–66 Concept of propagation delay.

Rise: 10% - 90%
FAll: 90%-10%

TTL Specifications

Table 5–8 details the meaning of each section of the data sheet of Figure 5–65.

Table 5–8 MEANING OF LOGIC CIRCUIT SPECIFICATIONS

Parameter	Meaning
RECOMMENDED OPERATING CONDITIONS	
Supply voltage V_{cc}	The minimum, nominal, and maximum allowed voltages that should be used to power the device.
High-level input voltage V_{IH}	The minimum amount of voltage that can be applied to a logic input and recognized as an input HIGH.
Low-level input voltage V_{IL}	The maximum amount of voltage that can be applied to a logic input and be recognized as an input LOW.
High-level output current I_{OH}	The current flowing from the logic output of the device when its output is a logic HIGH.
Low-level output current I_{OL}	The current flowing from the logic output of the device when its output is a logic LOW.
ELECTRICAL CHARACTERISTICS	
Input clamp voltage V_{IK}	This is not intended to be an operating condition. The manufacturer specifies this as a method of rating the input circuit of the logic device.
High-level output voltage V_{OH}	The voltage value at the logic output of the device when the output is HIGH.
Low-level output voltage V_{OL}	The voltage value at the logic output of the device when the output is LOW.
Maximum high input current I_I	This is the maximum amount of current that will flow into a logic input when that logic input has the maximum voltage specified applied to it.
High-level input current I_{IH}	The amount of current that flows to a logic input when that logic input has the minimum HIGH level voltage applied to it.
Low-level input current I_{IL}	The amount of current flowing out of a logic input when that logic input has the maximum LOW level voltage applied to it.
Output drive current I_O	The amount of current flowing into the logic output when the output is in its OFF state. This usually applies to devices that will operate devices such as digital displays.
Supply current	The amount of current that the device will take from power source when all outputs are HIGH and when all outputs are LOW.
SWITCHING CHARACTERISTICS	
Propagation delay time—LOW to HIGH T_{PLH}	The amount of time it takes for the output to respond to a LOW to HIGH change on the input.
Propagation delay time—HIGH to LOW T_{PHL}	The amount of time it takes for the output to respond to a HIGH to LOW change on the input.

Absolute Maximum Ratings (Notes 1 & 2)

Supply Voltage (V_{CC})	−0.5 to +7.0V
DC Input Voltage (V_{IN})	−1.5 to V_{CC}+1.5V
DC Output Voltage (V_{OUT})	−0.5 to V_{CC}+0.5V
Clamp Diode Current (I_{IK}, I_{OK})	±20 mA
DC Output Current, per pin (I_{OUT})	±25 mA
DC V_{CC} or GND Current, per pin (I_{CC})	±50 mA
Storage Temperature Range (T_{STG})	−65°C to +150°C
Power Dissipation (P_D) (Note 3)	500 mW
Lead Temp. (T_L) (Soldering 10 seconds)	260°C

Operating Conditions

	Min	Max	Units
Supply Voltage (V_{CC})	2	6	V
DC Input or Output Voltage (V_{IN}, V_{OUT})	0	V_{CC}	V
Operating Temp. Range (T_A)			
MM74HCU	−40	+85	°C
MM54HCU	−55	+125	°C

DC Electrical Characteristics (Note 4)

Symbol	Parameter	Conditions	V_{CC}	T_A=25°C		74HCU T_A=−40 to 85°C	54HCU T_A=−55 to 125°C	Units
				Typ	Guaranteed Limits			
V_{IH}	Minimum High Level Input Voltage		2.0V		1.7	1.7	1.7	V
			4.5V		3.6	3.6	3.6	V
			6.0V		4.8	4.8	4.8	V
V_{IL}	Maximum Low Level Input Voltage		2.0V		0.3	0.3	0.3	V
			4.5V		0.8	0.8	0.8	V
			6.0V		1.1	1.1	1.1	V
V_{OH}	Minimum High Level Output Voltage	$V_{IN}=V_{IL}$ $\|I_{OUT}\| \le 20\ \mu A$	2.0V	2.0	1.8	1.8	1.8	V
			4.5V	4.5	4.0	4.0	4.0	V
			6.0V	0.0	5.5	5.5	5.5	V
		V_{IN}=GND $\|I_{OUT}\| \le 4.0$ mA	4.5V	4.2	3.98	3.84	3.7	V
		$\|I_{OUT}\| \le 5.2$ mA	6.0V	5.7	5.48	5.34	5.2	V
V_{OL}	Maximum Low Level Output Voltage	$V_{IN}=V_{IH}$ $\|I_{OUT}\| \le 20\ \mu A$	2.0V	0	0.2	0.2	0.2	V
			4.5V	0	0.5	0.5	0.5	V
			6.0V	0	0.5	0.5	0.5	V
		$V_{IN}=V_{CC}$ $\|I_{OUT}\| \le 6.0$ mA	4.5V	0.2	0.26	0.33	0.4	V
		$\|I_{OUT}\| \le 7.8$ mA	6.0V	0.2	0.26	0.33	0.4	V
I_{IN}	Maximum Input Current	$V_{IN}=V_{CC}$ or GND	6.0V		±0.1	±1.0	±1.0	μA
I_{CC}	Maximum Quiescent Supply Current	$V_{IN}=V_{CC}$ or GND $I_{OUT}=0\ \mu A$	6.0V		2.0	20	40	μA

Note 1: Absolute Maximum Ratings are those values beyond which damage to the device may occur.

Note 2: Unless otherwise specified all voltages are referenced to ground.

Note 3: Power Dissipation temperature derating — plastic "N" package: −12 mW/°C from 65°C to 85°C; ceramic "J" package: −12 mW/°C from 100°C to 125°C.

Note 4: For a power supply of 5V ±10% the worst case output voltages (V_{OH}, and V_{OL}) occur for HC at 4.5V. Thus the 4.5V values should be used when designing with this supply. Worst case V_{IH} and V_{IL} occur at V_{CC}=5.5V and 4.5V respectively. (The V_{IH} value at 5.5V is 3.85V.) The worst case leakage current (I_{IN}, I_{CC}, and I_{OZ}) occur for CMOS at the higher voltage and so the 6.0V values should be used.

Figure 5–67 Reprinted with permission of National Semiconductor Corporation.

CMOS Specifications

The following example demonstrates the use of a CMOS specification sheet for a hex inverter. It also makes an important comparison between the CMOS and TTL hex inverters.

Example 5–29

Figure 5–67 shows a specification sheet for a CMOS IC hex inverter. Answer the following questions from this specification sheet (they apply to the 74HCU).

1. What is the range of power supply voltages this device may use?
2. For a supply voltage of 4.5 volts, what voltage reading would you expect to get at a logic output that was HIGH?
3. For a supply voltage of 6.0 volts, what is the amount of input voltage required to make an input HIGH?
4. For a supply voltage of 6.0 volts, how much electrical power does the device consume. How does this compare to the electrical power consumed by the TTL hex inverter described in Figure 5–65?

Solution

1. The range of power supply voltages is given under the section called OPERATING CONDITIONS. This particular CMOS may operate within a range of 2 volts to 6 volts.
2. This is under DC ELECTRICAL CHARACTERISTICS and is listed as V_{OH}. For a power supply value of 4.5 volts, the minimum voltage reading for a HIGH would be 4.0 volts.
3. For a supply voltage of 6.0 volts, the amount of input voltage required to make an input HIGH is V_{IH}, which is equal to 4.8 volts.
4. The amount of electrical power consumed by this device with a supply voltage of 6.0 volts is

$$P = I \times V$$
$$P = 20\ \mu\text{A} \times 6.0\ \text{V}$$
$$P = 120\ \mu\text{W}$$

This compares to (for a TTL hex inverter)

$$P = I \times V$$
$$P = 4.2\ \text{mA} \times 5\ \text{V}$$
$$P = 21\ \text{mW}$$

Thus, the TTL IC consumes:

$$\frac{21\ \text{mW}}{120\ \mu\text{W}} = 175 \text{ times more power then the CMOS IC!}$$

Conclusion

This section presented the TTL and CMOS data sheets. Here you had a chance to see what the different specifications meant and how to apply them. Knowing how to interpret these data sheets is an important skill for the digital technician and designer. Test your understanding of this section by trying the following review questions.

5–8 Review Questions

1. Is a logic device operated at the *absolute maximum ratings*? Explain.
2. Under which of the major parts of the data sheet would you find the operating temperature of the device?
3. Explain the purpose of the function table in the specification sheet.
4. What is meant by the switching characteristics of a digital device?
5. Where would you find the data in order to compute the *noise margin* for a HIGH level of a digital IC?

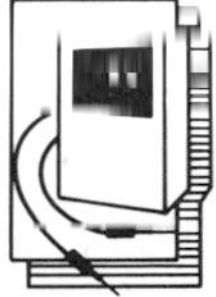

TROUBLESHOOTING SIMULATION

The troubleshooting simulation for this chapter found on your student disk gives you practice in troubleshooting logic networks consisting of more than one gate. You will see the effect of a faulty gate on an entire logic network. It will be your task to determine the most likely problem based upon your observation of the symptoms.

SUMMARY

- Logic networks can be broken down into *logic levels* for analysis purposes.
- Boolean expressions may be developed from gate networks by using *empty terms*.
- Boolean expressions may be developed from gate networks by first developing the terms from the deepest level first and then working toward the lowest level in the analysis.
- The *sum of products* Boolean expression consists of the ORing of terms that are ANDed.
- The *product of sums* Boolean expression consists of the ANDing of terms that are ORed.
- An SOP expression can be derived from a truth table by having a product term for every TRUE output condition.
- A POS expression can be derived from a truth table by having a sum term for every FALSE output condition.
- The method used for developing a Boolean expression from a truth table (SOP or POS) depends on the output conditions of the truth table. Generally, use the method that produces the fewest terms.
- Any SOP expression can be represented by two-level logic consisting of AND gates feeding a single OR gate.
- Any POS expression can be represented by two-level logic consisting of OR gates feeding a single AND gate.

- It is considered good practice to construct SOP or POS logic networks using NAND/NOR logic rather than AND/OR logic.
- It is convenient to represent a group of digital data by a set of parallel connectors called a *bus*.
- Two levels of NAND gates can be used to represent POS Boolean expressions.
- Two levels of NOR gates can be used to represent SOP Boolean expressions.
- A Karnaugh map (K-map) is a graphical method of simplifying logic networks.
- A logic network can be simplified with a K-map by considering adjoining *pairs, quads,* and *octets*.
- A K-map can be *rolled* with either or both edges touching in order to facilitate logic simplification.
- *Don't care* conditions can be treated as a 1 or a 0 in a K-map to facilitate the simplification of gate networks.
- Overlapping groups in K-mapping are adjacent cells that can be used more than once in the gate simplification process.
- POS terms may be used in K-map simplification by considering the cells containing 0's.
- There is a definite set of steps used in the design of gate logic networks. These steps are helpful in converting an idea to an actual logic network.
- A mnemonic code is used in programming to help recall specific instructions to the microprocessor.
- All microprocessors come with an instruction set that contains a unique mnemonic code for that microprocessor.
- There are different programming language levels, depending on how close the programming language is to the actual bit pattern used by the microprocessor.
- An assembler is a computer program that converts a program in mnemonics to the actual bit pattern used by the microprocessor.
- Assembly language programming consists of writing the program code in terms of mnemonics.
- Machine language is the actual bit pattern that is used by the microprocessor.
- Usually, hex code is also referred to as *machine language*.
- Logic gate specification sheets show the operating conditions for the specific logic gate.
- The specifications for *CMOS* differ considerably from those of *TTL*.

CHAPTER SELF-TEST

I. TRUE/FALSE

Answer the following questions true or false.

1. The Boolean expression $AB + CD$ represents an SOP expression.
2. A POS expression is represented by OR gates feeding into a single AND gate.

3. Three AND gates feeding into a single OR gate can all be replaced by four NOR gates.
4. Three OR gates feeding into a single AND gate can all be replaced by four AND gates.
5. Four NAND gates feeding a single NAND gate represent a POS Boolean expression.

II. MULTIPLE CHOICE

Answer the following by selecting the most correct answer.

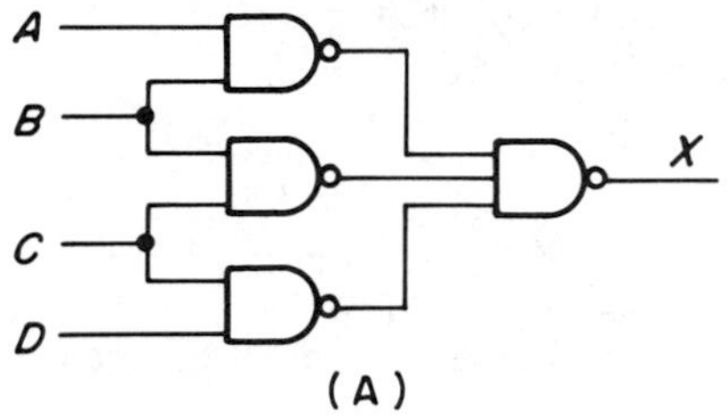

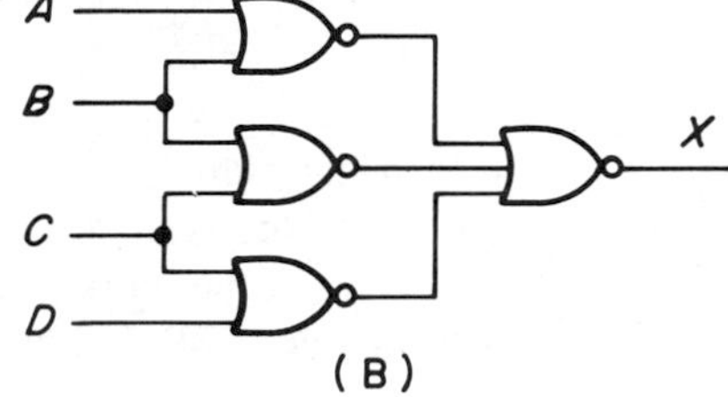

Figure 5–68

6. In the logic network of Figure 5–68(A), the empty term expression of the output of the first level is
   (A) $(\ \ )_1(\ \ )_2(\ \ )_3$
   (B) $\overline{(\ \ )_1}\,\overline{(\ \ )_2}\,\overline{(\ \ )_3}$
   (C) $(\ \ )_1 + (\ \ )_2 + (\ \ )_3$
   (D) None of the above
7. For the logic network of Figure 5–68(B), the empty term expression of the output of the first level is
   (A) $(\ \ )_1(\ \ )_2(\ \ )_3$
   (B) $\overline{(\ \ )_1(\ \ )_2(\ \ )_3}$
   (C) $\overline{(\ \ )_1 + (\ \ )_2 + (\ \ )_3}$
   (D) None of the above
8. For the logic networks of Figure 5–68:
   (A) Both produce an SOP expression.
   (B) Neither produce an SOP expression.
   (C) Figure 5–68(A) produces a POS while 5–68(B) produces an SOP.
   (D) Figure 5–68(A) produces an SOP while 5–68(B) produces a POS.
9. For the truth table

A	B	C	X
0	0	0	1
0	0	1	0
0	1	0	0
0	1	1	1
1	0	0	1
1	0	1	0
1	1	0	0
1	1	1	1

an SOP expression would be:
(A) $\overline{A}\,\overline{B}\,\overline{C} + \overline{A}BC + A\overline{B}\,\overline{C} + ABC$
(B) $ABC + A\overline{B}\,\overline{C} + \overline{A}BC + \overline{A}\,\overline{B}\,\overline{C}$
(C) $(\overline{A} + \overline{B} + \overline{C})(\overline{A} + B + C)(A + \overline{B} + \overline{C})(A + B + C)$
(D) $(A + B + C)(A + \overline{B} + \overline{C})(\overline{A} + B + C)(\overline{A} + \overline{B} + \overline{C})$

10. For the truth table of problem 9 the correct POS expression would be
(A) $\overline{A}\,\overline{B}\,\overline{C} + \overline{A}BC + A\overline{B}\,\overline{C} + ABC$
(B) $AB\overline{C} + A\overline{B}C + \overline{A}B\overline{C} + \overline{A}\,\overline{B}C$
(C) $(A + B + \overline{C})(A + \overline{B} + C)(\overline{A} + B + \overline{C})(\overline{A} + \overline{B} + C)$
(D) $(A + B + \overline{C})(\overline{A} + B + \overline{C})(A + \overline{B} + C)(A + B + \overline{C})$

III. FILL-IN

Fill in the blanks with the correct term(s).

11. The Boolean expression $AB + ACD + D$ is a/an ______ expression.
12. A logic network consisting of ______ levels represents a POS expression.
13. For an SOP expression, the number of inputs to each AND gate represents the number of ______ in each product term.
14. A group of wires in a digital system treated as a unit is called a ______.
15. When an SOP expression is converted to AND/OR logic, the first level will consist of a/an ______ gate.

IV. MATCHING

Using Figure 5–69, match the logic networks to the truth tables that represent them.

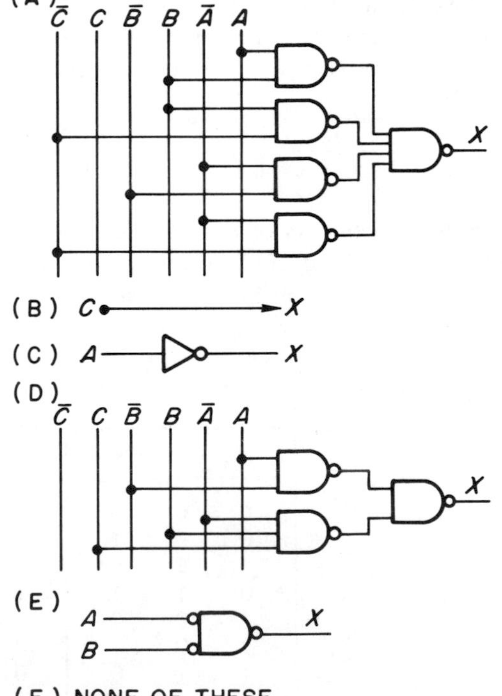

Figure 5–69

16.

A	*B*	*C*	*X*
0	0	0	0
0	0	1	1
0	1	0	0
0	1	1	1
1	0	0	0
1	0	1	1
1	1	0	0
1	1	1	1

17.

A	*B*	*X*
0	0	0
0	1	1
1	0	1
1	1	1

18.

A	*B*	*X*
0	0	1
0	1	1
1	0	0
1	1	0

19.

A	*B*	*C*	*X*
0	0	0	1
0	0	1	1
0	1	0	1
0	1	1	0
1	0	0	0
1	0	1	0
1	1	0	1
1	1	1	1

20.

A	*B*	*C*	*X*
0	0	0	0
0	0	1	0
0	1	0	0
0	1	1	1
1	0	0	1
1	0	1	1
1	1	0	0
1	1	1	0

V. OPEN-ENDED

Answer the following questions as indicated.

21. Construct the K-map for the following truth table, using POS.

A	*B*	*C*	*X*
0	0	0	0
0	0	1	0
0	1	0	1
0	1	1	1
1	0	0	0
1	0	1	0
1	1	0	1
1	1	1	1

22. Develop the Boolean POS expression from the K-map in problem 21.
23. Construct the logic network for the SOP expression of the K-map in problem 21, using NAND/NOR logic.
24. Construct the logic network for the POS expression of the K-map in problem 21, using NAND/NOR logic.
25. Develop the logic network using NAND/NOR logic with the fewest number of gates for the following truth table.

A	*B*	*C*	*D*	*X*
0	0	0	0	1
0	0	0	1	1
0	0	1	0	0
0	0	1	1	0
0	1	0	0	0
0	1	0	1	0
0	1	1	0	1
0	1	1	1	1
1	0	0	0	1
1	0	0	1	1
1	0	1	0	0
1	0	1	1	0
1	1	0	0	0
1	1	0	1	0
1	1	1	0	1
1	1	1	1	1

Answers to Chapter Self-Test

1] T 2] T 3] F 4] F 5] F 6] D 7] C 8] D 9] A 10] C
11] SOP 12] two 13] variables 14] bus 15] OR 16] B
17] E 18] C 19] A 20] D
21] $X = \overline{A}B\overline{C} + \overline{A}BC + AB\overline{C} + ABC$

	$\bar{C}$	C
$\bar{A}\bar{B}$	0	0
$\bar{A}B$	1	1
AB	1	1
$A\bar{B}$	0	0

22] $X = B$
23] $B \rightarrow X$ (no logic necessary)
24] $B \rightarrow X$ (no logic necessary)
25] See Figure 5–70.

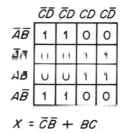

	$\bar{C}\bar{D}$	$\bar{C}D$	CD	$C\bar{D}$
$\bar{A}\bar{B}$	1	1	0	0
$\bar{A}B$	0	0	1	1
AB	0	0	1	1
$A\bar{B}$	1	1	0	0

$X = \bar{C}\bar{B} + BC$

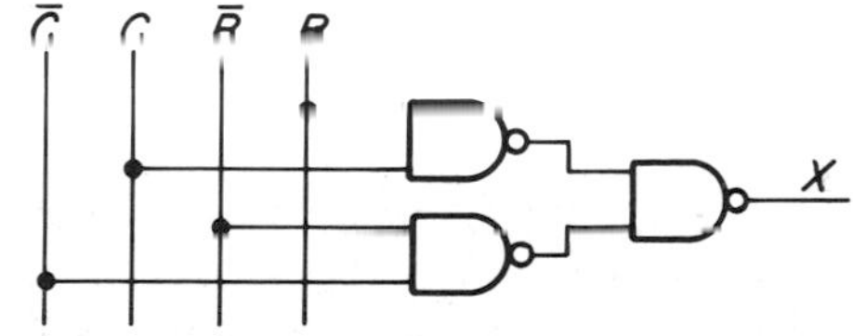

Figure 5–70

CHAPTER PROBLEMS

Basic Concepts

Section 5–1:

1. Determine the number of logic levels for the logic network of Figure 5–71(A).
2. For the logic network of Figure 5–71(B), determine the number of logic levels.
3. Use the method of *empty terms* to develop the Boolean expression for the logic network of Figure 5–71(A).
4. Develop the Boolean expression for the logic network of Figure 5–71(B), using the method of *empty terms*.

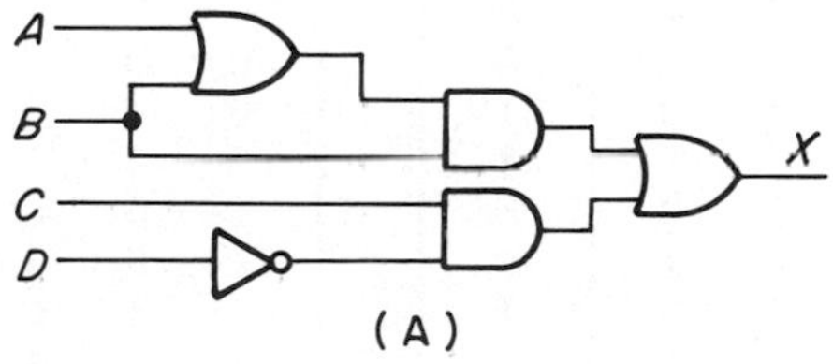

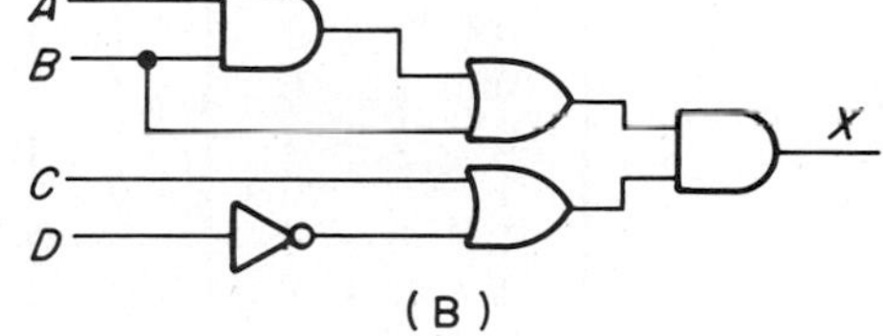

Figure 5–71

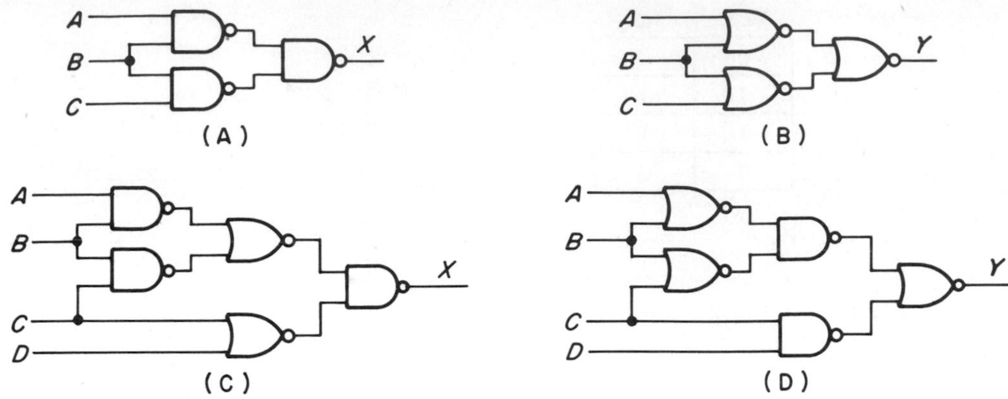

Figure 5–72

5. Develop the Boolean expression for the logic networks of Figure 5–72(A) and (B), using the method of empty terms.
6. For the logic networks of Figure 5–72(C) and (D), develop the Boolean expressions, using the empty terms method. Be sure to show all steps.
7. Develop the Boolean expression using the *gate-by-gate* technique for the logic network of Figure 5–72(A) and (B). Be sure to show all steps.
8. Using the logic network of Figure 5–72(C) and (D), develop the Boolean expression using the *gate-by-gate* technique. Show all steps.

Section 5–2:

9. Determine which of the Boolean expressions are SOP and which are POS.
   (A) $AB\overline{C} + BC + \overline{A}B$
   (B) $(A + B + \overline{C})(\overline{A} + B)C$
   (C) $A + \overline{A}B + CDE$
10. For the Boolean expressions given, determine which are POS and which are SOP expressions.
    (A) $(R + S + \overline{T})(A + C)\overline{B}$
    (B) $C(\overline{D} + \overline{E})(C + E)$
    (C) $X\overline{Y}\overline{Z} + \overline{Z} + \overline{Y}\overline{X}$
11. Develop the SOP Boolean expression for Truth Table 5–1.

Truth Table 5–1

A	B	C	X
0	0	0	1
0	0	1	1
0	1	0	1
0	1	1	1
1	0	0	0
1	0	1	0
1	1	0	0
1	1	1	0

Truth Table 5–2

A	B	C	X
0	0	0	0
0	0	1	0
0	1	0	1
0	1	1	1
1	0	0	0
1	0	1	0
1	1	0	1
1	1	1	1

Truth Table 5–3

A	B	C	X
0	0	0	1
0	0	1	1
0	1	0	0
0	1	1	0
1	0	0	1
1	0	1	0
1	1	0	1
1	1	1	1

12. Create the SOP expression for Truth Table 5–2.
13. Using Truth Table 5–3, develop the SOP Boolean expression.

Truth Table 5–4

A	*B*	*C*	*D*	*X*
0	0	0	0	1
0	0	0	1	1
0	0	1	0	0
0	0	1	1	0
0	1	0	0	0
0	1	0	1	1
0	1	1	0	1
0	1	1	1	1
1	0	0	0	0
1	0	0	1	0
1	0	1	0	1
1	0	1	1	1
1	1	0	0	1
1	1	0	1	1
1	1	1	0	0
1	1	1	1	1

Truth Table 5–5

A	*B*	*C*	*D*	*X*
0	0	0	0	0
0	0	0	1	0
0	0	1	0	1
0	0	1	1	1
0	1	0	0	1
0	1	0	1	0
0	1	1	0	0
0	1	1	1	0
1	0	0	0	1
1	0	0	1	1
1	0	1	0	1
1	0	1	1	1
1	1	0	0	0
1	1	0	1	0
1	1	1	0	0
1	1	1	1	0

14. Write the SOP expression for Truth Table 5–4.
15. Using Truth Table 5–5, develop the SOP expression.
16. Create the POS expression for Truth Tables 5–1 and 5–2.
17. Referring to Truth Table 5–3, develop the POS expression.
18. Write the POS expression for Truth Table 5–4.
19. Using Truth Table 5–5, develop the POS expression.
20. Using Boolean algebra, demonstrate that the SOP expressions for Truth Tables 5–1 and 5–2 are equivalent to the POS expressions for the same tables when $A = 1$, $B = 1$, and $C = 0$.
21. Show that the POS and SOP expressions for Truth Table 5–4 are equivalent, using Boolean algebra when $A = 1$, $B = 0$, $C = 1$, and $D = 0$.
22. For Truth Table 5–5, demonstrate, using Boolean algebra, that the POS and SOP expressions are equivalent when $A = 0$, $B = 1$, $C = 1$, and $D = 0$.

Section 5–3:

23. Convert Truth Table 5–1 into AND/OR logic using POS.
24. Convert Truth Table 5–1 into AND/OR logic using SOP.
25. Create the AND/OR logic network for Truth Table 5–2, using first the SOP and then the POS.
26. Using NAND/NOR logic, convert Truth Table 5–4 into a logic network, using POS.
27. For Truth Table 5–5, develop its logic network with NAND/NOR gates, using SOP.

28. Convert the following Boolean expressions to their equivalent AND/OR logic networks:
    (A) $A\overline{B} + \overline{A}B$
    (B) $(A + \overline{B})(B + A)$
    (C) $A\overline{B}\,\overline{C} + AB + \overline{B}C$
    (D) $(A + \overline{B} + C)(\overline{A} + B)C$
29. Using AND/OR logic, convert the following Boolean expressions to their equivalent logic networks.
    (A) $(A + B)(\overline{A} + B)$
    (B) $\overline{A}\,\overline{B}C + ABC + A\overline{B}\,\overline{C}$
    (C) $(A + \overline{B} + C)(A + B)C$
30. Convert the Boolean expressions of problem 29 to their equivalent NAND/NOR logic.
31. For the Boolean expressions in problem 28, convert them to equivalent NAND/NOR logic networks.
32. Implement Truth Tables 5–1 and 5–2 to their equivalent logic networks using NAND/NOR logic.
33. Develop the NAND/NOR logic for Truth Table 5–4.
34. For Truth Table 5–5, develop the equivalent NAND/NOR logic.

Section 5–4:

35. In Figure 5–73, determine if the K-maps correctly represent the corresponding truth tables.

(A)

A	B	C	X
0	0	0	0
0	0	1	1
0	1	0	0
0	1	1	0
1	0	0	0
1	0	1	0
1	1	0	1
1	1	1	0

	$\overline{B}\,\overline{C}$	$\overline{B}C$	BC	$B\overline{C}$
$\overline{A}$	0	1	0	0
A	0	0	0	1

(B)

A	B	C	X
0	0	0	0
0	0	1	1
0	1	0	0
0	1	1	0
1	0	0	0
1	0	1	0
1	1	0	1
1	1	1	0

	$\overline{C}$	C
$\overline{A}\,\overline{B}$	0	1
AB	1	0
$\overline{A}B$	0	0
$A\overline{B}$	0	0

(C)

A	B	X
0	0	0
0	1	1
1	0	1
1	1	0

	$\overline{B}$	B
$\overline{A}$	0	1
A	1	0

(D)

A	B	C	X
0	0	0	0
0	0	1	0
0	1	0	1
0	1	1	1
1	0	0	0
1	0	1	0
1	1	0	1
1	1	1	1

	$\overline{C}$	C
$\overline{A}\,\overline{B}$	0	0
$\overline{A}B$	1	1
AB	1	1
$A\overline{B}$	0	0

Figure 5–73

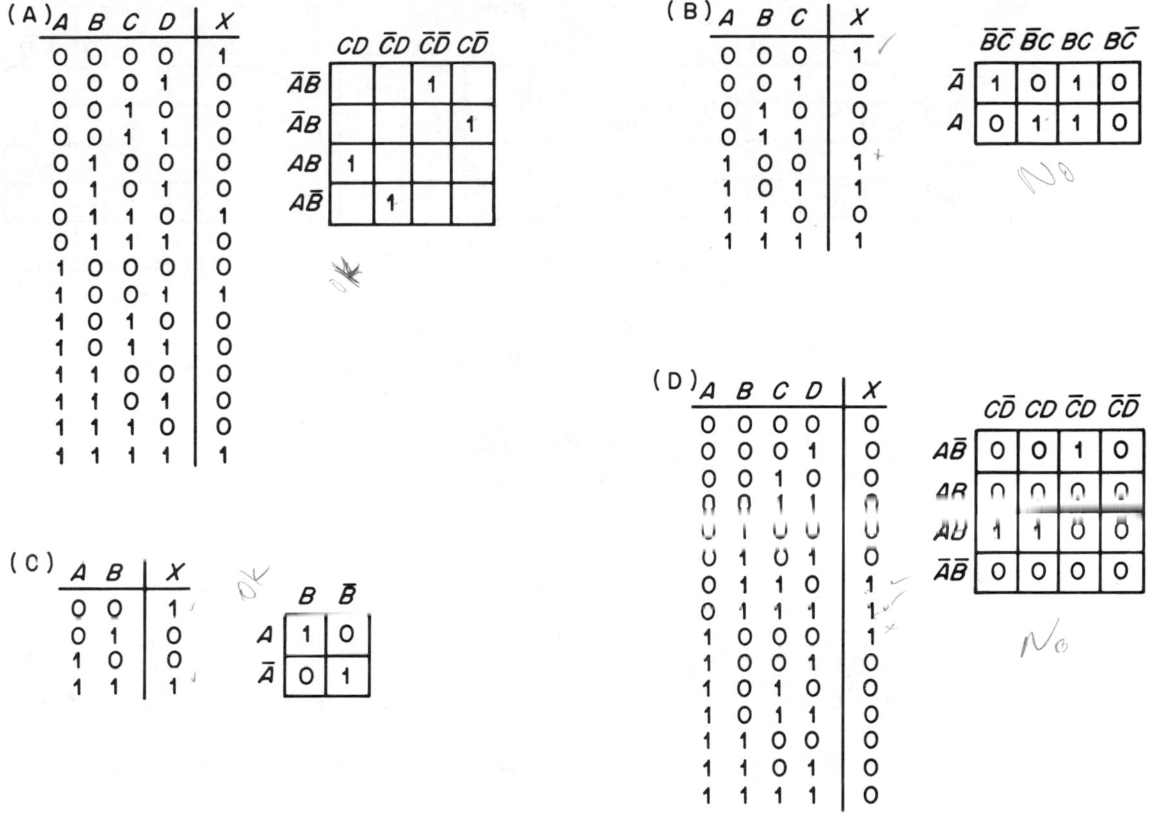

(A)

A	B	C	D	X
0	0	0	0	1
0	0	0	1	0
0	0	1	0	0
0	0	1	1	0
0	1	0	0	0
0	1	0	1	0
0	1	1	0	1
0	1	1	1	0
1	0	0	0	0
1	0	0	1	1
1	0	1	0	0
1	0	1	1	0
1	1	0	0	0
1	1	0	1	0
1	1	1	0	0
1	1	1	1	1

	CD	$\overline{C}D$	$\overline{C}\overline{D}$	$C\overline{D}$
$\overline{A}\overline{B}$			1	
$\overline{A}B$				1
AB	1			
$A\overline{B}$		1		

(B)

A	B	C	X
0	0	0	1
0	0	1	0
0	1	0	0
0	1	1	0
1	0	0	1
1	0	1	1
1	1	0	0
1	1	1	1

	$\overline{B}\overline{C}$	$\overline{B}C$	BC	$B\overline{C}$
$\overline{A}$	1	0	1	0
A	0	1	1	0

(C)

A	B	X
0	0	1
0	1	0
1	0	0
1	1	1

	B	$\overline{B}$
A	1	0
$\overline{A}$	0	1

(D)

A	B	C	D	X
0	0	0	0	0
0	0	0	1	0
0	0	1	0	0
0	0	1	1	0
0	1	0	0	0
0	1	0	1	0
0	1	1	0	1
0	1	1	1	1
1	0	0	0	1
1	0	0	1	0
1	0	1	0	0
1	0	1	1	0
1	1	0	0	0
1	1	0	1	0
1	1	1	1	0

	$C\overline{D}$	CD	$\overline{C}D$	$\overline{C}\overline{D}$
$A\overline{B}$	0	0	1	0
AB	0	0	0	0
AB	1	1	0	0
$\overline{A}\overline{B}$	0	0	0	0

Figure 5–74

36. Referring to Figure 5–74, determine if the K-maps correctly represent their corresponding truth tables.
37. Write the Boolean expression for K-maps 5–1 and 5–2.

K-map 5–1

	$\overline{C}\overline{D}$	$C\overline{D}$	CD	$\overline{C}D$
$\overline{A}\overline{B}$	1	0	0	0
$A\overline{B}$	1	0	0	0
AB	0	0	1	1
$\overline{A}B$	0	0	0	0

K-map 5–2

	CD	$C\overline{D}$	$\overline{C}\overline{D}$	$\overline{C}D$
AB	0	1	1	0
$A\overline{B}$	0	0	0	0
$\overline{A}\overline{B}$	0	0	0	0
$\overline{A}B$	0	1	1	0

38. Develop the Boolean expression for K-maps 5–3 and 5–4.

K-map 5–3

	$\bar{C}\bar{D}$	$C\bar{D}$	CD	$\bar{C}D$
$\bar{A}\bar{B}$	1	0	0	1
$A\bar{B}$	0	1	1	0
AB	0	1	1	0
$\bar{A}B$	1	0	0	1

K-map 5–4

	CD	$C\bar{D}$	$\bar{C}\bar{D}$	$\bar{C}D$
AB	0	1	1	0
$A\bar{B}$	1	0	0	1
$\bar{A}\bar{B}$	1	0	0	1
$\bar{A}B$	0	1	1	0

K-map 5–5

	$C\bar{D}$	CD	$\bar{C}D$	$\bar{C}\bar{D}$
$A\bar{B}$	X	X	X	X
AB	1	1	1	1
$\bar{A}B$	1	0	0	1
$\bar{A}\bar{B}$	1	0	0	1

39. Redraw the K-map 5–3 so that its terms (*AB* and *CD*) change in the same manner as those of K-Map 5–4.
40. Develop the Boolean expression, using POS for K-map 5–5.
41. Using the SOP expressions, develop the Boolean expression for K-map 5–5.

Section 5–5:

42. Which of the K-maps have (A) pairs, (B) quads?
43. Which of the K-maps have (A) overlapping groups, (B) redundant groups?
44. Develop the truth table for K-maps 5–1 and 5–2.
45. Using K-maps 5–3 and 5–5, develop the corresponding truth tables.
46. Using K-map simplification techniques, create the simplified logic network for Truth Tables 5–1 and 5–2. Use NAND/NOR logic for the final circuit.
47. Develop the equivalent NAND/NOR logic using K-map simplification techniques for Truth Tables 5–3 and 5–4.
48. Using POS terms, develop the equivalent NAND/NOR logic network for K-map 5–4.
49. Using POS terms, develop the equivalent NAND/NOR logic network for K-map 5–5.

Section 5–6:

50. Create a logic network for four variables that will activate an LED when any two or more variables are TRUE.
51. Design a logic network that will activate an alarm for the following conditions.
    1. Any of three doors open, and it's dark outside.
    2. If it's not dark outside, you don't care about the condition of the doors.

Applications

The following questions pertain to the 4-bit microprocessor presented in this chapter and its corresponding instruction set.

52. For Program 5–1, show the corresponding bit pattern you would expect to find in memory.

Program 5–1

Address	Operation
0	LDA B
2	LDB C
4	ADD
5	HLT

Program 5–2

Address	Operation
0	CLA
1	TAB
2	NTA
3	AND
4	HLT

Program 5–3

Address	Operation
0	LDB 3
	LDA 8
	ADD
	HLT

53. Indicate the bit pattern you would expect to find in memory for Program 5–2.
54. Complete the addresses in Program 5–3.
55. Indicate what the contents of the microprocessor registers would be after Program 5–1 is completed.
56. At the completion of Program 5–2, what would be the contents of the microprocessor registers?
57. Determine the contents of the microprocessor registers when Program 5–3 is executed.
58. Write an assembly language program that would accomplish the following: add the numbers 4 and 3.
59. Write an assembly language program that would subtract 3 from 5. (Recall how to subtract using the complement of a number.)
60. Write the assembly language program for Bit Pattern 5–1.

Bit Pattern 5–1

0	0	0	0	1
1	1	1	1	0
2	0	0	1	0
3	0	0	1	1
4	0	1	0	1
5	1	0	0	0
6	1	1	1	1

Bit Pattern 5–2

0	0	0	0	1
1	0	1	0	0
2	0	0	1	0
3	0	1	0	1
4	0	1	0	1
5	1	0	1	1
6	1	1	0	1
7	1	1	1	1
8	1	0	0	1

Bit Pattern 5–3

0	1	1	1	0
1	0	1	0	0
2	0	1	0	1
3	1	0	0	1
4	0	1	0	1
5	1	0	1	0
6	1	1	0	1
7	1	1	1	1
8	1	0	0	1

61. Develop the assembly language program for Bit Pattern 5–2.
62. Write the assembly language program for Bit Pattern 5–3.

63. Determine the contents of the microprocessor registers after the program represented by Bit Pattern 5–1 is executed.
64. Determine the contents of the microprocessor registers after the program represented by Bit Pattern 5–2 is executed.
65. For Bit Pattern 5–3, determine the contents of the microprocessor registers and the new memory bit pattern when the program is executed.

Troubleshooting

Refer to Figure 5–75 for the following questions.

66. What is the maximum supply voltage allowed?
67. Determine the fan-out of the chip.
68. Calculate the noise margin for a HIGH logic level.
69. Calculate the noise margin for a LOW logic level.
70. For three levels of this logic gate, how long would it take the output of the first level to respond to a change in the input of the third level from FALSE to TRUE?
71. Calculate the maximum power dissipation for this chip. Under what conditions does this happen?
72. For a logic system that consisted of 50 of these chips, what is the maximum amount of current necessary from the power source?
73. Using a CMOS data book, what would be the amount of delay for an equivalent logic circuit between input and output if three levels of this gate were used?
74. For the equivalent CMOS gate, what is the maximum power dissipation of the IC. Under what conditions does this happen?
75. For the equivalent CMOS gate, what would be the maximum current requirements from the power supply for 50 of these ICs in a logic system?

Analysis and Design

76. Write an assembly language program and develop the corresponding NAND/NOR logic fed by the accumulator that would replicate the first eight conditions of the following truth table:

A	*B*	*C*	*D*	*X*
0	0	0	0	1
0	0	0	1	1
0	0	1	0	0
0	0	1	1	0
0	1	0	0	0
0	1	0	1	1
0	1	1	0	1
0	1	1	1	0
1	0	0	0	1
1	0	0	1	0
1	0	1	0	0
1	0	1	1	0
1	1	0	0	1
1	1	0	1	0
1	1	1	0	0
1	1	1	1	1

DM54LS00/DM74LS00 Quad 2-Input NAND Gates

General Description

This device contains four independent gates each of which performs the logic NAND function.

Connection Diagram

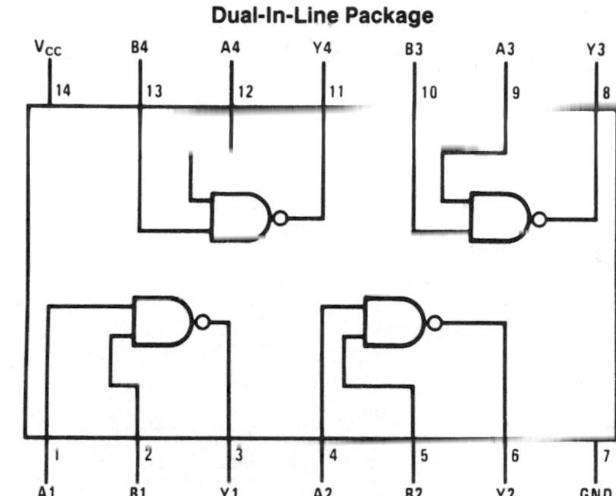

TL/F/6439–1

Order Number DM54LS00J, DM74LS00M or DM74LS00N
See NS Package Number J14A, M14A or N14A

Function Table

$Y = \overline{AB}$

Inputs		Output
A	B	Y
L	L	H
L	H	H
H	L	H
H	H	L

H = High Logic Level
L = Low Logic Level

Figure 5–75 *Reprinted* with permission of National Semiconductor Corporation. Continued on page 96.

Absolute Maximum Ratings (Note)

Specifications for Military/Aerospace products are not contained in this datasheet. Refer to the associated reliability electrical test specifications document.

Supply Voltage	7V
Input Voltage	7V
Operating Free Air Temperature Range	
DM54LS	−55°C to +125°C
DM74LS	0°C to +70°C
Storage Temperature Range	−65°C to +150°C

Note: *The "Absolute Maximum Ratings" are those values beyond which the safety of the device cannot be guaranteed. The device should not be operated at these limits. The parametric values defined in the "Electrical Characteristics" table are not guaranteed at the absolute maximum ratings. The "Recommended Operating Conditions" table will define the conditions for actual device operation.*

Recommended Operating Conditions

Symbol	Parameter	DM54LS00			DM74LS00			Units
		Min	Nom	Max	Min	Nom	Max	
V_{CC}	Supply Voltage	4.5	5	5.5	4.75	5	5.25	V
V_{IH}	High Level Input Voltage	2			2			V
V_{IL}	Low Level Input Voltage			0.7			0.8	V
I_{OH}	High Level Output Current			−0.4			−0.4	mA
I_{OL}	Low Level Output Current			4			8	mA
T_A	Free Air Operating Temperature	−55		125	0		70	°C

Electrical Characteristics over recommended operating free air temperature range (unless otherwise noted)

Symbol	Parameter	Conditions		Min	Typ (Note 1)	Max	Units
V_I	Input Clamp Voltage	V_{CC} = Min, I_I = −18 mA				−1.5	V
V_{OH}	High Level Output Voltage	V_{CC} = Min, I_{OH} = Max, V_{IL} = Max	DM54	2.5	3.4		V
			DM74	2.7	3.4		
V_{OL}	Low Level Output Voltage	V_{CC} = Min, I_{OL} = Max, V_{IH} = Min	DM54		0.25	0.4	V
			DM74		0.35	0.5	
		I_{OL} = 4 mA, V_{CC} = Min	DM74		0.25	0.4	
I_I	Input Current @ Max Input Voltage	V_{CC} = Max, V_I = 7V				0.1	mA
I_{IH}	High Level Input Current	V_{CC} = Max, V_I = 2.7V				20	μA
I_{IL}	Low Level Input Current	V_{CC} = Max, V_I = 0.4V				−0.36	mA
I_{OS}	Short Circuit Output Current	V_{CC} = Max (Note 2)	DM54	−20		−100	mA
			DM74	−20		−100	
I_{CCH}	Supply Current with Outputs High	V_{CC} = Max			0.8	1.6	mA
I_{CCL}	Supply Current with Outputs Low	V_{CC} = Max			2.4	4.4	mA

Switching Characteristics at V_{CC} = 5V and T_A = 25°C (See Section 1 for Test Waveforms and Output Load)

Symbol	Parameter	R_L = 2 kΩ				Units
		C_L = 15 pF		C_L = 50 pF		
		Min	Max	Min	Max	
t_{PLH}	Propagation Delay Time Low to High Level Output	3	10	4	15	ns
t_{PHL}	Propagation Delay Time High to Low Level Output	3	10	4	15	ns

Note 1: All typicals are at V_{CC} = 5V, T_A = 25°C.

Note 2: Not more than one output should be shorted at a time, and the duration should not exceed one second.

Figure 5–75 Continued

CHAPTER 6

Programmable Logic Devices

OBJECTIVES

After completing this chapter, you should be able to:

- ☐ Understand the general concepts of a programmable logic gate.
- ☐ Explain the difference between hardware and software.
- ☐ Determine what constitutes the simplest kind of computer.
- ☐ Explain how multiplexers work and where they are applied.
- ☐ Understand programming and logic structure of the programmable logic array (PLA).
- ☐ Describe the commonalities of programmable logic devices (PLD).
- ☐ Understand logic notation used for programmable logic devices.
- ☐ Determine what logic makes up decoders and how decoders are used in logic systems.
- ☐ Understand programming and logic structure of read-only memories (ROM).
- ☐ Explain how the ROM, PLA, and PAL are related and how they are different.
- ☐ Explain the differences and similarities between demultiplexers and encoders.
- ☐ Describe the operation and application of a CMOS bilateral switch.
- ☐ Describe an electronic diode and how it is used in programmable logic devices.
- ☐ Explain how sound is synthesized.
- ☐ Explain the bus orientation that is common to all microprocessors.
- ☐ Describe a general programming sequence that is common to all microprocessors.
- ☐ Explain the need and use of pull-up resistors in logic circuits.

INTRODUCTION

In the last chapter, you saw how to use gates to implement the requirements of any truth table. In this chapter, you will discover another way of doing exactly the same thing.

Here, you will see the real relationship between the concept of *software* and *hardware*. You will find that many practical logic problems require more than just one output to solve a particular problem.

This chapter presents many new ideas that depend upon what you have learned up to this point. The new ideas you will learn here took many years for those in the industry to understand and refine. These ideas are not difficult, they just required a new and important way of looking at logic circuits. You will find this an exciting and useful chapter. When you finish with this chapter, you will be a giant step closer to the mastery of digital systems. Let's get started.

KEY TERMS

Programmable Logic Gate (PLG)
Computer
Hardware
Software
Read-Only Memory (ROM)
Read-Write Memory
Multiplexer
Data Selector
Programmable Logic Array (PLA)
Diode
Product Line
Decoder
Mask Programmable ROM
Programmable Read-Only Memory (PROM)
Erasable Programmable Read-Only Memory (EPROM)
Demultiplexer
Encoder
Priority Encoder
Bilateral Switch
Waveform Synthesis
Music Synthesizers
Binary Counter
Voice Synthesizer
Address Register
Address Bus
Data Bus
Control Bus
Pull-up Resistor
Open-Collector Logic

6–1 ANOTHER WAY

Discussion

You have come a long way in your ability to take an idea and implement it into a real digital network using logic gates. In this section, you will discover another way of looking at the same thing. But this method promises to lay the foundation of what modern computers, robotics, and other practical automated systems use. You will find the material presented here fun to learn and very useful.

A Familiar Logic Network

Look at the logic network in Figure 6–1. As you can see, this is nothing more than your old friend, the SOP logic using AND/OR gates. There is a major difference—besides the

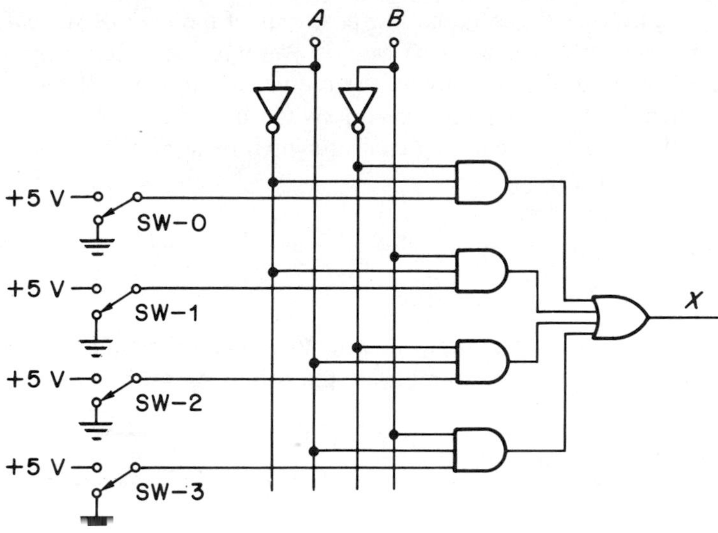

Figure 6–1 A prewired logic network.

standard logic inputs (*A* and *B* with their complements), there is another set of inputs connected to four switches (SW-0 through SW-3). As shown in the figure, these switches can have either a TRUE value or a FALSE value.

With all of the input switches at a FALSE as shown in Figure 6–1, no matter what the values of *A* and *B*, the output of the logic network will always be FALSE. This idea is illustrated in Figure 6–2.

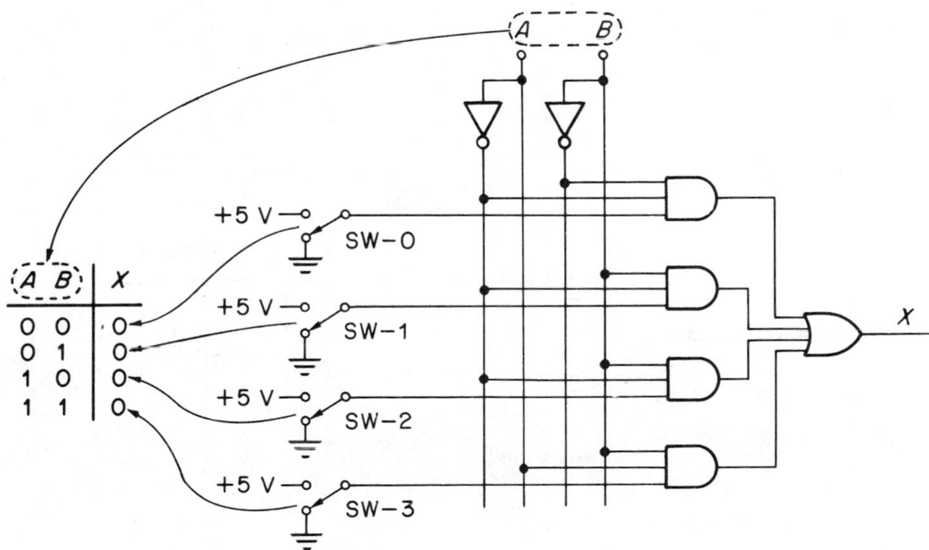

Figure 6–2 Developing a truth table.

Figure 6–2 suggests that the positions of the *control switches* can indicate the results of the truth table. This is the case as you will see. The positions of the control switches are identical to the outcome of the required truth table. If the output of the truth table is a 0 then the corresponding control switch must be LOW. If the output of the truth table is 1, then the corresponding control switch must be HIGH. This is illustrated by the following example.

Example 6–1

Using an AND/OR network with logic control switches, show how you would set the control switches to replicate the following truth tables:

(A)

A	*B*	*X*
0	0	0
0	1	1
1	0	1
1	1	0

(B)

A	*B*	*X*
0	0	1
0	1	0
1	0	0
1	1	1

Solution

Use the truth table output to indicate the setting of the control switches. When this is done, the output of the gate network will replicate the requirements of the truth table. The method is shown in Figure 6–3.

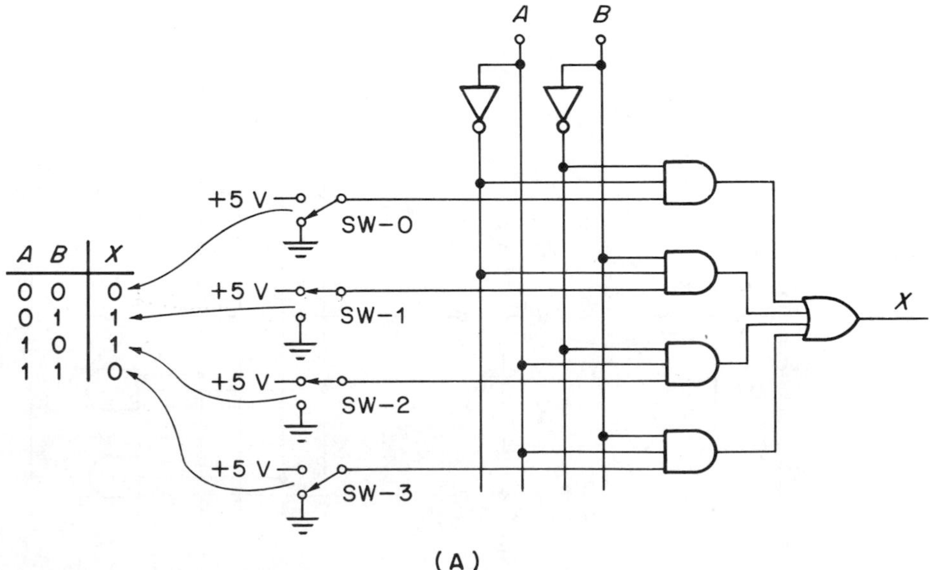

Figure 6–3

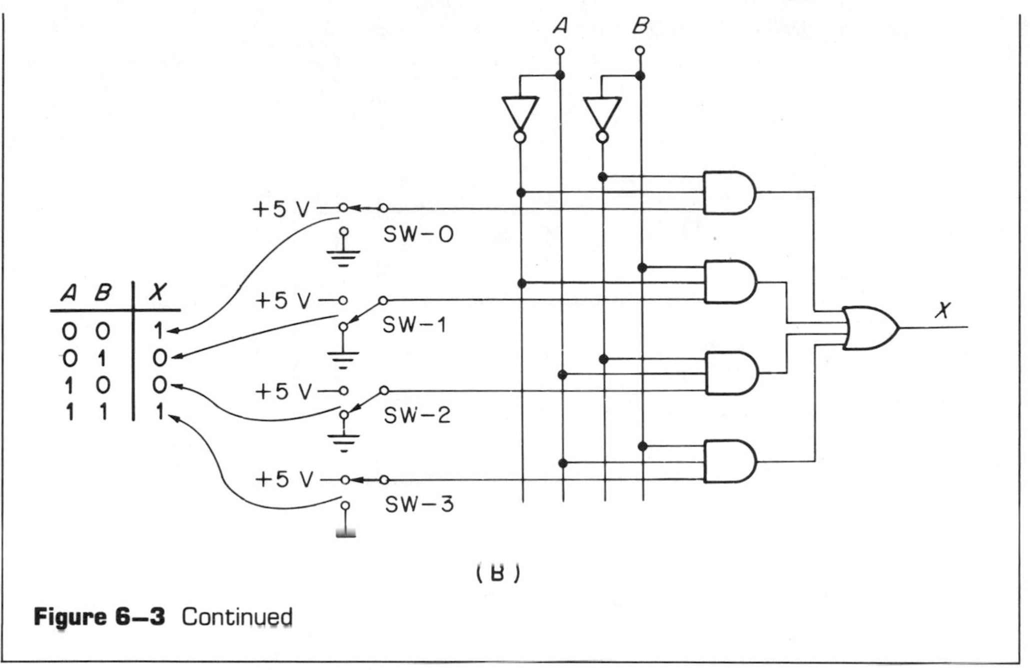

Figure 6–3 Continued

Observe the results of this example. Note that Table (A) is for an XOR gate. As shown in the solution, there are four possible conditions on the truth table and four control switches. In order to replicate the logic of the XOR gate, the switches need only copy the output pattern of the truth table. In this case, the truth table output is (starting from the top) 0110. The corresponding position of each switch is (again, starting from the top) LOW, HIGH, HIGH, LOW. The same procedure is used with Table (B). This is a truth table for the XNOR gate. The output of the truth table is (starting from the top) 1001. This pattern is replicated by the switches: HIGH, LOW, LOW, HIGH.

Creating Logic Gates

By using the logic combination with the control switches, you can replicate any logic gate you choose. This may at first seem like a lot of gates to replicate just one of them, but what is important here is the concept of what is happening. Once you have the concept of using a bit pattern (the positions of the control switches) to make any logic condition, then this can be extended to a whole world of practical applications.

Consider Figure 6–4. Note how the block diagram of the logic network can be thought of as an adjacent part that is used to determine the logic of the network. Now, any logic gate with two input variables can be replicated by this system by the mere setting of the control switches. The control switch settings for all of the basic logic gates are shown in Figure 6–5.

Programmable Logic Gate

The previous logic network can be thought of as a **programmable logic gate (PLG).**

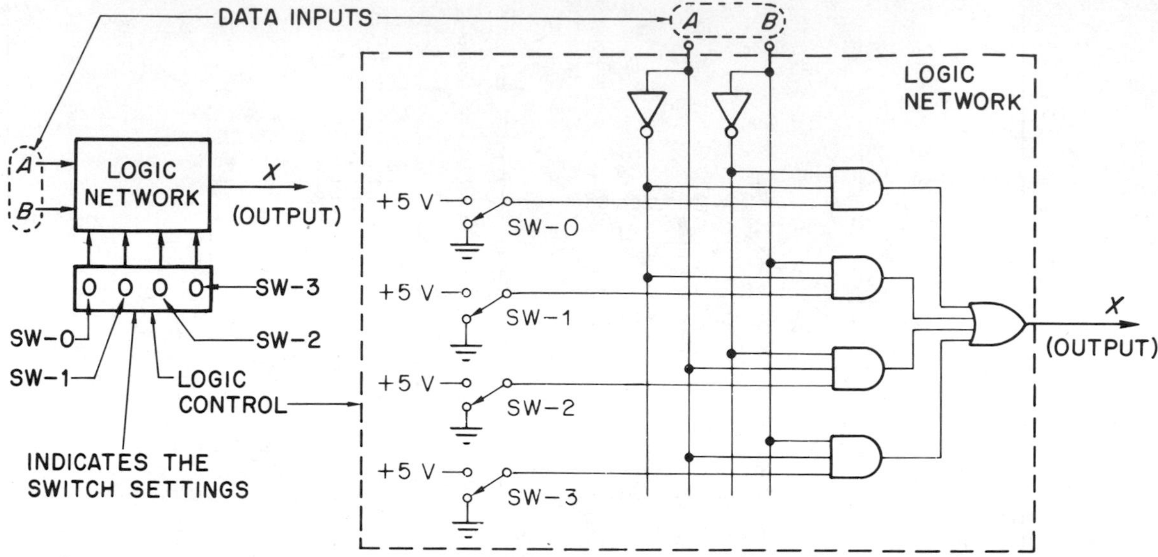

Figure 6–4 Another representation.

Programmable Logic Gate (PLG):
A logic network that can replicate any of the fundamental gates (AND, OR, NAND, NOR, XOR) with the setting of a bit pattern that determines the type of gate it will replicate.

You can consider the PLG as the most basic *programmable* device possible. Consider the illustration of Figure 6–6. As you can see from the figure, the PLG has all of the parts of a *computer*. The control switches are acting as *memory*. The settings of these switches act as the *instruction*. The logic network of AND/OR gates acts as the *processing* part. There is an *input* and *output*. The PLG is *programmed* by setting the bit pattern of the switches. Once programmed, the PLG "remembers" what kind of gate it is supposed to be. As an example, the bit pattern of 0001 can be thought of as the "instruction" for an AND gate. The major concepts presented are summarized in Table 6–1.

Viewing the PLG in the manner described by Table 6–1 presents the simplest model of a **computer.**

Computer:
A programmable processing machine consisting of an input, output, processing unit, and memory that is capable of accepting information, applying prescribed processes to the information, and supplying the results of that process.

With the PLG as a model, it now becomes easy to differentiate between the meanings of **hardware** and **software.** These concepts are illustrated in Figure 6–7.

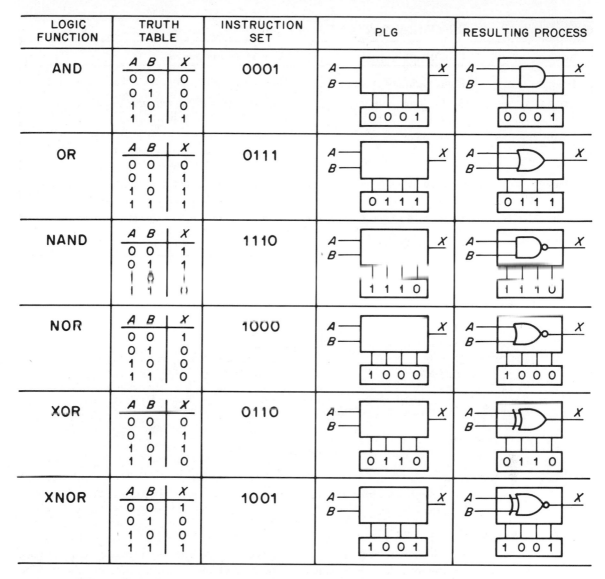

LOGIC FUNCTION	TRUTH TABLE	INSTRUCTION SET	PLG	RESULTING PROCESS
AND	A B | X 0 0 | 0 0 1 | 0 1 0 | 0 1 1 | 1	0001	A, B → X 0 0 0 1	A, B → X 0 0 0 1
OR	A B | X 0 0 | 0 0 1 | 1 1 0 | 1 1 1 | 1	0111	A, B → X 0 1 1 1	A, B → X 0 1 1 1
NAND	A B | X 0 0 | 1 0 1 | 1 1 0 | 1 1 1 | 0	1110	A, B → X 1 1 1 0	A, B → X 1 1 1 0
NOR	A B | X 0 0 | 1 0 1 | 0 1 0 | 0 1 1 | 0	1000	A, B → X 1 0 0 0	A, B → X 1 0 0 0
XOR	A B | X 0 0 | 0 0 1 | 1 1 0 | 1 1 1 | 0	0110	A, B → X 0 1 1 0	A, B → X 0 1 1 0
XNOR	A B | X 0 0 | 1 0 1 | 0 1 0 | 0 1 1 | 1	1001	A, B → X 1 0 0 1	A, B → X 1 0 0 1

Figure 6–5 Control switch settings for all logic gates.

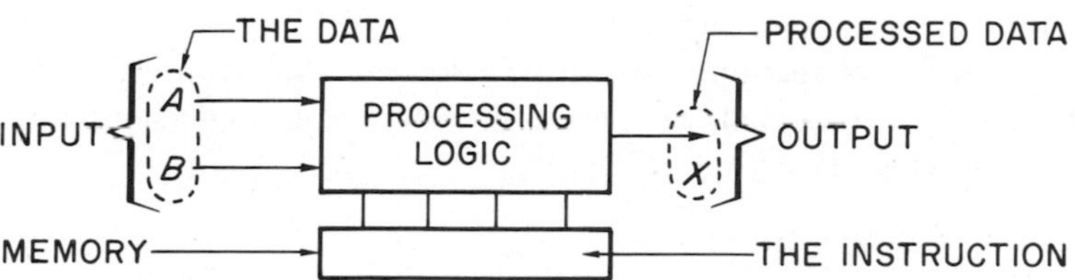

Figure 6–6 Another way of looking at the PLG.

Table 6–1 PLG NOMENCLATURE

Section	Consists of	What It Represents
Input	Input variables *A* and *B*	This is the input *data*
Output	Single logic output	The logic output that depends upon the type of instruction and value of data
Processing unit	Logic gates	The architecture that determines the relationship between the instruction and the results of the input data
Memory	Four switches (control switches)	Determines the effect the input data will have on the output
Instruction	Bit pattern of the switches	Documentation used to determine how the PLG will process its data

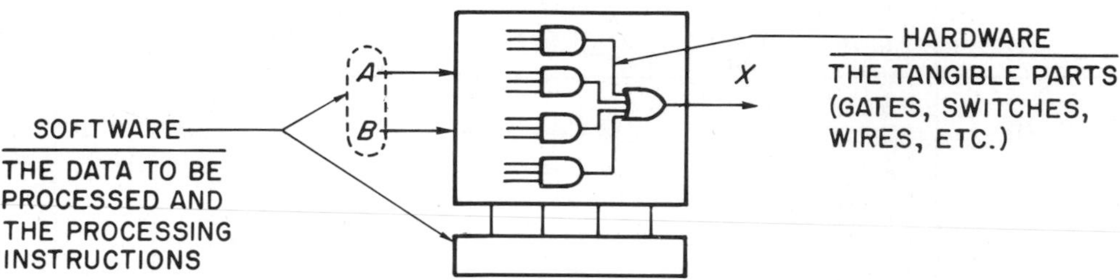

Figure 6–7 Hardware and software.

Hardware:
The tangible part of a logic system.

Software:
The nontangible part of a logic system that determines the process performed by the hardware.

Conclusion

This was a very important section. Here you discovered how to use a simple logic network to make a programmable device that contained all of the elements of a *computer*. The PLG is capable of accepting information, applying prescribed processes to the information, and supplying the results of the process. Just as important, you saw the relationship between hardware and software, that is, the way a bit pattern can determine how information is processed by logic networks.

In the next section, you will see some real hardware that can be used to implement the concepts presented here. For now, check your understanding of this section by trying the following review questions.

6–1 Review Questions

1. Describe the major parts of a computer.
2. Explain what makes up each of the following parts of a PLG: (A) memory, (B) processing circuit, (C) instruction.

3. What is the difference between *hardware* and *software*? How is this accomplished with the PLG?
4. Define what is meant by an *instruction,* a *process.*
5. Explain how a PLG replicates a computer.

6–2 THE MULTIPLEXER/DATA SELECTOR

Discussion

This section presents some real hardware that will act as a PLG. As you will see here, these ICs also serve other useful applications. Here, you will discover some of the available ICs that already contain combinations of AND/OR gates so that the amount of extra wiring on your part is held to a minimum.

The Real Thing

A multiplexer/data selector will accept several different inputs and allow only one of these inputs to appear on the output at a time. As shown in Figure 6–8, the device consists of *data inputs* and *address inputs* (sometimes referred to as *data select inputs*). Essentially the address inputs determine which data input will appear on the output of the device. As an example, if the address inputs are 000, then the output will reflect the condition of the D_0 input. If the address inputs are 111, then the output will reflect the condition of the D_7 input. These data inputs may consist of a series of logic pulses as well as a fixed logic level.

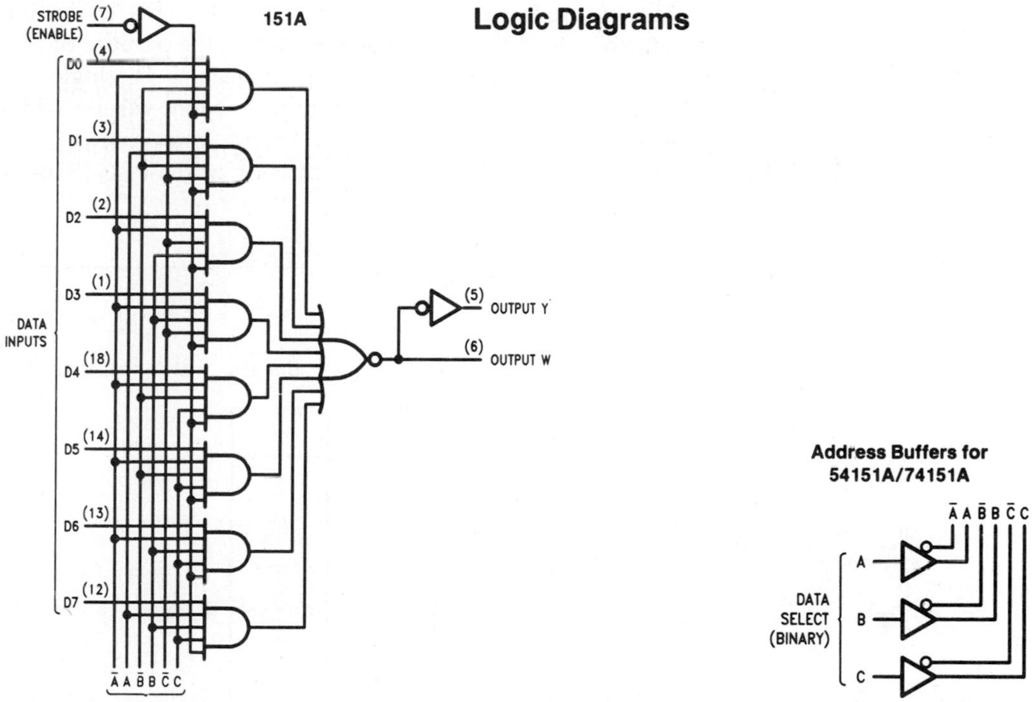

Figure 6–8 Logic diagram of eight-input selector. *Reprinted with permission of National Semiconductor Corporation.*

National holds no responsibility for any circuitry described. National reserves the right at any time to change without notice said circuitry.

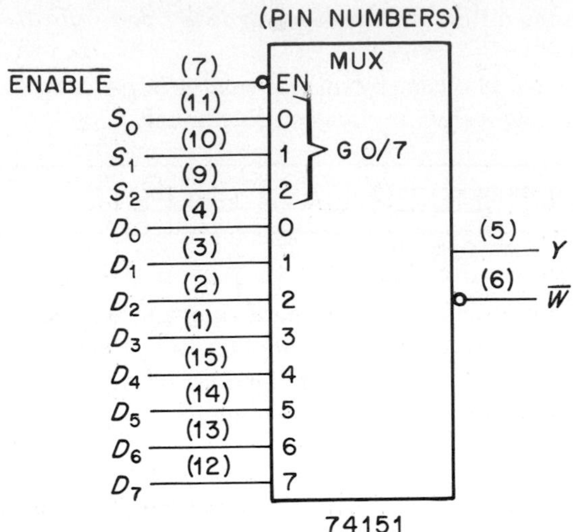

Figure 6–9 Logic symbol of 8-bit data selector.

Note from Figure 6–8 that the complement of the output is available (output W) as well as the uncomplemented output (output Y). The bottom of Figure 6–8 shows that for the 54151A and 74151A, that there are only three data select inputs because each input goes to a buffer/inverter so that the complemented and uncomplemented form of the inputs are internally available. As shown in Figure 6–8, the STROBE (ENABLE) input must be active (LOW) before the device is enabled. This input serves the purpose of allowing more than one of these devices to be used in the same digital system. The logic symbol of the 8-bit data selector/multiplexer is illustrated in Figure 6–9. More will be said about using the device in this manner.

This logic network can also be viewed as an extension of the PLG. This is illustrated in the following example.

Example 6–2

Using the 74151A, show how you would implement the following truth table:

C	B	A	X
0	0	0	1
0	0	1	0
0	1	0	1
0	1	1	1
1	0	0	0
1	0	1	1
1	1	0	0
1	1	1	1

Solution

The actual logic connections and the corresponding wiring diagram are shown in Figure 6–10.

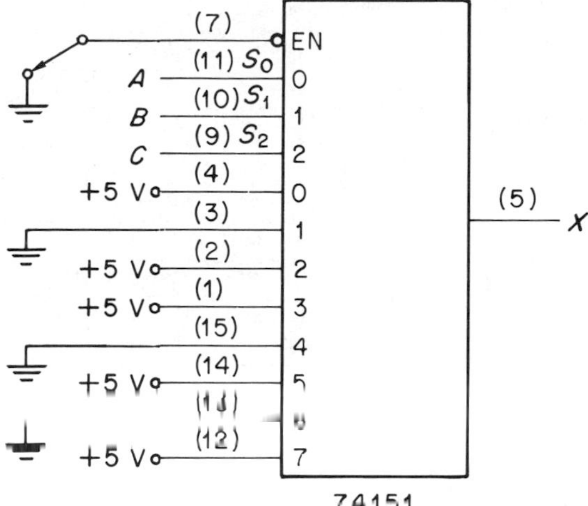

Figure 6–10

As shown in the previous example, you use the same approach as you did with the PLG. All that is necessary is to set the data switches in the same pattern as the required output of the truth table. The truth table output is (starting from the top) 10110101 and thus the data switches were set accordingly: HIGH, LOW, HIGH, HIGH, LOW, HIGH, LOW, HIGH.

Since the pattern of this truth table is permanently wired to this device, you can think of this as a memory. The input wiring can be thought of as a permanent instruction to cause the device to always treat the variable data (the inputs *A*, *B*, and *C*) exactly the same way. This kind of *permanent* memory is referred to as **read-only memory (ROM).**

Read Only Memory (ROM):
Memory that contains a set of permanent instructions that cannot be easily changed.

On the other hand, if you had wired the control inputs to the 74154A to a set of eight switches, you would have created a memory that is easy to change. This is shown in Figure 6–11.

When you change the switch settings to accommodate a specific three-input truth table, this is called *writing* to memory. When the device is using the instructions made by this bit pattern, it is said to be *reading* from memory. Thus, this kind of memory that is easy to write to as well as read from is called **read-write memory.**

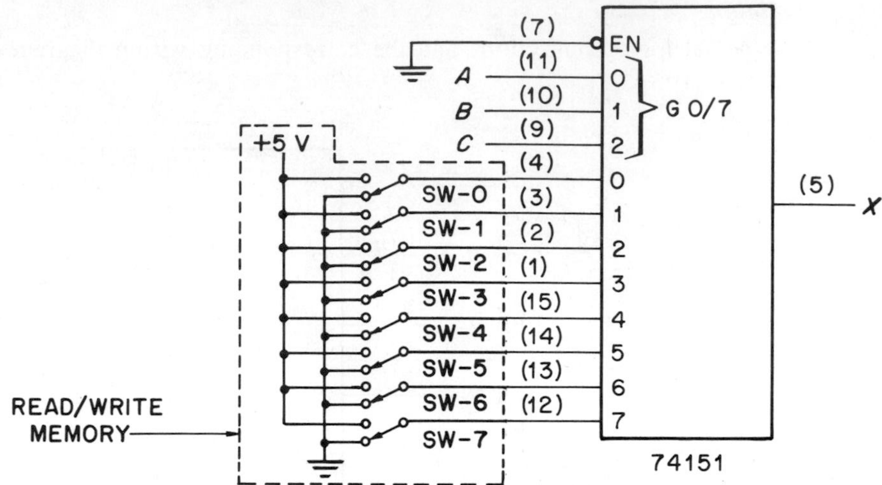

Figure 6–11 Creating READ/WRITE memory for the 74151.

Read-Write Memory:
Memory that can have its instructions easily changed.

You now have a device that can replicate *any three-input truth table*! All you need to do is to change the switch settings (reprogram it) for the desired logic output. There are other uses for this versatile device as you will see.

Multiplexer

The word *multiplex* means to take several different things and combine them into one. The 74151A is a digital **multiplexer.**

Multiplexer:
A logic network that is capable of taking several different inputs and transmitting the result on one output.

You can think of the multiplexer as the logic gate equivalent of a multiposition switch. This is illustrated in Figure 6–12. You can also see why it is called a **data selector.** As you can see, it could be used to select one of the input *data* lines and then transmit that information out along a single line.

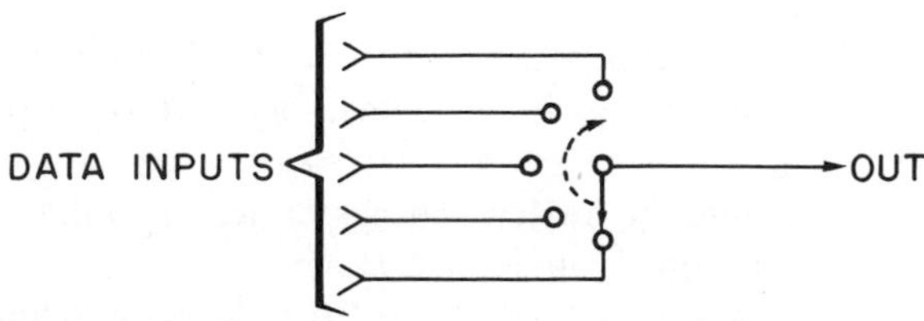

Figure 6–12 Equivalent circuit of a multiplexer/data selector.

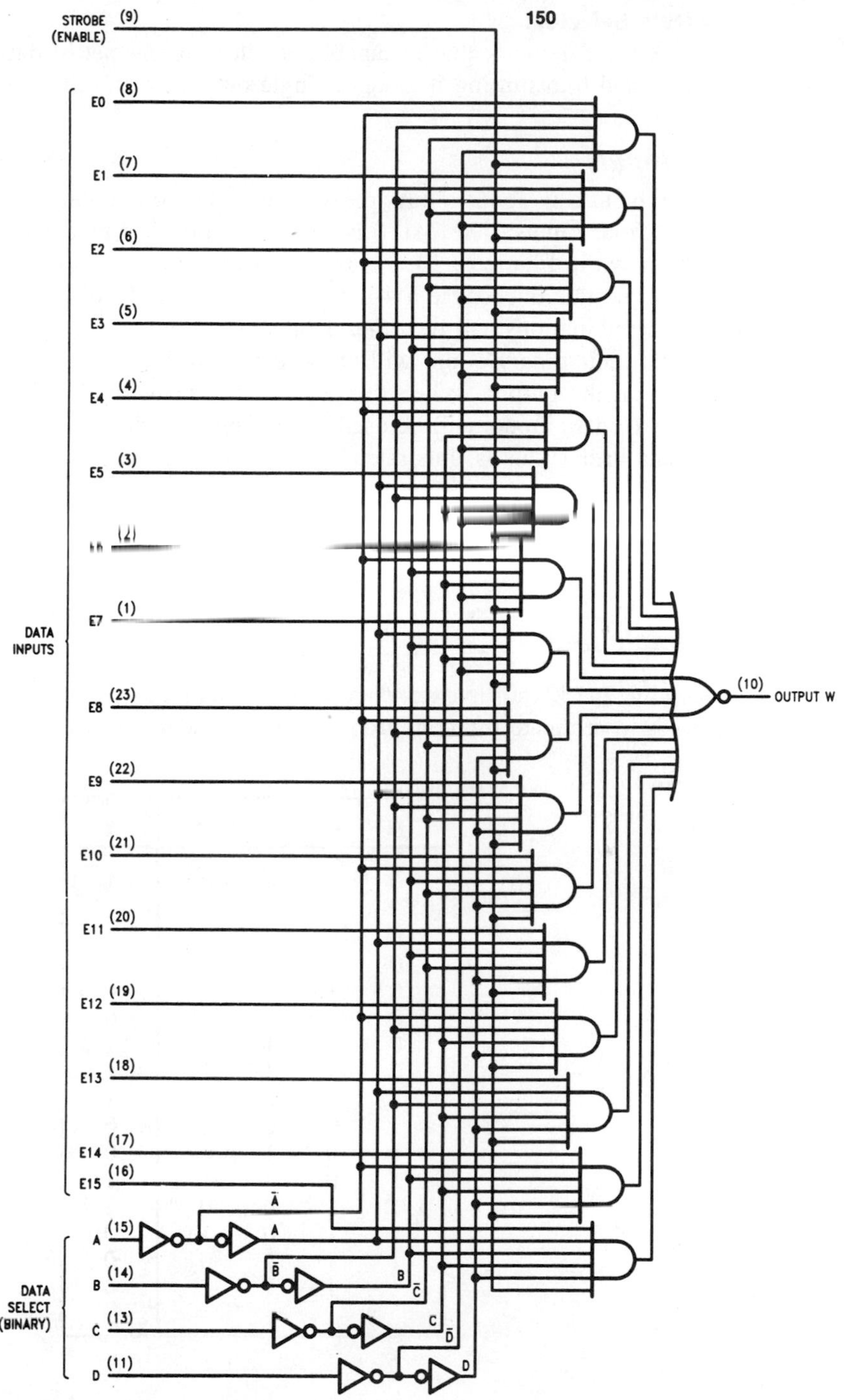

Figure 6–13 The 74150 multiplexer/data selector. *Reprinted with permission of National Semiconductor Corporation.*

National holds no responsibility for any circuitry described. National reserves the right at any time to change without notice said circuitry.

Data Selector:
A logic network that is capable of selecting one set of data from a group of data and transmitting it along a single output line.

More Inputs

You may be asking about a multiplexer/data selector that can be used to program a truth table with four inputs. The 74150 is available to do just that. Its connecting diagram and logic is shown in Figure 6–13. This device can replicate a 4-input truth table. The reason why it can do this is because a 4-input truth table has 16 different combinations. Each of these 16 combinations can be either 1 or 0. The device of Figure 6–13 has 4 data select inputs and their binary count will cause the condition of only one of the 16-data inputs to appear on the output. When this device is used to replicate the conditions of a truth table, the 16 data inputs will be used as memory and the 4 data select inputs will be used as the data itself. This is illustrated in the following example.

Example 6–3

Using the 74150 multiplexer/data selector, and read-write memory, show the switch settings to replicate the conditions of the following truth table.

D	*C*	*B*	*A*	*X*
0	0	0	0	1
0	0	0	1	1
0	0	1	0	1
0	0	1	1	0
0	1	0	0	0
0	1	0	1	0
0	1	1	0	1
0	1	1	1	1
1	0	0	0	1
1	0	0	1	0
1	0	1	0	0
1	0	1	1	1
1	1	0	0	1
1	1	0	1	0
1	1	1	0	0
1	1	1	1	0

Solution

See Figure 6–14. (Note the addition of the output inverter.)

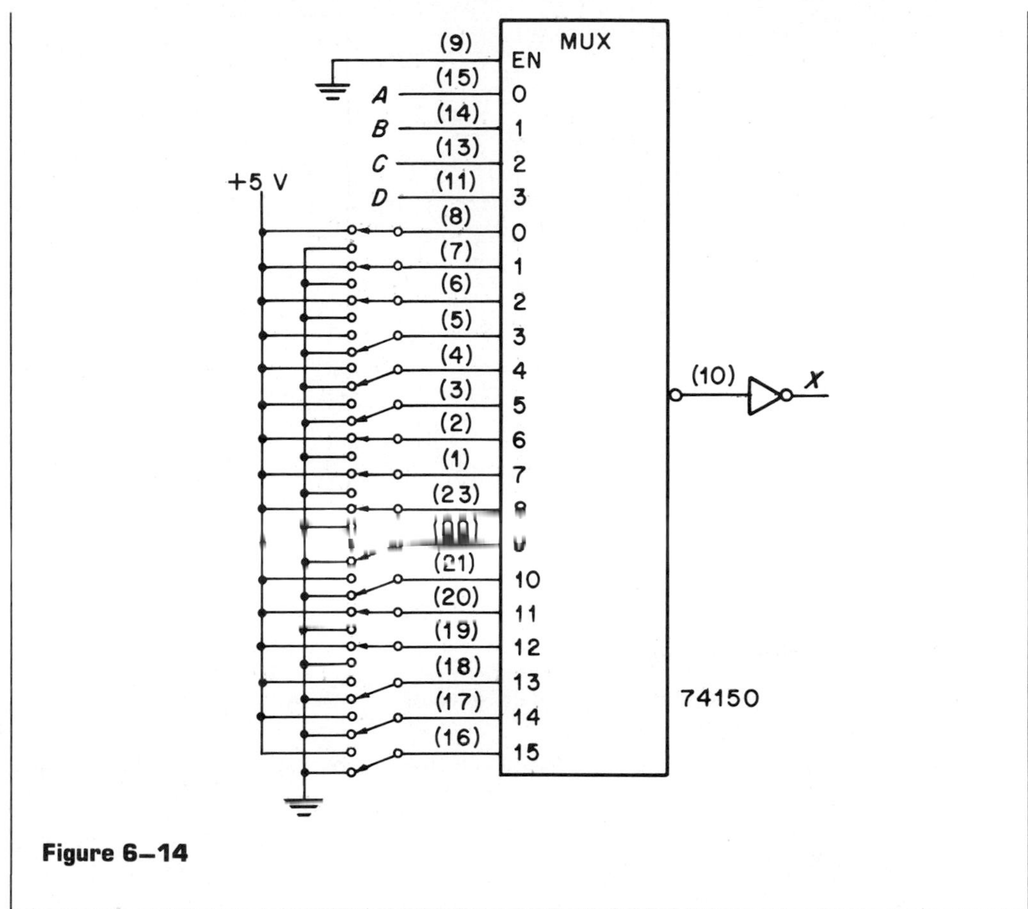

Figure 6–14

Folding

There is another way of implementing a four-input truth table by using a three-input data selector (instead of a four-input as in the previous example). The method of doing this is to just observe that the count of the three least significant bits repeat themselves in the truth table. They do this once when the MSB is 0 and again when the MSB is 1. Thus, there are essentially two logic conditions that can be forced upon the data inputs. One logic condition is for when the MSB is LOW, the other is when it is HIGH. This method is illustrated in Figure 6–15.

Conclusion

In this section, you were shown some applications of the multiplexer/data selector. Here you saw how to use this device as a programmable device that could replicate the conditions of any truth table. In the next section you will be introduced to another important programmable device that looks very much like this one. For now, test your understanding of this section by trying the following review questions.

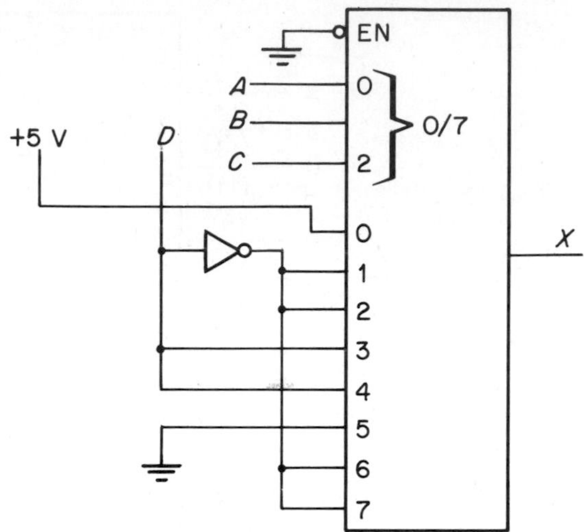

Figure 6–15 Expanding the multiplexer for a 32-row truth table.

6–2 Review Questions

1. Describe the operation of a data selector.
2. What is a digital multiplexer?
3. Explain how a multiplexer/data selector can be used to replicate the logic conditions of a truth table.
4. What is ROM?
5. What is read-write memory?

6–3 INTRODUCTION TO THE PLA

Discussion

This section presents an IC that can be programmed to represent a given logic condition. As you will see, it consists of a logic network you have worked with many times in the past.

Basic Idea

Look at the logic network of Figure 6–16. As you can see, it consists of your old friend the two-level AND/OR logic that represents the SOP Boolean expression. Note from the figure that every AND gate has an input to each line of the input bus. This means that no matter what the input combination of *A* and *B*, the output of every AND gate will always be FALSE because every AND gate will have at least one false input. Thus, the output of the OR gate will always be FALSE as well. The truth table of this apparently useless logic network follows.

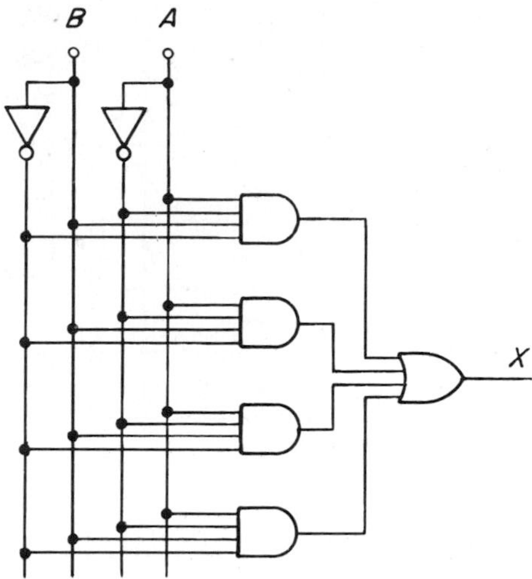

Figure 6–16 Basic AND/OR logic.

A	*B*	*X*
0	0	0
0	1	0
1	0	0
1	1	0

Despite the appearance of a logic network of little utility, there is a way to make this network very useful.

Making It Programmable

Assume that you could easily buy this network in an inexpensive chip. Further, assume that all of the connections to each of the AND gates were made by microscopic fuses. This arrangement is illustrated in Figure 6–17. These fuses are constructed in such a manner that they can electrically be blown to selectively create an open anywhere you want one. As an example, suppose you wanted this device to represent the logic function of an XOR gate:

A	*B*	*X*	
0	0	0	
0	1	1	$\rightarrow \overline{A}B$
1	0	1	$\rightarrow A\overline{B}$
1	1	0	

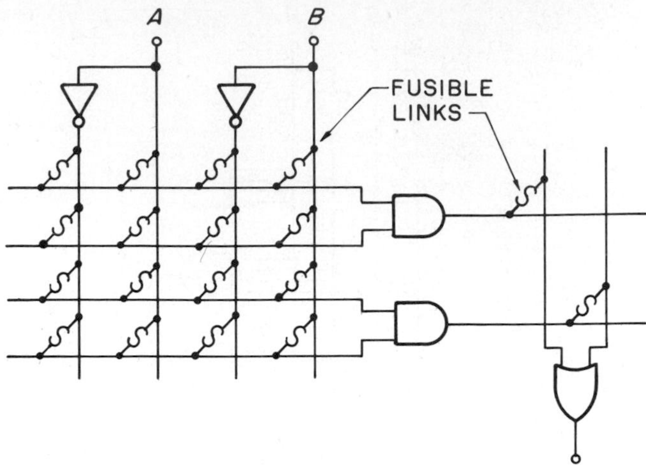

Figure 6–17 Fuses with gates.

The Boolean expression for the truth table becomes $\overline{A}B + A\overline{B}$. The fuses could be blown so that the final logic network appears as shown in Figure 6–18. Methods for physically blowing these fusible links are explained in Chapter 9. As shown in this figure, the output of the OR gate will be TRUE only when the conditions $\overline{A}B$ or $A\overline{B}$ are on the input lines. For all other combinations of the inputs, the output is FALSE. What you have done here is to *program* this AND/OR network to produce a desired result. Such an arrangement of a programmable AND/OR network is called a **programmable logic array (PLA).**

Programmable Logic Array (PLA):
A two-level AND/OR POS logic network where both the AND connections and the OR connections can be programmed.

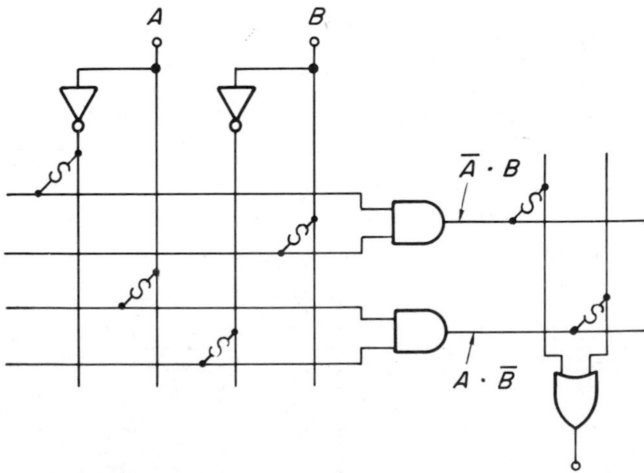

Figure 6–18 Logic to represent $\overline{A}B + A\overline{B}$.

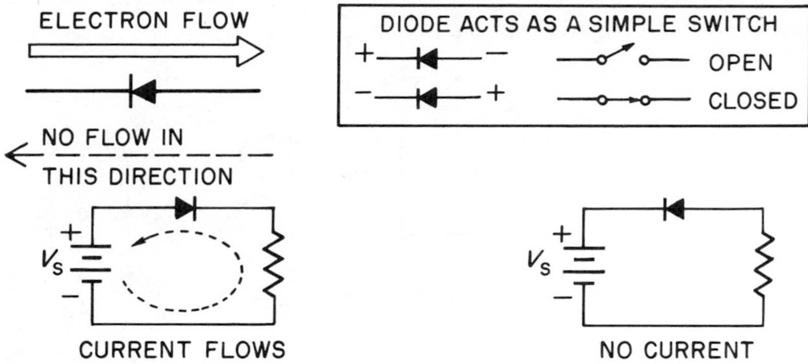

Figure 6–19 Diode symbol and operation.

The Diode

There is an important addition that is needed for the PLA. It is an electrical device called a diode.

> **Diode:**
> An electrical device that allows current flow in only one direction.

Figure 6–19 shows the schematic of a diode and the direction of electron flow within the device. As you can see, the diode only allows current to flow in one direction. Knowing this about diodes is necessary in order to see why they are needed in a digital memory.

Need for the Diode

Figure 6–20 shows why a diode is needed in a PLA. As you will see, each fuse link really contains a diode. These are necessary so that there is no conflict of logic levels on the programmable lines.

A Mini-PLA

The example PLA just presented had only one output. You will find, however, that PLAs do have many outputs. This makes them more versatile. As an example, consider the PLA

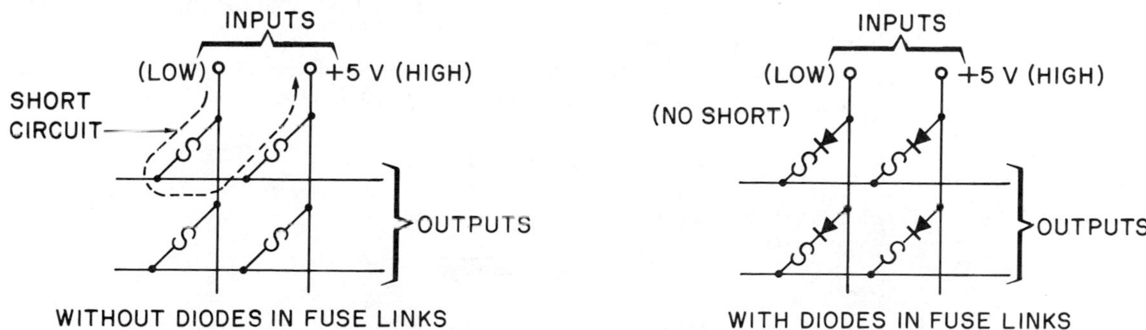

Figure 6–20 The need for a diode.

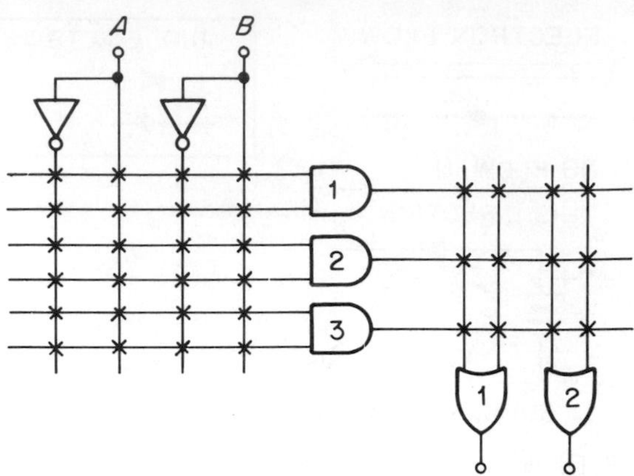

Figure 6–21 A mini-PLA.

of Figure 6–21. This mini-PLA has two inputs and two outputs. Note how the fuses are drawn—as a small x. When programmed, the blown fuse is represented as the absence of x.

Suppose you wished to program it so that one output represented a NAND gate and the other output represented a NOR gate. Recall the information presented in the last chapter on how to develop Boolean expressions from truth tables and then how to develop logic networks from these Boolean expressions. These skills can be applied here in programming the PLA. For example, suppose you wanted to program the mini PLA to have one of its output replicate a NAND gate. Since this is a standard gate, the Boolean expression is

$$X = \overline{AB}$$

The PLAs are SOP networks, so using DeMorgan's theorem you have

$$\overline{AB} = \overline{A} + \overline{B}$$

Since A must be inverted, then both the inverted A outputs are placed on the input to the first AND gate producing $\overline{A}$. The same thing is done with the inverted B outputs to the second AND gate producing $\overline{B}$. These two outputs then become the inputs to the first OR gate as shown in Figure 6–22.

Now suppose you wanted to program the other output of the PLA so that it would represent a NOR gate. The Boolean expression for a NOR gate is

$$Y = \overline{A + B}$$

Applying DeMorgan's theorem yields

$$\overline{A + B} = \overline{A} \cdot \overline{B}$$

Inverting both inputs to the third AND gate will produce the desired results. This is also shown in Figure 6–22.

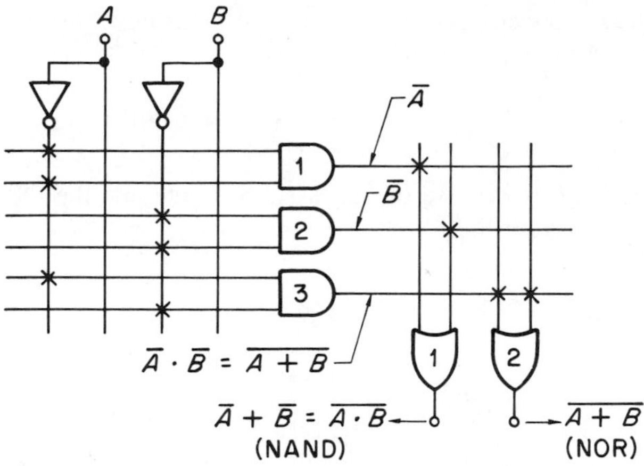

Figure 6–22 Programming the mini-PLA for logic function outputs.

Example 6–4

Program the mini-PLA shown in Figure 6–23 so that one output acts as an XOR function and the other as an AND function.

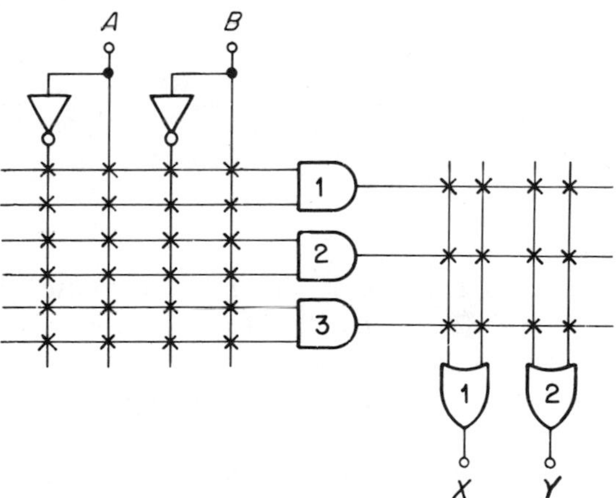

Figure 6–23

Solution

The solution uses the same method as described previously. For the XOR function write the Boolean expression:

$$X = \overline{A}B + A\overline{B}$$

From the previous SOP expression, you can see that there is one OR gate being fed by two AND gates. The first AND gate has an $\overline{A}$ term and a B term. The second

AND gate has an A term and a $\overline{B}$ term. This development is shown in Figure 6–24. For the AND expression:

$$Y = A \cdot B$$

This simply makes use of the last AND gate and the OR gate becomes a buffer as shown in Figure 6–24.

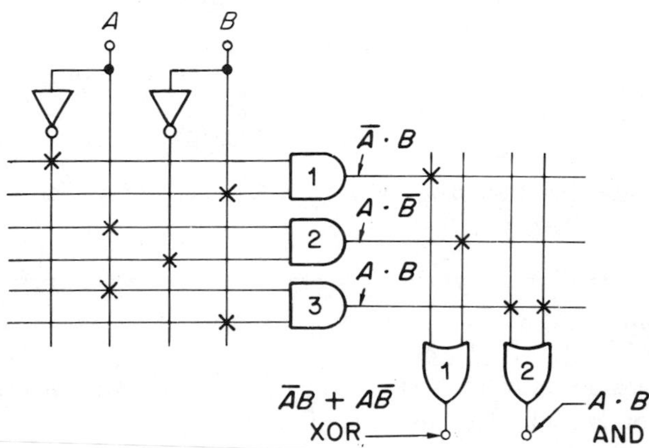

Figure 6–24

An Application

The application used here will demonstrate how to use a PLA to add 2 binary bits. First, recall the rules for the addition of 2 binary bits. This along with the desired PLA response is shown in Figure 6–25.

The process of developing this relationship is to convert the truth table output to a Boolean expression. As shown in Figure 6–26, the SUM output is the same as an XOR gate, while the CARRY output is the same as an AND gate. Thus, the Boolean expressions were

$$\text{SUM} = \overline{A}B + A\overline{B}$$

and

$$\text{CARRY} = A \cdot B$$

The previous discussion and example demonstrated how to implement these functions. Now, you can begin to see the usefulness of having more than one output from a PLA. What you have just seen is a powerful concept in digital systems. This is illustrated by the following example.

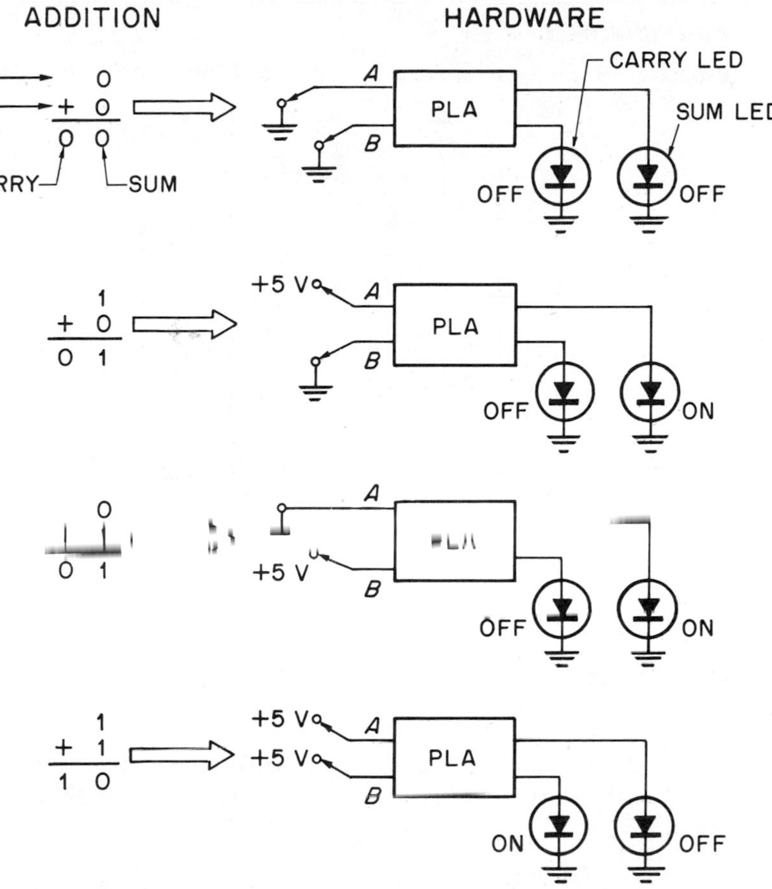

Figure 6–25 Requirements for adding 2 bits.

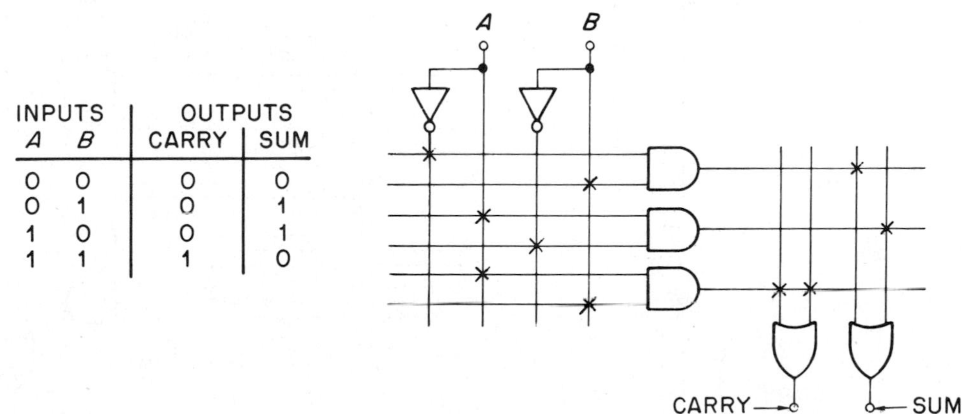

INPUTS		OUTPUTS	
A	B	CARRY	SUM
0	0	0	0
0	1	0	1
1	0	0	1
1	1	1	0

Figure 6–26 Programming the PLA to add two numbers.

Example 6–5

Program the PLA of Figure 6–27 so it will give the results of adding three bits.

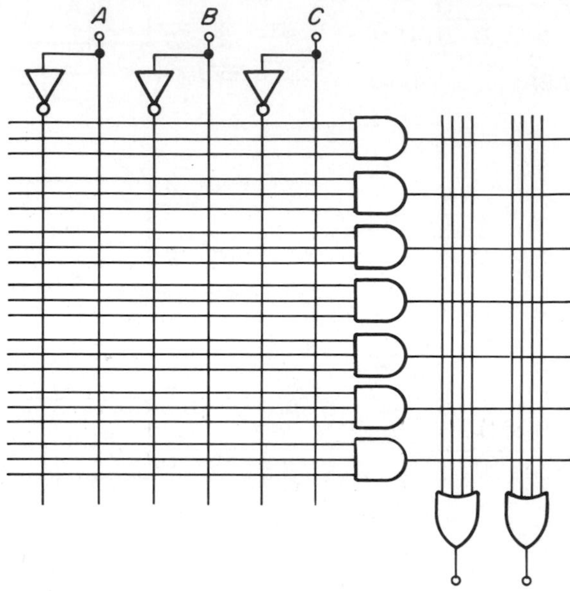

Figure 6–27

Solution

Again, develop the truth table for the relationship between the input and output of the PLA. Do this by referring to the rules for binary addition. This is all shown in Figure 6–28.

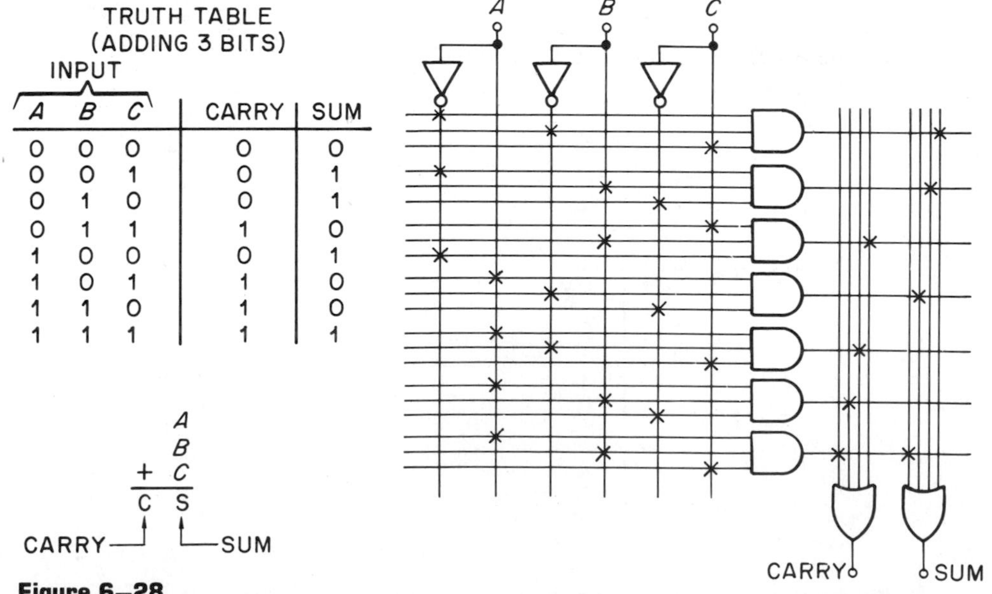

TRUTH TABLE
(ADDING 3 BITS)

INPUT				
A	*B*	*C*	CARRY	SUM
0	0	0	0	0
0	0	1	0	1
0	1	0	0	1
0	1	1	1	0
1	0	0	0	1
1	0	1	1	0
1	1	0	1	0
1	1	1	1	1

Figure 6–28

In this example, the third bit can be thought of as a *carry* bit from another addition problem. You will see more logic networks that add binary numbers, but for now, the important point is how to make a PLA replicate this function.

Developing Logic Statement from PLAs

When you work with a PLA, sometimes it will be necessary to translate a programmed one into a truth table so that you can follow an output waveform for a given input waveform. This process is illustrated by the following example.

Example 6–6

For the programmed PLA of Figure 6–29, determine the waveform from each output for the indicated input waveforms.

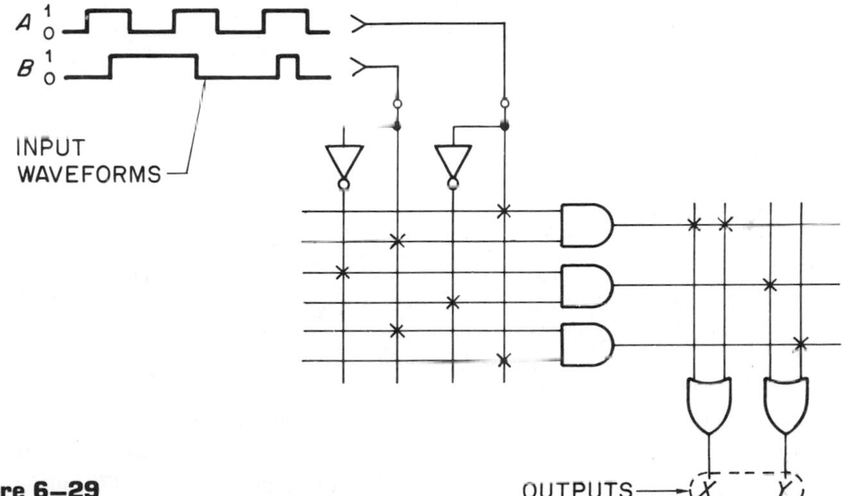

Figure 6–29

Solution

First develop a truth table to indicate the logic level of each output for all possible input conditions. Then refer to this table to determine the output waveform. This process is illustrated in Figure 6–30.

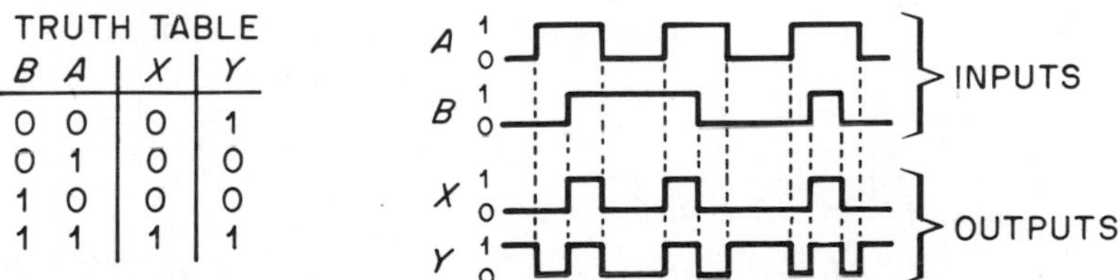

TRUTH TABLE

B	A	X	Y
0	0	0	1
0	1	0	0
1	0	0	0
1	1	1	1

NOTE: FROM THE TRUTH TABLE, *X* IS HIGH WHEN *A* AND *B* ARE HIGH. *Y* IS HIGH WHEN *A* AND *B* ARE THE SAME (BOTH HIGH OR BOTH LOW)

Figure 6–30

Conclusion

In this section you were introduced to the important concepts of a programmable logic device (PLD) known as a programmable logic array (PLA). These devices are finding wide applications in all phases of digital systems and are quickly replacing individual logic gates.

In the next section, you will learn about some very important PLD notation and then get a chance to meet a real PLA. Check your understanding of this section by trying the following review questions.

6–3 Review Questions

1. What is a PLA?
2. What is the logic output of an unprogrammed PLA?
3. What makes a PLA programmable?
4. Give an application of a PLA.
5. What is the purpose of a diode?

6–4 PLD NOTATION

Discussion

Because of the increased popularity of PLDs, a unique logic notation is used for them. This has evolved from many different methods used to represent these important logic networks. The notation presented here makes it easy to indicate how the PLD should be programmed.

Developing the Product Line

The structure of a PLD can be simplified so that it is easier to see how it can be programmed. As an example, consider another method of indicating a multiple-input AND gate. This is shown in Figure 6–31. The figure shows the PLD logic notation for a three-input AND gate. In this notation, the PLD notation for an AND gate is called the **product line.**

Product Line:
In PLD notation, a line that represents multiple inputs to an AND gate.

Observe that the three vertical lines in Figure 6–31 represent the input bus to the logic network. A closed (unprogrammed) line still has its fuse intact, but for a programmed line

Figure 6–31 PLD **AND** notation.

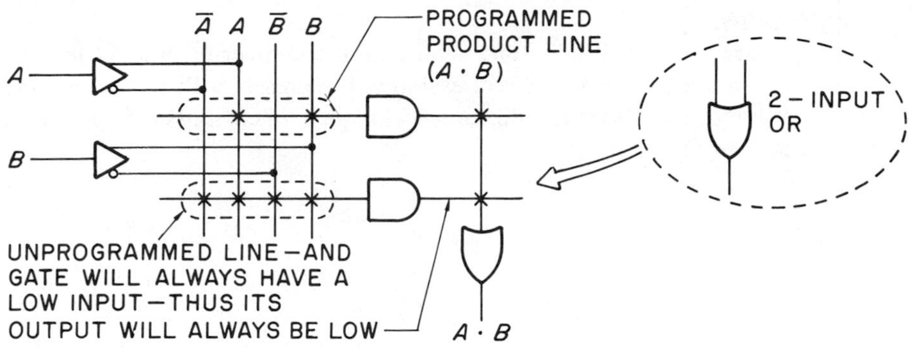

Figure 6–32 Programmed PLD to generate *AB*.

the fuse has blown. As an example of a programmed PLD using this notation, consider the PLD of Figure 6–32. As you can see, every input line is available to the product line of the PLD. This means that A, $\overline{A}$, B, $\overline{B}$ are available to each AND gate. This is made possible by the input drivers where one output acts as a buffer while the other acts as an INVERTER. Thus, both the input variable and its complement are available to the logic network.

The following example gives some practice in the use of PLD notation.

Example 6–7

Using the PLD of Figure 6–33, indicate how you would program it to create the following: (A) two-input XOR (B) three-input NAND (C) three-input NOR.

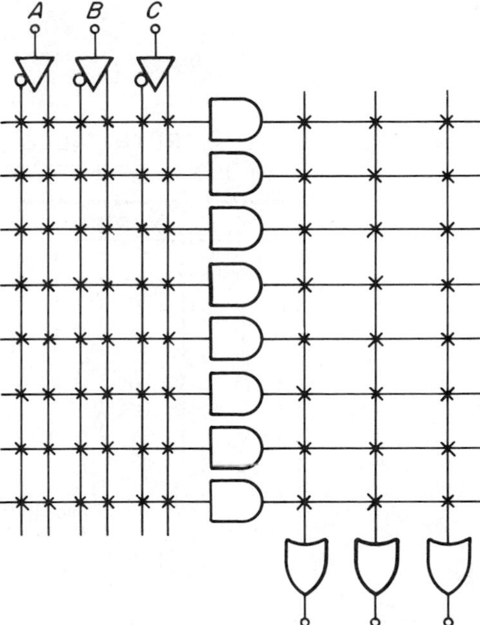

Figure 6–33

Solution

Construct the truth table and determine the number of product terms. Each product term in the final Boolean expression will represent a PLD product line that must be programmed. Use simplification techniques where possible. These steps are illustrated in Figure 6–34.

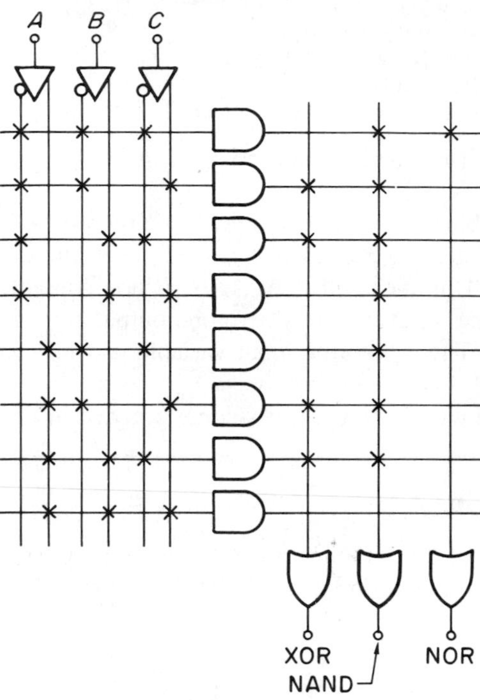

TRUTH TABLE

INPUTS			OUTPUTS		
A	*B*	*C*	XOR	NAND	NOR
0	0	0	0	1	1
0	0	1	1	1	0
0	1	0	1	1	0
0	1	1	0	1	0
1	0	0	0	1	0
1	0	1	1	1	0
1	1	0	1	1	0
1	1	1	0	0	0

USES INPUTS *B* & *C* ONLY

Figure 6–34

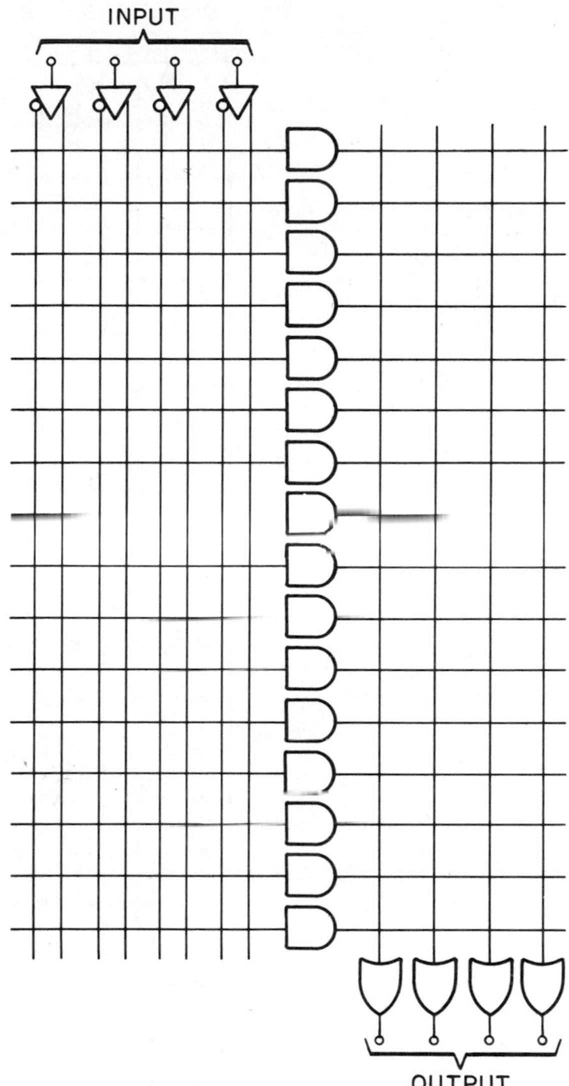

Figure 6–35 Typical PLA.

A Real PLA

The last section introduced you to one kind of PLD, the PLA. Since you now have an understanding of real PLD notation, the logic diagram of a PLA is shown in Figure 6–35.

Observe from this figure that there are 4 inputs, 16 product lines, and 4 outputs. This means that you could program any available logic sequence with four variables. As an example of the usefulness of this device, the PLA of Figure 6–36 has been programmed to convert the gray code into binary numbers.

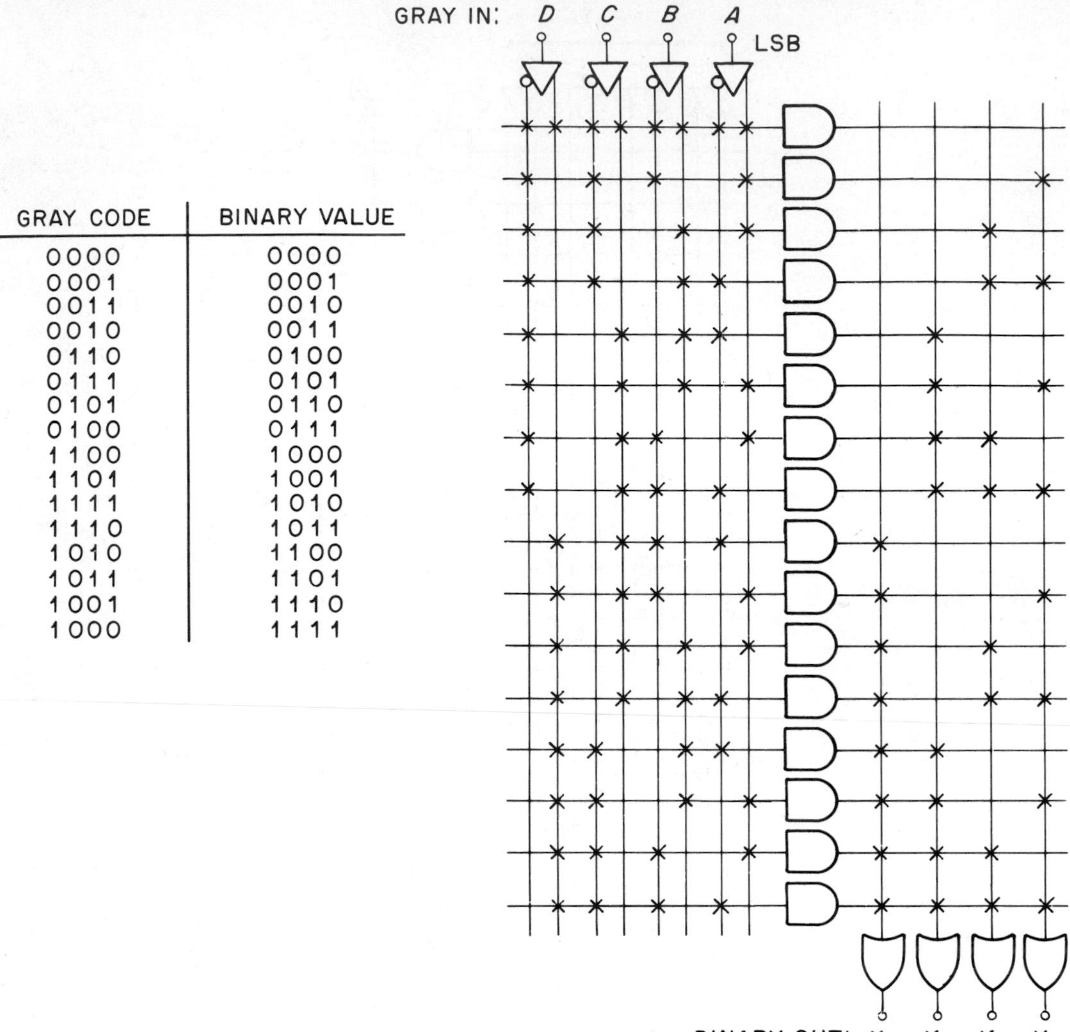

GRAY CODE	BINARY VALUE
0000	0000
0001	0001
0011	0010
0010	0011
0110	0100
0111	0101
0101	0110
0100	0111
1100	1000
1101	1001
1111	1010
1110	1011
1010	1100
1011	1101
1001	1110
1000	1111

Figure 6–36 Gray code to binary code conversion.

Design Example

Observe from Figure 6–36 that you could program a PLA so that any input bit pattern would produce a predictable bit pattern on the output. Consider another application of the PLA presented in the following example.

Example 6–8

You are required to program a PLA so that the binary input will cause a hexadecimal display on a seven-segment readout. Show the steps you would take to determine the programming required and the final program code of the PLA.

Solution

Refer to Figure 6–37.

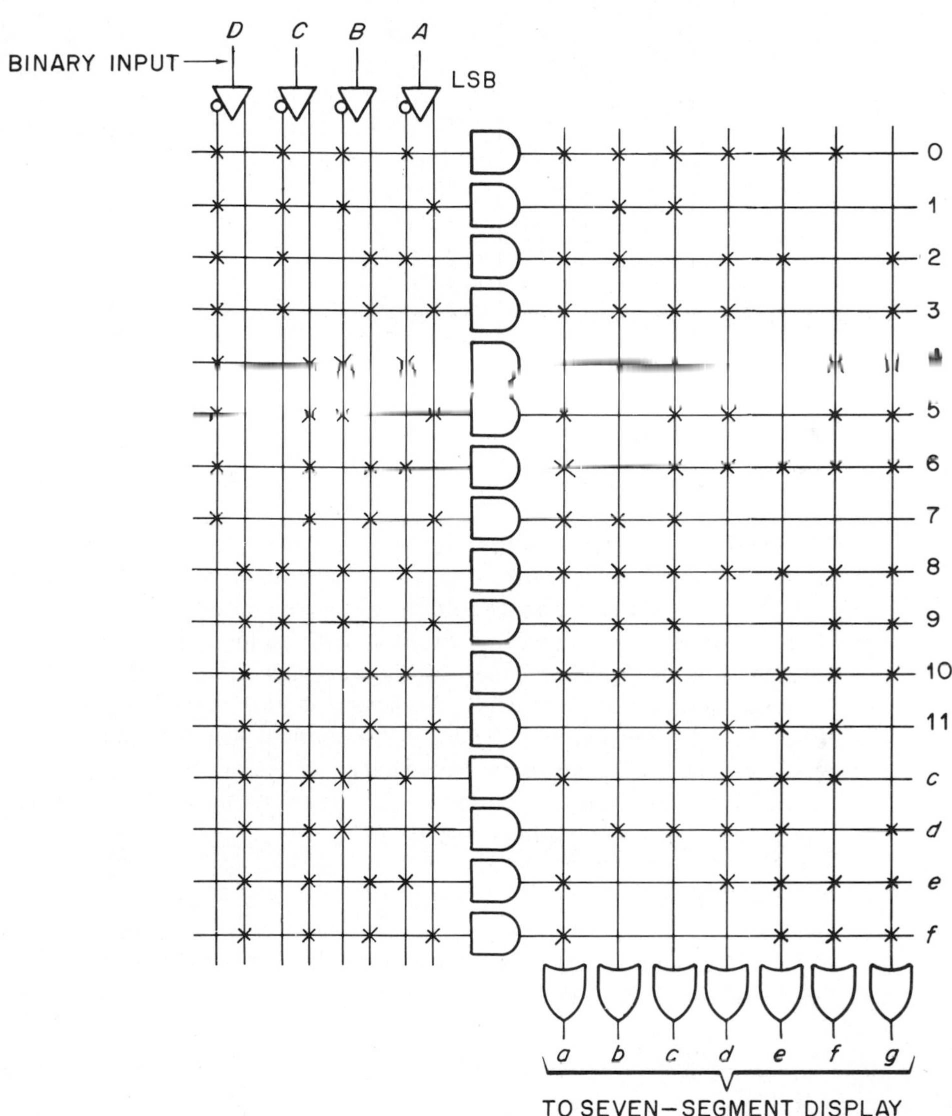

Figure 6–37

DECIMAL	HEX	BINARY	SEGMENTS ACTIVATED
0	0	0000	*a, b, c, d, e, f*
1	1	0001	*b, c*
2	2	0010	*a, b, g, e, d*
3	3	0011	*a, b, c, d, g*
4	4	0100	*b, c, f, g*
5	5	0101	*a, c, d, f, g*
6	6	0110	*a, c, d, e, f, g*
7	7	0111	*a, b, c*
8	8	1000	*a, b, c, d, e, f, g*
9	9	1001	*a, b, c, f, g*
10	A	1010	*a, b, c, e, f, g*
11	b	1011	*c, d, e, f*
12	C	1100	*d, e, f, a*
13	d	1101	*b, c, d, e, g*
14	E	1110	*d, e, g, f, a*
15	F	1111	*e, g, f, a*

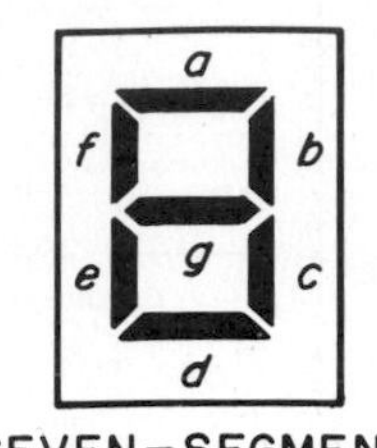

SEVEN-SEGMENT READOUT

Figure 6–37 Continued

Conclusion

In this section, you learned how to use PLD notation and, more specifically, how it applied to the PLA. Here, some important applications were presented along with a typical design problem. Hopefully, you can begin to see how these logic devices are quickly replacing discrete gates as the building blocks of digital systems. Using PLDs allows the designer to prototype digital systems quickly, because it is easy enough to change the PLD chip as design changes are required. Thus, it is important that the technician understand the nomenclature and operation of the new PLDs.

Check your understanding of this section by trying the following review questions.

6–4 Review Questions

1. What is PLD notation?
2. What is a *product line*?
3. How are the number of inputs to a product line indicated in PLD notation?
4. State what the logic value will be for an unprogrammed product line.

6–5 DECODERS

Discussion

There is a unique logic circuit arrangement that finds wide applications in digital systems. As you will soon see, this logic network is also used in programmable logic devices. Here, you will see the concepts behind this important logic network and some practical ICs that contain this powerful feature.

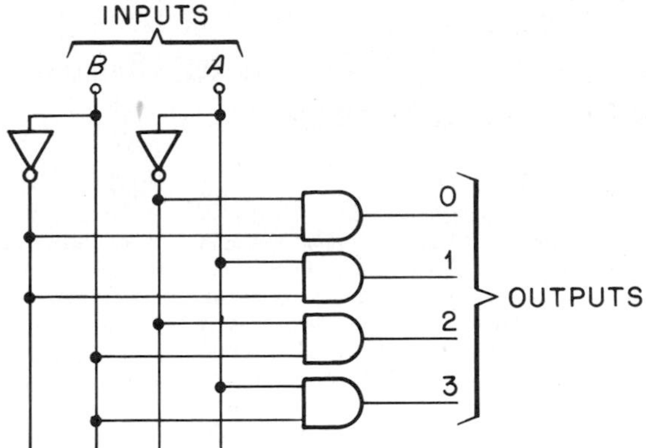

Figure 6–38 Basic decoder.

General Idea

Consider the logic network of Figure 6–38. Note from the figure that there are two input variables (*A* and *B*) and four outputs (0, 1, 2, and 3). Also note that only one output will be TRUE at a time—the one that is TRUE depends upon the binary value of the input. This is illustrated in the following function table.

Function Table						
Binary Value	**Inputs**		**Output**			
	B	***A***	**0**	**1**	**2**	**3**
0	0	0	1	0	0	0
1	0	1	0	1	0	0
2	1	0	0	0	1	0
3	1	1	0	0	0	1

As you can see from the function table, if you consider the binary values of the input variables, then their values correspond to the output that will be active. Such a logic system is called a **decoder.**

Decoder:
A logic network that will convert a specific bit pattern on its input to a specific output level.

The example decoder of Figure 6–38 is called a *2-line to 4-line* decoder. It is called this because the 2-bit binary count on the input is converted to the activation of one of four lines on the output.

Practical Decoders

Decoders are available on a single chip. Figure 6–39 illustrates a 4-line to 16-line decoder. Observe from this figure that the outputs are active LOW. This means that there will be

DM54154/DM74154 4-Line to 16-Line Decoders/Demultiplexers

General Description

Each or these 4-line-to-16-line decoders utilizes TTL circuitry to decode four binary-coded inputs into one of sixteen mutually exclusive outputs when both the strobe inputs, G1 and G2, are low. The demultiplexing function is performed by using the 4 input lines to address the output line, passing data from one of the strobe inputs with the other strobe input low. When either strobe input is high, all outputs are high. These demultiplexers are ideally suited for implementing high-performance memory decoders. All inputs are buffered and input clamping diodes are provided to minimize transmission-line effects and thereby simplify system design.

Features

- Decodes 4 binary-coded inputs into one of 16 mutually exclusive outputs
- Performs the demultiplexing function by distributing data from one input line to any one of 16 outputs
- Input clamping diodes simplify system design
- High fan-out, low-impedance, totem-pole outputs
- Typical propagation delay
  3 levels of logic 19 ns
  Strobe 18 ns
- Typical power dissipation 170 mW

Connection Diagram

Dual-In-Line Package

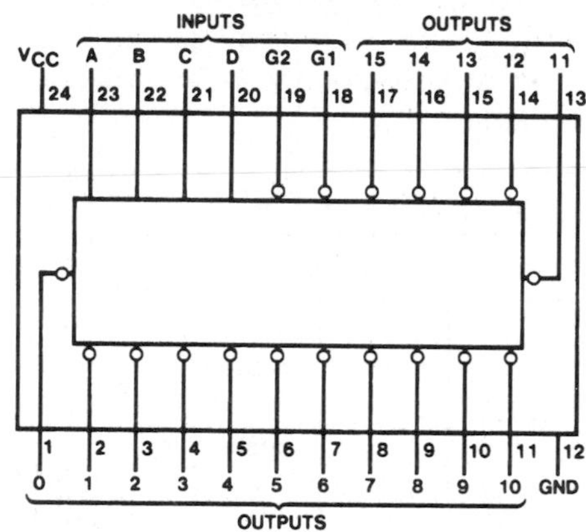

Order Number DM54154J or DM74154N
See NS Package Number J24A or N24A

Logic Diagram

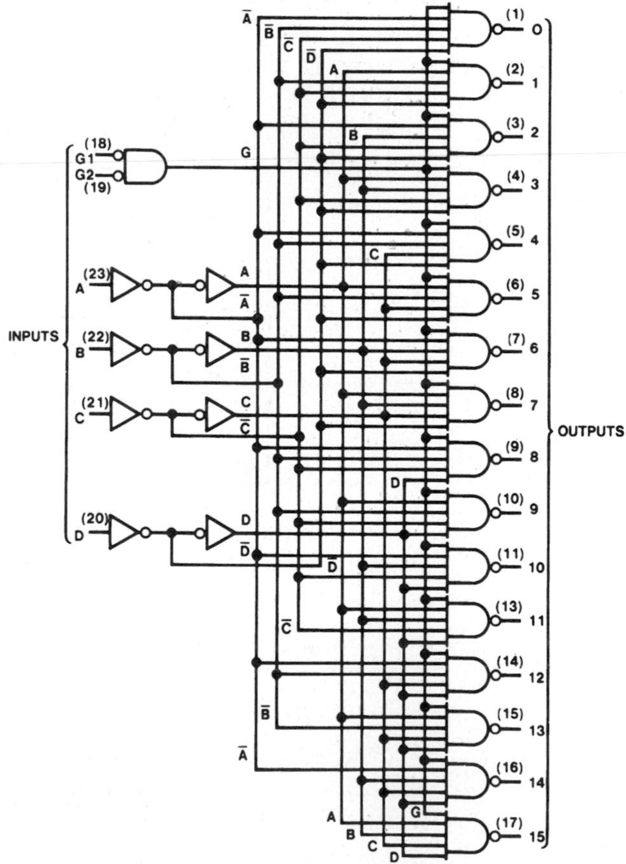

Figure 6–39 Practical 4-line to 16-line decoder. *Reprinted with permission of National Semiconductor Corporation.*

National holds no responsibility for any circuitry described. National reserves the right at any time to change without notice said circuitry.

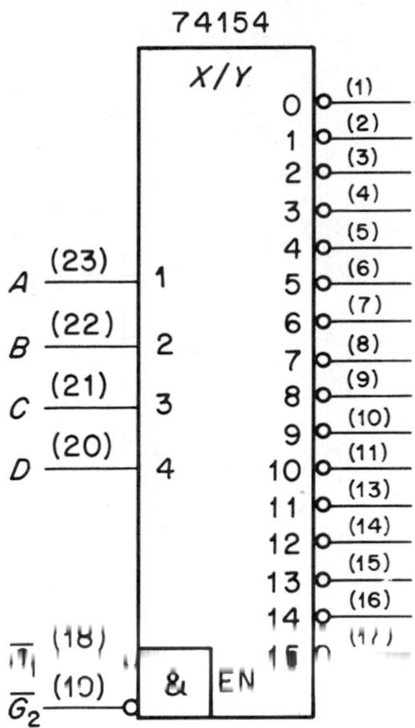

Figure 6–40 Logic symbol of 4-line to 16-line decoder.

only 1 of the 16 output lines LOW at a time according to the binary value of the four input lines (*A*, *B*, *C*, and *D*). Note that the NAND gate outputs indicate that each output is active LOW, and all the rest are HIGH.

The input to the decoder also contains two enabling inputs called G_1 and G_2. These must both be LOW in order for any of the 16 outputs to be active (note that their inputs indicate an active LOW). These enabling lines are used when more than one decoder is needed. The logic symbol for the 4- to 16-line decoder is shown in Figure 6–40. Observe from the logic symbol of the decoder that the enable lines are indicated as an AND gate with active LOW inputs. Note that all 16 outputs are indicated as active LOW.

The numbers in parentheses for the logic diagrams in Figure 6–40 represent the pin numbers of the chips, and the other numbers represent functional values. For example, the input pins *A*, *B*, *C*, and *D* are indicated by their binary weights (1, 2, 4, 8) and each output pin is identified by a decimal number that is the value of the binary count that will activate it (0 through 15).

Example 6–9

For the 74154 decoder, what will be the conditions of its output lines if the input lines are $G_1 = 0$, $G_2 = 0$, $D = 0$, $C = 1$, $B = 1$, $A = 0$?

Solution

Since both G_1 and G_2 are active (their inputs are active LOW), the device is enabled. The decimal equivalent of the input value of 0110_2 is 6_{10}. Thus, all output lines from 0 through 15 with the exception of line 6 will be HIGH. Line 6 will be LOW.

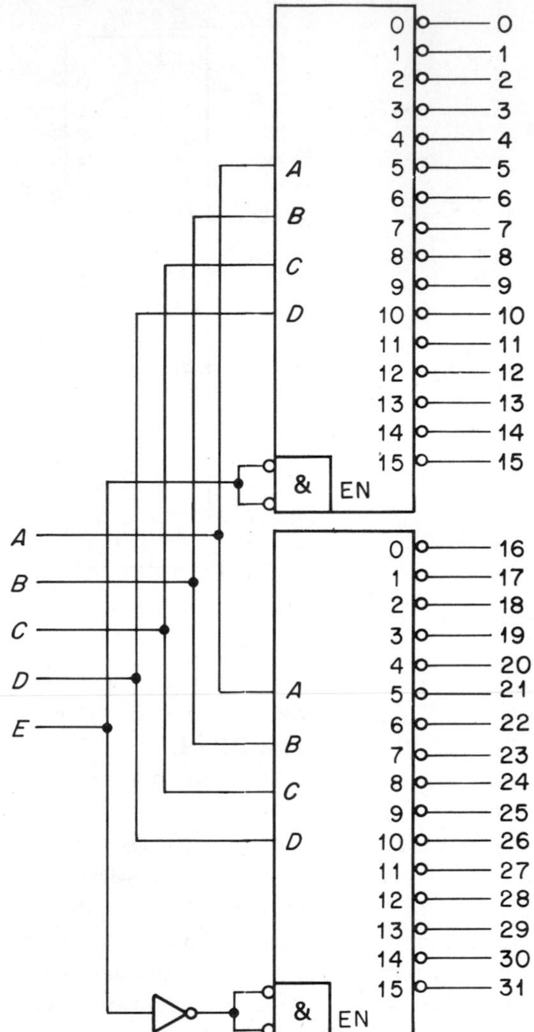

Figure 6–41 Using more than one decoder.

More Decoding

You can use more than one of these 4-line to 16-line decoders as shown in Figure 6–41. As you can see, this combination becomes a 5-line to 32-line decoder ($2^5 = 32$). When the most significant bit (E) is LOW, the value of the input binary count will be from 0 through 15, and thus only the first decoder is active. When the most significant bit is HIGH, the value of the input binary count will now be from 16 to 31, and the second decoder will be active.

An Application

Consider the decoder of Figure 6–42. Note that its output is connected to memory locations. From the operation of a decoder, you know that only one output at a time can be

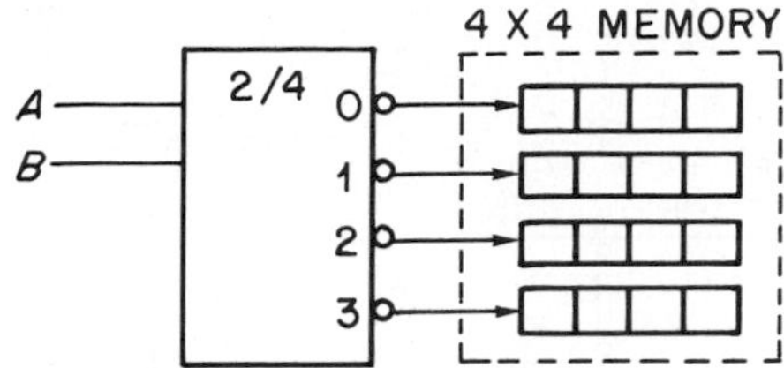

Figure 6–42 Decoder output connected to memory locations.

active. From Figure 6–42, you can see how this device would be used to select only one memory location in a digital system. Here, the binary count inputted to the decoder determines which memory location is selected and hence displayed on the seven-segment readouts. You can begin to see the importance of this IC in real digital systems.

Conclusion

This section presented the important concepts of a decoder. You will see this type of digital device used many times in programmable logic devices. In the next section, a very important and powerful application of the decoder is revealed. Check your understanding of this section by trying the following review questions.

6–5 Review Questions

1. What is a decoder?
2. Describe the action of a 2-line to 4-line decoder.
3. Explain the purpose of the 74154 IC.
4. State how two decoders can be used to increase the number of output lines.
5. What is the importance of using a decoder with memory?

6–6 READ-ONLY MEMORIES

Discussion

In this section, you will learn about one of the most important topics in the digital control of systems. Here you will see the fundamental logic network that makes your computer know what to do when you first turn it on, makes your automatic dialing telephone smart enough to know how to store phone numbers, and controls the logic networks in your digital recordings.

The Basic ROM

The section on PLAs showed you how to use these devices to create a logic array that added 2 binary bits and produced the correct answer for the sum and the carry. You can also do the same thing with a *read-only memory* (ROM). For example, look at the logic circuit of Figure 6–43.

As shown in this figure, the sum and carry outputs behave exactly the same way as if the *A* and *B* inputs represented the addition of 2 binary bits. Now, here is the reason for those diodes. Keep in mind that a diode will only allow current flow in one direction.

B	A	CARRY	SUM
0	0	0	0
0	1	0	1
1	0	0	1
1	1	1	0

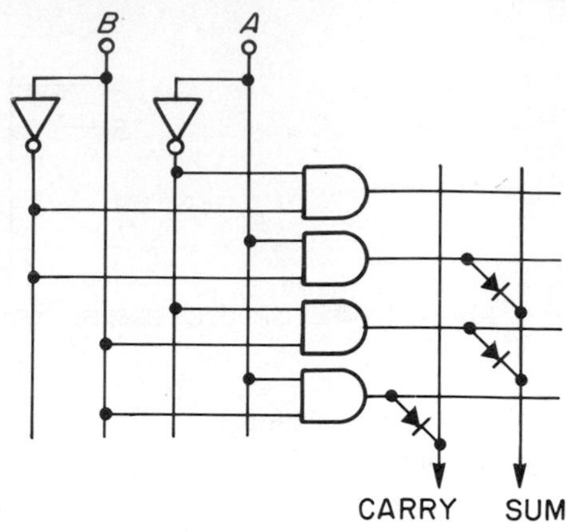

Figure 6–43 ROM for adding 2 binary bits.

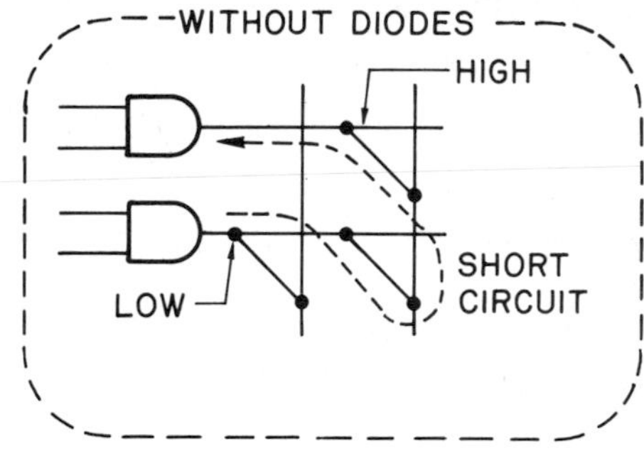

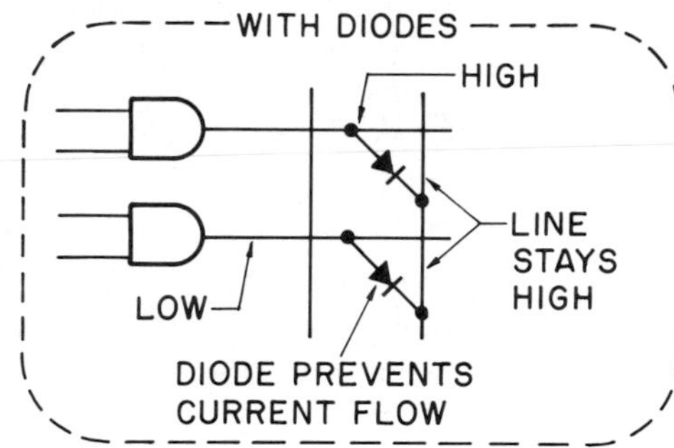

Figure 6–44 Reason for diodes in ROM.

Now look at Figure 6–44. Without the diodes, current would flow in directions not allowed and cause possible circuit damage to the controlling digital gates.

A Laboratory ROM

You could actually construct a ROM in the lab using diodes and switches. Such a ROM is shown in Figure 6–45. Note that each line on the ROM is controlled by a 4-line to 16-line decoder. This means that only one line of the ROM will be activated at a time and will correspond to the value of the binary count of the four input variables. Also note that there are four outputs of the ROM. You will see the reason why four outputs were selected.

By use of the switches, you can now program the ROM for any bit pattern you choose. Figure 6–46 shows the standard nomenclature used for an open or closed switch. Note that this is similar to the fuses used by PLAs. The difference here is that you can change the program in this ROM any time you wish by simply changing the setting of each switch.

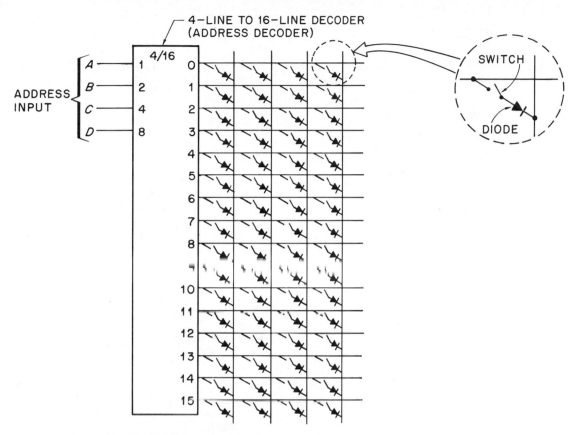

Figure 6–45 ROM construction from switches.

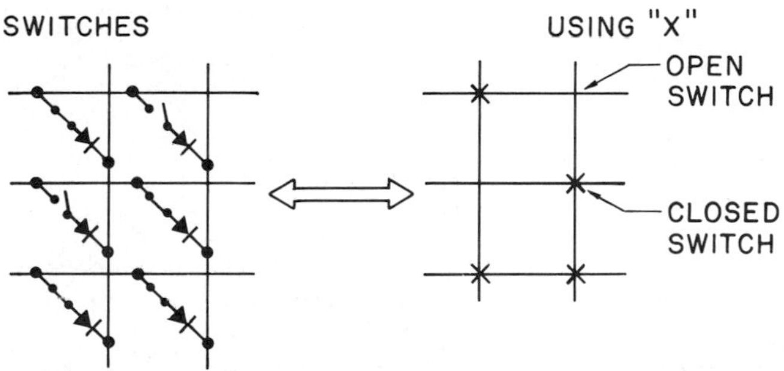

Figure 6–46 Nomenclature for open and closed switches.

A ROM Application

The section on PLAs showed how to simulate the action of any gate. Well, you can do the same thing with any ROM. This is illustrated in the following example.

Example 6–10

Show the bit pattern for the laboratory ROM that would simulate the logic of the following four gates (one for each output): NAND, NOR, XOR, AND.

Solution

Refer to Figure 6–47.

B	*A*	NAND	NOR	XOR	AND
0	0	1	1	0	0
0	1	1	0	1	0
1	0	1	0	1	0
1	1	0	0	0	1

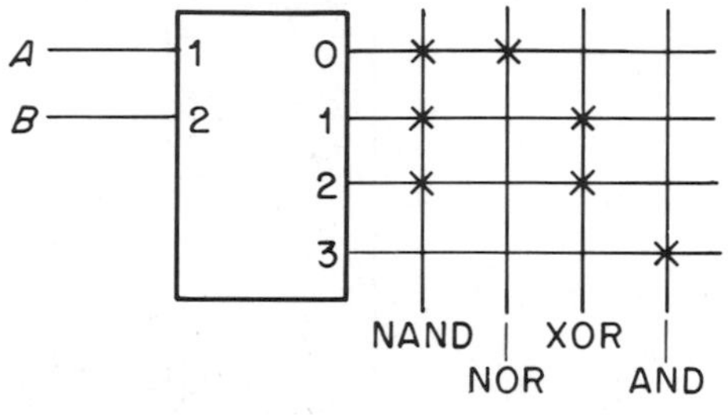

Figure 6–47

You could also program your laboratory ROM to do a binary to gray code conversion, or, with more outputs, a binary to seven segment read out conversion. The point here is that with the concept of the ROM, you have another powerful digital control network.

Diode OR Gate

Figure 6–48 shows how the diodes used in a ROM follow the function of an OR gate. Observe Figure 6–48(A). Here, neither diode conducts, therefore they both act as an open

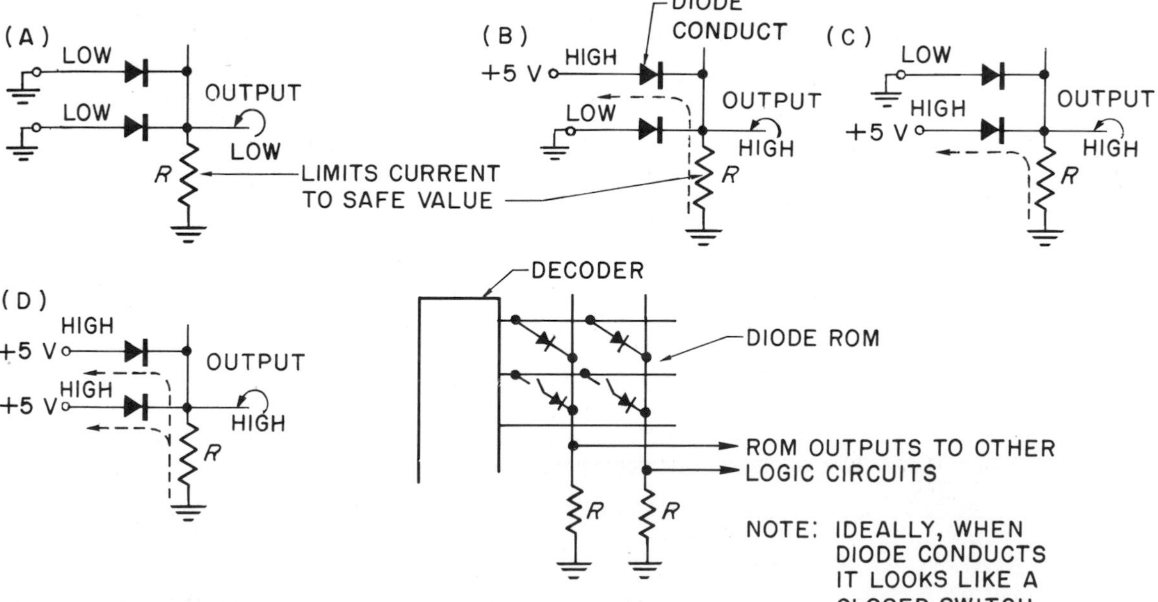

Figure 6–48 OR gate action of diodes.

switch. This causes the output to be LOW (no +5 V present at the output). In Figure 6–48(B), the top diode conducts and therefore acts as a closed switch. It makes no difference if the bottom diode is conducting or not, the output will appear HIGH (have +5 V). Figure 6–48(C) is similar to 6–48(B), just one diode is conducting (acting like a closed switch) this causes the output to again be HIGH. In the last case (Figure 6–48(D)) both diodes conduct and the output is HIGH. This action is identical to the action of an OR gate. Thus, the diode/resistor combination of a diode ROM replicates the OR gate function. Because of the OR gate action of diodes in a ROM, the diodes can be represented as OR gates.

Factory ROMs

ROMs are another form of PLDs. The difference between a ROM and a PLA is that in a ROM only the OR gates are programmable; the AND gate bit pattern is fixed. This is illustrated in Figure 6–49. Note that the fixed AND bit pattern is set up as a *decoder*. Thus the distinguishing feature of the ROM as a PLD is that the AND gate arrangement is preprogrammed as a decoder and the output of each decoder position (address) will be the bit pattern determined by the ROM user. Thus, the ROM is not as flexible as the PLA, but it is very useful in applications where the number of inputs is limited. Because the AND arrangement of the ROM is always fixed as a decoder, it is not possible to perform logic minimization between inputs and outputs, as it is with a PLA.

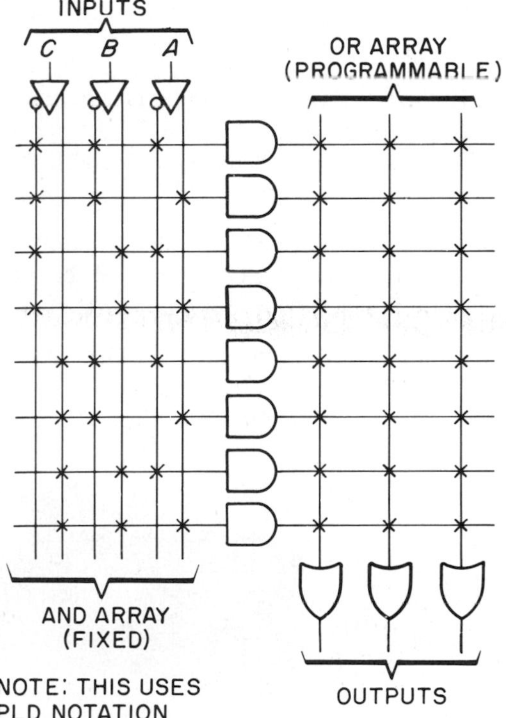

Figure 6–49 Programmable part of the ROM.

Types of ROMs

Many different types of ROMs are available as ICs. As an example, a ROM that is supplied from the factory with a built-in program is said to be **mask programmable ROM.**

Mask-Programmable ROM:
A ROM that is programmed by the manufacturer.

These ROMs are called mask programmable because in their manufacturing process a photographic *mask* is placed over the ROM circuits that replicates the desired bit pattern. Thus, during the manufacturing process, this mask imparts the desired bit pattern to each ROM. Mask-programmable ROMs are usually used when large quantities all with the same bit patterns are needed. This would be used for already developed digital systems that are mass produced.

User-Programmable ROMs

Another type of ROM is the *user-* or *field*-programmable ROM. This is referred to as **Programmable Read Only Memory (PROM).**

Programmable Read Only Memory (PROM):
A programmable ROM that can be programmed in the field by the user.

A PROM contains a fusible link similar to that used by a field-programmable PLA. These can be electrically programmed by the technician. These are used when small quantities are needed for prototypes or limited production runs. PROMs are cheaper in small quantities than factory programmed ROMs are. Hence, they are used in small quantities.

Erasable PROMs

An *erasable* PROM is a ROM that can be programmed in the field by the technician, and then the bit pattern can be again changed. These are more versatile than PROMs because the bit pattern is not permanent. This is an **Erasable Programmable Read Only Memory (EPROM).**

Erasable Programmable Read-Only Memory (EPROM):
Erasable programmable ROM. A ROM that can be programmed in the field by the user and can have its bit pattern changed again.

The EPROM comes in two different types. One can be erased electrically, and the other is erased by the application of ultraviolet light. Typical EPROMs are shown in Figure 6–50. You will learn more details about these devices in Chapter 9.

Figure 6–50 Typical EPROM. *Courtesy of Motorola. Inc.*

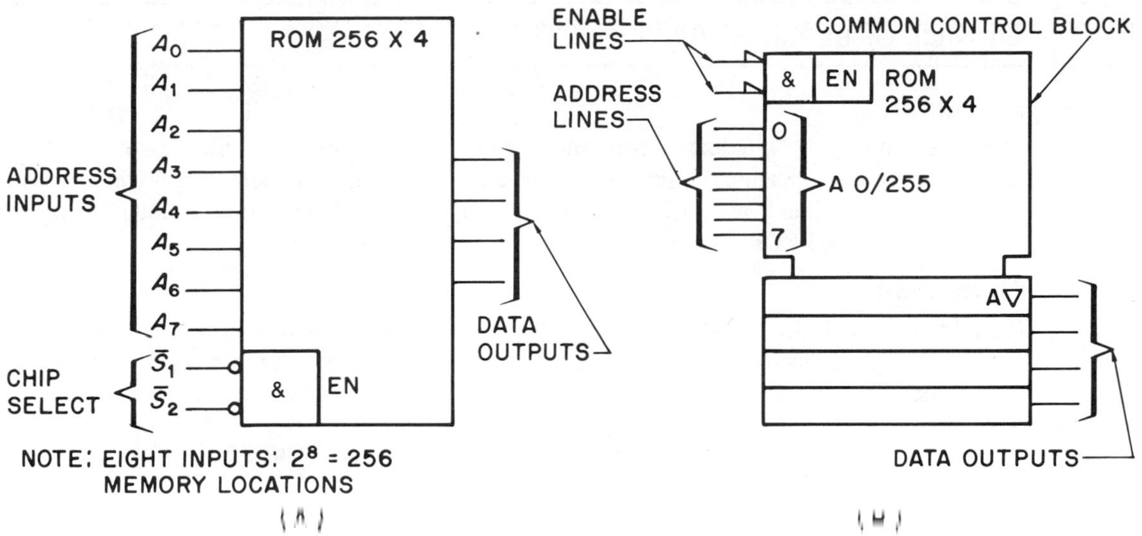

Figure 6–51 Logic symbol for typical ROM.

Some Real ROMs

The logic symbols for some typical ROMs are shown in Figure 6–51. Note that the first ROM shown in Figure 6–51(A) is a 256 × 4 ROM, which means it is 4 bits wide with 256 memory locations. This makes sense, since there are eight address inputs (2^8 = 256). Figure 6–51(B) illustrates the same ROM using the IEEE dependency notation. As you can see, the figure consists of a *common control block*. Below this is the body of the ROM that contains the data. The control block has 8 address lines (labeled 0 through 7) with a bracket pointing to the letter *A*. This means address dependency. Each data output is address dependent thus it also has the same *A* on its output (only the first one is shown, the rest are implied). Next to the *A* on the common control block is the address range values (0 to 255 for 256 address locations). Each data output has an inverted letter delta to indicate that it is a tri-state output. Again only the first output is labeled, the others are implied. You will see more of the use of this dependency notation in the chapter on memory and the appendix.

Conclusion

In this section, you saw how another important PLD, the ROM, is constructed. Here you were shown a simple laboratory ROM made from switches and diodes. You also saw the need for diodes in this type of ROM. As a PLD, you were given the differences between several different ROM types with their advantages and disadvantages. In the next section, you will be introduced to a third kind of PLD. For now, test your understanding of this section by trying the following review questions.

6–6 Review Questions

1. Why is a diode necessary in a ROM?
2. State the difference between a ROM and a PLA.
3. Can you use K-mapping techniques in designing ROM programs? Explain.
4. Explain the difference between a PROM and an EPROM.

6–7 DEMULTIPLEXERS AND ENCODERS

Discussion

In this section, you will learn about more useful and interesting logic networks. These devices serve a wide range of applications from keyboard inputs to selecting different kinds of data inputs. When you finish with this section, you will have a firm foundation in the different types of PLDs and their associated digital networks.

Basic Idea

Recall that a *multiplexer* had many different inputs that could each be connected to a single output. Well, a **demultiplexer** does just the opposite, as shown in Figure 6–52. This figure shows the selector switch representations of both types of digital networks.

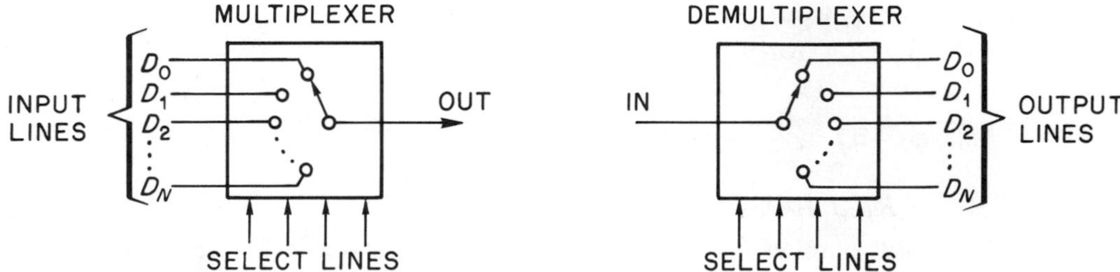

Figure 6–52 Comparison of multiplexer and demultiplexer.

Demultiplexer:
A digital network that converts the logic of one input line to one of several output lines, as determined by the binary value of its select lines.

The logic diagram of a demultiplexer is shown in Figure 6–53. Observe that the logical *A* and *B* inputs control which AND gates output will reflect the logic condition of the data input. For example, if *A* and *B* are both LOW, then the D_0 output level will be the same as the DATA IN level.

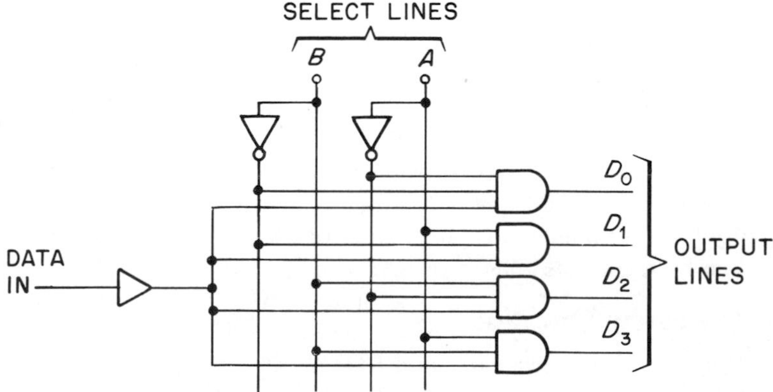

Figure 6–53 Logic diagram of 1-line to 4-line demultiplexer.

Example 6–11

For the waveform conditions applied to the 1- to 4-line demultiplexer of Figure 6–53, determine the resulting outputs. See waveforms of Figure 6–54.

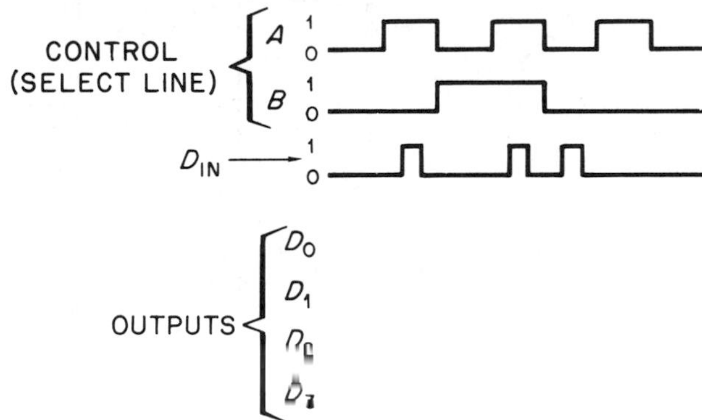

Figure 6–54

Solution

For a demultiplexer, only one output at a time is active, depending on the binary count of the controlling input. The solution is shown in Figure 6–55.

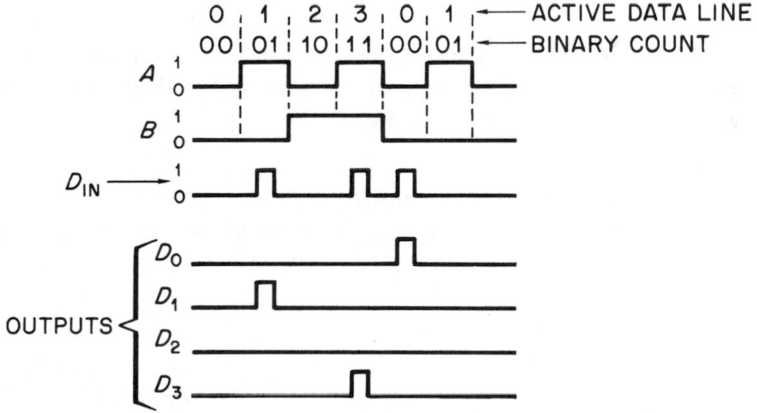

Figure 6–55

As shown in the previous example, the given input control signals for A and B represent a binary count (as shown above the waveforms in Figure 6–55). Thus the output lines D_0 through D_3 mirror the input pulses in sequence, starting with D_0 and progressing through D_3.

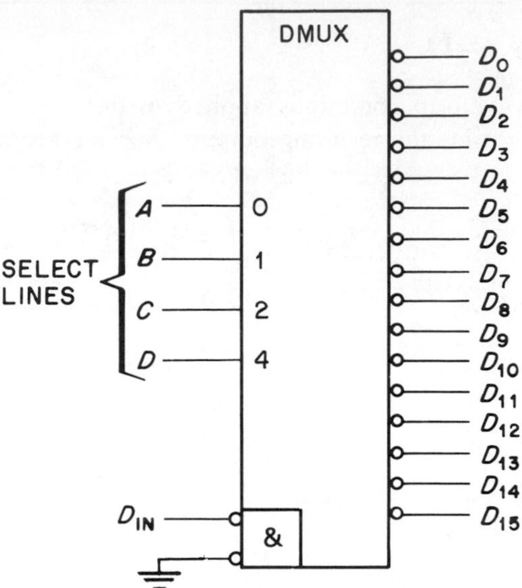

Figure 6–56 Logic symbol of the 74154 demultiplexer.

Real Demultiplexers

A real demultiplexer is shown in Figure 6–56. This demultiplexer was used as the 4-line to 16-line decoder presented in Section 6–5.

Encoders

Recall that a *decoder* converted a binary count into a single selected line, depending upon the binary value of the input. An **encoder** is the dual of a *decoder* and does just the opposite. This comparison is shown in Figure 6–57.

Encoder:

A logic network that converts information into a form of code.

The logic network of a 4-line to 2-line *encoder* is shown in Figure 6–58. Observe from the figure that if only one input line is HIGH at a time while the others are LOW, then

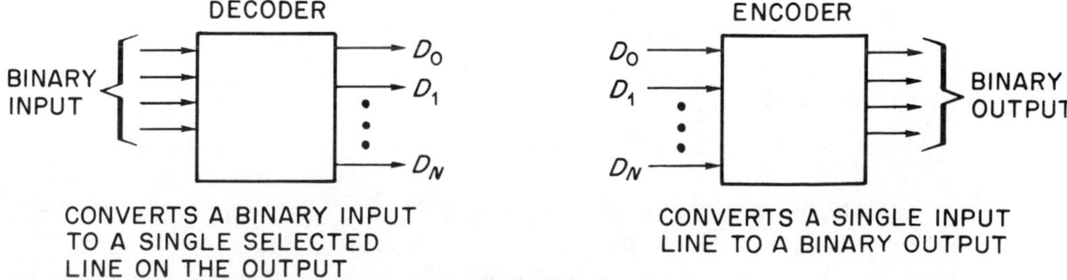

Figure 6–57 Comparison of decoder to encoder.

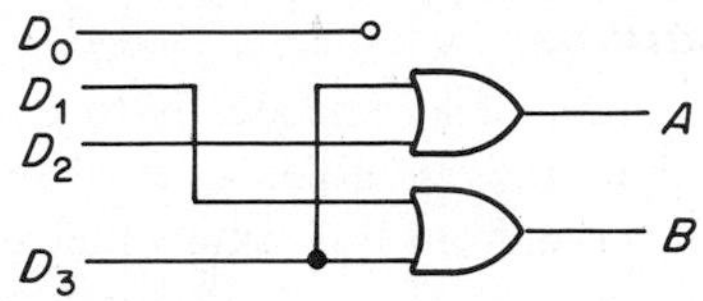

Figure 6–58 Four-line to 2-line encoder.

the output lines A and B will have a binary value equal to the indicated value of the line. This is shown in the following table.

HIGH Input Line	Output A	B
D_0	0	0
D_1	0	1
D_2	1	0
D_3	1	1

Example 6–12

For the encoder of Figure 6–58, determine the A and B outputs for the input waveforms of Figure 6–59.

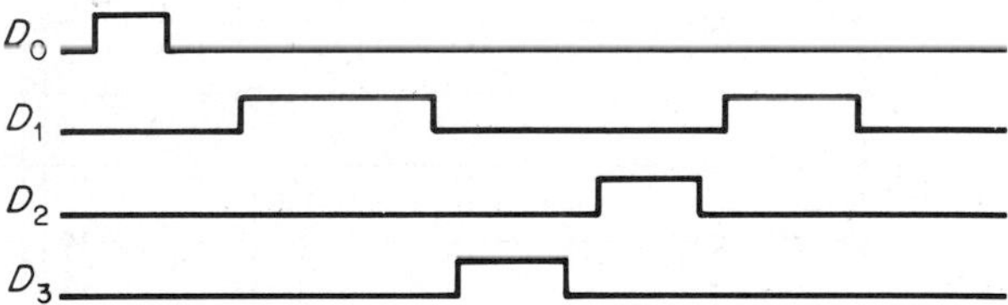

Figure 6–59

Solution

Observe how each of the input lines are connected to the OR gates of the encoder. From this observation, sketch the output waveforms. The results are shown in Figure 6–60.

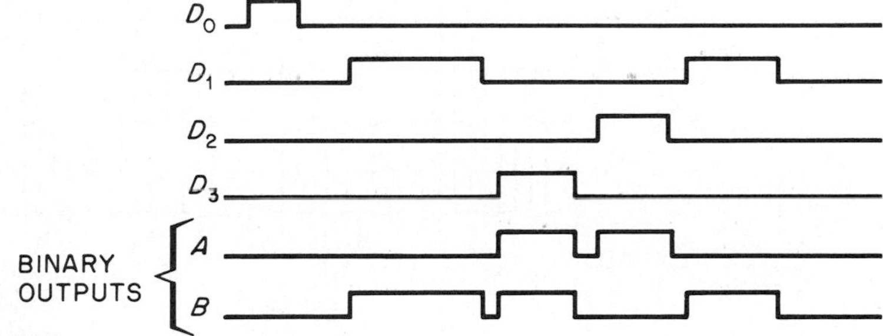

Figure 6–60

Encoder Applications

One common application of an encoder is to convert a switch input to its binary code. The logic network that converts a hex keypad to its equivalent binary value is shown in Figure 6–61(A). Observe from this figure that the binary output of each depressed key will appear on data lines D_0, D_1, D_2 and D_3. Assume that each hex key normally causes a LOW to appear on its output until depressed, when it will then cause its line to be HIGH. For example, if the "1" key is depressed, the data output line D_0 will be HIGH and all the other data output lines will be LOW, thus giving the binary value of 0001_2. However, if key "1" is released and key "7" is depressed, the first three data output lines will be HIGH, producing the binary value of $0111_2 = 7_{16}$. When the "F" key is depressed, all outputs will be HIGH: $1111_2 = F_{16} = 15_{10}$. Figure 6–61(B) shows a **priority encoder.**

Priority Encoder:
A special type of encoder that senses if two or more inputs are active and then gives an output corresponding to the highest value of the input.

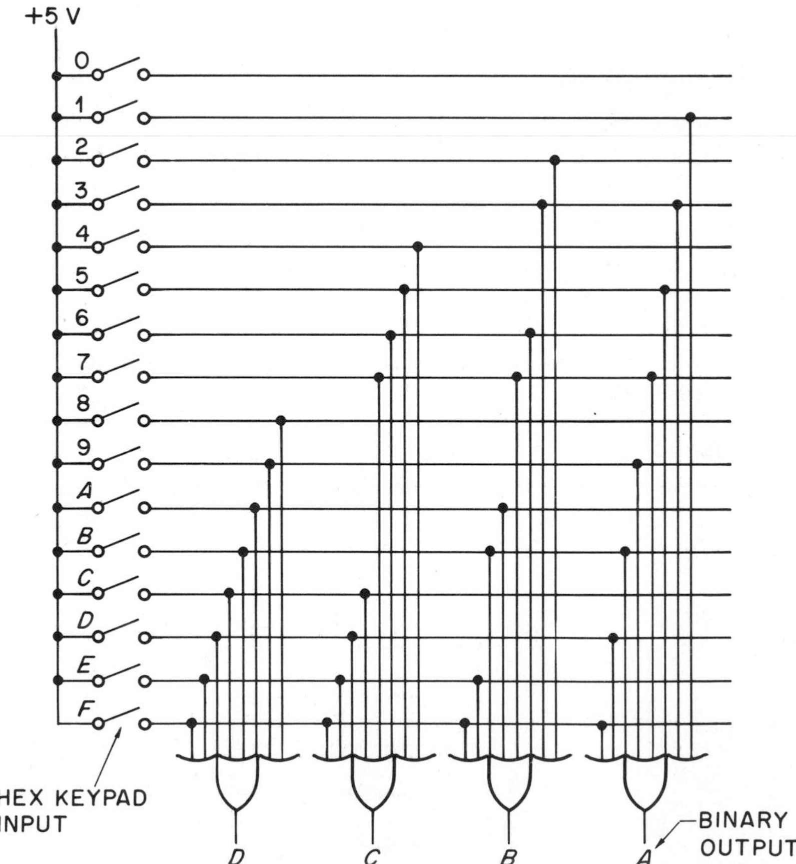

Figure 6–61A Hex keypad encoder and priority encoder.

DM54148/DM74148 Priority Encoders

General Description

This TTL encoder features priority decoding of the input data to ensure that only the highest-order data line is encoded. The DM54148 and DM74148 encode eight data lines to three-line (4-2-1) binary (octal). Cascading circuitry (enable input E1 and enable output E0) has been provided to allow octal expansion without the need for external circuitry. For all types, data inputs and outputs are active at the low logic level.

Features

- Encodes 8 data lines to 3-line binary (octal)
- Applications include:
  N-bit encoding
  Code converters and generators

Connection Diagram

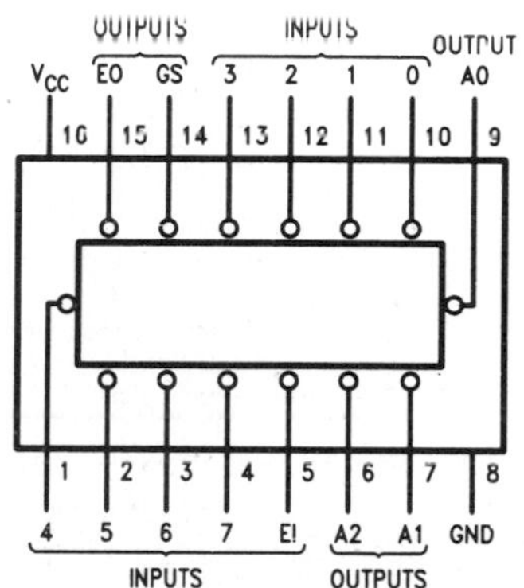

TL/F/6545-1

Order Number DM54148J or DM74148N
See NS Package Number J16A or N16A

Function Table

54148/74148

Inputs									Outputs				
E1	0	1	2	3	4	5	6	7	A2	A1	A0	GS	E0
H	X	X	X	X	X	X	X	X	H	H	H	H	H
L	H	H	H	H	H	H	H	H	H	H	H	H	L
L	X	X	X	X	X	X	X	L	L	L	L	L	H
L	X	X	X	X	X	X	L	H	L	L	H	L	H
L	X	X	X	X	X	L	H	H	L	H	L	L	H
L	X	X	X	X	L	H	H	H	L	H	H	L	H
L	X	X	X	L	H	H	H	H	H	L	L	L	H
L	X	X	L	H	H	H	H	H	H	L	H	L	H
L	X	L	H	H	H	H	H	H	H	H	L	L	H
L	L	H	H	H	H	H	H	H	H	H	H	L	H

H = High Logic Level, L = Low Logic Level, X = Don't Care

Figure 6–61B *Reprinted with permission of National Semiconductor Corporation.*
National holds no responsibility for any circuitry described. National reserves the right at any time to change without notice said circuitry.

The problem with the encoder of Figure 6–61(A) is that it can give an erroneous output code if more than one switch is depressed. The 74148 priority encoder will always give the highest value on the output for any combination of inputs. For example, look at the function table for this device. Note that all of the inputs and output are active LOW. There are 8 inputs which require three binary bits to represent these 8 lines (4-2-1). The ENABLE line must be LOW to activate the chip. When all inputs are HIGH (not active) the outputs (A_2, A_1, A_0) are all HIGH (meaning 000 since the outputs are inverted). When line 7 is LOW (active) it makes no difference what the other input lines are (marked by don't cares X), the output will represent a binary 7 by being all LOW. When line 6 is the largest value active line (meaning 6 is LOW and 7 is HIGH) then the output will represent a binary 6 (LOW, LOW, HIGH).

A Real Encoder

Figure 6–62 shows an IC used as a 16-key (hex) to binary *encoder*. For the 16-key encoder, the $\overline{KP}$ output is normally HIGH and will go LOW when any input key is pressed. Thus, this logic output is used to indicate when a key is pressed. If more than one input is HIGH at the same time (indicating more than one key is being pressed), then the $\overline{KRO}$ (key rollover) output will go LOW, indicating an invalid code. The KRO output could be constructed with a PLA, looking for more than one key being pushed at a time. Many other kinds of decoders are available, such as the 74147, which is a 10-key to BCD encoder.

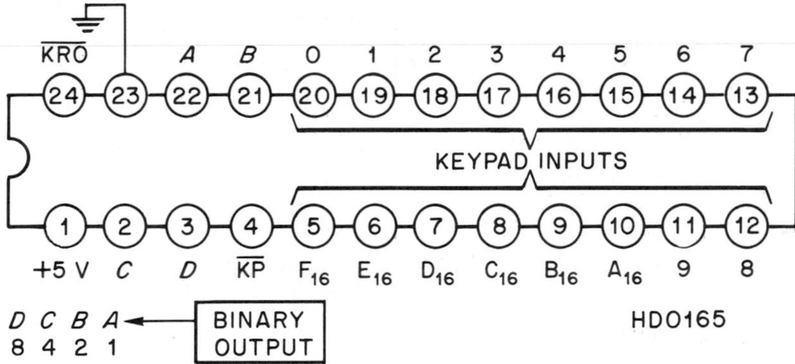

Figure 6–62 Keypad encoder.

Conclusion

This section presented two useful logic networks that were duals of others presented earlier. Here you saw an application of these devices and some real ICs that represented them. In the next section you will be introduced to an important CMOS device. There you will see how ROM can be used to synthesize sound. Check your progress in this section by trying the following review questions.

6–7 Review Questions

1. Describe the difference between a *multiplexer* and a *demultiplexer*.
2. What is a 1-line to 8-line demultiplexer?
3. State how a demultiplexer can be used.
4. Describe the difference between an *encoder* and a *decoder*.
5. Give an application of an *encoder*.

6–8 CMOS BILATERAL SWITCH

Discussion

In this section, you will see that, unlike TTL, CMOS can offer digital devices that operate as if they were mechanical ON/OFF switches. This feature provides great flexibility for the operation of digital systems.

Basic CMOS Switch

Figure 6–63 shows a basic **bilateral switch.**

Bilateral Switch:
A switch that can be operated in any direction. In reference to CMOS, a solid-state device that electrically behaves as such a switch.

This switch is called *bilateral* because data may be put into the switch from any direction. This is different from TTL, where there is a definite input and a definite output. As an example, consider the bilateral switch used as either a multiplexer or as a demultiplexer. See Figure 6–64. The reason for this versatility is that the CMOS bilateral switch, unlike anything in TTL, can be used in either direction.

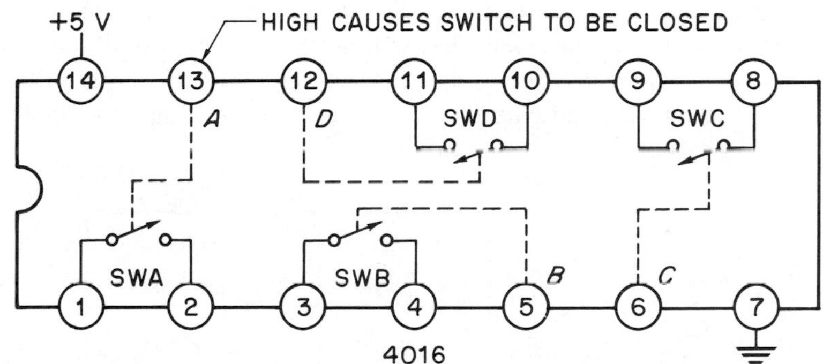

Figure 6–63 A basic CMOS bilateral switch.

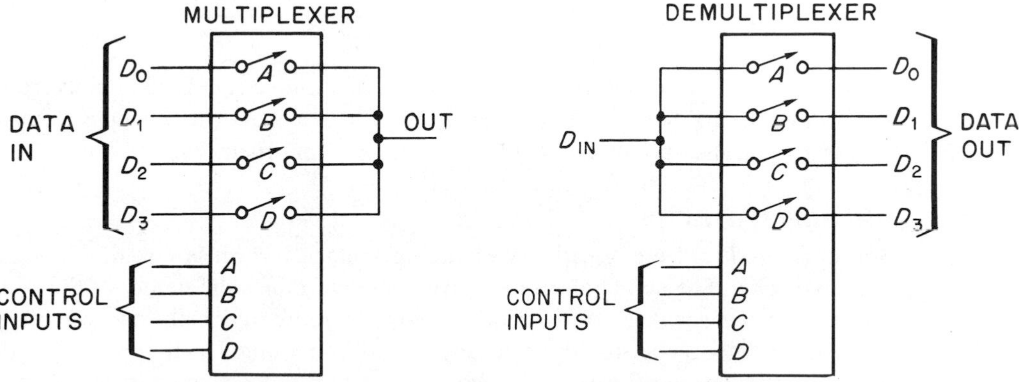

Figure 6–64 Bilateral switch as a multiplexer or demultiplexer.

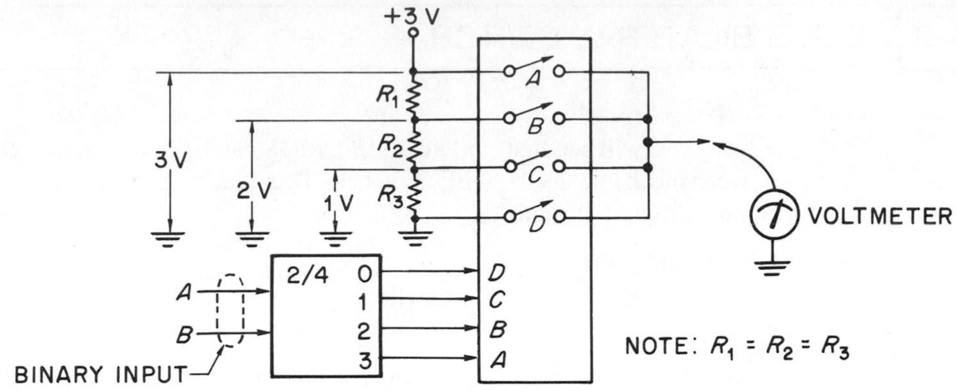

Figure 6–65 Digital to voltage level converter.

Digital to Voltage

A CMOS bilateral switch can also be used to change a binary value into a voltage level. For example, consider the logic circuit of Figure 6–65. Observe from this figure that only one CMOS switch will be closed at a time. The binary value of the input variable will determine which of these switches will be closed. Because of the resistive voltage divider, the closure of a switch will produce a unique voltage reading on the voltmeter. The voltmeter reading will indicate the binary value of the input variables as indicated by the following table.

Inputs		Output
A	*B*	Voltage (volts)
0	0	0
0	1	1
1	0	2
1	1	3

Such a device is referred to as a *digital-to-analog converter* (D to A converter). You will learn more about these important devices in Chapter 9. For now, focus your attention on a useful relationship between the bilateral switch and memory.

Making Waves

Sound is produced electrically by changing voltages within a circuit. Figure 6–66 gives the basic idea. The actual electrical waveform determines the nature of the sound that will come from the speaker. As a matter of fact, if you could make the waveform look any way you wanted, you could create any sound you wanted. The process of doing this is called **waveform synthesis** and is the fundamental process used in **music synthesizers.**

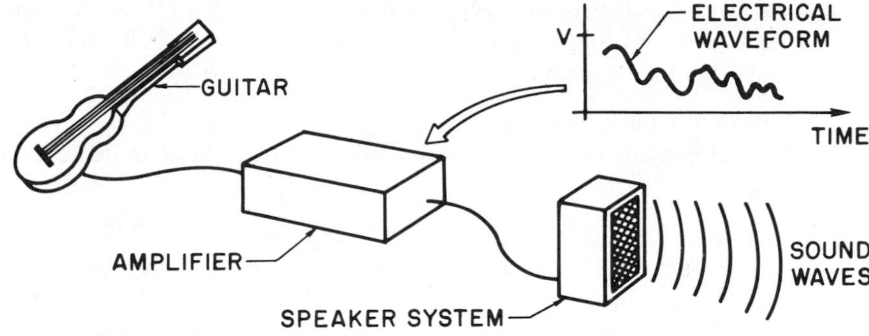

Figure 6–66 Basic idea of electrical waveform.

Waveform Synthesis:
The process of creating a desired waveform.

Music Synthesizer:
A device capable of creating various sounds selected by the user.

As an example, suppose a particular sound had the electrical waveform as shown in Figure 6–67. There is a way that such a sound could be created using a CMOS bilateral switch and a switch-programmable ROM. Such an arrangement is shown in Figure 6–68. The

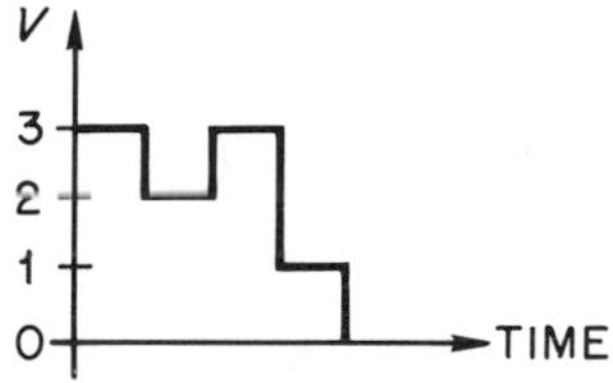

Figure 6–67 Waveform of a particular sound.

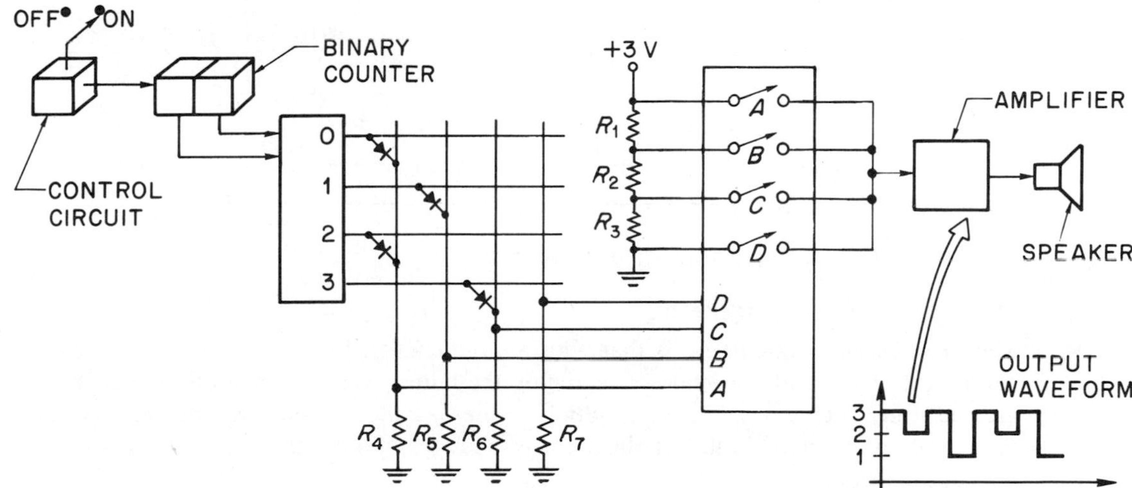

Figure 6–68 Basic sound synthesizer.

control circuit causes the **binary counter** to count from 00_2 to 11_2, and then repeat from 00_2, and so on, as long as power is applied to the circuit. The *control circuit* determines how fast the binary counter will count.

Binary Counter:
A logic device capable of sequentially incrementing or decrementing a binary value.

The output of the binary counter goes to the input of the 2-line to 4-line decoder. This causes the output of the decoder to sequentially address the diode ROM. The output of the ROM determines which switch of the bilateral switch will be closed. The amplifier operates the speaker to produce the actual sound of the synthesized wave. The complete process is summarized in Table 6–2.

Table 6–2 SUMMARY OF SOUND PRODUCTION

Binary Counter	Decoder Output	ROM Output	Bilateral Switch Closed	Output Voltage (volts)
00	Line 0	1000	A	3
01	Line 1	0100	B	2
10	Line 2	1000	A	3
11	Line 3	0010	C	1
00	Line 0	1000	A	3
01	Line 1	0100	B	2
⋮	⋮	⋮	⋮	⋮

The resulting waveform is shown in Figure 6–69.

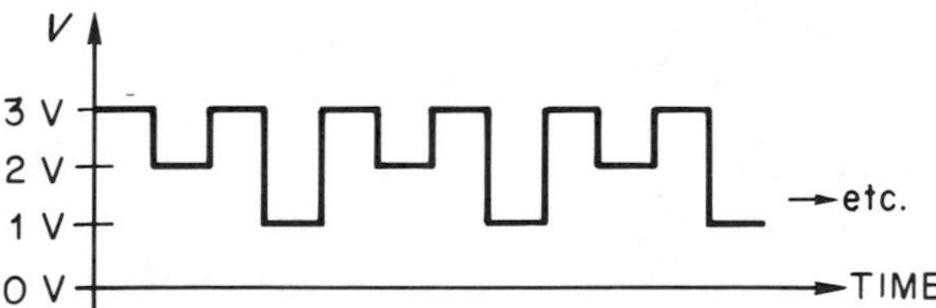

Figure 6–69 Resulting synthesized waveform.

Waveform Synthesizers

The principle illustrated here is that with a large enough binary counter and ROM, you can create any sound you choose, including the human voice. When the human voice is created using such a system, the resulting system is called a **voice synthesizer.** A microprocessor can be used in this application, as shown in Figure 6–70.

Voice Synthesizer:
A system for replicating the human voice by creating waveforms of the required form that emulate human sounds.

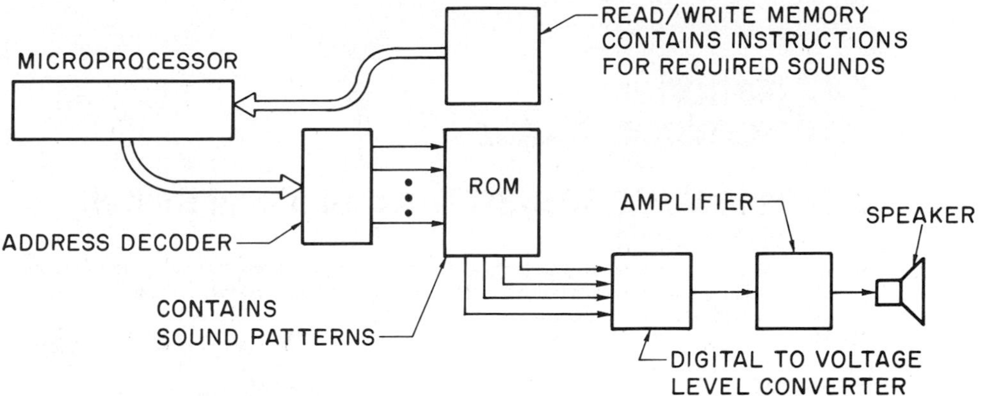

Figure 6–70 Fundamental voice synthesizer process.

Example 6–13

From the programmed ROM of Figure 6–71, construct the resulting output waveform.

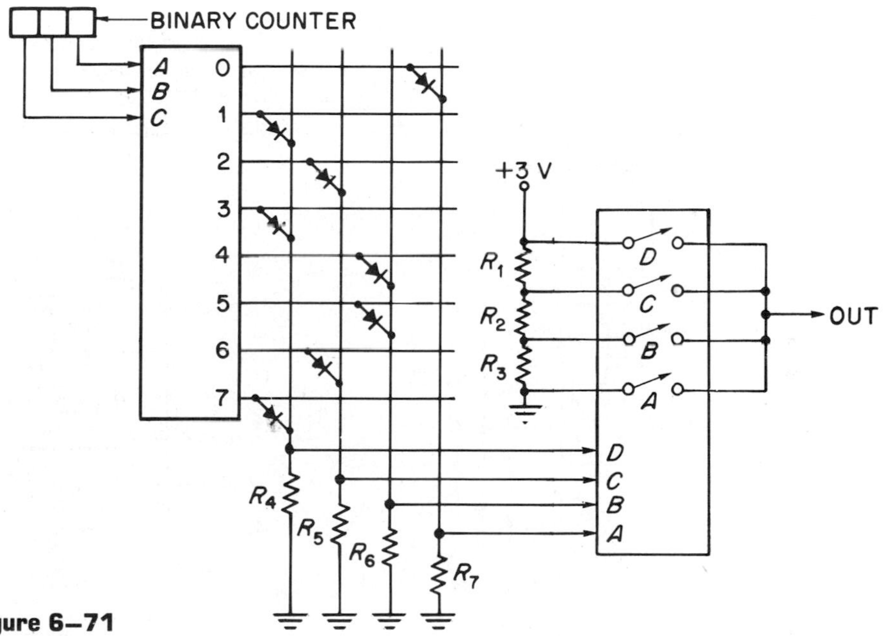

Figure 6–71

Solution

Refer to Figure 6–72.

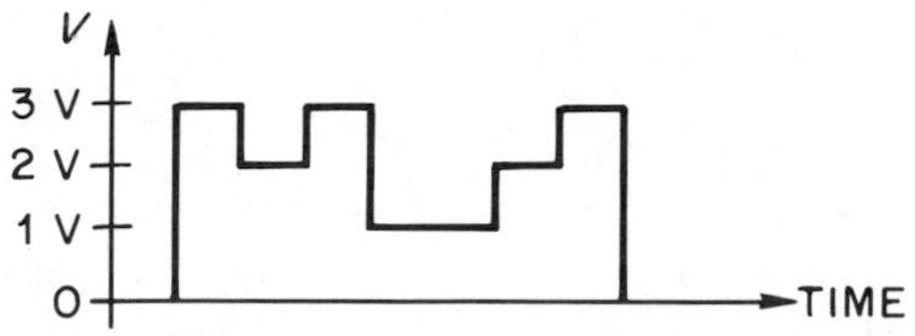

Figure 6–72

CD4016BM/CD4016BC Quad Bilateral Switch

General Description

The CD4016BM/CD4016BC is a quad bilateral switch intended for the transmission or multiplexing of analog or digital signals. It is pin-for-pin compatible with CD4066BM/CD4066BC.

Features

- Wide supply voltage range 3V to 15V
- Wide range of digital and analog switching $\pm 7.5\ V_{PEAK}$
- "ON" resistance for 15V operation 400Ω (typ.)
- Matched "ON" resistance over 15V signal input $\Delta R_{ON} = 10\Omega$ (typ.)
- High degree of linearity 0.4% distortion (typ.) @ $f_{IS} = 1$ kHz, $V_{IS} = 5\ V_{p\text{-}p}$, $V_{DD} - V_{SS} = 10V$, $R_L = 10$ kΩ
- Extremely low "OFF" switch leakage 0.1 nA (typ.) @ $V_{DD} - V_{SS} = 10V$ $T_A = 25°C$
- Extremely high control input impedance $10^{12}\Omega$ (typ.)
- Low crosstalk between switches −50 dB (typ.) @ $f_{IS} = 0.9$ MHz, $R_L = 1$ kΩ
- Frequency response, switch "ON" 40 MHz (typ.)

Applications

- Analog signal switching/multiplexing
  - Signal gating
  - Squelch control
  - Chopper
  - Modulator/Demodulator
  - Commutating switch
- Digital signal switching/multiplexing
- CMOS logic implementation
- Analog-to-digital/digital-to-analog conversion
- Digital control of frequency, impedance, phase, and analog-signal gain

4 Input Multiplexer

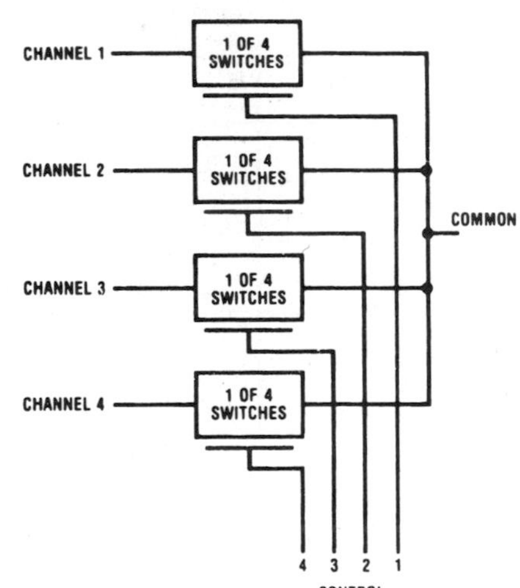

Dual-In-Line Package

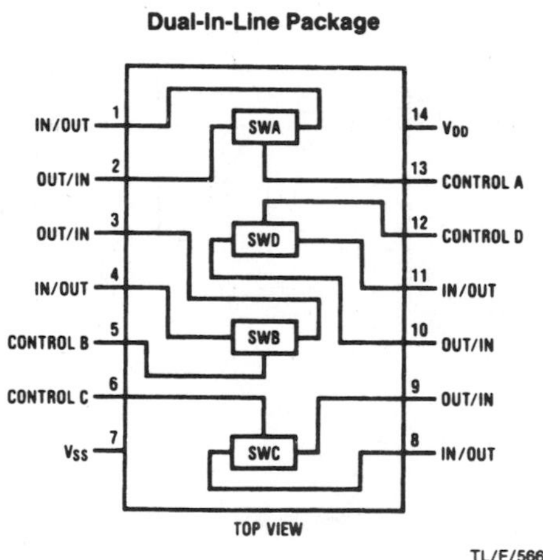

TL/F/5661-1

Figure 6–73 A real CMOS bilateral switch. *Reprinted with permission of National Semiconductor Corporation.*

The Real Thing

Figure 6–73 shows an actual data sheet for a CMOS bilateral switch. As listed in the figure under features, the switches are not perfect switches; they do exhibit some resistance ($\approx$400 Ω).

Conclusion

This section presented the versatility of the CMOS bilateral switch. Here, you were introduced to a very useful and powerful application—the synthesis of sound. In the next section, you will learn more about the commonalities of all microprocessors. Check your understanding of this section by trying the following review questions.

6–8 Review Questions

1. What is a *bilateral switch?*
2. Describe the basic operation of a CMOS bilateral switch.
3. State some of the applications of a CMOS bilateral switch.
4. What is a *sound synthesizer*? a *voice synthesizer*?

MICROPROCESSOR APPLICATION

Discussion

The hypothetical 4-bit microprocessor you have been working with up to this point will again be used as a model to present the important features that are common to all microprocessors. You are now at the point where you can see how this versatile device actually works with memory.

Another Register

The 4-bit microprocessor used up to this point has an *accumulator, B register,* and *data register*. These internal registers, or registers similar to them, are common to all microprocessors. Another register that is common to all microprocessors is the *binary counter*. Recall that this digital device was introduced in the last section. The binary counter for the 4-bit microprocessor contains 4 bits. It is designed in such a fashion that it will start at the count of 0000_2 when the control circuit of the microprocessor is first turned ON and will continue its count forward, 1 bit at a time, to 1111_2. This sequence is illustrated in Figure 6–74.

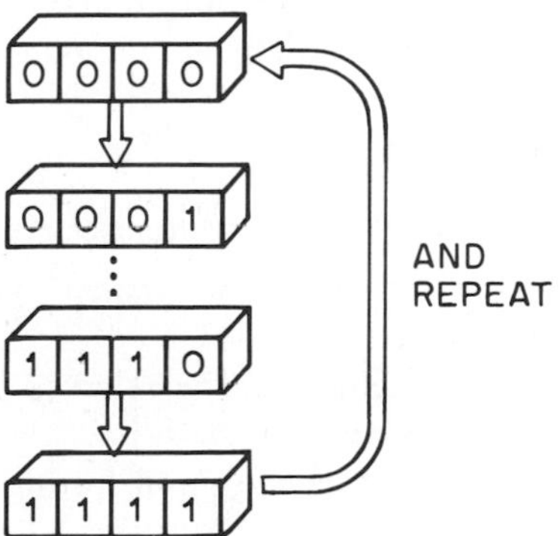

Figure 6–74 Four-bit binary counter sequence.

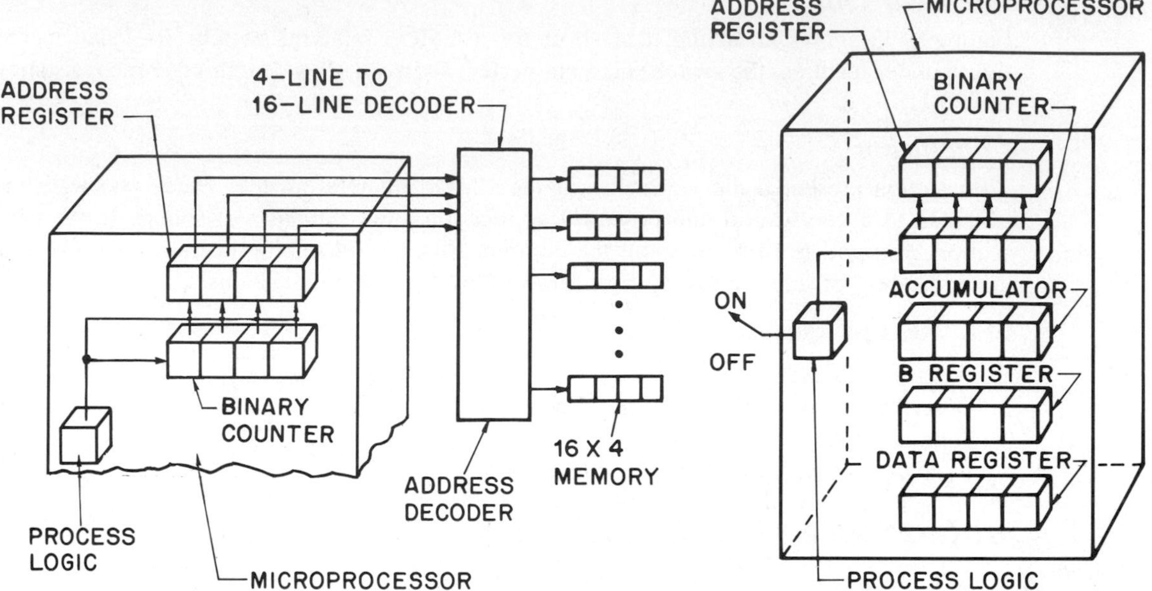

Figure 6–75 Addition of 4-bit binary counter to microprocessor.

The addition of the binary counter to the 4-bit microprocessor is shown in Figure 6–75. As you can see, the binary counter for the 4-bit microprocessor is controlled by the process logic. When the microprocessor is first turned on, the process logic causes the counter to start at 0000_2. Circuitry inside the process logic then causes the counter to start counting.

The process register causes the contents of the binary counter to be transferred to the **address register.**

Address Register:
A register inside the microprocessor that stores the binary number that represents the location (address) of data being processed by the microprocessor.

Notice from Figure 6–75 that the output of the address register goes to a 4-line to 16-line decoder. The output of the decoder in turn goes to a 16 × 4 switch-programmable ROM. Thus, the value of the binary number in the *address register* will determine the activated row in the ROM.

Microprocessor Sequence

The sequence used by the 4-bit microprocessor is as follows:

1. Transfer the contents of the binary counter into the address register.
2. Increment the binary counter.
3. Copy the contents of memory at the selected address into the data register.
4. Perform the required process.
5. Start the procedure over again by transferring the contents of the binary counter into the address register.

This process is shown in Figure 6–76.

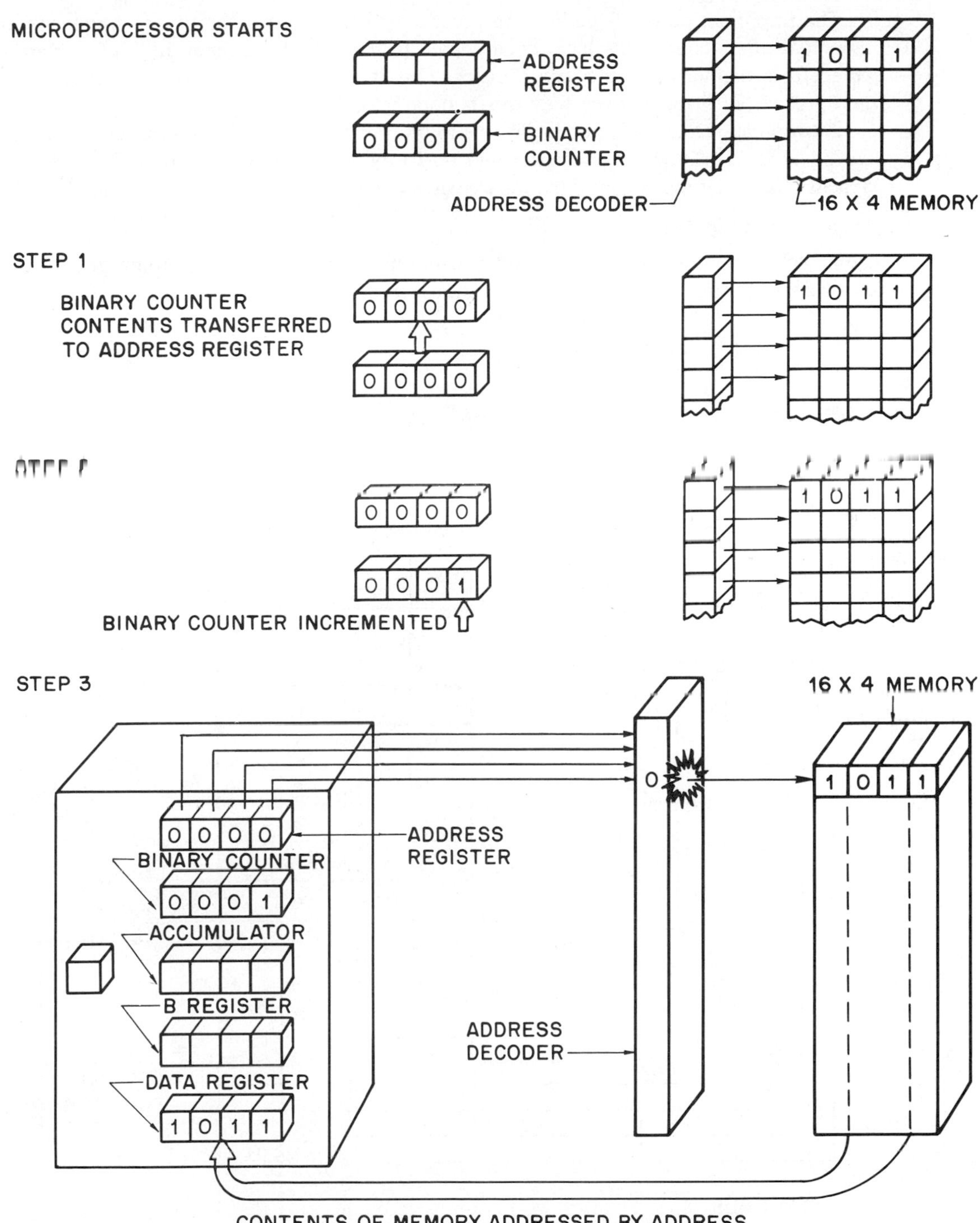

Figure 6–76 Microprocessor sequence.

The important point of this microprocessor sequence is that the binary counter is incremented *after* its contents have been copied to the address register and *before* data is copied from memory into the data register. This is an important step and is common to microprocessors. You will see why this is done later.

As you can see from Figure 6–76, the microprocessor is sequenced through memory, starting at the first memory location and on to the last memory location. This process happens because of the action of the binary counter along with the 4-line to 16-line decoder.

Programmable Memory

The switch-programmable ROM introduced in this chapter can be used to program the 4-bit microprocessor. This arrangement is shown in Figure 6–77.

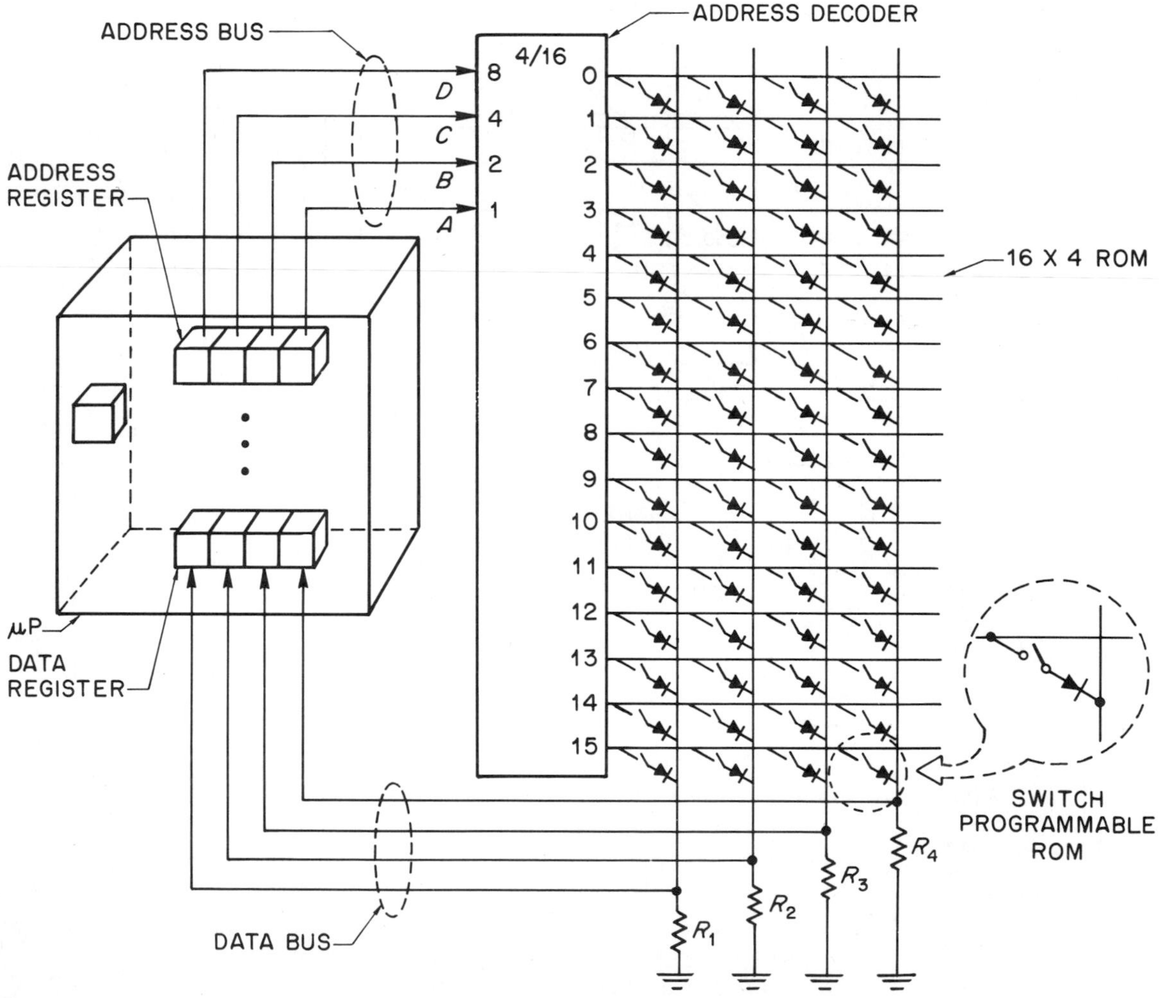

Figure 6–77 Relationship of ROM and 4-bit microprocessor.

As you can see, it is now easy to program the microprocessor. You only need to set the bit pattern in ROM corresponding to each desired instruction. For convenience, an abbreviated table for the 4-bit microprocessor instruction set follows.

4-BIT MICROPROCESSOR INSTRUCTION SET	
Mnemonic	**Hex Code**
NOP	0
LDA	1
LDB	2
TBA	3
TAB	4
NTA	5
ANA	6
ORA	7
XOR	8
TAM	9
TBM	A
INR	B
DCR	C
ADD	D
CLA	E
HLT	F

Because the memory that is being used is ROM, the microprocessor can only copy bit patterns *from* the memory.

Example 6–14

Indicate the bit pattern you would place into the programmable ROM in order to implement the following assembly language program:

Address	**Instruction**
0	LDA 5 ; Load accumulator with 5.
2	LDB 3 ; Load B reg with 3.
4	ADD ; Add 3 to 5.
5	TAB ; Copy answer to B reg.
6	HLT ; Stop processing.

Solution

Refer to the instruction set and convert the hex code into binary. Then set the binary bit pattern for each of the indicated addresses. The result is shown in Figure 6–78.

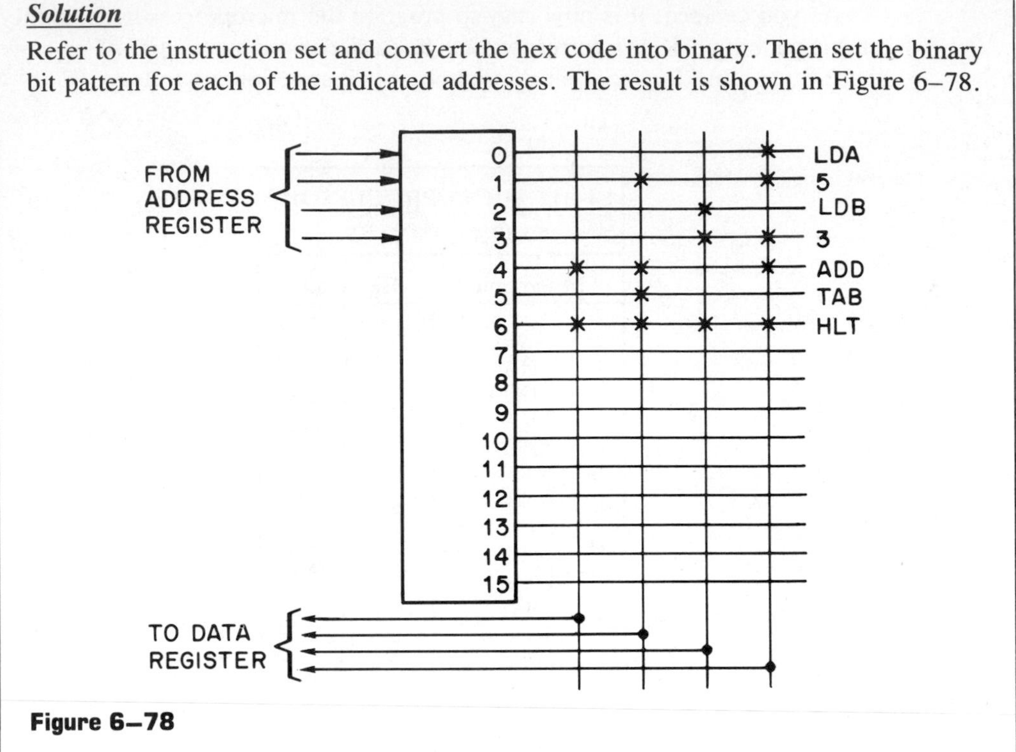

Figure 6–78

The Buses

The 4-bit microprocessor has four wires coming from its address register to the 4-line to 16-line decoder. Recall that a *bus* is a group of conductors treated as a unit. For connections from the address register to the 4-line to 16-line decoder, these conductors are called the **address bus.**

Address Bus:
The group of conductors from a microprocessor that carries the information for the activation of a specific memory or device location.

In like manner, the connections between the *data register* and memory are called the **data bus.**

Data Bus:
The group of conductors from a microprocessor that will carry the data transferred to and from memory or a device.

All microprocessors contain an address bus and a data bus, and all address buses are used to select a specific memory location or device, whereas all data buses are used to transfer information between the microprocessor and the selected memory location.

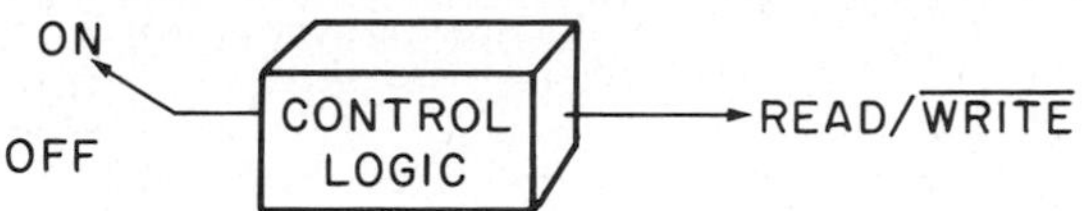

Figure 6–79 READ/WRITE line from control logic.

Outputting

You saw that by using a switch-programmable ROM, you could now actually enter a program for the 4-bit microprocessor. This arrangement provides a method for getting information *into* the system; however, there are no provisions for getting information *out of* the system (such as the answer to an addition problem).

To do this requires a visit to the *control logic* of the hypothetical 4-bit microprocessor. Here, another important control line, other than the ON/OFF control, will be added. Look at Figure 6–79.

Here is the idea behind the READ/$\overline{\text{WRITE}}$ line. The digital circuitry inside the control logic is such that when the microprocessor is copying information *from* memory into the data register, the READ/$\overline{\text{WRITE}}$ line will be HIGH. When the microprocessor is having information copied *from* the data register *into* memory, the READ/$\overline{\text{WRITE}}$ line will be LOW. This important relationship is shown in the following table.

READ/$\overline{\text{WRITE}}$	Process
HIGH (1)	Copy from memory
LOW (0)	Copy to memory

Now you can see the reason for the bar over the word $\overline{\text{WRITE}}$ on the READ/$\overline{\text{WRITE}}$ memory line. It indicates that when this line is LOW, the microprocessor is in the process of having data copied from its data register into memory.

Implementing Output

Figure 6–80 shows how the 4-bit microprocessor uses its READ/$\overline{\text{WRITE}}$ line to activate either the ROM or the seven-segment decoder. You can think of the seven-segment decoder as another memory location. The difference is that this "memory location" will only be activated when the READ/$\overline{\text{WRITE}}$ line from the microprocessor is LOW. Now whatever appears on the *data bus* will activate the seven-segment decoder and cause the seven-segment readout to display the hex value of the contents of the data register. The reason for this is that when the READ/$\overline{\text{WRITE}}$ line is LOW, the enable input of the PAL will be active, and no matter what is on the data bus at this time its hex value will be displayed by the seven-segment readout. The two instructions that cause the READ/$\overline{\text{WRITE}}$ line to go LOW are TAM (transfer the accumulator contents to memory) and TBM (transfer the B register contents to memory).

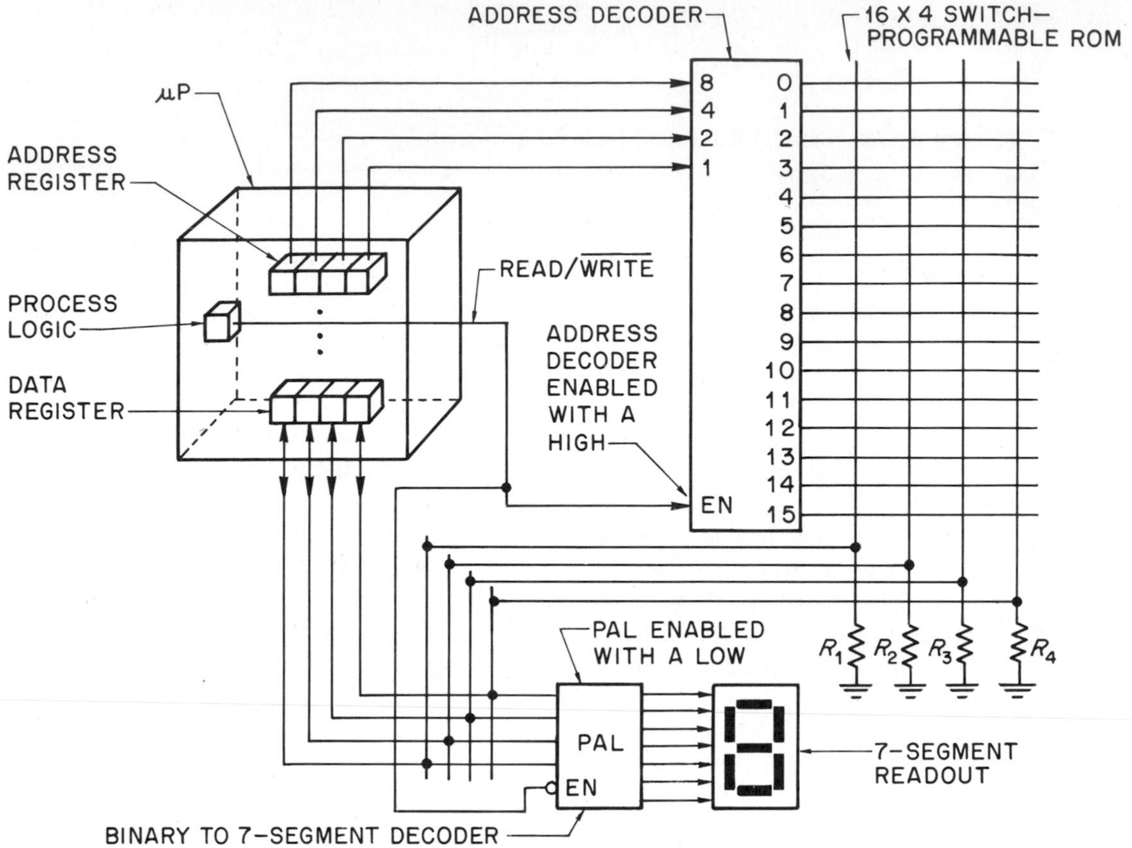

Figure 6–80 Adding an output.

Now the 4-bit microprocessor contains two control lines: the ON/OFF line and the READ/$\overline{\text{WRITE}}$ line. These two lines form a third bus called the **control bus.**

Control Bus:
A group of conductors from the microprocessor that carries information for or about the operation of the microprocessor.

Example 6–15

Write an assembly language program that will add the numbers 2 and 3, then display the results on the seven-segment readout. For each process, indicate the condition of the READ/$\overline{\text{WRITE}}$ line.

Solution

First, write the program:

Address	Instruction	
0	LDA 2	; Load accumulator with 2.
2	LDB 3	; Load B reg with 3.
4	ADD	; Add 2 + 3.
5	TAM	; Copy accumulator to memory.
6	HLT	; Stop processing.

When this program is processed, the action of the READ/WRITE line will be

READ/WRITE		Instruction/Data	
Read instruction	→ HIGH	LDA	← Instruction
Read data	→ HIGH	2	← Data
Read instruction	→ HIGH	LDB	← Instruction
Read data	→ HIGH	3	← Data
Read instruction	→ HIGH	ADD	← Instruction
Read instruction	→ HIGH	TAM	← Instruction
Write data	→ LOW	5	← Data (answer)
Read instruction	→ HIGH	HLT	← Instruction

The previous example shows that the READ/$\overline{\text{WRITE}}$ line will be LOW when the microprocessor is in the process of having the contents of the data register copied into memory. Recall that the instruction TAM actually copies the contents of the accumulator into the data register (the addition process caused the result to be in the accumulator). Once the data register has the answer, the TAM instruction causes the READ/$\overline{\text{WRITE}}$ line to go LOW at the same time the bit pattern of the data register is on the data bus. Thus, the seven-segment decoder becomes activated and displays the hex value of the answer.

Bus Oriented

All microprocessors use a bus-oriented system. The different buses along with the microprocessor power connections are illustrated in Figure 6–81. The word *microprocessor* is frequently abbreviated to μP (μ = the Greek letter mu, which, in metric notation, stands for micro). Table 6–3 summarizes the purpose of each bus.

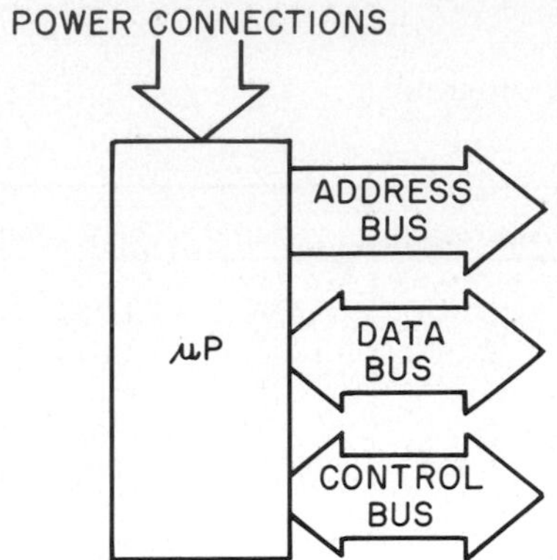

Figure 6–81 Microprocessor bus system.

Table 6–3 μP BUSES

Bus	Purpose
Power bus	Supplies the necessary voltages to power the microprocessor
Address bus	A one-way bus that contains the binary value of the address being used by the microprocessor
Data bus	A bidirectional bus that transfers information to and from the microprocessor
Control bus	A bus containing input and output lines for transferring information concerning the operation of the microprocessor

Conclusion

This section presented the concepts of interfacing memory and a device with the hypothetical 4-bit μP. Here you saw the meaning of a *bus-oriented* system. You also saw the operation and purpose of the READ/$\overline{\text{WRITE}}$ line. What you have learned here will give you the foundation needed for today's microprocessor technology. Test your understanding of this section by trying the following review questions.

6–9 Review Questions

1. What is a *bus* (as used in microprocessor terminology)?
2. Describe the purpose of an *address bus*.
3. Describe the purpose of a *data bus*.
4. State the purpose of a READ/$\overline{\text{WRITE}}$ line.
5. What is a *control bus?*

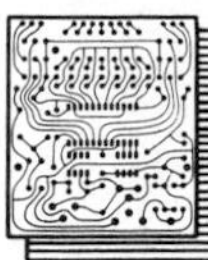

ELECTRONIC APPLICATION

Discussion

In this section, you will discover some important uses of resistors in digital systems. One use is to limit the amount of current in a circuit to a safe amount.

A Switch Circuit

Consider the SPDT switch of Figure 6–82. Observe from the figure that when the position of the switch is changed from LOW to HIGH, there is a time in the transition when the output from the switch is an open. For TTL logic, during the time that the switch is in transition, the open will appear as a HIGH (recall that for TTL, because of the construction of its internal circuitry, an open input appears as a HIGH). This action may not be desirable. In CMOS, an open input can cause the device to use an excessive amount of current from the power source, or even produce oscillations in the output.

A better arrangement exists for using a switch for controlling the inputs of logic levels. It uses the simplified kind of switch shown in Figure 6–83. Here, an SPST switch is used with a very serious problem. As you can see, connecting an SPST switch in this manner eliminates the problem of a momentary open. However, it creates a new problem of possibly blowing a fuse in the power source or causing a fire from the heat generated from excessive current flow. This arrangement can be used if a resistor is added to reduce the amount of current flow.

Adding a Resistor

The idea of an SPST switch to control logic levels is very appealing. Because of the simplicity of the switch, it can be made smaller and more economical, and it is more reliable.

If a resistor is used in conjunction with the switch, the current flow from the power source can now be kept to a save value. Such an arrangement is shown in Figure 6–84. When the switch is open, the output line has a voltage of +5 volts. When the switch is closed, the output voltage is now at ground. The resistor reduces the resulting amount of current from the +5-volt source to a safe value. If the value of the selected resistor were 1 000 ohms, then the resulting current flow would be only 0.005 ampere (5 milliamperes).

Using the resistor in this manner gives the ordinary resistor a special name in digital systems—a **pull-up resistor.**

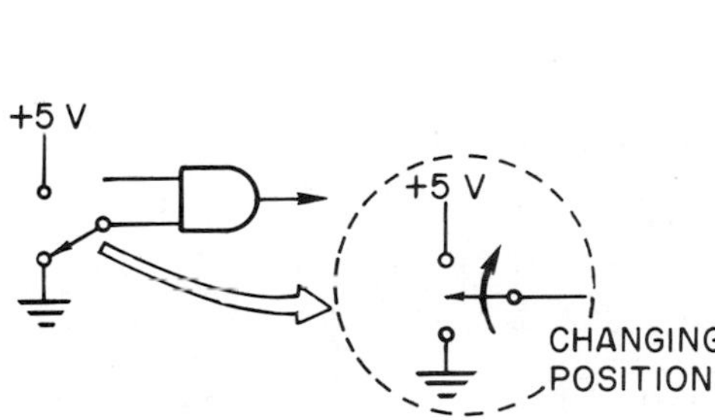

Figure 6–82 SPST switch connected to logic gate.

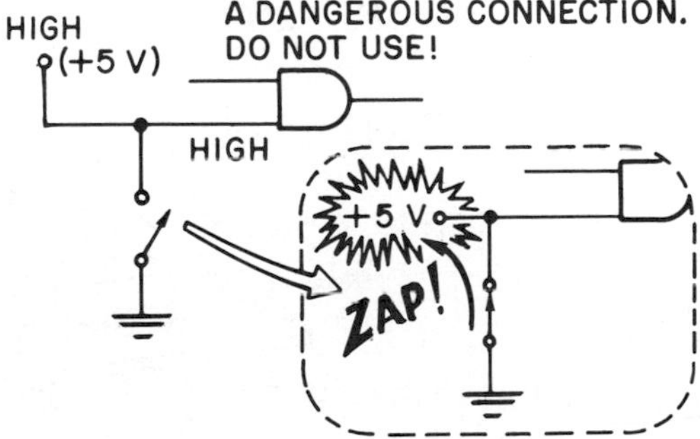

Figure 6–83 How *not* to use a SPST switch for a logic input.

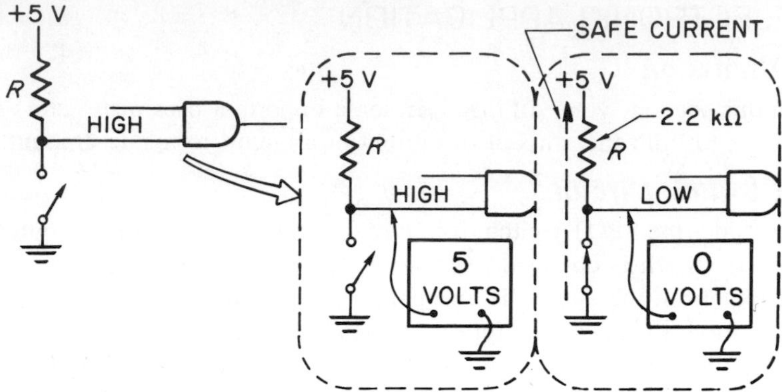

Figure 6–84 Correct use of SPST switch for logic input.

Pull-up Resistor:
In digital systems, a resistor connected to a voltage source in such a manner as to supply the voltage potential and also limit any resulting current flow.

Input Switches

This arrangement of using a pull-up resistor with a SPST switch opens even more possibilities. For example, now a simple PBNO or PBNC can be used to momentarily input a logic level. This arrangement is shown in Figure 6–85. When a PBNO is open, the output line will be HIGH and when it is depressed the output line will be LOW. This arrangement allows for the manufacture of printed circuit boards that make very inexpensive and reliable keypads. Such construction is shown in Figure 6–86.

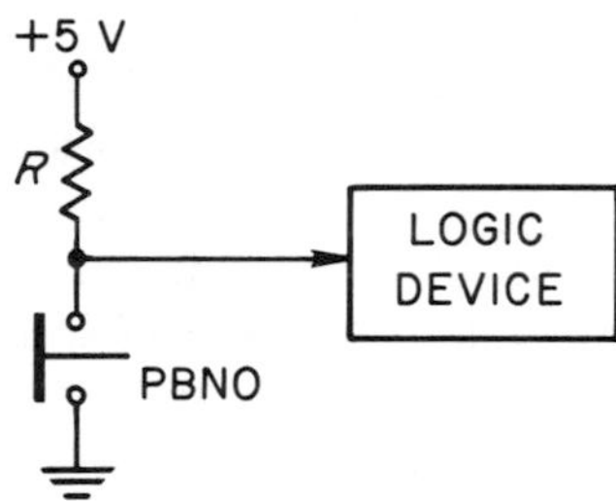

Figure 6–85 PBNO with pull-up resistor.

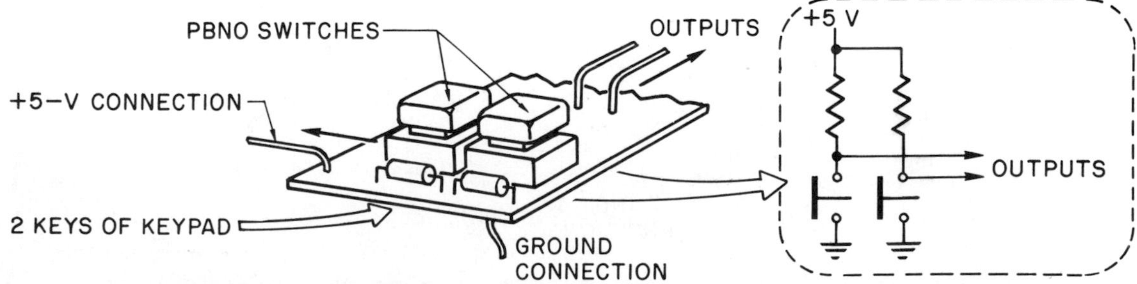

Figure 6–86 Printed circuit construction of keypad.

Now the input keys to a decoder need only be simple PBNO switches and the software within the programmable logic devices can be used to do the "complicated" things.

Open-Collector Outputs

Some logic devices come with what is called an **open-collector** output. This means that the internal circuit in the output of the logic device is not complete and requires the addition of a pull-up resistor.

> **Open-Collector Logic:**
> A logic device whose output requires the addition of a pull-up resistor for proper operation.

The advantage of such a system is that it allows the outputs of several logic circuits to be connected together with the pull-up resistor. This arrangement is shown in Figure 6–87.

Logic gates that do not have open-collector outputs must not have their outputs connected together. The reason is that the output lines from the gate originate from internal circuits that must have either a HIGH or a LOW output. Connecting these two outputs together can cause the chip to become permanently damaged from overheating.

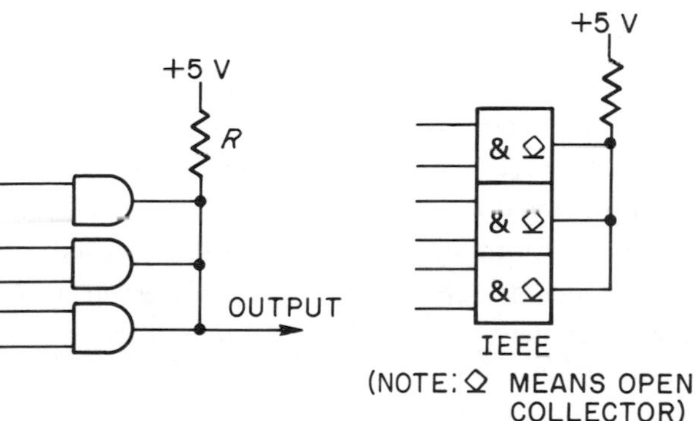

Figure 6–87 Connecting open-collector logic gates.

Conclusion

This section presented the use of resistors in digital circuits. Here you saw how to use a simple switch to control the input logic—this was made possible with the use of a resistor to limit current flow. The open-collector logic circuit was presented here along with one of the major reasons for its use.

Check your understanding of this section by trying the following review questions.

6–10 Review Questions

1. State one of the purposes of a resistor in logic circuits.
2. Explain why a resistor is needed when a PBNO switch is used to control a logic input.
3. What is a pull-up resistor?
4. Discuss the open-collector logic circuit. How can these be used together?

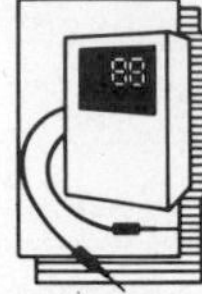

TROUBLESHOOTING SIMULATION

The troubleshooting simulation for this chapter simulates a programmable logic device. With this simulation program, you can practice your skills at programming these versatile devices. Then, once you have programmed the device, you may actually test it.

When you are ready, the computer will then put a troubleshooting problem in the PLD. It will then be up to you to test the PLD and determine the problem.

SUMMARY

- A programmable logic gate is a logic network that can be programmed to produce any standard logic.
- A programmable logic gate can be thought of as emulating a computer in that it contains all of the essential sections required by the definition of a computer.
- Hardware is the tangible part of a digital system.
- Software is the nontangible part of a digital system that determines the process performed by the hardware.
- A multiplexer/data selector can be used as a programmable logic gate.
- A multiplexer/data selector can be used to easily replicate a truth table.
- Read-only memory is memory that contains a set of permanent instructions that cannot be easily changed.
- Read-write memory is memory that can have its instructions easily changed.
- Another use of a multiplexer/data selector is to have one of several inputs outputted to a single output.
- More than one multiplexer can be used to expand the number of available inputs.
- A programmable logic array (PLA) consists of a two-level AND/OR logic array, where both the AND gate connections as well as the OR gate connections are programmable.
- A diode is an electrical device that causes current to flow in only one direction.
- A diode is used in programmable logic devices in order to prevent electrical shorts.
- A PLA can be programmed to simulate different logic functions, one for each output.
- One practical application of a PLA is to perform arithmetic operations on digital data.
- There is a logic notation used especially for programmable logic devices. The purpose of this notation is to make the programming of these devices easier.
- A decoder is a logic network where a specific bit pattern on its input will activate a single logic line on its output.
- The outputs of a decoder can be active HIGH or active LOW.
- Decoders can be combined to produce a larger number of decoded outputs.
- A read-only memory (ROM) is a special type of programmable logic device where the AND gate array is fixed and the OR gate array is programmable.
- An experimental ROM can be constructed in the laboratory with diodes and switches.
- A mask-programmable ROM is a ROM that is factory programmed.

- A PROM is a programmable ROM that can be programmed by the user in the lab.
- An EPROM is a programmable ROM that can be programmed as well as erased for a new program by the user.
- A demultiplexer can be thought of as the opposite of a multiplexer.
- A demultiplexer can take the logic level of one line on its input and route it to one of several lines on its output.
- An encoder is a logic network that converts information into a form of code.
- One common use of an encoder is to convert a keypad consisting of simple switches into a digital code.
- A bilateral switch is a switch that can be operated in any direction.
- A bilateral switch can be used as a digital to voltage level converter.
- A bilateral switch can be used to create synthesized waveforms.
- A binary counter is a digital device that is capable of producing a binary count.
- An address register is a register inside the microprocessor that stores the binary number that represents the address being used by the microprocessor.
- A microprocessor goes through a very ordered and predictable sequence of processes.
- All microprocessors use a bus-oriented system consisting of four buses: the power bus, the data bus, the address bus, and the control bus.
- A pull-up resistor in a digital system is used to supply the proper potential and limit the circuit current.
- Logic gates are available with open-collector outputs. These outputs must be connected to a supply voltage through a pull-up resistor and allow several outputs to be connected together.

CHAPTER SELF-TEST

I. TRUE/FALSE

Answer the following true or false.

1. A programmable logic gate can be programmed to simulate any standard gate.
2. Hardware can be thought of as the tangible parts and software as the bit patterns that control the routing of bits within the tangible parts.
3. A PLG cannot replicate a computer because it is too simple a logic network.

4. Read-write memory can easily have its instructions changed.
5. ROM is the kind of memory that cannot easily have its instructions changed.

II. MULTIPLE CHOICE

Answer the following questions by selecting the most correct answer.

6. A data selector has

   (A) several outputs and one input
   (B) several inputs and one output
   (C) several inputs and outputs
   (D) a variety of inputs and outputs

7. In a PLA the programmable part(s) is/are the
   (A) AND gate array
   (B) OR gate array
   (C) AND and OR gate array
   (D) none of the above
8. A diode is an electrical device that
   (A) can act as an electrical switch
   (B) allows voltage to go in only one direction
   (C) allows current to flow in only one direction
   (D) can do both (A) and (C)
9. In PLD notation, the product line is
   (A) the different PLDs stocked by the manufacturer
   (B) the line that takes the product of the input
   (C) the line that represents the programmable part
   (D) none of the above
10. A logic network that will convert a specific bit pattern on its input to a familiar form is called a/an
    (A) decoder
    (B) encoder
    (C) multiplexer
    (D) demultiplexer

III. MATCHING

Match the correct answers on the right to the statements on the left.

11. PAL	(A) AND gate array that is programmable
12. PLA	(B) OR gate array that is programmable
13. ROM	(C) Both AND and OR gate arrays that are programmable
14. PLD	(D) Can be programmed to represent any logic gate
15. PLG	(E) General notation used for any programmable logic device
	(F) None of these

IV. FILL-IN

Fill in the blanks with the most correct answer(s).

16. A ______ is a programmable ROM that can be programmed in the field.
17. An EPROM is an ______ programmable read-only memory.
18. A/An ______ is a logic network that converts information into a form of code.
19. A bilateral switch is a switch that can be operated in any ______.
20. Waveform ______ is a process of creating a desired waveform.

V. OPEN-ENDED

Answer the following questions as indicated.

21. Describe the bus system as used with microprocessors.
22. What is the relationship between the binary counter and the address register in the 4-bit microprocessor presented in this chapter?
23. When is the binary counter used in the 4-bit microprocessor incremented?
24. State the purpose of a pull-up resistor used with an SPST switch.
25. Describe what is meant by open-collector logic.

Answers to Chapter Self-Test

1] T 2] T 3] F 4] T 5] T 6] B 7] C 8] D 9] D
10] A 11] C 12] A 13] B 14] E 15] D 16] PROM
17] erasable 18] encoder 19] direction 20] synthesis
21] A microprocessor bus system consists of the power bus, address bus, data bus, and control bus.
22] The binary counter feeds the address register.
23] The binary counter is incremented before information is taken from memory.
24] The purpose of a pull-up resistor used with a SPST switch is to limit the amount of current flow.
25] Open-collector logic consists of logic devices that require a pull-up resistor for proper operation.

CHAPTER PROBLEMS

BASIC CONCEPTS

Section 6–1:

1. Describe the major parts of a computer.
2. Demonstrate how a PLG meets the criteria for being a computer.
3. Sketch a PLG and label its major sections.
4. Illustrate the difference between *hardware* and *software*.
5. Explain the difference between an instruction and a process.
6. Show how you would program a PLG to replicate the following:
   (A) AND gate (B) OR gate (C) XOR gate
7. Using a PLG, show how you would program it to replicate the following:
   (A) NAND gate (B) NOR gate (C) XNOR gate

Section 6–2:

8. Using the 74151A, show how you would implement the following truth tables:

(A)

A	*B*	*C*	*X*
0	0	0	1
0	0	1	1
0	1	0	1
0	1	1	0
1	0	0	0
1	0	1	1
1	1	0	1
1	1	1	1

(B)

A	*B*	*C*	*X*
0	0	0	0
0	0	1	0
0	1	0	0
0	1	1	1
1	0	0	1
1	0	1	1
1	1	0	0
1	1	1	0

9. Show how you would implement the following truth tables, using the 74151A:

(A)

A	*B*	*C*	*X*
0	0	0	1
0	0	1	0
0	1	0	0
0	1	1	0
1	0	0	0
1	0	1	1
1	1	0	0
1	1	1	1

(B)

A	*B*	*C*	*X*
0	0	0	1
0	0	1	1
0	1	0	0
0	1	1	0
1	0	0	1
1	0	1	1
1	1	0	0
1	1	1	0

10. Explain what is meant by a *data selector/multiplexer*.
11. Sketch a mechanical switch analogy of a data selector/multiplexer.
12. Illustrate how you would connect two 74150 multiplexers to create a 32-word input.
13. What is the difference between ROM and read-write memory?

Section 6–3:

14. Design a PLA that could be used to multiply two binary numbers ($0 \times 0 = 0$, $1 \times 0 = 0$, $0 \times 1 = 0$, $1 \times 1 = 1$).
15. Create a PLA whose output is double the value of the input. Do this for a 2-bit input number. This means there must be two inputs. Since the largest number on the output is 11_2 (doubled) $= 110_2$, there must be three outputs.
16. Design a PLA that will take a 3-bit binary number on the input and give you the two's complement on the output.
17. Construct a truth table for the PLA shown in Figure 6–88.

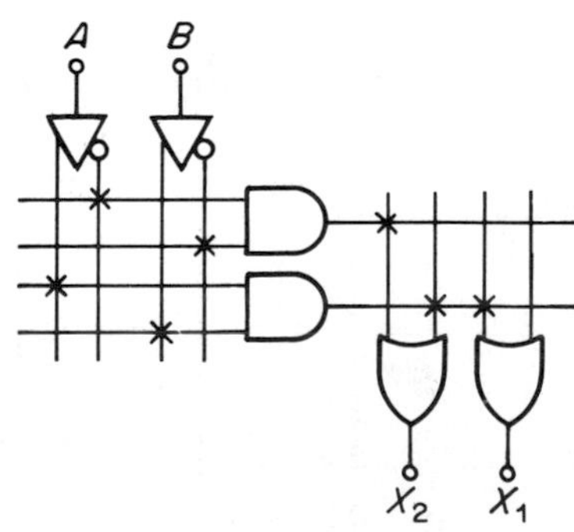

Figure 6–88

18. Construct the truth table for the PLA shown in Figure 6–89.

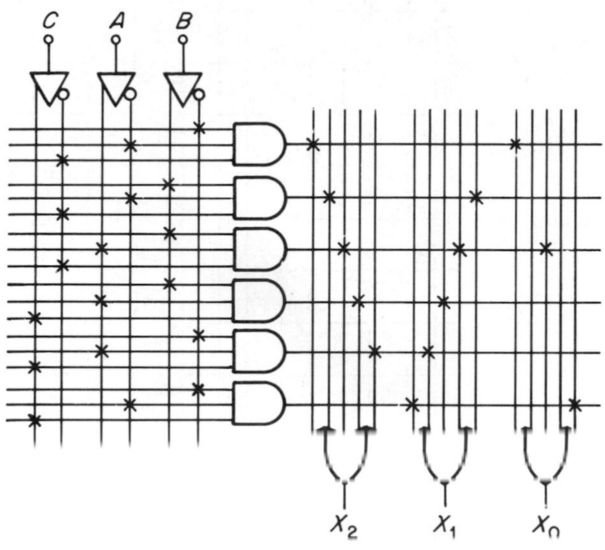

Figure 6–89

19. Assume that the waveforms indicated in Figure 6–90(A) are applied to the PLA of Figure 6–88. Sketch the resulting outputs.
20. The waveforms of Figure 6–90(B) are applied to the inputs of the PLA of Figure 6–89. Sketch the resulting output waveforms.

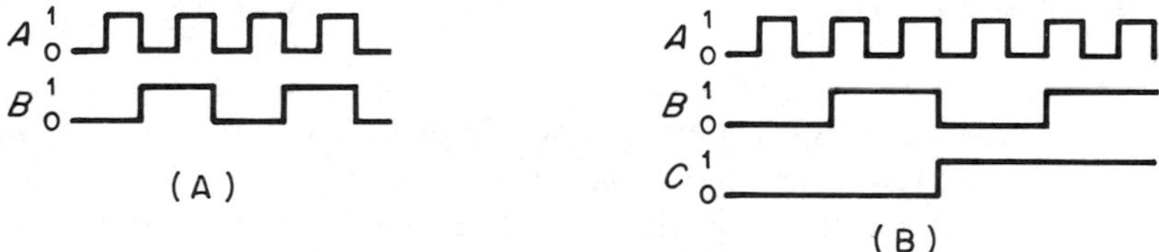

Figure 6–90

Section 6–4:

21. Using a 4-line input, 7-line output PLD, indicate how you would program it to replicate a gray code to seven-segment hex readout.
22. Using the PLD of Figure 6–91, illustrate how you would program it in order to get the two's complement of the input number.

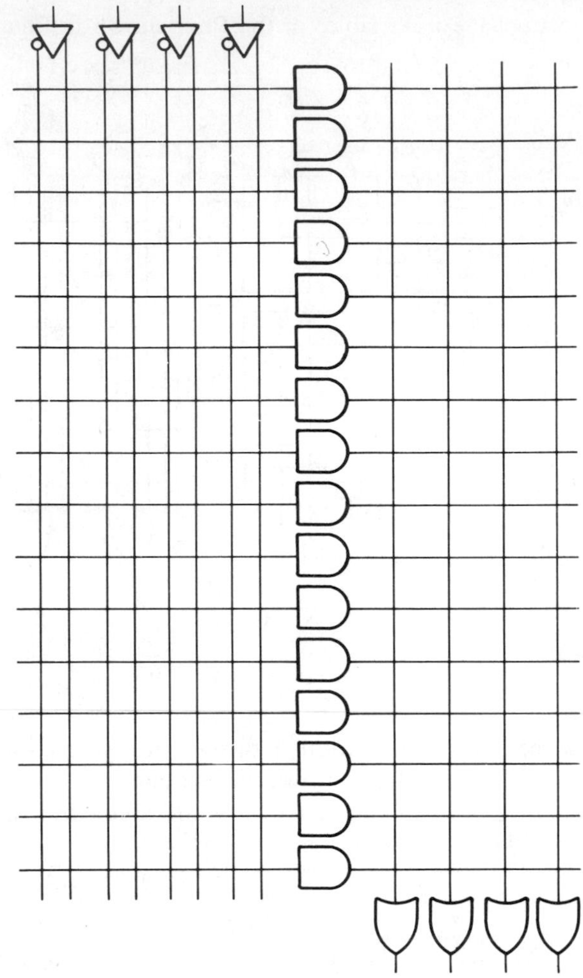

Figure 6–91

Section 6–5:

23. Indicate the bit pattern that would cause the output of a 4-line to 16-line decoder to activate every odd output line in sequence. Consider output line 1 to be an odd output and thus start the control sequence with 0001.
24. Do the same thing as required in question 23, but this time for all the even output lines.
25. Assume that the bit pattern of Figure 6–92(A) is applied to the input of a 4-line to 16-line decoder (with active LOW outputs). Indicate the output line conditions for D_1, D_5, D_{10}, and D_{15}.

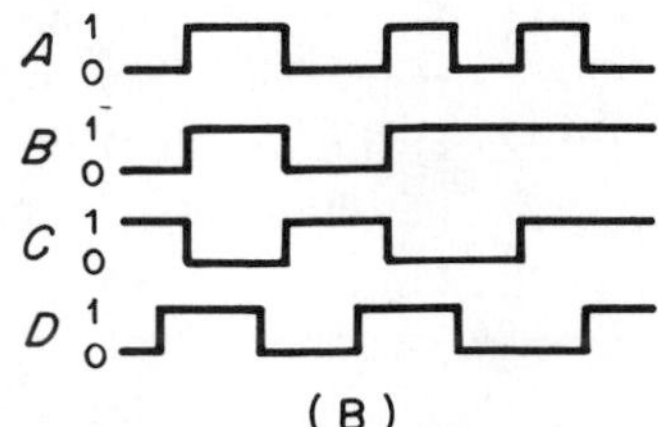

Figure 6–92 (A) (B)

26. For the bit pattern shown in Figure 6–92(B), indicate the condition of D_3, D_4, D_{11}, and D_{12} for a 4-line to 16-line decoder where each output line is active HIGH.

Section 6–6:

27. Explain why diodes are used in read-only memories.
28. What are the unique features of read-only memories when compared to other programmable logic devices?
29. Illustrate how you would construct a switch-programmable ROM that would multiply 2 binary bits.
30. Sketch a switch-programmable ROM that would (A) add two binary numbers of 2 bits each, (B) multiply two binary numbers of 2 bits each.

Section 6–7:

31. Explain the difference between a *demultiplexer* and an *encoder*.
32. Sketch the output conditions for each line of 1- to 4-line demultiplexer for the input waveforms shown in Figure 6–93(A).
33. For the input waveforms shown in Figure 6–93(B), determine the output conditions for a 1 line to 4 line demultiplexer.

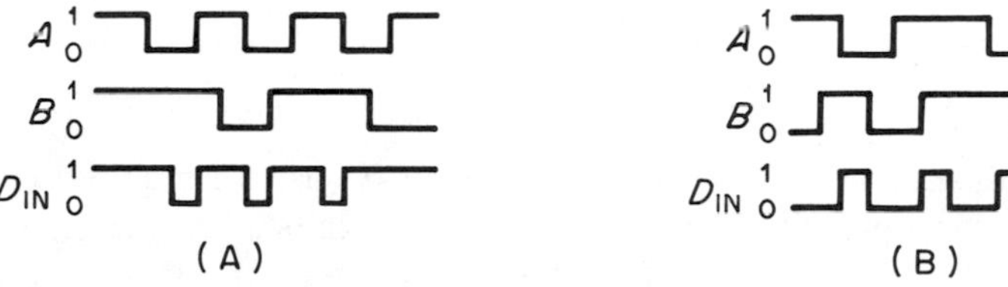

Figure 6–93

34. Sketch the output conditions for each line of a 4-line to 2-line encoder for the waveforms shown in Figure 6–94(A).

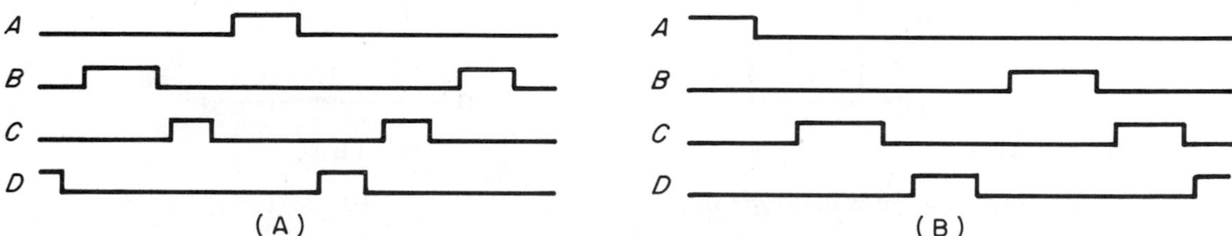

Figure 6–94

35. For the input waveforms shown in Figure 6–94(B), sketch the output conditions of a 4-line to 2-line encoder.
36. Design an encoder that would convert a decimal input to BCD.
37. Design an encoder that would convert an octal keypad to binary.
38. Design a 16-key encoder that converts each input to BCD.

Section 6–8:

39. Show how you would connect a bilateral switch to behave as the following gates: (A) OR (B) AND. Note: Do not use any other gates to accomplish this.

40. For the bilateral switch arrangement shown in Figure 6–95(A), determine the resulting waveform.
41. Sketch the resulting waveform for the bilateral switch for Figure 6–95(B).

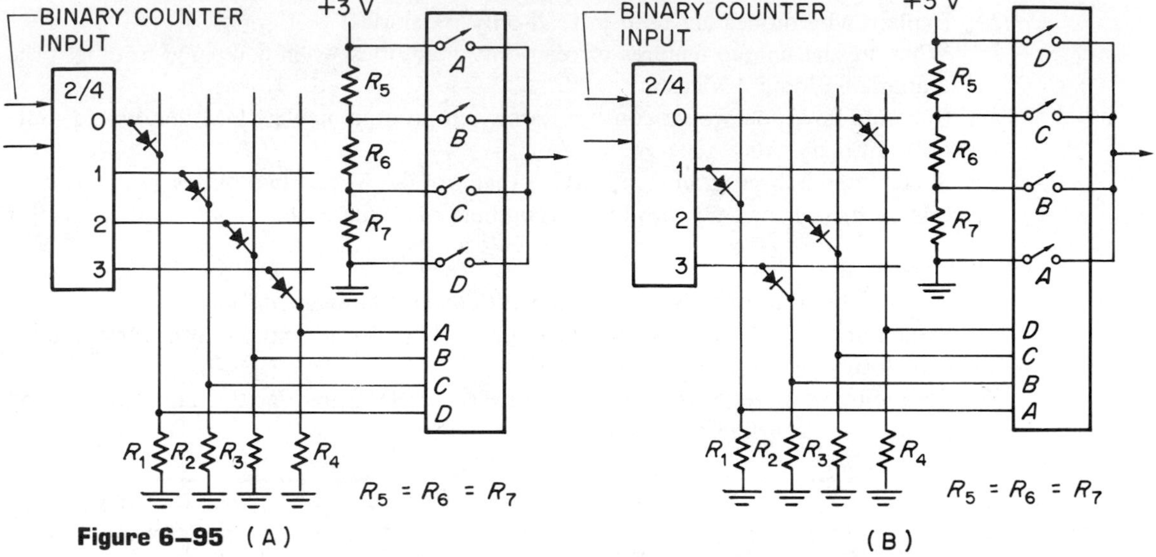

Figure 6–95 (A) (B)

42. Indicate the switch-programmable ROM pattern required of the circuits in Figure 6–95 in order to achieve the waveform indicated in Figure 6–96(A).
43. For the waveform of Figure 6–96(B), draw the ROM switch pattern required to produce it, using the logic circuits of Figure 6–95.

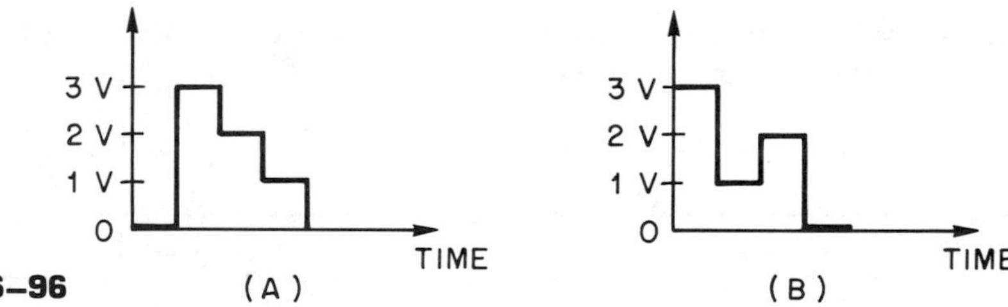

Figure 6–96 (A) (B)

Microprocessor Applications

44. State the sequence a microprocessor uses in relationship to ROM.
45. Sketch a 4-bit microprocessor and indicate its four different buses. Describe the purpose of each of the buses.
46. An 8-bit microprocessor has 8 data lines and 16 address lines. How many memory locations can it address? How many different instructions can it have?
47. Using the instruction set given in this chapter for the 4-bit microprocessor, convert the ROM bit pattern of Figure 6–97(A) to assembly language.
48. Convert the ROM bit pattern of Figure 6–97(B) to an assembly language program, using the instruction set for the 4-bit microprocessor used in this chapter.
49. Using the microprocessor circuit of Figure 6–80, write the assembly language program you would use to cause the seven-segment readout to sequence through the letters DAFFIE.

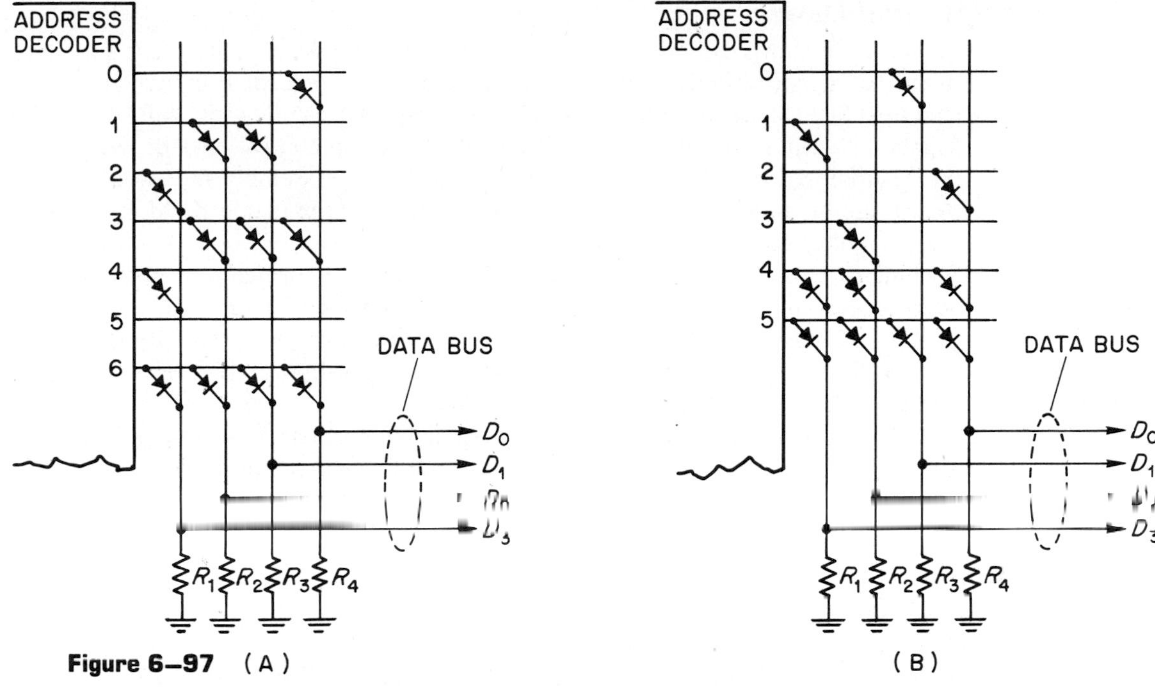

Figure 6–97 (A) (B)

50. Indicate the condition of the READ/$\overline{\text{WRITE}}$ line for each process from the program developed in problem 49.

Electronic Applications

51. Referring to Figure 6–98(A), indicate if there is a problem with any of the switch circuits. If so, state what the problem is.
52. Indicate if there is a problem with any of the logic circuits of Figure 6–98(B). State what the problem is.

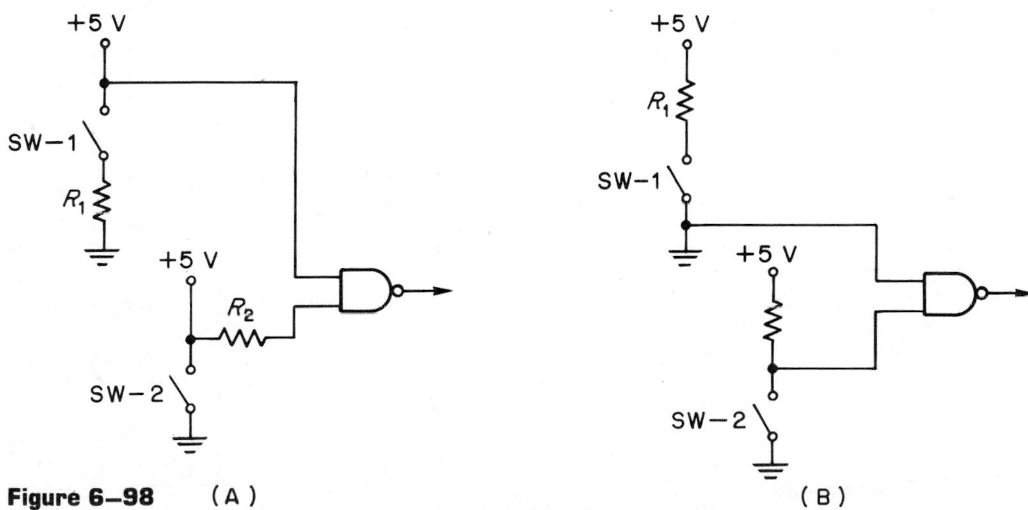

Figure 6–98 (A) (B)

Analysis and Design

53. Figure 6–99 shows an industrial controller. Create an assembly language program in the 16 memory locations of the 4-bit microprocessor presented in this chapter to start up a factory. Fit as much of the following sequence as you can. (A BIT tricky). Lights on—Alarms disarmed—Vault unlocked—Temp control on—Blower fans on—Shipping doors unlocked—Assembly line 1 on—Assembly line 2 on
54. Sketch the bit pattern you would put in the switch-programmable ROM to perform the operations required in problem 53.
55. Assume that you have access to a new 4-bit microprocessor that now contains five address lines instead of four. This means that it is capable of addressing a maximum of $2^5 = 32$ memory locations. Design an industrial control system similar to the one shown in Figure 6–99 that would now use two 4-line to 16-line decoders to operate a 32 × 4 switch-programmable ROM.
56. For the industrial controller system you designed in problem 55, write an assembly language program that would do all the operations required in problem 53.

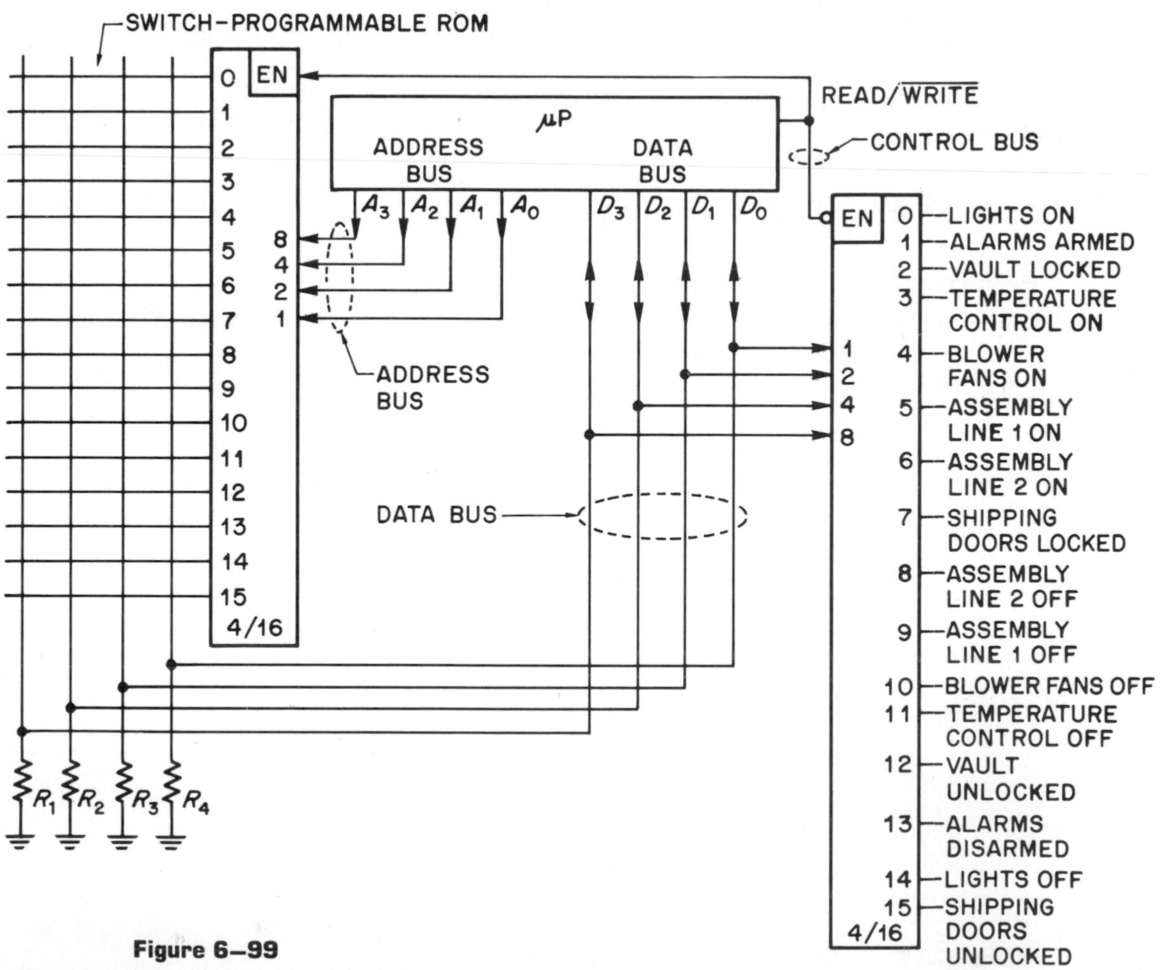

Figure 6–99

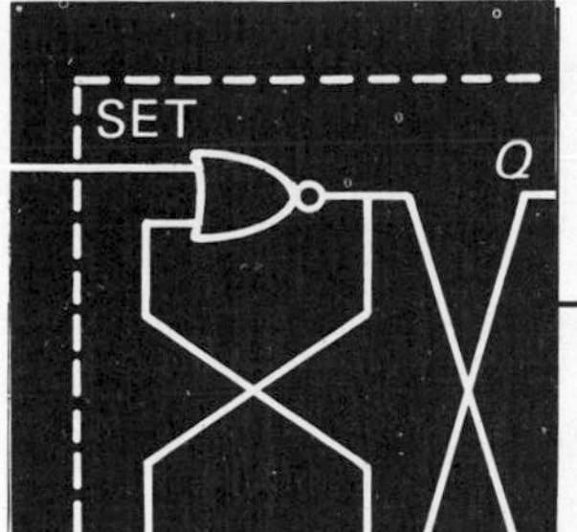

CHAPTER 7

Introduction to Latches and Flip–Flops

OBJECTIVES

After completing this chapter, you should be able to:

- ☐ Describe the concept of a storage element
- ☐ Know the logical construction of a latch.
- ☐ Present some applications for a latch.
- ☐ Understand the similarities and differences between different types of latches.
- ☐ Explain the operation of a gated latch.
- ☐ Realize the difference between latches and flip-flops.
- ☐ Predict the output of flip-flops for a given input.
- ☐ State the difference between the synchronous and asynchronous inputs of a flip-flop.
- ☐ Predict the output of flip-flops for given synchronous and asynchronous input waveforms.
- ☐ Discuss several applications for flip-flops.
- ☐ Recognize the different types of multivibrators.
- ☐ Describe the operation of a monostable multivibrator.
- ☐ Describe the operation of an astable multivibrator.
- ☐ Understand the operation and application of a TRI-STATE buffer.
- ☐ Know the logical construction of a simple industrial controller.
- ☐ Develop and interpret assembly language programs for a simple industrial controller.
- ☐ Explain how to make a clock generator from a popular IC.
- ☐ Explain the operation of a practical one-shot.
- ☐ Demonstrate how a Schmitt trigger operates and its application.

INTRODUCTION

This chapter presents the concept of a storage element. Here you will again see the versatile NAND and NOR gates used in yet another powerful application–the ability to remember a prior condition.

As you will see in this chapter, these digital storage elements may be constructed in many different ways, producing a rich variety of useful applications. Here you will meet the digital circuits which produce single pulses and trains of pulses. You will also see some of the fundamental secrets of automated systems and how to control them through hardware and programming.

The material here is important and is the backbone of the digital systems which produced the miracle of the digital revolution. Learn this material well. The next two chapters depend on it and so does the rest of your digital career.

KEY TERMS

Latch
Gated Latch
Gated *D* Latch
Flip-Flop
Clock
Dynamic Input Indicator
Multivibrator
Monostable Multivibrator
Trigger
Debouncer
Astable Multivibrator
Symmetry
Sync Pulse
Pulse-Triggered Master-Slave Flip-Flop
Data Lock-Out Flip-Flops
TRI-STATE
Addressing Modes
Immediate Addressing
Direct Addressing
Input/Output (I/O)
Robotics
Comparator
Saturation
Schmitt Trigger

7–1 STORAGE ELEMENTS

Discussion

This section presents a very simple idea and then demonstrates what can be done with it. You will see that the addition of this important device will complete all of the major building blocks required by digital systems.

A Storage Element

Recall the cups and straws used in Chapter 2 to illustrate binary numbers. These could be considered as storage elements that were either full or empty. Because each cup was only allowed two conditions (and nothing in between), it served well to illustrate the concept of binary numbers. There is a similar device in digital circuits called a **latch.**

Latch:
A logic circuit that maintains a given logic condition until changed by an external source.

Figure 7–1 Basic idea of a latch.

The basic idea of a latch is shown in Figure 7–1. Note from the figure that there are two inputs to the latch. One input is called the SET and the other is called the RESET. As shown in the figure, when the set is activated, the latch will store a 1 (HIGH). When the RESET is activated, the latch will store a 0 (LOW). The important thing to note is that the latch will remain in its previous condition *until* changed by the input lines. This

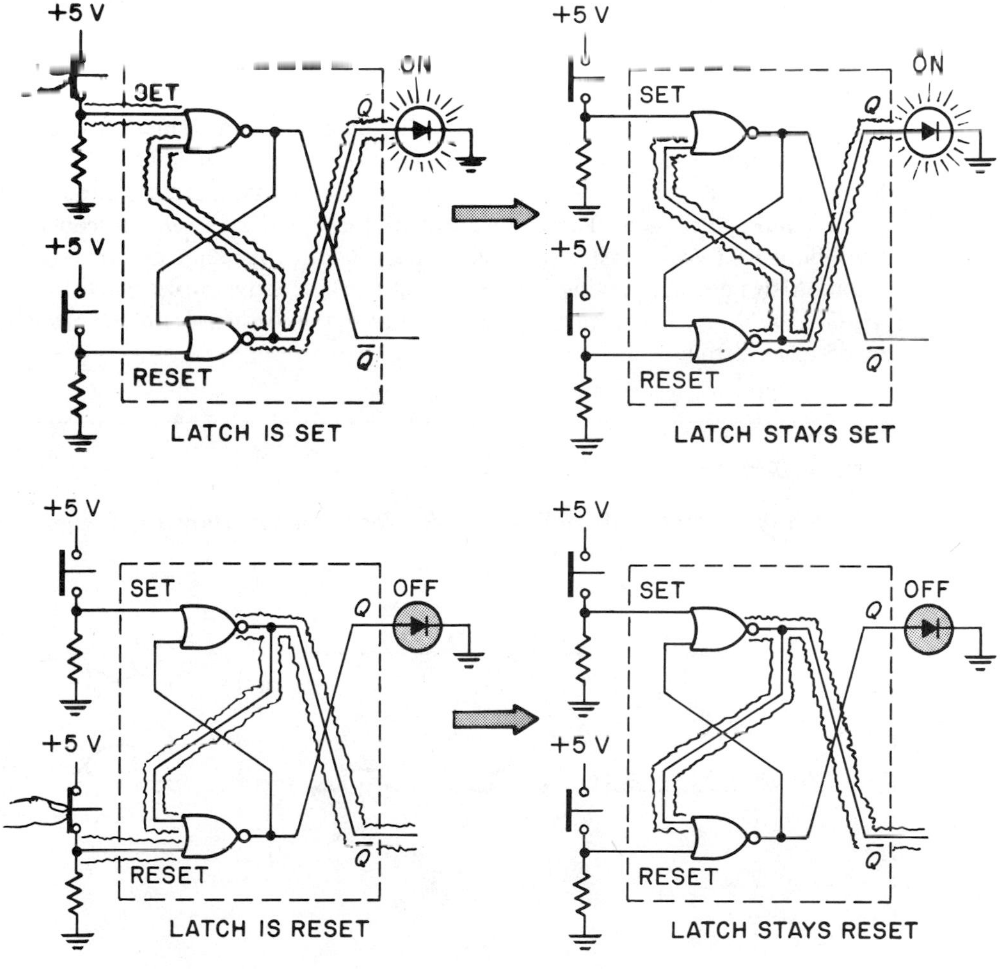

Figure 7–2 Constructing a latch with NOR gates.

Table 7–1 NOR GATE LATCH CONDITIONS			
SET	RESET	Q	$\overline{Q}$
1	0	1	0
0	1	0	1
0	0	No change in output.	
1	1	Invalid condition.	

means that if the latch were SET to a 1, it will retain this 1 until its RESET line is activated. Conversely, when the latch is RESET to a 0, it will retain this 0 until its SET line is active. It is because of this feature of *retaining* a HIGH or a LOW that it can be used as a digital storage element.

Latch Construction

A latch may be constructed from NOR gates. When this is done, the resulting latch will actually have two outputs. These outputs will always be complements (when one is HIGH, the other will be LOW). By convention, these outputs are called Q and $\overline{Q}$. Such construction is illustrated in Figure 7–2. Note from the figure that there are three valid conditions for the NOR gate latch. These are given in Table 7–1. As can be seen from the table, when both SET and RESET are inactive, the latch will retain its previously stored condition. Also note that having both SET and RESET active at the same time produces an *invalid condition*. Thus, unless otherwise noted, the control inputs to latches are not allowed to be active at the same time.

Example 7–1

Sketch the output of the following latch for the inputs shown in Figure 7–3.

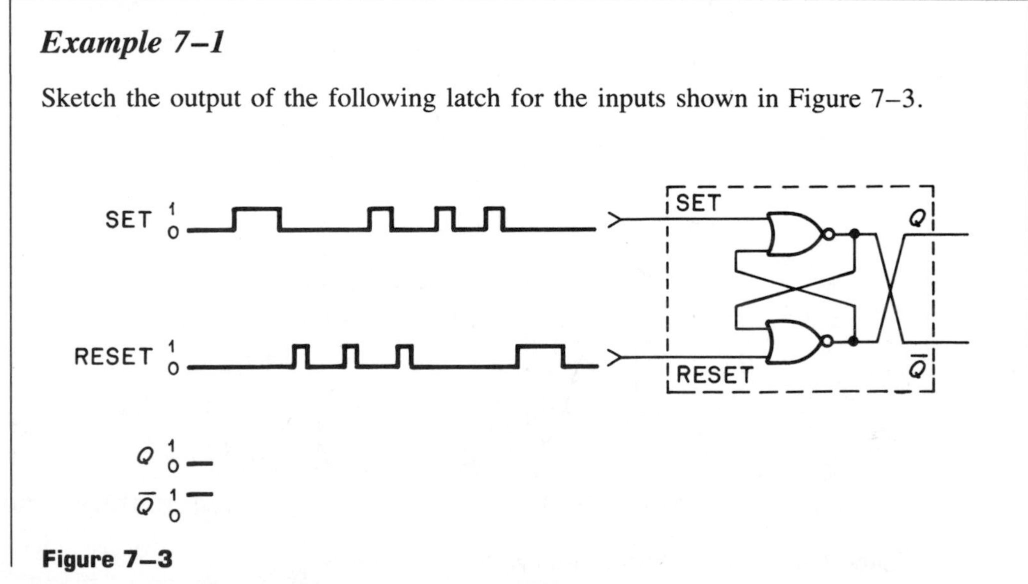

Figure 7–3

Referring to the operation of the latch as given in Table 7–1, the resultant outputs are shown in Figure 7–4.

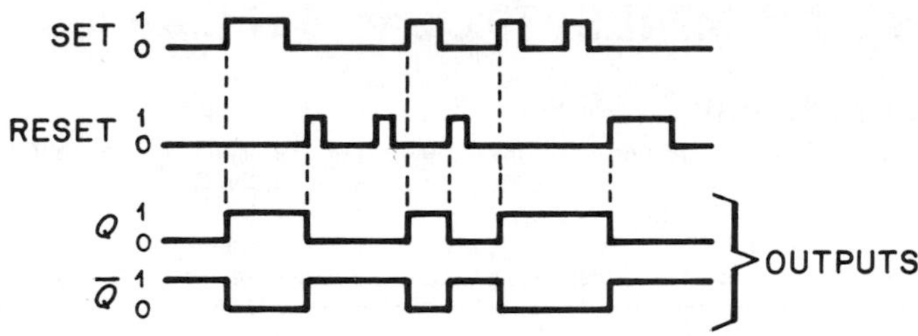

Figure 7–4

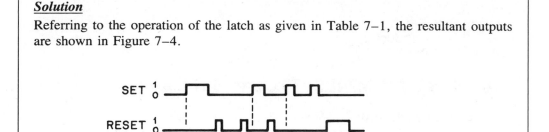

Latches may also be constructed from NAND gates. See Figure 7–5. Note from the figure that a LOW activates the input; otherwise, the NAND gate latch behaves the same as the NOR gate latch.

A Real Latch

Figure 7–6 shows the data sheet for a quad *S-R* (SET-RESET) latch. Observe the FUNCTION TABLE of the quad latch in Figure 7–6. Note that having both inputs active (both LOW) produces a HIGH on the *Q* output. However, there is a footnote that warns you that the output is not stable. Thus again, having both control inputs to the latch active at

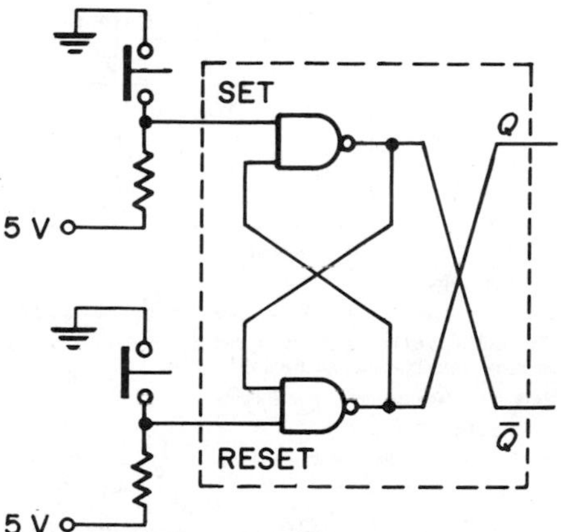

Figure 7–5 Constructing a latch with NAND gates.

DM54LS279/DM74LS279 Quad $\overline{S}$-$\overline{R}$ Latches

General Description

This device consists of four individual and independent Set-Reset Latches with active low inputs. Two of the four latches have an additonal $\overline{S}$ input ANDed with the primary $\overline{S}$ input. A low on any $\overline{S}$ input while the $\overline{R}$ input is high will be stored in the latch and appear on the corresponding Q output as a high. A low on the $\overline{R}$ input while the $\overline{S}$ input is high will clear the Q output to a low. Simultaneous transistion of the $\overline{R}$ and $\overline{S}$ inputs from low to high will cause the Q output to be indeterminate. Both inputs are voltage level triggered and are not affected by transition time of the input data.

Connection Diagram

Dual-In-Line Package

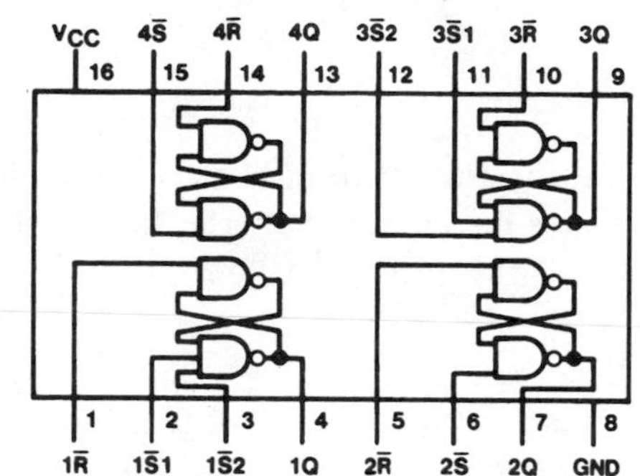

TL/F/6420–1

Order Number DM54LS279J, DM74LS279M or DM74LS279N
See NS Package Number J16A, M16A or N16A

Function Table

Inputs		Output
$\overline{S}$(1)	$\overline{R}$	Q
L	L	H*
L	H	H
H	L	L
H	H	Q_0

H = High Level

L = Low Level

Q_0 = The Level of Q before the indicated input conditions were established.

*This output level is pseudo stable; that is, it may not persist when the $\overline{S}$ and $\overline{R}$ inputs return to their inactive (high) level.

Note 1: For latches with double $\overline{S}$ inputs:

H = both $\overline{S}$ inputs high

L = one or both $\overline{S}$ inputs low

Figure 7–6 74LS279 quad *S-R* latch. *Reprinted* with permission of National Semiconductor Corporation.

National holds no responsibility for any circuitry described. National reserves the right at any time to change without notice said circuitry.

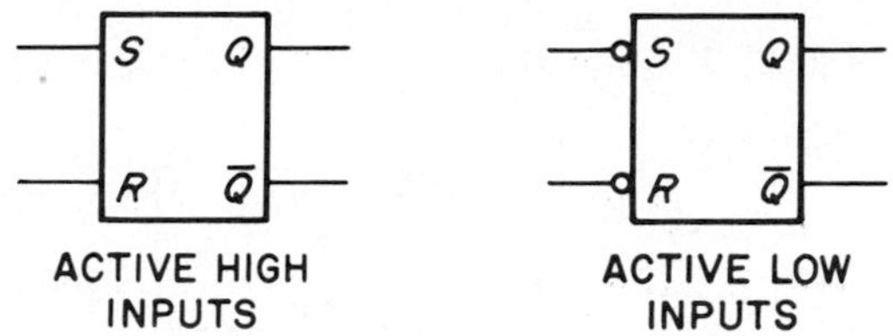

Figure 7–7 Logic diagram of an *S-R* latch.

the same time should be avoided. The logic diagrams of SET-RESET latches are shown in Figure 7–7.

Note from the diagram that both types of *S-R* latches are shown, the *S-R* active HIGH input and the *S-R* active LOW input. In both cases each latch has two complementary outputs referred to as Q and $\overline{Q}$.

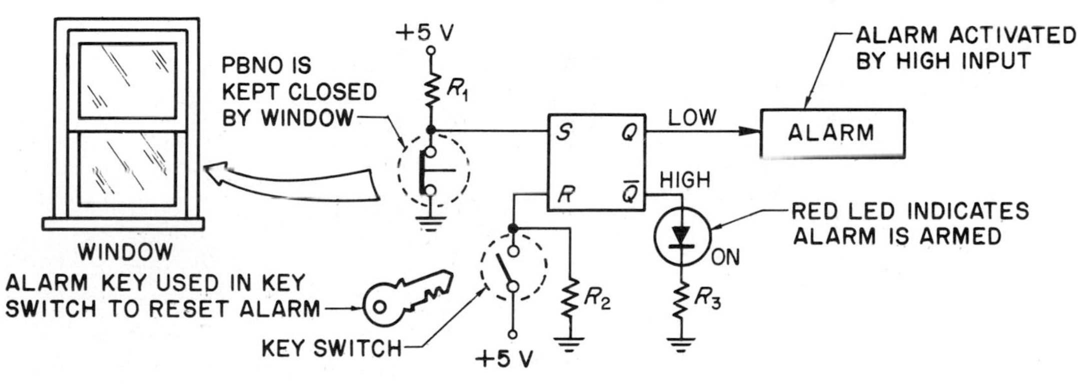

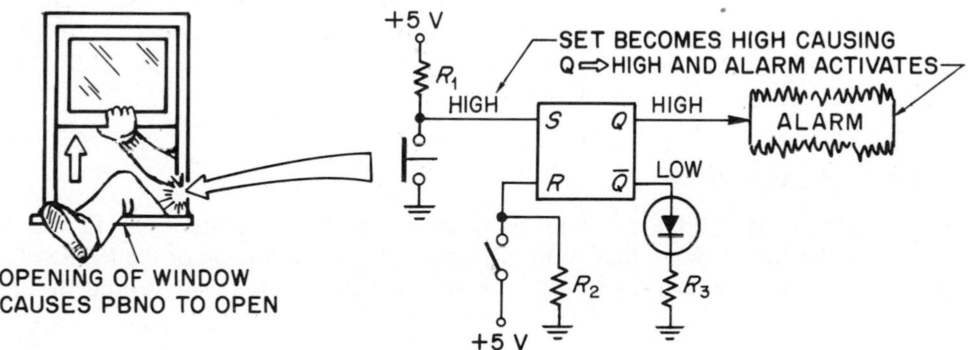

Figure 7–8 Alarm system using an *S-R* latch.

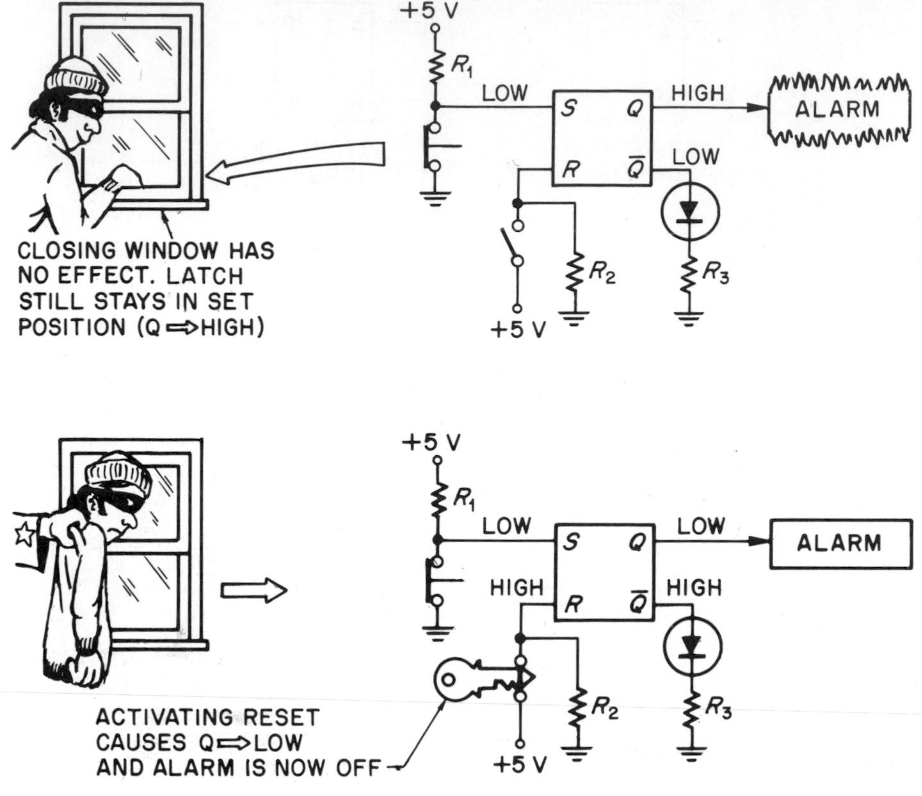

Figure 7–8 Continued

S-R Latch Application

An application of an *S-R* latch is illustrated in Figure 7–8. As shown in the figure, once the latch is set, it will remain set even though the SET input is deactivated (by the act of closing the window). The only way the latch may be reset is by having its RESET input activated. This may only be done with a key.

S-R Latch Comparison

A comparison of the operation of both types of *S-R* latches is illustrated in Figure 7–9. As shown in the figure, when the inputs are not active, the condition of the latch will not change. The operation of both types of *S-R* latches is demonstrated in the following example.

ACTIVE HIGH INPUTS

ACTIVE LOW INPUTS

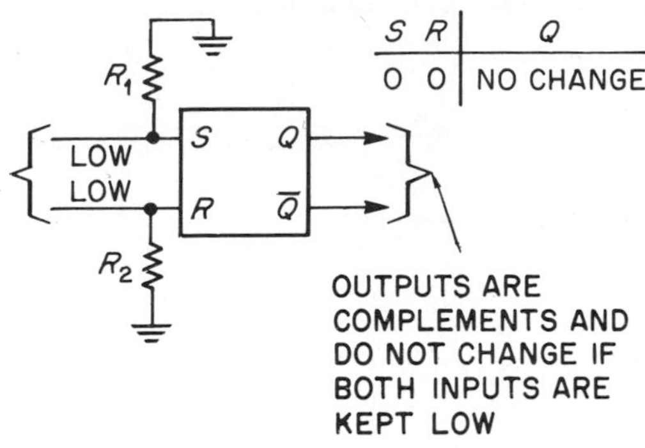

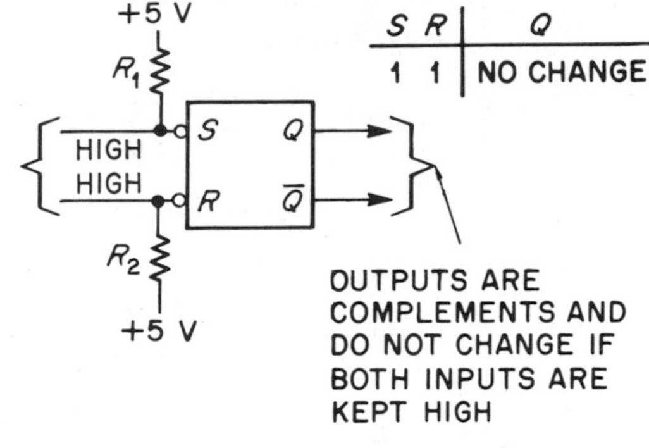

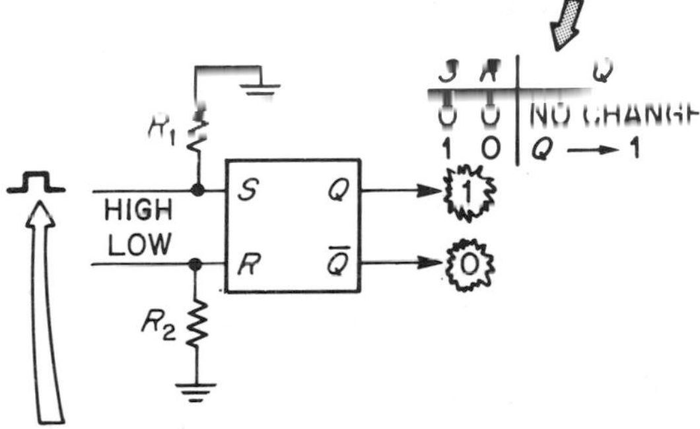

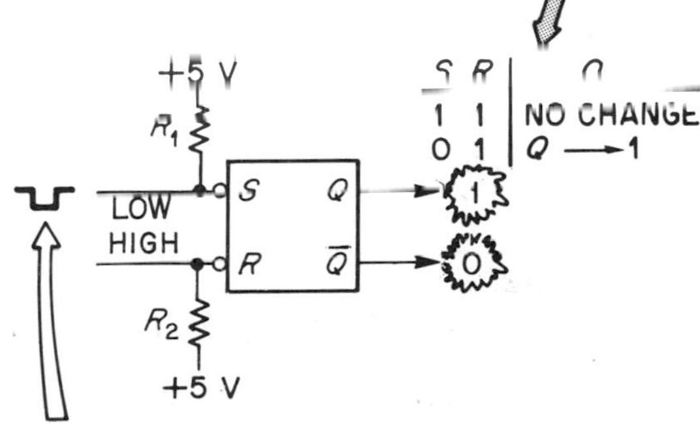

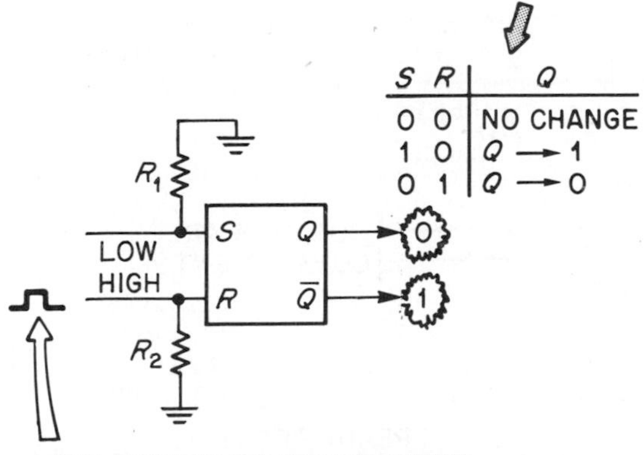

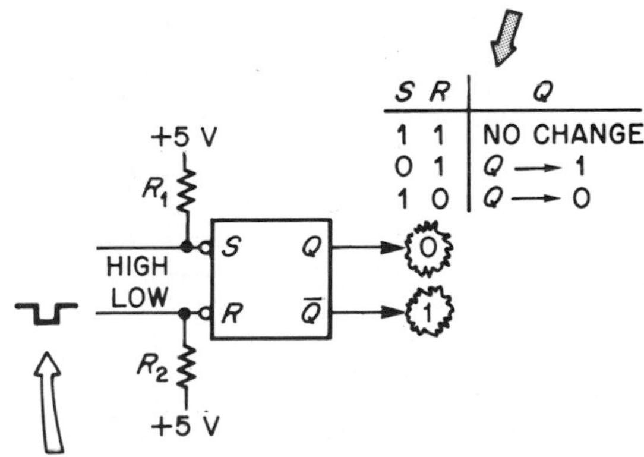

Figure 7–9 Comparison of both types of *S-R* latches.

Example 7–2

For the indicated input waveforms of Figure 7–10, determine the resulting output waveforms.

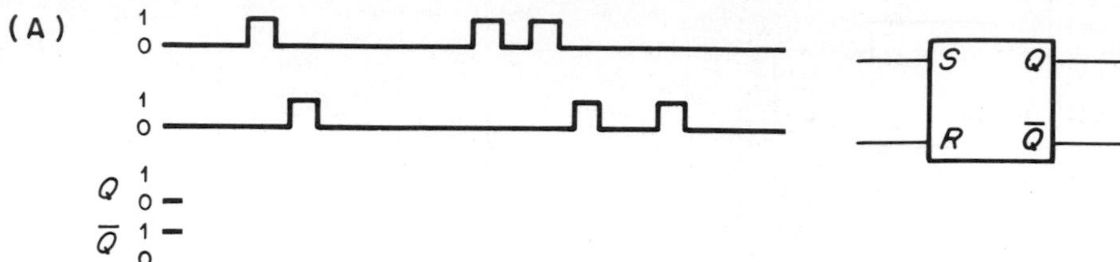

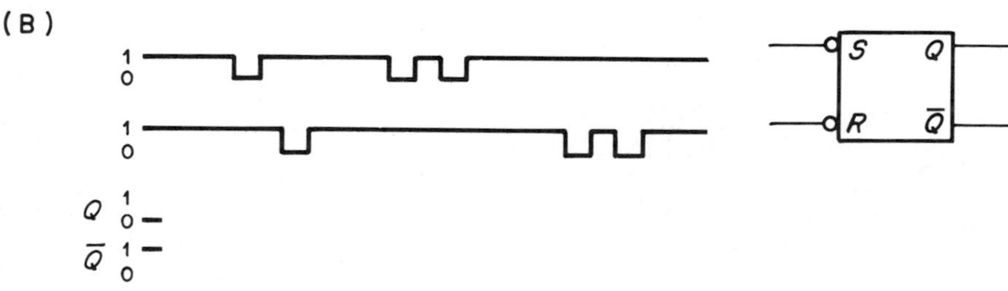

Figure 7–10

Solution

The solution is shown in Figure 7–11.

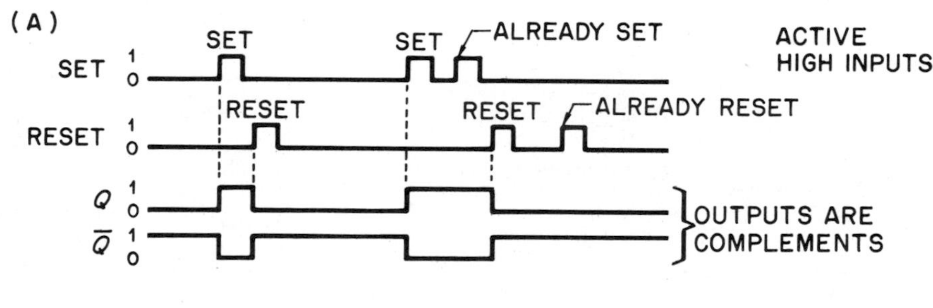

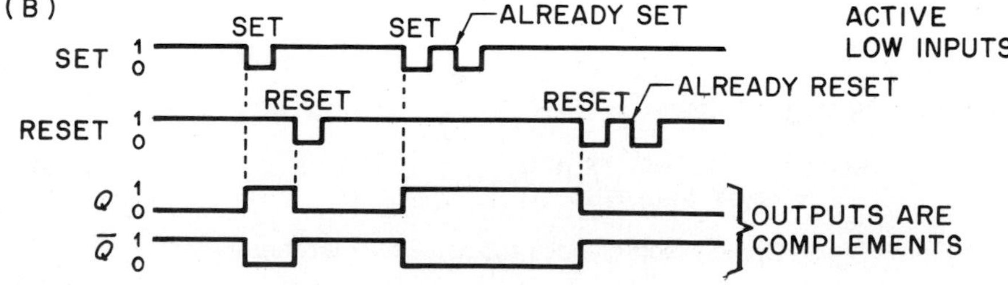

Figure 7–11

Conclusion

This section presented the fundamental concept behind all latches which is that they are used to store the condition of a binary bit. In the next section, you will be introduced to some of the many other uses of this versatile logic element.

Test your understanding of this section, by trying the following review questions.

7–1 Review Questions

1. Describe what a latch does.
2. State the purpose of the SET input.
3. State the purpose of the RESET input.
4. When are the inputs to a NAND gate latch active? To a NOR gate latch?
5. State what input condition is usually not allowed with a latch. Why?

7–2 TYPES OF LATCHES

Discussion

In this section you will be introduced to different types of latches. These are the next building block from the simple latches introduced in the last section.

Gated Latch

A **gated latch** is illustrated in Figure 7–12.

> **Gated Latch:**
> A SET-RESET latch with a third input called the ENABLE input. The ENABLE input must be active before the SET or RESET inputs have any affect on the latch.

Observe from the figure that this latch needs to have its ENABLE line active in order to process any input on its SET or RESET line. The function table for this latch is illustrated in Table 7–2. As you can see from the function table for the gated *S-R* latch, the SET and RESET lines perform the same function as they did for the ungated *S-R* latch. However, this time, the ENABLE input must be active (HIGH in this case). When the ENABLE line is inactive (LOW) it makes no difference what the conditions of the SET and RESET lines are. This is indicated by the letter *X* (meaning don't care) for these conditions.

The operation of the gated *S-R* latch is illustrated in the following example.

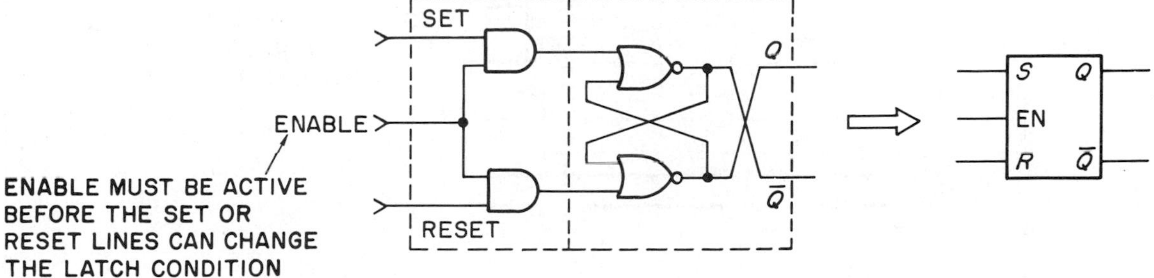

Figure 7–12 A gated latch.

Table 7–2	GATED S-R LATCH			
ENABLE	**SET**	**RESET**	Q	$\overline{Q}$
HIGH	LOW	LOW	No change from previous condition.	
	LOW	HIGH	0	1
	HIGH	LOW	1	0
	HIGH	HIGH	Invalid condition	
LOW	X	X	No change from previous condition.	

Example 7–3

For the input waveforms into the gated *S-R* latch of Figure 7–13, determine the resulting output.

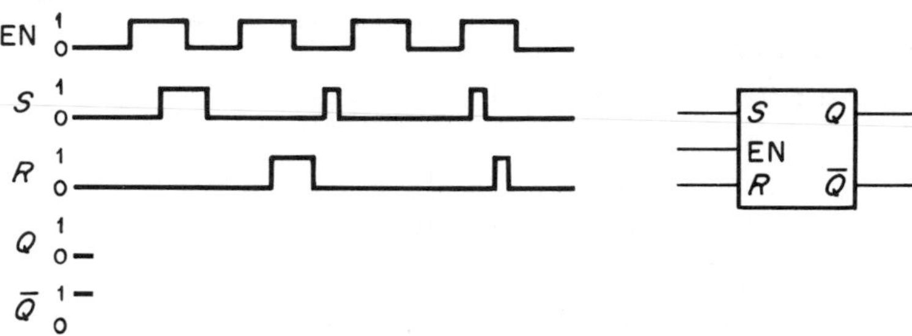

Figure 7–13

Solution

Refer to the function table for this latch (Table 7–2). The solution is given in Figure 7–14.

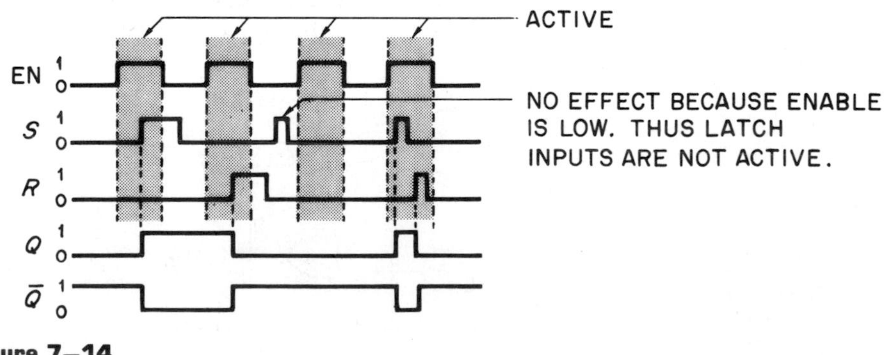

Figure 7–14

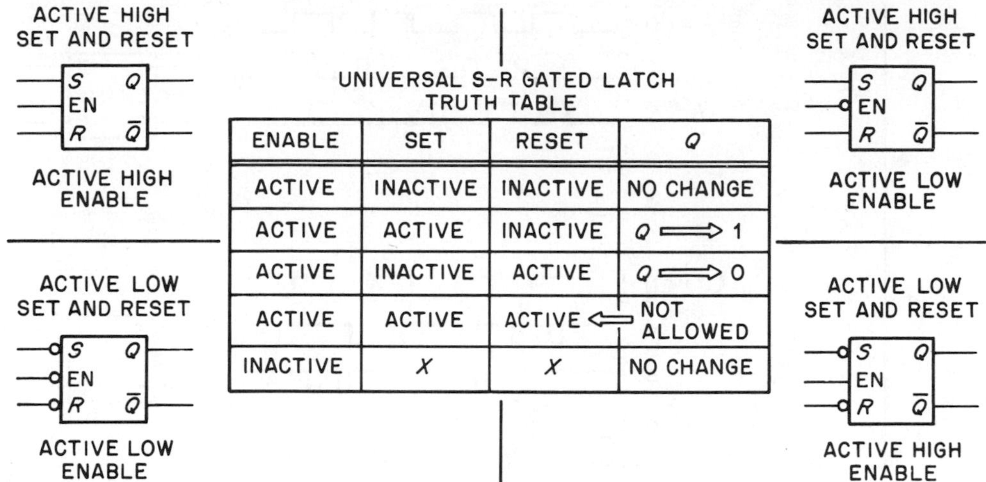

ENABLE	SET	RESET	Q
ACTIVE	INACTIVE	INACTIVE	NO CHANGE
ACTIVE	ACTIVE	INACTIVE	Q ⟹ 1
ACTIVE	INACTIVE	ACTIVE	Q ⟹ 0
ACTIVE	ACTIVE	ACTIVE ⟸	NOT ALLOWED
INACTIVE	X	X	NO CHANGE

Figure 7–15 Types of *S-R* latches.

Types of S-R Latches

Figure 7–15 shows various types of *S-R* latches. As shown in the figure, the latches may have a variety of different kinds of inputs. The important point to recognize is the universality of the truth table. If you analyze the latches in terms of *active* or *inactive* inputs, then their commonalities emerge. The following example illustrates.

Example 7–4

For the waveforms of Figure 7–16 with the corresponding type of gated latch, determine the resulting output waveforms.

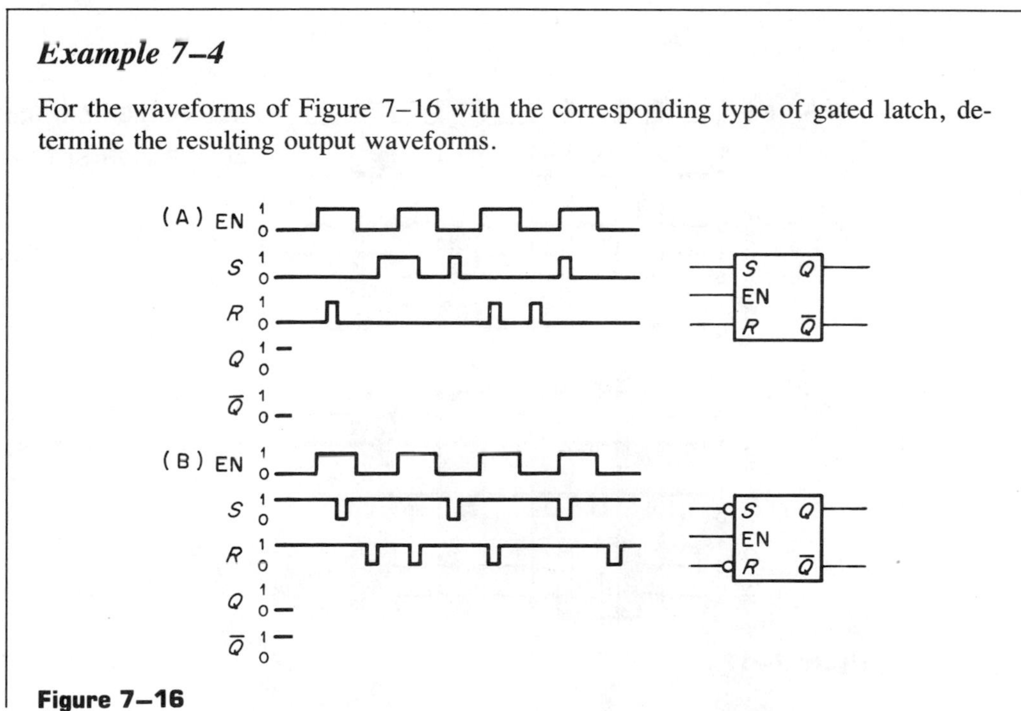

Figure 7–16

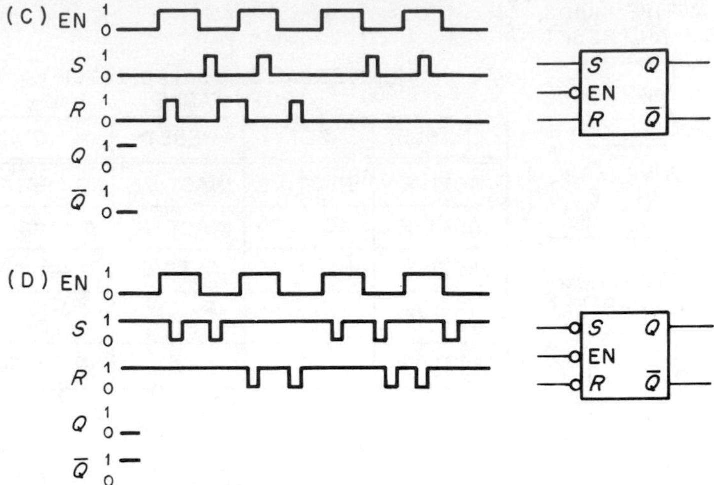

Figure 7–16 Continued

Solution

Keep in mind that the only time the SET and RESET inputs to the latch are active is when the ENABLE line is active. The resulting waveforms are shown in Figure 7–17.

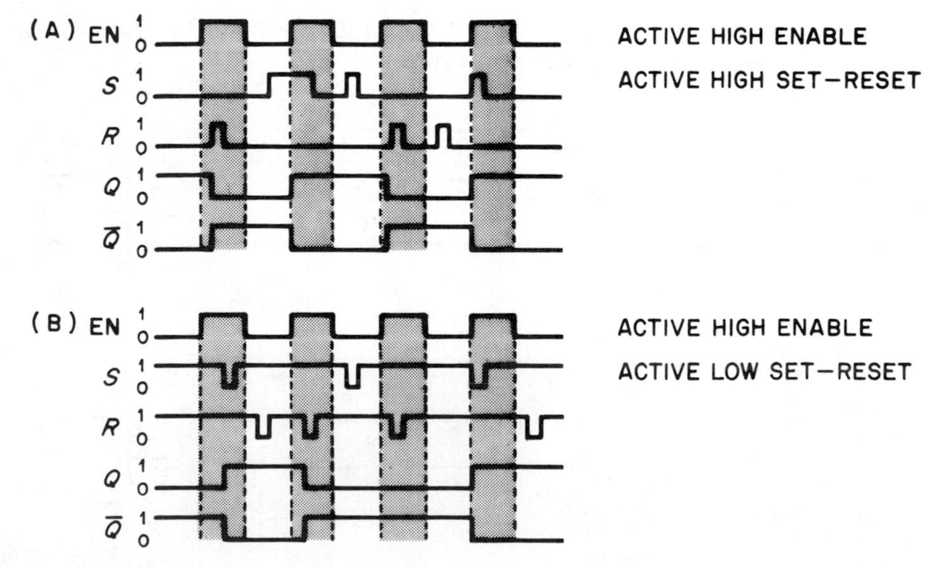

Figure 7–17

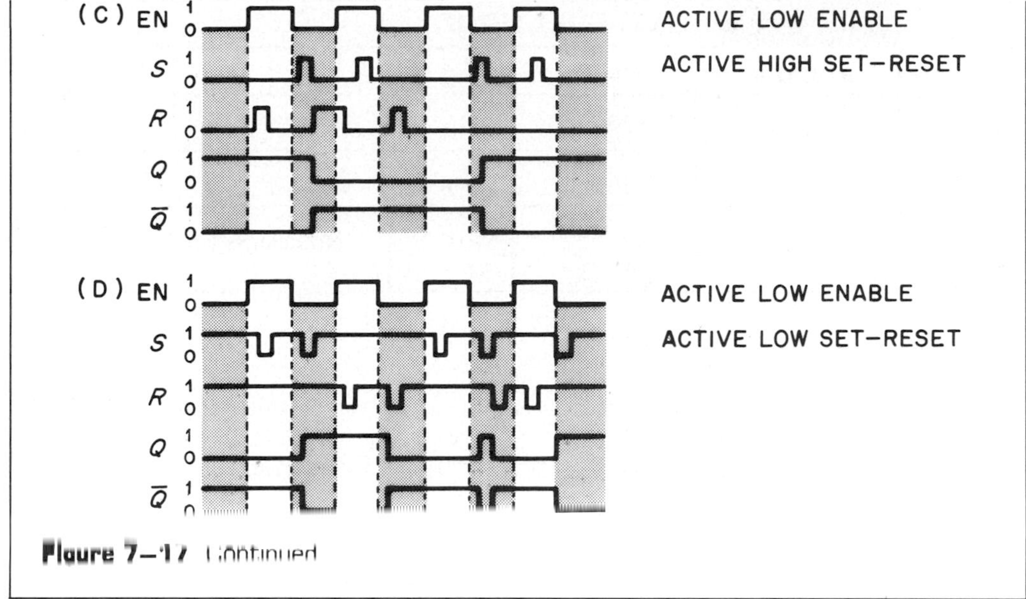

Figure 7–17 Continued

The Gated D Latch

Another type of latch is the **gated *D* latch.**

> **Gated *D* Latch:**
> A latch with two inputs called the ENABLE and *D* inputs. The logic level at the *D* input will determine the state of this latch when the ENABLE line is active.

The gated *D* latch is illustrated in Figure 7–18. Note from the figure, that it is now impossible for both the SET and RESET inputs to be active at the same time. The important point about the gated *D* latch is that the *Q* output will follow the *D* input when the ENABLE line is active. The function table for the gated *D* latch is shown in Table 7–3. As shown in the table, when the ENABLE line is inactive (LOW in this case), the state of the *D* input has no effect on the latch. Again, this condition is indicated by the don't care

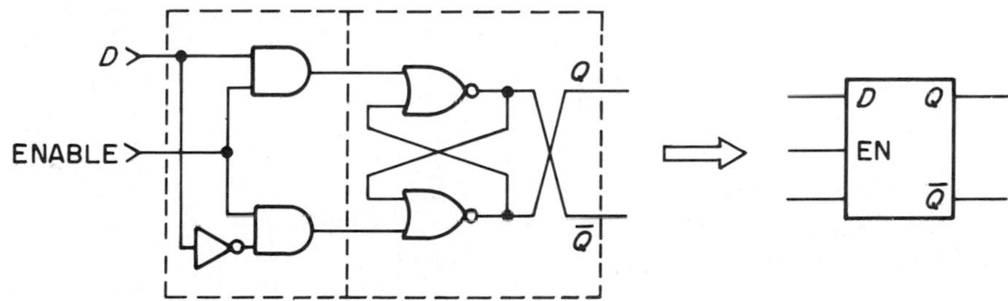

Figure 7–18 The gated *D* latch.

Table 7–3	GATED *D* LATCH		
ENABLE	***D* input**	***Q***	$\overline{Q}$
HIGH	LOW	0	1
	HIGH	1	0
LOW	*X*	*X*	No change from previous condition.

X. However, when the ENABLE line is active (HIGH) the output of the latch follows the input logic of the *D* input.

The operation of the *D* latch is illustrated by the following example.

Example 7–5

For the waveforms applied to the gated *D* latch of Figure 7–19, determine the resulting outputs.

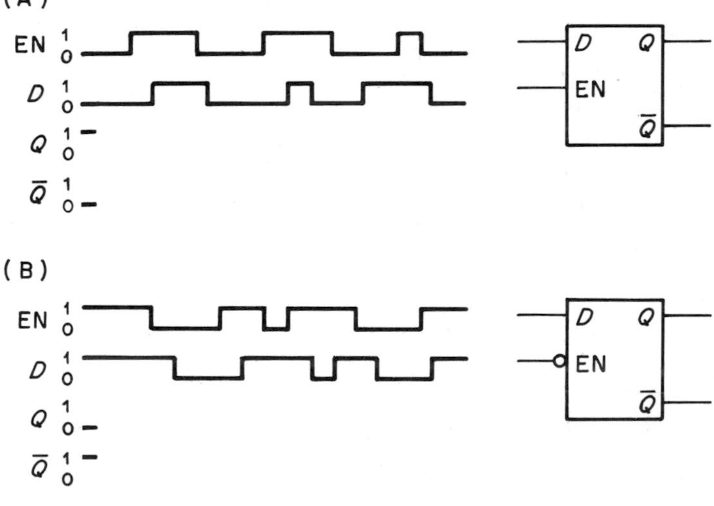

Figure 7–19

Solution

Refer to the function table for the gated D latch. The resulting outputs are shown in Figure 7–20.

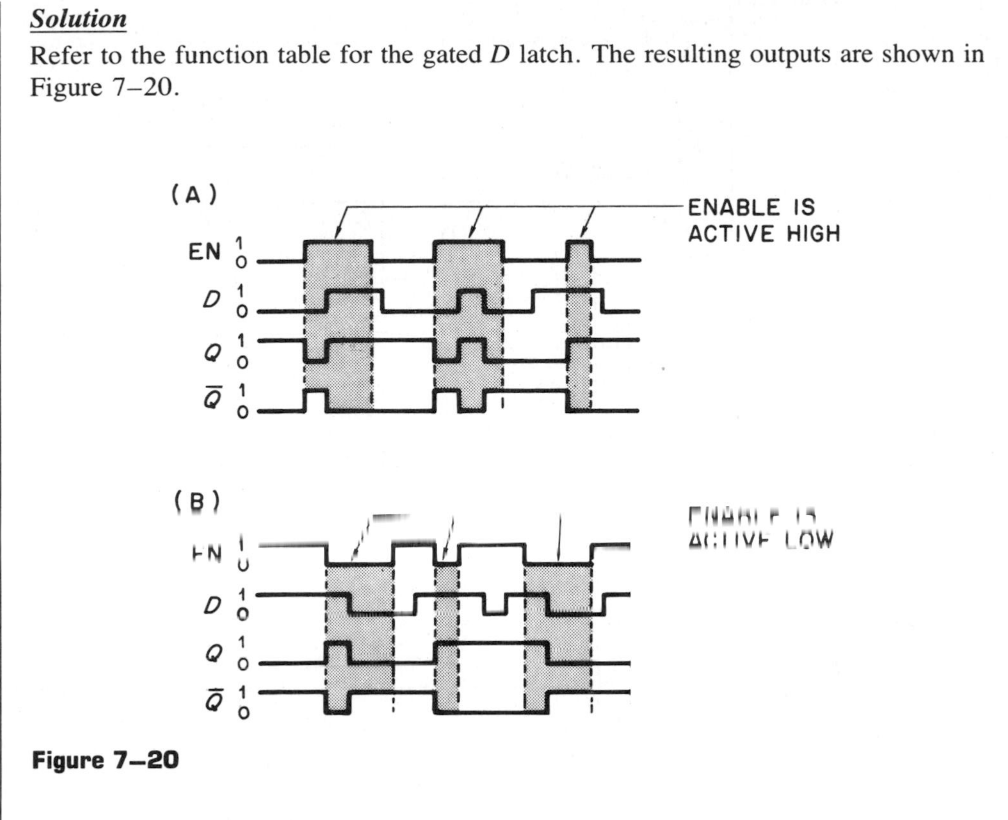

Figure 7–20

Gated Latch Applications

Figure 7–21 shows gated latches used as memory to store the bit patterns of a four bit word. Observe from the figure that the latches are arranged in rows of four each. This means they may store the bit pattern of a four-bit word (a nibble). The row in which the bit pattern will be stored is selected by the ADDRESS SELECTOR switch. The position of the switch determines which row of four latches will have their ENABLE lines active. When a key on the hex keypad is depressed, thc 16 to 4 line encoder will place the corresponding binary bit pattern on the data bus. The row of latches selected by the ADDRESS SELECTOR switch will then copy this bit pattern via their D inputs. Thus, the user of such a system may easily store a program into this memory by advancing the ADDRESS SWITCH and then using the hex keypad to enter the desired instruction or data into the memory.

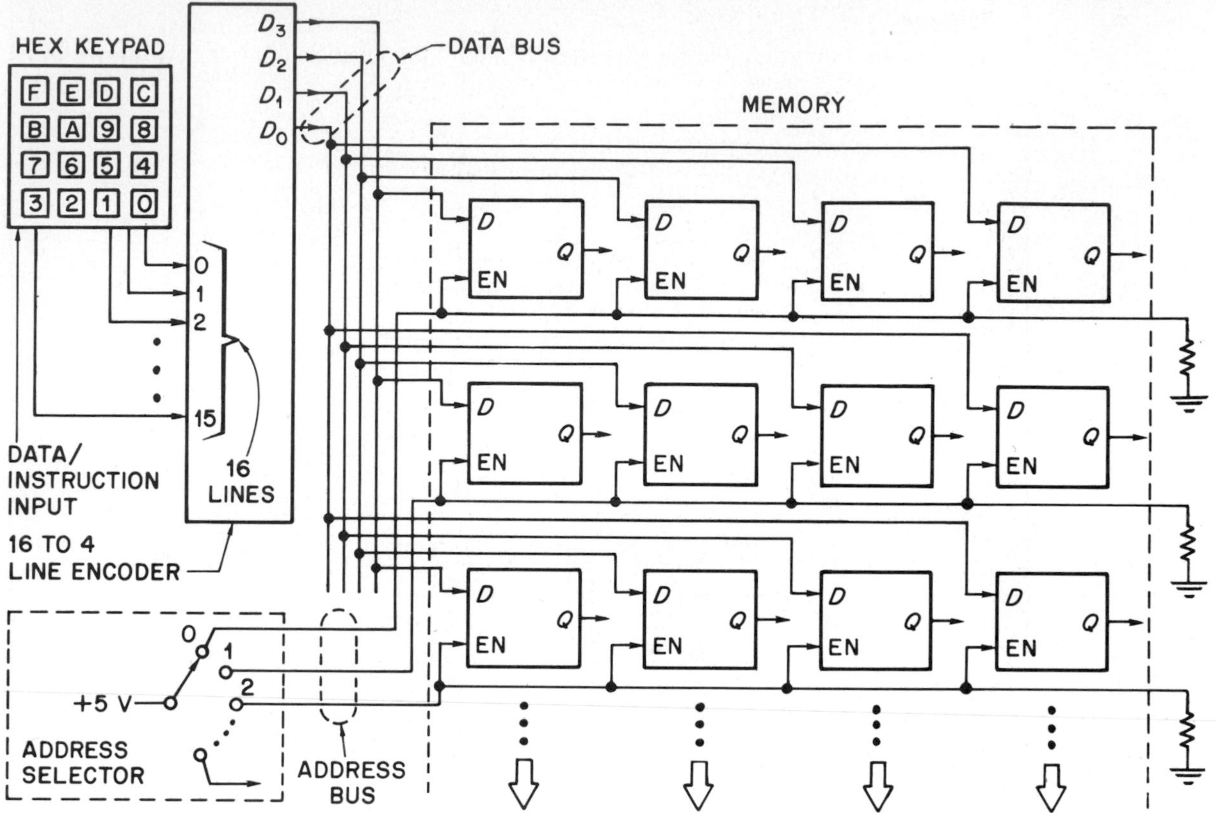

Figure 7–21 Gated latch used as memory.

Conclusion

This section presented the gated latch. Here you saw the gated *S-R* latch as well as the gated *D* latch. You were given an opportunity to see how these devices behaved under various waveform conditions. You also saw an application for the gated latch. Check your understanding of this section by trying the following review questions.

7–2 Review Questions

1. Describe a *gated S-R latch*.
2. State the purpose of an ENABLE input.
3. Describe a *gated D latch*.
4. Give an application of a gated latch.

7–3 WHAT ARE FLIP-FLOPS?

Discussion

The last two sections introduced you to the latch. There you saw that a latch had two stable states. As you will see in this section, the flip-flop is similar to the latch in that it

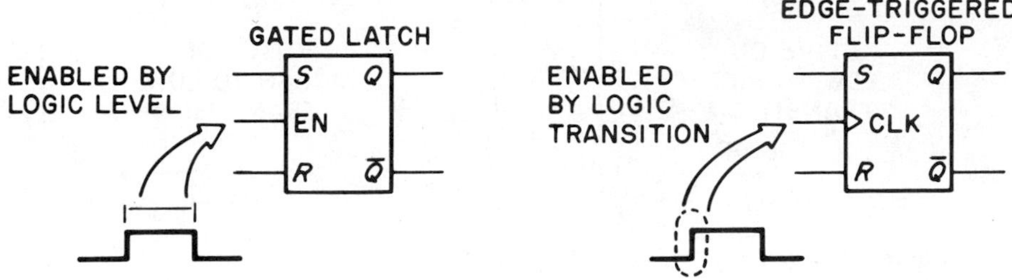

Figure 7–22 Similarities between latches and flip-flops.

also has two stable states, has two complementary outputs and similar control inputs. The difference is in how the flip-flop is *enabled*.

Basic Idea

The latch is similar to the **flip-flop** in several ways.

> **Flip-Flop:**
> A bistable digital device that is synchronized by an external source.

The similarities are illustrated in Figure 7–22. As shown in the figure, the difference between a latch and a flip-flop is that a flip-flop may have its output change *only at the edge* of its ENABLE signal. This ENABLE signal is commonly referred to as a **clock.**

> **Clock:**
> A series of digital pulses used in the synchronization of digital devices.

Figure 7–23 illustrates a logic network that creates a flip-flop from a simple gated latch. As shown in the figure, the slight delay inherent in the INVERTER allows the creation of a spike type pulse at the output of the AND gate. This short pulse in turn enables the input to the gated latch so that the SET and RESET inputs are now active.

The S-R Flip-Flop

Figure 7–24 illustrates the logic symbol of an *S-R* flip-flop. As you can see from the figure, there are two basic kinds of *edge triggered S-R* flip-flops. Those that are activated on the leading edge of the clock and those that are activated on the trailing edge of the clock.

Note the use of the triangle for the clock. This indicates that the synchronized inputs to the flip-flop are active only during a clock *transition*. The inversion bubble means

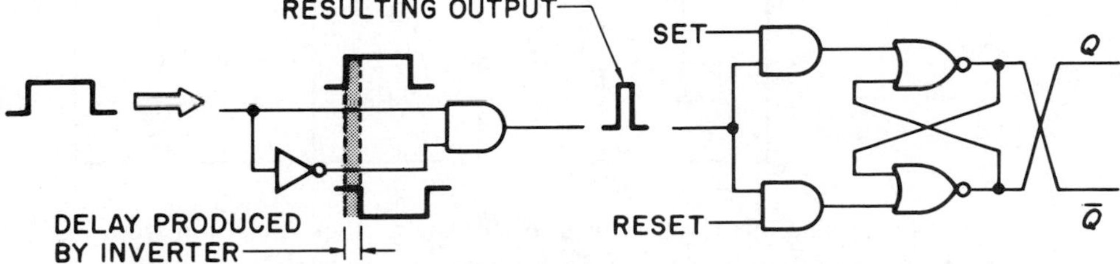

Figure 7–23 Logic network for developing an edge-triggered pulse.

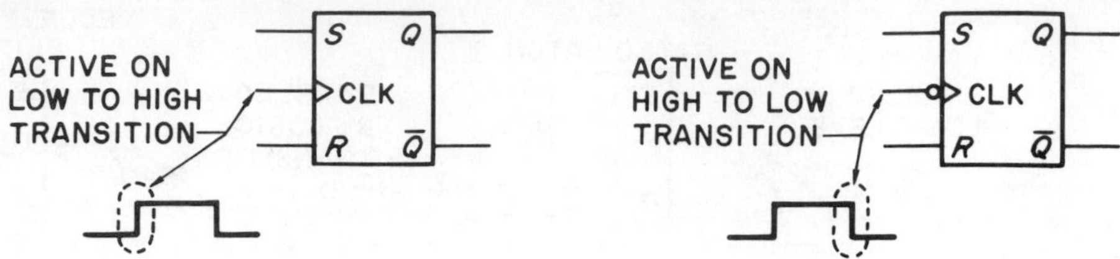

Figure 7–24 Logic symbol of an *S-R* flip-flop.

activation takes place on the HIGH to LOW transition, and the omission of the inversion bubble means activation on the LOW to HIGH transition of the clock. The triangle symbol used for the clock is referred to as the **dynamic input indicator.**

Dynamic Input Indicator:
The triangular symbol used on the clock input of a flip-flop to indicate that the device is synchronized by a logic transition of the clock. The edge-triggered flip-flop allows for greater timing accuracy over that of the latch.

Truth Tables

The truth tables for an edge-triggered *S-R* flip-flop follow.

LOW TO HIGH TRANSITION			
Clock	**SET**	**RESET**	***Q***
↑	0	0	No change
↑	1	0	$Q \rightarrow 1$
↑	0	1	$Q \rightarrow 0$
0	X	X	No change

HIGH TO LOW TRANSITION			
Clock	**SET**	**RESET**	***Q***
↓	0	0	No change
↓	1	0	$Q \rightarrow 1$
↓	0	1	$Q \rightarrow 0$
0	X	X	No change

Note the use of the up arrows to indicate a clock transition from LOW to HIGH and the down arrows to indicate a clock transition from HIGH to LOW.

Example 7–6

For the waveshapes indicated in Figure 7–25, determine the resultant output waveforms.

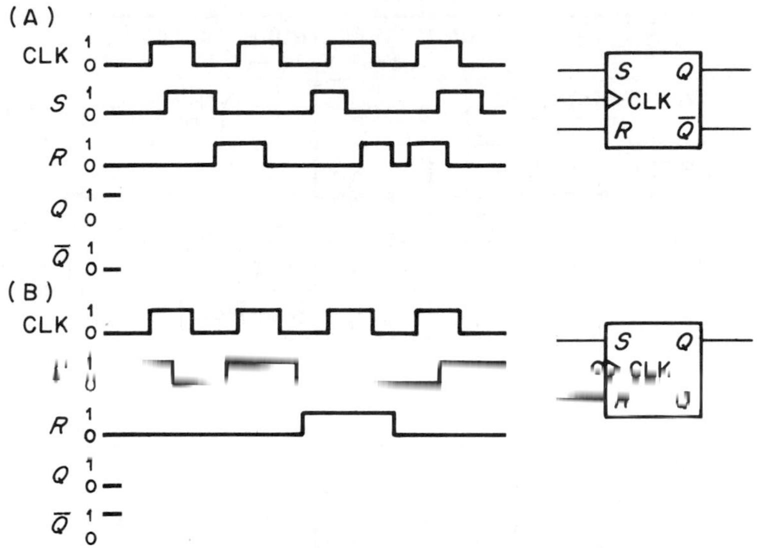

Figure 7–25

Solution

Refer to the truth tables for edge-triggered *S-R* flip-flops. The resulting outputs are shown in Figure 7–26.

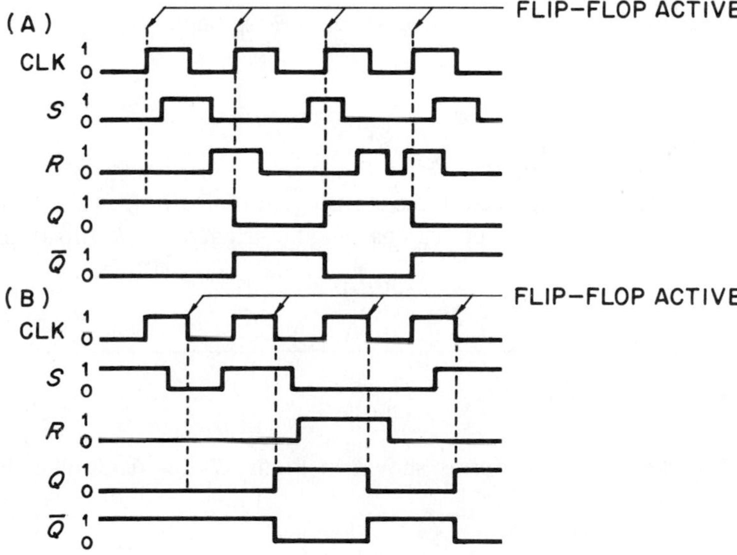

Figure 7–26

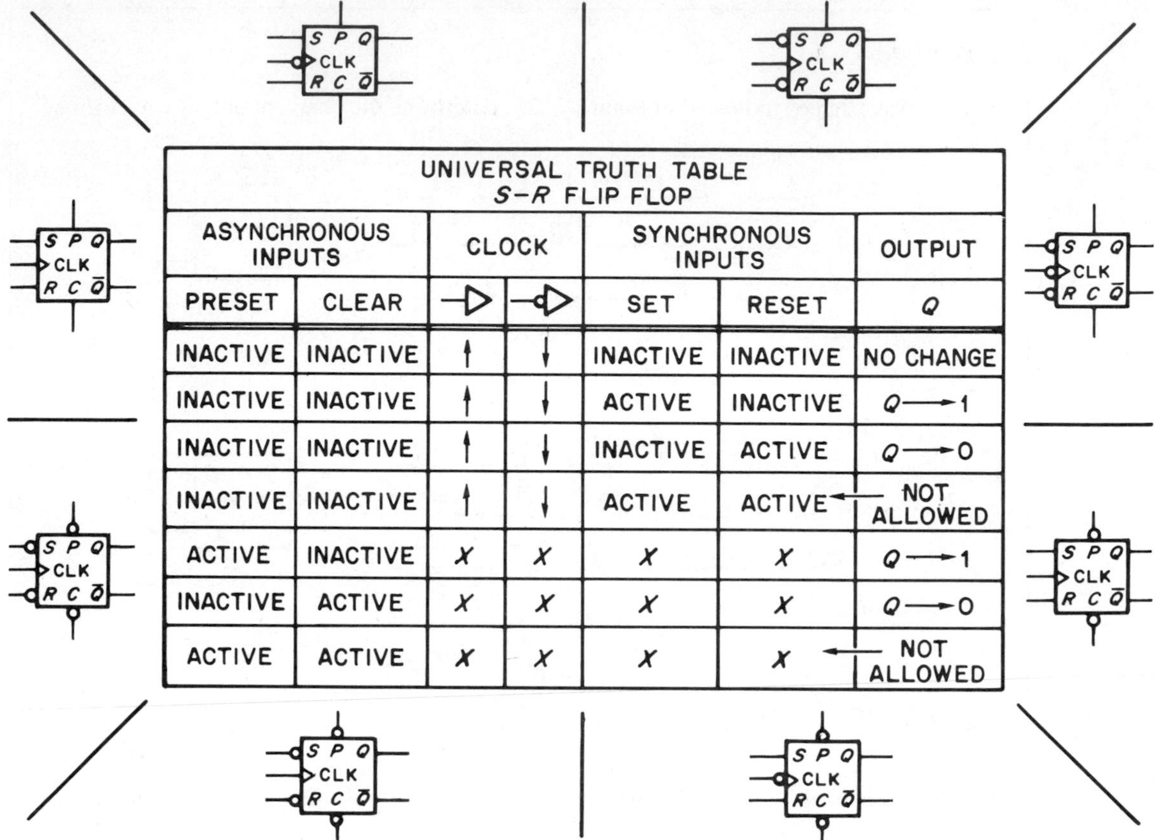

UNIVERSAL TRUTH TABLE S–R FLIP FLOP						
ASYNCHRONOUS INPUTS		CLOCK		SYNCHRONOUS INPUTS		OUTPUT
PRESET	CLEAR	▷ (rising edge)	○▷ (falling edge)	SET	RESET	Q
INACTIVE	INACTIVE	↑	↓	INACTIVE	INACTIVE	NO CHANGE
INACTIVE	INACTIVE	↑	↓	ACTIVE	INACTIVE	$Q \rightarrow 1$
INACTIVE	INACTIVE	↑	↓	INACTIVE	ACTIVE	$Q \rightarrow 0$
INACTIVE	INACTIVE	↑	↓	ACTIVE	ACTIVE	← NOT ALLOWED
ACTIVE	INACTIVE	X	X	X	X	$Q \rightarrow 1$
INACTIVE	ACTIVE	X	X	X	X	$Q \rightarrow 0$
ACTIVE	ACTIVE	X	X	X	X	← NOT ALLOWED

Figure 7–27 Various combinations of *S-R* flip-flops.

Asynchronous Inputs

As you may have guessed, the edge-triggered *S-R* flip-flop also comes with *asynchronous* inputs. The various combinations of edge-triggered *S-R* flip-flops are shown in Figure 7–27.

Example 7–7

For the input waveforms shown in Figure 7–28, determine the resultant flip-flop outputs.

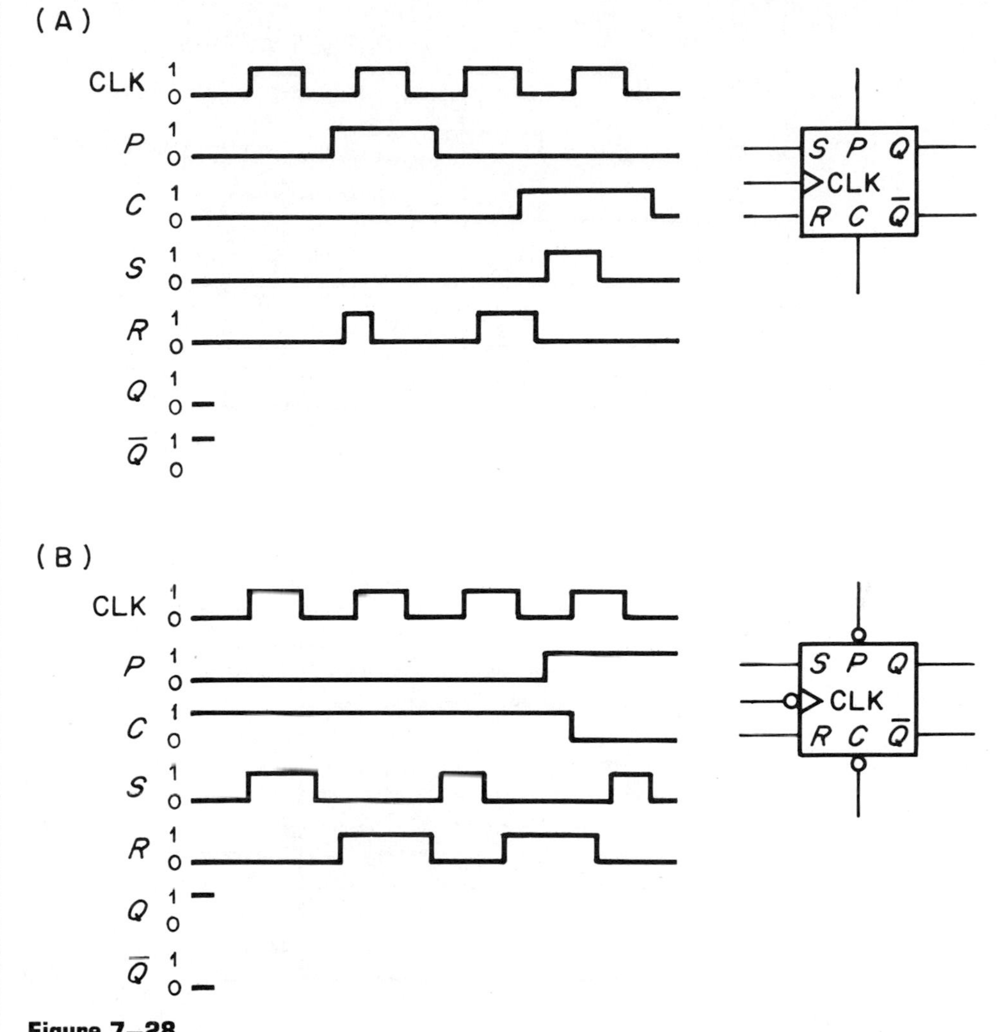

Figure 7–28

Solution

Refer to the universal truth table for the edge-triggered *S-R* flip-flop with PRESET and CLEAR inputs. The resulting waveforms are illustrated in Figure 7–29.

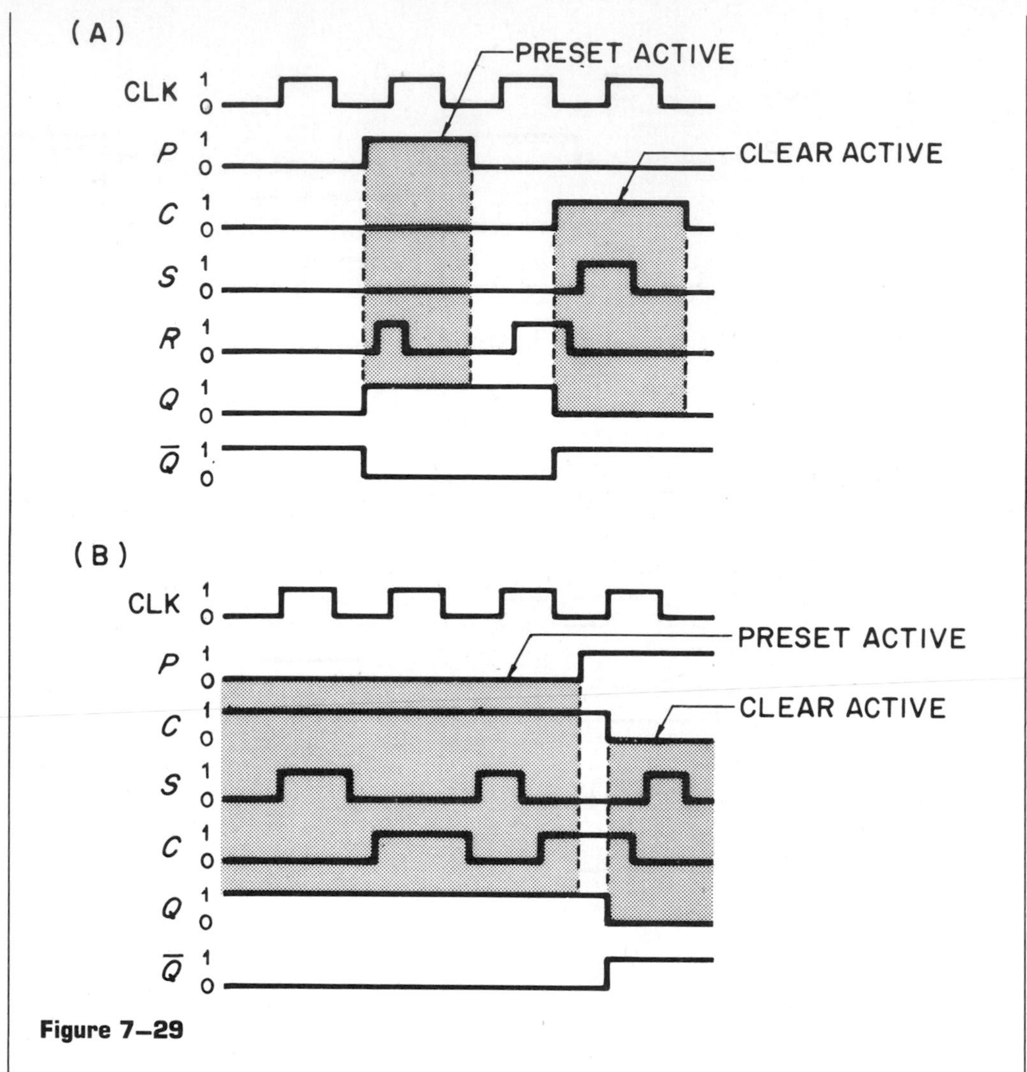

Figure 7–29

Edge-Triggered D Flip-Flops

Edge-triggered *D* flip-flops with asynchronous PRESET and CLEAR are also used in digital systems. Figure 7–30 illustrates some of the possible combinations. Observe that these types of flip-flops can be activated in the same manner as the edge-triggered *S-R* flip-flops with PRESET and CLEAR. The following example illustrates.

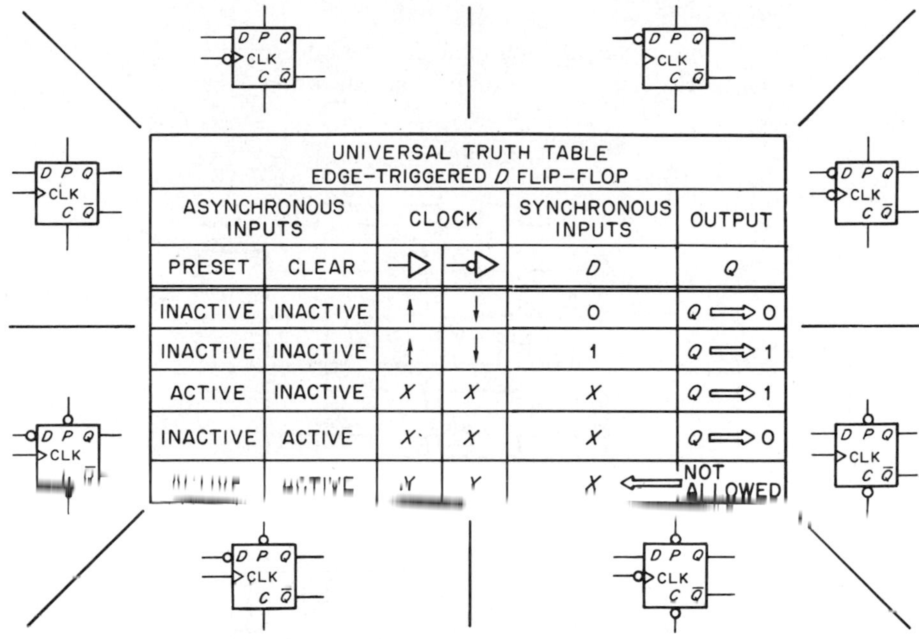

UNIVERSAL TRUTH TABLE
EDGE-TRIGGERED *D* FLIP-FLOP

ASYNCHRONOUS INPUTS		CLOCK		SYNCHRONOUS INPUTS	OUTPUT
PRESET	CLEAR	▷	○▷	*D*	*Q*
INACTIVE	INACTIVE	↑	↓	0	*Q* ⇒ 0
INACTIVE	INACTIVE	↑	↓	1	*Q* ⇒ 1
ACTIVE	INACTIVE	*X*	*X*	*X*	*Q* ⇒ 1
INACTIVE	ACTIVE	*X*	*X*	*X*	*Q* ⇒ 0
ACTIVE	ACTIVE	*X*	*X*	*X* ⇐	NOT ALLOWED

Figure 7–30 Variations of edge-triggered *D* flip-flops with PRESET and CLEAR.

Example 7–8

For the waveforms shown in Figure 7–31, determine the resultant outputs. Assume that the *Q* outputs of the flip-flops start out HIGH.

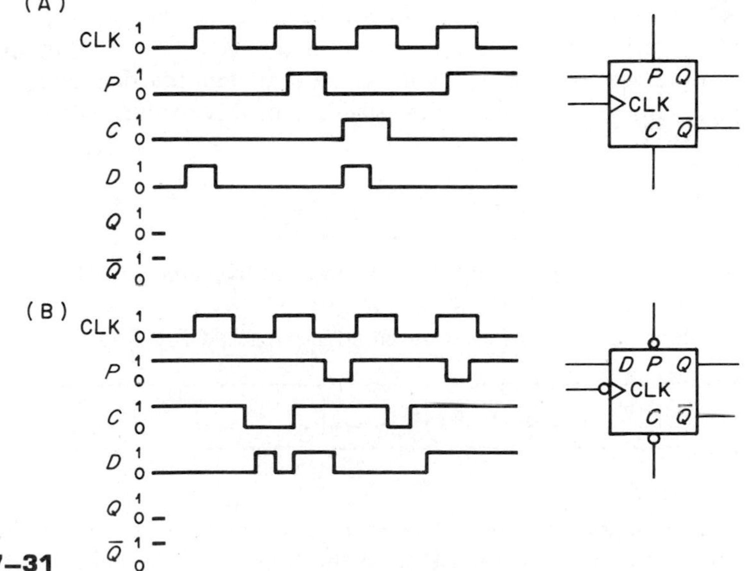

Figure 7–31

Solution

Refer to the universal truth table for the edge-triggered *D* flip-flop with PRESET and CLEAR. The resultant outputs are shown in Figure 7–32.

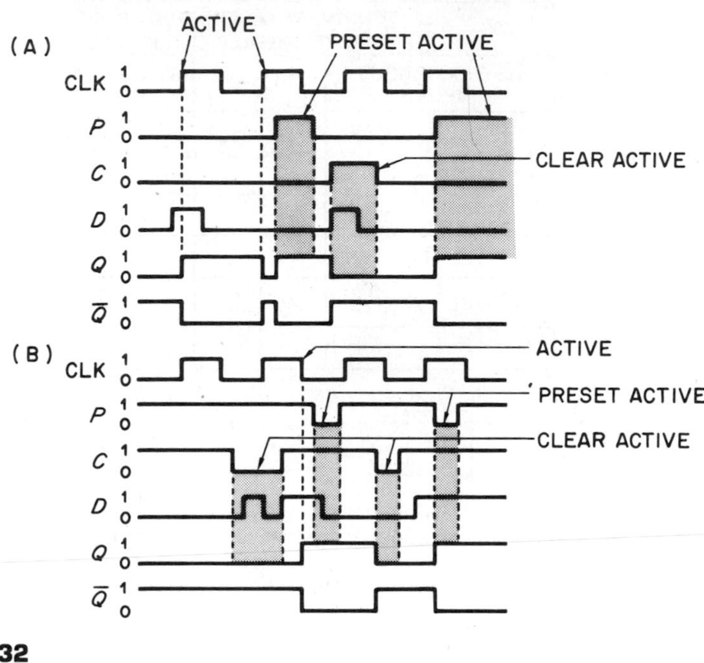

Figure 7–32

Conclusion

This section presented the important concepts of activating a flip-flop more precisely. The edge-triggering method is widely used in almost every flip-flop application.

In the next section, you will see an important practical addition to the versatile flip-flop. For now, check your understanding of this section by trying the following review questions.

7–3 Review Questions

1. What is a *latch?*
2. Briefly describe the operation of an edge-triggered flip-flop.
3. What is a *dynamic input indicator?*
4. Explain the reason for using an edge-triggered flip-flop.

7–4 WHY FLIP–FLOPS ARE USED

Discussion

The following chapters will show you the actual details of exactly how the digital systems presented here are constructed from flip-flops. It is the purpose of this section to familiarize you with the main ideas behind them without getting bogged down in details. If you have an idea of what flip-flops are used for, then it will make your understanding of how flip-flops work a lot easier. Knowing the purpose of a journey is a great aide in understanding the reason for the guided tours along the way.

The Binary Counter

In the microprocessor applications section of the last chapter, you were introduced to the binary counter. As you probably suspect by now, a binary counter consists of flip-flops connected in a special way that causes the bit pattern to change in the form of a binary count.

The basic idea of a binary counter is illustrated in Figure 7–33. In order for a binary

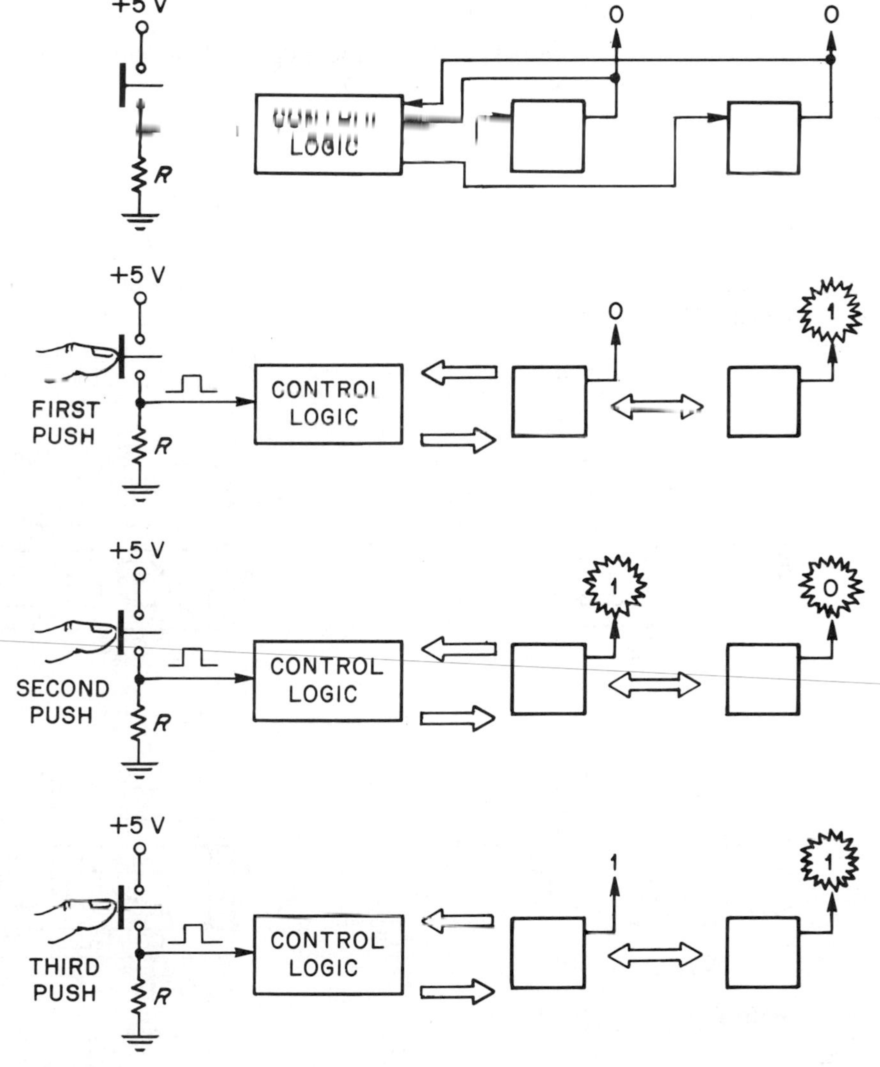

Figure 7–33 Basic idea of a binary counter.

counter to count, it needs something to count. For the counter shown in Figure 7–33, every time the count switch is depressed, the binary counter will be incremented by 1.

Example 7–9

What will be the contents of a 4-bit binary counter that always starts at 0000_2 if the number of times the count switch is depressed is (A) 5? (B) 10? (C) 16?

Solution

(A) The count will be 0101_2.

(B) The count will be 1010_2.

(C) The count will be 0000_2 (it will count up to 1111_2 and then start over again at the count of 16).

Registers

Recall that a *register* was nothing more than a group of flip-flops working together as a unit. Figure 7–34 shows some of the different methods of using registers for storing, modifying, or transferring binary bit patterns. Table 7–4 summarizes the different types of registers.

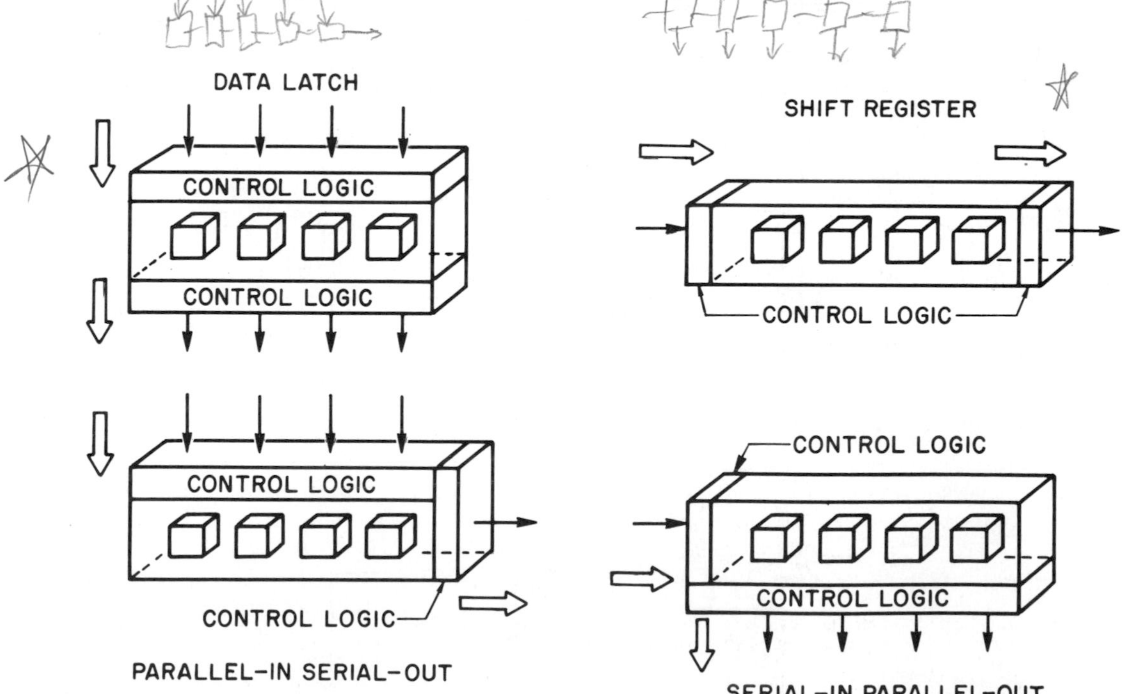

Figure 7–34 Methods of using registers.

Table 7–4 SUMMARY OF REGISTER TYPES

Type	Basic Operation	Application
Data latch	Parallel in—all data in at the same time. All data can be sensed at the same time.	Computer memory.
Shift Register	Bit patterns are transferred into the register 1 bit at a time.	Specialized digital circuits for control functions.
Parallel in–serial out	Bit patterns all in at the same time. Transferred out 1 bit at a time.	Converting digital information for transmission over a few wires such as telephone lines.*
Serial in–parallel out	Bit patterns in 1 bit at a time. Data sensed at the same time.	Receiving digital information from a telephone line.

*Digital pulses themselves are not sent directly over telephone lines. These pulses must first be converted into audio tones. You will see the details of this in Chapter 9.

Example 7–10

Identify each register type as shown in Figure 7–35.

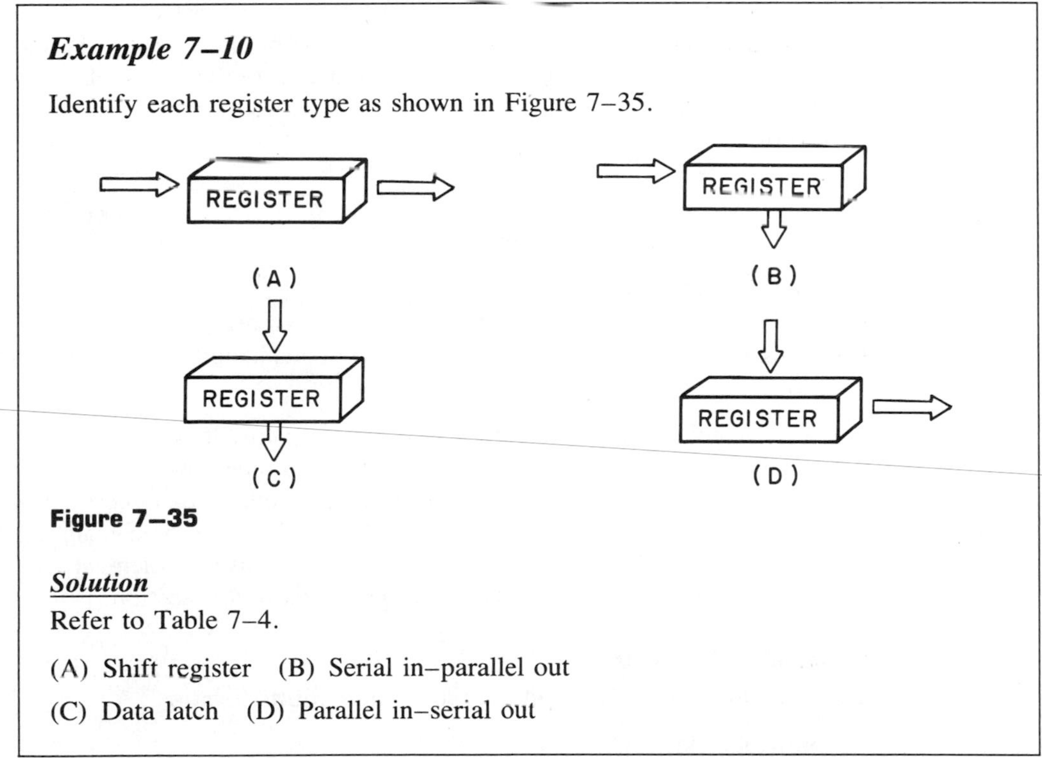

Figure 7–35

Solution

Refer to Table 7–4.

(A) Shift register (B) Serial in–parallel out

(C) Data latch (D) Parallel in–serial out

Conclusion

It was the purpose of this section to give you some idea of the use of flip-flops in registers. Registers are an important element in digital systems. They are so important, there is a whole chapter, Chapter 8, devoted to them—and, they are one of the major elements in the study of microprocessor technologies.

Check your understanding of this section by trying the following review questions.

7–4 Review Questions

1. What is a register?
2. Explain what a binary counter does.
3. State what kind of register could be used to transmit digital information from a computer over a telephone line.
4. What kind of a register would be used to receive digital information from a telephone line into a computer?

7–5 TYPES OF MULTIVIBRATORS

Discussion

In the previous sections you have studied two types of bistable elements: the latch and the flip-flop. As you will discover in this section, the bistable element is a special class of another more generalized logic circuit. There are two other logic elements that have their own special class on equal footing with the bistable elements. These are presented in this section.

Classification

Flip-flops are a class of **multivibrator.**

> **Multivibrator:**
> A two-state output circuit capable of acting as an oscillator by generating its own signal or maintaining one stable state or maintaining one of two stable states.

As you can see from the definition of a *multivibrator,* it is a very versatile circuit. The first multivibrators were constructed from vacuum tubes, were bulky, and were relatively expensive. Today, as you may expect, these useful circuits come in a variety of IC packages. These multivibrator circuits can be divided into three main categories, as illustrated in Figure 7–36. In the previous sections, the bistable element was presented. The monostable and the astable elements are presented in this section.

Monostable Multivibrator

A class of multivibrators is called the **monostable multivibrator.**

> **Monostable Multivibrator:**
> A multivibrator that will stay in one condition for a predetermined amount of time and then automatically go back to its prior condition.

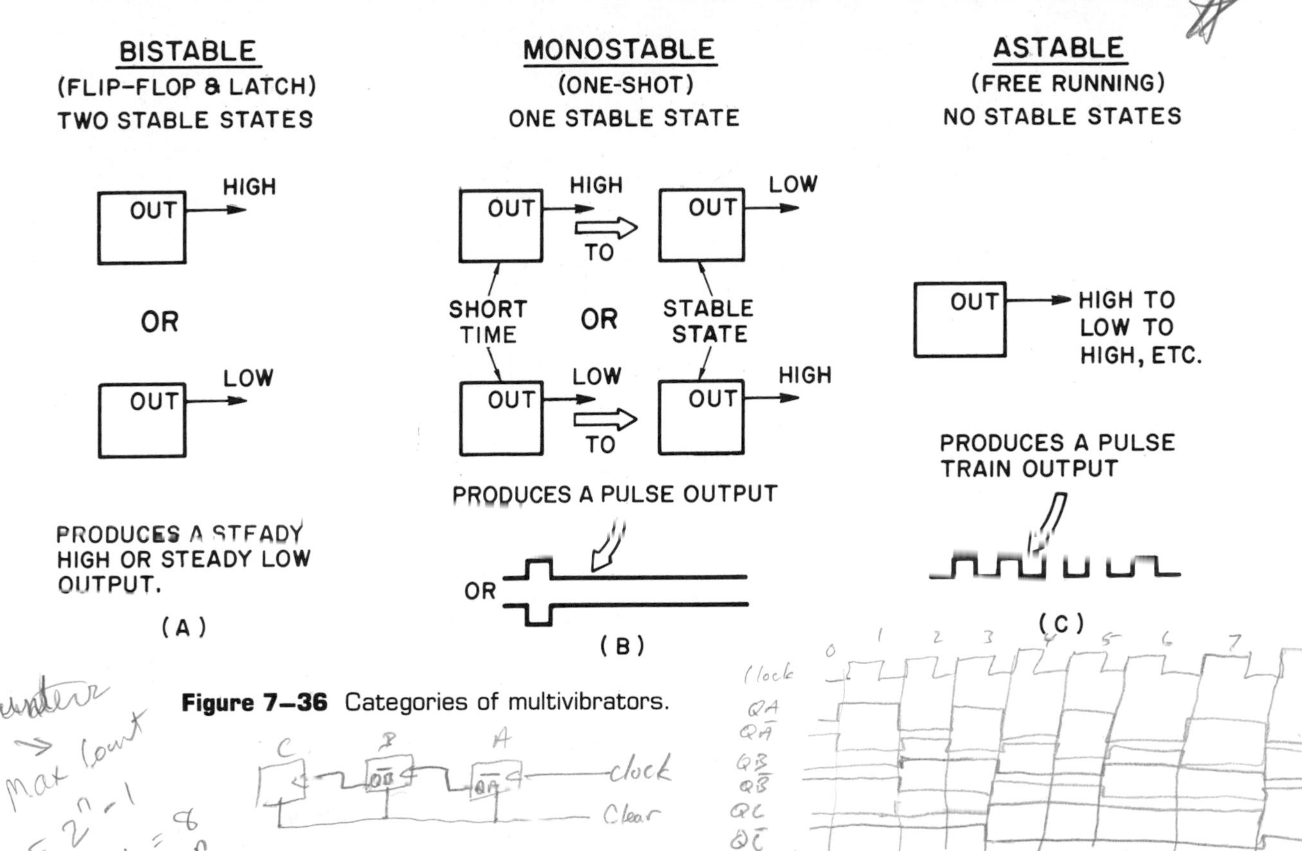

Figure 7–36 Categories of multivibrators.

The monostable multivibrator will produce an output pulse for a certain amount of time. This kind of multivibrator is often referred to as the *one-shot*. Figure 7–37 gives the basic idea of this type of digital circuit. As shown, the one-shot will only produce a change in its output for a certain amount of time. This change in output is caused by a brief change in the input, called a **trigger.**

Trigger:
A logic pulse of short duration used to start a digital process.

How long the device stays in its astable state depends upon the value of external circuit components. This is discussed in the electronic application section of this chapter.

Types of One-Shots

There are basically two types of one-shots: the retriggerable and the nonretriggerable. The operation of both of these types is illustrated in Figure 7–38. Observe that the retriggerable one-shot will produce a pulse every time it is triggered. In the case of the nonretriggerable one-shot, the effect of any trigger on its input is ignored as long as the device is in its ON time. This is illustrated by the following example.

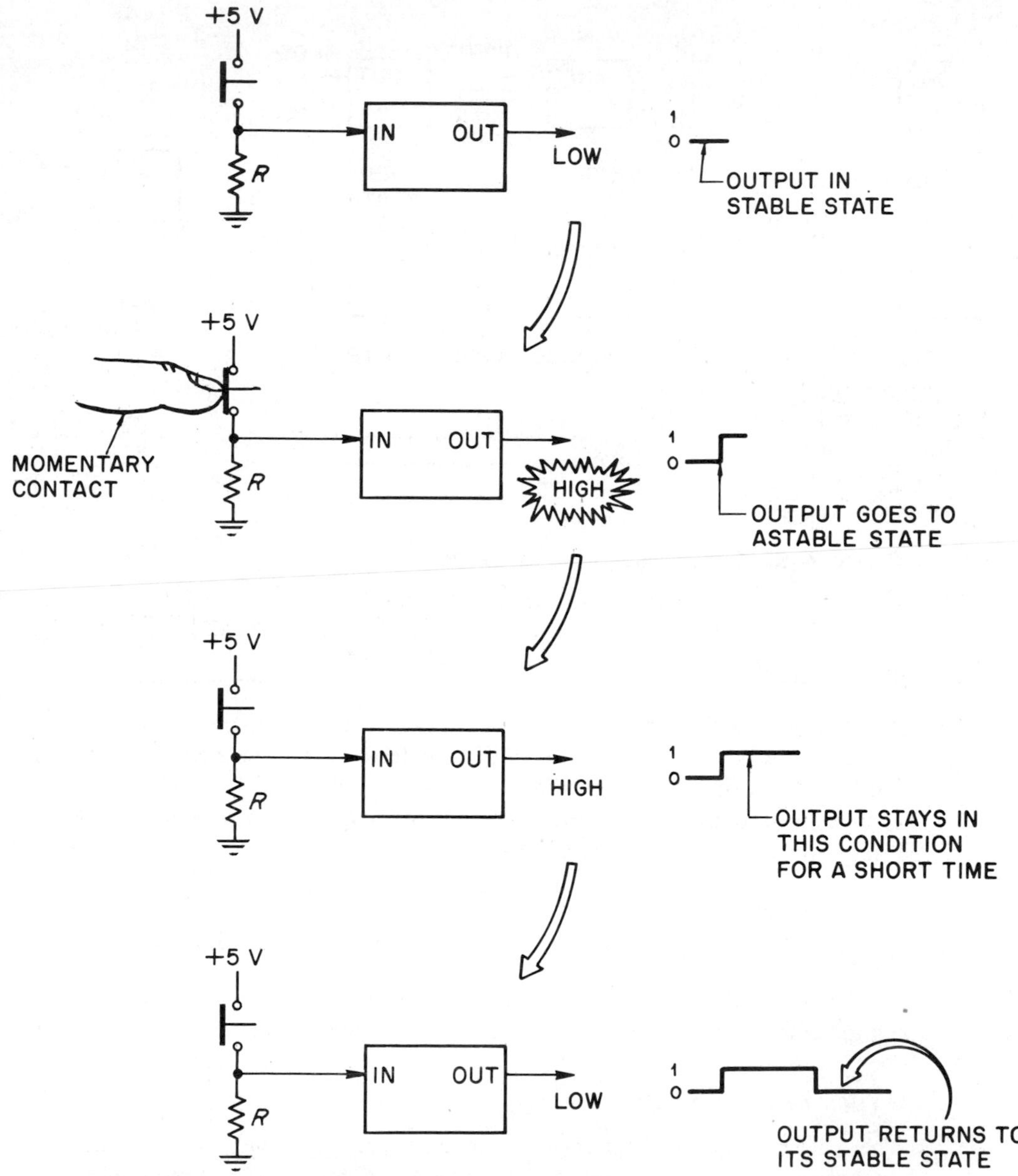

Figure 7–37 Basic idea of a one-shot.

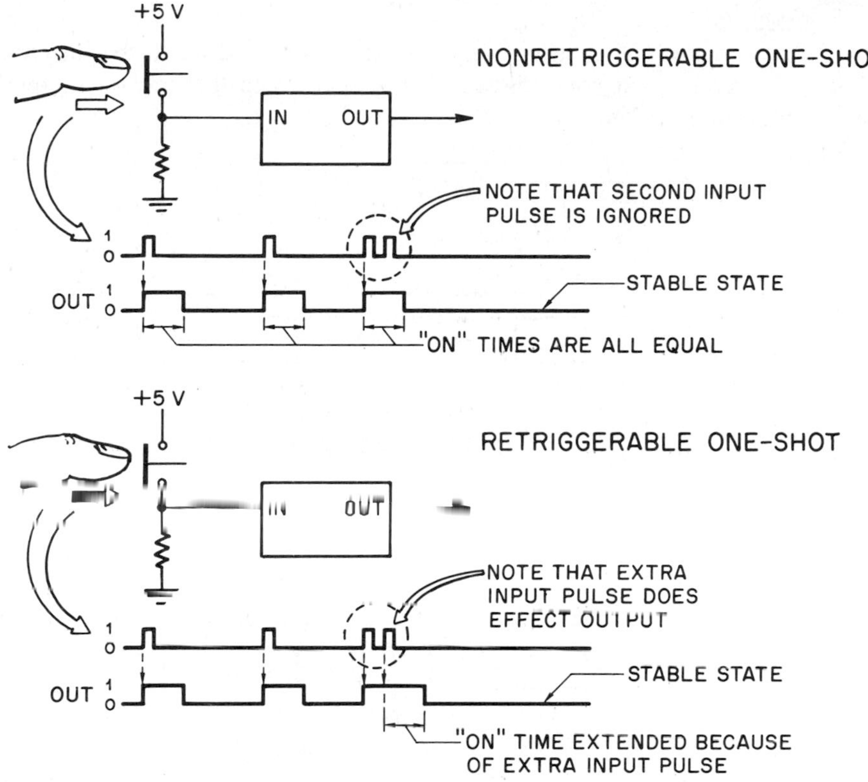

Figure 7–38 Action of a retriggerable and nonretriggerable one-shot.

Example 7–11

Sketch the output waveforms for the one-shots shown in Figure 7–39.

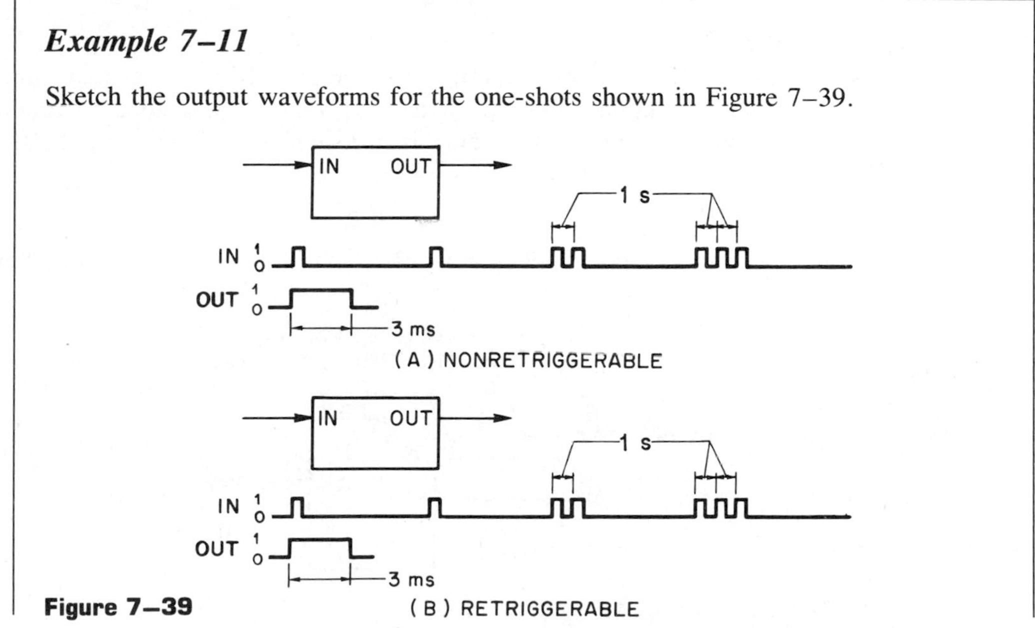

Figure 7–39

Solution

Recall that a one-shot has only one stable state and recall the difference between a retriggerable and nonretriggerable one-shot. With this in mind, Figure 7–40 illustrates the solution.

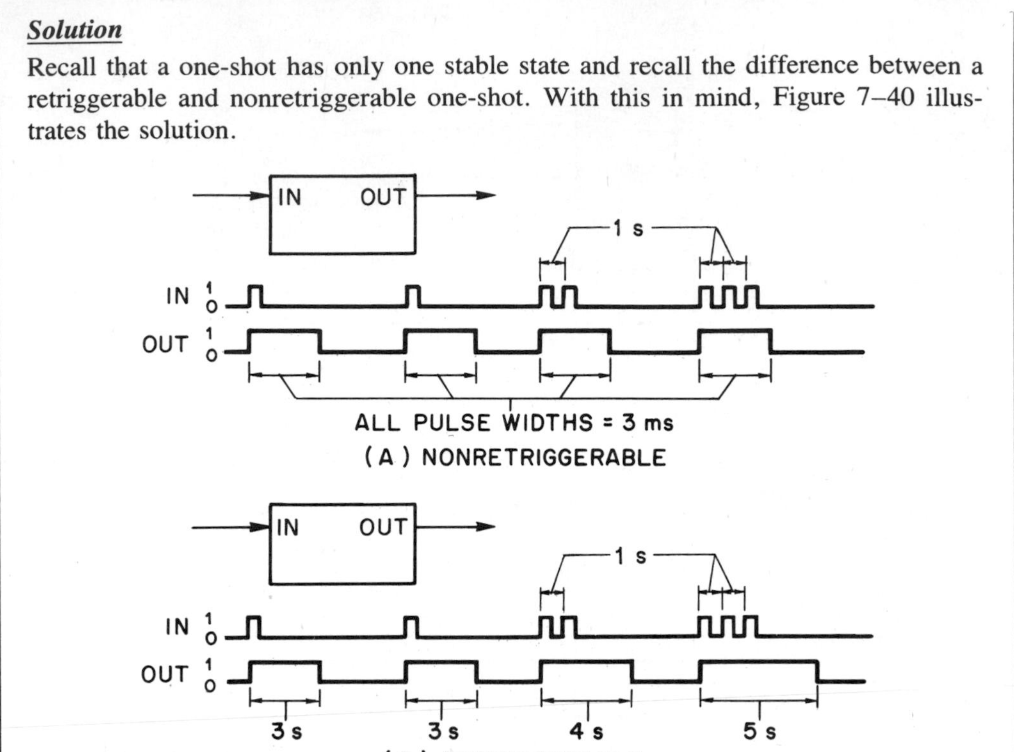

Figure 7–40

A Real One-Shot

Figure 7–41 shows the logic diagram of a real one-shot. Note that the amount of time a one-shot is on can be determined by the user. This is done by selecting appropriate values of circuit components. You will see how this is done in the electronic application section of this chapter.

One-Shot Application

There are many useful applications of a one-shot. One application is that of a **debouncer.**

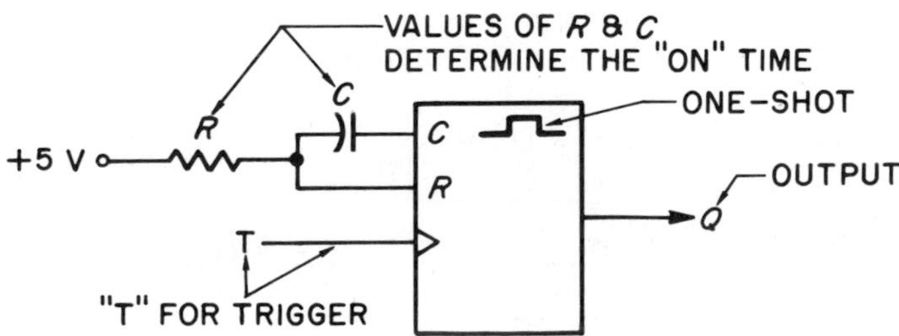

Figure 7–41 Logic diagram for a one-shot.

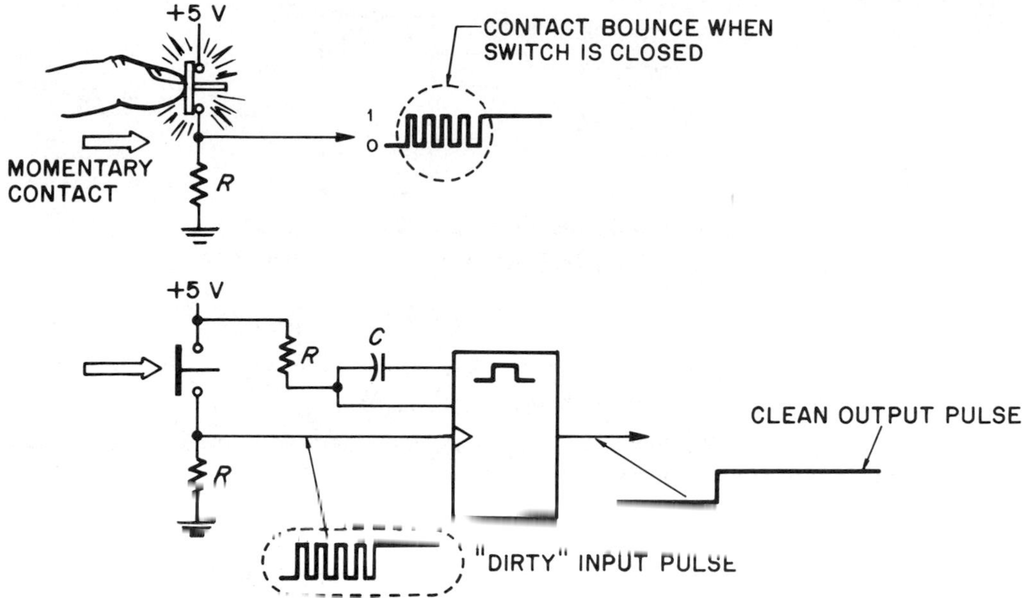

Figure 7–42 Use of one-shot for a switch debouncer.

Debouncer:
A logic circuit used to remove unwanted changes in logic levels usually caused by mechanical switch contacts.

When a contact is made in a mechanical switch, it is not immediate. There is a short period of time where the contacts actually "bounce," in part from the mechanical shock of being brought together. This can cause an unwanted rapid change in logic levels. For many digital circuits, such as the binary counter, this can be disastrous. Figure 7–42 illustrates the problem of contact bounce and the use of a one-shot as a switch debouncer.

Astable Multivibrators

The **astable multivibrator** is sometimes referred to as the free-running multivibrator.

Astable Multivibrator:
Sometimes called the "free-running" multivibrator— produces its own output logic waveform by constantly switching back and forth between HIGH and LOW.

The concept of such a multivibrator is shown in Figure 7–43. Note from the figure that the frequency of the output pulse as well as the **symmetry** of the output pulse can be controlled by the user of the device by selecting the proper values of resistors and capacitors.

Symmetry:
When used in digital, refers to the amount of pulse ON time versus the amount of pulse OFF time. A symmetrical pulse has the same amount of ON time as OFF time. A nonsymmetrical pulse does not have these equal times.

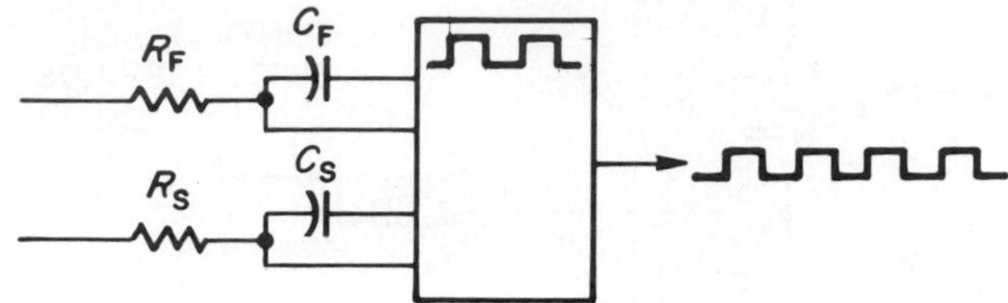

Figure 7–43 Basic idea of free-running multivibrator.

Free-Running Multivibrator Applications

These types of multivibrators play a very important role in practical digital systems. They are used by microprocessor systems to control the timing of all processes. As an example, this type of multivibrator produces a train of pulses used to increment the binary counter that feeds the address register. When these types of flip-flops are used to synchronize digital circuits, the resulting pulse waveform is referred to as a **sync pulse.**

Sync Pulse:
Abbreviation for synchronized. Pulses used to cause certain events to happen at a predetermined time and sequence.

Generally speaking, the *clock pulse* is used as the reference for all operations in a digital system. A *sync pulse* is usually considered as a specifically generated pulse within the same system. It may be generated by the clock pulse or another digital circuit. Figure 7–44 illustrates some applications of a free-running multivibrator.

Conclusion

This section presented the three types of flip-flops used in all digital systems. In the remaining sections of this chapter, you will study these flip-flops in necessary detail. What is important at this point is that you now know the major commonalities and differences of all flip-flops. Since you understand where you are going, you are now ready for the details.

Check your understanding of this section by trying the following review questions.

7–5 Review Questions

1. Describe what is meant by a *multivibrator*.
2. Name the three classifications of multivibrators, and describe what each one means.

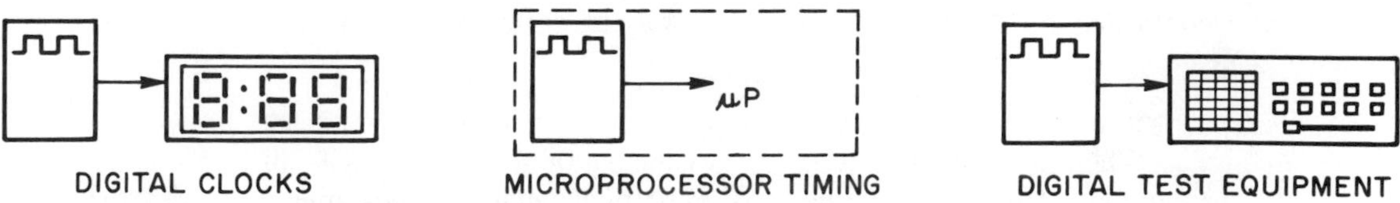

Figure 7–44 Applications of free-running multivibrators.

3. What is the purpose of a *switch debouncer?*
4. Describe the meaning of *symmetry*.
5. Explain what is meant by a *clock*.

7–6 MASTER-SLAVE TECHNIQUES

Discussion

This section concerns itself with the solution to a very practical problem. Here, the problem will first be presented and then the solution used by IC manufacturers will be shown.

A Problem

Recall from Section 7–4 that one of the applications of flip-flops is the ability to transfer digital information serially, 1 bit at a time, into a register. This kind of process is used as part of the process for transmission and reception of digital information over telephone lines. See Figure 7–45.

For the purpose of discussion, assume that the bit pattern you are attempting to serially input into a 4-bit register is 0101. Ideally, the transfer of data would be as shown in Figure 7–46. As you can see from the figure, it takes four clock pulses to load in the complete 4-bit word 0101. This kind of serial register can be constructed utilizing edge-triggered *D* flip-flops, as shown in Figure 7–47.

The basic idea of how the register is suppose to work is that the digital data (0101 in this case) is loaded into the first *D* flip-flop 1 bit at a time. The output of this flip-flop is connected to the control input of the next flip-flop, and so on down the line of flip-flops. The LOW to HIGH transition of each clock pulse is in step with each bit of the digital data. If you assume that the register starts out as 0000, then after the first clock pulse the input flip-flop will contain the least significant bit of the data word and the register will now contain 1000. At the next LOW to HIGH transition of the second clock pulse, the HIGH condition of the input flip-flop is felt on the *D* input of the second flip-flop. The second bit of the input word (LOW) is at the same time felt on the *D* input of the input flip-flop. When the clock transition occurs, the input flip-flop becomes LOW and the second flip-flop becomes HIGH. The word inside the register after the second clock pulse is now 0100.

This process continues, until after four clock LOW to HIGH transitions, the 4-bit word has been transferred. This is how it is suppose to work. In a practical sense, it may not always work. Figure 7–48 shows the potential problem.

Overcoming the Problem

The previous problem is so common in digital systems that manufacturers devised ways of constructing flip-flops that actually contained two flip-flops in one. This idea is now

Figure 7–45 Serial transmission and reception.

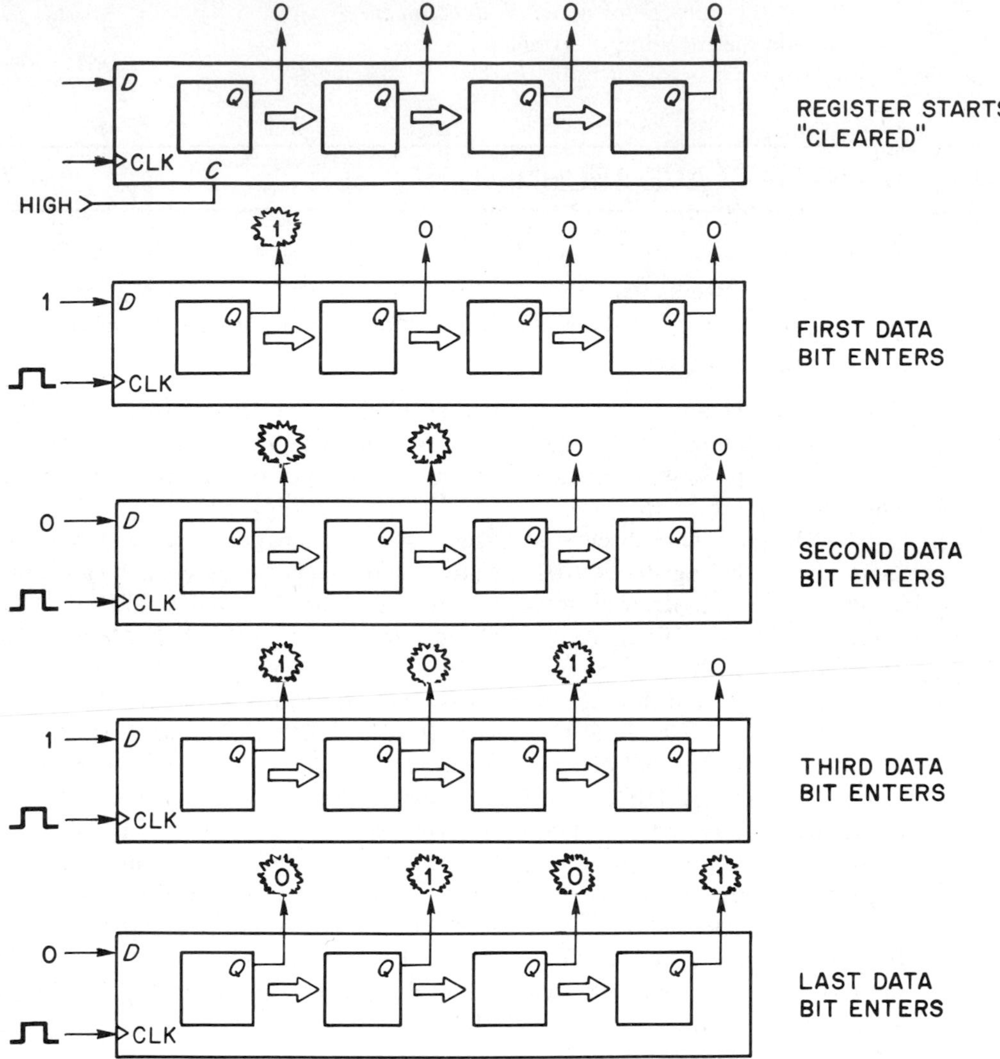

Figure 7–46 Serial loading example.

commonly used by all manufacturers of IC flip-flops. Figure 7–49 shows the basic idea. As shown in the figure, the idea is to keep the output of the flip-flop steady while it is being read. Remember, this was the problem with the serial input register: the output of a flip-flop could be changing at the same time it was trying to be read by a second flip-flop.

To overcome this problem, the level edge of the clock activates the input. Now the flip-flop is to "remember" what was at the input when activated. At the next clock level, the output will reflect what was "remembered" by the input.

This arrangement of flip-flops is called a **pulse-triggered master-slave flip-flop.**

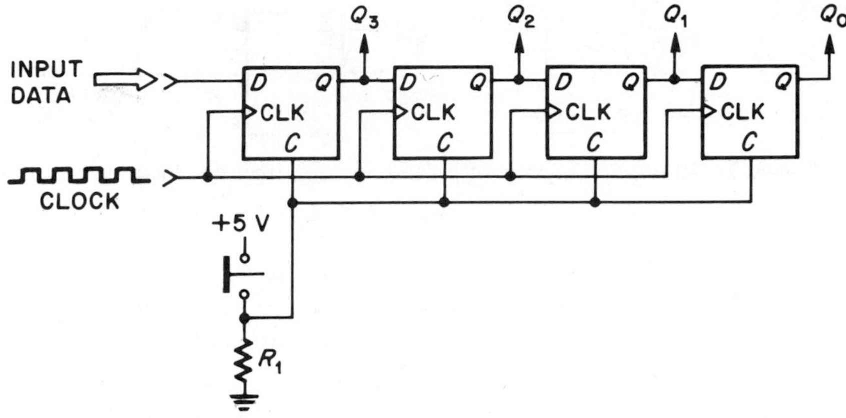

Figure 7–47 Utilizing D flip-flops for serial register.

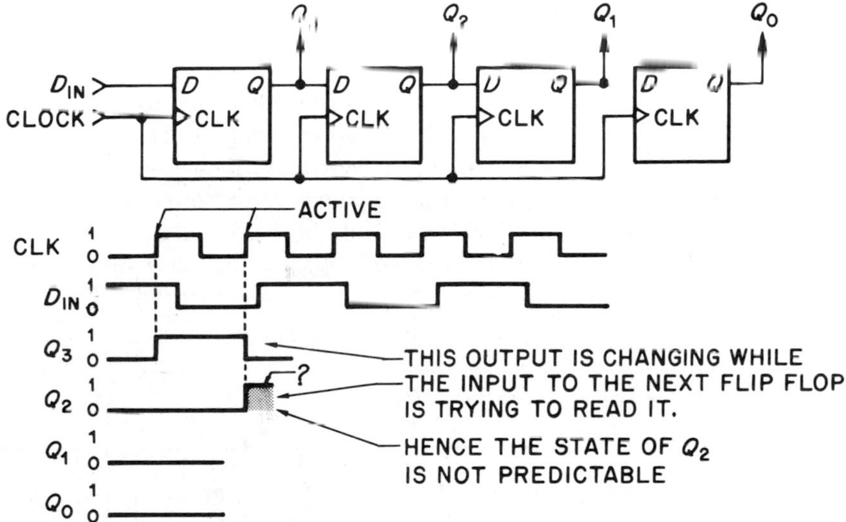

Figure 7–48 Potential register problem.

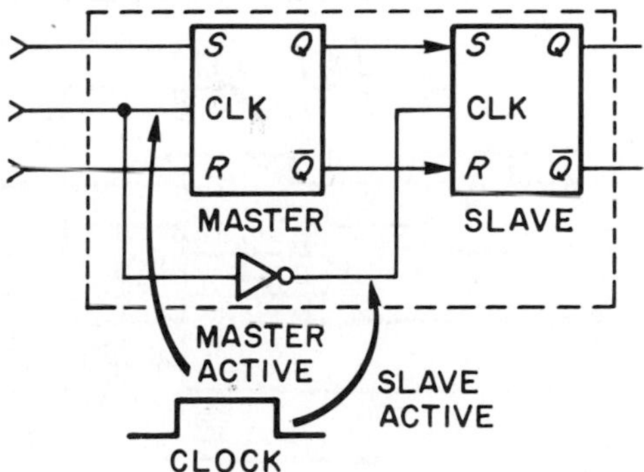

Figure 7–49 Basic idea of dual flip-flop construction.

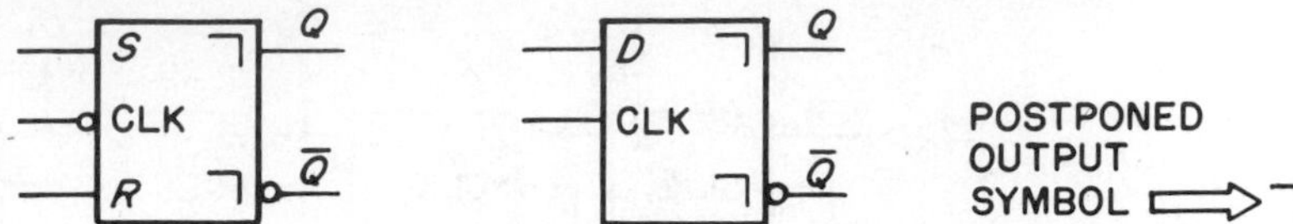

Figure 7–50 Logic diagram of pulse-triggered master-slave flip-flop.

Pulse-Triggered Master-Slave Flip-Flop:

A flip-flop whose inputs are active at one level of the clock and whose outputs will reflect the input condition only at a different level of the clock.

It is important that the control inputs to the pulse-triggered master-slave flip-flop do not change while the input is active, because this will cause the input to change. This flip-flop is another class of bistable multivibrator. The logic diagram is illustrated in Figure 7–50. Observe two features of the logic diagram: The use of the *postponed output* symbol at the outputs and the absence of the dynamic input indicator triangle.

By using a master-slave *D* flip-flop, you can now achieve a serial input register without having the problem of the output of a flip-flop trying to change while it is being read by the input of another. This is illustrated in Figure 7–51. As you can see, the outputs of the flip-flops are held steady while the control inputs are active. This happens on the active level of the clock pulse. The outputs become active when the inputs are inactive.

The master-slave flip-flop is available in the *S-R* flip-flop as well as in the *D* flip-flop. Usually conventional notation will not show the pulse sign on the output of the flip-flop of a master-slave flip-flop.

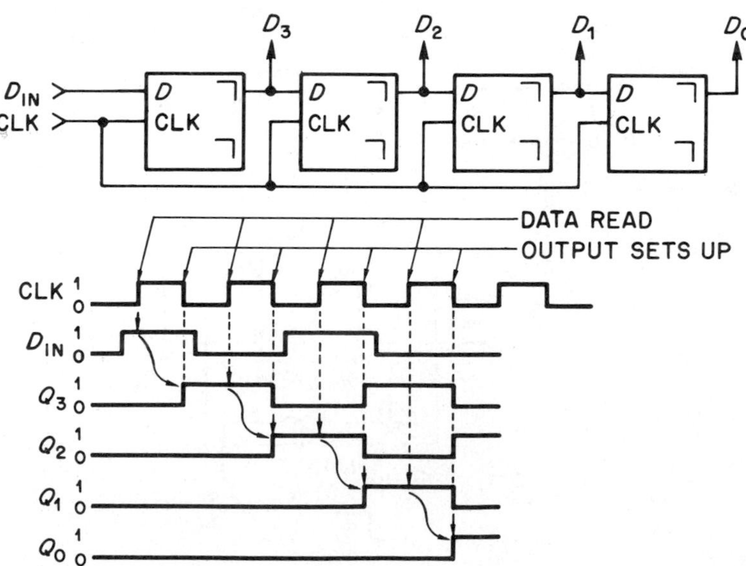

Figure 7–51 Action of master-slave flip-flops.

Example 7–12

For the input waveforms of Figure 7–52, determine the resultant waveforms.

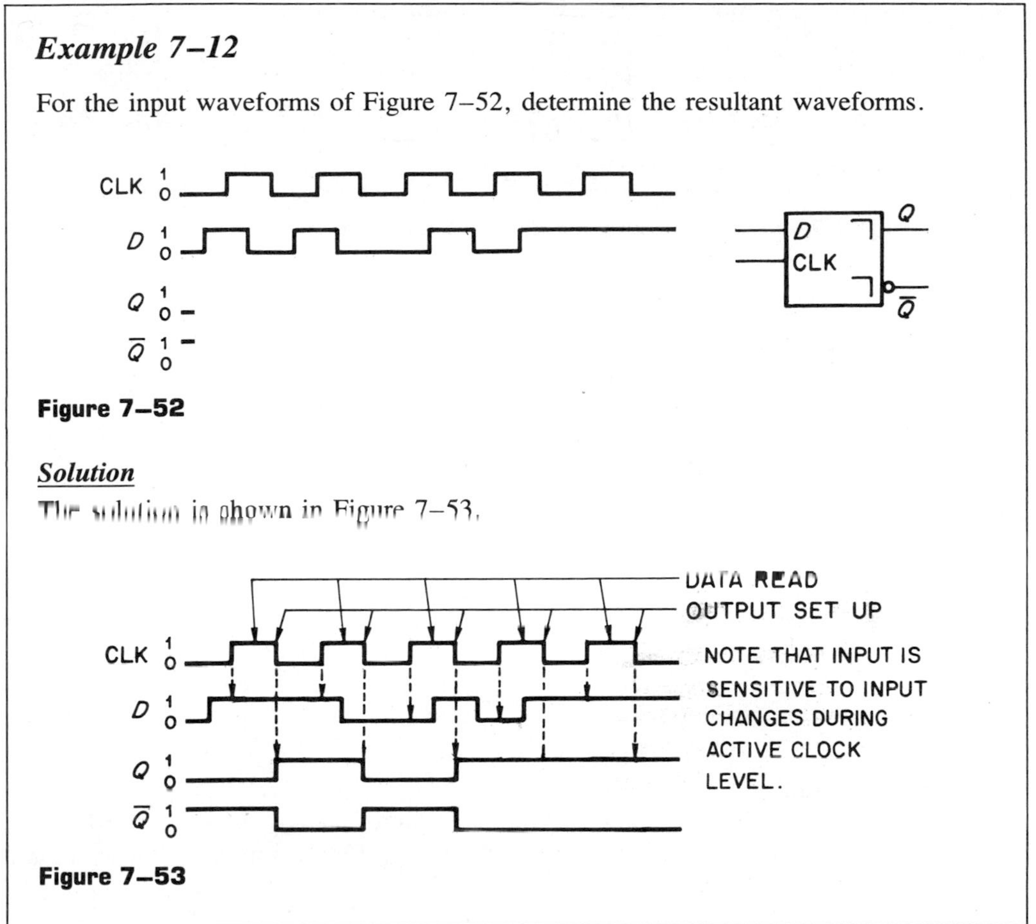

Figure 7–52

Solution

The solution is shown in Figure 7–53.

Figure 7–53

Data Lock-Out Flip-Flops

Data lock-out flip-flops are similar to the pulse-triggered master-slave flip-flops presented in the last section.

> **Data Lock-Out Flip-Flops:**
> A master-slave flip-flop where the input is sensitive to changes only during the clock transition.

The logical construction of a data lock-out flip-flop is shown in Figure 7–54. As you can see from the figure, the master portion is activated only during the clock transition. This means that input data may now change during the duration of the clock but its effect is "locked out." The slave portion of the flip-flop will reflect the condition of the master only after the next transition of the clock. Note the use of the postponed output symbol to indicate a master-slave relationship and the use of the dynamic input indicator to indicate edge triggering on the clock.

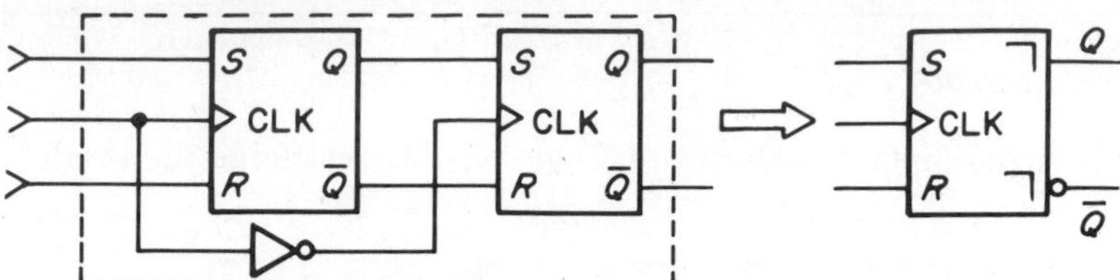

Figure 7–54 Logic diagram of data lock-out flip-flop.

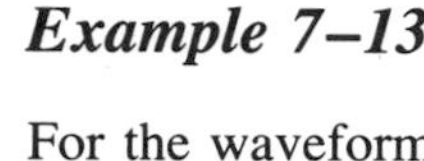

Example 7–13

For the waveforms shown in Figure 7–55, indicate the resulting outputs.

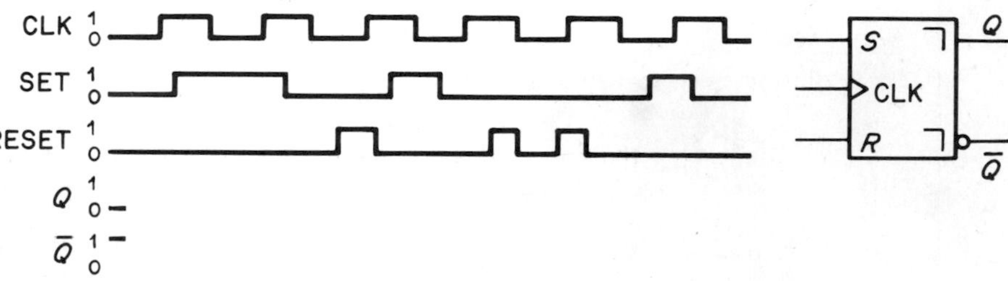

Figure 7–55

Solution

Keep in mind that the only time the input of the data lock out flip-flop is active is during the indicated transition of the clock. With this in mind, the solution is shown in Figure 7–56.

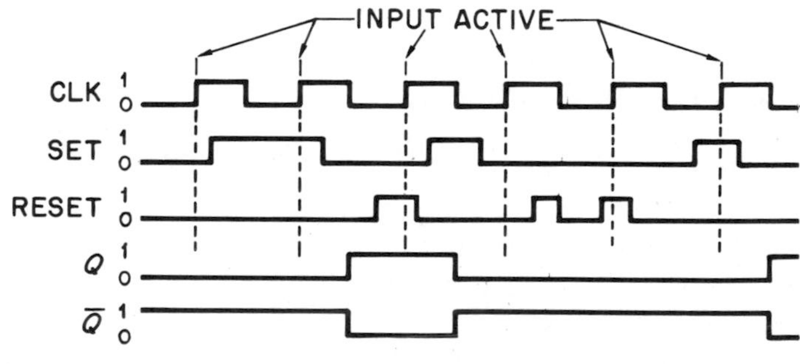

Figure 7–56

Conclusion

You have covered a wide variety of flip-flop designs in this chapter. In the next chapter, you will be introduced to another type of flip-flop that is used extensively in digital systems. In the next section, you will see some exciting applications of these versatile flip-flops. For now, test your understanding of this section by trying the following review questions.

7–6 Review Questions

1. Explain one of the problems in using a single-design flip-flop (one that is not master-slave).
2. Describe the basic construction of a master-slave flip-flop.
3. What are the timing relations in a master-slave flip-flop?
4. State the advantage of a master-slave flip-flop.

MICROPROCESSOR APPLICATION

Discussion

You are now ready to see an important development that took place in IC technology. This development allowed the use of a printed circuit board in the connection of hundreds of ICs and thousands of memory locations. This is effectively the way your personal computer is constructed. You will also be introduced to a new advanced version of a 4-bit hypothetical microprocessor called μP_2.

TRI-STATE Buffer

The first TRI-STATE device was developed by National Semiconductor Corporation. The term "TRI-STATE" is copyrighted by them. It is commonly used in the industry to describe this important development.

TRI-STATE:
A logic output that has three states: HIGH, LOW or not connected (sometimes called the high-impedance state).

Figure 7–57 shows the basic concept of a TRI-STATE buffer. What a TRI-STATE output accomplishes is the ability to make the output of a logic device electrically connect or disconnect from a circuit. To demonstrate the importance of this, consider the illustration in Figure 7–58. In this figure, the contents of register A or register B may be stored in register C. This is accomplished by selecting either register A or register B with the register select switch. This switch causes one group of TRI-STATE buffers to become active while the other group is disconnected. The active group of TRI-STATE buffers allows the contents of the corresponding register to appear on the data bus. This data may then be copied into register C where its contents will be displayed on the LEDs.

TRUTH TABLE

TRISTATE	IN	OUT
ACTIVE	1	1
ACTIVE	0	0
INACTIVE	X	OPEN

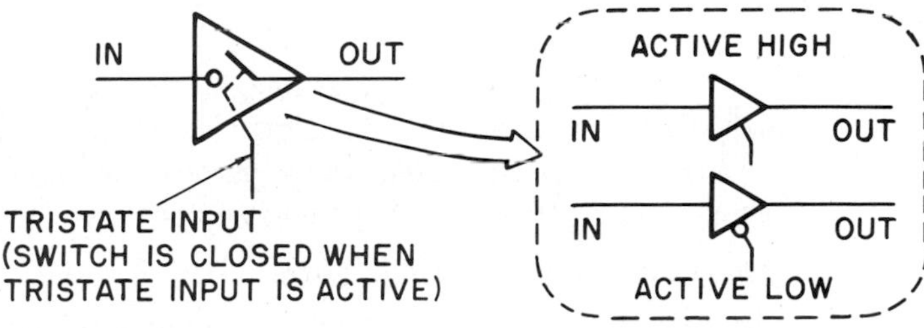

Figure 7–57 Basic concept of a TRI-STATE buffer.

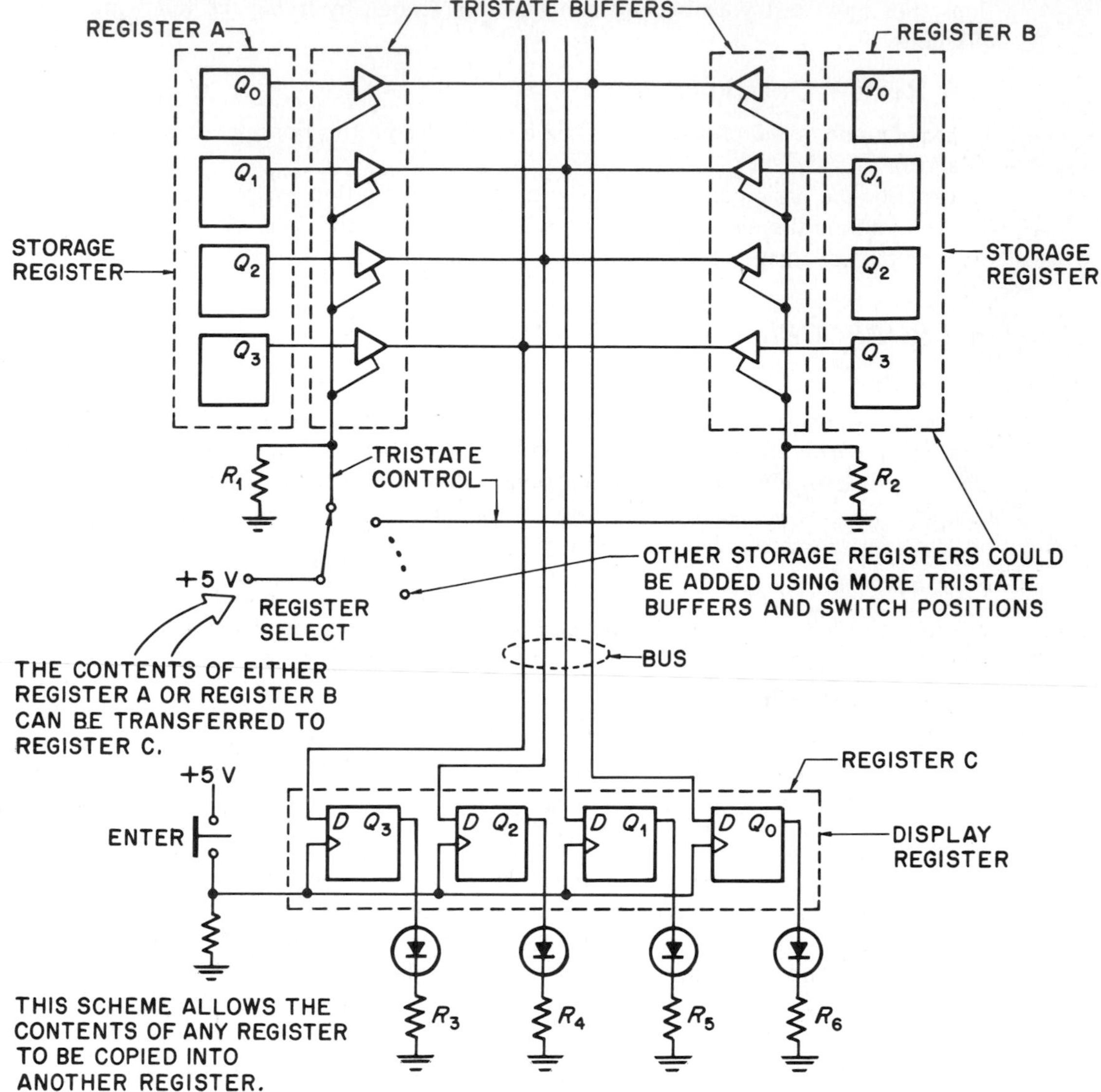

Figure 7–58 Application of TRI-STATE outputs.

TRI-STATE outputs are available in all logic circuits such as logic gates, flip-flops, and inverters. Figure 7–59 illustrates two types of TRI-STATE buffers.

As you will see in the following discussion, the use of TRI-STATE devices greatly increases the versatility of the microprocessor.

An Advanced 4-Bit Machine

The hypothetical 4-bit microprocessor you have been using up to this point served as a fine model. It helped you understand how the digital devices you were learning were

DM54LS126A/DM74LS126A Quad TRI-STATE® Buffers

General Description

This device contains four independent gates each of which performs a non-inverting buffer function. The outputs have the TRI-STATE feature. When enabled, the outputs exhibit the low impedance characteristics of a standard LS output with additional drive capability to permit the driving of bus lines without external resistors. When disabled, both the output transistors are turned off presenting a high-impedance state to the bus line. Thus the output will act neither as a significant load nor as a driver. To minimize the possibility that two outputs will attempt to take a common bus to opposite logic levels, the disable time is shorter than the enable time of the outputs.

Absolute Maximum Ratings (Note 1)

Supply Voltage	7V
Input Voltage	7V
Storage Temperature Range	−65°C to 150°C

Note 1: The "Absolute Maximum Ratings" are those values beyond which the safety of the device can not be guaranteed. The device should not be operated at these limits. The parametric values defined in the "Electrical Characteristics" table are not guaranteed at the absolute maximum ratings. The "Recommended Operating Conditions" table will define the conditions for actual device operation.

Connection Diagram

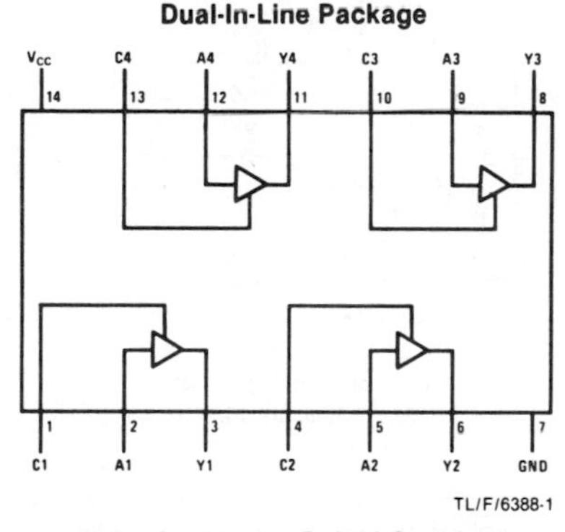

DM54LS126A (J) DM74LS126A (N)

Function Table

Y = A

Input		Output
A	C	Y
L	H	L
H	H	H
X	L	Hi-Z

H = High Logic Level
L = Low Logic Level
X = Either Low or High Logic Level
Hi-Z = TRI-STATE (Outputs are disabled)

Figure 7–59 Typical TRI-STATE outputs. *Reprinted* with permission of National Semiconductor Corporation.

DM54LS125A/DM74LS125A Quad TRI-STATE® Buffers

General Description

This device contains four independent gates each of which performs a non-inverting buffer function. The outputs have the TRI-STATE feature. When enabled, the outputs exhibit the low impedance characteristics of a standard LS output with additional drive capability to permit the driving of bus lines without external resistors. When disabled, both the output transistors are turned off presenting a high-impedance state to the bus line. Thus the output will act neither as a significant load nor as a driver. To minimize the possibility that two outputs will attempt to take a common bus to opposite logic levels, the disable time is shorter than the enable time of the outputs.

Absolute Maximum Ratings (Note 1)

Supply Voltage	7V
Input Voltage	7V
Storage Temperature Range	−65°C to 150°C

Note 1: The "Absolute Maximum Ratings" are those values beyond which the safety of the device can not be guaranteed. The device should not be operated at these limits. The parametric values defined in the "Electrical Characteristics" table are not guaranteed at the absolute maximum ratings. The "Recommended Operating Conditions" table will define the conditions for actual device operation.

Connection Diagram

Dual-In-Line Package

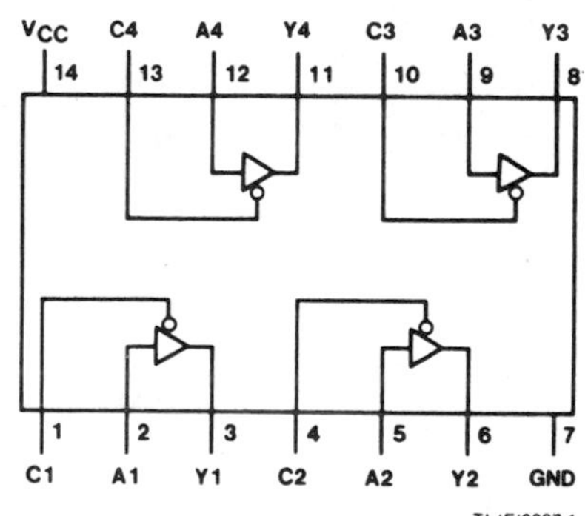

DM54LS125A (J) DM74LS125A (N)

Function Table

Y = A

Input		Output
A	**C**	**Y**
L	L	L
H	L	H
X	H	Hi-Z

H = High Logic Level
L = Low Logic Level
X = Either Low or High Logic Level
Hi-Z = TRI-STATE (Outputs are disabled)

Figure 7–59 Continued

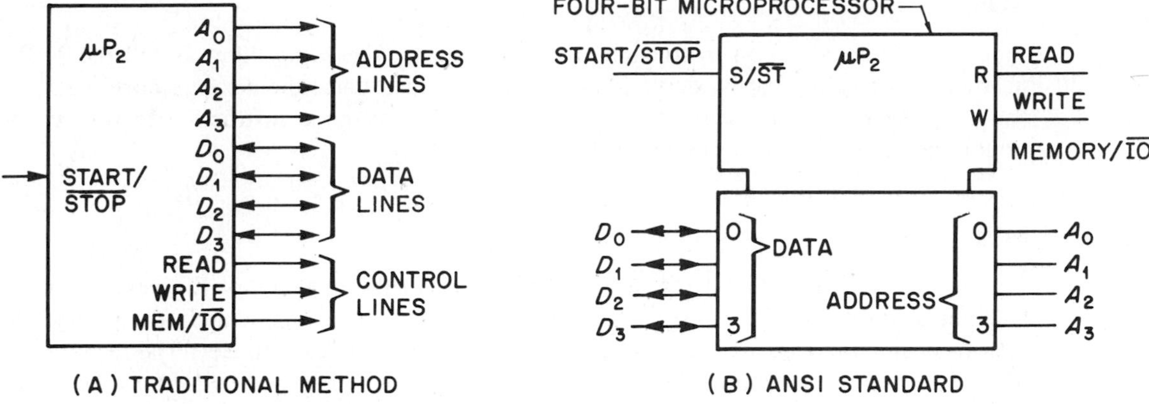

Figure 7–60 Logic diagram of μP_2.

applied to a microprocessor system. With your understanding of flip flops and TRI-STATE logic, you are now ready for a new microprocessor.

To distinguish between the first 4-bit microprocessor and this new 4-bit microprocessor, the terminology μP_2 will be used for the new 4-bit machine. From this point on, unless otherwise noted, any discussion of a hypothetical 4-bit microprocessor will be about the μP_2.

Figure 7–60 shows the logic diagram of the new 4-bit microprocessor. Notice a few new additions. First, there is a separate READ line and a separate WRITE line. There is also a new output line called MEM/$\overline{\text{IO}}$. Note from this figure that two different logic diagrams of the microprocessor are shown. Figure 7–60(A) is the traditional method of showing microprocessors while Figure 7–60(B) is the new ANSI standard. You should become familiar with both types.

The internal registers inside μP_2 are essentially the same as in μP_1; they are all 4-bit registers. The purpose and application of the separate READ and WRITE lines along with the new MEM/$\overline{\text{IO}}$ line will be presented shortly. For now, it's important to look at the new instruction set.

New Instruction Set

The bit patterns, mnemonic, and definitions of its first three instructions follow.

Operation Code	Mnemonic	Definition
0000	NOP	No operation
0001	MVI, A	Move immediate (accumulator)
0010	LDA	Load accumulator from memory
1111	HLT	Halt (stop processing)

Here is what each of these instructions means.

NOP

This is the No Operation instruction. It means exactly the same thing as did the NOP instruction for μP_1. That is, don't do anything more than go to the next memory location and read the next instruction. As you will see, this is a good instruction to use when troubleshooting a microprocessor-based system.

MVI

This means load an internal register immediately with the contents of the next memory location. For example, the instruction MVI A,5 means move immediately the number 5 into the accumulator. This is the same as the LDA instruction used in μP_1. The instruction MVI B,7 means move immediately the number 7 into the B register; and this is the same as LDB 7 used with μP_1. One of the reasons for changing the mnemonic here is so it can easily be distinguished from the LDA instruction.

LDA

This instruction means load the accumulator directly with the contents of a given memory location. For example, the instruction LDA 5 means to load the accumulator with the bit pattern contained in memory location 5. What the accumulator will contain will not necessarily be the number 5 but the number that was contained in memory location 5. If the bit pattern in memory location 5 happened to contain the number 3, then the instruction LDA 5 would cause the number 3 to be loaded into the accumulator.

Figure 7–61 shows the differences between the MVI and LDA instructions. As shown in the figure, μP_2 has two different ways of getting data from memory. It can get data from the memory location *immediately* following the location of the instruction, or it can get data from a different memory location at an address *directed* to by a part of the instruction. These two different methods of getting data into an internal register of the microprocessor are called **addressing modes.**

Addressing Modes:
The method used by the microprocessor for transferring information.

The two addressing modes illustrated in Figure 7–61 are called **immediate addressing** and **direct addressing.**

Immediate Addressing:
The process of transferring data located in the memory location immediately following the instruction into an internal register of the microprocessor.

Direct Addressing:
The process of transferring data between the microprocessor and the external system by designating the address for the location of the data.

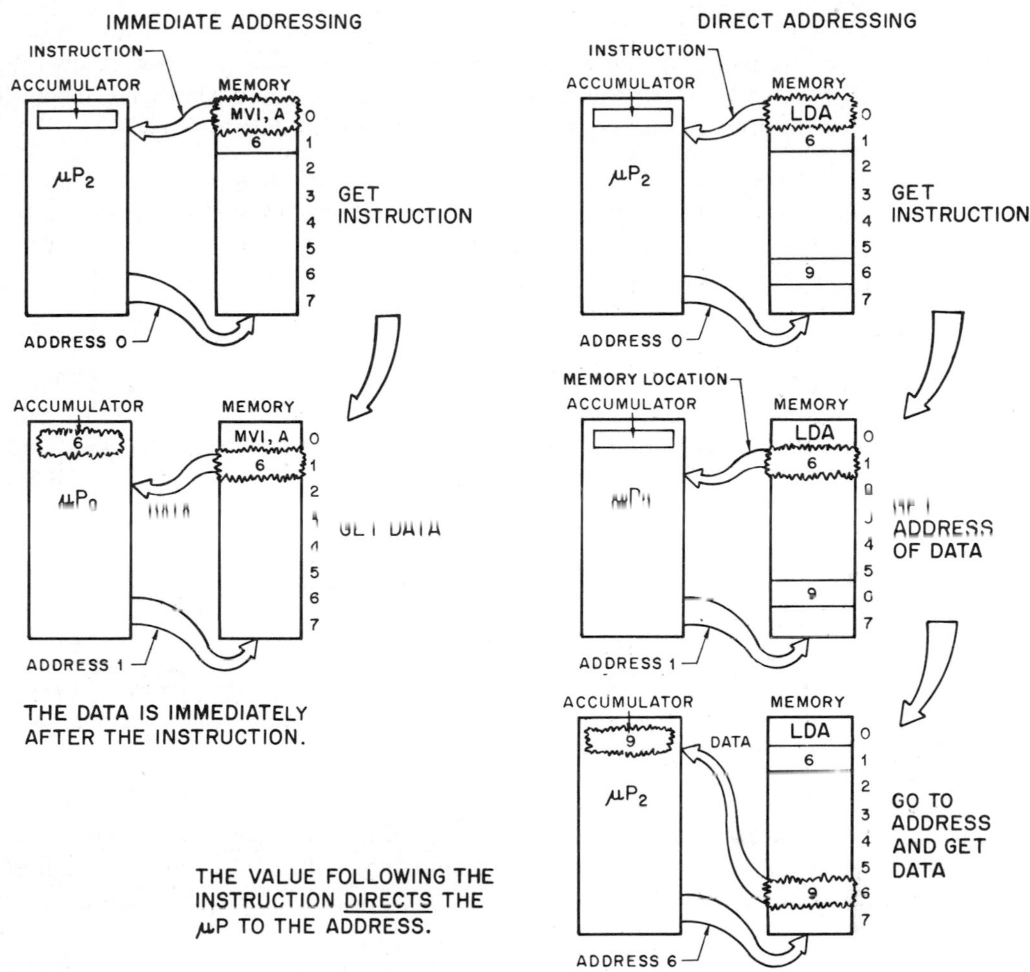

Figure 7–61 Differences between addressing modes.

Example 7–14

For the following assembly language programs, show how the bit patterns would appear in memory before and after program execution. Also state what the final contents of the accumulator would be after program execution.

PROGRAM A	
Address	**Instruction**
0	MVI A, 3
2	HLT
3	B

PROGRAM B	
Address	**Instruction**
0	LDA 3
2	HLT
3	B

Solution

The mnemonics are converted to their operation codes. Program A uses the *immediate* addressing mode, and program B uses the *direct* addressing mode. The distinction between the two programs is illustrated in Figure 7–62.

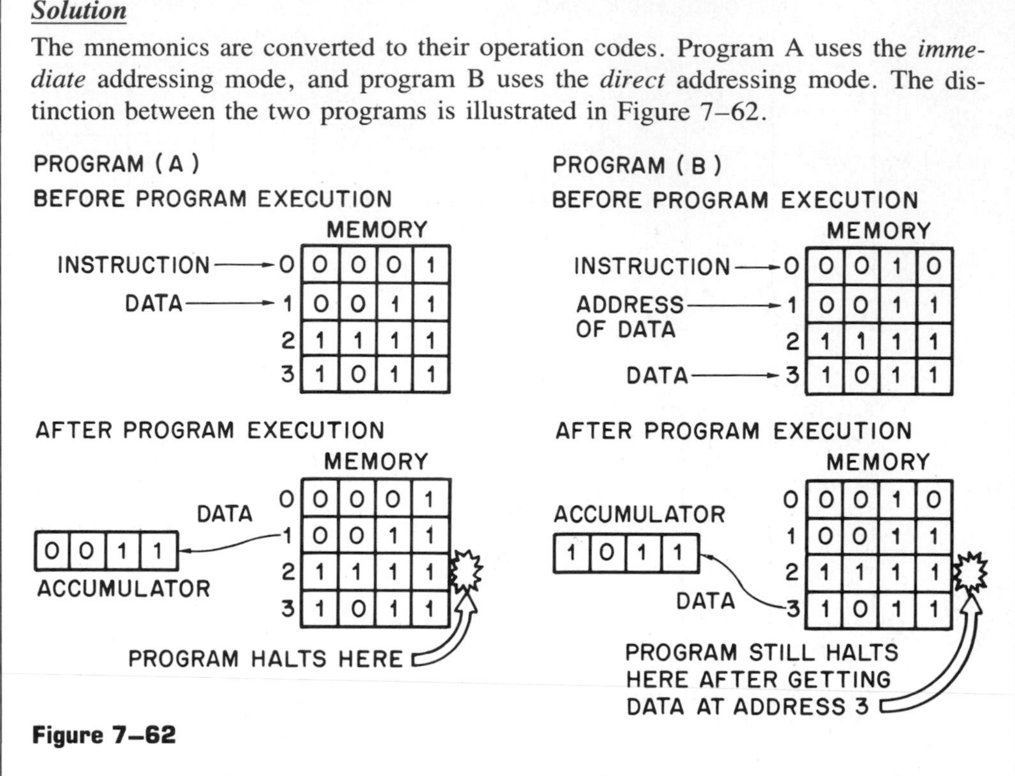

Figure 7–62

It is very important that you clearly understand the difference between the two programs illustrated in the previous example. The utilization of the two addressing modes will be used to show the application of TRI-STATE logic and flip-flops to microprocessor-based systems.

Input/Output

You have seen several examples of how a microprocessor-based digital system can output information to control external devices such as motors. Several of these examples were also included as end-of-chapter exercises. In the last chapter, you also saw how μP_1 could be used to output data to a seven-segment display.

In microprocessor-based digital systems there are methods for getting information from other sources, such as a hex keypad, on/off switch, telephone line, digital thermometer, digital pressure gauges, another computer, voice or vision recognition. The process of getting information into and out of a digital system is called **input/output** and is abbreviated *I/O*.

Input/Output (I/O):
The process of transferring information between a digital system and devices external to the digital system.

Figure 7–63 illustrates two IC chips that are used with the new microprocessor to perform I/O. The ENABLE line is tied to the clock input of each data latch in the buffer

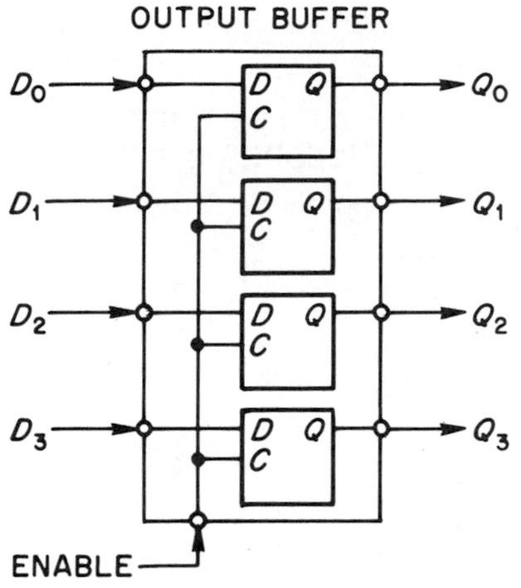

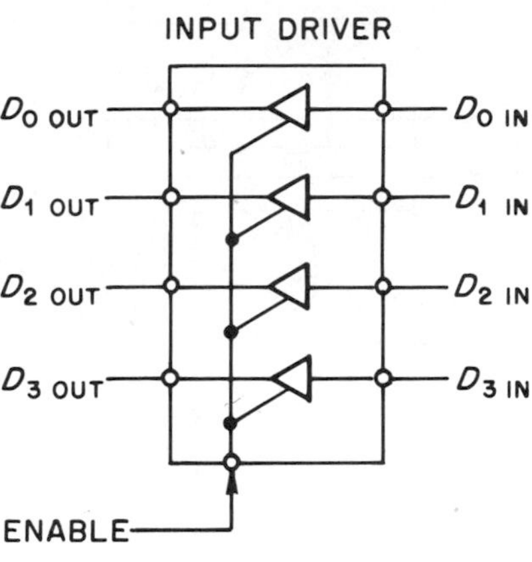

Figure 7–63 Output buffer and input driver

register. Thus, the bit pattern on the data lines will be stored in the buffer when the ENABLE line is HIGH. For the input driver, the ENABLE line controls the condition of the TRI-STATE buffers. When this line is LOW, the outputs of the input driver appear disconnected (in their high impedance state).

The purpose of the output buffer is to store a 4-bit word. Its operation is illustrated in Figure 7–64. As shown in the figure, the selector switch activates only one set of latches at a time. The position of the data input switches will then determine the bit pattern stored in the output buffer.

The purpose of the input driver is to transfer a 4-bit word from its input to its output when the ENABLE line is active. Its operation is illustrated in Figure 7–65. As shown in the figure, the position of the selector switch determines which input driver will have its TRI-STATE buffers active. The settings are transferred to the data bus where they will be displayed on the LEDs. This output buffer and input driver will open a whole world of applications for the digital devices you have been learning, all the way from music and voice synthesizers to the automated control of complex assembly lines. But first, two more new instructions need to be introduced.

Operation Code	Mnemonic	Definition
0000	NOP	No operation
0001	MVI, A	Move immediate (accumulator)
0010	LDA	Load accumulator from memory
0011	OUT	Output accumulator to a device
0100	IN	Input to accumulator from a device
1111	HLT	Halt (stop processing)

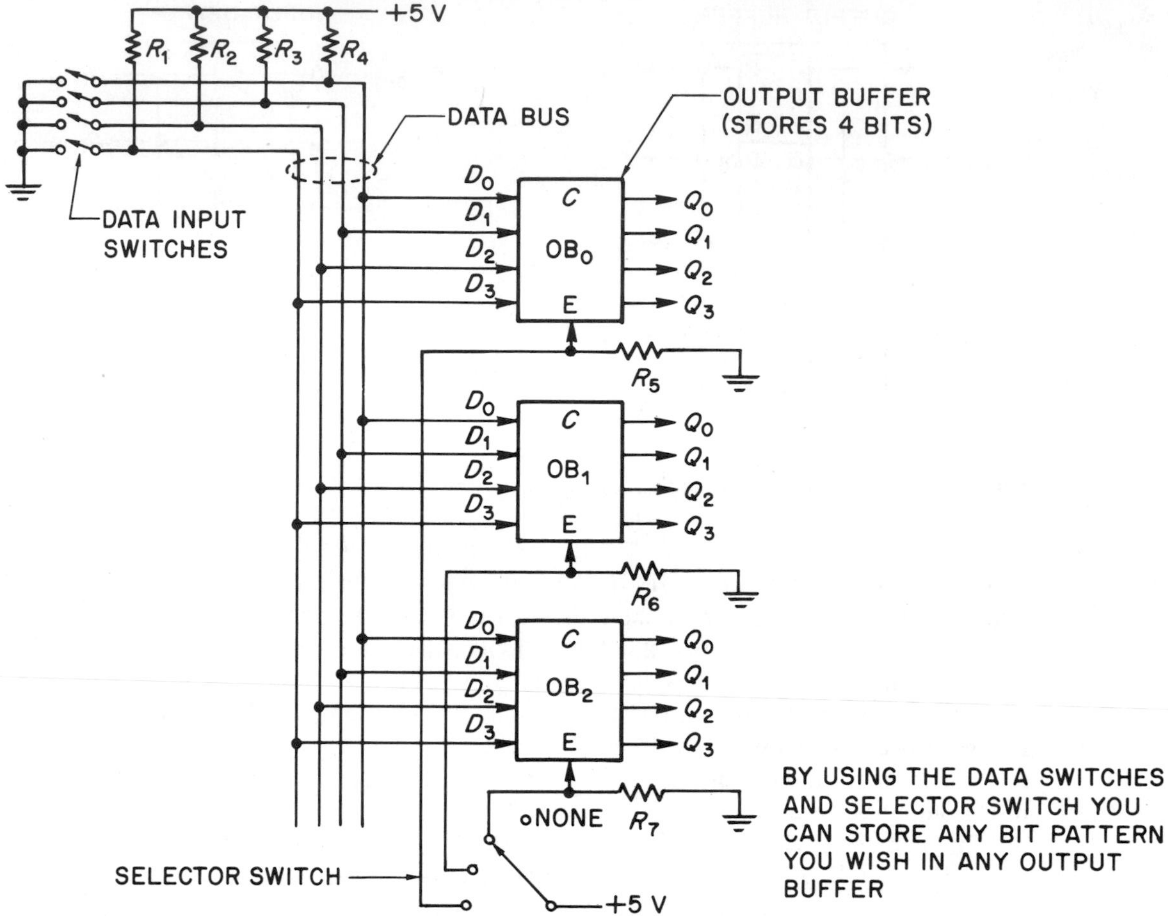

Figure 7–64 Operation of output buffer.

Some More Instructions

The two new instructions are OUT and IN. What both of these instructions do is cause the MEM/$\overline{\text{IO}}$ line to become LOW. They are the only instructions that will do that. All the other instructions will keep the MEM/$\overline{\text{IO}}$ line HIGH. There is one more thing each instruction does. The OUT instruction will cause the WRITE line to be HIGH while the IN instruction will cause the READ line to be HIGH. The addressing mode used by both of these instructions is direct, so IN 5 means input the data at I/O address 5 into the accumulator. OUT 5 means output the contents of the accumulator to I/O location 5. This is illustrated in the following example.

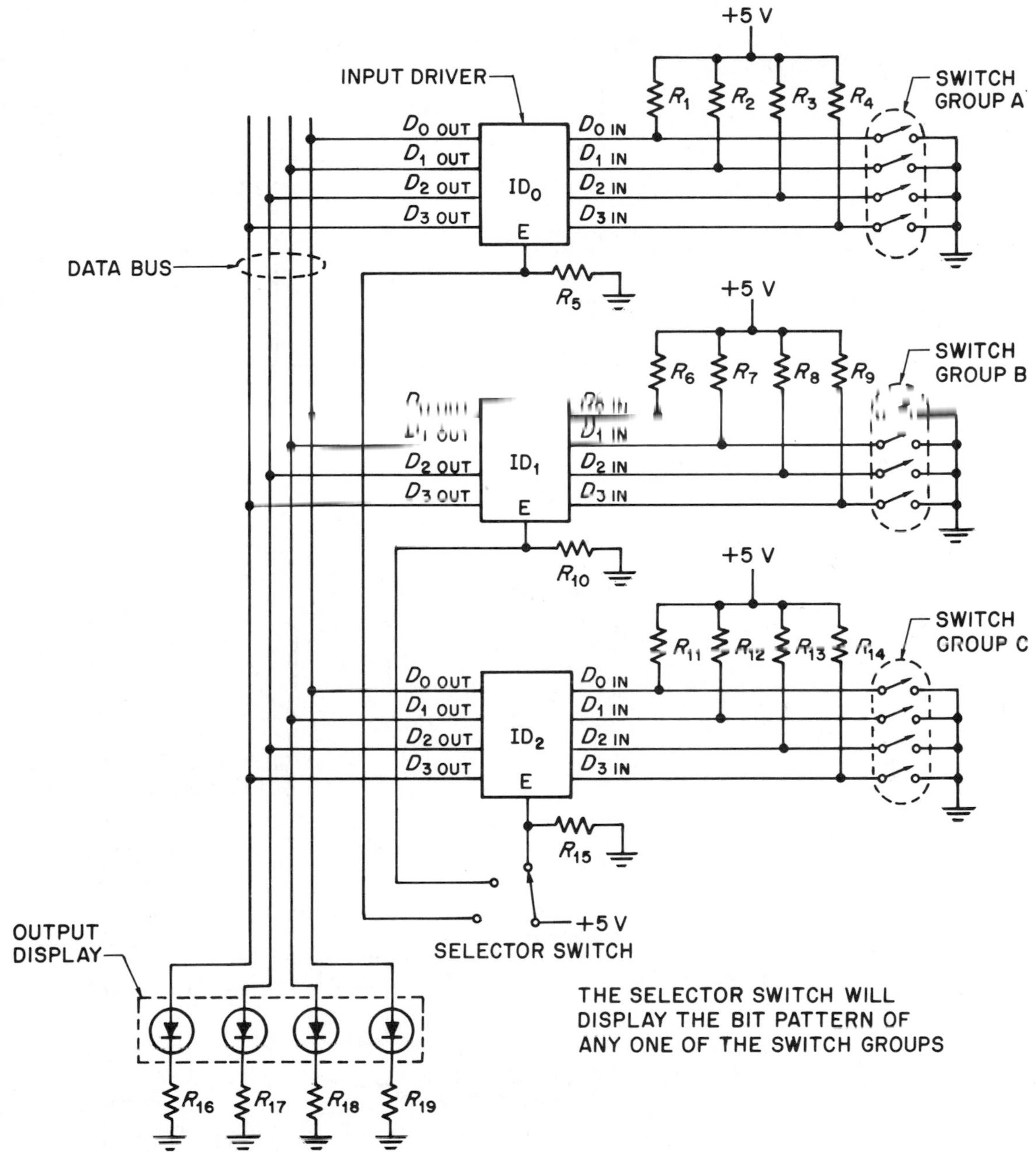

Figure 7–65 Operation of the input driver.

Example 7–15

For each of the following programs, indicate the condition of the READ, WRITE, and MEM/$\overline{IO}$ lines of the 4-bit microprocessor. Assume accumulator contents = 1011_2 and input port 3 = 0110_2.

PROGRAM A	
Address	**Instruction**
0	OUT 3
2	HLT

PROGRAM B	
Address	**Instruction**
0	IN 3
2	HLT

Solution

Keep in mind the effect that the IN and OUT instructions have on the control lines of the microprocessor. The solution is presented in Figure 7–66.

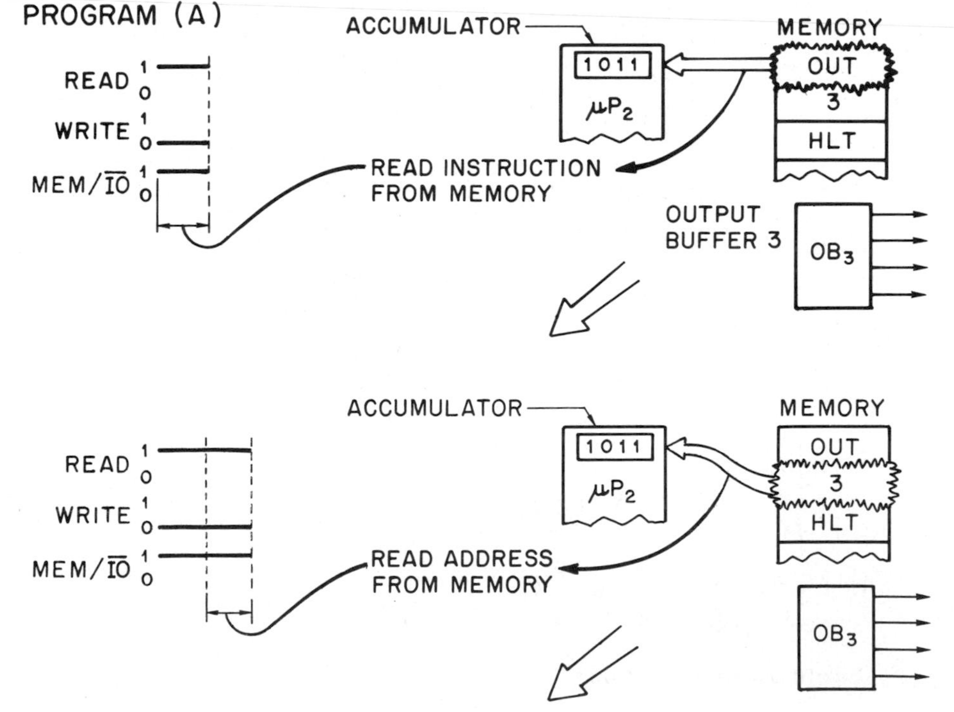

Figure 7–66

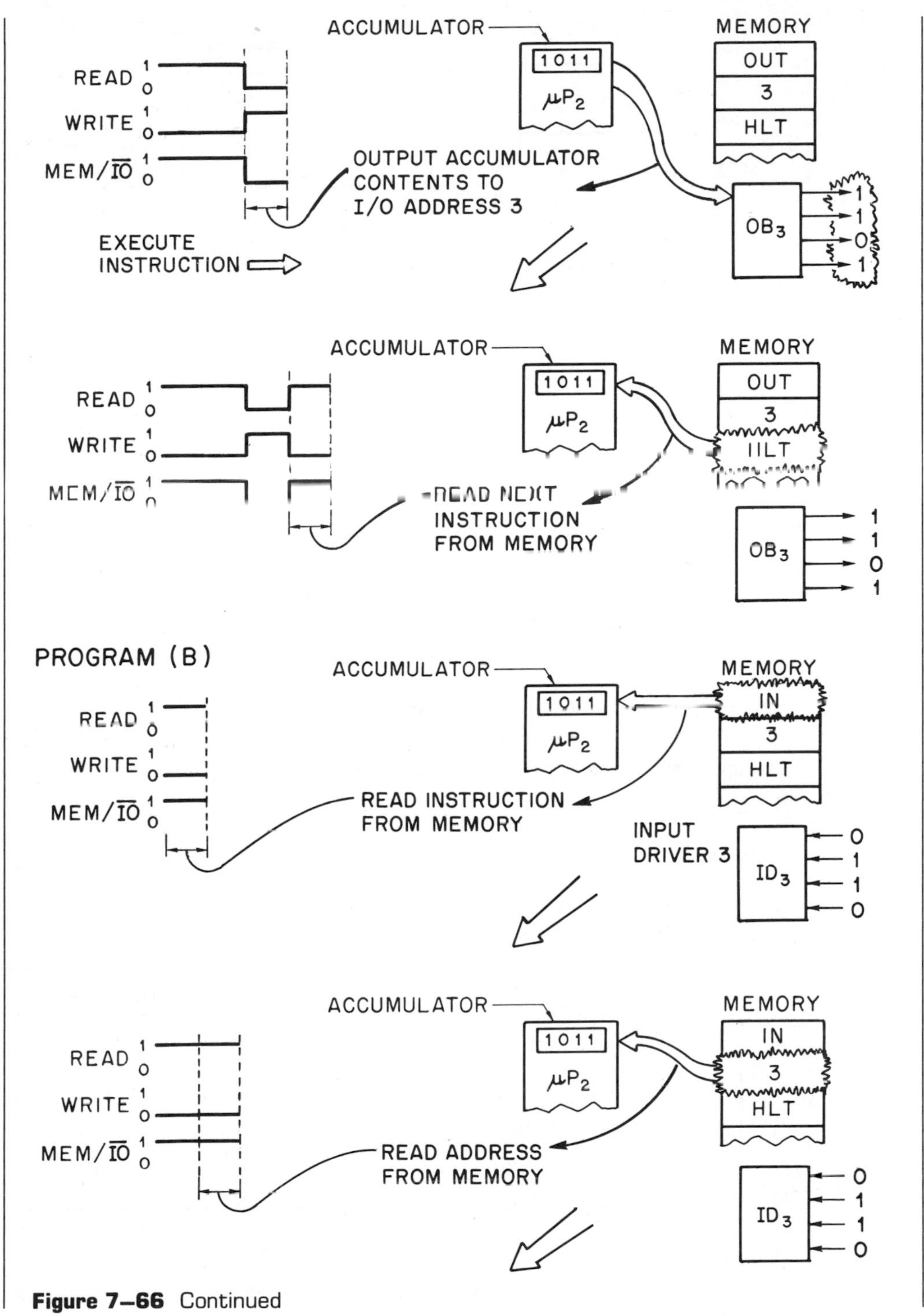

Figure 7–66 Continued

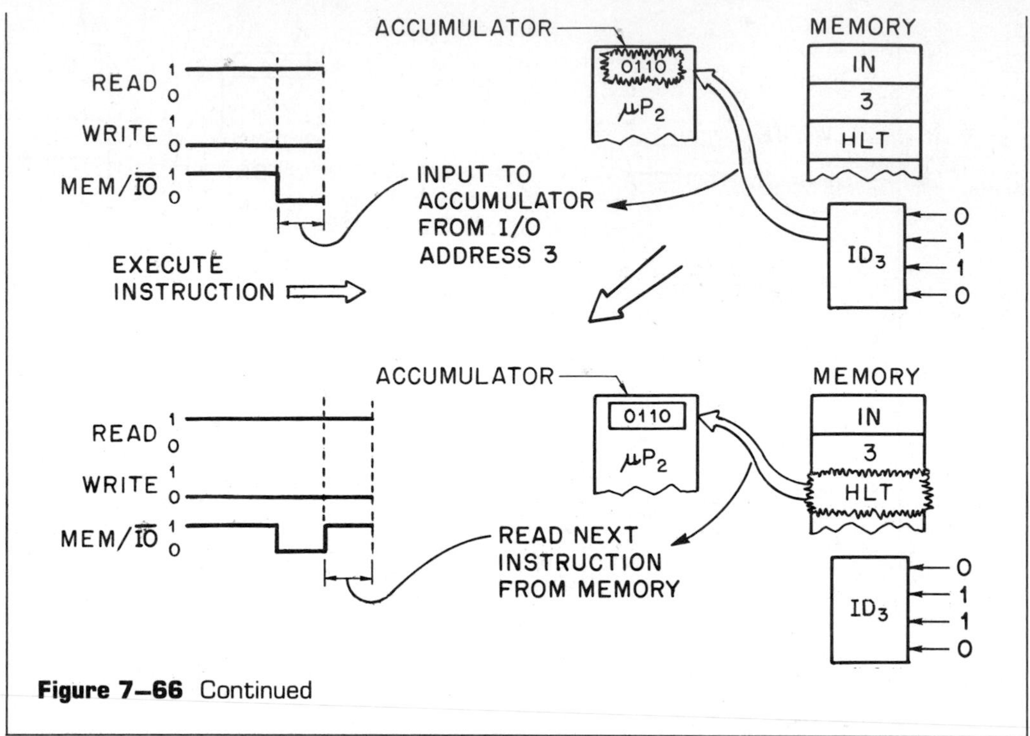

Figure 7–66 Continued

Note from Example 7–15, the contents of the accumulator were transferred to the output buffer in program A. In program B, the bit pattern on the input to the input driver were copied into the accumulator. Compare the previous results with the following example.

Example 7–16

For each of the following programs, indicate the condition of the READ, WRITE, and MEM/IO lines of the 4-bit microprocessor.

PROGRAM A	
Address	**Instruction**
0	MVI A, 3
2	HLT
3	B

PROGRAM B	
Address	**Instruction**
0	LDA 3
2	HLT
3	B

Solution

Recall that the only instructions that cause the MEM/$\overline{IO}$ line to go low are the IN and OUT instructions. Keeping this in mind and realizing that the MVI and LDA instructions are both READ instructions, refer to Figure 7–67 for the solution.

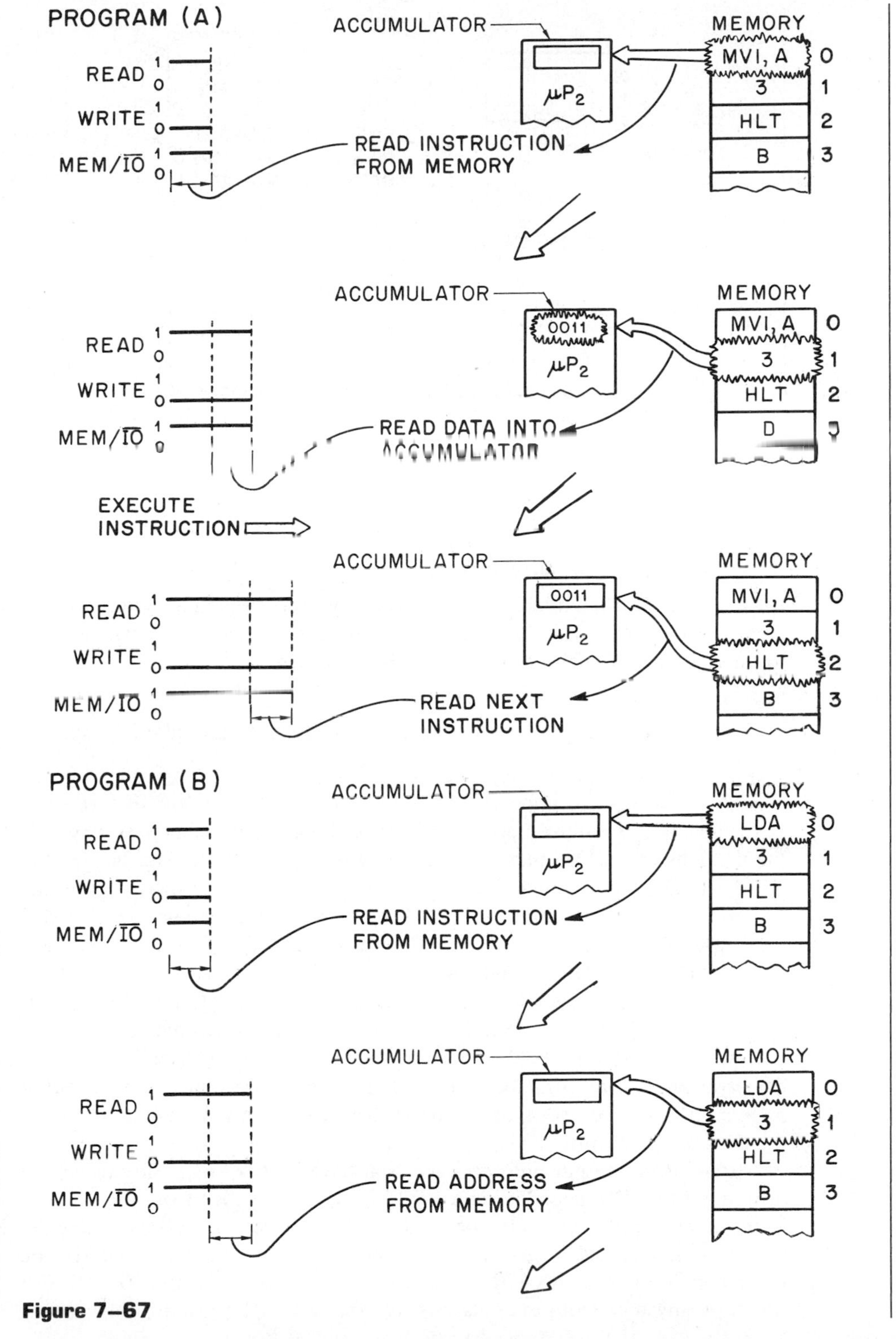
PROGRAM (A)
ACCUMULATOR
MEMORY
MVI, A
3
HLT
B
0
1
2
3
READ
WRITE
MEM/$\overline{IO}$
μP_2
READ INSTRUCTION FROM MEMORY
0011
READ DATA INTO ACCUMULATOR
D
EXECUTE INSTRUCTION
READ NEXT INSTRUCTION
PROGRAM (B)
LDA
READ ADDRESS FROM MEMORY

Figure 7–67

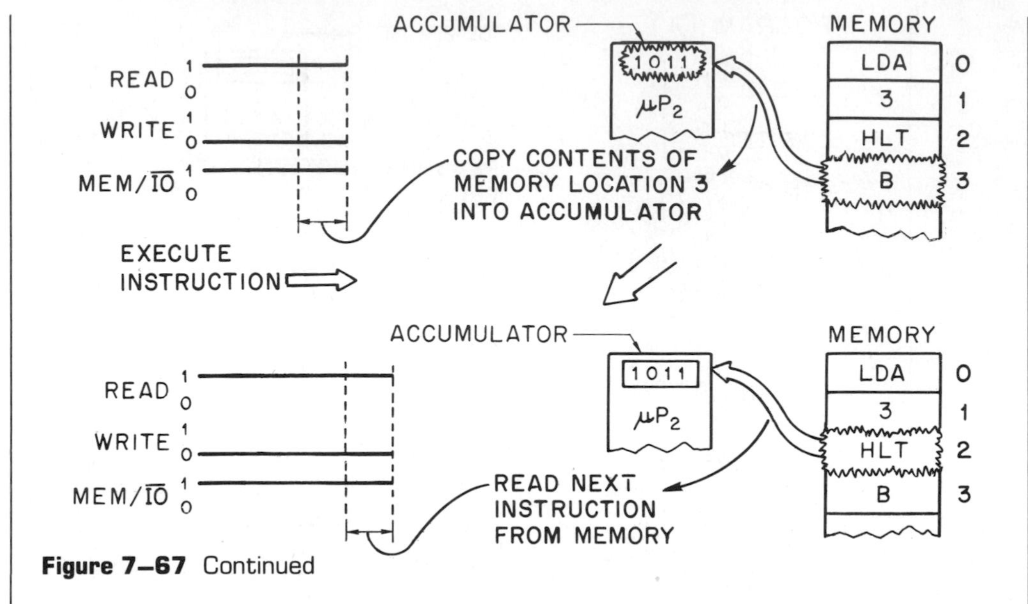

Figure 7–67 Continued

A Microprocessor-Based System

Figure 7–68 shows a microprocessor-based system utilizing a switch-programmable ROM with a 4 to 16 line decoder. What is involved in this system is typical of industrial controllers used in **robotics** or other automated systems.

> **Robotics:**
> The study of systems that can exercise control and judgment without human intervention.

There are several important features of this system. Observe that the ROM decoder is enabled by *two* control lines: READ and MEM/$\overline{IO}$. When both of these lines are HIGH, the contents of a ROM memory location will appear on the data bus. Both of these lines will be HIGH whenever the microprocessor is reading information from the ROM.

Now, observe the input drivers. They each have an enable line that will be active by the output of the 2 to 4 line decoder (called READ I/O). This decoder is active only when the READ line is HIGH and the MEM/$\overline{IO}$ line is LOW. Thus when the READ I/O decoder is active, the address line inputs (A_0, A_1) will determine which input driver will be selected (ID_0 through ID_3). Under any other conditions, the ENABLE line to each ID will not be active and the TRI-STATE buffers will be electrically disconnected from the data bus and no information will be transferred from their input to output. Using this system, only one or none of the input drivers can have its input information appear on the data bus at a time.

Look at the output buffers; they each have an ENABLE line that is controlled by a different 2 to 4 line decoder (called WRITE I/O). Observe from the figure that the output buffer ENABLE lines will be active only when the WRITE line is HIGH, the MEM/$\overline{IO}$ line is HIGH, and their corresponding address is on the address bus. Under any other conditions, the ENABLE line will not be active and the TRI-STATE buffers will not copy any data from the data bus into the *D* flip-flops. With this system, only one or

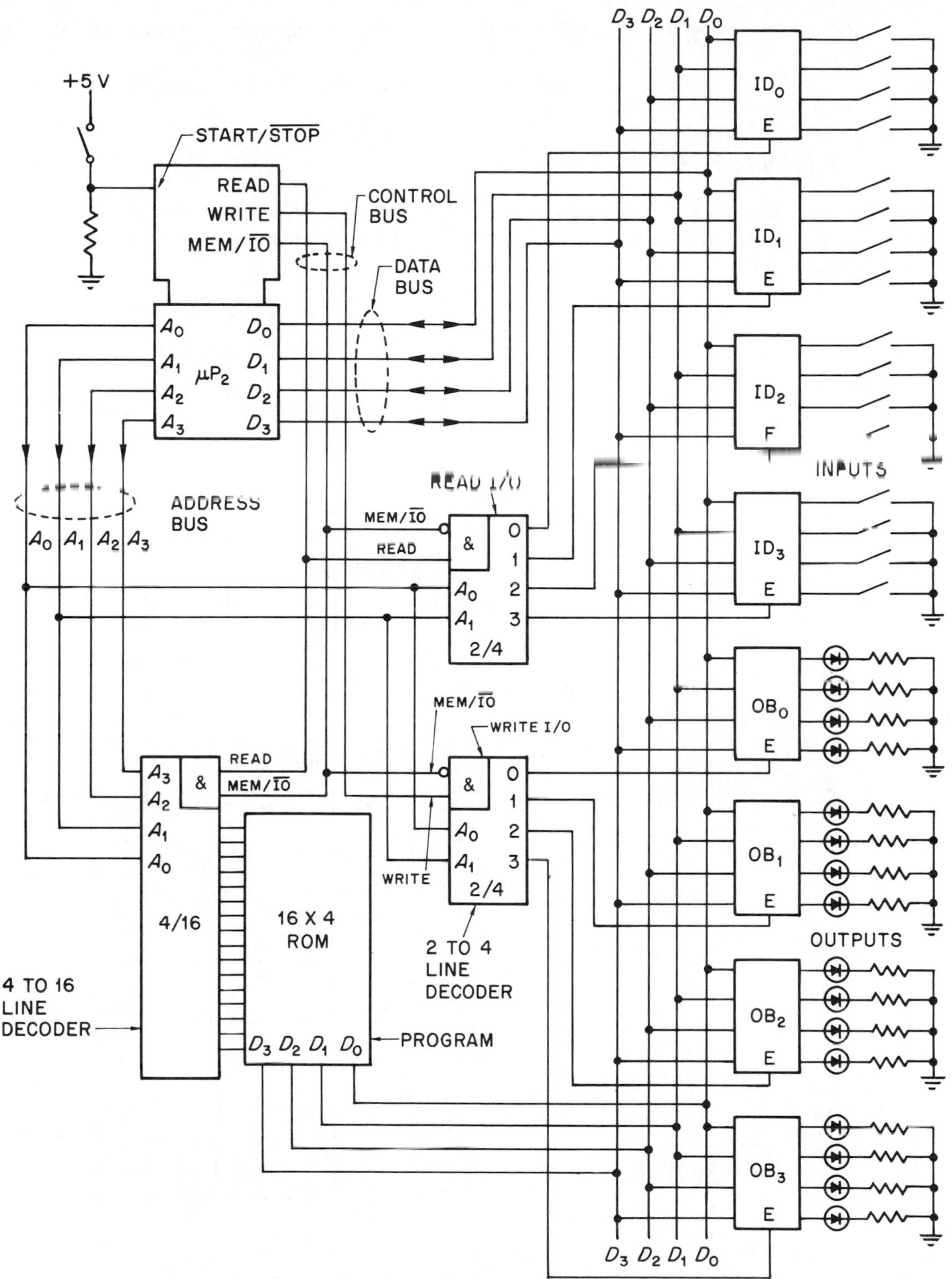

Figure 7–68 A microprocessor-based system.

none of the output buffers will have the bit pattern on the data bus copied into the flip-flops.

To demonstrate how the system works, look at the following example.

Example 7–17

For the initial conditions of switches and LEDs of the microprocessor system shown in Figure 7–69, determine the final LED conditions after the following program is executed.

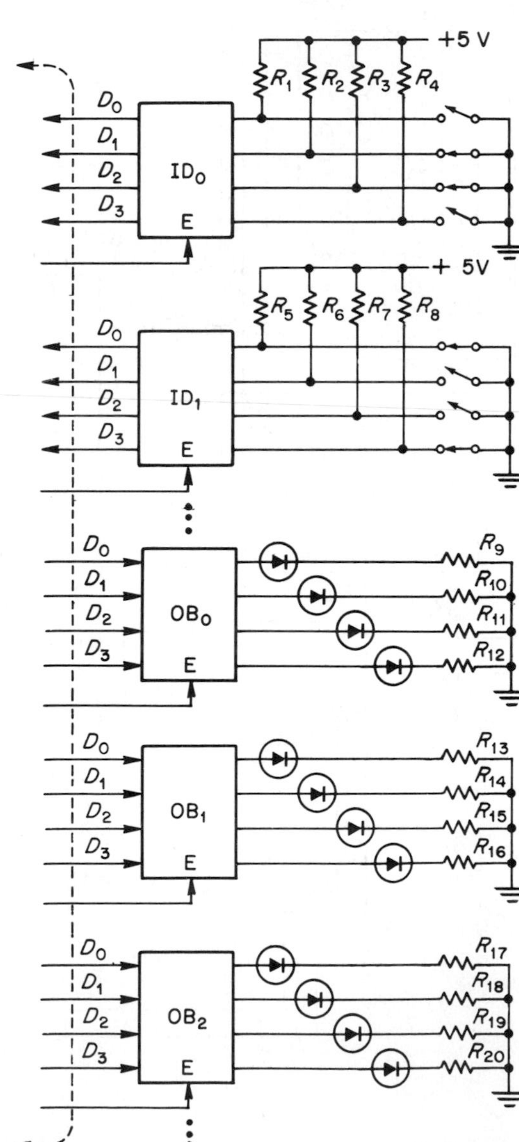

Figure 7–69

Address	Instruction
0	IN 0
2	OUT 0
4	IN 1
6	OUT 1
8	OUT 2
A	HLT

Solution

Keep in mind the effects that the IN and OUT instructions have on the control lines of the microprocessor. Use the address lines to determine which driver or buffer will be active. The process is illustrated in Figure 7–70.

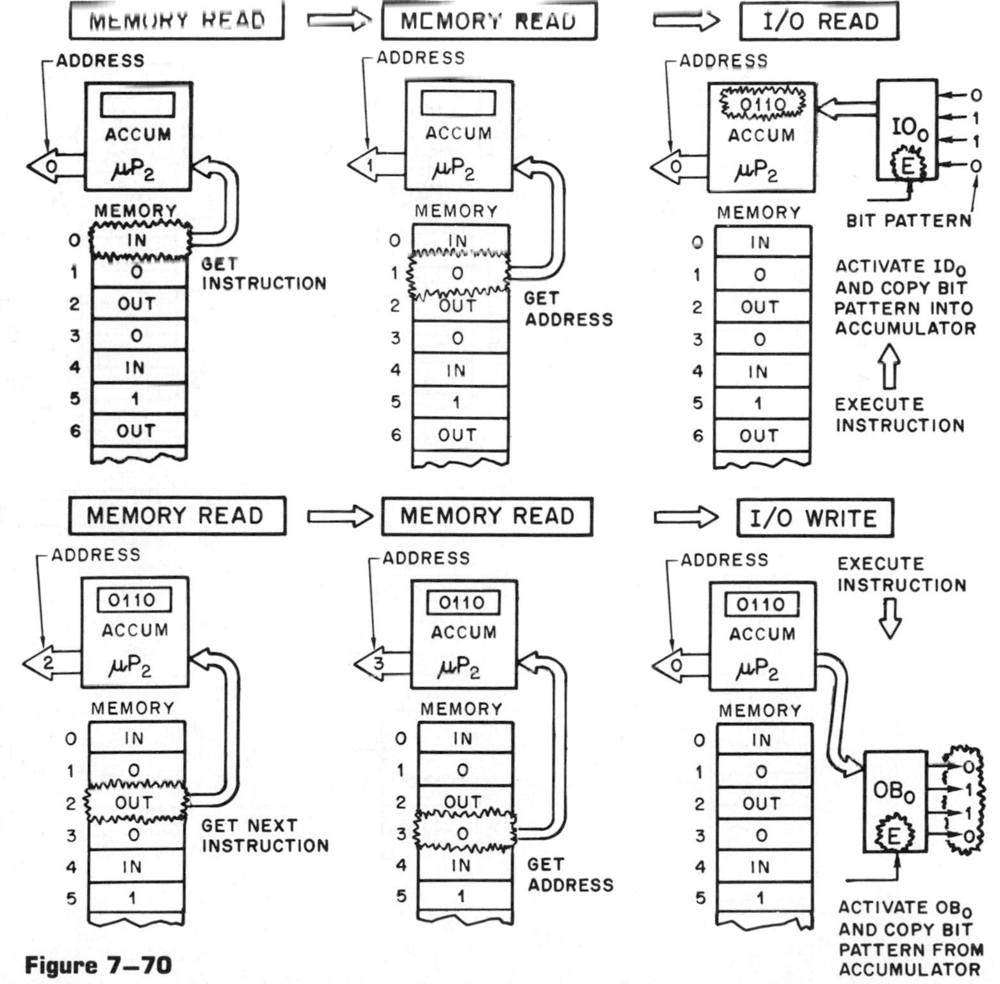

Figure 7–70

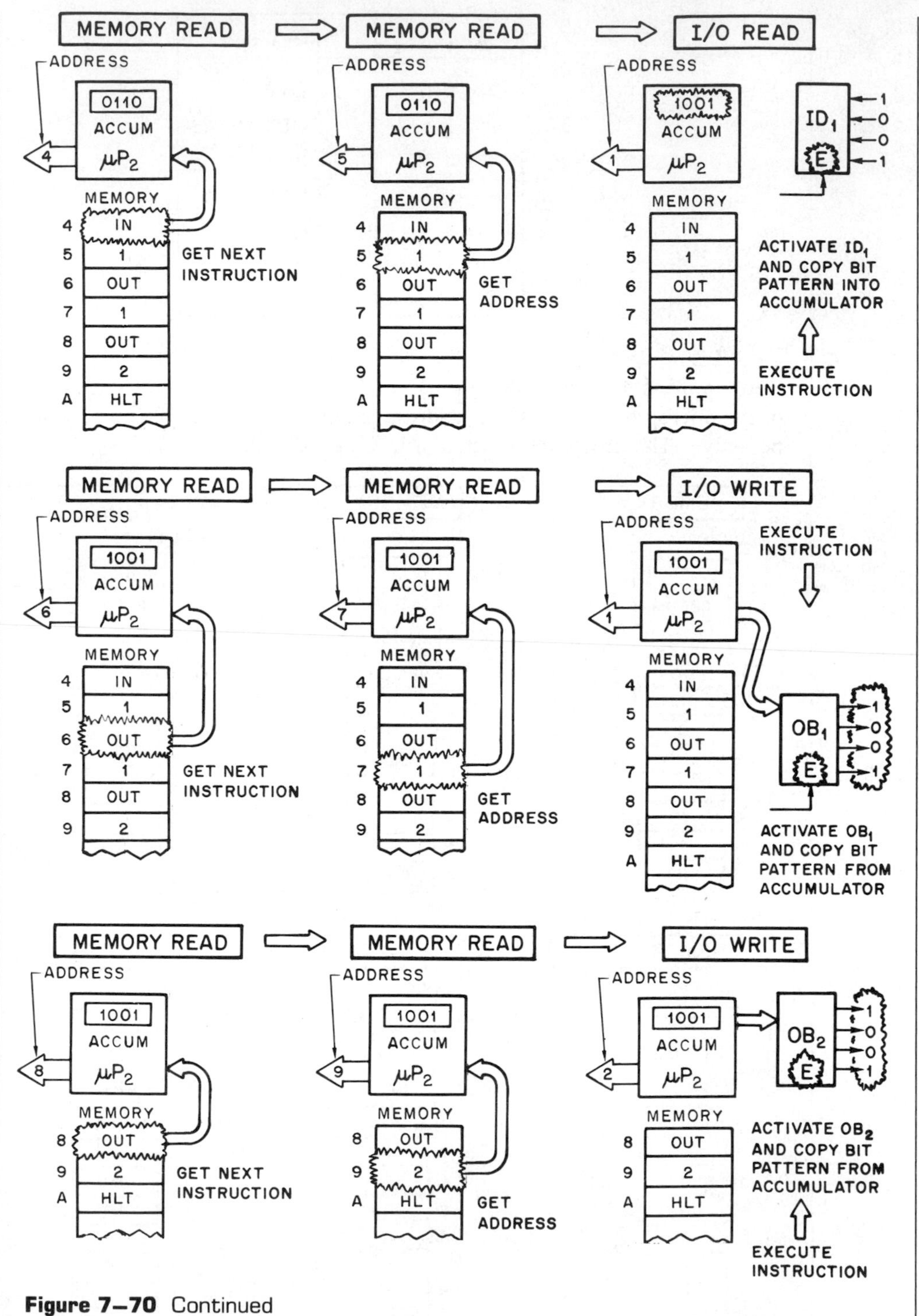

Figure 7–70 Continued

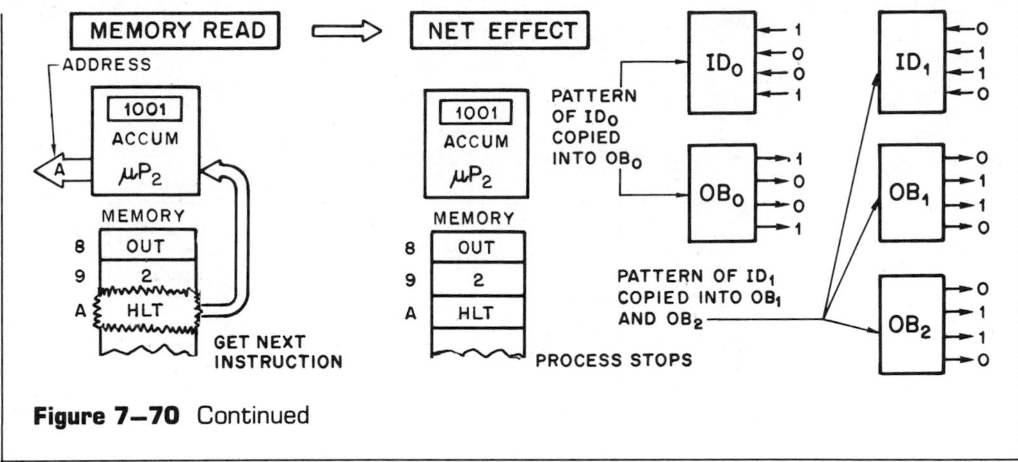

Figure 7–70 Continued

Such a microprocessor system can have other devices connected to its inputs and outputs. Consider the system in Figure 7–71.

The following table lists the I/O devices and their application:

INPUT/OUTPUT DEVICES FOR PROCESS CONTROLLER		
Device	**Input/Output**	**Function**
Driver 0	Simple switches	Inputs a bit pattern
Driver 1	Hex keypad encoder	Inputs a binary code
Driver 2	Photoresistors	Inputs light information
Driver 3	Shaft encoder	Inputs position information
Out Buffer 0	Individual LEDs	Displays a bit pattern
Out Buffer 1	Seven-segment out	Displays a hex number
Out Buffer 2	CMOS switch	Controls motor speed
Out Buffer 3	Individual relays	Controls various devices

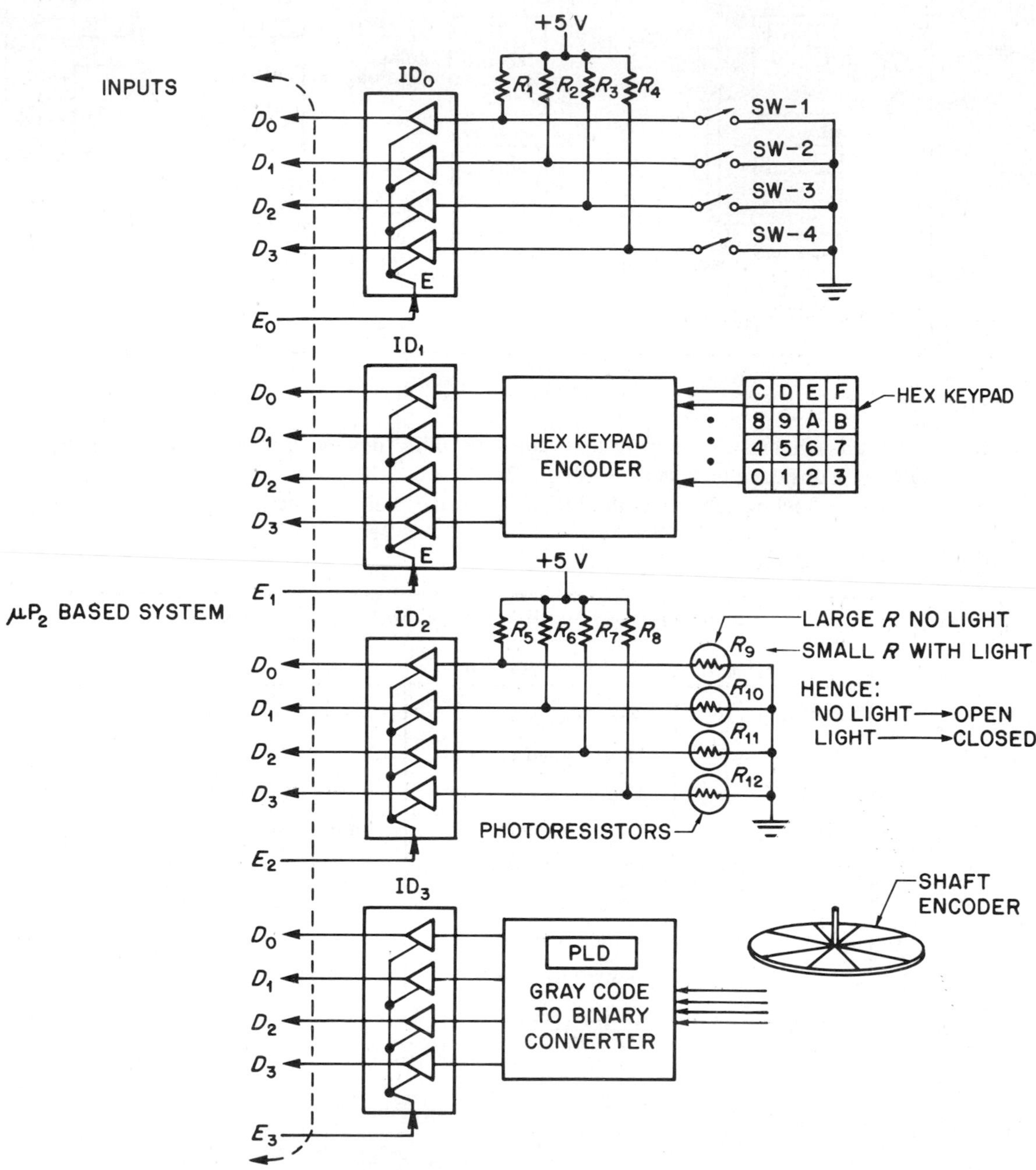

Figure 7–71 A process controller.

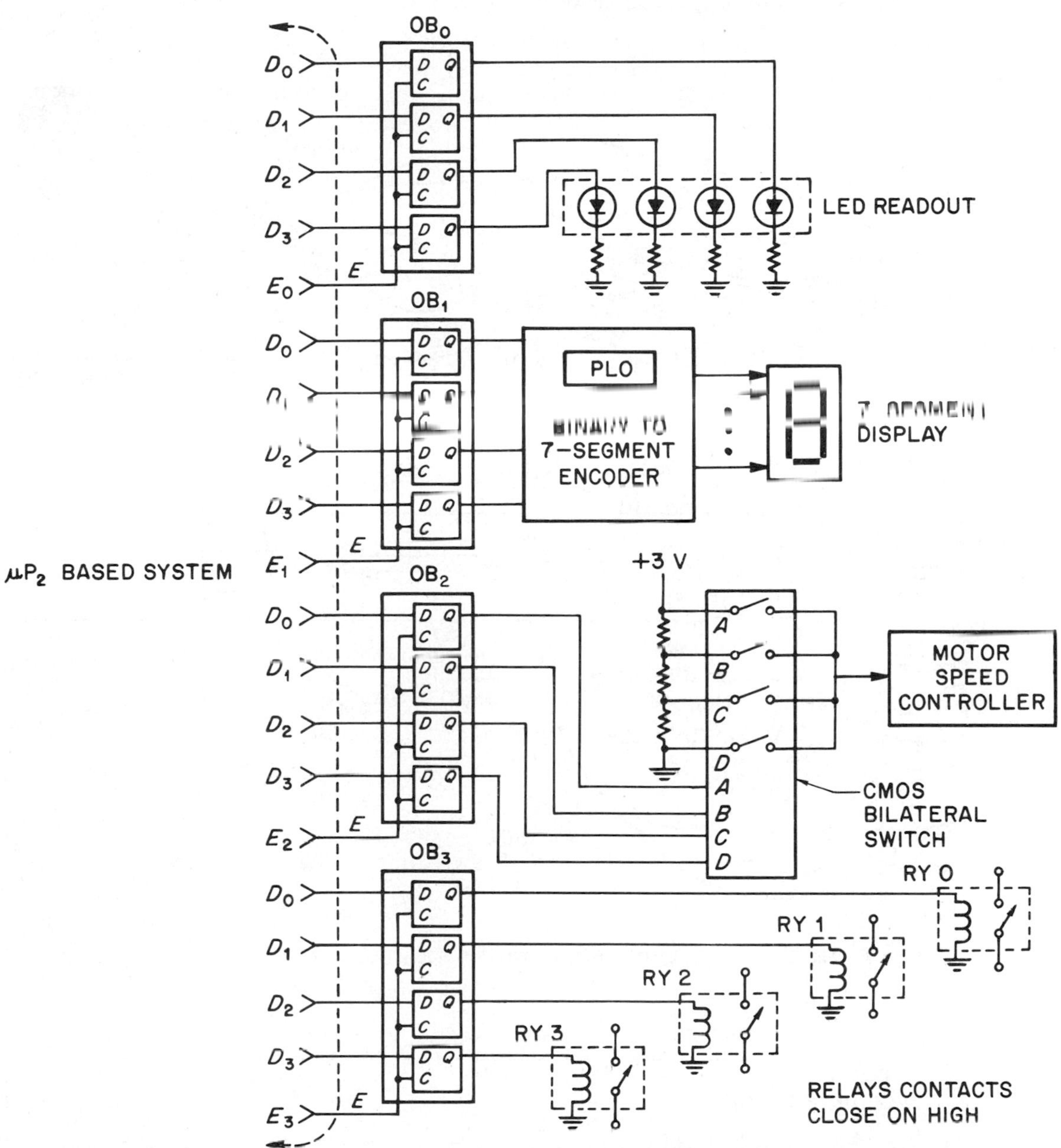

Figure 7–71 Continued

An Industrial Application

An application of a process controller is shown in Figure 7–72. The following examples illustrate some of the applications of this system.

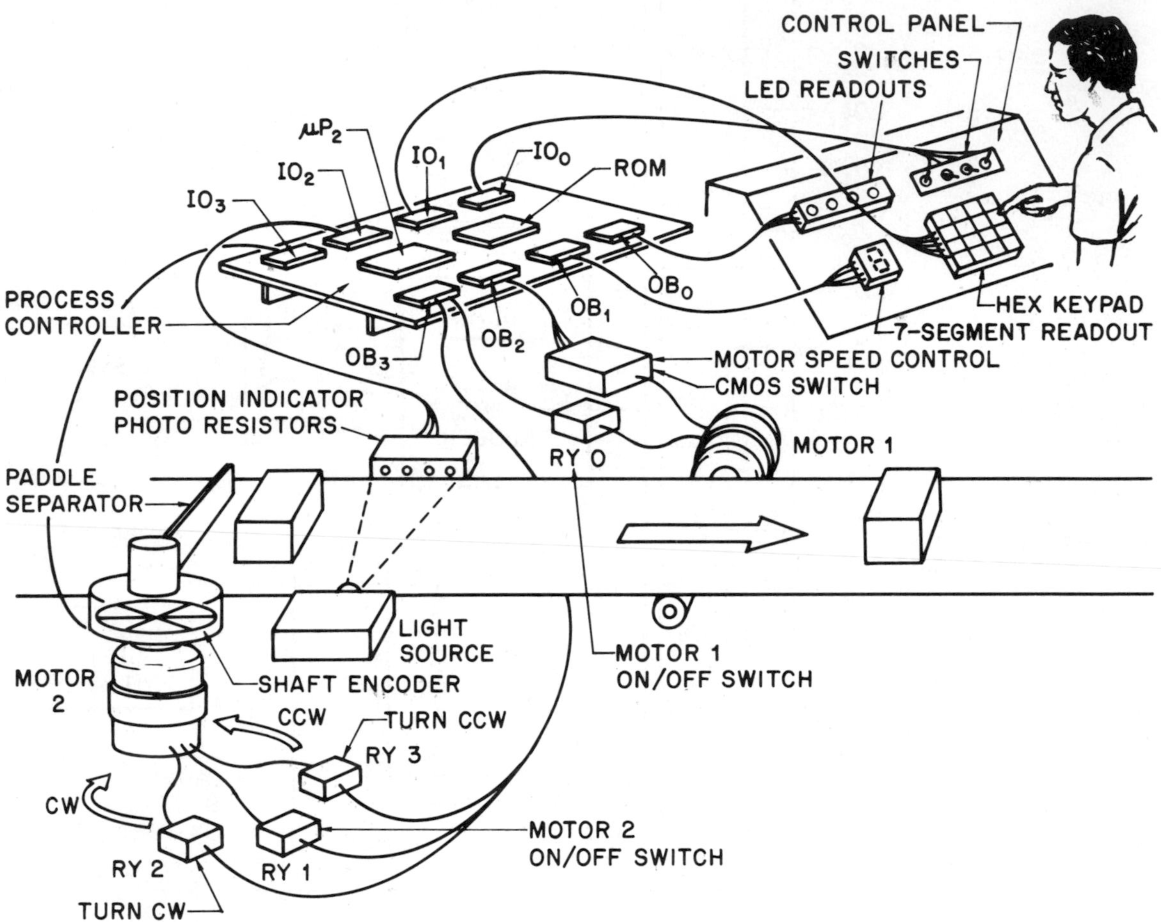

Figure 7–72 Illustration of process controller application.

> ### *Example 7–18*
>
> Using the system illustrated in Figures 7–71 and 7–72, write assembly language programs that would do the following:

(A) Cause the switch positions to appear on the seven-segment display.

(B) Cause a reading from the shaft encoder to appear on the seven-segment display.

(C) Cause an input from the hex keypad to control the speed of the motor.

(D) Cause the positions of the individual switches to appear on the LED readouts and control the individual relays.

Solution

The solution is shown in Figure 7–73.

(A)

ADDRESS	INSTRUCTION
0	IN 0
2	OUT 1
4	HLT

(B)

ADDRESS	INSTRUCTION
0	IN 3
2	OUT 1
4	HLT

(C)

ADDRESS	INSTRUCTION
0	IN 1
2	OUT 2
4	HLT

(D)

ADDRESS	INSTRUCTION
0	IN 0
2	OUT 1
4	OUT 3
5	HLT

Figure 7–73

Example 7–19

Using the system illustrated in Figures 7–71 and 7–72, write an assembly language program that would do the following:

(A) Cause motor 2 to rotate CCW and motor 1 to rotate at maximum speed (have maximum voltage applied).

(B) Cause motor 2 to rotate CW and motor 1 to be off.

Solution

The solution is shown in Figure 7–74.

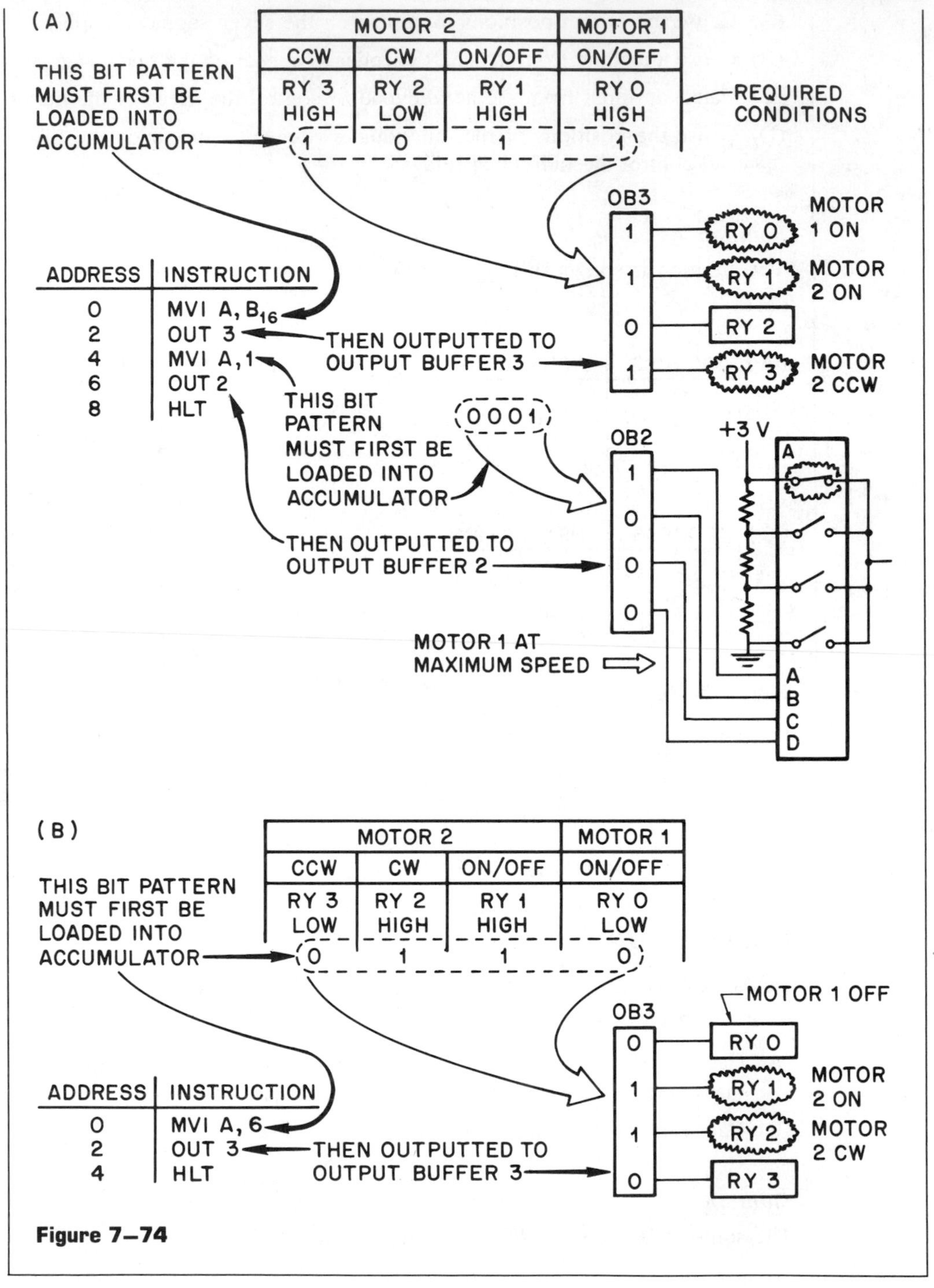
(A)
MOTOR 2
MOTOR 1
CCW
CW
ON/OFF
ON/OFF
THIS BIT PATTERN MUST FIRST BE LOADED INTO ACCUMULATOR
RY 3 HIGH
RY 2 LOW
RY 1 HIGH
RY 0 HIGH
REQUIRED CONDITIONS
1 0 1 1
OB3
MOTOR 1 ON
MOTOR 2 ON
MOTOR 2 CCW
RY 0
RY 1
RY 2
RY 3
ADDRESS
INSTRUCTION
0 MVI A, B16
2 OUT 3
4 MVI A, 1
6 OUT 2
8 HLT
THEN OUTPUTTED TO OUTPUT BUFFER 3
THIS BIT PATTERN MUST FIRST BE LOADED INTO ACCUMULATOR
0 0 0 1
OB2
+3 V
THEN OUTPUTTED TO OUTPUT BUFFER 2
MOTOR 1 AT MAXIMUM SPEED
A
B
C
D
(B)
MOTOR 2
MOTOR 1
CCW
CW
ON/OFF
ON/OFF
THIS BIT PATTERN MUST FIRST BE LOADED INTO ACCUMULATOR
RY 3 LOW
RY 2 HIGH
RY 1 HIGH
RY 0 LOW
0 1 1 0
MOTOR 1 OFF
OB3
RY 0
RY 1
RY 2
RY 3
MOTOR 2 ON
MOTOR 2 CW
ADDRESS
INSTRUCTION
0 MVI A, 6
2 OUT 3
4 HLT
THEN OUTPUTTED TO OUTPUT BUFFER 3

Figure 7–74

Example 7–20

Determine what each of the assembly language programs would cause the system of Figures 7–71 and 7–72 to do.

PROGRAM A	
Address	**Instruction**
0	IN 0
2	OUT 1
4	OUT 2
6	MVI A, 4
8	OUT 3
A	HLT

PROGRAM B	
Address	**Instruction**
0	MVI A, 2
2	OUT 0
4	OUT 3
6	IN 2
8	OUT 1
A	HLT

Solution

The solution is illustrated in Figure 7–75.

(A)

ADDRESS	INSTRUCTION	
0	IN 0 ←	INPUTS SWITCH PATTERN TO ACCUMULATOR
2	OUT 1 →	OUTPUTS SWITCH PATTERN FROM ACCUMULATOR TO 7-SEGMENT DISPLAY
4	OUT 2 →	OUTPUTS SWITCH PATTERN FROM ACCUMULATOR TO MOTOR SPEED CONTROL
6	MVI A,4 ←	PUTS THE BIT PATTERN 0100 INTO ACCUMULATOR
8	OUT 3 →	OUTPUTS BIT PATTERN 0100 IN ACCUMULATOR TO RELAYS
A	HLT →	STOPS ALL PROCESSING

(B)

ADDRESS	INSTRUCTION	
0	MVI A,2 ←	PUTS THE BIT PATTERN 0010 INTO ACCUMULATOR
2	OUT 0 →	OUTPUTS BIT PATTERN 0010 IN ACCUMULATOR TO LEDS
4	OUT 3 →	OUTPUTS BIT PATTERN 0010 IN ACCUMULATOR TO RELAYS
6	IN 2 →	INPUTS PHOTORESISTORS INTO ACCUMULATOR
8	OUT 1 →	OUTPUTS PHOTORESISTOR PATTERN IN ACCUMULATOR TO 7-SEGMENT DISPLAY
A	HLT →	STOPS ALL PROCESSING

Figure 7–75

Example 7–21

As a technician, you are responsible for the maintenance and repair of the industrial controller system of Figures 7–71 and 7–72. Determine what the system is supposed to do from the ROM bit patterns of Figure 7–76.

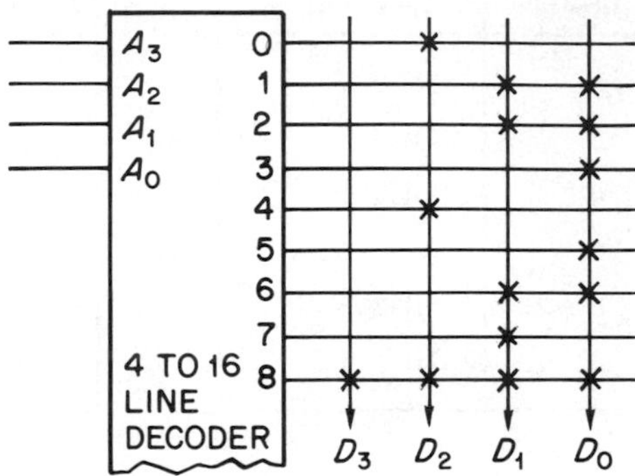

Figure 7–76

Solution

The solution is given in Figure 7–77.

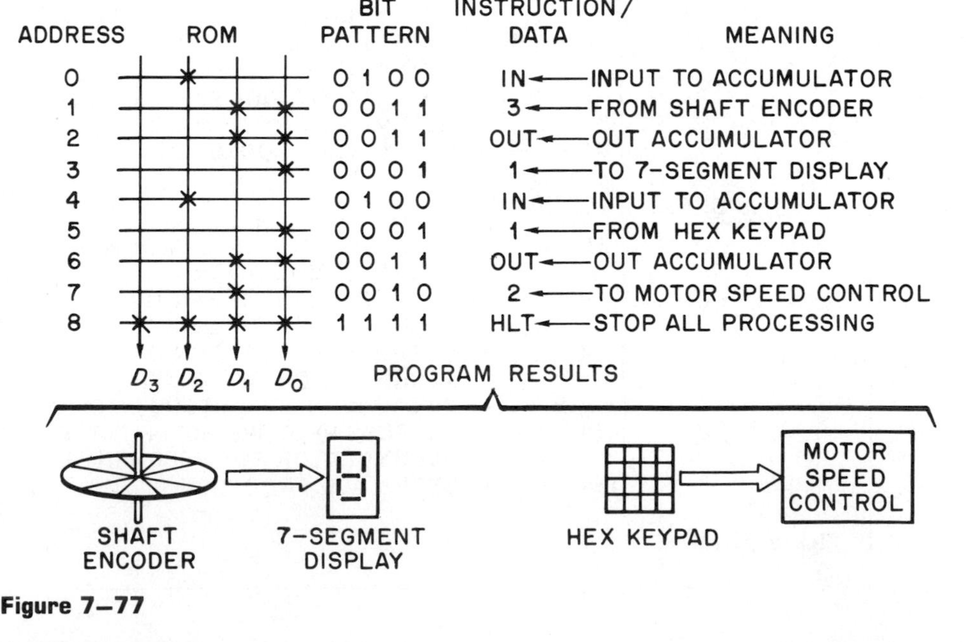

ADDRESS
ROM
BIT PATTERN
INSTRUCTION/ DATA
MEANING
0 0 1 0 0 IN INPUT TO ACCUMULATOR
1 0 0 1 1 3 FROM SHAFT ENCODER
2 0 0 1 1 OUT OUT ACCUMULATOR
3 0 0 0 1 1 TO 7-SEGMENT DISPLAY
4 0 1 0 0 IN INPUT TO ACCUMULATOR
5 0 0 0 1 1 FROM HEX KEYPAD
6 0 0 1 1 OUT OUT ACCUMULATOR
7 0 0 1 0 2 TO MOTOR SPEED CONTROL
8 1 1 1 1 HLT STOP ALL PROCESSING
D_3 D_2 D_1 D_0
PROGRAM RESULTS
SHAFT ENCODER
7-SEGMENT DISPLAY
HEX KEYPAD
MOTOR SPEED CONTROL

Figure 7–77

Conclusion

You have come a long way in your studies of digital systems. You are at the point where the applications for what you have learned are unveiling the secrets of the computer revolution. By now you can understand the heart of an industrial control system and perhaps gain a glimpse of a computer system. What is important is that you are seeing the *applications* of what you are learning. Knowing how to apply what you are learning makes the learning fun and helps you retain what you have learned. This is a far cry from memorization of apparently disjointed facts that are soon forgotten.

Test your understanding of this section by trying the following review questions.

7–7 Review Questions

1. Describe the operation of a TRI-STATE buffer.
2. What is an *addressing mode?*
3. What is the difference between *immediate addressing* and *direct addressing?*
4. Describe the effect the instructions IN and OUT have on the MEM/$\overline{\text{IO}}$, WRITE, and READ lines.
5. Explain the operation of the output buffer and input driver as used in the digital system presented in this section.

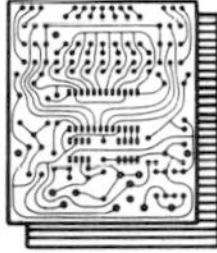

ELECTRONIC APPLICATION

Discussion

In this section, you will discover the electrical details of a most versatile IC: the 555 timer. Here you will also learn about the workings of other similar devices, including the one-shot. Once you have finished this section, you will have a solid foundation in the principles of pulse generation and reshaping techniques.

The 555 Timer

The 555 timer is a number for a popular IC that has a wide variety of practical applications. See Figure 7–78. One of its many applications is that of an astable flip-flop. Before starting on the specific application, it's important to know the basic building blocks that go to make up this versatile device. The first item you should know about is the comparator.

Figure 7–78 The 555 timer integrated circuit.

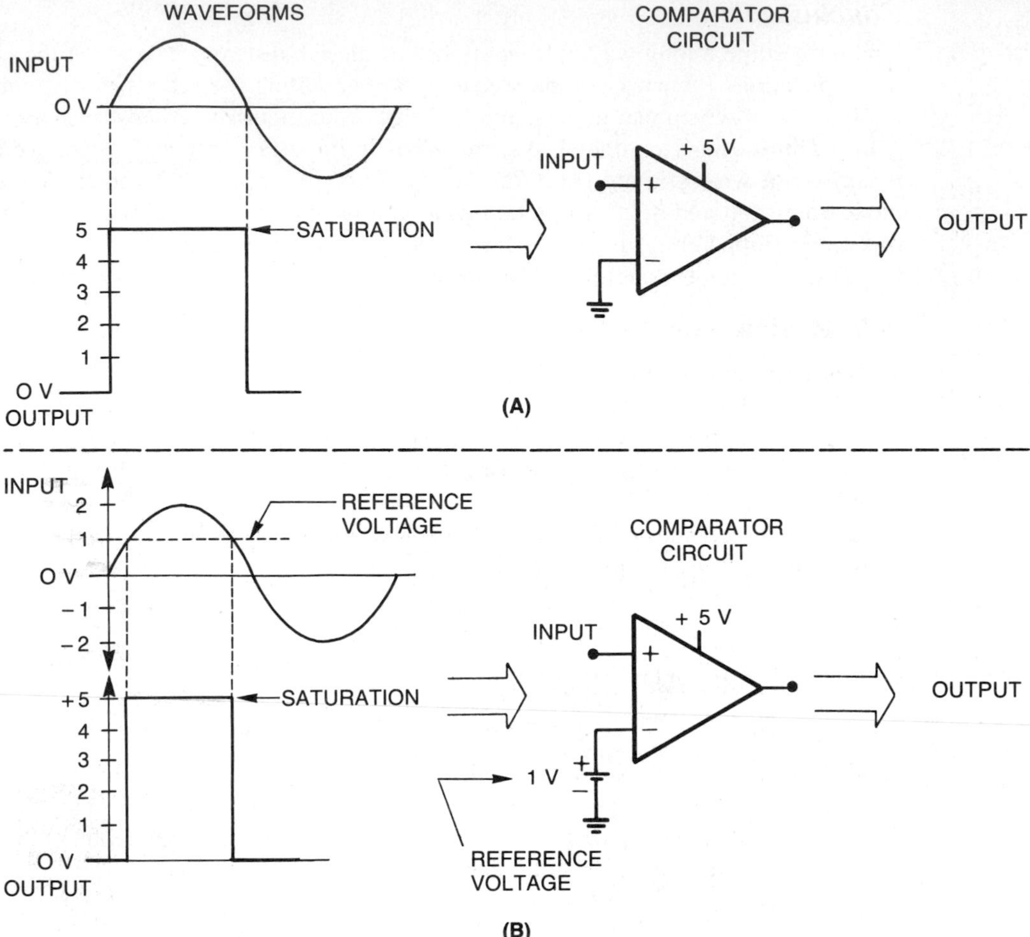

Figure 7–79 Operation of a comparator.

The Comparator

A **comparator** is a circuit that compares an input voltage to a reference voltage.

Comparator:
A circuit that compares an input voltage to a reference and produces an output voltage indicating the comparison.

Figure 7–79 shows the schematic and operation of a comparator. As you can see, the comparator has two inputs, one of which is used for the signal (called the *threshold input*), the other for the reference. This circuit will produce an output voltage of either +V or 0 V and nothing in between. That is, if the voltage at its (+) terminal is more than the voltage at its (−) terminal, then the output will be +V. This condition is called **saturation** because the output signal cannot get any larger than +V.

Saturation:

A circuit condition where an increase of the input signal no longer produces a change in the output.

Note from Figure 7–79 that if the reference terminal (− terminal) has a voltage of +1 V, then the output of the comparator will not saturate until the input signal is greater than +1 V. At values less than +1 V, the input signal keeps the comparator output at 0 V. Thus, for this discussion, the output of the comparator has two possible values: 0 V and +5 V.

Basic Idea

The basic idea behind the 555 timer is shown in Figure 7–80. The circuit consists of an *S-R* flip-flop. In the condition shown, with SW-1 at the ground position, the Q output of the flip-flop is +5 volts and the NPN transistor is saturated (looks like a short across capacitor C_1). For the purpose of these discussions, consider that the transistor will look like an open between its collector and emitter or a short between these same two terminals. What determines this is the voltage applied to its base. When this voltage is positive, the transistor will appear as a short; when this voltage is zero, the transistor will appear as an open.

Consider switch SW-1 as the starting switch. When it is momentarily switched to its +5-V position and then switched back to ground, +5 V is applied to the *RESET* input of the *R-S* latch, causing the Q output to go to 0 V, which causes the transistor to now look like an open between its collector and emitter terminals. At the same time, the $\overline{Q}$ output goes true (+5 V), as shown in Figure 7–80. Since the transistor now appears as an open, the capacitor C_1 begins to charge toward V_{CC}.

Note that the voltage across C_1 is applied to the (+) *threshold* input of the comparator. Also note that the reference input (−) of the comparator is slightly less than 5 V (because of the voltage divider of R_1 and R_2). As long as the voltage across the capacitor is less than this reference voltage, the output of the comparator will be 0 V. However, as soon as the capacitor charges up to a voltage greater than this reference voltage, the output of the comparator will go to +5 V. When this happens, the flip-flop is again set, and Q goes to +5 V. With Q at +5 V, the transistor is again made to look like a short and the capacitor will discharge. Thus the circuit will be right back to where it started before switch SW-1 was activated.

The net result of all this is that an output pulse was produced (like a one-shot). The width of this pulse was determined by the amount of time it took capacitor C_1 to charge. For the simple circuit shown, the time it takes this capacitor to charge is determined by the value of the capacitor and resistor R_3.

Complete Timer

Figure 7–81 shows the complete diagram of the 555 timer. This figure shows all of the control and output pins of the 555 timer with their descriptive names. To use the 555 timer as an astable flip-flop, connect external components as shown in Figure 7–82. Observe from this figure that pin 7 (the discharge output) of the 555 is connected to a voltage divider (R_1 and R_2) with an external capacitor. The *trigger* and *threshold* inputs (pins 6 and 2) are connected together and to the capacitor. The other inputs of the two comparators are internally connected to a voltage divider of three equal resistors.

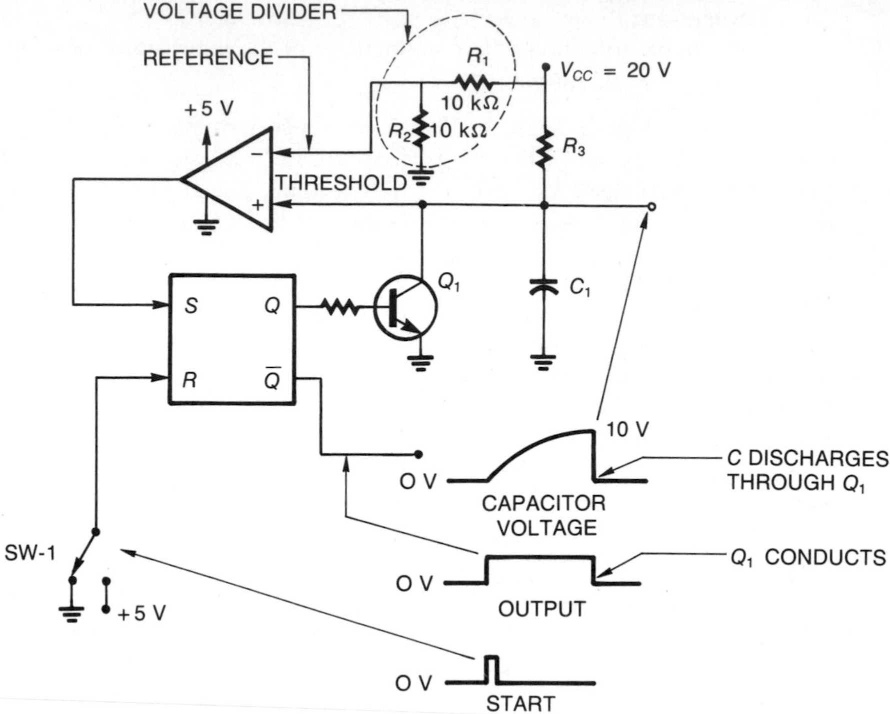

Figure 7–80 Basic idea of a 555 timer.

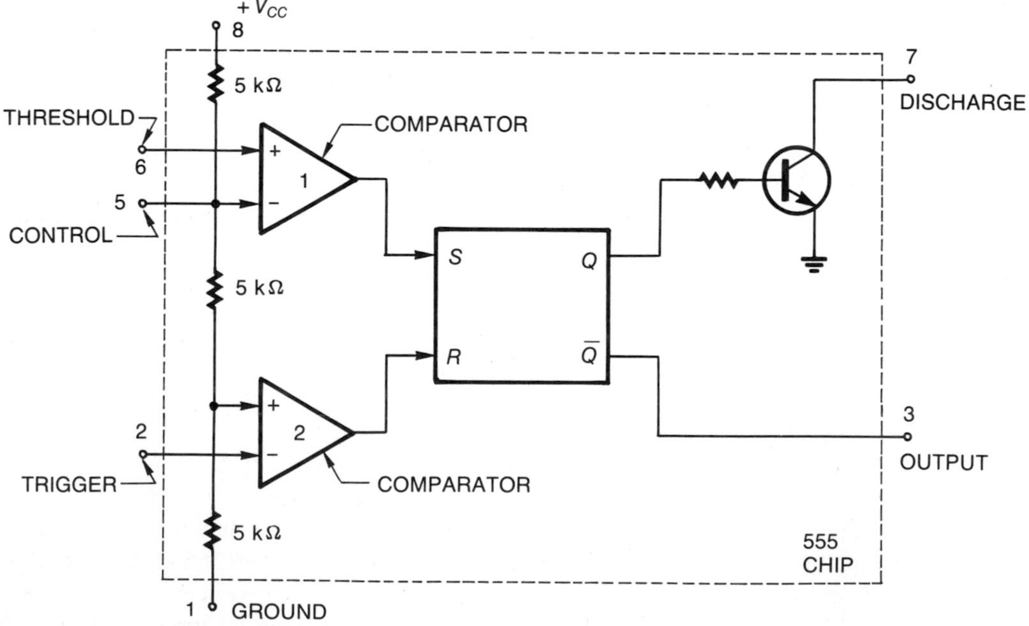

Figure 7–81 Complete functional diagram of 555 timer.

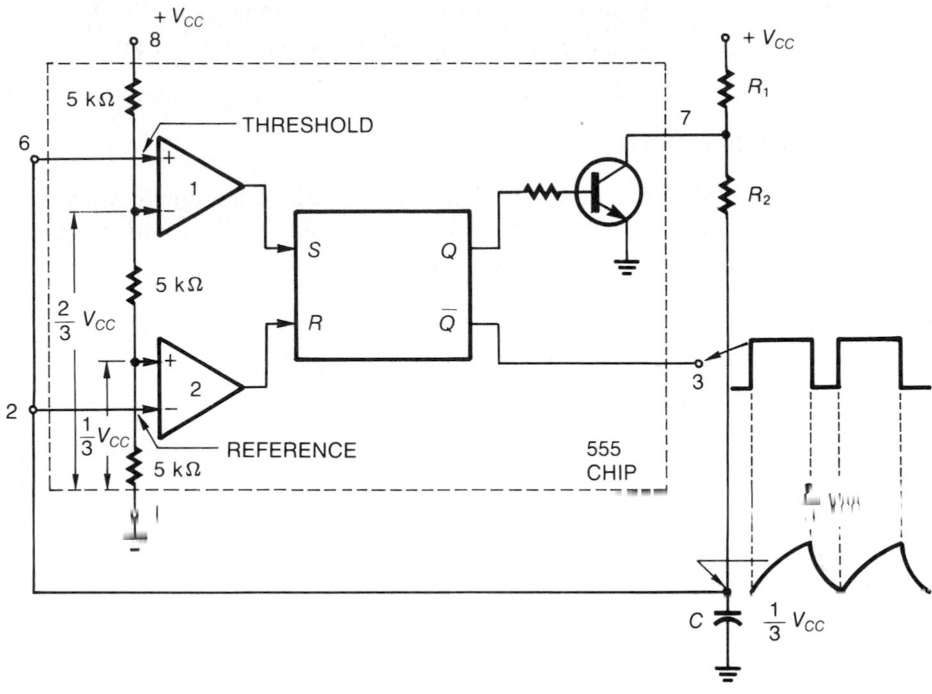

Figure 7–82 Connecting the 555 to produce its own waveform.

Assume that the capacitor starts at 0 V. This means that the threshold input to comparator 1 will be 0 V while its reference input is $^2/_3 V_{CC}$. Hence, its output will be zero volts. The reference input to comparator 2 will be 0 V while its threshold input is $^1/_3 V_{CC}$. When this happens, the reference input of comparator 2 will now be less positive than its control input *k* and the output of comparator 2 will go to +5 V, causing the *Q* output of the *RS* flip-flop to return to zero. This in turn will apply 0 V to the base of the transistor, again causing it to look like an open.

Creating Square Waves

The frequency of the square-wave output can be determined from the formula

$$f = \frac{1.44}{(R_1 + 2R_2)C}$$

where: f = frequency of oscillation in hertz
R_1, R_2 = values of external resistors in ohms
C = value of external capacitor in farads

The ON times and OFF times are determined by

$$T_{ON} = 0.693(R_1 + R_2)C$$
$$T_{OFF} = 0.693R_2C$$

where: T_{ON} = amount of time in seconds the output pulse is HIGH
T_{OFF} = amount of time in seconds the output pulse is LOW

Example 7–22

For the 555 timer shown in Figure 7–82, sketch the output waveform, indicating the frequency, the ON time, and the OFF time of the resulting waveform. (Let $R_1 = 3.3\ \text{k}\Omega$, $R_2 = 5\ \text{k}\Omega$, and $C = 1\ \mu\text{F}$.)

Solution

Using the relationships for the 555 timer, you have:

Output frequency:

$$f = \frac{1.44}{(R_1 + 2R_2)C}$$

$$f = \frac{1.44}{(3.3\ \text{k}\Omega + 2(5\ \text{k}\Omega))(1.0\ \mu\text{F})}$$

$$f = \frac{1.44}{13.3 \times 10^{-3}}$$

$$f = 108\ \text{Hz}$$

On time:

$$T_{ON} = 0.693(R_1 + R_2)C$$

$$T_{ON} = 0.693(3.3\ \text{k}\Omega + 5\ \text{k}\Omega)(1.0\ \mu\text{F})$$

$$T_{ON} = 0.693(8.3 \times 10^{-3})$$

$$T_{ON} = 5.75\ \text{ms}$$

Off time:

$$T_{OFF} = 0.693R_2C$$

$$T_{OFF} = 0.693(5\ \text{k}\Omega)(1.0\ \mu\text{F})$$

$$T_{OFF} = 0.693(5 \times 10^{-3})$$

$$T_{OFF} = 3.46\ \text{ms}$$

The resulting waveform is sketched in Figure 7–83.

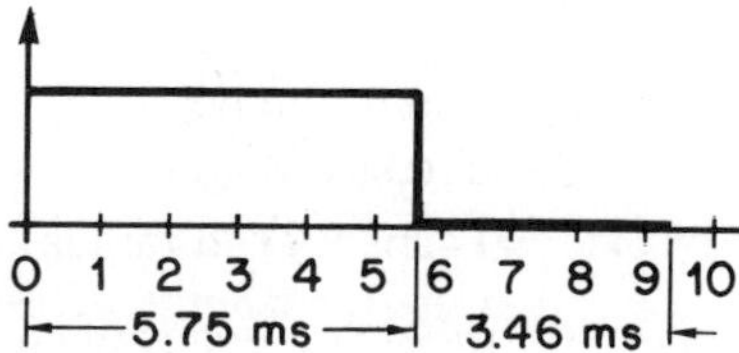

Figure 7–83

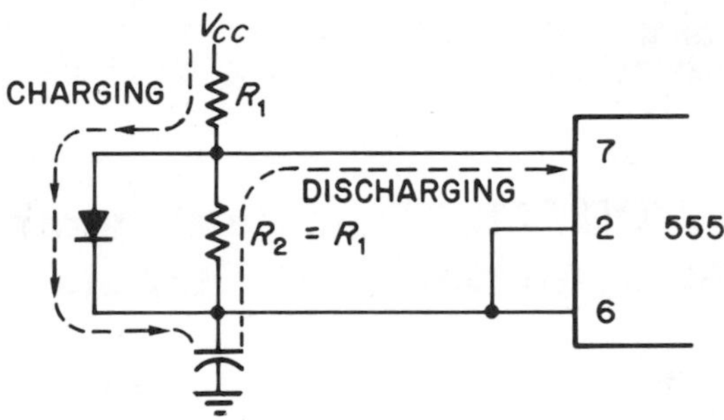

Figure 7–84 The 555 as a square-wave generator.

Creating True Square Waves

In order to get a true square wave from the output of the 555 timer, you need to add a diode, as shown in Figure 7–84. The diode is necessary in order to create a charge time that is equal to its discharge time. Now, if both resistors (R_1 and R_2) are set equal to each other, the ON time will be equal to the OFF time.

One-Shots

A one-shot is a particular class of multivibrator that is monostable—that is, it has only one stable state. A general-purpose one-shot is the 74123 dual retriggerable one-shot with clear inputs. The logic diagram is shown in Figure 7–85.

Either one of these one-shots may be triggered by a HIGH to LOW transition or by a LOW to HIGH transition. This depends upon the input to which you apply the input pulse. For example, if input A_1 is held LOW, then a LOW to HIGH transition on the B_1 input will cause the one-shot to trigger. If, on the other hand, input B_1 were held HIGH, then a LOW to HIGH transition on the A_1 input would cause triggering. The amount of time that the one-shot is in its unstable state is determined by the values of external resistors and capacitors. Figure 7–86 shows the one-shot with the required external components and the formula to determine the amount of time the one-shot will be in its unstable state.

The new logic symbol for the dual one-shot is shown in Figure 7–87. Note the pulse symbol used to indicate that it is a one-shot.

Schmitt Triggers

Another important element in a digital system is the **Schmitt trigger,** named after its inventor.

Schmitt Trigger:
A circuit that changes state abruptly when the input signal crosses a specified DC triggering level.

The symbol for a Schmitt trigger is shown in Figure 7–88 along with its basic principle of operation. The device will not trigger LOW until a lower threshold is reached and holds that LOW, despite any noise, until a much higher upper threshold triggers a HIGH output. Schmitt triggers are particularly useful in the transmission and reception of digital signals. They can be used to "clean up" noisy digital signals.

Figure 7–89 shows the 74132 quad two-input NAND gates with Schmitt trigger inputs.

DM54123/DM74123 Dual Retriggerable One-Shot with Clear and Complementary Outputs

General Description

The DM54/74123 is a dual retriggerable monostable multivibrator capable of generating output pulses from a few nano-seconds to extremely long duration up to 100% duty cycle. Each device has three inputs permitting the choice of either leading-edge or trailing edge triggering. Pin (A) is an active-low transition trigger input and pin (B) is an active-high transition trigger input. The clear (CLR) input terminates the output pulse at a predetermined time independent of the timing components.

National's '123 device features a unique logic realization not implemented by other manufacturers. The "Clear" input will not trigger the device, a design tailored for applications where it shall only terminate or reduce a timing pulse.

To obtain the best and trouble free operation from this device please read the operating rules as well as the NSC one-shot application notes carefully and observe recommendations.

Features

- DC triggered from active-high transition or active-low transition inputs
- Retriggerable to 100% duty cycle
- Direct reset terminates output pulse
- Compensated for V_{CC} and temperature variations
- DTL, TTL compatible
- Input clamp diodes

Functional Description

The basic output pulse width is determined by selection of an external resistor (R_X) and capacitor (C_X). Once triggered, the basic pulse width may be extended by retriggering the gated active-low transition or active-high transition inputs or be reduced by use of the active-low transition clear input. Retriggering to 100% duty cycle is possible by application of an input pulse train whose cycle time is shorter than the output cycle time such that a continuous "HIGH" logic state is maintained at the "Q" output.

Connection Diagram

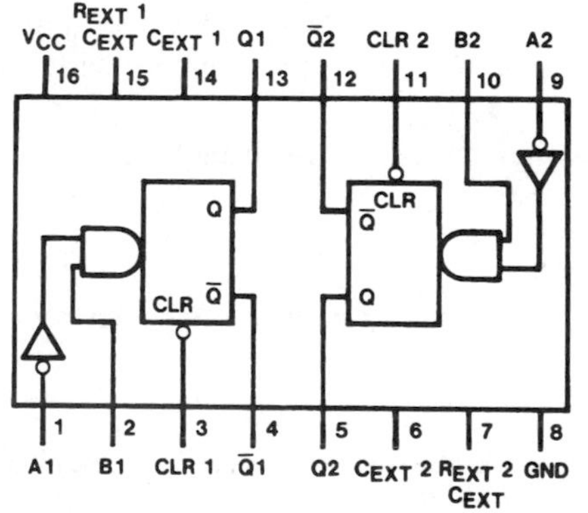

Order Number DM54123J or DM74123N
See NS Package Number J16A or N16A

Function Table

Inputs			Outputs	
CLEAR	A	B	Q	$\overline{Q}$
L	X	X	L	H
H	H	X	L	H
H	X	L	L	H
H	L	↑	⎍	⊔
H	↓	H	⎍	⊔

H = High Logic Level
L = Low Logic Level
X = Can Be Either Low or High
↑ = Positive Going Transition
↓ = Negative Going Transition
⎍ = A Positive Pulse
⊔ = A Negative Pulse

Figure 7–85 The 74123 one shot. *Reprinted* with permission of National Semiconductor Corporation.

National holds no responsibility for any circuitry described. National reserves the right at any time to change without notice said circuitry.

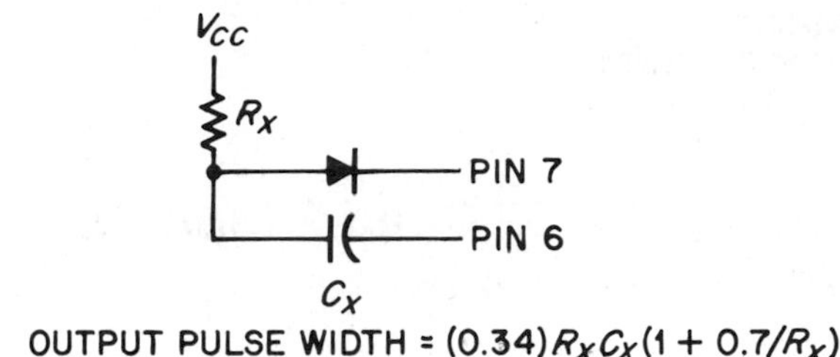

Figure 7–86 Components connected to a one-shot.

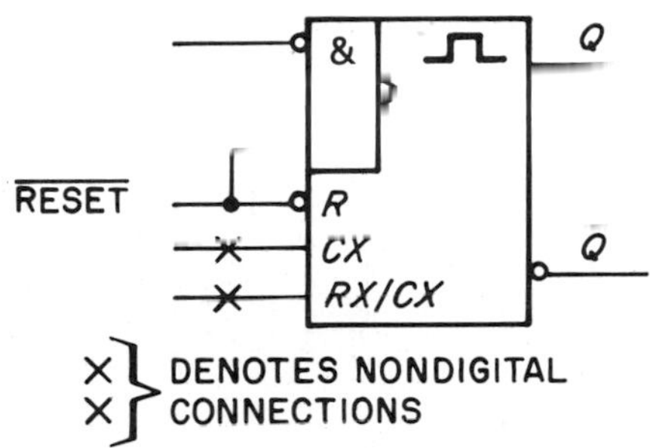

Figure 7–87 New logic symbol for a one-shot.

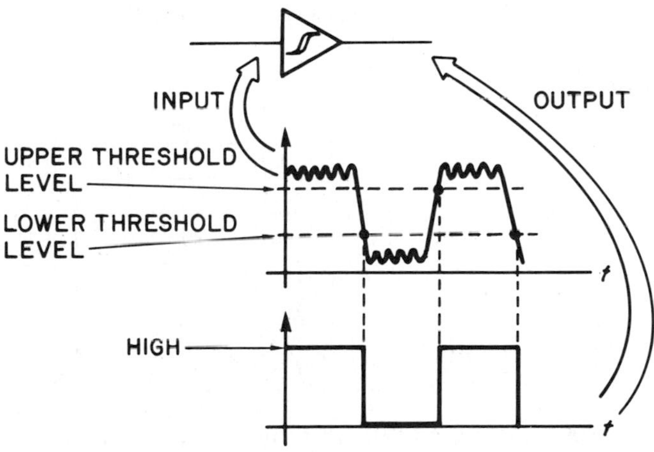

Figure 7–88 Schmitt-trigger symbol and application.

DM54132/DM74132 Quad 2-Input NAND Gates with Schmitt Trigger Inputs

General Description

This device contains four independent gates each of which performs the logic NAND function. Each input has hysteresis which increases the noise immunity and transforms a slowly changing input signal to a fast changing, jitter-free output.

Connection Diagram

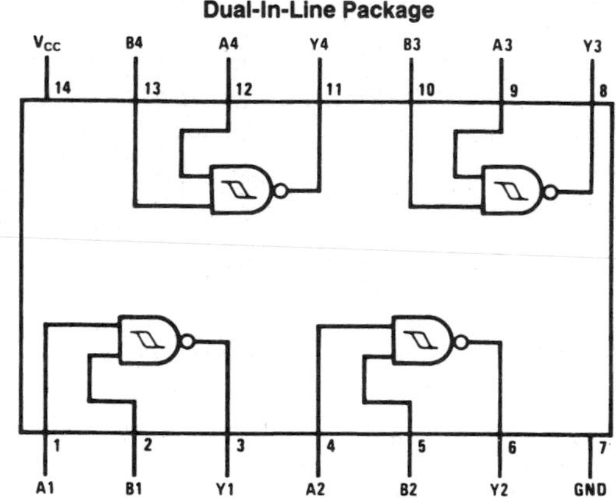

TL/F/6542–1

Order Number DM54132J or DM74132N
See NS Package Number J14A or N14A

Function Table

$Y = \overline{AB}$

Inputs		Output
A	B	Y
L	L	H
L	H	H
H	L	H
H	H	L

H = High Logic Level
L = Low Logic Level

Figure 7–89 74132 quad two-input NAND gates with Schmitt-trigger inputs. *Reprinted with permission of National Semiconductor Corporation.*

Conclusion

In this section, you were introduced to one of the most popular multivibrator circuits—the 555. Here you saw how to create a clock pulse generator using this versatile device. Along with this, you were introduced to the operation of a comparator. Here, you also saw a real one-shot and were introduced to the operation of a Schmitt trigger.

In the next chapter, you will be introduced to one of the most versatile of all flip-flops and see how to use it in a variety of applications. For now, check your understanding of this section by trying the following review questions.

7–8 Review Questions

1. What is one application of the 555 timer?
2. State the function of a comparator.
3. How many comparators are there in the 555 timer? Basically, what purpose do they serve?
4. What determines the frequency of the output of a 555 timer?
5. What determines how long a one shot will stay in its unstable state?
6. Briefly describe the operation of a Schmitt trigger.

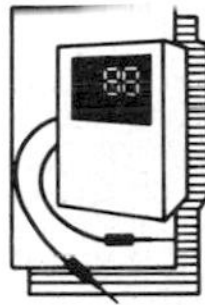

TROUBLESHOOTING SIMULATION

The troubleshooting simulation for this chapter gives you an opportunity to troubleshoot an *S-R* flip-flop. Here you will have a chance to first see the correct operating mode and then the computer will put in a problem when you are ready. This is an opportunity to develop troubleshooting skills in the analysis of bistable elements used in digital systems.

SUMMARY

- One of the simplest digital storage elements is called a latch.
- A latch may be logically constructed from NAND gates or NOR gates.
- The basic latch has two inputs called SET and RESET.
-  The basic latch has two outputs called Q and $\overline{Q}$. These outputs are complements of each other.
- A latch may have active HIGH inputs or active LOW inputs.
- A gated latch has an ENABLE input which must be active before the SET and RESET inputs can affect the latch.
- Another type of gated latch is the gated D latch.
- The gated D latch has two inputs: the ENABLE input and the D input.
- A flip-flop is a bistable logic device that is synchronized by an external source.
- A clock is a series of pulses used to synchronize digital systems.
- There are two basic kinds of edge-triggered flip-flops, those that are activated on the leading edge of the clock and those that are activated on the trailing edge.
- A dynamic input indicator is used to signify that a flip-flop is edge triggered.
- Flip-flops may have asynchronous inputs that override the synchronous inputs.

- Edge-triggered flip-flops with asynchronous PRESET and CLEAR are available.
- One application of a flip-flop is in the construction of a *binary counter*.
- A binary counter is a group of flip-flops connected in a way that will cause the conditions of the flip-flops to change so that a binary count appears.
- A register is a group of flip-flops treated as a unit.
- Registers are classified according to how digital information is put into and taken from them.
- Digital information may be put into a register all at once (parallel in) and taken out all at once (parallel out).
- Digital information may be put into a register 1 bit at a time (serial in) and taken out 1 bit at a time (serial out).
- Registers can be constructed so that any combination of serial and parallel data handling is available.
- Flip-flops are actually a class of multivibrators.
- A monostable multivibrator has only one stable state and will stay in its unstable state for a predetermined amount of time.
- A trigger can be thought of as a logic pulse with a short duration used to start (or stop) a digital process.
- A debouncer is a logic circuit that removes undesirable changes in logic levels usually due to the mechanical actions of switches.
- An astable multivibrator produces its own output logic pulses.
- Sync pulses are used to cause certain events to happen at a predetermined time.
- A clock pulse is the basic timing waveform used by a digital system.
- A special logic symbol called the *dynamic input indicator* is used on the clock input of a flip-flop to indicate that the flip-flop is edge triggered.
- A master-slave flip-flop consists essentially of two flip-flops in one: the input flip-flop (called the master), and the output flip-flop (called the slave).
- The basic operation of a pulse-triggered master-slave flip-flop is that the master is active only during the clock level of the clock pulse and the slave is active when the clock changes state.
- Use of a master-slave flip-flop prevents the outputs of a flip-flop from changing while its inputs are reading the output of a similar flip-flop.
- A data lock-out flip-flop is a master-slave arrangement where the input is active only on the leading edge of the clock. Data is transferred from the master to the slave after the clock changes state.
- A TRI-STATE logic device has three output states: HIGH, LOW, and disconnected (called the *high-impedance state*).
- TRI-STATE devices are used in bus-oriented systems.
- Addressing mode is the method used by the microprocessor for transferring information.
- Immediate addressing means that the data *immediately* follow the instruction.
- Direct addressing means that the data are located at a memory location whose address is given immediately after the instruction.

- I/O refers to the input/output interface between a digital system and the external environment accessible to it.
- One of the most popular digital timers is the 555 IC.
- A comparator is a circuit that compares an input voltage to a reference voltage and produces a HIGH or LOW output based upon this comparison.
- The frequency, on time, and off time of a 555 timer used as a clock generator is determined by the values of the external components selected by the user.
- The amount of time a one-shot stays in its unstable state depends on the values of the external components selected by the user.
- A Schmitt trigger is a device that has an abrupt change in its output when a certain DC voltage level is presented to its input.

CHAPTER SELF-TEST

I. TRUE/FALSE

Answer the following questions true or false.

1. A latch is used to store the logic condition of a bit.
2. An application for a latch is the storage of binary data in a read/write memory.
3. A latch has only one output, which is either HIGH or LOW.
4. A *register* is a group of logic gates working together as a unit.
5. A serial input register means that data are inputted 1 bit at a time.

II. MULTIPLE CHOICE

Answer the following questions by selecting the most correct answer.

6. A serial-in–parallel-out register means that digital data is transferred into the register

(A) 1 bit at a time and transferred out 1 bit at a time.
(B) all at once and transferred out 1 bit at a time.
(C) 1 bit at a time and transferred out all at once.
(D) all at once and transferred out all at once.

7. A multivibrator is an electrical circuit that

(A) can maintain one or two stable states.

(B) can maintain one stable state.
(C) can produce its own output signal.
(D) All of the above are correct.

8. A flip-flop will

(A) maintain one of two stable states.
(B) produce its own output square wave.
(C) maintain one stable state.
(D) flip or flop, depending on its condition.

9. A *monostable* multivibrator will

(A) maintain one of two stable states.
(B) produce its own output square wave.

(C) maintain one stable state.
(D) flip or flop, depending on its condition.

10. An *astable* multivibrator will
    (A) maintain one of two stable states.
    (B) produce its own output square wave.
    (C) maintain one stable state.
    (D) flip or flop, depending on its condition.

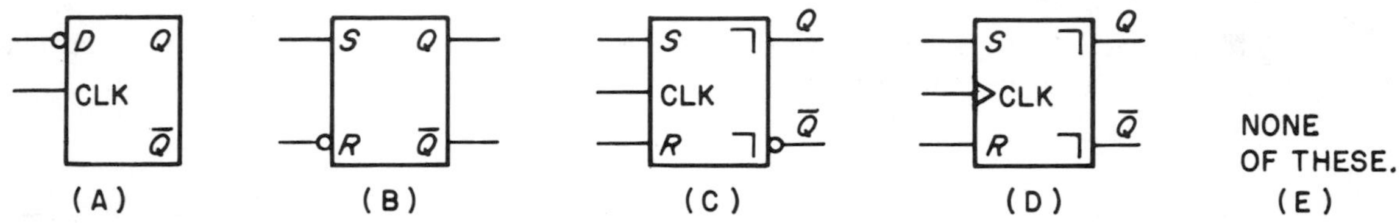

Figure 7–90

III. MATCHING

Match the correct flip-flop symbol in Figure 7–90 to the statement on the left.

11. *S-R* latch.
12. Level-triggered *S-R* latch.
13. Data lock-out master-slave flip-flop.
14. Pulse-triggered master-slave flip-flop.
15. Level-triggered *D* latch.

IV. FILL-IN

Fill in the blanks with the most correct answer(s).

16. Flip-flop inputs that are synchronized with the clock are called ______ inputs.
17. In a clocked flip-flop, those inputs that are independent of the clock are called ______ inputs.

18. Flip-flops whose inputs are active only during the edge of the clock are called ______ -triggered flip-flops.
19. The triangular symbol used on the clock input of a flip-flop to indicate synchronization on the clock ______ is called the dynamic input indicator.
20. A flip-flop whose inputs and outputs are not active at the same time is called a ______-______ flip-flop.

V. OPEN-ENDED

Answer the following questions as indicated.

21. Using the process controller presented in the microprocessor application section of this chapter, develop an assembly language program that would cause the position of the shaft encoder to appear on the seven-segment readout.
22. Using the microprocessor controller, write an assembly language program that would allow the speed of the motor to be controlled by the position of the input switches. The position of the switches must also be displayed on the LED readouts.

23, Using the microprocessor controller, determine what the following assembly language program will cause it to do.

Address	Instruction
0	MVI A, 6
2	OUT 3
4	OUT 0

24. Explain the operation of a TRI-STATE buffer.
25. Explain the operation of a Schmitt trigger.

Self-Test Answers

1] T 2] T 3] F 4] F 5] T 6] C 7] D 8] A 9] C A 10] B
11] B 12] E 13] D 14] C 15] A 16] [illegible]
17] asynchronous 18] edge 19] transition 20] master-slave

21]

Address	Instruction
0	IN 3
2	OUT 1
4	HLT

22]

Address	Instruction
0	IN 0
2	OUT 2
4	OUT 0
6	HLT

23] The bit pattern 0110 would be outputted to the control switches causing motor 2 to be ON and to rotate CW. The LEDs would display the bit pattern 0110.
24] A TRI-STATE buffer has three output states: HIGH, LOW, and open (called the high-impedance state).
25] A circuit that changes its output abruptly when the input signal crosses a specified DC triggering level.

CHAPTER PROBLEMS

Basic Concepts

Section 7–1:

1. Explain the main idea behind the concept of a latch.
2. State some applications of latches.
3. How many outputs does a latch have? What is always true about them?

Section 7–2:

4. For the input waveforms into the gated *S-R* latch of Figure 7–91(A), determine the resulting output.

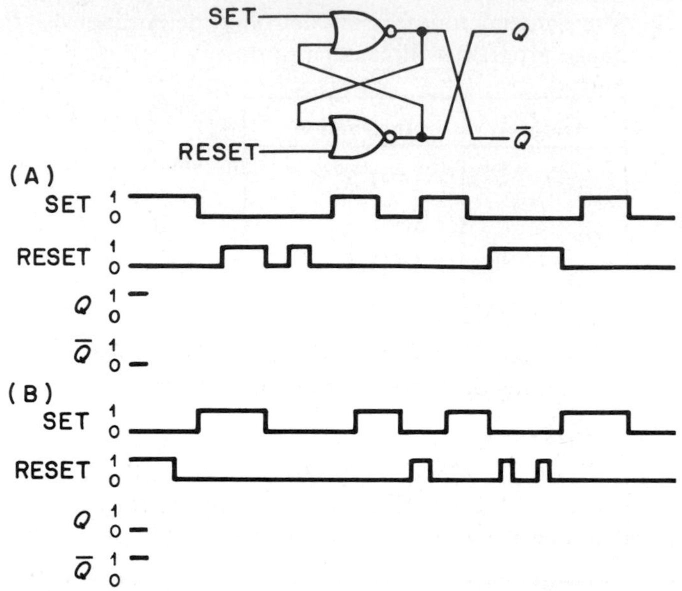

Figure 7–91

5. Determine the resulting output for the input waveforms of Figure 7–91(B).
6. Sketch the resulting output for the indicated latch for the input waveforms of Figure 7–92(A).
7. For the input waveforms and indicated latch of Figure 7–92(B), determine the resultant waveforms.

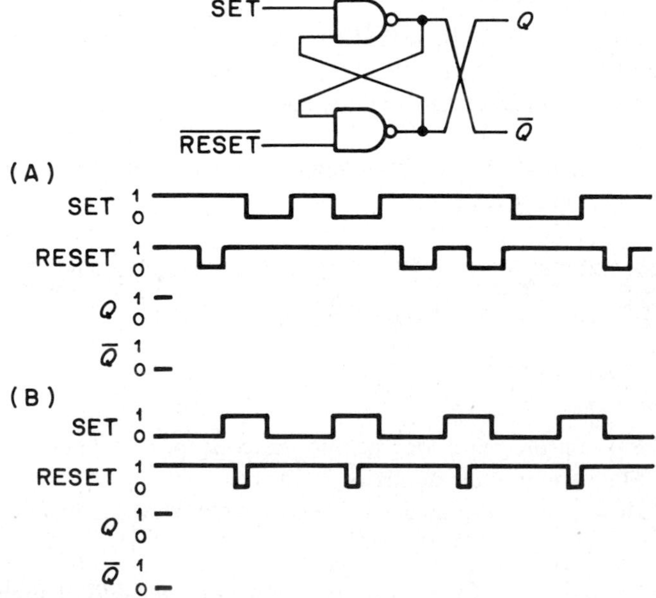

Figure 7–92

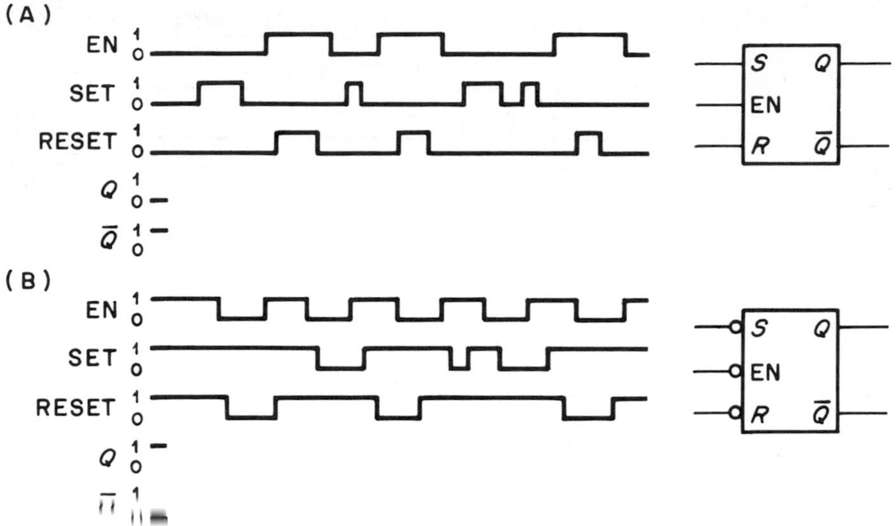

Figure 7–93

8. Determine the resultant waveforms for the gated *D* latch of Figure 7–93(A).
9. For the gated *D* latch of Figure 7–93(B), determine the resultant waveforms.

Section 7–3:

10. State the difference between latches and flip-flops.
11. Describe the operation of the flip-flops illustrated in Figure 7–94. Construct a truth table for each and describe each part in detail.

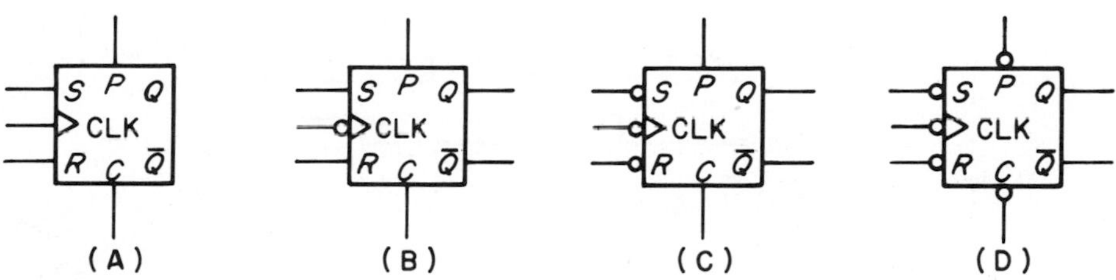

Figure 7–94

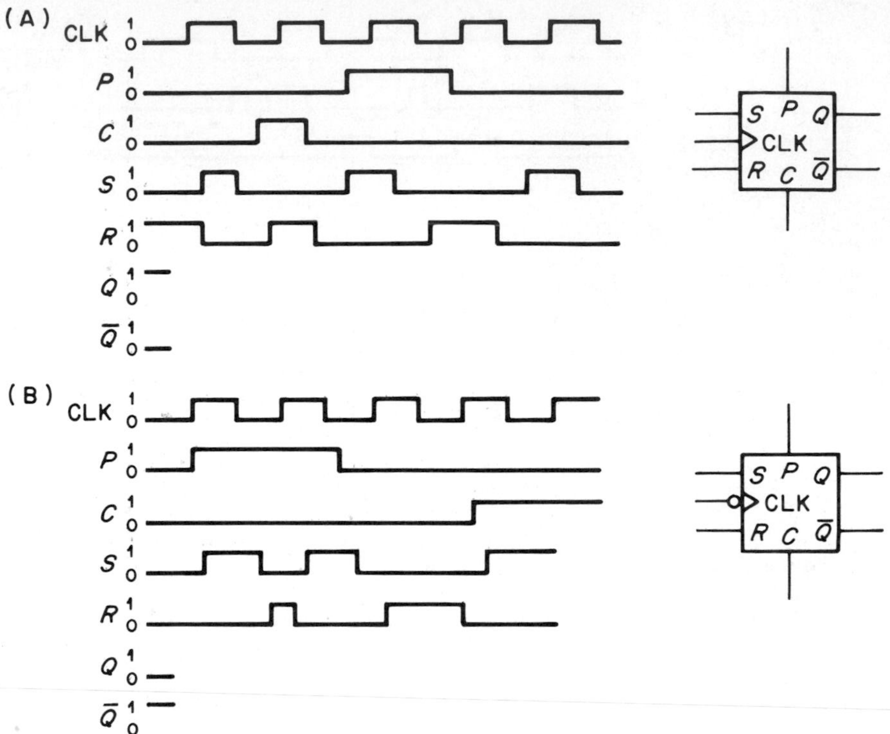

Figure 7–95

12. For the waveforms and flip-flops indicated in Figure 7–95, determine the resulting output waveforms.
13. For the waveforms and flip-flops of Figure 7–96, determine the resulting output waveforms and mark the Q waveform where it would be indeterminate.

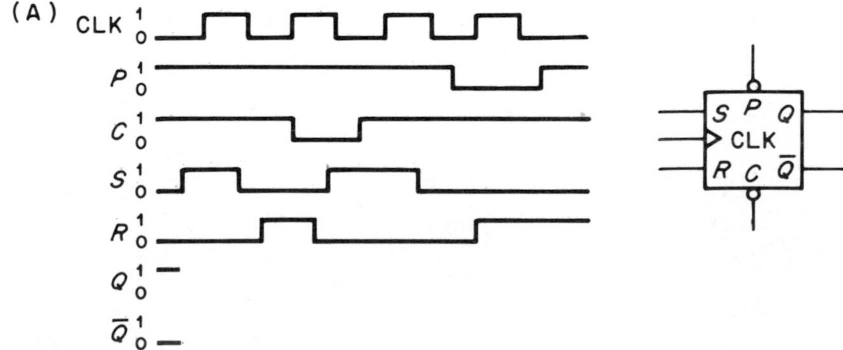

Figure 7–96

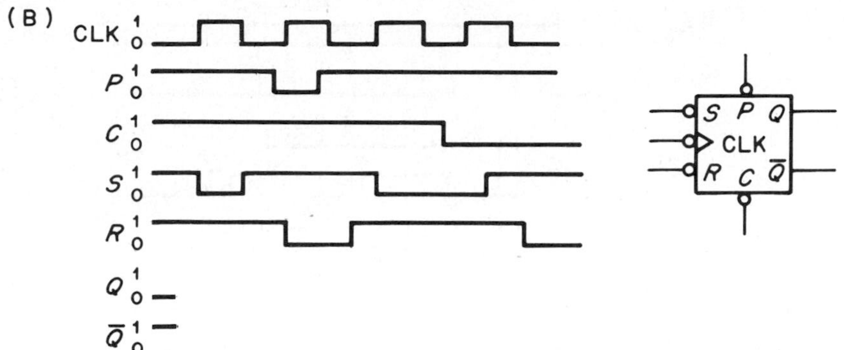

Figure 7–96 Continued

14. Describe the operation of each of the flip-flops shown in Figure 7–97. Construct a truth table for each and describe in detail each part of the table.

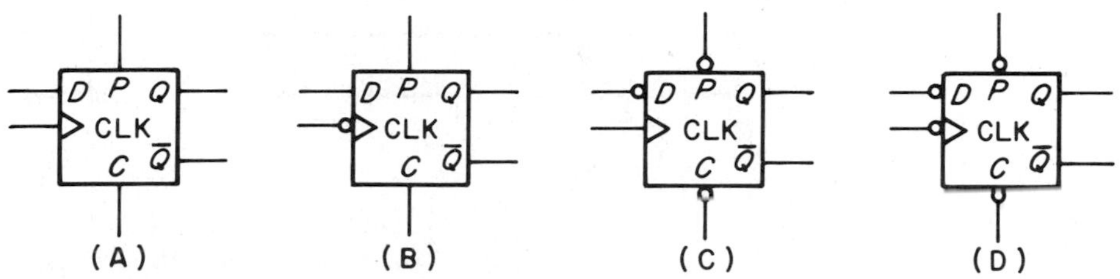

Figure 7–97

15. For the flip-flops and indicated input waveforms of Figure 7–98, determine the resulting outputs.

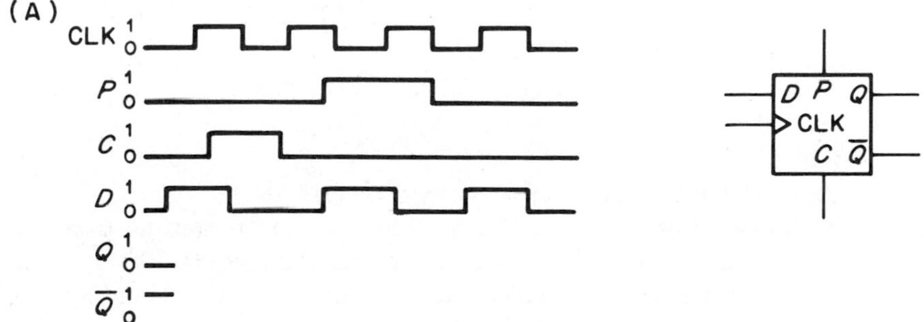

Figure 7–98

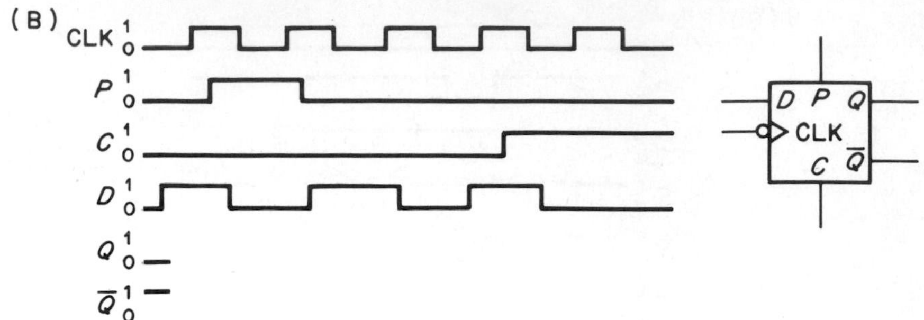

Figure 7–98 Continued

16. For the flip-flops and indicated input waveforms of Figure 7–99, determine the resulting outputs assuming all control input edges effectively lag the clock edges.

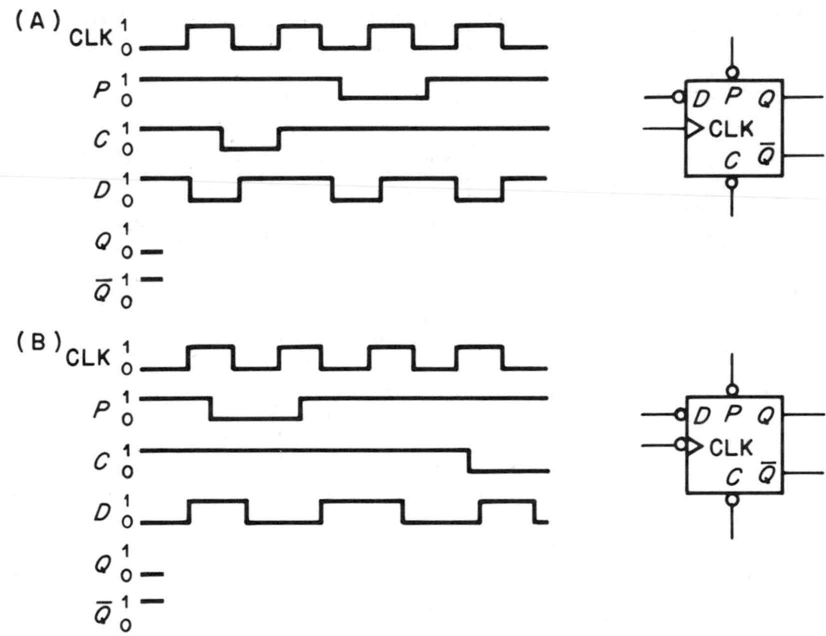

Figure 7–99

Section 7–4:

17. Describe the action of a binary counter.
18. Show all of the states of a 5-bit binary counter starting from the count of 00000_2.
19. For an 8-bit binary counter starting its count at 00000000_2, how many times must its input count switch be depressed in order for it to count up to its maximum value and then recycle back to 00000000_2?

20. Sketch a *data latch* register and indicate the direction of input and output data.
21. Explain the basic idea behind a shift register.
22. Describe the difference between a parallel-in–serial-out register and a serial-in–parallel-out register.
23. What kind of register would be used for transmitting information from a computer data bus over telephone lines?
24. What kind of register would be used for receiving information from a telephone line and inputting it to a computer data bus?

Section 7–5:

25. Describe the difference between *bistable*, *monostable*, and *astable* as they apply to multivibrator.
26. Sketch a switch debouncer. Explain why such a device is needed.
27. Explain the similarities and differences between a *sync pulse* and a *clock pulse*.
28. Describe the meaning of *symmetry*.

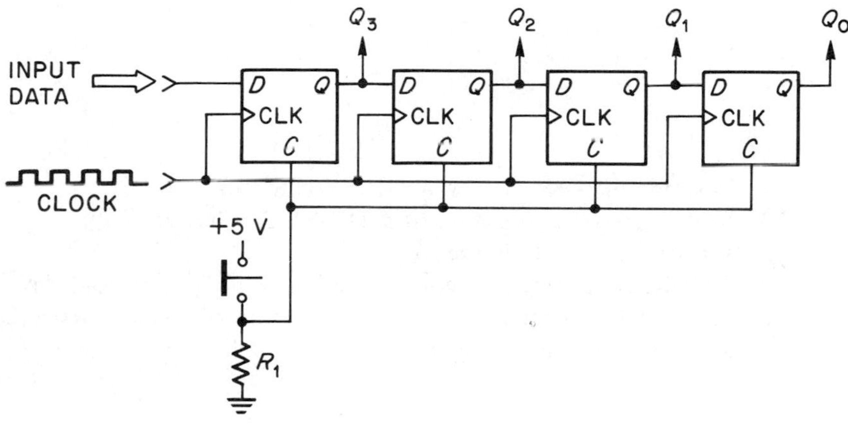

Figure 7–100

Section 7–6

29. Explain the operation of the register shown in Figure 7–100.
30. Demonstrate, by means of drawings, the potential problem with the register shown in Figure 7–100.
31. Explain the basic idea of a master-slave flip-flop.

32. For the waveforms shown in Figure 7–101(A), determine the output waveforms.
33. Determine the output waveforms for the input waveforms shown in Figure 7–101(B).

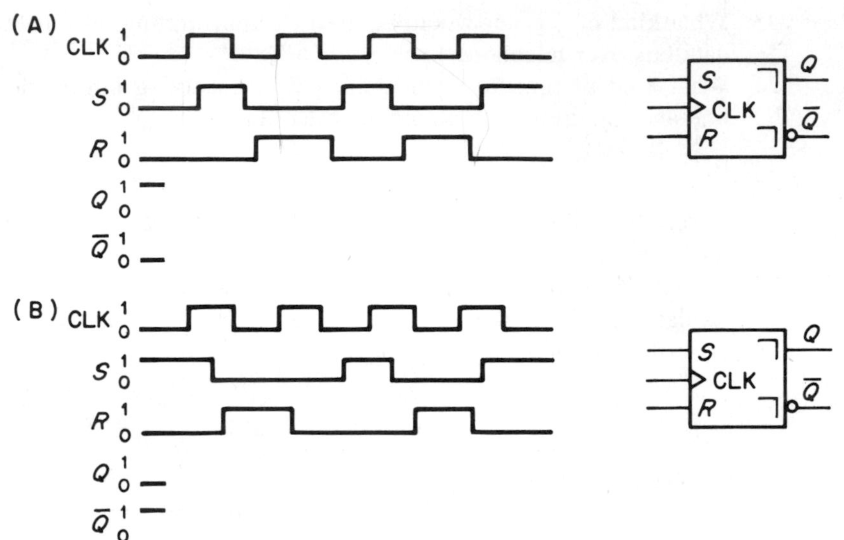

Figure 7–101

Applications

34. Describe the operation of a TRI-STATE buffer.
35. Sketch a logic application of a TRI-STATE buffer.
36. What is an *addressing mode?*
37. Explain the difference between *immediate addressing* and *direct addressing*.
38. State what the contents of the accumulator would be after execution of the following assembly language programs:

(A)

Address	Instruction
0	MVI A,5
2	NOP
3	NOP
4	HLT
5	8

(B)

Address	Instruction
0	LDA 5
2	NOP
3	NOP
4	HLT
5	8

39. Write an assembly language program that would (A) load the number 4 into the accumulator, (B) load the contents of memory location 4 into the accumulator.
40. For the following programs, indicate the condition of the READ, WRITE, and MEM/$\overline{\text{IO}}$ line.

(A)

Address	Instruction
0	MVI A,3
2	OUT 3
4	IN 2
5	HLT

(B)

Address	Instruction
0	LDA 5
2	OUT 3
4	HLT
5	9

41. For the microprocessor-based system presented in the microprocessor applications section of this chapter, write an assembly language program that would do the following:
(A) Display the hex keypad on the seven-segment readout.
(B) Cause the conditions of the photoresistors to control the speed of motor 1.
42. For the microprocessor-based system presented in the microprocessor applications section of this chapter, determine what the following assembly language programs would do:

(A)

Address	Instruction
0	IN 3
2	OUT 0
4	MVI A, 7
6	OUT 3
8	HLT

(B)

Address	Instruction
0	LDA 5
2	OUT 3
4	HLT
5	A

Electronic Application

43. Explain the operation of a comparator.

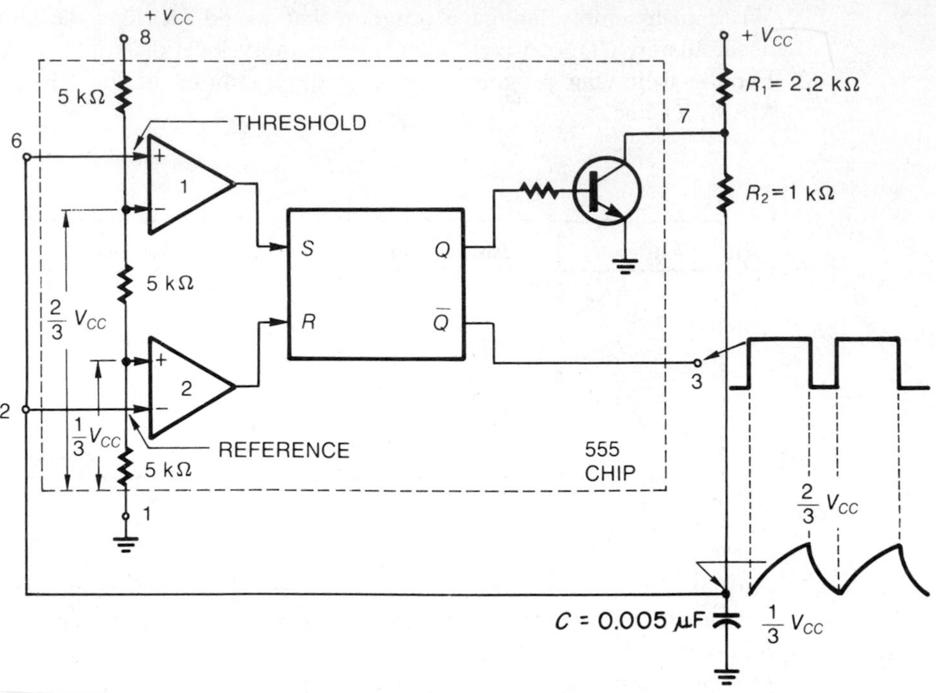

Figure 7–102

44. For the 555 timer in Figure 7–102 determine the output frequency, the pulse ON time and the pulse OFF time.
45. Explain the operation of a one-shot. What determines the amount of time the circuit will stay in its unstable state?
46. Describe the operation of a Schmitt trigger. State how this is a useful device.

Analysis and Design

47. Show how you would connect the industrial controller presented in the microprocessor application section so it would test the condition of the following ICs. You may use any devices you wish to connect to the input ports and output ports:
    (A) a quad inverter
    (B) an *RS* flip-flop
    (C) a 4-bit binary counter
    (D) a 4-bit shift register
    (E) a quad two-input NAND gate

CHAPTER 8

JK Flip-Flops, Counters, and Registers

OBJECTIVES

After completing this chapter, you should be able to:

- ☐ Explain the basic idea behind the *JK* flip-flop.
- ☐ Demonstrate practical applications of the *JK* flip-flop.
- ☐ Explain the fundamental properties of counters.
- ☐ Recognize the difference between synchronous and asynchronous counters.
- ☐ Use mod numbers and demonstrate their application to counters.
- ☐ Define the theory of operation of upcounters and downcounters.
- ☐ Demonstrate practical applications of counters and PLDs.
- ☐ Describe fundamental logic construction of registers.
- ☐ Describe the integration of PLDs with registers.
- ☐ Identify real IC flip-flops, counters, and registers.
- ☐ Explain the fundamentals of digital and analog comparators.
- ☐ Explain the basic principles of automated troubleshooting.
- ☐ Describe fundamentals of digital instrumentation.

INTRODUCTION

The focus of this chapter is on the analysis of digital processes. Here you see *how* registers are used in a complete digital system. Even though registers are made of flip-flops, you do not need to make registers in this manner. It is more economical to use a single chip that contains a complete register. The flip-flops introduced here is one of the most versatile of all.

As you know from the microprocessor application sections of the previous chapters, counters and registers are the necessary building blocks of all microprocessors and microprocessor-based systems. This chapter contains a lot of practical information that can be immediately applied to all kinds of digital systems.

Almost all of the information you have acquired up to this point will be used in this chapter. PLAs, logic gates, flip-flops, and number systems will be integrated here in a way that will lay the foundation for your understanding of real digital systems. The magic of robotics, computers, and other marvels of the information age is becoming a workable tool for your use. The foundations set forth in this chapter will serve you well.

KEY TERMS

JK Flip-Flop
Asynchronous Counter
Ripple Counter
Modulus
Mod Number
Programmable Counter
Decoding Glitch
Synchronous Counter
Ring Counter
State Machine
State Table
Microprogram
Shift Register
Machine Cycle
Microcode
Microinstruction
Dependency Notation
Overflow
Underflow
System Comparison
Magnitude Comparator
Difference Amplifier
Digital Margin Detector
Frequency Counter
Digital Voltmeter
VCO
Thermistor
Photoresistor

8–1 THE *JK* FLIP-FLOP

Discussion

In this section, you will be introduced to the last kind of flip-flop. This flip-flop was "saved" until now because it is the most versatile of all the flip-flops and requires a thorough understanding of the *S-R* flip-flop. Thus, you were given time to develop your skills with the *S-R* flip-flop presented in the last chapter.

What it Does

Recall that an *S-R* flip-flop had one state that was not allowed. This was the condition when both the *S* and *R* inputs were active. This represented a "wasted" state, and it could be applied to a useful condition. One such condition could be to cause the flip-flop to change state on each clock pulse. In order to distinguish this flip-flop from the *S-R* flip-flop, the *S* input is changed to *J* and the *R* input to *K*. This concept is illustrated in Figure 8–1.

As you can see, this modified *S-R* flip-flop now has four useful states instead of just three. To distinguish this modified *S-R* flip-flop from the standard *S-R* flip-flop, the identification letters used for the flip-flop inputs are labeled *J* (instead of *S*) and *K* (instead of *R*). This flip-flop is called a ***JK* flip-flop.**

***JK* Flip-Flop:**
A flip-flop with four useful input conditions: no change, set, reset, and toggle.

As you can see, the *JK* flip-flop is nothing more than the *S-R* flip-flop with an added feature—you can make it toggle (change states with each clock pulse).

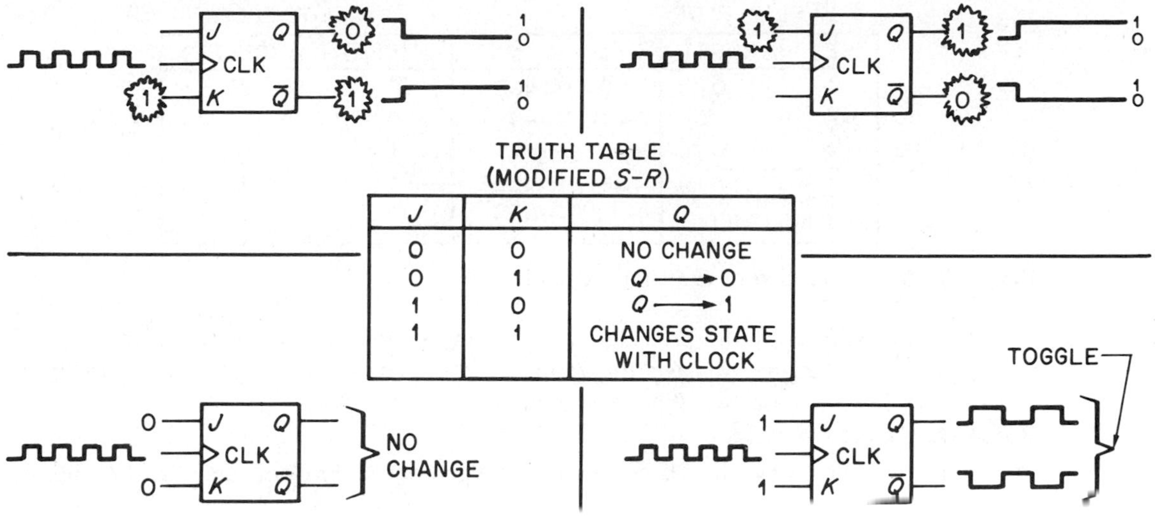

J	K	Q
0	0	NO CHANGE
0	1	Q ⟶ 0
1	0	Q ⟶ 1
1	1	CHANGES STATE WITH CLOCK

Figure 8–1 Using all flip-flop states.

Example 8–1

For the waveforms indicated in Figure 8–2, sketch the resulting outputs.

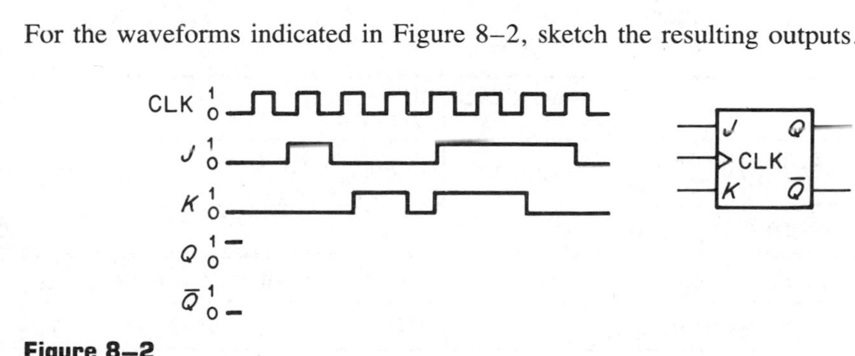

Figure 8–2

Solution

See Figure 8–3.

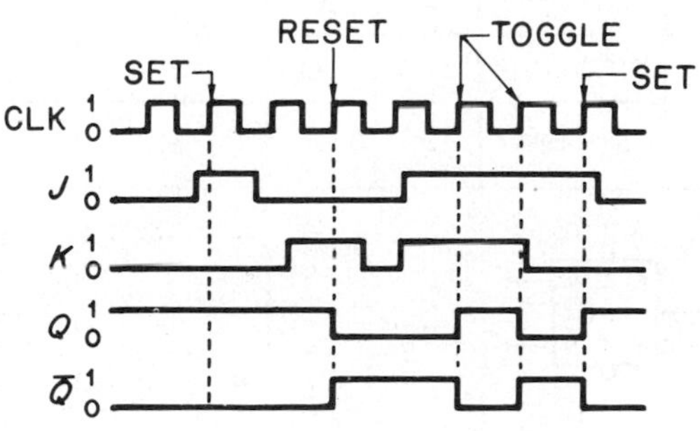

Figure 8–3

S-R FLIP-FLOP			*JK* FLIP-FLOP	
SET	RESET	Q $\bar{Q}$	*J* (SET)	*K* (RESET)
0	0	NO CHANGE	0	0
0	1	→ 0 → 1	0	1
1	0	→ 1 → 0	1	0
1 (NOT ALLOWED)	1 (NOT ALLOWED)	TOGGLE MODE (*JK* ONLY) ←	1	1

Figure 8–4 Comparison of *S-R* and *JK* flip-flops.

As demonstrated by the previous example, the only difference between the *JK* and *S-R* flip-flops is that the *JK* has a toggle mode while the *S-R* does not. A comparison of the truth tables for these two flip-flops is shown in Figure 8–4.

PRESET and CLEAR

As with other flip-flops, the *JK* comes with asynchronous *PRESET* and *CLEAR* inputs. Thus the *JK* flip-flop has great flexibility. Some of the possible variations of the *JK* flip-flop are illustrated in Figure 8–5.

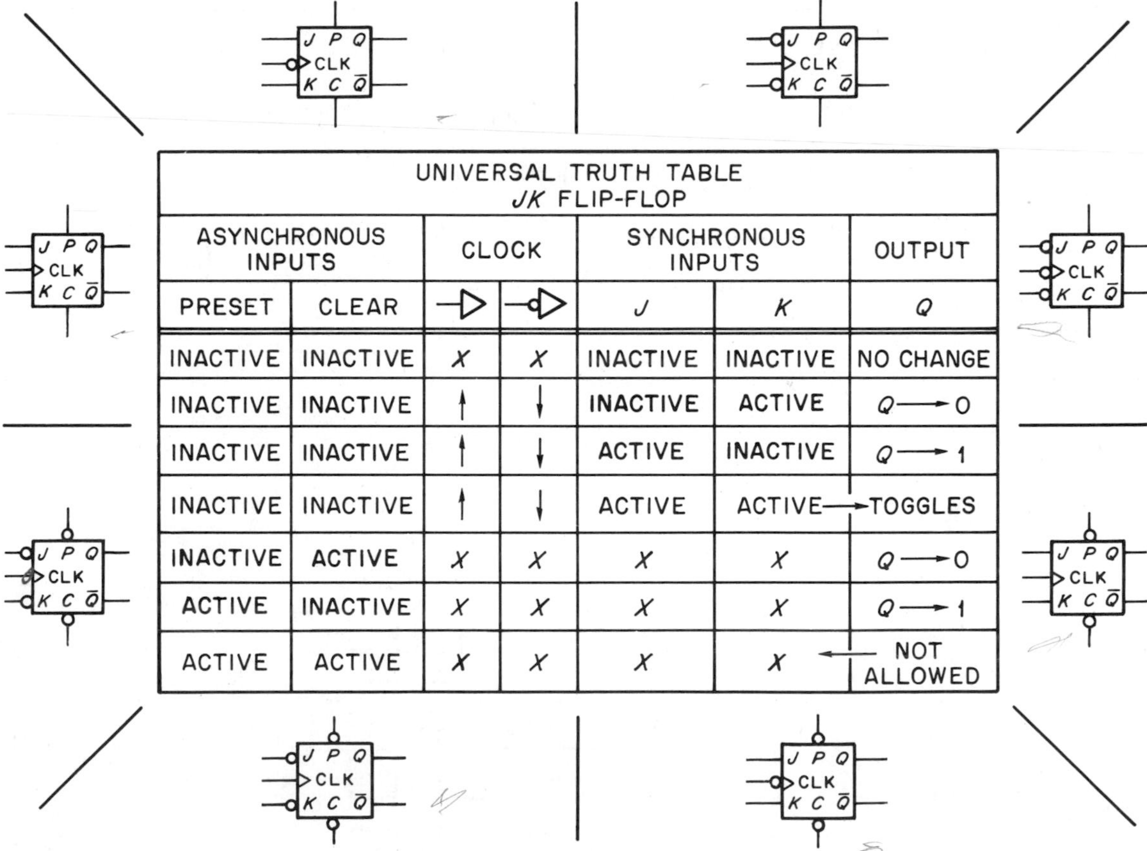

UNIVERSAL TRUTH TABLE *JK* FLIP-FLOP

ASYNCHRONOUS INPUTS		CLOCK		SYNCHRONOUS INPUTS		OUTPUT
PRESET	CLEAR	▷	o▷	*J*	*K*	*Q*
INACTIVE	INACTIVE	X	X	INACTIVE	INACTIVE	NO CHANGE
INACTIVE	INACTIVE	↑	↓	INACTIVE	ACTIVE	Q → 0
INACTIVE	INACTIVE	↑	↓	ACTIVE	INACTIVE	Q → 1
INACTIVE	INACTIVE	↑	↓	ACTIVE	ACTIVE	TOGGLES
INACTIVE	ACTIVE	X	X	X	X	Q → 0
ACTIVE	INACTIVE	X	X	X	X	Q → 1
ACTIVE	ACTIVE	X	X	X	X	NOT ALLOWED

Figure 8–5 Possible *JK* flip-flop variations.

Example 8–2

For the input waveforms shown in Figure 8–6, determine the resulting outputs.

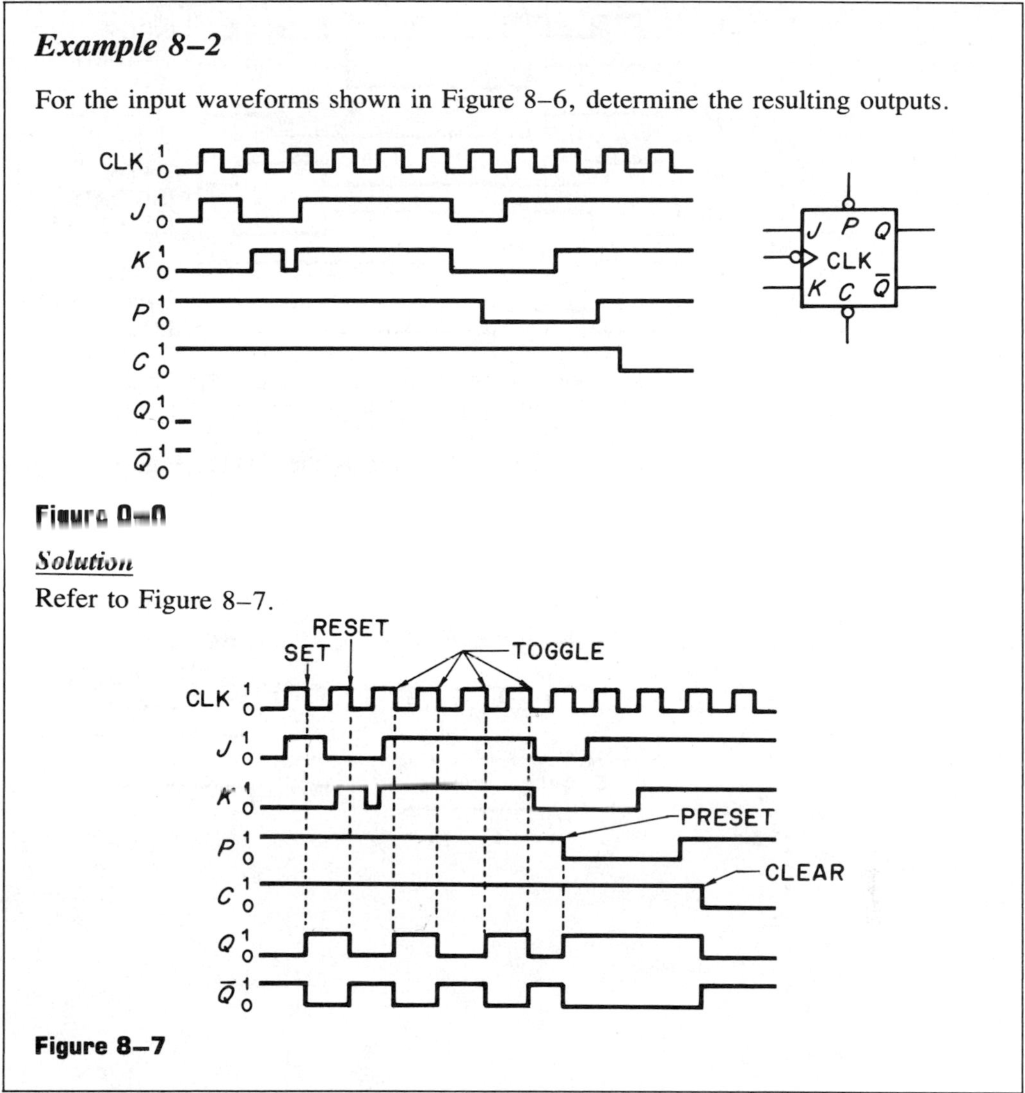

Figure 8–6

Solution

Refer to Figure 8–7.

Figure 8–7

Master-Slave JK Flip-Flops

JK flip-flops come in the master-slave configuration. There are two basic types: edge-triggered and level-triggered. The level-triggered master-slave has its inputs active during a clock pulse *level* and its output active when this level changes. Several of the "older" *JK* flip-flops have this feature. They are the 74H71, 7472, 7473, 7476, 7478, 74104, and 74105.

The other type of *JK* master-slave flip-flop is an edge-triggered flip-flop whose inputs are active only during the leading edge of the clock. Therefore, data may change on the input during a clock level with no effect. This type of flip-flop is referred to as a *data*

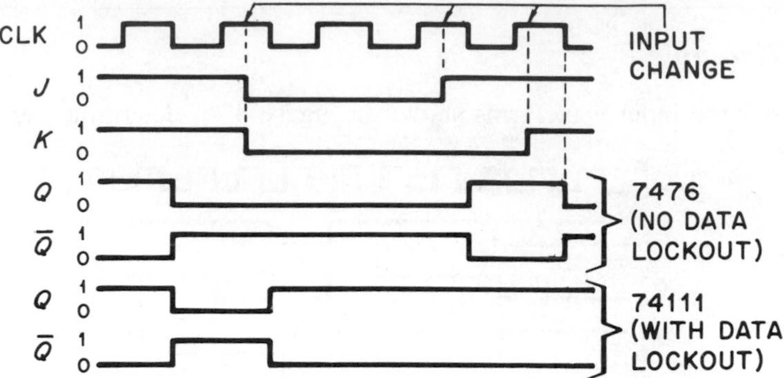

Figure 8–8 Comparison of two types of master-slave *JK* flip-flops. The 74111 flip-flop (with data lockout) is not affected by input change during clock levels.

lockout. An example of this kind of flip-flop is the 74111. Figure 8–8 illustrates the difference between the two.

Example 8–3

For the waveforms indicated in Figure 8–9, determine the resulting outputs.

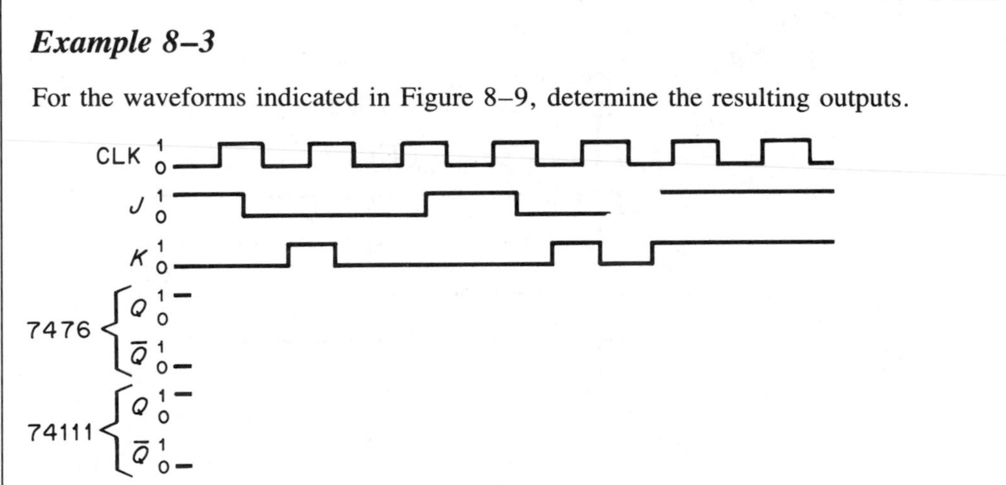

Figure 8–9

Solution

Refer to Figure 8–10. The 74111 is a *data lockout* flip-flop while the 7476 is not. As shown in the diagram, the *J* and *K* inputs of the 74111 are sensitive to level changes only during the leading edge (LOW to HIGH transition of the clock pulse). Thus, even though the *J* or *K* inputs may be changing after the leading clock edge, the data lock-out feature ignores them. On the other hand, the 7476 *JK* flip-flop does not have a data lock-out feature. Its *J* and *K* inputs are both active during the active *level* of the clock. Thus, as shown in the diagram, these inputs will respond to any level changes after the leading clock edge.

In both of these master-slave flip-flops, the output reflects the condition of the input at the trailing edge (HIGH to LOW transition) of the clock.

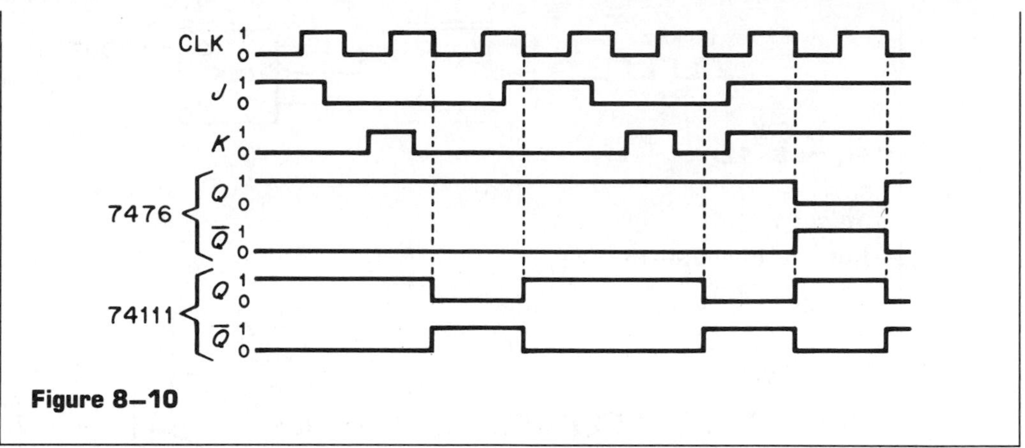

Figure 8–10

The 7476 is a *level-triggered* flip-flop, and the 74LS76A is an *edge-triggered* flip-flop. Thus it is important for the technician to always use the exact replacement or be sure to check the data sheets.

Frequency Division

One of the features of the *JK* flip-flop in the toggle mode (both *J* and *K* inputs are active) is that it will produce an output waveform that is exactly one half the frequency of the input waveform. This phenomena is illustrated in Figure 8–11. If this resulting output waveform is used as the clock input to a second *JK* flip-flop in the toggle mode, the resulting output will have a frequency of one quarter the input frequency. This is illustrated in the following example.

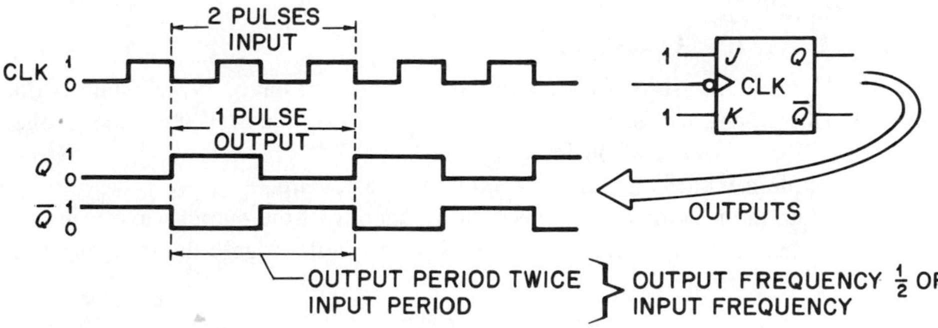

Figure 8–11 Frequency division with *JK* flip-flops.

Example 8–4

An input frequency of 12 kHz is applied to the *JK* flip-flop arrangement shown in Figure 8–12. Determine the resulting output frequency for each flip-flop and sketch the resulting waveform.

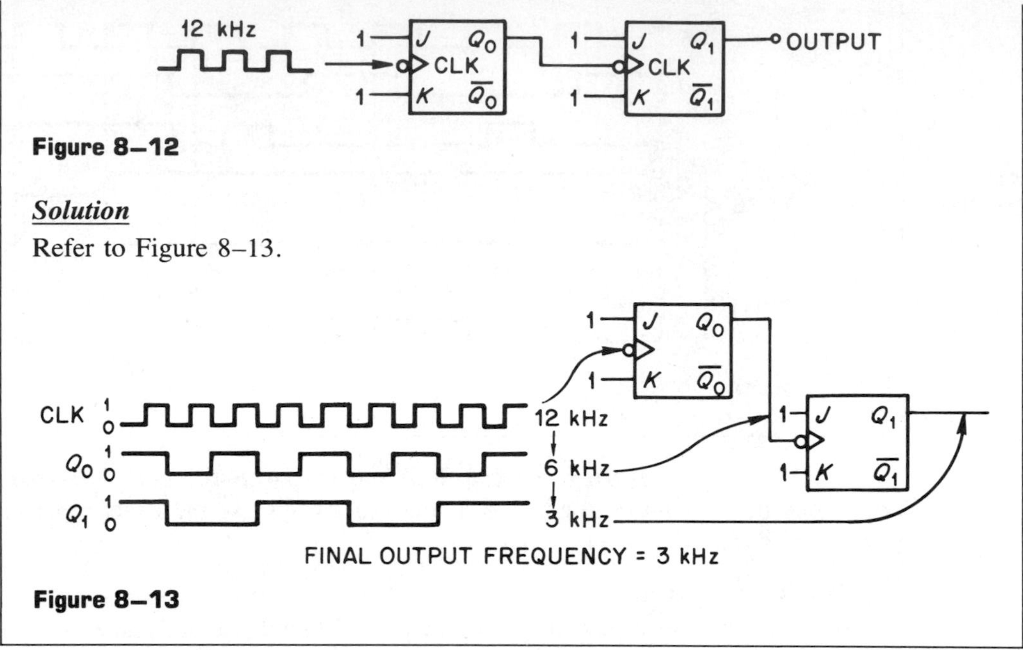

Figure 8–12

Solution

Refer to Figure 8–13.

Figure 8–13

Conclusion

This section presented the versatile *JK* flip-flop. Here you had an opportunity to work with this device. In the next section, you will see how this flip-flop is used in the construction of counters. Check your understanding of this section by trying the following review questions.

8–1 Review Questions

1. How many useful states does the *S-R* flip-flop have? Which state is not useful?
2. How many useful states does the *JK* flip-flop have? Which one is different from the *S-R* and what does it do?
3. Explain what is meant by a *data lockout* feature.
4. State the relationship between the frequency of the output waveform and the frequency of the clock waveform for a *JK* flip-flop in the toggle mode. What is this called?

8–2 ASYNCHRONOUS COUNTERS

Discussion

You have already been introduced to the concept of counters in the preceding chapter. In the microprocessor applications sections you saw the use of a counter in the 4-bit microprocessor. There a counter was used in order to supply sequential addresses to the address register. In this section you will see some of the most common ways of constructing counters.

Basic Idea

Perhaps the simplest kind of binary counter is one constructed from the *JK* flip-flop. You were really introduced to this in the last section, where the concept of frequency division was presented. Consider what a binary count looks like:

000	100
001	101
010	110
011	111

Observe from the binary count that the MSB changes state half the number of times as the next bit, and the next bit changes state half the number of times as the LSB. This concept is illustrated in Figure 8–14.

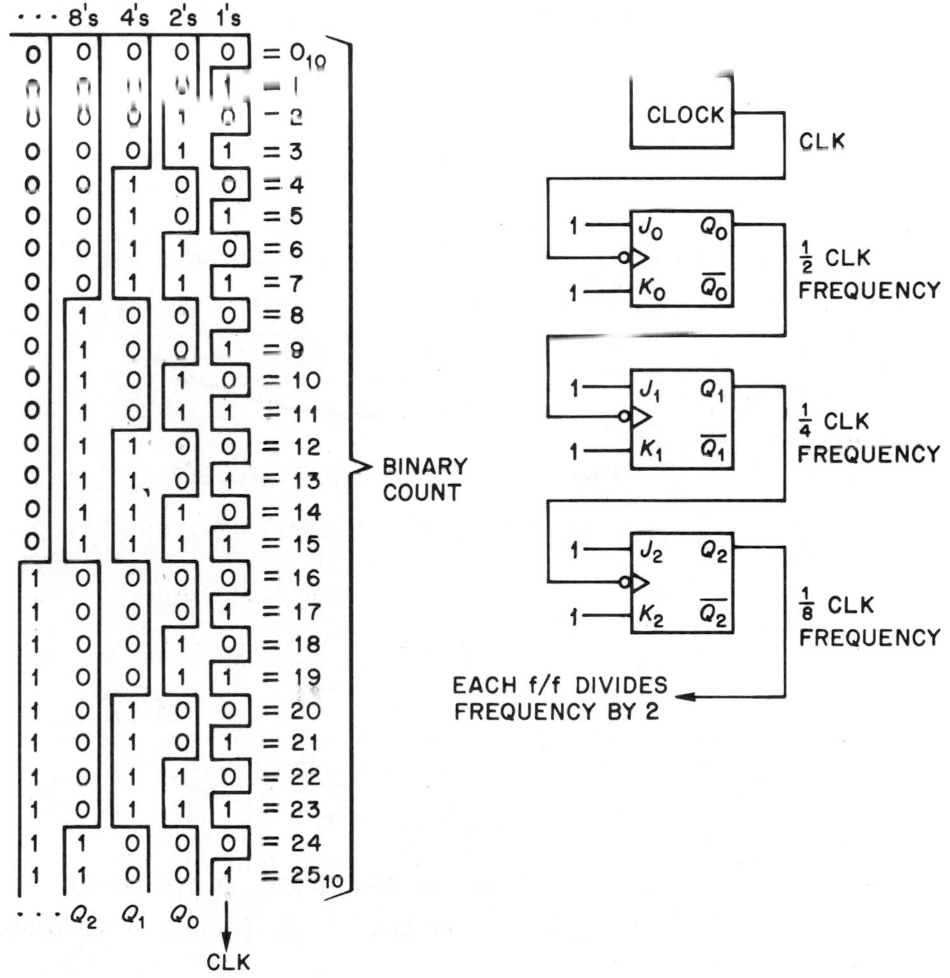

Figure 8–14 Concept of a binary counter from frequency division.

A detailed step-by-step analysis of what is actually happening is illustrated in Figure 8–15. From this figure, you can see that the 2-bit binary counter has four possible states ($2^2 = 4$): 00_2, 01_2, 10_2, and 11_2. The counter will then repeat the sequence. The following example illustrates the action of a 4-bit binary counter.

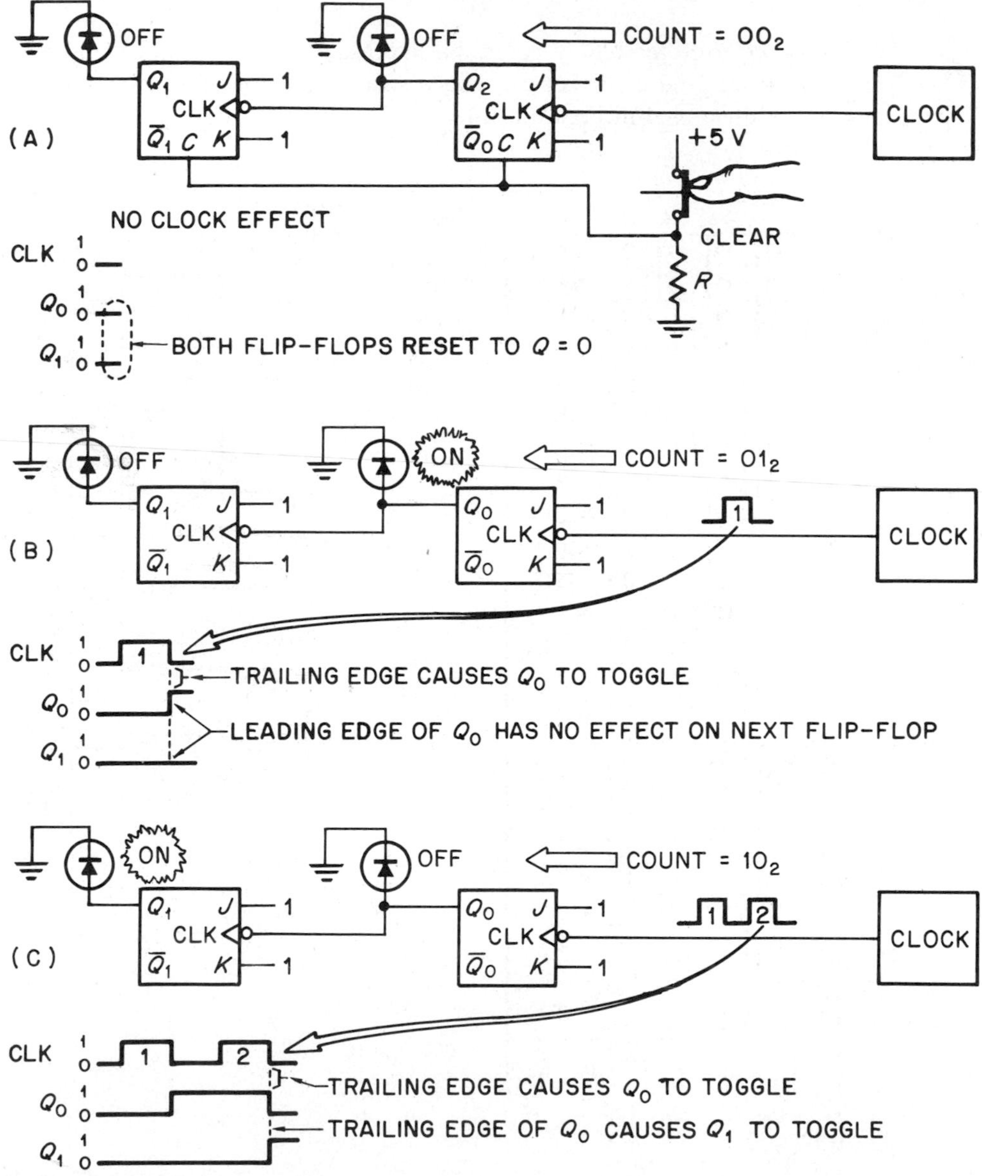

Figure 8–15 Sequence of a 2-bit binary counter.

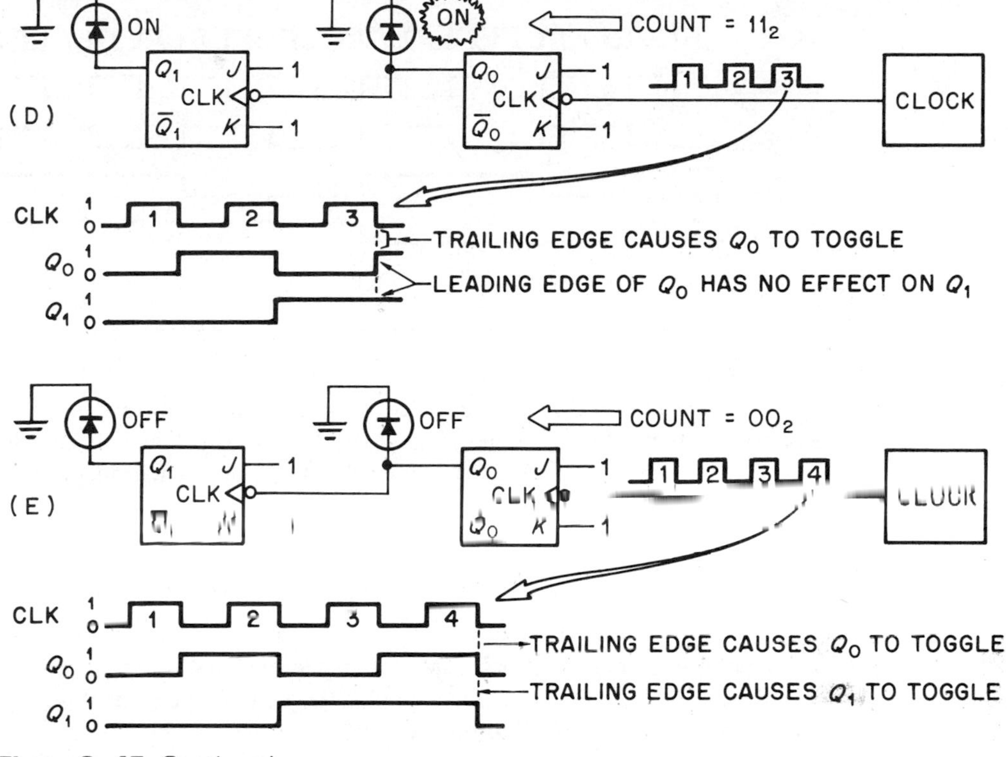

Figure 8–15 Continued

Example 8–5

Sketch the output waveforms you would expect to see from the 4-bit binary counter in Figure 8–16 for the first 16 clock pulses. Assume that the counter starts its count at 0000_2.

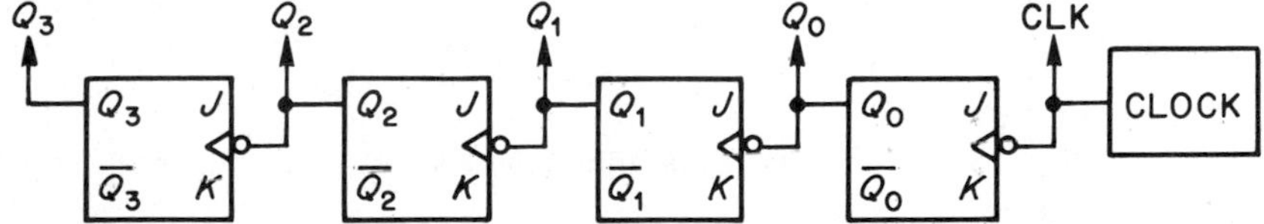

Figure 8–16

Solution

Sketch 16 clock pulses. Keep in mind that each of these flip-flops will toggle on the trailing edge (HIGH to LOW transition) of the clock pulse. The resultant waveform is illustrated in Figure 8–17.

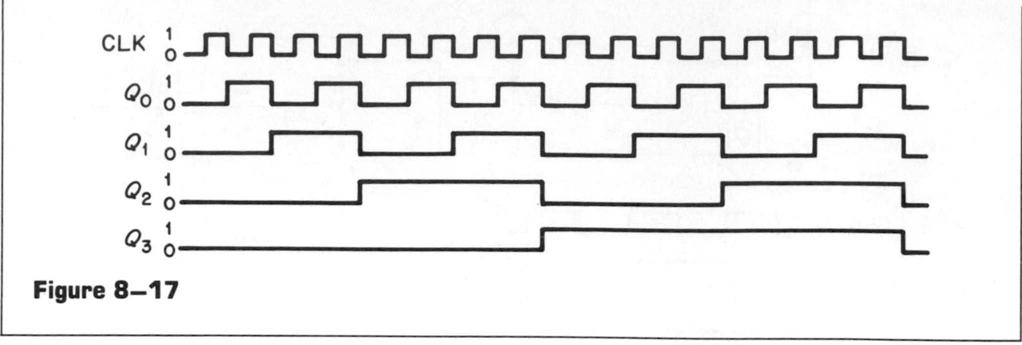

Figure 8–17

Counter Types

The binary counters presented here are called **asynchronous counters** because each flip-flop does not change by direct action of the clock. This means that only the first flip-flop is synchronized by the clock. All of the other flip-flops are synchronized by the output of the previous counter.

> **Asynchronous Counter:**
> A counter that does not have all of its flip-flops synchronized directly by the clock.

In an asynchronous counter, each flip-flop, except the first, is triggered by the output of the preceding flip-flop. Thus the count ripples through. Because of this, asynchronous counters are referred to as **ripple counters**.

> **Ripple Counter:**
> A counter where each flip-flop, except the first, is clocked by the preceding flip-flop.

Mod Number

Mod comes from the word **modulus**.

> **Modulus:**
> When applied to counters, refers to the maximum number of states the counter can assume.

The **mod number** of a counter is the number of states a counter will have under given conditions.

> **Mod Number:**
> Of a counter, refers to the number of states a counter will have under given conditions.

For example, a 2-bit binary counter is a mod 4 ($2^2 = 4$) counter and a 4-bit binary counter is a mod 16 ($2^4 = 16$) counter. Thus, the mod number of any counter may be determined by

$$\text{mod number} = 2^N$$

where, in a binary counter, N represents the number of flip-flops used.

Example 8–6

You are designing a microprocessor that must address at least 2 000 memory locations. How many flip-flops would be required in the counter for the address register?

Solution

This is easily solved by considering that $2^{10} = 1\ 024$, and $2^{11} = 2\ 048$. Thus 10 flip-flops would not be enough, but 11 would be. The extra counts would simply not be used.

Example 8–7

Most 8-bit microprocessors have 16 address lines. These address lines are fed from a binary counter with 16 flip-flops. With this information, determine the mod number of the 16 lines. Express your answer in K (1024).

Solution

The mod number would be $2^{16} = 65\ 536$.

In K, 65 536/1 024 = 64 K.

Up/Down Counters

In the previous presentation, only the Q outputs were considered. Suppose, in the previous counters, LEDs were connected to the $\overline{Q}$ outputs instead of the Q outputs. The results for a 4-bit counter are shown in the following table:

Clock	**Q Count**	**$Q_3\ Q_2\ Q_1\ Q_0$**	**$\overline{Q}_2\ \overline{Q}_2\ \overline{Q}_4\ \overline{Q}_0$**	**$\overline{Q}$ Count**
Start	0	0 0 0 0	1 1 1 1	15
1	1	0 0 0 1	1 1 1 0	14
2	2	0 0 1 0	1 1 0 1	13
3	3	0 0 1 1	1 1 0 0	12
4	4	0 1 0 0	1 0 1 1	11
5	5	0 1 0 1	1 0 1 0	10
6	6	0 1 1 0	1 0 0 1	9
7	7	0 1 1 1	1 0 0 0	8
8	8	1 0 0 0	0 1 1 1	7
9	9	1 0 0 1	0 1 1 0	6
10	10	1 0 1 0	0 1 0 1	5
11	11	1 0 1 1	0 1 0 0	4
12	12	1 1 0 0	0 0 1 1	3
13	13	1 1 0 1	0 0 1 0	2
14	14	1 1 1 0	0 0 0 1	1
15	15	1 1 1 1	0 0 0 0	0

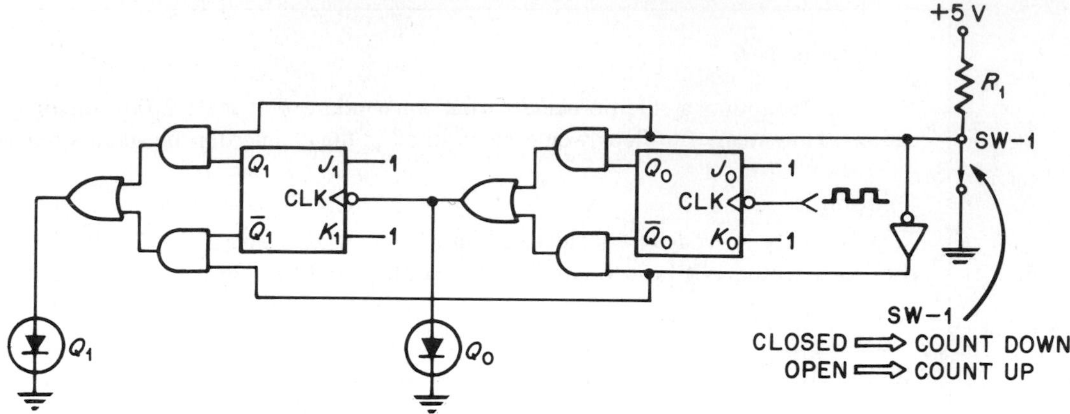

Figure 8–18 Logic diagram of up/down counter.

As shown in this table, while the Q outputs are counting up the $\overline{Q}$ outputs are counting *down*. If you used some gates, you could construct a binary counter that would count up or down. See Figure 8–18. As you can see from the figure, the count-up position of the switch enables the upper AND gate allowing the Q flip-flop output to control the clock input. Thus, the LEDs will indicate an up count. When the switch is in the countdown position lower AND gate will have the $\overline{Q}$ controlling the clock input. Now, the LEDs will indicate a down count.

Example 8–8

For a 2-bit binary counter, sketch the resulting output waveforms of the $\overline{Q}$ outputs for the first four clock pulses. Assume that the Q outputs start at 0.

Solution

See Figure 8–19.

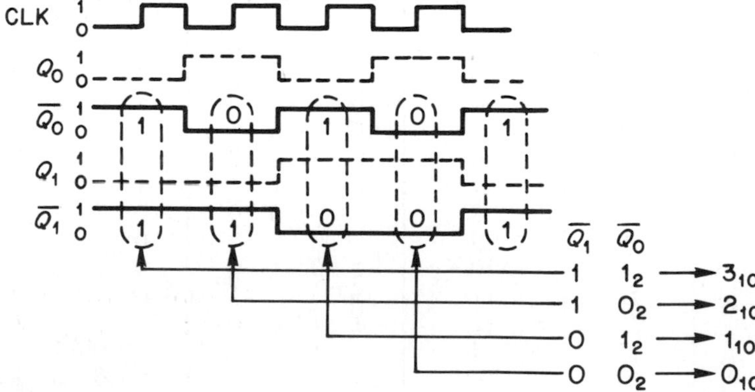

Figure 8–19

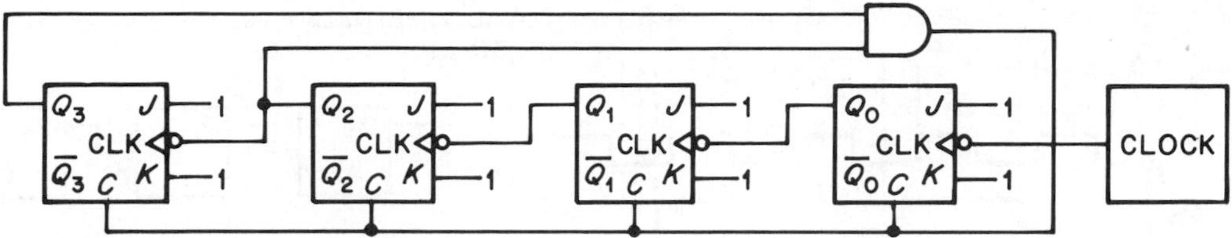

Figure 8–20 A mod-12 counter.

Controlling the Count

Occasionally you may want a counter to count less than its full capability. For example, a 4-bit counter is a mod-16 counter. But suppose you needed a counter that was a mod-12 counter. For example, this type of counter could be used in an automated egg packer where a dozen eggs are packaged in a single container.

You can reset any counter through the use of gates. This can be done at any count that is less than the mod number of the counter determined by the number of flip-flops it contains. Figure 8–20 illustrates a mod-12 counter. As shown in the figure, an AND gate is used to RESET the counter to 0000. This will happen the first time the Q outputs of the two most significant flip-flops are HIGH. This will be at the count of 1100, which is actually the thirteenth count. However, this pulse is of a very short duration compared to the "normal" counting pulses. Thus the mod-12 counter will have twelve "normal" counts of 0000_2 through and including 1011_2.

Programmable Counter:

You can construct a **programmable counter** that will achieve any required mod number that is equal to or less than its maximum mod number.

Programmable Counter:
A counter that has its mod number controlled by a changeable bit pattern.

A programmable counter that allows you to easily change the mod number is illustrated in Figure 8–21. As shown in this figure, the setting of the input switches determines when the counter will reset.

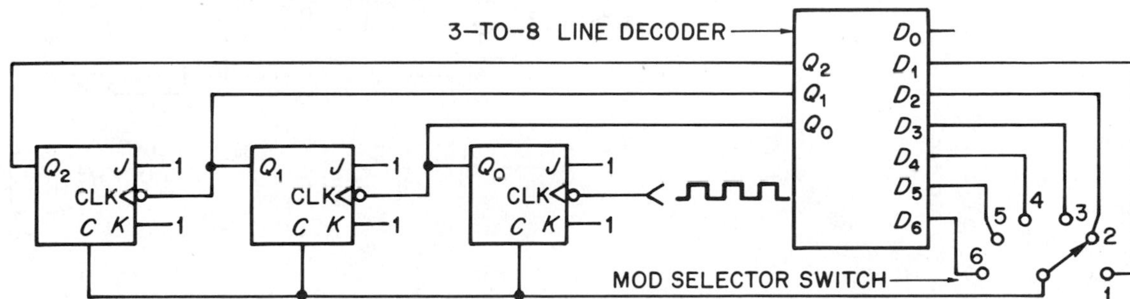

Figure 8–21 Programmable counter.

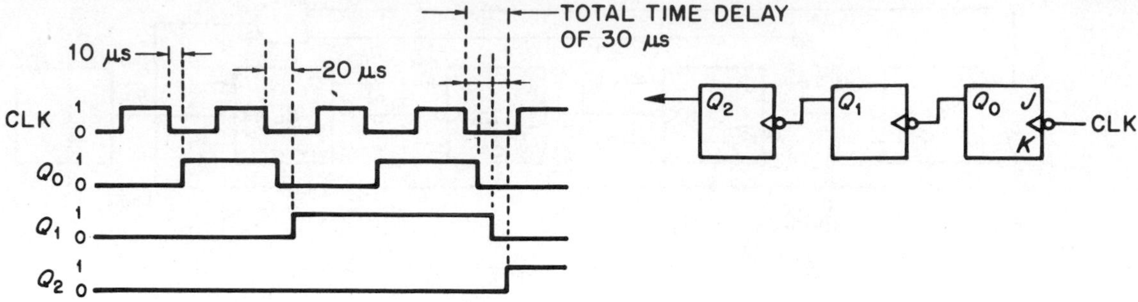

Figure 8–22 Potential time delay in asynchronous counters.

Problems with Asynchronous Counters

Asynchronous counters can present problems because it takes time for each flip-flop to change state. Suppose it took 10 μs for each flip-flop to respond to a change on its input. This would mean that the output waveform would not be what has been shown as the ideal because the next output count will be slightly behind the counter. This is illustrated in Figure 8–22. As you can see from the figure, the Q_2 waveform is 30 μs behind the trailing edge of the clock pulse and 20 μs behind the Q_0 output. If this were an 8-bit binary counter, then the difference between the generation of the most significant bit and the clock would be even greater (8×10 μs = 80 μs in this case).

The previous problem limits the clock frequency of the asynchronous counter. Another problem can occur if the output of the counter is being decoded. This problem is illustrated in the following example.

Example 8–9

For the 2-bit asynchronous counter shown in Figure 8–23, assume that each flip-flop has a 20-ns delay from the time the trailing edge of the clock arrives at the input to the time the flip-flop toggles. With this in mind, show exactly the output of each decoding line for the first five counts (assume the count starts at 0_2).

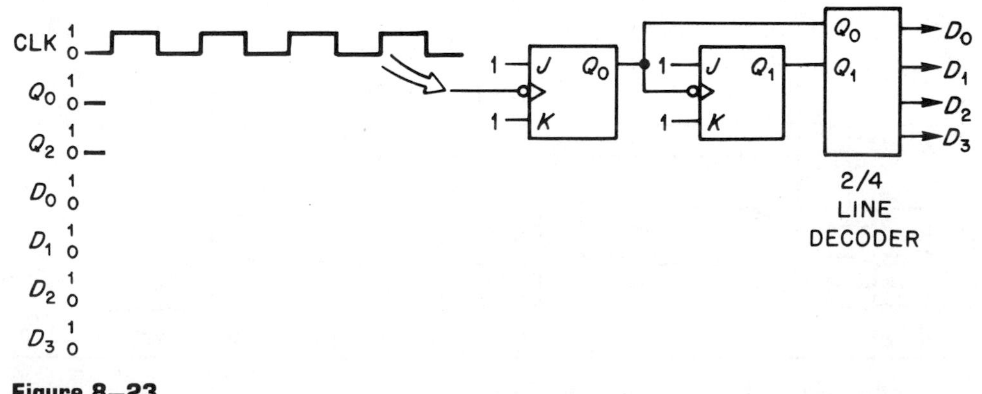

Figure 8–23

<u>***Solution***</u>

See Figure 8–24.

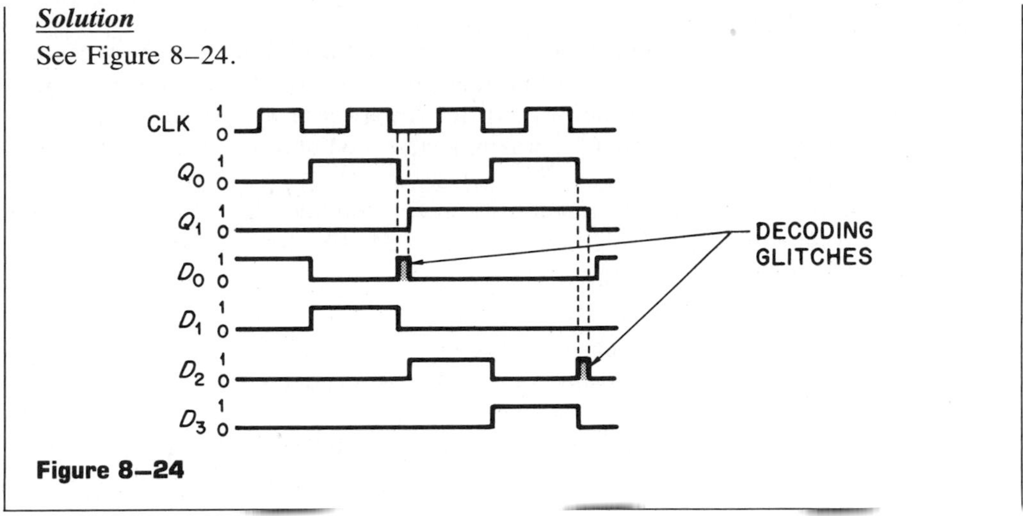

Figure 8–24

As demonstrated in Example 8–9, a **decoding glitch** can result from using an asynchronous counter.

Decoding Glitch:
An undesirable logic level of a short duration that is caused by time delays inherent in logic circuits.

Maximum Clock Frequency

The maximum clock frequency that can be tolerated by a ripple counter is equal to the reciprocal of the total delay. The formula for determining the maximum clock frequency is:

$$f_{max} = \frac{1}{(N \times t_d)}$$

Where:

f_{max} = Maximum clock frequency in hertz
N = Number of flip-flops in the counter
t_d = Time delay of flip-flop in seconds

The procedure for determining this is demonstrated in the following example.

Example 8–10

Determine the highest clock frequency for a mod-16 ripple counter if the delay of each flip-flop to a response on its input is 20 ns.

<u>***Solution***</u>

A mod-16 counter has four flip-flops. Thus:

$$f_{max} = \frac{1}{4 \times 20 \text{ ns}} = \frac{1}{80 \text{ ns}} = 12.5 \text{ MHz}$$

Conclusion

In this section, you saw the basic construction of asynchronous binary counters, and you learned the relationship among flip-flops used to construct the counters. You also learned how to make them count up or down for any given mod number.

The powerful concept of a programmable mod number, widely used in industry, was also presented.

In the next section, you will learn more about the binary counter. For now test your understanding of this section by trying the following review questions.

8–2 Review Questions

1. Explain what a binary counter does.
2. State how a binary counter acts as a *frequency divider*.
3. Describe what distinguishes an asynchronous counter.
4. What is the mod number of a counter?
5. Briefly describe the construction of a *programmable counter*.

8–3 SYNCHRONOUS AND RING COUNTERS

Discussion

In this section, you will be introduced to a counter where each flip-flop is affected directly by the clock.

Basic Idea

Consider the binary counter in Figure 8–25. The 3-bit counter shown is a **synchronous counter** because each flip-flop is fed directly by the clock pulse.

> **Synchronous Counter:**
> A counter where each flip-flop making up the counter is affected directly by the clock pulse.

Presettable Counters

You can devise counters that can load a bit pattern and then start their count from that pattern. There are several methods for doing this. One way is to load a single bit into a *JK* flip-flop, as shown in Figure 8–26. Whenever the LOAD input is placed in the LOW

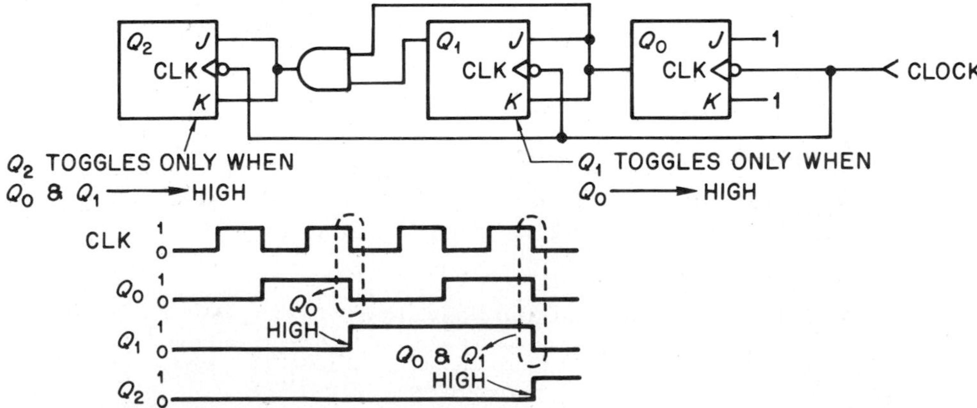

Figure 8–25 Basic idea of a synchronous counter.

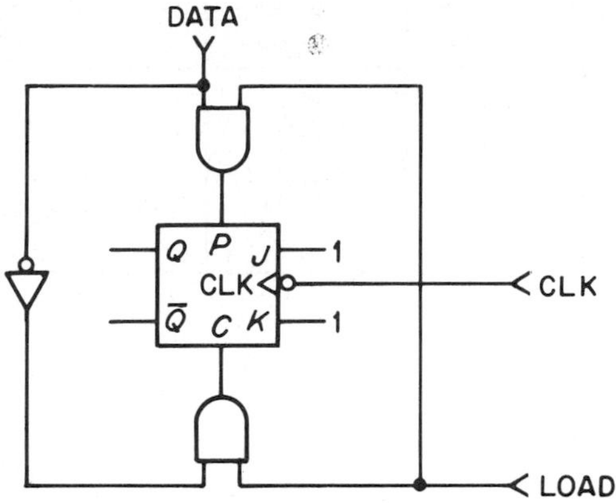

Figure 8–26 Loading a bit into a *JK* flip-flop.

position, the output of both AND gates will be LOW—thus deactivating both PRESET and CLEAR inputs. This condition allows the flip-flop to be in the *synchronous* mode of operation and will toggle for each clock pulse. However, when the LOAD input is HIGH, the flip-flop will be placed in *asynchronous* operation, and the condition of data bit D_0 will be copied into the *JK* flip-flop via its PRESET and CLEAR inputs.

This idea can be extended to as many flip-flops as you choose. A 4-bit synchronous counter with LOAD and CLEAR inputs is illustrated in Figure 8–27. The counter will continue a synchronous count as long as both the LOAD and CLEAR inputs are LOW. The following table illustrates the action of the counter.

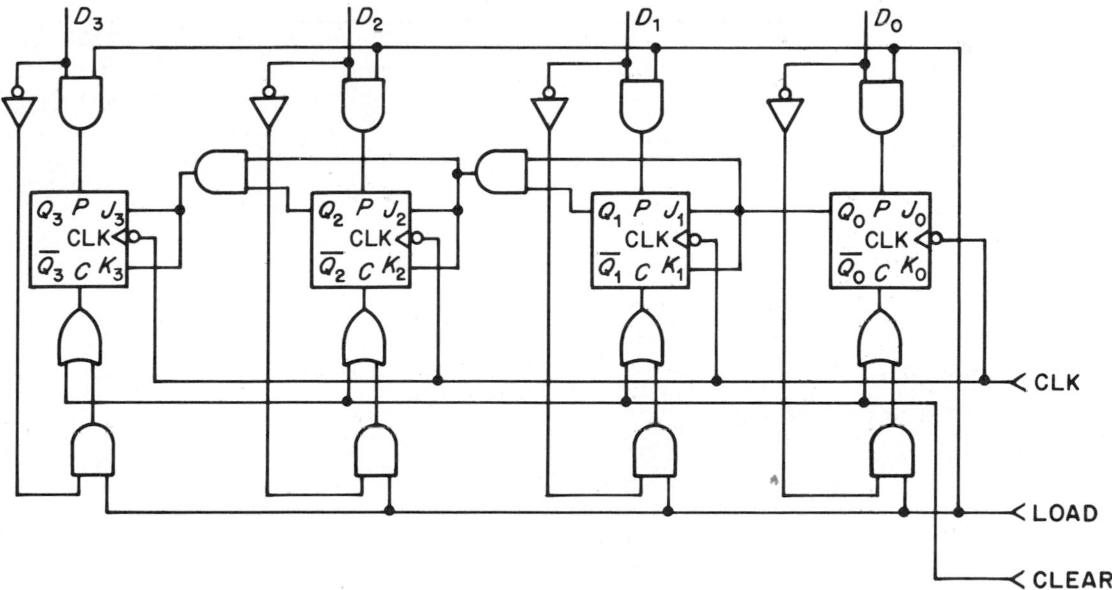

Figure 8–27 Synchronous counter with LOAD and CLEAR.

LOAD	CLEAR	Counter
LOW	LOW	Count
HIGH	LOW	Data word loaded into counter
LOW	HIGH	Counter set to 0000_2
HIGH	HIGH	Not allowed

Example 8–11

For the counter in Figure 8–28, determine the Q outputs for the indicated input waveforms.

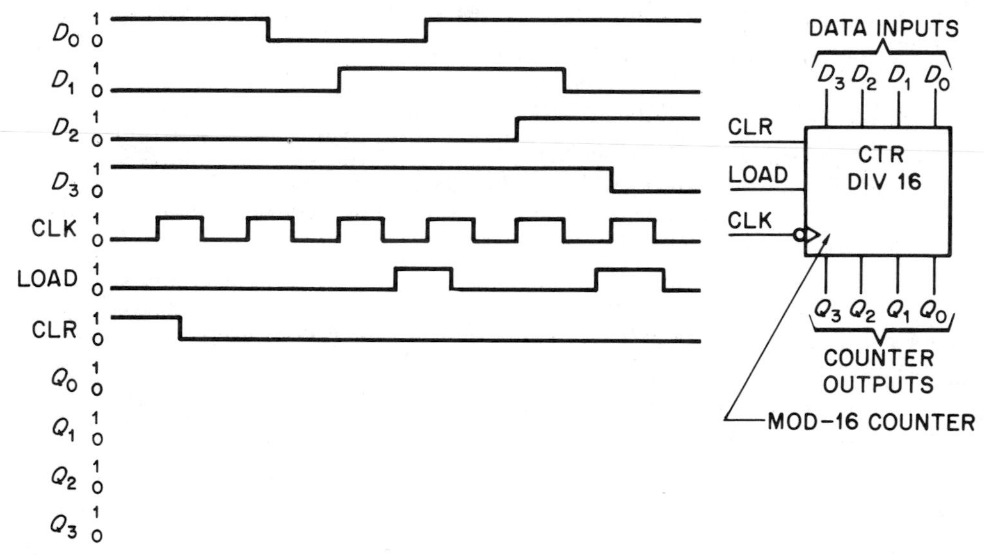

Figure 8–28

Solution

Refer to Figure 8–29.

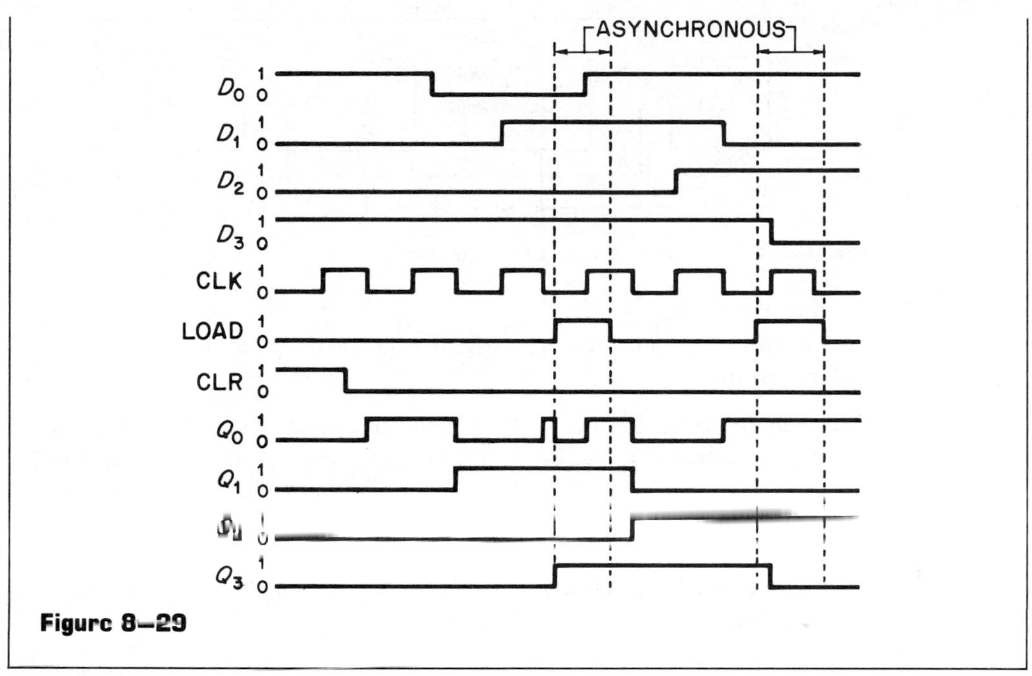

Figure 8–29

Ring Counters

The concept of a special type of counter called a **ring counter** is illustrated in Figure 8–30.

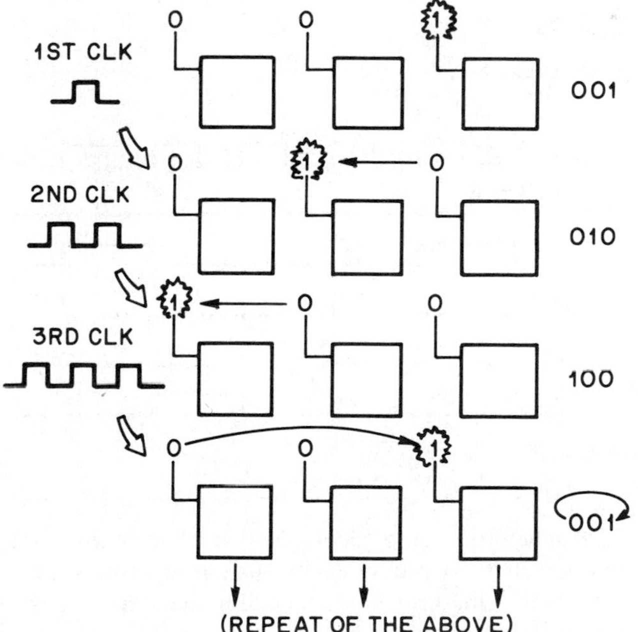

Figure 8–30 Concept of a ring counter.

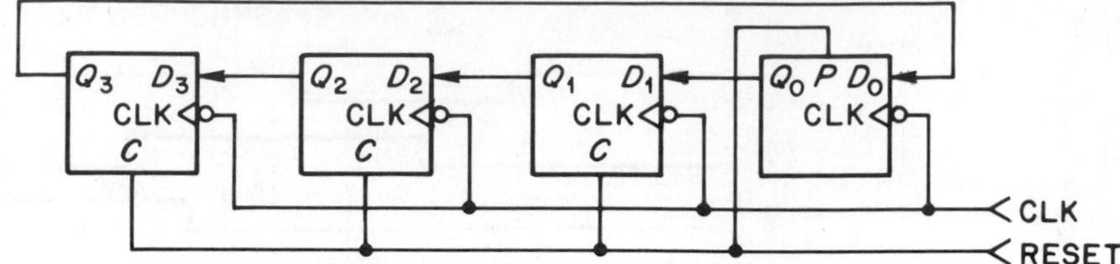

Figure 8–31 Construction of a ring counter.

Ring Counter:
A form of digital counter where a given bit pattern continually circulates through the counter.

As you can see from Figure 8–30, a ring counter is not like a binary counter. Instead what you really have is a serial shifting of a bit pattern within a register. The output of this register is connected back to the input. This causes the same bit pattern to be recycled through the counter. In the figure, a 1 is being recycled continually. Since the output is connected back to the input, this forms a "ring," hence the name *ring counter*. Such a counter can easily be constructed with *D*-type flip-flops. This arrangement is illustrated in Figure 8–31.

Example 8–12

For the ring counter of Figure 8–31, show the *Q* output waveforms for the first eight clock pulses after a RESET pulse.

Solution

Refer to Figure 8–32.

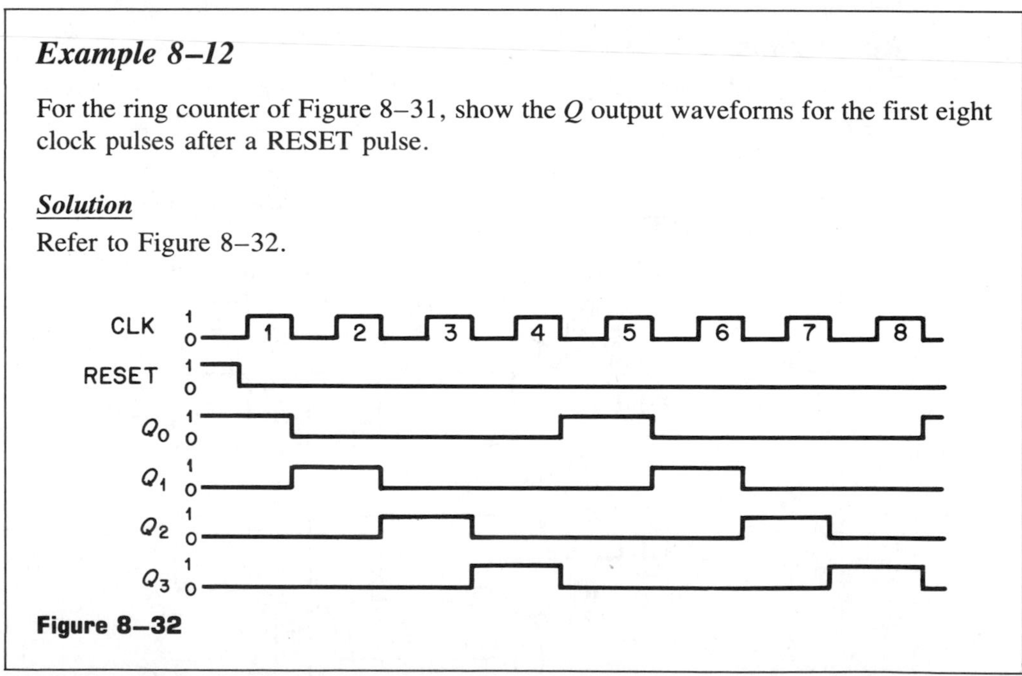

Figure 8–32

The advantage of a ring counter is that no decoding is necessary. For example, in a 4-bit ring counter the output of each flip-flop represents a unique state. The disadvantage of a ring counter is the limited number of unique states. For example, a 4-bit ring counter has only four unique states, whereas a 4-bit binary counter has 16 unique states.

Conclusion

This section presented the fundamental ideas behind synchronous counters. You were also introduced to the concept of a ring counter. In the next section you will discover how to design counter logic utilizing programmable logic devices. For now, check your understanding of this section by trying the following review questions.

8–3 Review Questions

1. Describe the difference between *synchronous* and *asynchronous* counters.
2. What is a *presettable* counter?
3. Describe the action of a *ring* counter.
4. What is the advantage of a *ring* counter? the disadvantage?

8–4 DESIGNING WITH COUNTERS AND PLDs

Discussion

In the last section, you worked primarily with counters that counted in a specific binary sequence. In this section, you will see how counters can be used to perform any sequence of events. Essentially, here you will see how to use your knowledge of programmable logic devices and counters to construct digital circuits that will perform a specific process.

Basic Idea

Figure 8–33 shows a 2-bit binary counter connected to a diode ROM. There are two pairs of LED readouts. One set, the counter readout, displays the exact binary count of the counter. The other pair, the ROM readout, displays the bit pattern output of the switch-

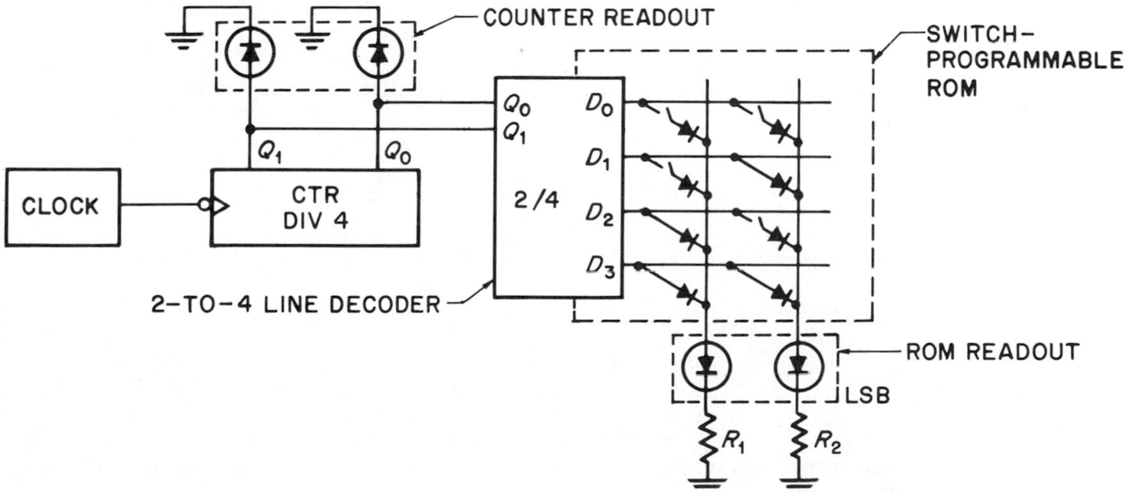

Figure 8–33 Binary counter connected to a diode ROM.

programmable ROM. For the system in Figure 8–33, the ROM is programmed to produce an up binary counter, as shown in the following table:

Binary Counter Output	ROM Output
0 0	0 0
0 1	0 1
1 0	1 0
1 1	1 1

A counter and a PLD (in this case a switch-programmable ROM) can produce any desired output sequence you wish. Consider the following example.

Example 8–13

Determine the output of the digital system shown in Figure 8–34.

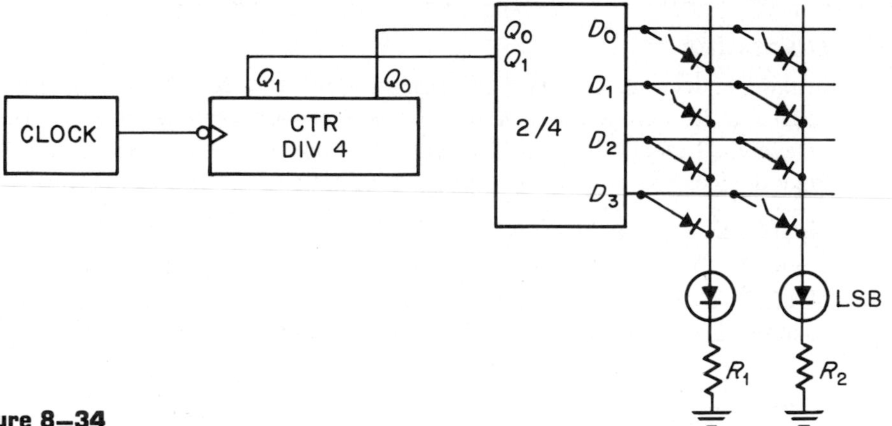

Figure 8–34

Solution

Since a closed switch represents a 1 and an open switch a 0, the following sequence is revealed:

Count	ROM Output
0	0 0
1	0 1
2	1 1
3	1 0
0	0 0
1	0 1
2	(etc.)

Upon close inspection of the output you can see that only one bit changes state at a time. This is the property of the gray code. Hence the output is that of a gray code counter.

State Machine

The previous system can be referred to as a **state machine**.

State Machine:
A sequential logic circuit exhibiting a specified sequence of events.

As you saw in previous examples, each state machine produced a unique sequence of events. In the first state machine, the events represented a binary upcounter. In the second, the events represented a gray code upcounter and a gray code downcounter in the last. Such a system can replicate the condition of any truth table. If you envision the count of the truth table as representing the sequence of a binary counter, then any sequence of HIGH/LOW events may be duplicated. This concept is illustrated in the following example.

Example 8–14

Design a state machine that will produce a 2-bit gray code downcounter.

Solution

First develop a *state table* that represents all of the unique conditions:

Count	State (Condition)
0	1 0
1	1 1
2	0 1
3	0 0
0	1 0
1	1 1
2	(etc.)

From this state table, you can program your ROM as shown in Figure 8–35.

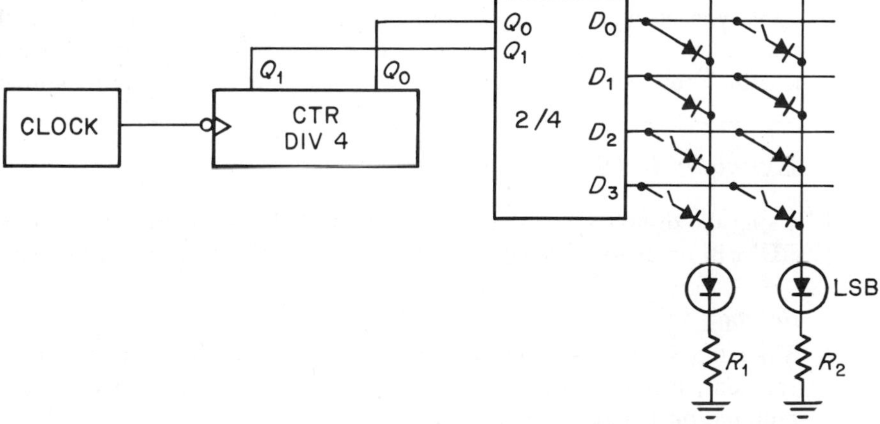

Figure 8–35

The ROM was programmed by duplicating the states presented on the given state table. Thus, on the first count, the state was 10. This is duplicated in the ROM by having switches on the D_0 line producing a 10 output on the first count. In a similar fashion, the switch pattern of each successive line of the ROM duplicates that of the state table. This pattern will repeat itself because the binary counter that sequences the system will reset itself back to zero and repeat its count.

State Table:
A list of the sequential states of a state machine.

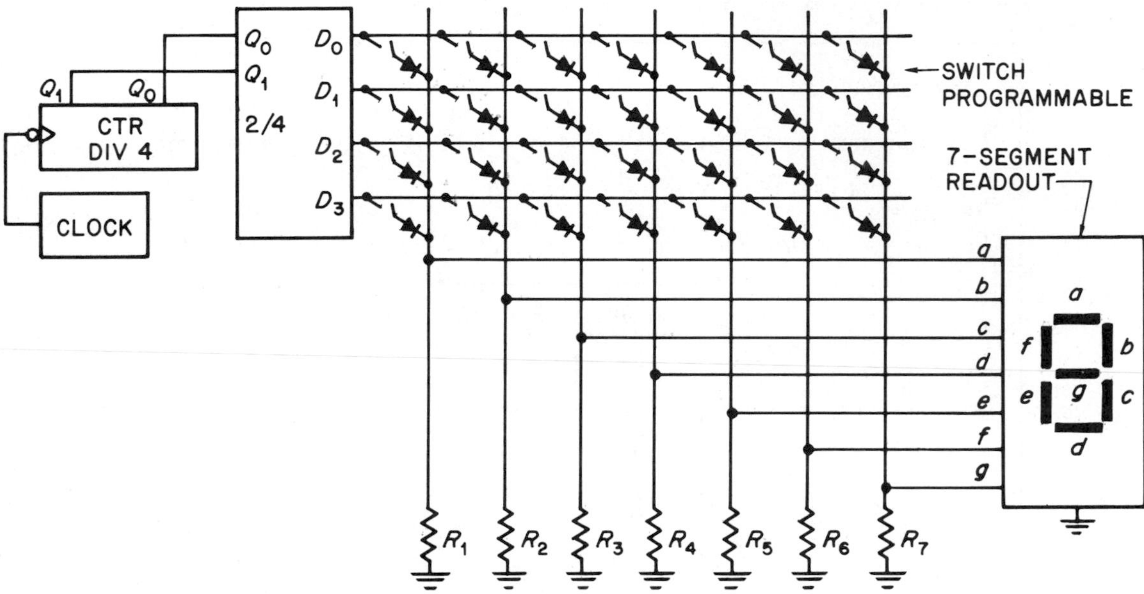

Figure 8–36 Creating a programmable readout.

Programmable Readout

To give you an idea of the versatility of a counter teamed up with a PLD, consider the digital system of Figure 8–36. You can cause the seven-segment display to display any repeatable sequence of four characters. Consider the following example:

Example 8–15

Using the digital system shown in Figure 8–36, show how you would program the ROM in order to produce the letter sequence HELP on the seven-segment readout.

Solution

Since the ROM has seven outputs and each output is connected directly to its own segment, it is only necessary to sketch the desired segments and then indicate the required bit pattern on a *state table*.

The solution is shown in Figure 8–37.

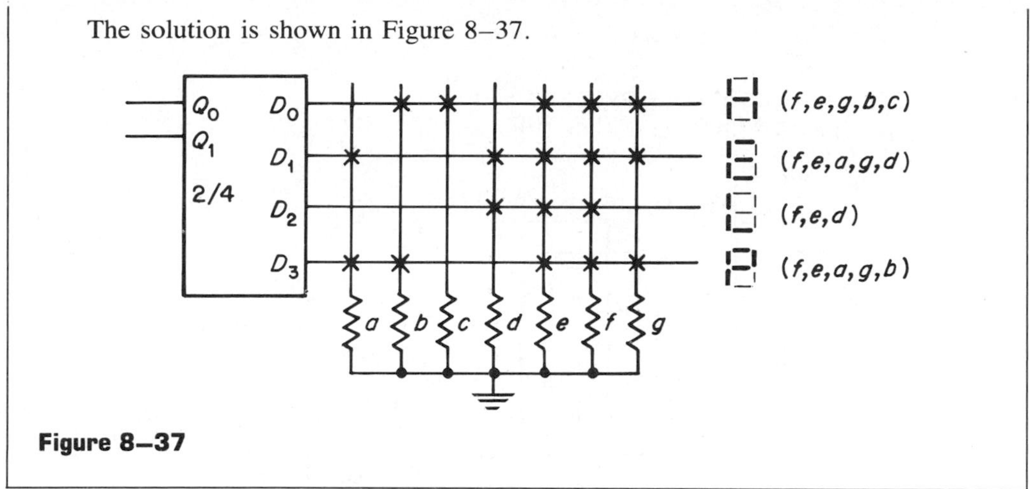

Figure 8–37

Developing Small Processors

Now that you have the basic idea of a single four-state machine, consider the system illustrated in Figure 8–38. Note that there are four different ROMs inside the section called the *processor*. Observe that the 2-to-4-line decoder for each ROM has an enable line. Thus, the ROM ENABLE line must be active for the ROM output to be active. Each ENABLE line comes from a 2-to-4-line decoder whose active output line is determined by the settings of the control switches (SW_0 and SW_1). Because of this, only one ROM will be active at a time. The results are shown in the following table.

Control Input [Instruction]	Output Result [Process]
0 0	Binary upcounter
0 1	Binary downcounter
1 0	Gray Code upcounter
1 1	Gray Code downcounter

This table shows the resulting process for a specific instruction. Thus, the state machine shown in Figure 8–38 can have four different small processes depending upon the selection of the CONTROL INPUT. Hence, what you really have is a small processor with an instruction set consisting of four instructions. By definition, this could be called a 2-bit microprocessor.

The concept of this machine is similar to that of all microprocessors. For example, each separate diode ROM can be thought of as a **microprogram**. What the 2-bit instruction is really doing is selecting the specific microprogram.

Microprogram:
The permanent instruction set inside the microprocessor used to perform a specific process.

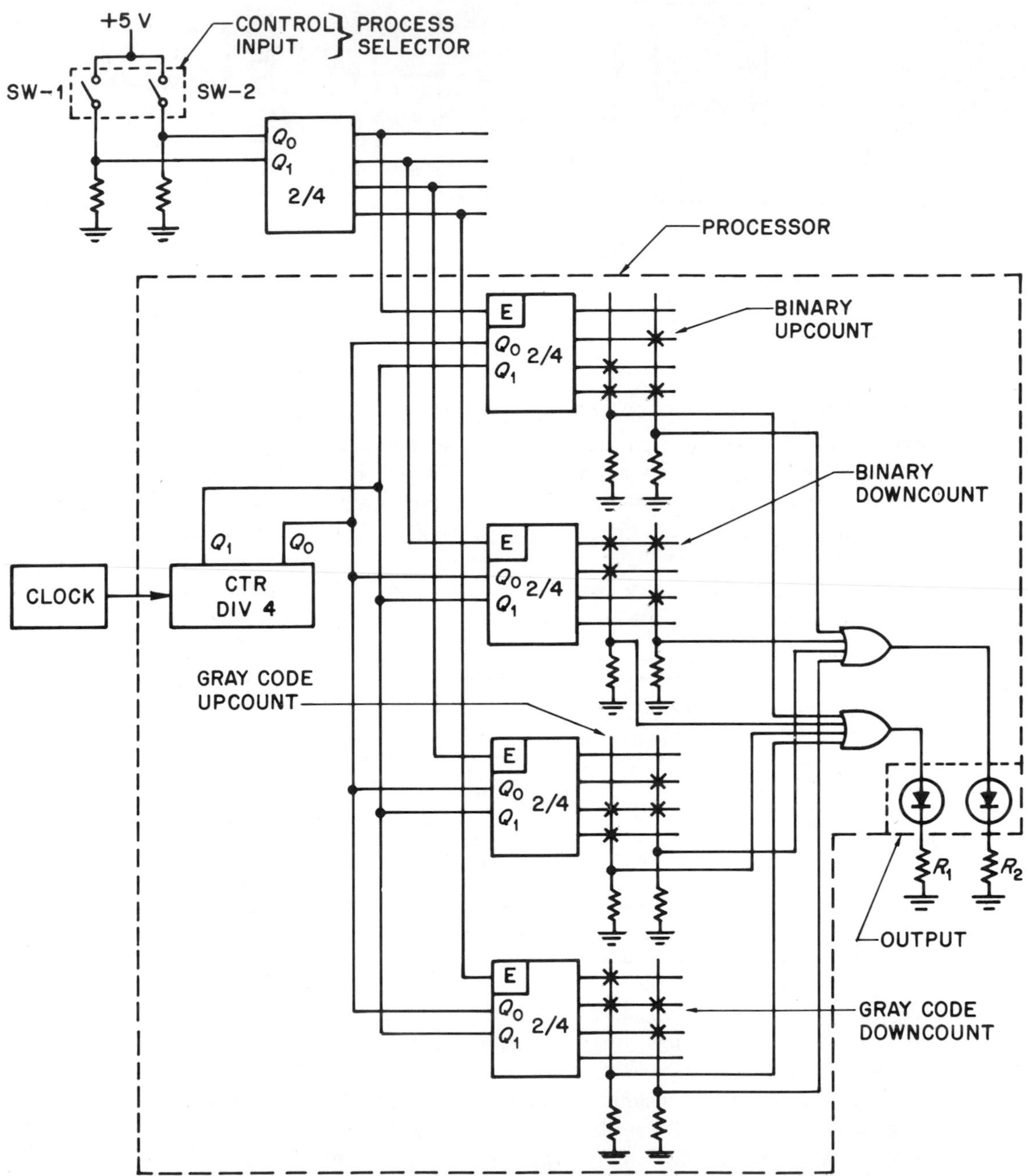

Figure 8–38 Process selector.

More Versatility

Using larger binary counters and PLDs gives you even more versatility in the design of a state machine. For example, consider the system of Figure 8–39. With the system shown, not only can you program the process (as before) with the switch-programmable ROM, but you can also change the mod number (the number of steps in the process) and how the process is to be sequenced.

For example, if the ROM were programmed as shown in the following table:

ADDRESS	CLR	UP	DOWN	ROM OUTPUTS			
0	0	1	0	0	0	0	0
1	0	0	0	0	0	0	1
2	0	0	0	0	0	1	1
3	0	0	0	0	0	1	0
4	0	0	0	0	1	1	0
5	0	0	1	0	1	1	1

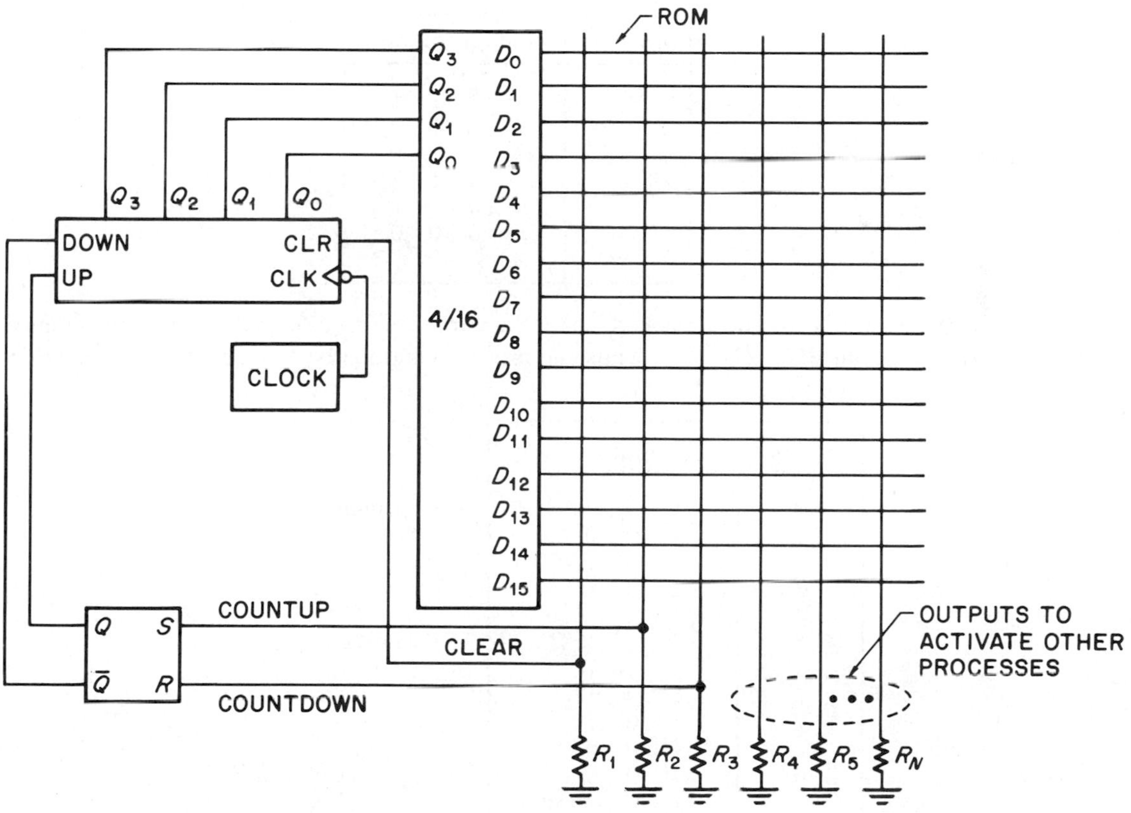

Figure 8–39 A larger state machine system.

you would have a mod-6 gray code up- to downcounter. This is what would happen. Assume that the binary counter starts at 0000_2. This means that address line 0 will be active. Here, the UP control line becomes active from the ROM. This in turn causes the control flip-flop to be SET ($Q \rightarrow 1$). Thus, the output of the ROM will be determined by the progression of *increasing* address changes (0, 1, 2, etc.). The state table will appear as

Count	Output Result
0	0 0 0 0
1	0 0 0 1
2	0 0 1 1
3	0 0 1 0
4	0 1 1 0
5	0 1 1 1

At the last count (5), the DOWN control line becomes active, thus resetting the control flip-flop ($Q \rightarrow 1$), and the binary counter will begin to count down from the count of 5. Thus the ROM outputs will now be as shown in the next state table.

Count	Output Result
5	0 1 1 1
4	0 1 1 0
3	0 0 1 0
2	0 0 1 1
1	0 0 0 1
0	0 0 0 0

Once address 0 is again reached, the UP control line becomes active, and the control flip-flop is again SET ($Q \rightarrow 1$). Thus the process continually repeats itself as shown in the complete state table:

Count	Output Result	
0	0 0 0 0	← upcount activated
1	0 0 0 1	
2	0 0 1 1	
3	0 0 1 0	
4	0 1 1 0	
5	0 1 1 1	← downcount activated
4	0 1 1 0	
3	0 0 1 0	
2	0 0 1 1	
1	0 0 0 1	
0	0 0 0 0	← upcount activated
	(Process repeats)	

Example 8–16

Using the digital system shown in Figure 8–39, program the ROM to create a 4-bit ring counter.

Solution

First develop a state table for the desired outcome.

Count	Output
0	0 0 0 1
1	0 0 1 0
2	0 1 0 0
3	1 0 0 0

The process repeats itselt at the count of 3. Thus, at the count of 4, the binary counter should be back to its condition at thc count of 0. Thc casicst way to do this is to have the counter RESET. Thus, the ROM should be programmed as follows:

ADDRESS	CLR	UP	DOWN	D_3	D_2	D_1	D_0
0	0	0	0	0	0	0	1
1	0	0	0	0	0	1	0
2	0	0	0	0	1	0	0
3	0	0	0	1	0	0	0
4	1	0	0	0	0	0	0

Conclusion

This section presented some powerful new concepts. Understanding how to create any sequence of bit patterns you choose is the foundation of microprocessor design.

In the next section you will be introduced to registers. With your background in counters and PLDs, an understanding of registers will allow you to build very sophisticated digital systems. Check your understanding of this section by trying thc following review questions.

8–4 Review Questions

1. State the advantage of using a binary counter with a PLD.
2. What is a *state machine*?
3. Describe a *state table*.
4. What is a *microprogram*?
5. Can a PLD be used to control the action of the binary counter that is selecting the address of the ROM? Explain.

8–5 REGISTERS

Discussion

Recall that a register was nothing more than a group of flip-flops treated as a unit. There are basically four different kinds of registers:

1. Parallel in–parallel out.
2. Serial in–serial out.
3. Parallel in–serial out.
4. Serial in–parallel out.

Combinations of these types are also available. In this section, each type will be presented.

Parallel In–Parallel Out

The most fundamental type of register is the parallel in–parallel out register. These registers are sometimes referred to as *buffer registers*. Figure 8–40 illustrates the logical construction of a 4-bit parallel-in–parallel-out register. The register in the figure allows a bit pattern to be transferred into the register when the LOAD line is HIGH. Note that the LOAD line connects directly to each of the clock inputs of the D latches.

For the same register two other control lines are used: the CLEAR and the output ENABLE. As you can see from Figure 8–40, the CLEAR line is connected to the CLEAR inputs of each latch. Thus, when the CLEAR line is active, each Q output of the latches in the register will be set to zero.

The output ENABLE line is connected to the ENABLE inputs of each of the output tristate buffers. Thus each output of this register has three possible conditions: HIGH, LOW, and high impedance.

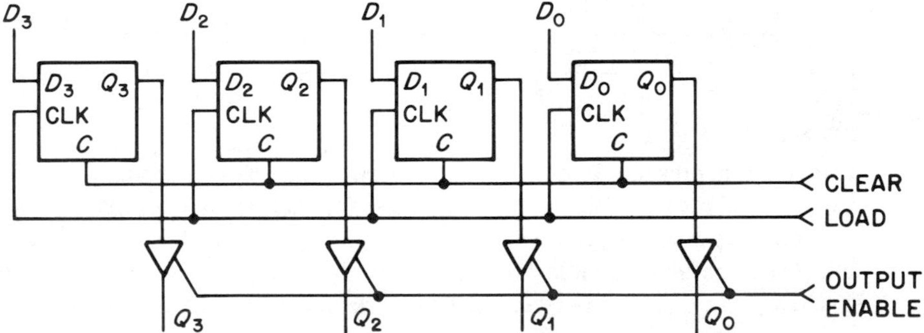

Figure 8–40 Logic construction of a parallel-in–parallel-out register.

> ### *Example 8–17*
>
> Determine the output conditions of the 4-bit register of Figure 8–40 for the waveforms shown in Figure 8–41.

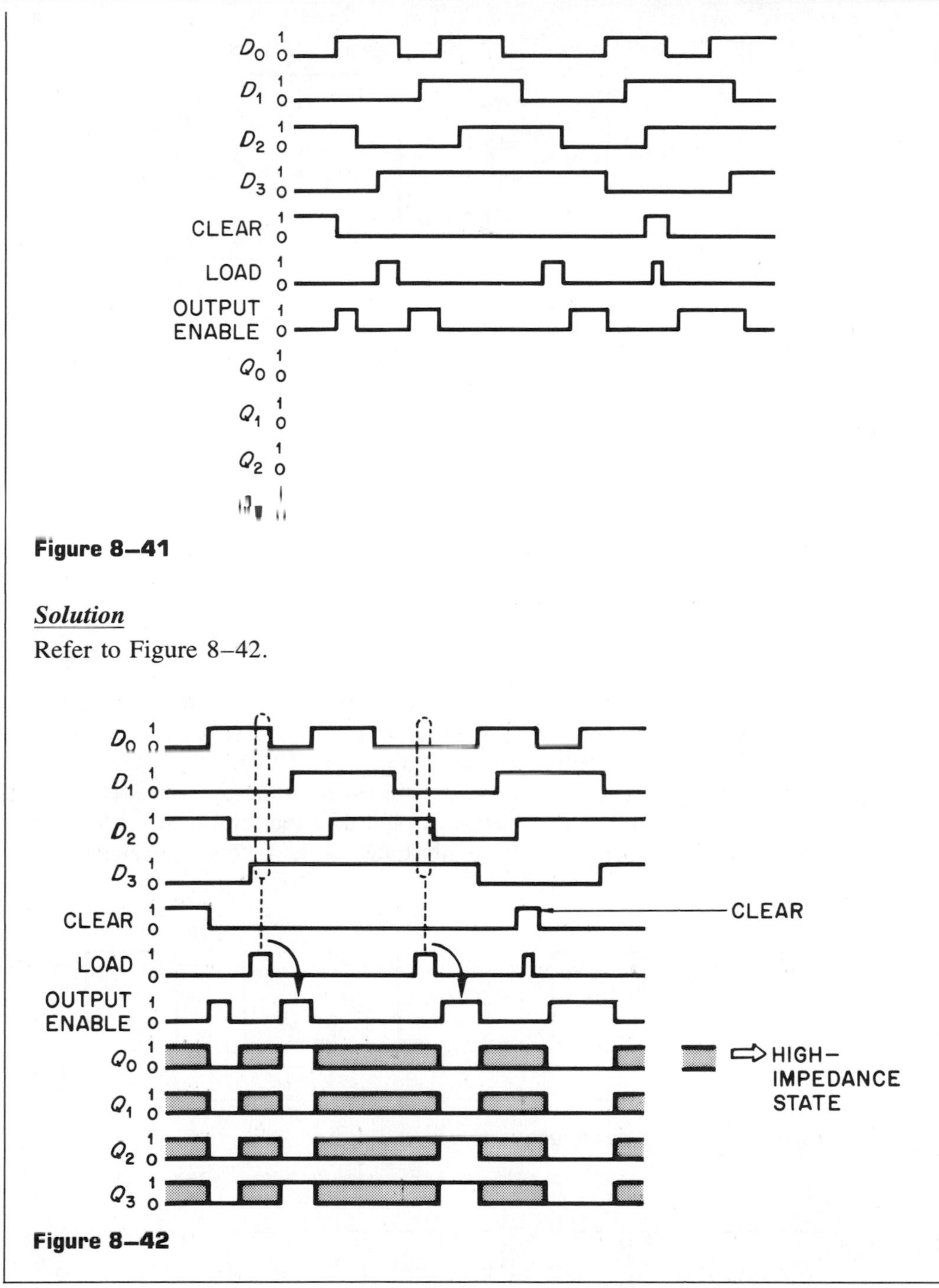

Figure 8–41

Solution

Refer to Figure 8–42.

Figure 8–42

For simplicity, the register of Figure 8–40 can be represented as shown in Figure 8–43. You will see many applications for this kind of register in digital systems. These registers are the backbone of microprocessors, memory, and most digital systems in general.

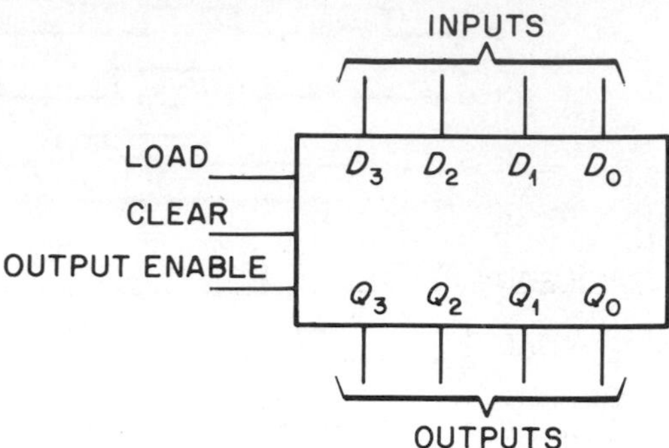

Figure 8–43 Logic diagram of 4-bit parallel-in–parallel-out register.

Serial In–Serial Out

A serial-in–serial-out register is similar to a ring counter without connecting the output back to the input. This type of register is sometimes referred to as a **shift register**.

> **Shift Register:**
> A type of register where the contents may be moved in a serial fashion within the register.

Figure 8–44 shows the logic construction of a shift register. Note that if the output of the shift register were connected to its input, you would have a ring counter. The example shift register has two control lines (a CLOCK and CLEAR), a single input and a single output. The contents of this register are shifted from left to right at each HIGH to LOW transition of the clock pulse. Thus, it would take four clock pulses to have a bit at the input felt at the output. This idea is illustrated in the following example.

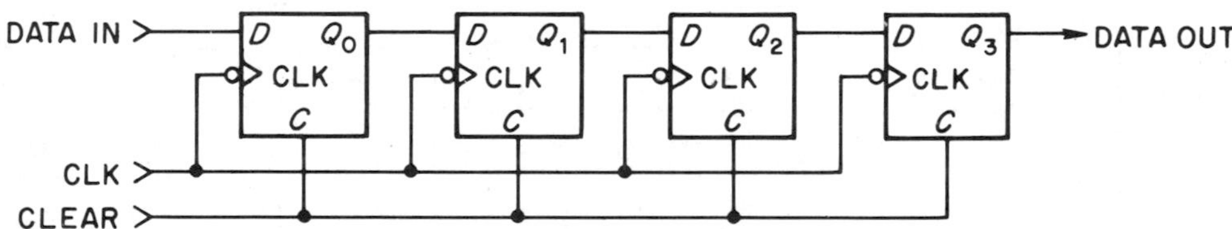

Figure 8–44 Logic construction of a shift register.

Example 8–18

Determine the output waveform of the 4-bit register shown in Figure 8–44 for the input waveforms of Figure 8–45.

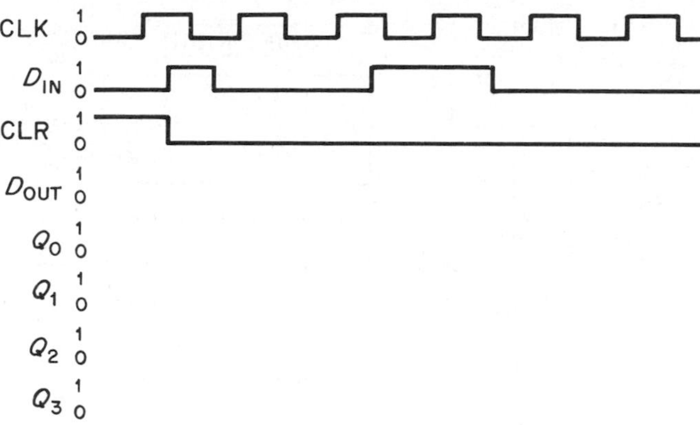

Figure 8–45

Solution

One method of arriving at the solution is to consider the condition of each flip-flop as shown in Figure 8–46. Note that the bit pattern of Q_0 is shifted to the other flip-flops within the register. This same pattern will eventually be presented on the output of the register.

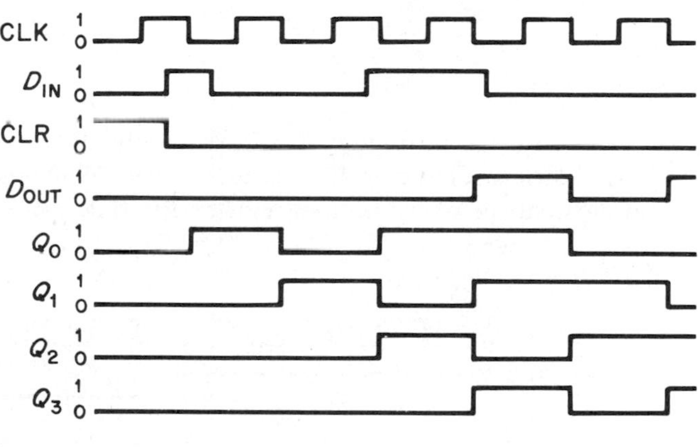

Figure 8–46

Shifting Either Way

Figure 8–47 shows two 4-bit shift registers connected to produce a system that will offer a shift right or shift left capability. The use of TRI-STATE buffers on the inputs and output of each register make this possible. As shown, when the $SL/\overline{SR}$ input is LOW, two tristate buffers are enabled. These TRI-STATE buffers cause the top shift register to become active while the bottom register effectively has its inputs and outputs disconnected. Thus, under these conditions, data appearing at the left of the figure are inputted serially into the top register and outputted to the right.

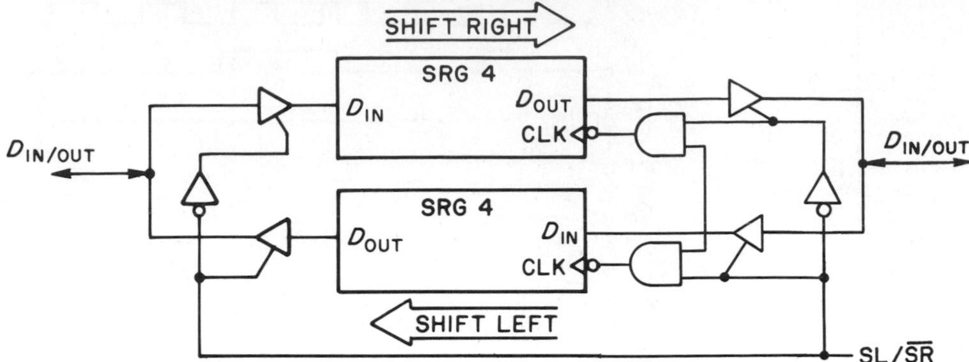

Figure 8–47 Shift RIGHT—shift LEFT register.

Just the opposite case takes place when the $SL/\overline{SR}$ line is HIGH. In this case, the other two TRI-STATE buffers become active, and data appearing at the right of the figure are inputted serially into the bottom register and outputted to the left. The following example illustrates the process.

Example 8–19

For the shift register of Figure 8–47, determine the resulting outputs for the waveforms shown in Figure 8–48. Assume that the contents of the top register are 1001 and the contents of the bottom register are 0110.

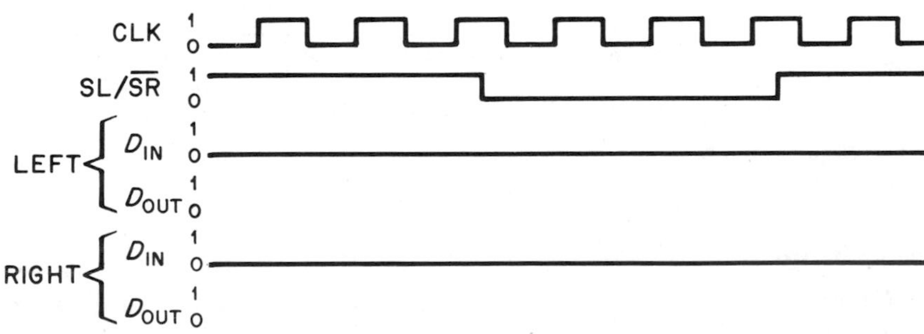

Figure 8–48

Solution

See Figure 8–49.

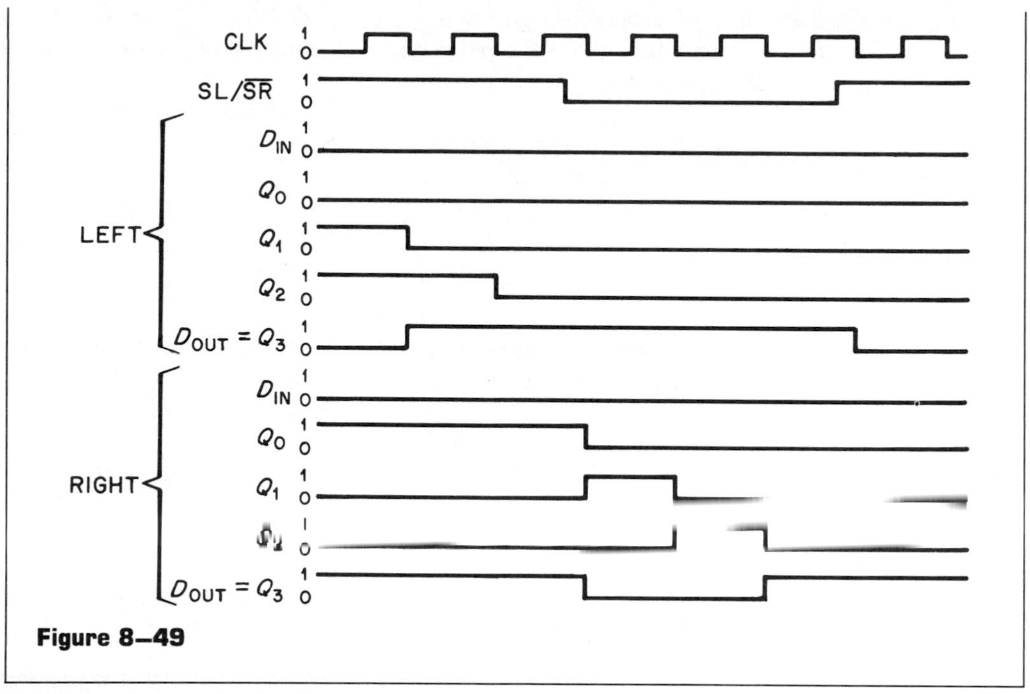

Figure 8–49

Parallel In–Serial Out

The logic construction of a parallel-in–serial-out register is shown in Figure 8–50. Note the similarity between the parallel-in–serial-out register and that of the presettable binary counter. The parallel-in–serial-out register has asynchronous parallel loaded inputs. When the LOAD line is HIGH, all of the AND gates will be active and the bit pattern present on the *D* input lines will be copied by the register. However, when the LOAD line is

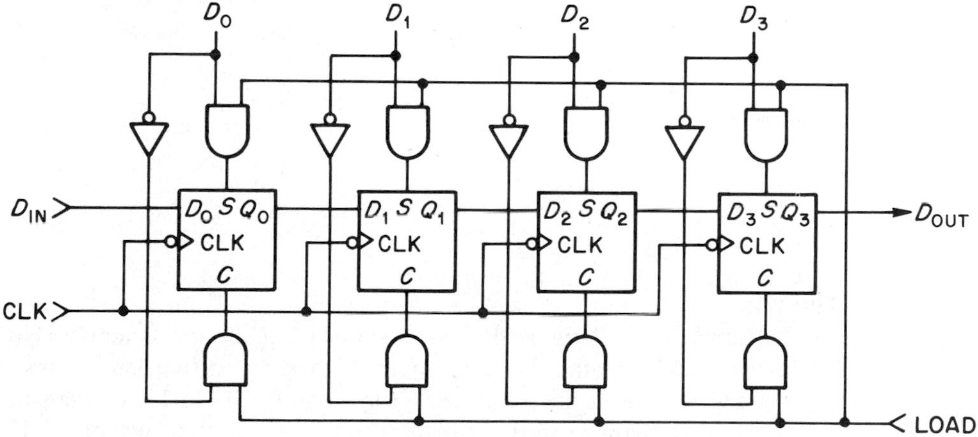

Figure 8–50 Parallel-in–serial-out register.

LOW, the output of each AND gate will be LOW and the register will then be in synchronous operation where the clock pulse will now shift the register contents to the right. Example 8–20 illustrates the process.

Example 8–20

For the parallel-in–serial-out register of Figure 8–50, determine the resulting output for the input waveforms shown in Figure 8–51.

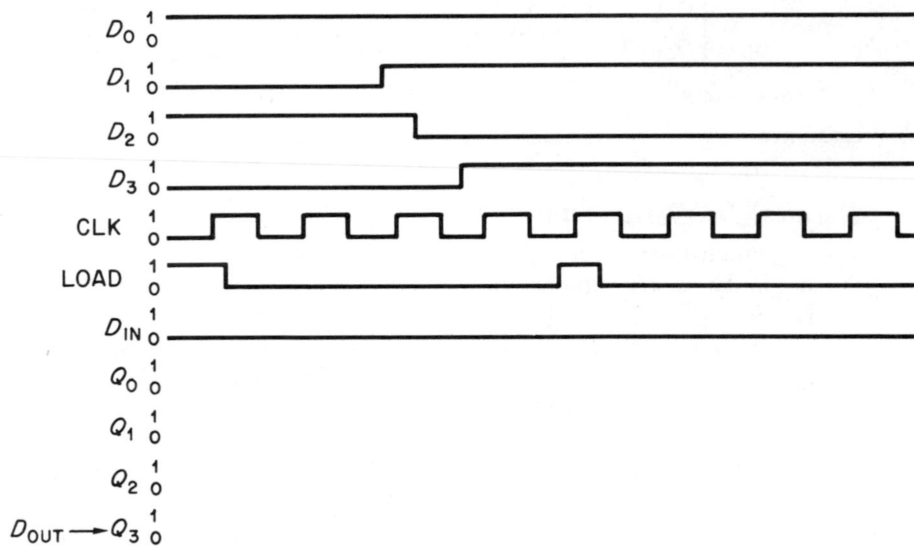

Figure 8–51

Solution

Keep in mind that as long as the asynchronous LOAD line is active (HIGH), then all flip-flops will maintain the bit pattern of their respective input lines. It is only after the LOAD line is inactive (LOW) that the HIGH to LOW transitions of the clock pulse will begin to shift the data. The resulting output waveform is shown in Figure 8–52.

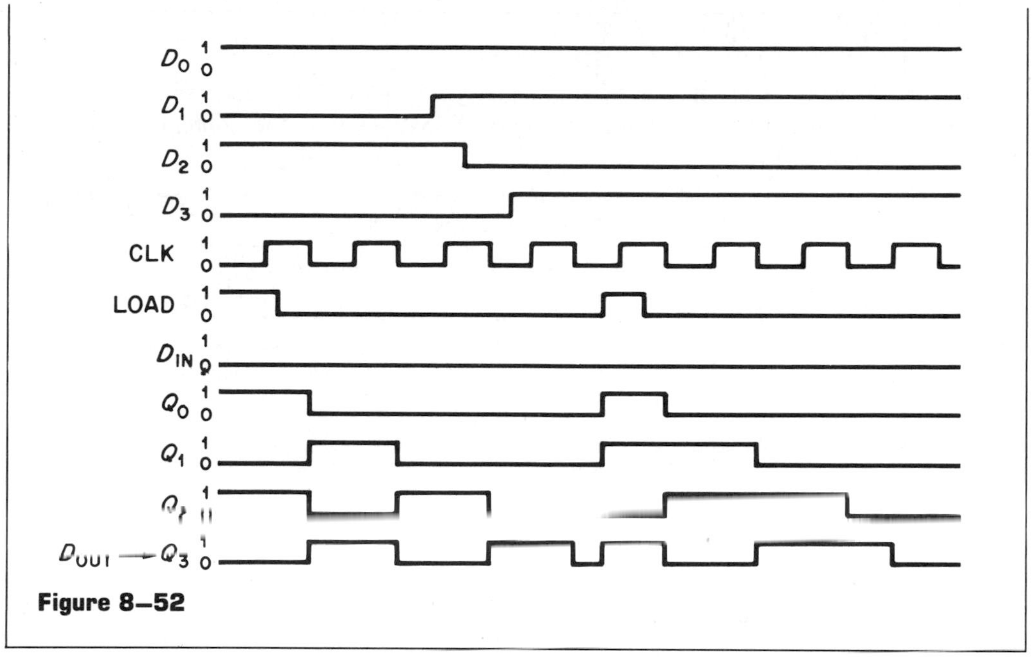

Figure 8–52

Serial In–Parallel Out

A serial-in–parallel-out register can use tristate buffers for its output lines. The logic construction of such a register is shown in Figure 8–53. As shown in the figure, the serial-in–parallel-out register is nothing more than a standard shift register with a TRI-STATE buffered output. Note the AND gate used with the clock pulse input. What happens here is that the clock pulses are inhibited when data of the register are outputted in parallel. This action prevents the register contents from shifting while the parallel outputs are active.

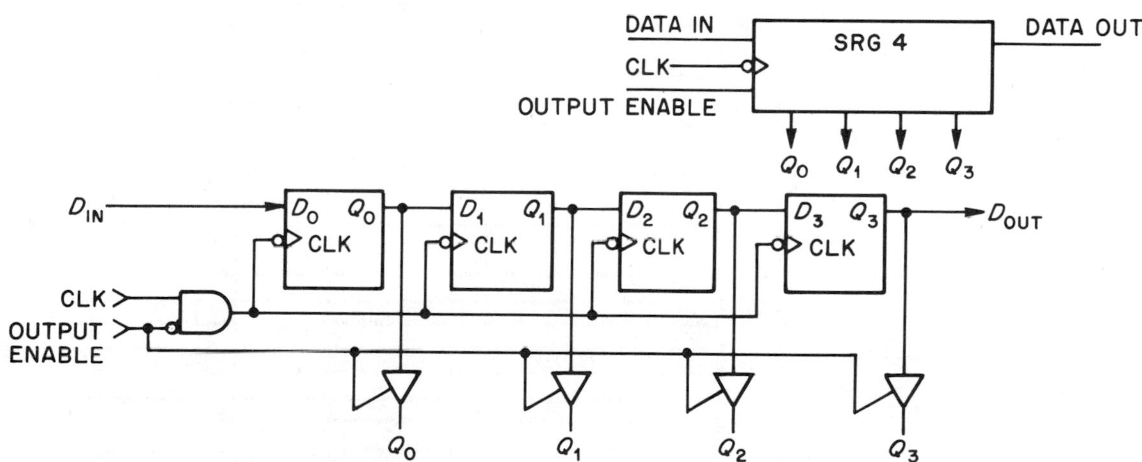

Figure 8–53 Construction of a serial-in–parallel-out register.

Example 8–21

For the serial-in–parallel-out register of Figure 8–53, determine the resulting output for the input waveforms of Figure 8–54. Assume the contents of the register start with 1010.

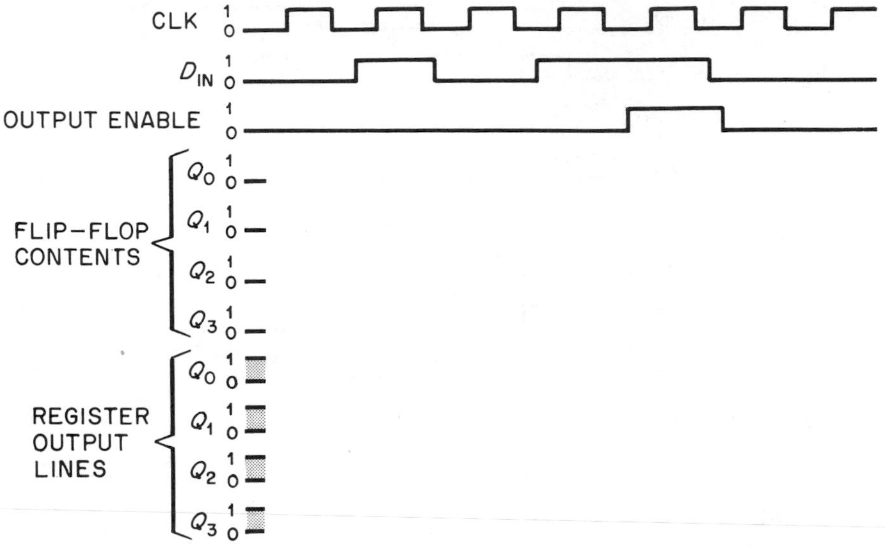

Figure 8–54

Solution

Keep in mind that the serial output will stay at the same logic level while the E_{OUT} line is active. The resulting waveforms are shown in Figure 8–55.

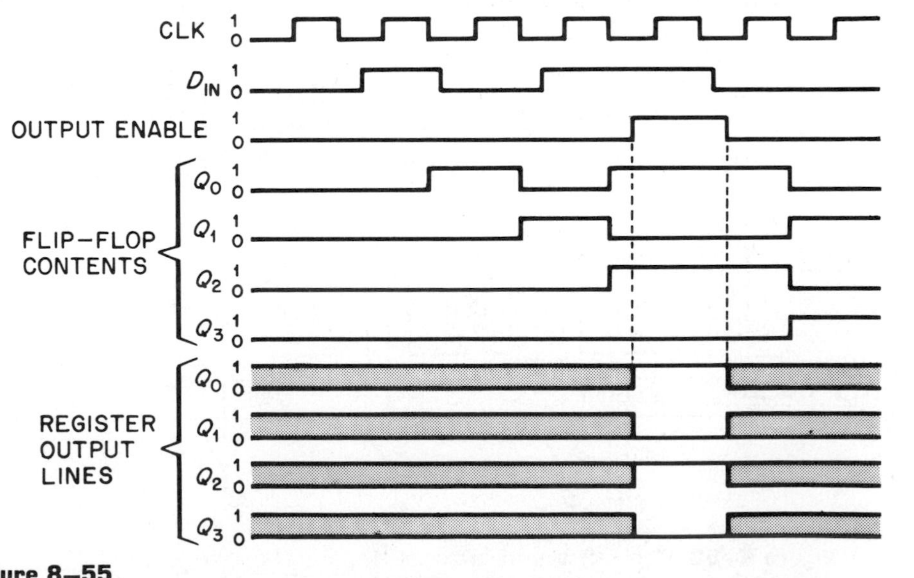

Figure 8–55

Register Applications

Registers are the backbone of most all digital systems. One of the most popular usages is in read/write memory. Since you have studied the serial loading and reading as well as the parallel loading and reading of registers in this section, an application of these characteristics will be presented.

In digital systems that employ busses, bit patterns are moved within the system in parallel (such as in a microcomputer). However, when connecting one digital system to another, it isn't practical to transmit digital information in parallel lines over long distances. As an example, when the telephone is used to transmit and receive digital information the data must be sent and received serially. Keep in mind that these digital pulses are converted to signals that are tones. It is these tones that are actually transmitted. You will see how this is done in the next chapter. For now, the focus is on the digital parts of such a system.

The basic idea of this concept is shown in Figure 8–56. There are two sets of digital interface units, one for each digital system. The basic function of the digital interface unit is to perform one of two major tasks:

[illegible]
1. Parallel load a bit pattern from its digital system and serially output this data.

Receiving:
2. Serially receive a bit pattern and load it in parallel to its digital system.

A logic diagram of such a unit is illustrated in Figure 8–57. As shown in the figure, there is a TRI-STATE bidirectional bus controller. This device allows the I/O bus to input data to or output data from the digital system. When the system is receiving information, the bus controller inputs data from the I/O bus. When the system is transmitting information, the bus controller outputs data to the I/O bus.

There are two control lines called output ENABLE and LOAD. These lines are not allowed to be active at the same time. When the output ENABLE line is active, the contents of the RECEIVE REGISTER are felt on the I/O bus. When the LOAD line is active, the contents of line of the I/O bus is transferred to the transmit register. The operation is as follows:

For the system to transmit data, the TRI-STATE bus controller is activated so that the data bus outputs the bit pattern to the I/O bus. The LOAD line is then made active, causing a parallel load of the bit pattern into the parallel-in–serial-out register. Then the LOAD line goes LOW causing the loaded bit pattern to be serially shifted out along the single data line.

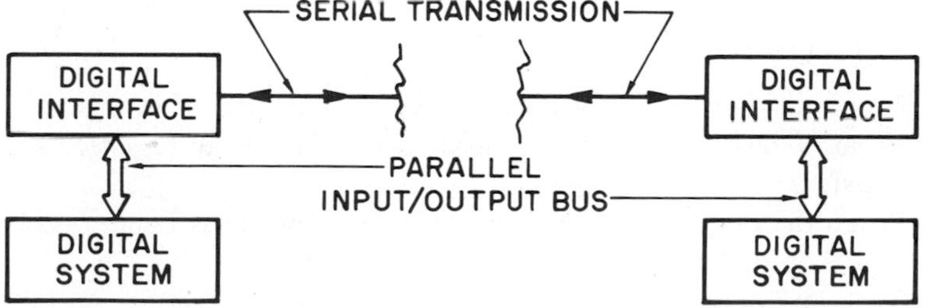

Figure 8–56 Basic concept of a serial data interface.

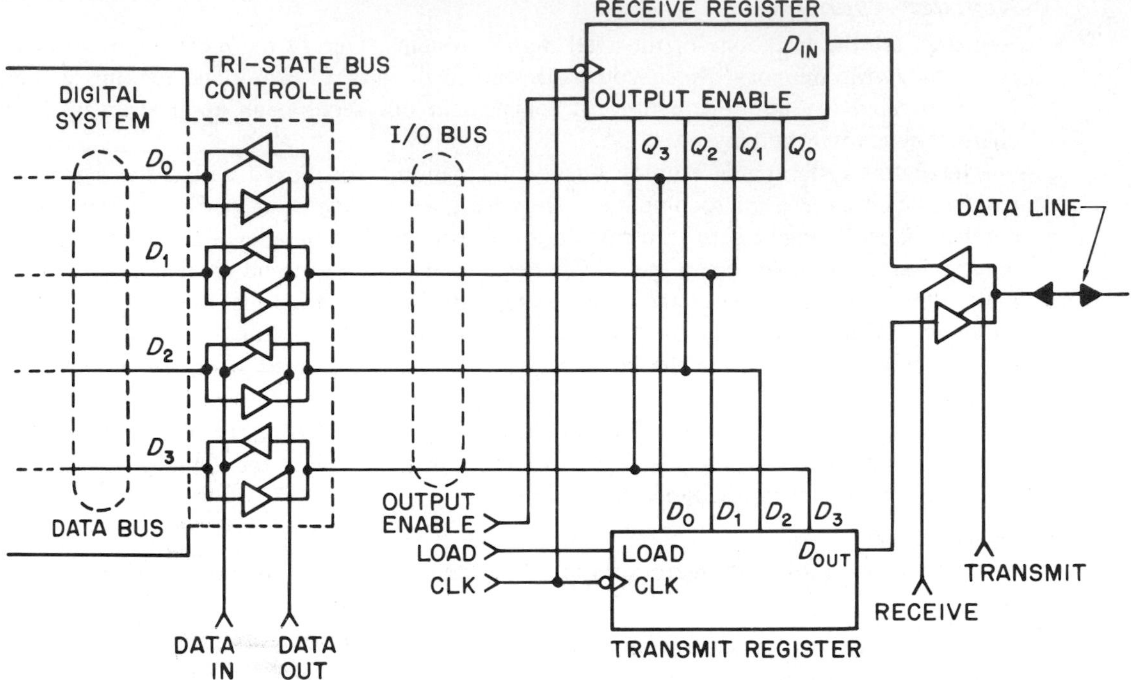

Figure 8–57 Basic concept of a data transmission system.

For the system to receive, the TRI-STATE bus controller is now activated so that the I/O bus will now act as an input to the digital system. However, there will not be any bit pattern on this bus as long as the output ENABLE line of the receiver register is not active. Thus, the receiver register first serially inputs data from the single data line and then its output ENABLE line becomes active. At this point, its data will now appear on the I/O bus.

The control of this unit is done by the digital system. As you will see in the chapter on digital interfacing, there are standard manufactured chips that perform this important process.

Conclusion

This section introduced you to the principles of registers. In the next section, you will see how registers can interact with each other using PLDs. Check your understanding of this section by trying the following review questions.

8–5 Review Questions

1. Name the four different kinds of registers presented in this section.
2. What is another name for a parallel-in–parallel-out register? For a serial-in–serial-out register?
3. When the output of a serial-in–serial-out register is connected back to the input, what is it then called?
4. State an application of a parallel-in–serial-out register.

8–6 REGISTERS AND PLDs

Introduction

This section presents some important concepts concerning the use of registers with PLDs. As you saw in the section on counters and PLDs, the use of the PLD gives the designer tremendous control over the counter. You will see that the same versatility is offered with the combination of registers and PLDs.

Register Connections

When the logic diagram of a register is used, it becomes easier to show the interconnections of more than one register. For example, look at Figure 8–58. As shown in this figure, two 4-bit registers have their inputs and outputs connected to a common bus. By doing this, the contents of one register may be copied into the other. Observe from the figure that there are six control switches. Each switch has the function shown in the following table.

Switch	Activates	Resulting Process
L_A	Load Reg A	Loads register A with the bit pattern on the bus.
L_B	Load Reg B	Loads register B with the bit pattern on the bus.
C_A	Clear Reg A	Sets the contents of Reg A to all zeros.
C_B	Clear Reg B	Sets the contents of Reg B to all zeros.
OE_A	Output Enable Reg A	The contents of Reg A will appear on the data bus.
OE_B	Output Enable Reg B	The contents of Reg B will appear on the data bus.

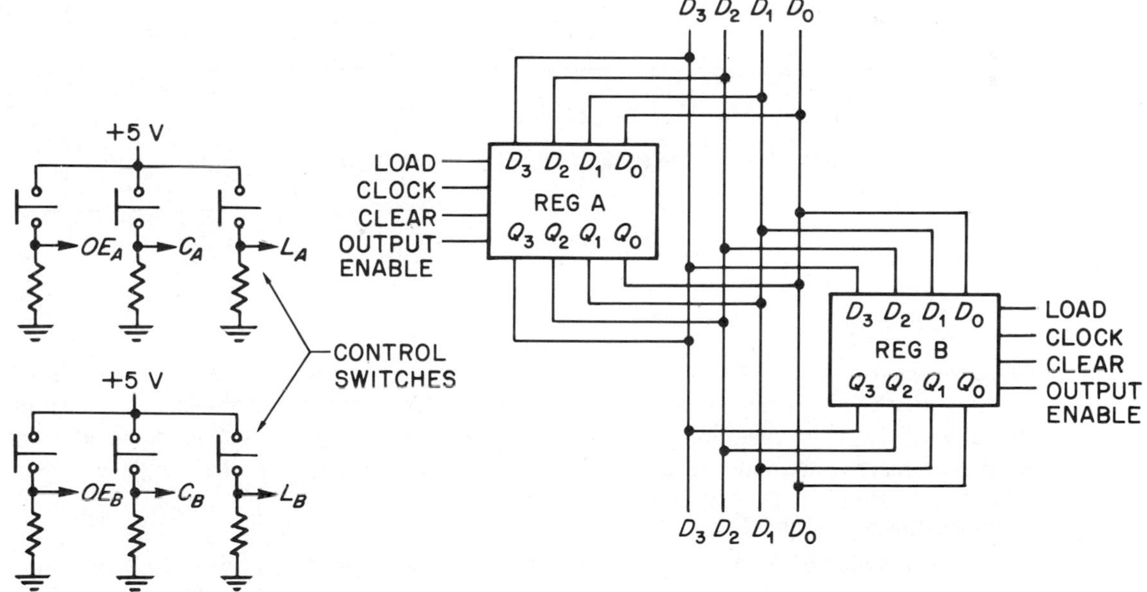

Figure 8–58 Connecting two registers.

It's important to note that only one enable line at a time is allowed to be active; otherwise there could be a bus conflict between the outputs of the two registers.

Example 8–22

Referring to the two register system shown in Figure 8–58, determine the final contents of each register for the following control switch sequence. Assume that the contents of each register are A = 1010 and B = 1111.

Control switch sequence: C_B
OE_A & L_B (both pressed at the same time)

Solution

Refer to the logic diagram and the previous table. The final conditions of each register are arrived at as follows:

Control Button	Resulting Process
C_B	B → 0000 (B register is cleared to zeros)
OE_A & L_B	A → B (Contents of the A register are copied into the B register. Thus both A and B registers contain 1010.)
C_A	A → 0000 (A register is cleared to zeros)

Thus, the final contents of the two registers are

A = 0000 and B = 1010

Another Way

Observe from Figure 8–59 that all of the logic connections shown mean the same thing. As shown, the simplified method of showing register interconnections leaves out some detail but makes the diagram easier to read. The important point is to understand what each type means, as industrial logic diagrams may employ any one of these.

Programming Registers

The control switches of the system in Figure 8–58 can be replaced by a PLD. By doing this, you can program the sequencing of register operations. Consider the following example.

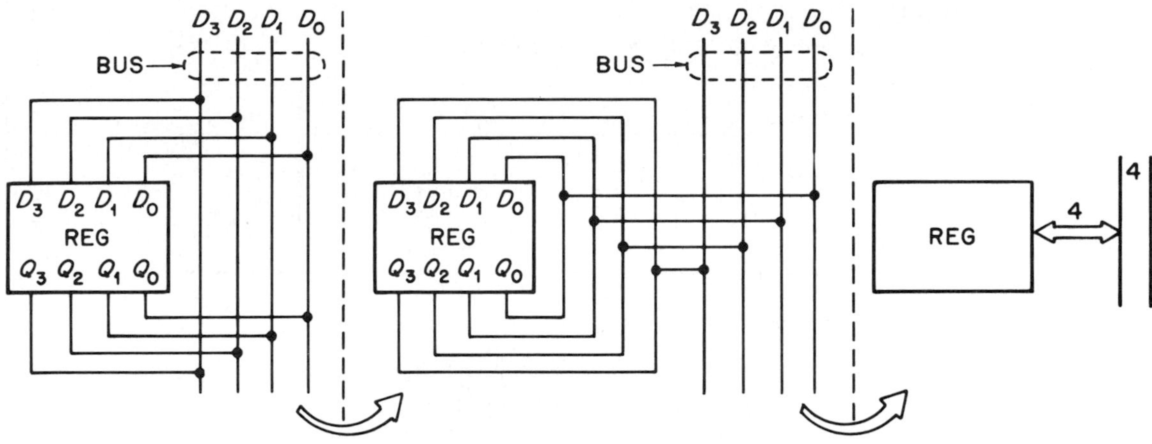

Figure 8–59 Different methods of representing registers connected to a bus.

Example 8–23

Determine the final contents of each register for the digital system shown in Figure 8–60. Assume each register starts with the following contents: A → 1011, B → 1001.

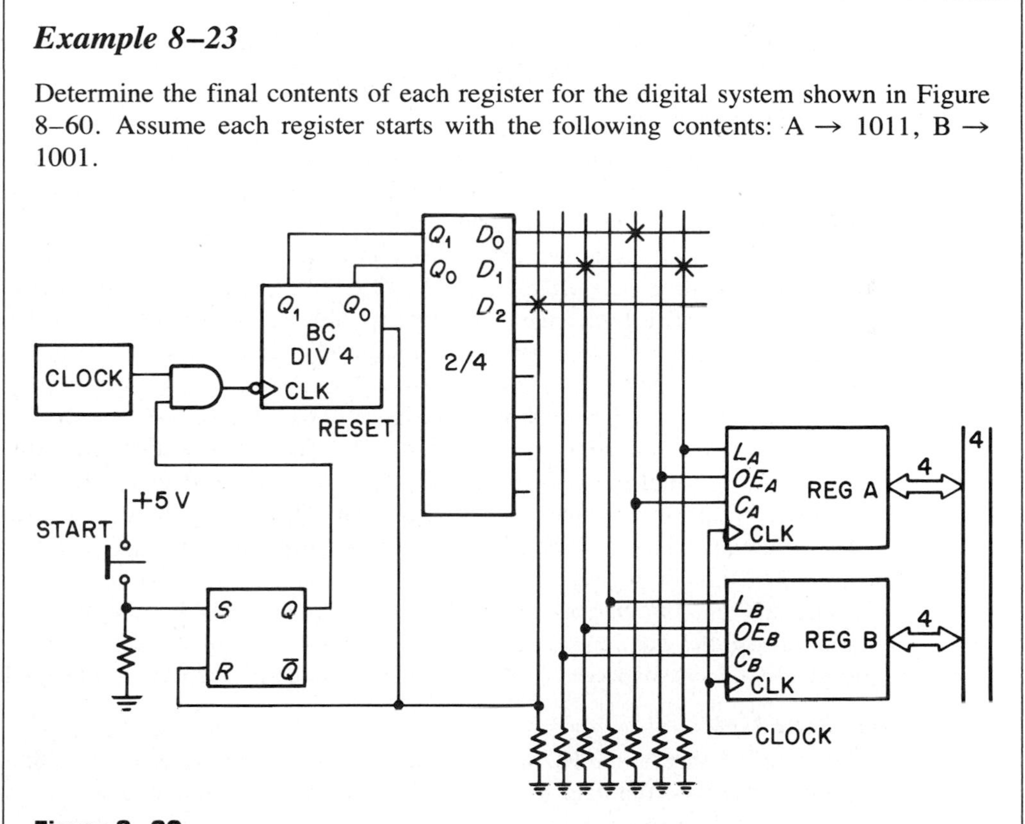

Figure 8–60

Solution

Develop a state table directly from the ROM coding:

Count	L_A	E_A	C_A	L_B	OE_B	C_B	Reset
0	0	0	1	0	0	0	0
1	1	0	0	0	1	0	0
2	0	0	0	0	0	0	1

From the state table you can see that

C_A: A → 0000 [clear the A register]

L_A & OE_B: B → A [copy B into A]

Thus, A = 1001, B = 1001.

From the previous example, since the binary counter is then reset to zero and the START/STOP flip-flop is RESET ($Q = 0$), the AND gate will now inhibit the arrival of the clock pulses until the START button is again depressed.

A Larger System

The concept of programming a PLD to present a specific sequence of operations between registers should begin to look suspiciously familiar to you. The same process takes place in a microprocessor. Recall that what a microprocessor does is copy a bit pattern from some register (contained in memory or from an input device), cause something to happen (a process), and then repeat the FETCH-EXECUTE cycle. Essentially then, the microprocessor is primarily transferring bit patterns between registers, and in the process of doing so it may perform other processes that modify in some way one or more of these bit patterns.

Consider the logic circuit shown in Figure 8–61. Figure 8–61 shows a 2-bit binary counter, a 2-to-4 line decoder with a diode ROM. This is used to control the action of a 2-bit serial-in–parallel-out register and a parallel-in–parallel-out register. Note from the figure that the shift line is controlled by the D_1 and D_2 count of the binary counter. During these two counts, data is shifted into the shift register (register A). At the count of D_3, the ENABLE line of register A and the LOAD line of register B are both activated. At this time, the data that has been shifted into register A is now copied into register B. The parallel-in–parallel-out register (register B) is connected to a pair of LEDs so that its condition is always displayed (note the output ENABLE line OE_B is always HIGH).

Machine Cycle

As Figure 8–61 shows, the ROM pattern will cause data to be shifted in to the 2-bit serial-in register, and then the data will be parallel outputted to the bus where its contents will be copied into the B register and displayed on the LEDs. A timing diagram of this process is shown in Figure 8–62.

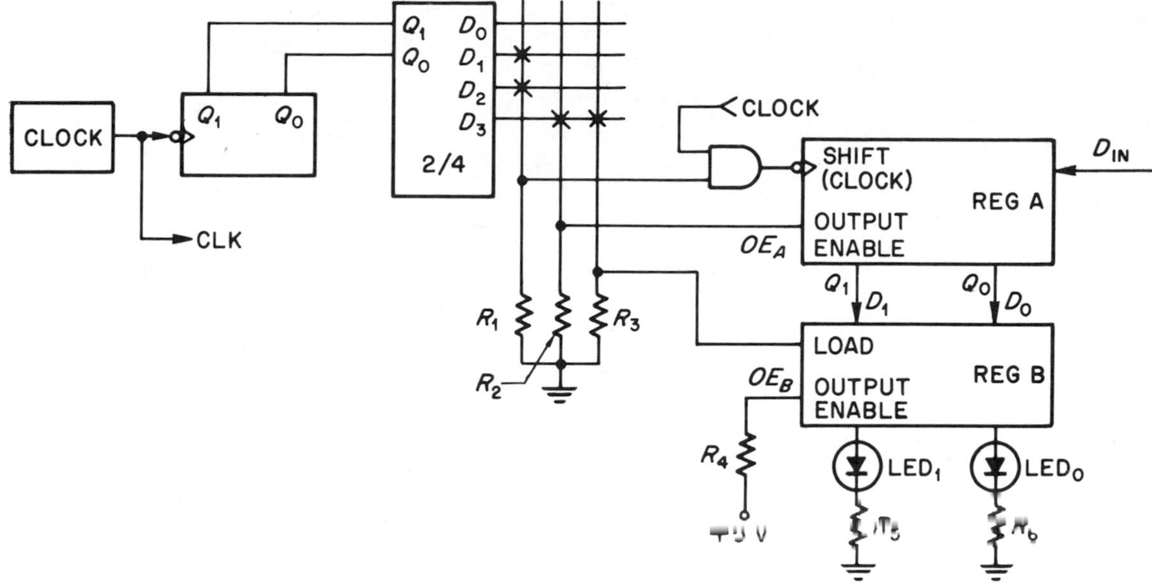

Figure 8–61 A 2-bit receiving process.

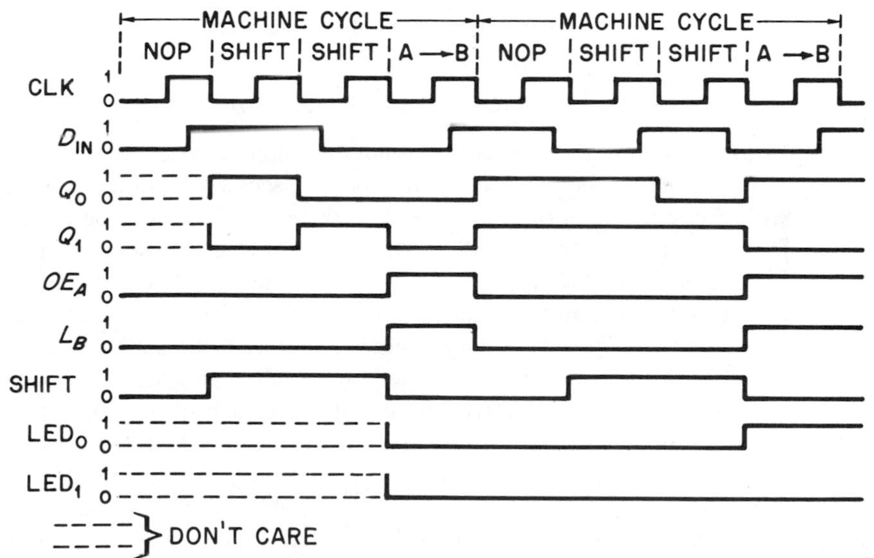

Figure 8–62 Timing diagram for 2-bit receiving process.

The timing diagram shows two complete machine cycles. Each machine cycle represents a series of processes. As indicated at the top of the diagram, the process is NOP (no operation), two consecutive shifts, and then to copying of data from register A to register B. This series of processes is repeated over and over again because of the action of the binary counter. It is important to note that the *type* of process in such a system is

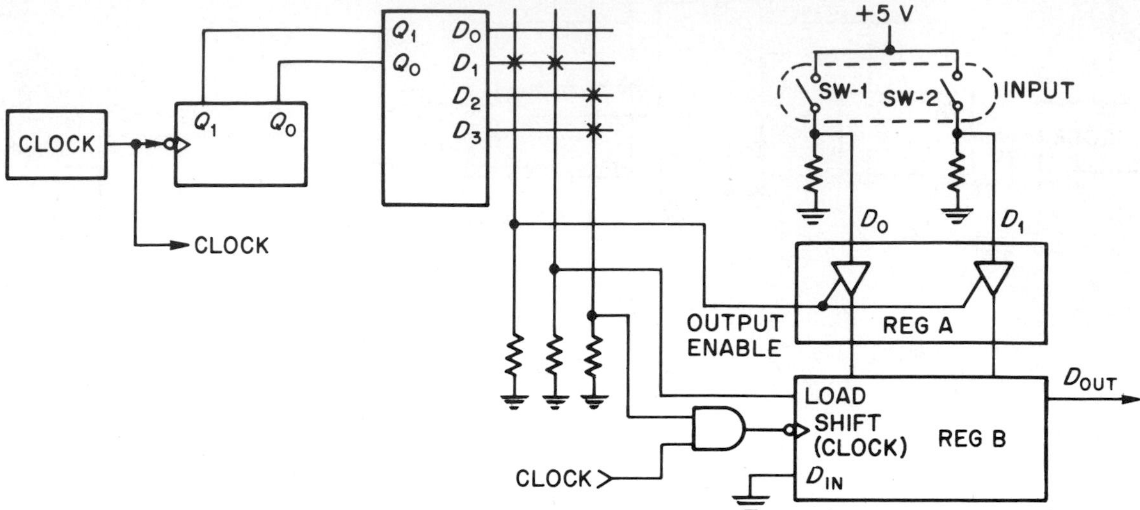

Figure 8–63 Two-bit transmitting process.

determined by the programming of the PLD (in this case a switch programmable ROM). Note from the timing diagram the indication of a **machine cycle**.

Machine Cycle:
The length of time required to complete a process.

To continue the presentation, consider the digital system shown in Figure 8–63. This system is similar to the 2-bit receiving process discussed previously. Again a binary counter controls the sequencing of a 2-to-4 line decoder, which selects the control lines of an input buffer (consisting of two TRI-STATE buffers) and a parallel-in–serial-out register. The bit pattern in the ROM will cause the state of the two input switches (SW-1 and SW-2) to be felt on the bus and loaded into the B register. This will take place when the binary counter is at 01. When the binary counter goes to 10, the shift input (S_B) is activated, and the contents of the B register are shifted out serially. This serial shifting process continues for the next count of 11. When the counter cycles back to 00, the process is ready to repeat. The timing diagram is shown in Figure 8–64. This timing diagram is similar to the last one presented for the input system. Here again two complete machine cycles are shown, each consisting of a series of predictable processes. As before, there are four processes, the first is NOP, followed by a LOAD then two consecutive shift pulses. Again the important point to make here is the use of a counter and a PLD in order to create a predictable and repeatable sequence of events.

Combining the Two

Both of the previous systems contained separate and unique processes. These two separate processes may be joined by a larger system that allows the user to select either of the processes. Such a system is illustrated in Figure 8–65. As shown, there are now two 2-bit binary counters. One is called the *program counter,* the other the *microprogram counter.*

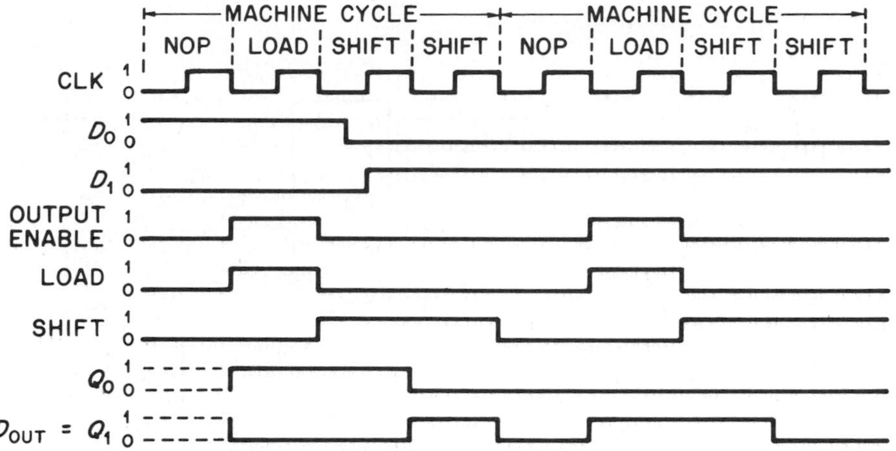

Figure 8–64 Timing diagram for 2-bit transmitting process.

Note the connection of the AND gate with inverted inputs between the two counters. The microprogram counter is controlled directly by the clock pulse, and the program counter is controlled by the transition of the microprogram counter from 11 to 00. This process is illustrated in Figure 8–66. As shown in the timing diagram, the program counter will increment after every fourth count of the microprogram counter.

This means that the program counter will activate one line of the switch-programmable ROM and the corresponding activated bit pattern will determine which process will be carried out for the next three clock pulses. As an example, observe how the switch-programmable ROM is programmed in Figure 8–65. Start with both binary counters at 00. Then, with each clock pulse the following will take place.

Clock Pulse Number	Micro-program Counter	Program Counter	Switch Program ROM Output	Process Activated by ROM	Micro-process
1	0 0	0 0	P_A	Receive	None
2	0 1	0 0	P_A	Receive	S_A
3	1 0	0 0	P_A	Receive	S_A
4	1 1	0 0	P_A	Receive	E_A, L_B
5	0 0	0 1	P_B	Transmit	None
6	0 1	0 1	P_B	Transmit	E_C, L_D
7	1 0	0 1	P_B	Transmit	S_D
8	1 1	0 1	P_B	Transmit	S_D
9	0 0	10 → 00	CLR → P_A	Clear	None
10	0 1	0 0	P_A	Receive	S_A
11	1 0	0 0	P_A	Receive	S_A
12	1 1	0 0	P_A	Receive	E_A, L_B
			⟨Process repeats⟩		

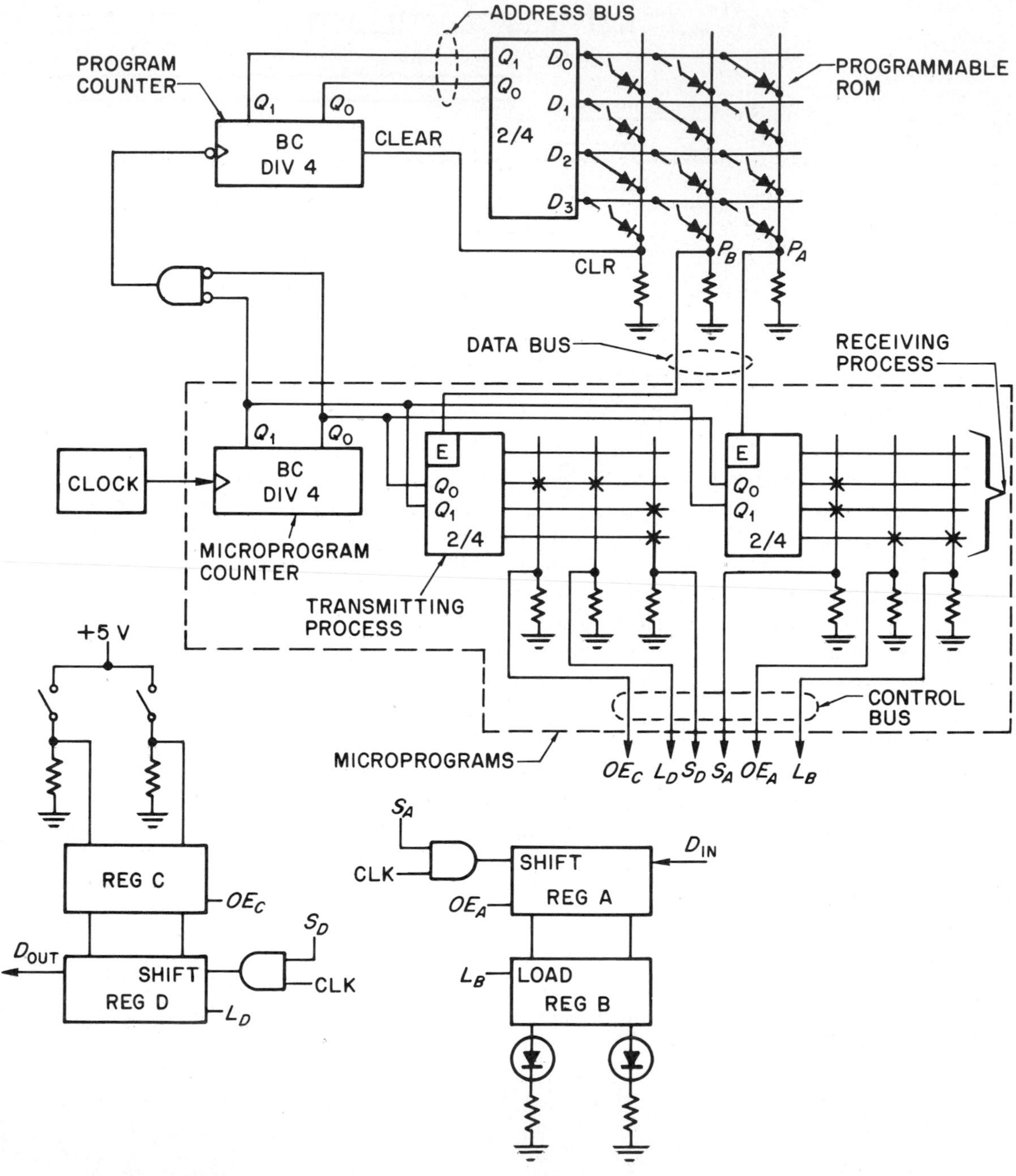

Figure 8–65 A larger process controlling smaller processes.

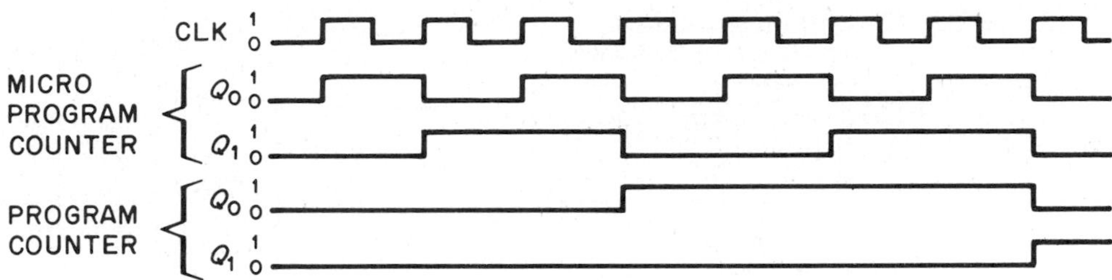

Figure 8–66 Timing relationships between the two counters.

As shown in this table, the two processes repeat. A timing diagram of this operation is shown in Figure 8–67. The whole system has only three processes: (1) transmit, (2) receive, and (3) clear the program counter. Any one of these processes may be selected by the bit pattern you put into the switch-programmable ROM. Thus, this system has three instructions, each of which selects a predictable small process.

Making Some Conclusions

The previous system presented the fundamental idea behind a microprocessor. Using microprocessor terminology, the bit pattern in the ROMs used to control each small process is called a **microcode** and is built into every microprocessor at the factory.

> **Microcode:**
> The instructions built into a microprocessor that determine a specific selectable process.

Each specific bit pattern within the microcode is called a **microinstruction**.

> **Microinstruction:**
> A single instruction in the microcode of a microprocessor.

Thus, when you program any microprocessor, you are really activating a specific microprogram that exists within the microprocessor. This powerful and useful concept is illustrated in Figure 8–68.

The user code consists of the bit patterns (instructions and data) the programmer puts into memory. In turn, during the FETCH, this bit pattern will determine which of the

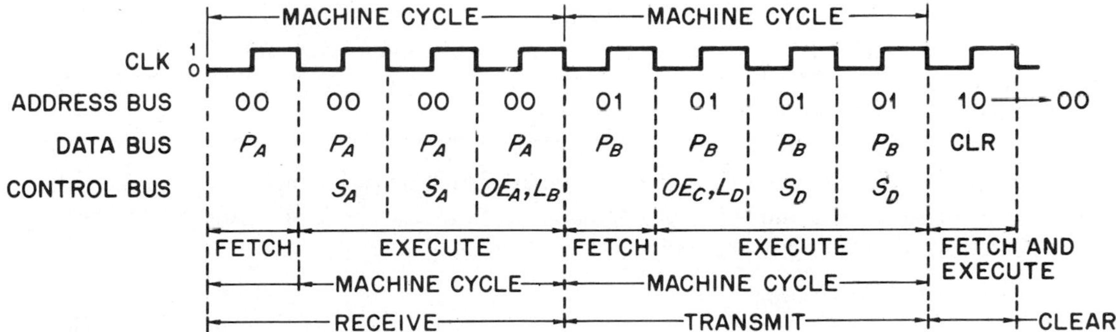

Figure 8–67 Processor timing diagram.

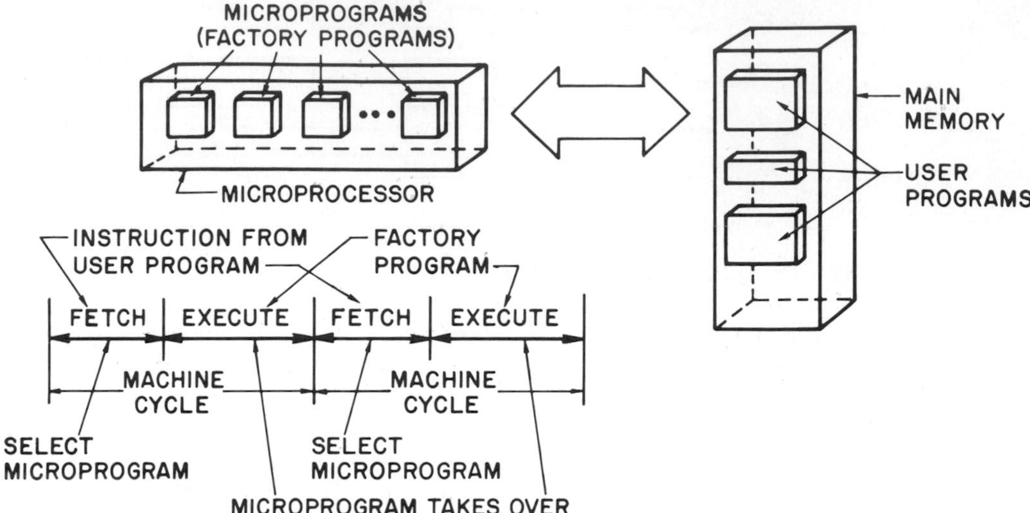

Figure 8–68 Concept of microcode and user code.

microinstructions will be activated. In the system presented the bit pattern was kept simple to convey this important relationship. In some real microprocessors the programmer has the option of selecting one of over 200 microprograms (processes) programmed into the device at the factory.

Conclusion

This section presented how registers and PLDs could be used together. When a counter was introduced with the system, the resultant mixture produced a microprocessor. This concept helped foster the invention of the modern microprocessor.

In the next section you will have an opportunity to see some real counters and registers used in industry. Check your understanding of this section by trying the following review questions.

8–6 Review Questions

1. State the advantage of connecting the inputs and outputs of registers to a common bus.
2. Define *machine cycle*.
3. Define *microcode* and *microinstruction*.
4. What is actually happening when you program a microprocessor?

8–7 REAL HARDWARE

In this section you will be introduced to some real ICs that make up *JK* flip-flops, counters, and registers. The fundamental operation of these devices is the same as the idealized models presented up to this point in the chapter.

Real JK Flip-Flops

Figure 8–69 shows a CMOS dual *JK* flip-flop, the DM7476. Observe from the truth table on the data sheet that the manufacturer states that when both $\overline{\text{PRE}}$ and $\overline{\text{CLR}}$ are active the

DM5476/DM7476 Dual Master-Slave J-K Flip-Flops with Clear, Preset, and Complementary Outputs

General Description

This device contains two independent positive pulse triggered J-K flip-flops with complementary outputs. The J and K data is processed by the flip-flop after a complete clock pulse. While the clock is low the slave is isolated from the master. On the positive transition of the clock, the data from the J and K inputs is transferred to the master. While the clock is high the J and K inputs are disabled. On the negative transition of the clock, the data from the master is transferred to the slave. The logic state of J and K inputs must not be allowed to change while the clock is high. The data is transferred to the outputs on the falling edge of the clock pulse. A low logic level on the preset or clear inputs will set or reset the outputs regardless of the logic levels of the other inputs.

Connection Diagram

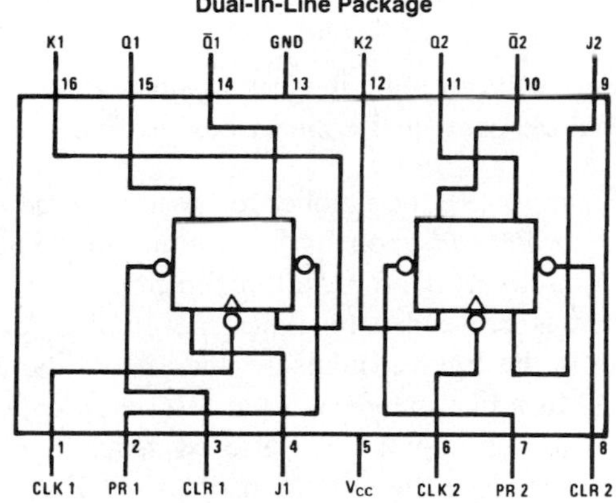

TL/F/6528-1

Order Number DM5476J or DM7476N
See NS Package Number J16A or N16A

Function Table

Inputs					Outputs	
PR	CLR	CLK	J	K	Q	$\bar{Q}$
L	H	X	X	X	H	L
H	L	X	X	X	L	H
L	L	X	X	X	H*	H*
H	H	⎍	L	L	Q_0	$\bar{Q}_0$
H	H	⎍	H	L	H	L
H	H	⎍	L	H	L	H
H	H	⎍	H	H	Toggle	

H = High Logic Level

L = Low Logic Level

X = Either Low or High Logic Level

⎍ = Positive pulse data. The J and K inputs must be held constant while the clock is high. Data is transfered to the outputs on the falling edge of the clock pulse.

* = This configuration is nonstable; that is, it will not persist when the preset and/or clear inputs return to their inactive (high) level.

Q_0 = The output logic level before the indicated input conditions were established.

Toggle = Each output changes to the complement of its previous level on each complete active high level clock pulse.

Figure 8–69 CMOS dual *JK* flip-flop. *Reprinted with permission of National Semiconductor Corporation.*

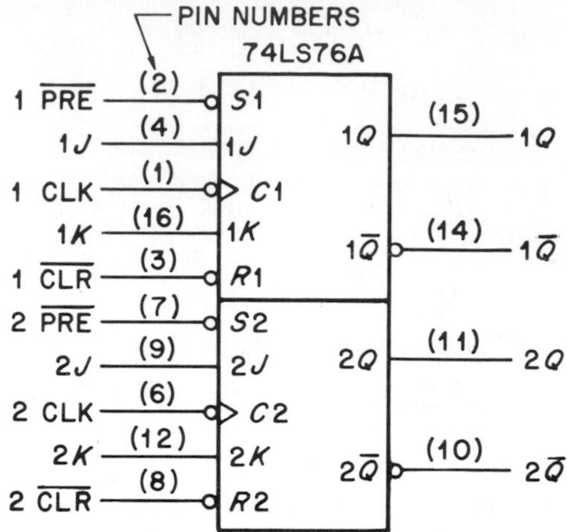

Figure 8–70 Logic symbol for *JK* flip-flop.

device is in an unstable condition where its output is not guaranteed when returned from this condition. The logic symbol for the 74LS76A *JK* flip-flop is shown in Figure 8–70. Note that each flip-flop can be independently controlled with connections for its own clock (*JK* inputs as well as PRE and CLR).

The logic diagram of Figure 8–70 uses **dependency notation.**

Dependency Notation:
A method of showing the relationship between logic devices inputs and outputs without showing all of the internal elements and connections involved.

As an example of the flip-flop notation in Figure 8–70, the number following a logic input indicates which of the other inputs or outputs *depend* upon its logic condition. For example, both clock inputs are followed by a number (a 1 for the top flip-flop and a 2 for the bottom). Any logic input or output that is preceeded by the same number is then dependent upon this clock input. As shown in the figure, inputs $1J$ and $1K$ are therefore *dependent* upon the condition of the CLK 1. In a like manner, $1Q$ and $\overline{1Q}$ are *dependent* upon the condition of the CLK 1. Note that the Q outputs must also be *dependent* upon the condition of $S1$ and $R1$. This makes sense, since the condition of the PRESET and CLEAR inputs does determine the condition of the Q outputs. You will see further use of this kind of notation used in this text. More information on dependency notation is presented in the appendix.

A Real Counter

Popular binary asynchronous counters are the 7490–7492 and 7493 series shown in Figure 8–71. The internal logic construction of each counter is illustrated in Figure 8–72. As shown in the data sheets for these counters, their logic diagrams are different. The difference is in the arrangement of the additional gating for each counter. The first, the 90A, is arranged so that the count cycle length is a divide by five. The 92A is arranged for a divide by six and the 93A for a divide by eight. In doing this, a variety of options is now

DM5490A/DM7490A, DM5492A/DM7492A, DM5493A/DM7493A Decade, Divide by 12, and Binary Counters

General Description

Each of these monolithic counters contains four master-slave flip-flops and additional gating to provide a divide-by-two counter and a three-stage binary counter for which the count cycle length is divide-by-five for the 90A, divide-by-six for the 92A and divide-by-eight for the 93A.

All of these counters have a gated zero reset and the 90A also has gated set-to-nine inputs for use in BCD nine's complement applications.

To use their maximum count length (decade, divide-by-twelve, or four-bit binary), the B input is connected to the Q_A output. The input count pulses are applied to input A and the outputs are as described in the appropriate truth table. A symmetrical divide-by-ten count can be obtained from the 90A counters by connecting the Q_D output to the A input and applying the input count to the B input which gives a divide-by-ten square wave at output Q_A.

Features

- Typical power dissipation
  - 90A 145 mW
  - 92A, 93A 130 mW
- Count frequency 42 MHz

Absolute Maximum Ratings (Note 1)

Supply Voltage	7V
Input Voltage	5.5V
Storage Temperature Range	−65°C to 150°C

Note 1: The "Absolute Maximum Ratings" are those values beyond which the safety of the device cannot be guaranteed. The device should not be operated at these limits. The parametric values defined in the "Electrical Characteristics" table are not guaranteed at the absolute maximum ratings. The "Recommended Operating Conditions" table will define the conditions for actual device operation.

Connection Diagrams

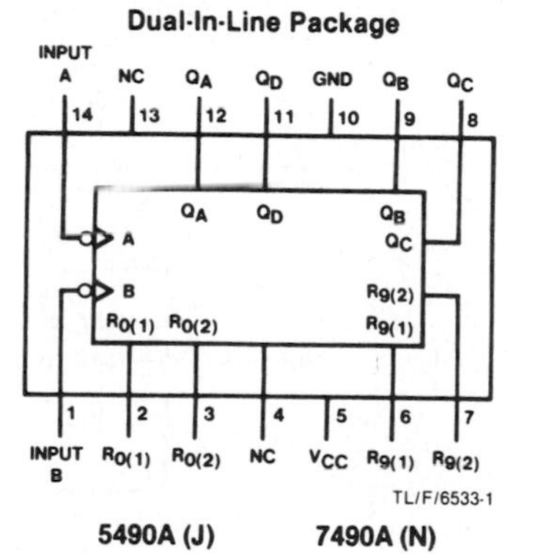

5490A (J)   7490A (N)

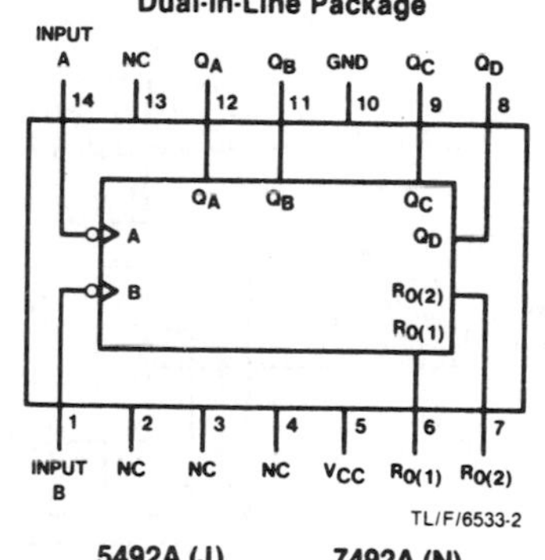

5492A (J)   7492A (N)

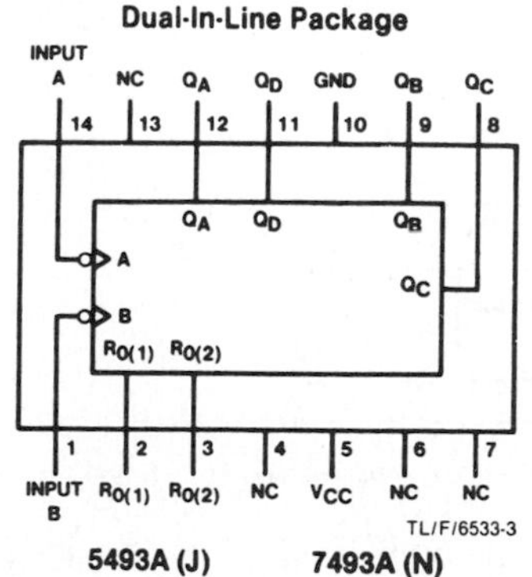

5493A (J)   7493A (N)

Figure 8–71 7490–7492 and 7493 series counters. *Reprinted with permission of National Semiconductor Corporation.*

Logic Diagram

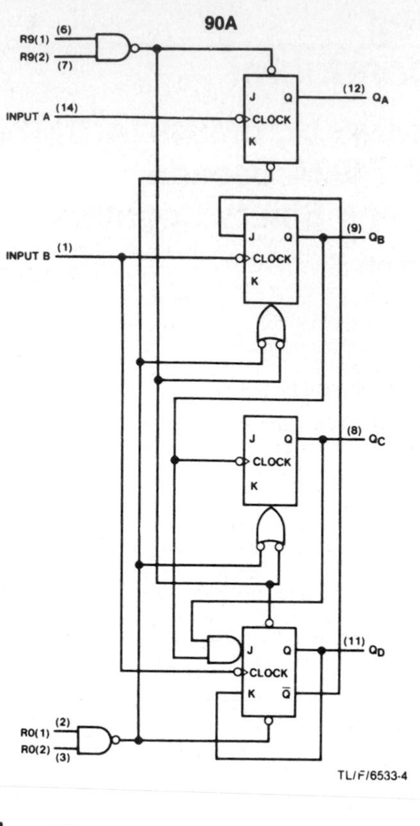

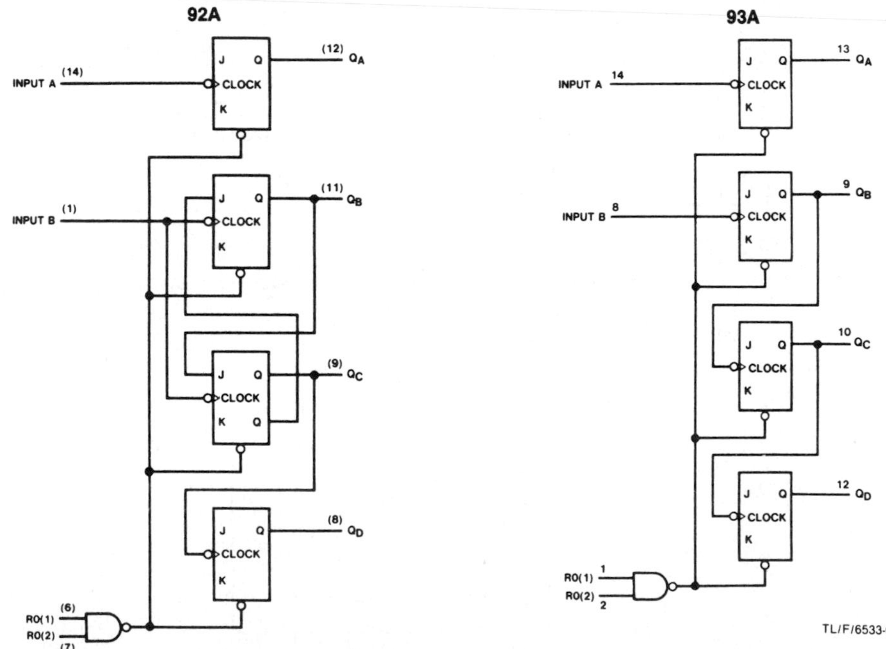

Figure 8–72 Internal logic construction of the 7490 series. *Reprinted with permission of National Semiconductor Corporation.*

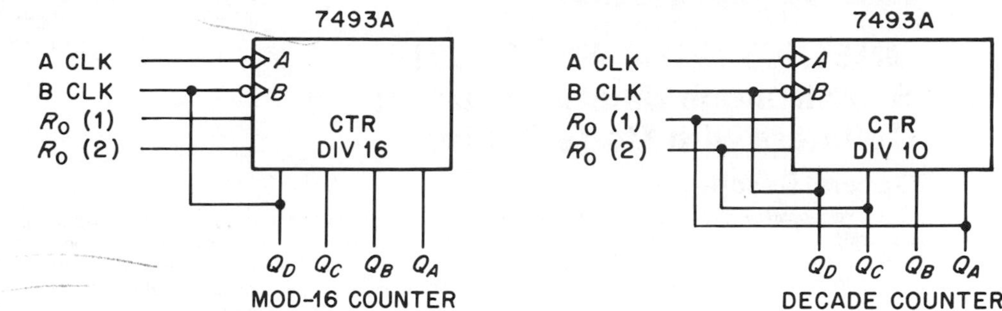

Figure 8–73 The 7493 as a mod-16 and decade counter.

available to the digital circuit designer. As an example of the versatility of these counters, the connections for the 7493 as a full 4 bit (mod 16) binary counter or as a decade counter are shown in Figure 8–73.

Real Synchronous Counters

A versatile synchronous counter is the 74HCT191 COMS binary up/down counter with a parallel loading feature. The data page for this counter is illustrated in Figure 8–74. The truth table on the data sheet shows that the counter can also act as a storage register when the enable input is HIGH. The logical construction of this device is illustrated in Figure 8–75.

The timing diagram showing the relationships of the inputs and outputs for this counter is in Figure 8–76. The double dashed lines of the timing diagram are *don't care* conditions. The LSB is Q_A, and the MSB is Q_O. The versatility of this counter includes a MAX/MIN output that produces an output pulse every time the counter count is 1111_2 (MAX) or 0000_2 (MIN). There is also an output called the RIPPLE clock that produces an output pulse at an **overflow** or **underflow** condition.

Overflow:
In a counter, when a maximum value count is reached and the counter counts forward, resetting itself to zero.

Underflow:
In a counter, when a minimum count of zero is reached and the counter counts backward, resetting itself to its maximum value count (all 1's).

The ripple clock output is used for cascading counters as shown in Figure 8–77. This process creates larger counters in groups of 4 bits each. Observe from the timing diagram that the only time the data inputs have an effect on the counter is when the LOAD line is active (LOW in this case). Also note that when ENABLE is not active (HIGH), the counting will stop.

National Semiconductor

PRELIMINARY

MM54HCT191/MM74HCT191 Synchronous Binary Up/Down Counters with Mode Control

General Description

These high speed synchronous counters utilize microCMOS technology, 3.0 micron silicon gate N-well CMOS. They possess the high noise immunity and low power consumption of CMOS technology, along with the speeds of low power Schottky TTL. These circuits are synchronous, reversible, up/down counters. The MM54HCT191/MM74HCT191 are 4-bit binary counters.

Synchronous operation is provided by having all flip-flops clocked simultaneously so that the outputs change simultaneously when so instructed by the steering logic. This mode of operation eliminates the output counting spikes normally associated with asynchronous (ripple clock) counters.

The outputs of the four master-slave flip-flops are triggered on a low-to-high level transition of the clock input, if the enable input is low. A high at the enable input inhibits counting. The direction of the count is determined by the level of the down/up input. When low, the counter counts up and when high, it counts down.

These counters are fully programmable; that is, the outputs may be preset to either level by placing a low on the load input and entering the desired data at the data inputs. The output will change independent of the level of the clock input. This feature allows the counters to be used as divide by N dividers by simply modifying the count length with the preset inputs.

Two outputs have been made available to perform the cascading function; ripple clock and maximum/minimum count. The latter output produces a high level output pulse with a duration approximately equal to one complete cycle of the clock when the counter overflows or underflows. The ripple clock output produces a low level output pulse equal in width to the low level portion of the clock input when an overflow or underflow condition exists. The counters can be easily cascaded by feeding the ripple clock output to the enable input of the succeeding counter if parallel clocking is used, or to the clock input if parallel enabling is used. The maximum/minimum count output can be used to accomplish look-ahead for high speed operation.

MM54HCT/MM74HCT devices are intended to interface between TTL and NMOS components and standard CMOS devices. These parts are also plug-in replacements for LS-TTL devices can be used to reduce power consumption in existing designs.

Features

- Level changes on Enable or Down/Up can be made regardless of the level of the clock.
- Low quiescent supply current: 80 μA maximum (74HCT Series)
- Low input current: 1 μA maximum
- TTL compatible inputs

Connection Diagram

Dual-In-Line Package

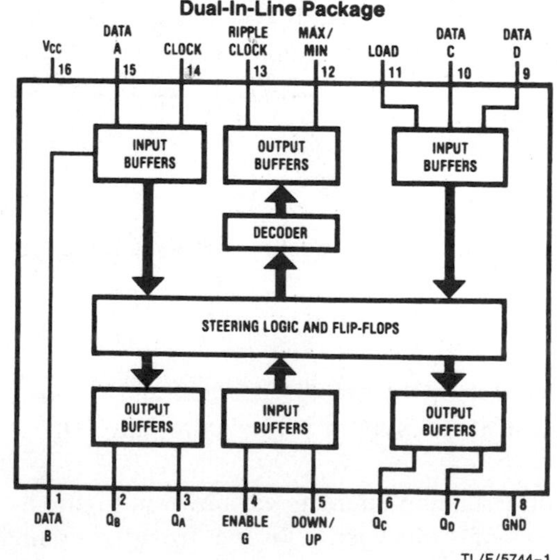

Truth Table

Load	Enable G	Down/Up	Clock	Function
H	L	L	↑	Count Up
H	L	H	↑	Count Down
L	X	X	X	Load
H	H	X	X	No Change

Order Number MM54HCT191J or MM74HCT191J, N
See NS Package J16A or N16E

Figure 8–74 The 74HCT191 synchronous counter. *Reprinted with permission of National Semiconductor Corporation.*

Logic Diagram

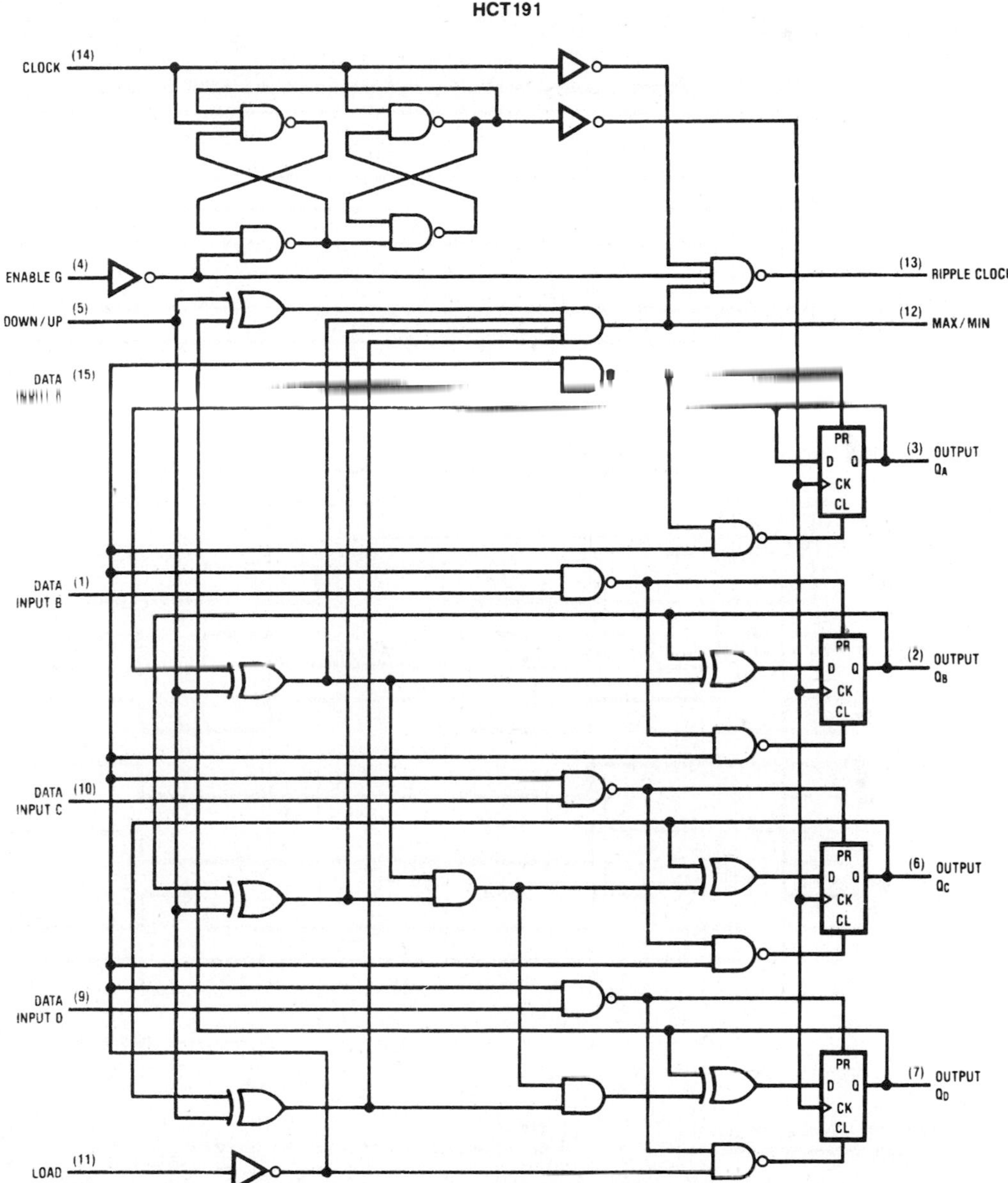

Figure 8–75 Logical construction of the 74HCT191. *Reprinted with permission of National Semiconductor Corporation.*

National holds no responsibility for any circuitry described. National reserves the right at any time to change without notice said circuitry.

Timing Diagram

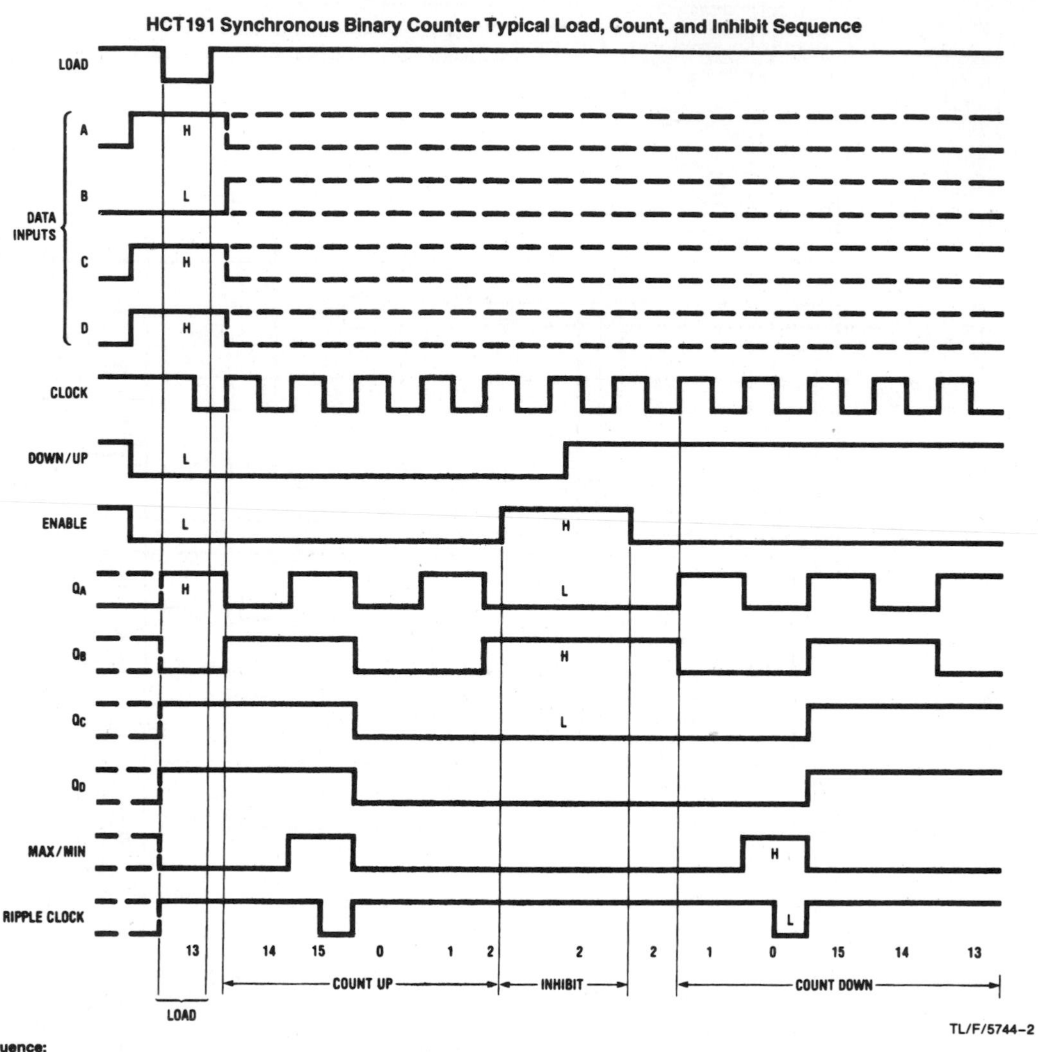

Sequence:

(1) Load (preset) to binary thirteen.

(2) Count up to fourteen, fifteen, zero, one, and two.

(3) Inhibit.

(4) Count down to one, zero, fifteen, fourteen, and thirteen.

Figure 8–76 The 74HCT191 timing diagram. *Reprinted with permission of National Semiconductor Corporation.*

National holds no responsibility for any circuitry described. National reserves the right at any time to change without notice said circuitry.

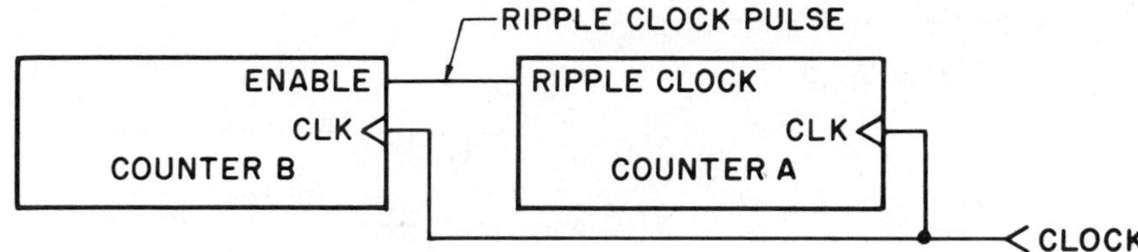

Figure 8–77 Cascading counters.

Example 8–24

Determine the output waveforms for a 74HCT191 counter if the waveforms of Figure 8–78 were applied. Indicate *don't care* conditions using the double dashed line notation.

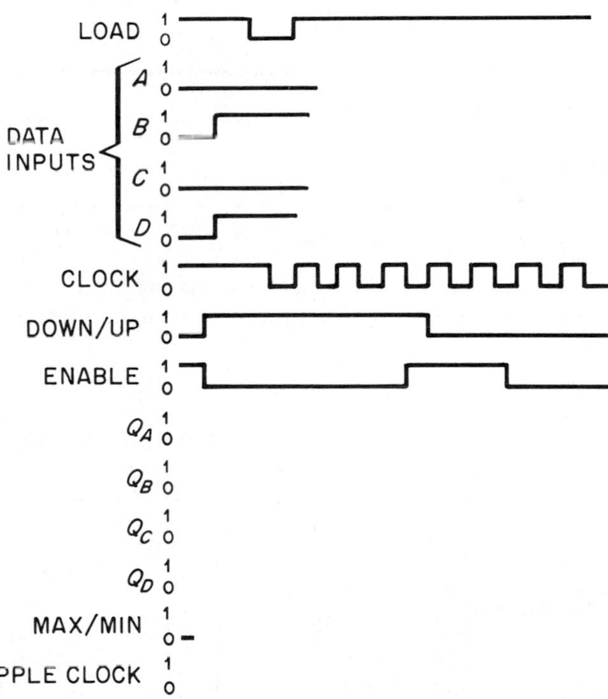

Figure 8–78

Solution

See Figure 8–79.

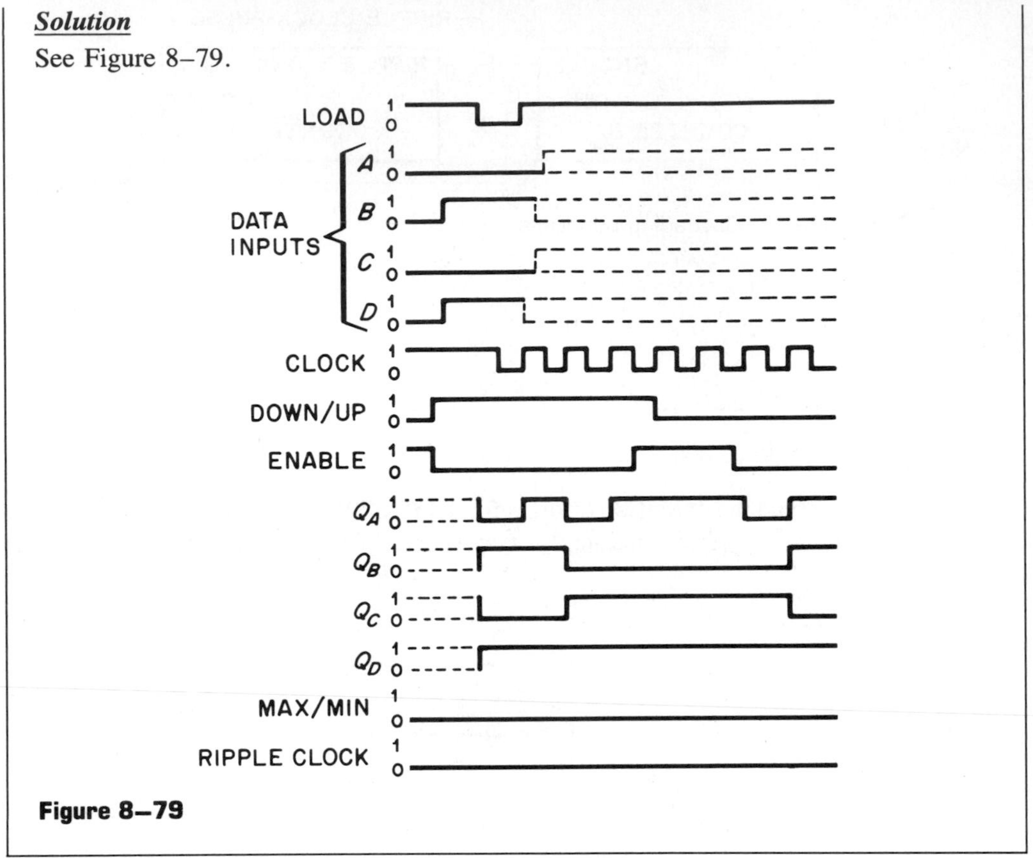

Figure 8–79

Hidden Counters

Many ICs use counters and are not called counters. One popular timer, the XR2240 programmable IC timer contains an 8-bit binary counter. See Figure 8–80. This device allows

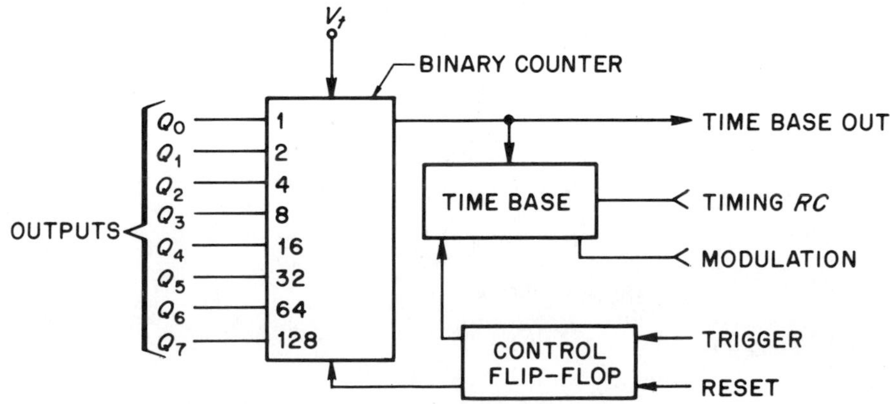

Figure 8–80 The XR2240 programmable IC timer.

255 different time periods with a single resistor-capacitor combination. The time base may be computed from

$$T = RC$$

where:

T = time base in seconds
R = timing resistor in ohms
C = timing capacitor in farads

Each successive output of the binary counter is related to the time base by

$Q_0 = 2^0 = 1T$
$Q_1 = 2^1 = 2T$
$Q_2 = 2^2 = 4T$
$Q_3 = 2^3 = 8T$
$Q_4 = 2^4 = 16T$
$Q_5 = 2^5 = 32T$
$Q_6 = 2^6 = 64T$
$Q_7 = 2^7 = 128T$

Example 8–25

The outputs of an XR2240 programmable timer are ANDed as shown in Figure 8–81. Determine the timing period for the indicated *RC* values.

(A) $R = 1\ \text{M}\Omega$, $C = 1\ \mu\text{F}$

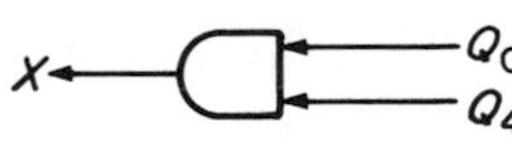

(B) $R = 1\ \text{M}\Omega$, $C = 5\ \mu\text{F}$

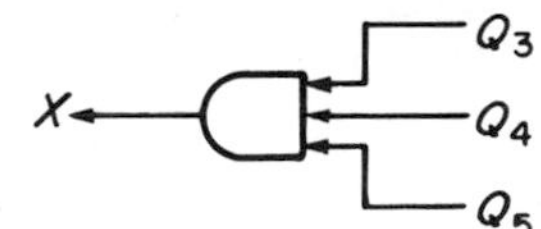

(C) $R = 1\ \text{M}\Omega$, $C = 500\ \mu\text{F}$

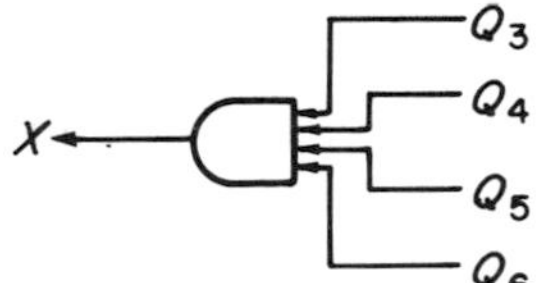

Figure 8–81

Solution

(A) $T = RC$
$T = 1\ \text{M}\Omega \times 1\ \mu\text{F}$
$T = 1\ \text{s}$

$Q_0 = 1\ \text{s}$
$Q_4 = 16 \times 1\ \text{s} = 16\ \text{s}$
Output time $= 1\ \text{s} + 16\ \text{s} = 17\ \text{s}$

(B) $T = RC$

$T = 1\ \text{M}\Omega \times 5\ \mu\text{F}$

$T = 5\ \text{s}$

$Q_3 = 8 \times 5\ \text{s} = 40\ \text{s}$

$Q_4 = 16 \times 5\ \text{s} = 80\ \text{s}$

$Q_5 = 32 \times 5\ \text{s} = 160\ \text{s}$

Output time $= 40 + 80 + 160 = 280$ s or 4 min 40 s

(C) $T = RC$

$T = 1\ \text{M}\Omega \times 500\ \mu\text{F}$

$T = 500\ \text{s}$

$Q_3 = 8 \times 500\ \text{s} = 4\ 000\ \text{s}$

$Q_4 = 16 \times 500\ \text{s} = 8\ 000\ \text{s}$

$Q_5 = 32 \times 500\ \text{s} = 16\ 000\ \text{s}$

$Q_6 = 64 \times 500\ \text{s} = 32\ 000\ \text{s}$

Output time $= 4\ 000 + 8\ 000 + 16\ 000 + 32\ 000$

$= 60\ 000\ \text{s} = 16$ h 40 min

Registers

A very flexible register is the 74194 shown in Figure 8–82. This device may be parallel loaded and shifted left or right. The device also has parallel data outputs. The timing diagram for this register is shown in Figure 8–83. The parallel inputs to this register are A, B, C and D. The parallel outputs are Q_3, Q_2, Q_1 and Q_0. Note that this may be CLEARED by applying a logic LOW at pin 1. The S_0 and S_1 inputs are the control inputs. These determine the mode of operation for the register. Serial data is fed through pins 2 or 7 depending upon the mode of operation, either shifting left or shifting right.

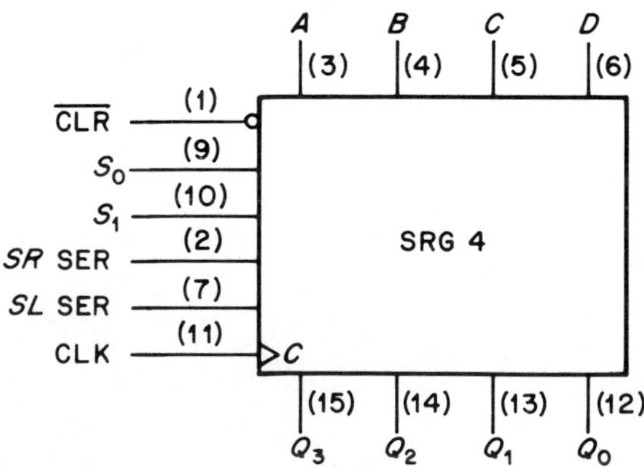

Figure 8–82 The 74194 register.

Timing Diagram

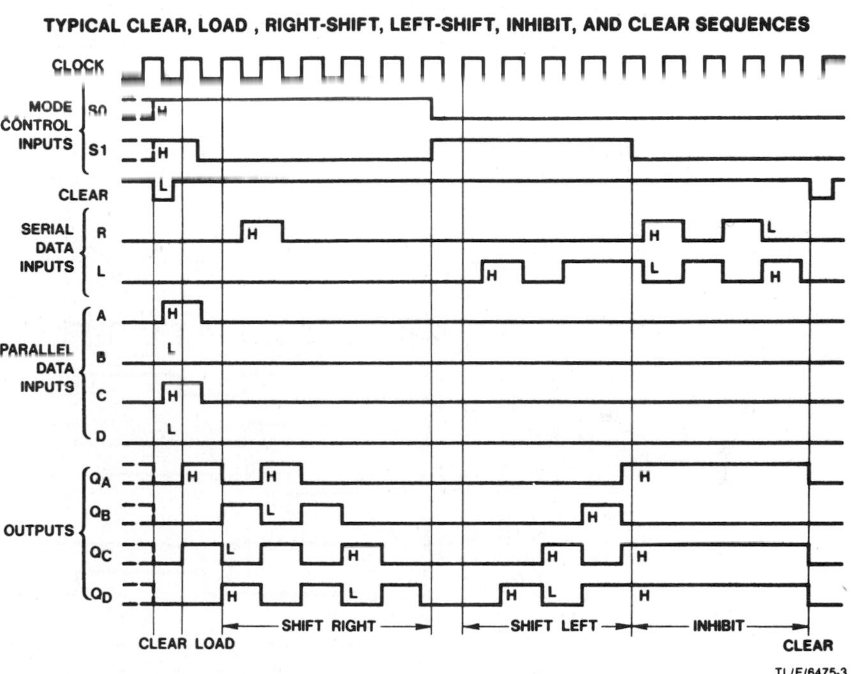

Figure 8–83 Timing diagram for the 74194 register. *Reprinted with permission of National Semiconductor Corporation.*

Observe from the timing diagram that the MODE control inputs (S_0, S_1), determine the type of operation as shown in the following table:

Mode Control		Result
S_0	S_1	Operation
0	0	Inhibit
0	1	Shift right
1	0	Shift left
1	1	Parallel load

Example 8–26

For the input waveforms of Figure 8–84, determine the resulting output waveforms for the 74194 shift register (be sure to indicate the *don't care* conditions).

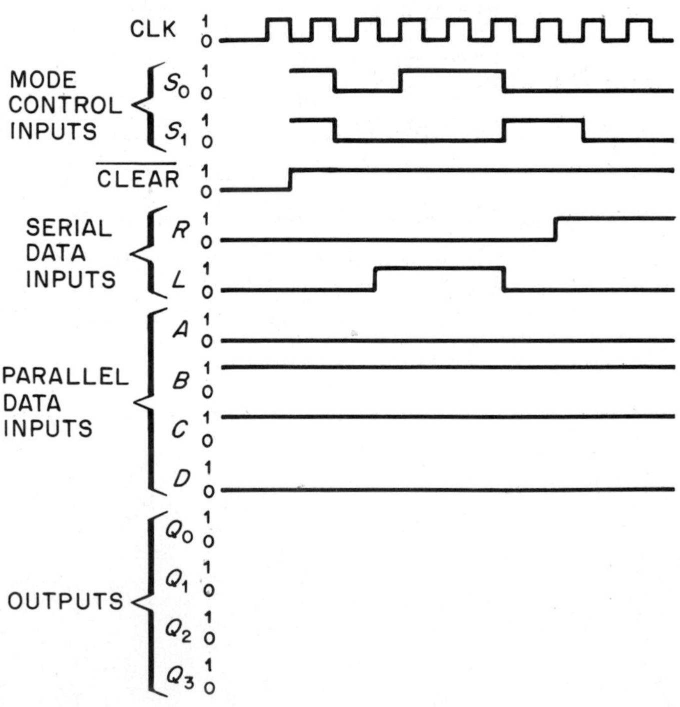

Figure 8–84

Solution

See Figure 8–85.

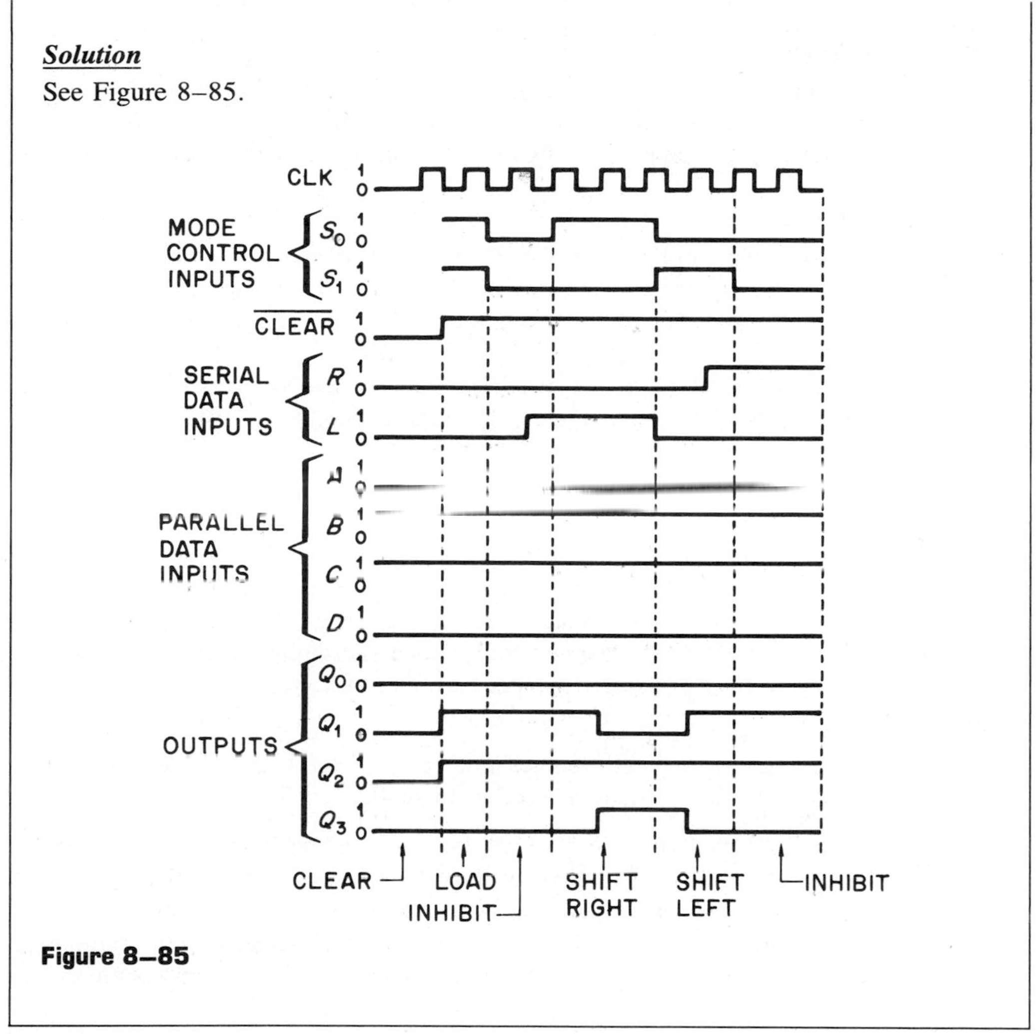

Figure 8–85

Conclusion

This completes your introduction to practical ICs using the digital devices presented in this chapter. The microprocessor applications section and the troubleshooting and instrumentation section will present many practical applications of these and other related digital devices. Check your understanding of this section by trying the following review questions.

8–7 Review Questions

1. For the 74HC76 flip-flop, describe some of its unique characteristics.
2. Describe the major differences between the 7490, 7492, and 7493 IC counters.
3. State some of the unique characteristics of the 74HCT191 counter.
4. Describe the major characteristics of the 74194 register.
5. State some of the unique characteristics of the XR2240 programmable timer.

MICROPROCESSOR APPLICATION

Discussion

This is an exciting section. It brings together all of the digital elements you have studied up to now and introduces some new ones in an important application. Here you will see the underlying principles of a digital system that does automated troubleshooting.

Basic Idea

Troubleshooting can be divided into two major approaches: batch troubleshooting where the same type of system is analyzed and compared to some given standard; and single-system troubleshooting where different kinds of systems or infrequent analysis of the same system is analyzed and compared to some standard. This section concerns itself with batch troubleshooting. This type of troubleshooting is usually encountered on assembly lines, by distributors who service a particular product line, or by large service centers where the same types of systems are frequently encountered.

Comparing Things

Suppose you had the task of testing 100 AND gates that were just purchased by your company. More than likely, these gates come in a standard IC package. With the knowledge you already possess in digital techniques, it would be a waste of time to test each of these ICs by hand (by using a logic probe and digital pulser). The following material presents a method of using a **system comparison** for doing batch troubleshooting.

System Comparison:
A method of batch troubleshooting where the system under test is compared to an identical system called the standard.

With this system of troubleshooting, one of your major tasks is to compare the device under test to some standard. For example, if you were to troubleshoot an AND gate by hand, you would compare the level of its output to various combinations of its inputs and see if this was the same as the truth table for an AND gate. This same process can be done with far greater accuracy and speed by using a digital **magnitude comparator**.

Magnitude Comparator:
A circuit used to compare the magnitude of two quantities in order to determine their relationship.

There are two basic kinds of comparators: a digital comparator and an analog comparator. Both types of comparators will be presented in this section. As you would suspect, when troubleshooting digital systems, digital comparators are used. The analog comparator, presented later in this section, requires some knowledge of electronics and may be omitted without loss of continuity in the text.

Digital Comparators

Recall the operation of the XOR gates. Their function is illustrated again in Figure 8–86. As shown, these gates can be used to compare one bit with another. This ability lends itself to a practical automated gate tester.

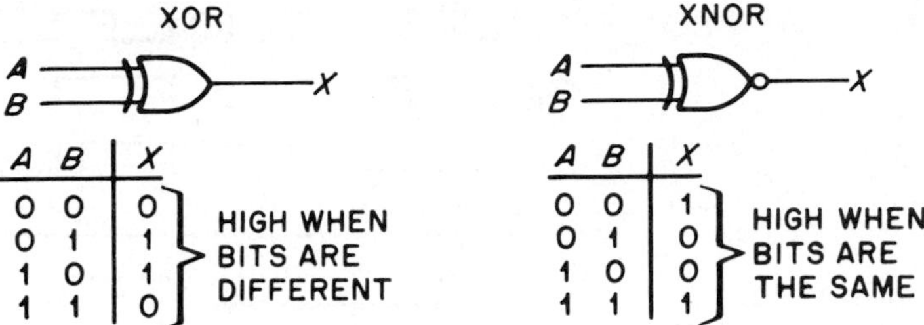

Figure 8–86 Operation of XOR and XNOR gates.

Automated Gate Tester

Consider the logic circuit shown in Figure 8–87. This system is comparing the logic output of one gate (called the *gate under test*) to another gate (called the *gate standard*). Essentially the binary counter is supplying all combinations of bit patterns to both AND gates, and the XOR gate is comparing the outputs of each gate. As long as the outputs of each AND gate are identical, the XOR gate output will be LOW, and the RESET input of the *test flip-flop* will not be activated. Thus, under these conditions, the green LED will remain ON. If, however, the output of the AND gates are not identical, the XOR gate output will go HIGH and cause the test flip-flop to be reset. This will turn on the red LED indicating that the test gate (G_T) does not match the gate standard (G_S). The red LED will remain ON until the test switch is again depressed. The buffers are used between the input leads of the gate under test and the gate standard to ensure electrical isolation between the inputs.

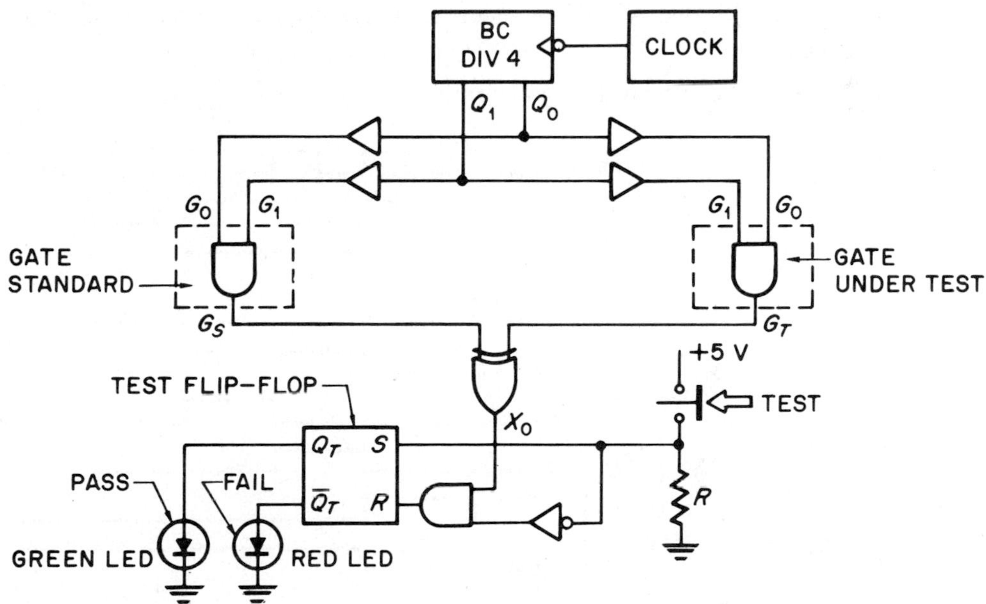

Figure 8–87 Automated logic circuit tester.

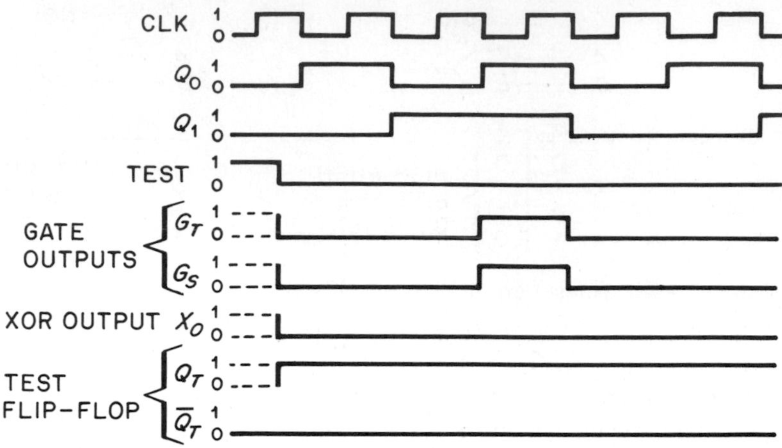

Figure 8–88 Automated tester timing diagram.

The basic operation of the system consists of inserting the gate to be tested, then tap the test button. This will cause the test flip-flop to be SET. If the clock frequency is 1 000 Hz, then the entire test time will take 4 ms. As previously stated, any HIGH output from the XOR gate will cause the flip-flop to be RESET ($Q \rightarrow 0$), and the red LED will light, turning off the green. Figure 8–88 shows the timing diagram for the system. The timing diagram assumes that both gates are identical.

Example 8–27

Sketch the timing diagram for the gate tester of Figure 8–87. Assume that the G_0 input lead to the AND gate under test is always stuck in the HIGH state.

Solution

Refer to the system timing diagram of Figure 8–89.

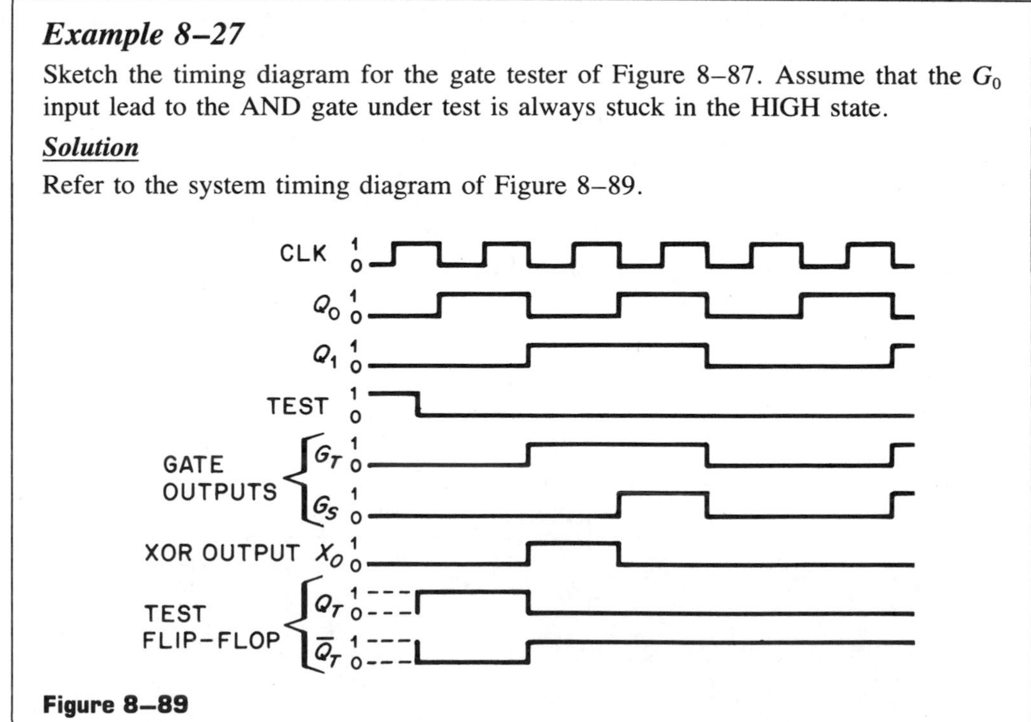

Figure 8–89

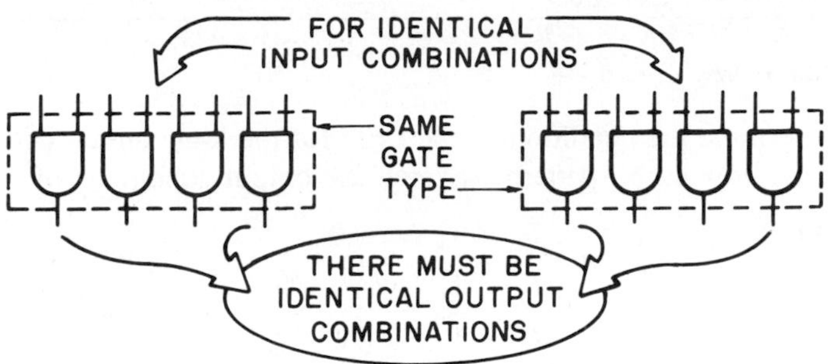

Figure 8–90 Concept of comparing multiple gate inputs.

Automated Testing of Other Gates

Any individual gate can be tested using the previous system as long as the gate standard is the same as the gate under test. For example, two OR gates can be compared in this manner, and so can NAND, NOR, XOR, and XNOR gates, including inverters. Individual flip-flops can also be tested by this method, though some circuit modification (such as the addition of a PLD) may be needed to accommodate all tests (this is left as an end-of-chapter exercise).

Multiple Inputs

The previous system is fine for testing only one gate at a time. However, practical ICs contain several gates—such as a quad two-input AND gate. Here there are a total of eight inputs and four outputs. If one gate is bad, for all practical purposes the entire IC should be rejected. Thus an automated troubleshooting system need not determine the bad gate or gates, only that an error did occur. This concept is shown in Figure 8–90.

Multiple-Input Digital Comparator

Consider the logic diagram of Figure 8–91. An understanding of how this works will lead you to the development of a real digital comparator.

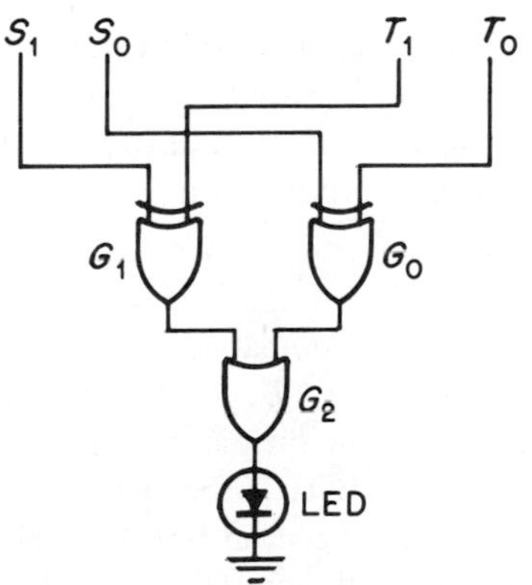

Figure 8–91 Logic diagram of comparison of two 2-bit states.

Example 8–28

Determine the condition of the LED for the logic circuit of Figure 8–91 for each of the following bit patterns. Sketch the output conditions of each XOR gate.

(A) $S_1 = 1, S_0 = 0; T_1 = 1, T_0 = 0$

(B) $S_1 = 1, S_0 = 1; T_1 = 1, T_0 = 0$

(C) $S_1 = 0, S_0 = 1; T_1 = 1, T_0 = 1$

Solution

Compare the inputs of each XOR gate. If they differ, then the gate output will be HIGH. See Figure 8–92.

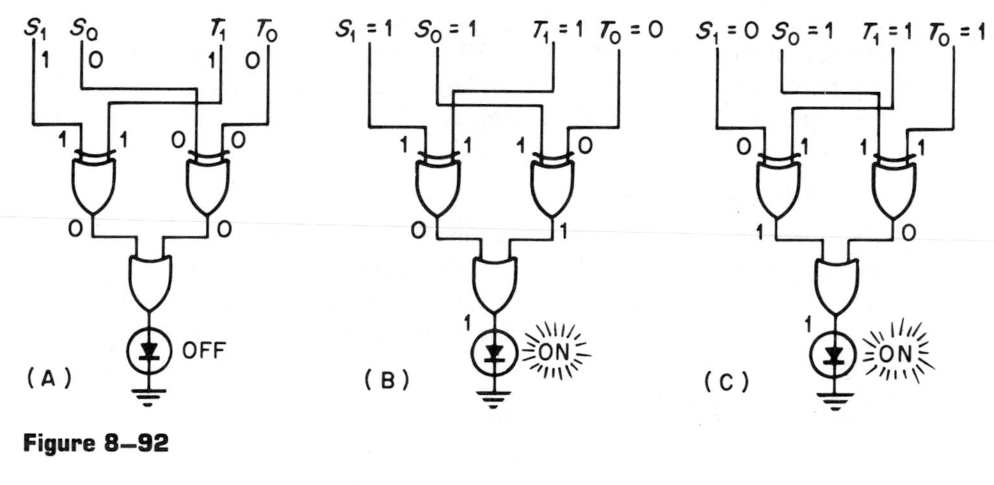

Figure 8–92

This concept can be extended so that the comparison of two 4-bit bit patterns can be made. Such an arrangement is shown in Figure 8–93. The 4-bit comparator has one output and two sets of 4-bit inputs. The output of this device will be HIGH whenever any of the bit patterns of the two sets are not equal.

A Real Comparator

The DM7485 4-bit magnitude comparator is shown in Figure 8–94. Observe that the DM7485 has three different outputs: $A < B$, $A = B$, and $A > B$. Thus not only will the output of the comparator be HIGH if the bit pattern of word A equals the bit pattern of word B, but the $A < B$ output will be HIGH if the binary value of the bit pattern for word A is less than that of B, and the $A > B$ output will be HIGH if the bit pattern for word A is greater than word B. The CASCADE inputs are used if more than one DM7485 is used—this would be done if you were comparing the bit pattern of two 8-bit words.

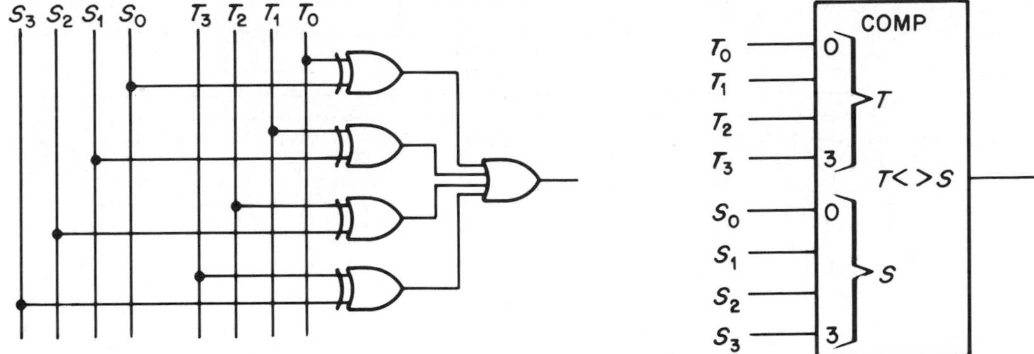

Figure 8–93 Diagram of a 4-bit comparator.

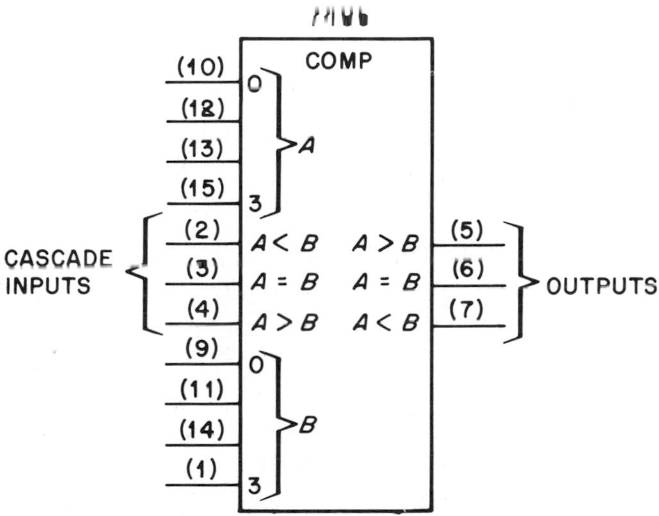

Figure 8–94 The DM7485 4-bit magnitude comparator.

A Practical Automated Digital IC Tester

Consider the automated digital IC tester of Figure 8–95. This tester will now compare two identical logic chips consisting of four gates with two inputs each. For the test setup shown, a known good quad two-input AND gate chip is being tested against an unknown one. A comparator is used, whose outputs are HIGH only when the bit patterns of each word are different. The practical DM7485 could be used for this with its $A = B$ output connected to an inverter. The operation of this system is the same as the previous one. Note that any failed test does not tell the user which gate is bad, or why, but simply that the IC under test did not compare well with the standard.

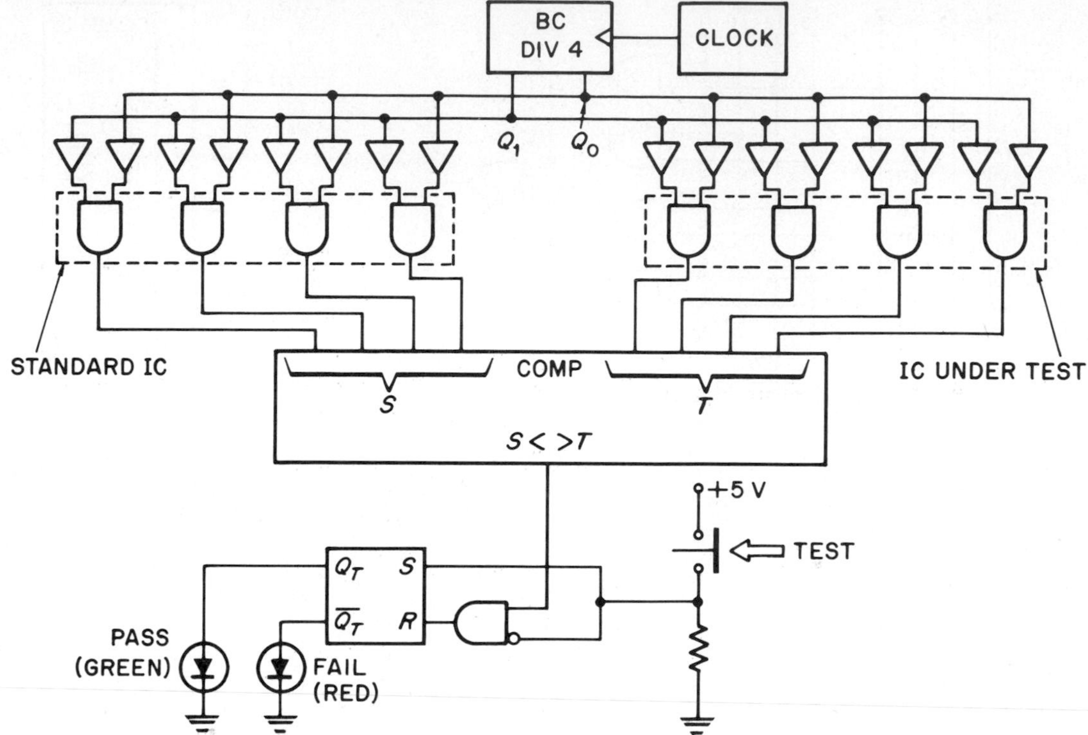

Figure 8–95 Practical digital IC tester.

Analog Testing

[Note: This portion contains some electronic analysis and may be omitted without loss of continuity.] Recall the comparator circuit in the 555 timer introduced in the last chapter. A close cousin of the comparator is the **difference amplifier**.

> **Difference Amplifier:**
> An amplifier whose output voltage is proportional to the difference of two input voltages.

The schematic of a difference amplifier is shown in Figure 8–96. The derivations for the relationship between the output voltage and input voltages may be found in any devices and circuit book or operational amplifier text, and is beyond the scope of this book. It is presented here without proof:

$$V_{\text{out}} = \frac{R_x}{R_y}(V_2 - V_1)$$

An analysis of this amplifier is presented in the following example.

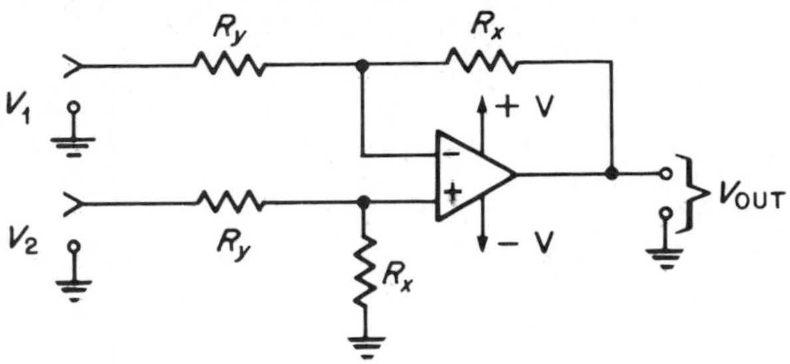

Figure 8–96 Schematic of a difference amplifier.

Example 8–29

For the difference amplifier of Figure 8–96, assume all resistors are 100 kΩ. Determine the output voltage of the difference amplifier for the following input voltages.

(A) $V_1 = 5$ V, $V_2 = 5.3$ V

(B) $V_1 = 3.8$ V, $V_2 = 2.7$ V

Solution

Using the formula for the difference amplifier, you have

$$\text{(A)}\quad V_{out} = \frac{R_x}{R_y}(V_2 - V_1)$$

$$V_{out} = \frac{100\text{ k}}{100\text{ k}}(5.3 - 5) = 1(0.3) = 0.3\ V$$

$$\text{(B)}\quad V_{out} = \frac{100\text{ k}}{100\text{ k}}(2.7 - 3.8) = 1(-1.1) = -1.1\ V$$

Automated Analog Troubleshooting

Consider the automated test circuit shown in Figure 8–97. This system utilizes two bilateral switches to compare the voltage readings of two voltage dividers. Like the digital system tester, this tester is comparing two circuits to see if they have identical characteristics. One is the standard, and the other is the one under test.

To operate this system, the technician steps through each test and observes the output of the meter. If the test circuit is identical to the standard, then the voltages into the inputs of the difference amplifier will be the same and its output will be zero. This will produce a zero reading on the meter. If, however, there is a difference between the two readings, the reading on the meter would change accordingly. The greater the voltage difference, the greater will be the reading on the meter. Thus, the technician must make a decision if the difference between the two circuits is acceptable.

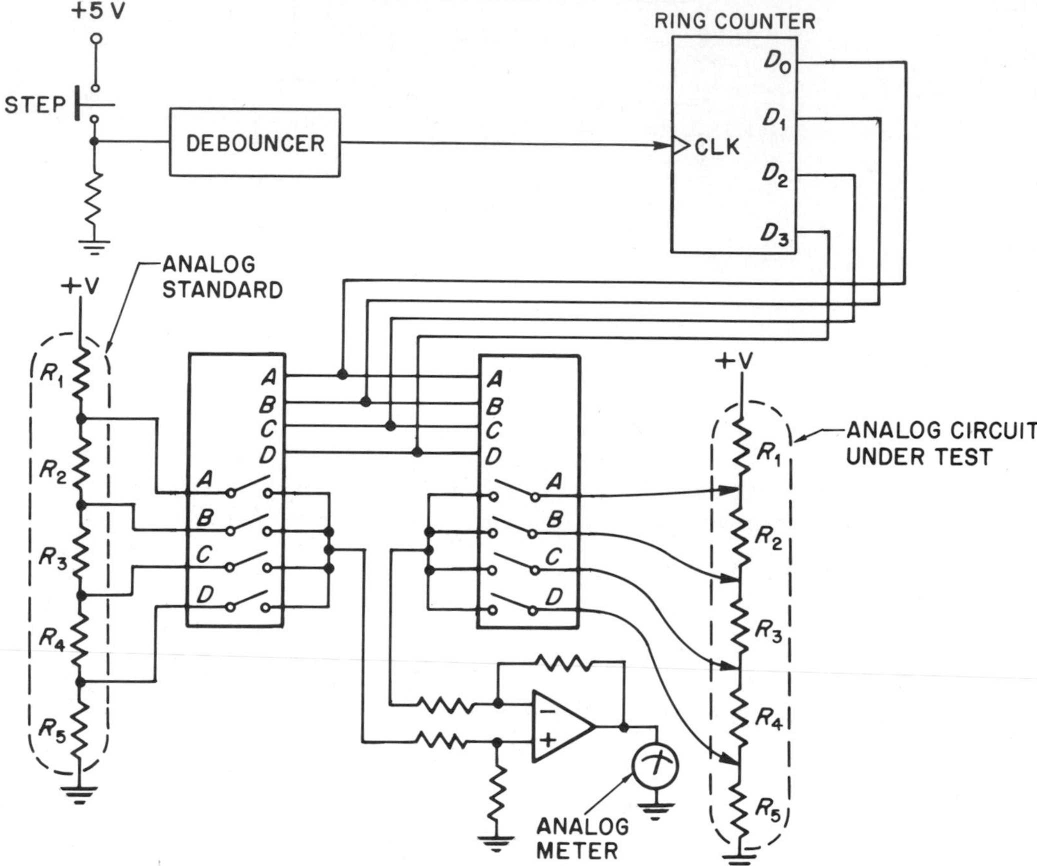

Figure 8–97 Semiautomated test circuit.

Automating the System

There are several methods for automating the above system. One method is to use comparators on the output of the difference amplifier. A brief review of the operation of an analog comparator is presented in Figure 8–98. Recall that a comparator was introduced in the last chapter. There you saw how it played a role in the operation of the 555 timer. Referring to Figure 8–98, you see that the output of the comparator will be one of two states, either HIGH or LOW. In both parts of the figure, the output state of the comparator is determined by the amount of input voltage (V_{in}) *compared* to a reference voltage (V_{Ref}).

Looking at Figure 8–98(A), this represents a *noninverting* comparator. This means that if V_{in} is more positive than V_{Ref}, the output will be HIGH. If V_{in} is less positive than V_{Ref}, the output will be LOW. Figure 8–98(B) shows an *inverting* comparator. This means that the condition of the output will be determined by how *negative* the reference voltage is compared to the standard. In this case, if V_{in} is more negative than V_{Ref}, the output will be HIGH; and, conversely, if V_{in} less negative than V_{Ref}, the output will be LOW. By

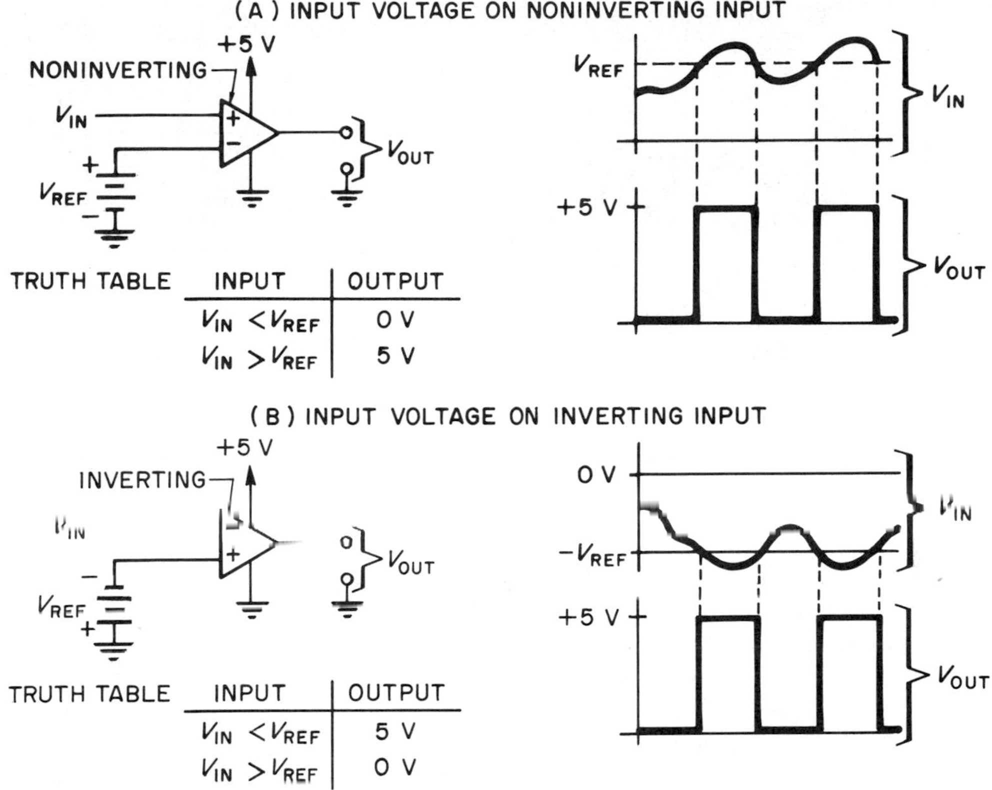

Figure 8–98 Operation of analog comparators.

combining both of these converters (the inverting and noninverting) a circuit may be created that will cause a HIGH output whenever an input voltage is *above* or *below* a specified voltage range. When this is done you have an analog-to-**digital margin detector**. See Figure 8–99.

Digital Margin Detector:

A circuit that produces a two-state output that depends upon a prescribed range of voltage inputs.

As you can see from the circuit in Figure 8–99, the heights of the margins are determined by the value of each reference voltage. When the analog input voltage goes above the + margin or below the − margin, the output of the gate will become HIGH.

The digital margin detector may now be combined with the difference amplifier to automate the analog tester. See Figure 8–100. What happens in this system, is that the DIFFERENCE AMPLIFIER will have an output equal to the voltage differences between the STANDARD and the TEST CIRCUIT. This voltage difference is fed into the DIGITAL MARGIN DETECTOR. The output of the DIGITAL MARGIN DETECTOR will be either HIGH or LOW effecting the state of the *S-R* flip-flop. If the DIFFERENCE

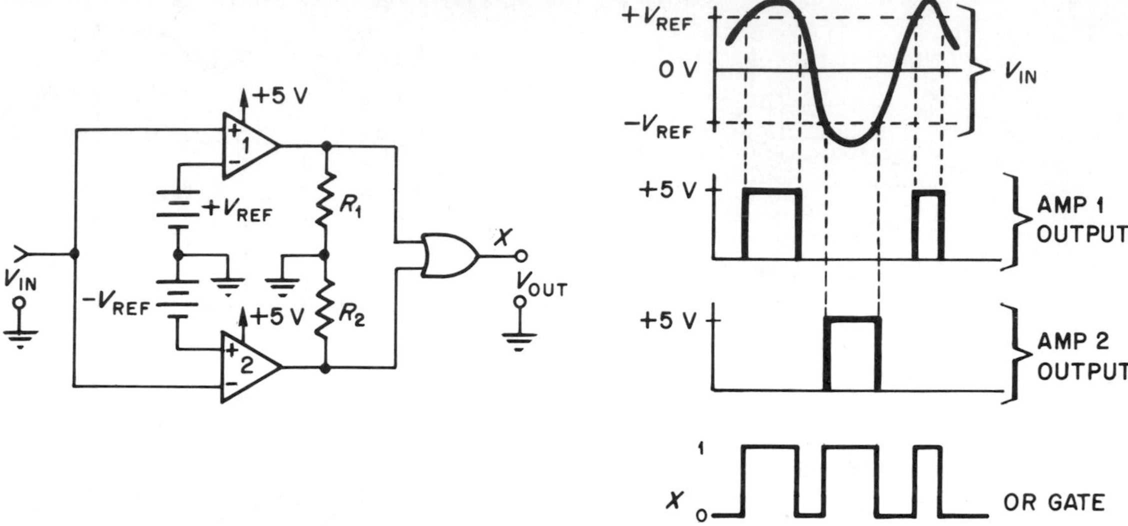

Figure 8–99 Digital margin detector and operation.

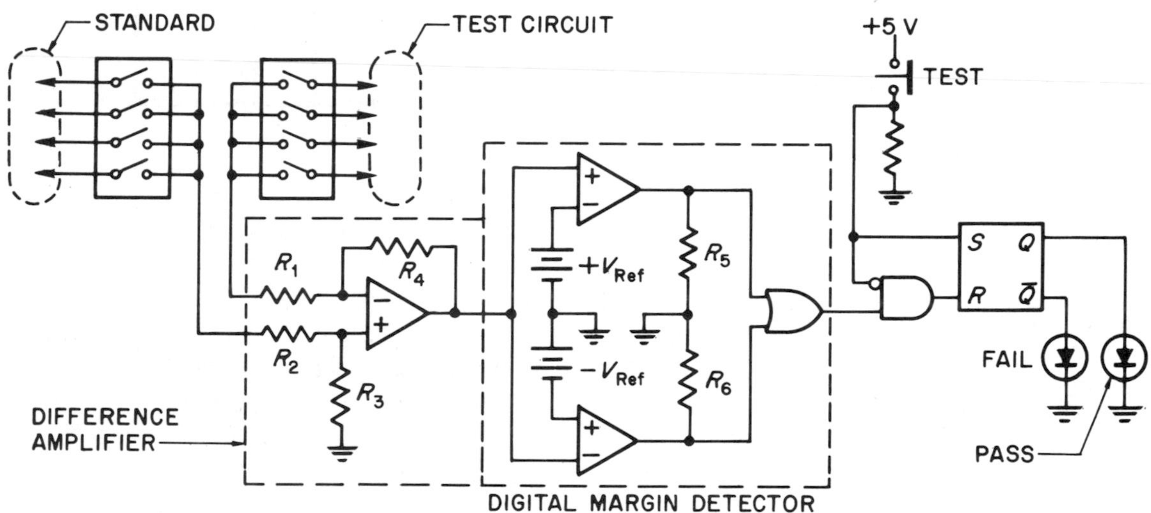

Figure 8–100 Combining difference amplifier with digital margin detector.

AMPLIFIER output is within the range of $+V_{Ref}$ and $-V_{Ref}$ then the output of the DIGITAL MARGIN DETECTOR will be LOW and the *S-R* flip-flop will remain SET (causing the PASS LED to stay ON). If, however, the output of the DIFFERENCE AMPLIFIER goes above $+V_{Ref}$, the output of the DIGITAL MARGIN DETECTOR will cause the *S-R* flip-flop to RESET and the FAIL LED will activate. This indicates that the voltage difference between the STANDARD and the TEST CIRCUIT was greater then allowed by the value of $+V_{Ref}$. In a like manner, if the ouput voltage of the DIFFERENCE AMPLIFIER is more negative than $-V_{Ref}$, then the output of the DIGITAL MARGIN DETECTOR will again be HIGH causing the *S-R* flip-flop to RESET and again the FAIL LED would activate.

The important point from all this is that the amount of variation allowed between the STANDARD and the TEST CIRCUIT may be set by adjusting the values of $+V_{Ref}$ and $-V_{Ref}$.

Completed System

Figure 8–101 shows the completed automated analog circuit tester. Note that this system has the ability to change the reference voltages from a PLD. Essentially, the system is comparing the DC voltage readings of two transistor amplifiers. This system is similar to

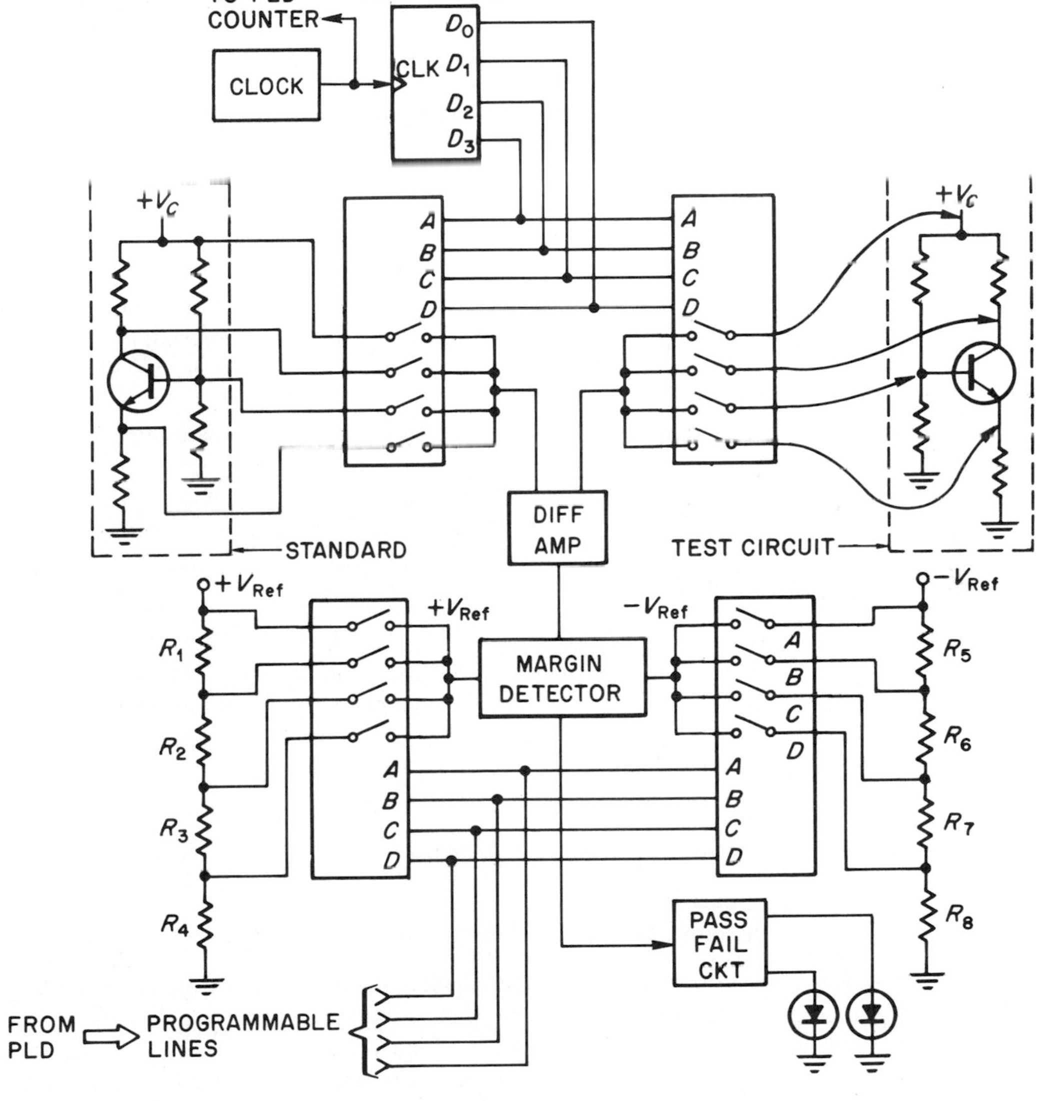

Figure 8–101 Completed automated analog/digital circuit tester.

the one in Figure 8–100. The main difference is that $+V_{Ref}$ and $-V_{Ref}$ are now programmable. This means that during the course of testing an analog circuit, the allowed margin for error may be changed. There is a reason for wanting to do this. In this figure, two transistor-amplifier circuits are being compared. One is the standard, the other the one being tested by being compared to the standard. When this is done, some voltages (such as the transistor collector voltage) may be allowed to have a voltage difference from the standard of + or −1 volt. In this case, $+V_{Ref} = +1$ V and $-V_{Ref} = -1$ V. However, for other voltage comparisons, the voltage range may be much smaller (such as comparing the base voltages). In this case the variation allowed may be no more than + or − 0.1 volt. In this case $+V_{Ref} = +0.1$ V and $-V_{Ref} = -0.1$ V. Thus, the PLD controlling the program lines of the bilateral switches for the reference voltages must reflect these requirements.

Conclusion

This section presented methods of automated troubleshooting utilizing some of the new information presented in this chapter. You will find that the techniques presented here are easily applied to microprocessor technology. Test your understanding of this section by trying the following review questions.

8–8 Review Questions

1. Name the two major approaches to troubleshooting. Describe the difference.
2. What is meant by *system comparison* as applied to troubleshooting?
3. Explain what is meant by a *magnitude comparator*. What are the kinds used for digital and analog comparisons?
4. Describe the fundamental gate used in a digital magnitude comparator, and an analog comparator.
5. Describe a *digital margin detector*.

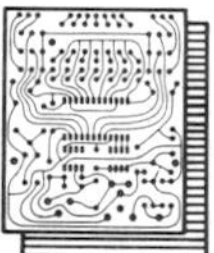

ELECTRONIC APPLICATION

Discussion

In the last section, you were introduced to a digital system that could perform batch troubleshooting. In this section, you will see how the digital building blocks presented up to this point are used to create powerful digital systems that are capable of responding to their external environment.

Frequency Counter

Fundamental to most digital instrumentation is the **frequency counter**.

Frequency Counter:
An instrument used to display the frequency of a periodic waveform.

The basic idea of a frequency counter is shown in Figure 8–102. As shown, the frequency counter consists of a one-shot, a binary counter, an AND gate and a readout system, such as a PLD, that converts a binary value into decimal for the seven-segment readouts. The ON time of the one-shot can be changed by the time base selector switch. As an example of the operation of this system, assume that the time base is set for 1 s.

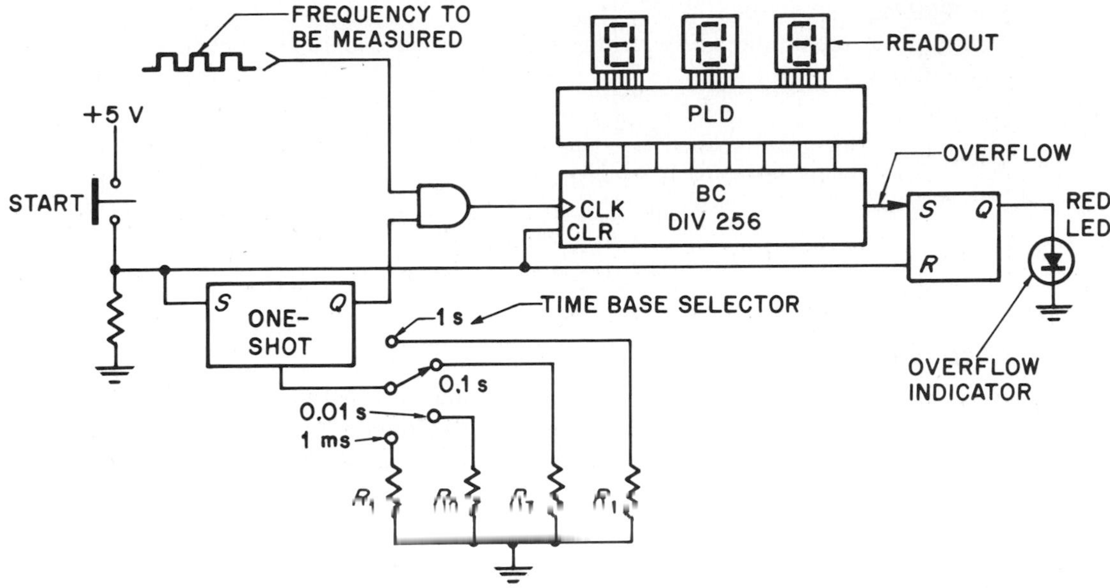

Figure 8–102 Basic frequency counter.

Assume that the frequency of the incoming signal is 10 Hz. When the START button is depressed, the counter is cleared to 0's; and when the button is released, the one-shot output goes HIGH for 1 s. This being the case, the binary counter will remain active for 1 s. Thus, if the input frequency is 10 Hz, the counter will have a binary count of 10 and the seven-segment readout will indicate a frequency of 10 Hz.

The overflow indicator is used to indicate that the frequency of the incoming signal was greater than the capacity of the binary counter. When this happens, the time base selector switch must be changed to cause the one-shot to be on for a shorter time.

The actual frequency being measured is related to the time base and count by

$$f = \frac{\text{Count}}{T}$$

where

f = frequency in Hz (cycles per second)
T = time base in seconds
Count = count of the binary counter

Example 8–30

The frequency counter shown in Figure 8–102 has the time base selector switch set to 10 ms. The START button is then depressed, and the seven-segment readouts display 23. From this, determine the frequency of the input waveform.

Solution

Using the relationship between the time base and count, you obtain the resulting frequency:

$$f = \frac{\text{Count}}{T}$$

$$f = \frac{23}{10 \text{ ms}} = \frac{23}{10 \times 10^{-3}} = 2.3 \times 10^3$$

$$f = 2.3 \text{ kHz}$$

Digital Voltmeter

A *voltmeter* is an instrument used to measure the electrical quantity of voltage. As you will see, the major building block of the **digital voltmeter** is the frequency counter. The major addition is that of a *voltage controlled oscillator* (**VCO**).

Digital Voltmeter:
An instrument used to measure voltage with a digital readout.

VCO:
Voltage controlled oscillator. An oscillator whose output frequency is determined by the value of an input voltage.

A 555 timer can be used as a VCO as shown in Figure 8–103. Recall from the last chapter, the 555 timer was introduced. When this timer was in the free-running mode, its

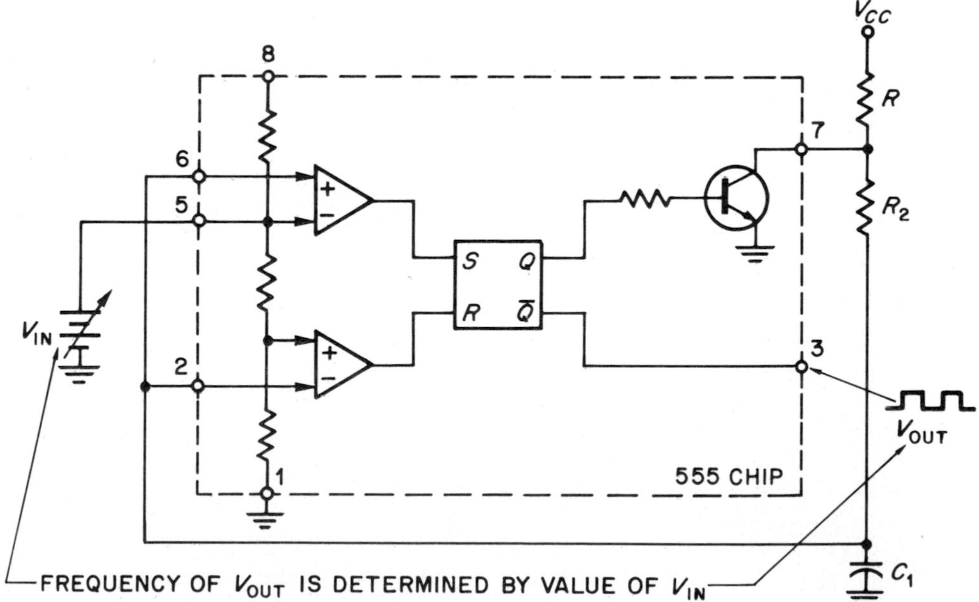

Figure 8–103 The 555 timer connected for VCO operation.

frequency was determined by the values of external resistors and a capacitor. The key to this operation was the charging of the capacitor (C_1). Note from the figure, that the capacitor is connected to pin 6 of the 555. This is the input to a comparator. It is the output of this comparator that determines when the *S-R* flip-flop will be SET, causing the capacitor to discharge. If the reference voltage to this comparator is now made variable (as with the addition of V_{in}) when this discharging takes place will depend upon the value of V_{in}. Thus, the frequency of this device will not only be determined by the values of the external resistors and capacitor, but also by the value of V_{in}. The smaller the input voltage, the higher the output frequency of the VCO. The important relationship here is that the frequency counter can now be used to measure the output frequency of the VCO. Thus, the count of the binary counter will be proportional to the input voltage of the VCO. This concept is shown in Figure 8–104. Note the use of a clock to automatically sample the readings. Thus, the user does not have to keep pressing a START button to take samples of the readings.

The PLD is programmed so that a smaller binary count indicates a larger voltage reading. This occurs because the larger the input voltage, the lower will be the resulting frequency from the VCO.

Digital Thermometer

The 555 timer can also be used in a digital thermometer. For example, when the 555 timer is used as a regular oscillator, its output frequency will depend upon the value of external resistors and a capacitor. If any of these values are changed, then the 555 timer output frequency will also change. Thus, if a temperature-sensitive device such as a **thermistor** is used as one of the frequency-determining components, then the frequency of the 555 timer will be a function of the ambient temperature.

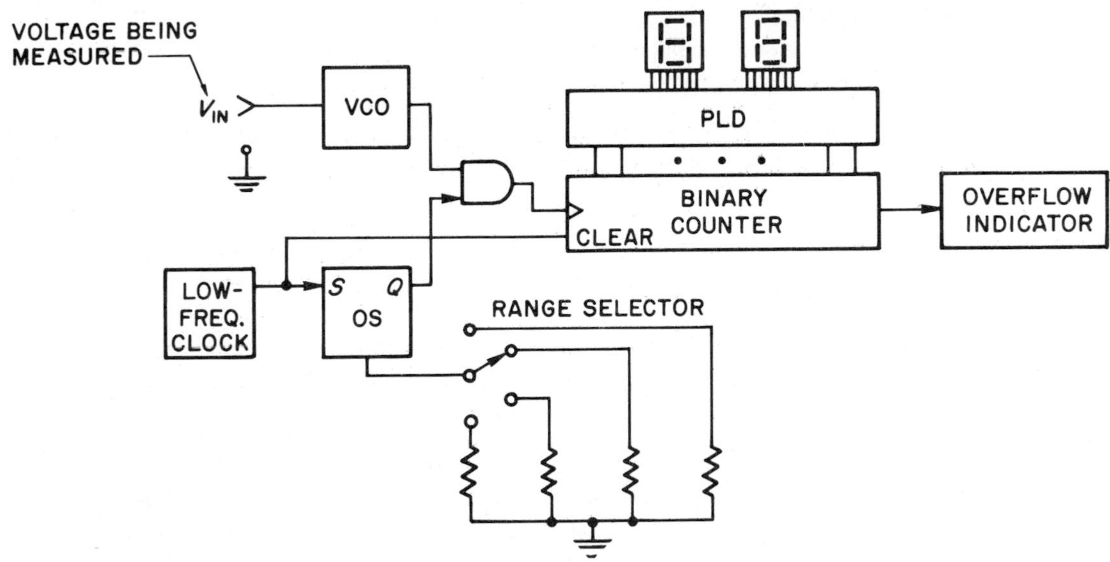

Figure 8–104 Basic digital voltmeter.

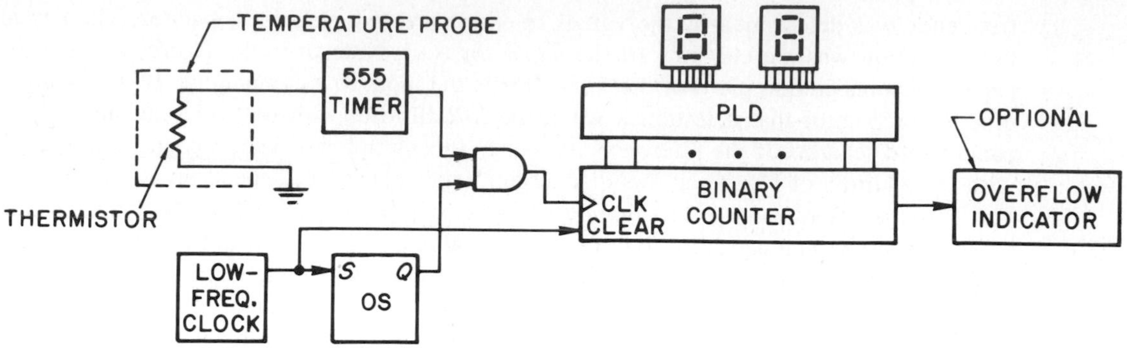

Figure 8–105 Logic construction of a digital thermometer.

Thermistor:
A device whose resistance value is related to its temperature and is developed for this particular application.

The logical construction of a digital thermometer is shown in Figure 8–105. Again, the concept of using a frequency counter is used in another application—a digital thermometer.

Resistive Devices

This principle of a changing resistance to an environmental change to control the output frequency of the 555 timer (or similar device) is fundamental to many different types of digital instruments. For example, a **photoresistor** can be used to construct a digital light meter using the same principle of a 555 timer and digital frequency counter.

Photoresistor:
A device whose resistance value is sensitive to the amount of light and is intended to be used for this affect.

Figure 8–106 illustrates several types of possible digital instruments where the resistance of a device is affected by some property of the environment.

Digital Sound Control

Consider the system shown in Figure 8–107. The system is activated by the microphone. The input frequency will determine the count of the binary counter. The output of the binary counter will then select an address from the ROM. The data stored at this address can represent an instruction to a microprocessor or directly control a small process itself. Thus the tone and duration of the signal can be used to remotely control the operation of another system.

Conclusion

This section presented many of the basic concepts used in digital instrumentation systems. In the last two sections of this chapter, you have covered many of the basic concepts that have revolutionized the electronics industry. Check your understanding of this section by trying the following review questions.

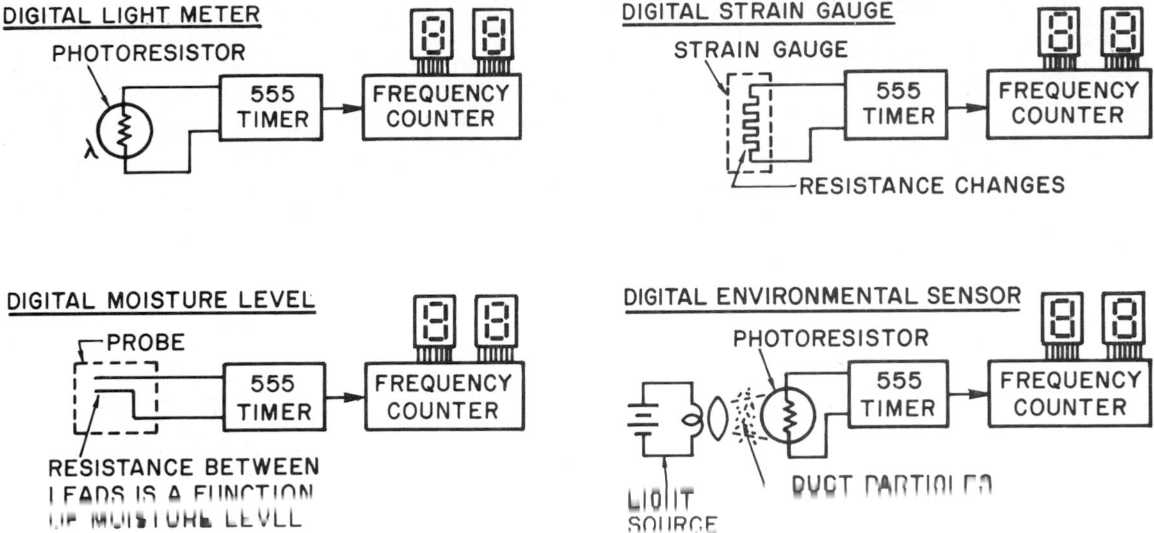

Figure 8–106 Various digital instruments utilizing the same principle.

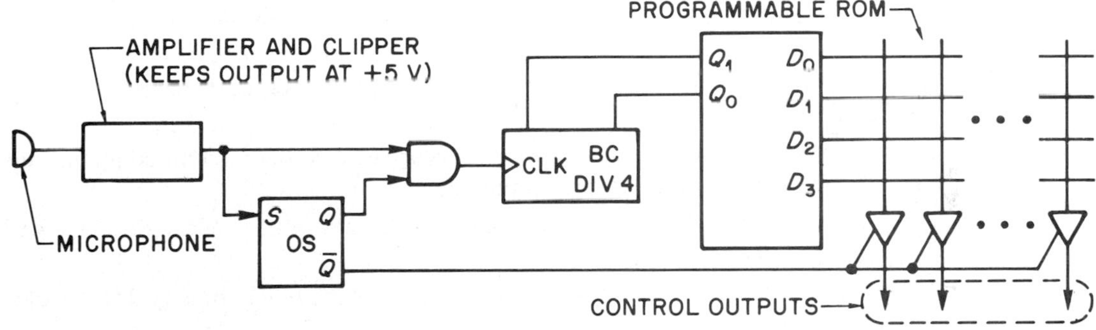

Figure 8–107 Digital sound control system.

8–9 Review Questions

1. Describe the basic operation of a digital frequency counter.
2. What is a VCO? Explain its function.
3. Describe the basic operation of a digital voltmeter.
4. Explain the basic operation of a digital thermometer.
5. State what a digital light meter, digital strain gauge, digital moisture detector, and digital sound detector have in common.

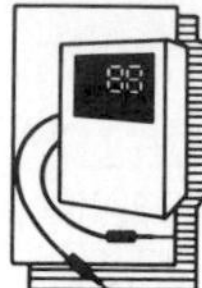

TROUBLESHOOTING SIMULATION

The troubleshooting simulation for this chapter simulates the troubleshooting of a real *JK* flip-flop. Here, you will have an opportunity to analyze the *JK* flip-flop. The program will allow you to test its condition without a hardware failure. Then, when you are ready, the computer will generate a hardware problem for you to troubleshoot. This is an important simulation program. Developing troubleshooting skills in this area serves as a building block for the remaining troubleshooting simulations for this book.

SUMMARY

- The *JK* flip-flop has one more state than the *S-R* flip-flop. This is the toggle state when both inputs are active.
- The *JK* flip-flop has four useful input conditions: no change, set, reset, and toggle.
- In the toggle mode the output waveform of a *JK* flip-flop is one half the frequency of the input clock.
- One of the simplest binary counters consists of a series of *JK* flip-flops all in the toggle mode, where the output of one is connected to the clock input of the next.
- An asynchronous counter is a counter that does not have all of its inputs synchronized directly by the clock.
- The mod number of a counter is the number of states a counter will have under a given condition.
- A programmable counter is a counter that can have its mod number controlled by a changeable bit pattern.
- The maximum counting frequency of an asynchronous counter is limited by the delay time of each flip-flop that makes up the counter.
- The time delays inherent in logic circuits can cause a brief, undesirable change in logic called a *decoding glitch*.
- A synchronous counter has each of its flip-flops controlled directly by the clock pulse.
- A presettable counter may have a binary value loaded into it before it begins a count.
- A digital counter that has the same bit pattern continuously circulating through it is called a *ring counter*.
- Combining a counter with a PLD produces a counter of great flexibility.
- A logic circuit exhibiting a predictable sequence of events is called a *state machine*.
- The list of sequential states of a state machine is called a *state table*.
- A digital system consisting of several small processes simulates the major concept of a microprocessor.
- A microprogram is a permanent instruction set inside a microprocessor used to perform a specific process.
- There are basically four different kinds of registers: parallel in–parallel out, serial in–serial out, parallel in–serial out, and serial in–parallel out.

- A register where the contents may be moved in a serial fashion within the register is called a serial register.
- Registers with a combination of serial in–parallel out and parallel in–serial out are utilized in computer interfacing.
- Combining registers with PLDs produces a system that allows the contents of different registers to be transferred between each other.
- The length of time required to complete a process is called a *machine cycle*.
- The instruction built into a microprocessor that determines a specific selectible process is called a *microcode*.
- A single instruction in the microcode of a microprocessor is called a *microinstruction*.
- Real hardware contains many extra features that may be used by the digital system designer and thus require familiarity on the part of the technician.
- Many IC devices utilize built-in counters for the overall operation of the device.
- Troubleshooting can be divided into two major approaches: batch troubleshooting and single-system troubleshooting.
- One method of automated troubleshooting is to compare the circuit under test to a known good circuit.
- Two major kinds of comparators are digital and analog.
- The fundamental building block of the digital comparator is the XOR gate. Its analog equivalent is the analog subtractor.
- An instrument used to display the frequency of a periodic waveform is called a *frequency counter*.
- The digital frequency counter is commonly found in most applications of digital instrumentation.
- The voltage controlled oscillator (VCO) produces an output frequency proportional to the value of an input voltage.
- The VCO is commonly found in digital instrumentation used to measure environmental conditions.

CHAPTER SELF-TEST

I. TRUE/FALSE

Answer the following questions true or false.

1. The *S-R* flip-flop has three useful states while the *JK* flip-flop has four.
2. When both inputs to a *JK* flip-flop are active, the flip-flop will be in the *toggle mode*.
3. The input clock frequency to a *JK* flip-flop in the toggle mode is half of the output frequency.
4. A binary counter may count forward but not backward.
5. A mod-5 counter means that its maximum count will have a maximum value of 5.

II. MULTIPLE CHOICE

Answer the following questions by selecting the most correct answer.

6. A counter where each flip-flop is not controlled directly by the clock is called
   (A) a synchronous counter.
   (B) an asynchronous counter.
   (C) a programmable counter.
   (D) a mod counter.
7. A counter where each flip-flop is controlled directly by the clock is called
   (A) a synchronous counter.
   (B) an asynchronous counter.
   (C) a programmable counter.
   (D) a mod counter.
8. The digital counter that contains a given bit pattern that continually circulates through the counter is called
   (A) a pattern counter.
   (B) an up/down counter.
   (C) a ring counter.
   (D) none of the above.
9. A counter whose starting value may be changed is called a
   (A) changeable counter.
   (B) presettable counter.
   (C) loadable counter.
   (D) preset counter.
10. A logic circuit that exhibits a predictable sequence of events is called a
    (A) microprocessor.
    (B) sequencer.
    (C) state machine.
    (D) programmable sequencer.

III. MATCHING

Match the flip-flop conditions in Figure 8–108 to the following statements.

11. $Q \rightarrow$ HIGH
12. Q toggles
13. $Q \rightarrow$ LOW
14. No change in Q
15. Not allowed

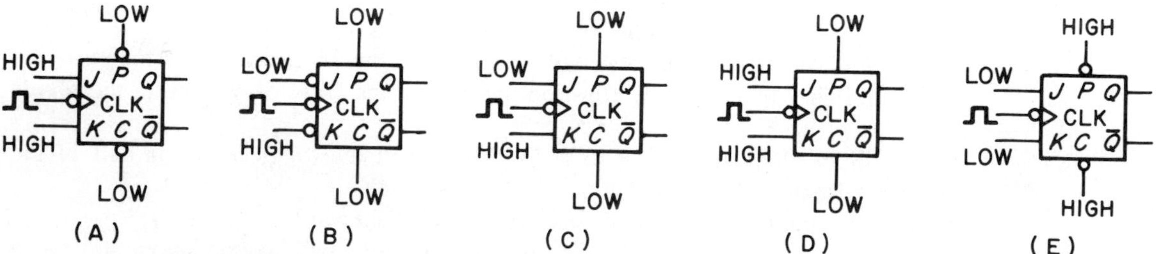

Figure 8–108

IV. FILL-IN

Fill in the blanks with the most correct answer(s).

16. A list of the sequential states of a state machine is called a ______ table.
17. The permanent instruction set inside the microcprocessor used to perform a specific process is called a ______.
18. A counter with a ______ mod number can be achieved by using a binary counter with a PLD.
19. A parallel-in–parallel-out register is sometimes referred to as a ______ register.
20. A register where the contents may be moved ______ is called a shift register.

V. OPEN-ENDED

Answer the following questions as indicated.

21. Sketch the logic diagram of a serial-in–parallel-out register.
22. Draw the logic diagram of a 2-bit serial transmitter/receiver that is parallel loaded and read.
23. Explain what the double dashed lines of a timing diagram indicate.
24. Sketch the logic diagram of an automated logic circuit tester.
25. Sketch the logic diagram of a digital voltmeter. Explain the purpose of each section.

Answers to Chapter Self-Test

1] T 2] T 3] F 4] F 5] F 6] B 7] A 8] C 9] B 10] C 11] B 12] D 13] C 14] E 15] A 16] state 17] microprogram 18] programmable 19] buffer 20] serially 21] Refer to Figure 8–53. 22] Refer to Figure 8–63. 23] A *don't care* condition. 24] Refer to Figure 8–87. 25] See Figure 8–104.

CHAPTER PROBLEMS

Basic Concepts

Section 8–1:

1. Explain the main difference between an *S-R* flip-flop and a *JK* flip-flop.
2. Construct the truth table for a *JK* flip-flop.
3. Sketch at least eight variations of a *JK* flip-flop.
4. For the input waveforms of Figure 8–109(A), determine the resulting outputs, assuming they are applied to a *JK* flip-flop.

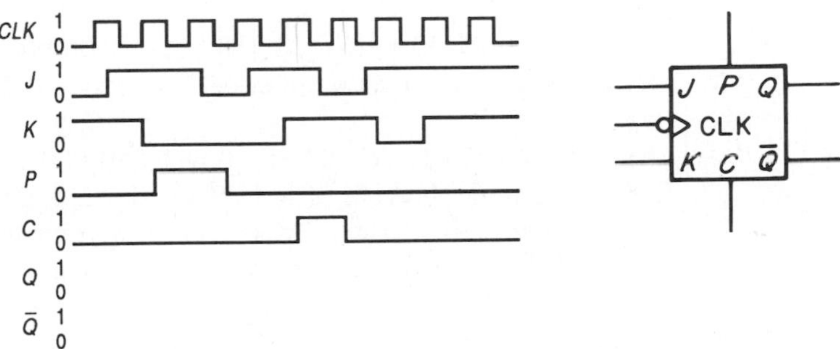

Figure 8–109(A)

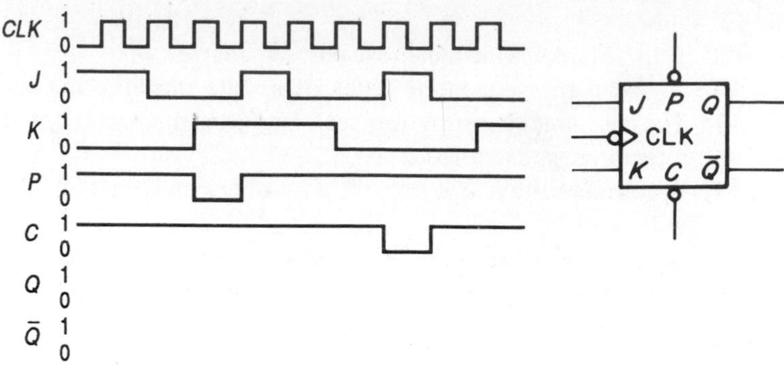

Figure 8–109(B)

5. If the input waveforms of Figure 8–109(B) were applied to a *JK* flip-flop, determine the resulting outputs.
6. Explain what is meant by frequency division when using a *JK* flip-flop. Give an example.
7. Sketch the resulting output waveforms for each of two *JK* flip-flops where the *Q* output of the first is connected to the clock input of the second.
8. For the flip-flops of problem 7, state the frequency relationships between the output of each flip-flop and the input clock pulse.

Section 8–2:

9. Explain the concept of a binary count from the standpoint of frequency division.
10. Sketch the output waveforms for each *JK* flip-flop of a 4-bit binary counter for each of the *Q* outputs.
11. For the binary counter in problem 10, sketch the output of each of the $\overline{Q}$ outputs.
12. (A) Explain the action of a *ripple counter*. (B) What is meant by an asynchronous counter?
13. Using the appropriate number of flip-flops, sketch a mod-8 counter.
14. Sketch a mod-32 counter using the appropriate number of flip-flops.
15. What are the minimum number of flip-flops required in order to have the following number of counts?
    (A) 1 000 (B) 333 (C) 12 169 (D) 1 000 000
16. Determine the minimum number of flip-flops in order to have the following number of counts.
    (A) 555 (B) 210 (C) 64 K (D) 1 500 000
17. Sketch the logic diagram of a 4-bit up/down counter.
18. Using logic gates, design a mod-10 (decade) counter.
19. Construct the timing diagram for the counter in problem 18.

20. Develop a *programmable counter* that will produce a mod-12 count.
21. Explain the major problem with an *asynchronous counter*.
22. Sketch the timing diagram of a 4-bit asynchronous counter if the delay in each flip-flop is 35 ns. (Do this for a full count.)
23. (A) What is a *decoding glitch*? (B) Decode the counter of problem 22 and show the occurrences of decoding glitches.
24. Calculate the highest clock frequency for the counter of problem 22.
25. Determine the highest clock frequency that can be used for an 8-bit binary counter if the delay time for each flip-flop is (A) 55 ns, (B) 10 ns.

Section 8–3:

26. Sketch the logic diagram of a 4-bit synchronous counter.
27. Define *synchronous* counter.
28. Sketch the logic diagram of an 8-bit *presettable* synchronous counter. Show the resulting timing diagram, including loading the counter with a given bit pattern.
29. For the counter of problem 28, show how you would add a feature to *clear* the counter to all 0's.
30. Sketch an 8-bit *ring* counter. Show the timing diagram of such a counter with one flip-flop HIGH for eight clock pulses.
31. Sketch a 4-bit ring counter that will start its count with the word 1010. Sketch the timing diagram.

Section 8–4:

32. Sketch a 4-bit binary counter connected to a switch-programmable ROM. Show the bit pattern you would use in order to implement a gray scale upcount for the 4-bit counter.
33. For the counter of problem 32, show the bit pattern you would use to cause the counter to count forward from 0000 to 1010 and back to 0101; then repeat within the range of 1010, 0101.
34. Using the counter of problem 32, sketch the bit pattern you would use to create a gray scale downcount.
35. Create a *state machine* that would give the user the option of getting a 4-bit gray scale of binary upcount.
36. Modify the state machine of problem 35 so that a user could have the option of getting a 4-bit gray scale or binary downcount.
37. Sketch the *state table* for the state machine of problem 35.
38. Develop the state table for the machine of problem 36.
39. Create a *programmable readout* using a state machine and a seven-segment LED that would continuously produce the letter sequence –HELLO–
40. Modify the state machine of problem 39 so it will continuously produce the letter sequence –DABBLE–
41. Explain what is meant by a *microprogram*. Give an example.

Section 8–5:

42. Describe the logic construction of a 4-bit parallel-in–parallel-out register.
43. Determine the output conditions of the 4-bit register for the input waveforms of Figure 8–110(A).

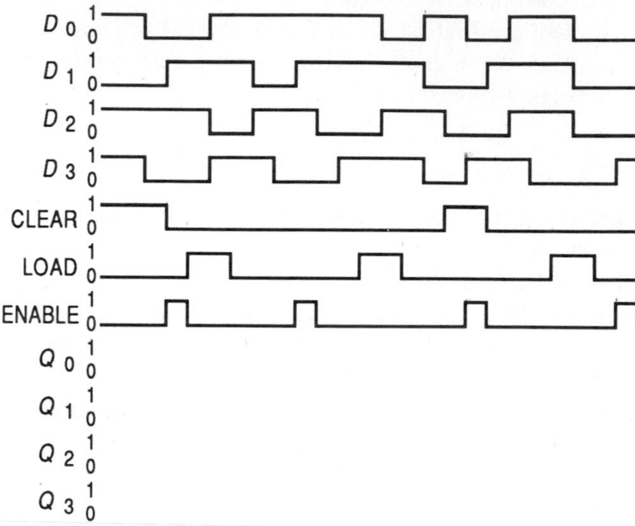

Figure 8–110(A)

44. The waveforms of Figure 8–110(B) are applied to the control inputs of a 4-bit register. Determine the resulting outputs.

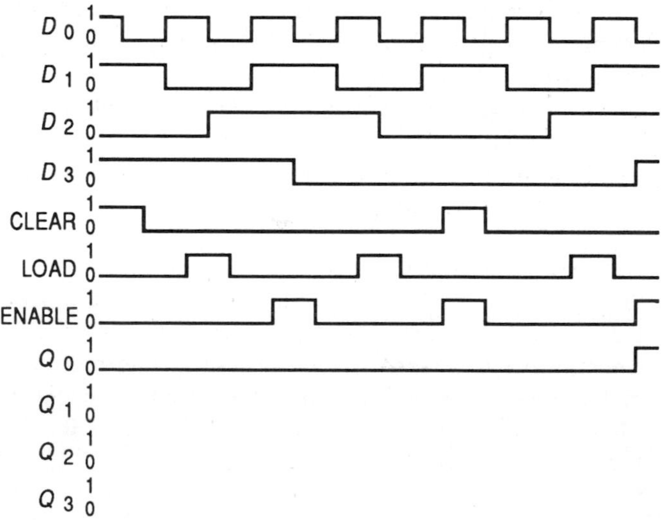

Figure 8–110(B)

45. Determine the output waveforms of a 4-bit shift register for the input waveforms shown in Figure 8–110(C).

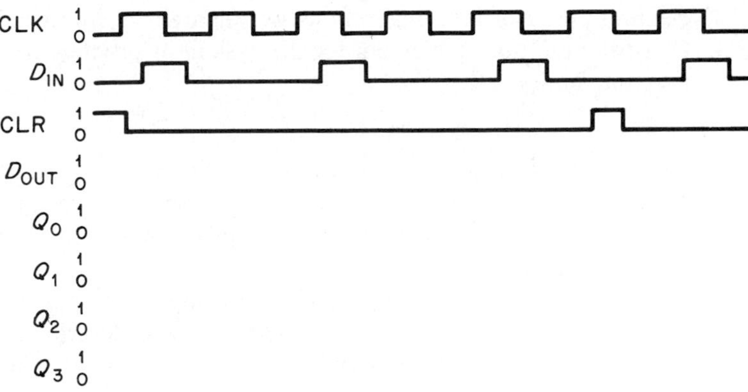

Figure 8–110(C)

46. For the input waveforms of Figure 8–110(D), determine the output waveforms for a 4-bit shift register.

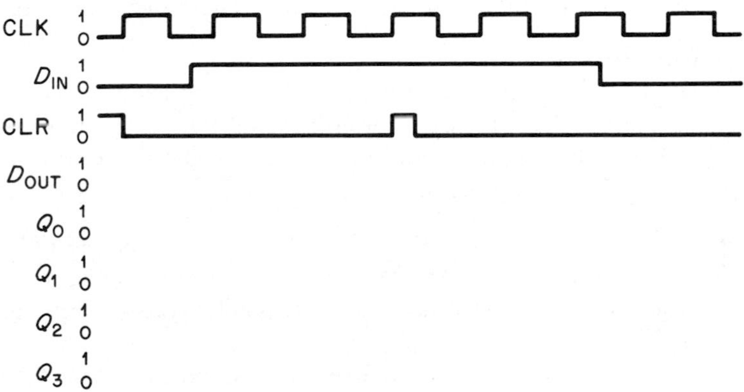

Figure 8–110(D)

47. Show the logic diagram of an 8-bit parallel-in–serial-out register.
48. Construct the timing diagram for the register of problem 47. (Show how a given word may be loaded.)
49. Show the logic diagram of an 8-bit serial-in–parallel-out register.
50. Construct the timing diagram for the register of problem 49. (Include the action of an output enable line.)

Section 8–6:

51. Illustrate how the inputs and outputs of two 4-bit registers may be connected to a common bus. (Label one register "A," the other register "B.")

52. Construct a timing diagram showing data being transferred from register "A" to register "B" as in problem 51.
53. Expand the system in problem 51 to include a PLD that will allow the option of clearing either or both registers as well as transferring data between them.
54. Construct the timing diagram for the system of problem 53. Indicate the length of each *machine cycle*.

Section 8–7:

[Note: The questions for this section require access to an IC data book.]

55. Using your IC data book, indicate the different *JK* flip-flops that are available.
56. For the *JK* flip-flops you indicated in problem 55, select two and describe their major differences.
57. Indicate the different types of registers available in your IC data book.
58. Select two of the registers from problem 57 and describe their major differences.
59. List the different types of counters available in your IC data book.
60. Take two of the registers from problem 59 and indicate their major differences.

Applications

61. (A) Explain the operation of a digital comparator. (B) Show its logic diagram.
62. (A) Explain the operation of an analog comparator. (B) Show its schematic diagram.
63. Devise a logic system that will automatically test a dual four-input NAND gate.
64. Show the timing relationships of the system of problem 63.

Instrumentation

65. Explain the operation of a digital frequency counter.
66. Construct the logic diagram of a digital frequency counter.
67. For the digital frequency counter of problem 66, create the timing diagram for a waveform of a given frequency.

Analysis and Design

68. Design a digital system that will test any of the standard TTL IC gates. This means that the system must accommodate quad two-input gates as well as an eight input NOR gate.
69. Design an automated tester that will test the condition of a single *JK* flip-flop with PRESET and CLEAR inputs. (Note: Make sure PRESET and CLEAR are not active at the same time.)
70. Modify the system of problem 69 so that it can automatically test a standard 4-bit register selected from an IC data book.

CHAPTER 9

Memories and Conversion

OBJECTIVES

After completing this chapter, you should be able to:

- ☐ Understand fundamental concepts common to all memory systems.
- ☐ Identify the different categories of digital memory.
- ☐ Identify the different classifications of digital memory.
- ☐ Apply memory terminology to various kinds of memory.
- ☐ Describe techniques used in electrical memory.
- ☐ Describe techniques used in magnetic memory.
- ☐ Describe techniques used in optical memory.
- ☐ Describe other memory techniques.
- ☐ Explain the operation of analog-to-digital converters.
- ☐ Explain the operation of digital-to-analog converters.
- ☐ Describe methods of connecting electrical memories and microprocessors.
- ☐ Read and interpret timing diagrams.
- ☐ Explain the principles of speech synthesizers.
- ☐ Describe methods of digital storage and retrieval of analog signals.

INTRODUCTION

This chapter presents the world as viewed by the microprocessor. Here you will discover the many different ways of exchanging digital information with the microprocessor. You already have a concept of memory. This was introduced in Chapter 2 when you learned about number systems. In this chapter, you will build upon this concept and broaden it to include many other kinds of digital storage techniques.

Recall from Chapter 1 the difference between digital and analog systems. Here you will see how these two systems can be made to work with each other. When this is done, a whole world of possibilities opens in the applications of digital systems. This is another important chapter. The information you learn here can be applied to advanced courses. It is also needed information to increase your troubleshooting and systems analysis ability.

KEY TERMS

Random Access
Sequential Access
Permanent Memory
Erasable Memory
Alterable Memory
Volatile Memory
Nonvolatile Memory
Write Operation
Read Operation
Interface
Port
ROM
Mask ROM
Look-up Table
Voice Synthesizers
Phonemes
PROM
EPROM
EEPROM
RAM
Static RAM
Zeropower RAM
Dynamic RAM
Refresh
Address Multiplexing
Burst Refresh
Distributed Refresh
Dynamic RAM Controller
First-In–First-Out (FIFO) Memory
CCD Memory
Magnetic Memory
Recording Format
MFM
Hard Disk
Bubble Memory
Compact Disk (CD)
Laser Diode
Coercivity
Kerr Effect
Analog-to-Digital (A/D) Converter
Digital-to-Analog (D/A) Converter
Flash Converter
Resolution
Stairstep Waveform
Stairstep A/D Converter
Conversion Time
Memory Map
High-Order Address
Low-Order Address
Partial Address Decoding
Full Address Decoding
I/O Ports
Memory-Mapped I/O
Isolated I/O
Bipolar Memories
MOS Memories
ECL
NMOS
Step Size

9–1 MEMORY TYPES

Discussion

You have already worked with digital memory that electrically stored digital information. In this section, you will see that there are four major kinds of digital memory. The kind of memory used depends on the application and on the physical characteristics of the medium. The important overview necessary to set the background for this chapter is presented here.

Basic Idea

Generally speaking, *digital memory* is any medium that stores digital data that can be accessed by some kind of digital processing system. Figure 9–1 illustrates the four mediums used by digital memory. Some of the mediums are a cross between two mediums. The point is that there are many different ways of storing digital information. The *storage* of digital information for any period of time is synonymous with *memory*. The simplest kind of memory may be depressing a single key on a keypad. The most complex kind of memory may be the digital patterns of a laser memory. Perhaps memory with the shortest duration is the electromagnetic wave that carries a digital signal to an earth satellite. The longest memory may be a laser-imprinted bar code on a sheet of stainless steel.

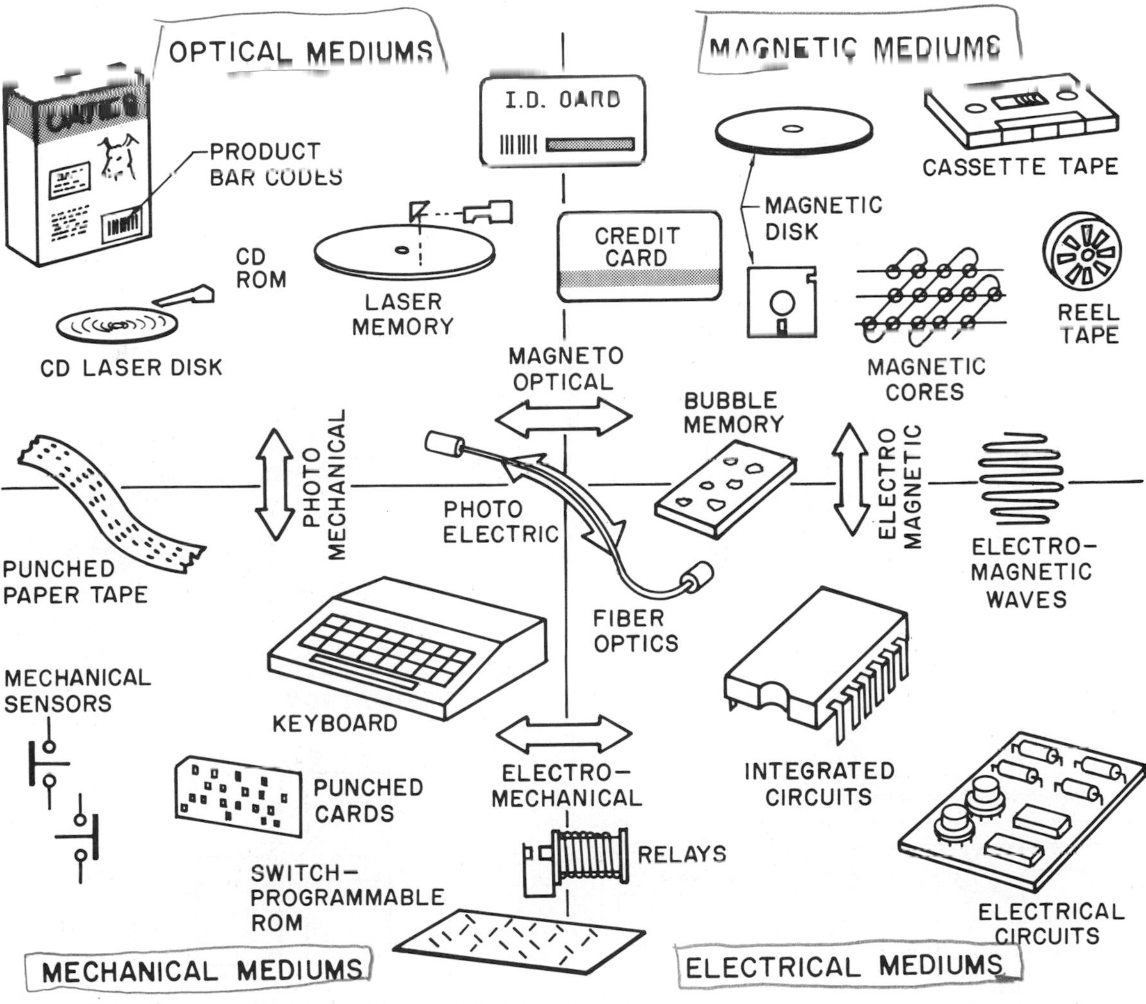

Figure 9–1 Four mediums used by digital memory.

Accessing Memory

There are two methods of accessing digital information in memory. One method is **random access,** the other is **sequential access.**

Random Access:
The amount of time it takes to begin to copy to or from a place in memory is the same.

Sequential Access:
The amount of time it takes to begin to copy to or from a place in memory depends upon the physical location in memory.

Figure 9–2 illustrates the concepts of randomly accessible memory and sequentially accessible memory and the concepts of sequential access and random access.

Altering Memory

Despite whether the memory is sequential or random, it may further be classified as **permanent memory, erasable memory,** or **alterable memory.**

Permanent Memory:
Digital information that cannot be changed without destroying the medium containing the digital information.

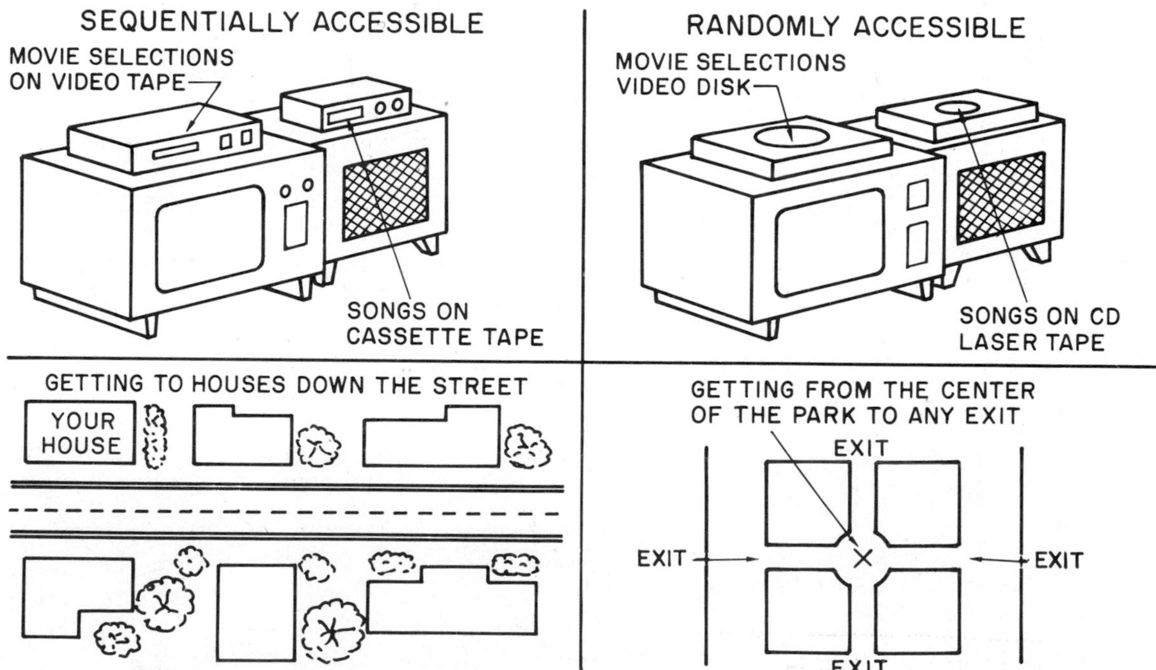

Figure 9–2 Types and concepts of sequential and random access.

Erasable Memory:
Digital information that can be deleted without destroying the medium containing the digital information.

Alterable Memory:
Any medium capable of having the storage of digital information changed.

An example of permanent memory would be the bar code on the package of a box of soap. This information cannot be changed without some disfiguration of the container. Erasable memory is digital data recorded on a cassette tape. Here the tape may be erased and not contain any digital information. The noise left on the tape could be interpreted as some kind of random digital pattern, but no "clean" digital pattern is left as in the case of a switch programmable ROM. Alterable memory would be a switch-programmable ROM. In this case no matter what the positions of the switches, there is *always* digital information in the memory (though it may not be useful).

Volatile and Nonvolatile Memories

Digital memory is generally placed in two broad areas called **volatile memory** and **nonvolatile memory.**

Volatile Memory:
Mediums that require an external source of energy to maintain the stored digital information.

Nonvolatile Memory:
Mediums that do not require an external source of energy to maintain the stored digital information

An example of volatile memory would be the storage of digital data used to display a number on the display screen of your pocket calculator. When you turn your calculator off (disconnect its source of energy) and then turn it back on, the display screen will no longer have the old number. An example of nonvolatile memory is the digital data contained on the bar code on a box of cereal, your bank checks or on the magnetic strip of your credit card.

Read or Write

The process of putting information into memory or getting information from memory is usually done by the microprocessor. Thus **write operation** and **read operation** refer to the microprocessor.

Write Operation:
The process of the microprocessor putting information into memory.

Read Operation:
The process of the microprocessor getting information from memory.

Interfacing

Most digital systems can read and write digital information to and from many different devices not usually thought of as memory. When the device is external to the digital system, it is **interfaced** to it by means of a **port.**

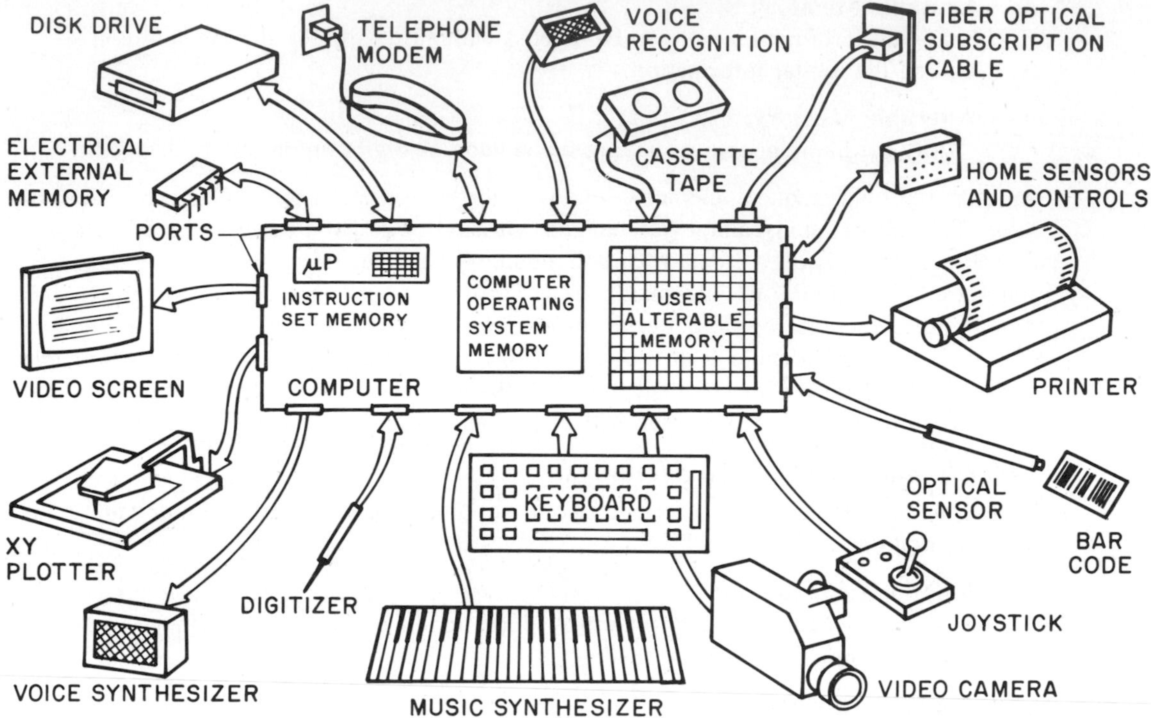

Figure 9–3 Computer memory and interfacing.

Interface:
The connection of a device to the digital system.

Port:
The connecting medium that allows communications between the digital system and an external device.

Figure 9–3 illustrates a computer with some of the more commonly found interface connections. Interface connections are used to convert between the +5 V and the 0 V binary environment of the computer and that of the device being interfaced. An example is the analog signal from the video camera, which must be converted to digital.

Conclusion

This section presented the four major kinds of digital memory. Here, important concepts and definitions were also presented. The following sections will present the details of what you have seen here. Check your understanding of this section by trying the following review questions.

9–1 Review Questions

1. What is digital memory?
2. Name the four major kinds of digital memory.

3. Give a specific example for each of the four different kinds of digital memory.
4. State the two different ways of accessing memory. Explain what they mean.
5. Describe the differences among *permanent, erasable* and *alterable* memories.

9–2 NONVOLATILE, NONALTERABLE, RANDOMLY ACCESSIBLE, ELECTRICAL MEMORY

Discussion

This section presents a specific type of digital memory. This memory does not require an external source of energy to maintain its information (nonvolatile), its bit pattern cannot be changed without destroying the medium (nonalterable), and it requires electrical energy to copy its bit pattern (electrical memory). It takes an equal amount of time to access any digital bit pattern (which means it is randomly accessible). As you will see, there are many applications for this type of memory.

ROM

Traditionally the type of memory called **ROM** referred to nonvolatile, nonalterable, electrical memory. The initials come from read only memory which meant that digital information could only be copied from it.

ROM:
Read-only memory—traditionally meant memory that could only have its stored bit pattern copied.

There are now several nonvolatile, alterable, electrical memories referred to as a type of ROM. The reason is that even though they are alterable, these devices are used more for bit pattern storage and less for altering the stored bit pattern.

Memory Nomenclature

The nomenclature used for most all electrical memory is illustrated in Figure 9–4. Much of the nomenclature has already been covered in the microprocessor applications sections of earlier chapters.

Device Construction

This memory device is permanently programmed during manufacturing. The process consists of placing a bit mask over the electrical memory cells used to create the bit pattern. Thus, this type of ROM is called a **mask ROM.**

Mask ROM:
A ROM that has its bit pattern put into it during the manufacturing process by the use of a bit mask that determines the bit pattern.

These memories are constructed by essentially having an electrical connection to represent a 1 and an open connection to represent a 0. The basic construction of a ROM is illustrated in Figure 9–5.

The manufacturer of such devices will charge a setup fee (around $1 000.00) for the design of a special mask. Once the mask is designed, the individual chips are produced.

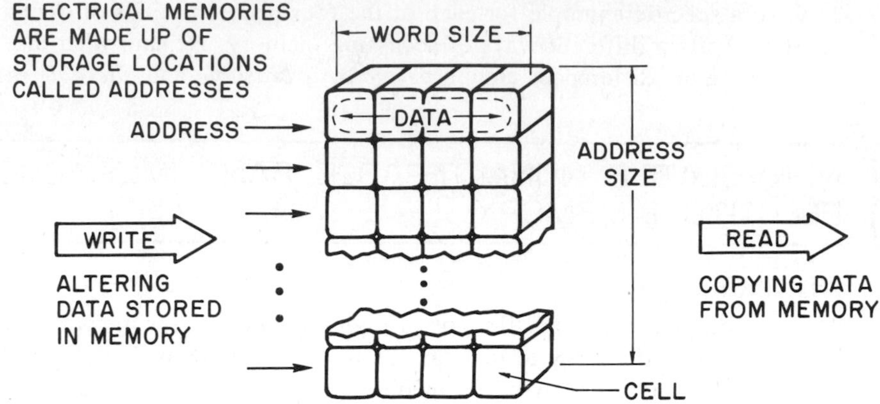

Figure 9–4 Nomenclature for electrical memory.

However, this process is only used when you are absolutely sure that your bit pattern works the way you intended it to work and you will be using this device in great quantities. Otherwise, there are more economical ways of producing such a device.

Some Real ROMs

Figure 9–6 illustrates the logic diagram of a real ROM IC. ROMs come in various sizes such as 32 × 8 and 512K × 8.

Look-Up Tables

One of the major uses of this type of electrical memory is as a **look-up table.**

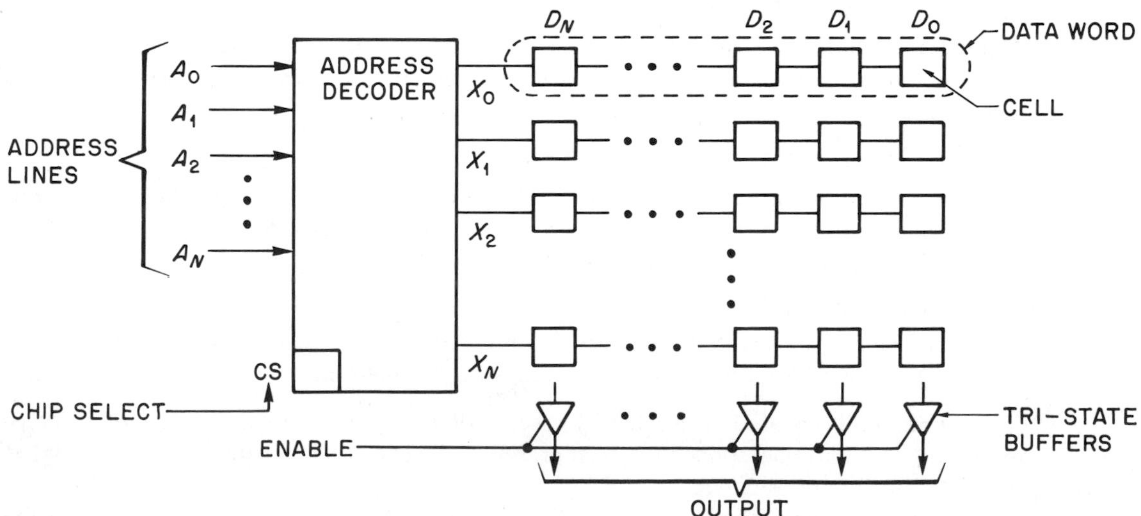

Figure 9–5 Internal construction of a ROM.

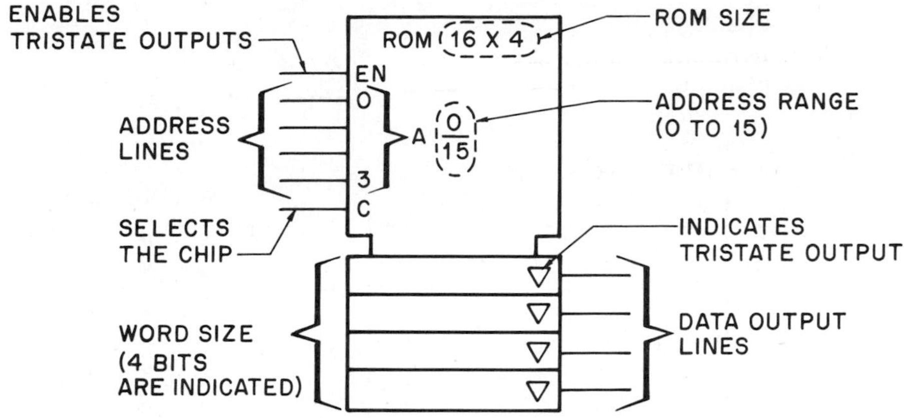

Figure 9–6 Logic diagram of electrical ROM.

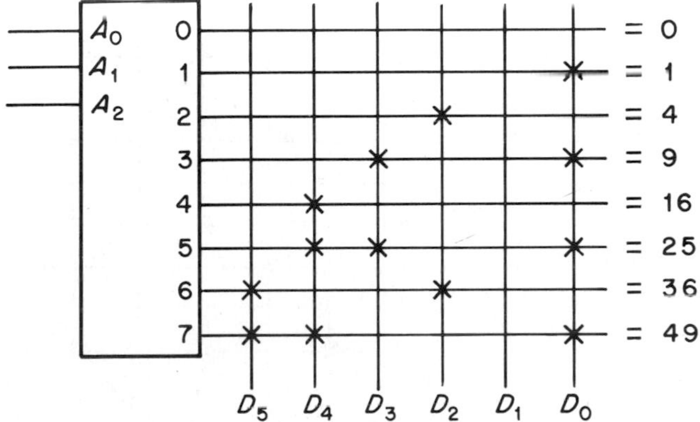

Figure 9–7 Basic idea of a look-up table.

Look-up Table:
A stored bit pattern where the address represents a value and the output represents a function of that value.

The basic idea behind a look-up table can be demonstrated by Figure 9–7. As shown in the figure, this device takes a number from 0 to 7 and produces the square of this number on the output. Keep in mind that all values in this circuit are binary numbers.

Example 9–1

Construct the bit pattern for a ROM look-up table that represents the function $y = 2x^2 - x$ for values of 0 to 7.

Solution

See Figure 9–8.

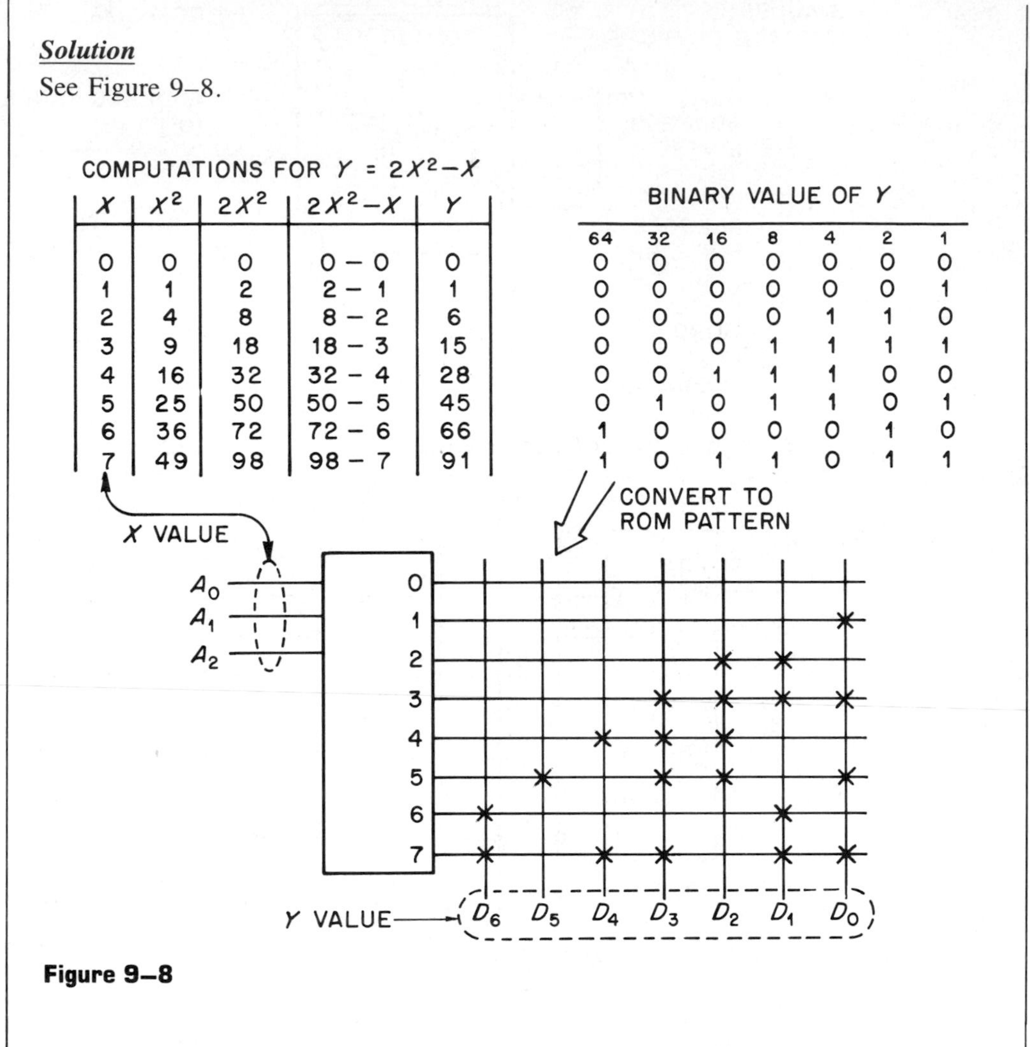

COMPUTATIONS FOR $Y = 2X^2 - X$

X	X^2	$2X^2$	$2X^2 - X$	Y
0	0	0	0 − 0	0
1	1	2	2 − 1	1
2	4	8	8 − 2	6
3	9	18	18 − 3	15
4	16	32	32 − 4	28
5	25	50	50 − 5	45
6	36	72	72 − 6	66
7	49	98	98 − 7	91

BINARY VALUE OF Y

64	32	16	8	4	2	1
0	0	0	0	0	0	0
0	0	0	0	0	0	1
0	0	0	0	1	1	0
0	0	0	1	1	1	1
0	0	1	1	1	0	0
0	1	0	1	1	0	1
1	0	0	0	0	1	0
1	0	1	1	0	1	1

Figure 9–8

Look-up tables can be constructed for converting giving the sine, cosine, tangent or any other trig relationship. Look-up tables can be constructed with any kind of digital memory.

Voice Synthesizers

Voice synthesizers are another common application of ROM.

Voice Synthesizers:
Digital systems that replicate the sound of the human voice.

Consider the voice synthesizer chip of Figure 9–9. In this system of speech synthesis, the speech processor contains a ROM that has a bit pattern that represents a unique **phoneme.**

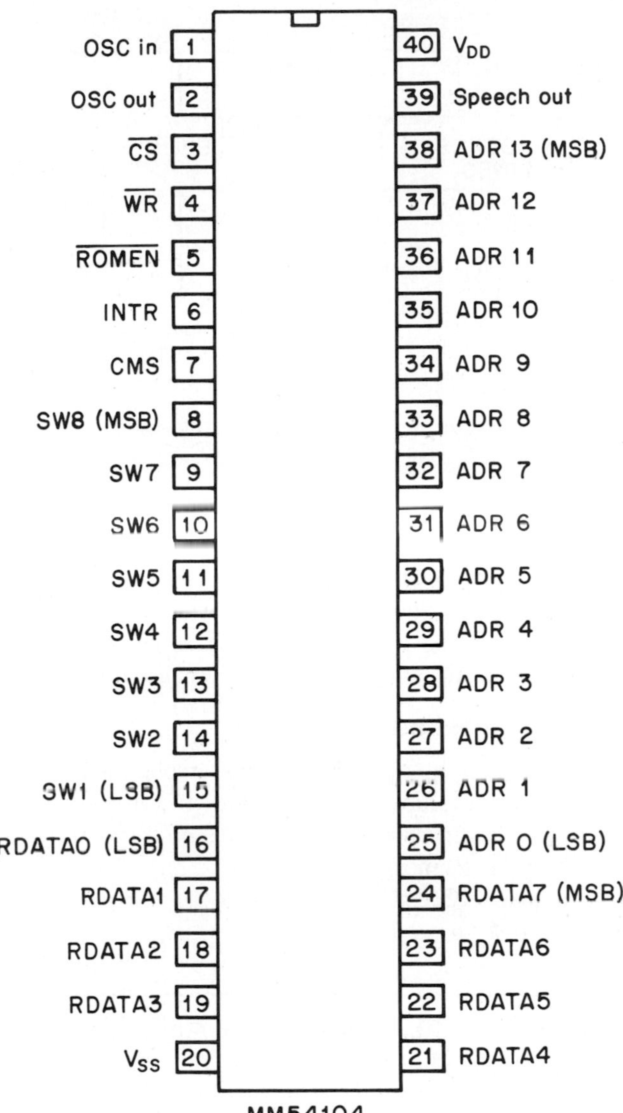

Figure 9–9 The MM54104 speech processor.

Phonemes:
The basic sounds that make up a language. English has about 60 phonemes.

Each address represents a unique phoneme. By progressively selecting the order of these addresses, any word can be created. For example, the word "hello" contains four phonemes: "h", "short e," "l", and "long o." The idea of how speech is constructed by such a system is illustrated in Figure 9–10.

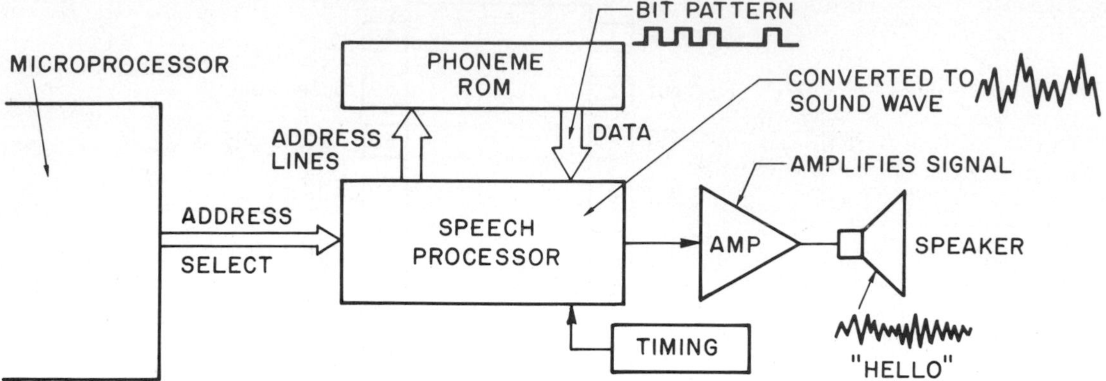

Figure 9–10 Construction of voice from phonemes.

Conclusion

This section presented the nonvolatile, nonalterable, randomly accessible, electrical memory commonly referred to as ROM. Here you saw the construction of the memory and some of its important applications. In the next section you will be introduced to nonvolatile, alterable, randomly accessible, electrical memory. For now, test your understanding of this section by trying the following review questions.

9–2 Review Questions

1. State what is meant by nonvolatile, nonalterable, randomly accessible, electrical memory.
2. Technically, what kind of memory does ROM refer to?
3. Explain the differences among the terms *address, data, cell,* and *word* in reference to electrical digital memory.
4. What is a *mask* ROM?
5. Describe a *look-up table*.

9–3 NONVOLATILE, ALTERABLE, RANDOMLY ACCESSIBLE, ELECTRICAL MEMORY

Discussion

This section presents another type of digital memory. As before, this memory does not require an external source of energy to maintain its information (it is therefore nonvolatile). However, its contents can be altered without damaging the medium that stores the information. The amount of time it takes to access any bit pattern is the same (meaning it is randomly accessible), and it does require electrical energy to copy its bit patterns (hence it is electrical memory).

Switch-Programmable ROM

Your switch-programmable read-only memory is a good example of this type of memory. Here, you altered the bit patterns by a set of switches. These switches did not require any external source of energy to maintain their digital patterns. However, it did require elec-

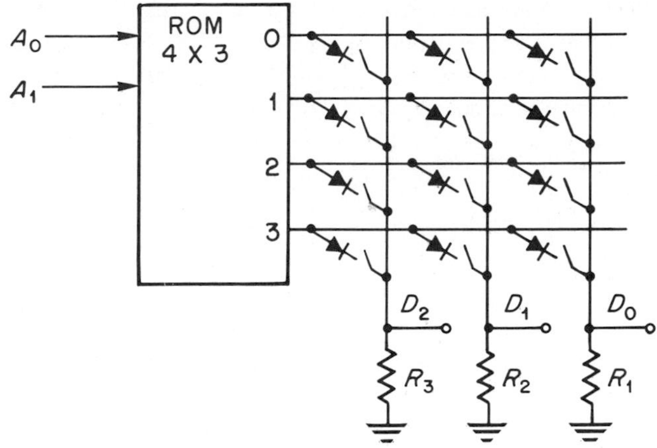

Figure 9–11 Typical switch-programmable ROM.

trical energy to cause the stored pattern to have some effect. Generally, the bit pattern was altered very few times compared to the number of times the bit pattern was used for a particular purpose. As a review, Figure 9–11 shows a typical switch-programmable memory.

PROM

The acronym **PROM** stands for programmable read-only memory.

> **PROM:**
> Nonvolatile, randomly accessible, electrical memory that may be altered only once by the user.

Think of a PROM as having the same internal construction as a switch-programmable ROM, but in place of the switches there are tiny miniature fuses (the same as with programmable logic devices). When you get the PROM from the factory, all of the fuses are intact (the device contains all 1's). You may then alter the contents by selectively blowing a fuse where you want to have a 0. However, once the fuse is blown, the process cannot be reversed. This concept is illustrated in Figure 9–12.

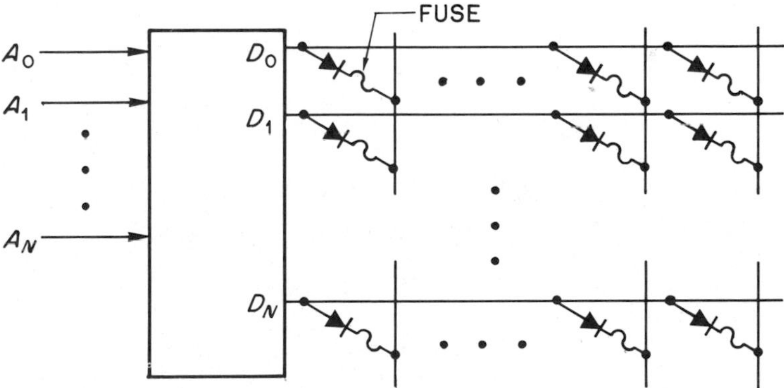

Figure 9–12 Concept of a PROM.

PROMs are useful when you need a small quantity of special ROMs. They are more economical than factory programmed ROMs. These devices are available in sizes that are equivalent to factory programmed ROMs. In small quantities, they are more economical than factory programmed ROMS. The process of programming a PROM is illustrated in Figure 9–13.

EPROM

The acronym **EPROM** stands for erasable programmable read-only memory.

EPROM:
A nonvolatile, alterable, randomly accessible electrical memory.

This type of memory is similar to the PROM but, instead of fusible links, uses special semiconductor devices that can be altered many times by the user. This type of memory is called *erasable,* meaning that its contents can be *altered*. It can never be erased in the sense that it never contains a bit pattern—it will always have a bit pattern as long as it is functional. However, it is a practice of industry to use the term *erasable* to mean *alterable* when talking about electrical memory.

These memories are programmed similar to a PROM. The technician uses instruments to help program PROMs and EPROMs. See Figure 9–14. The process of programming these devices is called *burning-in*.

The EPROM may have all of its cells set back to the same value by exposure to ultraviolet light (usually for about 30 minutes). A typical EPROM is shown in Figure 9–15.

EEPROMS

The acronym **EEPROM** stands for electrically erasable programmable read-only memory.

EEPROM:
Nonvolatile, electrically alterable, randomly accessible, electrical memory.

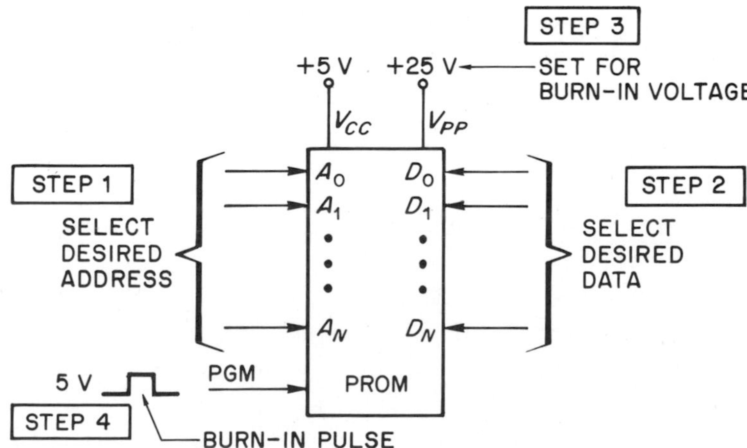

Figure 9–13 Programming a PROM.

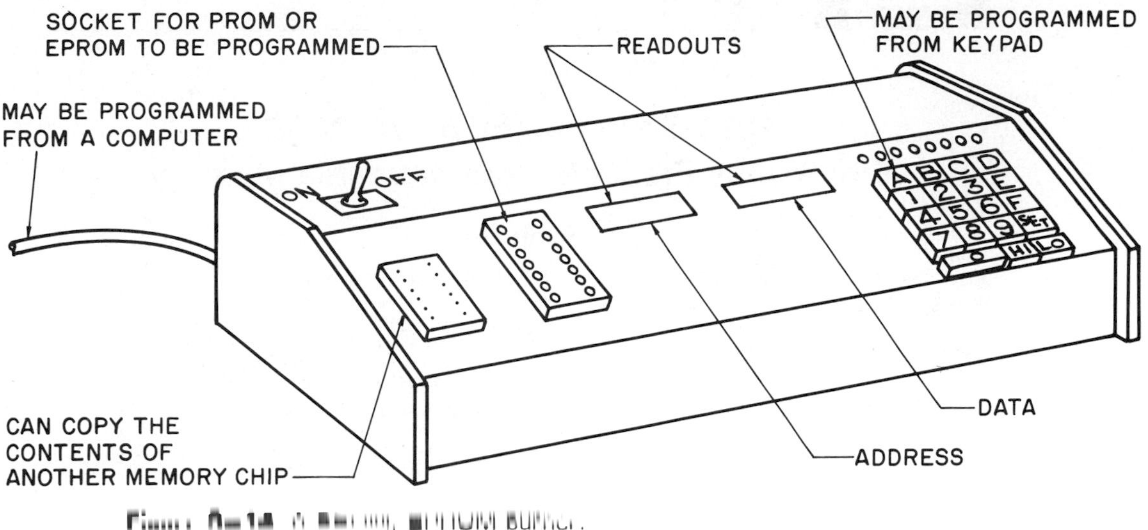

Figure 9–14 A PROM, EPROM burner.

The main difference between an EEPROM (sometimes referred to as E^2PROM) and an EPROM is that the device may be "erased" electrically instead of by ultraviolet light. Some manufacturers call this an EAROM (electrically alterable read-only memory). This term is more exact than the others, since you never really "erase" any bits, you simply alter them. Thus, EEPROMs may be programmed in their circuit of operation (unlike the EPROM, which must be removed to have new data written into it).

Figure 9-16 illustrates an EEPROM. This EEPROM is a serial, read/write, nonvolatile memory. This means that data are stored into and retrieved from this device one bit at a time. The advantage of doing this is that the pin count of the package is very small (8 pins). The disadvantage of this is that it takes more time to transfer data into and out of the memory. This is not a problem in most alarm systems and electronic locks.

The advantage of these memory devices is that they can be programmed quicker than the EPROM since they do not require long exposure to ultraviolet light to reset all bit patterns to the same value.

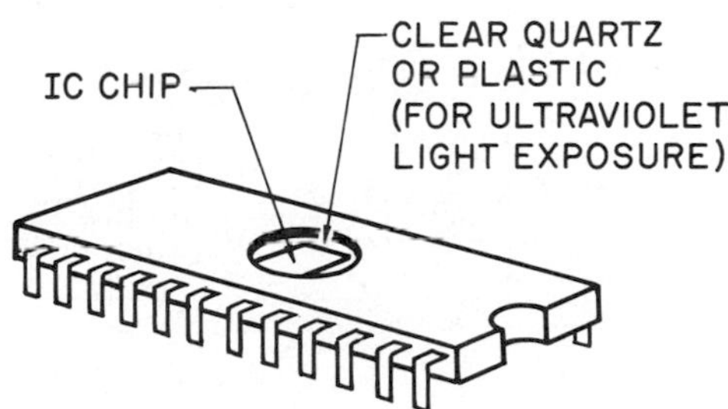

Figure 9–15 Typical EPROM.

INTERNATIONAL CMOS TECHNOLOGY, INC.

93C46
1024-Bit Serial (5V only) CMOS Electrically Erasable Programmable Memory

Features

- **ADVANCED CMOS E²PROM TECHNOLOGY**
- **READ/WRITE NON-VOLATILE MEMORY**
  - — Single 5V supply operation
  - — 1024 bits, 64 x 16 organization
  - — Easy to use yet versatile serial data interface
- **LOW POWER CONSUMPTION**
  - — 3mA Max Active
  - — 1mA Max Standby, TTL interface
  - — 100μA Max Standby, CMOS interface
- **SPECIAL FEATURES**
  - — Automatic write cycle time-out
  - — Ready/Busy status signal
  - — Software controlled write protection
- **IDEAL FOR LOW DENSITY DATA STORAGE**
  - — Low cost, space saving, 8-pin package
  - — Commercial, industrial, and military versions
  - — Interfaces with popular microcontrollers (ie., COP4XX, 8048, 8049, 8051, 8096, 6805, 6801, TMS1000, Z8) and standard microprocessors
- **APPLICATION VERSATILITY**
  - — Alarm Devices, Electronic Locks, Appliances, Terminals, Smart Cards, Satellite Receivers, Robotics, Meters, Telephones, Tuners, etc.
- **SUPER-SET COMPATIBILITY**
  - — National: NMC9306/COP494, NMC9346/COP495
  - — General Instrument: ER59256, ER5911
  - — NCR: 59306, 59308

General Description

The ICT 93C46 is a 1024-bit, 5V-only, serial, read/write, non-volatile memory device fabricated using an advanced CMOS EEPROM technology. Its 1024 bits of memory are organized into 64 registers of 16 bits each. Each register is individually addressable for serial read or write operations. A versatile serial interface, consisting of chip select, clock, data-in and data-out, can easily be controlled by popular microcontrollers (ie., COP4XX, 8048, 8049, 8051, 6805, 6801, TMS1000, Z8) or standard microprocessors.

Low power consumption, low cost, and space efficiency make the ICT 93C46 an ideal candidate for high volume, low density, data storage applications. Special features of the 93C46 include: automatic write time-out, ready/busy status signal, software controlled write protection and ultra-low standby power mode when deselected (CS low). Additionally 93C46 offers functional compatability with existing NMOS serial EEPROMs. The 93C46 is designed for applications requiring up to 10,000 erase/write cycles per register.

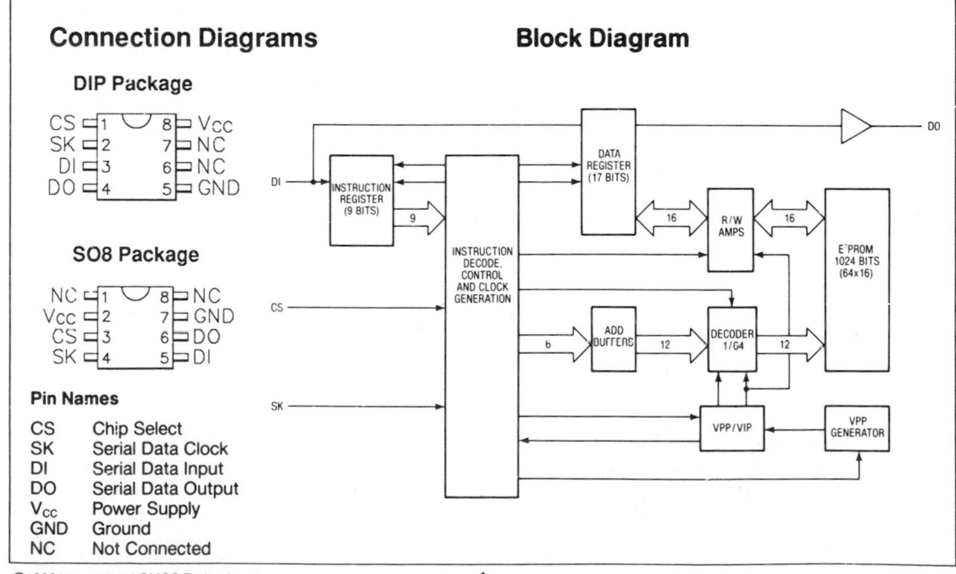

Figure 9–16 The 93C46 EEPROM. *Courtesy* International CMOS Technology, Inc.

Conclusion

This section presented electrical memory that was nonvolatile, alterable, and randomly accessible. In the next section you will be introduced to volatile, randomly accessible, electrical memory. For now, test your understanding of this section by trying the following review questions.

9–3 Review Questions

1. State what is meant by nonvolatile, alterable, randomly accessible, electrical memory.
2. What kind of electrical memory may be altered only once?
3. Explain how the above device is constructed.
4. What are the differences among PROMs, EPROMs and EEPROMs?
5. What is an EAROM?

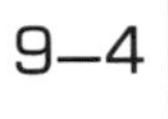

9–4 VOLATILE, RANDOMLY ACCESSIBLE, STATIC, ELECTRICAL MEMORY

Discussion

This section presents electrical memory that is *volatile*. That is, an external source of energy is required for it to maintain its programmed bit pattern. When the external source of energy is removed, the original bit pattern is lost and replaced by a purely random pattern. *Randomly accessible* means that it takes the same amount of time to access any piece of data. The memory is electrical because it requires electrical energy to store the required bit pattern and to retrieve any of its bit patterns.

Static RAM

Historically, this type of memory was referred to as **RAM,** randomly accessible memory. This name is a poor choice because all of the memory presented so far has been randomly accessible.

> **RAM:**
> Literally means randomly accessible memory, but is used to mean volatile, randomly accessible, electrical memory.

A more descriptive acronym for this type of memory would have been VARAEM, for volatile alterable randomly accessible electrical memory. However, it is still common practice for manufacturers to use RAM to mean this kind of memory. Sometimes RAM is referred to as READ/WRITE (R/W) memory.

There are two types of RAM: **static RAM** and dynamic RAM. Dynamic RAM is discussed in the next section.

> **Static RAM:**
> Volatile, randomly accessible, electrical memory that requires no further processing as long as an external source of energy is supplied to maintain its programmed bit pattern.

RAM Organization

A RAM consists of columns and rows of memory cells just as the nonvolatile memory does. The difference is that these cells can be altered just as easily as they can be copied.

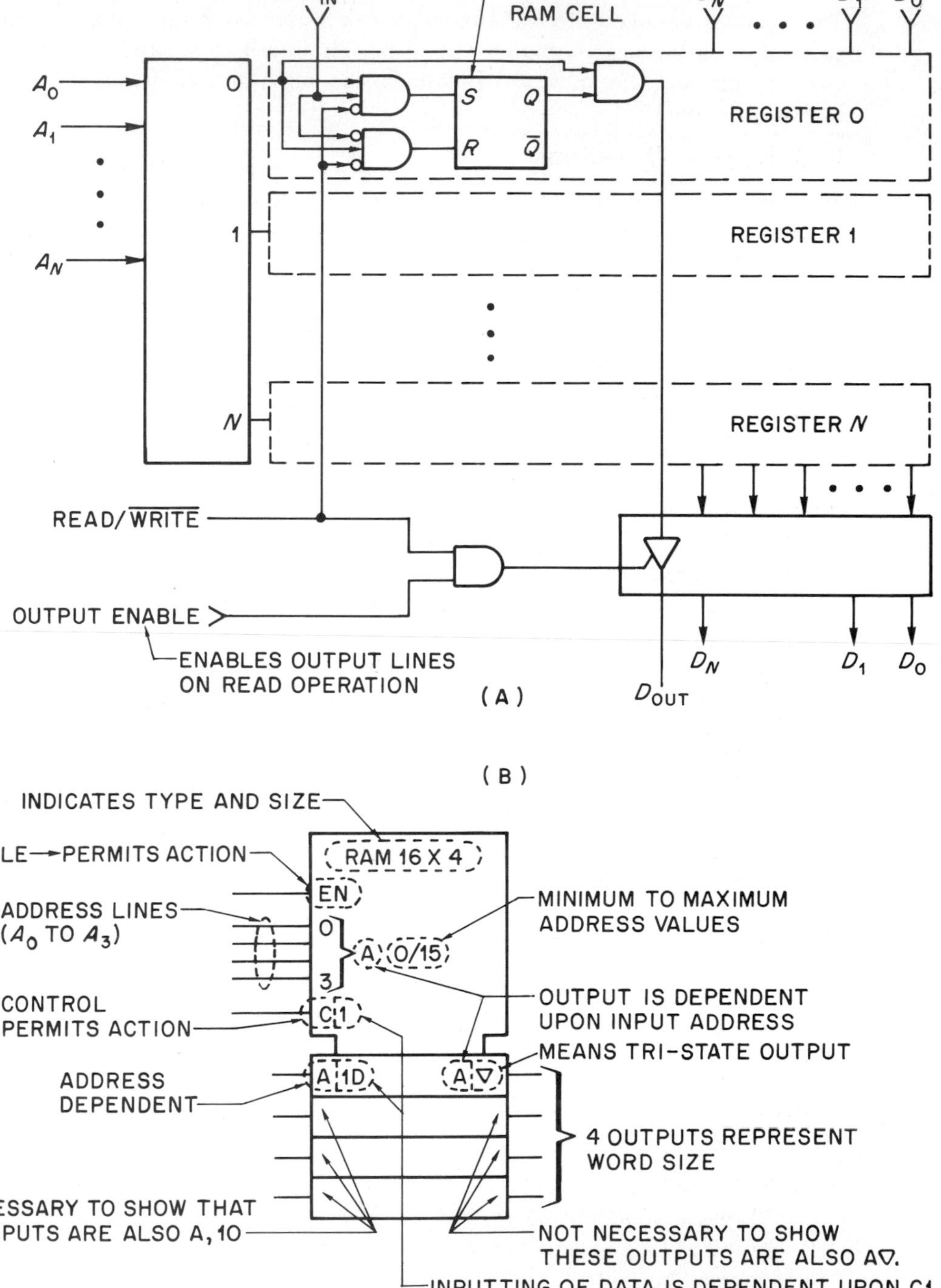

Figure 9–17 (A) Construction of a typical RAM. (B) Address dependency notation.

Essentially each cell is constructed from a flip-flop. Such construction is illustrated in Figure 9–17(A). The IEEE dependency notation for a typical RAM is shown in Figure 9–17(B).

These memory devices are used in digital systems where digital information is to be stored. For example, in your computer, any input you have from the keyboard is stored in RAM. Your screen displays are all stored in RAM. The programs you load into your computer from a disk are all stored in RAM. RAM is used whenever you require quick storage of digital information. The disadvantage of this kind of flexibility is that with the flip-flop as a storage element, the stored bit patterns are lost when power is no longer applied.

Real RAMs

Figure 9–18(A) shows a real RAM. The MK41H69 has twelve address lines. This means that it can address up to 4K of memory. The chip has four data lines. Note that since data may be written into or from the chip, these four lines are used as input or output lines. The timing diagram of Figure 9–18(B) shows the timing relationship of this chip.

As with ROMs, RAMs come in various sizes, depending upon the application.

Zeropower RAM

Because of the increased demand for nonvolatile memory that is easily altered, manufacturers have developed **zeropower RAM.**

> **Zeropower RAM:**
> Volatile, randomly accessible, static electrical memory with a built-in electrical energy source that allows bit patterns to be stored for long periods of time.

The data sheet for such a device is shown in Figure 9–19. This particular ZRAM (zeropower RAM) uses a type of technology (CMOS) which requires very little power. It uses only 5.5 mW of power to retain data. Its greatest power consumption is when active (when actively changing data). Its power consumption then jumps up to almost 100 times its standby power (440 mW). This extra power may be supplied by an external source thus preserving the life of the internal battery. This chip also contains a low internal battery warning. Under these conditions a back-up power source could be automatically engaged until data is transferred to another ZRAM. The built-in power source has a guaranteed life expectancy of 11 years. Thus, effectively, you have a nonvolatile, electrically alterable, randomly accessible, electrical memory that is just as easily altered as copied. Applications of this type of RAM could replace the magnetic digital memory, covered later in this chapter, in many applications.

Conclusion

This section presented the digital memory storage device commonly known as RAM. Here you were introduced to what is known as *static* RAM, meaning no further processing was required for it to maintain its bit pattern other than an external source of energy. Here, you also saw that what is commonly referred to as RAM is really volatile, randomly accessible, electrical memory.

In the next section you will be introduced to another type of RAM. For now, check your understanding of this section by trying the following review questions.

MK41H68/ MK41H69(N,P)-20/25/35

4K × 4 CMOS STATIC RAM

- 20, 25, AND 35 ns ADDRESS ACCESS TIME
- EQUAL ACCESS AND CYCLE TIMES
- 20-PIN, 300 MIL PLASTIC AND CERAMIC DIP
- ALL INPUTS AND OUTPUTS TTL COMPATIBLE, LOW CAPACITANCE, AND PROTECTED AGAINST STATIC DISCHARGE
- 50 μA CMOS STANDBY CURRENT (MK41H68)
- TTL STANDBY CURRENT UNAFFECTED BY ADDRESS ACTIVITY (MK41H68)
- HIGH SPEED CHIP SELECT (MK41H69)
- JEDEC STANDARD PINOUT

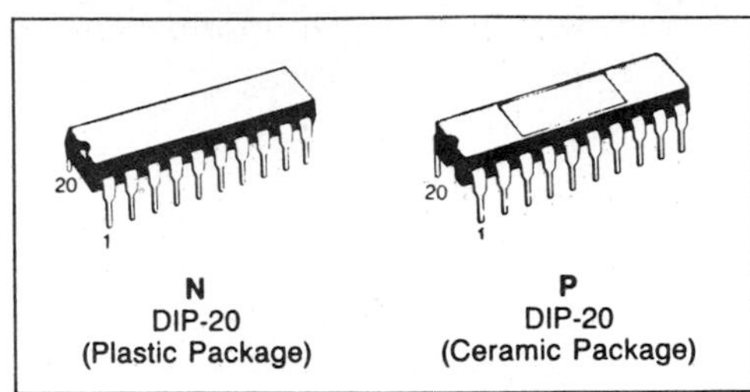

MK41H68 TRUTH TABLE

$\overline{CE}$	$\overline{WE}$	Mode	DQ	Power
H	X	Deselect	High Z	Standby
L	L	Write	D_{IN}	Active
L	H	Read	D_{OUT}	Active

MK41H69 TRUTH TABLE

$\overline{CS}$	$\overline{WE}$	Mode	DQ	Power
H	X	Deselect	High Z	Active
L	L	Write	D_{IN}	Active
L	H	Read	D_{OUT}	Active

X = Don't Care

DESCRIPTION

The MK41H68 and MK41H69 feature fully static operation requiring no external clocks or timing strobes, and equal address access and cycle times. Both require only a single +5V ± 10 percent power supply. Both devices are fully TTL compatible.

The MK41H68 has a Chip Enable power down feature which automatically reduces power dissipation when the $\overline{CE}$ pin is brought inactive (high).

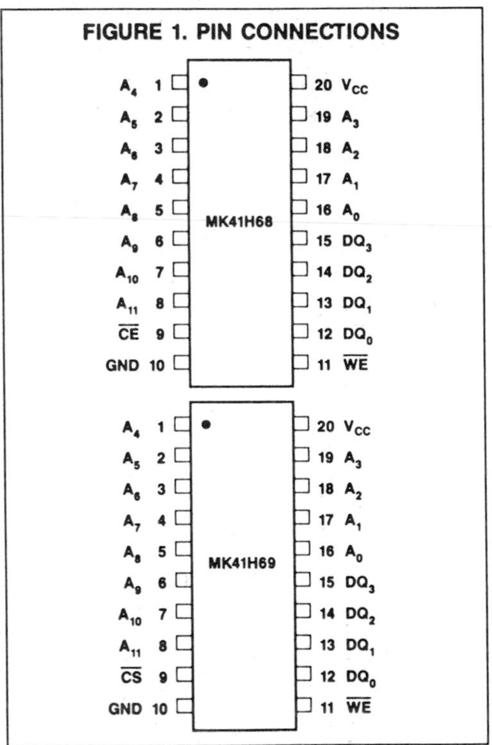

PIN NAMES

A_0 - A_{11} - Address	$\overline{WE}$ - Write Enable
DQ_0 - DQ_3 - Data I/O	GND - Ground
$\overline{CE}$ - Chip Enable (MK41H68)	V_{CC} - + 5 volts
$\overline{CS}$ - Chip Select (MK41H69)	

Figure 9–18(A) A real RAM. *Courtesy* of SGS-THOMSON Microelectronics.

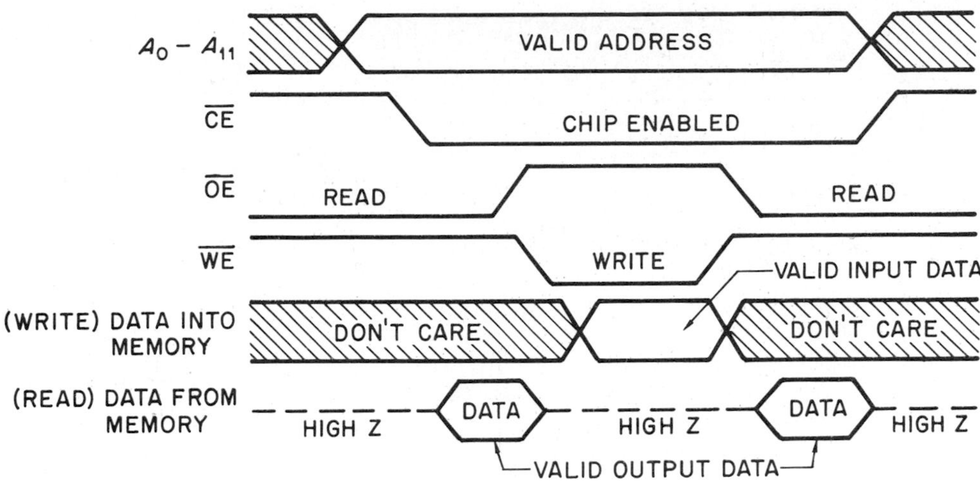

Figure 9–18(b) RAM timing diagram.

9–4 Review Questions

1. What is volatile, randomly accessible, static, electrical memory?
2. What kind of memory is normally meant by RAM? What do the letters RAM actually stand for?
3. Why is RAM used?
4. How do the data lines of RAM differ from the data lines of ROM?
5. Is any RAM nonvolatile? Explain.

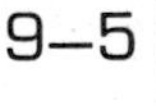

9–5 VOLATILE, RANDOMLY ACCESSIBLE, DYNAMIC, ELECTRICAL MEMORY

Discussion

The memory presented in this section is similar to that in the last section, except that it requires processing as well as a source of electrical energy to maintain its bit pattern.

Dynamic RAM

Dynamic RAM is similar to static RAM except that it requires an external process in order to maintain its programmed bit pattern.

> **Dynamic RAM:**
> Volatile, randomly accessible, electrical memory that tends to lose stored information over a period of time and requires "refreshing" in order to retain its bit pattern.

The reason for the process required by this type of memory is that the digital information is stored in cells that can accept tiny electrical charges. Using these small cells instead of flip-flops allows greater packaging densities, so you can get much more memory

MK48Z02/12(B) -12/15/20/25

2K × 8 ZEROPOWER™ RAM

- PREDICTED WORST CASE BATTERY LIFE OF 11 YEARS @ 70°C
- DATA RETENTION IN THE ABSENCE OF POWER
- DATA SECURITY PROVIDED BY AUTOMATIC WRITE PROTECTION DURING POWER FAILURE
- +5 VOLT ONLY READ/WRITE
- CONVENTIONAL SRAM WRITE CYCLES
- FULL CMOS-440 mW ACTIVE; 5.5 mW STANDBY
- 24-PIN DUAL IN LINE PACKAGE, JEDEC PINOUTS
- READ-CYCLE TIME EQUALS WRITE-CYCLE TIME
- LOW-BATTERY WARNING
- TWO POWER-FAIL DESELECT TRIP POINTS AVAILABLE
  MK48Z02 4.75V $\geq V_{PFD} \geq$ 4.50V
  MK48Z12 4.50V $\geq V_{PFD} \geq$ 4.20V

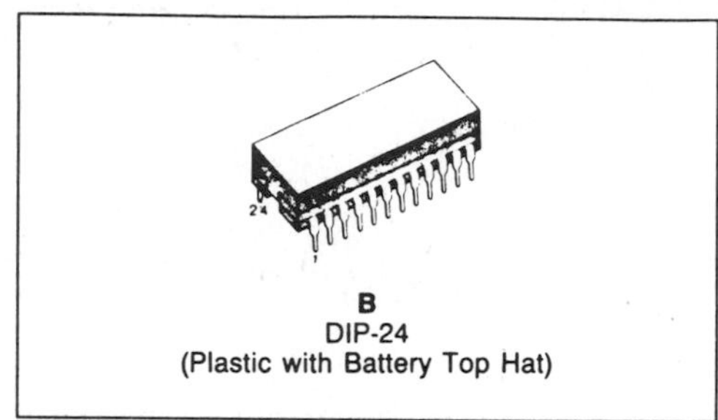

B
DIP-24
(Plastic with Battery Top Hat)

Part Number	Access Time	R/W Cycle Time
MK48ZX2-12	120 ns	120 ns
MK48ZX2-15	150 ns	150 ns
MK48ZX2-20	200 ns	200 ns
MK48ZX2-25	250 ns	250 ns

TRUTH TABLE (MK48Z02/12)

V_{CC}	$\overline{E}$	$\overline{G}$	$\overline{W}$	MODE	DQ
$<V_{CC}$ (Max) $>V_{CC}$ (Min)	V_{IH}	X	X	Deselect	High-Z
	V_{IL}	X	V_{IL}	Write	D_{IN}
	V_{IL}	V_{IL}	V_{IH}	Read	D_{OUT}
	V_{IL}	V_{IH}	V_{IH}	Read	High-Z
$<V_{PFD}$ (Min) $>V_{SO}$	X	X	X	Power-Fail Deselect	High-Z
$\leq V_{SO}$	X	X	X	Battery Back-up	High-Z

FIGURE 1. PIN CONNECTIONS

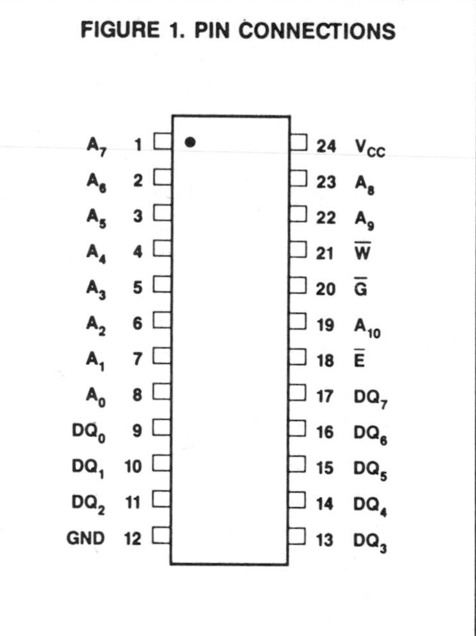

PIN NAMES

A_0 - A_{10}	Address Inputs	V_{CC}	System Power (+5 V)
$\overline{E}$	Chip Enable	$\overline{W}$	Write Enable
GND	Ground	$\overline{G}$	Output Enable
DQ_0—DQ_7	Data In/Data Out		

Figure 9–19 Zeropower RAM™. *Courtesy* of SGS-THOMSON Microelectronics.

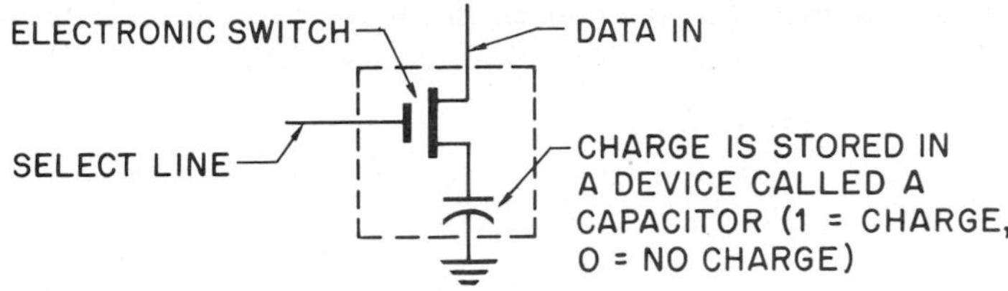

Figure 9–20 Basic idea of a dynamic RAM cell.

on a single chip than would be possible with static RAMs. The basic idea of a dynamic RAM cell is shown in Figure 9–20.

These small electrical charges will eventually leak off (discharge), and the data will be lost. Therefore, this type of memory device must be periodically recharged by a process called **refresh.** Typically this process needs to be done every 2 ms to 4 ms.

Refresh:
The process of maintaining the electrical cell charge in a dynamic RAM.

Sometimes this type of memory is referred to as DRAM, dynamic random access memory.

Refreshing the Cell

Figure 9–21 shows the basic process of refreshing a DRAM cell. Note that you may write a 1 or a 0 into the cell and read a 1 or a 0 from the cell. Also note that besides the ability to store the 1 or 0, a refresh operation is also needed. Here, the refresh will maintain a charge if one already existed and not put in a charge if one did not already exist. This charge is supplied by power from the external circuit.

Because of the requirements of the refresh operation of this memory, external circuity (which may be built into the chip) is required. This circuity must be synchronized with the digital system using this type of memory so a refresh does not happen at the same time a read or write operation would be taking place.

Basic DRAM Design

When using dynamic RAM, a large number of cells can be placed on a single chip due to the small size of the cells. Normally this would require many address lines to address this enormous amount of data. Under these conditions a large pin count would be required.

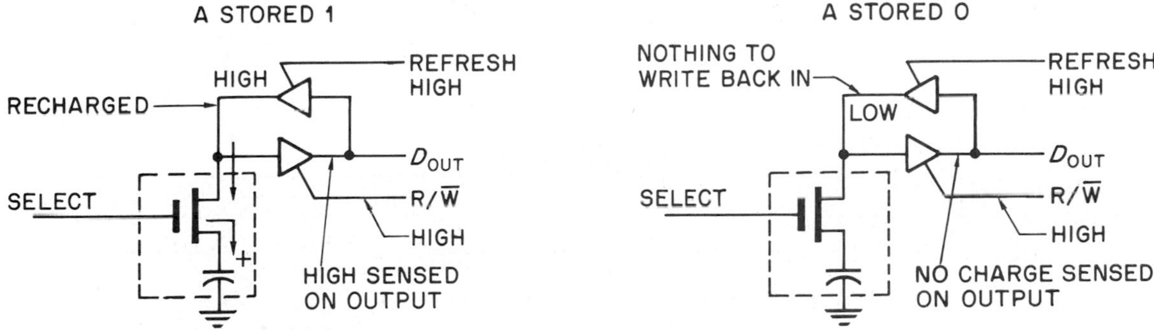

Figure 9–21 Basic process of refreshing a DRAM cell.

In order to reduce the pin count of the IC package, a different method of addressing the data is used. This method, called **address multiplexing,** is illustrated in Figure 9–22.

Address Multiplexing:

The process of splitting the complete address into two or more parts so that fewer address lines are required.

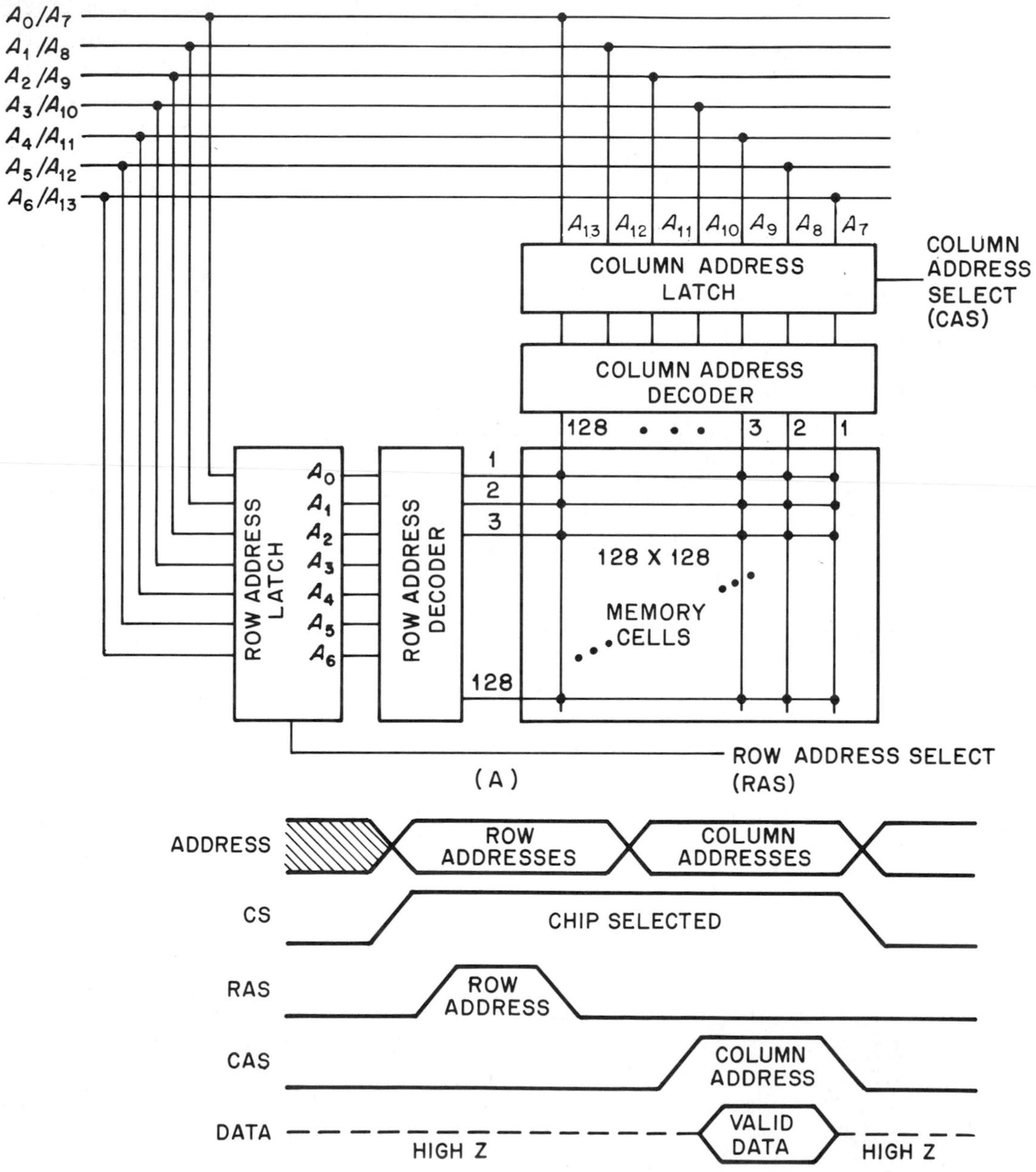

Figure 9–22 (A) Basic layout of a DRAM. (B) Address multiplexing timing diagram.

When the DRAM is being addressed, only the ROW ADDRESS LATCH or the COLUMN ADDRESS LATCH will be active, not both. When the ROW ADDRESS LATCH is active, the row address decoder selects the row of DRAM. During this time, the address lines are interpreted as address lines A_0 through A_6. Once the row has been selected, then the COLUMN ADDRESS LATCH becomes active and the column address decoder selects a memory column. In this manner 16,348 bits can be accessed by using only seven pins for address lines on the chip.

DRAM Refresh Circuitry

Figure 9–23 shows the basic construction of a typical refresh circuit for a DRAM. There are basically two methods used to refresh DRAM. One method is called **burst refresh;** the other is **distributed refresh.**

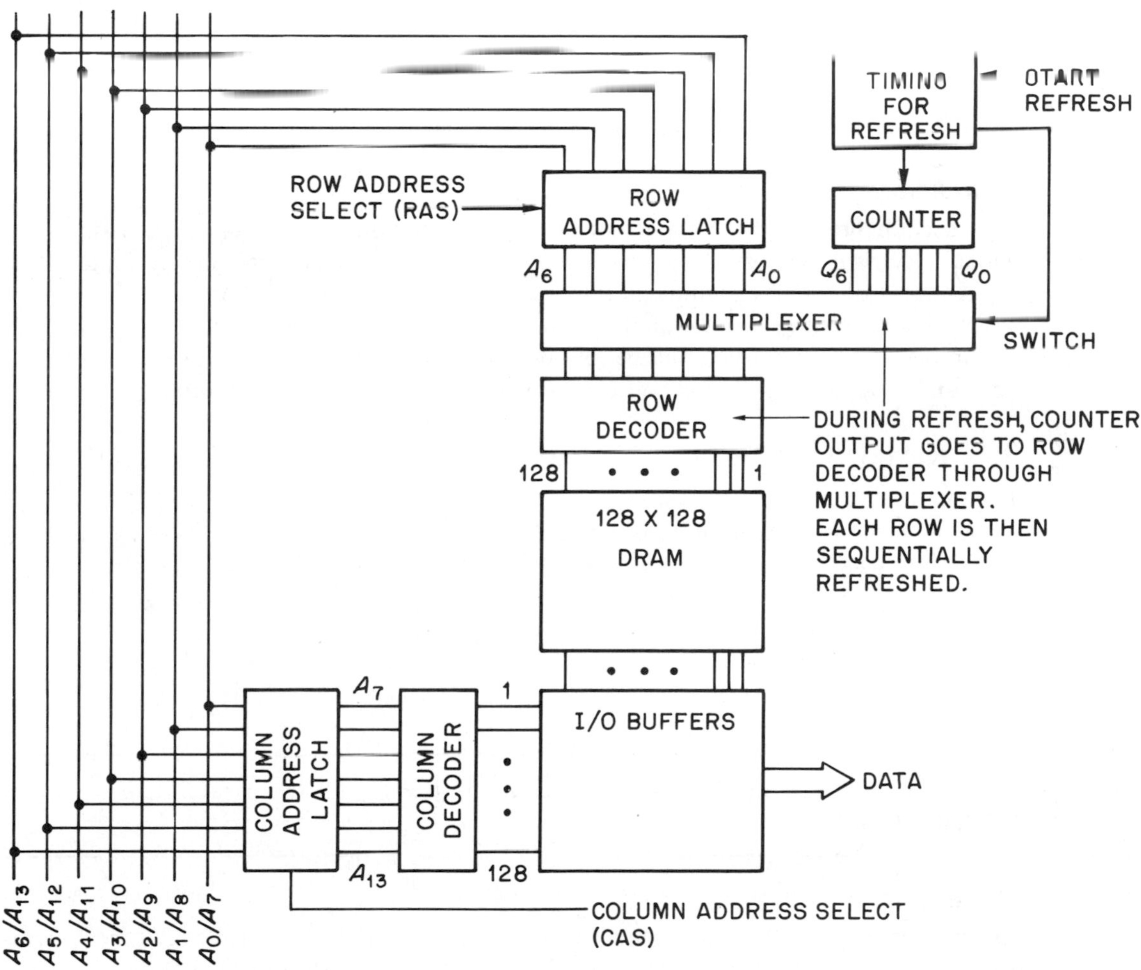

Figure 9–23 Basic refresh logic for DRAM.

Burst Refresh:
A method of refreshing DRAM where each cell row is sequentially refreshed.

Distributed Refresh:
A method of refreshing DRAM where cell rows are refreshed between read and write operations.

Essentially, a DRAM cell is refreshed everytime a READ operation is performed. However, the chip still needs all of its cells refreshed within the 2 ms to 4 ms time frame. This means that the row address multiplexer is used within this time frame to ensure that all of the DRAM cells receive their required refresh cycle. While this is being done, other circuitry is used to ensure that there is not a conflict with external attempts to read or write to the device. There are special integrated circuits available to perform the required refresh operation for DRAMs. These are referred to as **dynamic RAM controllers.**

Dynamic RAM Controller:
A special purpose integrated circuit used to coordinate the interfacing between a microprocessor and a dynamic RAM. This prevents a conflict between the addressing requirements of the RAM and that of the required RAM refreshing.

Conclusion

This section presented the fundamental concepts of volatile, randomly accessible, dynamic, electrical memory. Here you saw the difference between static and dynamic RAM. The concept of address multiplexing was also presented in this section. You saw the necessity for refreshing and the digital logic that could accomplish this task.

In the next section, you will be introduced to sequentially accessible, memories. Test your understanding of this section by trying the following review questions.

9–5 Review Questions

1. State the difference between *static* and *dynamic* RAM.
2. What is the advantage of dynamic RAM?
3. Explain what is meant by *refreshing* memory.
4. How are the number of pins on a DRAM package kept to a minimum?
5. Give the two different methods of refreshing DRAM.

9–6 VOLATILE, SEQUENTIALLY ACCESSIBLE, ELECTRICAL MEMORY

Discussion

This section presents the fundamentals of volatile, sequentially accessible, electrical memory. One of the main uses of this type of memory is in transferring digital data between digital systems.

First-In–First-Out Memories

The construction of a **first-in–first-out** (FIFO) memory is illustrated in Figure 9–24.

> **First-In–First-Out (FIFO) Memory:**
> Refers to a special arrangement of digital memory where the first bit of data written into the memory is the first bit of data read from the memory.

As you can see, this type of memory is constructed from shift registers. Essentially what happens is that data is input into the shift registers as a word and the data is extracted from the output of each shift register. This type of memory may be used in asynchronous operations where data are shifted out independently of data entry. This process requires two clocks and allows systems with different data rates to communicate. Figure 9–25 illustrates a real FIFO memory.

Note from the figure that the CMOS parallel FIFO comes in two different sizes: a 64 × 4 bit and a 64 × 5 bit FIFO memory. The differences are shown in the functional block diagram for the memory. D_4 is shown separately for types IDT72402 and IDT72404. These same two devices also contain an output enable pin (OE).

The INPUT READY line (IR) is used to indicate when the input is ready for new data. When HIGH it means that new data may be sent, and when LOW it indicates that the memory is full. The OUTPUT READY line (OR) is used to indicate when the output is ready to send data. When HIGH, it means that the device contains valid data and when LOW it indicates that the memory is empty.

This is an asynchronous device, meaning that two clocks are used, one for the input data and the other for the output data. Because of this, the device may be used to connect two digital systems of different operation clock frequencies.

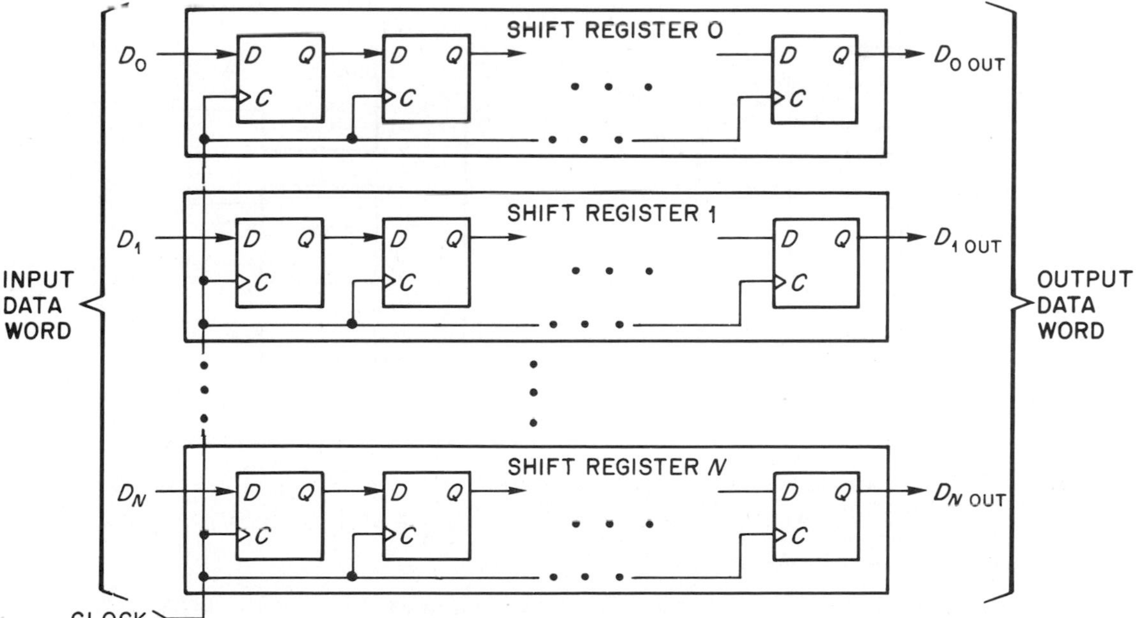

Figure 9–24 Basic construction of FIFO memory.

FEATURES:

- First-In/First-Out dual-port memory
- 64 x 4 organization (IDT72401/03)
- 64 x 5 organization (IDT72402/04)
- IDT72401/02 pin and functionally compatible with MMI67401/02
- RAM-based FIFO with low fall-through time
- Low power consumption
  - Active: 175mW (typ.)
- Maximum shift-rate – 45MHz
- High data output drive capability
- Asynchronous and simultaneous read and write
- Fully expandable by bit width
- Fully expandable by word depth at 35MHz
- IDT72403/04 have Output Enable pin to enable output data
- High-speed data communications applications
- High-performance CEMOS ™ technology
- Available in CERDIP, plastic DIP, LCC and SOIC
- Military product compliant to MIL-STD-883, Class B
- Standard Military Drawing# 5962-86846 is pending listing on this function. Refer to Section 2/page 2-4.

DESCRIPTION:

The IDT72401 and IDT72403 are asynchronous, high-performance First-In/First-Out memories organized 64 words by 4 bits. The IDT72402 and IDT72404 are asynchronous, high-performance First-In/First-Out memories organized as 64 words by 5 bits. The IDT72403 and IDT72404 also have an Output Enable ($\overline{OE}$) pin. The FIFOs accept 4-bit or 5-bit data at the data input (D_0–$D_{3,4}$). The stored data stack up on a first-in/first-out basis.

A Shift Out (SO) signal causes the data at the next to last word to be shifted to the output while all other data shifts down one location in the stack. The Input Ready (IR) signal acts like a flag to indicate when the input is ready for new data (IR = HIGH) or to signal when the FIFO is full (IR = LOW). The Input Ready signal can also be used to cascade multiple devices together. The Output Ready (OR) signal is a flag to indicate that the output contains valid data (OR = HIGH) or to indicate that the FIFO is empty (OR = LOW). The Output Ready signal can also be used to cascade multiple devices together.

Width expansion is accomplished by logically ANDing the Input Ready (IR) and Output Ready (OR) signals to form composite signals.

Depth expansion is accomplished by tying the data inputs of one device to the data outputs of the previous device. The Input Ready pin of the receiving device is connected to the Shift Out pin of the sending device and the Output Ready pin of the sending device is connected to the Shift In pin of the receiving device.

Reading and writing operations are completely asynchronous, allowing the FIFO to be used as a buffer between two digital machines of widely varying operating frequencies. The 45MHz speed makes these FIFOs ideal for high-speed communication and controller applications.

Military grade product is manufactured in compliance with the latest revision of MIL-STD-883, Class B.

FUNCTIONAL BLOCK DIAGRAM

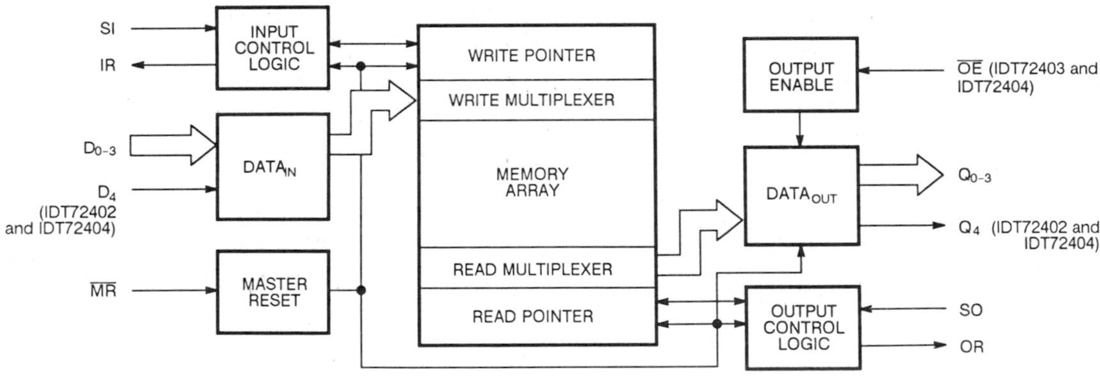

CEMOS is a trademark of Integrated Device Technology, Inc.

Figure 9–25 Real FIFO memory. *Courtesy* of Integrated Device Technology, Inc.

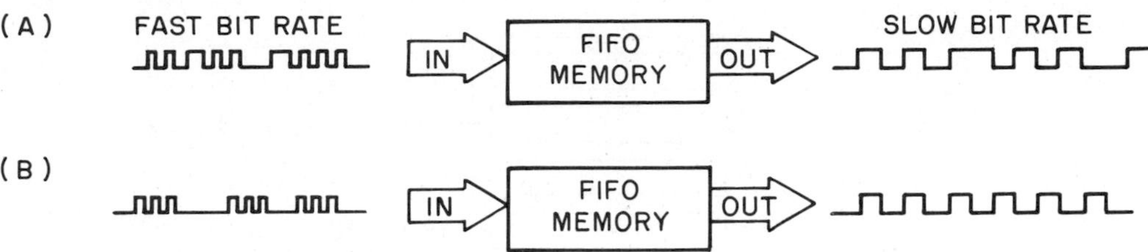

Figure 9–26 Application of FIFO memory.

Applications

One of the applications of this type of memory is the transfer of data from a computer to a printer. Since the computer can output data much faster than a printer can print the data, a FIFO memory can act as a buffer between the two. Figure 9–26(A) illustrates such an arrangement.

Another application of this type of memory is when digital information is received in irregular intervals and needs to be retransmitted at regular intervals. In this system, digital information can be temporarily stored and then transmitted at a uniform rate. This concept is illustrated in Figure 9–26(B)

Recirculating Memory

Recirculating memory is another form of volatile, sequential access, electrical memory. Essentially this type of memory is used where the same information needs to be referenced periodically. An example of this would be the display of characters on a monitor generated by a digital system. The basic idea of recirculating memory is illustrated in Figure 9–27. As shown in the figure, the recirculating memory consists of a shift register connected as a ring counter. The difference here is that data is being circulated with two different modes of operation. In one mode, the DATA IN line is activated, thus allowing new data to be placed into the register. In the other mode, the DATA IN line is inactive and the information inside the register is continuously cycled through. These kinds of memories may also be used to sample display readouts that present messages. An example of this is illustrated in Figure 9–28.

Charged-Coupled Devices

Another kind of volatile, sequentially accessible, electrical memory is the charged-coupled device **(CCD) memory.**

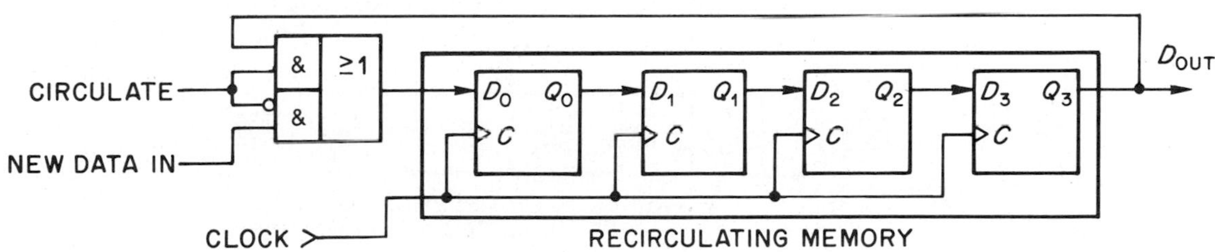

Figure 9–27 Basic idea of recirculating memory.

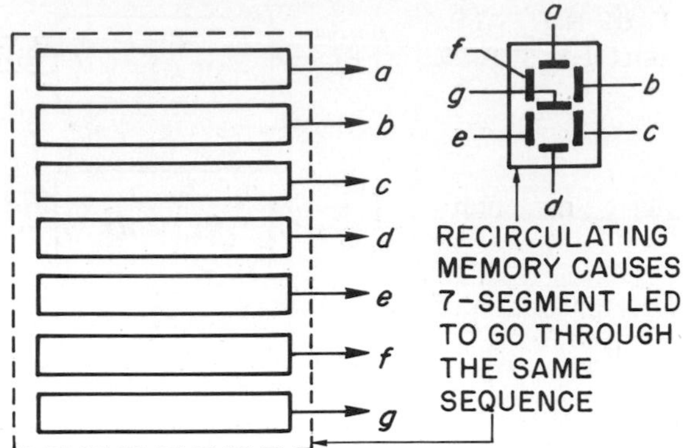

Figure 9–28 Using circulating memories for display readouts.

CCD Memory:

Charged-coupled device memory where charges are stored between tiny electrical plates fabricated in a high-density integrated circuit.

Figure 9–29 gives the basic construction of CCD memory. As shown in the figure, the charges move serially down the substrate, thus producing sequentially accessed data. These memories must also be refreshed as with the DRAM. Their main advantage is the high memory densities available. Their disadvantage is the slower access time involved since they are sequentially accessed.

Conclusion

This section presented sequentially accessed, volatile, electrical memory. Here you saw three methods used with this kind of memory: FIFO, recirculating, and CCD memories. You also learned of some of the important applications along with their advantages and disadvantages. In the next section, you will be introduced to magnetic memories. For now, test your understanding of this section by trying the following review questions.

9–6 Review Questions

1. Briefly describe what is meant by a volatile, randomly accessible, electrical memory.
2. What is meant by a FIFO memory?
3. Give an application of FIFO memory.
4. Describe the operation of recirculating memory.
5. What does CCD memory stand for? What is its main advantage?

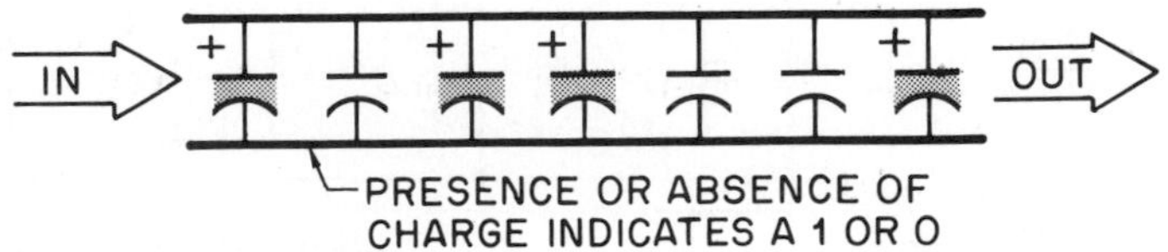

Figure 9–29 Basic idea of CCD memory.

9–7 NONVOLATILE, ERASABLE, RANDOMLY ACCESSIBLE, MAGNETIC MEMORY

Discussion

This is the first of two sections on **magnetic memory.**

> **Magnetic Memory:**
> Digital memory that uses the patterns of a magnetic field to represent stored digital information.

This section introduces randomly accessible, magnetic memory; the next section introduces sequentially accessible, magnetic memory.

Advantages of Magnetic Memory

The advantage of magnetic memory is that it is nonvolatile and erasable. This means that data may be altered or completely removed (leaving no digital pattern at all) as well as copied from the medium. It is nonvolatile because a magnetic field does not require an external source of energy. Thus magnetic material may be written to as well as read from.

Disadvantages of Magnetic Memory

The main disadvantages of magnetic memory are slow speed, high cost, and low reliability. Because most magnetic memories require some mechanical means of recording and reading digital information, this raises the cost and reduces the reliability and access time when compared to nonmechanical systems (such as electrical memory). Zeropower RAMs offer a promising alternative to magnetic memory since they require no moving parts and have a much faster access time.

Magnetic Disk

The most common method of making magnetic memory randomly accessible is through the use of a magnetic disk. The construction of a typical magnetic disk is illustrated in Figure 9–30.

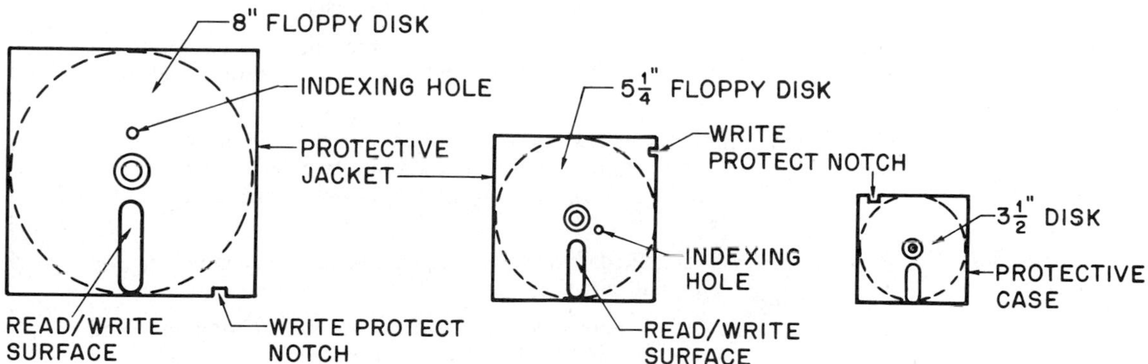

Figure 9–30 Construction of typical magnetic disks.

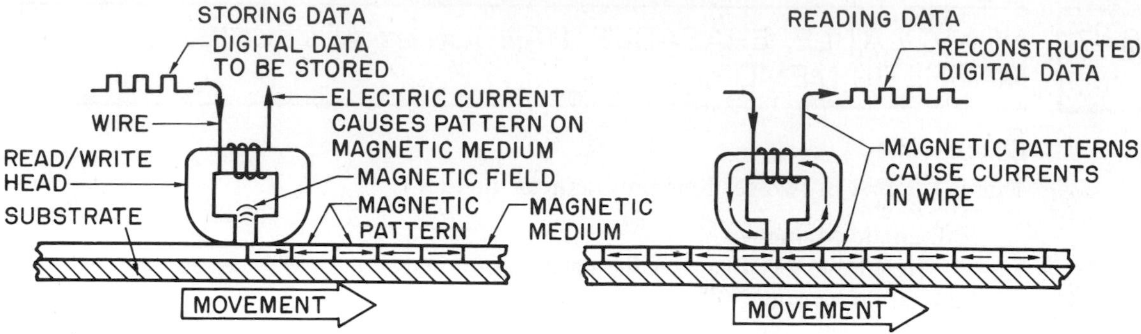

Figure 9–31 Storing and copying digital magnetic data.

The basic concept of how digital information is stored and copied from the disk is shown in Figure 9–31. The placement of digital information on a magnetic surface essentially amounts to impressing a magnetization pattern on a medium. There are several different ways of doing this, called **formats,** as outlined in Table 9–1.

Recording Format:
The method used to magnetically store data.

Most disk systems employ modified frequency modulation (**MFM**) to store digital data. Effectively this allows twice as much data to be stored on the disk as older systems. The frequency range used in this process is from 250 kHz to 500 kHz.

Table 9–1 METHODS OF RECORDING DIGITAL INFORMATION

Format	Method
Kansas City Standard	Uses two different frequencies where 2.4 kHz is used to represent a 1 and 1.2 kHz is used to represent a 0.
Manchester	A form of phase incoding where a HIGH to LOW transition represents a 0 and no transition represents a 1.
RZ Standard (Return to Zero)	Essentially a fixed-width pulse represents a 1, and no pulse represents a 0. However, there is always a return 0 after a 1 occurs.
NRZ Standard (Nonreturn to Zero)	Same format as the RZ standard, but there is no automatic return to 0 after each 1. If there are two or more 1's in succession, then the waveform will not return to zero until a 0 occurs.
FM	Frequency modulation where changes in frequency represent the digital data.
MFM	Modified frequency modulation is the most commonly used method for storing digital data on disks.

MFM:
Modified frequency modulation is a method of storing digital information on a disk. This method produces twice the density of the information as with standard frequency modulation techniques.

Types of Disk

Magnetic disks are available in three different packaging styles. These are listed in Table 9–2.

Hard Disk

A **hard disk** is one of the many types of magnetic disk storage of digital information.

Hard Disk:
A disk storage system of one or more nonremovable disks protected in a permanently sealed case.

Hard-disk systems contain one or more hard disks and can be used in all different computers. Microcomputers use the $5^1/_4$-inch hard-disk system where the storage capacity is 10 to 70 megabytes. In larger computer systems, the hard disk can supply storage capacities in the billions of bytes.

Floppy Disk

The operation of a floppy disk is essentially the same as that of a hard magnetic disk. Essentially the flexible disk spins inside its protective jacket. Digital information is stored on the disk in concentric tracks divided into pie-shaped sectors. See Figure 9–32. Data is semi randomly accessible because the action of a movable arm that holds the read/write head makes it random but the reading of the tracks makes it sequential. Thus data may quickly be picked up or copied from any sector of the disk.

Table 9–2 MAGNETIC DISK PACKAGE STYLES

Type	Construction	Application
Hard Disk	Aluminum substrate with a magnetic surface. Available as single disk or large *disk packs* where there are several magnetic disks.	Large computer systems for reading and writing large amounts of data
Floppy Disk	Flexible Mylar substrate with magnetic surface. Come in two sizes: 8″ and $5^1/_4$″.	Small computer systems for reading and writing data
Miniature $3^1/_2$″	Mylar substrate with magnetic surface.	Small computer systems for reading and writing data

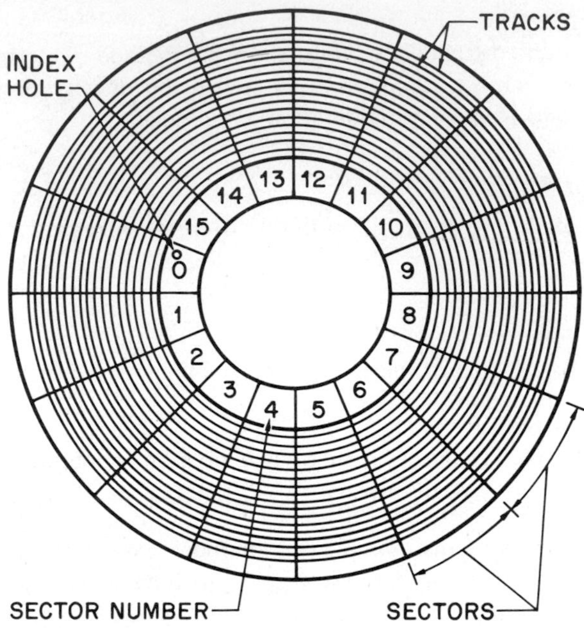

Figure 9–32 Floppy disk sectoring.

Care of Magnetic Disks

To get an idea of the importance of proper care for magnetic disks, consider the relative sizes of particles in a hard-disk unit. This is shown in Figure 9–33.

The following points must be observed when handling magnetic disks of any type:

- Never touch the surface.
- Do not bend the disk.
- Do not use sharp objects on the disk or its protective jacket (such as ballpoint pens or pencils).

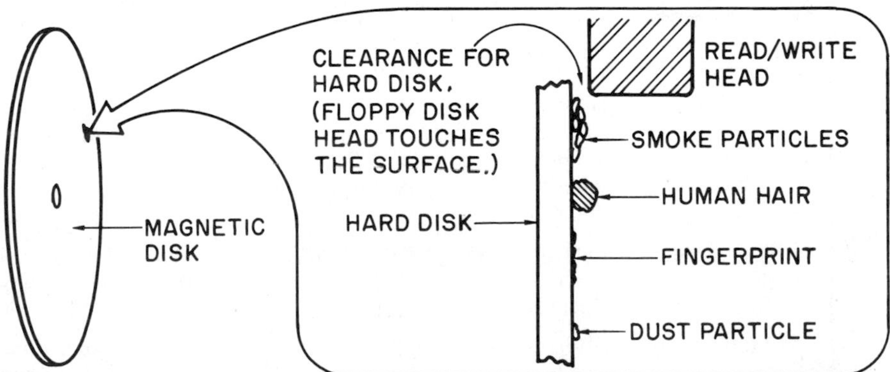

Figure 9–33 Disk contamination particles.

- Do not expose magnetic surface to direct sunlight.
- Keep disk between temperatures of 50° to 125°F (10° to 52°C).
- Do not expose the disk to magnetic fields or the stored data could be destroyed.
- Hard disks are kept in air-tight containers and should never be opened except in a "clean-room" environment.

Conclusion

This section introduced you to nonvolatile, erasable, randomly accessible, magnetic memory. Here you were introduced to the basic magnetic disk, as well as the principles involved in the reading and writing of digital information on the disk. In the next section, you will be introduced to sequentially accessible, magnetic memory. For now, test your understanding of this section by trying the following review questions.

9–7 Review Questions

1. What is meant by *magnetic memory*?
2. Give some advantages and disadvantages of magnetic memory.
3. Describe the basic construction of a magnetic disk.
4. Name the most common types of magnetic disks in current use.
5. What precautions should be taken with magnetic disks?

NONVOLATILE, ERASABLE, SEQUENTIALLY ACCESSIBLE, MAGNETIC MEMORY

Discussion

In the last section you learned about randomly accessible magnetic memory. This section introduces you to sequentially accessible magnetic memory.

Magnetic Tape

Magnetic tape is constructed from Mylar and coated with a magnetic material of iron oxide. The basic principle of recording with a magnetic tape is the same as illustrated in Figure 9–31.

Magnetic tape is by nature sequentially accessible. This means that the amount of time to access information on the tape is not always the same. To get required data, all information preceding the required data must be read. The advantage of using magnetic tapes is that they can store large quantities of digital information inexpensively. Because of this feature, they are often used for backup storage of digital information.

Magnetic tape storage comes in two basic sizes, $^1/_2$ inch and $^1/_4$ inch. For large computer systems, the $^1/_2$-inch tape is used. The use of $^1/_4$-inch tape allows you to use a high-quality audio cassette recorder to store your digital information from a personal computer.

Video Cassette Recorders

Video cassette recorders are used to back up a digital storage system. Large quantities of digital information can be stored and accessed. Figure 9–34 illustrates the recording and playback concepts used with these devices.

Video cassette recorders are for personal use and small business systems where it is necessary to store large amounts of data economically. The disadvantage of these systems is the long access time for getting information.

Magnetic Bubble Memory

Another form of nonvolatile, alterable, sequentially accessible, magnetic memory is **bubble memory.**

Bubble Memory:
Nonvolatile memory where data are represented by the presence or absence of magnetized areas (called bubbles) formed on a thin piece of garnet.

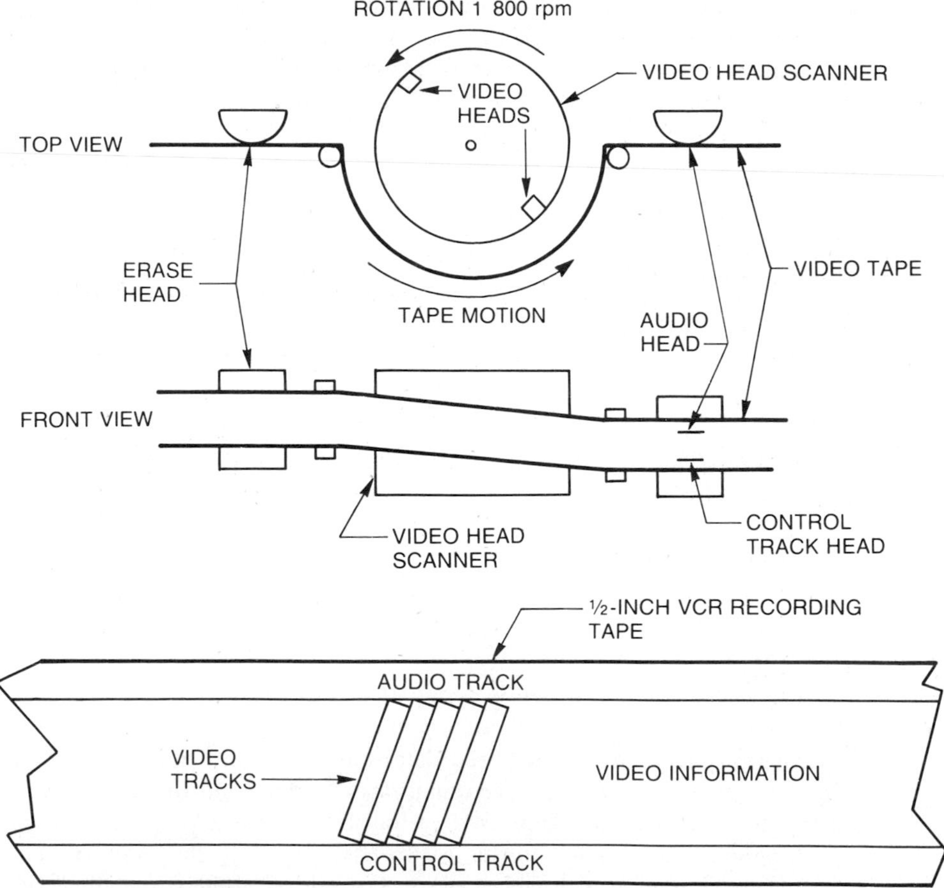

Figure 9–34 Recording method of video cassette recorders.

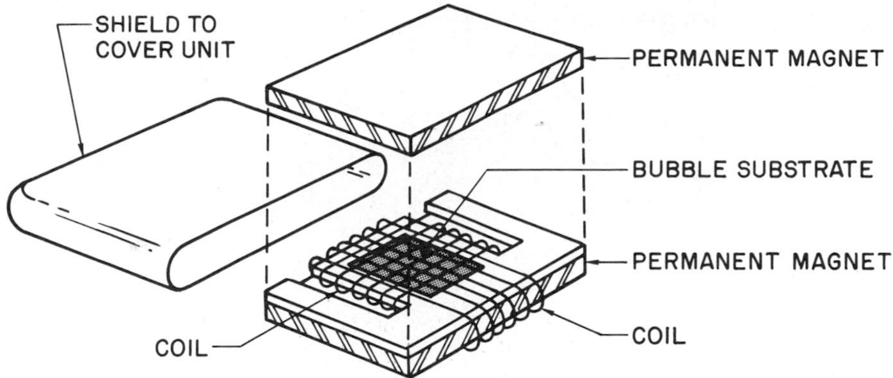

Figure 9–35 Basic construction of a magnetic bubble memory unit.

Magnetic bubble memory (MBM) comes in a unit constructed as shown in Figure 9–35. The magnetic bubble memory unit contains a list of complex elements that are necessary to generate, transfer, replicate and detect digital data. Essentially the unit consists of a thin magnetic film of *garnet*. The internal magnetic field of garnet will narrow into small magnetic domains when a perpendicular magnetic field is applied. If the strength of this magnetic field is slowly increased, the size of these magnetic domains will decrease even further. At a point in this process, these magnetic field domains will contract into small areas called *bubbles*. This process is illustrated in Figure 9–36.

If a rotating magnetic field is applied to these bubbles, they will move within the medium. If this process is carefully controlled, the presence or absence of a bubble can be used to represent digital information. In this process, data are sequentially accessed as these magnetic bubbles are moved within the system. A simplified magnetic bubble memory system is illustrated in Figure 9–37. The magnetic bubbles containing the digital data are stored in loops. These minor loops have their bubbles magnetically moved and then transferred to the main reading loop. This transfer process requires that the minor loops still retain the original information. Thus, a replication of the original bubble is preserved as each "read" bubble is split in two. The replicated bubble is then passed on to the major loop where the presence or absence of a bubble represents the required digital data.

When information is written into the bubble memory, the existing bubbles must be annihilated. This is accomplished by reading the minor loop without replication of any

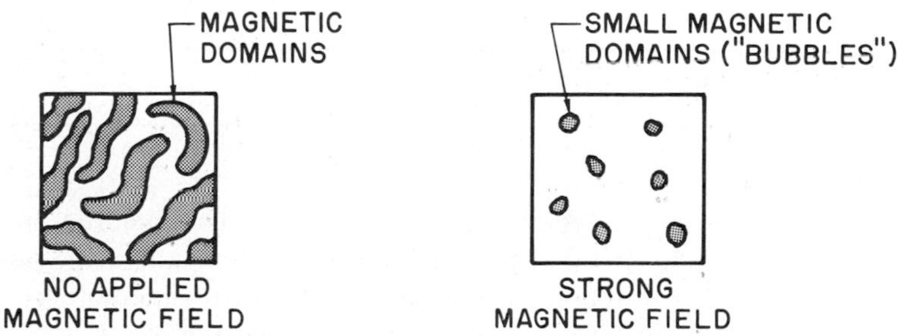

Figure 9–36 Action of magnetic field in garnet.

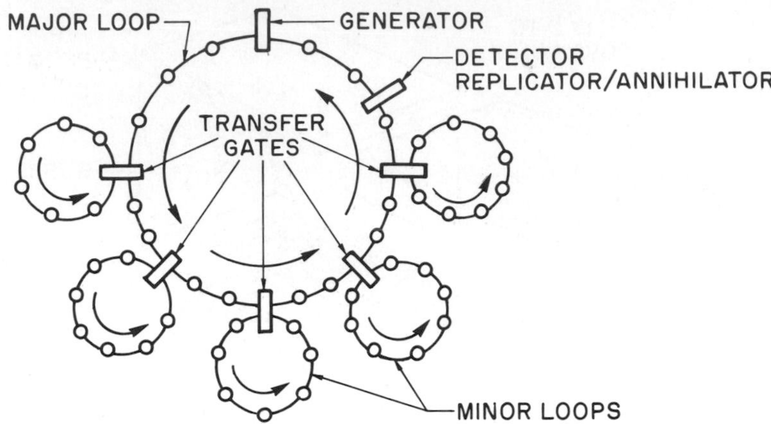

Figure 9–37 Simplified magnetic bubble memory system.

bubbles. This process causes the loop to contain all 0's. Next, new digital data are entered 1 bit at a time and stored in the appropriate loop.

The main disadvantages of MBM are its cost and long access time compared to electrical memory. Its main advantage is that it is a nonvolatile memory with write and read operations.

Conclusion

This section presented nonvolatile, erasable, sequentially accessible, magnetic memory. Here you saw how magnetic tape and magnetic bubble memory are used to store digital data. The next section introduces you to nonvolatile, nonalterable, randomly accessible, optical memory. Check your understanding of this section by trying the following review questions.

9–8 Review Questions

1. Describe the kind of digital storage produced by magnetic tape.
2. Name the three basic methods used to store digital data on magnetic tape.
3. How does the video cassette recorder differ from the audio cassette recorder?
4. What is bubble memory?
5. What are the advantages and disadvantages of bubble memory?

9–9 OPTICAL MEMORIES

Discussion

This section introduces you to optical memories. The material presented here is divided into two parts. The first part presents nonvolatile, nonalterable, randomly accessible, optical memory. The second part introduces nonvolatile, alterable, randomly accessible, optical memory. Both of these memories play an important role in the storage of large amounts of digital information.

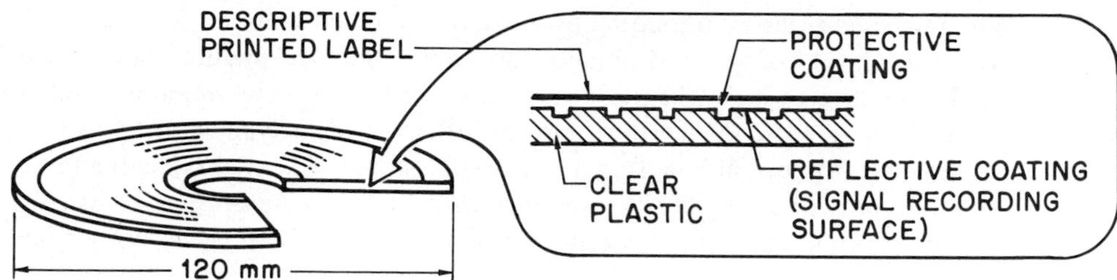

Figure 9–38 Basic structure of a compact disk.

CD ROM

The **compact disk (CD)** has earned a quick popularity for the recording of music. This same technology is also used to store digital information where the absence or presence of reflected light imparts stored information.

> **Compact Disk (CD):**
> An optically reflective disk that contains digital information stored as surface changes that change the reflection of a laser beam focused on the rotating disk.

The basic structure of a compact disk is shown in Figure 9–38. The disk is called a CD ROM because it is nonalterable. Larger disks are available for storing video (television) images and large amounts of digital information. The advantage of these disks is their ability to hold large amounts of information. The playback time of such a disk is around 74 minutes. Figure 9–39 illustrates a typical optical system used to read the information on a CD ROM.

A typical CD ROM is capable of storing 628 M bytes of information. This is equivalent to the information that can be stored on over 1 200 standard $5^1/_4''$ floppy disks or about 150 000 printed pages (well over 200 books of this size). Besides holding digital data for processing, it can have accompanying detailed pictures and high-quality voice.

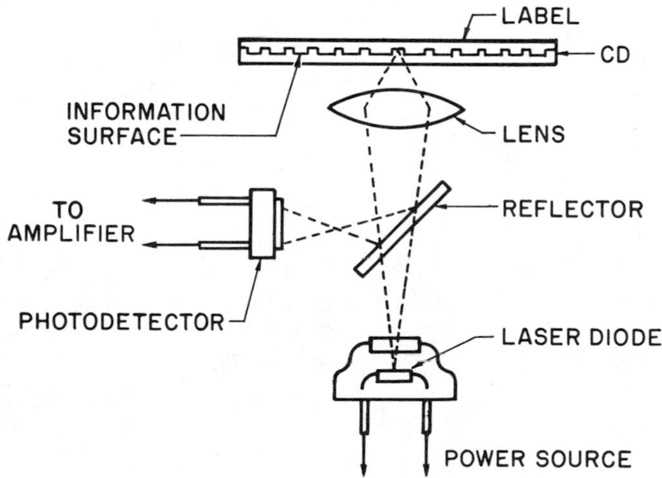

Figure 9–39 CD ROM optical system.

When used as a part of a training exercise or reference program, the CD ROM can provide over 8 hours of computer simulation programs, detailed pictures, and voice instruction.

In quantities, each CD ROM costs about $10.00 to manufacture (and stores as much information as $150.00 worth of microfilm). Another advantage of CD ROM is that it is not sensitive to normal handling and is easily stored. The main disadvantage of this memory medium is its relatively slow access time, about the same as a floppy disk drive. Thus for applications that require a fast data base search, CD ROM is not a viable contender.

Laser Memory

Emerging technology promises optical storage that is nonvolatile and alterable. One system uses a **laser diode** to change the optical properties of a specially prepared disk.

Laser Diode:
A solid-state device capable of emitting laser light when electrical energy is applied to it.

This scheme is illustrated in Figure 9–40. Essentially, the medium is magnetically sensitive to the heat from the laser. If just the right amount of heat is applied, the **coercivity** of the material drops enough to allow a weak external magnetic field to cause a change in direction of the magnetic field at that point.

Coercivity:
A measure of the magnetic field that must be applied in order to cause the magnetization in the medium to reverse direction.

To copy the written information, the same laser beam is again used, but this time at a much lower power (to not disturb the magnetic patterns). An effect known as the **Kerr effect** is sensed by the receiving lenses.

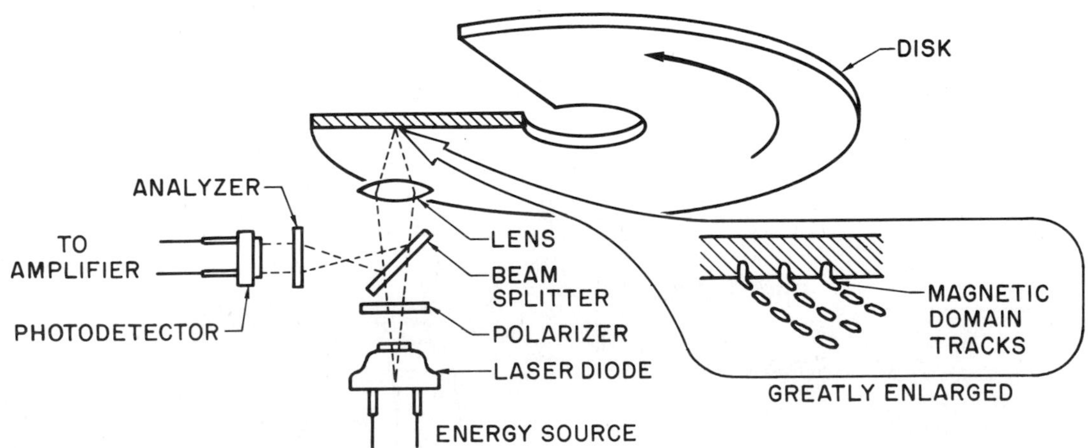

Figure 9–40 Laser memory.

Kerr Effect:
The effect on the polarization of a light beam when it is reflected off of a magnetic medium.

The major advantage of this type of system is the high storage capability (over 10 million characters per square centimeter). The disadvantage is the slow access time to write and read data when compared to hard disks.

Read/Write CD-ROM

Tandy Corporation has a unique technology called THOR® that allows recording, playback, erasure and re-recording of compact disks. This THOR-CD technology uses a laser beam to record and playback digital information on a CD-compatible optical disk. This technology can now be applied to music, voice and video as well as digital information. The system is compatible with existing CD audio and CD-ROM playback units. Someday, the floppy disk that accompanies this book may be replaced by an optical CD-ROM. With this, you will be able to see full color pictures of real hardware and actual interactive voice that will take you step by step in teaching you how to troubleshoot a system.

Conclusion

This section introduced you to the two major types of optical storage of digital information. You were introduced to the CD ROM, which can hold vast amounts of information but cannot be altered. The laser disk is a close kin to the CD ROM, has an even higher bit capacity, and can be altered. Both of these devices suffered from slow access time.

In the next section you will see how analog information can be converted into digital information for the purpose of storage and processing. Test your understanding of this section by trying the following review questions.

9–9 Review Questions

1. What is meant by a *compact disk*?
2. What is a CD ROM?
3. Name some of the advantages of CD ROM. Name a disadvantage.
4. What kind of memory is a laser disk?
5. Name some of the advantages of laser memory. Name a disadvantage.

9–10 ANALOG AND DIGITAL CONVERSION TECHNIQUES

Discussion

This section presents methods of converting from analog information to digital and from digital information to analog. Recall from the first chapter that analog information represented continuously changing information where digital information was represented by ON or OFF conditions. Before a digital system can process analog information, it must first be converted to a digital form. This process is done with a device called an **analog-to-digital (A/D) converter**.

Analog-to-Digital (A/D) Converter:
An electronic device that converts analog information into digital information.

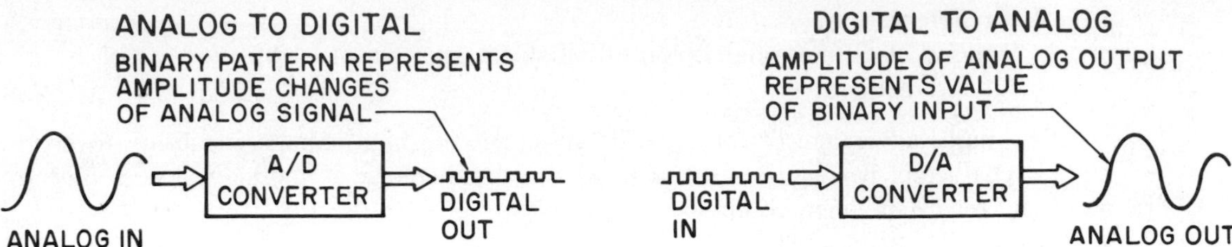

Figure 9–41 Basic concept of A/D and D/A conversion.

TRUTH TABLE		
INPUT	D_1	D_0
< 1 V	0	0
≥ 1 V AND < 2 V	0	1
≥ 2 V AND < 3 V	1	0
≥ 3 V	1	1

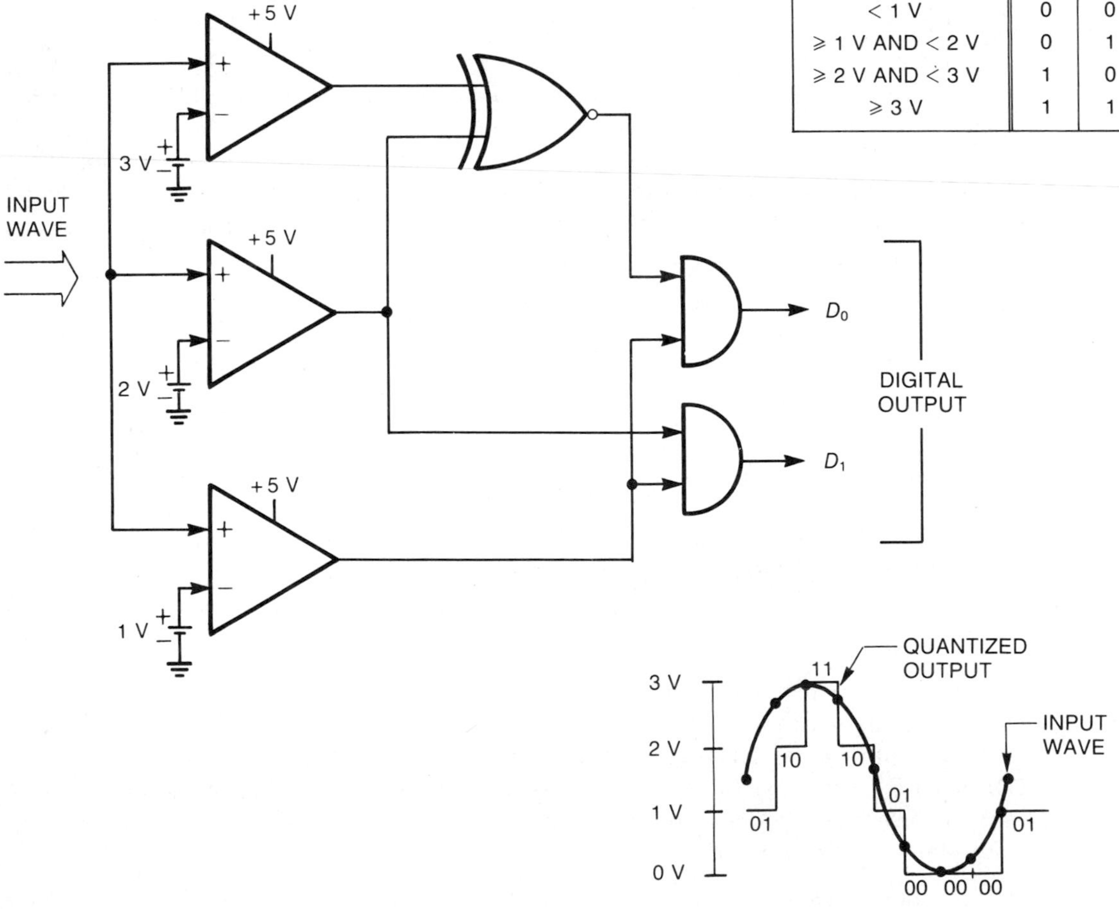

Figure 9–42 Flash converter.

Conversely, when digital information must be converted back to analog information, the device used is called a **digital-to-analog (D/A) converter**.

Digital-To-Analog (D/A) Converter:
An electronic device that converts digital information into analog information

The basic concept of this process is illustrated in Figure 9–41.

The Flash Converter

A very simple method of converting analog information into digital information is through the use of a **flash converter**. Such a system is illustrated in Figure 9–42.

Flash Converter:
An analog-to-digital converter that uses differential amplifiers and digital logic to produce a binary output that is equivalent to the magnitude of the analog input.

This type of analog to digital converter is referred to as a flash converter because the input signal is quickly converted into a binary code (done in a "flash"). This type of converter is sometimes referred to as a parallel or simultaneous converter. As shown in Figure 9–42, the circuit consists of three different amplifiers that have either a 0 V (LOW) or a +5 V (HIGH) output, depending on the level of the input signal (which in this case is a sine wave varying between 0 V and +3 V). The outputs of these differential amplifiers operate the logic circuity consisting of an EXOR gate and two AND gates. Recall that the output of a differential amplifier connected to a +5-V power source will be HIGH any time its + input is larger than the value of the reference voltage at its − input.

Example 9–2

Using the flash converter of Figure 9–42, indicate the resulting binary output for one cycle of each sine wave in Figure 9–43.

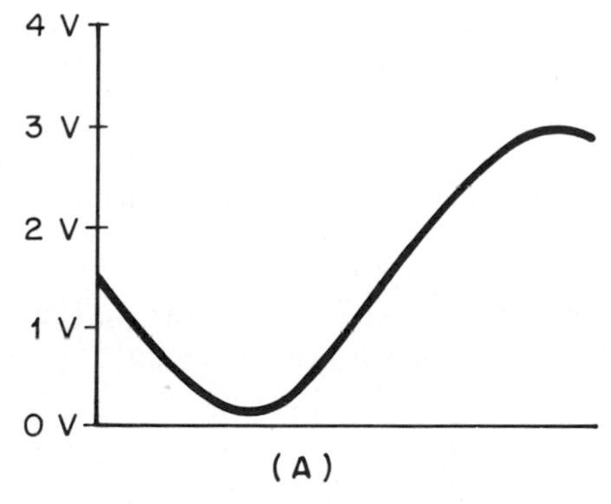

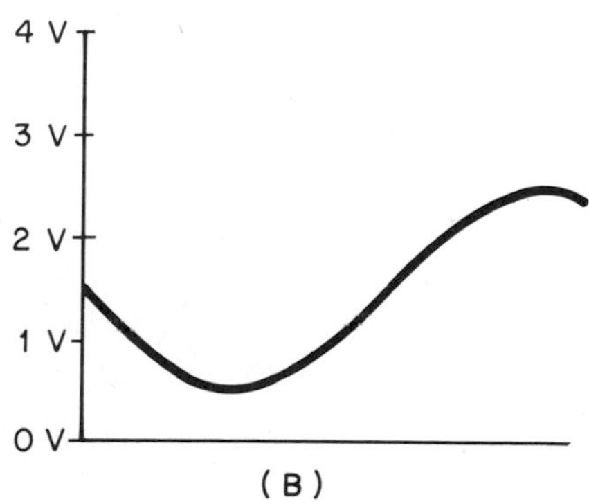

Figure 9–43

Solution

Refer to Figure 9–44.

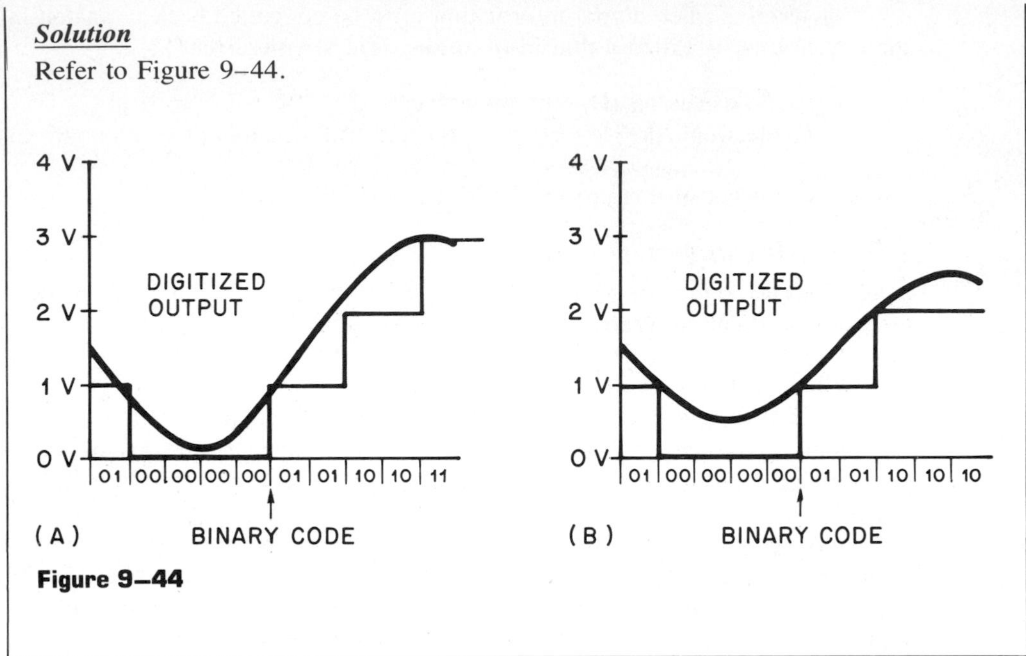

Figure 9–44

This example shows that it is difficult to distinguish between the waveforms because the **resolution** of the example flash converter is very poor.

Resolution:
The ability of an A/D converter to accurately convert small changes in an analog signal to binary data.

Real Flash Converters

Figure 9–45 shows the data sheet of a real flash converter. As you can see from the specification sheet, flash converters are available that can produce a high degree of resolution. Note from the figure of the A/D converter that there are 64 inputs with a 6-bit output. This allows good resolution with a fast conversion time. The device contains an overflow indicator on the output to indicate if the value of the analog input equals (or exceeds) the value of the reference voltage (V_{REF}). Note that all of the resistors used for the voltage divider are internal to the device and no external components are necessary.

Digital-to-Analog Converter

One method of converting digital information into analog is shown in Figure 9–46. The digital-to-analog converter consists of the following major sections:

- A 4-to-16-line decoder.
- A resistor voltage divider (each resistor has 1 V across it).
- A bilateral switch consisting of 16 switches and 16 control lines.

PIN NAME	DESCRIPTION
$-V_S$	Negative supply terminal, nominally -5.2V.
ANALOG GROUND	Analog ground return. All grounds should be connected together near the the AD9000.
$V_{HYSTERESIS}$	The hysteresis control voltage varies the comparator hysteresis from 15mV to 50mV, for a change of 0V to $+3$V at the hysteresis control pin.
ENCODE	The ENCODE pin controls the conversion cycle. Encode is rising edge sensitive and should be driven with a 50% duty-cycle waveform under normal conditions.
$-V_{REF}$	The most negative reference voltage for the internal resistor ladder.
ANALOG INPUT	Analog input pin.
$+V_S$	Positive supply terminal, nominally $+5.0$V.
$+V_{REF}$	Most positive reference voltage of the internal resistor ladder.
BIT 6 (LSB)	One of six digital outputs. BIT 6 (LSB) is the least-significant-bit of the digital output.
BIT 5 – BIT 2	One of six digital outputs.
BIT 1 (MSB)	One of six digital outputs. BIT1 (MSB) is the most-significant-bit of the digital output.
OVERFLOW	Overflow data output. Logic high indicates an input overvoltage ($A_{IN} \geq +V_{REF}$).
DIGITAL GROUND	Digital ground return. All grounds should be connected together near the AD9000.

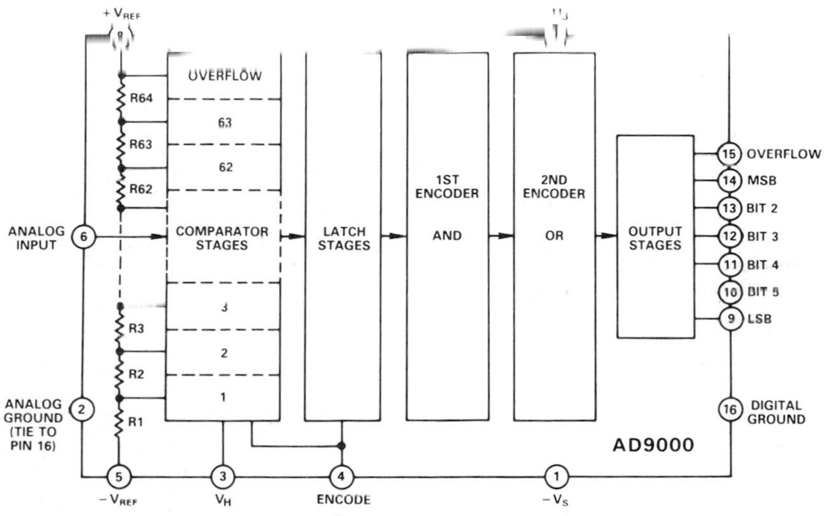

AD9000 Functional Block Diagram

Figure 9–45 A real flash converter. *Courtesy* of Analog Devices, Inc.

As an example of how the system works, consider the binary input to be 0101_2 (5_{10}). This means that line X_5 on the output of the decoder will be active. This will activate control line F on the input to the bilateral switch causing switch F to close. The value of the voltage on the analog output will be equal to the voltage drop across $R_{15} + R_{14} + R_{13} + R_{12} + R_{11}$. Since each resistor drops 1 V, the total voltage across these five resistors will be 1 V + 1 V + 1 V + 1 V + 1 V = 5 V. Thus, in this case, the value of the analog output voltage is equal to the binary value of the digital input.

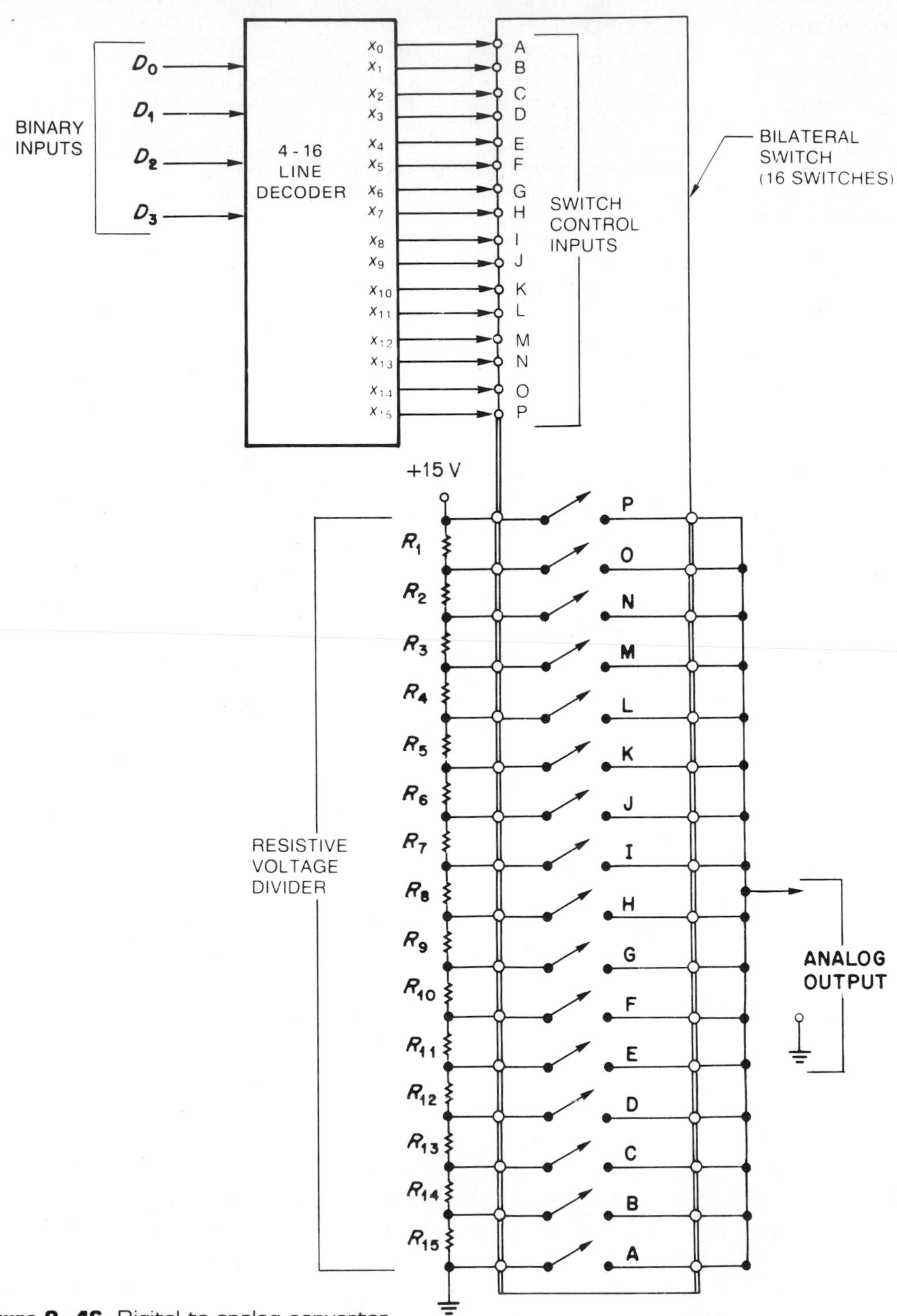

Figure 9–46 Digital-to-analog converter.

Example 9–3

For the D/A converter of Figure 9–46 assume a 4-bit binary counter is connected to the input of the 4-to-16 line decoder. Sketch the resulting output analog waveform assuming the binary counter starts its count at 0000 and cycles through two complete counts.

Solution

See Figure 9–47.

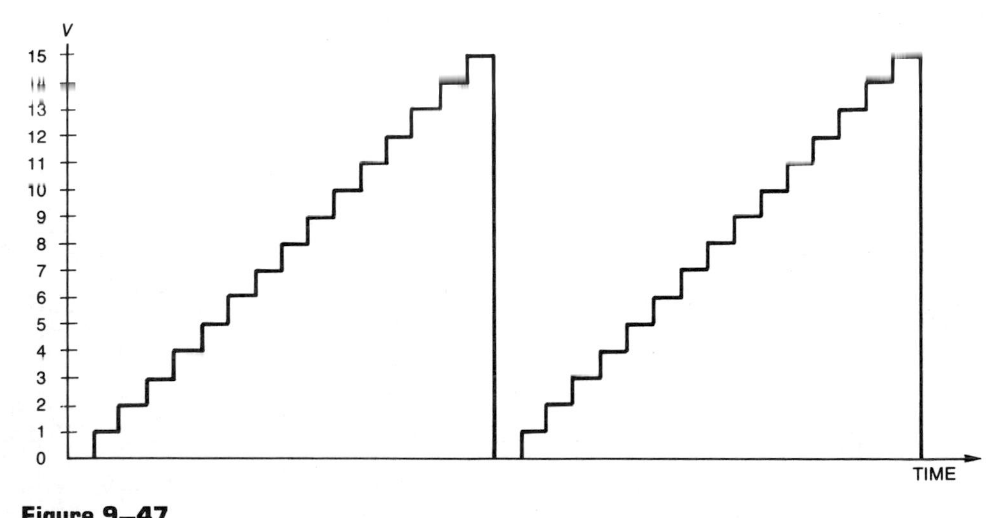

Figure 9–47

The resulting output waveform for Example 9–3 is called a **stairstep waveform**.

Stairstep Waveform:
A waveform where the changes in amplitude are of equal increments, thus producing a silhouette similar to a flight of stairs.

Stairstep A/D Converter

Another method of converting an analog signal to a digital signal is by a **stairstep A/D converter**.

Stairstep A/D Converter:

An analog-to-digital converter that uses a waveform with abrupt changes in its output, called *steps*. Each step represents an increase in the value of the analog signal and a corresponding increase in the binary value representing the analog input.

Figure 9–48 shows a stairstep A/D converter. Its operation can be explained as follows:

1. Assume that the binary counter starts at zero ($0000\ 0000_2$) and that the output of the comparator is zero.
2. An analog signal voltage is applied to the analog input of the comparator. If this voltage is greater than zero, it will cause the output of the comparator to switch to a HIGH output, which in turn enables the AND gate and allows the clock pulses to cause the binary counter to begin counting.

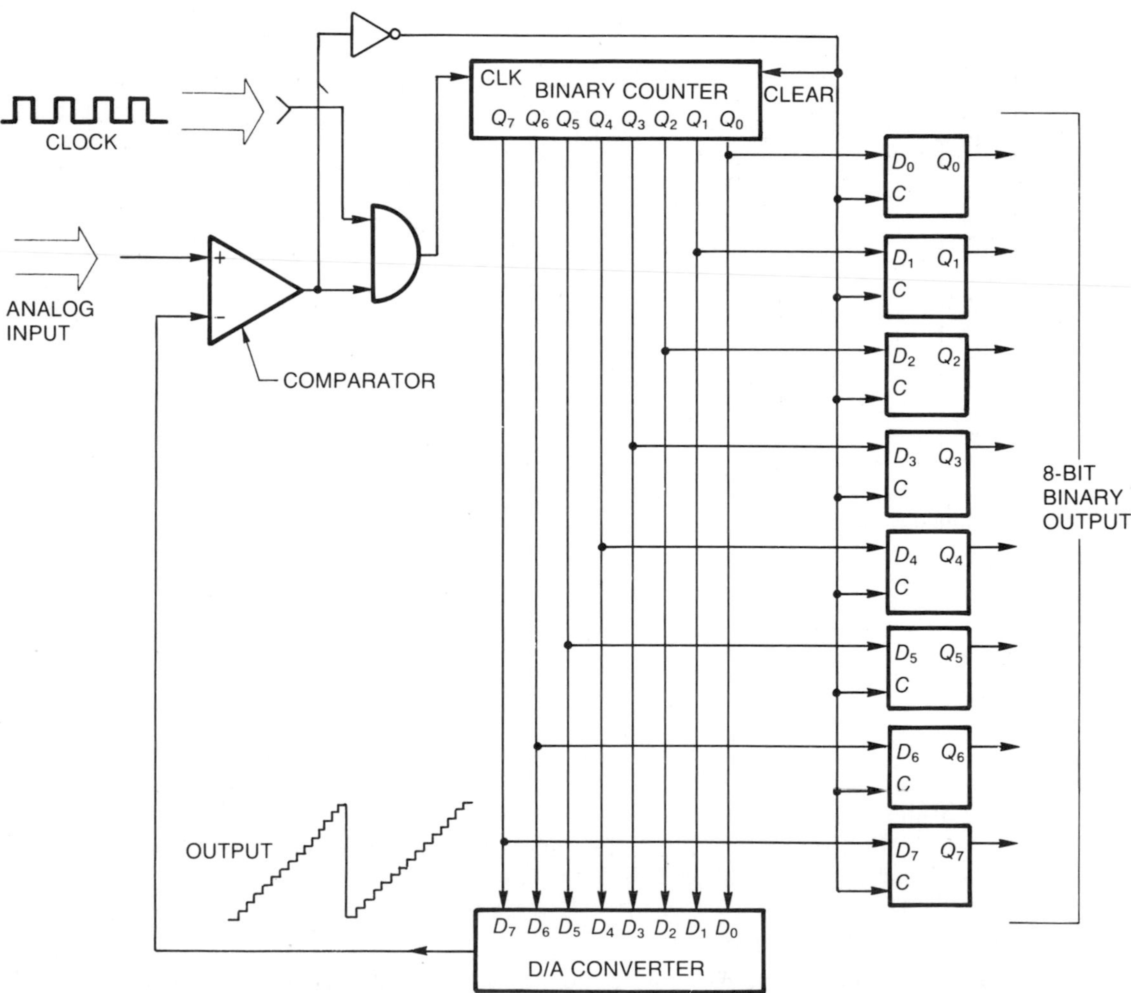

Figure 9–48 Stairstep A/D converter.

3. The binary counter causes a stairstep output from the D/A converter. This is fed to the − output of the comparator.
4. The counter continues its count until the stairstep voltage is larger than the input analog signal. As soon as this happens, the output of the comparator goes to zero.
5. When the output of the comparator goes to zero, the AND gate inhibits any more clock pulses from incrementing the binary counter. At the same time, the count of the binary counter is stored in the data flip-flops.
6. The value stored in the data flip-flops now represents the binary value of the analog signal at the time the signal was sampled.

The foregoing process is illustrated in Figure 9–49, which shows that this method is slower than the flash converter, because the larger the value of the input voltage, the longer the binary counter must count. In the worst case, the counter would have to go through 256 counts. As shown in the figure, for each sample of the input signal, the counter must start from zero up to the point where the stairstep reference voltage reaches a value larger than the input analog signal. Hence, the **conversion time** depends upon the value of the analog signal.

> **Conversion Time:**
> The amount of time it takes to convert an analog voltage reading into a digital value.

Specifically, the conversion time is 2^N where N is the number of clock pulses.

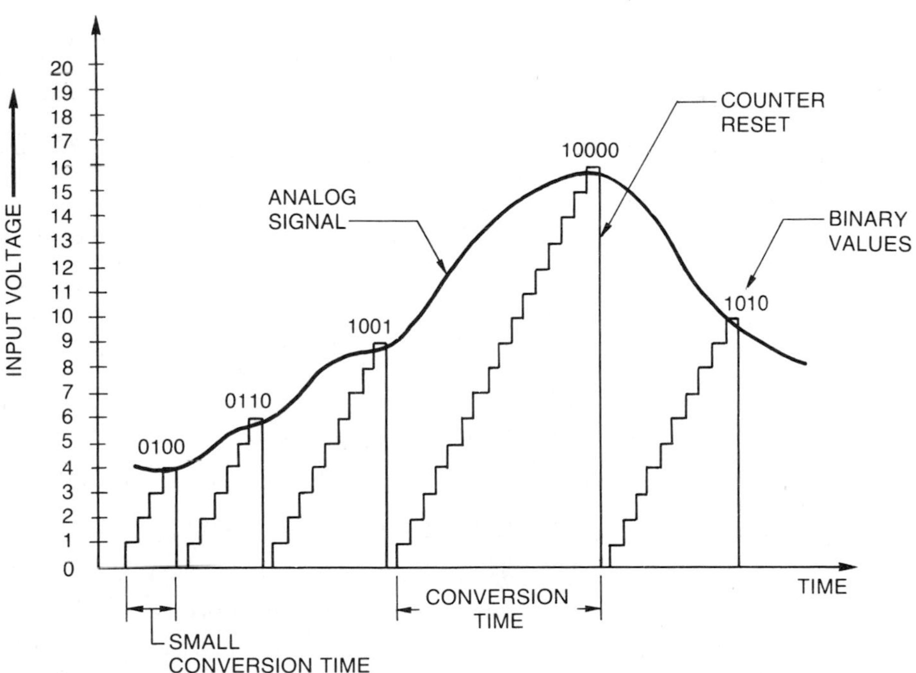

Figure 9–49 Example of stairstep A/D converter.

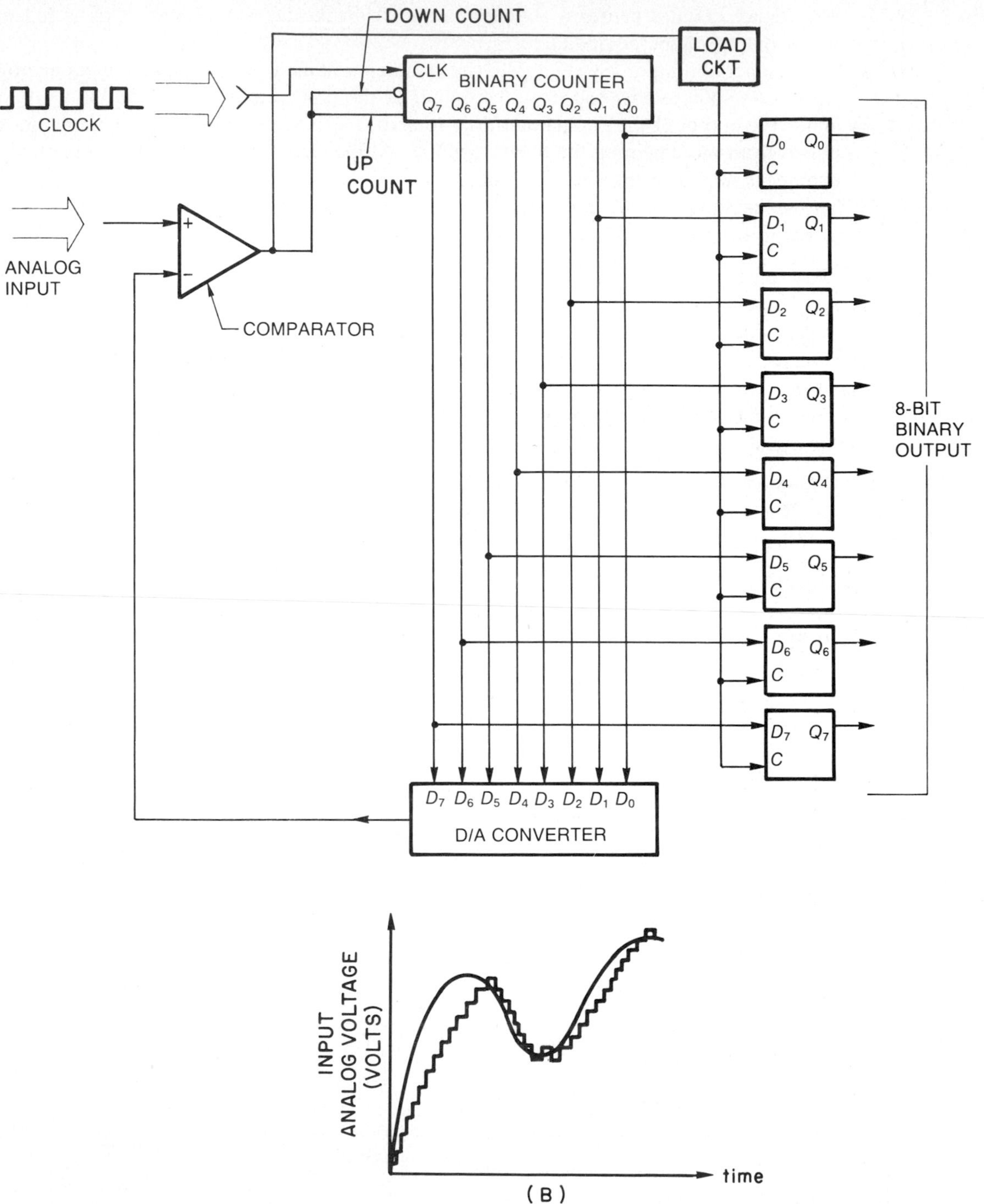

Figure 9–50 (A) Converting to a tracking A/D converter. (B) Waveform of tracking A/D converter.

Tracking A/D Converter

The stairstep A/D converter can be improved with a slight modification to its control logic. Instead of clearing the binary counter to zero, the counter can be made to count down to a new analog value as well as count up to a new analog value. The only changes that need to be made are shown in Figure 9–50(A).

Essentially the tracking A/D converter works exactly the same way as the stairstep A/D converter. The main difference is that a decrease in the value of the analog signal causes the binary counter to do a DOWN count. This produces a decreasing stairstep. The resulting waveform is shown in Figure 9–50(B).

Real D/A and A/D Converters

Figure 9–51 illustrates a real D/A converter. Figure 9–52 illustrates a real A/D converter.

Dual Slope A/D Converter

Another type of analog-to-digital converter is the dual slope A/D converter. The advantage of this type of converter is that it will ignore rapid variations of the input signal due to noise. A simplified block diagram of such an A/D converter is shown in Figure 9–53.

The mechanical switches shown in the figure would actually be electrical switches operated by the computer or controlling device. The basic operation of the system is outlined below.

1. A START pulse clears the binary counter and at the same time causes S_2 to close, discharging the capacitor C_1. S_2 then opens and S_1 is connected to the analog input voltage.
2. The capacitor C_1 starts to charge toward the value of the input analog voltage. This will cause the output of the comparator to go to +5 V (HIGH), thus enabling the AND gate input.
3. At this time, the output of the control logic along with the output from the comparator will cause the AND gate to allow each clock pulse to pass and the binary counter will begin its count. This binary count will continue until an overflow occurs. When this happens, an overflow signal is sent to the control logic.
4. The control logic will now cause switch S_1 to switch its connection to the reference voltage ($-V_{REF}$). Note that this voltage is opposite in polarity to the analog input voltage.
5. The capacitor C_1 has charged up to the value of the analog input voltage. This capacitor will now discharge at a linear rate and at the same time, the binary counter will begin its count again from zero. As soon as the capacitor has discharged to zero volts, the comparator output will become zero and cause the AND gate to stop clock pulses from going to the binary counter.
6. When the capacitor has discharged to zero, the control logic will then generate an EOC (end of conversion) pulse meaning that the value in the binary counter is now proportional to the value of the analog input voltage.

The conversion time for this type of converter is 2^{N+1} clock cycles. Thus for an 8-bit converter a total of 512 pulses are required to make a full-scale conversion.

Conclusion

This section presented the basic concepts of A/D and D/A conversion. Here you saw what was involved in both processes and some example circuits of how this process took

DAC0800/DAC0801/DAC0802 8-Bit Digital-to-Analog Converters

General Description

The DAC0800 series are monolithic 8-bit high-speed current-output digital-to-analog converters (DAC) featuring typical settling times of 100 ns. When used as a multiplying DAC, monotonic performance over a 40 to 1 reference current range is possible. The DAC0800 series also features high compliance complementary current outputs to allow differential output voltages of 20 Vp-p with simple resistor loads as shown in *Figure 1*. The reference-to-full-scale current matching of better than ±1 LSB eliminates the need for full-scale trims in most applications while the nonlinearities of better than ±0.1% over temperature minimizes system error accumulations.

The noise immune inputs of the DAC0800 series will accept TTL levels with the logic threshold pin, V_{LC}, grounded. Changing the V_{LC} potential will allow direct interface to other logic families. The performance and characteristics of the device are essentially unchanged over the full ±4.5V to ±18V power supply range; power dissipation is only 33 mW with ±5V supplies and is independent of the logic input states.

The DAC0800, DAC0802, DAC0800C, DAC0801C and DAC0802C are a direct replacement for the DAC-08, DAC-08A, DAC-08C, DAC-08E and DAC-08H, respectively.

Features

- Fast settling output current 100 ns
- Full scale error ±1 LSB
- Nonlinearity over temperature ±0.1%
- Full scale current drift ±10 ppm/°C
- High output compliance −10V to +18V
- Complementary current outputs
- Interface directly with TTL, CMOS, PMOS and others
- 2 quadrant wide range multiplying capability
- Wide power supply range ±4.5V to ±18V
- Low power consumption 33 mW at ±5V
- Low cost

Typical Applications

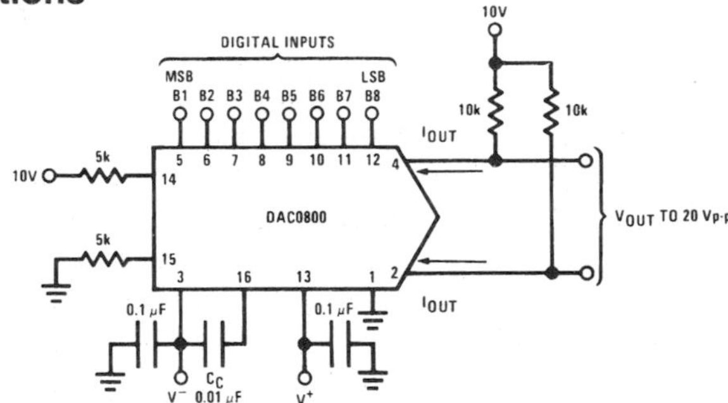

FIGURE 1. ±20 V_{P-P} Output Digital-to-Analog Converter (Note 4)

Ordering Information

Non-Linearity	Temperature Range	Order Numbers				
		J Package (J16A)*		N Package (N16A)*		SO Package (M16A)
±0.1% FS	$-55°C \le T_A \le +125°C$	DAC0802LJ	DAC-08AQ			
±0.1% FS	$0°C \le T_A \le +70°C$	DAC0802LCJ	DAC-08HQ	DAC0802LCN	DAC-08HP	DAC0802LCM
±0.19% FS	$-55°C \le T_A \le +125°C$	DAC0800LJ	DAC-08Q			
±0.19% FS	$0°C \le T_A \le +70°C$	DAC0800LCJ	DAC-08EQ	DAC0800LCN	DAC-08EP	DAC0800LCM
±0.39% FS	$0°C \le T_A \le +70°C$	DAC0801LCJ	DAC-08CQ	DAC0801LCN	DAC-08CP	DAC0801LCM

*Devices may be ordered by using either order number.

Figure 9–51 A real D/A converter. *Reprinted* with permission of National Semiconductor Corporation.

ADC0800 8-Bit A/D Converter

General Description

The ADC0800 is an 8-bit monolithic A/D converter using P-channel ion-implanted MOS technology. It contains a high input impedance comparator, 256 series resistors and analog switches, control logic and output latches. Conversion is performed using a successive approximation technique where the unknown analog voltage is compared to the resistor tie points using analog switches. When the appropriate tie point voltage matches the unknown voltage, conversion is complete and the digital outputs contain an 8-bit complementary binary word corresponding to the unknown. The binary output is TRI-STATE® to permit bussing on common data lines.

The ADC0800PD is specified over −55°C to +125°C and [illegible]

Features

- Low cost
- ±5V, 10V input ranges
- No missing codes
- Ratiometric conversion
- TRI-STATE outputs
- Fast — $T_C = 50\ \mu s$
- Contains output latches
- TTL compatible
- Supply voltages — 5 V_{DC} and −12 V_{DC}
- Resolution — 8 bits
- Linearity — ±1 LSB
- Conversion speed — 40 clock periods
- Clock range — 50 to 800 kHz

Block Diagram

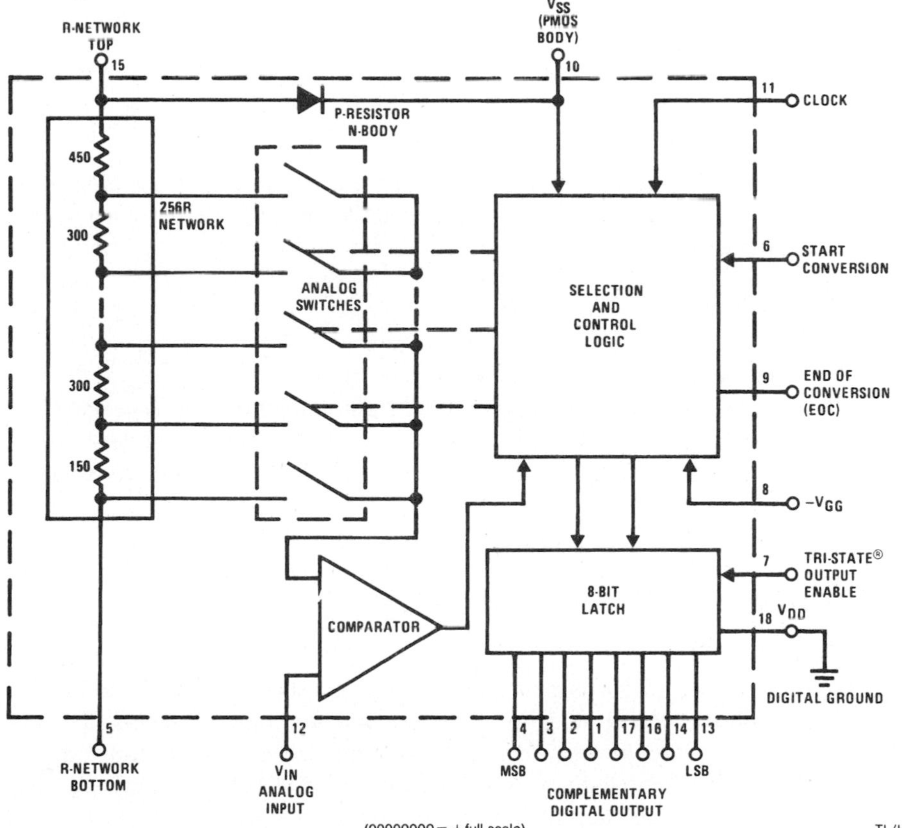

Figure 9–52 A real A/D converter. *Reprinted* with permission of National Semiconductor Corporation.

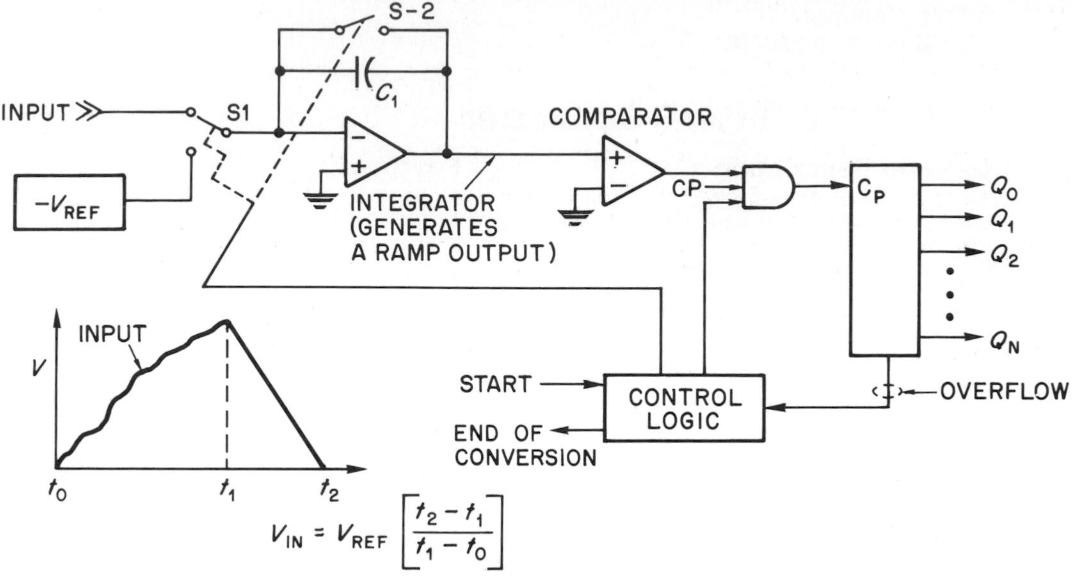

Figure 9–53 Dual slope A/D converter.

place. There is an electrical analysis of these converters in the electrical applications section of this chapter. Check your understanding of this section by trying the following review questions.

9–10 Review Questions

1. What is an A/D converter? What does it do?
2. What is a D/A converter? What does it do?
3. Briefly describe the operation of a flash converter.
4. What is the major disadvantage of a staircase A/D converter?
5. State the major difference between a tracking A/D converter and a staircase A/D converter.

MICROPROCESSOR APPLICATION

Discussion

This section introduces real 8-bit microprocessors. You will see how memory chips are used with the microprocessor to create complete digital systems. The material in this section sets the stage for the microprocessor and related courses to follow. This is an exciting section that brings together much of the material you have learned up to this point. The microprocessor section in Chapter 7 presented a complete system that interfaced input and output control with a microprocessor. Here you will see these same techniques used with memory.

An 8-Bit Microprocessor

Figure 9–54 illustrates the logic diagram of an 8-bit microprocessor. The figure shows only one control line (the R/W). Actually, several control lines are not shown because they will not be used in the discussion to follow. The next chapter will present all control lines along with the most popular microprocessors.

This microprocessor is called an 8-bit microprocessor because it has eight data lines (D_0 through D_7). This means that it has the capability of $2^8 = 256$ different instructions, which could result in 256 different processes built into its internal ROM. Note that there are 16 address lines (A_0 through A_{15}). These address lines mean that the 8-bit microprocessor has the capability of addressing $2^{16} = 65\,536$ different memory locations (65 536/1 024 = 64K of memory). This 8-bit microprocessor has many more capabilities than the 4-bit one you have been studying. However, the important point is that they both use the same principles of operation. What you learned about the 4-bit microprocessor can be applied here. The main difference is there are just more bits in the 8-bit processor.

Memory Map

While you will be using memory (RAM and ROM) with the microprocessor, it is important to establish a method of easily visualizing what parts of memory are activated by different addresses. One method of doing this is by the use of a **memory map**.

Memory Map:
A listing showing the contents of all memory locations.

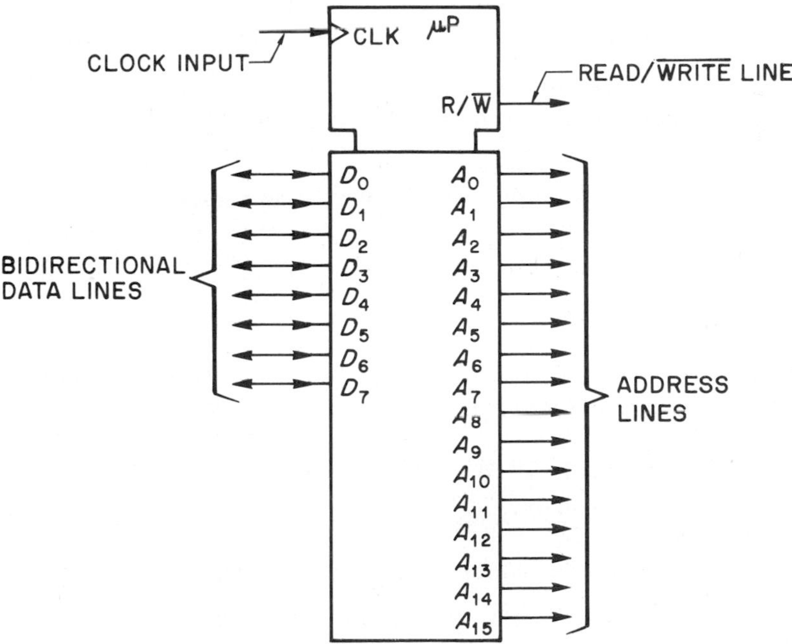

Figure 9–54 Logic symbol of an 8-bit microprocessor.

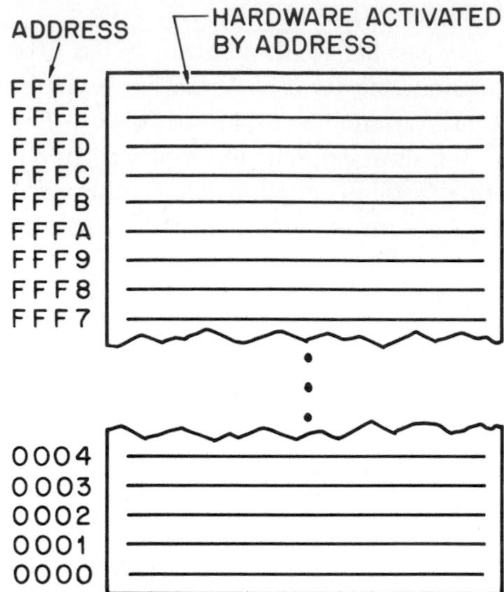

Figure 9–55 Memory map of 8-bit microprocessor.

Recall that you were first introduced to the concept of memory mapping in the microprocessor application section of Chapter 2. The microprocessor application section in Chapter 4 demonstrated how information could be stored in memory, while the microprocessor section of Chapter 6 showed how a bus system is used. Figure 9–55 presents the memory map of the 8-bit microprocessor. The purpose of the memory map is to ensure that the contents of all the possible addresses are accounted for.

High-Order Address:
The most significant 8 bits of a 16-bit address line.

Low-Order Address:
The least significant 8 bits of a 16-bit address line.

Example 9–4

Express the high order address and the low order address of each of the following addresses in hexadecimal as well as the complete hex address.
(A) 0000 0001 1011 1111 (B) 1001 0000 1100 0111
(C) 1111 1111 1110 1110

Solution

For the high order address, use the 8 most significant bits, $A_8 - A_{15}$:

(A) 0000 0001 = 01H (B) 1001 0000 = 90H

(C) 1111 1111 = FFH

For the low order address, use the 8 least significant bits, $A_0 - A_7$:

(A) 1011 1111 = BFH (B) 1100 0111 = C7H

(C) 1110 1110 = EEH

For the complete address, simply combine the two:

(A) 0000 0001 1011 1111 = 01BFH

(B) 1001 0000 1100 0111 = 90C7H

(C) 1111 1111 1110 1110 = FFEEH

(Don't forget that the H following the numbers means that they are hexadecimal numbers to the base 16.)

Memory Map Applications

Consider the 16 × 8 ROM chip connected to the 8-bit microprocessor shown in Figure 9–56. The logic notation used is such to emphasize the weight of each data bit and each control bit. Otherwise dependency notation is used in this and following figures in this section. As you can see from the memory map, the only address lines that affect this chip

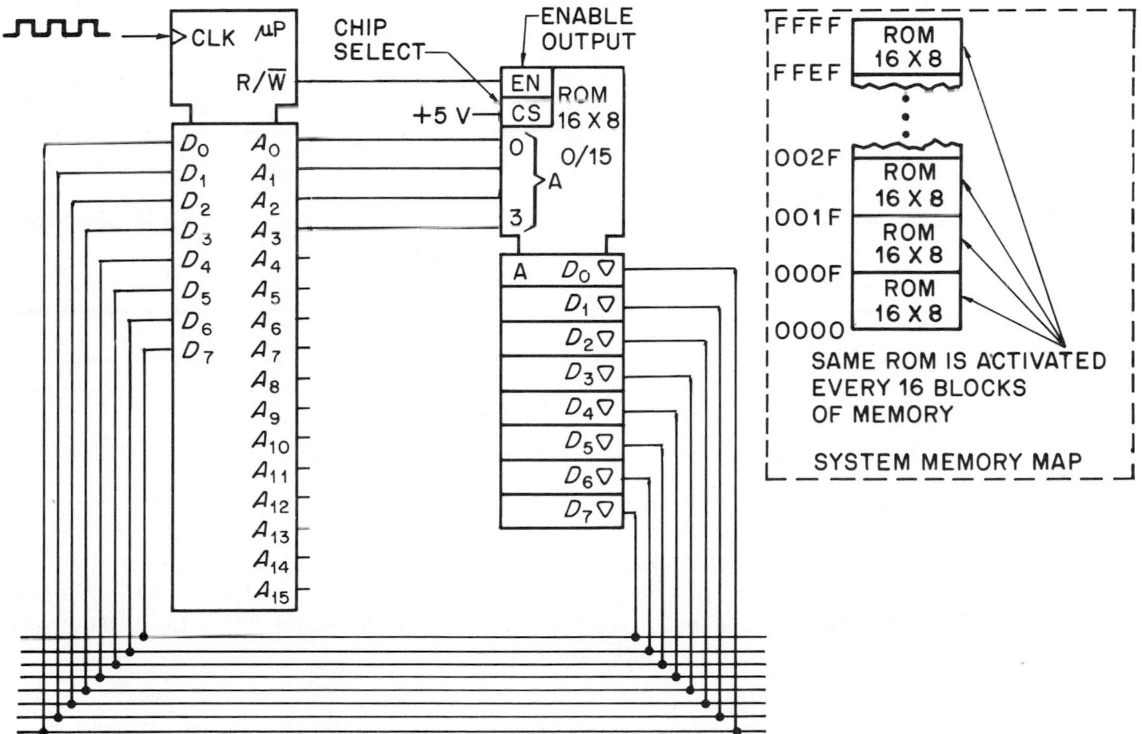

Figure 9–56 Microprocessor and ROM with resulting memory map.

are A_0 to A_3 (which results in $2^4 = 16$ memory locations in the ROM). The ROM memory is *not affected* by the status of address lines A_4 to A_{15}. Thus, address 0 of the ROM chip will be activated at any of the following address combinations:

A_{15}	A_{14}	A_{13}	A_{12}	A_{11}	A_{10}	A_9	A_8	A_7	A_6	A_5	A_4	A_3	A_2	A_1	A_0	
0	0	0	0	0	0	0	0	0	0	0	0	0	0	0	0	= 0000H
0	0	0	0	0	0	0	0	0	0	0	1	0	0	0	0	= 0010H
0	0	0	0	0	0	0	0	0	0	1	0	0	0	0	0	= 0020H
	⋮				⋮				⋮				⋮			
1	1	1	1	1	1	1	1	1	1	1	1	0	0	0	0	= FFF0H

And, by the same token, address 15H of the ROM is always activated at any of these address combinations:

A_{15}	A_{14}	A_{13}	A_{12}	A_{11}	A_{10}	A_9	A_8	A_7	A_6	A_5	A_4	A_3	A_2	A_1	A_0	
0	0	0	0	0	0	0	0	0	0	0	0	1	1	1	1	= 00FH
0	0	0	0	0	0	0	0	0	0	0	1	1	1	1	1	= 001FH
0	0	0	0	0	0	0	0	0	0	1	0	1	1	1	1	= 002FH
	⋮				⋮				⋮				⋮			
1	1	1	1	1	1	1	1	1	1	1	1	1	1	1	1	= FFFFH

In other words, the ROM chip doesn't care what values address lines A_{15} to A_8 assume. Thus the address decoding can be expressed as

A_{15}	A_{14}	A_{13}	A_{12}	A_{11}	A_{10}	A_9	A_8	A_7	A_6	A_5	A_4	A_3	A_2	A_1	A_0	
X	X	X	X	X	X	X	X	X	X	X	X	0	0	0	0	= XXX0H
	⋮				⋮				⋮				⋮			
X	X	X	X	X	X	X	X	X	X	X	X	1	1	1	1	= XXXFH

Recall that the X means "don't care." Because address lines A_{15} to A_8 were not used to decode the ROM chip, this chip can be addressed by many different address combinations. This type of decoding is called **partial address decoding**. Note the system memory map in Figure 9–56 graphically illustrates this concept.

Partial Address Decoding:
The results achieved when all address lines are not decoded.

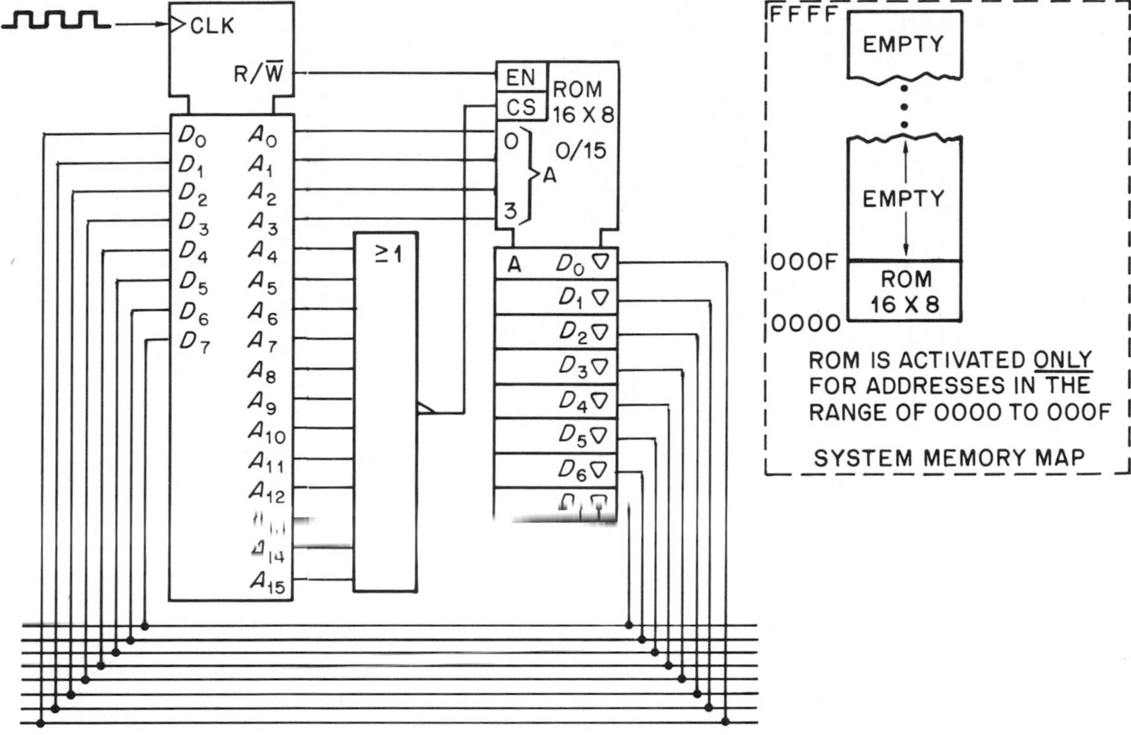

Figure 9–57 Full address decoding with ROM.

Figure 9–57 shows the same ROM connected to the same microprocessor, but this time, **full address decoding** is used.

Full Address Decoding:
The results achieved when all address lines are decoded.

Now look at the resulting memory map in Figure 9–57. You can see that because of the logic gate, the address selected on the ROM does now depend upon the value of *all* the address lines.

Example 9–5

Construct the memory map for the ROM and RAM chips shown connected to the 8-bit microprocessor of Figure 9–58.

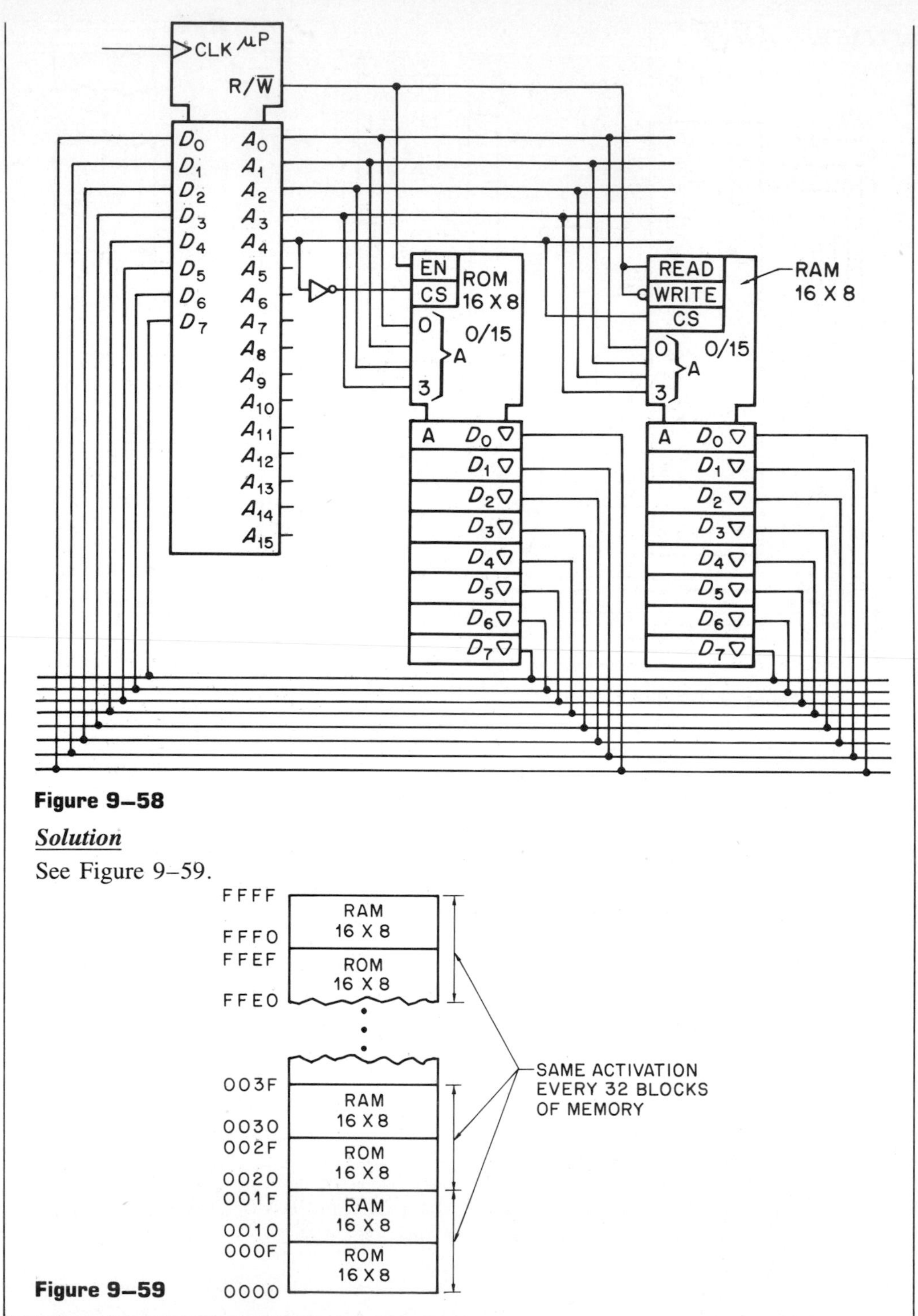

Figure 9–58

Solution

See Figure 9–59.

Figure 9–59

Example 9–6

Modify the circuit of Figure 9–58 so that there is full decoding. Draw the resultant memory map.

Solution

See Figure 9–60.

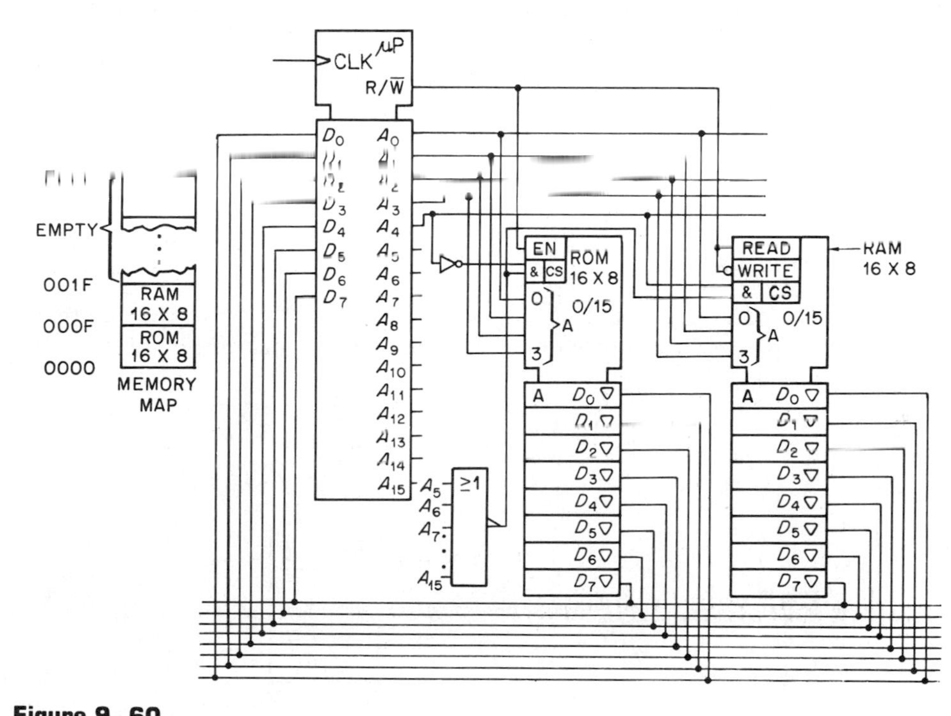

Figure 9–60

Expanding the Word Size

Figure 9–61 shows the 8-bit microprocessor using 1K × 1 RAM chips. All 8 chips are addressed at the same time. This allows the 8 data bits to move as a group. Memory chips come in various sizes, with word sizes of 1, 2, 4, and 8. It is therefore important to know how to use these chips in order to make a word size that is compatible with the word size of the microprocessor.

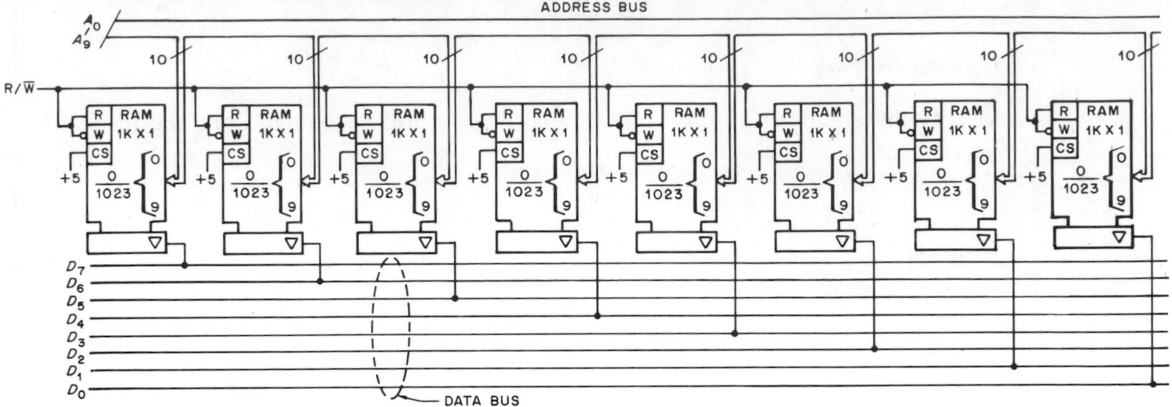

Figure 9–61 Connecting 1 K × 1 RAM chips.

Example 9–7

Show how you would use 1K × 4 ROM and 1K × 4 RAM chips with an 8-bit microprocessor in order to achieve the memory map of Figure 9–62 (8-bit word).

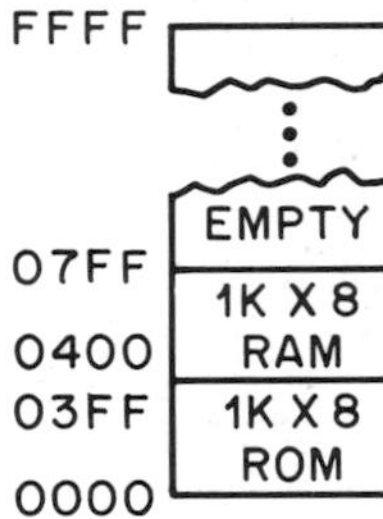

Figure 9–62

Solution

See Figure 9–63. Note that the ROM chips are addressed at the same time. Also note that one ROM chip has its four data lines connected to D_0–D_3 of the data bus while the other ROM chip has its four data lines connected to D_4–D_7 of the same data bus. This allows the two ROM chips to act as a single 8-bit word ROM chip. The same thing may be said about the two 4-bit word RAM chips. Both RAM chips respond to the same address and with their connections to the data bus now behave as if they were a single RAM chip with an 8-bit word size.

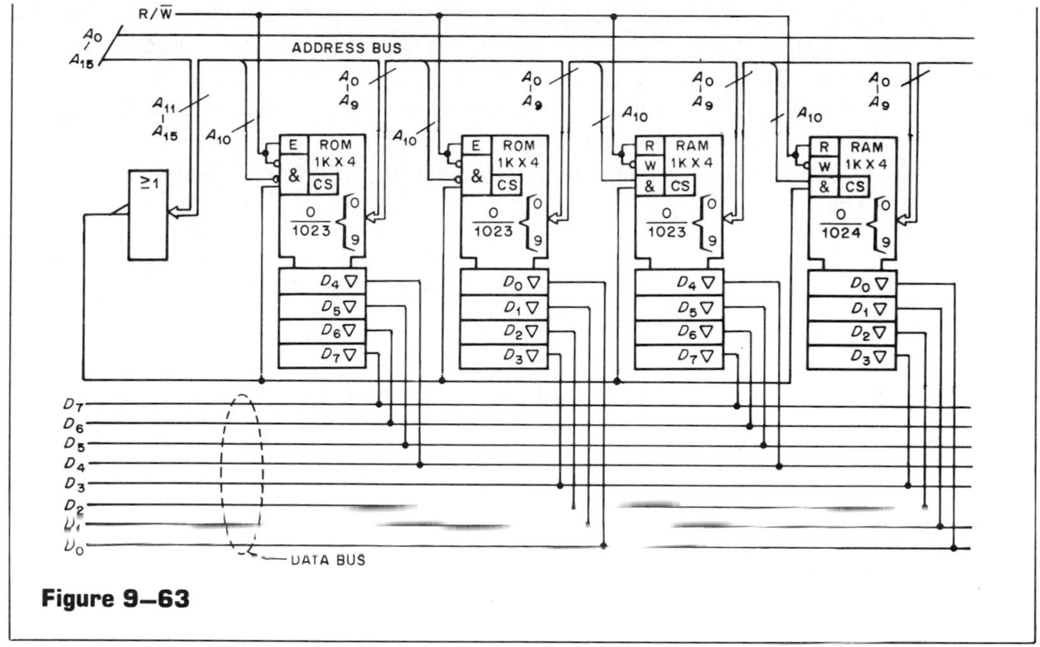

Figure 9–63

Expanding Memory Size

Not only can the word size of memory be expanded but so can its addressing size. This is illustrated in Figure 9–64, where two 1K × 8 RAM chips are connected in such a way

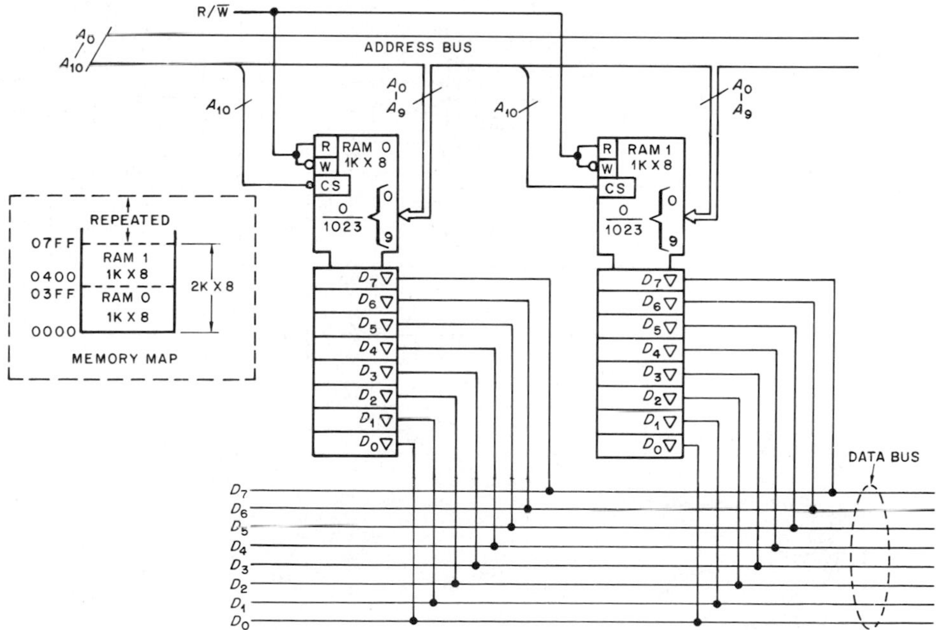

Figure 9–64 Expanding the memory size.

as to produce a total of 2K × 8 RAM. This is accomplished by having the address of one RAM start right after the ending address of the other. As shown in the figure, the memory map indicates that one RAM is addressed from 0000 to 03FF, while the other is addressed from 0400 to 07FF. This is the same result as if you had one single RAM chip whose address size allowed it to go from 0000 to 07FF. In effect, you have taken two 1K × 8 RAM chips, and by sequentially addressing them, have effectively created a single 2K × 8 RAM.

Example 9–8

Construct the memory map of the microprocessor system of Figure 9–65.

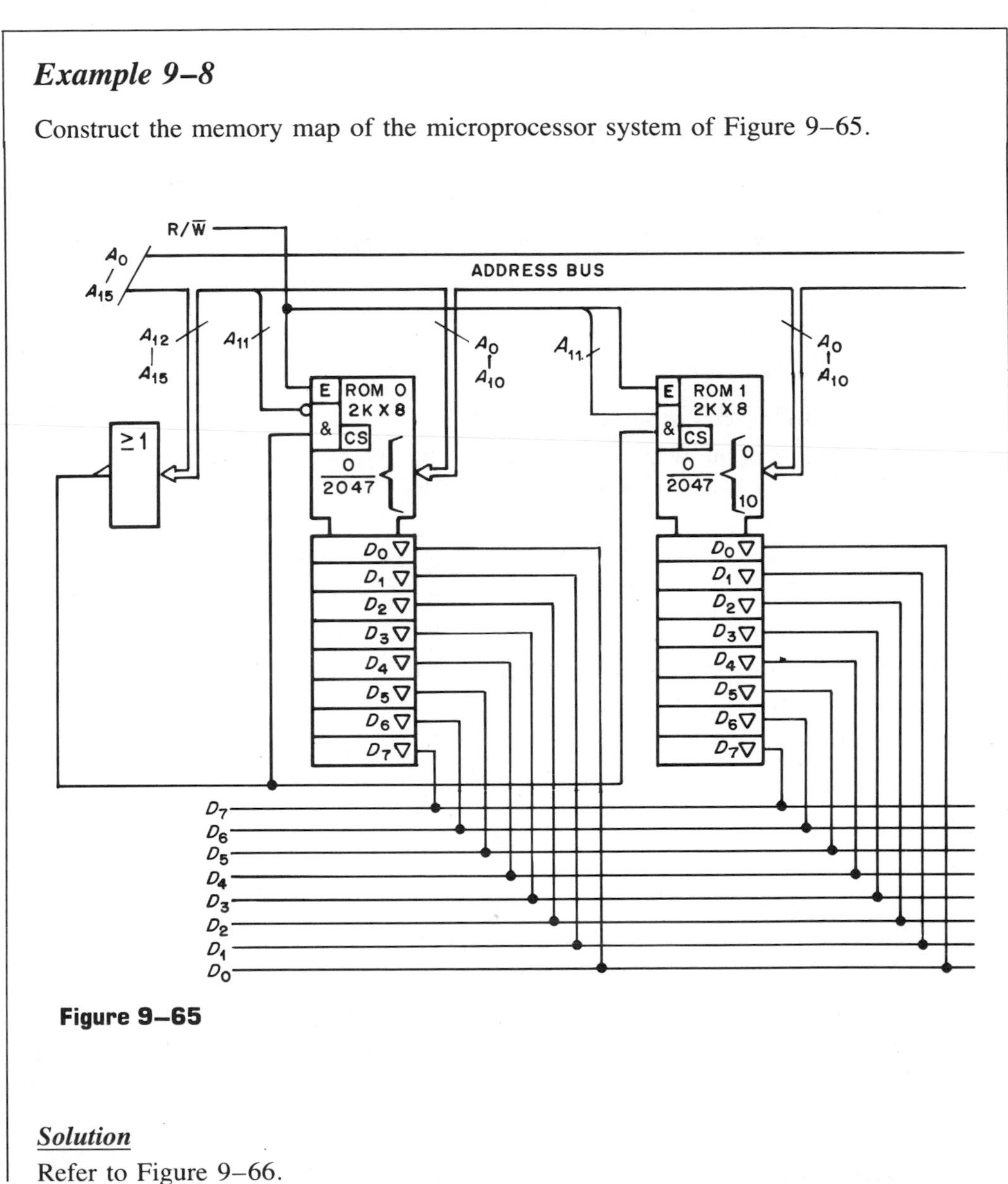

Figure 9–65

Solution

Refer to Figure 9–66.

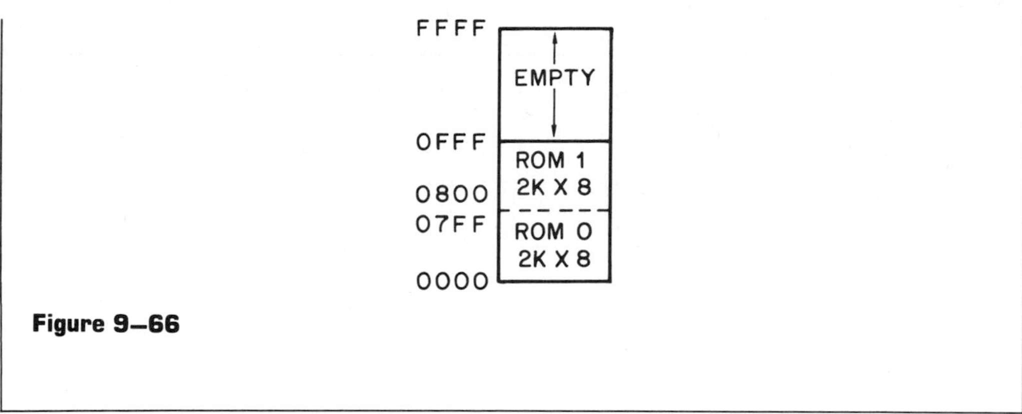

Figure 9–66

Memory I/O

Recall that I/O stood for input/output. Memory I/O simply means connecting the microprocessor system to some external I/O devices. The connections to these devices are called **I/O ports**.

> **I/O Ports:**
> A reference to any device that reads or writes data to anything external to the microprocessor, its associated memory, and control system.

Figure 9–67 illustrates an 8-bit microprocessor connected to two I/O ports. One port is an A/D converter, and the other is a D/A converter. Memory consists of a 1K × 8 ROM and a total of 8K × 8 RAM. As you can see from the memory map, the I/O ports are treated the same as any other memory locations. This is called **memory-mapped I/O**.

> **Memory-Mapped I/O:**
> Input/output treated as a regular memory location.

Some microprocessors do not use memory-mapped I/O. That is, I/O ports are treated separately from ROM or RAM memory. This process is called **isolated I/O**. You will be introduced to this type of microprocessor in the next chapter.

> **Isolated I/O:**
> Input/output treated differently from regular memory locations.

Memory Timing Diagrams

Figure 9–68 shows the timing relationship for a write and a read operation using static RAM. Note from the figure that both timing diagrams show the relationship between the address lines, R/W line, CE (CHIP ENABLE) line, and the data lines of the RAM. Observe that there is a certain amount of time required for RAM to respond to input changes. This is referred to as RAM *access* time. These timing relationships are discussed in detail in the electronic application section of this chapter. Figure 9–69 illustrates the timing relationships for write and read operations with dynamic RAM. Note that the address must first be latched into the RAM before a read or write operation.

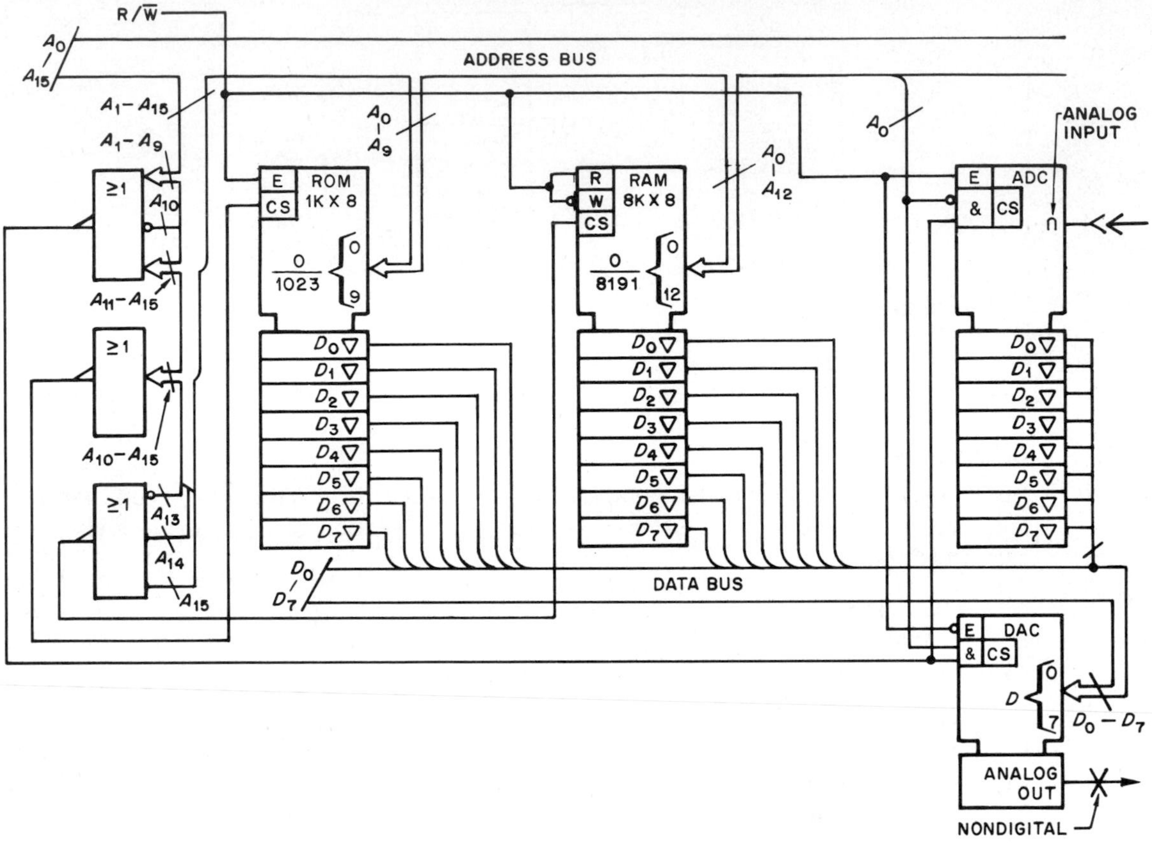

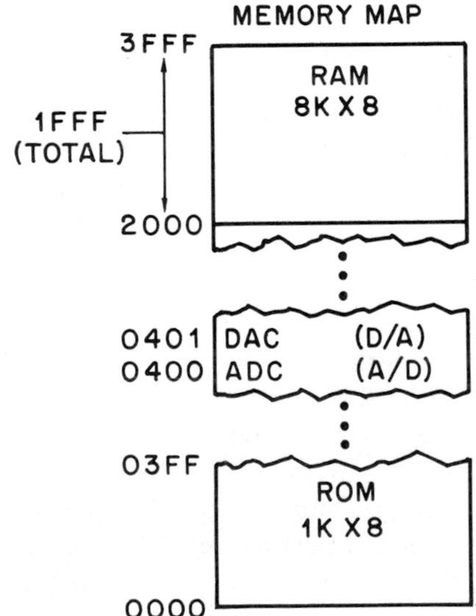

Figure 9–67 Microprocessor connected to I/O ports.

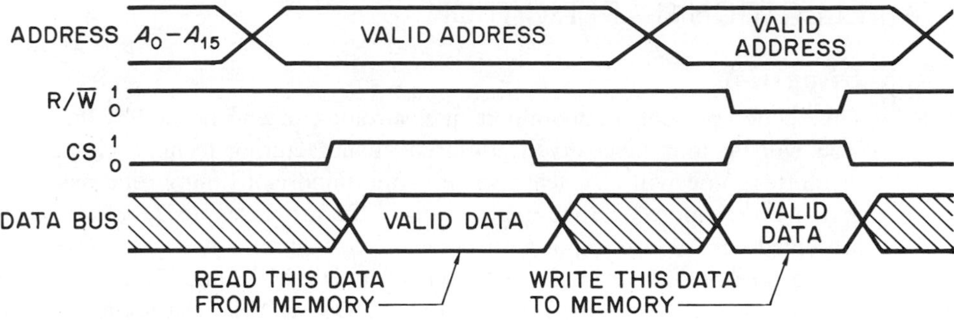

Figure 9–68 Memory timing relationships with static RAM.

Conclusion

This was an important section. You were introduced to your first 8-bit microprocessor. You also saw the various ways of connecting memory with the microprocessor as well as how to represent these connections with a memory map. Timing diagrams were also introduced. In the next section, you will have the opportunity to learn some of the important electrical specifications and differences in memory. There you will also learn many of the important details concerning D/A and A/D converters. Check your understanding of this section by trying the following review questions.

9–11 Review Questions

1. How many address lines and data lines does the microprocessor presented in this section contain? What is the significance of this?
2. What is a *memory map*?
3. What is the difference between a *high-order* address and a *low-order* address?
4. Explain the difference between *partial* address decoding and *full* address decoding.
5. What is *memory-mapped* I/O?

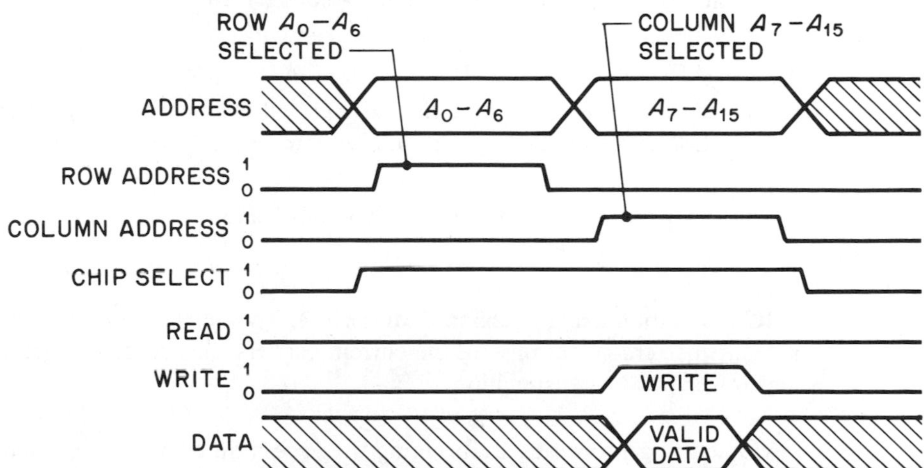

Figure 9–69 Memory timing relationships with dynamic RAM.

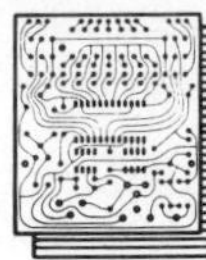

ELECTRONIC APPLICATION

Discussion

This section presents many important electronic concepts needed by the service technician. You will be introduced to the different manufacturing technologies used for electronic memories. You will also learn some of the important timing relationships used by these memories. A more detailed explanation of the processes of D/A and A/D conversion is presented here. More important technical information concerning these important devices is also presented. Interfacing techniques between MOS and TTL are presented here. The section concludes with the presentation of a DVM. This instrument helps merge much of the information you have studied up to this point.

Memory Technologies

Electrical memories can be divided into two major groups: **bipolar memory** and **MOS memory.**

Bipolar Memories:
Electrical memories manufactured from bipolar transistors.

MOS Memories
Electrical memories manufactured from field-effect transistors.

Essentially bipolar transistors are a type of semiconductor device that requires an input current for operation. This results in a larger power consumption from the input signal. On the other hand, field effect transistors require only an input voltage for operation. This results in a great reduction of power required from the input signal. Bipolar memories are fabricated on a single silicon wafer chip and must use a variety of electronic components such as resistors, diodes, and bipolar transistors. Because of this, the packaging density of this technique is low compared to MOS. Another form of bipolar technology is **ECL.**

ECL:
Emitter-coupled logic where the basic circuit is essentially a differential amplifier. Sometimes called current mode logic.

MOS memories primarily use **NMOS** in their fabrication.

NMOS:
IC fabrication using N-channel MOSFETs. This results in high package densities. Since free electrons are the current carriers, this results in greater speed and lower power consumption.

New manufacturing processes are bringing the access times of NMOS memories close to that of the bipolar memories. The CMOS memories present very low power consumption

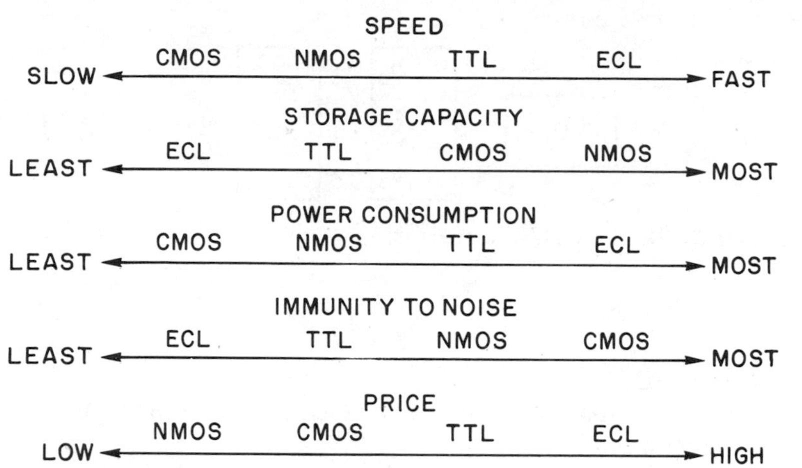

Figure 9–70 Memory technology characteristics.

along with high package densities. Figure 9–70 illustrates some of the important relationships between the technologies.

Timing Specifications

Figure 9–71 illustrates a typical memory timing diagram. The timing relationships between the microprocessor and memory are very critical. The design of the circuits that take these relationships into account are beyond the scope of this book. However, as a technician, you should realize that memory requires time to respond to requests from the microprocessor.

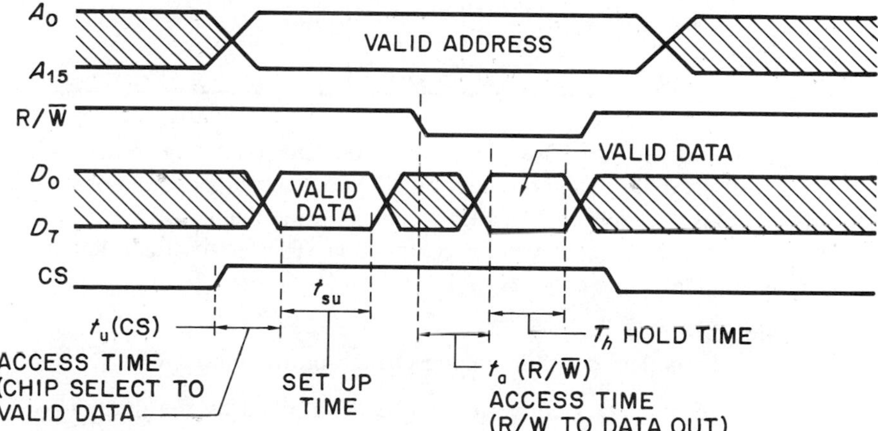

Figure 9–71 Typical memory timing diagram.

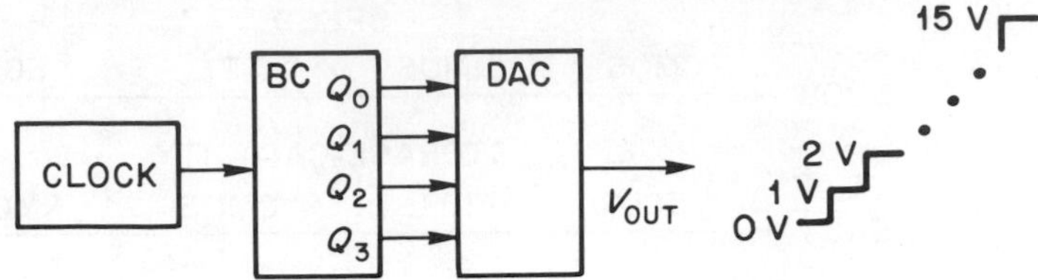

Figure 9–72 4-bit D/A converter.

D/A Analysis

Consider the 4-bit D/A converter of Figure 9–72. The output voltage for the 4-bit converter follows:

Binary Counter				Analog Voltage Out (V)
D_3	D_2	D_1	D_0	
0	0	0	0	0
0	0	0	1	1
0	0	1	0	2
0	0	1	1	3
0	1	0	0	4
0	1	0	1	5
0	1	1	0	6
0	1	1	1	7
1	0	0	0	8
1	0	0	1	9
1	0	1	0	10
1	0	1	1	11
1	1	0	0	12
1	1	0	1	13
1	1	1	0	14
1	1	1	1	15

As can be seen from this relationship, the LSB (D_0) contributes a change in the analog output voltage of 1 V, and the next bit (D_1) contributes a change of 2 V; bit D_2 contributes a change of 4 V, and the MSB (D_3) contributes a change of 8 V. The point is, that each data bit doubles the output voltage contribution. That is, the final output voltage is weighted by powers of 2. The **step size** is 1 V.

Step Size:
The smallest possible analog change on the output of a D/A converter.

The step size of a D/A converter can be determined by the relation:

$$\text{Step size} = \frac{\text{Rated maximum output level}}{2^N}$$

where:

Rated maximum output level = maximum value of analog output voltage.
N = number of digital bits on the input of the D/A converter.

Example 9–9

Determine the step size for the following D/A converters:

(A) 8-bit D/A, maximum analog output voltage = 10 V.

(B) 12-bit D/A, maximum analog output voltage = 10 V.

Solution

Using the relationship for step size, you have

(A) Step size $= (10 \text{ V})/2^8 = (10 \text{ V})/256 = 0.039 \text{ V}$

(B) Step size $= (10 \text{ V})/2^{12} = (10 \text{ V})/4\,096 = 0.002\,4 \text{ V}$

As you can see from Example 9–9, the greater the number of binary inputs, the smaller will be the step size. Example 9–10 illustrates the use of the step size.

Example 9–10

For the A/D converters of Example 9–9, determine the value of the analog output voltage for the following binary input values.

(A) (8-bit A/D converter): 1011 0011

(B) (12-bit A/D converter): 0001 1011 0101

Solution

Keep in mind that each step size is weighted by a power of 2.

(A) 1011 0011

(step size = 0.039 V) = 179_{10}

The final output voltage would be step size × input value.

$$e_o = 0.039 \times 179 = 6.98 \text{ V}$$

(B) 0001 1011 0101

(step size = 0.002 4 V) = 437_{10}

The final output voltage would be step size × input value.

$$e_o = 0.002\,4 \times 437 = 1.05 \text{ V}$$

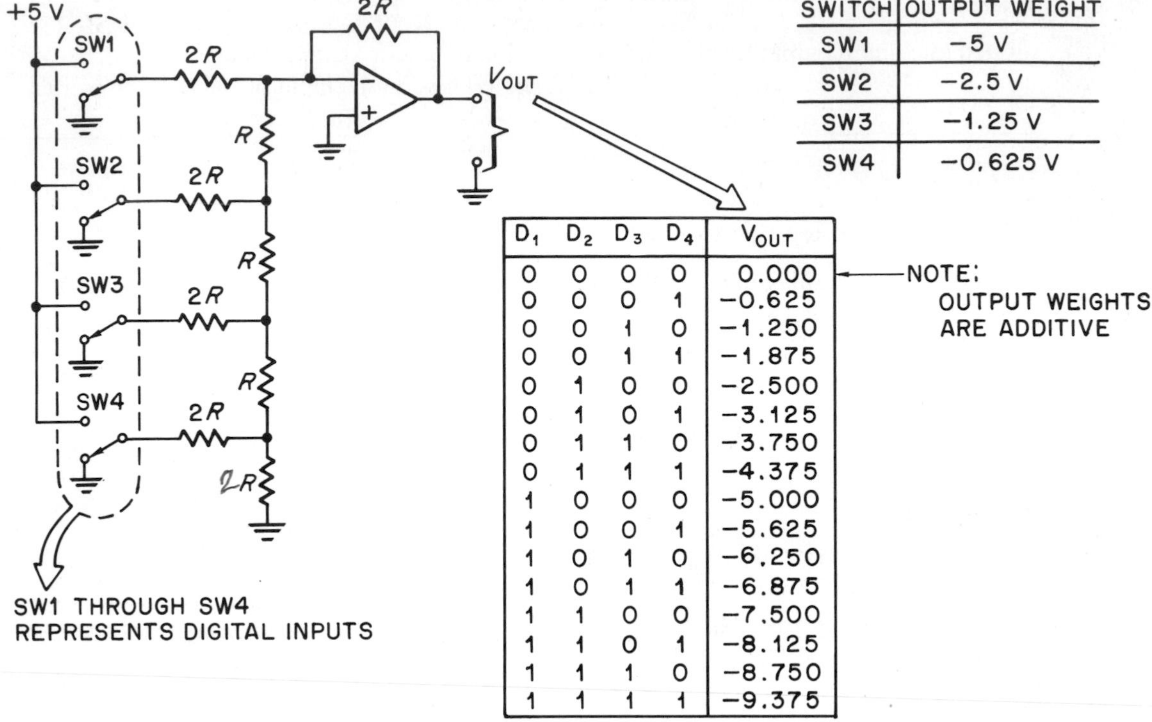

SWITCH	OUTPUT WEIGHT
SW1	−5 V
SW2	−2.5 V
SW3	−1.25 V
SW4	−0.625 V

D_1	D_2	D_3	D_4	V_{OUT}
0	0	0	0	0.000
0	0	0	1	−0.625
0	0	1	0	−1.250
0	0	1	1	−1.875
0	1	0	0	−2.500
0	1	0	1	−3.125
0	1	1	0	−3.750
0	1	1	1	−4.375
1	0	0	0	−5.000
1	0	0	1	−5.625
1	0	1	0	−6.250
1	0	1	1	−6.875
1	1	0	0	−7.500
1	1	0	1	−8.125
1	1	1	0	−8.750
1	1	1	1	−9.375

Figure 9–73 R/2R ladder network method for D/A converter.

$V_{out} = \left[\frac{1}{2} V_{IN} + \frac{1}{4} V_{IN} + \frac{1}{8} V_{IN} + \frac{1}{16} V_{IN}\right]$

Ladder Networks

Perhaps one of the most popular methods of constructing a D/A converter is an R/2R ladder network. This network along with the method of operation is illustrated in Figure 9–73.

> ### *Example 9–11*
>
> Determine the output voltage of the D/A R/2R ladder network of Figure 9–73 for a digital input of 0110.
>
> ***Solution***
>
> Referring to each of the analog weights, the second and third bits are HIGH, thus the final output voltage is
>
> $$0.625 \times 6 = -3.75 \text{ V}$$
>
> This answer checks with the result of the truth table on the same figure.

Table 9–3 IMPORTANT ANALOG AND DIGITAL CONVERSION TERMINOLOGY

Term	Meaning
Conversion Time	The time it takes for a complete measurement by an A/D or D/A converter.
Accuracy	Comparison of the actual output with the computed output. Expressed as a percentage of the maximum output voltage.
Gain Error (Full Scale Error)	The difference between the input that will produce a maximum output and the actual input that produces a full-scale output.
Resolution	For a D/A converter, the reciprocal of the number of discrete steps in the D/A output. Usually expressed as a percentage.
Linearity	Worst case deviation from the line between the endpoints (zero and full scale). Is expressed as a percentage of full scale or in fractions of an LSB.
Monotonicity	A monotonic function has a slope whose sign does not change. This means that the output always changes in the same direction for an increasing input.
Offset Error (Zero Error)	The output voltage that exists when the input is set to an ideal zero volts. For an A/D converter, this is the difference between the ideal input voltage ($^1/_2$ LSB) and the actual input voltage that is needed to make the transition from zero to 1 LSB. Many converters allow nulling of the offset with an external potentiometer.
Settling Time	The time from a change in input code until a D/A converter's output signal remains within $\pm^1/_2$ LSB (or some other specified tolerance) of the final value.

Analog and Digital Conversion Terminology

Table 9–3 lists some of the more important analog and digital conversion terminology.

D/A Applications

Figure 9–74 shows an application of a D/A converter with a ROM used as a sine wave look-up table. Here is the marriage of the ROM and the D/A converter. Essentially, this circuit allows the generation of very low frequency sine waves. Good-quality sine waves at very low frequencies are difficult to achieve by other methods. As pointed out in the figure, any periodic function can be duplicated by simply changing the contents of the ROM look-up table. Conceivably such a system could employ the large storage capabilities of a CD ROM and then reproduce the sound of any musical instrument as well as those that don't yet exist.

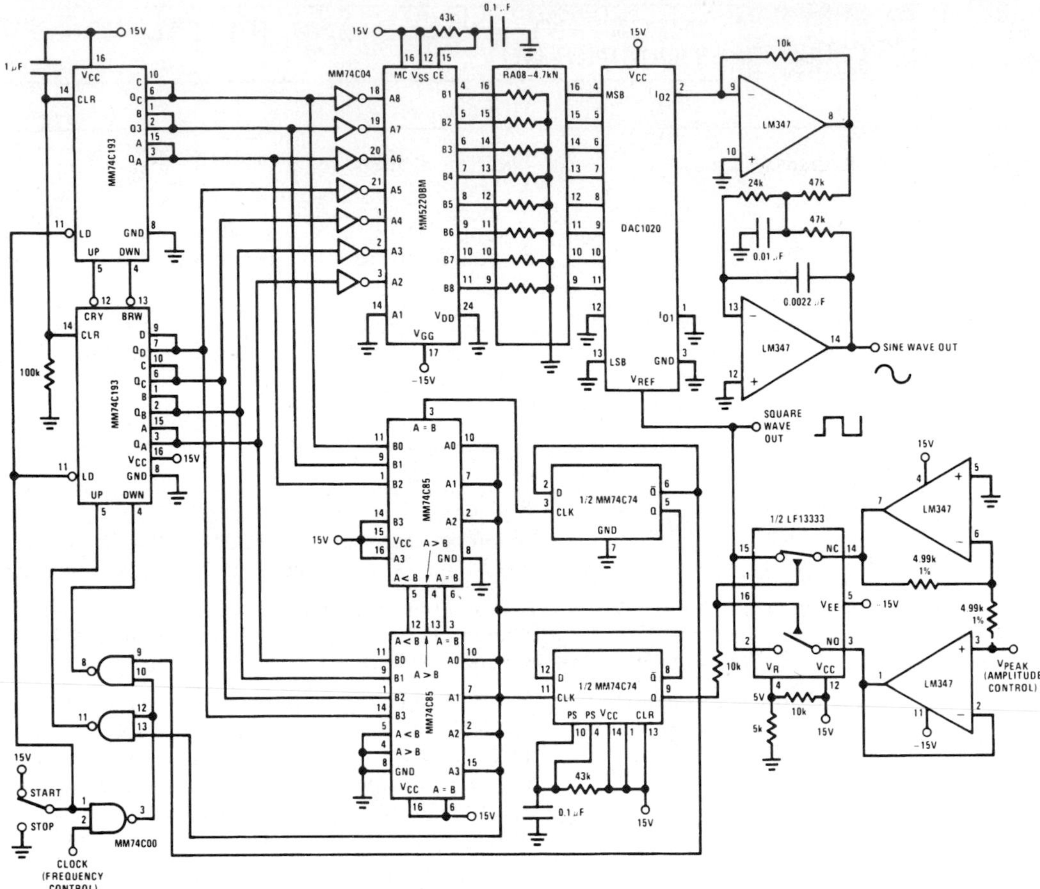

TL/H/5689-8

- Output frequency $= \frac{f_{CLK}}{512}$; $f_{MAX} \cong 2$ kHz
- Output voltage range = 0V − 10V peak
- THD < 0.2%
- Excellent amplitude and frequency stability with temperature
- Low pass filter shown has a 1 kHz corner (for output frequencies below 10 Hz, filter corner should be reduced)
- Any periodic function can be implemented by modifying the contents of the look up table ROM
- No start up problems

Figure 9–74 ROM and D/A sine wave generator. *Reprinted* with permission of National Semiconductor Corporation.

Other Types of A/D Converters

Another method of converting analog voltages to digital voltages is by using a voltage-to-frequency converter. This method was illustrated in the last chapter. Another popular method is by a process called *successive approximation.*

Successive Approximation A/D Converters

A successive approximation A/D converter is the fastest type of converter outside of the A/D "flash" converter. Basically what this type of converter does is to take the measured analog input voltage and compare it to $^1/_2$ of a reference voltage. If the input voltage is larger than this reference voltage then it is compared to $^3/_4$ of the same reference voltage. This type of comparison keeps on going until the fractional part of the reference voltage is equal (within the bit accuracy of the converter) to the incoming voltage.

The reason why this type of system is faster than others can be seen by using the analogy of a guessing game. Suppose someone says that they are thinking of a number from 1 to 12 and you are to guess it in as few tries as possible. One method would be to start at 1 and ask "is it one?", and if not, then ask "is it two?" and so on. A more efficient method, which results in fewer questions for many different "games" is to start in the middle and ask "is it greater than 6?" If the answer is yes then ask "is it greater than 9?" In this manner, you are taking half of what is left and testing it against the real value. The successive approximation A/D converter does exactly the same thing.

Look at Figure 9–75. It consists of an A/D converter with a reference voltage (V_{REF}), a comparator, some control logic, a shift register and output latches. There are two modes of operation to consider. One is when the input data is larger than $^1/_2$ of the reference voltage and the other is when it is less than $^1/_2$ the reference voltage. Both of these cases are illustrated in Figure 9–76. Consider first the case when V_{IN} is smaller than $^1/_2$ V_{REF}. See Figure 9–76(A).

Assume that $V_{REF} = 6$ V and $V_{in} = 2$ V. The shift register and data latches start at 0. On the first clock pulse, the MSB of the shift register is set to a 1 (giving a 1000 0000). Due to the internal circuitry of the D/A converter, this will produce an output voltage of $^1/_2$ $V_{REF} = 3$ V. This is applied to the − input of the comparator. This represents a value that is larger than the input voltage V_{IN}. Thus the comparator output will be LOW causing this bit to set to zero. On the next clock, the next most significant bit of the shift register is set to a 1 producing 0100 000. Since this bit represents half of the value of the previous

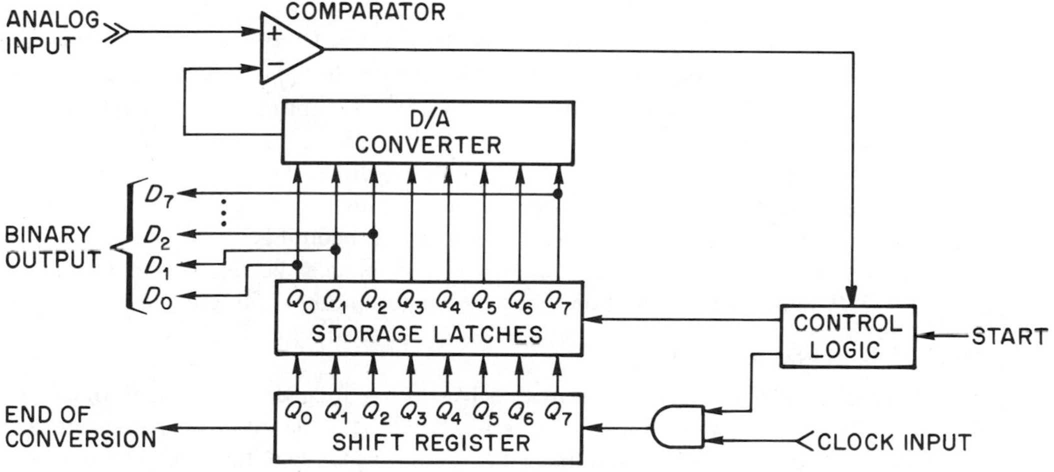

Figure 9–75 Successive approximation A/D converter.

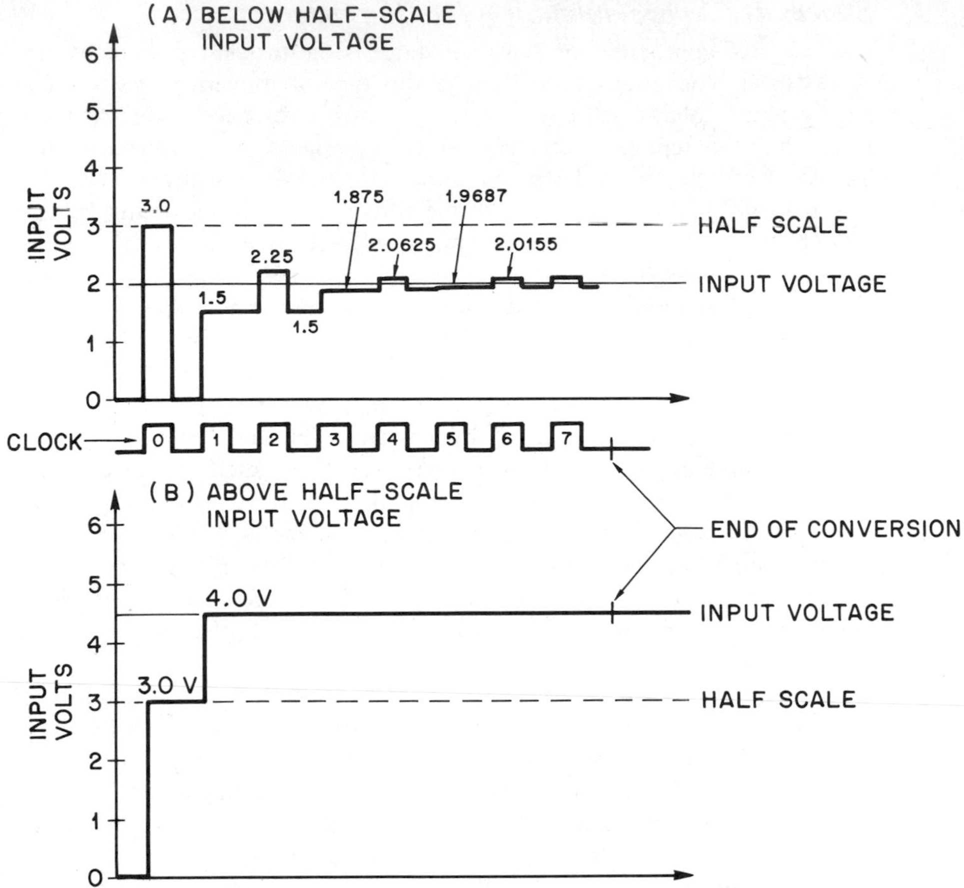

Figure 9–76 Successive approximation A/D converter waveforms.

bit, the new voltage to be compared is $^1/_2$ of 3 V or 1.5 V. When this is applied to the comparator the output of the comparator remains HIGH and this bit will not be set to zero.

On the next clock, the third MSB is set HIGH (representing 0.75 V) and the number is 0110 0000 which represents 1.5 V + 0.75 V = 2.25 V. This process continues until the ninth clock pulse when the shift register will overflow. This overflow condition represents the end of the conversion process. The resulting value represents the closest approximation within the accuracy of the system of the incoming analog voltage V_{IN}. This process required only $N + 1$ clock pulses where N equals the number of clock pulses. Figure 9–76(B) shows the condition when the analog input V_{IN} is greater than $^1/_2$ V_{REF}.

Digital Waveform Recorder

A method of recording analog information by storing it in digital form is illustrated in Figure 9–77. This system utilizes a total of 4K × 8 RAM. Essentially, analog information (such as voice or music) is taken from the analog input and then converted to digital information via an A/D converter. The digital data is then sequentially stored in the RAM.

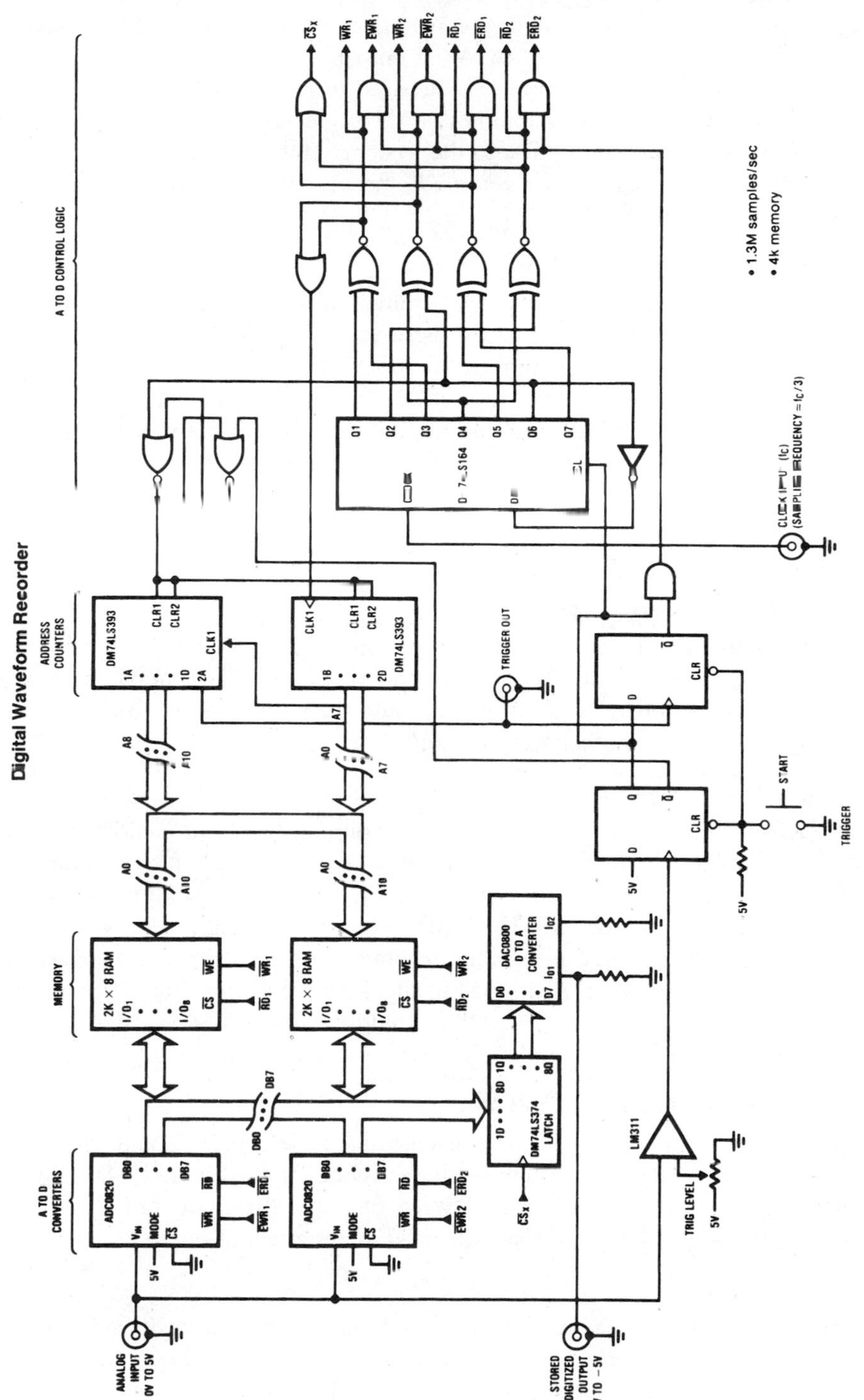

Figure 9–77 Digital waveform recorder. *Reprinted* with permission of National Semiconductor Corporation.
National holds no responsibility for any circuitry described. National reserves the right at any time to change without notice said circuitry.

For the system to play back the stored information, the foregoing process is essentially reversed. The digital data in RAM is sequentially sent to a D/A converter where the reconstructed analog signal is set to the output.

The principle in this circuit can be used to record voice or music and play it back. The advantage of such a system is that there are no moving parts or recording disks or tapes to wear out and add noise to the system. This type of system is used for telephone answering devices where the "recorded" message made by you is digitized and stored in RAM.

Digital Filtering

An example of digital filtering is the removal of noise present on old phonograph recordings. When the sound produced by these recordings is digitized and stored in RAM, it can then be processed by a computer. The digital data representing the recording (along with all of the noise from the old record) can now be operated upon by a computer program called a *digital filter*. Essentially the digital filter is a mathematical formula that tends to smooth out wide random variations (this is the noise). The resulting "clean" digital information can now be imprinted on a CD recording disk to be played back through your stereo system. This same technique is also applied to video, radio, and satellite communications as well as telephone.

Voice Recognition System

The previous system can also be used as part of a voice recognition system. Once your voice is digitized, the resulting digital pattern can now be compared to other previously stored digital patterns. A close match could then evoke a specific process such as controlling a machine or other such system.

DVM on a Chip

Recall from the last chapter that a digital voltmeter took an analog voltage and, using a VCO and a binary counter, converted it to a digital readout. The chip in Figure 9–78 takes an analog signal and converts it to a digital readout. With this chip, a DVM can be constructed as in Figure 9–79. Note that the seven-segment readouts are multiplexed. This means that only one seven-segment readout is on at a time. However, the DVM chip switches so fast between them that it appears to the user that all of the seven-segment readouts are on at the same time. This multiplexing system is used in order to conserve power and output lines from the chip to the seven-segment readouts.

Interfacing TTL and CMOS

Since many of the digital systems presented here use both TTL and CMOS, it is important to know how to interface between these technologies. Figure 9–80 illustrates the most common methods of interfacing between the two. Many of the 74 HC series chips are simply "plug-in-and-go" with no special TTL interfacing required.

Conclusion

This section presented much valuable information. You were introduced to the different technologies used in electrical memory. Also included were more important details on analog and digital conversion techniques as well as several current applications. You also saw how a DVM can be created with a single chip as the main component. Interfacing

ADD3501 3½ Digit DVM with Multiplexed 7-Segment Output

General Description

The ADD3501 monolithic DVM circuit is manufactured using standard complementary MOS (CMOS) technology. A pulse modulation analog-to-digital conversion technique is used and requires no external precision components. In addition, this technique allows the use of a reference voltage that is the same polarity as the input voltage.

One 5V (TTL) power supply is required. Operating with an isolated supply allows the conversion of positive as well as negative voltages. The sign of the input voltage is automatically determined and output on the sign pin. If the power supply is not isolated, only one polarity of voltage may be converted.

The conversion rate is set by an internal oscillator. The frequency of the oscillator can be set by an external RC network or the oscillator can be driven from an external frequency source. When using the external RC network, a square wave output is available. It is important to note that great care has been taken to synchronize digit multiplexing with the A/D conversion timing to eliminate noise due to power supply transients.

The ADD3501 has been designed to drive 7-segment multiplexed LED displays directly with the aid of external digit buffers and segment resistors. Under condition of overrange, the overflow output will go high and the display will read +OFL or −OFL, depending on whether the input voltage is positive or negative. In addition to this, the most significant digit is blanked when zero.

A start conversion input and a conversion complete output are included on all 4 versions of this product.

Features

- Operates from single 5V supply
- Converts 0V to ±1.999V
- Multiplexed 7-segment
- Drives segments directly
- No external precision component necessary
- Accuracy specified over temperature
- Medium speed - 200ms/conversion
- Internal clock set with RC network or driven externally
- Overrange indicated by +OFL or −OFL display reading and OFLO output
- Analog inputs in applications shown can withstand ±200 Volts

Applications

- Low cost digital power supply readouts
- Low cost digital multimeters
- Low cost digital panel meters
- Eliminate analog multiplexing by using remote A/D converters
- Convert analog transducers (temperature, pressure, displacement, etc.) to digital transducers

Connection Diagram

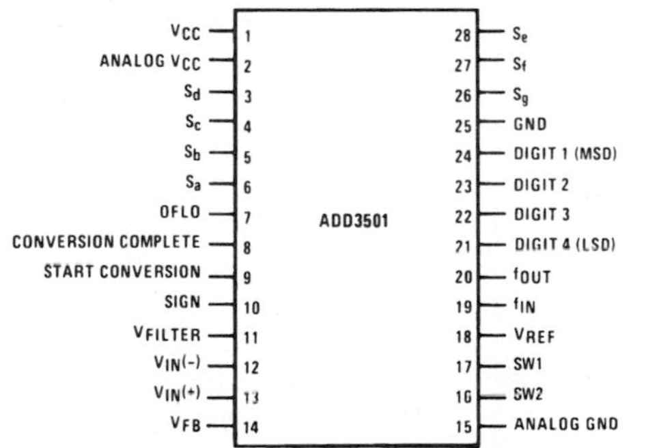

Order Number ADD3501CCN
See NS Package Number N28B

Figure 9–78 DVM chip. *Reprinted* with permission of National Semiconductor Corporation.

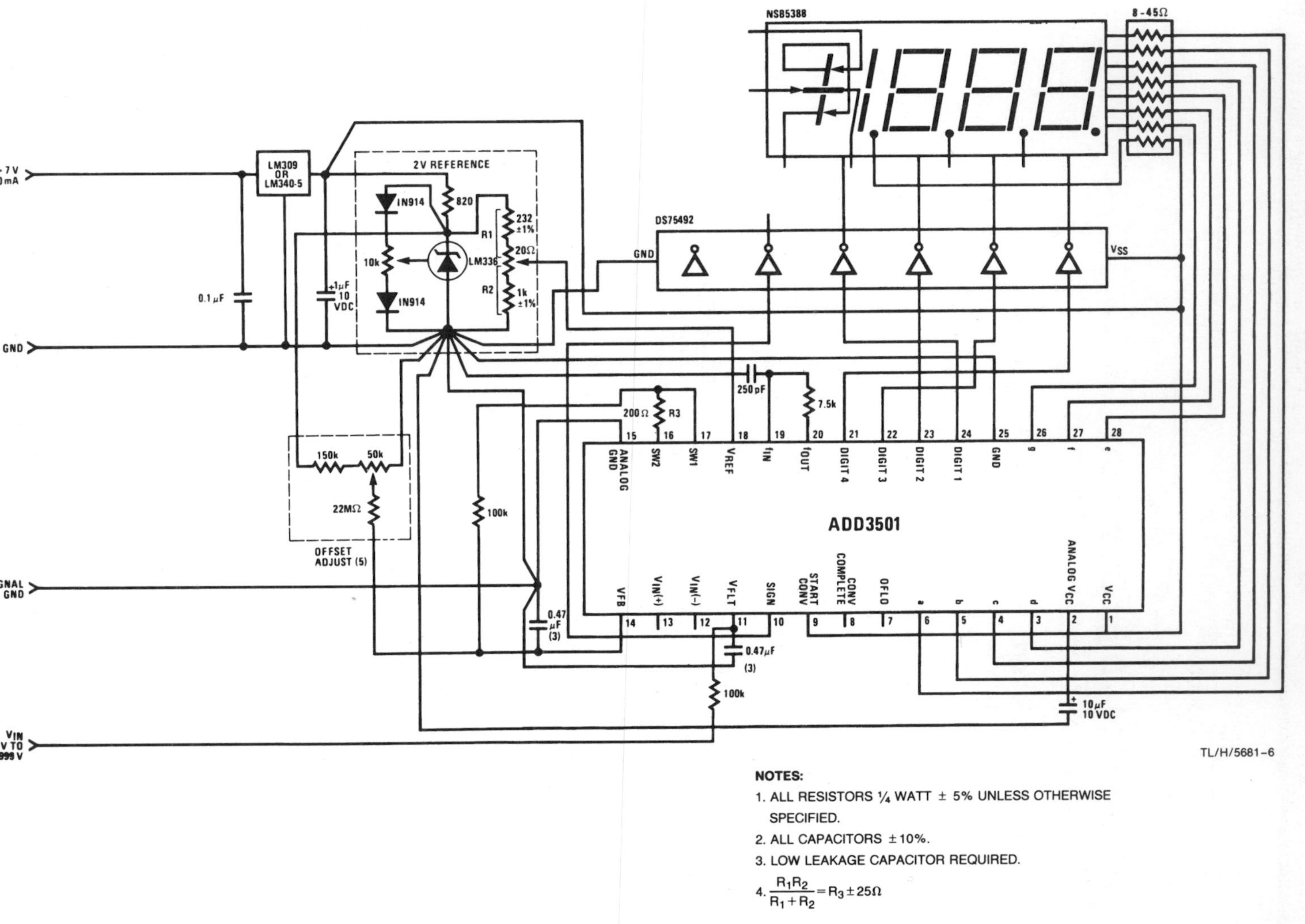

NOTES:

1. ALL RESISTORS ¼ WATT ± 5% UNLESS OTHERWISE SPECIFIED.
2. ALL CAPACITORS ± 10%.
3. LOW LEAKAGE CAPACITOR REQUIRED.
4. $\frac{R_1R_2}{R_1+R_2} = R_3 \pm 25\Omega$

Figure 9–79 DVM construction. *Reprinted* with permission of National Semiconductor Corporation.
National holds no responsibility for any circuitry described. National reserves the right at any time to change without notice said circuitry.

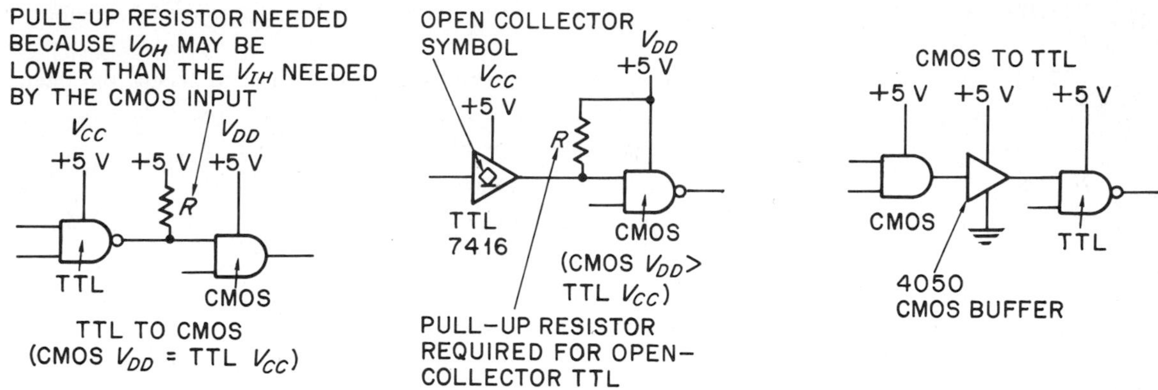

Figure 9–80 Interfacing between CMOS and TTL.

between CMOS and TTL was also presented. Check your understanding of this section by trying the following review questions.

9–12 Review Questions

1. Name the two major technologies used in the manufacturing of electronic memories.
2. Which of the popular electronic memory technologies has the greatest speed? Which uses the least amount of power?
3. What factors influence the step size of a D/A converter?
4. State the most popular method of constructing D/A converters.
5. Give the major advantage of a successive approximation A/D converter.

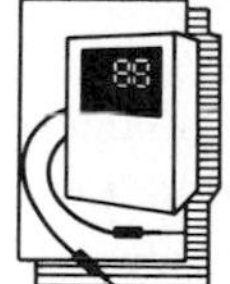

TROUBLESHOOTING SIMULATION

Discussion

The troubleshooting simulation program for this chapter provides an opportunity for the analysis and troubleshooting of registers. Here you will be able to test a simulated register under a variety of conditions. As with other troubleshooting simulation programs on your disk, there is a demonstration mode and a test mode. The demonstration mode familiarizes you with the register and the test mode will place a problem in the register for you to find.

This is an important troubleshooting simulation program. Here you have an opportunity not afforded to many students. You can spend as much time with a simulated register as you wish. The ability of the computer to generate a variety of problems and to assist you in discovering the answer is a powerful learning tool. Take advantage of it.

SUMMARY

- Digital memory is any medium that stores digital data.
- There are four mediums used by digital memory: electrical, magnetic, mechanical, and optical.

- Random-access is memory that takes the same amount of time to access any memory location.
- Sequentially accessed memory takes different amounts of time to access the data, depending upon the location of the data.
- Digital memory may be nonalterable, alterable, and/or erasable.
- Digital memory may be volatile or nonvolatile.
- Volatile memory requires an external source of energy to maintain its stored information; nonvolatile does not require the external energy.
- A write operation is the process of altering the contents of memory, and a read operation is the process of copying the contents of memory.
- ROM literally means read-only memory, which traditionally meant that it was memory that could only have its contents copied.
- A mask-programmable ROM has its bit patterns placed into it during the manufacturing process.
- A look-up table is a stored bit pattern where the address represents a value and the output represents a function of that value.
- A voice synthesizer is a digital system that replicates the human voice.
- The basic sounds that make up a language are called phonemes.
- An electronic memory that may be altered only once by the user is a PROM.
- An electronic memory that is nonvolatile and may be altered more than once by the user is an EPROM.
- EPROMs are altered by first exposing them to ultraviolet light and then electrically programming them.
- An EEPROM may be altered solely through the use of electricity.
- RAM literally means randomly accessible memory, but is used to mean volatile, randomly accessible, electrical memory.
- Static RAM does not require any processing to maintain its information. It does however, require the use of an external source of energy to maintain this information.
- Dynamic RAM does require processing to maintain its information as well as an external source of energy. The processing is called refreshing.
- The advantage of dynamic RAM over that of static RAM is that dynamic RAM has a much higher packaging density.
- Zeropower RAM has a built-in source of electrical energy that allows it to store information for very long periods of time (over 10 years).
- The process of splitting the complete address into two or more parts is called address multiplexing. This process is used in order to reduce the pinout count on high-density RAM chips.
- Sequentially accessed electrical memory is of a type called first-in–first-out (FIFO) memory.
- Charged-coupled device (CCD) memory uses tiny packets of electrical charge to sequentially store binary data. This type of memory has a very high package density.
- Digital memory that uses the patterns of a magnetic field to represent stored digital information is called magnetic memory.

- The main advantage of magnetic memory is that it is alterable and does not require an external source of energy in order to maintain its information.
- The disadvantage of magnetic memory is its slow access time compared to electrical memory.
- Magnetic tape, video and audio cassette tapes, and magnetic bubble memory (MBM) are all examples of nonvolatile, sequentially accessible, magnetic memory.
- A compact disk (CD) is an optically reflective disk that contains digital information stored as surface changes that cause a focused laser beam to be deflected in one of two patterns.
- The Kerr effect is the effect that the polarization of a light beam has when reflected off of a magnetic medium.
- The Kerr effect makes laser memory possible.
- An analog-to-digital converter is an electrical device that converts analog information into digital information.
- A digital-to-analog converter is an electrical device that converts digital information into analog information.
- A listing showing the contents of all memory locations accessed by a microprocessor is called a memory map.
- When all address lines are not decoded, the same physical memory location can be accessed with many different addresses.
- Memory address size and memory word size may be expanded with a microprocessor.
- Memory-mapped I/O is input/output treated as a regular memory location.
- ECL memories possess the highest access speed of any current electrical memories.
- NMOS memories have the greatest storage capacity of any electronic memories.
- CMOS memories have the lowest power consumption of any electronic memories.
- The R/2R ladder network is the most commonly used digital-to-analog conversion technique.
- The successive approximation method is most commonly used with analog-to-digital converters.
- A digital waveform recorder can be used as part of a voice recognition system.

CHAPTER SELF-TEST

I. TRUE/FALSE

Answer the following questions true or false.

1. Any medium that stores digital data can be referred to as digital memory.
2. ROM is not random-access memory.
3. RAM is random-access memory.
4. Sequential access memory requires an equal amount of time to access its data.
5. The process of altering data in memory is called a read operation.

II. MULTIPLE CHOICE

Answer the following questions by selecting the most correct answer.

6. ROM is
   (A) not randomly accessible
   (B) is volatile
   (C) requires an external source of energy to store its information
   (D) None of the above are correct.
7. A look-up table
   (A) can be used to store trig functions
   (B) is used to store a mathematical relationship
   (C) may be used to express the relationship between two quantities
   (D) All of the above are correct.
8. The basis of voice synthesizers is a look-up table that contains a digital pattern of
   (A) words
   (B) phonemes
   (C) vowels
   (D) consonants.
9. A switch-programmable ROM is
   (A) nonvolatile
   (B) alterable
   (C) randomly accessible
   (D) All of the above are correct.
10. A PROM
    (A) is programmed when it is manufactured
    (B) can be altered more than once
    (C) can be altered only once by the user
    (D) cannot be altered

III. Matching

Match the item(s) on the right to the statements on the left.

11. ROM	(A) Volatile
12. RAM	(B) Alterable
13. PROM	(C) Randomly accessible
14. EPROM	(D) Nonvolatile
15. EEPROM	(E) Sequentially accessible

IV. Fill-In

Fill in the blanks with the most correct answer(s)

16. Memory that requires a process in order to maintain its contents is called ______ RAM.
17. CCD memory is ______ accessible.
18. The method of magnetically recording digital data that uses two different frequencies is called the ______ ______ (two words) standard.
19. Magnetic ______ would be an example of nonvolatile, sequentially accessible, magnetic memory.
20. Bubble memory is ______ accessible.

V. OPEN-ENDED

Answer the following questions as indicated.

21. What kind of memory is laser disk?
22. State the difference between an analog-to-digital converter and a digital-to-analog converter.
23. What is an analog-to-digital converter that uses differential amplifiers called?
24. Explain what a memory map does.
25. What is the most common type of A/D converter technique used? Why?

Answers to Chapter Self-Test

1] T 2] F 3] T 4] F 5] F 6] D 7] D 8] B 9] D 10] C
11] C, D 12] A, B, C 13] B, C, D 14] B, C, D 15] B, C, D
16] dynamic 17] sequentially 18] Kansas City 19] tape
20] sequentially 21] nonvolatile, alterable, randomly accessible, optical memory
22] An analog-to-digital converter converts analog information into digital. A digital-to-analog converter converts digital information into analog.
23] A flash converter. 24] A memory map shows the contents of all memory locations addressed by a microprocessor.
25] The successive approximation technique because of its fast conversion time.

CHAPTER PROBLEMS

Basic Concepts

Section 9–1:

1. What is digital memory?
2. Describe the four types of digital memories.
3. Give an example of each method of storing digital memory.
4. How many different ways can memory be accessed? Explain what the different ways mean.
5. Describe the differences between alterable memory, permanent memory, and erasable memory.
6. Explain the meaning of each of the following as they apply to digital memory: (A) volatile (B) nonvolatile (C) alterable (D) sequentially accessible (E) randomly accessible
7. Give an example of each form of digital memory in problem 6.

Section 9–2:

8. Define nonvolatile memory and volatile memory. Which of the two are randomly accessible? Explain.
9. Describe the construction of a mask-programmable ROM. What is an application of such a device?
10. Explain the action of a look-up table. State two cases where one would be used.
11. Construct a look-up table for the area of a square if the side is known (use whole numbers only).
12. Construct a look-up table for the area of a circle if the radius is known (use whole numbers only).
13. Explain the operation of a voice synthesizer in terms of using a look-up table.

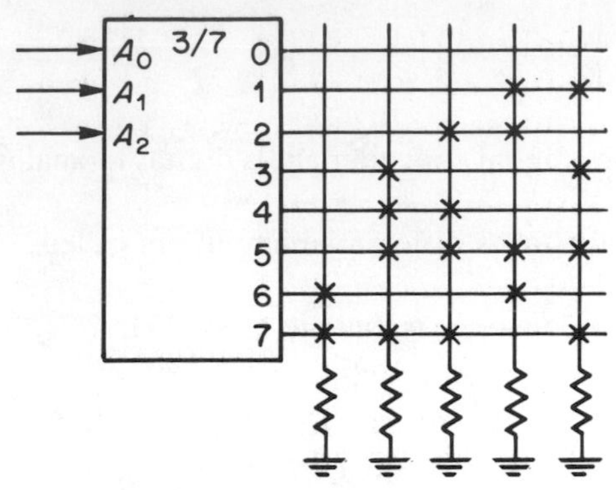

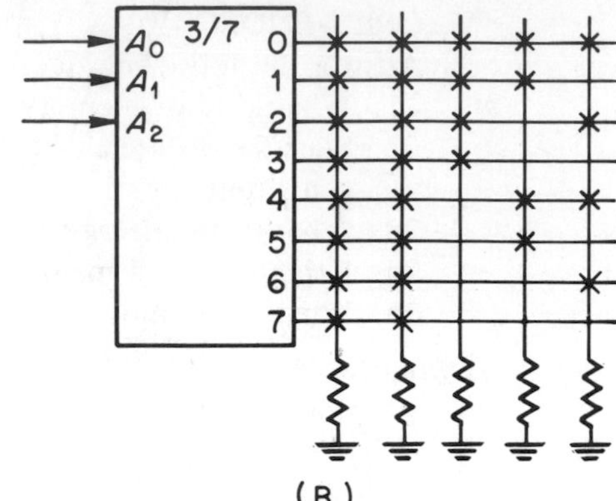

Figure 9–81 (A) (B)

Section 9–3:

14. Write the functional relationship between the address and the data output for the switch-programmable ROM shown in Figure 9–81(A).
15. Write the functional relationship between the address and the data output for the switch-programmable ROM shown in Figure 9–81(B).
16. State the differences between the following types of memories:
    (A) PROM (B) EPROM (C) EEPROM
17. State the similarities between the following types of memories:
    (A) PROM (B) EPROM (C) EEPROM
18. Sketch the bit pattern of an EPROM that would create a look-up table for the relationship $y = 2x^2$ for values of x from 0 to 7.
19. Sketch the bit pattern of an EEPROM that would create a look-up table that would relate the value of a digit to its ASCII code. Where could such a look-up table be used?
20. Describe how memory in an EPROM may be altered.
21. Describe how memory in an EEPROM may be altered.

Section 9–4:

22. What does the acronym RAM stand for, and what does it actually mean?
23. Construct the logic diagram of an 8 × 4 RAM. Why must electrical energy be applied in order for this type of memory to maintain its contents?
24. How many memory locations does the RAM in Figure 9–82 have and what is its word size?
25. Describe the purpose of each pin of the RAM in Figure 9–82.
26. Explain the operation of a zeropower RAM. What is its advantage? Give an example.

Section 9–5:

27. What is dynamic RAM? How does this differ from static RAM?
28. Describe the sequence of operation for the dynamic RAM in Figure 9–83.
29. What type of addressing is used by the dynamic RAM of Figure 9–83?
30. Describe the purpose of each pin for the RAM of Figure 9–83.

Section 9–6:

31. Describe the operation of FIFO memory.
32. Sketch the logic diagram of a 4-bit FIFO memory that will produce a delay of eight clock pulses.

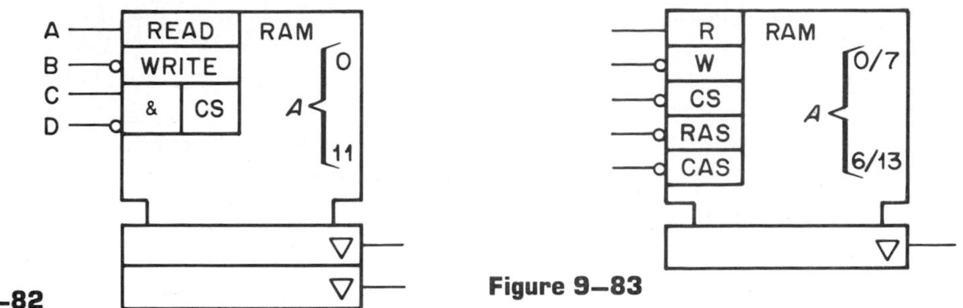

Figure 9–82

Figure 9–83

33. State an application of FIFO memory.
34. Sketch the logic diagram of a circulating memory that would cause the world HELLO to be continuously displayed on a 7-segment readout.
35. Briefly explain the operation of a CCD memory.

Section 9–7

36. List several advantages and disadvantages of magnetic memory.
37. Name the different recording techniques shown in Figure 9–84.
38. Describe the operation of a floppy disk.
39. List the important points for the care of a floppy disk.
40. State why a floppy disk may be described as sequentially accessible memory.
41. Briefly describe MFM.

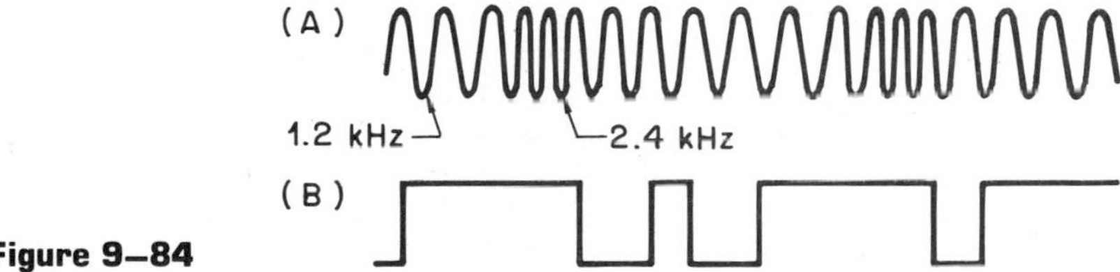

Figure 9–84

Section 9–8:

42. List some of the sequentially accessible magnetic memories.
43. Briefly describe the recording method used in a video cassette recorder.
44. What is bubble memory? What constitutes a magnetic bubble?
45. What are the advantages and disadvantages of MBM?

Section 9–9:

46. What is a compact disk?
47. What is CD ROM? What are its advantages and disadvantages?
48. Describe the operation of laser memory. List some of its advantages and disadvantages.
49. Which of the optical memories in this section are nonalterable? alterable? Explain.
50. What is *coercivity?* What is the *Kerr effect?*

Section 9–10:

51. Explain the basic properties of a digital-to-analog converter and of an analog-to-digital converter.
52. State the operation of a flash converter. Why is it given such a name?
53. Using the flash converter of Figure 9–42, determine the resulting output for the waveform shown in Figure 9–84(A) if it varies from 0 to 3 V.

54. For the D/A converter of Figure 9–47, assume that a 4-bit binary counter is connected to the input of the 4-to-16-line decoder. Sketch the resulting waveform. Assume that the binary counter is a downcounter and starts its count at 1111. Show this for two complete cycles.
55. Explain the operation of a stairstep A/D converter.
56. State the difference between a stairstep A/D converter and a tracking A/D converter.

Applications

57. Describe the major characteristics of an 8-bit microprocessor.
58. Express the higher-order address, the lower-order address, and the complete address in hex for the following 16 bit addresses:
    (A) 0000 1010 1100 0011 (B) 1100 0101 1110 1111
59. Express the following hex addresses in binary:
    (A) FBA0H (B) 3CD2H (C) 50AEH
60. Construct the memory map of the microprocessor system of Figure 9–67.
61. Construct a microprocessor system for the memory map in Figure 9–85.

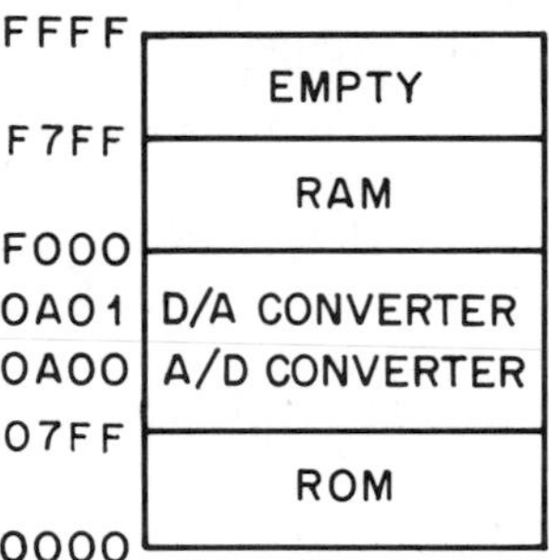

Figure 9–85

Troubleshooting

62. Which type of electrical memory has
    (A) the highest speed? (B) the lowest cost? (C) the highest storage capacity?
63. Determine the step size for the following A/D converters:
    (A) 8-bit, maximum analog output voltage = 5 V
    (B) 12-bit, maximum analog output voltage = 5 V
64. Determine the step size for the following A/D converters:
    (A) 12-bit, maximum analog output voltage = 12 V
    (B) 16-bit, maximum analog output voltage = 10 V
65. For the A/D converters in question 64, determine the value of the analog voltage for the following bit patterns:
    (A) 1100 0110 1001 (B) 1011 0011 1010 1111
66. For the A/D converters in question 63, determine the value of the analog voltage for the following bit patterns:
    (A) 0110 1001 (B) 1110 0111 1010
67. Briefly describe the operation of an R/2R ladder network A/D converter.
68. Explain the operation of a successive approximation A/D converter.
69. State some of the applications of a digital waveform recorder.
70. Explain how you would interface between CMOS and TTL and between TTL and CMOS.

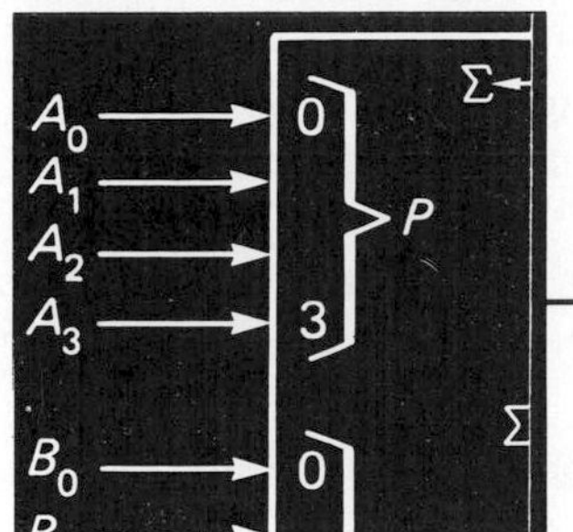

CHAPTER 10

Microprocessor Units and Interfacing

OBJECTIVES

After completing this chapter, you should be able to:

- ☐ Describe methods used to represent signed binary numbers.
- ☐ Use various techniques in working with signed binary numbers.
- ☐ Identify digital networks used for arithmetic processes.
- ☐ Understand the requirements for digital communications.
- ☐ Describe various methods used in digital communications.
- ☐ Describe methods used to interface digital systems.
- ☐ Describe various techniques used by microprocessors for manipulating data.
- ☐ Identify the similarities and differences between major microprocessor types.
- ☐ Understand microprocessor programming techniques.
- ☐ Explain advanced troubleshooting systems and methods.

INTRODUCTION

This is the last chapter of the text. For you it should represent a new beginning. This chapter represents the bridge between the end of the digital course and the beginning of your microprocessor course. Here is the culmination of all the material you have studied up to this point.

The material presented here will take you into the realm of logic networks that perform digital computations. Then you will see how digital systems can communicate with each other. In this, the devices as well as the important methods are presented. This is your introduction to real microprocessors. You will see the popular 8-bit, 16-bit, and 32-bit machines. The microprocessor application section will show you what it takes to program a real microprocessor. With this knowledge, you will be well prepared for any future microprocessor course. Finally, in the electrical application section, you will be introduced to advanced troubleshooting techniques. There you will learn about state-of-the-art systems and techniques used in today's digital industry.

KEY TERMS

Direct Method
Half Adder
Full Adder
Overflow
ALU
Modulate
Amplitude Modulation
Frequency Modulation
Phase Modulation
Laser
Simplex Mode
Half-Duplex Mode
Full-Duplex Mode
Modem
Direct Memory Access (DMA)
DMA Controller
Peripheral Interface Adapter (PIA)
Baud Rate
Bits Per Second
UART
ACIA
Acoustic Coupler
Register Architecture
Index Register
Stack Pointer
Stack
Status Flags
Multiplexing
Segmentation Registers
Label
Interrupt

10–1 BINARY NUMBERS REVISITED

Discussion

This section presents the background you will need in order to understand how logic networks perform the arithmetic processes of addition, subtraction, multiplication, and division. In order for such a system to be practical, a method of handling signed binary numbers must be used. This means that any logic network capable of performing practical arithmetic processes must be able to represent negative as well as positive numbers.

Representing Signed Numbers

Remember that whatever a logic network does, it can only do it in terms of 1's and 0's (HIGH or LOW); there is no other condition possible. This means that a minus sign or plus sign must be represented by a single binary bit. To get the general idea, suppose you and I agreed to the following conditions for representing signed binary numbers—call it the **direct method.**

Direct Method:
A method of representing signed binary numbers where the MSB represents the sign of the binary number. If it is 1, the number is negative; if it is 0, the number is positive.

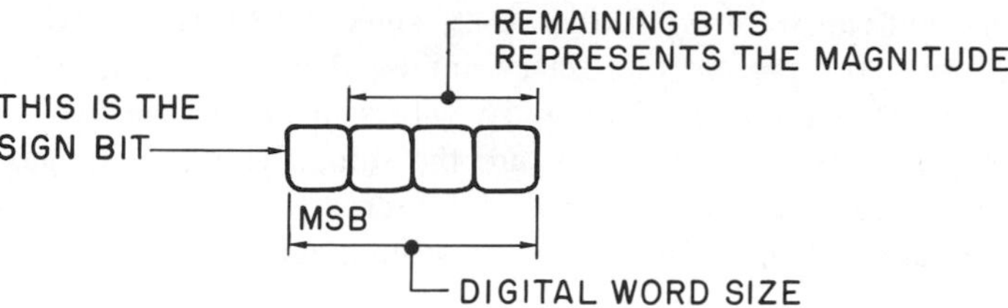

Figure 10–1 Concept of the sign bit.

These are the conditions:

1. All binary numbers will have only 4 bits.
2. The MSB will be the *sign bit* representing the sign of the binary number.
3. If the MSB is a 1, the binary number will be considered negative. If the MSB is 0, the binary number will be considered positive.

This concept is illustrated in Figure 10–1.

This method will of course limit the range of numbers that can be represented. All of the positive and negative numbers which can be represented using this system are shown in Table 10–1.

Table 10–1 DIRECT METHOD OF BINARY SIGN BIT NOTATION

Decimal Number	Binary Representation	Hexadecimal Representation
0	0 0 0 0	0H
1	0 0 0 1	1H
2	0 0 1 0	2H
3	0 0 1 1	3H
4	0 1 0 0	4H
5	0 1 0 1	5H
6	0 1 1 0	6H
7	0 1 1 1	7H
−0	1 0 0 0	8H
−1	1 0 0 1	9H
−2	1 0 1 0	AH
−3	1 0 1 1	BH
−4	1 1 0 0	CH
−5	1 1 0 1	DH
−6	1 1 1 0	EH
−7	1 1 1 1	FH

As you can see from Table 10–1, since the MSB is used as the sign bit, this leaves only 3 bits to represent the binary values. Thus, the range of numbers that can be represented is from −7 to +7 (still 16 values including the two representations for 0). For all practical reasons, −0 and 0 are the same. This system has a negative as well as a positive representation of this value. Observe from Table 10–1 the resulting hexadecimal representation. Under this agreed-to system, any hex value larger than 7H represents a negative number!

Example 10–1

Using the direct method of signed binary number representation, determine the decimal value, including the sign, of the following hex numbers.

(A) 3H (B) 8H (C) EH (D) AH

Solution

First convert the hex numbers to their binary equivalents. Then, checking the value of the sign bit, convert the binary number to its equivalent signed decimal value.

(A) 3H = 0011 = +3

(B) 8H = 1000 = −0 (same value as +0)

(C) EH = 1110 = −6

(D) AH = 1010 = −2

Example 10–2

Using the direct method, convert the following signed decimal numbers to their equivalent hex representations.

(A) 4 (B) −4 (C) 0 (D) −7

Solution

First convert the decimal numbers to their signed binary values. Then convert the resulting binary number to its hex representation.

(A) 4 = 0100 = 4H

(B) −4 = 1100 = CH

(C) 0 = 0000 = 0H (or −0 = 1000 = 8H)

(D) −7 = 1111 = FH

A Basic Problem

The direct method is one way of representing signed binary numbers. However, it isn't used in digital systems. The reason why is because it doesn't give the correct answer when signed numbers are added, subtracted, multiplied, or divided. It was presented to you here because it quickly shows how signed binary numbers can use the MSB to represent the sign and the resulting effect this has on the hexadecimal representation of negative decimal numbers. Now that you have these concepts, they can be extended to a system that is actually used in practical digital systems.

A Practical Method

Recall from Chapter 2 that you could use the two's complement method for subtracting binary numbers. Thus, using this method to perform

$$\begin{array}{r} 5 \\ \underline{-3} \end{array}$$

you first converted both numbers to their binary equivalents:

$$\begin{array}{r} 0101 \\ -0011 \end{array}$$

Then you took the two's complement of the subtrahend and added it to the minuend, disregarding the carry:

$$\begin{array}{r} 0100 \\ \underline{+1101} \\ \text{(ignore carry)} \rightarrow 1\ 0010 \end{array} \leftarrow \text{(final answer is 2)}$$

Another way of looking at what was just done is to say that you were adding two numbers, a +5 and a −3. In other words, expressing −3 in the two's complement form is similar to representing a negative number using the MSB again as the sign bit. This method is illustrated in Table 10–2.

As before, the MSB is the sign bit. The range of numbers is from −8 to +7 (using only 4 bits). The difference is that, this time, there is only one representation for −0 and +0, and that is 0 (recall that the previous method had two values for representing 0). You can still represent 16 values which now go from −8 to +7 (and includes a 0). The advantage of using the two's complement method of representing signed binary numbers is that when you do arithmetic operations, you get correct answers.

Note from Table 10–2 that any time the decimal value is negative, the binary representation is then converted to its two's complement form. This always results in the MSB being a 1. Thus, as before, when the MSB is a 1, for this system, it means two things:

1. A negative value is being represented.
2. The binary number is in the two's complement form and must be uncomplemented in order to represent its true value.

This means that if this is the agreed-to system for representing signed binary numbers, then 1101 represents a negative number. But, before its value can be determined, it must be uncomplemented (recall, from Chapter 2 that the complement of a complement was the original number). As an example, taking the two's complement of 1101 yields 0011. Thus 1101 represents a −3. This can be verified in Table 10–2.

Table 10–2 TWO'S COMPLEMENT METHOD OF REPRESENTING SIGNED BINARY NUMBERS

Decimal Number	Binary Representation	Hexadecimal Representation
0	0 0 0 0	0H
1	0 0 0 1	1H
2	0 0 1 0	2H
3	0 0 1 1	3H
4	0 1 0 0	4H
5	0 1 0 1	5H
6	0 1 1 0	6H
7	0 1 1 1	7H
−0	0 0 0 0 = 0 0 0 0	0H
−1	0 0 0 1 = 1 1 1 1	FH
−2	0 0 1 0 = 1 1 1 0	EH
−3	0 0 1 1 = 1 1 0 1	DH
−4	0 1 0 0 = 1 1 0 0	CH
−5	0 1 0 1 = 1 0 1 1	BH
−6	0 1 1 0 = 1 0 1 0	AH
−7	0 1 1 1 = 1 0 0 1	9H
−8	1 0 0 0 = 1 0 0 0	8H

One's Complement System

Recall from Chapter 2 that you could also use the one's complement system for subtracting binary numbers. This means you could also use this system for representing signed binary numbers. However, using the one's complement system results in again having two values for zero, as shown in Table 10–3.

Because the one's complement system of representing signed binary numbers has two values for zero, it is not popular in digital systems. The two's complement method is the standard notation used to represent signed binary numbers in most digital systems.

Example 10–3

The following hexadecimal numbers represent the two's complement method for representing signed binary numbers. Convert each of them to its equivalent signed decimal value.

(A) 3H (B) DH (C) 8H (D) FH

Solution

First convert the hexadecimal value to its equivalent binary representation. Then, check the MSB of the result. If it is 1, then the binary number is negative and in the two's complement form. This means it must be uncomplemented before its absolute value can be determined. Convert the resulting binary values to their decimal equivalents remembering to carry the sign to the decimal value.

(A) 3H = 0011 = 3

(B) DH = 1101 [MSB = 1, hence number must be uncomplemented]
1101 → 0011 = −3

(C) AH = 1010 [MSB = 1, number must be uncomplemented]
1010 → 0110 = −6

(D) FH = 1111 [MSB = 1, must uncomplement]
1111 → 0001 = −1

Table 10–3 ONE'S COMPLEMENT NOTATION OF REPRESENTING SIGNED BINARY NUMBERS

Decimal Number	Binary Representation	Hexadecimal Representation
0	0 0 0 0	0H
1	0 0 0 1	1H
2	0 0 1 0	2H
3	0 0 1 1	3H
4	0 1 0 0	4H
5	0 1 0 1	5H
6	0 1 1 0	6H
7	0 1 1 1	7H
−0	0 0 0 0 = 1 1 1 1	FH
−1	0 0 0 1 = 1 1 1 0	EH
−2	0 0 1 0 = 1 1 0 1	DH
−3	0 0 1 1 = 1 1 0 0	CH
−4	0 1 0 0 = 1 0 1 1	BH
−5	0 1 0 1 = 1 0 1 0	AH
−6	0 1 1 0 = 1 0 0 1	9H
−7	0 1 1 1 = 1 0 0 0	8H

Example 10–4

Using two's complement notation for signed binary numbers, express the following signed decimal numbers in hexadecimal.

(A) −3 (B) +3 (C) 0 (D) −5

Solution

Convert all the decimal numbers to their binary equivalents. If the number is negative, then take its two's complement. Convert all results to hex notation.

(A) −3 → 0011 [Number is negative, so take two's complement]
0011 → 1101 = DH

(B) +3 = 0011 = 3H

(C) 0 = 0000 = OH

(D) −5 → 0101 [Take two's complement]
0101 → 1011 = BH

Representing Larger Numbers

Practical digital systems use binary numbers that are larger than 4 bits. Usually 8 or 16 bits are used. An 8-bit system using two's complement notation for the representation of signed binary numbers (where the MSB is used for the sign) means that the range of signed numbers that can be represented is from −128 to +127. In a 16-bit system, the range is −32 768 to +32 767. Basically, with any standard size binary number, the setting of the MSB to HIGH may be used to indicate a negative number. With the two's complement system, the − is one larger than the + representation.

Example 10–5

Using two's complement notation for representing signed binary numbers, convert the following signed decimal numbers to their equivalent hexadecimal values. Use an 8-bit notation system.

(A) +3 (B) 52 (C) −3 (D) −52

Solution

As before, convert the decimal number to its equivalent binary value. Use eight binary places, and if the decimal number was negative take the two's complement of the resulting binary number. Convert the results to hex notation.

(A) +3 = 0000 0011 = 03H

(B) 52 = 0011 0100 = 34H

(C) −3 = 0000 0011 [Take two's complement]
0000 0011 → 1111 1101 → FDH

(D) −52 = 0011 0100 [Take two's complement]
0011 0100 → 1100 1100 → CCH

Example 10–6

Using two's complement notation for representing signed binary numbers, convert the following signed decimal numbers to their equivalent hexadecimal values. Use a 16-bit notation system.

(A) +1 (B) −1 (C) −515

Solution

Use the same procedure as in the previous example. The only difference is that all binary representations are to have 16 bits.

(A) +1 = 0000 0000 0000 0001 = 0001H

(B) −1 = 0000 0000 0000 0001 [Negative number, take two's complement]
0000 0000 0000 0001 → 1111 1111 1111 1111 = FFFFH

(C) −515 = 0000 0010 0000 0011 [Take two's complement]
0000 0010 0000 0011 → 1111 1101 1111 1101 = FDFDH

Example 10–7

Each of the following hex numbers represents a signed binary value using the two's complement method for signed binary representation. Convert each of them to its equivalent signed decimal number.

(A) 3CH (B) F (C) F035H

Solution

First convert the hex number to its binary equivalent. If the MSB of the resulting binary is 1, then it is representing a negative number in its two's complement form. In this case, it must be uncomplemented. Take the resulting binary values and convert them to their equivalent decimal values, remembering to carry through with the correct sign.

(A) 3CH = 0011 1100 = +60

(B) F = 1111 [A negative number, must uncomplement]
1111 → 0001 = −1

(C) F035 = 1111 0000 0011 0101 [Must uncomplement]
1111 0000 0011 0101 → 0000 1111 1100 1011 = −4043

Conclusion

This section presented the concept of how signed binary numbers can be represented. Here you saw several methods. The two's complement method of representing signed binary values was chosen as the preferred method because it has only one value for zero and it produces correct arithmetic results.

In the next section, you will perform arithmetic operations on signed numbers much the same way as done by practical digital systems. Check your understanding of this section by trying the following review questions.

10–1 Review Questions

1. Explain why a bit of a binary number must be used to represent its sign.
2. The direct method of representing signed binary numbers wasn't practical. Explain why.
3. What is the main difference that results in using the one's complement method of representing signed binary numbers over that of the two's complement method?
4. In using the methods employed here for representing signed binary numbers, what general statement can you make about any hex number whose MSB is 8 or larger?

10–2 HARDWARE ARITHMETIC

Discussion

In the last section, you were introduced to methods of representing signed numbers in digital systems. In this section you will be introduced to concepts of how digital hardware is used to implement digital arithmetic. What you learned in the last section will be used here. This material will prepare you for understanding the operation of real hardware.

Adder Logic

Recall that in Chapter 6 you saw how to program a PLD that would add 2 binary bits. This figure is shown again in Figure 10–2. Such a logic circuit is called a **half adder.**

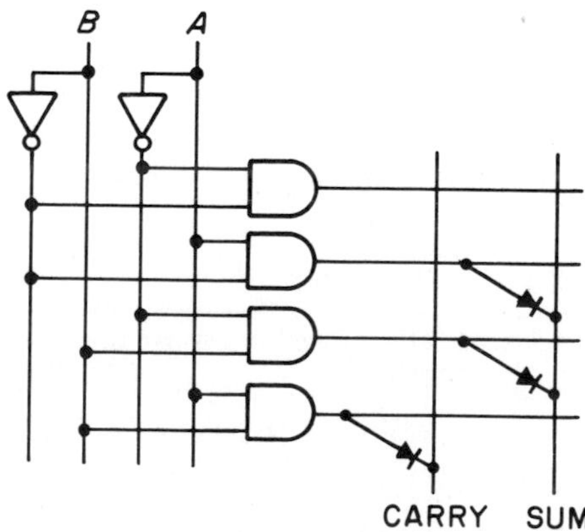

Figure 10–2 ROM for adding two binary numbers.

Half Adder:

Logic network with two inputs, the sum of which is represented by two outputs, the sum and the carry.

Also in Chapter 6, you saw how to construct a PLD that would add 3 bits (2 bits and a carry). The logic network from that chapter are shown again in Figure 10–3. Such a logic network is called a **full adder.**

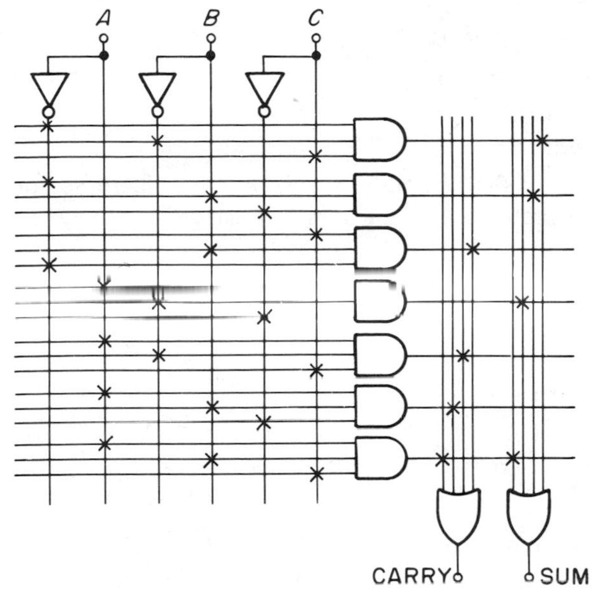

Figure 10–3 PLD full adder.

Full Adder:

Logic network with three inputs, the sum of which is represented by two outputs, the sum and the carry.

Using a Full Adder

The logic symbol of a full adder with its corresponding truth table are illustrated in Figure 10–4. Check your understanding of the operation of a full adder by following the next example.

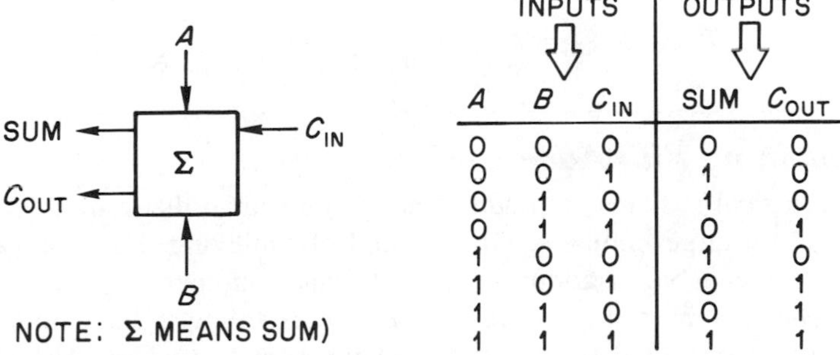

A	B	C_{IN}	SUM	C_{OUT}
0	0	0	0	0
0	0	1	1	0
0	1	0	1	0
0	1	1	0	1
1	0	0	1	0
1	0	1	0	1
1	1	0	0	1
1	1	1	1	1

Figure 10–4 Full adder and truth table.

Example 10–8

For the given inputs to a full adder, determine the resulting outputs (the sum and carry outputs):

(A) $A = 0,\ B = 1,\ C_{in} = 0$

(B) $A = 1,\ B = 0,\ C_{in} = 0$

(C) $A = 1,\ B = 0,\ C_{in} = 1$

(D) $A = 1,\ B = 1,\ C_{in} = 1$

Solution

Treat all of the full adder inputs as 3-bit binary addition problems. From this, determine the output of the sum and carry.

(A) $A = 0,\ B = 1,\ C_{in} = 0$

$$\begin{array}{rr} & 0 \leftarrow C_{in} \\ & 0 \leftarrow A \\ & +1 \leftarrow B \\ \hline C_{out} \rightarrow & 0\ 1 \leftarrow S_{out} \end{array}$$

Thus $S_{out} = 1,\ C_{out} = 0$

(B) $A = 1,\ B = 0,\ C_{in} = 0$

$$\begin{array}{rr} & 0 \leftarrow C_{in} \\ & 1 \leftarrow A \\ & +0 \leftarrow B \\ \hline C_{out} \rightarrow & 0\ 1 \leftarrow S_{out} \end{array}$$

Thus $S_{out} = 1,\ C_{out} = 0$

(C) $A = 1,\ B = 0,\ C_{in} = 1$

$$\begin{array}{rr} & 1 \leftarrow C_{in} \\ & 0 \leftarrow B \\ & +1 \leftarrow A \\ \hline C_{out} \rightarrow & 1\ 1 \leftarrow S_{out} \end{array}$$

(D) $A = 1,\ B = 1,\ C_{in} = 1$

$$\begin{array}{rr} & 1 \leftarrow C_{in} \\ & 1 \leftarrow A \\ & +1 \leftarrow B \\ \hline C_{out} \rightarrow & 1\ 1 \leftarrow S_{out} \end{array}$$

Thus $S_{out} = 1,\ C_{out} = 1$

Combining Full Adders

For practical systems, full adders are combined so that arithmetic operations may be performed on large numbers. One method of achieving this is shown in Figure 10–5. This system is capable of adding two 4-bit binary numbers.

The system in Figure 10–5 operates as follows. Each individual full adder has its carryouts connected to the carryin of the next full adder. The C_{in} of the first full adder (FA_0) is connected so that it is always LOW. This is done because the LSB of any addition

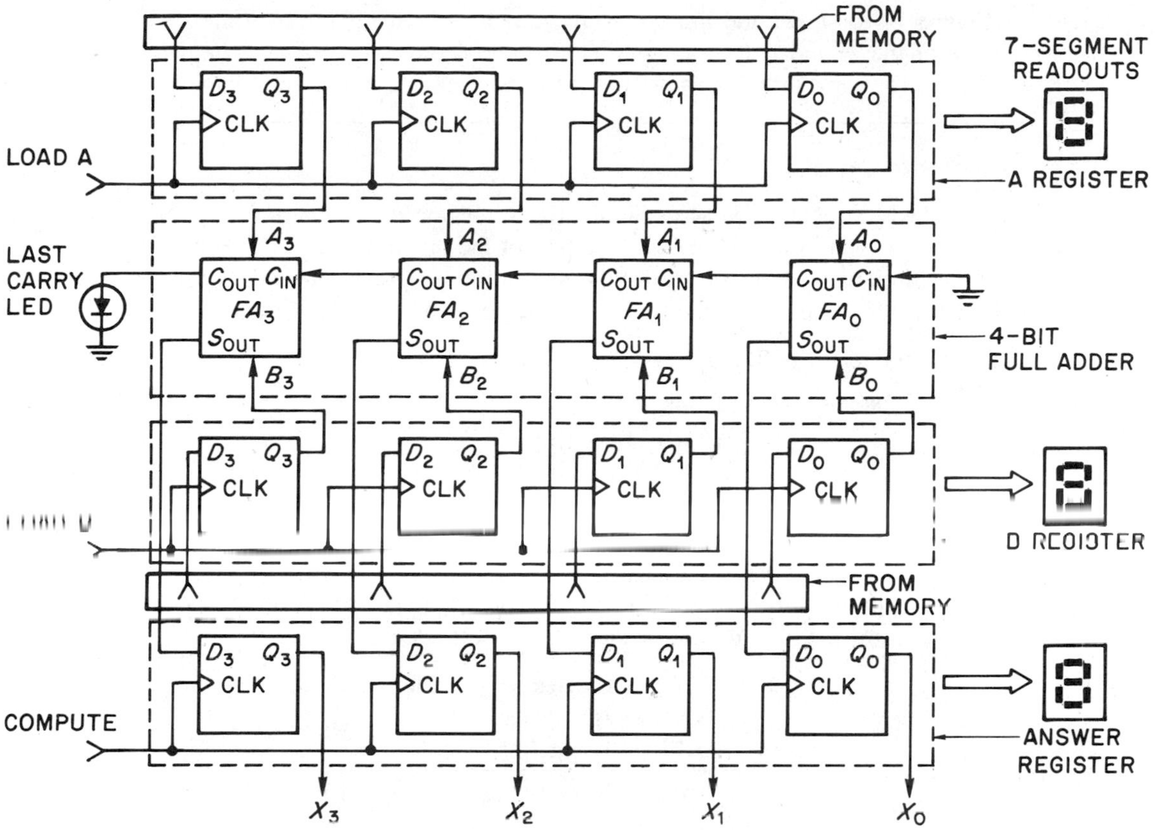

Figure 10–5 Combining full adders and registers.

never has a carry (there is nothing to carry from). The last carryout ($C_{out\ 3}$) is connected to an LED. This LED is used to indicate if the sum produced a value that was too large to fit in the answer register. The two numbers to be added are stored in register A (the augend) and register B (the addend). The sum is stored in the answer register. Each of the three registers has a separate input for activating its flip-flop clocks. The A and B registers each have a load input. When these inputs are activated, the numbers to be added (stored in memory) are transferred to these registers. The answer register has a compute input which will transfer the sum outputs from the full adders to the flip-flops within this register.

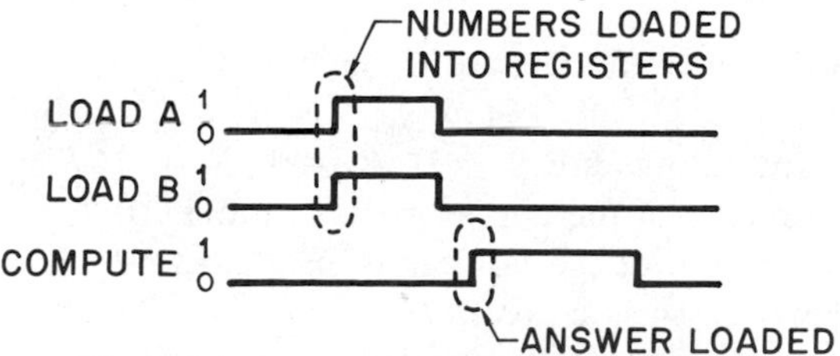

Figure 10–6 Timing relationships for full adder with registers.

Figure 10–6 shows the timing sequence that is used for the operation of adding the contents of the A and B registers and storing the results in the answer register.

The following examples show the operating details of this logic network.

Example 10–9

For the adder logic network of Figure 10–5 indicate the status of each carry output, each sum output, the carry LED condition of the bits in the answer register and each seven-segment readout for the following additions. Comment on the correctness of the absolute value verses the two's complement value of the output.

(A) A register = 0001, B register = 0011

(B) A register = 0111, B register = 0111

(C) A register = 1100, B register = 0111

Solution

Add each of the binary numbers as you would in normal binary addition. Keep track of all of the carries between terms; these carries represent the condition of each of the carry outputs for the corresponding full adder. Each sum from your computation will represent the sum output of each of the full adders.

(A)

		[hex readouts]
011	← Carries	
0001	← A register	1
+0011	← B register	+3
0100	← Answer reg.	4

Thus the carry outputs are $C_{out\,0} = 1$, $C_{out\,1} = 1$
$C_{out\,2} = 0$, $C_{out\,3} = 0$
The sum outputs are $S_0 = 0$, $S_1 = 0$, $S_2 = 1$, $S_3 = 0$
The contents of the answer register are 0100
The condition of the last carry LED is LOW
The answer 4 is correct (1 + 3 = 4)
This answer is correct because there was no carry overflow into the sign bit or the last carry LED.

(B)

		[hex readouts]
111	← Carries	
0111	← A register	7
+0111	← B register	+7
1110	← Answer reg.	E

Thus the carry outputs are $C_{out\,0} = 1$, $C_{out\,1} = 1$
$C_{out\,2} = 1$, $C_{out\,3} = 0$
The sum outputs are $S_0 = 0$, $S_1 = 1$, $S_2 = 1$, $S_3 = 1$
The contents of the answer register are 1110
The condition of the last carry LED is LOW
The answer 14 is correct (7 + 7 = 14)
This answer is correct provided it is understood that two's complement notation is not being used.

(C)

1100	← Carries	[hex readouts]
1100	← A register	C
+0111	← B register	+7
0011	← Answer reg.	3

Thus the carry outputs are $C_{out\,0} = 0$, $C_{out\,1} = 0$
$C_{out\,2} = 1$, $C_{out\,3} = 1$
The sum outputs are $S_0 = 1$, $S_1 = 1$, $S_2 = 0$, $S_3 = 0$
The contents of the answer register are 0011
The condition of the last carry LED is HIGH
The answer 3 is not correct (12 + 7 = 19)
The HIGH condition of the last carry LED indicates that the answer in the answer register is not correct.

Example 10–9 showed the operation of the full adder network under three conditions: when the numbers to be added and the answer left the MSB of all registers = 0; when the numbers to be added caused the MSB of the answer register to be 1; and when the numbers to be added produced an incorrect answer in the answer register. The first and second cases both produced a correct answer. However, if you wanted to represent signed binary numbers and use the two's complement notation, then the resulting answer indicates that you had a negative answer when adding two positive numbers (the MSB is 1 in the answer register, indicating a negative number in two's complement form). Thus under these conditions the answer would not be correct because with the MSB HIGH the value 1110 would indicate a −2 which is not the sum of 7 + 7. In the third case, when the last carry LED was active, it indicated an **overflow** condition. This meant that the answer in the answer register was not correct because an overflow had occurred.

Overflow:
A condition indicating that the result of an arithmetic operation is too large to fit in the number of bits in the answer register.

As you can see from this presentation, there are several practical conditions that must be considered for the analysis of a 4-bit full adder. These are the same conditions that would apply to an 8- or 16-bit full adder.

In order to facilitate the following presentation, a simplified logic diagram of the 4-bit full adder will be used. See Figure 10–7.

Adding Signed Binary Numbers

Figure 10–8 illustrates the four combinations that must be considered when developing logic hardware for arithmetic operations. From the analysis shown in Figure 10–8, a truth table can be constructed for each condition of the output. A PLD can then be used to replicate the table. By using this PLD along with a digital comparator, you can develop a logic network that would indicate when an overflow condition had occurred. This is shown in Figure 10–9. The actual programming of the PLD is left as an end-of-chapter problem assignment.

When an overflow does occur, the programmer must make corrections for this in the program. Overflow corrections are usually made by software. The next examples present problems that account for all four conditions stated in Figure 10–8.

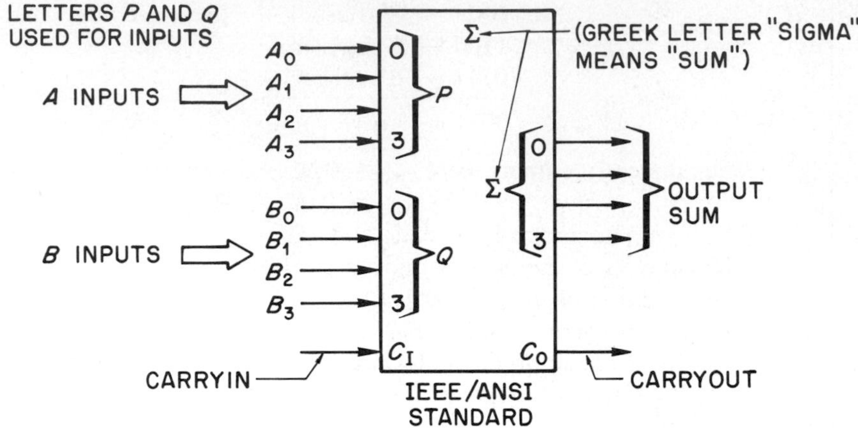

Figure 10–7 Logic diagram of 4-bit full adder.

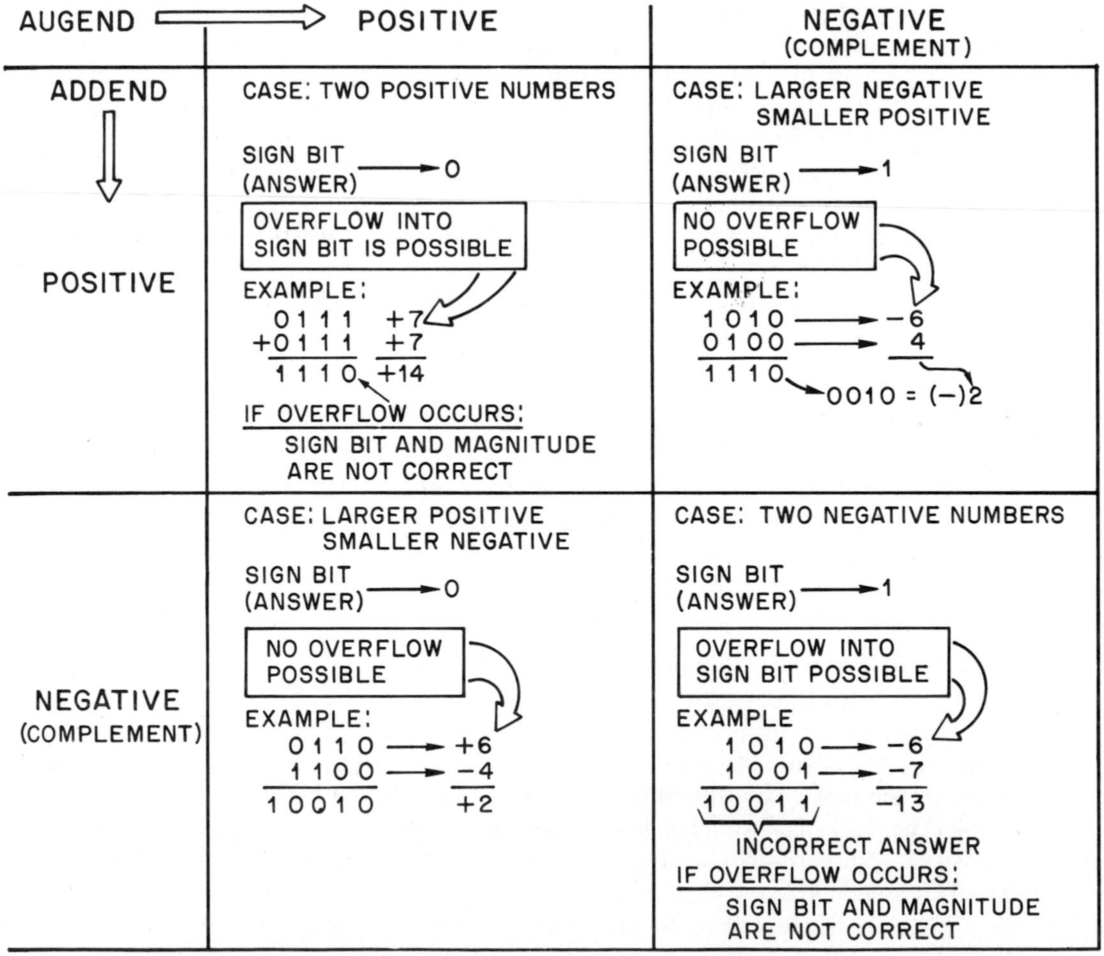

AUGEND ⟹ / ADDEND ⇓	POSITIVE	NEGATIVE (COMPLEMENT)
POSITIVE	CASE: TWO POSITIVE NUMBERS SIGN BIT (ANSWER) → 0 OVERFLOW INTO SIGN BIT IS POSSIBLE EXAMPLE: 0111 +7 +0111 +7 1110 +14 IF OVERFLOW OCCURS: SIGN BIT AND MAGNITUDE ARE NOT CORRECT	CASE: LARGER NEGATIVE SMALLER POSITIVE SIGN BIT (ANSWER) → 1 NO OVERFLOW POSSIBLE EXAMPLE: 1010 → −6 0100 → 4 1110 → 0010 = (−)2
NEGATIVE (COMPLEMENT)	CASE: LARGER POSITIVE SMALLER NEGATIVE SIGN BIT (ANSWER) → 0 NO OVERFLOW POSSIBLE EXAMPLE: 0110 → +6 1100 → −4 10010 +2	CASE: TWO NEGATIVE NUMBERS SIGN BIT (ANSWER) → 1 OVERFLOW INTO SIGN BIT POSSIBLE EXAMPLE 1010 → −6 1001 → −7 10011 −13 INCORRECT ANSWER IF OVERFLOW OCCURS: SIGN BIT AND MAGNITUDE ARE NOT CORRECT

Figure 10–8 Possible combinations of signed binary numbers.

MSB CONDITION			RESULT	
B	*A*	ANSWER	*A* > *B*	*B* > *A*
0	0	0	O.K.	O.K.
0	0	1	NO	NO
0	1	0	NO	O.K.
0	1	1	O.K.	NO
1	0	0	O.K.	NO
1	0	1	NO	O.K.
1	1	0	NO	NO
1	1	1	O.K.	O.K.

NO ⟶ INCORRECT RESULT
O.K. ⟶ CORRECT RESULT

Figure 10–9 Logic overflow analysis.

Example 10–10

For the following problems, write the expected decimal answer and indicate the hex representations of each term including the solution (use 4-bit binary words input to a full adder). State if an overflow occurs and indicate what this means. Use two's complement notation when necessary.

(A) +5 +6 = (B) +2 +3 = (C) −6 −5 = (D) −3 +6 =

(E) −2 −4 = (F) +3 −6 =

Solution

Convert each number to its binary equivalent. If the original number was negative, take the two's complement. Add the resulting binary values. Test the sign bit to see if an overflow has occurred. Refer to the diagram in Figure 10–8.

(A) **Decimal**	**Binary**	**Hex**
+5	0101	5
+6	0110	6
+11	1011	B

Resulting binary number indicates a negative value (sign bit is set to 1). Thus overflow has occurred, and for two's complement notation the answer is not correct because of overflow into the sign bit.

(B) **Decimal**	**Binary**	**Hex**
+2	0010	2
+3	0011	3
+5	0101	5

Resulting binary does not produce an overflow (correct sign bit is preserved). Thus the answer is correct.

(C) Decimal	Binary		Two's Complement	Hex
−6	0110	→	1010	A
−5	0101	→	1011	B
−11			1 0101	5

Resulting binary number produces an overflow into the sign bit (the sign bit should indicate a negative number, but it is set to 0, meaning positive). Thus the sign and magnitude of the answer are not correct.

(D) Decimal	Binary		Two's Complement	Hex
−3	0011	→	1101	D
+6	0110		0110	6
+3			1 0011	3

No overflow is possible under these conditions (see diagram in Figure 10–8). Sign bit and magnitude of the answer are correct.

(E) Decimal	Binary		Two's Complement		Hex
−2	0010	→	1110		E
−4	0100	→	1100		C
−6			1 1010	→ (−)0110	−6

Both numbers are negative, and resulting answer is negative. Thus there is no overflow of the sign bit. Since resulting answer has MSB as a 1, it is in the two's complement form and must be uncomplemented. Resulting answer indicates a −6, which is correct.

(F) Decimal	Binary		Two's Complement		Hex
+3	0011		0011		3
−6	0110	→	1010		A
−3			1101	→ (−)0011	D

No overflow is possible under these conditions (see diagram in Figure 10–8). Sign bit indicates a negative answer, therefore it must be uncomplemented (indicating a magnitude of 3). The resulting answer is correct.

Working with Larger Numbers

In practical systems, 8-bit and larger binary numbers are used. The 4-bit numbers were used here to present the important ideas. The following examples illustrate the use of arithmetic processes with larger binary numbers.

Example 10–11

Using 8-bit binary numbers, express the following problems in hex notation including the answer. Indicate if there is an overflow and, if so, what this means.

(A) +12 +25 = (B) −57 −93 = (C) +100 −55 =

Solution

Convert each number to its binary equivalent. If the original number was negative, then take the two's complement. Add the resulting binary values. Then check to see if an overflow exists.

(A) Decimal	Binary	Hex
+12	0000 1100	0C
+25	0001 1001	19
+37	0010 0101	25

Both numbers were positive, and the sign bit in the answer indicates a positive value. Thus, no overflow into the sign bit took place, and the answer is correct.

(B) Decimal	Binary		Two's Complement	Hex
−57	0011 1001	→	1100 0111	C7
−93	0101 1101	→	1010 0001	A1
−150			1 0110 1000	68

Both numbers were negative; thus the sign bit of the answer should also have been negative. Since the sign bit indicates a positive answer, overflow into the sign bit has occurred, and the sign and magnitude of the answer are not correct.

(C) Decimal	Binary		Two's Complement	Hex
+100	0110 0100		0110 0100	64
−55	0011 0111	→	1100 1001	C9
+45			1 0010 1101	2D

No overflow is possible when the two numbers have different signs. Thus the sign bit and magnitude of the answer are correct.

Example 10–12

Perform the indicated operations for the following hex numbers. Give the answer you would actually get using hardware arithmetic and state if the answer indicates an overflow condition. The number of binary bits used is indicated by the number of hex values used in the problem.

(A) 0E + 21 = (B) −3E + 7F = (C) 3ECA − 6495 =

Solution

First convert the given hex number to its binary equivalent. If the original value was negative, then take the two's complement of the number. Perform the addition on the resulting binary number. Check the sign bit for overflow and comment on the result accordingly.

(A)

Hex	Binary		
0E	0000 1110		
+21	0010 0001		
	0010 1111	→	2F

Both numbers were positive, and the answer should be positive. Since there was no overflow into the sign bit, the answer is correct.

(B)

Hex	Binary		Two's Complement		
−3E	0011 1110	→	1100 0010		
+7F	0111 1111		0111 1111		
			1 0100 0001	→	41

Recall from Figure 10–8 that there cannot be an overflow when a + and − value are combined. Hence the sign bit and magnitude of the answer are correct.

(C)

Hex	Binary		Two's Complement
3ECA	0011 1110 1100 1010		0011 1110 1100 1010
−6495	0110 0100 1001 0101	→	1001 1011 0110 1011
			1101 1010 0011 0101
			DA35

Again, no overflow is possible under these conditions. The indicated answer is negative (sign bit is = 1). Therefore the two's complement must be taken in order to obtain the correct magnitude for the number. This would be

0010 0101 1100 1011

which is $-96\ 75_{10}$.

Conclusion

This section presented the important concepts of performing signed arithmetic operations using the limitations of digital hardware. Armed with this knowledge, you will see how to perform multiplication and division using software. In the next section you will be introduced to real arithmetic hardware used to perform the operations presented here. Check your understanding of this section by trying the following review questions.

10–2 Review Questions

1. State the difference between a half adder and a full adder.
2. State what an overflow condition indicates.

3. Which combinations of arithmetic addition can produce answers that have an overflow into the sign bit?
4. Which combinations of arithmetic addition will always produce the correct answer?

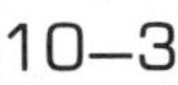

10–3 ARITHMETIC HARDWARE

Discussion

This section presents real hardware that will perform the arithmetic presented in the last section. Here, you will also see the hardware that performs logic operations as well. This section will prepare you for the arithmetic and logic hardware used in microprocessors.

The Accumulator

The arithmetic processing unit that was used in the last section contained three registers. Two registers were used to hold the numbers to be added, and a third register was used to store the answer. Real hardware does not do this. Instead two registers are used. One of these registers, called the *accumulator*, is used to store part of the problem as well as the answer. This concept is shown in Figure 10–10.

There are hardware devices used to perform the functions illustrated in Figure 10–10.

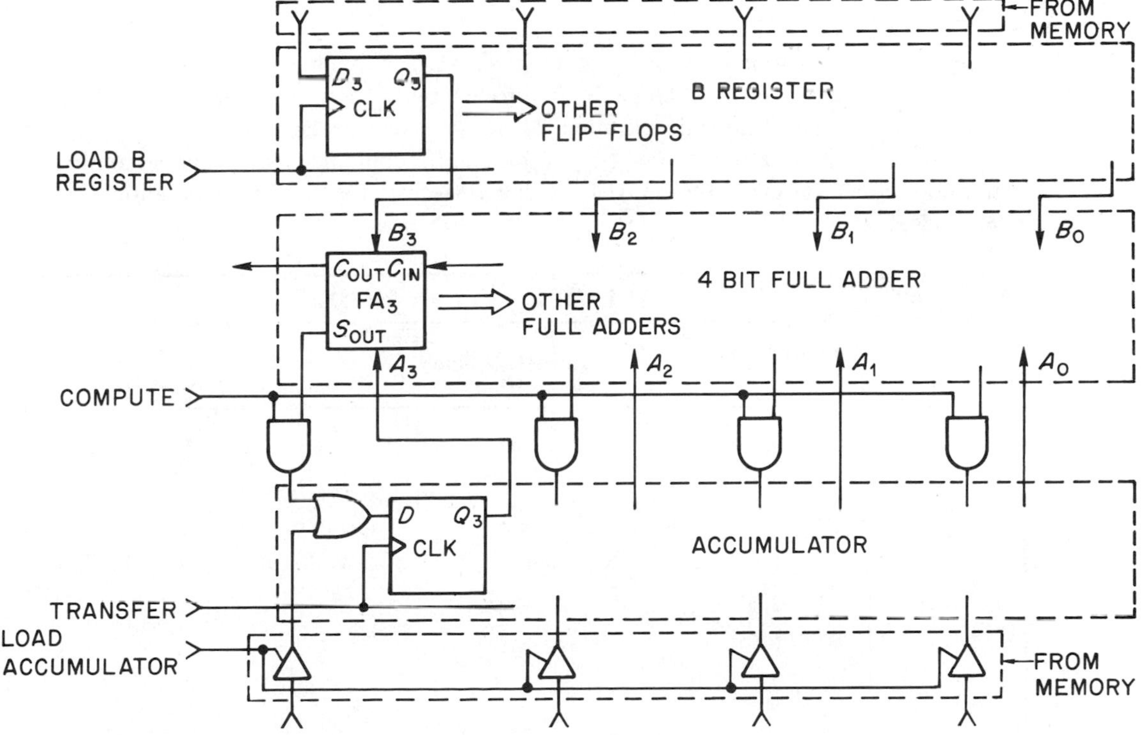

Figure 10–10 Concept of an accumulator.

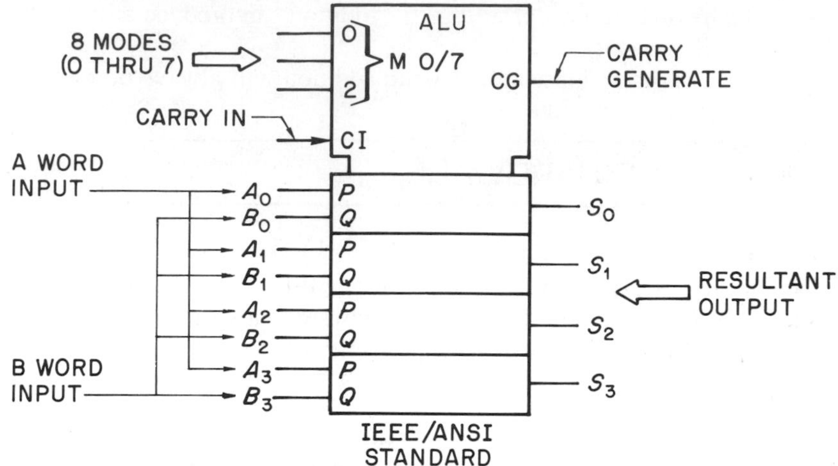

Figure 10–11 Typical block diagram of an ALU.

The ALU

An **ALU** is an arithmetic logic unit.

> **ALU:**
> Arithmetic logic unit. A hardware device that performs arithmetic and logic functions.

Figure 10–11 shows a block diagram of a typical ALU. Note that the figure uses the ANSI standard. The carry generate output is controlled by a carry out from the MSB of the answer produced in an arithmetic operation. There is no carry generate from a logic operation.

Most ALUs have a mode control. The bit pattern contained on the mode control will determine the action taken by the ALU. Table 10–4 lists typical ALU operations for various modes.

Table 10–4 TYPICAL ALU OPERATION MODES

Mode Control Inputs			Arithmetic/Logic	
S_2	S_1	S_0	Operation	Meaning of Operation
0	0	0	Clear	Accumulator set to 0
0	0	1	B − A	Subtraction
0	1	0	A − B	Subtraction
0	1	1	A + B	Addition
1	0	0	A ⊕ B	Exclusive OR
1	0	1	A + B	OR
1	1	0	A × B	AND
1	1	1	Preset	Accumulator set to 1's

Example 10–13

The ALU presented in this section has the following bit patterns on its mode control lines and the indicated values in its corresponding registers. State the contents of both registers after the indicated process is completed. Indicate if the process will cause an overflow condition. (This will happen if there is a carry from the MSB.)

(A) B = 3C, A = 41, mode = 6 (B) B = 7F, A = 3A, mode = 4

Solution

First convert all of the values into binary numbers. Then referring to the table for the various modes, perform the indicated arithmetic or logic operation. Check for an overflow condition.

(A) B = 3C = 0011 1100
A = 41 = 0100 0001 Mode = 6 = 110 = AND
0000 0000

Final register contents: B = 3C, A = 00, no overflow

(B) B = 7F = 0111 1111
A = 3A = 0011 1010 Mode = 4 = 100 = Exclusive OR
0100 0101

Final register contents: B = 7F, A = 45, no overflow

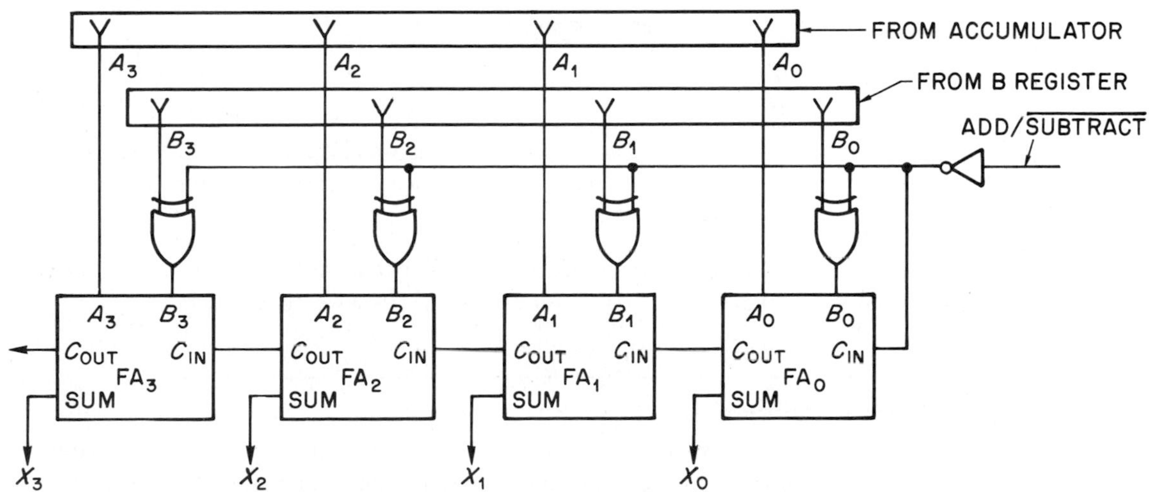

Figure 10–12 Logic for performing addition and subtraction.

Adder-Subtractor

Logic that can be implemented to actually perform addition and subtraction is shown in Figure 10–12. As you can see from the figure, subtraction of word B from word A is performed by using the two's complement method. This is accomplished by simply inverting the number to be subtracted (which is equivalent to taking the one's complement) and, at the same time, making the carryin HIGH. Making the carryin HIGH is the same as adding 1 to the one's complement—this process produces the two's complement.

Example 10–14

For the addition and subtraction logic shown in Figure 10–12, indicate the final contents of each register after the indicated process has been performed. Also state the status of the carryin and the carryout.

(A) $B = 3$, $A = 2$, add/$\overline{\text{subtract}}$ = HIGH

(B) $B = 7$, $A = 5$, add/$\overline{\text{subtract}}$ = LOW

Solution

First convert all values to their equivalent binary. Then perform the indicated operation using two's complement arithmetic for subtraction operations.

(A) $B = 3 =$ 0010
$A = 2 =$ 0101 add/$\overline{\text{subtract}}$ = HIGH = addition

1010 [Carryin bit = 0 because this was an addition problem and carryout = 0 because there was no carryout of the MSB of the answer.]

(B) $B = 7 =$ 0111
$A = 5 =$ 0101 add/$\overline{\text{subtract}}$ = LOW = subtraction

Using the two's complement system, you have

1001
+0101

1110 [Carryin bit = 1 because this was a subtraction problem and carryout = 0 because there was no carryout of the LSB of the answer.]

Final answer of 1110 indicates a negative number that must be uncomplemented to get the correct magnitude: 1110 → 0010 = (−)2. Answer is correct: $5 - 7 = -2$.

Binary-Coded Decimal Logic

Recall BCD notation from Chapter 2. BCD is a digital code that allowed direct translation between decimal numbers and a digital code. Thus the decimal number 532 was represented as $0101\ 0011\ 0010_{BCD}$. This meant that the largest BCD value allowed in 4 bits was 1001 (representing decimal 9), because to represent 10_{10} you would use $0001\ 0000_{BCD}$.

This system must be preserved when adding BCD numbers. The rules for adding BCD coded numbers are as follows:

1. Use normal binary addition to add two BCD values.
2. If the sum of any group of four is 9 or less, no carry is generated and the answer is correct.
3. If the sum is greater than 9, a carry is generated and the resulting sum must have its magnitude corrected by adding a 6 (0110). Adding 6 to any value over 9, represented by only 4 bits, will ensure that the result is less than 9.

The following example demonstrates the process.

Example 10–15

Add the following BCD numbers. Convert both the problem and the answer to decimal and check your answer.

(A) 0110 + 0010 (B) 0001 1001 + 0111 0100

Solution

(A)
```
  0110 =  6
 +0010 = +2
 ----------
  1000 =  8   Answer is correct
```

(B)
```
  0001 1001 =    19
 +0111 0100 =   +74
 ------------------
  1000 1101 = Sum > 9

  1000 1101
       0110   → add 6 and carry a 1
  ---------
  1001 0011   = 93 Answer is correct
```

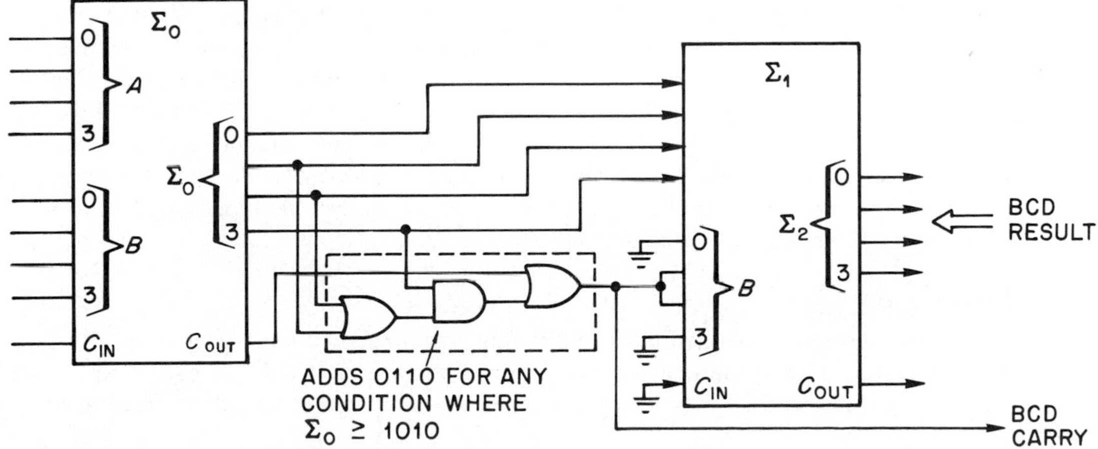

Figure 10–13 Hardware for BCD addition.

BCD Hardware

Figure 10–13 shows a logic circuit capable of performing BCD addition. Note the logic used to correct any sum that is greater than 9. Essentially this logic adds 6 to the resulting sum when it is 9 or more and produces a carry. This is demonstrated in the following example.

Example 10–16

For the BCD hardware shown in Figure 10–13, indicate what the contents of each register will be and the status of the carry bit after the following additions are performed.

(A) $B = 4$, $A = 5$ (B) $B = 6$, $A = 7$

Solution

(A)
$$\begin{array}{rr} B = 4 = & 0100 \\ A = 5 = & \underline{0101} \\ & 1001 \end{array}$$

Final register contents: $B = 4$, $A = 9$, no carry

(B)
$$\begin{array}{rrl} B = 6 = & 0110 & \\ A = 7 = & \underline{0111} & \\ & 1101 & \\ & \underline{0110} & \leftarrow 6 \text{ must be added} \\ & 1\,0011 & \text{Answer is } 13\ (6 + 7 = 13) \end{array}$$

Final register contents: $B = 6$, $A = 3$, carry generated

Cascading Adders

Binary and BCD adders may be cascaded as shown in Figure 10–14. Theoretically, this process of cascading adders can be kept up indefinitely. However, there is a practical problem. The carries generated cause a delay in the process. This means that with a large enough number, the adding would fall behind the carry and thus produce incorrect answers. One way around this is to slow down the addition process. Doing this, however, is not satisfactory for high-speed digital systems where many calculations must be done in very short times. A method of preventing this is called *look-ahead carry*. An extra logic circuit, with a minimum number of gates for speed, analyzes the input words. From this a final carry is generated before the full adder adds.

Conclusion

This section presented hardware for performing arithmetic and logic operations. Here you saw the actions of an arithmetic logic unit as well as logic networks for performing binary and BCD arithmetic. Test your understanding of this section by trying the following review questions.

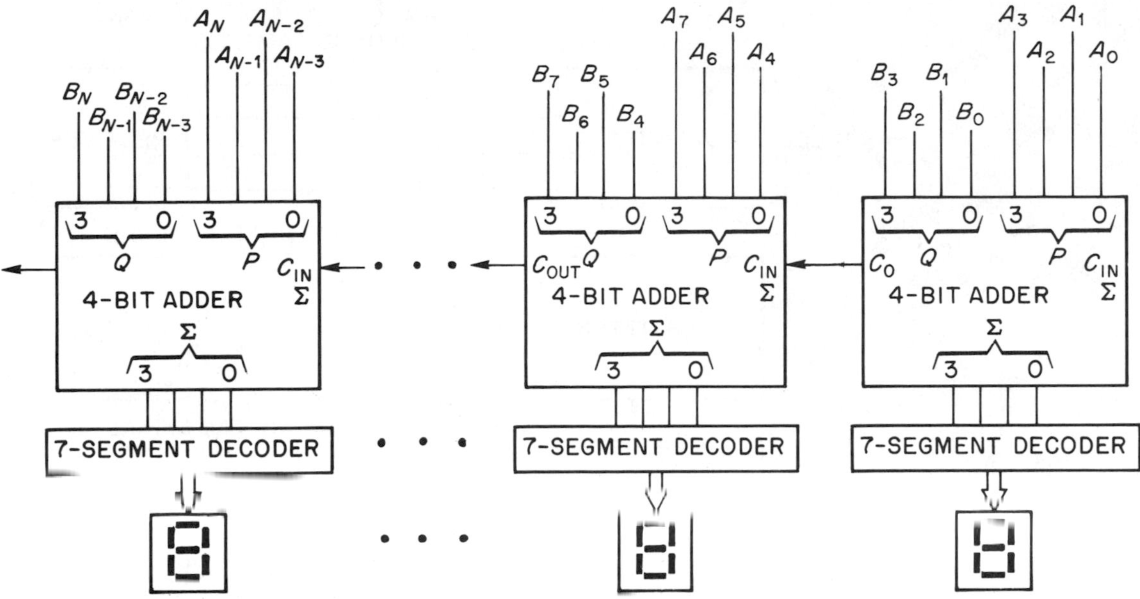

Figure 10–14 Cascading binary and BCD adders.

10–3 Review Questions

1. State the function of the accumulator.
2. What is an ALU?
3. Describe the basic operation of an adder-subtractor.
4. Explain how BCD addition differs from binary addition.
5. What does cascading adders mean?

10–4 DIGITAL COMMUNICATION MEDIUMS AND METHODS

Discussion

This section presents information concerning different methods of connecting one digital system to another. Here you will see the various mediums used to accomplish communications between computers and other digital systems. You will also see some of the advantages and disadvantages offered by these various communication methods.

Methods

As you have seen in previous chapters, there are basically two methods of transmitting digital information from one system to another. One is by *parallel* transmission, the other by *serial* transmission. The serial transmission method may be broken down into *synchronous* and *asynchronous* serial transmission. All of these methods are shown in Figure 10–15.

Parallel transmission is used for sending digital information over short distances, usually 1 meter or less. This is done because of the signal loss due to lengthy parallel trans-

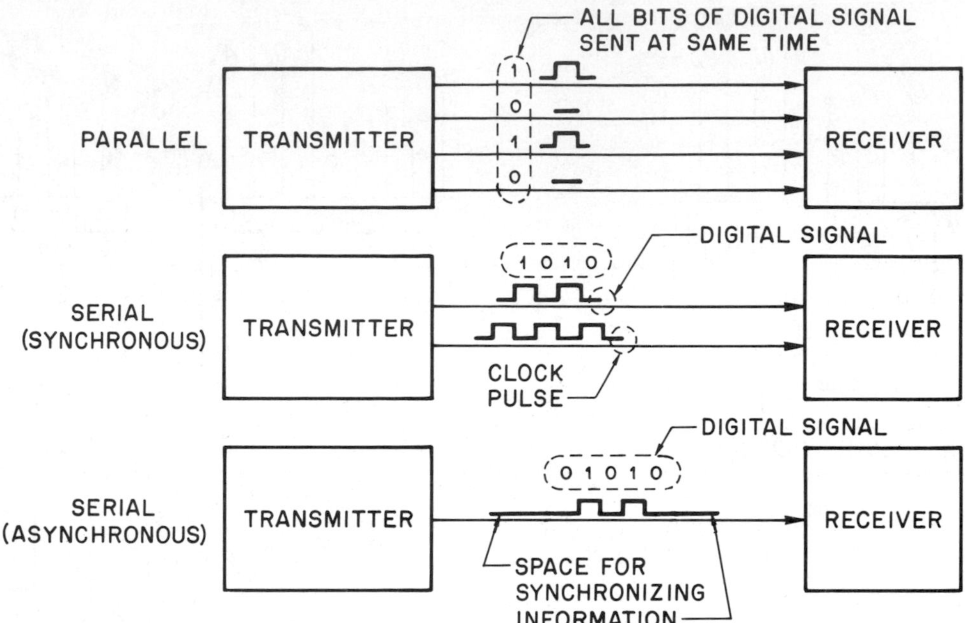

Figure 10–15 Methods of digital transmission.

mission lines as well as cost. Long-distance transmission is accomplished by serial transmission of data.

Communication Mediums

Many types of mediums can be used for the transmission of digital information. The most direct method is by a wired connection between the two systems. Several standard conditions that are used for direct connection are illustrated in Figure 10–16.

Space Transmission

Another medium used for the transmission of digital information is space. This technique is referred to as electromagnetic radiation (commonly called *radio waves*). Space is used

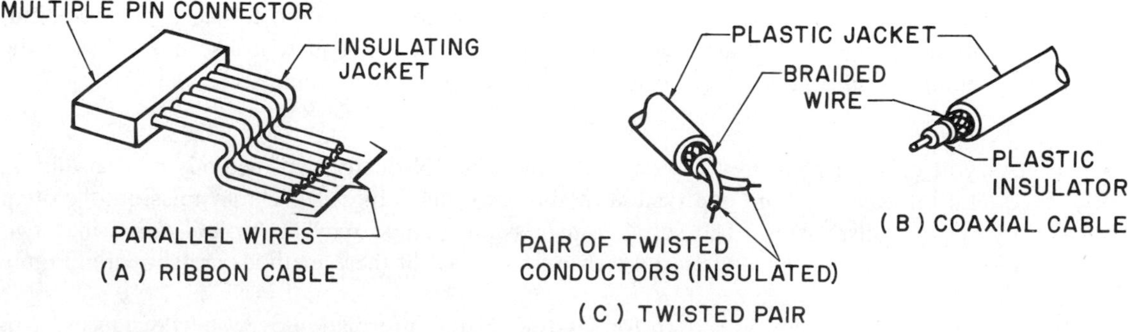

Figure 10–16 Different types of conductors.

as a communications medium when physical connections between transmitter and receiver are not possible nor practical. An example of this would be digital communication with a satellite.

The details of exactly how an electromagnetic wave is created and generated through space is beyond the scope of this book. For a detailed discussion, consult a communications textbook (such as *Electronic Communications*, T. Adamson, Delmar Publishers, 1988).

The basic idea behind electromagnetic radiation is that a radio wave propagates through space at or near the speed of light, and can be envisioned as a sine wave. Digital information can then **modulate** this wave in one of three ways:

1. By causing the *amplitude* of the carrier to change. This is called **amplitude modulation** (AM).
2. By causing the *frequency* of the carrier to change. This is called **frequency modulation** (FM).
3. By causing the *phase* of the carrier to change. This is called **phase modulation** (PM).

Modulate:
To change or modify. In communications the act of having information change a measurable characteristic of an electromagnetic wave (called the *carrier*).

Amplitude Modulation:
A process of having information change the strength (amplitude) of the carrier.

Frequency Modulation:
A process of having information change the frequency of the carrier.

Phase Modulation:
A process of having information change the phase of the carrier.

All three modulation techniques are shown in Figure 10–17.

Fiber Optics

Another method of connecting digital systems is through the use of light transmission in a medium that directs the path of the light. Fundamentally, the digital information can cause the intensity of light transmission in a glass fiber to change. This concept is illustrated in Figure 10–18.

The advantages of using fiber optics for the transmission of digital information are

1. Immunity from electromagnetic interference
2. Immunity from interception by external means (privacy)
3. Inexpensive and abundant materials (silicon is the main ingredient for the construction of optical fibers and is also the most abundant element on the earth's surface)
4. Resistance to corrosion and oxidation
5. Immune to atmospheric changes
6. Wide bandwidth (the ability to include many different transmissions at the same time)

Practical fiber-optic cables are being installed throughout the United States and in many other countries. These cables are designed to handle voice, video (TV), and computer (digital) data all at the same time. A typical fiber-optic link between San Francisco

(A) AMPLITUDE MODULATION (AM)

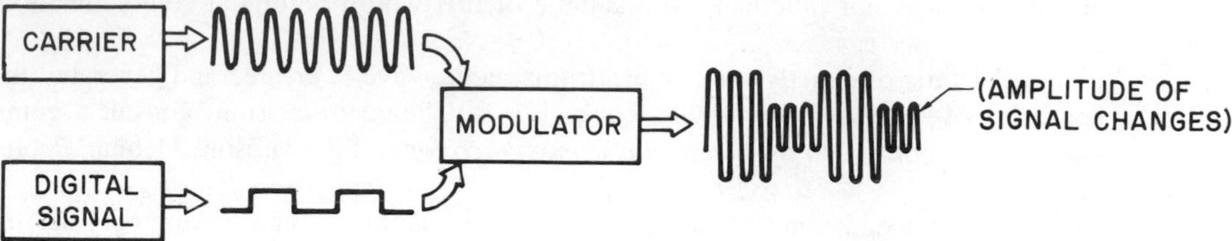

(B) FREQUENCY MODULATION (FM)

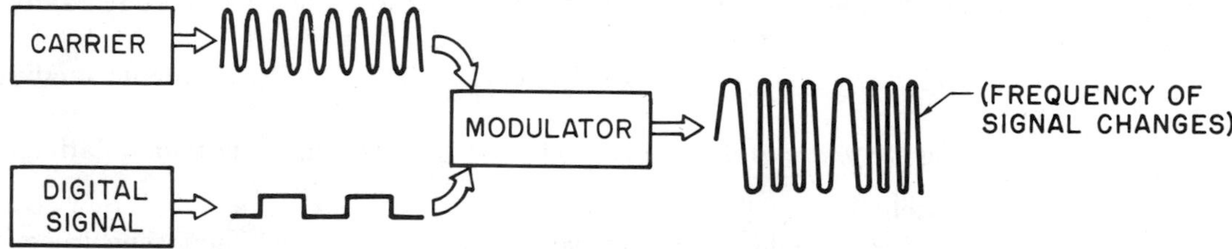

(C) PHASE MODULATION (PM)

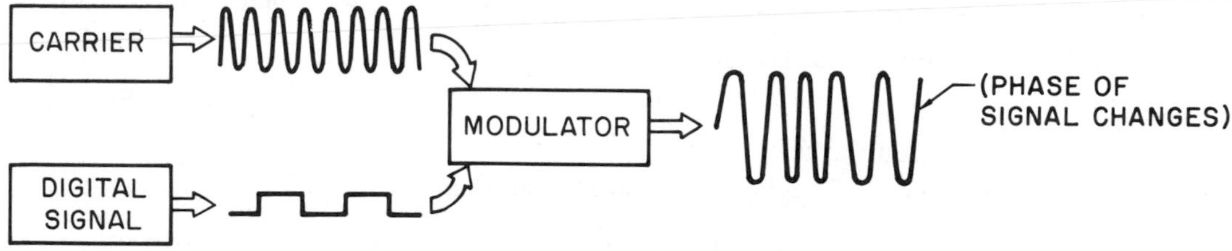

Figure 10–17 Various methods of carrier modulation.

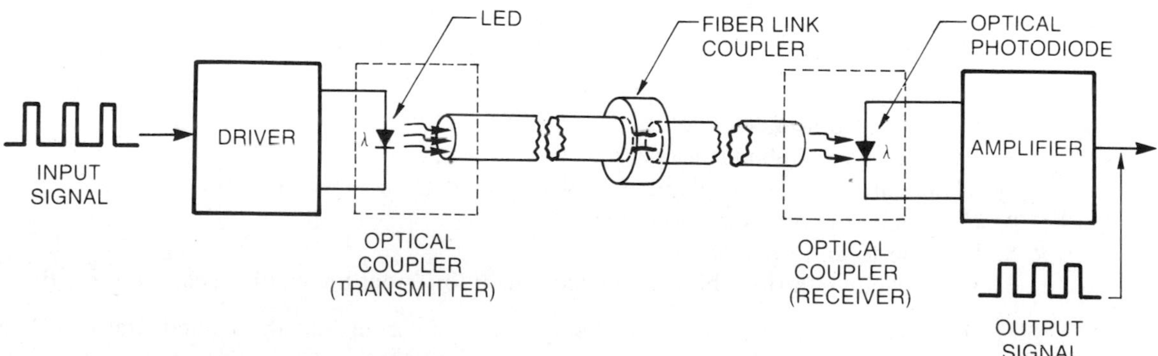

Figure 10–18 Fiber-optic connection.

and San Diego contains over 78,000 fiber-kilometers of lightwave circuits. The system can handle three signals of 90 megabits per second over the same fiber. This gives more than 240,000 digital channels at 64,000 bits per second in a cable with 144 optical fibers. Such systems will soon be commonplace in all households. It will be accessed by a wall receptacle that is no more complicated than the simple telephone connection it will replace. Thus, libraries, schools, museums, banks, news services, and other sources of information will be able to be interactively accessed directly from your home.

Laser Communications

The **laser** is another medium for linking two or more digital systems.

Laser:
Light amplification by stimulated emission of radiation.

The advantage of using laser light is that it will provide line-of-sight transmission without any physical connections. The difference between laser light and radio waves is that the laser can be easily directed at a specific target and thus affords the privacy that undirected radio waves cannot provide.

Another method of utilizing laser light is with a fiber-optic cable. This system can now be used to transmit digital information over long distances, and line-of-sight transmission is no longer necessary (the fiber-optic cable now acts as a light guide). Such an arrangement is illustrated in Figure 10–19.

Types of Transmission

There are three basic transmission types used for digital information: **simplex mode, half-duplex mode,** and **full-duplex mode.**

Simplex Mode:
A mode of transmission where data are sent in only one direction, from transmitter to receiver.

Half-Duplex Mode
A mode of transmission where two-way communication can take place between two systems, but not at the same time.

Full-Duplex Mode
A mode of transmission where two systems may transmit and receive data at the same time.

These various methods of transmission are shown in Figure 10–20.

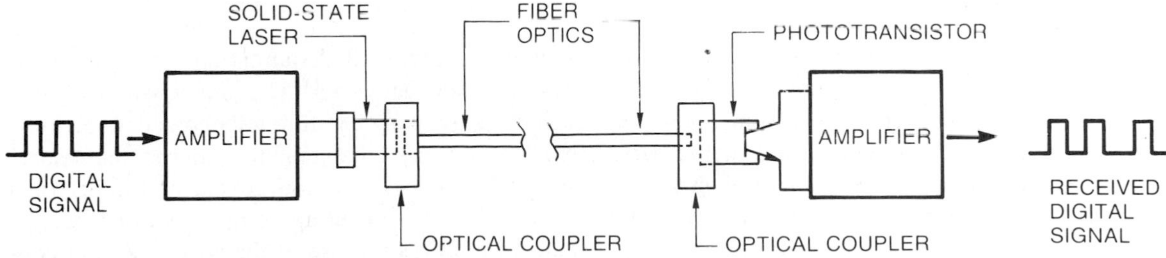

Figure 10–19 Laser—fiber-optic link.

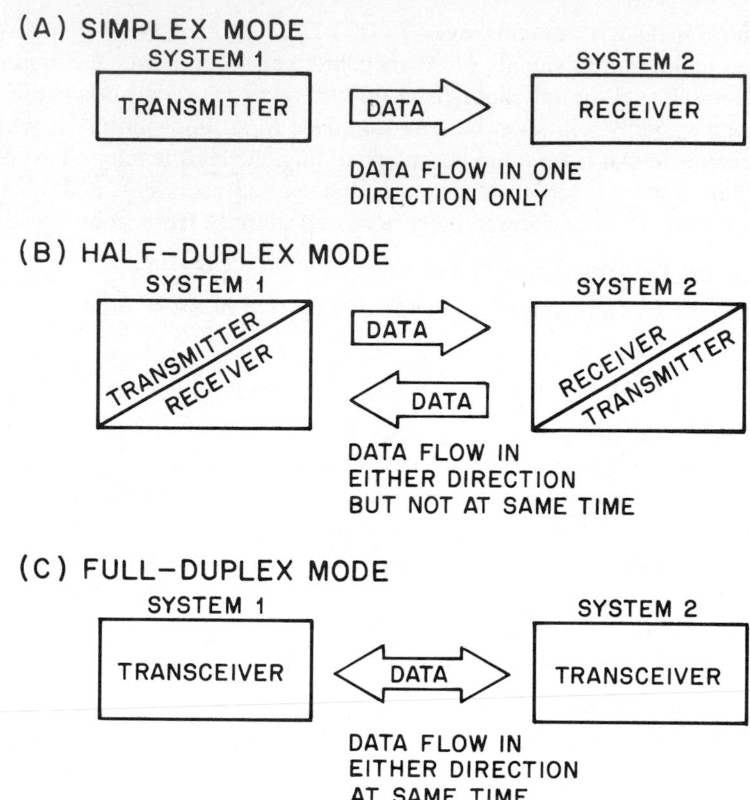

Figure 10–20 Various operational modes.

Interfacing Standards

There are basically three different systems used for making wired electrical connections between digital systems. These are *recommended standards* (RS) published by the Electronic Industries Association (EIA). They are called

1. RS-232C
2. 20-mA current loop
3. RS-422, RS-423, and RS-449

RS-232C

The RS-232C interface standard (RS stands for recommended standards) was first developed for data communications on public telephone networks. The intent was to allow different systems to communicate with each other by using existing telephone connections. This standard was used in the early 1960s (before the advent of the microcomputer) to encourage the use of time-shared computers. However, with the advent of the personal computer, this connection standard was also used for connecting computers to printers as well as other in-house terminals. This standard requires the use of the special 25-pin connector shown in Figure 10–21. The RS-232C standard connections are shown in Figure 10–22.

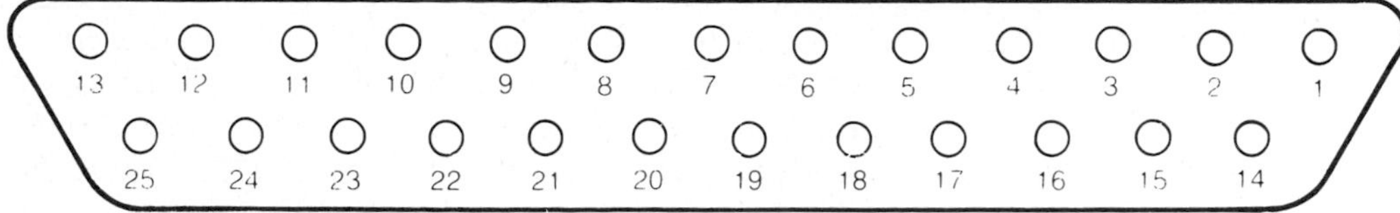

Figure 10–21 Pin diagram of 25-pin connector used for interfacing.

The standard classifies all computer equipment and peripherals in one of two categories:

1. Data terminal equipment (DTE): printers and terminals.
2. Data communications equipment (DCE): **modems**.

Modem:

Stands for *mo*dulator/*dem*odulator. A modulator converts digital information into transmission signals. A demodulator converts the signal to the original digital form upon reception.

RS-232C SIGNAL SPECIFICATIONS		
Pin No.	**Type of Signal**	**Purpose of Signal**
1	Ground	Protective ground
2	Data	Transmitted data
3	Data	Received data
4	Control	Request to Send
5	Control	Clear to Send
6	Control	Data Set Ready
7	Ground	Signal ground
8	Control	Received signal detector
9		Reserved for testing
10		Reserved for testing
11		No assignment
12	Control	Secondary received signal Detector
13	Control	Secondary Clear to Send
14	Data	Secondary transmission data
15	Timing	Transmission signal timing
16	Data	Secondary received data
17	Timing	Receiver signal timing
18		No assignment
19	Control	Secondary Request to Send
20	Control	Data Terminal Ready
21	Control	Signal quality detector
22	Control	Ring indicator
23	Control	Data signal rate selector
24	Timing	Transmitting signal timing
25		No assignment

Figure 10–22 RS-232C signal specifications.

Note that the term *microcomputer* is not in either category. This is because the standard was set before microcomputers became popular. The microcomputer can be viewed as a DTE when interfaced to a modem or as a DCE when connected to a printer. It is important to understand that, from the table, pin 2 *transmitted data* is from the DTE to the DCE. With pin 3, received data is from the DCE to the DTE. Thus, the definitions are from the point of view of the DTE.

With the RS-232C interface specifications, there are four types of connecting lines:

1. Data signals
2. Control signals
3. Timing signals
4. Grounds

The voltage levels for signals are −5 V to −15 V for ON (called a mark) condition and +5 V to +15 V for an OFF (called a space) condition. For any piece of equipment using the RS-232C standard, all pins do not have to be used.

20-mA Current Interface

This system uses a current of 20 mA to represent a HIGH and no current to represent a LOW. This is different from other operations where a voltage is used to represent the logic level. This interface was originally intended for use with teletype equipment, where a minimum current was necessary to close a relay contact. Even though this equipment is now outdated, the 20-mA current loop finds application where interference from external electrical noise may be a problem or when electrical isolation between systems is required. Figure 10–23 shows a typical application of the current loop.

Other Standards: RS-422, RS-423, and RS-499

Other interface standards are used to overcome some of the limitations of the RS-232C. The RS-422 allows for higher signal rates over longer distances than the RS-232C. The mechanical connections for the RS-422 are provided by the RS-499 standard, which requires a 37-pin connector.

In order to be compatible with other systems, the RS-423 provides a link designed to connect both the RS-232C and the RS-422 links. Doing this provides a method for allowing systems to use both the older 232C and newer 422 standards. Figure 10–24 presents the required signals for the RS-499 system.

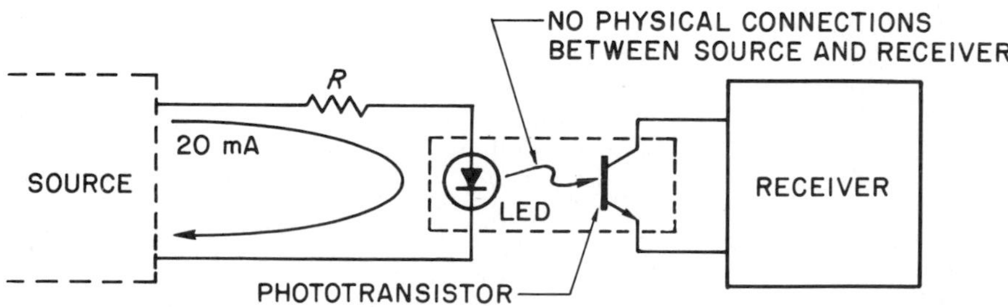

Figure 10–23 20-mA current loop application.

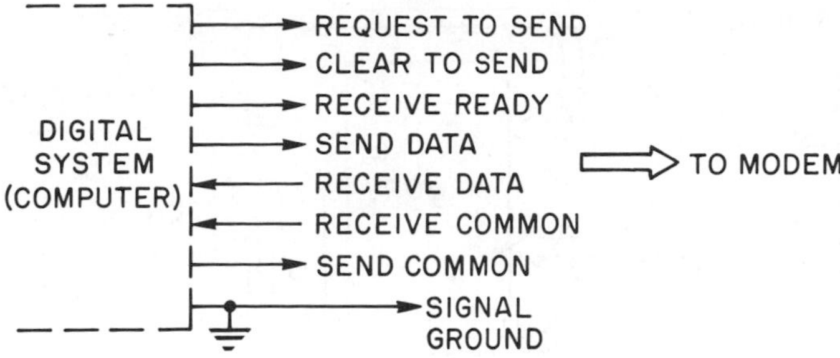

Figure 10–24 Required RS-499 signals.

Conclusion

This section presented information on the different mediums and methods of connecting one digital system to another. In the next section, you will see the hardware used to achieve the necessary interfacing between these systems. Check your understanding of this section by trying the following review questions.

10–4 Review Questions

1. When is parallel transmission of digital information used? When is serial transmission used?
2. What are the three ways for transmitting a digital signal on a radio wave?
3. Name some advantages of fiber optics for transmitting digital information.
4. What does laser stand for?
5. What does RS-232C mean?

10–5 INTERFACING HARDWARE

Discussion

This section presents specific methods and hardware for interfacing a digital system (such as a computer) with external devices (such as telephones and printers). Here you will see how the data used by one system can be shared with data from another system. Because of the need to share information with other systems, this is one of the fastest-growing areas in digital systems.

Direct Memory Access (DMA)

Direct memory access (DMA) is a method of entering data directly into a digital system while bypassing the microprocessor.

> **Direct Memory Access (DMA):**
> An I/O capable of transferring data to and from memory without intervention of the microprocessor.

Usually the microprocessor is used in the transfer of data between a microprocessor-based digital system and I/O. However, the process of transferring large blocks of data is a repetitious task that can be done more quickly by a special device called a **DMA controller.**

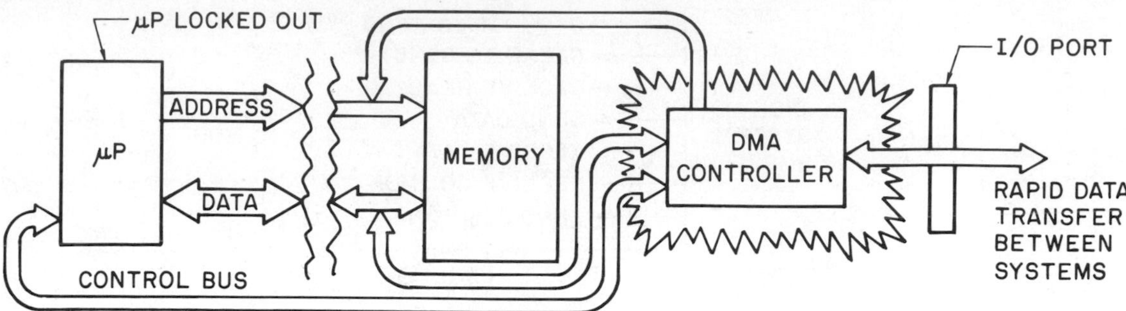

Figure 10–25 Basic idea of a DMA controller.

DMA Controller:
A direct memory access device used to control the I/O transfer of data without microprocessor intervention.

Figure 10–25 shows the basic idea of a DMA controller.

Since the purpose of the DMA controller is to transfer data between microprocessor-based digital systems, it must have complete access to the address and data bus in the same manner the microprocessor does. It must also have some way of letting the microprocessor know that this is going to happen so that there is no bus conflict while this process is taking place. The DMA controller must also know when the microprocessor will allow it to use the system bus for data transfer. Thus, the DMA controller must also have access to some of the microprocessor control lines. The block diagram of a typical DMA controller is shown in Figure 10–26.

The DMA controller has three registers that interface with the I/O port. The address register stores the initial value of the address of the data to be transferred. The command register holds instructions concerning how the data are to be transferred, and the count register contains information about the size of the block of memory to be transferred. Transfer then begins between the two systems. The μP is locked out of this transfer process.

Peripheral Interface Adapter (PIA)

The **peripheral interface adapter (PIA)** is a device that is used as a parallel data bus interface.

Peripheral Interface Adapter (PIA):
A device used to perform parallel I/O.

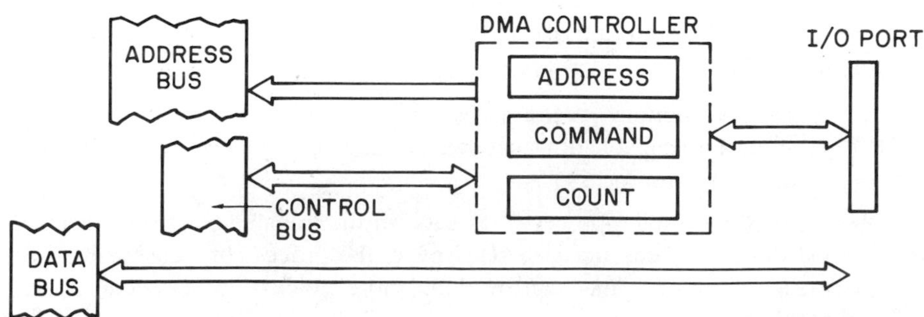

Figure 10–26 Block diagram of typical DMA controller.

The basic idea behind the PIA is to provide tristate registers that are capable of storing digital data between two separate digital systems. This concept is illustrated in Figure 10–27.

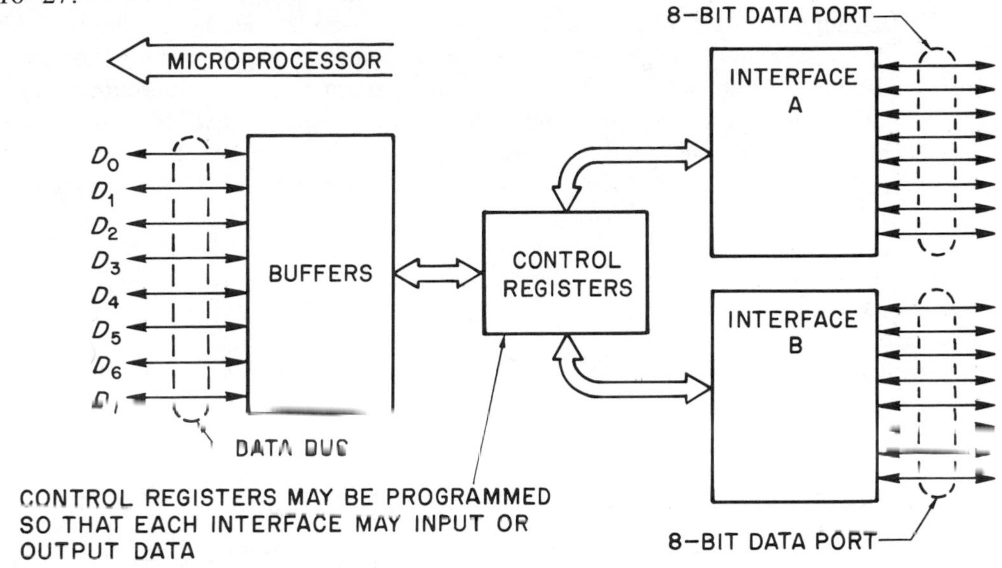

Figure 10–27 Basic idea of a PIA.

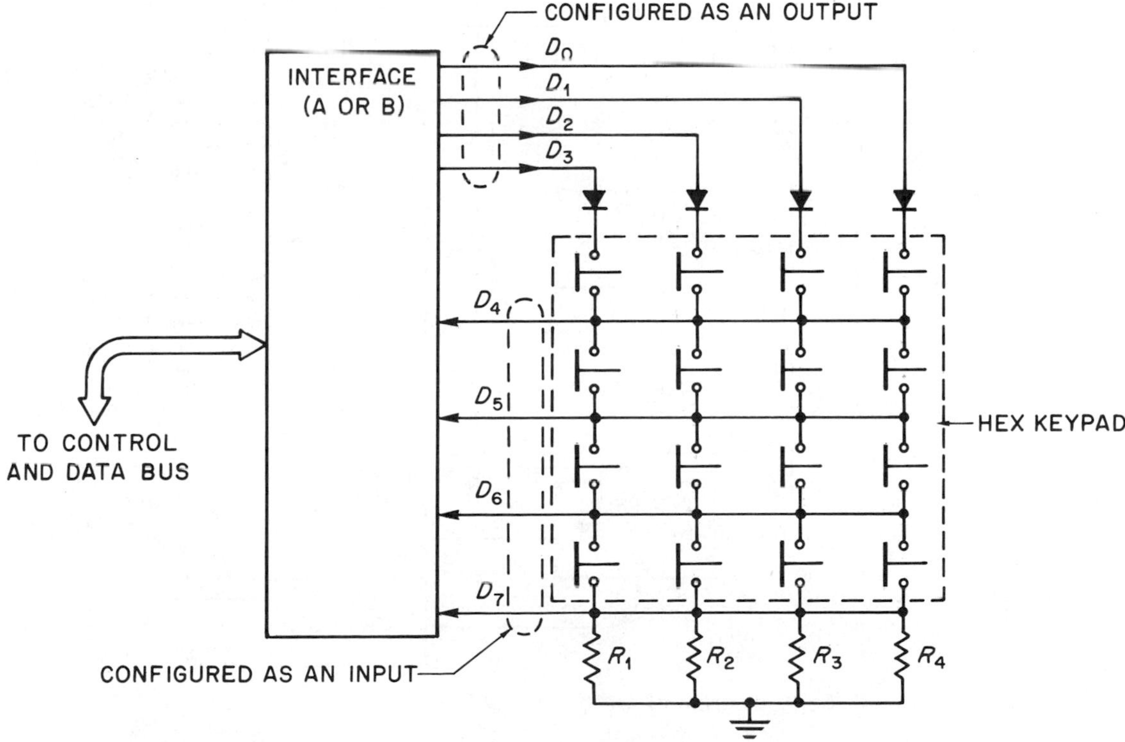

Figure 10–28 Using the PIA to read a hex keypad.

Unlike the DMA, the PIA works in conjunction with the microprocessor. As shown in Figure 10–27, the PIA contains more than one register. Depending upon the type of PIA, each of these 8-bit registers may be software configured to operate as an input or as an output register. Figure 10–28 shows a PIA used to read the condition of a hex keypad.

Under this configuration, 4 bits of an 8-bit port register are used as an output while the other 4 bits are used as an input. Bit patterns from the microprocessor activate one key column at a time while each row is read into the 4-input bits. Thus, by checking the bit pattern for each column scanned, the microprocessor can determine if a key has been pressed and, if so, which one. The logic diagram of a 6520 PIA is shown in Figure 10–29.

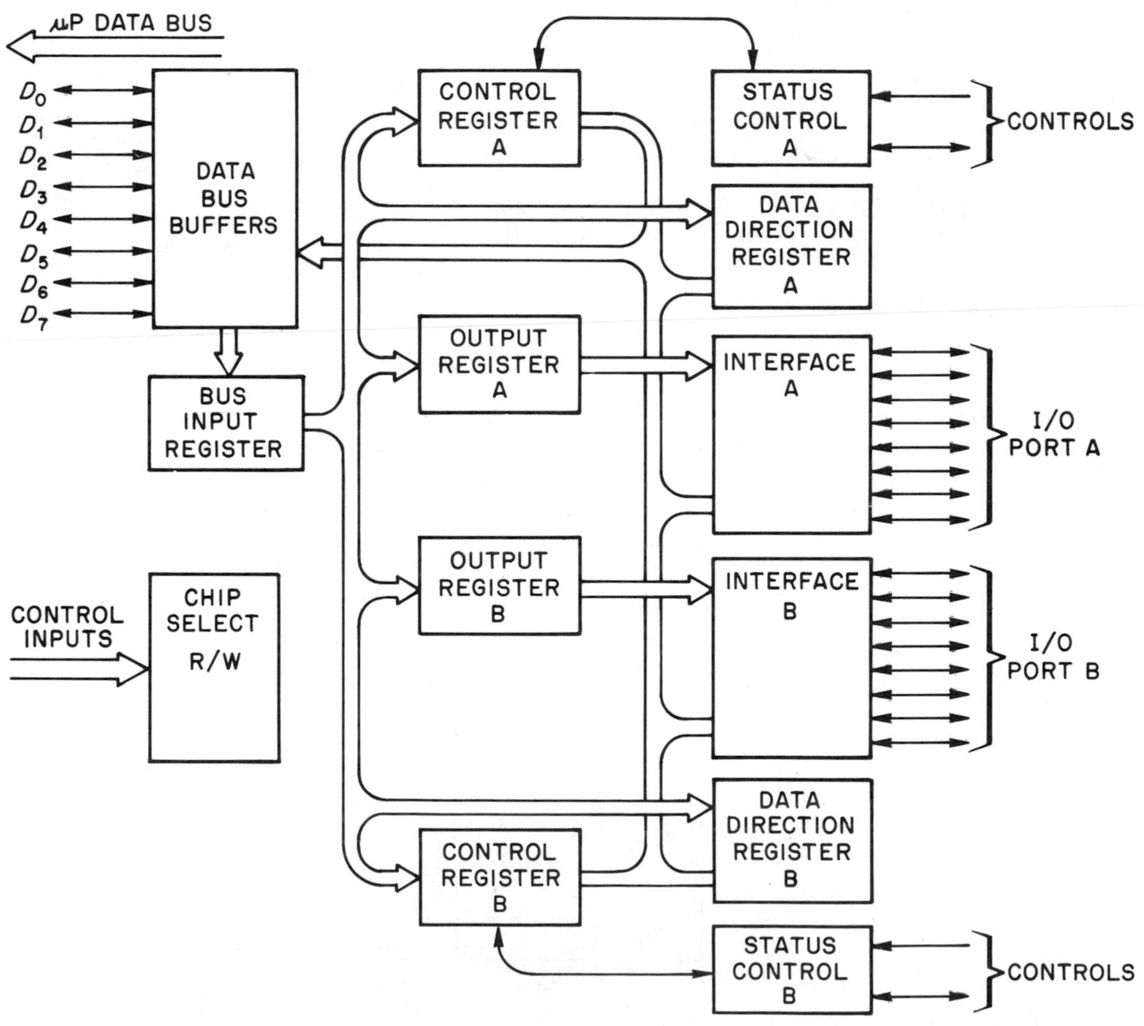

Figure 10–29 Logic diagram of the 6520 PIA.

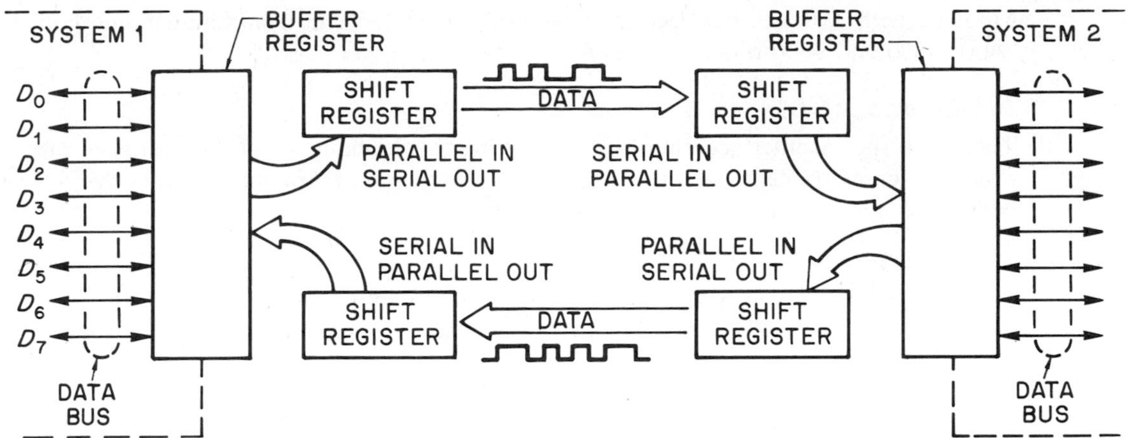

Figure 10–30 Concept of serial data transmission.

Serial Interfacing

Recall that the other method of transferring data between microprocessor-based digital systems was by serial transfer. A microprocessor-based digital system is basically a *parallel* machine. If data is to be sent between systems in a serial manner, a hardware device is needed in order to make the transition from parallel to serial and from serial to parallel. This concept is illustrated in Figure 10–30. You were introduced to this concept in Chapter 7 when the discussion of shift registers was presented.

There are two methods for transferring serial data between digital systems: asynchronous and synchronous. Essentially, serial data is transferred between systems in 8-bit words. In an asynchronous system, each word is treated as an individual transmission. This means that each word may be transmitted at nonfixed times. However, the word itself is transmitted at a known clock rate. Thus in asynchronous transmission the word itself is sent synchronously. Before proceeding further in this discussion, it is important that you understand a term used in serial transmission speeds.

Baud Rate

The **baud rate** is a method of measuring the transmission speed of digital data.

Baud Rate:
The rate, expressed in number of times per second, that a signal in a communication channel varies or makes a transition between states.

Baud rate is not necessarily the same as **bits per second**.

Bits Per Second:
The rate that data bits are transmitted each second in a communication channel.

Modems with 110 bits per second are considered slow. The near future points toward 2400, 4800 and 9600 bits per second.

UART and ACIAs

There is a special interface chip for transmitting asychronous serial data between microprocessor-based digital systems. It is called a *universal asynchronous receiver-transmitter* or **UART**

> **UART:**
> Acronym for universal asynchronous receiver-transmitter. Works with the microprocessor to send asynchronous data serially between microprocessor-based digital systems.

Some manufacturers call this device an *asynchronous communications interface adapter* or **ACIA**.

> **ACIA:**
> Acronym for asynchronous communications interface adapter. A UART device.

A block diagram of the UART (or ACIA) is shown in Figure 10–31.

The major format for asynchronous serial transmission is illustrated in Figure 10–32. There you can see that the actual 8 data bits to be transmitted are enclosed by a *start bit* and *stop bit(s)*. The diagram shows the ASCII code for the letter E being transmitted (45_{16}). In this method, the LSB is transmitted first, thus converting the hex value to binary: 45_{16} = 0100 0101. The number is then arranged so that the LSB is first, yielding 1010 0010.

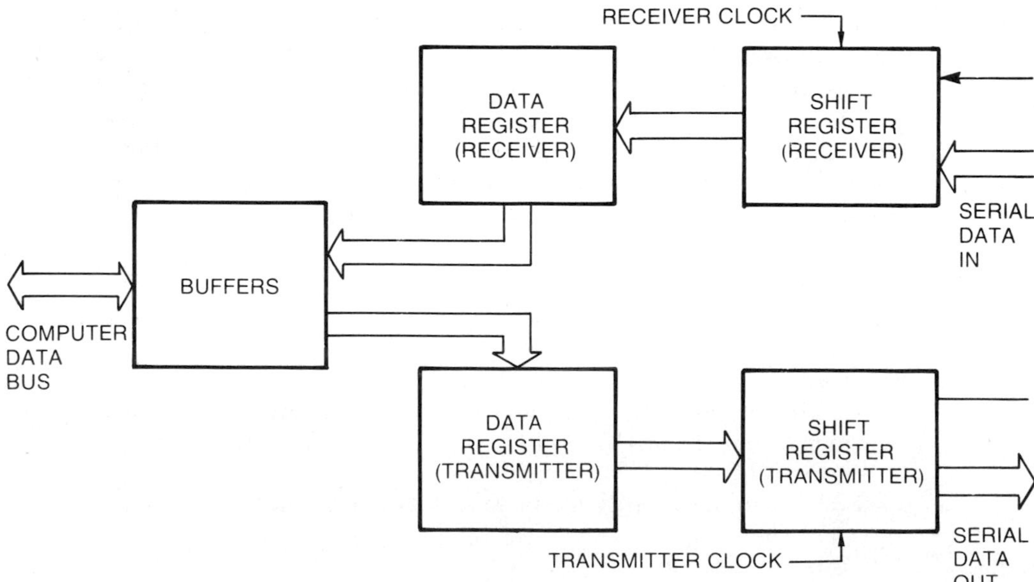

Figure 10–31 Block diagram of the UART.

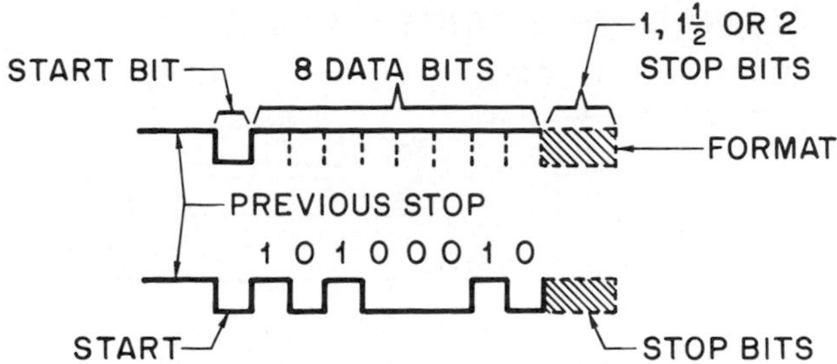

Figure 10–32 Typical asynchronous serial transmission format.

When not transmitting, the line is *idle* and stays at the HIGH state. When ready to transmit, the line then goes LOW. After the data are received, stop bits signal the end of the transmission of a single word. However, these start and stop bits also serve to synchronize the clock at the receiver to that of the transmitter.

The method employed in synchronizing these two clocks is to have the receiver clock operate at a rate that is higher than the baud rate of the received data. Usually the clock rate is 16 times greater than the baud rate. The clock usually works with a binary counter. When the start bit is first received, the binary counter is cleared and the clock counts for 8 counts. It is then at the middle of the start bit, clears the counter again, and then counts for 16 bits, and makes a reading. This process repeats itself, reading the middle of a received bit every 16 clock counts. This process is illustrated in Figure 10–33.

The reason for using a clock rate that is higher than the received baud rate is that this allows for a margin of error in the frequency of the receiver clock. If the receiver clock were exactly at the frequency of the received data, then any slight drift in the clock (since no clock is perfect) could cause the receiver to miss a bit.

Telephone Modems

A *modem* is a hardware device that can be used to transmit and receive data. It was presented in the last section. Basically a telephone modem is an interface between the computer serial I/O port and the telephone. The purpose of this is to transmit digital information using standard telephone lines. Older systems uses a device called an **acoustic coupler**.

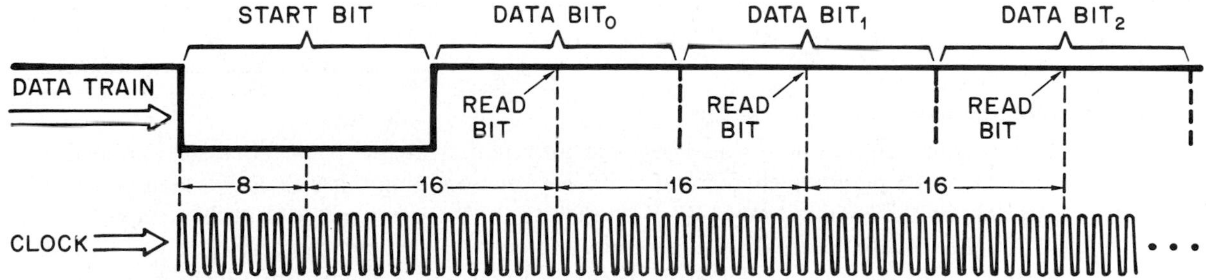

Figure 10–33 Clock format for resynchronization.

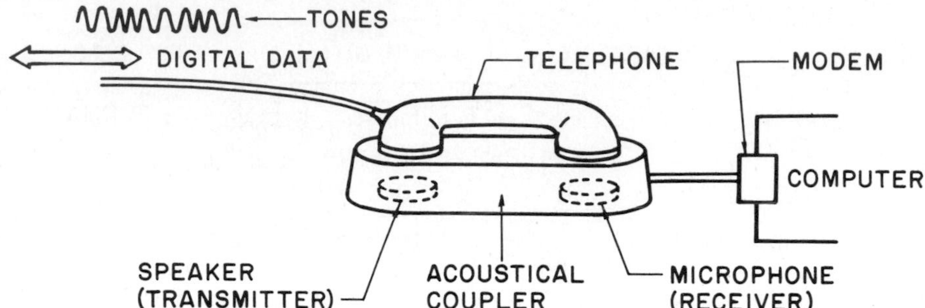

Figure 10–34 Pictorial of acoustic telephone interface.

Acoustic Coupler:
A device that converts electrical information into sound and sound information into electrical signals.

A pictorial of an acoustic telephone interface is shown in Figure 10–34. The advantage of a telephone interface is that digital information can be transmitted between computer terminals using a minimum investment in new hardware. This is because a standard telephone can be used at both the transmitter and the receiver with the telephone company's wiring between them. Most modern systems use an interface so that the computer can connect directly to the telephone jack. This eliminates the need for the acoustic coupler.

Conclusion

This section presented the most popular hardware used in interfacing digital systems. Here you saw how data was transferred directly, in parallel as well as serially. Test your understanding of this section by trying the following review questions.

10–5 Review Questions

1. What is direct memory access?
2. What is a PIA?
3. Give the main requirements for serially interfacing microprocessor-based digital systems.
4. What is baud rate?
5. What is a UART? an ACIA?

10–6 8-BIT MICROPROCESSORS

Discussion

This section requires that you know the information in the microprocessor application sections of the previous chapters. If you have been omitting those sections, then you may skip this section without any loss of continuity for the other material in this chapter.

In this section you will visit a variety of very real microprocessors. Here you will be introduced to several popular 8-bit machines. You will see what they have in common and their major differences.

The 6502 μP

The 6502 is the microprocessor used in the microprocessor application section of this chapter to introduce you to real machine and assembly language programming. In this section you will see its **register architecture.**

Register Architecure:
As pertaining to a microprocessor, a diagram of the internal registers of interest to the programmer.

Figure 10–35 shows the register architecture of a 6502 microprocessor. Note that the 6502 μP has an *accumulator*. This does effectively the same thing as the accumulator used in the ALU presented in this section and the accumulators of the μP presented in the microprocessor application sections. The results of arithmetic and logical operations are kept here. Also operations, such as shifting data to the left or right, may be used here.

This microprocessor contains a 16-bit *program counter*. Again, the program counter does the same thing as the program counters (address registers) of the microprocessors presented in previous sections. Since the program counter is 16 bits, it is *capable of addressing* $2^{16} = 65{,}536$ memory locations where each memory location contains 8 bits of data (the size of the accumulator).

There are two **index registers:** index X and index Y. These are general purpose registers similar to the B register used in the 4-bit μP presented in earlier chapters. However, here they have some added features.

Index Register:
An internal microprocessor register that may be used for process counting or modifying the address in the program counter.

As a programmer, you can place any 8-bit binary value into the index register and then decrement or increment it. Thus you can use it to keep track of how many times a process

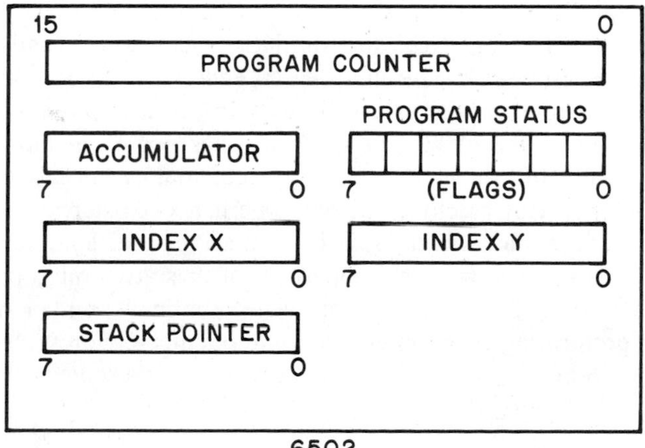

Figure 10–35 Architecture of 6502.

is repeated. The other use of this register is for using different *address modes*. You will see how to do this in the microprocessor application section.

The **stack pointer** is another useful internal register.

Stack Pointer:

An internal microprocessor register that contains the address for specially stored data that is not a part of the data for the main program.

You will learn some of the important details of this important register in the microprocessor application section. Basically what the register does is store an address. This address is usually a location in memory that is separate from the memory locations used to store the main program. This special section is referred to as **the stack.** There is nothing physically different about this memory location, it is distinguished only by software.

The Stack:

A location in memory usually different from the location of the main program. The address of this location is kept in an internal microprocessor register called the stack pointer.

The **status flags** (sometimes called the *condition code*) register contains useful information concerning the internal processes of the μP.

Status Flags:

A microprocessor register where each bit represents the results of some previous internal register process or the condition of the bit will have an affect on such a process.

As an example of the use of the status flags register, one of the bits indicates if an arithmetic overflow has occurred. Another bit tells if there was a carryout of the MSB. Still another bit, if set HIGH, will cause the μP to perform BCD operations. You will see more details concerning this useful register in the microprocessor application section. The logic diagram (using the ANSI standard) of the 6502 μP is shown in Figure 10–36.

The 6800 μP

Figure 10–37 shows the register architecture of the 6800 μP. The 6800 is another popular 8-bit microprocessor. Note the similarities between it and the 6502. The 6800 has a 16-bit program counter, meaning it can address up to 65,536 different memory locations that each contain an 8-bit word. It has a 16-bit *stack pointer*. This is larger than the stack pointer used by the 6502. The 6800 also has *two accumulators:* A and B. Each of these do essentially the same thing as any accumulator, but having two adds to the power of this machine. This machine has only one index register, called index register X. Like the 6502, the 6800 has a *status* register that serves the same purpose as it did for the 6502.

As you can see from a comparison of these two microprocessors, the design philosophy is essentially the same—that is, a minimum number of internal registers with the idea of performing much interaction with memory. You will see a big difference between these two 8-bit machines and the next two to be presented. The logic diagram of the 6800 is shown in Figure 10–38.

The 8085 μP

The register architecture of the 8085 μP is shown in Figure 10–39. As you can see, there are many more internal registers used in this microprocessor than there are in the 6502

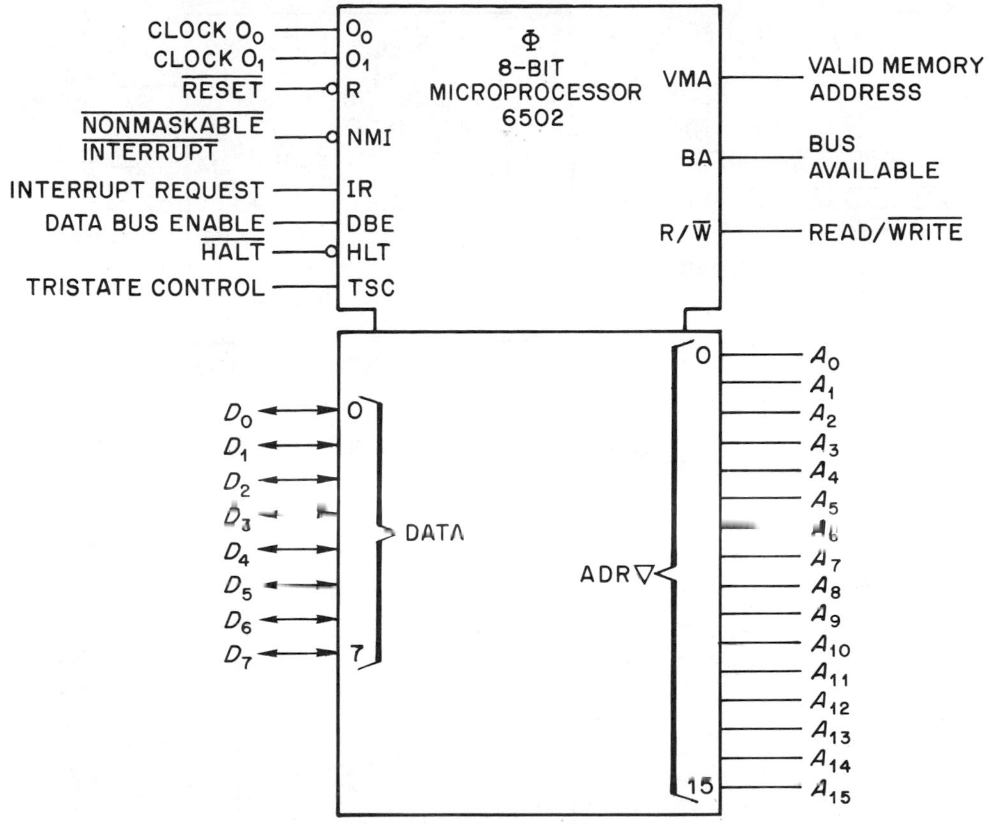

Figure 10–36 Logic diagram of 6502 μP.

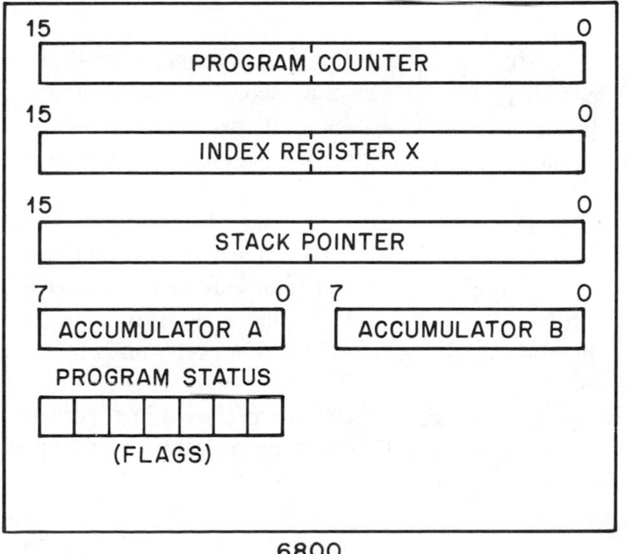

Figure 10–37 Architecture of the 6800 μP.

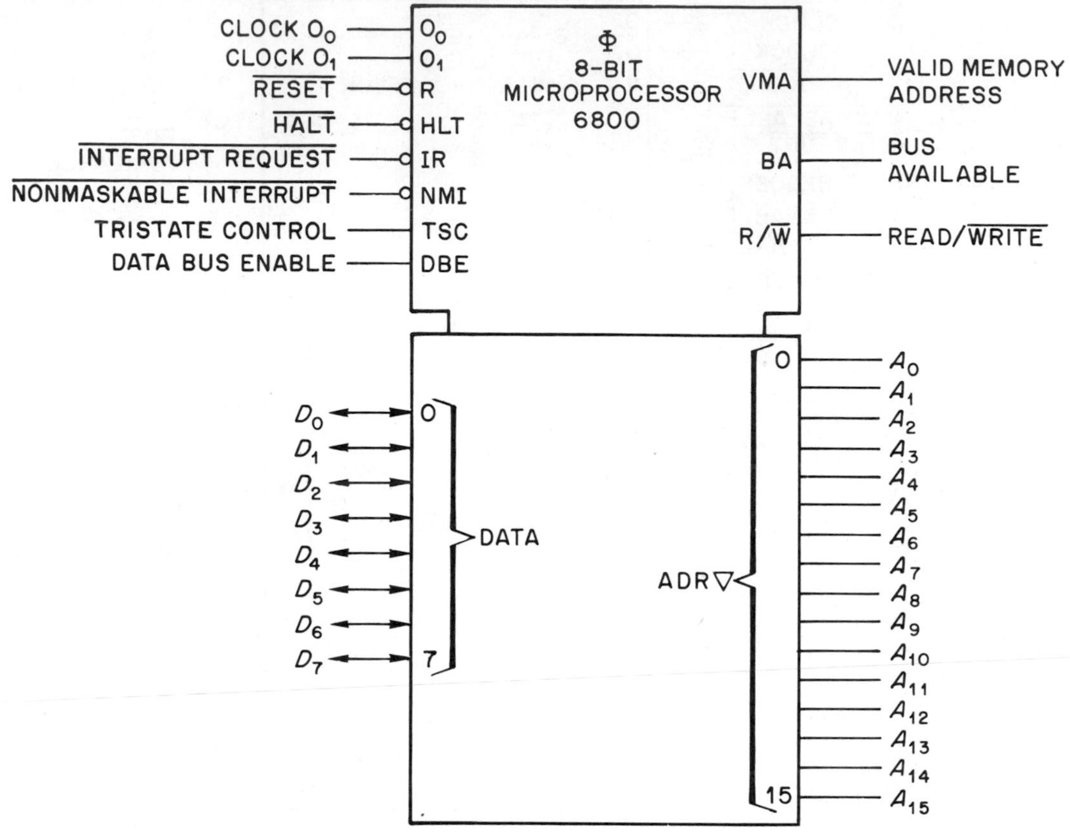

Figure 10–38 Logic diagram of the 6800 μP.

and 6800. The 8085 has a 16-bit program counter that allows it to address 65,636 different memory locations each with an 8-bit word. It also contains an accumulator that does essentially the same thing as all accumulators. It has a status flag register that is similar in function to that of other microprocessors. It also contains a 16-bit stack point similar in function to the 16-bit stack pointer of the 6800.

The difference comes in the other internal registers. These are broken into 8-bit pairs: B and C registers, D and E registers, and H and L registers. These registers can be used as separate 8-bit registers, or each pair can be treated as a single 16-bit register. These registers can be used as 8- or 16-bit index registers. The H and L registers may be used to affect the addressing modes of the 8085 (H stands for *high-order byte* and L for *low-order byte*).

The logic diagram of the 8085 μP is shown in Figure 10–40. Note that some address and data lines are shared. This is done to hold down chip pin count. More will be said about this later.

The Z80 μP

The register architecture of the Z80 μP is shown in Figure 10–41. Look closely at the Z80. It essentially contains the registers of two 8085 microprocessors. It has two accu-

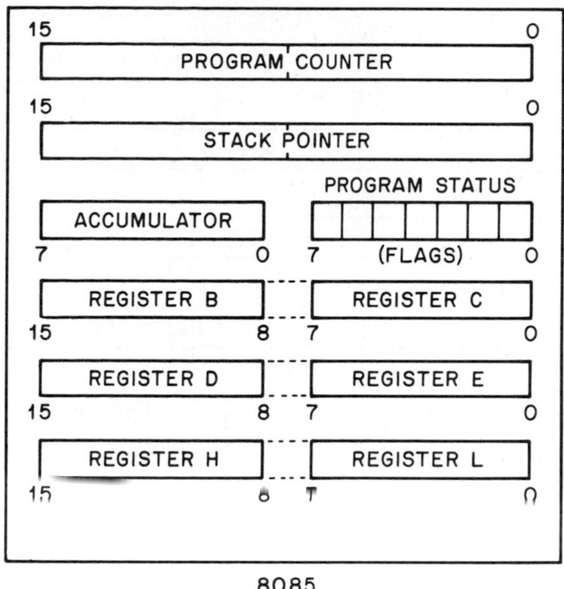

Figure 10–39 Architecture of the 8085 μP.

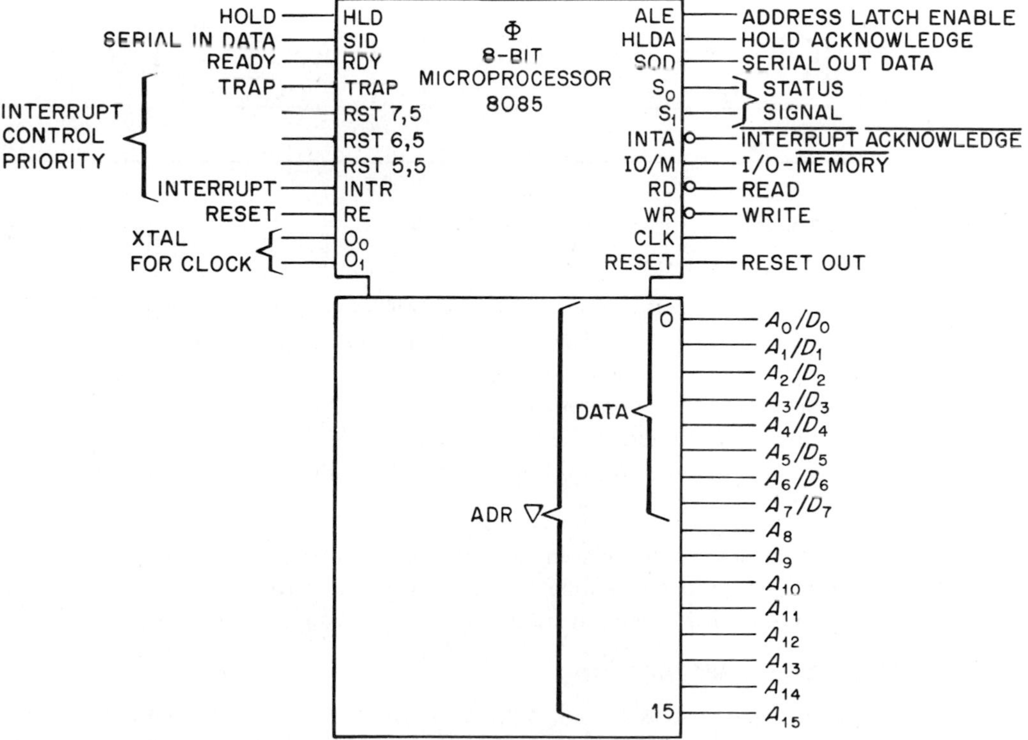

Figure 10–40 Logic diagram of the 8085 μP.

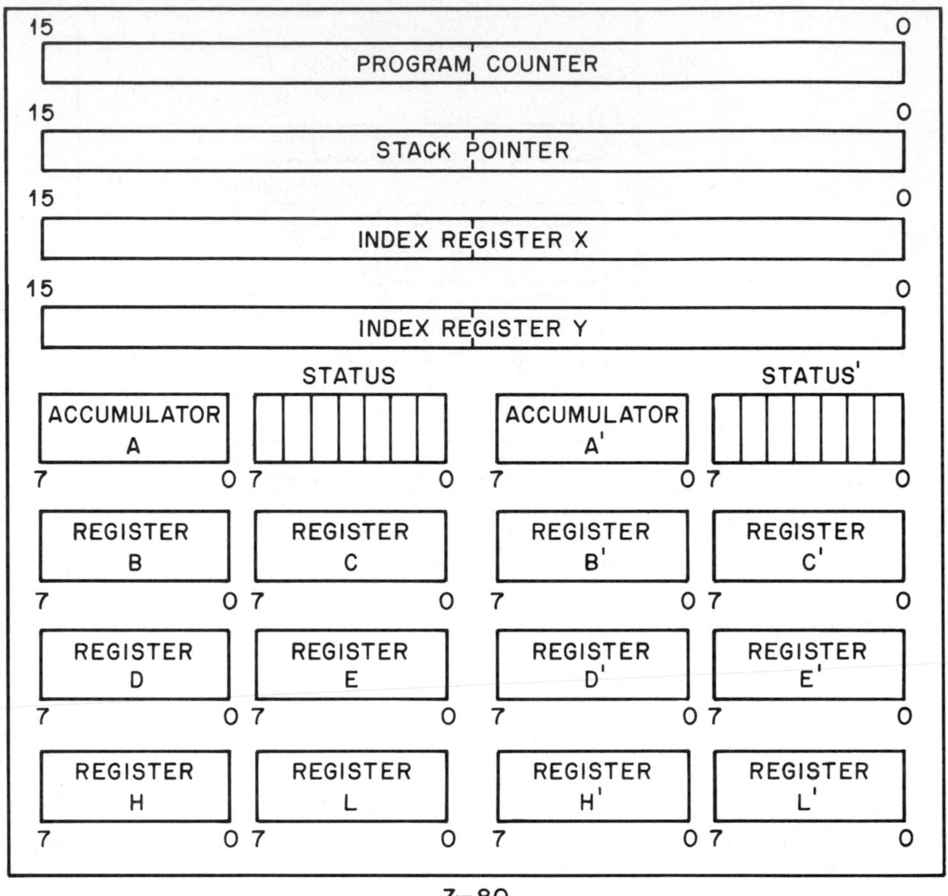

Figure 10–41 Architecture of the Z80 μP.

mulators (called A and A′), two status flag registers, and duplicate sets of the same internal register pairs (BC, DE, and HL pairs). It also has a 16 bit program counter, 16-bit stack pointer, and two 16-bit index registers called index register X and index register Y. It too is capable of addressing 65,536 different memory locations, each containing an 8-bit word. This duplication of internal registers makes the Z80 a very powerful 8-bit microprocessor. The logic diagram of the Z80 microprocessor is illustrated in Figure 10–42.

Conclusion

This section described the four most popular 8-bit microprocessors. In the next section, you will be introduced to some of the most popular 16-bit machines. For now, test your understanding of this section by trying the following review questions.

10–6 Review Questions

1. Name two registers that all 8-bit microprocessors have in common.
2. How many memory locations can all of these 8-bit microprocessors address? What is the word size?

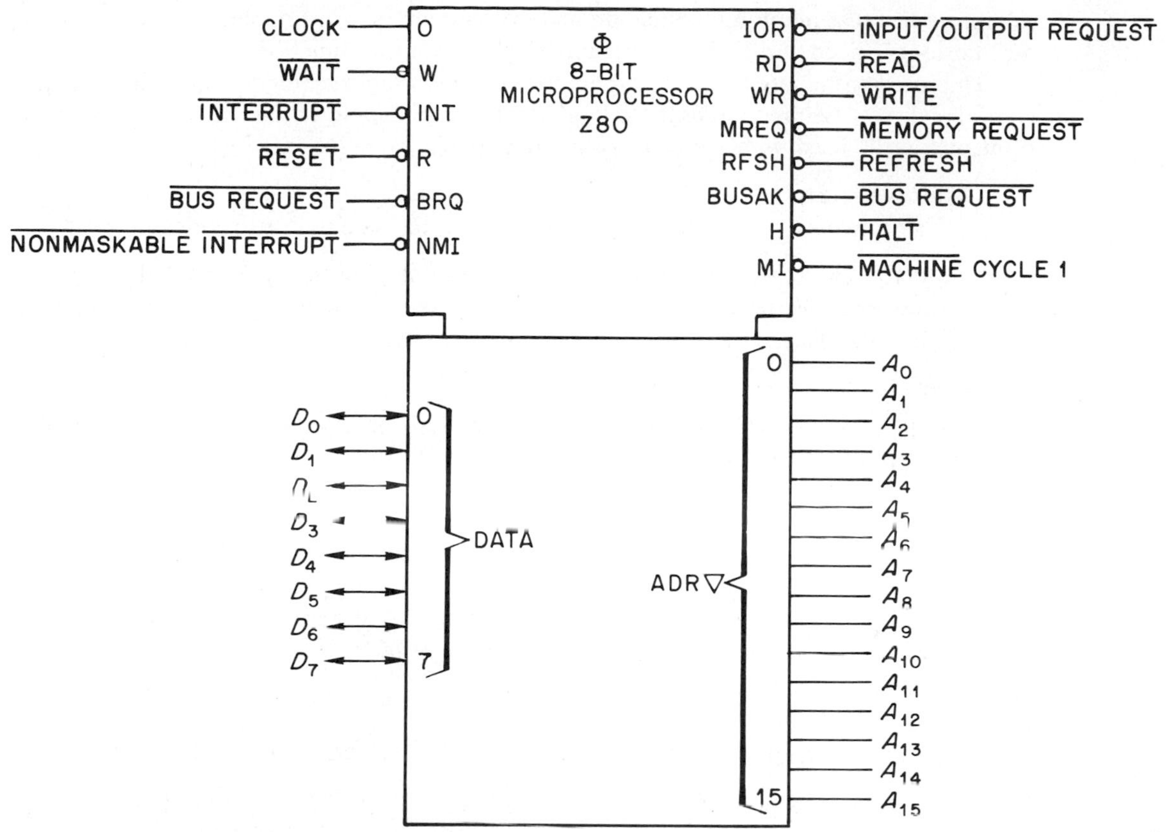

Figure 10–42 Logic diagram of the Z80 μP.

3. State the purpose of the *stack pointer*.
4. Which of the four microprocessors have similar architecture?
5. Which of these 8-bit microprocessors has the least number of internal registers? Which one has the most?

10–7 16-BIT AND 32-BIT MICROPROCESSORS

Discussion

This section presents some popular 16-bit microprocessors. When you compare these machines to the 8-bit microprocessors, you will find that they have much in common.

Overview of 16-Bit Machines

The main difference between the 8-bit microprocessors and the 16-bit machines is that the 16-bit machines are capable of taking data from a 16-bit data bus and can address more memory. Their internal registers are larger and their instruction set contains more instructions. This allows for more microprograms to be contained in their internal ROM. Thus, microroutines that allow multiplication and division are now available, and the programmer no longer needs to develop programs to perform these functions.

The increase in data bus size and ability to address more memory require a larger pin count for these machines. In order to hold down this pin count on the IC, a process called **multiplexing** is used with the address and data bus in some of these μP.

Multiplexing:
Using two or more different signals through the same electrical element.

Figure 10–43 shows the general concept of multiplexing the data and address buses. At one time some of the lines from the μP act as address lines, and at other times they act as data lines. This same type of process is used with the 8-bit 8085 μP.

The 8088 μP

The 8088 μP utilizes an 8-bit data bus, but internally behaves like a 16-bit machine. The architecture of the 8088 is shown in Figure 10–44. Note that the first eight data and address lines are multiplexed in order to hold the pin count to 40.

You may be asking, why produce a 16-bit machine with only an 8-bit data bus? The manufacturer (Intel) did this to make it compatible with the older 8-bit processors, the 8080 and the 8085. This allowed manufacturers of digital equipment who used the older μP to switch to the newer machine with a minimum of system hardware changes. The newer 8085 would operate programs written for the 8080.

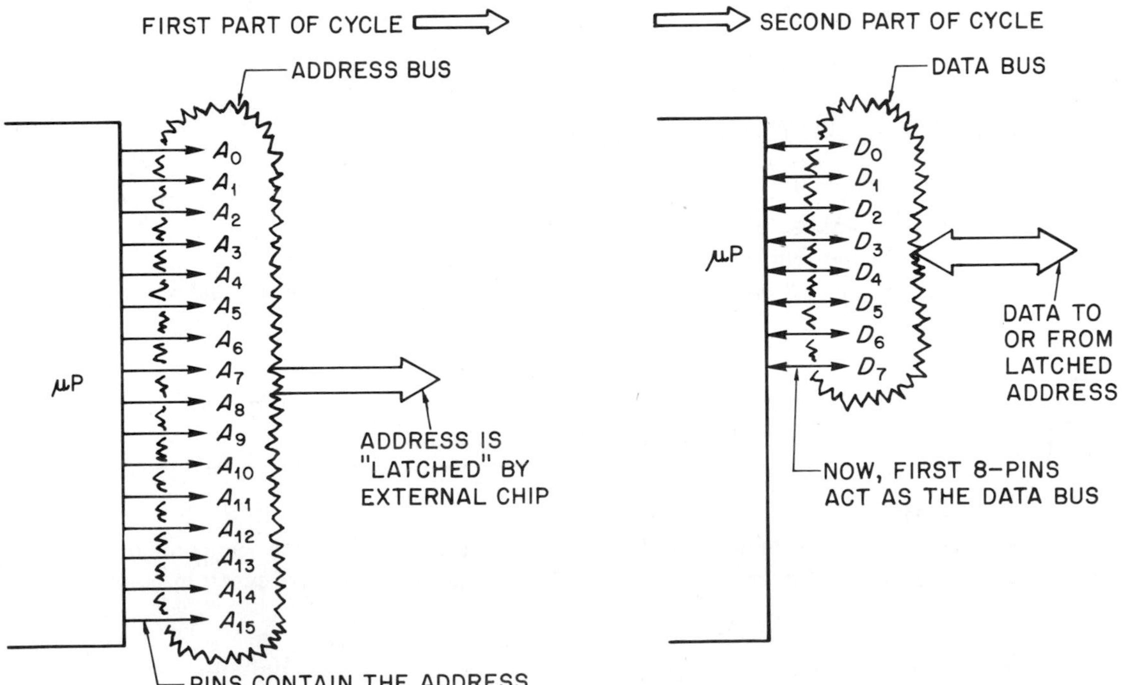

Figure 10–43 Multiplexing the data and address buses.

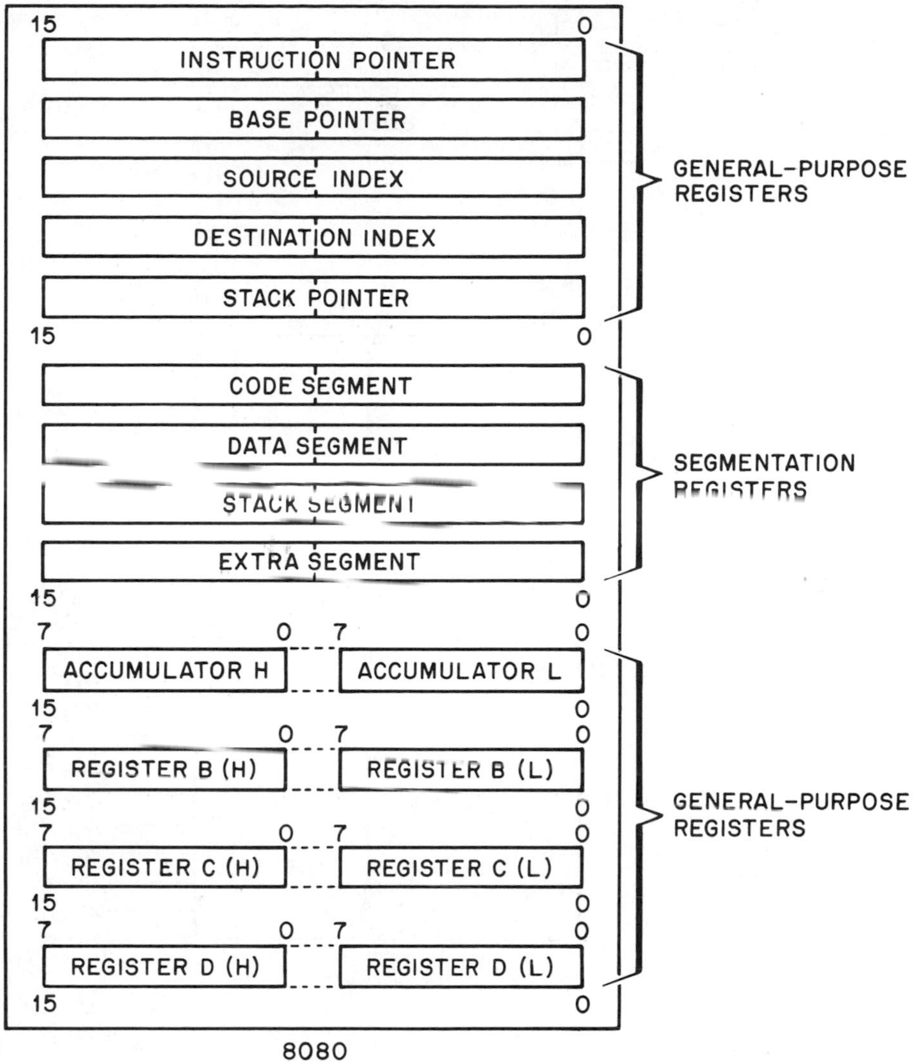

Figure 10–44 Architecture of the 8088 μP.

The logic diagram of the 8088 is illustrated in Figure 10–45. Note that the *general-purpose registers* are similar to those of the 8-bit 8085 μP. There are the accumulator and the register pairs. This machine has two 8-bit accumulators that may be software combined into a single 16-bit accumulator. There are, however, some major differences. You still see the stack pointer but there are other registers in the same group with the stack pointer. The *base pointer* register can be used to keep the address of large groups of data. The *source index* and the *destination index* registers are also used for addressing data in order to make processes more efficient. The *instruction pointer* is actually the program counter (address register), and it contains the address of the next instruction.

The **segmentation registers** are all new; they are not found in the older 8-bit machines.

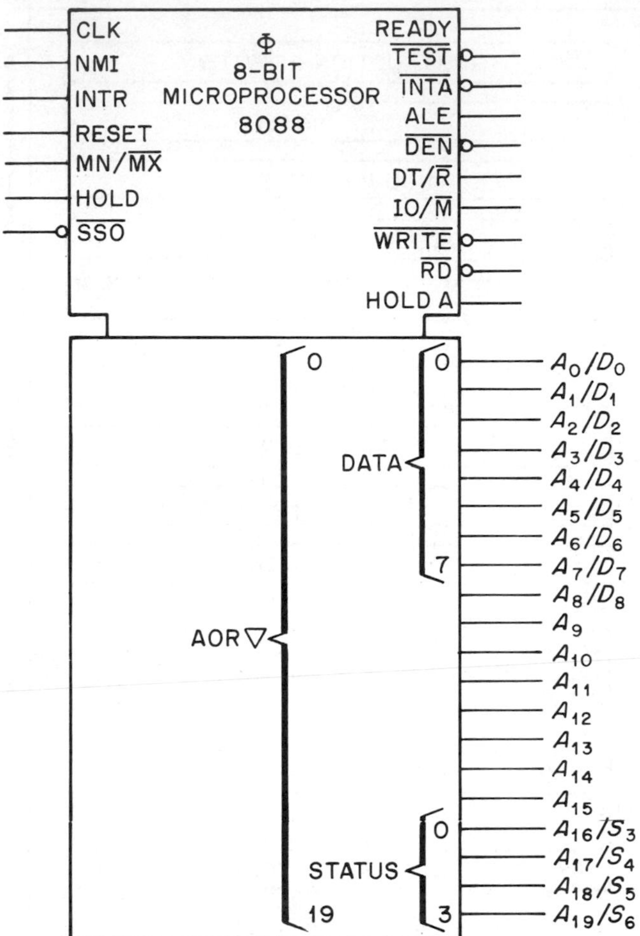

Figure 10–45 Logic diagram of the 8088 μP.

Segmentation Registers:
Internal storage for the address of a section of memory called a segment.

The reason for the segmentation registers is to overcome the problem of having address registers that are only 16 bits wide feeding a 20-line address bus to handle 1,048,576 (1M) separate 8-bit-wide memory locations. In order to address this much memory, it would normally require 20 address bits ($2^{20} = 1{,}048{,}576$). The way this is done with the 8088 is through a process called *segmentation*. Basically what happens is that the 16-bit address stored in the segmentation register is shifted to the left 4 bits (binary value of 16). This identifies the beginning of a *segment* of memory. Once this segment has been identified, it is then treated as a 64K block of continuous memory. The reason for having four segmentation registers is to divide the memory efficently. Thus, data are kept separate from instructions, which are kept separate from required "housekeeping" data. This same concept is used in the next *real* 16-bit machine.

The 8086 μP

The 8086 μP is a real 16-bit machine. Its logic diagram is shown in Figure 10–46. The main difference between the 8086 and the 8088 μP is the size of the data bus. With the 8086, data can be taken in with 16-byte chunks. This allows the 8086 to operate at a higher speed than the 8088 because it does not have to address two separate 8-bit memory locations in order to fill its 16-bit registers. From a programming standpoint, the internal register architecture of this machine is nearly identical to that of the 8088.

The 68000 μP

The 68000 μP is another real 16-bit machine. Its logic diagram is shown in Figure 10–47. As you can see from the diagram, this machine has a 16-bit data bus and a 24-bit address bus. This gives it the capability of addressing up to 16 Mbytes without having to use segmentation. It can handle data in 8-, 16-, or 32-bit values because its internal architecture is 32 bits instead of 16. The 68010 is an improved version of the 68000.

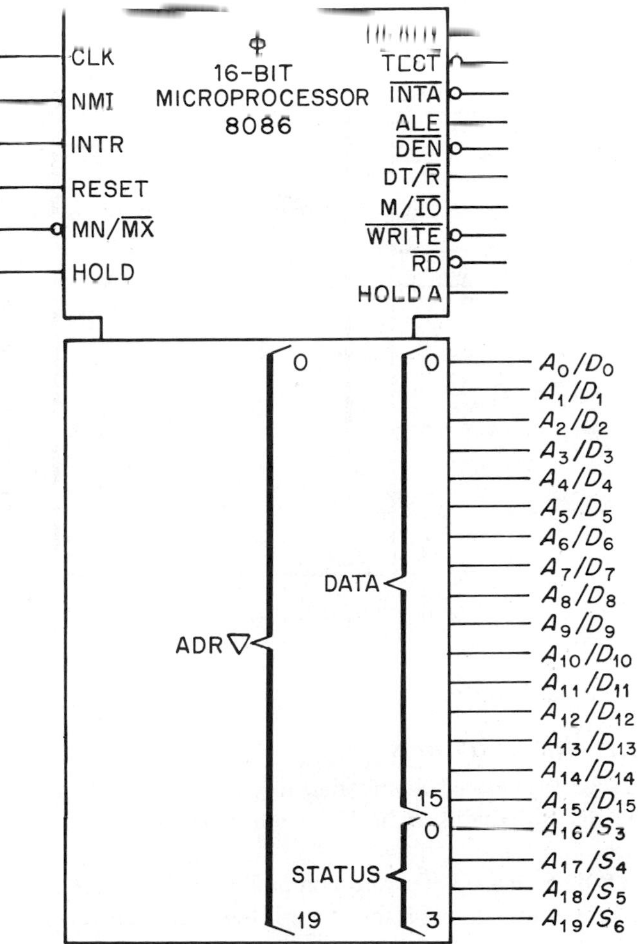

Figure 10–46 Logic diagram of the 8086 μP.

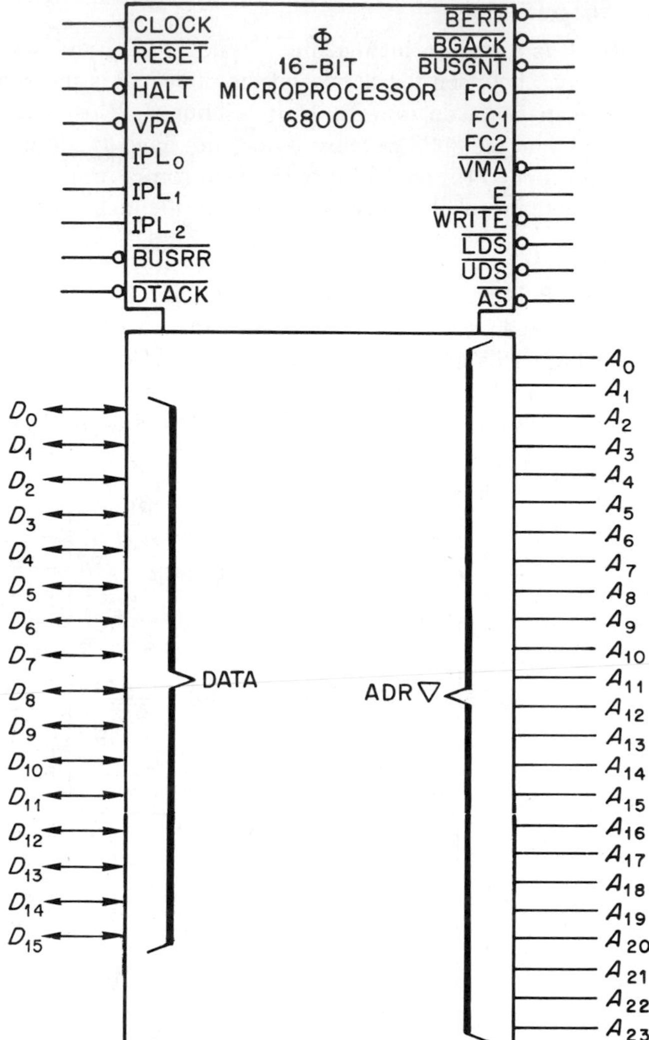

Figure 10–47 Logic diagram of 68000 μP.

Other 16-Bit Machines

The NS16032 (National Semiconductor) also has a 32-bit internal architecture with a 16-bit data bus. It contains eight 32-bit internal registers and is capable of addressing up to 16 Mbytes.

The 80286 (Intel) is upward-compatible with the 8086. Its internal architecture is similar in that it contains 16-bit registers. It has eight general-purpose registers and can handle memory up to 1000 Mbytes (1 Gbyte).

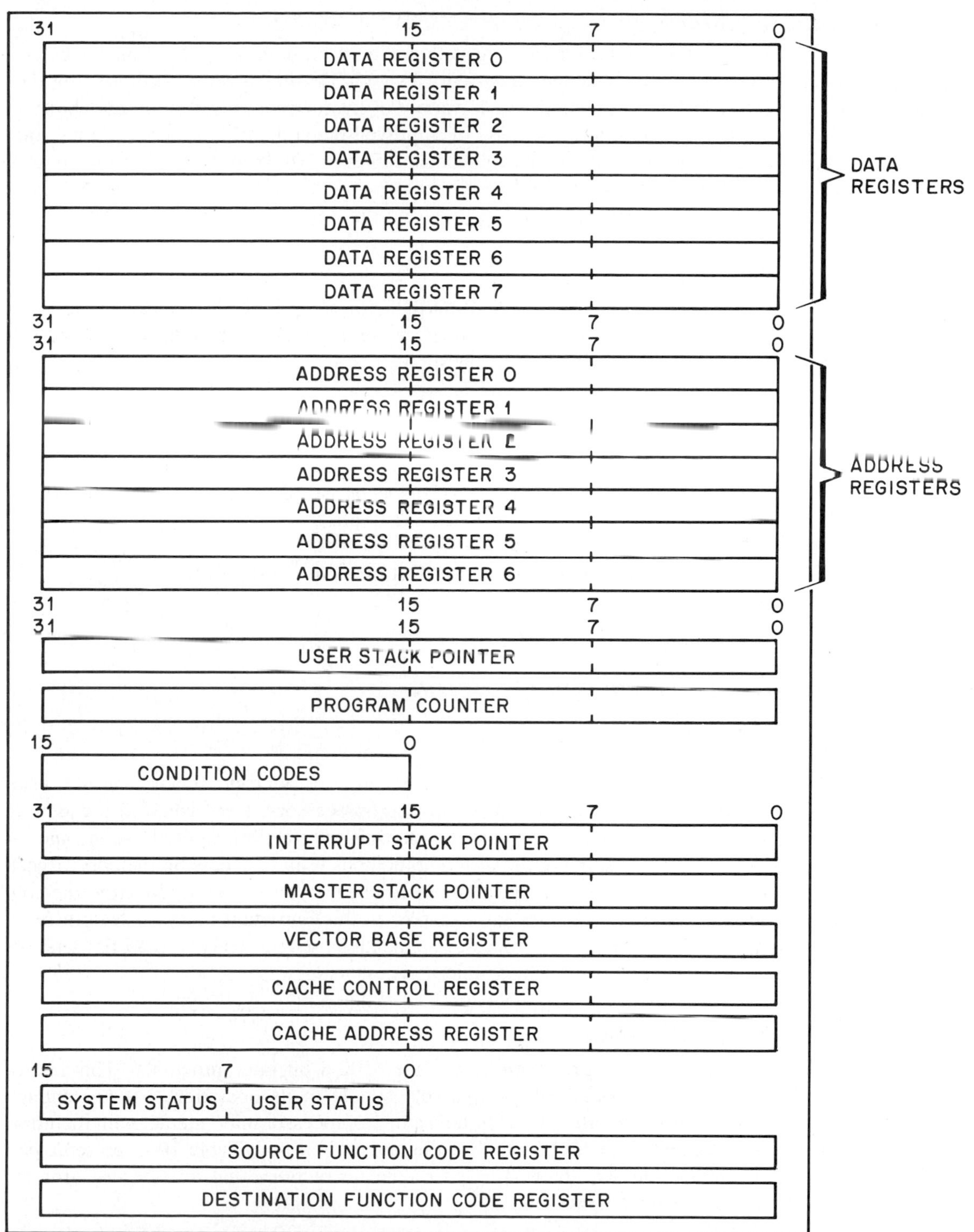

Figure 10–48 Architecture of the MC68020 32-bit μP.

32-Bit Machines

A 32-bit μP employs 32-bit internal registers and a 32-bit data bus. These machines also support larger addressing (theoretically 4 Gbytes with 32 address bits). With a 32-bit data bus, they are capable of handling a wide variety of instructions, resulting in the ability to handle complex microroutines. Because of improved manufacturing processes, these machines can also handle higher clock frequencies. Figure 10–48 shows the architecture of what one of these machines looks like. Other popular 32-bit machines are the NS32032 (National Semiconductor), the 80386 (Intel), and the Z80000 (Zilog).

Conclusion

This section introduced you to 16- and 32-bit μP. You saw the similarities and differences between these machines and their 8-bit predecessors. Check your understanding of this section by trying the following review questions.

10–7 Review Questions

1. State the main difference between a 16-bit microprocessor and an 8-bit microprocessor.
2. What term is used to describe the use of the same μP pins for addressing and data?
3. Is the 8088 μP a 16-bit or an 8-bit machine? Explain.
4. State the purpose of segmentation registers.
5. What is the main difference between a 32-bit μP and a 16-bit μP?

MICROPROCESSOR APPLICATION

Discussion

This section presents a real microprocessor. By using this microprocessor, you will learn practical fundamentals that are similar to all microprocessors. Here you will see how to perform basic arithmetic operations such as addition, subtraction, multiplication, and division. You will also see some of the different methods used in accessing memory. Recall from the microprocessor application section of Chapter 7 you were introduced to the concept of addressing modes. This section expands on this important practical feature of all microprocessors. This final microprocessor applications section ends with a brief analysis of future trends in the development of microprocessors.

A Real 8-Bit Microprocessor

The microprocessor that is presented here is the 6502 8-bit microprocessor. This microprocessor was selected because it is relatively simple in its operation, yet has enough complexity to develop important ideas of how a microprocessor may interact with memory. Figure 10–35 presented the architecture of this microprocessor. Table 10–5 presents part of the 6502 instruction set. It contains all of the instructions that will be used for the programs presented in this section.

Table 10–5 6502 INSTRUCTION SET [PARTIAL]

Mnemonic	Meaning	Hex Code
ADC	Add with carry	[65]
ASL	Arithmetic shift left	[0A]
BCC	Branch if carry clear	[90]
BNE	Branch if not equal to zero	[D0]
BRK	Break	[00]
CLC	Clear carry flag	[18]
DEX	Decrement index register X by 1	[CA]
INC	Increment memory by 1	[E6]
LDA	Load accumulator	[A5] Direct [A9] Immediate
ROL	Rotate left through carry	[26]
SBC	Subtract with borrow	[E5]
SEC	Set the carry flag	[38]
STA	Store accumulator	[85] Direct
TAX	Transfer accumulator to index register (X)	[AA]

The Status Register

Some mention of the status register was introduced in this chapter. However, the details of the 6502 status register will be presented here. Figure 10–49 illustrates the status register of the 6502 microprocessor. Each bit on this register is called a flag.

Table 10–6 summarizes the purpose of each of the flags in the status register. As you will see, the flags are affected by various microprocessor instructions.

An Introductory Program

One of the simplest processes a microprocessor can perform is to copy the contents of one of its internal registers into another internal register. There is no interaction with memory. This is presented here to get you introduced to the 6502 instructions. There are two instructions that will be used in this program: TAX and BRK. The TAX instruction does the actual transfer of data between the accumulator and index register X. For the

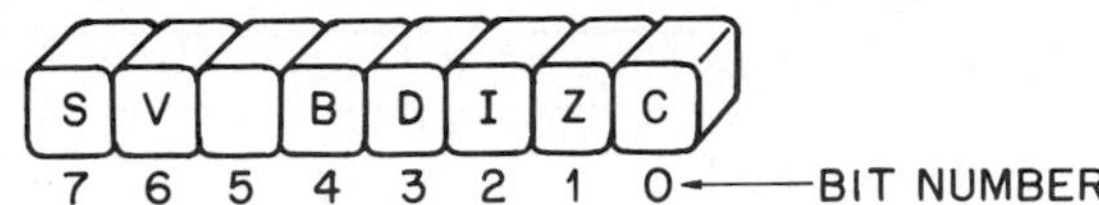

Figure 10–49 6502 status register.

Table 10–6 6502 STATUS FLAGS

Flag	Name	Purpose
C	Carry flag	Holds the carry from the most significant bit from any operation. Also used in a shift or rotate instruction.
Z	Zero status flag	Is set to 1 when any arithmetic or logical operation produces a result of zero. It is set to 0 when any arithmetic or logical operation produces a nonzero result.
I	Interrupt mask flag	When set to a 1, interrupts are disabled. When set to a 0, interrupts are enabled.
D	Decimal mode flag	When set to a 1, causes the add and subtract instructions to act in the BCD mode.
B	Software interrupt flag	Flag is set to 1 when the BRK command is used. A software interrupt.
V	Overflow status flag	Is set to a 1 if an overflow occurs in any arithmetic operation. Is set to a 0 if no overflow occurs in an arithmetic operation.
S	Sign status flag	Flag is set to value of the MSB for any arithmetic or logic operation. A 1 indicates a negative number, and a 0 a positive number if the two's complement notation is used.

purpose of this section, the BRK instruction will be used to end all programs and stop further processing. The first example program is as follows:

Problem: Transfer the contents of the accumulator to index register X.

Assembly Program:

```
TAX    ; Transfer accumulator to index X
BRK    ; Stop execution
```

Machine Code:

Memory Location [Hex]	Memory Contents [Hex]	Instruction /Operand
0000	AA	TAX
0001	00	BRK

Figure 10–50 illustrates the process of what took place in this program.

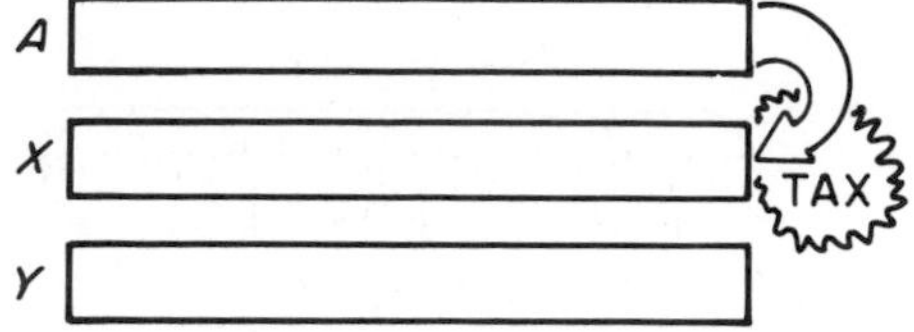

Figure 10–50 Transfer of accumulator to index register.

Interaction With Memory

The next program is the one of the most basic interactions with memory a microprocessor can do. It takes the contents of one memory location and copies it into another memory location. One common method of doing this is to load the accumulator from the memory location whose contents are to be copied. Then store the contents of the accumulator into the memory location you wish to copy to. The following program illustrates the technique. Note the use of the dollar sign before a number. In 6502 assembly language programming it indicates that the number following the dollar sign is a hexadecimal number. Thus $\$23 = 23_{16}$.

Problem: Move the contents of memory location 0030 to memory location 0035.

Assembly Program:

```
LDA   $30   ; Get data from memory
STA   $35   ; Transfer data to new location
BRK         ; Stop execution
```

Machine Code:

Memory Location [Hex]	Memory Contents [Hex]	Instruction /Operand
0000	A5	LDA
0001	30	$30
0002	85	STA
0003	35	$35
0004	00	BRK

Figure 10–51 shows this process.

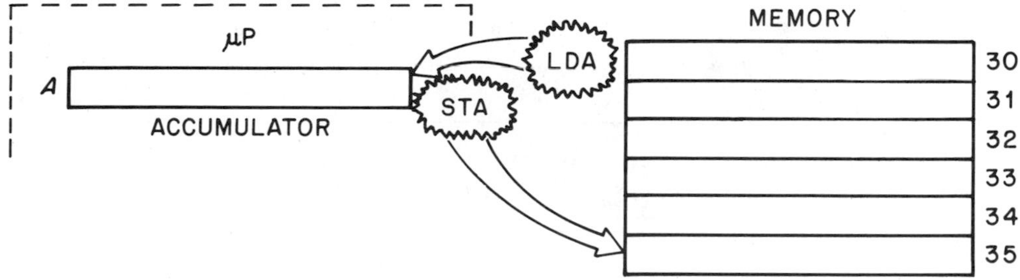

Figure 10–51 Copying data from one memory location to another.

Adding Two Numbers (8-Bit Addition)

The only add instruction in the 6502 is ADC, which will cause the contents of the accumulator to be added to a number in memory along with the status of the carry flag. This process is shown in Figure 10–52.

When performing an addition, make sure that the carry flag is cleared (set to 0); otherwise the answer could be 1 larger than it should be. The process of adding two numbers is illustrated in the following program:

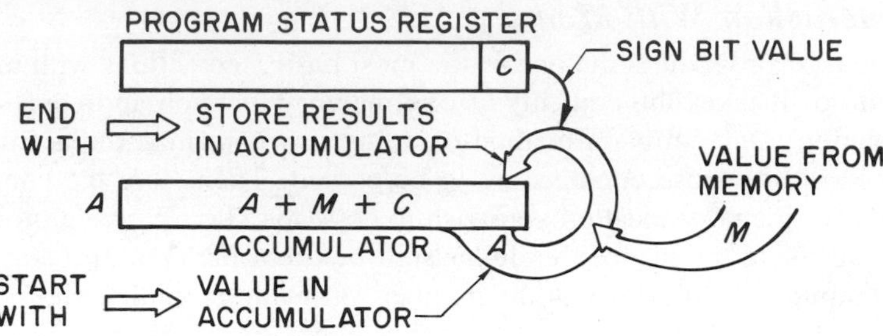

Figure 10–52 Add-with-carry operation.

Problem: Add the contents of memory locations 0030 and 0031. Store the result in memory location 0040.

Assembly Program:

```
CLC              ; Clear the carry flag
LDA     $30      ; Get first number
ADC     $31      ; Add memory contents to first number
STA     $40      ; Store the answer
BRK              ; Stop execution
```

Machine Code:

Memory Location [Hex]	Memory Contents [Hex]	Instruction /Operand
0000	18	CLC
0001	A5	LDA
0002	30	$30
0003	65	ADC
0004	31	$31
0005	85	STA
0006	40	$40
0007	00	BRK

In the assembly language program, the semicolon is used to indicate that a comment is to follow. Comments are ignored by the assembler. The details of this process are illustrated in Figure 10–53.

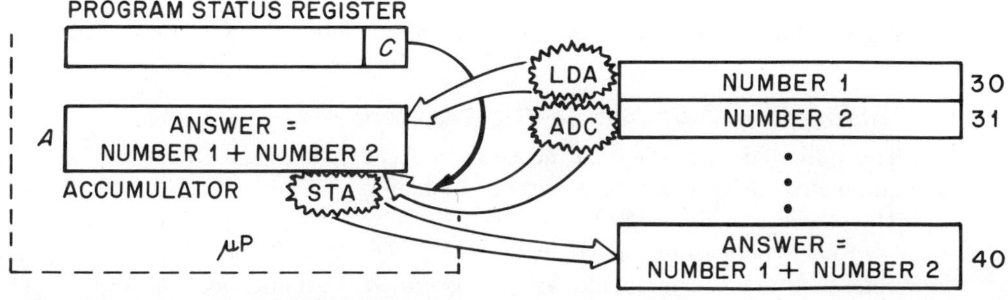

Figure 10–53 Details of 8-bit addition.

In the previous problem, an overflow may occur into the carry bit (thus setting the carry flag to 1). Programs can be created to account for such occurrences. Another use of the carry bit is in subtraction.

Subtraction with Borrow

Figure 10–54 presents the idea of how this instruction works. This instruction for the 6502 differs from most other microprocessors in that the carry flag is set to 1 if *no borrow is required*. If a borrow is required, the carry flag will be set to 0. It is therefore important that this flag be set to 1 (indicating no borrow) before any subtraction is started.

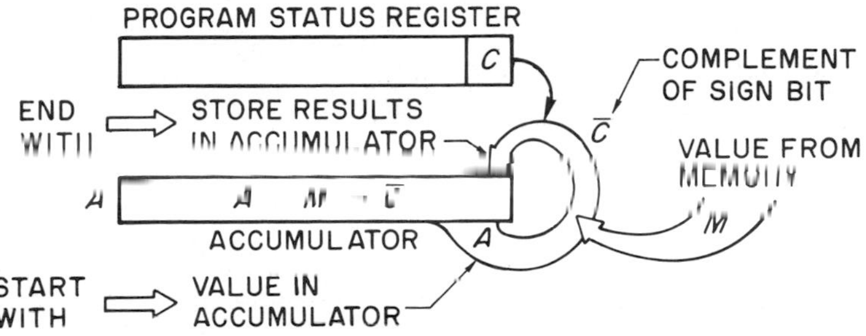

Figure 10–54 Operation of the subtract instruction.

More Addition (16-Bit)

More precision can be achieved in arithmetic operations as demonstrated by the following program that causes the 8-bit microprocessor to add two 16-bit numbers. This concept can be extended so that any multiple of 8 bits can be processed in the same manner. Again the carry flag must be cleared to start with. However, this time, the contents of this flag will be utilized in the program. Note that two consecutive 8-bit memory locations are used for the 16-bit numbers.

Problem: Add the 16-bit number in memory locations 0030 and 0031 to the 16-bit number in memory locations 0032 and 0033. Store the answer in memory locations 0040 and 0041. Make sure the least significant bit of the answer is in memory location 0041.

Assembly Program:

```
CLC          ; Clear the carry flag
LDA   $31    ; Get the least significant bit
ADC   $33    ; Add the least significant bits
STA   $41    ; Store least significant bit
LDA   $30    ; Get the most significant bit
ADC   $32    ; Add the most significant bit + carry
STA   $40    ; Store the most significant bit
BRK          ; Stop execution
```

Machine Code:

Memory Location [Hex]	Memory Contents [Hex]	Instruction /Operand
0000	18	CLC
0001	A5	LDA
0002	31	$31
0003	65	ADC
0004	33	$33
0005	85	STA
0006	41	$41
0007	A5	LDA
0008	30	$30
0009	65	ADC
000A	32	$32
000B	85	STA
000C	40	$40
000D	00	BRK

Figure 10–55 illustrates the details of this program.

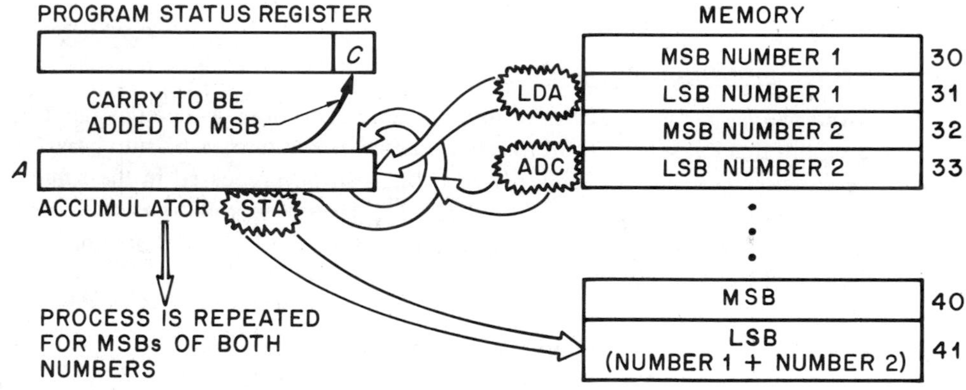

Figure 10–55 Details of 16-bit addition.

Shifting and Rotating Instructions

There are some important instructions that work on just the accumulator. You need to understand what these do in order to see how you can use this 8-bit machine to multiply and divide. The first one to be investigated is a *rotate* instruction. This process is illustrated in Figure 10–56.

There are other methods of altering data. One is called the *shift* instruction. This process is illustrated in Figure 10–57.

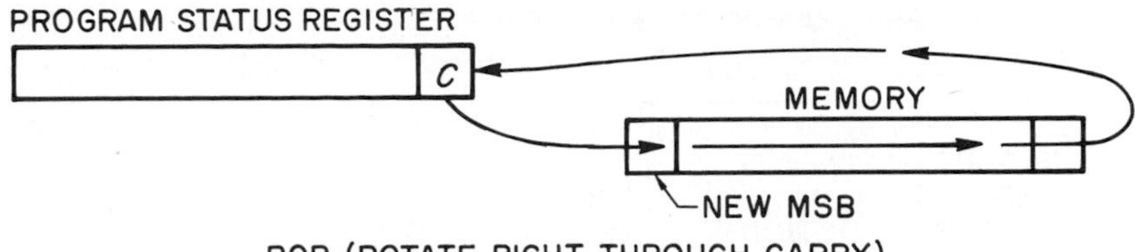

Figure 10–56 Process of a rotate instruction.

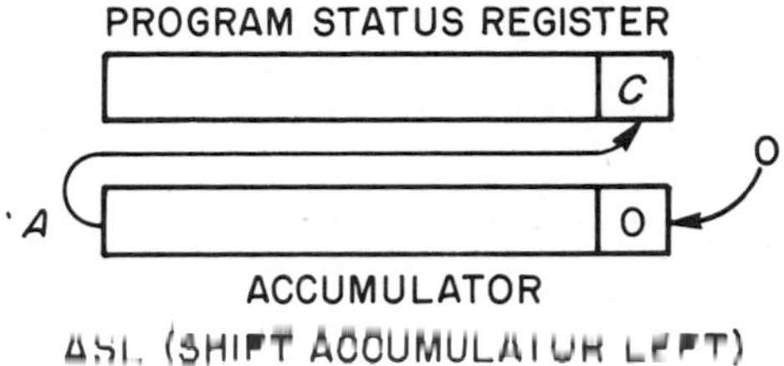

Figure 10–57 Process of a shift instruction.

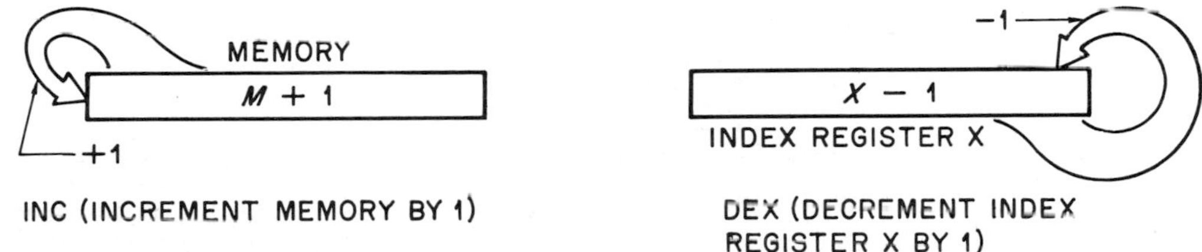

Figure 10–58 Concept of incrementing and decrementing.

Incrementing and Decrementing

Registers and memory locations can have their values changed by various instructions in the microprocessor set. This is illustrated in Figure 10–58. These processes may be used in conjunction with the following branch instructions.

A Branching Instruction

A branching instruction will cause the microprocessor to go to another memory location for further instructions if a certain condition is met. The general idea of one type of branching instruction is illustrated in Figure 10–59. As you will see, it is the branch instruction that gives the microprocessor the ability to repeat processes. This will happen when the number used for the branch is negative (two's complement notation). When this happens, the program will branch backward. Doing this causes a repeat of a prior part of the program.

Using Labels

Assembly language programs were first introduced in the microprocessor application section of Chapter 7. There you learned the basic concepts of mnemonics. Most assemblers utilize four fields for programming, as shown in Table 10–7.

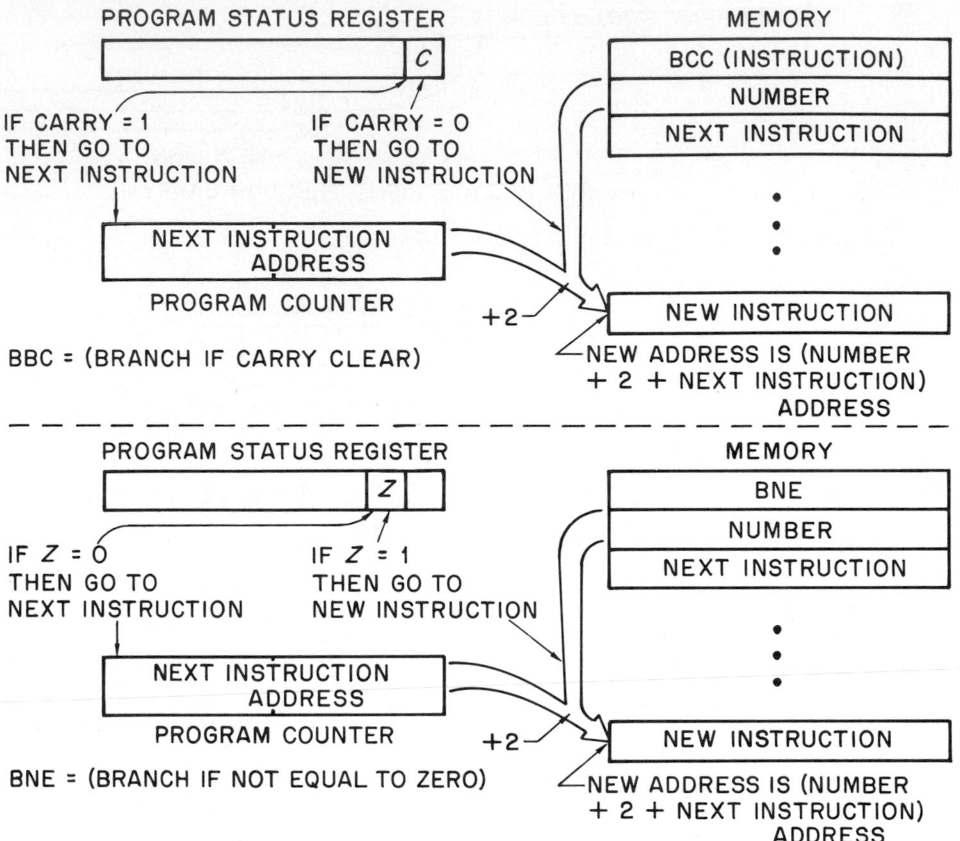

Figure 10–59 General idea of a branching instruction [BCC and BNE].

In the previous excerpt, the word BEGIN is a user-defined **label.**

Label:
In assembly language programming. The first field of the program that can be used to represent an address.

Table 10–7 FIELDS USED IN AN ASSEMBLY LANGUAGE PROGRAM

Label Field	Op Code or Mnemonic Field	Operand or Address Field	Comment Field
BEGIN	LDA	$31	; A COMMENT HERE
	STA	$55	; ANOTHER COMMENT
	⋮		
	BRK		; END IT ALL

The advantage of using a label as an address in an assembly language program is that you may subsequently use the same label in the address field of another instruction. The following program illustrates the use of labels in an assembly language program.

Multiplication

Multiplication can be performed using an 8-bit microprocessor. However, one of the advantages of 16- and 32-bit microprocessors is that they have built in multiplication instructions. The 8-bit ones do not.

The basic idea of how to multiply two binary numbers was presented in Chapter 2. There you were shown how to do this by repeated addition. Another method of multiplying binary numbers is by using the same method of multiplication you were taught in grade school. The only difference here is that since you will only be working with 1's and 0's, the multiplication table is much easier ($1 \times 1 = 1$ and $1 \times 0 = 0$). Consider, for example, multiplying 3×5 in binary:

```
     0011
   × 0101
  -------
     0011    ← Multiply by 1
    0000     ← Multiply by 0, shift left
   0011      ← Multiply by 1, shift left
  0000       ← Multiply by 0, shift left
  -------
  0001111    ← Add
```

Thus you get $3 \times 5 = 15$, as you should. This same process can be performed by an 8-bit microprocessor, as shown in Figure 10–60.

The following program illustrates the process of multiplying two numbers. The pound sign (#) in front of a number means that the microprocessor is to use the *immediate* addressing mode. Recall from the microprocessor applications section of Chapter 7 that this means the number represented will be treated as data and not as a memory location. Thus LDA #5 means load the number 5 into the accumulator (the resulting hex code for the immediate mode of LDA is also different).

Problem: Multiply an 8-bit unsigned number in memory location 30 by an 8-bit unsigned number in memory location 31. Store the answer in memory locations 40 and 41. Be sure that the most significant bits are stored in memory location 40.

Assembly Program:

```
        LDA    #0       ; Make LSB's of product a 0
        STA    $40      ; Make MSB's of product a 0
        LDX    #8       ; 8 bits in the multiplier
HERE    ASL    A        ; Shift accumulator left (clears carry)
        ROL    $40      ; Rotate most significant bit left
        ASL    $31      ; Put MSB of multiplier into carry flag
        BCC    NOADD    ; If carry flag is 0 do not add
        CLC             ; Clear the carry flag
        ADC    $30      ; Add multiplicand to the product
```

```
        BCC    NOADD    ; If carry clear, then no overflow
        INC    $40      ; There was overflow so add 1 to MSBs
NOADD   DEX             ; Decrease value in X register by 1
        BNE    HERE     ; If X register not 0 then repeat process
        STA    $41      ; Process completed, store LSBs in memory
        BRK             ; Stop execution
```

Machine Code:

	Memory Location [Hex]	Memory Contents [Hex]	Instruction /Operand
	0000	A9	LDA
	0001	00	#0
	0002	85	STA
	0003	40	$40
	0004	A2	LDX
	0005	08	#8
HERE	0006	0A	ASL
	0007	26	ROL
	0008	40	$40
	0009	06	ASL
	000A	31	$31
	000B	90	BCC
	000C	07	NOADD
	000D	18	CLC
	000E	65	ADC
	000F	30	$30
	0010	90	BCC
	0011	02	NOADD
	0012	E6	INC
	0013	40	$40
NOADD	0014	CA	DEX
	0015	D0	BNE
	0016	EF	HERE
	0017	85	STA
	0018	41	$41
	0019	00	BRK

Note the BNE instruction in memory location 0015. It has an operand equal to EF_{16}. Recall that this value is added to the current address + 2, and the program counter then uses this value as the next address. In this case

```
Current address →  0015
 Add the offset →  00EF
                   ----
                   0104   ← Disregard the carry
        Add 2 →       2
                   ----
                   0006   ← Address of next instruction
```

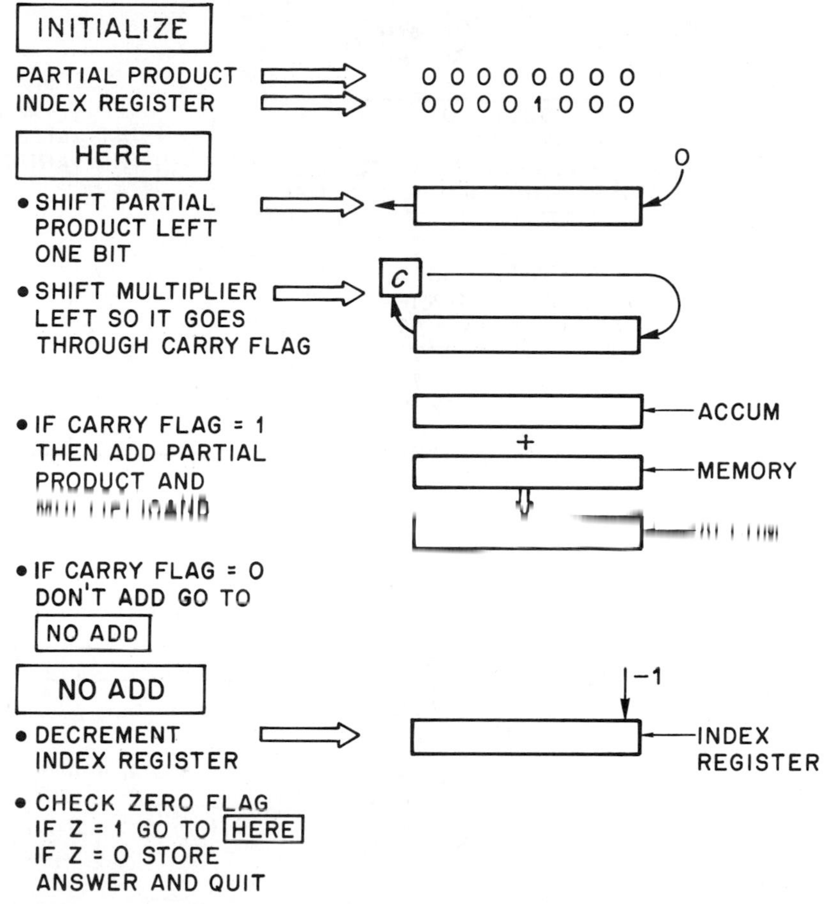

Figure 10–60 Basic idea of multiplication.

The address 0006 is the exact address where the label HERE is located. An analysis of this program is illustrated in Figure 10–61.

Interrupts

An **interrupt** is a method of quickly stopping the process of the microprocessor and have it pay attention to another detail.

> **Interrupt:**
> The process of stopping the microprocessor from its current operation and causing it to take its next instruction from a different memory location.

Interrupts are used for servicing alarms, power failures, or other external events that require immediate attention. The basic idea of an interrupt is illustrated in Figure 10–62. The 6502 has two interrupt inputs: an active LOW maskable interrupt (IRQ) and an active LOW nonmaskable interrupt (NMI). When these are activated, several important steps take place:

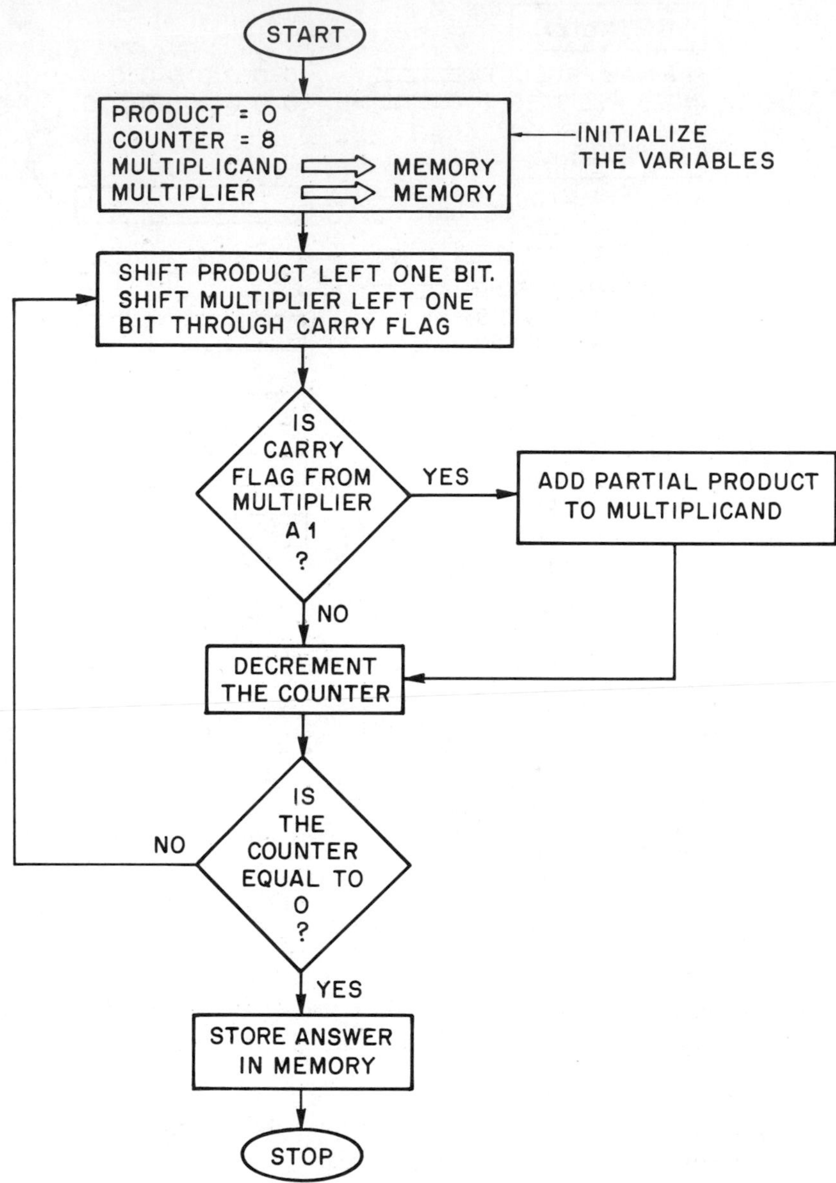

Figure 10–61 Analysis of multiplication program.

1. The contents of the program counter and the status register are saved in a memory location whose address is contained in the internal register called the stack pointer. These memory locations are referred to as the *stack*.
2. A new address is then copied into the program counter (address register). The value of this address is contained at memory locations FFFE and FFFF for an IRQ and at address FFFA and FFFB for a NMI.

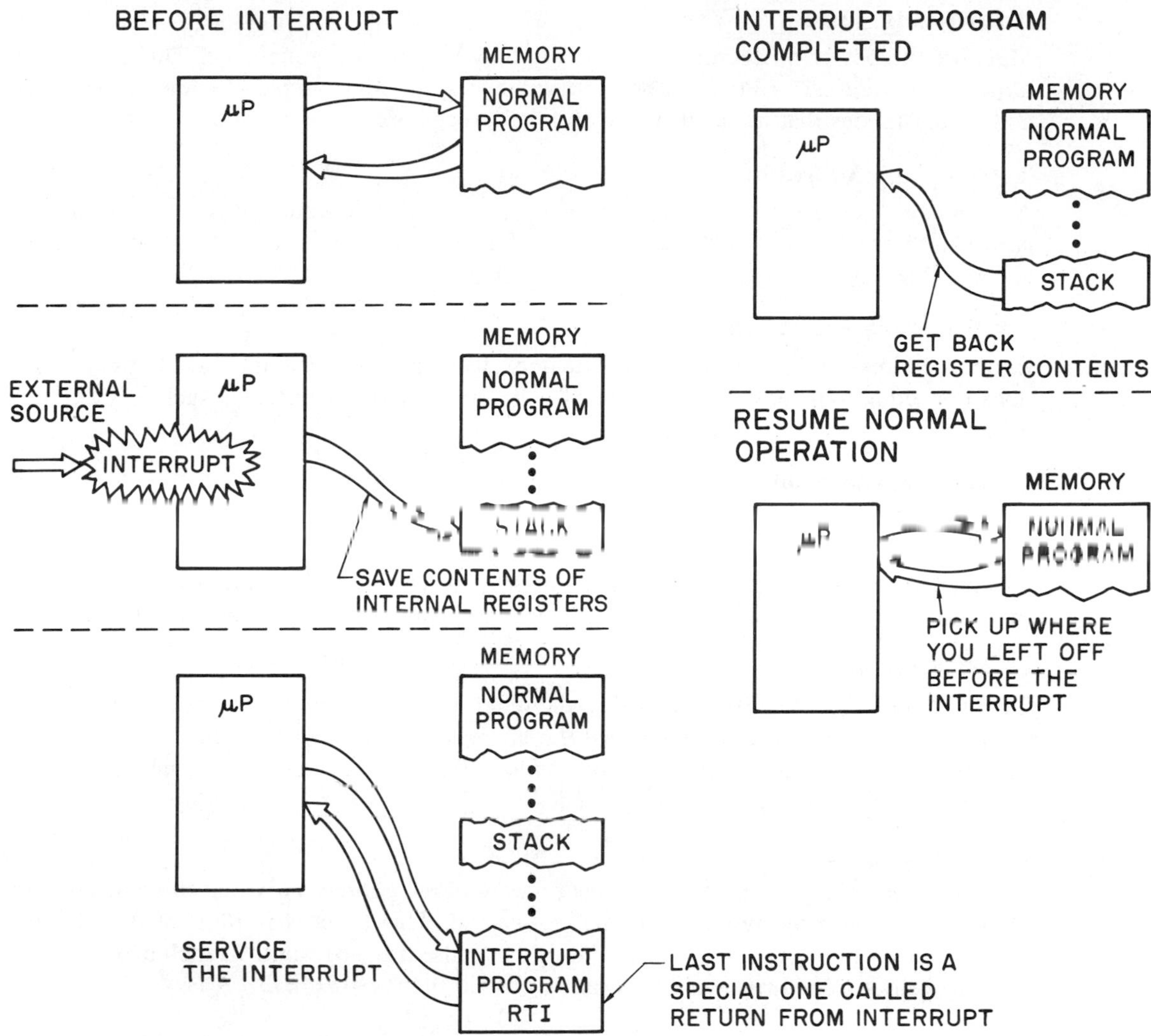

Figure 10–62 Basic process of an interrupt.

3. The microprocessor will then get its next instruction from the new address in its program counter. This section of memory contains the program called the *interrupt service routine*.
4. The last instruction in the interrupt service should be an RTI (return from interrupt). This causes the old address to be pulled from the stack and loaded back into the program counter, thus allowing the microprocessor to pick up where it left off before the interrupt occurred. It also loads back the original condition of the status register.

There is much more to interrupts than presented here. The purpose of this presentation was to give you an idea of what an interrupt was and one of the purposes of the stack pointer.

Addressing Modes

Much of the power of microprocessors is in how they relate to memory. This relation is termed their *addressing modes*. The 6502 will be used as an example of some of the many addressing modes that are available with microprocessors.

Accumulator Addressing

Accumulator addressing is perhaps the simplest of the addressing modes, being any instruction that acts solely on the accumulator: for example, the rotate and shift instructions such as ROL and ASL.

Implied (Inherent) Addressing

Implied addressing applies to any instruction other than the accumulator addressing that does not require an address to execute. This would include instructions such as CLC and TAX.

Immediate Addressing

Immediate addressing was presented in the microprocessor application section of Chapter 7. It means that the operand is present in the byte *immediately* following the first byte of the instruction. An example of this is LDA #5, which means load the accumulator with the number 5. The op code for LDA = A9. This mode is not used for STA.

Direct Addressing

Direct addressing was also presented in the microprocessor application section of Chapter 7. It means that the operand identifies the address in memory where the data is stored. An example is LDA $5, which now means load the accumulator with the number that is stored at memory location 5. The op codes for LDA = A5 and STA = 85.

Indirect Addressing

Indirect addressing causes the operand to represent the address at which the new address to be loaded onto the program counter is located. This process is illustrated in Figure 10–63. This mode is applied to instructions that cause the program to branch to a different area in memory for further instructions.

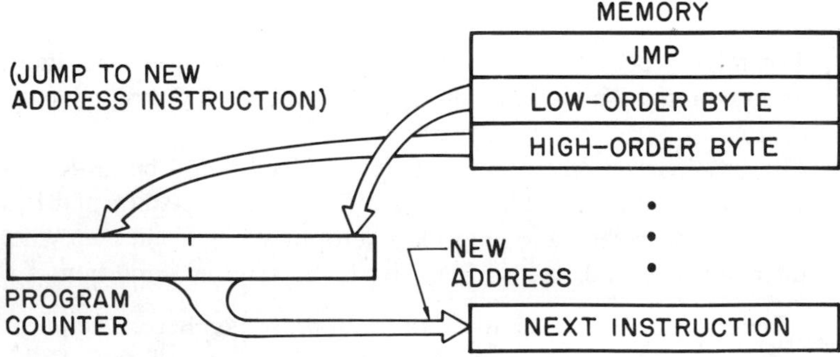

Figure 10–63 Operation of indirect addressing.

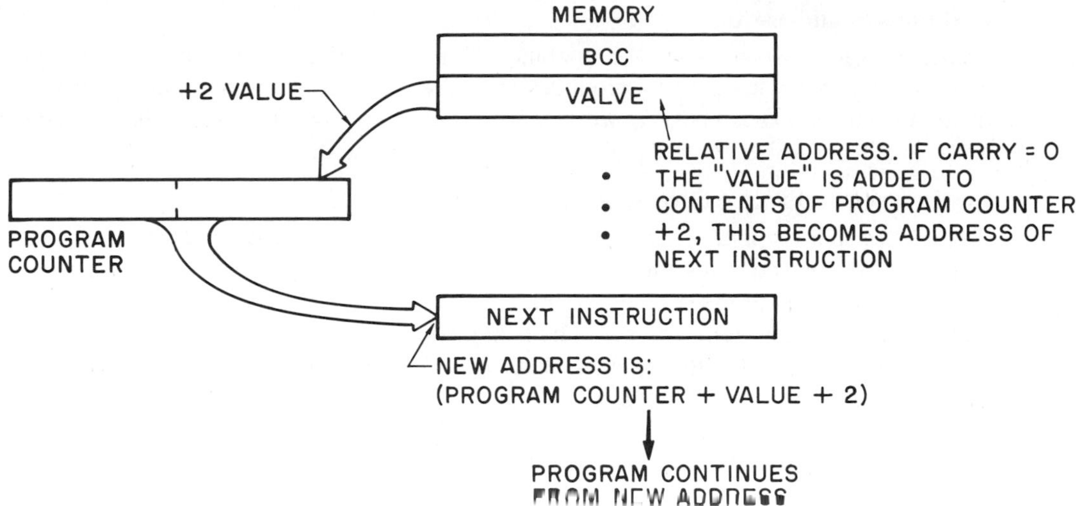

Figure 10–64 Operation of relative addressing.

Relative Addressing

In relative addressing the operand of the instruction represents a number to be added to the program counter. After the addition, the next instruction is fetched from this new memory location. Figure 10–64 presents the concept of this type of addressing. Note that a negative number will cause a branch backward in the program.

Indexed Addressing

In indexed addressing the contents of the X or Y index register are added to the operand of the instruction. This process is illustrated in Figure 10–65. The op codes for LDA = B5 and for STA = 95. Both are indexed by the X register.

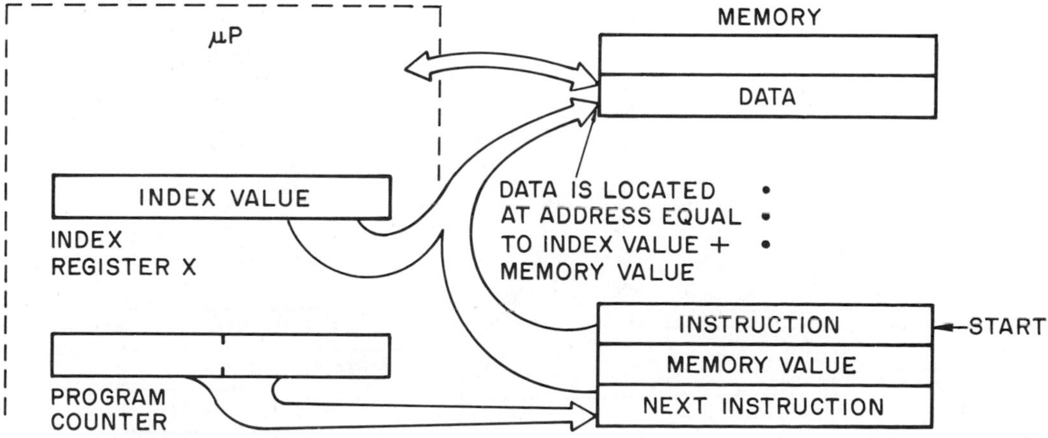

Figure 10–65 Operation of indexed addressing.

Preindexed Addressing

Preindexed addressing causes the operand of the instruction to be added to the contents of the X register. This process produces an address that contains the address of the next instruction to be executed. The process is illustrated in Figure 10–66. The op codes for LDA = A1 and for STA = 81. Both are indexed by the X register.

Postindexed Indirect Addressing

Postindexed indirect addressing is perhaps the most complex of the 6502 addressing modes. Here the operand contains the address where the address of a value is contained. This value is then added to the contents of the Y index register. The resulting number will then be loaded into the program counter. The next instruction will be fetched from this location. Figure 10–67 illustrates the process. The op codes for LDA = B1 and for STA = 91. Both are indexed by the Y register.

Zero Page Addressing

Zero page addressing simplifies the writing of programs, since all of the addressing will be done where the high-order byte is zero (00XX). Thus, the high-order byte never has to be specified. All of the programs in this section used this form of addressing. The op codes for LDA = A5 and STA = 85.

The Hex Code

A complete instruction set for any microprocessor will have all of the different hex codes for all of the addressing modes. For example, the hex codes for the 6502 given at the beginning of this section contained for the most part the hex codes for only zero page addressing. Table 10–8 presents all of the hex codes for the different modes that can be used by the LDA (load accumulator) instruction.

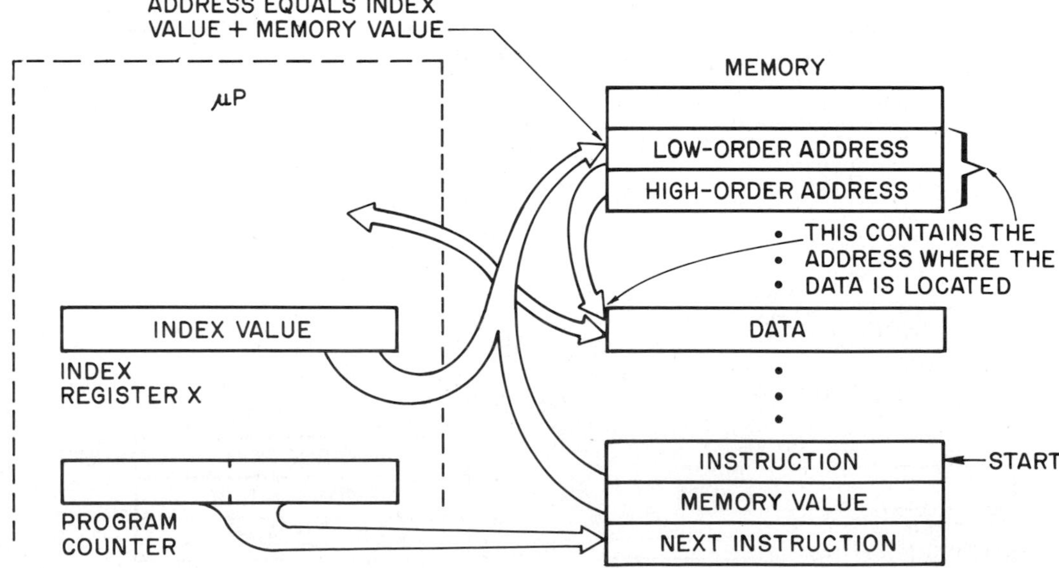

Figure 10–66 Operation of preindexed addressing.

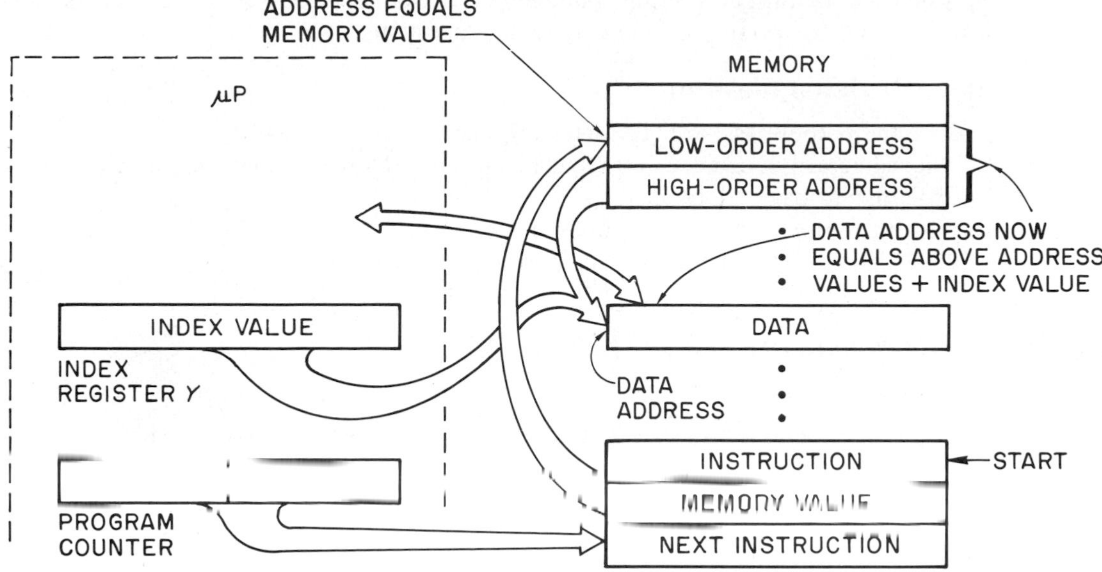

Figure 10–67 Process of postindexed indirect addressing.

As you can see, what first appears to be a very simple instruction can take on a variety of meanings, depending upon the mode of addressing selected by the programmer. You will learn much more about addressing modes in future microprocessor courses. They were presented here so you could get a general idea of the meaning of an addressing mode and appreciate the rich variety of programming options they offer.

Conclusion

This section presented the 6502 microprocessor. You were introduced to many important concepts used by all microprocessors. This section was intended as a springboard for con-

Table 10–8 HEX CODES FOR LDA IN DIFFERENT ADDRESSING MODES

Hex	Addressing Mode
A1	Indirect, preindexed with X index register
A5	Zero page direct addressing
A9	Immediate addressing
AD	Direct addressing
B1	Indirect, postindexed with Y index register
B5	Zero page indexed with X index register
B9	Absolute indexed with Y index register
BD	Absolute indexed with X index register

cepts that will be further developed in microprocessor courses. Check your understanding of this section by trying the following review questions.

10–8 Review Questions

1. State the purpose of the 6502 status register.
2. What addressing mode does not require an address and works only with the accumulator?
3. Explain zero page addressing.
4. Which addressing modes use the X index register?
5. What is the stack? What is the purpose of the stack pointer?

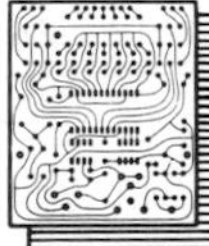

ELECTRONIC APPLICATION

Discussion

This section presents some advanced troubleshooting concepts. It also offers some practical suggestions to very basic troubleshooting ideas. In Chapter 8 you were introduced to a troubleshooting technique based on comparing the operation of one system with another. An elaboration of that technique is presented here.

Background Information

The very nature of a digital system lends itself to the act of troubleshooting. Digital systems can perform repetitive tasks, they can do specific sequences based upon certain predefined conditions, and these predefined conditions can depend upon measurements made from the external environment. Figure 10–68 illustrates some of the external measurements that can be made by a digital system. These external measurements are not limited to just electrical measurements but can simulate almost all of the human senses.

A Troubleshooting System

Consider the troubleshooting chart in Figure 10–69 for troubleshooting a hypothetical robot. Observe that the value of a specific measurement determines what the next measurement or action/decision should be. Many manufacturers provide these troubleshooting charts for systems they produce. It is this kind of system—next action based upon a prior measurement—that a digital system can replicate.

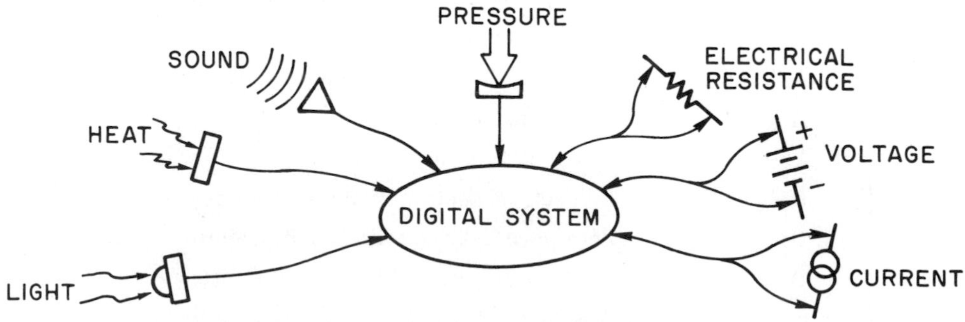

Figure 10–68 External measurements for a digital system.

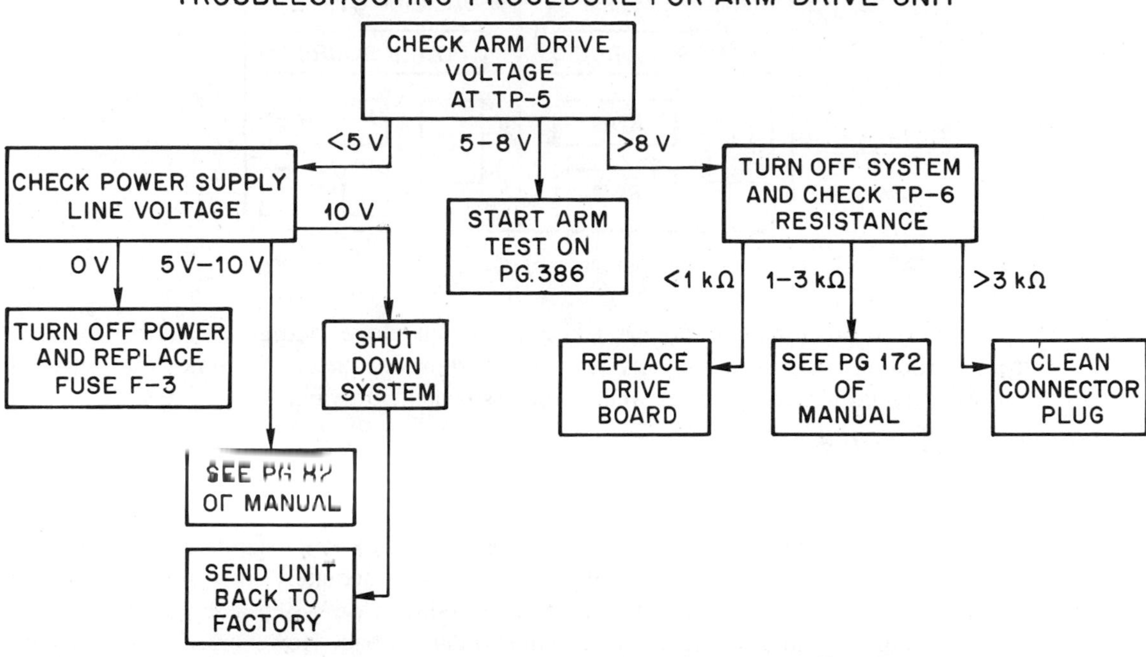

Figure 10–69 Troubleshooting chart for troubleshooting a hypothetical robot.

Computer-Aided Troubleshooting (CAT)

The troubleshooting chart for the hypothetical robot is an excellent candidate for computer-aided troubleshooting. The basic concept of such a system is illustrated in Figure 10–70. Measurement sensors interface with the robot to give input to the computer. A program inside the computer emulates the troubleshooting chart. The computer then determines what part is malfunctioning, orders the part from stock, maintains the stock inventory,

Figure 10–70 Basic concept of CAT.

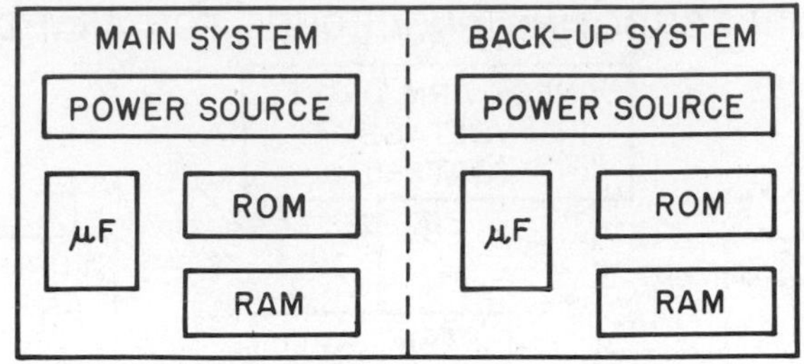

Figure 10–71 Concept of a parallel system.

makes out the bill for the customer, addresses the envelope, and gets ready for the next troubleshooting assignment. However, such a system does need a technician to repair it and to maintain the computer as well as the associated software.

Parallel Systems

The concept of a *parallel system* is illustrated in Figure 10–71. In a parallel system, there is a replication of every part and every section. The system is designed so that there are separate monitors that compare the system operation to a known standard. If one part of the system malfunctions, the *redundant* part or section immediately takes over. The system operator is then notified that a replacement is needed. Such a system is by no means foolproof. A monitor can malfunction, or a standard can go bad. However, such systems are used by the military and space systems as well as banking automatic tellers to increase overall system reliability.

General-Purpose Interface Bus (GPIB)

A bus system that is frequently used to interconnect automated testing equipment is the IEEE-488 standard bus. The basic connection scheme of such a system is illustrated in Figure 10–72.

The system can accommodate three basic classes of devices: *talker, listener,* and *controller*. The talker is an information-producing device (such as a temperature probe). A listener is a display device (such as a seven-segment readout). The controller is a device that controls the interaction between the talkers and listeners (such as a digital system).

The IEEE-488 bus consists of 24 wires connected to a standard connector. This connector along with its associated functions is illustrated in Figure 10–73.

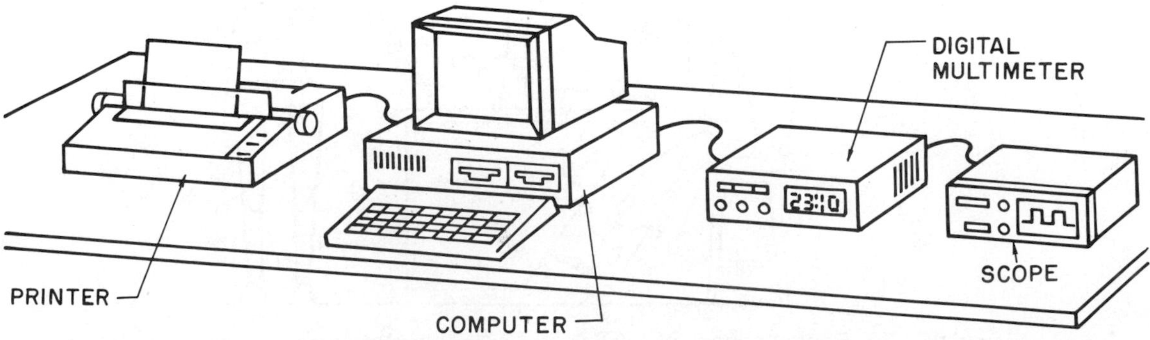

Figure 10–72 The IEEE-488 bus system.

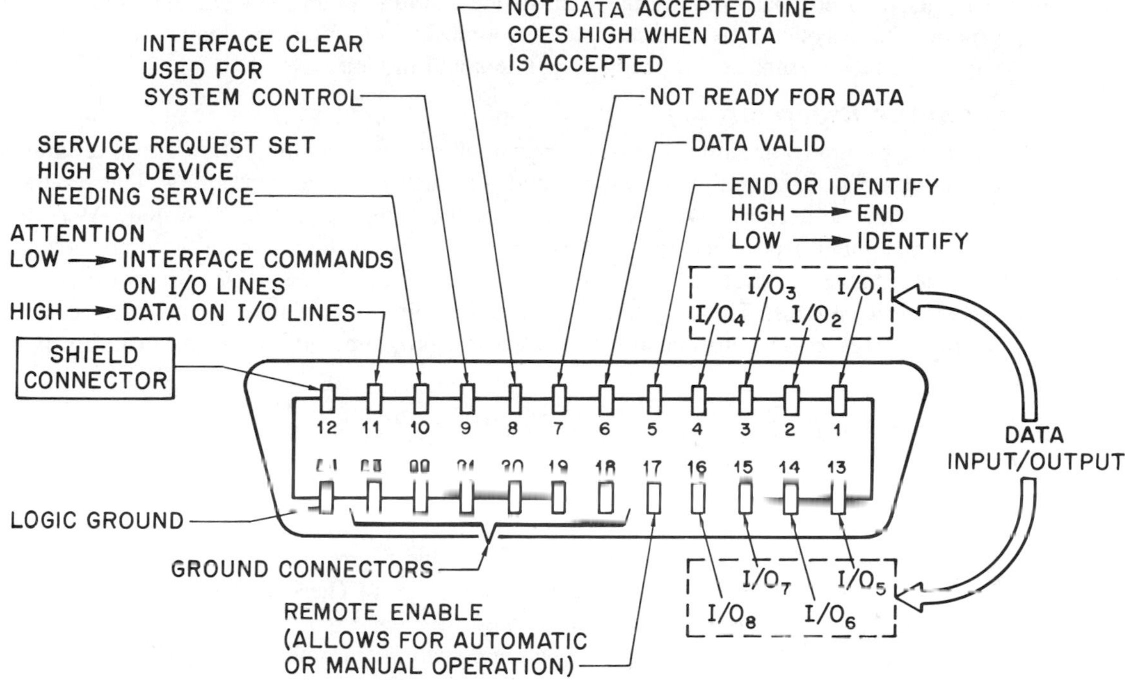

Figure 10–73 IEEE-488 standard connector.

Teachable Troubleshooters

Another approach to troubleshooting digital equipment is to place a troubleshooting instrument where the microprocessor chip is connected. This is illustrated in Figure 10–74.

The programmable microsystem troubleshooter (PMT) is first connected in the manner shown in Figure 10–74 to a known good system. It can then be put into a "self-programming" mode called the LEARN mode. While in the LEARN mode, it steps through all possible memory and I/O addresses. It does this to determine if there is a component at that address and how the component responds to control commands. From this, the PMT develops a complete system profile. Essentially it will completely "map" the entire system. This information can be stored for later use. When the system does malfunction, the PMT

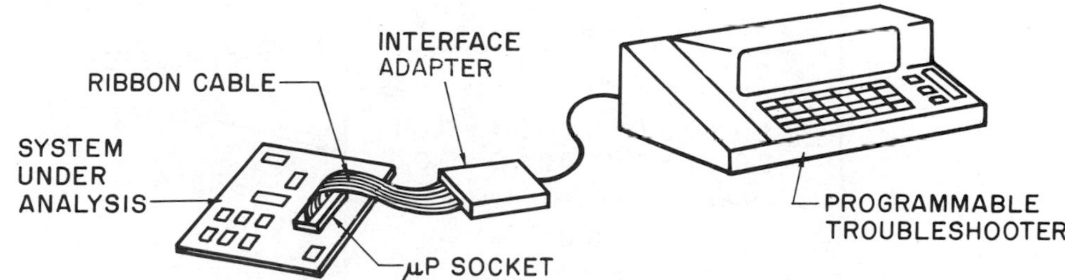

Figure 10–74 Replacing the microprocessor chip with a programmable microsystem troubleshooter.

can be used to compare its original data to the malfunctioning system and localize the problem. These systems are also capable of troubleshooting interface connections between computers and external devices such as printers and modems.

Diagnostic Programs

Diagnostic programs are simply computer programs to test the computer. These programs are available for large computer systems and for small personal computers. The program interacts with the user by asking which parts of the computer are to be tested. Available tests usually range from checking the keys on the keyboard to checking the color settings on the monitor. Such programs will also check the system RAM, ROM, disk drives, and other support devices. These systems are severely limited in that the computer must be performing well enough to operate the diagnostic program; otherwise the program is of little use.

Logic Analyzer

A *logic analyzer* is similar to an oscilloscope in that it can display waveforms. It differs from the oscilloscope in that it can display multiple waveforms. This allows you to observe the important timing relationships on a digital bus. The basic idea of a logic analyzer is illustrated in Figure 10–75. As you can see from the figure, the logic analyzer can also show the digital information in different display modes. These include binary, octal or hexadecimal as well as pulse waveform displays.

There are several advantages to using a logic analyzer in troubleshooting. The analyzer contains a memory that can store information about previously read signals. It can also be programmed to respond to only a certain type of signal or to a specific combination of signals. Thus if you were looking for a glitch you could program it to display pulses of only a certain duration or less, and longer, normal pulses, would not be displayed. You could also program the instrument to show you the action of the control bus or data bus only during certain addresses put out by the microprocessor.

Some logic analyzers are available that will translate the binary patterns of a bus into the mnemonics of the original instruction set. This capability is especially useful in troubleshooting new systems where the software as well as the hardware is still under development. Again, the instrument may be programmed to respond to only certain instructions or a range of addresses or to make a comparison to another program.

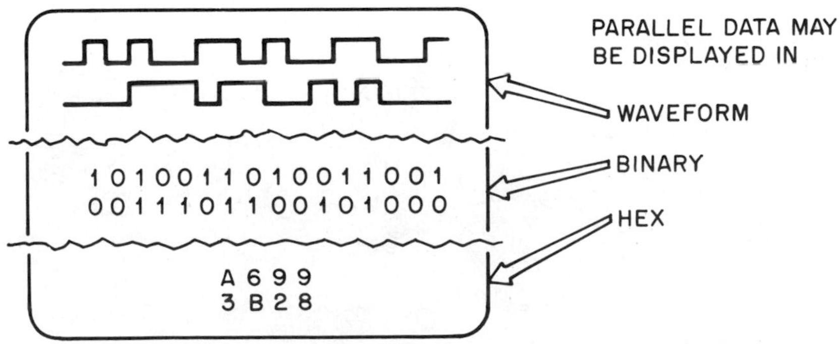

Figure 10–75 Basic idea of a logic analyzer.

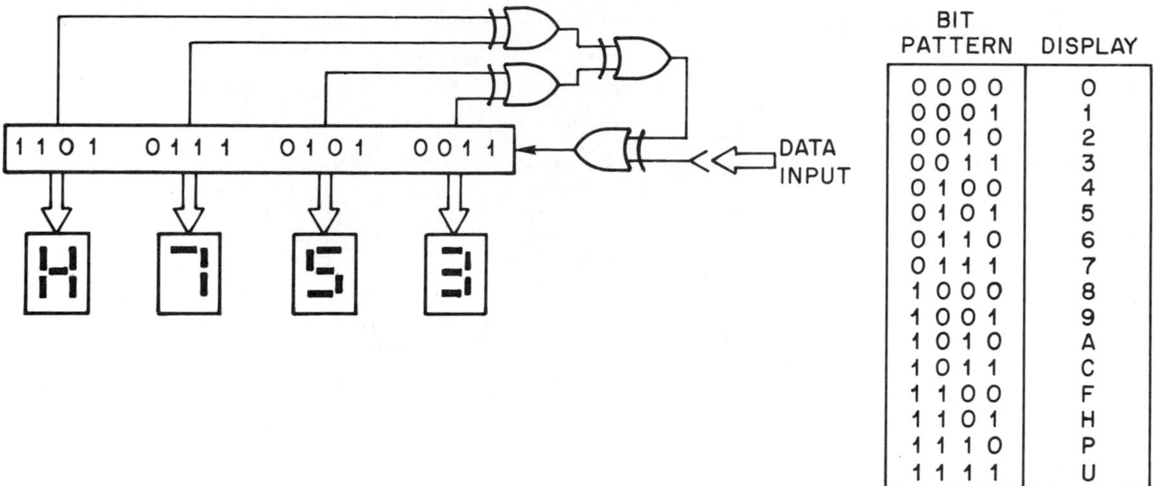

BIT PATTERN	DISPLAY
0000	0
0001	1
0010	2
0011	3
0100	4
0101	5
0110	6
0111	7
1000	8
1001	9
1010	A
1011	C
1100	F
1101	H
1110	P
1111	U

Figure 10–76 Design idea of a signature analyzer operation

Signature Analyzers

A *signature analyzer* costs less than the logic analyzer. It is used to troubleshoot digital systems by using a single probe. The instrument works by analyzing a sequence of bits. These bits could be on a particular line of a data bus or address bus. Such a specified sequence of pulses will always be the same under identical conditions. This is demonstrated by the pulse stream in Figure 10–76. A unique bit stream will produce a unique seven-segment display. This particular display is called the *signature* for that point and with that particular pulse stream. Any deviation from the pulse stream caused by a system error will result in a different signature. A typical signature analyzer is shown in Figure 10–77. Note that the 7-segment display is not hex. It is part of a display pattern unique to the signature.

Figure 10–77 Typical signature analyzer. *Courtesy* of Hewlett-Packard Company

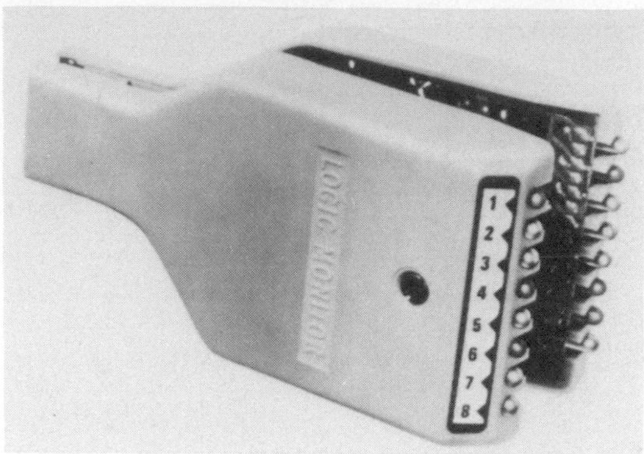

Figure 10–78 The logic clip. *Courtesy* of Global Specialties

Logic Clip

An inexpensive logic analyzer is the *logic clip*. Logic clips come in various sizes. A 16-pin logic clip is shown in Figure 10–78. The clip will indicate the logic condition of each pin on the IC. Thus, you can observe the logic level relationships of all pins on the chip in question. This device is usually level sensitive and does not respond to rapidly changing pulses.

Current Probe

A current probe is shown in Figure 10–79. This instrument can detect a changing or pulsating current in a wire. The current must be changing because the probe responds to the changing magnetic field produced by a changing current much the same way as a radio

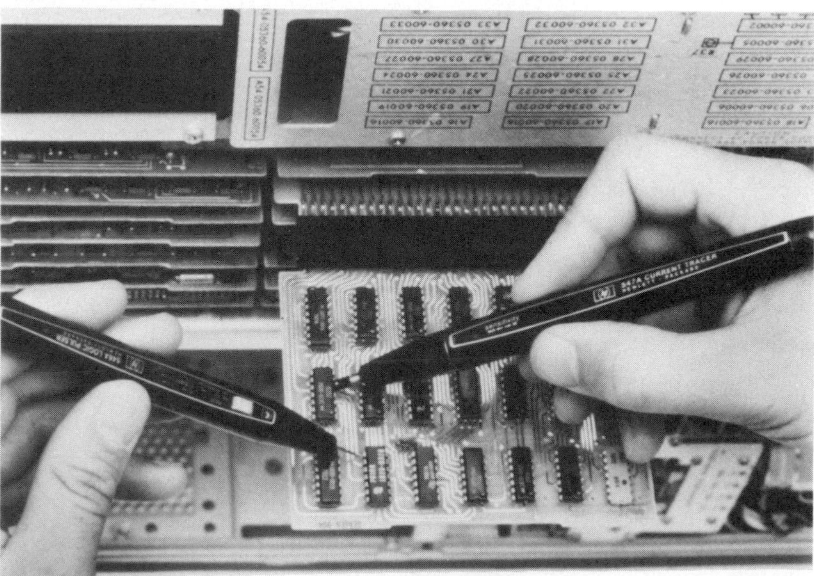

Figure 10–79 Logic current tracer. *Courtesy* of Hewlett-Packard Company

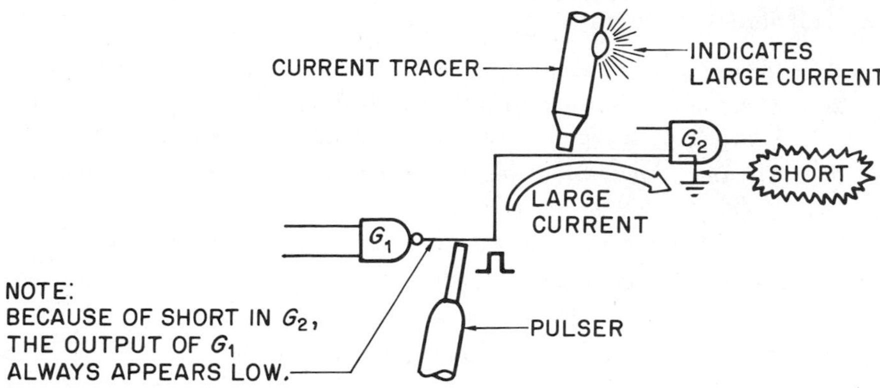

Figure 10–80 Use of the current tracer to find a short.

antenna responds to a radio wave (which consists, in part, of a changing magnetic field). The usefulness of this instrument is in tracing current paths in logic networks. This is especially helpful when it comes to isolating shorts in digital ICs. To cause a changing logic level, a logic pulser is used in conjunction with the current tracer. This is illustrated in Figure 10–80. The current tracer is useful in detecting faulty digital points that are *stuck LOW*. This means that since the device is shorted, it will force the output of a logic device attached to it to go LOW.

Conclusion

This section presented various diagnostic troubleshooting equipment. The systems used in advanced troubleshooting techniques are constantly improving, and, as a technician, you are sure to encounter many of them in the field. Check your understanding of this section by trying the following review questions.

10–9 Review Questions

1. State briefly the items needed to perform computer-aided troubleshooting.
2. What is the purpose of the IEEE-488 bus?
3. State one of the main disadvantages of diagnostic software.
4. What is the main advantage of a logic analyzer?
5. Give the main use of a current tracer.

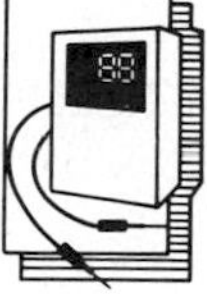

TROUBLESHOOTING SIMULATION

Discussion

The troubleshooting simulation for this chapter will provide you with valuable experience in troubleshooting software. The program presents a working 4-bit microprocessor, instruction set and RAM. The demonstration mode takes you step-by-step through the operation and addressing of this device. The test mode helps you sharpen your software diagnostic skills. This is an exciting experience and sets the foundation for your future microprocessor courses.

SUMMARY

- In signed binary notation, the MSB is used to represent the sign of the binary number.
- The convention for signed binary notation is if the sign bit is a 1, then the number is negative. If the sign bit is 0, the number is positive.
- The one's or two's complement notation may be used to represent signed binary numbers.
- In the one's complement system, there are two values represented for zero.
- In the two's complement system, there is only one value for zero.
- Half adders have two inputs and two outputs.
- A full adder has three inputs and two outputs.
- Full adders may be cascaded in order to accommodate the addition of larger numbers.
- Overflows may occur when adding signed binary numbers.
- Overflows are not possible when the signs of the numbers to be added are different.
- An accumulator is a register that starts out by containing one of the values for an arithmetic or logic operation and ends up with the answer.
- An arithmetic logic unit is a logic device capable of performing arithmetic and logic operations.
- Binary-coded decimal logic must take into account the requirements of adding BCD numbers.
- There are two methods for transmitting digital information: in parallel or serially.
- Serial transmission may be performed synchronously or asynchronously.
- Digital transmission may be done through space using radio waves as a carrier.
- The modulation techniques for space transmission are AM, FM, and PM.
- Fiber optics utilizes glass fibers to direct the transmission of light.
- Laser communications take advantage of the line-of-sight transmission produced by laser light.
- The three basic transmission types that can be used for digital information are the simplex mode, half-duplex mode, and full-duplex mode.
- There are various recommended standards for wired interfacing between digital systems.
- A telephone modem is a device that allows for the transmission and reception of digital information using a standard telephone.
- Direct memory access is a method of transferring digital information to and from memory without intervention from the microprocessor. This results in much faster rates of data transfer.
- A peripheral interface adapter (PIA) is a device used for the parallel transfer of digital information.
- The baud rate is the rate at which serial bits are transferred.
- The UART and ACIA are digital devices that allow for the serial transmission of digital information.
- The device used for converting digital information into sound for telephone transmission and back from sound to digital information is an acoustic coupler.
- Microprocessors contain various internal registers that are accessible by the programmer.

- The stack is a location in memory usually used to store information needed by the microprocessor. This memory location is usually separate from the locations used for the main program.
- The status flags are individual bits in the internal register that is sometimes called the condition code register. These bits are used to indicate the result of a previous operation or to influence a pending operation.
- Eight-bit microprocessors have internal registers that are 8 bits wide and access data through an 8-bit-wide interface.
- Sixteen-bit microprocessors have internal registers that are 16 bits wide and may access data through an 8- or 16-bit-wide interface.
- Thirty-two-bit microprocessors have internal registers that are 32 bits wide and may, depending on the type of machine, access data through an 8-, 16-, or 32-bit interface.
- Real microprocessors have many different addressing modes available for the same type of process.
- In assembly language programming, a label may be used to represent an address. This is for the convenience of the programmer.
- An interrupt is the process of causing the microprocessor to stop its current process and to begin a different process.
- Computer-aided troubleshooting utilizes the decision-making and interfacing abilities of the computer to diagnose digital systems.
- Parallel digital systems use redundant digital systems and subsystems to back up existing systems in case of failure.
- The general-purpose interface bus (GPIB) is a standard used to interface automated testing equipment.
- Teachable troubleshooters are digital troubleshooting systems that can "map" a known good system and then use this information to troubleshoot the system.
- Diagnostic programs are software used to test the operation of a computer system that must be in good enough condition to operate the software.
- A logic analyzer is capable of displaying many digital signals at the same time in a variety of formats.
- Signature analysis is a method of troubleshooting that analyzes a pulse train to an unexpected pattern.

CHAPTER SELF-TEST

I. TRUE/FALSE

Mark the following statements true or false.

1. When representing signed binary numbers, the MSB is used as the sign bit.
2. The direct method of representing −5 is 1101.
3. The hex number 7F represents a negative number.
4. A half adder can only add half a binary number.
5. Adding two binary numbers with different signs will never produce an overflow.

II. MULTIPLE CHOICE

Answer the following questions by selecting the most correct answer.

6. Which of the following additions of signed binary numbers could produce an overflow condition?
   (A) Two positive numbers
   (B) Two negative numbers
   (C) Two values of different signs
   (D) Both A and B are true.
7. An accumulator is a register that
   (A) will only contain the results of an operation
   (B) keeps the values of one of the operands
   (C) holds the instructions for the process to take place
   (D) None of the above are correct.
8. An ALU can perform
   (A) Addition and subtraction
   (B) Logical OR and AND
   (C) Exclusive OR
   (D) All of the above are correct.
9. In BCD addition, if the value of one digit becomes larger than 9,
   (A) you must add 6 to the value
   (B) discard the answer because it is wrong
   (C) convert to regular binary notation
   (D) None of the above are correct.
10. The difference between serial and parallel data transmission is
    (A) serial transmission is faster
    (B) parallel transmission requires fewer wires
    (C) parallel transmission is only asynchronous
    (D) None of the above are correct.

III. MATCHING

Match the answers on the right to the statements on the left.

11. Stores the result of an arithmetic or logic process
12. Holds the address of a special location in memory used for microprocessor information
13. Can indicate if an overflow occurred
14. Holds the address of the next instruction
15. May be used for counting the number of times through a loop

(A) Index register
(B) Stack pointer
(C) Status register
(D) Accumulator
(E) Program counter
(F) None of these

IV. FILL-IN

Fill in the blanks with the most correct answer(s).

16. The modulation technique that causes the strength of the carrier to change is called _______ modulation.
17. The modulation technique that causes the frequency of the carrier to change is called _______ modulation.

18. The modulation technique that causes the phase of the carrier to change is called _______ modulation.
19. The term *laser* stands for light amplification by _______ emission.
20. The mode of transmission where data may be transmitted and received at the same time is called _______-_______ mode.

IV. OPEN-ENDED

Answer the following questions as indicated.

21. State the main advantage of DMA.
22. What is the difference between a UART and an ACIA?
23. Is the 8088 a 16-bit microprocessor? Explain.
24. State the main difference between a logic analyzer and a signature analyzer.
25. Briefly explain the operation of a teachable troubleshooter.

Answers to Chapter Self-Test

1] T 2] [illegible] 3] [illegible] 4] F 5] T 6] D 7] D 8] D 9] A 10] D
11] D 12] B 13] C 14] E 15] A 16] amplitude
17] frequency 18] phase 19] stimulated 20] full-duplex
21] Increased speed in transferring data.
22] The only difference is in the name.
23] The 8088 has 16-bit internal registers and an 8-bit data interface bus.
24] The logic analyzer can observer pulses in parallel; the signature analyzer cannot.
25] A teachable troubleshooter can map a known good system and then use this information to troubleshoot the same system.

CHAPTER PROBLEMS

Basic Concepts

Section 10–1

1. Using the direct method for signed binary notation, determine the decimal value, including the sign, of the following hex numbers.
   (A) 24H (B) 8AH (C) 90DH (D) E3FH
2. Using the direct method of signed binary notation, determine the 8-bit hex representation of the following signed decimal numbers.
   (A) −5 (B) +25 (C) −52 (D) −127
3. Assume that the following hex numbers represent the one's complement of signed binary numbers. Convert each one to the equivalent signed decimal value.
   (A) 5EH (B) 40H (C) C4H (D) AC0H
4. Using the one's complement system for signed binary numbers, convert the following signed decimal numbers to their hexadecimal equivalent. Use an 8-bit representation.
   (A) −5 (B) +23 (C) −15 (D) −88

5. Using the two's complement system for representing signed binary numbers, convert the following hex numbers to their equivalent signed decimal values.
   (A) 5EH (B) FFH (C) EC0H (D) 9A5H
6. Using the two's complement system for representing signed binary numbers, convert the following signed decimal numbers to hex representation. Use an 8-bit representation.
   (A) −5 (B) +32 (C) −12 (D) −256
7. Repeat problem 6, but this time use a 16-bit representation.

Section 10–2

8. Given the following inputs to a half adder, determine the resulting outputs (sum and carry)
   (A) $A = 0, B = 1$ (B) $A = 1, B = 0$ (C) $A = 1, B = 1$
9. Given the following inputs to a full adder, determine the resulting outputs (sum and carry)
   (A) $A = 1, B = 0, C = 1$ (B) $A = 1, B = 1, C = 0$
   (C) $A = 1, B = 1, C = 1$
10. State what combinations of inputs could produce the following outputs of a full adder:
    (A) $S = 1, C = 0$ (B) $S = 1, C = 1$ (C) $S = 0, C = 1$
    (D) $S = 0, C = 0$
11. For the adder logic network of Figure 10–5 in Section 10–2, indicate the status of each carry output, each sum output, the carry LED, the answer register, and each seven-segment readout for the following additions. State if a carryout occurs.
    (A) A register = 0011, B register = 0010
    (B) A register = 1010, B register = 0101
    (C) A register = 1001, B register = 1111
12. Using two's complement notation, give the hex 8 bit representations of each of the following terms, including the solution. Use signed binary numbers with two's complement notation. Indicate the correctness of the resulting answer.
    (A) +2 +3 = (B) −5 +7 = (C) +2 −6 = (D) −3 −4 =
13. Using 8-bit binary numbers, express the following problems in hex notation including the answer. Indicate if there is an overflow and what it means.
    (A) +10 +15 = (B) −12 +18 = (C) −15 +4 = (D) −3 −12 =
14. Using signed binary notation with two's complementation, give the resulting answers in hex notation. Indicate the number of bits required for each of the following problems. State if an overflow condition exists and what this means.
    (A) 0A + 4B = (B) −5C + 20 = (C) −5F −7D = (D) 2DA −4CD =

Section 10–3

15. For the following problem, refer to Figure 10–81. (A) Which pins are the inputs to the accumulator? (B) From which pins would the answer be taken?
16. Referring to Figure 10–81. (A) Which pins determine the process to be performed by the ALU? (B) State the purpose of pin 13.

DM54S381/DM74S381 Arithmetic Logic Unit/Function Generator

General Description

The 'S381 is a Schottky TTL arithmetic logic unit (ALU)/function generator that performs eight binary arithmetic/logic operations on two 4-bit words as shown in the function table. These operations are selected by the three function-select lines (S0. S1, S2). A full carry look-ahead circuit is provided for fast, simultaneous carry generation by means of two cascade outputs ($\overline{P}$ and $\overline{G}$) for the four bits in the package. The method of cascading 54S182/74S182 look-ahead carry generators with these ALU's to provide multi-level full carry look-ahead is illustrated under typical applications data for the 'S182. The typical addition times shown illustrate the short delay time required for addition of longer words when full look-ahead is employed. The exclusive-OR, AND, or OR function of two Boolean variables is provided without the use of external circuitry. Also, the outputs can be either cleared (low) or preset (high) as desired.

Features

- A fully parallel 4-Bit ALU in 20-pin package for 0.300-inch row spacing
- Ideally suited for high-density economical processors
- Parallel inputs and outputs and full look-ahead provide system flexibility
- Arithmetic and logic operations selected specifically to simplify system implementation:
  - A minus B
  - B minus A
  - A plus B
  - and five other functions
- Schottky-clamped for high performance
  - 16-bit add time . . . 26 ns typ using look-ahead
  - 32-bit add time . . . 34 ns typ using look-ahead

Connection Diagram

Dual-In-Line Package

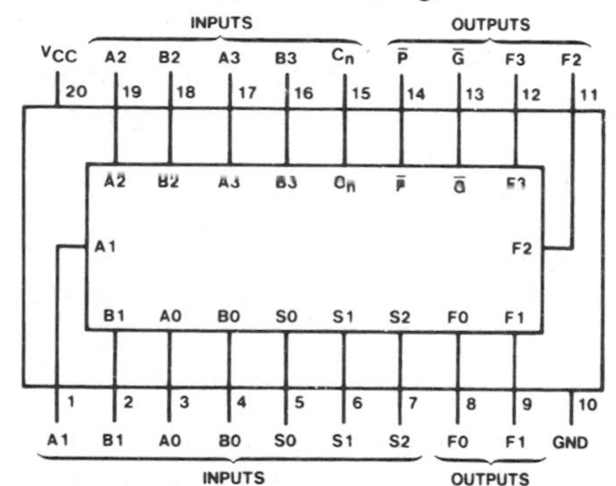

TL/F/6487-1

Order Number DM54S381J or DM74S381N
See NS Package Number J20A or N20A

Pin Designations

Designation	Pin Nos.	Function
A3, A2, A1, A0	17, 19, 1, 3	Word A Inputs
B3, B2, B1, B0	16, 18, 2, 4	Word B Inputs
S2, S1, S0	7, 6, 5	Function-Select Inputs
C_n	15	Carry Input for Addition, Inverted Carry Input for Subtraction
F3, F2, F1, F0	12, 11, 9, 8	Function Outputs
$\overline{P}$	14	Inverted Carry Propagate Output
$\overline{G}$	13	Inverted Carry Generated Output
V_{CC}	20	Supply Voltage
GND	10	Ground

Function Table

Selection			Arithmetic/Logic Operation
S2	S1	S0	
L	L	L	CLEAR
L	L	H	B MINUS A
L	H	L	A MINUS B
L	H	H	A PLUS B
H	L	L	A ⊕ B
H	L	H	A + B
H	H	L	AB
H	H	H	PRESET

H = high level, L = low level

Figure 10–81 *Reprinted* with permission of National Semiconductor Corporation

17. For the ALU in Figure 10–81, state the final contents of both registers after the indicated process is completed. (For correct mode, refer to selection on the Function Table.)
(A) B = FF, A = 01, mode = 7 (B) B = FF, A = AE, mode = 6
18. For the adder-subtractor unit shown in Figure 10–82, indicate the final contents of each register after the indicated process has been performed. Also state the status of the internal carryin and the carryout.
(A) $B = 6$, $A = 5$, Add/subtract = LOW
(B) $B = A$, $A = 3$, Add/subtract = HIGH

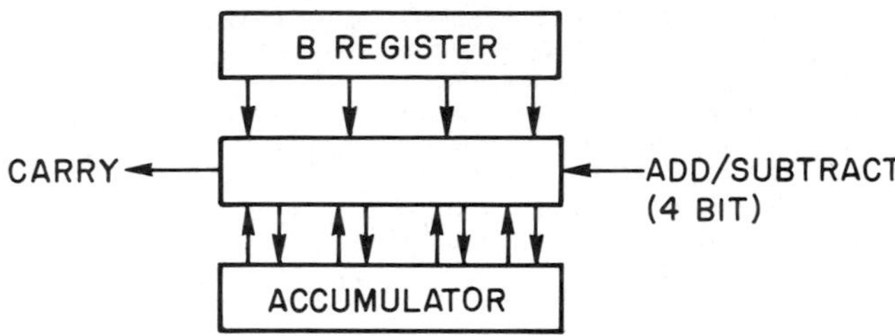

Figure 10–82

19. For the adder-subtractor unit of Figure 10–82, indicate the final contents of each register after the indicated process has been performed. Also state the status of the carryin and the carryout.
(A) $B = 7$, $A = 3$, Add/subtract = HIGH
(B) $B = $ C, $A = D$, Add/subtract = LOW
20. Add the following BCD numbers. Convert both the problem and the answer to decimal.
(A) 0110 + 0111 = (B) 0010 1001 + 0101 0110 =
(C) 1001 + 1001 =
21. Referring to the BCD hardware of Figure 10–83, indicate what the contents of each register will be and the status of the carry bit after the following additions are performed.
(A) $B = 9$, $A = 3$ (B) $B = 7$, $A = 1$
(C) $B = 0$, $A = 0$ (D) $B = 9$, $A = 9$

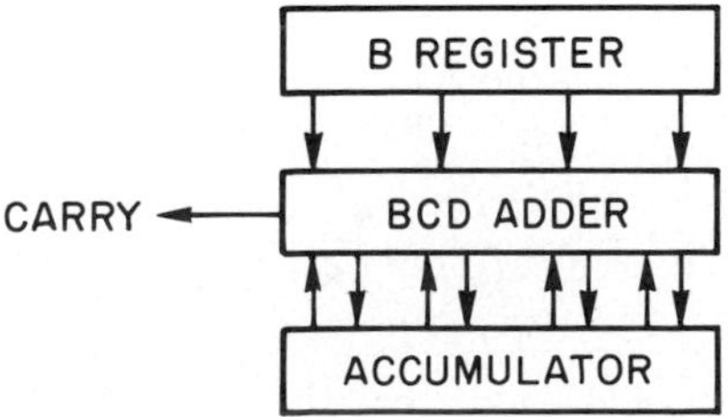

Figure 10–83

Section 10–4

22. Describe the two methods of digital transmission. Which one is the fastest?
23. Explain the difference between synchronous and asynchronous transmission of digital data.
24. Sketch a diagram of the most common types of connectors used in the transmission of digital information.
25. State what the word *modulate* means. Name three ways modulation can occur.
26. Sketch the waveforms illustrating the three different types of modulation.
27. Briefly describe the method of digital transmission used with fiber optics.
28. List some of the advantages of fiber-optic transmission.
29. State what the term *laser* means.
30. Discuss the three interfacing standards presented in this chapter.
31. State the differences between the transmission modes of problem 30.
32. Describe the main characteristics of the following wired interface standards.
    (A) RS-232C (B) 20-mA current loop (C) RS-422
33. Briefly describe the meaning of the term *modem*.
34. State the purpose of the RS 499 standard

Section 10–5

35. Explain the process of direct memory access.
36. What is the advantage of DMA?
37. Describe the operation of a DMA controller.
38. Explain the purpose and the operation of a PIA.
39. State some of the advantages of using a PIA.
40. Briefly explain what is meant by baud rate.
41. (A) Describe the operation of a UART. (B) How does it differ from an ACIA?
42. Sketch and explain a typical transmission format used for asynchronous serial transmission.
43. For the asynchronous digital format described in Section 10.5, sketch the resulting pulse train for each of the following characters.
    (A) A (B) ! (C) 9
44. What is the purpose of an acoustic coupler?
45. Briefly describe the operation of a telephone modem.

Section 10–6

46. (A) What is meant by the register architecture of a microprocessor.
    (B) Give an example.
47. (A) Sketch the register architecture of the 6502 8-bit microprocessor.
    (B) Explain the purpose of each register.
48. Which internal microprocessor register is used most frequently to:
    (A) Store an answer
    (B) Indicate an overflow
    (C) Count the times a process is repeated
    (D) Hold the address of a special place in memory
49. State the major differences between the 6502 and 6800 μP.
50. (A) Sketch the architecture of the 8085 μP. (B) State the purpose of each register.

51. (A) Sketch the architecture of the Z80 μP. (B) State the purpose of each register.
52. (A) Discuss the major differences between the 8085 and the Z80 microprocessors. (B) What are the major commonalities?

Section 10–7

53. Describe the main difference between an 8-bit microprocessor and a 16-bit microprocessor.
54. Explain how multiplexing is used in order to keep the number of pins needed by the microprocessor chip to a minimum.
55. (A) Describe the 8088 μP. (B) Sketch its logic diagram.
56. Sketch the architecture of the 8088 μP. Describe the purpose of at least three of its general purpose registers.
57. State the purpose of a segmentation register. Briefly describe how the process of segmentation works.
58. Compare the 8088 and the 8086 μP.
59. Briefly describe the major differences between a 16-bit μP and a 32-bit μP.
60. (A) Sketch the architecture of the MC6820 μP. (B) Describe the purpose of at least three of its internal registers.

Applications

61. Sketch the status register of the 6502 μP and explain the purpose of each of the flags in this register.
62. Convert the following assembly language program into machine code starting at location 0000H:

Assembly Program:

```
LDA    #0 ;
STA    $35
INC    $35
LDA    $35
STA    $36
BRK
```

63. Convert the following machine code into an assembly language program:

Machine Code:

Memory Location [Hex]	Memory Contents [Hex]	Instruction [Mnemonic]
0000	A9	
0001	00	
0002	85	
0003	40	
0004	00	

64. State the various addressing modes that are used in the program of problem 62.
65. State the various addressing modes that are used in the program of problem 63.
66. Write an assembly language program for the 6502 that would subtract two numbers.
67. Write an assembly language program for the 6502 that sets all of the memory locations from 40 to and including 50 all to zeros. Try to do this with a minimum of code.
68. State how you would write the previous program using relative addressing.
69. State how you would write the previous program using indexed addressing.
70. State how you would write the previous program using preindexed addressing.

Electronic Applications

71. (A) Explain the operation of CAT.
    (B) What are the minimum system requirements?
72. Describe the concept of a parallel system for decreasing the likelihood of system failure.
73. (A) What is a GPIB? (B) State its purpose.
74. Describe the concept of a teachable troubleshooter.
75. (A) Explain the use of a diagnostic program for troubleshooting a computer system.
    (B) What is the major disadvantage of the use of a diagnostic program?
76. (A) Explain the operation of a logic analyzer.
    (B) What is the advantage of such an instrument?
77. (A) Describe the basic idea of a signature analyzer.
    (B) What is its advantage over that of the logic analyzer?
    (C) What is its disadvantage?
78. (A) Describe the operation of a logic clip.
    (B) What is its main advantage?
    (C) Its disadvantage?
79. (A) Explain the operation of a current probe.
    (B) What other troubleshooting instrument is usually used with it?

Analysis and Design

80. Design a PLD that will indicate an overflow condition for a 3-bit binary adder.
81. (A) Using the 6502 μP, design a system that would transfer information from an 8-bit parallel input port to read/write memory.
    (B) Draw the resulting memory map.
82. Write an assembly language program to accomplish the task in problem 81.
83. Using the 6502 μP, design a zero page system (high order byte = 00H) that would transfer information from a memory location selected by a external hex keypad to an 8-bit parallel output port. Draw the resulting memory map.
84. Write an assembly language program to accomplish the task in problem 83.

GLOSSARY

Accumulator. A register inside a microprocessor. It is usually used for the intermediate storage of a word or the copying of a word to or from memory or another device external to the microprocessor. It is also used to temporarily store the result of an arithmetic, logical, or transfer process.
ACIA. Acronym for asynchronous communications interface adapter. A UART device.
Acoustic Coupler. A device that converts electrical information into sound and sound information into electrical signals.
Active Logic Level. The logic state (HIGH or LOW) used to activate a logic or other device.
Address. A label, name, or number identifying a location or unit where information is stored.
Address Bus. The group of conductors from a microprocessor that carries the information for the activation of a specific memory or device location.
Address Multiplexing. The process of splitting the complete address into two or more parts so that fewer address lines are required.
Address Register. A register inside the microprocessor that stores the binary number that represents the location (address) of data being processed by the microprocessor.
Addressing Modes. The method used by the microprocessor for transferring information.
Alphanumeric Codes. Codes that can represent letters of the alphabet, numerals, and other symbols such as punctuation marks and mathematical symbols.
Alterable Memory. Any medium capable of having the storage of digital information changed.
ALU. Arithmetic logic unit. A hardware device that performs arithmetic and logic functions.
American Standard Code for Information Interchange (ASCII). Pronounced "Askee." Standard data transmission code used to achieve compatibility between data devices. Has a total of 128 code combinations.
Amplitude Modulation. A process of having information change the strength (amplitude) of the carrier.
Analog. In the context of this book, refers to devices that have a range of possible conditions.
Analog-to-Digital (A/D) Converter. An electronic device that converts analog information into digital information.
Assembler. A computer program that operates on the symbolic input of instructions and translates them to bit patterns that are understood by the microprocessor.
Assembly Language. Programming instructions using mnemonics.
Astable Multivibrator. Sometimes called the "free-running" multivibrator—produces its own output logic waveform by constantly switching back and forth between HIGH and LOW.
Asynchronous Counter. A counter that does not have all of its flip-flop synchronized directly by the clock.
Axes. The lines of a graph used to indicate the magnitude of the variables represented by the graph. A two-dimensional graph uses two mutually perpendicular lines as the axes. The vertical line is referred to as the *Y*-axis and the horizontal as the *X*-axis.

Base. Number of characters used in each position of a number system. Sometimes called the radix.
Base Subscript. For identification of the base of the number. Unless indicated otherwise, the base of a number is assumed to be 10. Base subscript notation of a base 10 number, for example, is 123_{10}, and for a base 2 number 1010_2.
Baud Rate. The rate at which serial data bits are transferred.
Bilateral Switch. A switch that can be operated in any direction. In reference to CMOS, a solid-state device that electrically behaves as such a switch.
Binary Counter. A logic device capable of sequentially incrementing or decrementing a binary value.
Binary Numbers. A base 2 number system consisting of two symbols (1 and 0) to represent all numerical values. This system is useful for giving numerical value to the two-state (ON/OFF) condition of digital systems.
Bipolar Memories. Electrical memories manufactured from bipolar transistors.
Bit. Abbreviation for binary digit—a unit of data in binary notation. Each of the 1's and 0's used in binary is called a bit.
Black Box. A term used in electronics to denote a

complex circuit where the circuit details are not necessary for input and output analysis. The importance is placed on what the circuit does as a part of a total system.
Boolean Addition. Boolean representation of the OR function.
Boolean Algebra. A two-value algebra similar in form to ordinary algebra.
Boolean Multiplication. Boolean representation of the AND function.
Bubble Memory. Nonvolatile memory where data are represented by the presence or absence of magnetized areas (called bubbles) formed on a thin piece of garnet.
Bubbled (Input/Output). A method of showing a logic inversion.
Burst Refresh. A method of refreshing DRAM where each cell row is sequentially refreshed.
Bus. A group of wires in a digital system, treated as a unit.
Byte. Generally considered to be an 8-bit word.

CCD Memory. Charged-coupled device memory where charges are stored between tiny electrical plates fabricated in a high-density integrated circuit.
Chip (IC). An integrated circuit where complete circuits are manufactured on a single silicon substrate and housed in a single package.
Clock. A series of digital pulses used in the synchronization of digital devices.
CMOS. Complementary metal-oxide semiconductor.
Coercivity. A measure of the magnetic field that must be applied in order to cause the magnetization in the medium to reverse direction.
Compact Disk (CD). An optically reflective disk that contains digital information stored as surface changes that change the reflection of a laser beam focused on the rotating disk.
Comparator. A circuit that compares an input voltage to a reference and produces an output voltage indicating the comparison.
Complement. A reversion of the digital state. The complement of "OFF" is "ON," and the complement of "ON" is "OFF." Thus, the complement of 1 is 0, and the complement of 0 is 1.
Computer. A programmable processing machine consisting of an input, output, processing unit, and memory that is capable of accepting information, applying prescribed processes to the information, and supplying the results of that process.
Control Bus. A group of conductors from the microprocessor that carries information for or about the operation of the microprocessor.
Conventional Current. Current flow where the direction is taken to be from the positive voltage potential to the negative. This is opposite to the direction of electron flow.
Conversion Time. The amount of time it takes to convert an analog voltage reading into a digital value.
Current Sink. A device is said to act as a current sink when it accepts current flow into it from an external source.
Current Source. A device is said to act as a current source when it delivers current flow to an external connection.

Data. A general term that identifies basic elements of information that can be processed or produced by a digital system.
Data Bus. The group of conductors from a microprocessor that will carry the data transferred to and from memory or a device.
Data Lock-Out Flip-Flops. A master-slave flip-flop where the input is sensitive to changes only during the clock transition.
Data Selector. A logic network that is capable of selecting one set of data from a group of data and transmitting it along a single output line.
Debouncer. A logic circuit used to remove unwanted changes in logic levels usually caused by mechanical switch contacts.
Decoder. A logic network that will convert a specific bit pattern on its input to a specific output level.
Decoding Glitch. An undesirable logic level of a short duration that is caused by time delays inherent in logic circuits.
De Morgan's First Theorem. The Boolean relationship showing that a NOR gate is logically equivalent to an AND gate with inverted inputs.
De Morgan's Second Theorem. A NAND gate is logically the same as an OR gate with inverted inputs.
Demultiplexer. A digital network that converts the logic of one input line to one of several output lines, as determined by the binary value of its select lines.
Difference Amplifier. An amplifier whose output voltage is proportional to the difference of two input voltages.
Digital. In the context of this book, refers to devices that have discrete conditions.
Digital Integrated Circuit. An electronic circuit

where all of the components are produced on a single substance (usually silicon) and the circuit responds to a two-level (TRUE/FALSE or HIGH/LOW) electrical signal.

Digital Margin Detector. A circuit that produces a two-state output that depends upon a prescribed range of voltage inputs.

Digital-To-Analog (D/A) Converter. An electronic device that converts digital information into analog information.

Digital Voltmeter. An instrument used to measure voltage with a digital readout.

Diode. An electrical device that allows current flow in only one direction.

Direct Addressing. The process of transferring data between the microprocessor and the external system by designating the address for the location of the data.

Direct Memory Access (DMA). An I/O capable of transferring data to and from memory without intervention of the microprocessor.

Direct Method. A method of representing signed binary numbers where the MSB represents the sign of the binary number. If it is 1, the number is negative; if it is 0, the number is positive.

Distributed Refresh. A method of refreshing DRAM where cell rows are refreshed between read and write operations.

DMA Controller. A device used to control the I/O transfer of data without microprocessor intervention.

Don't Cares. A logic condition whose outcome has no effect on the rest of the system.

Dynamic Input Indicator. The triangular symbol used on the clock input of a flip-flop to indicate that the device is synchronized by a logic transition of the clock. The edge-triggered flip-flop allows for greater timing accuracy over that of the latch.

Dynamic RAM. Volatile, randomly accessible, electrical memory that tends to lose stored information over a period of time and requires "refreshing" in order to retain its bit pattern.

Dynamic RAM Controller. A special purpose integrated circuit used to coordinate the interfacing between a microprocessor and a dynamic RAM. This prevents a conflict between the addressing requirements of the RAM and that of the required RAM refreshing.

ECL. Emitter-coupled logic where the basic circuit is essentially a differential amplifier. Sometimes called current mode logic.

EEPROM. Nonvolatile, electrically alterable, randomly accessible, electrical memory.

8421 Code. The place value of each bit corresponds to the place value of a binary number.

$$2^3 2^2 2^1 2^0 = 8421$$

Electrical Noise. Undesirable and unpredictable electrical signals caused by natural phenomena and manufactured equipment.

Empty Term. Part of a Boolean expression used in the evaluation of a larger Boolean expression where the contents of the smaller expression have not yet been evaluated.

Encoder. A logic network that converts information into a form of code.

EPROM. A nonvolatile, alterable, randomly accessible electrical memory.

Erasable Memory. Digital information that can be deleted without destroying the medium containing the digital information.

Erasable Programmable Read-Only Memory (EPROM). Erasable programmable ROM. A ROM that can be programmed in the field by the user and can have its bit pattern changed again.

Even Parity. Creating a word with an extra bit so that the number of 1's in the word adds up to an even number.

Exclusive NOR Gate. A two-input logic gate where the output is TRUE only if the logic inputs are the same (both TRUE or both FALSE). Otherwise the gate output is FALSE.

Exclusive OR Gate. A two-input gate in which the output is TRUE when only one input is TRUE, and FALSE when both inputs are TRUE or both inputs are FALSE.

Fan-Out. The maximum number of logic devices that can be controlled by the output of a single logic device.

First-In–First-Out (FIFO) Memory. Refers to a special arrangement of digital memory where the first bit of data written into the memory is the first bit of data read from the memory.

Flash Converter. An analog-to-digital converter that uses differential amplifiers and digital logic to produce a binary output that is equivalent to the magnitude of the analog input.

Flip-Flop. A bistable digital device that is synchronized by an external source.

Frequency Counter. An instrument used to display the frequency of a periodic waveform.

Frequency Modulation. A process of having information change the frequency of the carrier.
Full Adder. Logic network with three inputs, the sum of which is represented by two outputs, the sum and the carry.
Full Address Decoding. The results when all address lines are decoded.
Full-Duplex Mode. A mode of transmission where two systems may transmit and receive data at the same time.

Gated *D* Latch. A latch with two inputs called the ENABLE and *D* inputs. The logic level at the *D* input will determine the state of this latch when the ENABLE line is active.
Gated Latch. A SET-RESET latch with a third input called the ENABLE input. The ENABLE input must be active before the SET or RESET inputs have any affect on the latch.
Graph. A picture showing the relationship between variables. In electronics a graph can show the relationship between an electrical variable and time.

Half Adder. Logic network with two inputs, the sum of which is represented by two outputs, the sum and the carry.
Half-Duplex Mode. A mode of transmission where two-way communication can take place between two systems, but not at the same time.
Hard Disk. A hard-disk storage system of one or more nonremovable disks protected in a permanently sealed case.
Hardware. The tangible part of a logic system.
Hexadecimal Number System. A number system with a base of 16. Thus, 16 symbols are used: 0–9, A, B, C, D, E, and F, which are sometimes referred to as hex numbers.
High-Order Address. The most significant 8 bits of a 16-bit address.
High-Order Nibble. The first 4 bits, including the MSB, of an 8-bit word.

IC Family. A group of integrated circuits manufactured using specific materials and processes.
Immediate Addressing. The process of transferring data located in the memory location immediately following the instruction into an internal register of the microprocessor.
Indeterminate Range. A range of voltage values for a digital IC that indicates the circuit is not functioning according to the manufacturer's specifications.
Index Register. An internal microprocessor register that may be used for process counting or modifying the address in the program counter.
Input/Output (I/O). The process of transferring information between a digital system and devices external to the digital system.
Input Profile. The specified range of logic HIGH and logic LOW input voltages for a digital circuit.
Instruction. A bit pattern that causes the microprocessor to perform a predictable process.
Instruction Set. A listing of the bit patterns that will cause the microprocessor to perform a specific process.
Instrumentation. The use of equipment to measure electrical and logical quantities for the purpose of keeping these quantities within prescribed limits to ensure proper system operation.
Interface. The connection of a device to the digital system.
Interrupt. The process of stopping the microprocessor from its current operation and causing it to take its next instruction from a different memory location.
I/O Ports. A reference to any device that reads or writes data to anything external to the microprocessor, its associated memory and control system.
Isolated I/O. Input/output treated differently from regular memory locations.

JK Flip-Flop. A flip-flop with four useful input conditions: no change, set, reset and toggle.

Karnaugh Map. A graphical method of simplifying logic networks.
Kerr Effect. The effect on the polarization of a light beam when it is reflected off of a magnetic medium.
Keypad. A small keyboard (or section of a keyboard) containing a smaller number of keys than a typewriter keyboard. It serves as one of the simplest input devices to a computer.

Label. In assembly language programming. The first field of the program that can be used to represent an address.
Language Level. How close the programming code is to the actual bit pattern used by the microprocessor. The closer the programming code is to that of the microprocessor, the lower the level of the programming language level.

Laser. Light amplification by stimulated emission of radiation.
Laser Diode. A solid-state device capable of emitting laser light when electrical energy is applied to it.
Latch. A logic circuit that maintains a given logic condition until changed by an external source.
Leading Edge. For a positive going pulse, it is the LOW-to-HIGH transition. For a negative going pulse, it is the HIGH-to-LOW transition.
Least Significant Bit (LSB). In a binary number, it is the significant bit contributing the smallest quantity of the value of the number. For example, in 101_2, the rightmost 1 is the LSB.
Light Emitting Diode (LED). A solid-state device that emits light when the correct amount and polarity of electrical potential are applied.
Load. A device or devices that consume electrical energy from a given source.
Logic Diagram. The graphical representation of one or more logic functions and their interconnections.
Logic Gate. An electrical circuit that simulates a logic function.
Logic Levels. The sectioning of a logic circuit for analysis purposes, where the first level is the output gate, the second level consists of the gates feeding the output gate, the third level is the logic gates feeding the second level, and so on.
Logic Symbol. A drawing that represents a logic function.
Look-up Table. A stored bit pattern where the address represents a value and the output represents a function of that value.
Low-Order Address. The least significant 8 bits of a 16-bit address.
Low-Order Nibble. The last 4 bits, including the LSB, of an 8-bit word.

Machine Cycle. The length of time required to complete a process.
Machine Language. The actual bit pattern used by the microprocessor. Writing a program using this bit pattern is called machine language programming.
Magnetic Memory. Digital memory that uses the patterns of a magnetic field to represent stored digital information.
Magnitude Comparator. A circuit used to compare the magnitude of two quantities in order to determine their relationship.
Mask-Programmable ROM. A ROM that is programmed by the manufacturer.
Mask ROM. A ROM that has its bit pattern put into it during the manufacturing process by the use of a bit mask that determines the bit pattern.
Memory Map. A listing showing the contents of all memory locations.
Memory-Mapped I/O. Input/output treated as a regular memory location.
Memory Size. An indication of how bit patterns are accessed in a two-dimensional matrix. The notation is $A \times D$, where A gives the number of addresses (rows) and D gives the number of bits at each address (column or word size).
MFM. Modified frequency modulation is a method of storing digital information on a disk. This method produces twice the density of disk information as with standard frequency modulation techniques.
Microcode. The instructions built into a microprocessor that determine a specific selectable process.
Microcomputer. A computer that uses a microprocessor as its processing unit.
Microinstruction. A single instruction in the microcode of a microprocessor.
Microprocessor. A single integrated circuit chip containing a fixed set of processes such as arithmetic, logic, and bit manipulation.
Microprogram. The permanent instruction set inside the microprocessor used to perform a specific process.
Mnemonic Code. A technique to assist the human memory. A mnemonic code resembles the word that describes the process that will be performed by the microprocessor.
MOD Number. Of a counter, refers to the number of states a counter will have under given conditions.
Modem. Stands for *mo*dulator/*dem*odulator. A modulator converts digital information into transmission signals. A demodulator converts the signal to the original digital form upon reception.
Modulate. To change or modify. In communications the act of having information change a measurable characteristic of an electromagnetic wave (called the *carrier*).
Modulus. When applied to counters refers to the maximum number of states the counter can assume.
Monostable Multivibrator. A multivibrator that will stay in one condition for a predetermined amount of time and then automatically go back to its prior condition.

MOS Memories. Electrical memories manufactured from field-effect transistors.
Most Significant Bit (MSB). In a binary number, it is the significant bit contributing the largest quantity of the value of the number. For example, in 101_2, the leftmost 1 is the MSB.
Multiplexer. A logic network that is capable of taking several different inputs and transmitting the result on one output.
Multiplexing. Using two or more different signals through the same electrical element.
Multivibrator. A two-state output circuit capable of acting as an oscillator by generating its own signal or maintaining one stable state or maintaining one of two stable states.
Music Synthesizers. A device capable of creating various sounds selected by the user.

NAND Gate. A logic circuit formed by inverting the output of an AND gate. The output is low only when all inputs are high.
Nibble. Generally considered to be a 4-bit word.
NMOS. IC fabrication using N-channel MOSFET's. This results in high package densities. Since free electrons are the current carriers, this results in greater speed and lower power consumption.
Noise Margin. The voltage difference between the input and output profiles of a logic circuit.
Nonperiodic. A waveform that is not periodic.
Nonvolatile Memory. Mediums that do not require an external source of energy to maintain the stored digital information.
NOR Gate. A logic gate formed by inverting the output of an OR gate. The output is HIGH when all inputs are LOW.
NOT Function. Complement or inversion of a variable. NOT TRUE is FALSE. NOT FALSE is TRUE.

Octet Simplification. Treating a group of eight horizontally and/or vertically adjacent cells as a single term.
Odd Parity. Creating a word with an extra bit so that the number of 1's in the word adds up to an odd number.
One's Complement Notation. A number in binary derived from another binary number resulting from the complement of each bit.
Open-Collector Logic. A logic device whose output requires the addition of a pull-up resistor for proper operation.
Output Profile. The specific range of logic HIGH and logic LOW output voltages for a logic device.
Overflow. A condition indicating that the result of an arithmetic operation is not correct.

Pair Simplification. In K-map, looking for cells that are either vertically or horizontally adjacent for the purpose of eliminating one variable from the final Boolean expression represented by the K-map.
Parity. Relates to the maintenance of a count by keeping an odd or even number of binary 1's in a computer word in order to be able to check that word's correctness.
Parity Bit. An extra bit added to a word so that an accidental change of a single bit in the word can be detected.
Partial Address Decoding. The results achieved when all address lines are not decoded.
Periodic. A waveform that continually repeats itself.
Peripheral Interface Adapter (PIA). A device used to perform parallel I/O.
Permanent Memory. Digital information that cannot be changed without destroying the medium containing the digital information.
Phase Modulation. A process of having information change the phase of the carrier.
Phonemes. The basic sounds that make up a language. English has about 60 phonemes.
Photoresistor. A device whose resistance value is sensitive to the amount of light and is intended to be used for this affect.
Place Value. The representation of quantities by a positional value system. For example, the symbol 3 represents different values, depending upon its placement with respect to the decimal point: 3.0, 30.0, 0.3.
Port. The connecting medium that allows communications between the digital system and an external device.
Positional Number Systems. A systematic method for representing values, where any value may be represented as a sequence of multiples of successive powers of a given base. For example, in decimal notation:

$$3241 = 3 \times 10^3 + 2 \times 10^2 + 4 \times 10^1 + 1 \times 10^0$$

Powers of 10. The power of 10 is expressed as an exponent of the base 10. The exponent indicates the number of decimal places to the left or right of the decimal place.

Priority Encoder. A special type of encoder that senses if two or more inputs are active and then gives an output corresponding to the highest value of the input.
Product Line. In PLD notation, a line that represents multiple inputs to an AND gate.
Product of Sums. A Boolean expression that consists of the ANDing of terms that are ORed.
Programmable Counter. A counter that has its MOD number controlled by a changeable bit pattern.
Programmable Logic Array (PLA). A two-level AND/OR POS logic network, where both the AND connections and the OR connections can be programmed.
Programmable Logic Gate (PLG). A logic network that can replicate any of the fundamental gates (AND, OR, NAND, NOR, XOR) with the setting of a bit pattern that determines the type of gate it will replicate.
Programmable Read Only Memory (PROM). A programmable ROM that can be programmed in the field by the user.
PROM. Nonvolatile, randomly accessible, electrical memory that may be altered only once by the user.
Pull-up Resistor. In digital systems, a resistor connected to a voltage source in such a manner as to supply the voltage potential and also limit any resulting current flow.
Pulse. A momentary change in an electrical characteristic of a circuit—usually voltage or current.
Pulse-Triggered Master-Slave Flip-Flop. A flip-flop whose inputs are active at one level of the clock and whose outputs will reflect the input condition only at a different level of the clock.

Quad Simplification. Treating a group of four vertically or horizontally adjacent cells of a K-map as a single Boolean term.

RAM. Literally means randomly accessible memory but is used to mean volatile, randomly accessible, electrical memory.
Random Access. The amount of time it takes to begin to copy to or from a place in memory is the same.
Read-Only Memory (ROM). Memory that contains a set of permanent instructions that cannot be easily changed.
Read Operation. The process of the microprocessor getting information from memory.
Read-Write Memory. Memory that can have its instructions easily changed.
Recording Format. The method used to magnetically store data.
Redundant Group. In K-mapping, a group of cells that are already used by other groups in logic network simplification.
Refresh. The process of maintaining the electrical cell charge in a dynamic RAM.
Register. A storage place for one or more computer words.
Register Architecture. As pertaining to a microprocessor, a diagram of the internal registers of interest to the programmer.
Resolution. The ability of an A/D converter to accurately convert small changes in an analog signal to binary data.
Ring Counter. A form of digital counter where a given bit pattern continually circulates through the counter.
Ripple Counter. A counter where each flip-flop, with the exception of the first, is clocked by the preceding flip-flop.
Robotics. The study of systems that can exercise control and judgment without human intervention.
ROM. Read-only memory—traditionally meant memory that could only have its stored bit pattern copied.

Saturation. A circuit condition where an increase of the input signal no longer produces a change in the output.
Schmitt Trigger. A circuit that changes state abruptly when the input signal crosses a specified DC triggering level.
Segmentation Registers. Internal storage for the address of a section of memory called a segment.
Sequential Access. The amount of time it takes to begin to copy to or from a place in memory depends upon the physical location in memory.
Shift Register. A type of register where the contents may be moved in a serial fashion within the register.
Simplex Mode. A mode of transmission where data are sent in only one direction from transmitter to receiver.
Software. The nontangible part of a logic system that determines the process performed by the hardware.
Square Wave. A periodic waveform where the ON time is equal to the OFF time.
Stack. A location in memory usually different from the location of the main program. The address of this

location is kept in an internal microprocessor register called the stack pointer.

Stack Pointer. An internal microprocessor register that contains the address for specially stored data that is not a part of the data for the main program.

Stairstep A/D Converter. An analog-to-digital converter that uses a waveform with abrupt changes in its output, called *steps*. Each step represents an increase in the value of the analog signal and a corresponding increase in the binary value representing the analog input.

Stairstep Waveform. A waveform where the changes in amplitude are of equal increments, thus producing a silhouette similar to a flight of stairs.

State Machine. A logic circuit exhibiting a predictable sequence of events.

State Table. A list of the sequential states of a state machine.

Static RAM. Volatile, randomly accessible, electrical memory that requires no further processing as long as an external source of energy is supplied to maintain its programmed bit pattern.

Status Flags. A microprocessor register where each bit represents the results of some previous register process or the condition of the bit will have an affect on such a process.

Step Size. The smallest possible analog change on the output of a D/A converter.

Stored Program Concept. The concept used for programming digital systems where the instructions are included in memory with the data.

Subfamilies. A group of ICs from the same family with similar electrical characteristics.

Sum of Products. A Boolean expression that consists of the ORing of terms that are ANDed.

Symmetry. When used in digital, refers to the amount of pulse ON time versus the amount of pulse OFF time. A symmetrical pulse has the same amount of ON time as OFF time. A nonsymmetrical pulse does not have these equal times.

Sync Pulse. Abbreviation for synchronized. Pulses used to cause certain events to happen at a predetermined time and sequence.

Synchronous Counter. A counter where each flip-flop making up the counter is affected directly by the clock pulse.

System Comparison. A method of batch troubleshooting where the system under test is compared to an identical system called the standard.

Thermistor. A device whose resistance value is related to its temperature and is developed for this particular application.

Trailing Edge. For a positive going pulse, it is the HIGH-to-LOW transition. For a negative going pulse, it is the LOW-to-HIGH transition.

Trigger. A logic pulse of short duration used to start a digital process.

TRI-STATE. A logic output that has three states: HIGH, LOW or not connected (sometimes called the high-impedance state).

Troubleshooting. The act of locating and repairing in an electrical circuit, a fault that causes improper operation of the circuit.

Truth Table. A listing showing all possible input combinations of a digital circuit and the resultant output(s). Usually done in terms of any two-state symbolism such as 1's and 0's.

TTL. Transistor-transistor logic.

Two's Complement Notation. A value arrived at by taking the one's complement and adding 1 to the result.

UART. Acronym for universal asynchronous receiver-transmitter. Works with the microprocessor to send asynchronous data serially between microprocessor-based digital systems.

Underflow. In a counter, when a minimum count of zero is reached and the counter counts backwards, resetting itself to its maximum value count (all 1s).

Unit Load (UL). A gate input represents a unit load to a gate output within the same family.

VCO. Voltage controlled oscillator. An oscillator whose output frequency is determined by the value of an input voltage.

Voice Synthesizer. A system for replicating the human voice by creating waveforms of the required form that emulate human sounds.

Volatile Memory. Mediums that require an external source of energy to maintain the stored digital information.

Waveform. The graph of an electrical characteristic of a circuit. Usually the vertical axis represents voltage or current, and the horizontal axis is time.

Waveform Synthesis. The process of creating a desired waveform.

Word. Set of bit patterns that occupies a specific storage location and is treated by the digital system as a unit and moved around the system as such.

Worse-Case Conditions. Circuit parameters (voltage, current, etc.) are measured under the worst conditions of voltage and temperature.

Write Operation. The process of the microprocessor putting information into memory.

Zero-power RAM. Volatile, randomly accessible, static electrical memory with a built-in electrical energy source that allows bit patterns to be stored for long periods of time.

APPENDIX A
Logic Wheel

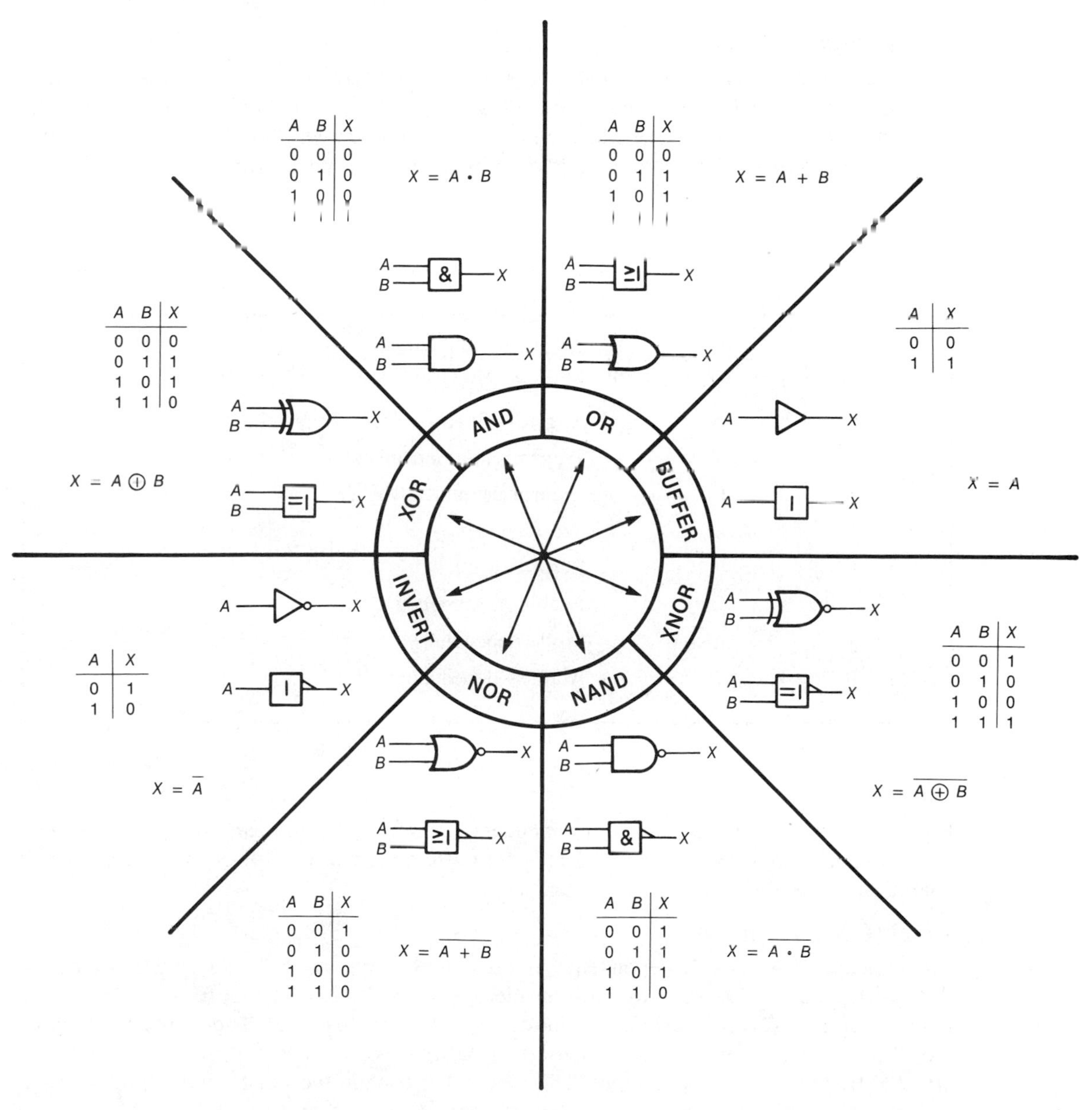

APPENDIX B
Dependency Notation

Discussion

A symbolism called *dependency notation* was developed by the International Electrotechnical Commission (IEC). The purpose of this new symbolism is to show the relationship of each *input* of a digital circuit to each *output* without showing the internal logic of the circuit. The types of dependency notation presented are listed in Table B-1.

Table B-1 DEPENDENCY TYPES

Symbol	Meaning
G	AND gate dependency
V	OR gate dependency
N	Negate—Exclusive OR dependency
X	Transmission dependency
C	Control dependency
S	Set dependency
R	Reset dependency
EN	Enable dependency
M	Mode dependency
A	Address dependency

Basic Idea

Figure B–1 shows the basic idea of dependency notation. As you can see from the figure the *common control block* is used to indicate those lines that control or indicate a control condition of the device.

AND/OR Dependency

The concept of AND/OR dependency is illustrated in Figure B–2. To give you an idea of how dependency notation works, consider the 16 × 4 RAM of Figure B–3. As shown in the figure, the C input designates control (by a clock in this case). The number following the C notation (C1 in this case) is used to indicate what other parts of the device are affected by the state of the C input. The inputs that contain the same number prefix (such

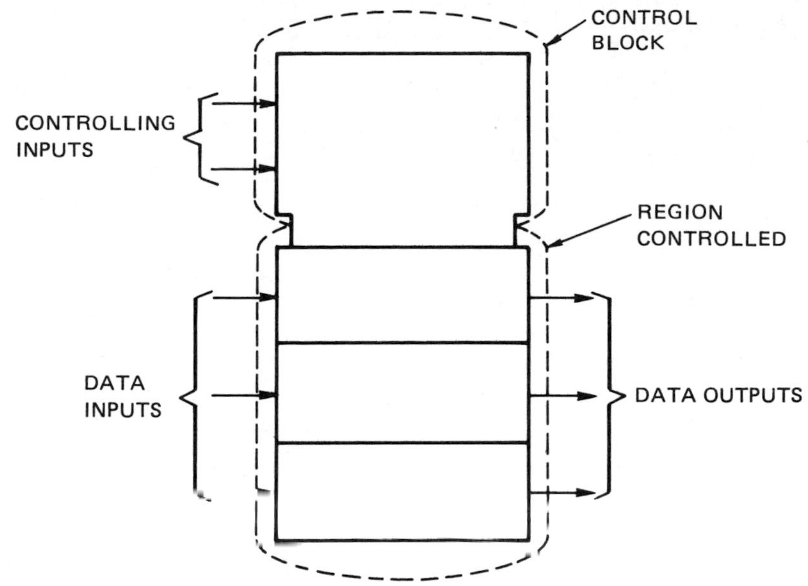

Figure B–1 Basic idea behind dependency notation.

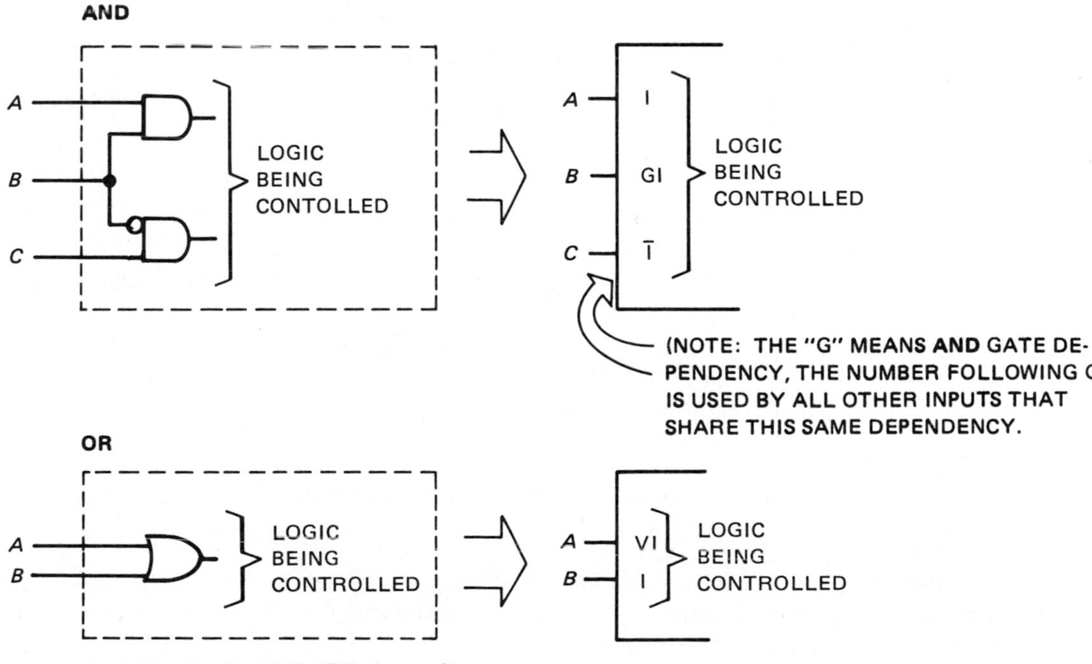

Figure B–2 AND/OR dependency.

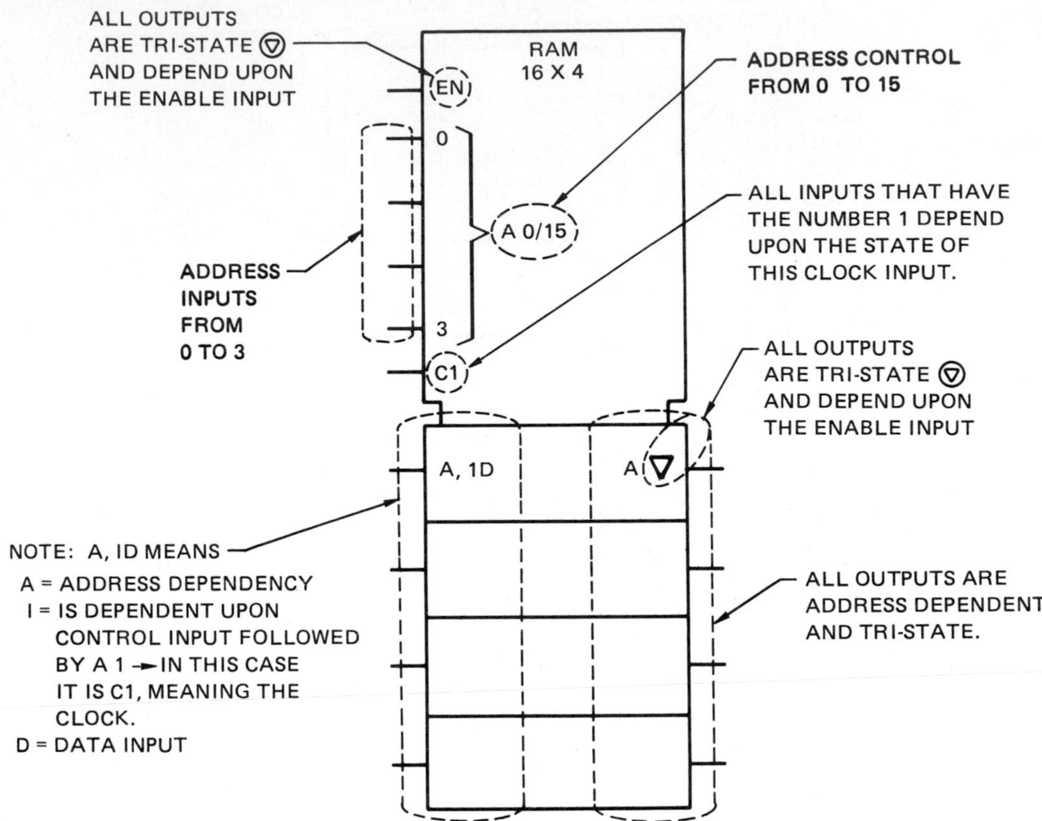
ALL OUTPUTS
ARE TRI-STATE
AND DEPEND UPON
THE ENABLE INPUT
RAM
16 X 4
EN
ADDRESS CONTROL
FROM 0 TO 15
0
A 0/15
ALL INPUTS THAT HAVE
THE NUMBER 1 DEPEND
UPON THE STATE OF
THIS CLOCK INPUT.
ADDRESS
INPUTS
FROM
0 TO 3
3
C1
ALL OUTPUTS
ARE TRI-STATE
AND DEPEND UPON
THE ENABLE INPUT
A, 1D
A
NOTE: A, ID MEANS
A = ADDRESS DEPENDENCY
I = IS DEPENDENT UPON
CONTROL INPUT FOLLOWED
BY A 1 → IN THIS CASE
IT IS C1, MEANING THE
CLOCK.
D = DATA INPUT
ALL OUTPUTS ARE
ADDRESS DEPENDENT
AND TRI-STATE.

Figure B–3 Dependency notation for a 16 × 4 RAM.

as 1A, 1D) are all controlled by the state of the C input (the D stands for data). The A input indicates address dependency. Address dependency is indicated on the input to the latches by the letter A. Note that only the first latch symbol is labeled. It is assumed that the others will be the same. The G input indicates AND dependency meaning that the device will be enabled only when both of these inputs are active.

More Examples

More examples of this type of dependency notation are illustrated in Figure B–4.

Conclusion

It is the idea of dependency notation to simplify the understanding of the relationship between the inputs and outputs of various integrated circuits. Some manufacturers have adopted this standard while others have yet to do so. Both this method and the more traditional one should be understood.

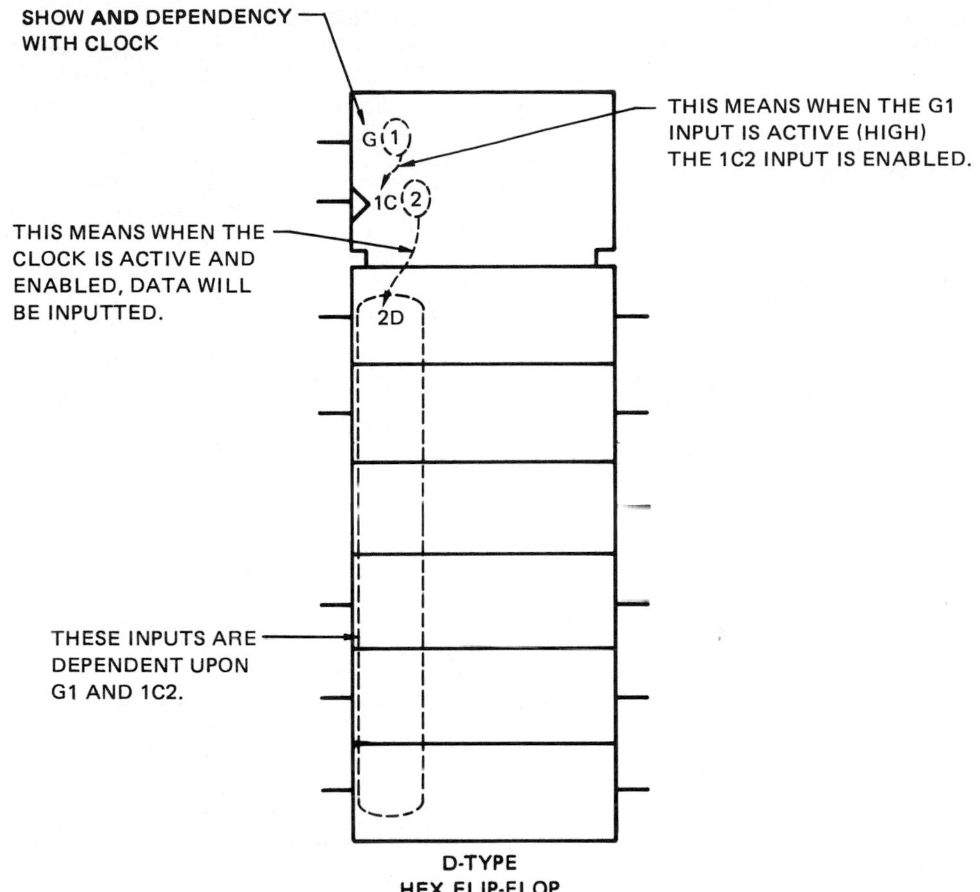
SHOW AND DEPENDENCY WITH CLOCK
THIS MEANS WHEN THE G1 INPUT IS ACTIVE (HIGH) THE 1C2 INPUT IS ENABLED.
G1
1C2
THIS MEANS WHEN THE CLOCK IS ACTIVE AND ENABLED, DATA WILL BE INPUTTED.
2D
THESE INPUTS ARE DEPENDENT UPON G1 AND 1C2.
D-TYPE HEX FLIP-FLOP

Figure B–4 More examples of dependency notation.

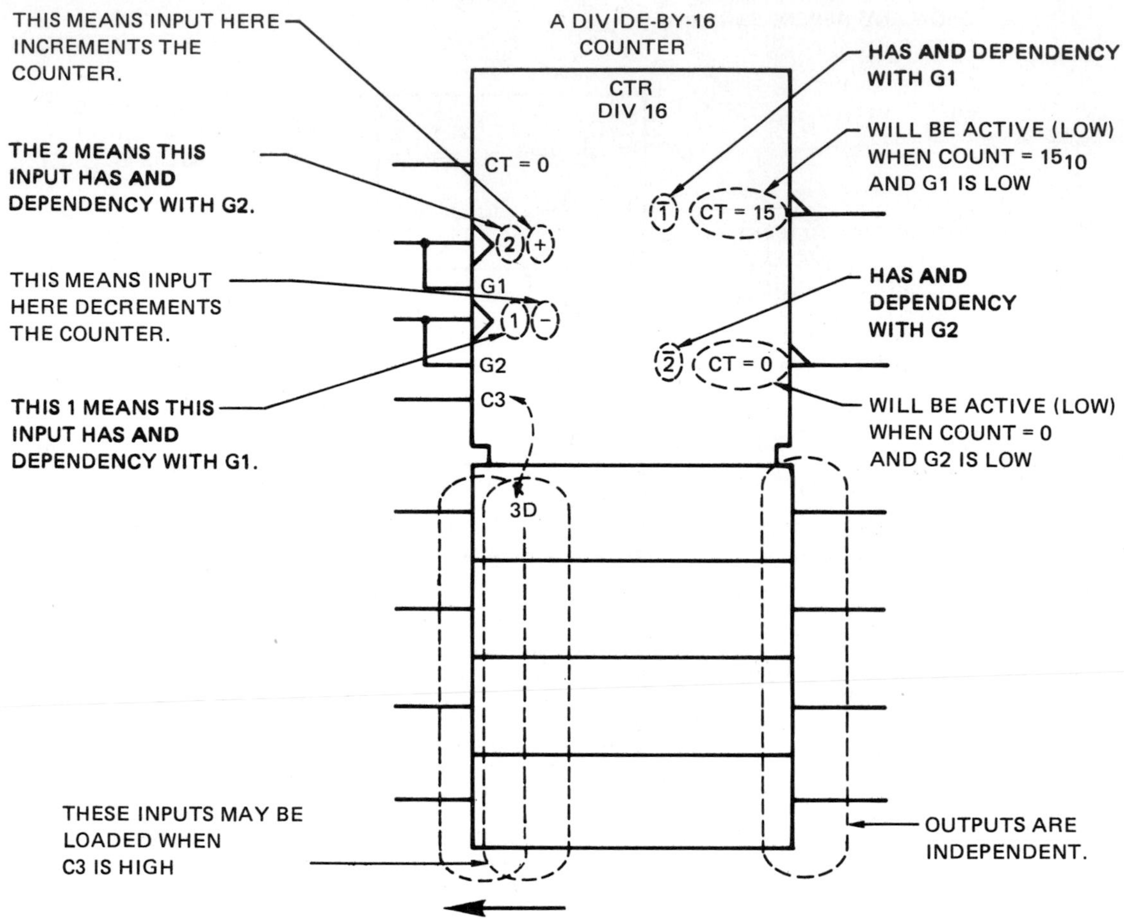
THIS MEANS INPUT HERE INCREMENTS THE COUNTER.
A DIVIDE-BY-16 COUNTER
HAS AND DEPENDENCY WITH G1
CTR DIV 16
WILL BE ACTIVE (LOW) WHEN COUNT = 15₁₀ AND G1 IS LOW
THE 2 MEANS THIS INPUT HAS AND DEPENDENCY WITH G2.
CT = 0
1
CT = 15
2
+
G1
THIS MEANS INPUT HERE DECREMENTS THE COUNTER.
1
−
HAS AND DEPENDENCY WITH G2
G2
2
CT = 0
THIS 1 MEANS THIS INPUT HAS AND DEPENDENCY WITH G1.
C3
WILL BE ACTIVE (LOW) WHEN COUNT = 0 AND G2 IS LOW
3D
THESE INPUTS MAY BE LOADED WHEN C3 IS HIGH
OUTPUTS ARE INDEPENDENT.

Figure B–4 Continued

APPENDIX C
How to Use the Student Disks

Introduction

Two student disks are available to support the text. They are $5^1/_4''$ floppy disks that will work with an IBM PC or true compatible with at least 360K of memory. The computer you use does not need graphics capability. A color monitor is desirable but not necessary.

Disk Purpose

The disks contain 30 different programs divided into 3 sets for each of the 10 chapters in the book. The two disks are identified as Disk A and Disk B. Disk A contains the programs for chapters 1–5 and Disk B contains the programs for chapters 6–10. Each chapter set consists of

- Chapter problems consisting of multiple-choice questions
- Test feature with interactive demonstration and test modes
- Troubleshooting simulation with demonstration and test modes

The purpose of the disks is to provide another learning/teaching environment for you and your instructor. The programs on the disks may be used as homework assignments, self-tests, examinations or any other self-paced learning activity.

Using Your Disk Package

When you first use either disk it will ask for your name, student number, class section number and date. When using Disk B, the programs are loaded into the computer. Disk B is then replaced by Disk A. All of this information will be permanently recorded on Disk A and cannot be changed. Thereafter, each time you use your disks your name will appear on the screen. Additionally, Disk A keeps a permanent record of the following information:

- Number of times you used your disk package.
- Amount of time you used your disk package.
- Number of times you attempted a testing session.
- Number of times you completed a testing session.
- Percent score of the last testing session completed.
- Running average score for each test.
- Date you first used your disk package.
- Date you last used your disk package.

Your instructor may then choose to periodically collect your disk and extract this information for the purpose of helping you in your course progress. Only Disk A needs to be collected since it contains all the information from both disks. Your instructor has a special disk that analyzes all of the information on your disk. This may be done to generate an individual student assessment or a class analysis to see your standing in relation to others in the class. When you use your disks, you are not working alone. By being able to analyze your progress with the material on your disk, your instructor can give you specific recommendations designed to help your learning process.

What is on Your Disks

After you enter your name, student number, section number and date on your disk, you will see a menu consisting of the number and name of each chapter in your DIGITAL textbook. What you will see is shown in Figure C–1. Chapters 1–5 are on Disk A and chapters 6–10 are on Disk B.

Each chapter contains three separate programs. That is, when a chapter is selected, a new menu screen will appear. This screen is shown in Figure C–2. This is a sample of the program set for Chapter 3. Like all the other chapter selections, it contains three separate programs, each of which contains material presented in that chapter of your DIGITAL book. For Chapter 3, there are the CHAPTER PROBLEMS which consists of multiple-choice questions keyed to each section of the book. There is also the TEST FEATURE which for this chapter allows you to analyze the action of logic gates. The last selection is the TROUBLESHOOTING SIMULATION which, for this chapter, gives you the opportunity to develop your troubleshooting skills in the analysis of typical logic gate problems.

Chapter Problems

The screen for a typical test question is shown in Figure C–3. As shown in the figure, the number of questions in the test and the number of questions completed are displayed

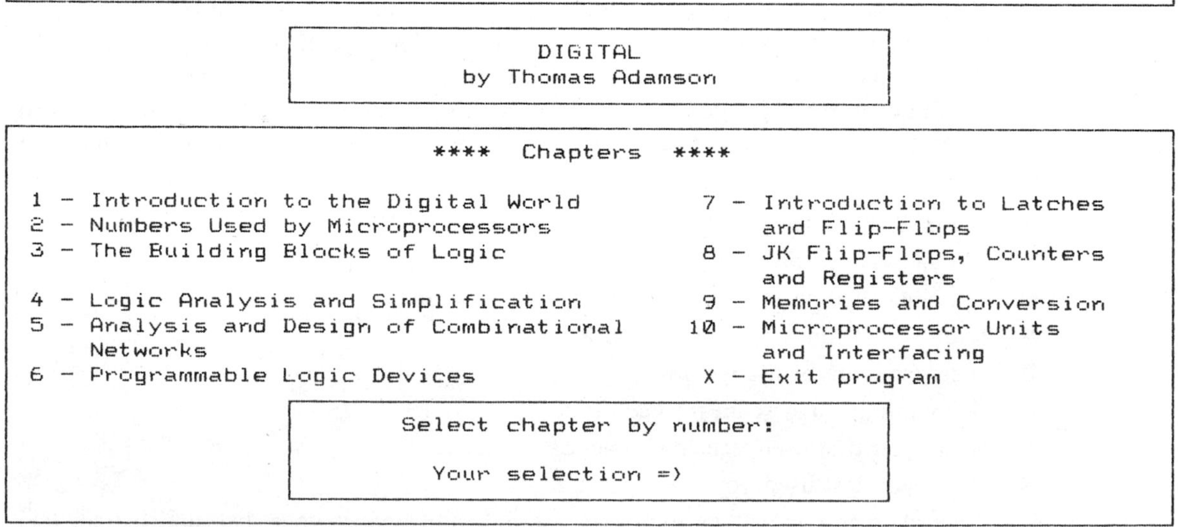

Figure C–1 Student disk menu.

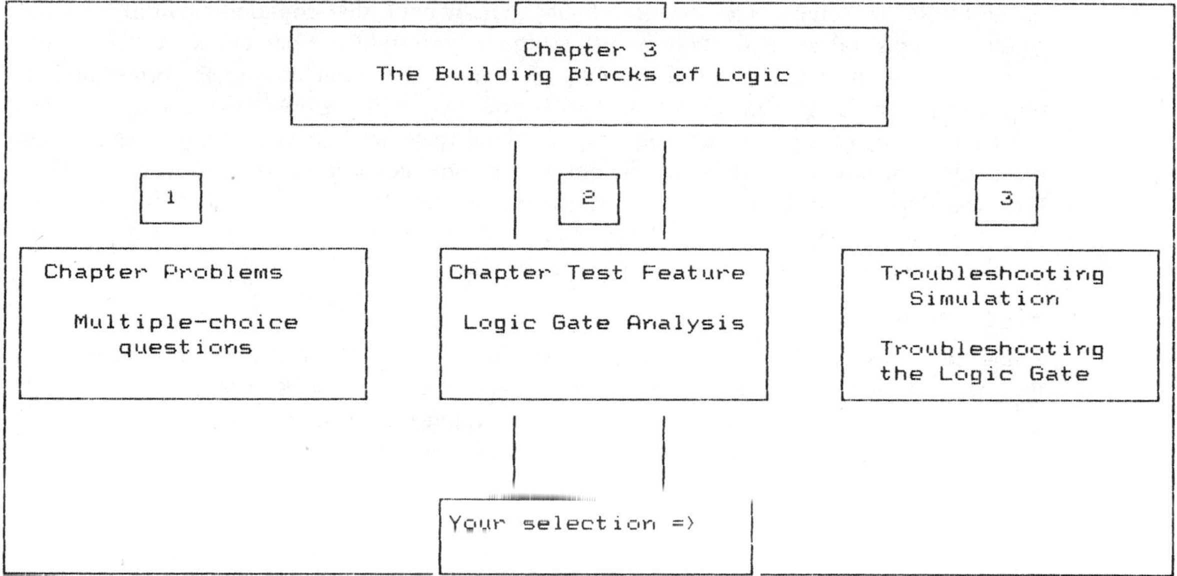

Figure C–2 Chapter menu.

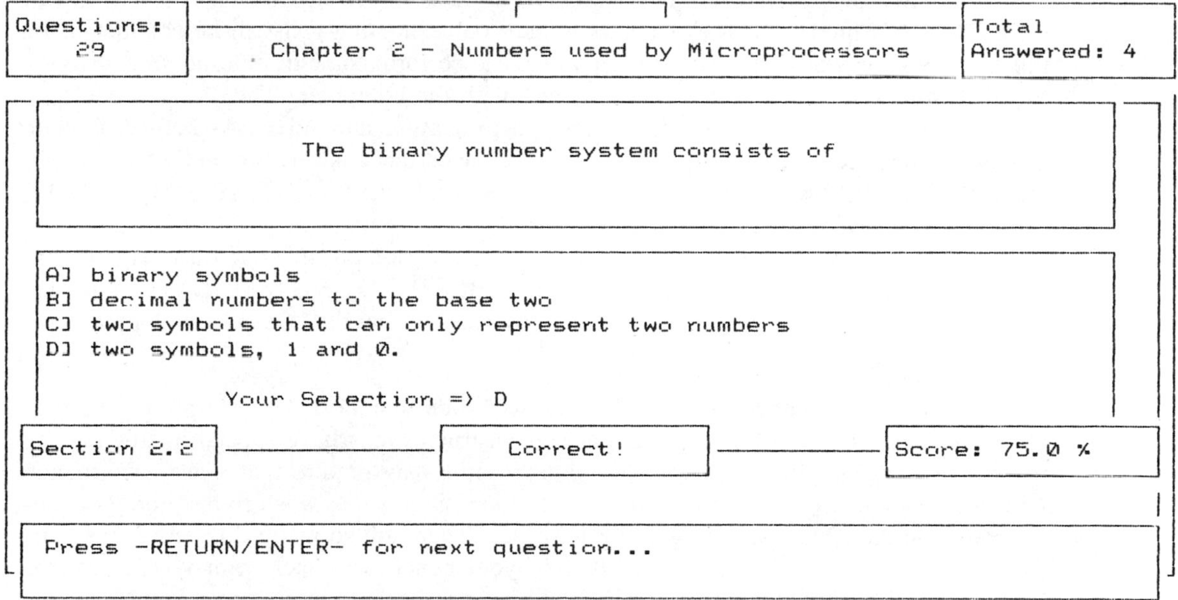

Figure C–3 Typical test questions from multiple-choice test.

at the top of the screen. The section number in your book that contains material relevant to the question appears at the bottom left. At the bottom right is a percent score that shows your progress up to this point in the test. You are told if your answer is correct and if not, you will be given the correct answer. If you take the same test again, you will find the answers rearranged so that you may continue to benefit from a "fresh" test. (This rearrangement is completely random, only your computer knows the arrangement. Any time you start one of these tests, this information is recorded on your disk. It is suggested that you not try any of these until you receive direction from your instructor as to how these tests will be used as part of your course grade.

Test Feature

The TEST FEATURE program contains two parts. One is the demonstration mode and the other the test mode. The demonstration mode presents instructions on how the test mode will be conducted. It also presents information that helps you review the chapter material. You may freely use the demonstration mode as often as you wish because nothing you do here is ever recorded on your disk. The other part is the TEST mode. If you enter this, that fact will be recorded on your disk. Figure C–4 illustrates a typical screen from the DEMONSTRATION and TEST mode for the TEST FEATURE of Chapter 3.

As shown in the figure, the TEST gives you a running percent score. As with all tests, if your response to a question is not correct, you will be shown the correct response. Again, the sequence of specific data used in the test is randomly generated every time you use it. Thus you always have the benefit of a "fresh" test.

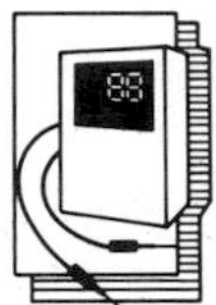

TROUBLESHOOTING SIMULATION

The TROUBLESHOOTING SIMULATION is divided into two parts. There is the DEMONSTRATION mode which explains how the troubleshooting test will be conducted. In this mode you may also be given the opportunity to perform some troubleshooting analysis to make sure you understand how to interact with the computer. The TEST mode will then present troubleshooting problems for you to analyze and solve. As before, if your analysis is incorrect, the computer will show you the correct answer. Figure C–5 illustrates the DEMONSTRATION and TEST screen for the TROUBLESHOOTING SIMULATION from Chapter 3 of your text.

Each time you try the TEST mode, this fact is recorded on your disk. Again, the sequence of specific problems used in the TROUBLESHOOTING SIMULATION is randomly generated each time you use the disk so you will have a "fresh" problem set.

Conclusion

It is strongly recommended that you do not use your student disks until after your first class meeting. The reason for this is that your instructor may have special instructions as to how your disks will be used in the course. Some may require that your performance with the disks be used as part of your course grade. Others may wish to have only certain programs on the disks used for grading purposes. There are so many different ways your disks can be used in a class setting. It is to your benefit to check with your instructor before using your disks for the first time. You have a valuable learning tool at your disposal.

```
This is a buffer...

          i                   The output is always the same as the input...
Ø —| A    X |—Ø
                                  X = A

                                  Ø = Ø

This is an inverter...

          1
1 —| A    X |\_Ø              The output is always opposite the input...

                                  X = Ā

                                  Ø = 1

                    Press any key to continue...
```

Figure C–4A Typical screen from the demonstration mode for the test feature.

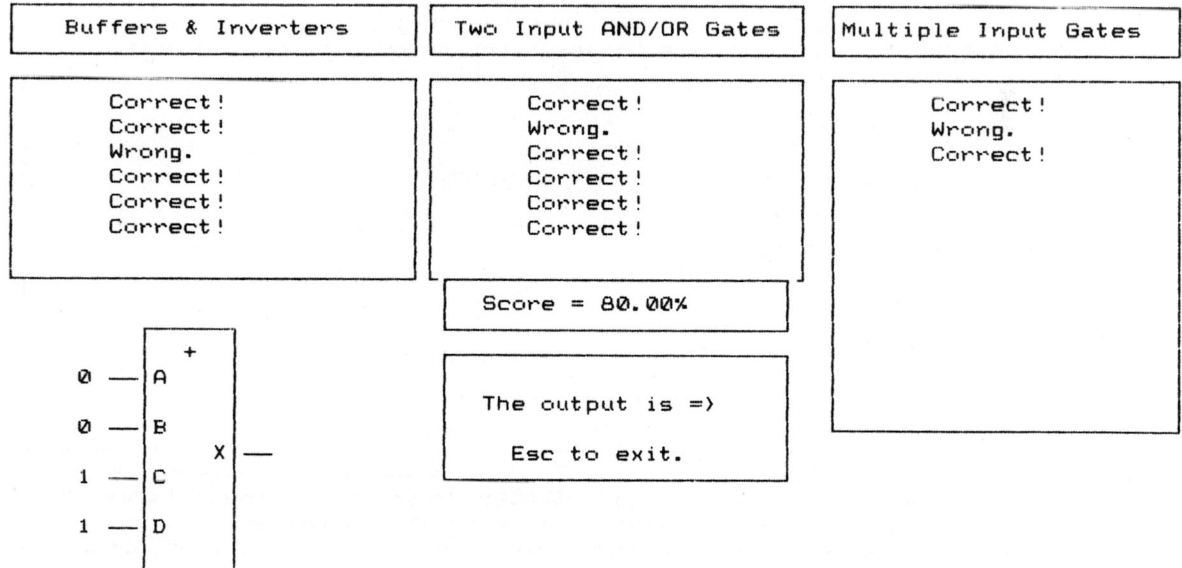

Figure C–4B Typical screen from the test mode for the test feature.

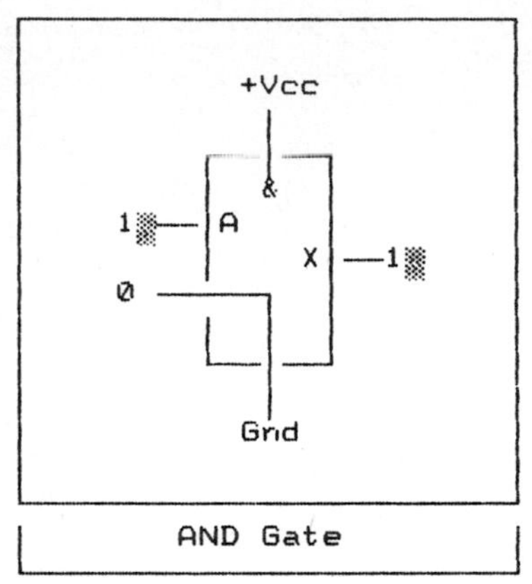

An Internal Short

Gates can experience an internal short this can happen between any input and +Vcc or ground.

In the case of an internal short in an AND gate between an input and ground, the output will not respond to changes on the input.

This is shown in the demonstration.

*** Press spacebar to continue or Esc to exit ***

Figure C–5A Typical screen from the demonstration mode for the troubleshooting simulation.

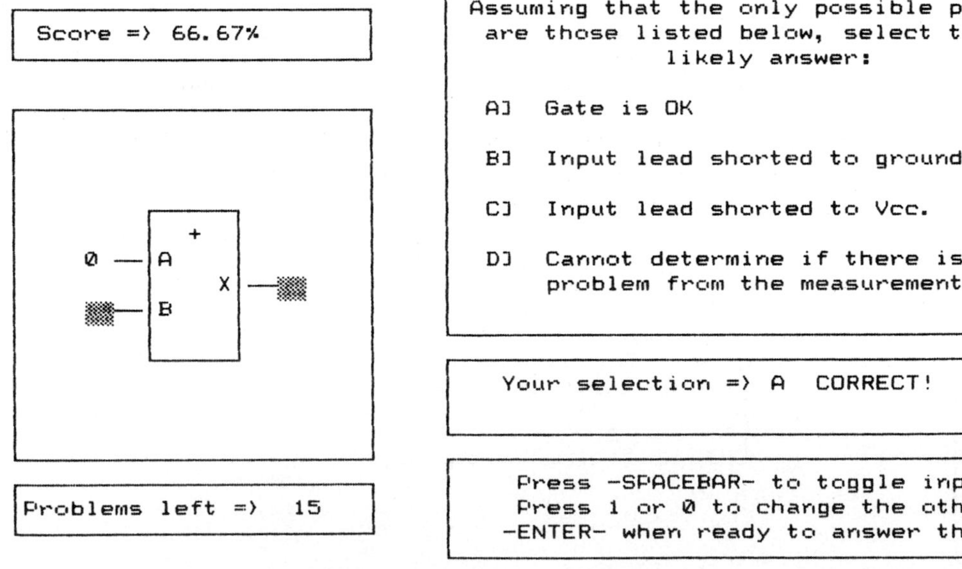

Score => 66.67%
+
0
A
X
B
Problems left => 15
Assuming that the only possible problems are those listed below, select the most likely answer:
A] Gate is OK
B] Input lead shorted to ground.
C] Input lead shorted to Vcc.
D] Cannot determine if there is a problem from the measurement.
Your selection => A CORRECT!
Press -SPACEBAR- to toggle input level.
Press 1 or 0 to change the other input.
-ENTER- when ready to answer the problem.
*** Press spacebar to continue or Esc to exit ***

Figure C–5B Typical screen from the test mode for the troubleshooting simulation.

INSTRUCTIONS

Instructions for using the IBM FORMATTED STUDENT DISKS

- Insert your System DOS disk into Drive A.
- Switch on computer and the monitor (after several seconds the disk drive light will come on, indicating the disk is being read, then it will go off).
- Enter the date and the time (if necessary), then the DOS prompt (A>) will appear.
- Remove the DOS disk and replace it with the STUDENT DISK. (Disk A for chapters 1–5 and Disk B for chapters 6–10.)
- Then type in START and press enter/return key.
- Follow the instructions on the screen.

APPENDIX D
Answers

Review Questions Chapter 1

1–1:

1. The main purpose of this book is to prepare you for a course in microprocessor technology.
2. Three of the subject areas that require the information in this book are microprocessors, robotics and industrial automation. Some of the other areas are computer maintenance and repair, digital communications, computer interfacing and biomedical electronics.
3. It is not necessary to know electronics to use this book because of the extensive use of *chips* in digital systems.
4. You need to know the cause and effects of input and output conditions of the chip.

1–2:

1. Every section of each chapter contains a *discussion*, *conclusion* and *review questions*.
2. The material contained at the beginning of each chapter includes *chapter objectives,* an *introduction*, and a list of *key terms*.
3. The three key sections contained in each chapter include *microprocessor application*, *electronic application*, and *troubleshooting simulation* sections. The microprocessor application sections contain material that will prepare you for a microprocessor course. The electronic application sections require a knowledge of electronic circuits and contain background material necessary for the troubleshooting of digital systems. The troubleshooting simulation section presents an overview of the material on the disk used with this text.
4. The end of chapter problems are arranged according to each section. There is also an *application* section, a *troubleshooting* or *instrumentation* section as well as an *analysis and design* section.

1–3:

1. The change that has taken place concerning the people who need to know the material in this book has caused other areas of technology other than the electronic major to study this material.
2. The purpose of the electronic application section is to present background information necessary for the maintenance and repair of digital systems.
3. The purpose of the microprocessor application sections is to prepare you for the advanced topics now required by most microprocessor courses.

1–4:

1. A computer, robot and automated assembly all use the same kind of "thinking" elements.
2. The other skills you could acquire as an electronics major are laboratory skills through a coordinated laboratory manual available with this book.
3. This book can help for future courses and employment. Courses in microprocessor technology and related fields require the knowledge available here. Learning features contained in this book are also a valuable tool for helping you obtain that first job.

1–5:

1. A discrete operation is an ON/OFF light switch. Continuous operation is a light dimmer control.
2. In reference to this book, digital refers to devices that have discrete conditions, while analog refers to a full range of possible conditions.
3. An example of a digital device is a digital watch and an analog device is a sun dial.

1–6:

1. The main event that prompted the design of a microprocessor was the need of a Japanese calcu-

lator company to have a special calculator designed in a compact manner.
2. Ted Hoff and Federico Faggin of Intel Corporation, designed the first microprocessor.
3. The first microcomputers were produced by Ed Roberts founder of MITS.
4. The major philosophy of Steve Jobs concerning the marketing of microcomputers was that these systems could be used by the general public for personal as well as business use.
5. One of the major differences between today's microcomputers and advanced machines is that of parallel processing where thousands of programs are executed at the same time.

1–7:

1. Troubleshooting can be defined as the act of locating and repairing a fault that causes improper operation in an electrical circuit. Instrumentation is the use of equipment to measure electrical quantities for the purpose of keeping these quantities within prescribed limits to ensure proper system operation.
2. The easiest category to troubleshoot is *complete failure*. The most difficult is *intermittent fault*.
3. You must be sure that a system fault exists.
4. A good habit when making measurements is to record them.

Chapter Problems (odd-numbered)

1-1. To prepare for other courses and any career using microprocessors

1-3. Many electronic circuits on a small piece of silicon, housed in a single package.

1-5. Each related topic presented in a chapter is in its own section. Each section starts with a discussion overview and ends with a conclusion and review.

1-7. (A) The summary can provide an overview before reading the chapter. The summary can also be read as a review before job interviews, end of chapter problems, or chapter quizzes and tests.
(B) The self test provides immediate feedback to enhance your performance on the same kinds of questions or in similar situations.

1-9. The questions are grouped by sections for easy reference. Each is listed numerically followed by an Applications set and an Analysis and Design set of questions.

1-11. Concurrent with a DC/AC course, including all sections in the text.

1-13. Electronics Majors preparing for job entry need as much lab time as possible.

1-15. Made of essentially the same "computing" parts.

1-17. Methods of design, construction and measurement for simplifying or troubleshooting digital systems.

1-19. Microprocessors, Robotics, Industrial Automation, Computer Maintenance, Digital Communication, Computer Interfacing, Biomedical Electronics and many others.

1-21. ON-OFF switches

1-23. Discrete operations change in large measurable steps. Continuous change so gradually, the accuracy of your measuring tools determines the smallest steps you can measure.

1-25. Automatically adjusting room light according to sensors in room.

1-27. Clocks, temperature probes, music and testing instruments, entertainment devices, and so on.

1-29. The Busicon and Intel corporations, and the individuals,Ted Hoff and Fredrico Faggin

1-31. The APPLE was specifically targeted at the general public.

1-33. Locating and repairing a fault that causes improper operation of an electrical circuit.

1-35 (A) Complete System (B) Intermittent
(C) The complete system fault is often an easily diagnosed fault of the main power circuits or connections. The intermittent fault is difficult because symptoms may seem to have no pattern or reason, making it impossible to isolate.

1-37 (1) A decision is made to measure, some system behavior, logic or voltage level.
(2) A procedure to measure is decided upon.
(3) The observation or measure is recorded.
(4) The technician analyzes and compares the observation or measurement.
(5) The technician acts, depending on the discrepancy or similarities observed to replace, repair or make further measurements.

Review Questions Chapter 2

2–1:

1. There are ten symbols used in the "everyday" number system.
2. The base of the "everyday" number system is ten.
3. Place value means the representation of quantities by a positional value system. For example, the value of the symbol 5 depends upon its placement with respect to the decimal point: 5.0, 50.0, and 0.5.
4. $6 \times 10^3 + 2 \times 10^2 + 5 \times 10^1 + 8 \times 10^0 + 2 \times 10^{-1}$.

2–2:

1. A binary number is a number system to the base two. 101_2 represents 5_{10}.
2. A binary number can represent any quantity.
3. Binary values less than one are represented by their position to the right of the binary point: 0.1 is less than one.

2–3:

1. A binary number can be converted to decimal by taking the sum of the weight of each bit.
2. The term "weight" means the place value of the symbol.
3. A decimal number may be converted to binary by repeating division by two or by observing the sum of the binary weights that equal the digital number.

2–4:

1. A solid state device that emits light when the correct potential and polarity are applied.
2. The ON and OFF pattern of lights can be interpreted as the 1 and 0 pattern of a binary number. In the same manner, the ON and OFF pattern of switches can be interpreted as the 1 and 0 pattern of a binary number.
3. A small keyboard that serves as an input device to a digital system.
4. BCD is binary coded decimal.
5. The advantage of BCD is its easy conversion between decimal and binary, the disadvantage is that six possible combinations of a four bit binary number are not used.

2–5:

1. A hex keypad consists of sixteen keys and is in the hexadecimal number system.
2. The advantage of a hex keypad is that it allows for easy conversion to a four bit binary number and all of the number combinations are used by the binary number.
3. The hexadecimal number system is a base sixteen number system that uses the symbols 0–9 plus the characters A, B, C, D, E, and F.
4. To convert from hex to binary simply convert each hex digit to its four bit binary equivalent. To convert from binary to hex, break the binary number up into groups of four bits and then convert each group to its hex equivalent.
5. To convert from decimal to hex you may do so by repeated division of 16. To convert from hex to decimal use the sum of the individual hexa-

decimal weights while converting each hex character to its decimal equivalent. (A pocket scientific calculator can also be used to make the conversions.)

2–6:

1. One disadvantage of the binary number system is in the transition from one consecutive value to the other more than one bit can be changing at a time. This may produce erroneous readings in some automated sensing instruments.
2. The disadvantage stated for binary numbers can be eliminated by using a number code of 1's and 0's where only one bit changes at a time in a sequential count.
3. The major advantage of the Gray code is that only one bit changes at a time in a sequential count.
4. Decimal numbers are converted to excess-3 code by first adding three to each decimal digit then converting the result directly to a four bit binary group.

2–7:

1. An alphanumeric code is one that can represent letters of the alphabet, numerals and other symbols such as punctuation marks and mathematical symbols.
2. ASCII stands for the *American Standard Code for Information Interchange*.
3. The ASCII code is an alphanumeric code and as such can represent the symbols of that code including system transmission codes.
4. The ASCII code is a seven bit code where all of the decimal digits start with 011 and then their BCD equivalent. The capitol letters start with 100 and then the BCD count of 0001. The lowercase letters start with 110 and then the binary count of 0001.

2–8:

1. The binary count is 11000_2, 11001_2, 11010_2, 11011_2.
2. The rules for binary addition are $0 + 0 = 0$, $1 + 0 = 1$, $0 + 1 = 1$, $1 + 1 = 10_2$ and $1 + 1 + 1 = 11_2$.
3. Complement means to change the digital state. Thus the complement of 1 is 0 and the complement of 0 is a 1.
4. The 1's complement of a binary number is found by complementing each bit of the number. The 2's complement is found by adding one to the 1's complement.
5. The hexadecimal count is 7_{16}, 8_{16}, 9_{16}, A_{16}, B_{16}, C_{16}, D_{16}, E_{16}, F_{16}, and 10_{16}.

2–9:

1. The notation used to indicate how digital memory is organized is $A \times D$ where A is a number that states the number of rows and D is a number that gives the number of columns.
2. An address is the row number and the *data* is the bit pattern stored at a given row number.
3. To use the hexadecimal loader, press the hex value of the desired address and then the address button. Then enter the hex value of the high order nibble, press the HI button. Next, enter the hex value of the low order nibble and press the LOW button.
4. Even parity uses a parity bit to ensure that the sum of the 1's in a given word is an even number. Odd parity uses a parity bit to ensure that the sum of the 1's in a given word is an odd number.

2–10:

1. The most common method of electrically representing a 1 is by +5 volts and the most common method of electrically representing a 0 is by 0 volts or ground.
2. A PBNO switch is a push-button switch that is normally opened. A PBNC is a push-button switch that is normally closed.
3. A prototype board allows parts to be easily and quickly inserted for testing of circuits.
4. Wire wrapping requires a special tool to wrap connecting wires around wire-wrapping terminals. This should be used when a more permanent construction is required.

Chapter Problems (odd-numbered)

2-1. (A) 2: 10's, 5: 1's. (B) 3: 100's, 7: 10's, 9: 1's. (C) 3: 1000's, 9: 100's, 8:10's, 7: 1's.

(D) 2: 1,000,000's, 0: 100,000's, 5: 10,000's, 7: 1,000's, 0: 100's, 0: 10's, 8: 1's, 9: tenths, 7: hundredth's.

(E) 3: 1's, 1: tenth, 4: hundredths, 1: thousandth, 5: ten thousandths, 9: hundred thousandths.

2-3. (A) 304 (B) 0.005 (C) 4830.25

2-5. (A) 101101_2 (B) 1101100_2

2-7. (A) 1000_2 (B) 10010_2 (C) 1010110_2 (D) 1001000111_2 (E) 1111100111_2 (F) 11000001011100_2

2-9. (A) 2.5_{10} (B) 1.25_{10} (C) 13.6875_{10} (D) 83.34375_{10} (E) 1334.2890625_{10} (F) $14{,}229.066650390625_{10}$

2-11. (A) 1010_2 (B) 0011_2 (C) 100100_2 (D) 10010110_2

2-13. (A) $0011\ 1100_2$ (B) $0100\ 1010_2$ (C) $0110\ 1101_2$ (D) $1111\ 1111\ 1111\ 1111_2$

2-15. (A) 4_{10} (B) 44_{10} (C) 973_{10} (D) 4210_{10} (E) $85{,}794_{10}$

2-17. (A) 0H (B) AH (C) 1AH (D) 5DH (E) 249H (F) 14D97H (G) 59D7H

2-19. (A) 1H (B) 10H (C) 100H (D) 400H (E) 4000H (F) 10000H

2-21. (A) 011_2 (B) 100_2 (C) 001_2 (D) 101_2 (E) 110_2

2-23. Students initials in ASCII (varies)

2-25. (Start text)
STX You did a good job!
ETB (End transmission)

2-27. (A) 101_2
(B) 1110_2
(C) 11111_2
(D) 1100101_2
(E) 1000000_2
(F) 10011010_2

2-29.

(A)
```
  101
 - 10
 +101
 ----
 1010
  + 1
 ----
   11
```

(B)
```
  1100
  - 11
+ 1100
------
 11000
   + 1
------
  1001
```

(C)
```
  1010
- 1001
+  110
------
 10000
   + 1
------
     1
```

(D)
```
 101101
-  10011
+101100
-------
1011001
 +    1
-------
  11010
```

(E)
```
 1011011011
 -    11111
+1111100000
-----------
11010111011
 +        1
-----------
 1010111100
```

2-31.

(A)
```
  101
 - 11
+ 101
-----
1 010
```

(B)
```
  1010
  - 101
+ 1011
------
1 0101
```

(C)
```
  1111
- 1000
+ 1000
------
1 0111
```

(D)
```
  101101
-   1101
+ 110011
--------
1 100000
```

(E)
```
 1111 0000
  - 1 0110
+ 11101010
----------
1 11011010
```

2-33.

(A) 100 ÷ 10
```
+ 110
-----
1 010
+ 110
-----
1 000 : twice
```

(B) 1000 ÷ 100
```
+ 1100
------
1 0100
+ 1100
------
1 0000 : twice
```

(C) 1010 ÷ 101
```
+ 1011
------
1 0101
+ 1011
------
1 0000 : twice
```

(D) 1100 ÷ 10
```
+ 1110
------
1 1010
+ 1110
------
1 1000
+ 1110
------
1 0110
+ 1110
------
1 0100
+ 1110
------
1 0010
+ 1110
------
1 0000 : six times
```

2-35. (A) 7H (B) CH (C) 11H
(D) 22H (E) 116H (F) FF34H

2-37. (A) 8 x 4 (B) 2K x 8 (C) 4K x8 (D) 1M x 16

2-41. [0] —> [ADD-RESS] —> [4] —> [HI] —> [D] —> [LOW]

The addresses and HEX data listed at the left in problem 40 would be entered one at a time through number 14 just as address 0 has been entered above.

2-43. Student answers will vary as the HEX loader is used to enter their initials with even and odd parity.

2-45. (A) L_1 has no current limiting resistor and will soon be a memory.
(B) This could cause a short circuit, excess current will flow when the switch closes.
(C) This should work but the resistor must be large enough to limit current to a safe level.
(D) This circuit should also work correctly if the resistors are the correct size.

2-47. For the part to turn through 360º with a separate reading every 22.5º there must be 16 readings. 16 separate states requires a 4 bit code. To eliminate false radings as the shaft sensor turned the gray code should be used and converted to digital or decimal data.

RECOMENDATIONS: Gray code circle with 4 fingers to read the position of the shaft.
TEST: Rotate and check that no momentary inaccuracies occur. Display converted gray code as decimal values on a seven segment display.

2-39.
(A) 32
(B)

0	00000010
1	10100101
2	01000000
3	00111100
4	10111101
5	11111010
6	00000100
7	01010111
8	00111000
9	10011010
10	11101111
11	11101101
12	01110010
13	00011010
14	10111101
15	01001100
16	01010010
17	10000000
18	00100001
19	10001010
20	11000100
21	11111111
22	00100011
23	01010101
24	00000000
25	00101010
26	10111100
27	11010000
28	11111010
29	10111011
30	00001000
31	10011100

Review Questions Chapter 3

3–1:

1. A graph for representing the conditions of a digital circuit can be constructed using a vertical axis that represents the ON and OFF condition of the circuit and a horizontal axis that represents time measured in seconds.
2. The horizontal axis of the digital graph always represents *time* measured in seconds.
3. Other terms used for ON and OFF are 1 and 0; +5 volts and 0 volts; HIGH and LOW.
4. A digital pulse represents a transition from the normal condition of the digital circuit. The leading edge represents the part of the pulse that changes from the normal circuit condition and the trailing edge represents the part of the pulse that changes back to the normal circuit condition.
5. A digital waveform is a representation of a series of pulses. A periodic waveform is one that repeats itself. A nonperiodic waveform does not repeat itself.

3–2:

1. The AND function means that the only time a result is HIGH is when all inputs are HIGH.
2. Connect the switches, LED and resistor in such a way that both switches must be closed before the

LED will light (such an arrangement is called a series circuit).

3. A truth table indicates the output of a logic circuit for all possible combinations of the inputs. A truth table is constructed by doing a binary count of all the input lines and indicating the corresponding result on the output.
4. An AND gate is a logic circuit capable of performing the AND function.
5. An IC that consists of four AND gates where each gate has two inputs each is called a quad 2-input AND gate.

3–3:

1. The OR function means that the output will be HIGH when any input is HIGH.
2. An electrical circuit that demonstrates the OR function would have its two switches connected to the power source and LED in such a way that if either or both switches were closed, the LED would be activated. Such a connection of the switches is called a parallel circuit.
3. An OR gate is a logic circuit that produces the OR function.
4. The truth table for a two input OR gate will have four combinations of the two inputs and the only time the output will be 0 is when both inputs are 0.
5. An IC that consists of four OR gates with two inputs each is called a *quad 2-input OR gate*.

3–4:

1. An INVERTER function changes the logic level of the input, that is, if the input is HIGH, then the output will be LOW and if the input is LOW, then the output will be HIGH.
2. A circuit that demonstrates the INVERTER function consists of one switch wired to a voltage source and LED in such a way that when the switch is closed, the LED goes OFF and when the switch is open, the LED remains ON. The precaution is to make sure that the LED resistor is not shorted by the closed switch or the voltage source could become damaged due to excessive current flow.
3. An INVERTER circuit is a circuit that replicates the INVERTER function.
4. The truth table for an INVERTER has one input and one output. There are only two conditions possible for the input: 0 and 1. The condition of the output will be the opposite of that of the input.
5. An IC chip with six INVERTERS is called a *hex inverter*.

3–5:

1. The two values that are allowed in Boolean algebra are TRUE and FALSE.
2. The Boolean NOT statement means the opposite state, that is NOT TRUE is FALSE and NOT FALSE is TRUE.
3. Boolean multiplication is the same as the AND function, where all inputs must be TRUE before the output is TRUE.
4. Boolean addition is the same as the OR function where if any input is TRUE, then the output will be TRUE.
5. Boolean algebra is useful in the analysis of digital circuits because its two-state answers (TRUE and FALSE) can be directly related to the two-state condition of digital circuits.

3–6:

1. An existing logic circuit can be analyzed by developing its truth table and Boolean expression.
2. In a truth table, the term YES is a 1 and the term NO is a 0.
3. This is an OR gate.
4. This is an INVERTED AND gate.
5. You ensure that all possible combinations of input conditions of a digital system have been considered by doing a binary count using all of the inputs.

3–7:

1. The purpose of a microprocessor is to perform a process on binary words.
2. The accumulator is the most used register in the microprocessor.
3. The main difference between a 4-bit and an 8-bit microprocessor is the word size.
4. The NOT process of a microprocessor consists of complementing each bit of the word in the accumulator.
5. The AND process consists of ANDing the bit pairs formed by the accumulator and the B register and storing the result back in the accumulator. The OR process consists of ORing the bit pairs of the accumulator and B register and storing the results back in the accumulator.

3–8:

1. A digital IC is a complete digital circuit in a chip.
2. An IC family uses the same materials and man-

ufacturing process. A subfamily is the same as a family but has different electrical characteristics.

3. The main advantages of TTL are its historical ease of handling, low cost and availability.

4. The main advantages of CMOS are low power consumption.

5. Two disadvantages of CMOS are slower switching speed and sensitivity to static electricity.

Chapter Problems (odd-numbered)

3-1. High for 9 seconds, Low from then on.

3-3. It stays Low for 6 seconds, then stays High.

3-5. Vertical: Volts, State, Condition
Horizontal: Time

3-7. on / off — seconds

3-9. (A) Periodic (C) Periodic

(B)

PB-a	PB-b	PB-c	LED
off	off	off	Low
off	off	on	Low
off	on	off	Low
off	on	on	Low
on	off	off	Low
on	off	on	Low
on	on	off	Low
on	on	on	High

3-11. (A)

PBa	PB-a	LED
off	off	Low
off	on	Low
on	off	Low
on	on	High

3-13.

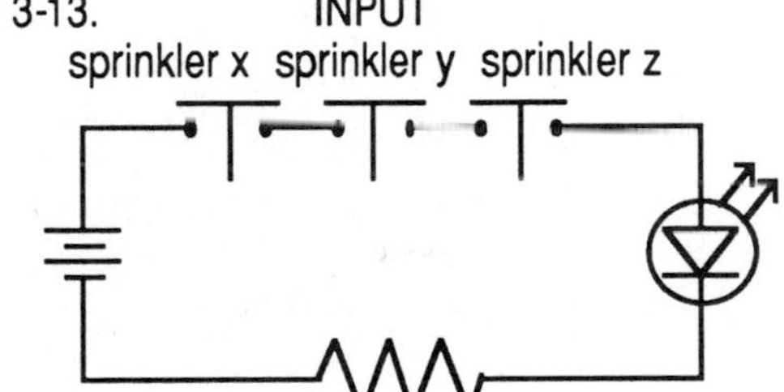

3-15.

PB-a	PB-b	PB-c	LED
off	off	off	Low
off	off	on	Low
off	on	off	Low
off	on	on	Low
on	off	off	Low
on	off	on	Low
on	on	off	Low
on	on	on	High

3-17.

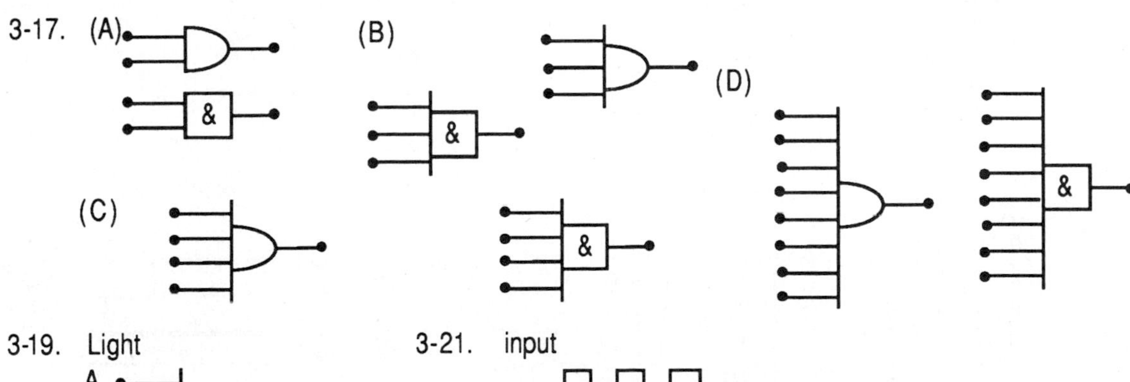

3-19. Light

A
B
C

3-21. input

+5V
(pass)
control (stop)
output

3-23. (A) high / low

(B) high / low

(C) high / low

(D) high / low

3-25. (A) high / low

high / low

(B) high / low

high / low

high / low

(C) high / low

3-27. When any Push Button switch is on the LED will also be on.

3-29. Lights off only if all are off

A

B

C

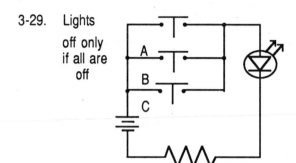

3-31.

PB-x	PB-y	PB-z	LED
off	off	off	Low
off	off	on	High
off	on	off	High
off	on	on	High
on	off	off	High
on	off	on	High
on	on	off	High
on	on	on	High

3-33. (A)

A	B	X
0	0	0
0	1	0
1	0	0
1	1	1

(B)

A	B	C	X
0	0	0	0
0	0	1	0
0	1	0	0
0	1	1	0
1	0	0	0
1	0	1	0
1	1	0	0
1	1	1	1

(C)

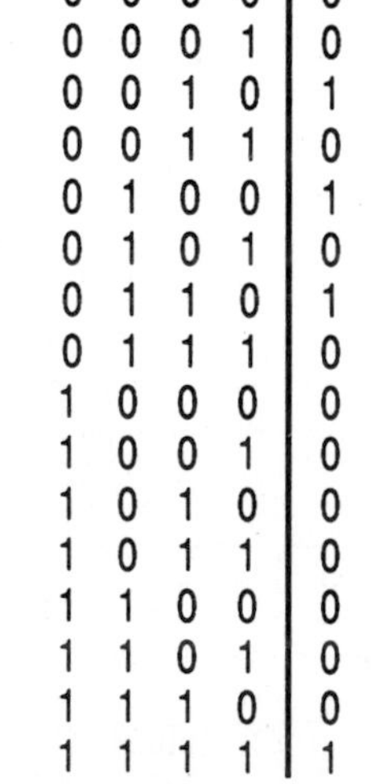

A	B	C	D	X
0	0	0	0	0
0	0	0	1	0
0	0	1	0	1
0	0	1	1	0
0	1	0	0	1
0	1	0	1	0
0	1	1	0	1
0	1	1	1	0
1	0	0	0	0
1	0	0	1	0
1	0	1	0	0
1	0	1	1	0
1	1	0	0	0
1	1	0	1	0
1	1	1	0	0
1	1	1	1	1

3-35. Sprinkler

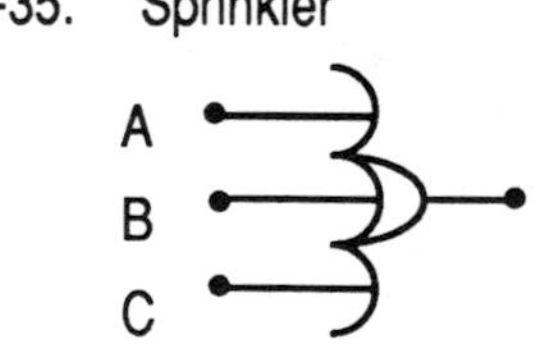

3-37.

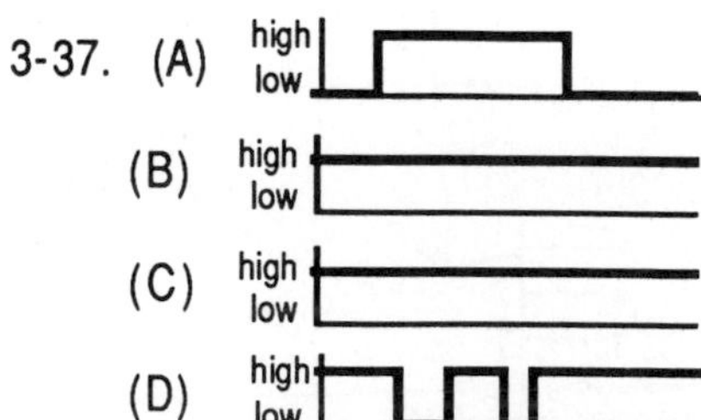

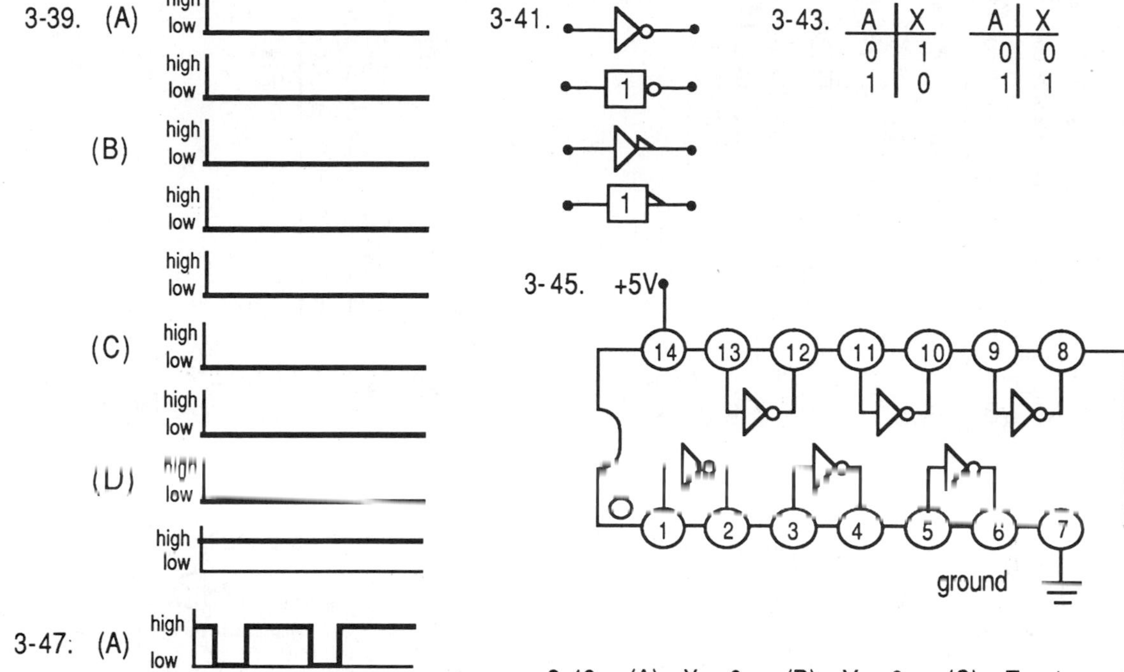

3-47: (A) high low

(B) high low

(C) high low

(D) high low

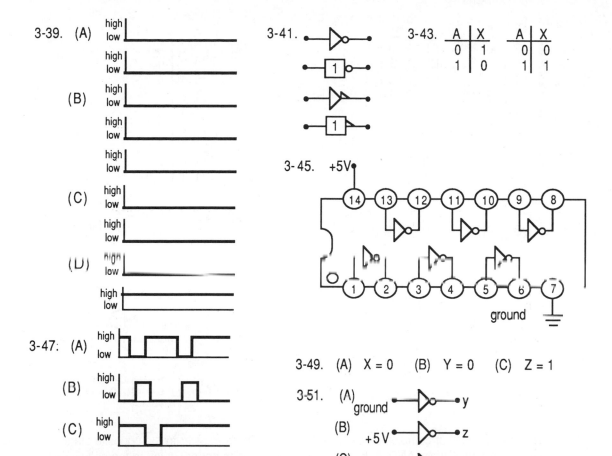

3-49. (A) X = 0 (B) Y = 0 (C) Z = 1

3-51. (A) ground → y

(B) +5V → z

(C) +5V → x

3-53. The BOOLEAN operation that uses the multiplication symbol AND's the logic values. If they are all true the output is true. If all 1's are ANDed a 1 is output. Any 0 or false input or term will cause a false output when using the multiplication symbol. The addition symbol is used to signify the OR operation. 0 or false conditions do not cause a false evaluation as long as there is at least one true or 1 input or term in the expression being OR'd.

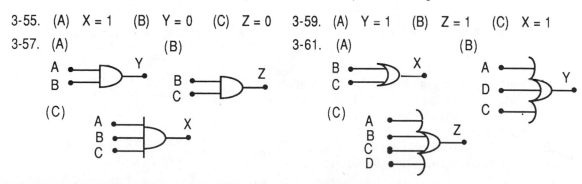

3-55. (A) X = 1 (B) Y = 0 (C) Z = 0

3-57. (A) A, B → Y (B) B, C → Z (C) A, B, C → X

3-59. (A) Y = 1 (B) Z = 1 (C) X = 1

3-61. (A) B, C → X (B) A, D, C → Y (C) A, B, C, D → Z

3-63. for (L)

(A) $\bar{A} + B = X$

(B)

A	B	X
0	0	1
0	1	0
1	0	1
1	1	1

(C) X = 0

for (M)

(A) $\overline{\bar{A}\ B} = X$

(B)

A	B	X
0	0	1
0	1	0
1	0	1
1	1	1

(C) X = 0

for (N)

(A) $\overline{\bar{A} + B} = X$

(B)

A	B	X
0	0	0
0	1	0
1	0	1
1	1	0

(C) X = 0

for (O)

(A) A B C = X

(B)

A	B	C	X
0	0	0	0
0	0	1	0
0	1	0	0
0	1	1	0
1	0	0	0
1	0	1	0
1	1	0	0
1	1	1	1

(C) Y = 0

3-65.

W	W	C	OK
0	0	0	0
0	0	1	0
0	1	0	0
0	1	1	0
1	0	0	0
1	0	1	0
1	1	0	0
1	1	1	1

3-67. (A) t1 t2 t3 Main

(B)

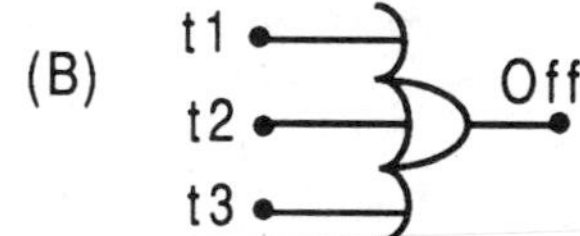

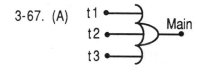

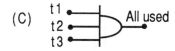

(C) t1 t2 t3 All used

3-69. (A) The microprocessor complements each bit of a word in it's register to the opposite state.
(B) Two word patterns are ANDed, bit by bit, and the resulting pattern is left in the accumulator.
(C) Two word patterns are OR'd, bit by bit, and the result is also in the accumulator.

3-71. (A) FFH (B) ABH (C) 98H

3-73. (A) EEH (B) 20H (C) 34H

3-75. (A) FH (B) 0H (C) 3CH

3-77. (A) 07H (B) FFH (C) BDH

3-79. 7 a b c d e f g 1 / 8 h i j k l m n 14

(A)

8 a b c d e f g h 1 / 9 i j k l m n o p 16

(B)

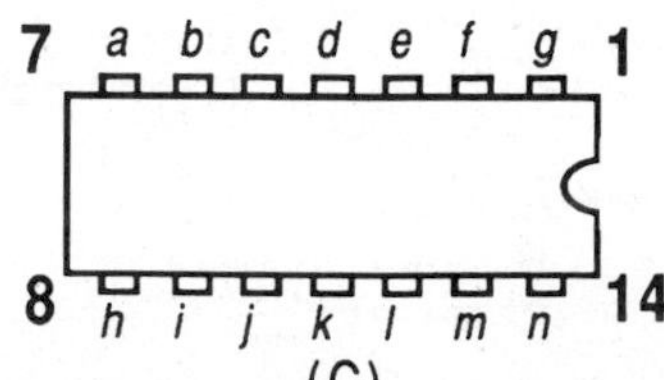

(C)

3-81. (A) The OR gate is not allowing the +5V to affect the circuit. Check the socket, if it is connecting the correct voltages to the the input pins, replace the chip.
(B) The AND gate should not hold the output HIGH with a LOW input. Check the socket or chip.

3-83.

	OPERATION / START;	register B	Accumulator	
	OPERATION / START;	0101	1010	All motors OFF
■	NOT the Accumulator	0101	0101	All motors ON
	NOT register B	1010	0101	
■	AND the Accumulator and B	1010	0000	Even motors OFF
■	OR the Accumulator and B	1010	1010	All motors OFF
	NOT register B	0101	1010	
■	OR the Accumulator	0101	1111	Odd motors OFF
	NOT register B	1010	1111	
■	AND the Accumulator and B	1010	1010	All motors OFF

Review Questions Chapter 4

4–1:

1. The logical construction of a NOR gate is an OR gate followed by an INVERTER.
2. The only time the output of a NOR gate is TRUE is when all inputs are FALSE.
3. The Boolean expression for a three input NOR gate is $X = \overline{A + B + C}$.
4. No, $\overline{A + B}$ is not equal to $\overline{A} + \overline{B}$.
5. The expressions $\overline{A + B}$ and $\overline{A} + \overline{B}$ can be demonstrated not equal by developing a truth table for each one and showing that the results are different for identical input conditions.

4–2:

1. The logical construction of a NAND gate is an AND gate followed by an INVERTER.
2. The only time the output of a NAND gate is LOW is when all inputs are HIGH.
3. The Boolean expression for a 3-input NAND gate is:

$$X = \overline{A \cdot B \cdot C}$$

4. No, $\overline{A \cdot B}$ is not equal to $\overline{A} \cdot \overline{B}$.
5. The expressions $\overline{A \cdot B}$ and $\overline{A} \cdot \overline{B}$ can be demonstrated not equal by developing a truth table for each one and show that the results are different for identical input conditions.

4–3:

1. The product of NOT A and NOT B equals NOT A OR B.
2. Develop truth tables for a NOR gate and an AND gate with inverted inputs. Then show that the results are the same for identical input conditions.
3. The equivalent logic circuit of an AND gate with four inverted inputs is a four-input NOR gate.

4–4:

1. The product of NOT A OR NOT B is equal to the quantity NOT A and B.
2. Develop a truth table for a NAND gate and an OR gate with inverted inputs. Show that the results of both tables are the same for identical inputs.
3. The equivalent logic circuit of an OR gate with inverted inputs is a NAND gate.

4–5:

1. Precedence of operations means the order in which logical operations are performed.
2. The cummulative laws state that is makes no difference in which order variables are ORed or ANDed.
3. The associative laws state that it makes no difference how variables are ORed or ANDed, the results will be the same.
4. The output of an AND gate will always be equal to the input when all inputs are connected to the same variable ($AA = A$).
5. An odd number of inverters produces an inversion of the input, and even number of inverters does not produce an inversion of the input.

4–6:

1. The OR function will have a FALSE output only when all inputs are FALSE. The XOR will have a TRUE output only when its inputs are different.

2. The output of an XOR gate will be LOW when its inputs are the same.
3. XOR gates find applications in parity generation and checking as well as arithmetic circuits in digital computers.
4. The basic logic construction of an XNOR gate consists of an XOR gate followed by an INVERTER.
5. The logic symbol for the XOR process is a + sign with a circle around it.

4–7:

1. The primary use for alternative forms of logic symbols is an attempt to make logical analysis of digital systems easier.
2. An active logic level is the level that causes an event of interest to take place.
3. Active HIGH implies that an event of interest will take place at a HIGH, while active LOW implies the opposite.
4. A generalization that can be made about standard gate symbols and their equivalents is if the original uses an AND symbol, then its equivalent will use an OR symbol, and if the original uses an OR symbol, its equivalent will use an AND symbol.

4–8:

1. C 2. D 3. E 4. B 5. A

4–9:

1. To create an INVERTER from NAND gates, connect the inputs of a NAND gate together. The results is an INVERTER.
2. The algebraic proof that you can create an OR gate from NAND gates is: $X = \overline{\overline{A}\,\overline{B}} = \overline{\overline{A}} + \overline{\overline{B}} = A + B$.
3. The equivalent gate that requires the most NAND gates is the NOR. It requires four NAND gates.
4. De Morgan's theorem is used in gate simplification anytime you convert a standard NAND/NOR gate to its equivalent or do the reverse process.

4–10:

1. To create an INVERTER from NOR gates, connect the inputs of a NOR gate together. The results is an INVERTER.
2. The algebraic proof that you can create an AND gate from NOR gates is $X = \overline{\overline{A} + \overline{B}} = \overline{\overline{A}} \cdot \overline{\overline{B}} = A \cdot B$.
3. The equivalent gate that requires the most NOR gates is the NAND. It requires four NOR gates.
4. Using NAND gates or NOR gates, you can construct any basic logic circuit you choose. This can be demonstrated by Boolean algebra or gate simplification techniques.

4–11:

1. The microprocessor got its name from the fact that it is constructed to perform a predefined set of small processes.
2. The instruction set documents the process that will be performed by the microprocessor by a specific bit pattern.
3. The stored program concept is the idea of having both *instructions* and *data* in memory.
4. The first thing the microprocessor does when it is turned on is to get a copy of the bit pattern in the first location in memory and interpret it as an *instruction*.
5. The difference between an *instruction* and *data* is that an *instruction* is a bit pattern that will cause the microprocessor to do a specific process while *data* is a bit pattern that will take place in the process. They both appear the same in memory. You cannot tell the difference between an *instruction* or *data* by its bit pattern. It is the sequence by which they are put into memory that determines the difference.

4–12:

1. The logic levels for TTL circuits are not exactly +5 volts for a logic HIGH and 0 volts for a logic LOW. Because of practical considerations, these logic levels will vary from a low of +2 volts for an input HIGH minimum to 0.8 volt for an input LOW maximum.
2. An input and output profile represents the manufacturer's guaranteed maximum and minimum acceptable voltages for a guaranteed logic LOW and logic HIGH.
3. The input and output profiles for logic circuits are not the same. The reason for this is to create a buffer region for noise called the noise margin. This reduces the chance of logic levels being affected by noise voltages.
4. Current sourcing is when the device in question supplies current to an external device. Current sinking is when the device accepts current from an external device. The minus sign for current in-

dicates that the conventional current is flowing out of the device.

5. Fan-out means the number of logic devices that can be logically operated by another logic device. For reliable logic operation, it is important that the fan-out limits of a logic device not be exceeded.

Chapter Problems (odd-numbered)

4-1. (A) 0 (B) 1 (C) 0

4-3.

A	B	C	X
0	0	0	0
0	0	1	0
0	1	0	1
0	1	1	0
1	0	0	1
1	0	1	0
1	1	0	1
1	1	1	0

4-5. (A) $\overline{A + B} = X$

(B) $\overline{A + B + C} = X$

(C) $\overline{A + B + C + D} = X$

4-7.

(L)

A	B	X
0	0	1
0	1	0
1	0	0
1	1	0

(M)

A	B	X
0	0	1
0	1	0
1	0	0
1	1	0

(N)

A	B	X
0	0	1
0	1	1
1	0	1
1	1	0

(O)

A	B	X
0	0	0
0	1	0
1	0	0
1	1	1

4-9. (A) 1 (B) 1 (C) 1

4-11.

A	B	C	X
0	0	0	1
0	0	1	0
0	1	0	1
0	1	1	0
1	0	0	1
1	0	1	0
1	1	0	1
1	1	1	1

4-13. (A) A B = X (B) A B C = X (C) A B C D = X

4-15.

(L)

A	B	X
0	0	1
0	1	1
1	0	1
1	1	0

(M)

A	B	X
0	0	1
0	1	1
1	0	1
1	1	0

(N)

A	B	X
0	0	1
0	1	0
1	0	0
1	1	0

(O)

A	B	X
0	0	0
0	1	1
1	0	1
1	1	1

4-17. $\overline{A + B} = Y$

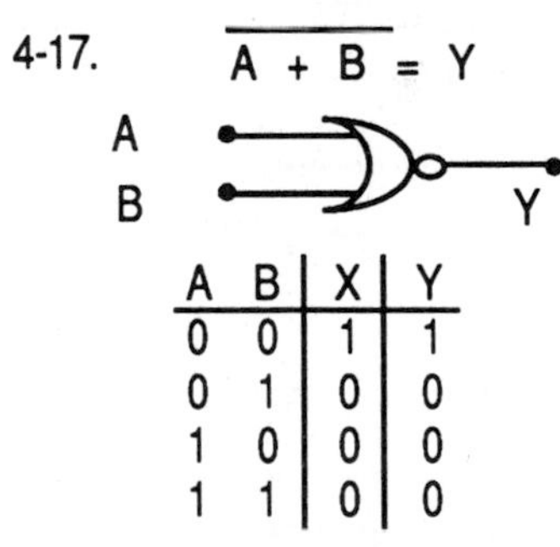

A	B	X	Y
0	0	1	1
0	1	0	0
1	0	0	0
1	1	0	0

$\overline{A}\ \overline{B} = X$

A
B
X

4-19. (A) $\overline{A + B}$ (B) $\overline{A + B + C}$ (C) $\overline{A + B + C + D}$

4-21. A NOR gate is logically equivalent to an OR gate with inverted inputs.

4-23. (A) Order BOOLEAN operations are done in an expression.

(B) For the expression: $A + B\ C + \overline{(D + E)}$

Do A + B and $\overline{(D + E)}$ first,

AND each with C and F then finally OR

4-25. $(A + B)(B + C) = X$

A	B	C	X
0	0	0	0
0	0	1	0
0	1	0	1
0	1	1	1
1	0	0	0
1	0	1	1
1	1	0	1
1	1	1	1

4-27. (A) $(AB)C = X$

$ABC = X$

A B C → X

(B) $(AB) + A = X$

$A(B + 1) = X$

$A(1) = X$

$A = X$

A —— X

4-29 (C)

A	B	B+(BA)	B+A
0	0	0	0
0	1	1	1
1	0	1	1
1	1	1	1

(D)

A	B	C	(AB)+C	(A+B) C
0	0	0	1	1
0	0	1	1	1
0	1	0	1	1
0	1	1	0	0
1	0	0	1	1
1	0	1	0	0
1	1	0	1	1
1	1	1	0	0

4-31. (A) X = A
(B) Y = A
(C) X = 1
(D) Y = 0

4-33
(A) A —— X
(B) A —— Y
(C) +5V —— X
(D) ⏚ —— Y

4-35. (C) $\overline{(AB) + (BC)} = X$

$\overline{B(A + C)} = X$

A, C → ≥1; B → & (inverted output) → X

4-35 (D)

$A + (\bar{A}B) = X$

$\overline{A + (\bar{A}B)} = \bar{X}$

$\bar{A} \cdot \overline{(\bar{A}B)} = \bar{X}$

$\bar{A} \cdot (A + \bar{B}) = \bar{X}$

$\bar{A}A + \bar{A}\bar{B} = \bar{X}$

$0 + \bar{A}\bar{B} = \bar{X}$

$\bar{A}\bar{B} = \bar{X}$

$\bar{\bar{A}}\,\bar{\bar{B}} = \bar{\bar{X}}$

$A + B = X$

A, B → ≥1 → X

4-37. Odd parity

4-39.

A	B	C	D	X
0	0	0	0	1
0	0	0	1	0
0	0	1	0	0
0	0	1	1	1
0	1	0	0	0
0	1	0	1	1
0	1	1	0	1
0	1	1	1	0
1	0	0	0	0
1	0	0	1	1
1	0	1	0	1
1	0	1	1	0
1	1	0	0	1
1	1	0	1	0
1	1	1	0	0
1	1	1	1	1

4-41. 5 bit odd parity word

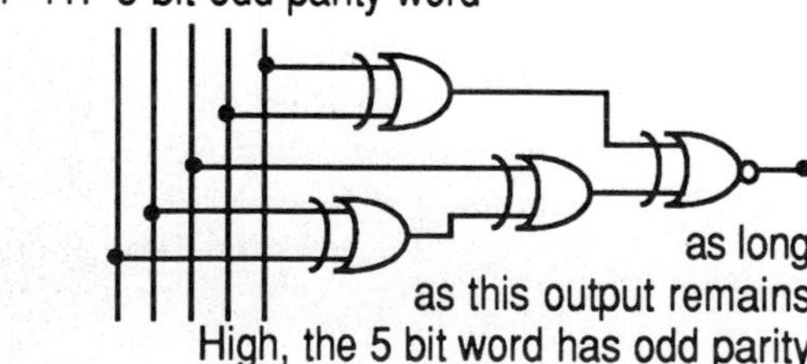

as long as this output remains High, the 5 bit word has odd parity

4-43. (A)

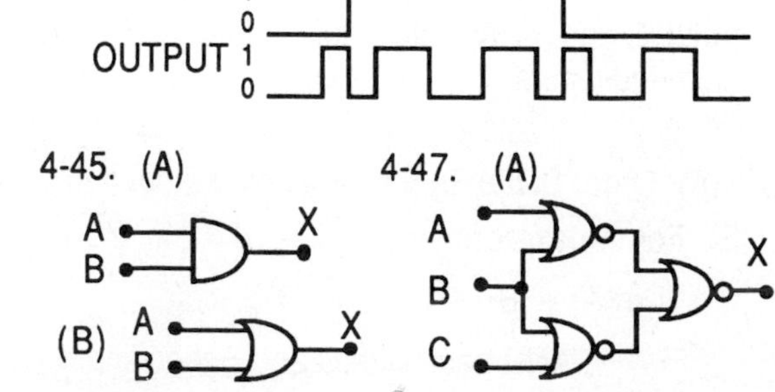

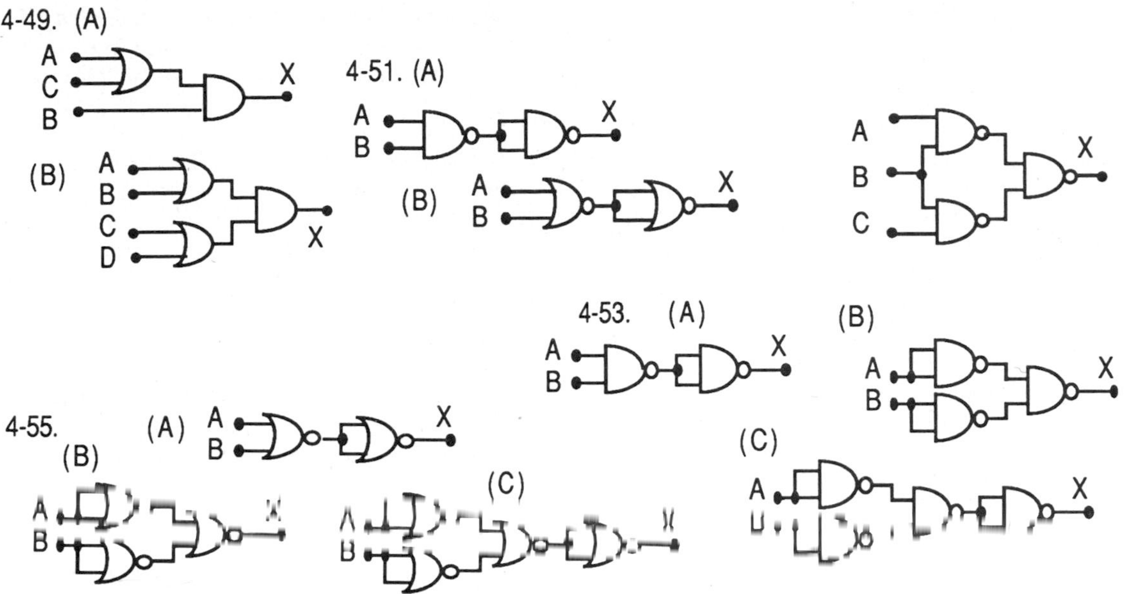

4-57 (A) The accumulator stores the result of logic operations. (B) B-register stores word that can be used with the accumulator for logic operations. (C) Data register stores word that controls process logic. (D) Process logic copies words and does logic on them.

4-59. The microprocessor will "GET" the first instruction and move it to the data regiser. Depending on the contents, the process logic "Tells" the microprocessor what to do next.

4-61. 0101 NOT Accumulator
1100 Decrease Acc. by 1
1011 Increase Acc. by 1
1110 ZERO the Accumulator

4-63. 1001 Acc. » Mem. 1010 B reg. » Mem.

4-65. (A)

A	1001
B	1010
D	1111

0	1000
1	1101
2	1001
3	1001
4	1111

(B)

0	0101
1	0110
2	1101
3	1100
4	1001
5	0011
6	1010
7	1010
8	1111

4-67
(A) High
(B) Problem
(C) Low
(D) High

A	0011
B	1010
D	1111

4-71.

0	1110
1	0001
2	0101
3	1110
4	0001
5	1010
6	0001
7	0001
8	0001
9	0011
A	0001
B	0111
C	0001
D	1111
E	1111

4-69.

		standard	Schottkey	Low power	Low power Schottkey
(A)	INVERTER	-200uA	-250uA	-50uA	-100uA
	each NAND	40uA	50uA	10uA	20uA
(B)	INVERTER	8mA	10mA	0.9mA	1.8mA
	each NAND	-1.6mA	-2mA	-.18mA	-.36mA

Review Questions Chapter 5

5–1:

1. A logic level is a section of gates and their connections that all feed toward the same input. It is used in network analysis where the first level is the output gate, the second level is those logic gates feeding into it and so forth.
2. The method of analyzing a logic circuit by empty terms is to start at the output gate (first level) and write the Boolean expression that would represent its output. Since the expressions to the input of this gate are not yet known, parentheses are used with no information inside them—hence the phrase "empty terms."
3. The Boolean expression for Figure 5–15(A) is $\overline{(A + B)(B + C)}$.
4. The Boolean expression for Figure 5–15(B) is $\overline{AB + BC}$.

5–2:

1. The term *sum of products* is a Boolean expression that is the ORing of variables that are ANDed.
2. The term *product of sums* is a Boolean expression that is the ANDing of variables that are ORed.
3. A *sum of products* expression is developed from a truth table by ORing the product terms from each TRUE condition of the truth table.
4. A *product of sums* expression is developed from a truth table by ANDing the sum of the complement of terms from each FALSE condition of the truth table.
5. When developing a Boolean expression from a truth table, use either the SOP or POS depending upon which expression will present the least number of terms.

5–3:

1. A POS Boolean expression is implemented with two-level logic that consists of an OR gate at the first level fed by AND gates or single variables.
2. An SOP Boolean expression is implemented with two-level logic that consists of an AND gate at the first level fed by OR gates or single variables.
3. A POS Boolean expression will produce two-level NAND logic.
4. An SOP Boolean expression will produce two-level NOR logic.
5. Since any logic network can be represented by a truth table, than any logic network can be represented by a POS or SOP Boolean expression. This results in the ability to represent any logic network with all NAND gates or all NOR gates.

5–4:

1. K Mapping is a graphical method of reducing the number of gates in a gate network.
2. They are as follows: 1] All combinations of the variables are represented by a box inside the K Map. 2] Along the sides of a K Map only one variable at a time is allowed to change from its complemented to uncomplemented form or from its uncomplemented to complemented form. 3] For a given variable combination that is TRUE, that cell inside the K Map representing the variable combination is given a 1, while for a variable combination that is FALSE, its corresponding cell is given a 0. 4] The K Map contains the same information as a truth table.
3. *Pair*, *quad* and *octet* simplification refers to treating *two*, *four* or *eight* adjacent cells as a single SOP term resulting in a reduction of the number of terms in the final SOP expression.
4. The process of *folding* the K Map means that the sides of the K Map can be considered as adjacent for the purpose of simplifying adjoining cells.
5. Overlapping cells in a K Map refer to cells that may be shared by two or more pairs, quads or octets for the purpose of K Map simplification.

5–5:

1. A redundant group is a group of adjacent cells that are already used by other groups in logic simplification.
2. A *don't care* condition is a logic condition whose input has no effect on the desired results.
3. *Don't cares* can be used in K Map simplification by treating them as either 1's or 0's depending upon which value causes the least number of terms in the final Boolean expression.
4. To develop POS terms from a K Map use the complement of each term in a sum from each 0 in a K Map cell.
5. The form that requires less steps depends upon the number of K Map cells with 1's compared to those with 0's.

5–6:

1. The steps used in the design of a digital logic system are as follows: 1] State the problem in writ-

ing. 2] Assure there is a need for the design. 3] Assure gate logic is the best solution. 4] Reduce problem to truth table. 5] Use K-Map. 6] Use Boolean algebra to minimize gates. 7] Implement with reliable and economical components.

2. No, gates are seldom used in the implementation of a logic design. Most commonly this is done with software or programmable logic.
3. Boolean algebra is used in order to ensure that the number of gates used are minimum.
4. When developing a K-Map use either the SOP or POS whichever one produces the easiest solution to the K-Map.
5. The kind of gates that are used for the implementation of the problem are NAND/NOR gates. These are used because they are reliable, economical and easily obtainable.

5–7:

1. Mnemonics are used because they help serve as a "memory jog" to indicate the meaning of the instruction.
2. Any example from Table 5–7 will do. One example is LDA which means load the accumulator from memory.
3. An *assembly language* program contains mnemonics which must eventually be reduced to the bit patterns understood by the microprocessor. *Machine language* is the actual bit pattern that is used for instructions by the microprocessor.
4. The meaning of the program is

0	LDA 7	; Load Accumulator with 7_{16}
2	LDB 3	; Load B register with 3_{16}
4	ADD	; Add Accum and B register
5	HLT	; Stop further processing

5. The Accumulator will contain $7_{16} + 3_{16} = A_{16} = 1010_2$, the B register will contain $3_{16} = 0011_2$.

5–8:

1. A logic device is never operated at the *absolute maximum ratings*. These are simply the values which, if exceeded, will damage the device.
2. For the data sheets shown in this section, the operating temperature of the device is under the section called Absolute Maximum Ratings.
3. The purpose of the Function Table in a specification sheet is to show the logic relations between the input and output of the device.
4. The switching characteristics state the amount of time it takes the output to respond to a given change on the input.
5. In the section of the data sheet that specified the minimum high level input voltage and the maximum high level output voltage.

Chapter Problems (odd-numbered)

5-1. 3 logic levels

5-3. $X = (\quad)_1 + (\quad)_2$

$X = [(\quad)_3 (B)] + [(C)(\quad)_4]$

$X = [(A+B)(B)] + [(C)(D)]$

5-5. (A) $X = \overline{(\quad)_1 (\quad)_2}$

$X = \overline{(\overline{AB})(\overline{BC})}$

(B) $Y = \overline{(\quad)_1 + (\quad)_2}$

$Y = \overline{(\overline{A+B}) + (\overline{B+C})}$

5-7. (A) 1st level: $\overline{AB}$, $\overline{BC}$ — 2nd level: $\overline{(\overline{AB})(\overline{BC})}$

(B) 1st level: $\overline{A+B}$, $\overline{B+C}$ — 2nd level: $\overline{(\overline{A+B}) + (\overline{B+C})}$

5-9. (A) SOP (B) POS (C) SOP

5-11. $X = (\overline{A}\,\overline{B}\,\overline{C}) + (\overline{A}\,\overline{B}\,C) + (\overline{A}\,B\,\overline{C}) + (\overline{A}\,B\,C)$

5-13. $X = (\overline{A}\,\overline{B}\,\overline{C}) + (\overline{A}\,\overline{B}\,C) + (A\,\overline{B}\,\overline{C}) + (A\,B\,\overline{C}) + (A\,B\,C)$

5-15. $X = (\overline{A}\,\overline{B}\,C\,\overline{D}) + (\overline{A}\,\overline{B}\,C\,D) + (\overline{A}\,B\,\overline{C}\,\overline{D}) + (A\,\overline{B}\,\overline{C}\,\overline{D}) + (A\,\overline{B}\,\overline{C}\,D)$
$+ (A\,\overline{B}\,C\,\overline{D}) + (A\,\overline{B}\,C\,D)$

5-17. $X = (A + \overline{B} + C)(A + \overline{B} + \overline{C})(\overline{A} + B + \overline{C})$

5-19. $X = (A + B + C + D)(A + B + C + \overline{D})(A + \overline{B} + C + \overline{D})(A + \overline{B} + \overline{C} + D)(A + \overline{B} + \overline{C} + \overline{D})$
$(\overline{A} + \overline{B} + C + D)(\overline{A} + \overline{B} + C + \overline{D})(\overline{A} + \overline{B} + \overline{C} + D)(\overline{A} + \overline{B} + \overline{C} + \overline{D})$

5-21. from table 5-4 A=1, B=0, C=1, D=0

POS $(A+B+\overline{C}+D)\ (A+B+\overline{C}+\overline{D})\ (A+\overline{B}+C+D)\ (\overline{A}+B+C+D)\ (\overline{A}+B+C+\overline{D})\ (\overline{A}+\overline{B}+\overline{C}+D) = X$

$(1+0+\overline{1}+0)\ (1+0+\overline{1}+\overline{0})\ (1+\overline{0}+1+0)\ (\overline{1}+0+1+0)\ (\overline{1}+0+1+\overline{0})\ (\overline{1}+\overline{0}+\overline{1}+0) = X$

$(1+0+0+0)\ (1+0+0+1)\ (1+1+1+0)\ (0+0+1+0)\ (0+0+1+1)\ (0+1+0+0) = X$

$(1)\ (1)\ (1)\ (1)\ (1)\ (1) = \underline{1}$

SOP:

$(\overline{A}\overline{B}\overline{C}\overline{D})+(\overline{A}\overline{B}\overline{C}D)+(\overline{A}B\overline{C}D)+(\overline{A}BC\overline{D})+(\overline{A}BCD)+(A\overline{B}C\overline{D})+(A\overline{B}CD)+(AB\overline{C}\overline{D})+(AB\overline{C}D)+(ABCD) = X$

$(\overline{1}\overline{0}\overline{1}\overline{0})+(\overline{1}\overline{0}\overline{1}0)+(\overline{1}0\overline{1}0)+(\overline{1}01\overline{0})+(\overline{1}010)+(1\overline{0}1\overline{0})+(1\overline{0}10)+(10\overline{1}\overline{0})+(10\overline{1}0)+(1010) = X$

$(0101)+(0100)+(0000)+(0011)+(0010)+(1111)+(1110)+(1001)+(1000)+(1010) = X$

$(0)+(0)+(0)+(0)+(0)+(\underline{1})+(0)+(0)+(0)+(0) = \underline{1}$

5-23.

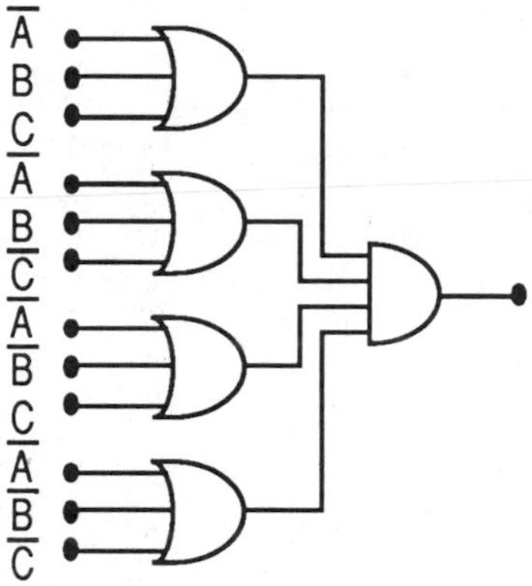

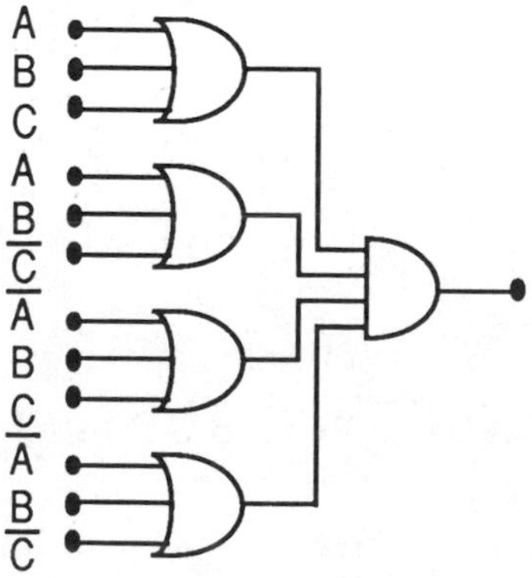

5-27. SOP from #15

5-25. SOP from #12

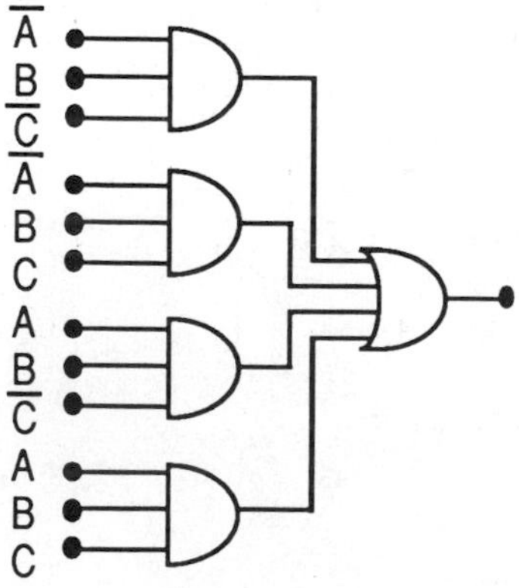

POS from #16

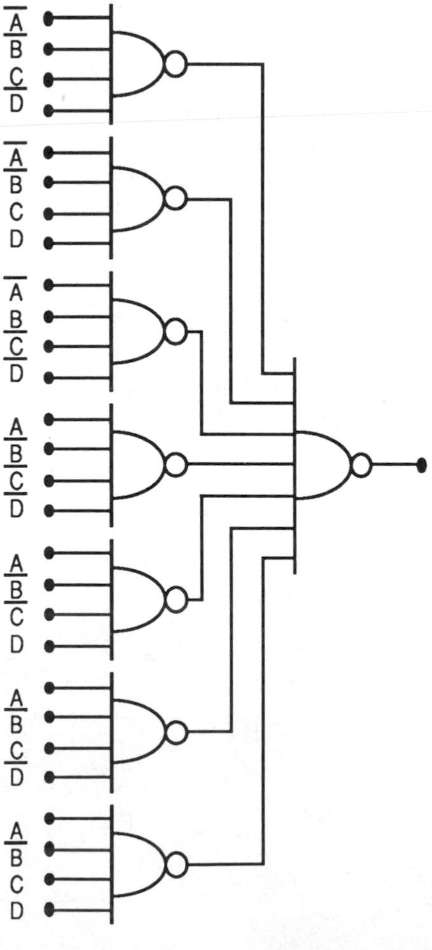

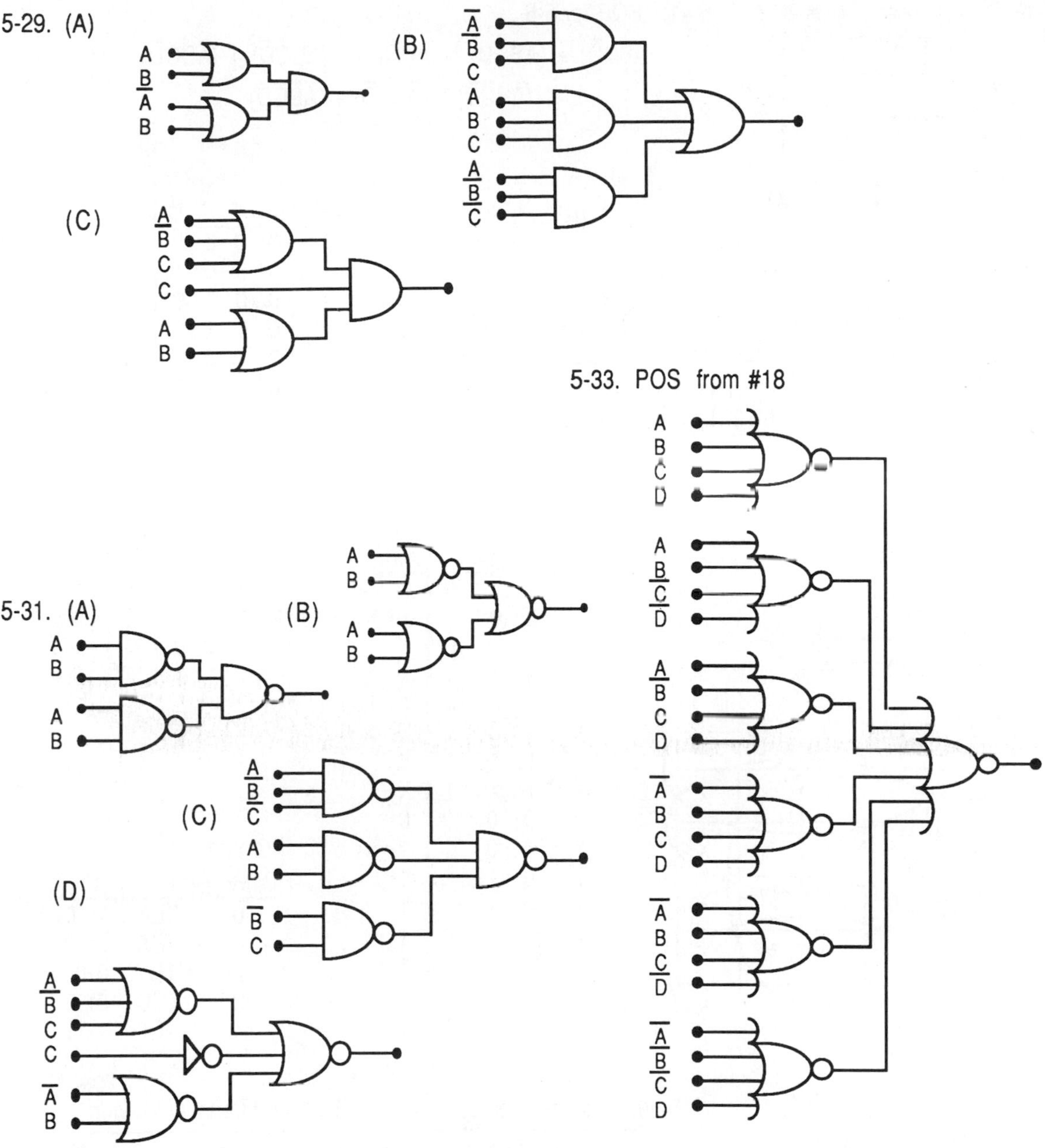

5-35. (A) OK (B) OK (C) OK (D) OK

5-37. from K-MAP 5-1 $(\overline{B}\,\overline{C}\,\overline{D}) + (ABD)$

from K-MAP 5-2 $B\overline{D}$

5-39.

	CD	$C\overline{D}$	$\overline{C}\overline{D}$	$\overline{C}D$
AB	1	0	0	1
$A\overline{B}$	0	1	1	0
$\overline{A}\overline{B}$	0	1	1	0
$\overline{A}B$	1	0	0	1

5-41. SOP for 5-5

$$(ABC\overline{D}) + (ABCD) + (AB\overline{C}D) + (AB\overline{C}\overline{D}) + (\overline{A}BC\overline{D}) + (\overline{A}B\overline{C}\overline{D}) + (\overline{A}\overline{B}C\overline{D}) + (\overline{A}\overline{B}\overline{C}\overline{D})$$

5-43. (A) 5-5

(B) 5-5

5-45.

A	B	C	D	5-3	5-4
0	0	0	0	1	0
0	0	0	1	1	1
0	0	1	0	0	0
0	0	1	1	0	1
0	1	0	0	1	1
0	1	0	1	1	0
0	1	1	0	0	1
0	1	1	1	0	0
1	0	0	0	0	0
1	0	0	1	0	1
1	0	1	0	1	0
1	0	1	1	1	1
1	1	0	0	0	1
1	1	0	1	0	0
1	1	1	0	1	1
1	1	1	1	1	0

5-47. from #38 (5-3)

5-49. from #40

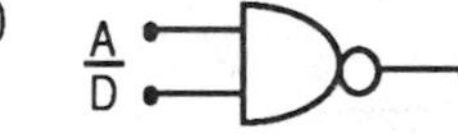

5-51. D1 + D2 + D3 = Alarm

	$da\overline{D1}$	$daD1$	$\overline{da}D1$	$\overline{da}\overline{D1}$
$D2D3$	1	1	X	X
$D2\overline{D3}$	1	1	X	X
$\overline{D2}\overline{D3}$	0	1	X	X
$\overline{D2}D3$	1	1	X	X

dark	D1	D2	D3	Alarm
0	0	0	0	X
0	0	0	1	X
0	0	1	0	X
0	0	1	1	X
0	1	0	0	X
0	1	0	1	X
0	1	1	0	X
0	1	1	1	X
1	0	0	0	0
1	0	0	1	1
1	0	1	0	1
1	0	1	1	1
1	1	0	0	1
1	1	0	1	1
1	1	1	0	1
1	1	1	1	1

53. Program 5-2

Address		Bits
0	CLA	1110
1	TBA	0100
2	NTA	0101
3	AND	0110
4	HLT	1111

5-55. Data Register: 1111
Accumulator: 0111
B-Register: 1100

5-57. Data Register: 1111
Accumulator: 1011
B-Register: 0011

5-59.

Address	Operation
0	LDB 5
2	LDA 3
4	NTA
5	INR
6	ADD
7	HLT

5-61.

Address	Operation
0	LDA 4
2	LDB 5
4	NTA
5	INR
6	ADD
7	HLT
8	TAM

5-63. Data Register: 1111
Accumulator: 0010
B-Register: 0011

5-65
Data Register: 1111
Accumulator: 1111
B-Register: 0000

Address	Operation	Bits
0	CLA	1110
1	TAB	0100
2	NTA	0101
3	TAM (4)	1001
		1111
5	TBM (6)	1010
		0000
7	HLT	1111
8	(TAM) ?	1001

5-67. 20

5-69. 0.45V

5-71. 21.1 mW all outputs Low, Vcc at 5.25 V

5-73. This should exceed 45 nsec. in worst case conditions depending on the type chosen.

5-75. Again, much less than the 200 mA above, depending on the type chosen.

Review Questions Chapter 6

6–1:

1. The major parts of a computer are the input, the output, processing circuit and memory.
2. (A) Memory is made up of control switches. (B) The processing circuit consists of logic gates. (C) Instruction is the bit pattern made by the control switches.
3. Hardware is the tangible part of a logic system. Software is the nontangible part that determines the process the hardware will use on the input data. For the PLG, the hardware is the gates, switches and connecting wires, while the software is the bit pattern of the switches.
4. An *instruction* determines the process that the processing circuits will use on the input data. A *process* is that which determines the relationship between the input data and the results of the output.
5. A PLG replicates a computer because it contains all of the essential elements that are used to define a computer. It is a programmable logic device that consists of an input, output, processing circuit and memory.

6–2:

1. A data selector is capable of selecting one line of data (from several) and transmitting it along one output.
2. A digital multiplexer can be thought of as the equivalent of a rotary switch that can select one of several logic levels to a single output.
3. A Multiplexer/Data Selector can be used to replicate the logic conditions of a truth table by using its *data inputs* to control the logic level of the output and have the *data select inputs* represent the data of the truth table.
4. ROM stands for *read only memory*. It is memory that contains instructions that are not easy to change.
5. Read-write memory is memory that can easily have its instructions changed.

6–3:

1. A PLA is a programmable logic array consisting of AND/OR gates that represent a POS expression. Both the AND gate and OR gate lines are programmable.
2. The logic output of an unprogrammed PLA is always the same no matter what the input.
3. A PLA is programmable because of the microscopic fuses contained within the device. These can be selectively 'blown' to program the device.
4. One application of a PLA is to perform arithmetic or logic functions.
5. The purpose of a diode is to cause current to flow in one direction only.

6–4:

1. PLD notation is a method of representing a programmable logic device so that it is easy to see the bit pattern of the device.
2. In PLD notation, a product line is a single line that represents multiple inputs to an AND gate.
3. The number of input lines are indicated on a product line by the number of vertical bus lines cross-

ing the product line with an indication, such as a small x, that the fuse connection is not blown.

4. The logic value of an unprogrammed product line is FALSE.

6–5:

1. A decoder is a logic network that takes a specific bit pattern on its input and converts it to a specific logic condition on its output.
2. In a 2-line to 4-line decoder, only one output line is active at a time depending upon binary value of the two input lines.
3. The 74154 is a 4-line to 16-line decoder with two active LOW enable lines, four input lines and sixteen output lines only one of which will be active LOW depending upon the conditions of the input.
4. Two decoders can be used to increase the number of output lines by having the most significant bit of the input control the enable lines of each decoder so that only one unit is enabled at a time.
5. A decoder is the logic device that enables the logic system to select only one memory location at a time.

6–6:

1. Diodes are necessary in a ROM in order to prevent the two opposite logic outputs from being electrically connected to each other. If this happened, electrical damage to the system would result.
2. In a ROM, the AND gate connections are preprogrammed as a decoder and only the OR connections are programmable. In a PLA both the AND and OR combinations are programmable.
3. Unlike the PLA, gate minimization techniques are useless in a ROM because all of the AND connections are programmed at the factory.
4. A PROM is a programmable ROM when once programmed, its bit pattern remains fixed. An EPROM is an erasable programmable ROM whose bit pattern may again be changed.

6–7:

1. A *multiplexer* takes one of several inputs and channels it to a single output. A *demultiplexer* takes one input and channels it to several outputs.
2. A 1-line to 8-line demultiplexer means a logic device that takes one input and can channel it to one of eight outputs.
3. A *demultiplexer* can also be used as a *decoder*.
4. A *decoder* takes coded information and converts it to a familiar form while an *encoder* takes familiar information and converts it to a coded form.
5. An application of an encoder is to convert a keypad input into binary code.

6–8:

1. A *bilateral switch* is a switch that can be operated in either direction.
2. A CMOS bilateral switch is a solid state device that electrically appears as a bilateral switch.
3. A CMOS bilateral switch can be used as an *encoder* or a *decoder* as well as a digital-to-analog converter.
4. A *sound synthesizer* is capable of generating sound through the creation of desired waveforms. A *voice synthesizer* is a sound synthesizer that generates waveforms that sound like the human voice.

6–9:

1. A *bus* is a group of conductors treated as a unit.
2. An *address bus* is a group of conductors that contains information about a specific memory location or device.
3. A *data bus* is a group of conductors from a microprocessor that contains data being transferred to or from memory or a device.
4. The purpose of the READ/$\overline{\text{WRITE}}$ line is to active memory or devices for the purpose of copying information *from* them (READ) or *to* them ($\overline{\text{WRITE}}$).
5. A *control bus* is a group of conductors from a microprocessor that contains information for or about the microprocessor.

6–10:

1. One of the purposes of a resistor in logic circuits is to limit the amount of current flow from the 5-volt source.
2. A resistor is needed when a PBNO switch is used to control a logic input in order to limit the resulting current to a safe value.
3. An ordinary resistor is called a pull-up resistor when it is connected in such a manner as to produce an output logic level and reduce any resulting current to a safe level.
4. An open collector logic circuit requires a pull-up resistor to be connected to the +5-volt source for proper operation. The advantage of such a circuit is that the outputs of several gates can be connected together.

Chapter Problems (odd-numbered)

6-1. Computers have MEMORY, which stores information. Some INPUT allows information to be placed in memory, a PROCESSOR applies certain rules to the information, and some OUTPUT supplies the results of the processing.

6-3.

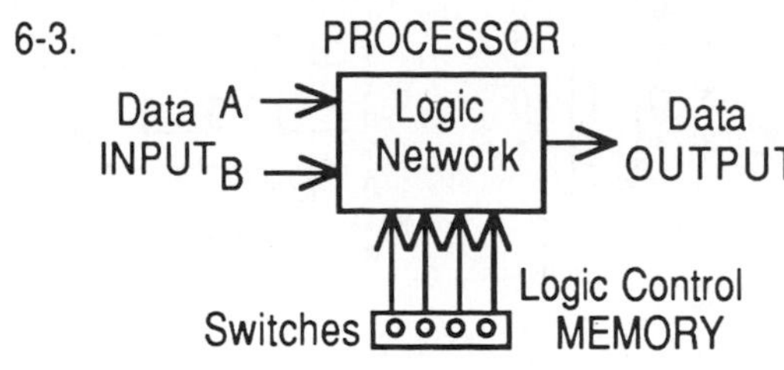

6-5. The instruction is the setting of the switches. This determines what gate the PLG will be like. The process is what the PLG does to the INPUT to create the OUTPUT.

6-7.

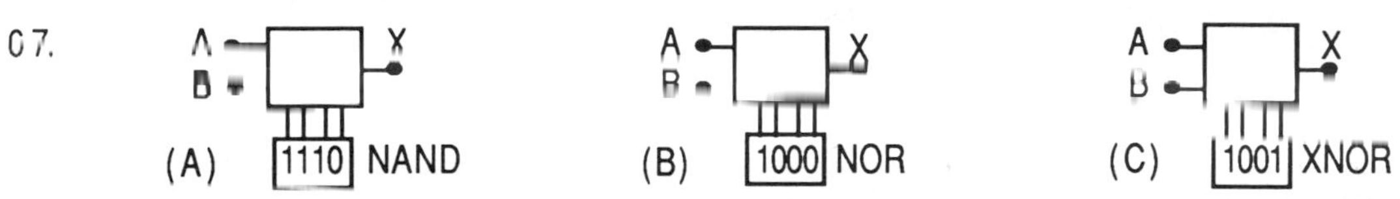

6-9.

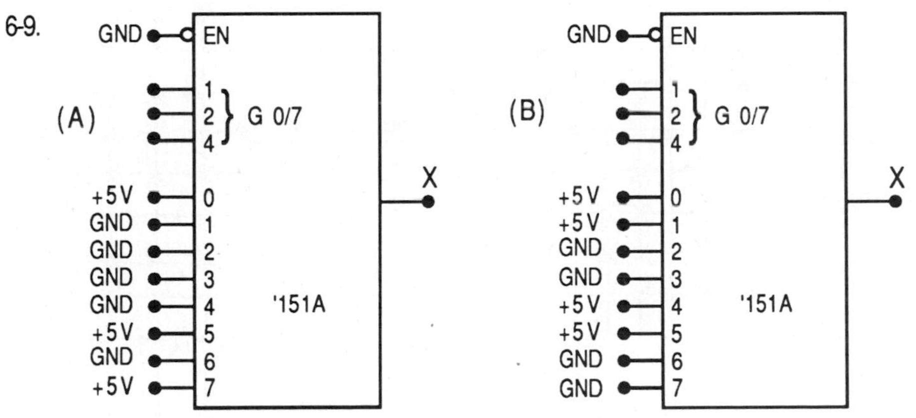

6-11.

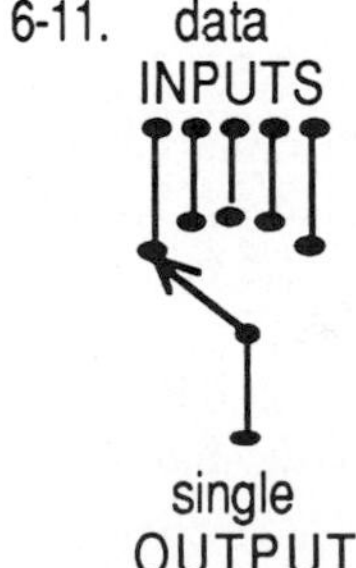

6-13. ROM is permanent. READ-WRITE memory is not. R/W can be easily changed.

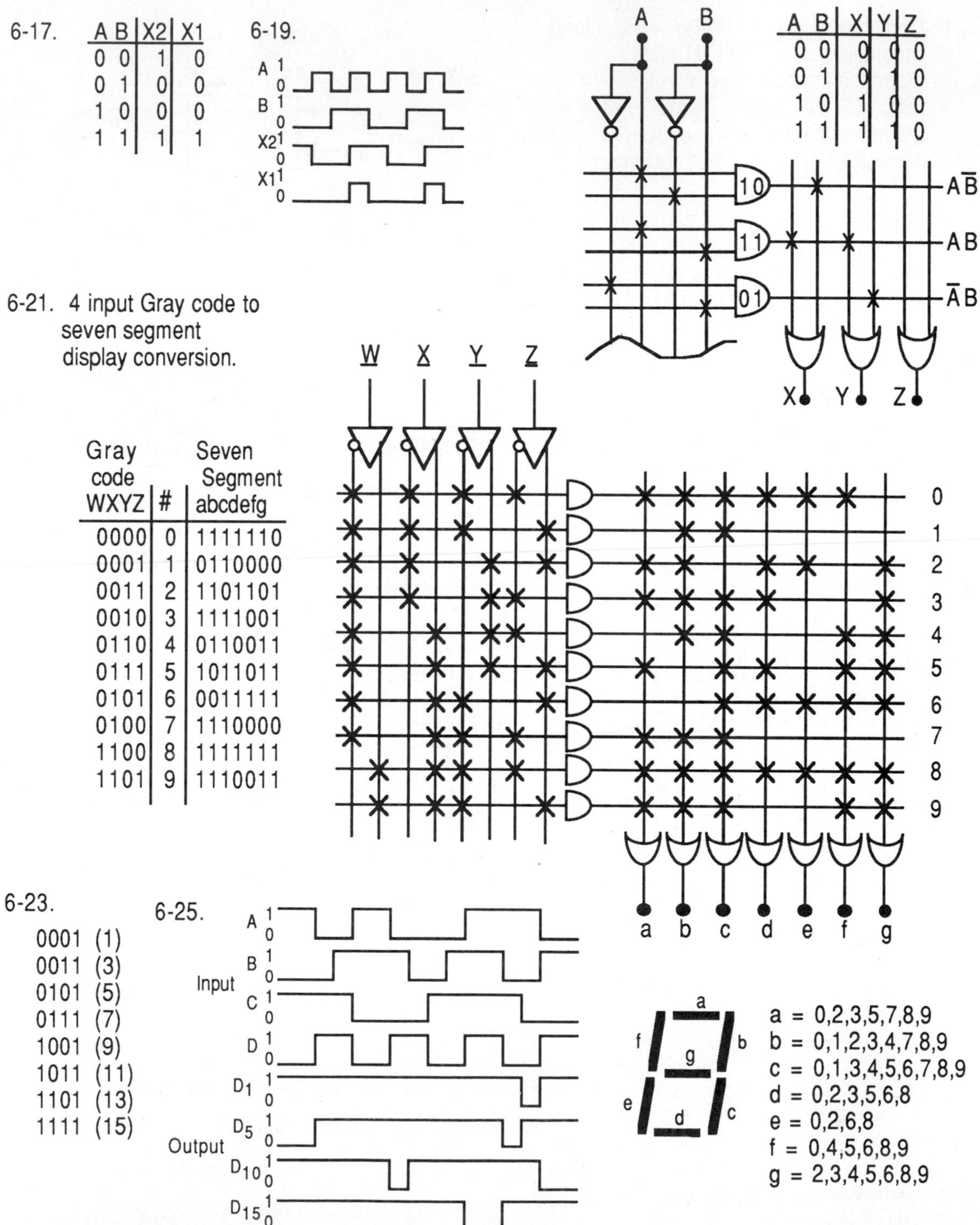

6-15.

A	B	X	Y	Z
0	0	0	0	0
0	1	0	1	0
1	0	1	0	0
1	1	1	1	0

6-17.

A	B	X2	X1
0	0	1	0
0	1	0	0
1	0	0	0
1	1	1	1

6-21. 4 input Gray code to seven segment display conversion.

Gray code WXYZ	#	Seven Segment abcdefg
0000	0	1111110
0001	1	0110000
0011	2	1101101
0010	3	1111001
0110	4	0110011
0111	5	1011011
0101	6	0011111
0100	7	1110000
1100	8	1111111
1101	9	1110011

a = 0,2,3,5,7,8,9
b = 0,1,2,3,4,7,8,9
c = 0,1,3,4,5,6,7,8,9
d = 0,2,3,5,6,8
e = 0,2,6,8
f = 0,4,5,6,8,9
g = 2,3,4,5,6,8,9

6-23.

0001 (1)
0011 (3)
0101 (5)
0111 (7)
1001 (9)
1011 (11)
1101 (13)
1111 (15)

6-27. A diode is used to prevent an electrical short circuit inside the ROM if a 0 and 1 input or output are both connected to the same line at the same time.

6-29. 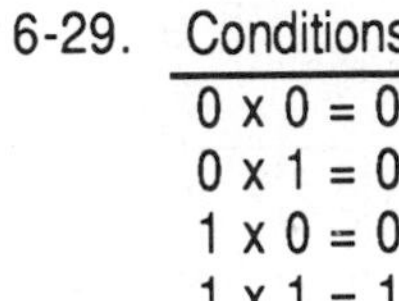

Conditions
0 x 0 = 0
0 x 1 = 0
1 x 0 = 0
1 x 1 = 1

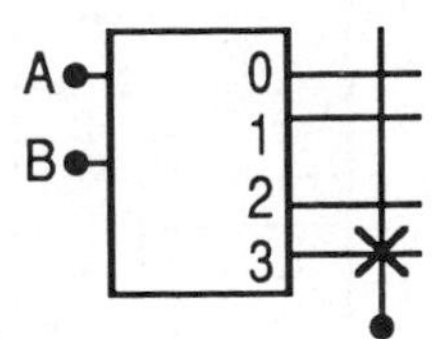

6-31 A demultiplexor routes one lineto different outputs. An encoder converts a pattern of inputs to a pattern of outputs. The demultiplexor is an addressable AND gate array, while theencoder is just the OR gate output with many possible inputs.

6-33. 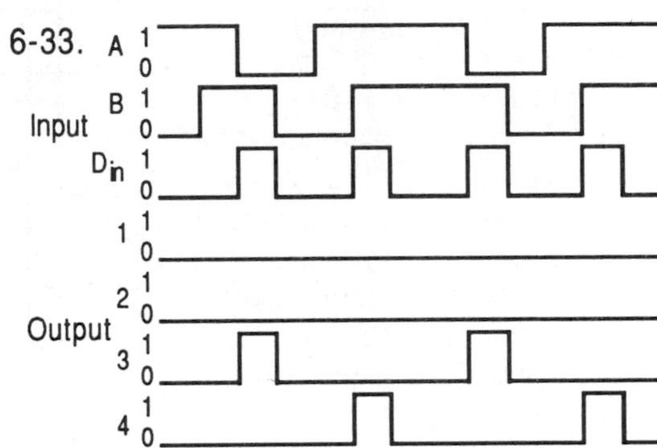

6-35. 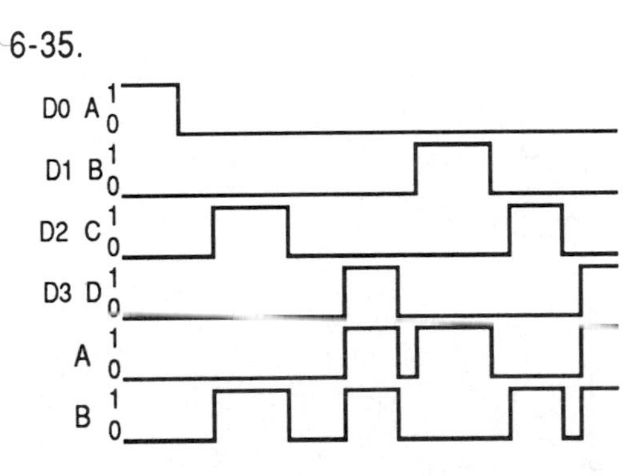

6-37. 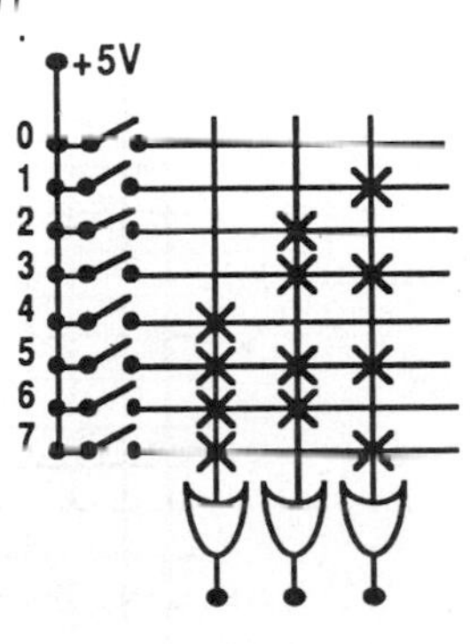

6-39. 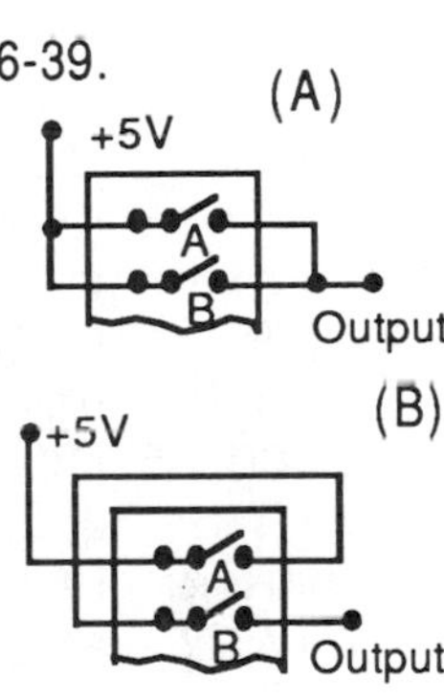

6-41. 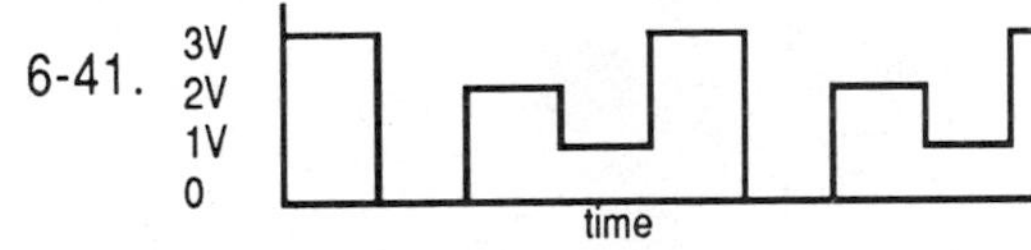

6-43. 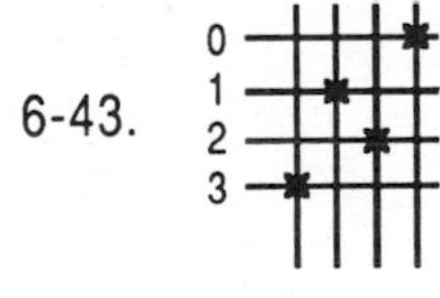

6-45.

Address Bus
μP
Data Bus
Control Bus
Power Bus

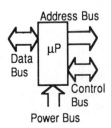

Address Bus: activates a specific address or device.
Data Bus: transfers bit patterns to and from μP.
Control Bus: provides signals that coordinate operation of μP and other parts of the system.
Power Bus: provides correct voltage and enough current for system operation.

6-47.

0 LDA 6
2 XOR
3 ORA
4 XOR
5 NOP
6 HLT

6-51. S1 will always read High. S2 will cause a short circuit.

6-49.

0	LDA	D	(1101)
2	TAM		(D)
3	LDB	A	(1010)
5	TBM		(A)
6	ORA		(1111)
7	TAM		(F)
8	TAM		(F)
9	LDA	1	(0001)
B	TAM		(1)
C	NTA		(1110)
D	TAM		(E)
E	HLT		

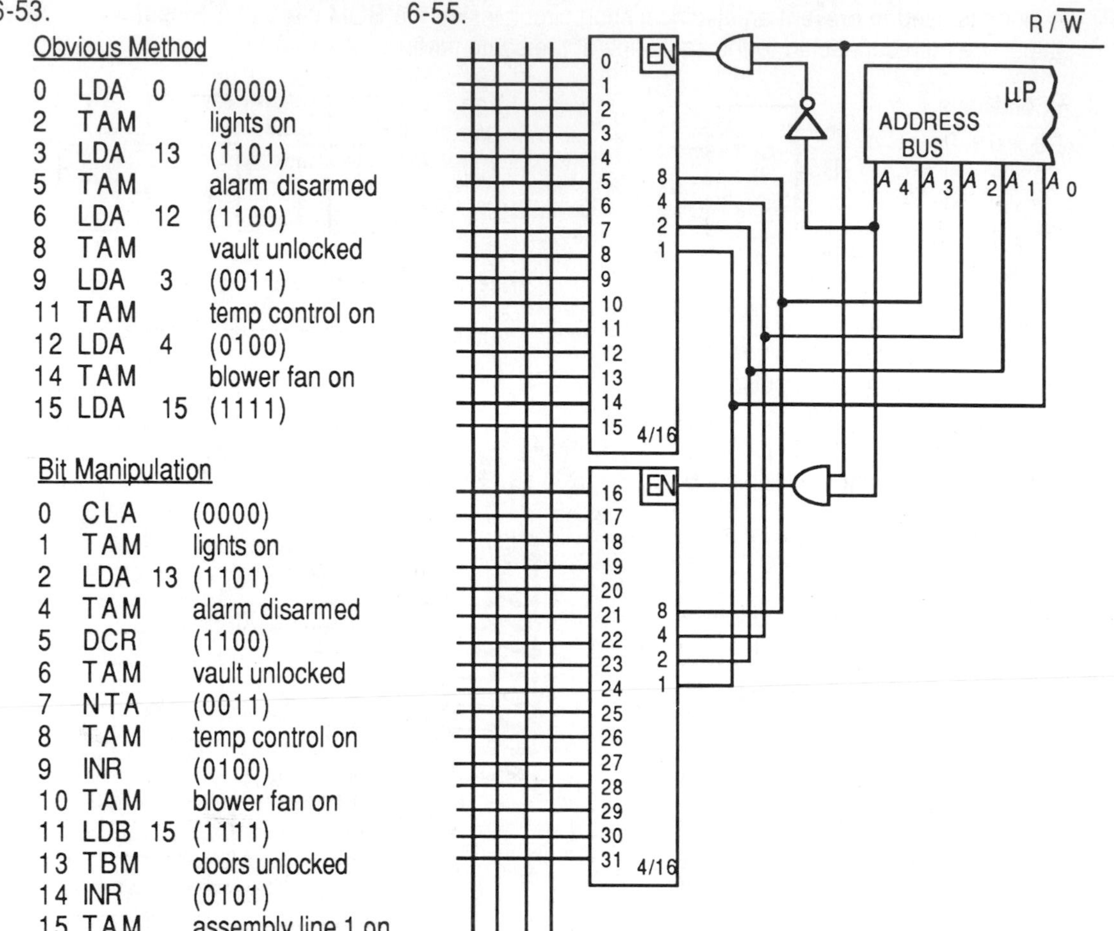
6-53.

Obvious Method

```
0  LDA  0   (0000)
2  TAM      lights on
3  LDA  13  (1101)
5  TAM      alarm disarmed
6  LDA  12  (1100)
8  TAM      vault unlocked
9  LDA  3   (0011)
11 TAM      temp control on
12 LDA  4   (0100)
14 TAM      blower fan on
15 LDA  15  (1111)
```

Bit Manipulation

```
0  CLA      (0000)
1  TAM      lights on
2  LDA  13  (1101)
4  TAM      alarm disarmed
5  DCR      (1100)
6  TAM      vault unlocked
7  NTA      (0011)
8  TAM      temp control on
9  INR      (0100)
10 TAM      blower fan on
11 LDB  15  (1111)
13 TBM      doors unlocked
14 INR      (0101)
15 TAM      assembly line 1 on
```

6-55.

Review Questions Chapter 7

7–1:

1. A latch stores a HIGH or LOW condition.
2. The purpose of the SET input is to cause $Q \rightarrow 1$ when SET is activated and remain in that condition until changed by the RESET input.
3. The purpose of the RESET input is to cause $Q \rightarrow 0$ when RESET is activated and remain in that condition until changed by the SET input.
4. The inputs to a NAND gate latch are active HIGH. The inputs to a NOR gate latch are active LOW.
5. The condition not usually allowed for a latch is to have both SET and RESET inputs active at the same time. The reason for this is that the output will be either unpredictable or unstable.

7–2:

1. A gated latch is an *S-R* latch with an *enable* input.
2. The purpose of an *enable* input is to control when the *S* and *R* inputs may effect the condition of the latch.
3. A gated-*D* latch has two inputs, one is an enable input and the other for data.
4. An application of a gated latch is a memory.
5. The term *read* means to copy the stored bit pattern from memory and the term *write* means to change the contents of the memory.

7–3:

1. A *latch* is a level-triggered flip-flop.
2. An edge-triggered flip-flop is activated on a logic

transition of the clock pulse, either HIGH to LOW or LOW to HIGH.

3. A *dynamic input indicator* is a triangular symbol on the clock input of a flip-flop used to indicate an edge-triggered flip-flop.
4. An edge-triggered flip-flop provides for more precise timing of digital systems using flip-flops.

7–4:

1. A register is a group of flip-flops treated as a unit.
2. A binary counter is a register that can be incremented or decremented.
3. A parallel-in serial-out shift register could be used to transmit digital information from a computer over a telephone line.
4. A serial-in parallel-out shift register could be used to receive digital information from a telephone line.

7–5:

1. A multivibrator is a two-state output circuit capable of acting as an oscillator by generating its own signal or maintaining one stable state or maintaining one of two stable states.
2. The three classifications of multivibrators are bistable multivibrator—has one of two stable states; monostable multivibrator has only one stable state and astable multivibrator which does not have a stable state.
3. The purpose of a switch debouncer is to remove the unwanted momentary level changes brought about by the mechanical action of switch closure.
4. Symmetry is a measurement of the ON time to the OFF time of a train of pulses.
5. A clock is used as the basic timing signal in a digital system.

7–6:

1. One of the problems in using a non master-slave flip-flop is that the output of a flip-flop can be in the process of changing at the same time this output is trying to be read by the input of another flip-flop.
2. The basic construction of a master-slave flip-flop consists of two flip-flops in one, where the output of the first flip-flop feeds to the input of the second flip-flop.
3. The timing relations in a master-slave flip-flop are such that the master is activated on the leading edge of the clock while the slave is active on the trailing edge of the clock.
4. The advantage of a master-slave constructed flip-flop is that the output is stable while the input is active and the input is stable while the output is active. Thus when the input of one flip-flop is being fed by the output of another, the output will not change while it is being read.

7–7:

1. A TRI-STATE buffer has three output conditions: HIGH and LOW and disconnected from the circuit.
2. An *addressing mode* is the method used by the microprocessor for transferring information.
3. Immediate addressing means that the data is in the next memory location *immediately* following the instruction while direct addressing means that the data will be in the memory location *directed* by the instruction.
4. The IN instruction will cause the MEM/$\overline{\text{IO}}$ and WRITE lines to both be LOW and the READ line to be HIGH. The OUT instruction will cause the MEM/$\overline{\text{IO}}$ and the READ lines to be LOW and the WRITE line to be HIGH.
5. The output buffer will have the bit pattern on its input lines loaded into its flip-flops when its enable line is active. The input driver will have the bit pattern of its input lines appear on the output lines when its enable line is active.

7–8:

1. The 555 timer can be used as an astable flip-flop.
2. A comparator compares an input voltage to a reference voltage. Its output will be either HIGH or LOW depending upon the relationship of the input voltage to the reference voltage.
3. There are two comparators in the 555 timer. One causes the *S-R* flip-flop to SET while the other causes it to RESET.
4. The output frequency of a 555 is determined by the value of external components (two resistors and a capacitor).
5. The amount of time a one-shot stays in its unstable state is determined by the values of its external components (a resistor and a capacitor).
6. A Schmitt trigger is a device that will change its output state abruptly when the input signal reaches a specified DC level.

Chapter Problems (odd-numbered)

7-1. A Latch can store the condition of a single bit of information until changed by an external source.

7-3. A Latch has two outputs. The outputs must always be in opposite states.

7-5.

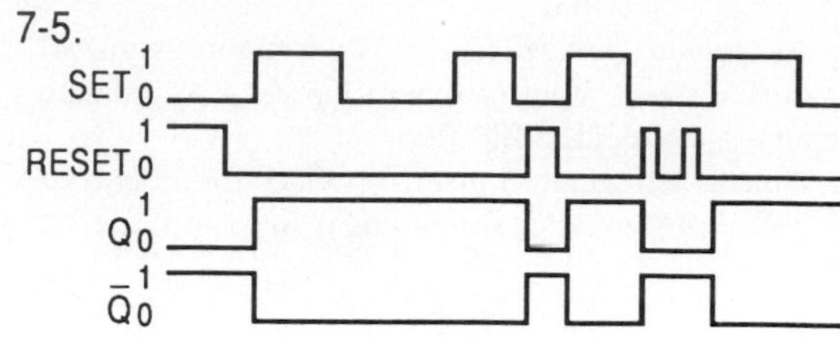

7-7.

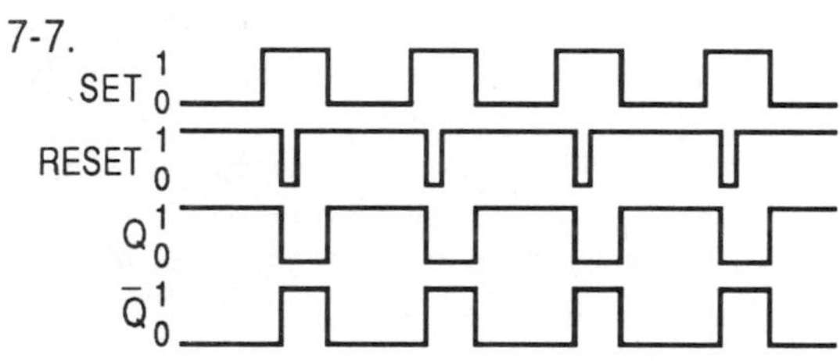

7-9.

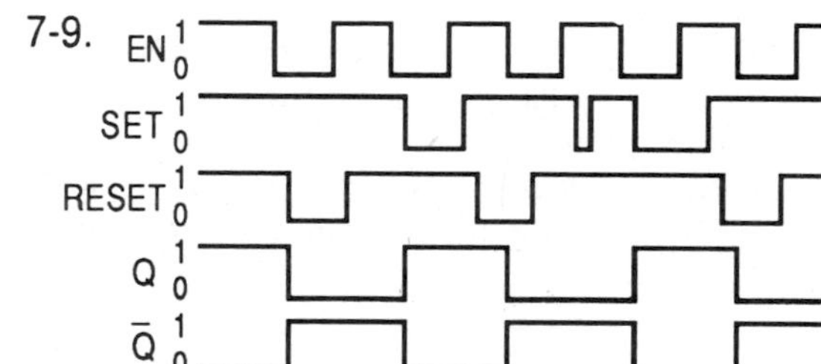

7-11. Each flip flop will Set or Reset Q anytime P or Q is active without regard to the state of the clock. If S, or R and the CLK are simultaneously active, it will Set or Reset.

(A)

P	C	CLK	S	R	Q
0	0	Î	0	0	n/c
0	0	Î	0	1	0
0	0	Î	1	0	1
0	0	Î	1	1	n/a
0	0	!	X	X	n/c
0	1	X	X	X	0
1	0	X	X	X	1
1	1	X	X	X	n/a

(B)

P	C	CLK	S	R	Q
0	0	!	0	0	n/c
0	0	!	0	1	0
0	0	!	1	0	1
0	0	!	1	1	n/a
0	0	Î	X	X	n/c
0	1	X	X	X	0
1	0	X	X	X	1
1	1	X	X	X	n/a

(C)

P	C	CLK	S	R	Q
0	0	!	1	1	n/c
0	0	!	1	0	0
0	0	!	0	1	1
0	0	!	0	0	n/a
0	0	Î	X	X	n/c
0	1	X	X	X	0
1	0	X	X	X	1
1	1	X	X	X	n/a

(D)

P	C	CLK	S	R	Q
1	1	!	1	1	n/c
1	1	!	1	0	0
1	1	!	0	1	1
1	1	!	0	0	n/a
1	1	Î	X	X	n/c
1	0	X	X	X	0
0	1	X	X	X	1
0	0	X	X	X	n/a

In the chart above, n/a refers to a not allowed input condition and n/c means no change in Q. The S, R, P and C inputs are level active. In figure (A) and (B) they are all active high. In (C) P and C are active high while S and R are active low and in figure (D) all inputs are active low. When P is active the Q output goes high. When C is active the Q output goes low. IF both P and C are inactive, an active S input can set Q to high and an active R input can reset Q to low if the CLK makes an active transition. CLK is active on the leading edge in (A) and on the trailing edge in (B), (C), and (D). .Q and Q are outputs, and are always in opposite states. The pairs of asynchronous inputs P and C or synchronous inputs S and R can not be active simultaneously.

7-13.

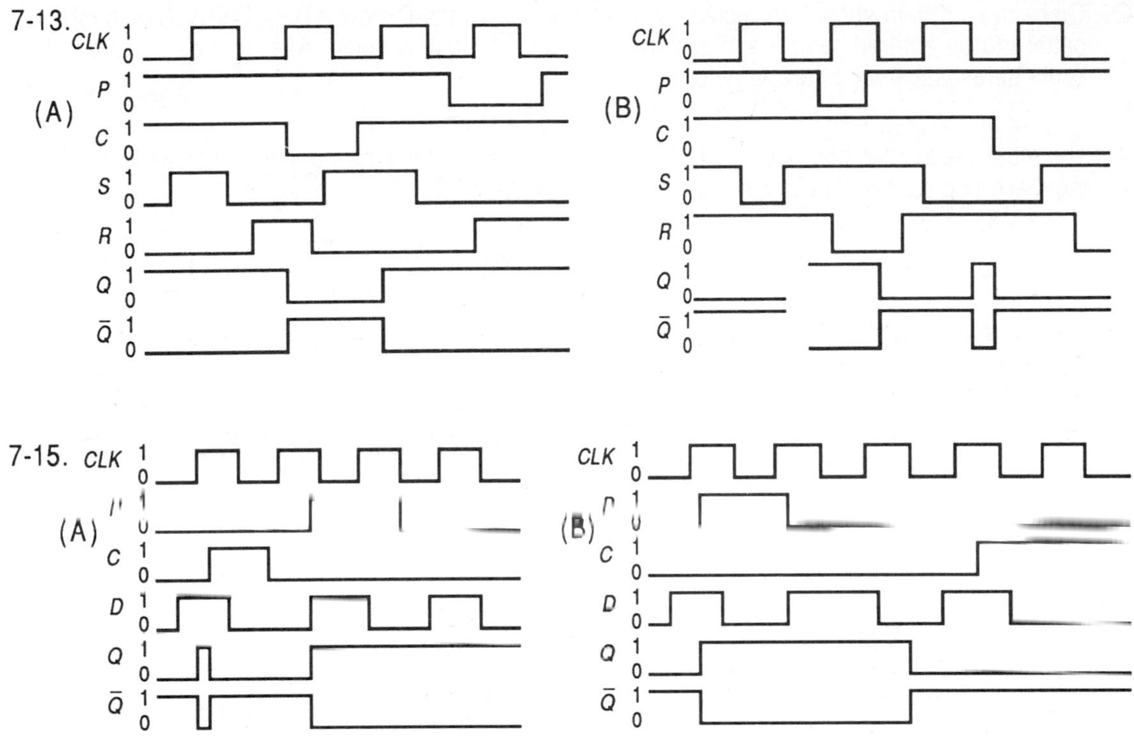

7-17. A binary counter is a group of flip flops connected so that on each CLK pulse, the pattern of 1's and 0's changes as designed. A binary count appears, in order, on the flip flop outputs.

7-19. 65536 times

7-21. Bit patterns are transferred into the register 1 bit at a time.

7-23. Parallel-in-serial-out

7-25. A bistable multivibrator will maintain either state, depending on it's control inputs. The monostable type of multivibrator will maintain one state, and switch to the other, unstable state for a designed period of time. The astable multivibrator will not maintain a single state, but switches back and forth at a designed rate, creating an "oscillating" digital signal.

7-27. Both the sync and clock pulses vary their logic levels in a predetirmined pattern to activate the computer system. The clock is used as a reference for all operations in the computer. Sync pulses are additional signals (often generated from the clock) that time events in the system.

7-29. On each CLK data should be copied by the Q output from the D input. This should allow a bit pattern to be entered serially and and shift right through the flip flops. After four clock pulses it could be available as a parallel output.

7-31. The master-slave flip flop has two flip flops in one. The input is active during the clock level and the slave copies the data to the output a half clock cycle later.

7-33.

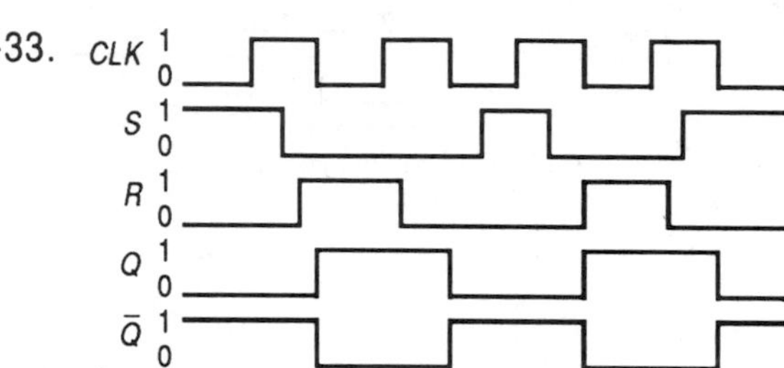

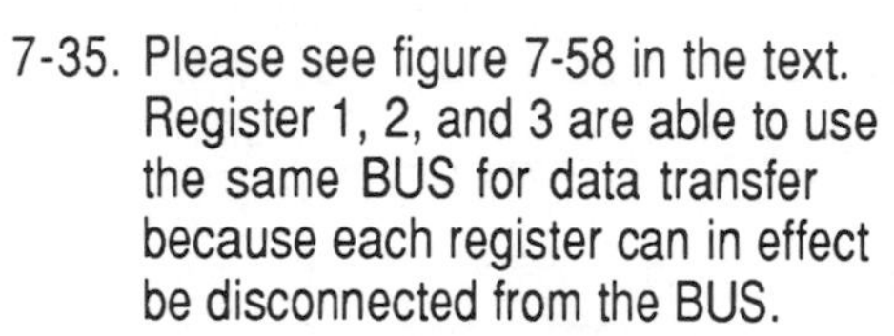

7-35. Please see figure 7-58 in the text. Register 1, 2, and 3 are able to use the same BUS for data transfer because each register can in effect be disconnected from the BUS.

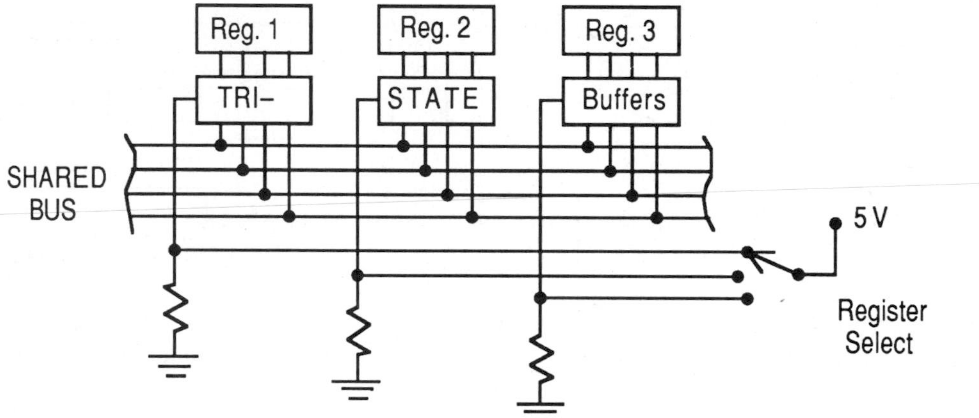

7-37. Immediate means the data immediately follows the instruction. Direct means the microprocessor can be "directed" to the correct memory location address that holds the data. The number following the instruction is the address that "directs" the μP to find the data.

7-39. (A)
```
0 MVI A,4
2 HLT
```
(B)
```
0 LDA 4
2 HLT
```

7-41. (A)
```
0 IN 1
2 OUT 1
4 HLT
```
(B)
```
0 IN 2
2 OUT 2
4 HLT
```

7-43. A comparator compares the levels of two voltages, and outputs a HIGH or LOW based on which is larger.

7-45. The one-shot has one stable state but will output the unstable state when triggered for a set time. The length of the unstable pulse is set by the values of external parts chosen by the designer or technician.

7-47.

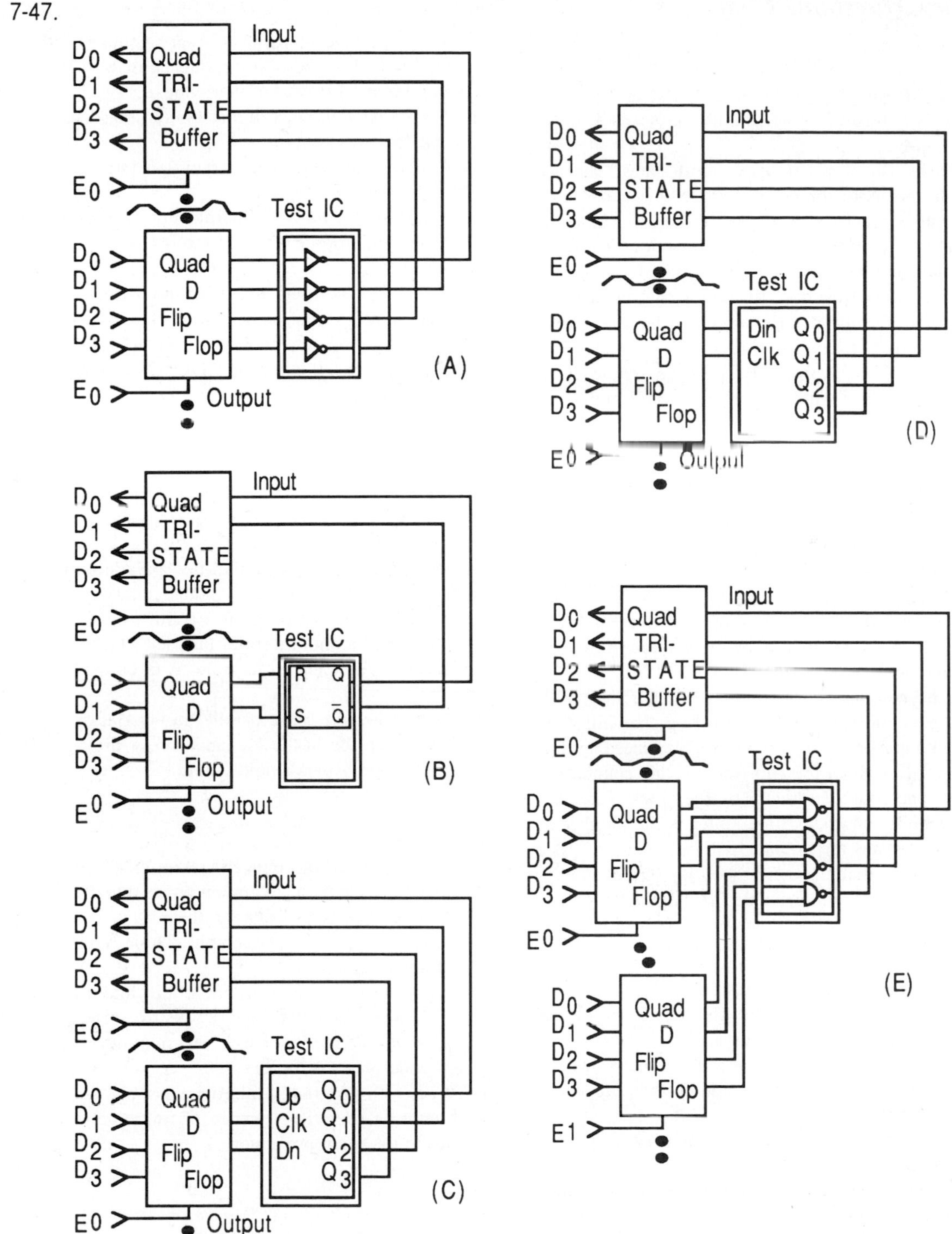

Input
Quad
TRI-
STATE
Buffer
Test IC
Quad
D
Flip
Flop
Output
(A)
(B)
(C)
(D)
(E)
R
S
Q
Up
Clk
Dn
Din
Clk
Q0
Q1
Q2
Q3
E0
E1
D0
D1
D2
D3

Review Questions Chapter 8

8–1:

1. The *S-R* flip-flop has three useful states. The one that is not useful is when both the *R* and *S* inputs are active.
2. The *JK* flip-flop has four useful states. The one state that is different from the *S-R* is when both inputs are active. This condition will cause the output of the *JK* to toggle with each clock pulse.
3. A data lock-out feature refers to an edge-triggered master-slave flip-flop. It means that changes of the data during a clock *level* will have no affect on the output.
4. The output waveform has half the frequency as the input clock when the *JK* flip-flop is in the toggle mode. This is called frequency division.

8–2:

1. A binary counter is capable of displaying a bit pattern that represents a sequence representative of counting in binary.
2. A binary counter acts as a frequency divider because the output waveform of each flip-flop is exactly half the frequency of the waveform applied to its clock input.
3. An asynchronous counter is a counter that, with the exception of the first flip-flop, has all of its clock inputs from the output of the preceeding flip-flop. Thus all of the flip-flops of this counter are not synchronized directly by the clock pulse.
4. The MOD number of a counter is the number of distinct states the counter will assume.
5. A programmable counter has its MOD number controlled by a changeable bit pattern.

8–3:

1. A *synchronous* counter has each of its flip-flops controlled directly by the clock pulse while an *asynchronous* counter does not.
2. A *presettable* counter is a counter that can have a bit pattern loaded into it and continue its count from there.
3. A *ring* counter continually circulates a bit pattern through it.
4. The advantage of a *ring* counter is that no decoding is necessary to take advantage of its unique states. The disadvantage of one is that it has less unique states than a binary counter.

8–4:

1. The advantage of using a binary counter with a PLD is that the resulting sequence of bit patterns may be any bit pattern you choose.
2. A *state machine* is a digital system that has a set sequence of events.
3. A *state table* shows the condition (state) of the output for a given count.
4. A *microprogram* is the permanent instruction set inside a microprocessor used to perform a specific process.
5. A PLD can be used to control the action of a binary counter. This can be done by having the PLD determine when the counter will be CLEARED or placed in an UP count or a DOWN count.

8–5:

1. The four different kinds of registers presented in this section were: 1] parallel-in—parallel-out. 2] serial-in—serial-out. 3] parallel-in—serial-out. 4] serial-in—parallel-out.
2. Another name for a parallel-in—parallel-out register is a buffer register. Another name for a serial-in—serial-out register is a shift register.
3. When the output of a serial-in—serial-out register is connected back to its input it is called a *ring counter*.
4. An application for a parallel-in—serial-out register would be in the transmission of digital data from a bus over a single output.

8–6:

1. The advantage of connecting the inputs and outputs of registers to a common bus is that data may then be transferred between them.
2. A machine cycle is the length of time required to complete a process.
3. A *microcode* is the instruction built into a microprocessor that determines a specific process. A *microinstruction* is a single instruction in the microcode of a microprocessor.
4. When you program a microprocessor you are activating a specific sequence of microinstructions inside the microprocessor.

8–7:

1. The 74HC76 is a dual *JK* flip-flop that has separate inputs and outputs for each flip-flop.

2. The 7490 is a decode counter, the 7492 a divide-by-12 counter and the 7493 a full 4-bit binary counter.
3. The 74HCT191 counter can be parallel loaded and has a parallel output. It also can be used as a storage register.
4. The 74194 may be parallel loaded, shifted to the left or right and used as a storage register.
5. The XR2240 programmable timer contains an 8-bit binary counter that allows multiples of the time base (in powers of 2) to be used.

8–8:

1. The two major approaches to troubleshooting are batch troubleshooting and single system troubleshooting. In batch troubleshooting, several systems of the same type are being analyzed. In single system troubleshooting, different systems are analyzed or the same system is analyzed on a not frequent basis.
2. System comparison as applied to troubleshooting is a method of batch troubleshooting where the system under test is compared to an identical system called a standard.
3. A *magnitude comparator* is a circuit used to compare two quantities in order to determine their relationship. There are two types, a digital comparator and an analog comparator.
4. The fundamental gate used in a digital magnitude comparator is the XOR gate. In an analog comparator, the basic circuit is a difference amplifier.
5. A *digital margin detector* is a circuit that will produce a two state output that depends upon a prescribed range of voltage inputs.

8–9:

1. The basic operation of a frequency counter consists of having the frequency to be counted serve as a clock for a binary counter for a known amount of time.
2. A VCO is a voltage controlled oscillator. Its function is to have an output frequency that is proportional to an input voltage.
3. The basic operation of a digital voltmeter is to have a frequency counter read the frequency of a VCO where the input voltage of the VCO is the voltage to be measured.
4. A digital thermometer consists of an oscillator whose frequency is determined by a temperature sensitive element. The output frequency of the oscillator is then read by a frequency counter whose output is converted to a temperature reading through a PLD.
5. They all have in common a device that is sensitive to the measurement being conducted and this device in turn determines the frequency of an oscillator. The resulting oscillator frequency is read by a digital counter and the result displayed through a PLD to a readout.

Chapter Problems (odd-numbered)

8-1. The JK flip flop has the added ability to toggle on each clock pulse if both control inputs are active.

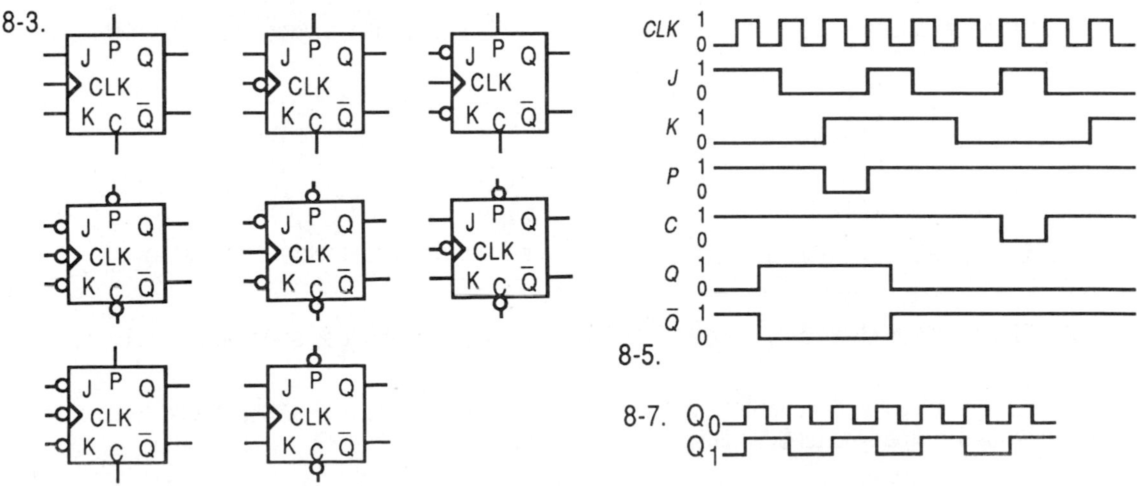

8-9. The pattern is apparent when you list and study the binary numbers in a vertical chart. The 1's position toggles every count, and each successive higher power position toggles half as often as the next lower position. Each position toggles at half the frequency of the next column. This counting pattern could thus be exactly simulated by flip flops connected to toggle on the preceeding flip flop's output as they divide the frequency by two.

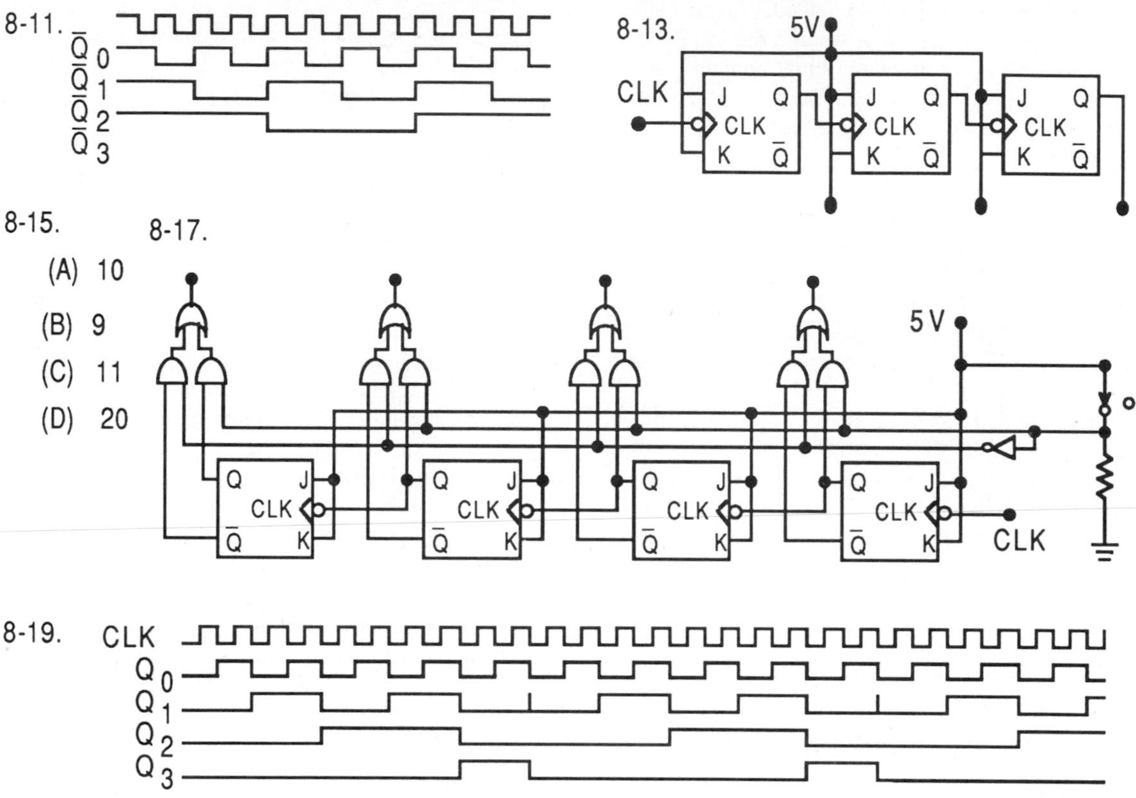

8-15.
(A) 10
(B) 9
(C) 11
(D) 20

8-21. The maximum counting frequency is limited by the total delay of all flip flops in the counter.

8-23.

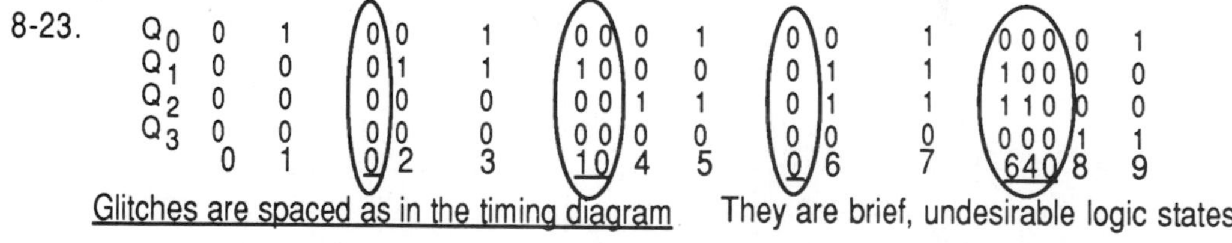

Glitches are spaced as in the timing diagram They are brief, undesirable logic states

8-25. 2.27 MHz, 12.5 MHz 8-27. A counter where every flip-flop changes state on the clock edge.

8-29. Refer to figure 8-27 in the chapter. The 4 flip-flop diagram includes a clear signal and need only be extended correctly to all 8 flip-flops in your sketch.

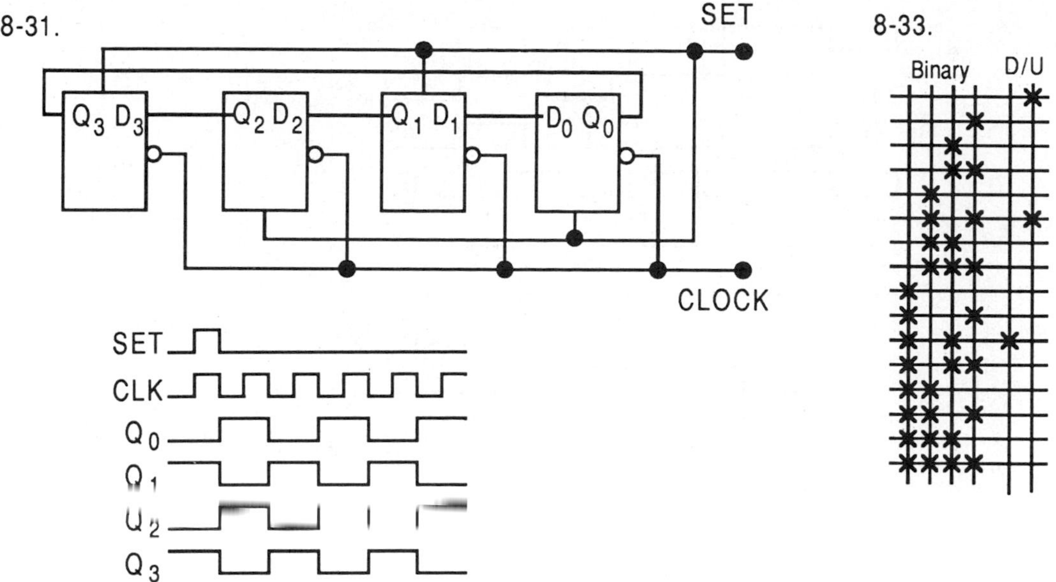

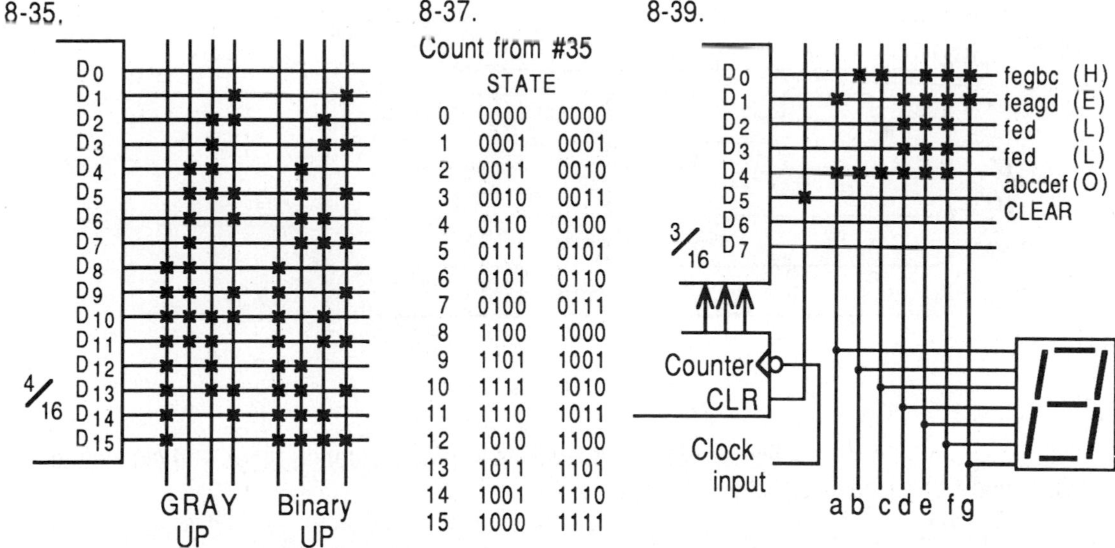

8-37.
Count from #35

STATE		
0	0000	0000
1	0001	0001
2	0011	0010
3	0010	0011
4	0110	0100
5	0111	0101
6	0101	0110
7	0100	0111
8	1100	1000
9	1101	1001
10	1111	1010
11	1110	1011
12	1010	1100
13	1011	1101
14	1001	1110
15	1000	1111

8-41. The set of instructions built into a microprocessor so it can do one of many different jobs. Example: the ring counter from the chapter in five steps including a reset on count 5.

8-43.

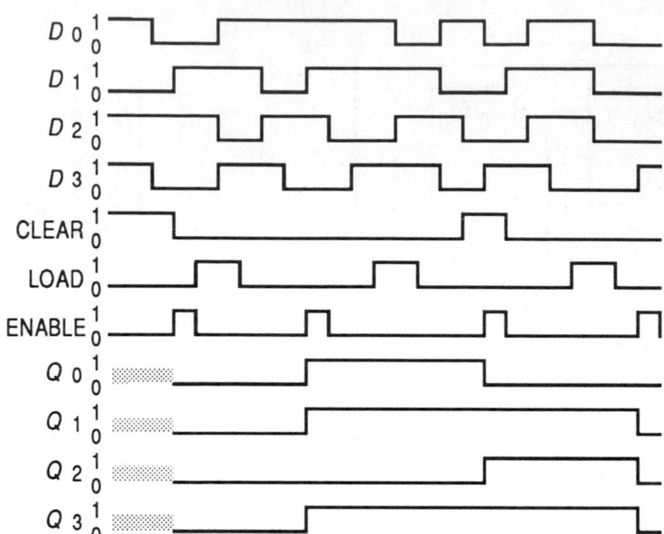

8-45.

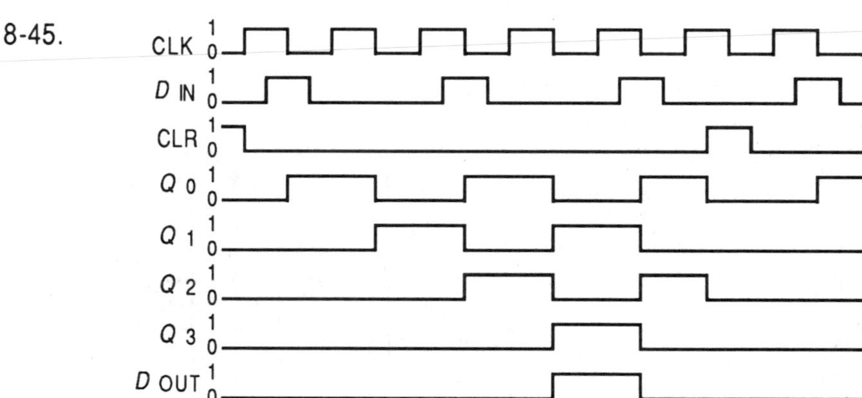

8-47. Refer to figure 8-50 in the chapter. Sketch 8 flip-flops with data steering gates and extend the LOAD and CLK signals correctly.

8-49. Refer to figure 8-53 in the chapter. The duplicate sketch requires 8 flip-flops with individual TRI-STATE outputs connected by one ENABLE line.

8-51. Sketch figure 8-58 from chapter. Control inputs are not required for question.

8-53.

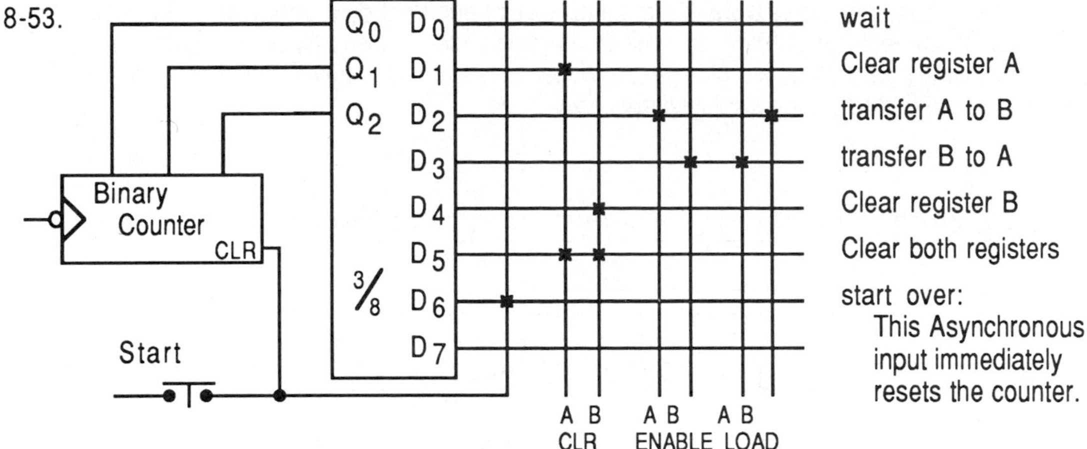

8-55. 74-(L) (S) (HC) etc. -73, -76, -107, etc. 4027, 4006 etc. are some possibilities.

8-57. Shift: 74-95, 96, 164, 165; 40-14, 15, 21, 35, etc. are possiblities
Buffers, Memory, Latch, 74-40, 89,173, etc.

8-59. 74-90, -92, -191, -192, etc. 40-17, -18, -29, -40, etc. are possibilities.

8-61. Each BIT's logic level is compared to the similar level in another word. A real comparator is internally gated to tell if two words are identical, and which is the larger of the two.

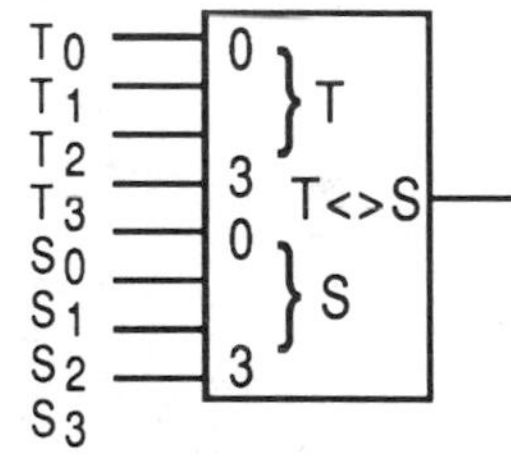

8-63.

Refer to figure 8-87 in the chapter. One possibility would be to replace the 2 bit counter, buffers, test gates, and XOR with the hardware at the right;

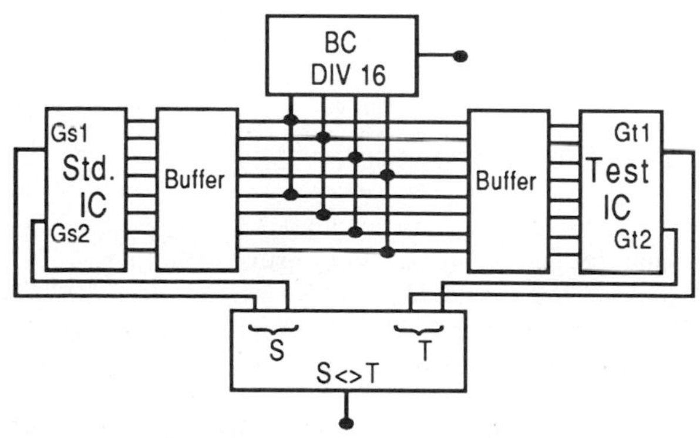

8-65. A frequency counter samples a waveform for an exact period set by a one shot. During this time interval, the test frequency is connected to a binary clock. If the clock overflows the one shot is switched to a quicker time period. When a valid count is taken, the count and the one shot period determine the displayed frequency.

8-67.

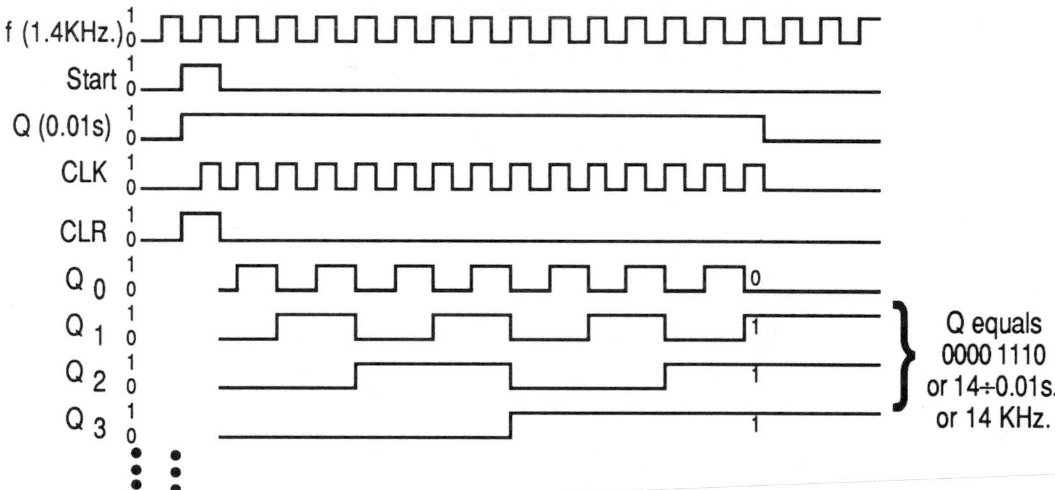

8-69. One possibility based on figure 8-87 in the chapter.

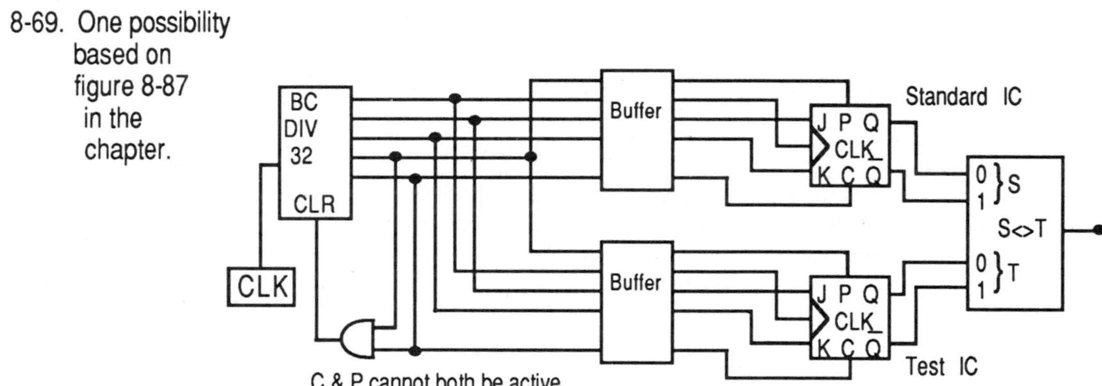

Review Questions Chapter 9

9–1:

1. Digital memory is any medium capable of storing digital information.
2. The four major kinds of digital memory are magnetic, electrical, mechanical and optical.
3. A specific example for each kind of digital memory would be: Magnetic—cassette tape; Electrical—switch programmable ROM; Mechanical—punched card; Optical—bar code.
4. The two different ways of accessing memory are *random access* which means that it takes the same amount of time to get any piece of digital information and *sequential access* which means the amount of time it takes to get information depends upon its location in the storage medium.
5. *Permanent* memory cannot have its digital infor-

mation altered without destroying the medium on which it is contained. *Erasable* memory can have all digital information removed and *alterable* memory can have new digital information stored in it.

9–2:

1. Nonvolatile, nonalterable, randomly accessible electrical memory means digital memory that does not require an external source of energy to store its bit patterns, cannot have its bit patterns changed without destruction of the storage medium, takes an equal amount of time to access any one bit pattern and requires electrical energy to make a copy of its bit patterns.
2. ROM refers to any type of digital memory that does not require an external source of energy to store its information and cannot have its contents altered without destroying the storage medium. It literally means Read Only Memory meaning memory that can only have its bit patterns copied.
3. A *cell* is the smallest unit in electrical memory capable of storing one bit of data. A *word* is a group of bits treated as a unit (one piece of information). An *address* is a number that represents the location in memory of the *word*. *Data* is the information stored in the memory.
4. A *mask* ROM is a ROM that has its bit patterns put into it at the factory by using a mask that contains the desired bit patterns during the manufacturing process.
5. A *look up table* is a ROM where the address represents an independent variable and the resulting word a dependent variable. The resulting output is thus some function of the numerical value of the address.

9–3:

1. Nonvolatile means the memory does not require an external source of energy to store its bit pattern. Alterable means its bit pattern may be changed by the user. Randomly accessible means that it takes the same amount of time to access any piece of data stored in the device. Electrical means that it requires electrical energy to copy the bit patterns.
2. The kind of electrical memory that may be altered only once is a PROM.
3. A PROM is constructed in the same manner as a switch programmable ROM but the difference is that each switch is replaced by a fuse. The device comes from the factory with all of its fuses intact (every cell is a 1). Its contents are programmed by electrically blowing a fuse in order to make the selected cell a 0.
4. A PROM may have its contents altered only once, an EPROM may have its contents altered between exposure to ultraviolet light and an EEPROM may have its contents altered after an electrical pulse has been applied to bring its cells all to the same logic level.
5. An EAROM is the same as an EEPROM. It means Electrically Alterable Read Only Memory.

9–4:

1. Volatile means that it requires an external source of energy to maintain its bit pattern, randomly accessible means that it requires the same amount of time to get any bit pattern, static means it requires no further processing to maintain its bit pattern and it is electrical memory because it requires electrical energy to retrieve its data.
2. RAM means volatile memory that may be altered (WRITTEN to) as easily as it is copied (READ from). The letters RAM actually stand for Randomly Accessible Memory.
3. RAM is used in digital systems where quick storage of digital data is required. In these systems it is desirable to have data altered as quickly as it can be copied.
4. The data lines of RAM are bidirectional, where the data lines of ROM are output lines only (data may only be copied).
5. Zero power RAM acts as nonvolatile RAM at least for the first 11 years of its life. This is accomplished by having a built in power source.

9–5:

1. Static RAM does not require any processing to maintain its programmed bit pattern. Dynamic RAM does require a process to maintain its programmed bit pattern.
2. The advantage of dynamic RAM is the ability to get higher storage density resulting in more memory in a smaller package.
3. *Refreshing* memory is the process of maintaining the correct bit pattern in dynamic memory by maintaining the correct cell charge.
4. The number of pins on a DRAM package are kept to a minimum by address multiplexing.

5. The two different methods of refreshing DRAM are burst refreshing and distributed refreshing.

9–6:

1. A volatile, randomly accessible, electrical memory requires an external source of energy to maintain its stored data, has a variable access time and requires electrical energy to copy its data.
2. A FIFO memory stands for first-in–first-out memory. This means that the first piece of data entered into the memory will be the first piece of data read from the memory.
3. An application of FIFO memory would be the transferring of digital information from a high speed computer to a slow speed printer.
4. Recirculating memory utilizes a shift register connected as a ring counter to continually shift its stored data.
5. CCD memory stands for charge coupled device. Its main advantage is high packaging density.

9–7:

1. Magnetic memory is digital memory that uses patterns of a magnetic field to store digital information.
2. The advantages of magnetic memory are its nonvolatility and ease to WRITE or READ data. The disadvantages are slow speed and expense and unreliability of mechanical devices as compared to electrical memory which does not require any mechanical devices for its READ and WRITE operations.
3. The basic construction of a magnetic disk consists of a hard or flexible substrate with a magnetic material coating.
4. The most common types of magnetic disks are the hard disk, $5^1/_4$ inch floppy disks and small $3^1/_2$ diskette.
5. The precautions that should be used when with magnetic disks are as follows: Don't contaminate the surface; don't attempt to bend the surface; do not use sharp objects on the disk or its protective jacket; do not expose to direct sunlight; do not expose to temperature extremes; and do not expose to magnetic fields.

9–8:

1. The kind of storage produced by magnetic tape is nonvolatile, erasable, sequentially accessible, magnetic memory.
2. The three basic methods of storing digital data on tape are audio cassette tape, video cassette tape and large reel tape used in large computer systems.
3. The video cassette recorder has a rotating playback/record head that moves across the video tape. The audio cassette recorder has a stationary playback/record head.
4. Bubble memory is magnetic memory where digital information is represented by the presence or absence of small movable magnetic areas called "bubbles."
5. The advantage of bubble memory is that it is nonvolatile READ/WRITE memory. The disadvantage is that it is sequentially accessible which requires a longer access time over that of electrical memory.

9–9:

1. A compact disk is an optically reflective disk that contains digital information stored as surface changes that change the reflection of a laser beam focused on the rotating surface.
2. A CD ROM is a compact disk that stores digital data which can be used for digital processing.
3. Some of the advantages of CD ROM are its high storage density and the ability to carry visual as well as audio information. Its main disadvantage is slow access time.
4. A laser disk memory is a nonvolatile, randomly accessible, alterable, optical memory.
5. Some of the advantages of laser memory are high storage density plus the ability to alter information. Its major disadvantage is slow access time compared to other types of existing memories.

9–10:

1. An A/D converter stands for analog-to-digital converter. It converts analog information into digital information.
2. A D/A converter stands for digital-to-analog converter. It converts digital information into analog information.
3. A flash converter uses differential amplifiers and associated digital logic to convert an analog signal into a digital code.
4. The major disadvantage of a staircase A/D converter is that the greater the amplitude of the analog signal the longer will be the conversion time.
5. The major difference between a tracking A/D converter and a staircase A/D converter is that a tracking A/D converter does have to reset back

to zero before each sampling of the analog waveform.

9–11:

1. The microprocessor presented in this section contained 16 address lines and 8 data lines. The significance of this is that it can address 2^{16} memory locations and has the capability of having 2^8 different processes.
2. A *memory map* is a listing showing what is contained in all memory locations.
3. A *high order* address represents the first 8 most significant address bits of a 16 bit address line and a *low order* address represents the first 8 least significant address bits of a 16 bit address line.
4. *Partial* address decoding means that some of the address lines of the microprocessor are not decoded. *Full* address decoding means that all of the microprocessor address lines are decoded.
5. *Memory mapped* I/O means that an input/output port is treated the same as any other memory location by the microprocessor.

9–12:

1. The two major technologies used in the manufacturing of electrical memories is bipolar and MOS.
2. The electrical memory technology that has the greatest speed is ECL and the one that uses the least amount of power is CMOS.
3. The factors that influence the step size of a D/A converter are the value of the rated maximum output voltage and the number of digital inputs.
4. The most popular way of constructing D/A converters is using the R/2R ladder network.
5. The major advantage of a successive approximation A/D converter is a reduction of conversion time.

Chapter Problems (odd-numbered)

9-1. Any medium that stores digital data.

9-3. Electrical: I.C.'s, Magnetic: Disk, Mechanical: Punch Cards, Optical: CD Rom, and refer to figure 9-

9-5. Alterable can be changed, Permanent can not be changed without destroying the medium itself. Erasable can have all digital information removed.

9-7. (A) Calculator numbers. (B) Calculator key operations. (C) see A. (D) Tape drive. (E) I.C.'s

9-9. A silicon chip is masked during manufacturing so that certain areas that might be any value are permanently set to 1's or 0's. This contains the operations that must always occur in a system. A monitor, wake-up, math functions, or operating system could be placed in ROM.

9-11.

length	area	ROM
0	0	000000
1	1	000001
2	4	000010
3	9	001001
4	16	010000
5	25	011001
6	36	100100
	etc.	

9-13. Each address in a ROM can represent a sound. Accessing the ROM in correct order and with certain timing, speech is electronically synthesized.

9-15. Y = 32 – X

9-17. All three are user programmable at least once, randon access, and non-volatile. (B) and (C) can also be altered by the user repeatedly.

9-19. Keyboard

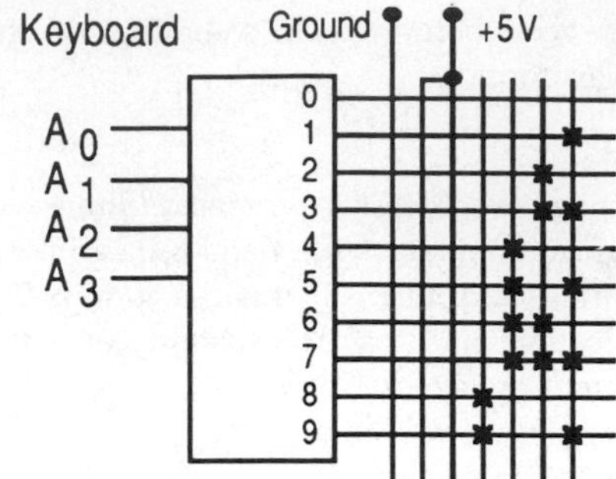

9-21. The pattern can be erased with a special voltage on a pin. It is then "burned in" like the EPROM.

9-23. See figure 9-17 in the chapter. The bit pattern is a charge, it is dissipated when turned off.

9-25. READ: connects stored bit pattern to the output. WRITE: stores an external bit pattern. CS: must be active for READ or WRITE to work, allows use of multiple chips.

9-27. Needs external refresh of the stored bit pattern as well as a constant energy source to maintain data. Static RAM does not need external refresh.

9-29. Address multiplexing is used to reduce pin count.

9-31. The first bit or the first word written into memory is also the first that can be read.

9-33. Printer buffer.

9-35. 1's and 0's are stored serially as charges in a solid material.

9-37. (A) Kansas city standard (B) Non-return to Zero Standard

9-39. Do not touch, bend, heat, magnetize, or write on the surface of the disk.

9-41. A type of magnetic recording format where changes in frequency represent data.

9-43. The video signal is recorded or read on an angle as the tape passes the head.

9-45. It is non-volatile, alterable, but also expensive and slow.

9-47. The non-alterable form of a laser disk is a CD-ROM. It is non-volatile, inexpensive, can store large amounts of data but is slow, mechanical and not alterable.

9-49. New CD's and Laser are alterable, CD-ROMs are not because the reflective patterns are fixed when they are manufactured.

9-51. D/A converts digital pulses or words into varying degrees of voltage or current.
A/D samples varying degrees of voltage or current and changes them to digital steps.

9-53.

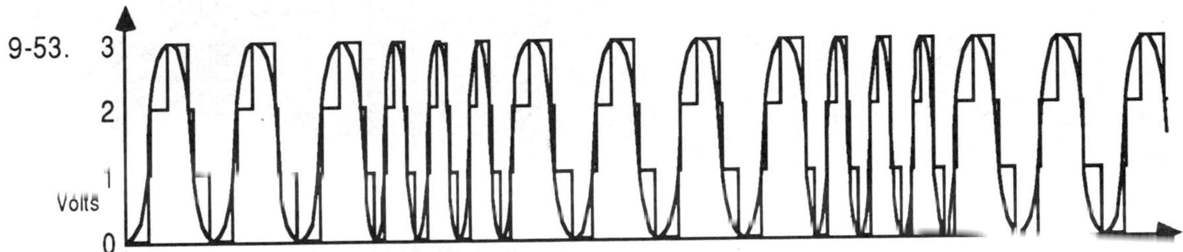

Note, the output immediately tracks the input as it reaches the next sampled level.

9-55. The input voltage is fed to a differential amplifier. The amplifier constantly compares the input to a voltage generated by a digital counter. When they match the count is latched as the data. The count is reset to 0 and a new count and comparison begins.

9-57. An 8-bit μP generally has 8 data, 16 address, CLK and R/$\overline{W}$ lines in order to operate. The 8 data bits allow as many as 256 operations.

9-59. (A) 1111 1011 1010 0000
(B) 0011 1100 1101 0010
(C) 0101 0000 1010 1110

9-61. See the next page

9-63. (A) 0.01953V (B) 0.0012207V

9-65. (A) 9.31V (B) 8.27V

9-67. An R/2R ladder network divides the input voltage in half as each section is added to the resistor ladder. As each switch connects another section, the voltage division by 2 exactly mirrors the binary weight of the switch. The output is thus directly proportional to the binary value actuating the switches.

9-69. A digital waveform recorder could be used for voice, music, to take and play back messages, for digital filtering, voice recognition, etc.

9-61.

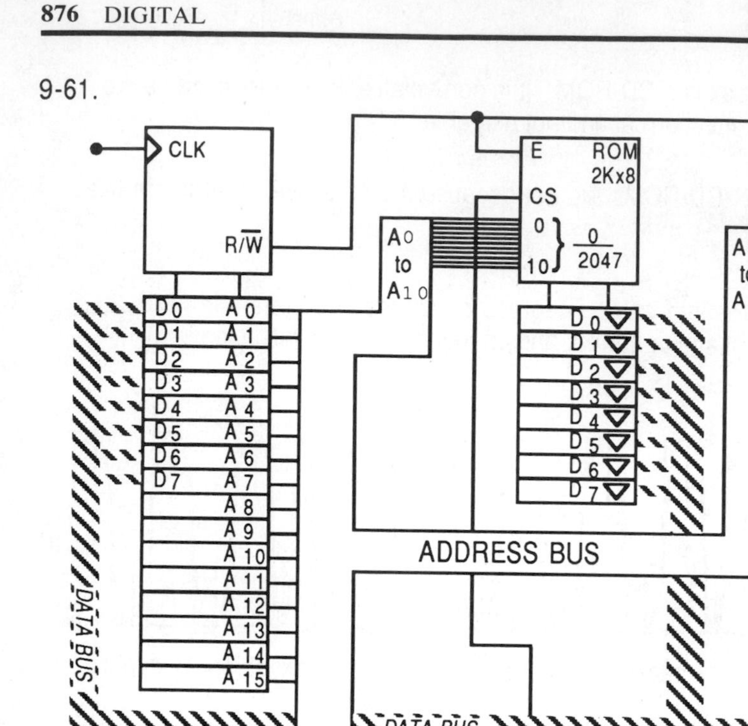
CLK
R/W
E
ROM
2Kx8
CS
0
10
0
2047
A0
to
A10
R
W
RAM
2Kx8
CS
D0
D1
D2
D3
D4
D5
D6
D7
A0
A1
A2
A3
A4
A5
A6
A7
A8
A9
A10
A11
A12
A13
A14
A15
DATA BUS
ADDRESS BUS
DATA BUS
ADDRESS BUS
A15
A14
A13
A12
A11
&
RAM
≥1
ROM
E
ADC
CS
ANALOG
INPUT
ADDRESS BUS
A15
A14
A13
A12
A10
A8
A7
A6
A5
A4
A3
A2
≥1
A11
A9
A1
A/D
D/A
E
DAC
CS
D
ANALOG
OUT
DATA BUS

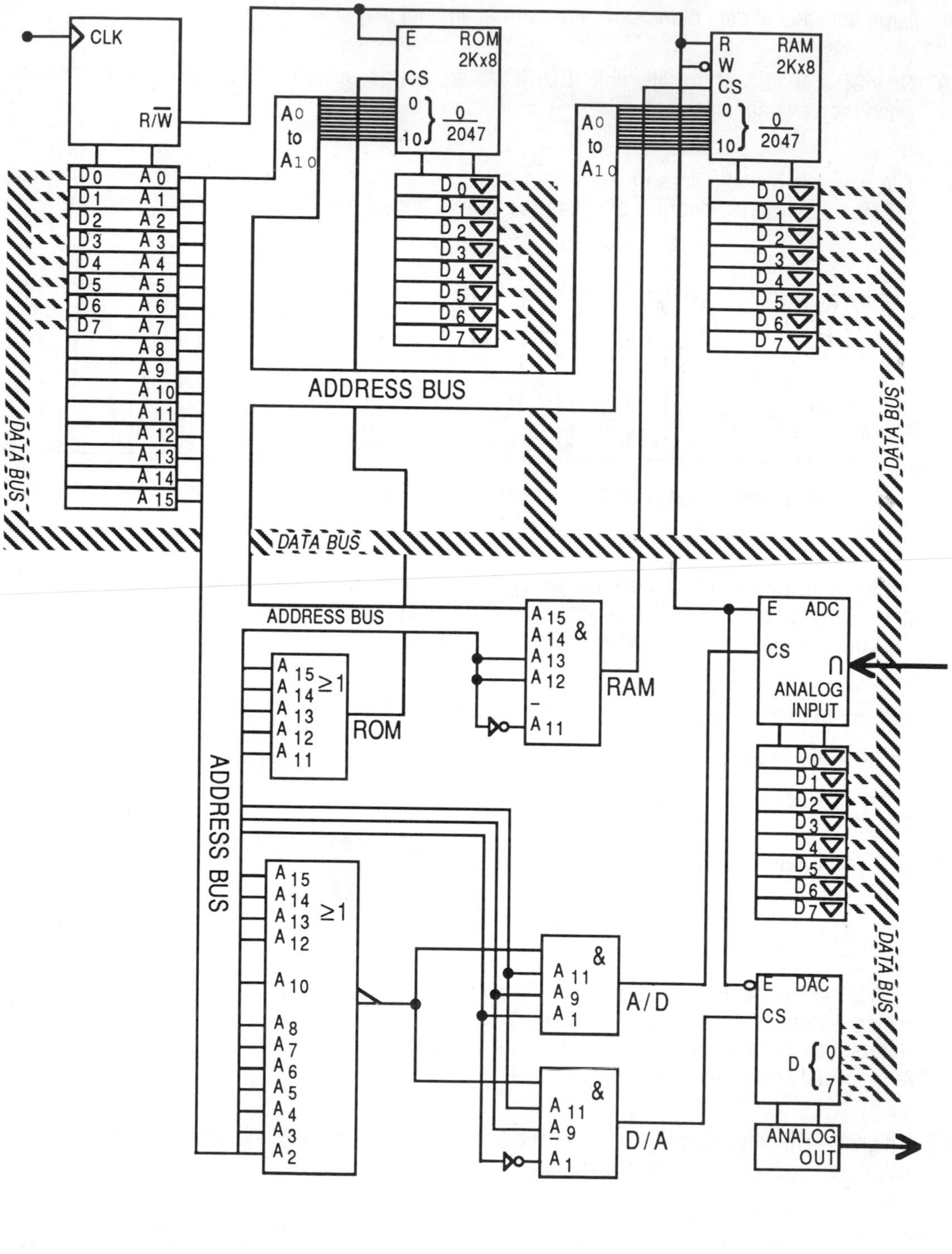

Review Questions Chapter 10

10–1:

1. The reason why a bit of a binary number must be used to represent its sign is because digital systems have only two states. Thus any symbol must be represented by an interpretation of this two-state system.
2. The direct method of representing signed binary numbers wasn't practical because when used in arithmetic computations, it would not produce correct results.
3. The one's complement method produces two values for zero where the two's complement method does not.
4. When using the methods presented here for signed binary number representation, any hex number whose MSD is 8 or greater is representing a negative number.

10–2:

1. A half adder has two inputs, while a full adder has three inputs (one is for the carry in).
2. An overflow condition indicates that the sign bit and magnitude of the answer is not correct.
3. An overflow into the sign bit can be produced by adding two positive or two negative numbers.
4. A correct answer will always be produced when the signs of the two numbers are different.

10–3:

1. The *accumulator* is a register that stores part of the problem and after exercising a logic or arithmetic operation, it will contain the answer.
2. An *ALU* is a digital device called an arithmetic logic unit. It performs arithmetic and logic operations.
3. An adder-subtractor operates as a full adder and takes the two's complement when performing a subtraction process.
4. In BCD, no number larger than 9 is allowed to be represented by four bits. Thus, unlike regular binary addition, any sum of a four bit group that results in a value of 9 or greater must be adjusted. This adjustment is accomplished by adding 6 to the answer and producing a carry.
5. Cascading adders means connecting them in such a manner so that larger numbers can be added. This is accomplished by having the carry out of one adder connected to the carry in of the next.

10–4:

1. Parallel transmission of digital information is used for short distances. Serial transmission is used for longer distances.
2. Three different ways of transmitting a digital signal on a radio wave are amplitude modulation, frequency modulation and phase modulation.
3. Some of the advantages for using fiber optics in the transmission of digital information are as follows:
   a. Immunity from electromagentic interference.
   b. Immunity from interception by external means (privacy).
   c. Inexpensive and abundant materials (silicon is the main ingredient for the construction of optical fibers and is the also the most abundant element on the earth's surface).
   d. Resistance to corrosion and oxidation
   e. Immune to atmospheric changes.
   f. Wide bandwidth (the ability to include many different transmissions at the same time).
4. The term laser stands for light amplification by stimulated emission of radiation.
5. RS-232-C means a recommended standard for interconnecting digital systems.

10–5:

1. *Direct memory access* means transferring data to or from memory without the use of the microprocessor.
2. A *PIA* is a peripheral interface adapter for transmitting data in parallel.
3. The main requirements for serially interfacing microprocessor based digital systems is to convert parallel data into serial data for transmission. The serially received data must then be converted back to parallel data.
4. *Baud rate* is the rate at which data bits are transferred.
5. A *UART* is a universal asynchronous receiver-transmitter. An ACIA is an asynchronous communications interface adapter.

10–6:

1. Two registers that all 8-bit microprocessors have in common are the *accumulator* and the *program counter*. The other internal registers that they have in common are the *status flag* and *stack pointer*. They also have registers that can be used as *index* registers.

2. Each of these 8-bit microprocessors can address 65,536 different memory locations. The word size for each is 8 bits.
3. The *stack pointer* is an internal microprocessor register that is used to hold the address of a location in memory referred to as the *stack*.
4. The 6502 and 6800 have similar architecture. So do the 8085 and the Z80.
5. The 6502 has the least number of internal registers. The Z80 has the most.

10–7:

1. The main difference between a 16-bit microprocessor and an 8-bit microprocessor is that a 16-bit machine can take data in 16-bytes and has more addressing capability.
2. The term multiplexing is used to describe the use of the same μP pins for address and data information.
3. The 8088 is internally a 16-bit machine, but externally it has an 8-bit data interface bus.
4. The purpose of segmentation registers is to provide a method of having a 16-bit address register support a 20-line address-bus interface.
5. A 32-bit μP can interface directly with a 32-line data bus, while a 16-bit μP cannot.

10–8:

1. The purpose of the status register is to show the conditions of various status flags. These status flags are used to indicate the results of previous processes and to control other processes.
2. The addressing mode that does not require an address and works only with the accumulator is called accumulator addressing.
3. Zero page addressing means dealing with instructions and data that are all located in memory where the higher order address is 00.
4. The X index register is used by the indexed addressing and the preindexed indirect addressing modes.
5. The stack is a location in memory where the contents of the internal registers of the microprocessor are usually stored. The purpose of the stack pointer is to contain the address of the stack.

10–9:

1. The items needed to perform computer aided troubleshooting are a computer, interface systems for measuring the system under test, and a program that performs the actual testing.
2. The main purpose of the IEEE-488 bus is to interconnect automated testing equipment.
3. The main disadvantage of diagnostic software is that the system it is testing must be functioning well enough to run the software.
4. There are several advantages inherent in logic analysis. One major advantage is the ability of this device to display several waveforms together based upon a predetermined set of conditions.
5. The main use of a current tracer is to locate a logic short.

Chapter Problems (odd-numbered)

10-1. (A) 36 (B) −10 (C) −269 (D) −1599

10-3. (A) 94 (B) 64 (C) −59 (D) −1343

10-5. (A) 94 (B) −1 (C) −320 (D) −1627

10-7. (A) FFFBH (B) 0020H (C) FFF4H (D) FF00H

10-9. (A) S=0, C=1 (B) S=0, C=1 (C) S=1, C=1

10-11.

	Each Cout 3 2 1 0	Each Sout 3 2 1 0	Carry LED	Answer Register	7-segment display A	7-segment display B	7-segment Answer
(A)	0 0 1 0	0 1 0 1	0	0101	3	2	5
(B)	0 0 0 0	1 1 1 1	0	1111	A	5	F
(C)	1 1 1 1	1 0 0 0	1	1000	9	F	8

10-13. (A) 0AH + 0FH = 19H, correct, no overflow
(B) F4H + 12H = 06H, correct, no overflow
(C) F1H + 0FH = F5, correct, no overflow
(D) FDH + F4H = F1H, correct, no overflow

10-15. (A) Data input – 1, 2, 3, 4; 15; 16, 17, 18,
(B) Data output – 8, 9, 11, 12; 13; 14

10-17. (A) B = FFH A = FFH
(B) B = FFH A = AEH

10-19. (A) A/S is HIGH so carry in is LOW, B = 7H, A = AH, carry out is LOW.
(B) A/S is LOW so carry in is HIGH, B = CH, A = FH, carry out is LOW.

10-21. (A) B = 9, A = 2, carry =1
(B) B = 7, A = 8, carry = 0
(C) B =0, A = 0, carry =0
(D) B = 9, A = 8, carry =1

10-23. Synchronous transmission requires a shared clock signal for reference. Asynchronous starts and stops the data with a synchronizing signal in the same medium.

10-25. Modulate means to change or modify. In electronics we use Amplitude, Frequency or Phase modulation to store and transmit informatiom

10-27. Light intensity or frequency is modulated and focused into the fiber optic tube. The reciever must convert the pulses back into information.

10-29. Light Amplification by Stimulated Emission of Radiation.

10-31. Simplex is only one directional. Half duplex can go in both directions, but not simultaneously. Full duplex can be simultaneously bidirectional allowing efficient system communications.

10-33. Modulator/Demodulator translates digital signals to and from tones for telephone transmission.

10-35. A special μP optimized to transfer information can be included in a system so I/O can occur rapidly. Some systems include the ability for a separate μP to take over the Address, Data and control lines when necessary for this efficient transfer to occur.

10-37. The main μP sets up the address, size, and any special handling instructions in the DMA controller registers, then gives DMA total control of memory.

10-39. The PIA provides several registers that can very easily allow input or output connections.

10-41. (A) The μP loads each data word into the UART. The UART frames start, synchronizing, and stop bits around the word, then sends them one at a time.
(B) They are the same.

10-43. (A) STOP 1 0 0 0 0 0 1 0 STOP
START

(B) STOP 1 0 0 0 0 1 0 0 STOP
START

(C) STOP 1 0 0 1 1 1 0 0 STOP
START

10-45. Digital information is translated into sound and sent serially over the telephone. At the receiving end the reverse operation occurs.

10-47. (A) Sketch figure 10-35 from the chapter.
(B) Program Counter points at the current/next instruction. Accumulator does arithmetic, logic and moves data. Status keeps track of the operation of the μP and signals the mode, errors, and results of calculations. Indexes keep track of counts or modify the adressimg mode being used. Stack points at specially stored data.

10-49. The 6800 has a larger Stack Pointer, 2 Acumulators, but only one Index Register.

10-51. (A) Sketch should match figure 10-41 in the chapter.
(B) A 16 bit Program Counter, Stack Pointer and 2 Index registers allow efficient 64K operation. Two 8 bit Accumulators each having its own status register and two sets of six 8 bit or three 16 bit dual purpose Index Registers allow maximum program flexibility.

10-53. A 16 bit μP uses 16 bit data. It usually addresses more memory. The internal registers are therefore larger, and the instruction set includes more operations.

10-55. (A) The 8088 uses an 8 bit data bus but can perform internal 16 bit logic and arithmetic. It can also address 1 Megabyte of memory.
(B) Sketch should match figure 10-45 in the chapter.

10-57. These registers allow the 16 bit address registers to be used over and over in 64K segments, so 20 address lines can manipulate a Megabyte of memory and still be somewhat compatible with the older 8 bit μP. The 4 left bits of the segment register become the high address bits of the 20 bit address so 64K blocks can be quickly and individually addressed.

10-59. A 32 bit µP has a 32 bit data bus and can address as much as 4 Gigabytes. The internal registers, operations and instructions are increased over a 16 bit µP.

10-61. Sketch should exactly match figure 10-49 in the chapter.

CARRY; holds most significant bit for arithmetic, used to test bit values.
ZERO; sets to 1 when the last operation resulted in a 0 in a register.
INTERUPT; enables (0) or disables (1) interupt signals to redirect the operation of the µP.
DECIMAL; tells µP to use HEX (0) or BCD (1) arithmetic.
BREAK; sets to (1) if a software interupt has occurred.
OVERFLOW; sets to a (1) if an overflow occurs in the accumulator.
SIGN; sets to (0) for positive, (1) for negative results or numbers in the registers.

10-63

```
LDA   #0
STA   $40
BRK
```

10-65.

```
LDA   immediate
STA   zero page direct
BRK   implied
```

10-67.

```
LDA  #$00
LDX  #$11
DEX
STA  $40,X
BNE  $FB
BRK
```

10-69. You can use the X register as an index to move a table of 0's from zero page locations to the memory from $40 to $50. This might be a larger program, but it could operate faster!

10-71. (A) Measuring sensors are attached to a computer and a system under test. A program in the computer tests the system in a systematic way and signals its results or takes some action.
(B) Sensors, Computer, Program, Technician.

10-73. (A) A General Purpose Interface Bus (B) Used to interconnect automated testing equipment.

10-75. (A) A program used in a system to test itself.
(B) The system must operate fairly well for a diagnostic program to run.

10-77. (A) A particular point should have a certain pattern of electrical activity in a correctly operating system. (B) Biggest advantage is that it is inexpensive.
(C) Biggest disadvantage is that it only shows one signal.

10-79. (A) Placed near a circuit trace or lead, a current probe can sense the changing current flow nearby. (B) A logic pulser is also used to inject changing logic levels.

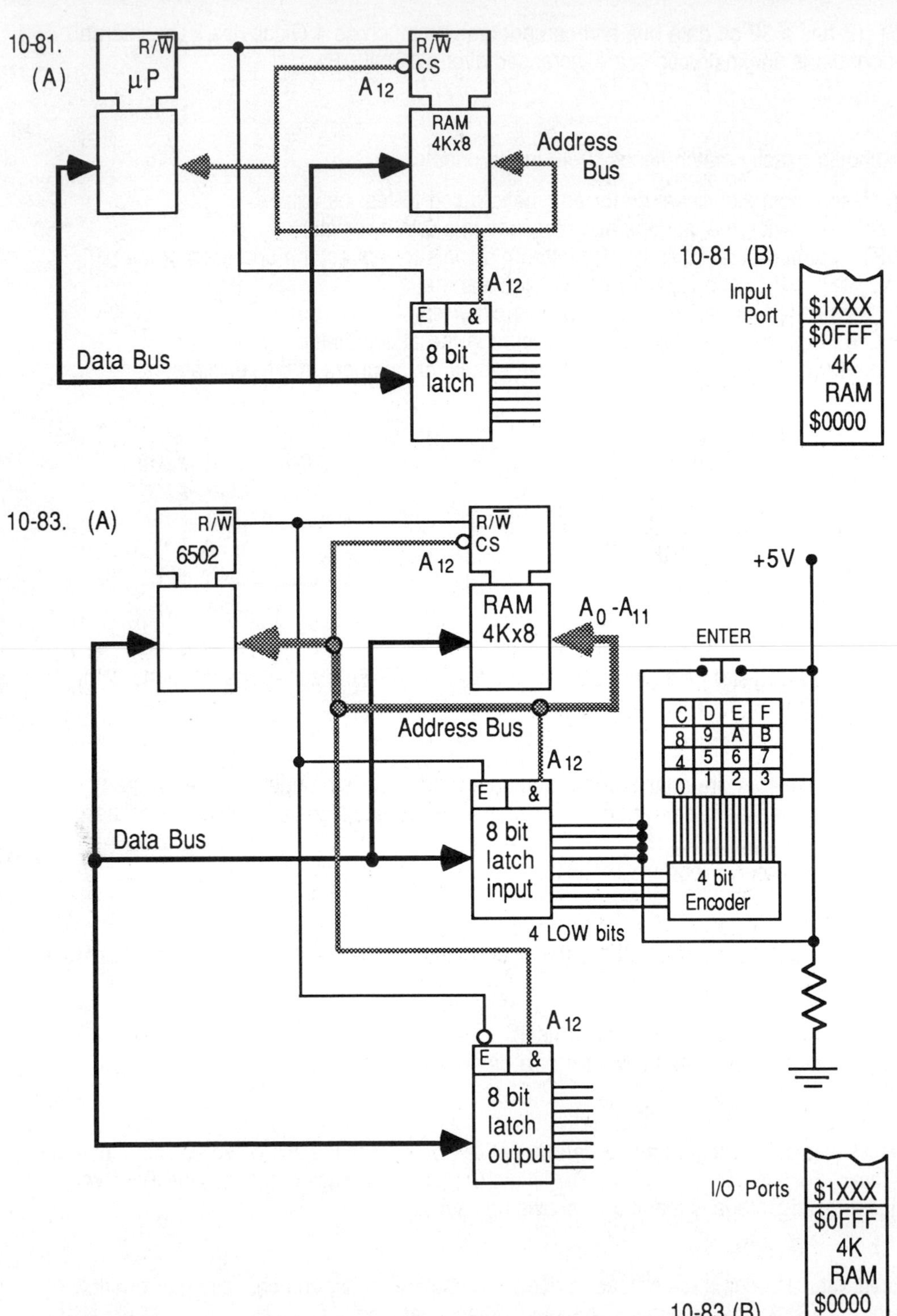
10-81.
(A)
R/W̄
μP
R/W̄
CS
A 12
RAM
4Kx8
Address
Bus
A 12
E
&
Data Bus
8 bit
latch
10-81 (B)
Input
Port
$1XXX
$0FFF
4K
RAM
$0000
10-83. (A)
R/W̄
6502
R/W̄
CS
A 12
RAM
4Kx8
A0 -A11
+5V
ENTER
C D E F
8 9 A B
4 5 6 7
0 1 2 3
Address Bus
A 12
E
&
Data Bus
8 bit
latch
input
4 bit
Encoder
4 LOW bits
A 12
E
&
8 bit
latch
output
I/O Ports
$1XXX
$0FFF
4K
RAM
$0000
10-83 (B)

INDEX